MOLECULAR BIOLOGY

Organism	Gene name	Example	Mutant allele	Example	Protein name	Example
Bacteria	Three lowercase letters, followed by upper case letter, all italicized	*recA*	Same as gene name, followed by allele number (can have non-integer allele designations such as 'am' or 'ts' for amber- and temperature-sensitive mutants, respectively)	*recA11*	Same as gene name except first letter is upper case and gene name is not italicized	RecA
Saccharomyces cerevisiae	Letters (all uppercase if dominant, all lowercase if recessive) followed by an Arabic number, all italicized	*URA3*	Same as gene name followed by a hyphen and an Arabic number (can have additional information about how mutant was generated)	*ura3-52*	Uppercase first letter, followed by lowercase letters and number, not italicized	Ura3
Schizosaccharomyces pombe	Three lowercase letters followed by a number and superscript +, all italicized	*cdc2$^+$*	Same as gene name, followed by allele number (but no superscript +)	*cdc2-5*	Same as gene name except first letter is uppercase and gene name is not italicized	Cdc2
Caenorhabditis elegans	Three to four lowercase letters, followed by a hyphen and a number, all italicized	*dpy-5*	Same as gene name, followed by an allele name (one or two letters followed by a number) in parentheses	*dpy-5(e61)*	Same as gene name except all uppercase letters and gene name is not italicized	DPY-5
Drosophila melanogaster	Can be any word lowercase italicized (most genes also have a shorter unique symbol)	*dacapo (dap)*	Same as gene name followed by a superscript number(s) or letter(s) (for dominant mutants, the gene name is followed by a superscript D)	*dacapo^4*, *dacapo^D*	Same as gene name except first letter is uppercase and gene name is not italicized	Dacapo
Mus musculus	Usually three to five letters and Arabic numbers (maximum ten characters) begin with an uppercase letter (not a number), followed by lowercase letters and numbers, all italicized	*Grid2*	Same as the gene with the original mutant symbol added as a superscript to the gene symbol	*Grid2^{ho}*	Same as gene name except all uppercase letters and gene name is not italicized	GRID2
Homo sapiens	Maximum six characters: all uppercase letters or by a combination of uppercase letters and Arabic numbers, all italicized	*ATM*	Sequence variants are described by the specific sequence change in the DNA with sequence change, insertion, and deletions having specific nomenclature	c.1636C4G (p.Leu546-Val) (this example corresponds to a C to G change at position 1636 of the *ATM* coding sequence	Same as gene name except not italicized	ATM

Nomenclature table. Note that the names of some genes and proteins that have become accepted in the literature, such as the human Rb and p53 proteins, do not follow the conventions listed in this table.

MOLECULAR BIOLOGY

Principles of Genome Function

Nancy L Craig

Orna Cohen-Fix

Rachel Green

Carol W Greider

Gisela Storz

Cynthia Wolberger

OXFORD
UNIVERSITY PRESS

OXFORD

UNIVERSITY PRESS

Great Clarendon Street, Oxford OX2 6DP

Oxford University Press is a department of the University of Oxford.
It furthers the University's objective of excellence in research, scholarship,
and education by publishing worldwide in

Oxford New York

Auckland Cape Town Dar es Salaam Hong Kong Karachi
Kuala Lumpur Madrid Melbourne Mexico City Nairobi
New Delhi Shanghai Taipei Toronto

With offices in

Argentina Austria Brazil Chile Czech Republic France Greece
Guatemala Hungary Italy Japan Poland Portugal Singapore
South Korea Switzerland Thailand Turkey Ukraine Vietnam

Oxford is a registered trade mark of Oxford University Press
in the UK and in certain other countries

Published in the United States
by Oxford University Press Inc., New York

British Library Cataloguing in Publication Data

Data available

Library of Congress Cataloging in Publication Data

Data available

Typeset by MPS Limited, A Macmillan Company
Printed in the United States of America on acid-free paper.

ISBN 978-0-19-956206-0
ISBN 978-0-19-956205-3

10 9 8 7 6 5 4 3 2

To our teachers

ABOUT THE AUTHORS OF MOLECULAR BIOLOGY

Nancy L Craig received an A.B. in Biology and Chemistry from Bryn Mawr College in 1973 and a Ph.D. in Biochemistry in 1980 at Cornell University in Ithaca, New York, where she worked on DNA repair with Jeff Roberts. She then worked on phage lambda recombination as a postdoctoral fellow with Howard Nash at the National Institutes of Health. She joined the faculty of Department of Microbiology and Immunology at the University of California, San Francisco in 1984 and began her work on transposable elements. She joined the Department of Molecular Biology and Genetics at the Johns Hopkins University School of Medicine in 1991, where she is currently a Professor and a Howard Hughes Medical Institute Investigator, as well as the recipient of the Johns Hopkins University Alumni Association Excellent in Teaching Award. Nancy Craig is a Fellow of the American Academy of Microbiology, the American Academy of Arts and Sciences and the American Association for the Advancement of Science, and was elected to the National Academy of Sciences.

Orna Cohen-Fix received a B.A. from the Tel Aviv University in 1987 and a Ph.D. in Biochemistry with Zvi Livneh at the Weizmann Institute of Science in 1994. She did a post-doctoral fellowship with Doug Koshland at the Carnegie Institution of Washington in Baltimore, studying the regulation of chromosome segregation. In 1998, she moved to the National Institute of Diabetes and Digestive and Kidney Diseases in Bethesda, where she is now a Senior Investigator. Her research focuses on cell cycle regulation and nuclear architecture, using budding yeast and *C. elegans* as model organisms. She is also the Co-Director of the NIH/Johns Hopkins University Graduate Partnership Program. She is a recipient of a Presidential Early Career Award for Scientists and Engineers, and an Association of Women in Science Mentoring Award for her work on promoting the retention of women in science.

Rachel Green received a B.S. in chemistry from the University of Michigan in 1986 and a Ph.D. in Biological Chemistry from Harvard University in 1992, where she worked with Jack Szostak studying catalytic RNA. She then did postdoctoral work in the laboratory of Harry Noller at the University of California, Santa Cruz, studying the role played by the ribosomal RNAs in the function of the ribosome. She is currently a Professor in the Department of Molecular Biology and Genetics at The Johns Hopkins University School of Medicine and an Investigator of the Howard Hughes Medical Institute. Her work continues to focus on the mechanism and regulation of translation in bacteria and eukaryotes. She is the recipient of a Johns Hopkins University School of Medicine Graduate Teaching Award.

Carol W Greider received a B.A. from the University of California at Santa Barbara in 1983. In 1987, she received her Ph.D. from the University of California at Berkeley, where she and her advisor, Elizabeth Blackburn, discovered telomerase, the enzyme that maintains telomere length. In 1988, she went to Cold Spring Harbor Laboratory as an independent Fellow and remained as a Staff Scientist until 1997, when she moved to The Johns Hopkins University School of Medicine. She is currently a Professor and Director of the Department of Molecular Biology and Genetics and her work focuses on telomerase and the role of telomeres in chromosome stability and cancer. She is a member of the National Academy of Sciences and is the recipient of numerous awards, including the Gairdner Foundation International Award, the Louisa Gross Horwitz Prize, and the Lasker Award for Basic Medical Research. In 2009, she was awarded the Nobel Prize in Physiology or Medicine together with Elizabeth Blackburn and Jack Szostak for the discovery of telomerase.

Gisela Storz graduated from the University of Colorado at Boulder in 1984 with a B.A. in Biochemistry and received a Ph.D. in Biochemistry in 1988 from the University of California at Berkeley, where she worked for Bruce Ames. After postdoctoral fellowships with Sankar Adhya at the National Cancer Institute and Fred Ausubel at Harvard Medical School, she moved to the National Institute of Child Health and Human Development in Bethesda, where she is now a Senior Investigator. Her research is focused on understanding gene regulation in response to environmental stress as well as elucidating the functions of small regulatory RNAs. She is a fellow of the American Academy of Microbiology and received the American Society for Microbiology Eli Lilly Award.

Cynthia Wolberger received her A.B. in Physics from Cornell University in 1979 and a Ph.D. in Biophysics from Harvard University in 1987, where she worked with Stephen Harrison and Mark Ptashne on the structure of the phage 434 cro repressor bound to DNA. She went on to study the structures of eukaryotic protein–DNA complexes as a postdoctoral fellow, first in the laboratory of Robert Stroud and the University of California, San Francisco and then in the laboratory of Carl Pabo at The Johns Hopkins University School of Medicine, where she is now Professor of Biophysics and Biophysical Chemistry and an Investigator of the Howard Hughes Medical Institute. Her research focuses on the structural and biochemical mechanisms underlying transcriptional regulation and ubiquitin-mediated signalling. She is a Fellow of the American Association for the Advancement of Science.

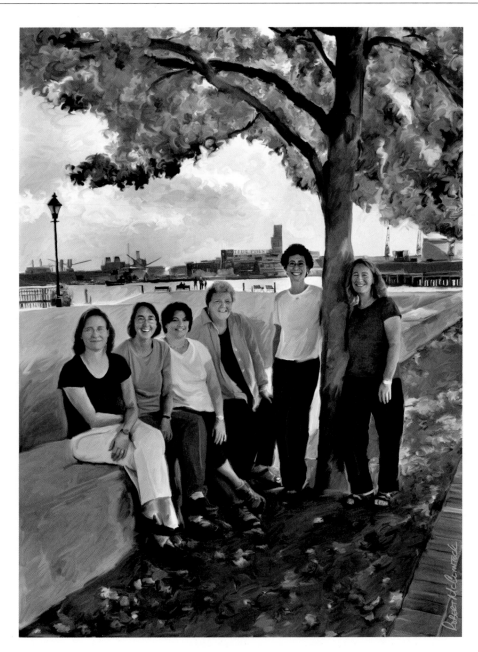

Molecular Biologists of Fells Point, Baltimore: (L–R) Rachel Green, Gisela (Gigi) Storz, Orna Cohen-Fix, Nancy Craig, Cynthia Wolberger and Carol Greider. The photo-digital illustration was created by Robert McClintock a Fells Point artist.

PREFACE

A new approach to molecular biology for the twenty-first century

Molecular Biology: Principles of Genome Function offers a fresh, distinctive approach to the teaching of molecular biology. It is an approach that reflects the challenge of teaching a subject that is in many ways unrecognizable from the molecular biology of the twentieth century – a discipline in which our understanding has advanced immeasurably, but about which many intriguing questions remain to be answered. Among the students being taught today are the molecular biologists of tomorrow; these individuals will be in a position to ask fascinating questions about fields whose complexity and sophistication become more apparent with each year that passes.

We have written the book with several guiding themes in mind, all of which focus on providing a faithful depiction of molecular biology in the twenty-first century, and on communicating this reality to students in a way that will engage and motivate, rather than overwhelm and intimidate.

A focus on the underlying principles

Arguably one of the biggest challenges facing instructors and students of molecular biology today is the vast amount of information encapsulated by the field. It is impossible for an instructor to convey every last detail (and equally impossible for students to absorb everything that there is to know). Indeed, we believe that, in order to understand the main concepts of molecular biology and to appreciate their exquisite complexity, it is not necessary to delve into every fine detail. Therefore, our approach focuses on communicating the *principles* of the subject.

We believe it is better for students to truly understand the foundational principles rather than simply learn a series of facts. To this end, we do not try to be exhaustive in our coverage. In the digital age in which we live, it is easier than ever before for students to gather a vast amount of information on a particular topic of interest. This information is of little value, however, if the student lacks a conceptual framework within which to make sense of all the information to which they are exposed.

By focusing on key principles, we seek to equip students with a conceptual framework, which we believe will be invaluable to them during their later careers.

An emphasis on commonalities

Until relatively recently, much more was known about the molecular components and processes of bacterial systems than of their archaeal and eukaryotic counterparts. In recent years, however, our understanding of archaeal and eukaryotic systems has increased enormously. With this increased understanding has come the realization that bacterial, archaeal, and eukaryotic systems exhibit many commonalities – commonalities that point to the common ancestry of the three kingdoms of life.

Throughout this book, therefore, our emphasis is on the *common features* of bacterial, archaeal, and eukaryotic systems. Differences do exist, of course – an inevitable outcome of evolutionary processes generating biological diversity. However, we have strived where possible to present a single view of key topics based on conserved processes and components. We have then discussed key differences between bacterial processes and their archaeal and eukaryotic counterparts where they exist, and where they have helped to further our understanding.

We recognize that some may feel that the processes occurring in bacteria, and in eukaryotes and archaea, are best taught separately. However, our focus on principles – and on constructing an overarching conceptual framework – leads us strongly to believe that an emphasis on commonalities is a valuable educational approach.

Integration of key themes and concepts

One of the most startling realizations of recent years has been the widespread importance of certain molecular phenomena, such as chromatin modification, or RNA silencing, which have impacts on genome function in ways far more diverse than had previously been recognized. Rather than examining each of these phenomena in isolation, our approach reflects their diverse impacts by presenting them in the various contexts in which they function. Therefore, you will not see a separate chapter on such topics; instead, the impact of these basic processes on genome function is woven throughout the book.

⊙ Look at the *Map of key themes* in the Online Resource Center at www.oxfordtextbooks.co.uk/orc/craig/ for an illustration of how topics such as regulation are blended throughout the book.

Similarly, while many books deliberately separate the regulatory control of basic processes from the processes themselves, we have chosen to put them together. As more is learned about how regulation takes place, it becomes increasingly apparent that regulation is typically nothing more than the alteration of the core process, so that an alternative, but related, pathway is chosen. Regulation simply acts on the core mechanistic features of the process, and so it makes sense to present them side by side.

We believe this overall approach reflects the reality of molecular biology, and helps students to appreciate molecular biology as a unified discipline, with many components and phenomena acting in concert, rather than as a series of isolated topics.

A demonstration of how we know what we know

At heart, molecular biology is an experimental science. Our understanding of the field is increased through the accumulation of experimental evidence, which leads to the gradual emergence of key ideas and paradigms. Therefore, a central element to the understanding of molecular biology is an appreciation of the approaches taken to yield the information from which concepts and principles are deduced.

However, as instructors, we face a potential conflict: a mass of experimental evidence can often be overwhelming for students, and can make it more challenging for them to grasp the central ideas and paradigms that the experimental evidence has allowed us to elucidate. On the other hand, ignoring the experimental evidence deprives students from fully understanding the fundamental aspects of molecular biology (and, indeed, of science in general). In response to this seeming conflict, our approach has been for the main body of the text to focus on the communication of key concepts, free from the layer of complexity that experimental evidence might introduce.

The 'experimental approach' panels

In recognizing the central importance of experimental evidence to furthering our understanding of molecular biology, we have complemented our coverage of key concepts in the main body of the text with separate panels entitled 'Experimental approach', which branch off from the text in a clearly signposted way. These panels describe pieces of research that have been undertaken and which have been particularly valuable in elucidating difference aspects of molecular biology.

Importantly, experimental research represents an ongoing journey of discovery, where the experimental approaches adopted develop as much as our understanding of the field. Uniquely, therefore, the experimental approach panels present, wherever possible, two approaches – one 'classic' and one 'contemporary'. Although all approaches have revealed valuable insights, regardless of whether they could be considered classic or contemporary, we believe that coupling the approaches in this way has additional educational value in terms of showing how both experimentation and the knowledge gained from such experimentation can evolve with time.

In addition to the experimental approach panels, further support for encouraging students to engage with experimental evidence is provided by an online Journal Club, as described more fully below.

The methods used in molecular biology

Many of the experimental approach panels (and the research work featured in the Journal Club papers) draw on certain laboratory techniques, which are used in different contexts throughout molecular biology research. The final chapter of this book, 'Tools and techniques in molecular biology', provides an overview of the basic techniques that are exploited during the course of much experimental work in molecular biology. Rather than describing general methods in detail within the experimental approach panels, we have directed the reader to appropriate coverage in Chapter 16, where they can learn more about the methodological tools that are at a molecular biologist's disposal, and how these tools work and what they can tell us.

NLC
OCF
RG
CWG
GS
CW
Baltimore, Maryland, July 2010

LEARNING FROM THIS BOOK

Beyond the overall approach adopted in writing this book, which we believe will make it a valuable teaching and learning resource, it includes a number of other features to help students get the most out of their molecular biology studies.

The experimental approach panels

As noted previously, molecular biology is an experimental science. To help you gain an understanding of how some of the key molecular processes and components described in this book were characterized, without overburdening the main text with lots of experimental detail, virtually every chapter features 'experimental approach' panels. These panels describe pieces of research that have been undertaken, and which have been particularly valuable in elucidating different aspects of molecular biology.

Further reading

Each chapter ends with a list of further reading materials, typically review articles, which we feel would make a good next step when looking to explore in more detail the topics covered in the book. Each further reading list is divided into chapter sections, to help you pinpoint articles that are of relevance to the particular topic you are interested in.

Glossary

Molecular biology, like many scientific disciplines, has its own particular vocabulary, and descriptions of molecular processes and procedures feature terms that may at first glance be unfamiliar. We have compiled an extensive glossary of all of the key terms featured in the book; we hope this will be of value as you master the language of the subject.

Cross-references

As we note previously, molecular biology comprises a range of interconnected topics, not a series of discrete, isolated ones. To help you make the connection between the topics presented in the book, and see how these topics come together to give a rounded picture of molecular biology, each chapter features numerous cross-references to other chapters in the book.

PDB codes

Many of the molecular structures that appear throughout this book have been generated from data deposited in the Protein Databank (PDB). Each entry in the PDB is assigned a unique code; this code can be used to retrieve the data related to the entry in question, which often includes crystallographic data, and onscreen renderings of molecular structures in three dimensions. The PDB codes relating to many of the molecular structures in the book are given in the relevant figure legends. Visit the PDB website (http://www.rcsb.org/pdb/home/home.do) and enter the PDB codes related to molecules of interest to retrieve data related to those molecules for yourself.

 ## Online Resource Center

Molecular Biology: Principles of Genome Function does not end with this printed book. Instead, additional resources for both instructors and their students are available in the book's Online Resource Center.

Go to http://www.oxfordtextbooks.co.uk/orc/craig/

For instructors

Electronic artwork

Figures from the book are available to download, for use in lectures.

Journal Club

Most chapters in the book are accompanied by an online Journal Club, which features suggested research papers and discussion questions linked to topics featured in the chapters. Understanding the details presented in primary literature articles can often be challenging; the purpose of the Journal Club is to guide students through some selected papers in a structured way, to build their confidence in reading and critically evaluating the work of others.

For students

New and noteworthy

This is a note of key highlights from the field of molecular biology since the book's publication, and is updated for the start of each semester.

ACKNOWLEDGMENTS

Many people made this textbook possible thanks to their advice and support.

We were fortunate to have worked under the guidance of two gifted individuals. The project was begun with Miranda Robertson at New Science Press, who convinced us that we could write a new textbook and helped show us the way. We benefited from her tremendous vision, advice, and insistence on clarity, as well as from her many visits to work with us in Baltimore and Washington. It was Jonathan Crowe at Oxford University Press who ushered us across the finish line by providing superb editorial advice, while teaching us how to work ever more effectively. He made outstanding contributions to the writing and organization of this book, and we are grateful for his efforts in helping us bring this project to completion.

We also benefited from assistance by Philip Meneely (Haverford College) and Brendan Cormack (Johns Hopkins School of Medicine), who helped to write and commented on multiple sections.

Our work was supported by many others. Matthew McClements is responsible for the beautiful illustrations while Lore Leighton did the graphics rendering of macromolecules. At New Science Press, Karen Freeland, Joanna Miles, and Gina Lyons helped to keep us on track and organized despite ourselves, while Eleanor Lawrence and Matthew Day lent their excellent editing skills to the project. At Oxford University Press, Bethan Lee and Marionne Cronin provided organizational and editorial support, while Lotika Singha did an excellent job of copy editing. Our administrative assistants, particularly Patti Kodeck, helped to organize our meetings and carve out time for us to work on the book.

Our colleagues at Johns Hopkins and at the National Institutes of Health were generous, answering innumerable questions and providing comments on the text and figures. We are also grateful to the scientists outside our respective institutions for responding to our many emails and phone calls asking for information and clarifications. We owe special thanks to our students, postdoctoral fellows and laboratory staff, who gladly provided us with input while being patient when the writing took us away from the laboratory.

We extend our most profound thanks to our families for their support and encouragement throughout this project, which took far more time and work than any of us had imagined. We are grateful to our children, Rachel and Joshua Adams, Charles and Gwendolyn Comfort, Eric, Toby and Noel Cormack, Tal and Jonathan Fix, and Ella, Toby, and Felix Wu for their forbearance. We especially thank our spouses, Jeffrey Adams, Nathaniel Comfort, Brendan Cormack, Alan Fix, Helen Lee McComas, and Carl Wu, for cheering us on while picking up the slack at home.

The authors and publisher extend their sincere thanks to the following individuals whose constructive comments improved the drafts of this book immeasurably.

Angelika Amon, MIT
John Atkins, The University of Utah
Brenda Bass, The University of Utah
Joel Belasco, New York University School of Medicine
Stephen D. Bell, Massachusetts Institute of Technology
Andrew Bell, University of Illinois
James Berger, University of California
Doug Black, University of California
Michael Botchan, University of California
Chris Burge, Massachusetts Institute of Technology
Sean Burgess, University of California
Douglas Burks, Wilmington College
Mike Carey, University of California
Mike Chandler, Laboratoire de Microbiologie et Génétique
Moléculaire, Toulouse
Cris Cheney, Pomona College
Mitchell Chernin, Bucknell University
Christopher Cole, University of Minnesota
Joan Conaway, Stowers Institute for Medical Research
Victor Corces, Emory University
Jeffrey Corden, Johns Hopkins University School of
Medicine
Don Court, Center for Cancer Research
Patrick Cramer, Gene Center Munich
James Dahlberg, University of Wisconsin-Madison
Seth Darst, The Rockefeller University
Stephanie Dellis, College of Charleston
Tom Dever, National Institutes of Health
Shery Dolhopf, Alverno College
Robert Drewell, Harvey Mudd College
Richard Ebright, Rutgers, The State University of New Jersey
Tom Eickbush, University of Rochester
Gary Felsenfeld, National Institutes of Health
Paul Foglesong, University of the Incarnate Word
Chris Francklyn, The University of Vermont
John Geiser, Western Michigan University
Raymond Gesteland, The University of Utah
Grace Gill, Tufts University
Rick Gourse, University of Wisconsin-Madison
Paula Grabowski, University of Pittsburgh
Nigel Grindley, Yale University
Carol Gross, University of California
Jim Haber, Brandeis University
Charlotte Hammond, Quinnipiac University
Phil Hanawalt, Stanford University
Frank Healy, Trinity University
Robert Heath, Kent State University
Tamara Hendrickson, Wayne State University
Alan Hinnebusch, National Institutes of Health
Ann Hochschild, Harvard Medical School
Paul Huber, University of Notre Dame

Dottie Hutter-Lobo, Monmouth University
Richard Jackson, University of Leeds
Steve Jackson, Gurdon Institute, University of Cambridge
Rudolph Jaenisch, Massachusetts Institute of Technology
Lisa Johansen, University of Colorado, Denver
Jim Kadonaga, University of California, San Diego
Roland Kanaar, Erasmus Medical Center
Gary Karpen, University of California
Mikhail Kashlev, Center for Cancer Research
Paul Kaufman, University of Massachusetts Medical School
Tom Kelly, Memorial Sloan-Kettering Institute
Bob Kingston, Harvard University
Christopher Korey, College of Charleston
Roger Kornberg, Stanford University
Doug Koshland, University of California
Justin Kumar, Indiana University
John Kuriyan, University of California
Josef Kurtz, Emmanuel College
Michael Lichten, Center for Cancer Research
Michael Lieber, USC Norris Comprehensive
Cancer Center
Lasse Lindahl, University of Maryland
Gary Linquester, Rhodes College
John Lis, Cornell University
Bob Lloyd, University of Nottingham
John Lorsch, John Hopkins University School of Medicine
Sue Lovett, Brandeis University
Karolin Luger, Colorado State University
Mitch McVey, Tufts University
Peter B. Moore, Yale University
Margaret Olney, Saint Martin's University
Jennifer Osterhage, Hanover College
Anthony Ouellette, Jacksonville University
William J Patrie, Shippensburg University
Tanya Paull, University of Texas at Austin
Stephen Pelsue, University of Southern Maine
Eric Phizicky, IBM Life Sciences Discovery Center
Marie Pizzorno-Simpson, Bucknell University
Anna Marie Pyle, Yale University
Venki Ramakrishnan, MRC Laboratory of Molecular
Biology
John Rebers, Northern Michigan University
Margaret Richey, Centre College
Timothy J. Richmond, Swiss Federal Institute of Technology
Zurich
Don Rio, University of California
Jeff Roberts, Cornell University
Peter Sarnow, Stanford University
Paul Schimmel, The Scripps Research Institute
Rey Sia, The College at Brockport – SUNY
Erik Sontheimer, Northwestern University

Forrest Spencer, John Hopkins University School of Medicine

Stephen Spiro, University of Texas at Dallas

Kevin Sullivan, National University of Ireland, Galway

Ben Szaro, State University of New York, Albany

Song Tan, Pennsylvania State University

David Tollervery, University of Edinburgh

Deborly Wade, Central Baptist College

Gabriel Waksman, University College London

Tracy Ware, Salem State College

Jon Warner, Albert Einstein College of Medicine, Yeshiva University

Alan Weiner, Washington University

Stephen West, London Research Institute

Jamie Williamson, The Scripps Research Institute

John Wilson, Michigan State University

Fred Winston, Harvard Medical School

Kate Leslie Wright, Rochester Institute of Technology

Stephen Wright, Carson-Newman College

Ramakrishna Wusirika, Michigan Tech University

Philip Zamore, University of Massachusetts Medical School

Ana Zimmerman, College of Charleston

Figure acknowledgments

We are grateful to the following for permission to reproduce or draw on copyright material.

Chapter 1

Figure 1.28 from Meneely, P. (2009). *Advanced Genetic Analysis: Genes, Genomes, and Networks in Eukaryotes.* Reproduced with permission by Oxford University Press.

Chapter 2

Figure 2.54 reproduced with permission from Wolberger, C., Vershon, A.K., Liu, B., Johnson, A.D., and Pabo, C.O. (1991). Crystal structure of a MATα2 homeodomain-operator complex suggests a general model for homeodomain-DNA interactions. *Cell* **67**: 517–528.

Chapter 4

Figure 4.2 from http://jama.ama-assn.org/cgi/content/extract/281/7/600-a

Figure 4.13 reproduced from Cosgrove S., Boeke, Jef D., and Wolberger, C. (2004). Regulated nucleosome mobility and the histone code. *Nature Structural and Molecular Biology* **11**: 1037–1043.

Figure 4.21 adapted from Becker, P., and Horz, W. (2002). ATP-dependent nucleosome remodelling. *Annual Review of Biochemistry* **71**: 247–273.

Figure 4.26a reproduced from Lippman, Z., and Martienssen, R. (2004). The role of RNA interference in heterochromatic silencing. *Nature* **431**: 364–370.

Figure 4.36c reproduced from the Laboratory for Cell Biology and Genetics, Rockefeller University, New York. Titia de Lange (http://delangelab.rockefeller.edu).

Figure 4.42 reproduced from Chubb, J. R. and Bickmore, W. A. (2003). Considering nuclear compartmentalization in the light of nuclear dynamics. *Cell* **112**: 403–406.

Chapter 5

Figure 5.12 adapted from Zheng, N., Schulman, B.A., Song, L., *et al.* (2002). Structure of the Cul1-Rbx1-Skp1-F boxSkp2 SCF ubiquitin ligase complex. *Nature* **416**: 703–709.

Chapter 6

Figure 6.7 reproduced from Johnson, S.J., and Beese, L.S. (2004). Structures of mismatch replication errors observed in a DNA polymerase. *Cell* **116**: 803–816.

Figure 6.11 reproduced from Enemark, E.J., and Joshua-Tor, L. (2006). Mechanism of DNA translocation in a replicative hexameric helicase. *Nature* **442**: 270–275.

Figure 6.12 adapted from Enemark, E.J., and Joshua-Tor, L. (2008). On helicases and other motor proteins. *Current Opinion in Structural Biology* **18**: 243–257.

Figure 6.14 from Corbett K.D., and Berger, J.M. (2004). Structure, molecular mechanisms, and evolutionary relationships in DNA topoisomerases. *Annual Review of Biophysics and Biomolecular Structure* **33**: 95–118.

Figure 6.16 from O'Donnell M., Jeruzalmi D., and Kuriyan, J. (2001). Clamp loader structure predicts the architecture of DNA polymerase III holoenzyme and RFC. *Current Biology* **11**: R935–R946.

Figure 6.17 from Bowman G. D., et al (2004). Structural analysis of a eukaryotic sliding DNA clamp-clamp loader complex. *Nature* **429**: 724–730.

Figure 6.20 from Erzberger, J.P., Mott, M.L., and Berger, J.M. (2006). Structural basis for ATP-dependent DnaA assembly and replication-origin remodeling. *Nature Structural and Molecular Biology* **13**: 676–683.

Chapter 7

Figure 7.3 adapted from Kireeva, N., Lakonishok, M., Kireev., I., Hirano, T., and Belmont, A. (2004). Visualization of early chromosome condensation: a hierarchical folding, axial glue model of chromosome structure. *Journal of Cell Biology* **166**: 775–785.

Figure 7.7 adapted from Li, H., et al. (2002). Microtubule structure at 8 Å resolution. *Structure* **10**: 1317–1328.

Figure 7.11b from McGill, M., Highfield, D.P., Monahan, T.M., and Brinkley, B.R. (1976). Effects of nucleic acid specific dyes on centrioles of mammalian cells. *Journal of Ultrastructure Research* **57**: 43–53.

Figure 7.16 adapted from Lampson, M.A., Renduchitala, K., Khodjakov, A., and Kapoor, T.M. (2004). Correcting improper chromosome–spindle attachments during cell division. *Nature Cell Biology* **6**: 232–237.

Figure 7.23 adapted from Anderson, D.J., and Hetzer, M.W. (2008). Reshaping of the endoplasmic reticulum limits the rate for nuclear envelope formation. *Journal of Cellular Biochemistry* **182**: 911–924.

Figure 7.27b from O'Connor, C. (2008). Meiosis, genetic recombination, and sexual reproduction. *Nature Education* 1 (1).

Figure 7.28 from Schmekel, K., and Daneholt, B. (1998). Evidence for close contact between recombination nodules and the central element of the synaptonemal complex. *Chromosome Research* **6**: 155–159.

Chapter 8

Figure 8.7 reproduced from Murakami, K.T., and Darst, S.A. (2003). Bacterial RNA polymerases: the wholo story. *Current Opinion in Structural Biology* **13**: 31.

Figure 8.15 adapted from Cramer, P.,et al. (2002). Multisubunit RNA polymerases. *Current Opinion in Structural Biology* **12**: 89–97.

Figure 8.17 reproduced from http://ssrl.slac.stanford.edu/research/highlights_ archive/pol-ii.html. With permission from R. Koinberg.

Figure 8.18 reproduced from Murakami, K.T., and Darst, S.A. (2003). Bacterial RNA polymerases: the wholo story. *Current Opinion in Structural Biology* **13**: 31.

Figure 8.22 from Cramer, P. (2004). RNA polymerase II structure: from core to functional complexes. *Current Opinion in Genetics and Devlopment* **14**: 218.

Figure 8.27b from Skordalakes, E., and Berger, J. (2003). Structure of the Rho Transcription Terminator. *Cell* **114**: 135, figure 3b.

Figure 8.31 from Spector, D.L., and Gasser, S.M. (2003). A molecular dissection of nuclear function conference on the dynamic nucleus: questions and implications. *EMBO reports* **4**: 18–23.

Figure 8.33 reproduced from Saunders, A. Core, L.J., and Lis, J.T. (2006). Breaking barriers to transcription elongation. *Nature Reviews Molecular Cell Biology* **7**: 557–567.

Figure 8.46a-c from Heldwein, E. E., and Brennan, R.G. (2001). Crystal structure of the transcription activator BmrR bound to DNA and a drug. *Nature* **409**: 328.

Figure 8.48 from Ekaterina, E., Heldwein, Z., and Brennan, R.G. (2001). Crystal structure of the transcription activator BmrR bound to DNA and a drug. *Nature* **409**: 378–382.

Figure 8.57 from Panne, D., Maniatis, T., and Harrison, S.C. (2007). An Atomic Model of the Interferon-β Enhanceosome. *Cell* **129**: 1111–1123.

Figure 8.70 from Bienvenu, T., and Chelly, J. (2006). Molecular genetics of Rett syndrome: when DNA methylation goes unrecognized. *Nature Reviews Genetics* **7**: 415–426.

Chapter 9

Figure 9.17 from Roberts, R.G., (1995). Dystrophin, its gene, and the dystrophinopathies. *Advanced Genetics* **33**: 177–231.

Figure 9.33 adapted from Lau, N.C., Lim, L.P., Weinstein, E.G., and Bartel, D.P. (2001). An Abundant Class of Tiny RNAs with Probable Regulatory Roles in *Caenorhabditis elegans*. *Science* **294**: 858–862, Figure 1D.

Chapter 10

Figure 10.17 from Miller, O.L., Hamkalo, B.A., and Thomas, C.A. (1970). Visualization of bacterial genes in action. *Science*, **169**: 392–395. Reprinted with permission from AAAS.

Figure 10.29b from Parnell, M.K., and Strobel, S.A. (2003). HIS & HERS, magic magnesium and the ballet of protein synthesis. *Current Opinion in Chemical Biology* **7**: 528–33.

Figure 10.30b from Valle et al, (2003). Locking and Unlocking of Ribosomal Motions. *Cell* **114**: 123–124.

Figure 10.31 from Youngman, E.M., McDonald, M.E., and Green, R. (2008). Peptide Release on the Ribosome: Mechanism and Implications for Translational Control. *Annual Review of Microbiology* **62**: 353–373.

Figure 10.61 from Spahn, M.T., Kieft, J.S., Grassucci, R.A. *et al* (2001). Hepatitis C Virus IRES RNA-Induced Changes in the Conformation of the 40S Ribosomal Subunit. *Science* **291**: 1959–1962. Reprinted with permission from AAAS.

Chapter 11

Figure 11.1 adapted from Deuerling, E., and Bukau, B. (2004). Chaperone-Assisted Folding of Newly Synthesized Proteins in the Cytosol. *Critical Reviews in Biochemistry and Molecular Biology* **39**: 261–277.

Figure 11.2 adapted from Walsh, C.T. (2006). *Posttranslational Modification of Proteins: Expanding Nature's Inventory*. Roberts and Company Publishers.

Figure 11.3 adapted from Thorpe, C., and Coppock, D.L. (2007). Generating Disulfides in Multicellular Organisms: Emerging Roles for a New Flavoprotein Family. *Journal of Biological Chemistry* **282**: 13929.

Figure 11.12 adapted from Hall, T.M., Porter, J.A., Young, K.E., Koonin, E.V., Beach, P.A., and Leahy, D.J. (1997). Crystal structure of a Hedgehog autoprocessing domain: homology between Hedgehog and self-splicing proteins. *Cell* **91**: 8–97.

Figure 11.17 adapted from Walsh, C.T. (2006). *Posttranslational Modification of Proteins: Expanding Nature's Inventory*. Roberts and Company Publishers.

Figure 11.29 adapted from Pickart, C.M., and Fushman, D. (2004). Polyubiquitin chains: polymeric protein signals. *Current Opinion in Chemical Biology* **8**: 610–616.

Chapter 12

Figure 12.14 from Yang, W. (2005). Portraits of a Y-family DNA polymerase. *FEBS Letters* **79**: 868–872.

Figure 12.15 from Bunting, K.A., Roe, S.M., and Pearl, L.H. (2003). Structural basis for recruitment of translesion DNA polymerase Pol IV/DinB to the β-clamp. *EMBO Journal* **22**: 5883–5892.

Figure 12.18 from Voloshin, O.N., Ramirez, B.E., Bax, A., et al. A model for the abrogation of the SOS response by an SOS protein: a negatively charged helix in DinI mimics DNA in its interaction with RecA. *Genes and Development* **15**: 415–427.

Figure 12.20 from Morgan, D. O. (2007). *The Cell Cycle: Principles of Control,* New Science Press.

Figure 12.21 from Wyman, C., Warmerdam, D.O., Kanaar, R. (2008). From DNA End Chemistry to Cell-Cycle Response: The Importance of Structure, Even When It's Broken. *Molecular Cell,* pp 5–6.

Figure 12.22 from Crimpich, K.A., and Cortez, D. (2008). ATR: an essential regulator of genome integrity. *Nature Reviews Molecular Cell Biology* **9**: 616–627.

Figure 12.23 from Crimpich, K.A., and Cortez, D. (2008). ATR: an essential regulator of genome integrity. *Nature Reviews Molecular Cell Biology* **9**: 616–627.

Chapter 13

Figure 13.8 from Bernstein, K.A., and Rothstein, R. (2009). At loose ends: resecting a double-strand break. *Cell* **137**: 807–809.

Chapter 14

Figure 14.1 from Belancio, V.P., Hedges, D.J., and Deininger, P. (2008). Mammalian non-LTR retrotransposons: For better or worse, in sickness and in health. *Genome Research* 2008 **18**: 343–358.

Figure 14.2 from Belancio, V.P., Hedges, D.J., and Deininger, P. (2008). Mammalian non-LTR retrotransposons: For better or worse, in sickness and in health. *Genome Research* **18**: 343–358.

Figure 14.6a from Feschotte and Pritham, Computational analysis and paleogenomics of interspersed repeats in eukaryotes. In: Stojanovic, N., ed. *Computational Genomics: Current Methods.* Wyndmonham, Norfolk, UK: Horizon Bioscience, 2007.

Figure 14.6b adapted from Pritham, E.J. (2009). Transposable elements and factors influencing their success in eukaryotes. *Journal of Heredity* **100**: 648–655.

Figure 14.11 from Steiniger-Whitem M., Rayment, I., and Reznikoff, W.S. (2004). Structure/function insights into Tn5 transposition. *Current Opinion in Structural Biology* **14**: 50–57.

Figure 14.13 from Hickman, A.B., Chandler, M., Dyda, F. (2010). Integrating prokaryotes and eukaryotes: DNA transposases in light of structure. *Critical Reviews In Biochemistry and Molecular Biology* **45**: 50–69.

Figure 14.30 from Deininger, P.L., and Batzer, M.A. (1999). Alu repeats and human disease. *Molecular Genetics and Metabolism* **67**: 183–193.

Figure 14.32 from Beauregard, A., Curcio, M.J., and Belfort, M. (2008). The Take and Give Between Retrotransposable Elements and their Hosts. *Annual Review of Genetics* **42**: 587–617.

Figure 14.33b from Halling, S.M., Simons, R.W., Way, J.C., Walsh, R.B., and Kleckner, N. (1982). DNA sequence organization of IS*10*-right of Tn*10* and comparison with IS*10*-left. *Proceedings of the National Academies of Science U S A* **79**: 2608–2612.

Figure 14.37 adapted from Brennecke, J., Aravin, A., Stark, A., Dus, M., Kellis, M., Sachidanandam, R., Hannon, G. (2007). Discrete Small RNA-Generating Loci as Master Regulators of Transposon Activity in *Drosophila*. *Cell* **128**: 1089–1103.

Figure 14.42 adapted from Grindley, N.D.F., Whihteson, K.L., and Rice, P.A. (2006). Mechanisms of site-specific recombination. *Annual Review of Biochemistry* **75**: 567–605.

Figure 14.45 and 14.47 from Li *et al.* (2005). Structure of a synaptic γδ-resolvase tetramer covalently linked to two cleaved DNAs. *Science* **309**: 1210–1215.

Figure 14.48 and Figure 14.49 adapted from Chen, Y., and Rice, P.A. (2003). New insight into site-specific recombination from FLP recombinase-DNA structures. *Annual Review of Biophysics and Biomolecular Structures* **32**: 135–159.

Figure 14.50a from Ghosh, K., Guo, F., and Van Duyne, G. (2007). DNA: replication, repair, recombination, and chromosome dynamics: *Journal of Biological Chemistry* **282**: 24004–240016.

Figure 14.50b from Gopaul, D.N., Guo, F., and Van Duyne (1998). Structure of the Holliday junction intermediate in Cre-loxP site-specific recombination. *EMBO Journal* **17**: 4175–4187.

Figure 14.51 adapted from Val M.-E., Kennedy S.P., El Karoui M., Bonné L., Chevalier F., et al. (2008) FtsK-Dependent Dimer Resolution on Multiple Chromosomes in the Pathogen *Vibrio cholerae*. PLoS Genet 4(9): e1000201. doi:10.1371/journal.pgen.1000201.

Figure 14.53 adapted from Grindley, N.D.F., Whiteson, K.L., and Rice, P.A. (2006). Mechanisms of Site-Specific Recombination. *Annual Review of Biochemistry* **75**: 567–605.

Figure 14.55 from van Duyne, G.D. (2005). Lambda integrase: armed for recombination. *Current Biology* **15**: R658–660.

Chapter 15

Figure 15.1 adapted from Gregory, R.T., (2005). Synergy between sequence and size in large-scale genomics. *Nature Reviews Genetics* **6**: 699–708.

Figure 15.3 a from Boone, C., Bussey, H., and Andrews, B. (2007). Exploring genetic interactions and networks with yeast. *Nature Reviews Genetics* **8**: 437–449.

Figure 15.6 from Breier, A.M., and Grossman, A.D. (2009). Dynamic association of the replication initiator and transcription factor DnaA with the *Bacillus subtilis* chromosome during replication stress. *Journal of Bacteriology* **191**: 486–493.

Figure 15.11 from Santani, A., Raudsepp, T., and Chowdhary, B.P. (2002). Interstitial telomeric sites and NORs in Hartmann's zebra (*Equus zebra hartmannae*) chromosomes. *Chromosome Research* **10**: 527–534.

Figure 15.14 from Turner, J.M.A., Mahadevaiah, S.K., Fernandez-Capetillo, O., Nussenzweig, A., Xu, X., Deng, C-X., and Burgoyne, P.S. (2004). Silencing of unsynapsed meiotic chromosomes in the mouse. *Nature Genetics* **37**: 41–47.

Figure 15.20 from Waterston *et al.* (2002). Initial sequencing and comparative analysis of the mouse genome. *Nature* **420**: 520–562.

Figure 15.27 from Aceland *et al.*, *Proceedings of the National Academy of Sciences of the U S A* 1998. **95**: 3048–3053.

Figure 15.28 from Sutter *et al.* (2007). A Single *IGF1* Allele Is a Major Determinant of Small Size in Dogs. *Science* **316**: 112–115.

Chapter 16

Figure 16.21, 16.22, 16.23, 16.24 from Meneely, P. (2009). *Advanced Genetic Analysis: Genes, Genomes, and Networks in Eukaryotes.* Oxford: Oxford University Press.

Figure 16.26 from Harm, W. (1980). *Biological Effects of Ultraviolet Radiation.* Cambridge: Cambridge University Press.

Figure 16.37 from Karr, T.L. (2008). Application of proteomics to ecology and population biology. *Heredity* **100**: 200–206.

Figure 16.38a from Barrick, J.E., Sudarsan, N., Weinberg, Z., Ruzzo, W.L., and Breaker, R.R. (2005). 6S RNA is a widespread regulator of eubacterial RNA polymerase that resembles an open promoter. *RNA* **11**: 774–784.

Figure 16.38b from Bousquet-Antonelli, C., Henry, Y., Gélugne, J-P., Caizergues-Ferrer, M., and Kiss, T. (1997). A small nucleolar RNP protein is required for pseudouridylation of eukaryotic ribosomal RNAs. *EMBO Journal* **16**: 4770–4776.

Figure 16.41 from Shevchenko, A., Wilm, M., Vorm, O., and Mann, M. (1996). Mass Spectrometric Sequencing of Proteins from Silver-Stained Polyacrylamide Gels. *Analytical Chemistry* **68**: 850–858.

Figure 16.44 adapted from Meisner and Johnson, *Methods* 2008; **45**: 133–141.

Figure 16.46 from MacVille, M., Schrock E., Padilla-Nash, H., Keck, C., Ghadimi, B.M., Zimonjic, D., Popescu, N., and Ried, T. (1999). Comprehensive and definitive molecular cytogenetic characterization of HeLa cells by spectral karyotyping. *Cancer Research* **59**:141–150.

Figure 16.47 from Shinawi, M., and Cheung, S.W. (2008). *Drug Discovery Today* **13**: 760.

Figure 16.49a from Opdyke, J. A., Kang, J.-G., and Storz, G. (2004). GadY, a small-RNA regulator of acid response genes in *Escherichia coli. Journal of Bacteriology.* **186**: 6698-6705.

Figure 16.50 adapted from Spellman, P.T., Sherlock, G., Zhang, M.Q., Iyer, V.R., Anders, K., Eisen, M.B., Brown, P.O., Botstein, D., and Futcher, B. (1998). Comprehensive identification of cell cycle-regulated genes of the yeast *Saccharomyces cerevisiae* by microarray hybridization. *Molecular Biology of the Cell* **9**: 3273–3297.

Figure 16.51 from Brodigan, T.M., Liu, J., Park, M. Kipreos, E.T., and Krause, M. (2003). Cyclin E expression during development in *C. elegans*. *Developmental Biology* **254**: 102–115.

Figure 16.57b adapted from Tartaglia, L.A., Gimeno, C.J., Storz, G., and Ames, B. (1992) Multidegenerate DNA recognition by the OxyR transcriptional regulator. *Journal of Biological Chemistry* **267**: 2038–2045.

Figure 16.58 adapted from Labib, K., Kearsey, S.E., and Diffley, J.F.X. (2001). MCM2-7 Proteins Are Essential Components of Prereplicative Complexes that Accumulate Cooperatively in the Nucleus during G1-phase and Are Required to Establish, But Not Maintain, the S-phase Checkpoint. *Molecular Biology of the Cell* **12**: 3658–3667.

Figure 16.60 from Cooper, M. (2002). Optical biosensors in drug discovery. *Nature Reviews Drug Discovery* **1**: 515–528.

Figure 16.61 from Elowitz, M.B., Levine, A.J., Siggia, E.D., and Swain, P.S. (2002*).* Stochastic Gene Expression in a Single Cell. *Science* **297**: 1183.

Figure 16.62a from Inoué, S., and Oldenbourg, R. (1998). Microtubule dynamics in mitotic spindle displayed by polarized light microscopy. *Molecular Biology of the Cell*, **9**:1603–1607.

Figure 16.64b from http://www.tsienlab.ucsd.edu/images.htm

Figure 16.64c reproduced with permission from Franz, C, Askjaer, P., Antonin, W., *et al The EMBO Journal* **24**: 3519–3531 (29 September 2005) doi 10. 1038/sj. emboj. 7600825

Figure 16.65 adapted from Damelin, M., and Silver, P.A. (2002). In situ analysis of spatial relationships between proteins of the nuclear pore complex. *Biophysical Journal* **83**: 3626–3636.

Figure 16.68 adapted from Hoog, J.L., Schwartz, C., Noon, A.T., O'Toole, E.T. Mastronarde, D.N., McIntosh, J.R., and Antony, C. (2007). Organization of interphase microtubules in fission yeast analyzed by electron tomography. *Developmental Cell* **12**: 349–361.

Figure 16.69 adapted from Ushiki, T., and Hoshi, O. (2008). Atomic force microscopy for imaging human metaphase chromosomes. *Chromosome Research* **16**: 383–396.

Figure 16.71b from Hickman, A.B., Perez, Z.N., Zhou, L., Musingarimi, P., Ghirlando, R., Hinshaw, J., Craig, N.L., and Dyda, F. (2005). Molecular architecture of a eukaryotic DNA transposase. *Nature Structural and Molecular Biology* **12**: 715–721.

Figure 16.73a from Frank, J. (2002). Single particle imaging of macromolecules by cryo-electron microscopy. *Annual Review of Biophysics and Biomolecular Structure* **31**: 303–319.

Figure 16.73b from Schaffitzel, C., Oswald, M., Berger, I., Ishikawa, T., Abrahams, J.P., Koerten, H.K., Koning, R.I., and Ban, N. (2006). Structure of the *E. coli* signal recognition particle bound to a translating ribosome. *Nature* **444**: 503–506.

Figure 16.75 and 16.76 from Mardis, E.R. (2008). Next-generation DNA sequencing methods. *Annual Review of Genomics and Human Genetics* **9**: 387–402.

Figure 16.79 adapted from Strachan and Read (2004). *Human Molecular Genetics*, 3rd edition, Garland Science.

Figure 16.82 from Freedman, M.L., *et al.* (2006). Admixture mapping identifies 8q24 as a prostate cancer risk locus in African-American men. *Proceedings of the National Academies of Science USA* **103**: 14068–14073.

While we have made every endeavour to seek permission from copyright holders, some omissions may have occurred. If so, please contact the publisher, who will be glad to rectify any such omission at the earliest possible opportunity.

CONTENTS IN OVERVIEW

CONTENTS IN FULL

ABBREVIATIONS

4E-BP	eIF4E-binding protein
A	adenosine
AAA+	ATPases associated with a variety of cellular activities
AAV	adeno-associated virus
aCGH	array comparative genomic hybridization
ADAR	adenosine deaminase that acts on RNA
ADP	adenosine diphosphate
AFM	atomic force microscopy
ALT	alternative lengthening of telomeres
AMP	adenosine monophosphate
APC	anaphase-promoting complex
ARE	AU-rich element
Asn-tRNA^{Asn}	asparagine-bound tRNA species
ATM	ataxia telangiectasia mutated
ATP	adenosine triphosphate
ATR	ATM related
ATRIP	ATR-interacting protein
attB	*attachment site bacteria*
attL	*attachment site left*
attP	*attachment site phage*
attR	*attachment site right*
BAC	bacterial artificial chromosome
bHLH	basic region-helix-loop-helix
BIR	break-induced replication
BLAST	Basic Local Alignment Sequence Tool
bp	base pairs
BRCA	breast cancer (protein)
bZIP	basic region-leucine zipper
C	cytidine
C	carbon
Ca	calcium
CAK	Cdk-activating kinase
CAK	Cdk-activating kinase
cAMP	cyclic adenosine monophosphate
CAP	catabolite activator protein
CAR	cohesin-associated region
CBP	CREB-binding protein
CDE	centromere DNA element
Cdk	cyclin-dependent kinase
CENP-A	centromere protein-A
CFP	cyan fluorescent protein
ChIP	chromatin immunoprecipitation
CKI	cyclin kinase inhibitor
CMCT	1-cyclohexyl-(2-morpholinoethylcarbodiimide metho-p-toluene sulfonate
CNV	copy number variant
CPE	cytoplasmic polyadenylation element

CRP	cAMP receptor protein
CsCl	cesium chloride
CSSR	conservative site-specific recombination
CTD	C-terminal domain
CTP	cytidine triphosphate
D	diversity (segement)
Da	dalton
DAPI	4′-6′-diamidino-2-phenylindole
DARS	DnaA reactivating sequences
datA	DnaA titrator
DIC	differential interference contrast
DNA	deoxyribonucleic acid
DPE	downstream promoter element
dsRBD	double-stranded RNA-binding domain
ECS	editing site complementary sequence
eEFsec	C-terminal extension
EFG	elongation factor
EGF	epidermal growth factor
eIF2	eukaryotic initiator
EJC	exon junction complex
EMS	ethyl methanesulfonate
EMSA	electrophoretic mobility shift assay
ENU	ethylnitrosourea
ES	embryonic stem (cells)
ESE	exonic splicing enhancer
FACT	facilitates chromatin transcription
FAD	flavin adenine dinucleotide
FIS	factor for inversion stimulation
FISH	fluorescence *in situ* hybridization
FLIP	fluorescence loss in photobleaching
FMN	flavin mononucleotide
FRAP	fluorescence recovery after photobleaching
FRET	fluorescence resonance energy transfer
G	guanosine
G0	G zero
G1	gap phase 1
G2	gap phase 2
GAP	GTPase-activating protein
GC	gas chromatography
GDP	guanosine diphosphate
GEF	guanine-nucleotide exchange factor
GFP	green fluorescent protein
GGR	global genomic repair
Gln-tRNA^{Gln}	glutamine-bound tRNA species
GlyRS	glycyl-tRNA synthetase
GST	glutathione-S-transferase
GTP	guanosine triphosphate
GWAS	genome-wide association studies
H	hydrogen
HAT	histone acetyltransferase
HDAC	histone deacetylase

hGH	human growth hormone
HisRS	histidyl-tRNA synthetase
HIV	human immunodeficiency virus
HLA	human leukocyte antigen
HMT	histone methyltransferase
HPLC	high pressure or high performance liquid chromatography
HSF	heat-shock factor
IEF gel	isoelectric focusing
IF	initiation factor
IGF	insulin-like growth factor
IHF	integration host factor
IleRS	isoleucyl-tRNA synthetase
INR	initiator element
Int	integrase
iPS	induced pluripotent stem
IRE	iron response element
IRES	internal ribosome entry site
IRIF	ionizing-radiation–induced foci
IRP	iron-response element binding protein
IS	insertion sequence
ISE	intronic splicing enhancer
ITAF	IRES-transacting factors
J	joining
K_a	equilibrium association constant
K_d	equilibrium dissociation constant
K_{eq}	equilibrium constant
KH	Hn RNP K homology
LC	liquid chromatography
LINE	long interspersed element
LNA	locked nucleic acid
LOD	log of the odds
LTR	long terminal repeat
LUCA	last universal common ancestor
M	mitotic
MALDI	matrix-assisted laser desorption/ionization,
MASV	mobile element associated structural variant
Mbp	million base pairs
miRNA	microRNA
MRN	MRE II-RAD50-NBS1
mRNA	messenger RNA
MRX	Mrell-Rad50-Xrs2
MuLV	murine leukemia virus
N	nitrogen
NaCl	sodium chloride
NAD	nicotinamide adenosine dinucleotide
NADP	nicotinamide adenosine dinucleotide phosphate
NER	nucleotide excision repair
NextGen	Next Generation
NF-κB	nuclear factor-κB
NHEJ	non-homologous end joining
NMD	nonsense-mediated mRNA decay

NMD	nonsense mediated decay
NMR	nuclear magnetic resonance spectroscopy
NOR	nucleolar organizer region
NSD	non-stop decay
NTD	N-terminal domain
NURF	nucleosome-remodeling factor
O	oxygen
OMIM	Online Mendelian Inheritance in Man
ORC	origin recognition complex
ORF	open reading frame
ori	origins of replication
PABP	poly(A)-binding protein
PAGE	polyacrylamide gel electrophoresis
PALB2	partner and localizer of BRCA2
PAZ	PIWI/Argonaute/Zwille
PCAF	CBP-associated factor
PCNA	proliferative cell nuclear antigen
PCR	polymerase chain reaction
PIKK	phosphoinositide-3-kinase–related protein kinase
piRNAs	Piwi-interacting RNAs
PITC	phenylisothiocyanate
PNPase	polynucleotide phosphorylase
poly(A)	polyadenosine
pre-RNA	precursor RNA
PSF	polyadenylation specificity factor
PTB	polypyrimidine tract binding protein
PTC	peptidyl transferase center
PTC	premature termination codons
PTGS	post-transcriptional gene silencing
qRT-PCR	quantitative reverse transcription-polymerase chain reaction
RACE	rapid amplification of complementary DNA ends
RBD	RNA-binding domain
rDNA	ribosomal RNA
RdRp	RNA-dependent RNA polymerase
RF	release factor
RFC	replication factor C
RIDA	regulatory inactivation of DnaA
RISC	RNA-induced silencing complex
RNA	ribonucleic acid
RNAi	RNA interference
RNP	ribonucleoprotein
RPA	replication protein A
RRM	RNA recognition motif
RSS	recombination signal sequence
rut	*r*ho *ut*ilization
SDS	sodium dodecyl sulfate
SDSA	synthesis-dependent strand annealing pathway
SDS-PAGE	sodium dodecyl sulfate-polyacrylamide gel electrophoresis
SECIS	selenocysteine insertion sequence
SEM	scanning electron microscope
SepRS	specialized class II synthetase

SILAC	stable isotope labeling with amino acids in culture
SINE	short interspersed element
siRNA	small inhibitory RNA
SKY	spectral karyotyping
SL RNA	spliced leader RNA
SMC	structural maintenance of chromatin
snoRNA	small nucleolar RNA
SNP	single nucleotide polymorphism
snRNP	small nuclear ribonucleoprotein
SPR	surface plasmon resonance
SRF	serum response factor
SSB protein	single-stranded binding protein
SSLP	simple sequence length polymorphism
STS	sequence tagged sites
T	thymidine
TAF	TBP-associated factor
TBP	TATA binding protein
TCR	transcription-coupled repair
TEM	transmission electron microscopy
ter	terminus region
TERT	telomerase reverse transcriptase
TIR	Terminal Inverted Repeat
TLC	thin-layer chromatography
TPRT	target-primed reverse transcription
tRNA	transfer RNA
U	uridine
UAS_G	Upstream Activating Sequence for Gal
UCE	upstream control element
uORF	upstream open reading frame
UTR	untranslated region
UV	ultraviolet
V	variable
YAC	yeast artificial chromosome
YFP	yellow fluorescent protein
γ-TuRC	γ-tubulin ring complex

GENOMES AND THE FLOW OF BIOLOGICAL INFORMATION

INTRODUCTION

The biological world is a maze of interconnections on a multitude of scales: atoms join to form molecules – molecules cluster to form cells – cells interact to form tissues – and tissues aggregate to form an organism. The interplay continues beyond the organism, as organisms form populations, populations inhabit ecosystems, and ecosystems unite to form the world around us.

Through all the layers of connectivity runs a common thread: the communication and onward passage of information – from cell to cell, from organism to organism and, ultimately, from generation to generation. This information is stored, at the most fundamental level, in each living cell in our body. But how does this information – no more than a static repository of data – come alive to direct the processes that constitute life?

The answer lies in the concerted action of molecular components, which cooperate in a series of ingenious processes to bring the information deposited in each of us, in our **genome**, to life. These components and processes lie at the heart of molecular biology, and are the key players in the chapters that follow.

Before embarking on our journey to discover these components and processes, however, we begin by considering some of the fundamental biological themes on which the study of molecular biology relies. You may be familiar with some of these themes, and the concepts and ideas introduced, from earlier studies. If so, look on this chapter as a refresher before later chapters lead you into more unfamiliar – but ultimately fascinating – topics.

1.1 THE ROOTS OF BIOLOGY

The diversity of life is unified by some common themes

Life on earth is remarkably diverse, ranging from the relatively simple unicellular organisms, such as bacteria, to more complex multicellular organisms, such as plants and humans. Despite their diversity at the macroscopic level, where distinctive features of organisms are readily apparent with the naked eye (Figure 1.1), the core molecular features of all organisms are remarkably similar. Indeed, we can think of life on Earth as being unified by a number of key themes.

- First, a living organism must be distinct in some way from its environment, that is, it must be defined by a physical barrier, which serves to separate 'organism' from 'environment'. This separation allows the internal environment of an organism to be carefully regulated, and ultimately distinguishes 'self' from 'non-self'.

- Second, a living organism must be able to store information in a stable way, and also have a way of using this information to determine its characteristic features – its structural composition, how it functions, and so on.

- Third, a living organism must be able to reliably replicate and pass information from one generation to the next. **Replication** lies not only at the heart of the propagation of a **species** from generation to generation, but also at the heart of the growth and repair of a multicellular organism within a single generation. Without replication, not only could an organism not reproduce to yield

Figure 1.1 The life forms on earth are diverse and interconnected. (a) A monarch butterfly on a milkweed plant (Louise Murray/Science Photo Library), (b) bacteria on the scales of an insect, (c) a California Alligator Lizard carrying multiple ticks (Dante Fenolio/Science Photo Library), (d) a clownfish surrounded by the tentacles of a sea anemone (Andrew J. Martinez/Science Photo Library), (e) an oxpecker bird sitting on an African buffalo (Dr P. Marazzi/Science Photo Library), and (f) a fungus on a tree (Ron Bass/Science Photo Library).

offspring, but a multicellular organism could not develop from an embryonic state to a fully formed individual. The processes of replication and transmission must remain faithful to the entity being replicated, just as a recipe – itself an information store – ensures that a given dish can be prepared in a consistent way time after time.

- Finally, living organisms require a source of energy from their surroundings to grow and reproduce; this energy is used to drive the biological processes that keep an organism alive. There are substantial differences in how energy is harvested by various organisms, for example, plants get their energy from the sun whereas humans get energy from food. Nevertheless, the core molecular mechanisms used by cells in all organisms to grow and propagate are remarkably similar (that is, they are highly conserved). For example, the same metabolic cycles are used in all organisms to transform sugars into the energy currency of the cell, adenosine triphosphate (ATP).

In modern-day organisms, the physical barrier that separates individuals from their surroundings (and so defines 'self') is a lipid-based membrane, while a molecular species known as **nucleic acid** stores the biological information, and allows for ready replication of this information (thus conferring 'self' identity). These two basic molecular features come together – the lipid membrane surrounding the nucleic acid – to form the cell, which is the building block of organismal life (Figure 1.2). Some organisms are composed of a single cell (unicellular) whereas others are formed of many cells (multicellular). However, the cell is not the only fundamental building block of life; common building blocks exist at the molecular level too.

Figure 1.2 All cells share the same basic components. False-colored scanning electron micrograph showing chains of *Streptococcus* bacteria (orange) and two human cells: a macrophage (gray) and a lymphocyte (pink). Both of the human cells are part of the immune response, and act together to eliminate the bacteria. Despite their different sizes, all three types of cell share the basic components of a cell: a membrane that isolates each individual cell; a genome that is the 'blueprint' of each cell; a means to divide and reproduce; and a means by which to utilize energy.
Electron micrograph courtesy of James A Sullivan.

Living organisms are constructed from common molecular building blocks

It is thought that the entire diversity of modern-day life derives from a common original life form often referred to as the last universal common ancestor (LUCA) or **progenote**, whose molecular components have been conserved from generation to generation, and from species to species, as different organisms have evolved. In support of this, when we compare modern organisms, we find that the core building blocks of all organisms are the same.

The four basic classes of molecules from which living organisms are constructed are (Figure 1.3):

- nucleic acids
- proteins
- lipids
- carbohydrates.

These molecules reflect a further unifying theme – of large biological molecules being constructed from smaller, repeated subunits.

Nucleic acids are the informational molecules and typically have a simple, linear structure composed of the monomer building blocks (the nucleotides) that are the instructions for life, and thus for the propagation of life. The information stored in

Figure 1.3 All cells are composed of four basic classes of molecules. (a) Nucleic acids (DNA and RNA) store information and serve as the interpreters of these instructions for life. (b) Proteins carry out most of the structural and functional tasks. (PDB 1A00). (c) 'Water-fearing' lipids form the membranes that comprise the physical barrier to the outside of the cell. (d) Carbohydrates provide energy and often adorn extracellular proteins.

these molecules directs the synthesis of proteins (a 'coding' role), and, in so doing, specifies the structure and function of the cell, as noted below. In addition, nucleic acids play essential non-coding roles within the cell as functional components of molecular machines and by regulating how the information they contain is used.

Proteins are also linear polymers, composed of distinct building blocks known as **amino acids** that typically fold into complex three-dimensional structures. Proteins are often referred to as the workhorses of the cell, carrying out most of the critical structural and functional tasks.

Lipids are 'water-fearing' molecules that self-assemble to form the membranes that present a physical barrier to the exterior milieu, thus defining self. Lipids are also constructed from repeated subunits – multiple fatty acid molecules join together to form the lipid membrane.

Carbohydrates are 'water-loving' molecules that play a wide range of roles in the cell, from providing energy to increasing the solubility of otherwise insoluble proteins. Carbohydrates continue to reflect the theme of large molecules being composed of smaller subunits: they are polymeric molecules constructed from simple sugar monomers.

➔ We learn more about the structure and chemical nature of these four classes of biological molecules in Chapter 2.

From these four components are built the macromolecular machines that are responsible for all cellular transactions – from copying the biological information stored in nucleic acids, to producing functional proteins, to harvesting energy from the environment.

The flow of biological information supports the maintenance of life

The title of this chapter refers to the flow of biological information – a phenomenon that is made possible through the components and processes that comprise 'molecular biology'. As we have mentioned above, storage and passage of information is one of the unifying characteristics of living organisms. But why is the flow of biological information necessary? Ultimately, a flow of information is needed to ensure that the offspring of an organism remains faithful to its parents in terms of the characteristics it exhibits. In other words, a parent must somehow act as the blueprint for its offspring. Without such a flow of information, a cat could not reliably give birth to another cat, or a human could not reliably give birth to another organism that was recognizably human.

The maintenance of life relies on the process of reproduction – both in the context of the growth of a specific tissue through rounds of cell division, and in the development of new offspring from a parent. In both cases, for the offspring to be a faithful reproduction of its parents, there must be a means of storing information (and using it), and a mechanism through which this information can be copied and inherited by the next generation.

Heritable information is stored in a simple polymer of monomer building blocks

Information in all cells is stored in a repeating polymer structure called deoxyribonucleic acid (DNA), which is easily copied and transmitted to subsequent generations through processes we will explore later in the book. DNA is a type of nucleic acid, and is composed of four building blocks known as **nucleotides**: guanosine (G), adenosine (A), thymidine (T), and cytidine (C). These nucleotides are strung together to form an extended linear polymer. The resulting strings of letters specify our genetic identity: they encode all functional molecules that must be synthesized for the cell and ultimately the organism to survive and propagate.

A particularly important chemical feature of these four distinct nucleotides is that the faces of A and T fit together, as do those of G and C, that is their shapes are complementary. When two different nucleic acids have complementary sequences – that is, where A on one strand is mirrored by T on another, and where C on one strand is mirrored by G on the other – the close association of these pairs of nucleotides allows the two sequences to form extended double-stranded structures with a characteristic double-helical conformation (Figure 1.4). This double-helical structure is the basic form in which the genetic information is stored.

Each DNA strand in a double helix has directionality: one end of the strand is denoted 3′ and the o ther 5′. When two complementary strands come together they do so in an antiparallel fashion – that is, the 3′ end of one strand aligns with the 5′ end of the other, and vice versa. (In essence, the two strands lie 'head to toe'.)

Duplication of the information stored in DNA is facilitated by the simple complementary shape correspondence – a complementary, fully informational strand can be made by separating the two strands and using the sequence of nucleotides in each of the two original strands to direct the synthesis of two new strands, as depicted in Figure 1.5. Such strand replication follows the same rules of complementarity described above: an A in one of the original strands directs the addition of a T in a new strand (and vice versa), and a C in the original strand directs the addition of a G in the new strand (and vice versa). It is important to

Figure 1.4 The DNA double helix is a repeating polymer that stores information in cells. DNA is composed of the nucleotide building blocks guanosine (G), adenosine (A), thymidine (T), and cytidine (C). The distinct feature that G pairs with C and A pairs with T allows a strand of DNA to be copied.

original 'parental' double strand

two doubled-stranded daughter strands are produced, each a full copy of the original parental strand

Figure 1.5 The process of replication is common to all organisms. The complementary nature of the two DNA strands allows the DNA to be copied.

note the word 'complementary' above: when a nucleic acid strand is replicated, the strand produced as a result is not identical to the original strand, but is complementary to it.

DNA is a very stable nucleic acid, making it the nearly ubiquitous material for the storage of information in modern biology. However, the chemically similar but less stable nucleic acid, ribonucleic acid (RNA), is occasionally used. (This does not mean that RNA is a mere bit-player on the biological stage: we will see throughout this book the myriad functional roles of RNA in the workings of an organism.)

1.2 THE GENOME: A WORKING BLUEPRINT FOR LIFE

A linear sequence of nucleotides in DNA comprises the genetic makeup of an organism, called the **genome** of an organism. This repository stores all information needed to specify cellular function and can be considered a blueprint for life. In the past decade, researchers across the world have determined the complete genome sequence of numerous organisms, from thousands of bacterial species to the human genome sequence. These approaches have generated great excitement in all areas of biology and medicine as analysis of these sequences has greatly increased our understanding of function and evolution.

Transmission and maintenance of the genome is essential for life

As we note above, reproduction by all organisms depends on their ability to make a copy of their blueprints for life to hand down to their progeny. At the cellular level, the copying of the genetic information (replication), the separation (or '**segregation**') of the copies, and their subsequent transfer into the cells of the next generation is required even if the recipient is to remain part of the same organism – for example, a new skin cell, or muscle cell. We refer to this copying and segregation process as the transmission of the genome. In molecular terms, this means that the nucleic acid, typically DNA, must be copied in its entirety, without many mistakes being made, in order to maintain the integrity of the genome – that is, the accuracy of the information contained within it. The process of segregation must also happen correctly, so that each daughter cell receives a full complement of DNA.

The molecular processes involved in transmission and maintenance of the genome are highly conserved from bacteria to human beings. DNA is first copied or replicated by core machinery that separates the complementary strands of the double helix and simultaneously makes copies of each individual strand. The molecular machine, or enzyme, responsible for synthesizing the new DNA copies is known as **DNA polymerase**, and as we shall see, is very highly conserved. Following this DNA replication process, the duplicated copies of the DNA are brought together to ensure that when the actual physical division of the cell occurs, the genome will be accurately partitioned (or segregated) between the two progeny cells (Figure 1.6). Throughout each of these processes, there are elaborate mechanisms in place to ensure that replication and segregation are extremely accurate.

The genome is composed of genes and intergenic regions

The genome – the total DNA content of the cell – contains the instructions for cellular life. Within the genome, much of the critical information is found in discrete regions referred to as **genes**. A gene is typically defined as a region of DNA that controls a discrete, hereditary characteristic, and as such usually specifies the production of a functional product (a protein or RNA molecule). Different types of sequence are typically found within a given gene: some of the sequence directly encodes the product of the gene, whereas other sequences within a gene are important for specifying how and when to make the product. These regions are often referred to as coding and regulatory regions, respectively. Genes can vary in length from fewer than 100 nucleotides for certain functional RNAs, for example, to 2.4 million nucleotides for the human dystrophin gene (which is defective in muscular dystrophies).

It is worth noting that we give a 'typical' definition of a gene above. In reality, giving a clear-cut definition for a gene – something that has never been as straightforward as it might appear – is becoming increasingly difficult. Why is this? As more genomes are sequenced and studied in detail, we are finding that far more of the genome appears to be expressed (that is, actively 'used' in some way) than previously anticipated. So, we may need to broaden our view of what a 'gene' really is to reflect the ever-expanding proportion of the genome that is active – beyond those areas that simply code for a single functional product.

Going further, recent innovations are allowing for breathtaking amounts of sequence information to be rapidly generated and this new information coupled with functional analysis is changing our view of biology. We are now asking questions such as 'How discrete an element is a gene?' and 'Are all RNA transcripts worthy of being called a gene?' It is an exciting time to be a molecular biologist and this book sets the stage for understanding new discoveries and how they impact our understanding of evolution and biology as a whole.

The genome is physically organized into chromosomes

The large amount of DNA that constitutes the genome must be packaged in an organized way that facilitates the various key events in the life of a cell, including the replication of the DNA and the expression of the genes. The individual genes are arranged in a linear array, either densely or less densely packed with intergenic DNA spacers (segments of DNA between the genes themselves). All of this genomic DNA is packaged into one or more functional units called **chromosomes**. Some organisms have multiple, independent linear chromosomes, while others have a single, circular chromosome.

In eukaryotic chromosomes, the DNA is wrapped tightly around packing proteins (referred to as histones) to assume a highly ordered three-dimensional structure (Figure 1.7). This highly compacted form of the DNA is optimal for storage as it is

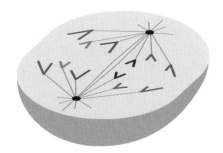

Figure 1.6 Replicated chromosomes are partitioned into two progeny cells by a process called segregation. To ensure accurate partitioning, the replicated chromosomes (having the same shape and color in this figure) are first brought together before they are pulled away into the progeny cells.

➲ We learn about replication in more detail in Chapter 6 and about chromosomal segregation in Chapter 7.

➲ Chromosomal features are the focus of Chapter 4.

DNA double helix

histone

chromosome

Figure 1.7 DNA is packaged into chromosomes. The double-stranded DNA strands are wound around packing proteins (histones in eukaryotes) and then further wound to assume a highly ordered and compact three-dimensional structure.

very space efficient. However, highly compacted DNA does not facilitate information retrieval. Instead, specialized mechanisms drive the opening of this tightly wound package to allow access to the DNA strands by various molecular components as needed to mediate gene expression. In addition to chromosomes, some cells have **plasmids**, which are extra-chromosomal pieces of DNA that can also carry important genetic information.

Given that the genome contains all of the genes of an organism, how then are the genes assembled? Are the genes placed in a particular order or in a particular location on the chromosome? Are they densely or sparsely distributed? What can we learn about an organism by examining overall DNA composition and structure? As we shall see, the overall composition of genomes varies widely across biology. For example, in small genomes, such as a typical bacterium (*Escherichia coli* has 4.6 million base pairs (Mbp) of DNA), the genes are rather tightly packed together, sometimes even overlapping one another to minimize overall genome size. By contrast, the human genome comprises 3200 Mbp of DNA and is composed predominantly (98%) of intergenic regions that do not appear to contain gene units. The composition of these intergenic regions is not uniform nor is the distribution of genes throughout the genome. As more genomes have been sequenced, and more studies of **gene expression** are performed, it has become very clear that we still have much to learn about the impact of overall genome structure on the fitness of the organism.

Gene number and arrangement vary greatly among organisms

The number of genes in an organism ranges widely – from 500 to 800 in some bacteria to a grossly estimated 30 000 in humans. Although in general, single-celled bacteria have fewer genes than multicellular organisms, as shown in Figure 1.8, there is no simple correlation between genome size, gene number, and organism complexity. As an example, although at the organismal level, zebrafish and lungfish appear to be very similar, the genome sizes are substantially different: the zebrafish has a genome of 2000 Mbp while that of the lungfish is 65 times larger with 130 000 Mbp. Furthermore, the nematode *Caenorhabditis elegans* and the fruit fly *Drosophila melanogaster* are certainly less complicated than humans, but possess only slightly fewer genes (around 20 000 in *C. elegans*, and around 15 000 in *D. melanogaster*). What are some potential explanations for this conundrum?

First, it is clear that the number of different products (RNAs or proteins) that can be generated from a given gene is large. So, even though the genome of a given organism may feature a seemingly modest number of genes, the same genome may, in fact, encode a huge range of different functional products – more, perhaps, than a different organism that possesses a larger number of genes.

Further, it is clear that the density of gene packaging varies widely from organism to organism, such that some genomes feature relatively more intergenic regions than others. But why is this important? Intergenic regions were previously thought to be little more than 'junk', acting solely as 'spacers' between genes. However, we now think that, far from being junk, many intergenic regions play critical roles in regulating gene expression. For example, we shall learn in Chapter 14 that much

Approximate genome size and gene number for representative organisms		
element	gene number	genome size
Mycoplasma genitalium	~480 genes	580 000 bp
Escherichia coli	4600 genes	4 600 000 bp
Saccharomyces cerevisiae	~ 6000 genes	12 100 000 bp
Schizosaccharomyces pombe	~ 5000 genes	14 000 000 bp
worm (*Caenorhabditis elegans*)	~23 000 genes	98 000 000 bp
fruit fly (*Drosophila melanogaster*)	~27 000 genes	130 000 000 bp
duckweed (*Arabidopsis thaliana*)	~29 000 genes	157 000 000 bp
zebrafish (*Danio rerio*)	~13 000 genes	2 000 000 000 bp
human (*Homo sapiens*)	~32 000 genes	3 200 000 000 bp
marbled lungfish (*Protopterus aethiopicus*)		130 000 000 000 bp
amoeba (*Amoeba dubia*)		670 000 000 000 bp

Figure 1.8 Approximate gene number and genome size for representative organisms.

of the DNA of certain organisms is composed of **transposable elements** that can (or once could) move from place to place within a genome. While this DNA was for some time thought to be functionally unimportant, new information on its critical function in evolution and in gene expression is emerging. Further, despite earlier expectations that a majority of the human genome is functionally silent, recent studies indicate that a significant fraction of the human genome is actively being transcribed.

Therefore, it is becoming clear that the value of the genome, in terms of governing the processes that underpin biological complexity, comes from more than just the genes from which a genome is composed. Instead, it comes from a combination of gene number, the production of multiple gene products from the information carried in a single gene, and the regulatory activity of the genome's intergenic regions.

Molecular conservation is highlighted by DNA sequencing

We noted at the start of this chapter how the diversity of life is unified by some common themes, with commonalities extending to the molecular building blocks used to construct living organisms, and the biological processes through which life is maintained. Direct proof of the remarkable extent of molecular conservation in all cells comes from analysis of the genome of an organism.

With the relatively recent advance of whole-genome sequencing has come new understanding of the extent to which commonalities exist between different organisms, as well as of the incredible complexity of modern-day life. This new understanding has come from the examination and comparison of whole-genome sequences from different organisms, from which similarities in genome sequence have been revealed. Such commonalities in genome sequence correspond, at least in part, to commonalities in the components and processes that operate at the molecular level in the many organisms studied (applying the logic that, if two genomes have the same (or very similar) sequences, then what results from their expression (for example, proteins) must also be similar). The new information revealed by genome sequencing has allowed us to expand on earlier views originally based on systematic analysis of **model organisms**, key reactions, and specific molecular features.

In this book we will focus on one particular area of cellular metabolism – on understanding key molecular mechanisms critical to the storage and copying of genetic information (the genome) and the expression of this information in its functional form.

Throughout the book, emphasis will be placed on the commonalities of these mechanisms in all organisms, but we will also highlight instructive differences. The impact of whole-genome sequencing on our understanding of molecular evolution and function cannot be overstated, and promises to continue to reveal fundamental surprises about both biological commonality and diversity on earth.

1.3 BRINGING GENES TO LIFE: GENE EXPRESSION

We have now seen how genetic information is captured by the sequence of nucleotides in nucleic acids. But how is this information interpreted by the organism in which it is stored? How is it converted from a static sequence of nucleotides into dynamic biological activity, termed gene expression? It is now that we begin our exploration of molecular biology to discover the molecular machinery and processes that bring the genome to life.

Genes encoded by DNA must be 'expressed'

While the blueprint of the cell is encoded in the DNA, the implementation of the blueprint is accomplished through the 'expression' of the genes into their functional forms. The gene products consist of two types of macromolecule, RNA and protein. These two molecules are principal structural and functional components of the cell, and are responsible for reactions that harvest energy, synthesize cellular components, and facilitate the import and export of materials. In short, these molecules implement cellular function.

In order for the individual genes to be expressed, the packets of information need to be transformed into the specified products. It is this transformation that lies at the heart of this book, and is a process we will gradually unravel in the chapters that follow. Genomes are the blueprints for life, but it is through molecular biology – the array of molecular components and processes – that this blueprint is used to direct how life at the level of a given organism proceeds.

The first stage in the expression of all genes is the synthesis of a copy of the DNA region of interest in the form of RNA through a process called **transcription**. Once the RNA copy of the gene has been synthesized, it is either 'translated' into the corresponding protein sequence or used directly, such that the RNA molecule produced by transcription has its own specific biological function, as illustrated in Figure 1.9. Let us now consider each of these processes (transcription and **translation**) in a little more detail.

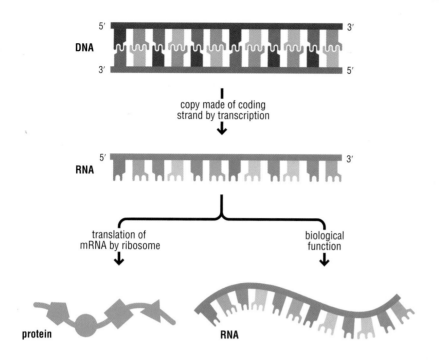

Figure 1.9 Two functions of RNA. RNA copies of DNA can serve as messenger RNAs (mRNAs) to direct the synthesis of a protein (left) or they can have intrinsic functions (right).

Transcription is the first step in gene expression for the production of both non-coding RNA and proteins

The process of copying the DNA sequence of interest into the corresponding RNA is referred to as transcription. The molecular machine that is responsible for carrying out transcription is called **RNA polymerase**, and is closely related to the DNA polymerase that is responsible for replicating the genome. The start point and end point of transcription along a particular DNA template strand are specified by sequences within the DNA template that are interpreted by RNA polymerase as a signal to start or stop.

⊃ The molecular details of the transcription process are described in greater detail in Chapter 8.

So, the RNA polymerase catalyzes the synthesis of an RNA strand by joining together nucleotide monomers (single nucleotides) in a sequence dictated by the DNA template that the RNA polymerase is 'reading'. The process of joining together nucleotides to form a chain-like polymer is an example of a polymerization reaction. How are nucleotides added in the correct sequence to the growing RNA strand by RNA polymerase? The physical basis for the polymerization reaction is the complementarity in shape between a nucleotide on the template DNA and an incoming nucleotide that is destined for the growing RNA chain, as depicted in Figure 1.10. This complementarity mirrors that seen for the process of DNA replication, with one important difference: as in DNA replication, a C on the DNA strand directs the addition of a G on the growing RNA strand and vice versa; a T on the DNA strand directs the addition of an A on the growing RNA strand, but an A on the DNA strand directs the addition of a U (uridine, a nucleotide unique to RNA) on the growing RNA strand.

The initial product of a transcription event is referred to as the primary transcript. For some genes, the final functional gene product is an RNA species derived directly from the primary RNA transcript; such RNA species often possess regulatory functions. For other genes, the primary RNA product is an intermediate that will be used to direct the synthesis of a protein product (called translation, as

Figure 1.10 The process of transcription is common to all organisms. The complementary nature of the two DNA strands allows the DNA to be copied into a complementary strand of RNA. When DNA is copied into RNA, a uridine (rather than a thymidine) is incorporated opposite A.

⊙ The post-transcriptional RNA processing events are described in greater detail in Chapter 9.

calmodulin protein   hemaglutinin protein

Figure 1.11 Proteins vary considerably in shape, size and function. Proteins can range in size from only a few amino acids to thousands of amino acids and have roles in every cellular process, ranging from roles as signaling molecules to receptors and structural scaffolds in the cell. In the examples shown here, the 148 amino acid calmodulin protein (PDB 1UP5) binds calcium (indicated by yellow balls), and the complex formed by the three proteins comprising the hemagglutinin protein (PDB 3HMG) mediates binding of influenza virus to the mammalian host cell and ultimately membrane fusion and entry.

⊙ We explore the features of the genetic code more fully in Chapter 10.

discussed below). These two classes are commonly referred to as non-coding and coding RNAs, respectively; the key distinction is that non-coding RNAs are not translated to give a protein product, whereas coding RNAs are.

This versatility of RNA is thought to have been central to the origin of life, and is still key to the functioning of modern-day cells. These ideas will be discussed more thoroughly in Chapters 2, 9, and 10. For both non-coding and coding RNAs, the primary transcript RNA is typically subjected to a series of events that trim away and modify the RNA in a refinement process broadly referred to as RNA processing, to yield the final RNA molecule.

The RNA sequence is next 'translated' to make functional proteins

Many genes do not encode a functional RNA but instead encode a protein. Proteins are composed of linear strings of 20 distinct building blocks, the amino acids, which are encoded by the sequence of nucleotides comprising the RNA transcript template. Proteins built from this diverse set of building blocks (relative to the four quite similar nucleotide building blocks of DNA and RNA) can form a wide range of structures that in turn perform quite diverse roles in the cell (Figure 1.11).

The nucleic acid code for protein synthesis is composed of the four different nucleotides, which are assembled in long linear arrays that are 'read' three nucleotides at a time to specify which of the 20 different amino acids should be incorporated at a given point in the protein, a process termed translation (Figure 1.12). Proteins are not directly synthesized from the DNA copy of the gene, but are instead synthesized from the transcribed RNA copy of the gene, which acts as an intermediate in the overall synthesis pathway. Such RNAs that function as coding molecules to direct protein synthesis are called messenger RNAs (mRNAs). Distinctive sequences in the mRNA specify the start and stop sites for protein synthesis, while the successive triplet nucleotide sequences in between encode the linear sequence of amino acids that comprise the final protein product.

The term 'translation' makes sense when one understands that this is the process by which the sequence of nucleotides within a nucleic acid is 'translated' into the sequence of amino acids in a protein. The code that correlates triplet nucleotides with the encoded amino acid is referred to as the genetic code (Figure 1.13). The nucleotide triplets are called codons. We see, for example, that the codon UUU is translated as phenylalanine whereas AUG is translated as methionine. All codons dictate a particular outcome, either the incorporation of an amino acid or the halting of translation (a 'stop' site), and so we say that the code is non-ambiguous.

Translation is carried out by an elaborate macromolecular machine known as the **ribosome**, which comprises both RNA and protein molecules. At the core of the process is a functional RNA called transfer RNA (**tRNA**), that acts as a physical link between the coding information (in the form of a triplet nucleotide sequence that reads the codons) and the appropriate amino acid. Figure 1.12 shows how the tRNA functions as the actual decoder that makes translation possible: in essence, a population of tRNAs sequentially 'reads' the codons in a processive march along the mRNA template, recruiting amino acids to the growing protein chain in the order specified by the mRNA template. These incoming amino acids are then chemically linked together as the march proceeds. The ribosome is the structural scaffold that facilitates this molecular march.

Most proteins that emerge from the ribosome are not fully functional. Different portions of the protein product may be removed (cleaved or spliced out) and additional moieties may be added. These processes are collectively referred to as post-translational processing.

The functional repertoire of the cell is much more complex than the genome

A direct consequence of the fact that RNA and protein products can be modified following their initial synthesis, is that the complete set of functional RNAs and proteins is far more extensive than the set that is simply encoded by the genome. While the human genome encodes an estimated number of 30 000 genes, for example, the number of different human gene products is estimated in the range of one million and it is likely that this is an underestimate. Moreover, we are beginning to understand that certain classes of molecule have escaped notice for a variety of reasons, including low expression levels and sizes that are difficult to identify. However, recent technical advances are leading to more systematic approaches to gene identification. There has been a recent explosion in known gene products exemplified by the discovery of several new classes of small, functional RNA. These provide a compelling example of how the overall gene count is likely to remain in flux for the foreseeable future. These issues will be discussed in more detail in Chapter 15. It is the exquisite control of the production of an increasingly large number of cellular components that allows for the incredible complexity that we see in the living world.

Figure 1.12 The process of translation is common to all organisms. The nucleic acids in RNA are read three at a time to specify which amino acid should be incorporated.

⊙ The process of translation is explored in more detail in Chapter 10 and post-translational processing is described in Chapter 11.

1st position (5′ end)	2nd position				3rd position (3′ end)
	U	**C**	**A**	**G**	
U	Phe	Ser	Tyr	Cys	U
	Phe	Ser	Tyr	Cys	C
	Leu	Ser	stop	stop	A
	Leu	Ser	stop	Trp	G
C	Leu	Pro	His	Arg	U
	Leu	Pro	His	Arg	C
	Leu	Pro	Gln	Arg	A
	Leu	Pro	Gln	Arg	G
A	Ile	Thr	Asn	Ser	U
	Ile	Thr	Asn	Ser	C
	Ile	Thr	Lys	Arg	A
	Met	Thr	Lys	Arg	G
G	Val	Ala	Asp	Gly	U
	Val	Ala	Asp	Gly	C
	Val	Ala	Glu	Gly	A
	Val	Ala	Glu	Gly	G

Amino acids	Abbreviations		Codons			
Alanine	Ala	**A**	GCA	GCC	GCG	GCU
Cysteine	Cys	**C**	UGC	UGU		
Aspartic acid	Asp	**D**	GAC	GAU		
Glutamic acid	Glu	**E**	GAA	GAG		
Phenylalanine	Phe	**F**	UUC	UUU		
Glycine	Gly	**G**	GGA	GGC	GGG	GGU
Histidine	His	**H**	CAC	CAU		
Isoleucine	Ile	**I**	AUA	AUC	AUU	
Lysine	Lys	**K**	AAA	AAG		
Leucine	Leu	**L**	UUA	UUG	CUA	CUC
Methionine	Met	**M**	AUG			
Asparagine	Asn	**N**	AAC	AAU		
Proline	Pro	**P**	CCA	CCC	CCG	CCU
Glutamine	Gln	**Q**	CAA	CAG		
Arginine	Arg	**R**	AGA	AGG	CGA	CGC
Serine	Ser	**S**	AGC	AGU	UCA	UCC
Threonine	Thr	**T**	ACA	ACC	ACG	ACU
Valine	Val	**V**	GUA	GUC	GUG	GUU
Tryptophan	Trp	**W**	UGG			
Tyrosine	Tyr	**Y**	UAC	UAU		

Figure 1.13 The genetic code.

Gene expression is regulated in both time and space

Every step in the overall process of gene expression – be it transcription, RNA processing, translation, or protein modification – represents an opportunity for regulation. Each gene can be expressed when and where the product is specifically needed by the organism, a process broadly referred to as **gene regulation**. As we shall see throughout this book, organisms commit tremendous cellular resources to gene regulation.

Differences in cellular behavior or identity can be brought about by differences in when or at what level a given gene product is expressed. For example, different tissues are defined by the different cells from which they are formed, and different cells are defined by the particular elements of the genome – the particular subset of genes – they express. Therefore, a liver cell will express a different set of genes than that from a nerve cell; these differences in gene expression mean that the two cells are biologically distinct.

A view of the impact of spatial regulation of single genes on the correct development of an organism is provided by the segmentation of the *Drosophila* embryo during early development. Correct segmentation of the organism depends on the localized expression of several genes including *even-skipped*. Figure 1.14 shows how the expression of *even-skipped* is localized in discrete bands of cells along the length of the embryo, laying down the pattern for the segmentation of the embryo later in development. When this localized expression is disrupted, the patterning of the embryo fails to occur as it should.

1.4 CELLULAR INFRASTRUCTURE AND GENE EXPRESSION

The cell is far from being an empty vessel in which molecular components tumble around freely. Instead, it is compartmentalized into distinct structures, each with its own function, which act in concert to mediate the processes essential to life. In this section we learn in broad strokes how the process of gene expression – from retrieval of information from the genome to its expression as gene products – is accommodated within the cell, and explore the cellular compartments in which these processes take place.

Figure 1.14 Expression of *even-skipped* gene in a *Drosophila* embryo. The *even-skipped* gene is expressed in discrete bands of cells along the length of the embryo at an early stage of *Drosophila* development and helps to lay down the pattern for the segmentation of the fly.
Figure © Sabrina Desbordes.

An internal membrane-bound compartment in the cell contains the genome in eukaryotes

At the beginning of this chapter, we discussed how the cellular membrane is key to life because it defines self and non-self. The membrane encapsulates the genetic information critical for self-propagation together with the functioning molecules of the cell. In addition to the exterior membrane shared by all organisms, more complex organisms have additional membrane-bound compartments in their interior. Many unicellular species, including members of the bacterial and the archaeal kingdoms, do not have internal membranes within their cells; because of this distinction they are referred to as **prokaryotes** (meaning 'before the kernel' or pre-nuclear). It should be noted, however, that these cells are not completely without compartmentalization. For example, the chromosomal DNA is sequestered in a region called the **nucleoid** (Figure 1.15).

By contrast, higher-complexity organisms (including those that are unicellular and multicellular) contain an internal membrane-bound structure called the **nucleus** where the genetic information is contained. The name eukaryote literally translates as 'well kernel', and defines a kingdom of organisms having a nuclear compartment that contains their genetic information. The contents of the cell lying outside the nucleus are collectively referred to as the **cytoplasm**.

Eukaryotic cells feature multiple membrane-bound compartments with specialized functions

While the nucleus is the defining compartment within eukaryotic cells, other membrane-bound compartments play equally important roles in cellular life. These compartments are generally referred to as organelles; the key organelles in eukaryotic cells are illustrated in Figure 1.16. The **mitochondria** are known as the powerhouses of the cell and contain the macromolecular machinery responsible for deriving chemical energy from food precursors. **Chloroplasts**

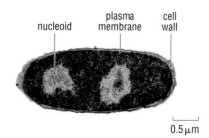

Figure 1.15 Electron micrograph of an *E. coli* cell. The nucleoid containing the DNA and the plasma membrane and cell wall surrounding the cell are labeled. From Sinauer *et al.* Biozentrum, University of Basel. Science Photo Library. Photo Researchers, Inc.

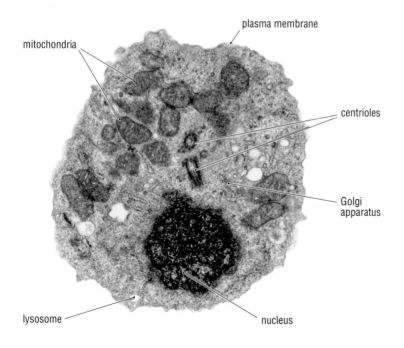

Figure 1.16 Electron micrograph of a human lymphocyte cell. This is the same type of cell that is shown in pink in Figure 1.2. The nucleus, Golgi apparatus, centrioles, mitochondria, lysosomes, and plasma membrane are labeled. From Dr Gopal Murti/Visuals Unlimited.

are organelles found in plant cells that house the machinery for harvesting energy from sunlight for the production of sugars within the cell. Both mitochondria and chloroplasts contain DNA unique to these organelles that encodes certain components critical to their function. This feature led scientists to hypothesize that these organelles originated as **endosymbionts**, organisms living within another cell that eventually became wholly dependent on their host for existence.

Other well-characterized membrane-bound compartments in the eukaryotic cell include the **endoplasmic reticulum** and the **Golgi apparatus**, which are essential for the production of proteins that are inserted into the membrane or secreted from the cell. Certain other organelles, which will not be discussed further in this book, are those devoted to the metabolic and catabolic needs of the cell. These include the lysosome and the peroxisome.

In addition to the membrane-bound compartments that we describe here, the eukaryotic cell also depends on other subcellular localization mechanisms both in the nuclear and cytoplasmic regions of the cell. These regions, which generally were first visualized by microscopy, include structures denoted P-bodies, stress granules, and neuronal particles in the cytoplasm, as well as the nucleolus and Cajal bodies in the nucleus.

What advantages does compartmentalization confer upon a cell? Why does it help biological processes occur in the way that they do?

Compartmentalization is important for facilitating chemical reactions

If we look at biological diversity, we see that overall cell size increases in the eukaryotic lineage when compared with simpler bacterial cells. All cellular reactions depend on interactions between molecules; by localizing various components together, their effective cellular concentrations are increased and so reactions can proceed more efficiently. These principles will be described in more detail in Chapter 2.

In addition to the chemical benefits of increased concentrations of cellular components, cells also benefit from the level of organization that accompanies compartmentalization. Just as we organize our own lives and make them more efficient by compartmentalizing our possessions – we store food items together in the kitchen ready for the preparation of meals, we gather books, class notes, and laptop on a desk ready for study – by co-localizing related processes with shared substrates, intermediates or macromolecular components, cellular efficiency is increased. Cell biological approaches provide striking images of detailed subcellular localization patterns for various cellular components that highlight the potential of such compartmentalization (Figure 1.17).

Compartmentalization is crucial for regulating gene expression in eukaryotic cells

The natural consequence of the fact that the DNA in eukaryotic cells is housed in a membrane-bound compartment, the nucleus, is that there has developed a large amount of regulatory control of gene expression based on this physical

Figure 1.17 Subcellular localization allows for more efficient reactions and regulation. An RNA polymerase (light blue) is localized to specific regions in the nucleus (stained dark blue with dye that binds to DNA) of a eukaryotic cell, while ribosomes (a ribosomal protein is stained green) are localized to the cytoplasm surrounding the nucleus (blue). This separates transcription and translation, providing an opportunity for regulation.

separation from the rest of the cell. Transcription of the DNA (to make RNA) takes place in the nucleus where the genome is housed, whereas translation of the RNA (to make proteins) takes place in the cytoplasm. This difference in location offers an immediate opportunity for regulation: just as you can regulate movement from one room to another in a house by locking (or unlocking) the door between them, so the physical barrier presented by the nuclear membrane provides an opportunity to regulate passage from the nucleus to the cytoplasm (and vice versa), such that passage can only occur if certain conditions are met.

For example, we know that in eukaryotic cells, transcripts that are produced in the nucleus are significantly modified following transcription (RNA processing), but that this modification occurs before the transcript is exported from the nucleus to the cytoplasm. In fact, transcripts cannot be exported from the nucleus unless they have been fully processed. This feature thus provides the cell with an opportunity for quality control – the blocking of movement of the transcript from the nucleus to the cytoplasm across the nuclear membrane – which prevents inadequately processed mRNAs from encountering the translational machinery. In bacteria, where there is no nucleus to sequester the genetic information, transcription and translation are spatially coupled, leading to distinct forms of regulation in these organisms.

An additional complexity faced by gene expression is the production of proteins destined for different locations – for example, for the cytoplasm, for insertion into the cellular membranes or beyond the cell itself. How can proteins be targeted to their correct locations?

As we will discover, proteins that are destined for the membrane or for export from the cell have physical properties that are incompatible with passage through the membrane without specific assistance. In bacteria, where there are no internal organelles, proteins destined for secretion are directly targeted to the extracellular membrane as they are translated. Eukaryotic cells appear to have elaborated on this plan, but instead of directly targeting proteins to the exterior of the cell, translation occurs on the membrane of an extensive organelle system known as the endoplasmic reticulum. This organelle is really a network of tubules where membrane and secreted proteins are translated, folded, modified and then sent on their way.

The final stop in the maturation of many proteins from genes is the Golgi apparatus, an organelle that modifies the nearly completed protein with sugars before packaging the protein for final export. Proteins are sent to the Golgi apparatus from the endoplasmic reticulum, and are ultimately sent to the cell membrane for export.

While this book will not focus on the cell biological aspects of gene expression, it is important to remember the critical role played by cellular structure in all biological processes.

1.5 EXPRESSION OF THE GENOME

In Section 1.2 we explored the expression of individual genes. However, genes do not operate in isolation. Just as individual humans in a community interact, and influence the behavior of others, so genes operate in networks, where the activity

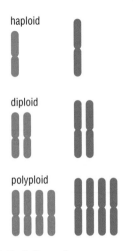

haploid

diploid

polyploid

Figure 1.18 Cells can have one or more copies of their chromosomes. Haploid cells only contain a single copy of the two distinct red and blue chromosomes and diploid cells have two copies. Some cells are polyploid and have multiple copies.

of one gene can quite often influence the activity of another. In this section we look beyond the expression of individual genes to ask how whole communities of genes – that is, genomes – are expressed.

The output of the genome is dictated by the interplay of genes

The expression of the information in a genome determines the physical properties of an organism. The visual features and properties of an organism are referred to as the **phenotype**, while the collective DNA sequence that determines the phenotype is referred to as the **genotype** of the organism. The output of the genome, the phenotype, is dictated by which genes are expressed, at what times, and at what levels. Each organism is thus the product of a combinatorial process dictated by a great many genes working together.

An additional complexity is that organisms can carry one or more copies of the individual genes (Figure 1.18). For example, yeast can survive indefinitely as a **haploid** organism, carrying a single copy of its genome. More complex multicellular organisms by contrast typically have two copies of their genome and are referred to as **diploid**. Some organisms have more than two versions of each of their genes and are said to have higher ploidy – for example, the frog species *Xenopus laevis*, commonly used for biochemical experiments, is **tetraploid** – it has four copies of each gene.

The importance of ploidy is that, in a haploid organism, a single version of a gene is expressed whereas in a diploid organism there is typically a mixed population of gene products derived from the expression of two different copies of the gene. Why can this be important? If one copy of a gene is defective, such that a functional gene product is not produced when the gene is expressed (or is produced in reduced amounts), a diploid organism has the advantage of a second copy of the gene to compensate for the defective one. By contrast, a haploid organism does not have the benefit of a 'back-up' gene, putting it at a greater risk of experiencing phenotypic defects if its only copy of a given gene is mutated.

Genetic analysis has been essential in deciphering the natural function of genes

We see above how different genes are expressed to produce different RNA or protein products, and that it is the concerted action of all the genes in a genome that directs the development and function of an organism. But how do we know what individual genes actually do? How can we say that gene *x* drives process *y*?

Much of our understanding of biology comes from what are referred to as genetic approaches, which probe the natural function of genes by comparing so called **wild-type** (normal) properties, the typical or most common form, with non-wild-type (mutant) ones. In classical genetic approaches, scientists observed organisms under various environmental conditions following the imposition of DNA damage, looking for visible phenotypic changes. That is, they looked to see the consequence of different genes being rendered inactive or altered. When such mutant (variant) organisms were found, then the specific changes in the genome that specified the difference between mutant and wild-type were pursued in order to provide information on potential gene function. So, for example, if a wild-type fruit fly had red eyes, but a defect in gene *a* produced a mutant with white eyes, one could deduce that the function of gene *a* was to determine eye color in some way (Figure 1.19).

This general approach is typically referred to as '**forward genetics**'. For example, Figure 1.20 shows a single mutant *Arabidopsis thaliana* plant seedling surrounded by a group of wild-type plants. In the example shown, the phenotype of the mutant plant is an elongated stem while the genotype turns out to be a **mutation** in a gene encoding a blue light photoreceptor. A specific change in the gene for the blue photoreceptor is manifested as this obvious change in the growth properties of the plant. We will sometimes refer to these two different versions of the gene as the wild-type and mutant **alleles** of the gene. An allele is thus a version of a gene.

With a more complete understanding of genomes and their complete composition, and with sophisticated tools of gene manipulation, it has become possible to specifically disrupt a gene of interest in some organisms and to study the phenotypic consequences. This powerful approach is typically referred to as **reverse genetics** and has been used with wide success to decipher gene function.

Genotypic changes have phenotypic consequences

In forward genetic studies, the phenotype of the mutant organism is typically tracked by the scientist with the ultimate goal being to identify the specific changes in the DNA (mutation) that resulted in the phenotype in question. This type of analysis in principle establishes causality in that a certain change in the DNA results in the physical properties being followed.

Mutations that cause differences from the wild-type organism are typically categorized as recessive or dominant. These terms take on meaning when one remembers that organisms do not always have just a single copy of a given gene. If, in a diploid, the phenotype of the mutant gene product is masked by the presence of the wild-type version, then the mutation is said to be recessive. A common example of a **recessive mutation** is a change in an enzyme that reduces its activity, but which is compensated for by the wild-type version that provides sufficient activity. By contrast, if a mutant gene product has obvious phenotypic consequences when the wild-type product is present, then the mutation is said to be dominant. A common example of a **dominant mutation** might be a change in a structural protein that disrupts higher-order protein structure, such that filament formation is globally disrupted by the presence of the mutant protein. Figure 1.21 shows schematically how recessive and dominant mutations impact mouse coat color.

What is the physical nature of the mutations that can cause such phenotypic changes? The mutations can be the result of very small changes in the gene (i.e. the alteration of a nucleotide), the rearrangement of large regions of the chromosome, the insertion of new segments of DNA or the **deletion** or excision of small or large sections of the genome (Figure 1.22). These changes can result in direct changes to the gene product encoded by a given gene, or in disruptions of overall gene expression (where the actual identity of the gene product is not changed). Mutations that eliminate the function of a gene are termed **null** or **loss of function mutations**. These types of mutation typically are associated either with a change in the product that completely eliminates function or a large disruption in the DNA sequence that eliminates gene expression.

Figure 1.19 *Drosophila* **mutant with a defective pigmentation gene.** Wild-type fruit flies have red eyes (left), while mutant flies carrying a defective *white* gene have white eyes (right). Dr Jeremy Burgess/Science Photo Library.

Figure 1.20 *Arabidopsis* **mutant unable to sense blue light.** The stem (hypocotyl) extension of wild-type *Arabidopsis* seedlings is inhibited by blue light. The growth of a mutant carrying a defective blue light receptor *cry* gene is not inhibited by blue light so it stands out among the wild-type seedlings.

Figure 1.21 Effects of dominant and recessive coat color mutations in mouse. In the cross between one parent (gray) contributing the wild-type gene and another parent (pink) contributing a gene with a dominant mutation, all offspring will have the phenotype of the parent carrying the dominant mutation. In the cross between one parent (gray) contributing the wild-type gene and another parent (blue) contributing a gene with a recessive mutation, all offspring will have the phenotype of the parent carrying the wild-type gene.

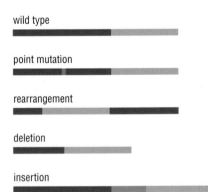

Figure 1.22 Types of mutation. The sequence of DNA can be mutated by a single nucleotide change (point mutation, red), shuffling of different regions of DNA here indicated by the two shades of gray (rearrangement), excision of DNA (deletion) or insertion of DNA (blue).

Single nucleotide changes in the DNA sequence can directly result in changes in the identity of the encoded gene product (either RNA or protein). In protein coding genes a change of one nucleotide can be sufficient to change a codon from one which encodes a particular amino acid to one encoding a different one. Such changes in amino acid identity are referred to as **missense mutations** (Figure 1.23). If the amino acid substitution results in gene function being completely lost the missense mutation can be thought of as a null mutation. By contrast, other substitutions may result in the production of partially functional gene products.

In some cases, there can be a change in the DNA sequence that introduces a premature stop signal such that only truncated species of the protein product are made. Such changes are typically referred to as **nonsense mutations** (Figure 1.23). Sometimes there are changes in the DNA sequence that do not result in obvious phenotypic changes. Such changes are typically referred to as **silent mutations** (Figure 1.23).

The origin of disease can often be traced to mutations in the organism

In multicellular organisms, there are two general types of cell:

- **somatic cells** – these constitute the building blocks of the body of an organism and simply transmit their genetic information to daughter cells in the same organism

- **germline cells** – these are involved in the production of the progeny and transmit the genetic information to the next generation of the organism.

Consequently, mutations found in the germline cells typically affect the progeny of the organism whereas those found in somatic cells affect the organism itself.

One common consequence of gene mutation is the onset of disease, whereby normal biological function is disrupted in some way. Some diseases are the consequence of variation in a single gene and are said to be **monogenic**. A well-known example of a monogenic disease is phenylketonuria, in which a mutation in a gene encoding a specific liver enzyme can lead to severe mental retardation. In other cases, disease states are thought to result from variations in multiple genes and are said to be **polygenic** (though the term 'multifactorial' is also commonly used). Some of the most prevalent diseases, such as heart disease, Alzheimer's, arthritis, and diabetes are thought to be polygenic.

The presence of a specific mutation does not always lead to a disease state, however. For example, mutations in genes called *BRCA1* and *BRCA2* are associated with breast cancer, but not all individuals carrying these mutations develop cancer during their lifetime. The **penetrance** of a given mutation is defined as the percentage of individuals carrying a particular mutant genotype that exhibit the mutant phenotype. The risk of developing a mutant phenotype when carrying a mutant genotype is determined by both the genetic makeup of an organism and environmental influences, such as diet and exposure to different environmental conditions.

The relative extent to which the genotype (nature) and the environment (nurture) influence the phenotype associated with a specific mutation has long been debated and undoubtedly will continue to be a subject of controversy. In the example of the *BRCA1* and *BRCA2* mutations, other genetic factors clearly influence whether individuals are more or less susceptible to developing cancer, but environmental factors such as diet and smoking can also affect the penetrance of these mutations. As we shall see in Chapter 15, whole-genome sequencing is providing critical information for the genetic mapping of complex diseases such as cancer.

1.6 EVOLUTION OF THE GENOME AND THE TREE OF LIFE

wild type	**ATG**	**AAT**	**ATT**	**CGA**	**GAT**
	Met	Asn	Ile	Arg	Asp
missense	**ATG**	**AAT**	**ACT**	**CGA**	**GAT**
	Met	Asn	Thr	Arg	Asp
nonsense	**ATG**	**AAT**	**ATT**	**TGA**	**GAT**
	Met	Asn	Ile	stop	Asp
silent	**ATG**	**AAT**	**ATC**	**CGA**	**GAT**
	Met	Asn	Ile	Arg	Asp

Figure 1.23 Multiple types of point mutation. Point mutations can lead to the incorporation of the wrong amino acid (missense), premature termination (nonsense), or not affect the sequence (silent) of the translated protein.

The genome of a particular species is not a static, unchanging entity. If it were, there would not be the remarkable variety of life in the world around us. Instead, genomes are dynamic: they undergo gradual, incremental change over time, which results in progressive change in the organism whose development and function they are orchestrating. The progressive change in an organism's phenotype over many generations, underpinned by a gradual change in that organism's genotype, is described by the process of evolution, as we now discuss.

The evolution and diversification of life is gradual

The argument in favor of a common ancestor (LUCA, see above) is based most simply on the fundamental similarities in the molecular features of all organisms, for example, the usage of nucleic acids to store genetic information and the genetic code that specifies the amino acid sequence of proteins. It is highly unlikely that similar mechanisms would have evolved independently in many different organisms. Instead, it is far more likely that these common mechanisms were passed down from our ancestors into each new species as they evolved.

If common features have been conserved between species, how did the great diversity of life on earth derive from the last common ancestor? If we see such similarities between species, how could the differences that delineate different species arise at all? Evolution is the underlying principle of modern biology that explains this conundrum. It is the process by which organisms gradually change over successive generations through a process of random change or mutation in the genome, coupled with natural selection, the outcome of competition. Let us consider a hypothetical example to illustrate this process (Figure 1.24).

Imagine a population that shares a common genome: all members of the population have the same genotype (genomic composition) and, hence, the same phenotype. This genome can only be passed on to future generations if an organism

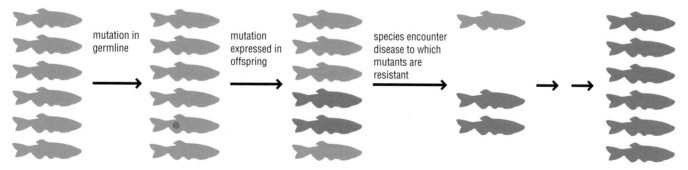

Figure 1.24 Ever-larger populations of individuals carrying specific DNA changes due to selection. Initially all the fish have the same genotype. Then a random mutation arises in the germline cells of one of the fish. By chance, the offspring (pink) of this fish have increased resistance to a particular disease. Since the offspring are more likely to survive, they are also more likely to reproduce, leading to an increase in the number of fish carrying the DNA change.

survives long enough to reproduce. In this case, however, all of this population will be equally healthy (because they have the same genome) and so all should be equally likely to reproduce, and have their genome transmitted to the next generation.

Now, imagine that one or two of the population develop a mutation in their germline cells. This mutation would not affect them directly (because it is in their germline and not their somatic cell line). However, the mutation *will* be expressed when the genome is transmitted to the next generation.

Let us assume the mutation confers upon the individual carrying it a particular resistance to disease. The next generation will therefore contain a minority of individuals who show unusually high disease resistance. If the population as a whole is then exposed to a disease in the surrounding environment, it is those individuals who carry the mutated (unusually effective) disease-resistance gene who are most likely to survive – and are therefore most likely to reproduce. Consequently, there is an increased chance of this mutated gene being passed on to the *next* generation.

Over time, environmental influences, such as disease, will continue to 'select for' the mutant individuals and, over time, an ever-larger proportion of the population will contain the mutant gene, until it becomes the norm. This is the process of natural selection in action. As a consequence of natural selection, as generations come and go, we see our original species evolving into one that is more disease-resistant than it was before. Notice how evolution is not a rapid process – it is a gradual change in genotype which, if beneficial to the population, becomes increasingly established, driving a concomitant shift in the observed phenotype for the population.

So we see how mutations that result in increased success for the organism will be propagated or selected; such selection eventually leads to the emergence of a reproductively isolated group of organisms or species. A species is defined as a group of organisms capable of interbreeding and producing fertile offspring.

All organisms can be grouped into three domains of life based on sequence comparisons

We have learned that the collection of genes in an organism, and their expression, determines the phenotype and the identity of an organism. By comparing the sequences of specific genes or whole genomes from different organisms, we can learn something about the likelihood that a given property will be similar among different organisms, which tells us how closely these organisms are related in evolutionary terms.

As shown in an alignment of the sequences of the same region of the ribosomal RNA genes from many different organisms (Figure 1.25), the sequences for individual genes can be very similar in certain regions while varying more extensively in others. The extent of relatedness can be quantitated and these results then represented in a branched model called a **phylogenetic tree**. An example of a phylogenetic tree based on the comparison of ribosomal RNA genes is shown in Figure 1.26. In these models, the length of the line reflects the relatedness of the gene at one end of the line to that at the other end, and thus by extension, the relatedness of these organisms. Therefore, an organism that is separated from its most recent ancestor by a short line is more closely related to its ancestor than a second organism that is separated from the same ancestor by a longer line.

T. acidophilum	GGCCUUUAGUACGAGAGGAACAAGGG
E. coli	UGCUCCUAGUACGAGAGGACCGGAGU
B. subtilis	UGUCCUUAGUACGAGAGGACCGGGAU
E. histolytica	ACAACUCAGUACGAGAGGAACCGUUG
S. cerevisiae	UGAACUUAGUACGAGAGGAACAGUUC
C. elegans	CCUGCUUAGUACGAGAGGAACAGCGG
D. melanogaster	CCUGCGUAGUACGAGAGGAACCGCAG
R. norvegicus	GGUGCUCAGUACGAGAGGAACCGCAC
O. sativa	UCAACCUAGUACGAGAGGAACCGUUG
A. thaliana	UCAACCUAGUACGAGAGGAACCGUUG
H. sapiens	CCUGCUCAGUACGAGAGGAACCGCAG

Figure 1.25 Similarity between sequences from distantly related organisms. The alignment of sequences corresponding to ribosomal RNAs from distantly related organisms shows regions of similarity and variation.

Trees constructed from comparisons of different genes can differ, and so phylogenetic trees are typically based on comparisons of multiple genes. The precision of such classification approaches far exceeds more traditional methods of classification that depended on visually apparent physical features such as whether or not cells have nuclei.

Sequence comparisons based on many genes indicate that there are three distinct groups, or branches, of the tree into which all of organismal life can be split. These broad groups are referred to as the **three domains of life** and include the bacteria, archaea, and eukaryotes.

- The bacterial domain encompasses the largest number of organisms and continues to grow as the environment is explored. These organisms are single celled and have minimal subcellular compartmentalization.

- Archaeal organisms share features with organisms in the bacterial domain in that they are unicellular and do not have a nuclear structure. However, core archaeal proteins typically show more similarity to the corresponding proteins found in eukaryotic organisms. Thus, early classification identified these organisms as bacterial, based on their physical features, but sequence comparisons allowed for the identification of this separate domain of life as recently as 1977.

- Eukaryotic organisms are distinguished by more extensive compartmentalization than found in either bacteria or archaea, including the nucleus that holds their DNA, and by possessing a cytoskeleton that helps to structure and organize

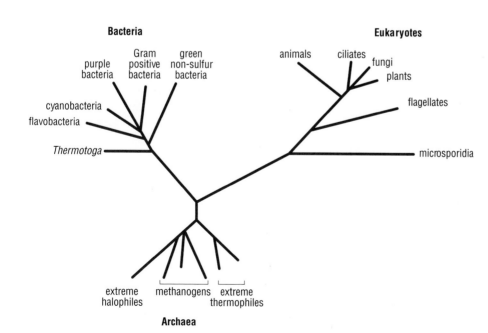

Figure 1.26 Relatedness of organisms depicted by a phylogenetic tree. The tree shows the relatedness of the sequences encoding ribosomal RNAs from bacteria, archaea, and eukaryotes. The length of each of the lines corresponds to the relatedness of the genes. From Woese, *Microbiological Reviews* 1987; **51**:221–271.

Figure 1.27 Commonly studied fungi *S. cerevisiae* **and** *S. pombe*. As seen in the micrograph, *S. cerevisiae* is oval shaped and divides by budding, whereas *S. pombe* is rod shaped and divides by fission. From Morgan D.O. 2007 *The Cell Cycle: Principles of Control.*

their larger cell volume. All multicellular organisms, including animals, plants, and fungi, are eukaryotic, but not all eukaryotes are multicellular: certain fungi such as yeast, algae, and slime molds are single-celled eukaryotes.

This tree of life is an organizing principle for biological studies and we will see below how biologists have chosen a range of organisms covering some of this diversity to focus on explorations of molecular function.

Studies of model organisms have been key to understanding many biological processes

The study of organisms that have features of agricultural or medical importance, yet are easy to manipulate under laboratory conditions, has been critical to the understanding of the many processes in biology. These organisms, often referred to as model organisms, are typically directly evaluated and manipulated, subjected to mutational studies, or are used to prepare samples for biochemical analysis.

Two commonly studied bacteria, *E. coli* and *Bacillus subtilis*, belong to two distinct groups of the bacterial domain referred to as Gram-negative and Gram-positive, respectively, based on their differential staining with a particular violet dye (Gram stain). Two commonly studied fungi belonging to the eukaryotic domain are *Saccharomyces cerevisiae* and *Schizosaccharomyces pombe* (Figure 1.27). While both are fungi, these two organisms exhibit significant differences in their biology – including differences in something as essential as their mode of cell division, where budding is seen in *S. cerevisiae* and fission in *S. pombe*. Careful comparison of these genomes suggests that *S. pombe* is more closely related to higher eukaryotes such as humans.

Multicellular eukaryotes that are commonly studied are the fruit fly *D. melanogaster*, the worm *C. elegans*, the frog *X. laevis*, the zebrafish *Danio rerio* and the plant *A. thaliana*. The predominant mammalian model organism is the mouse *Mus musculus* (Figure 1.28). But what gives an organism its 'model' status? Most of the model organisms were chosen for practical reasons: they have advantages in terms of their biological composition, development, or behavior that make them particularly amenable to study. For example, many of the organisms have

Figure 1.28 Commonly studied multicellular eukaryotes. Model multicellular eukaryotes commonly studied in biology include the fruit fly *D. melanogaster* (Hermann Eisenbeiss/Science Photo Library), the worm *C. elegans* (Sinclair Stammers/Science Photo Library), the frog *X. laevis* (Zigmund Leszcynski/Photolibrary Group.com), the zebrafish *D. rerio* (Oxford Scientific (OSF) Photolibrary Group), the plant *A. thaliana* (Dr Jeremy Burgess/Science Photo Library), and the mouse *M. musculus* (Oxford Scientific (OSF)/Photolibrary Group).

fast generation times: the average generation time of a fruit fly is two weeks, while an *Arabidopsis* plant is able to produce seed a month after germination. All of the model organisms mentioned above are also relatively small and can thus be cultivated under typical laboratory conditions.

In addition, each of the organisms mentioned above has different strengths and weaknesses regarding the types of analysis that can be readily performed. For example, *X. laevis* is a key organism for studying development in a biochemically tractable system since the developing oocyte is of large enough size to permit direct manipulation. However, genetic approaches are not very well developed in this system. By contrast, *D. melanogaster*, *C. elegans*, and *S. cerevisiae* are probably best known for their potential as genetically tractable organisms where unbiased genetic screens are readily performed.

It bears remembering that model organisms are simply that – they allow us to build models (or informed estimates) of how biological systems operate but do not allow us to formulate exhaustive descriptions of exactly what is going on in a single investigation. It is only by successive rounds of investigation, sometimes in different model organisms, each perhaps only revealing a small new insight, that our understanding is incrementally increased. As these insights build up, so we refine our understanding, and build gradually more detailed, and more refined, best guesses of the reality of life. Equally importantly, model organisms do not always provide a definitive indication of biological processes throughout the kingdoms of life; it may not be possible to extrapolate an observation made in one organism to a seemingly similar organism.

The study of viruses has provided tremendous insights into biological function

Viruses are particles composed of a nucleic acid-based genome surrounded by a protein coat. Although viruses contain their own genomes, they cannot replicate on their own. Instead, viruses replicate within another organism where they usurp components of the host cell machinery to replicate or express their own genomes (Figure 1.29). Viruses have been found to be associated with all organisms and are present in an astounding variety and number. For example, it is estimated that 10 000 000 **bacteriophages** (the term for viruses that infect bacteria) are present in 1 mL of coastal seawater.

Viruses use quite varied mechanisms to usurp the host cell machinery. Therefore, studies of the mechanisms used by viruses to infect and replicate in their host cells have time and again provided fundamental insights into basic cellular functions, and the strategies and mechanisms used by organisms to bring their genomes to life. So we can learn much even from entities that we would not classify as model organisms.

The molecular components of different organisms are described using different nomenclature rules

As a consequence of the large body of work carried out with the model organisms mentioned above, many of the examples cited in the subsequent chapters will be from studies involving them. In citing such examples we need to be aware that, despite the common molecular components and processes that operate in

Figure 1.29 Electron micrograph of a virus-infected cell. The picture illustrates the relative size of the crystalline-like virus particles (VC) relative to the eukaryotic cell.

different organisms, the ways in which these components are described varies. In particular, the conventions used for naming genes, proteins, and mutant alleles in different organisms – the nomenclature used – vary from organism to organism. For example, in some organisms, a wild-type gene name is written in uppercase letters, whereas the mutant gene name is written in lowercase letters. Sometimes, the gene name is in italics, and the name of the protein resulting from the expression of that gene is written in roman (non-italic) type. Throughout this book, we observe the most widely accepted nomenclature rules for each organism, and these are summarized in the table on page ii at the front of the book.

Also, we need to be aware that homologous genes and proteins in different organisms – that is, genes and proteins with conserved sequences (and, hence, conserved functions) – may have different names. For example, the eukaryotic homolog of the bacterial sliding clamp β-protein involved in DNA replication is called PCNA (proliferative cell nuclear antigen, a name that derived from how the factory was originally identified). Such differences can make it more challenging to see patterns and similarities, but they simply reflect the rich history and diversity of research that underpins our current knowledge of the field.

We have now completed our brief tour of biological systems, and the journey from genetic information to the living organism. We have seen how our genome directs our development and function, and how changes in our genome propagate changes in our physical character as evolutionary forces are brought to bear. We have also had our first glimpse of the molecular mechanisms that retrieve the information from our genome, and, by transforming information into activity, bring the genome to life.

In the next two chapters we move from the level of the cell to the level of atoms and molecules, to consider the chemical infrastructure of life, before beginning our more detailed exploration of molecular biology, and the principles of genome function.

✱ SUMMARY

The unifying features of life

- The diversity of life is unified by some common themes: a living organism must be distinct from its environment, have a way of storing biological information, be able to replicate and transmit the information to subsequent generations, and have a source of energy from the environment to grow and reproduce.

- The molecular building blocks of all modern organisms, nucleic acids, proteins, lipids, and carbohydrates, are conserved and are constructed from small, repeating subunits.

The blueprint of life: the genome

- Biological information is typically stored in the nucleic acid DNA and is referred to as the genome.

- A gene is typically defined as a region of DNA that controls a discrete hereditary characteristic. The number and arrangement of genes within a genome varies from species to species; gene number is a poor indicator of biological complexity.

Gene expression

- The first stage of gene expression is transcription, in which an RNA copy of the gene is synthesized. The product of transcription may be a functional RNA, which does not encode a protein (a non-coding RNA), or a messenger RNA (mRNA), which does encode a protein (a coding RNA).

- mRNAs are translated to produce a protein product in a process mediated by the ribosome. During translation, the nucleotide sequence of the mRNA is deciphered in three-base triplets called codons; each codon represents a particular amino acid (or tells the ribosome to stop translation).

- A single gene may encode more than one RNA or protein product such that the estimated 30 000 human genes are estimated to encode more than one million gene products.

- The regulation of gene expression in both time and space is extensive and occurs at every level of expression.

The cellular basis of gene expression

- Eukaryotic cells feature membrane-bound compartments called organelles, such as the nucleus which contains the genome. Bacteria and archaea lack organelles, though the genome is sequestered in a region of the cell called the nucleoid.

- Subcellular compartmentalization provides opportunities to increase the efficiency of cellular events and to regulate biological processes.

Expression of whole genomes

- The sequence of an organism's genome is its genotype. The overall expression of an organism's genome determines its phenotype (its physical features and properties).

- Null or loss of function mutations eliminate the function of a gene; missense mutations change the identity of a single amino acid within a protein product; and nonsense mutations result in the premature termination of translation to give a truncated protein product.

Evolution of the genome

- All modern-day life forms evolved from a single common ancestor.

- Random mutations that confer advantageous traits upon an organism will be selected for – that is, there will be an increased likelihood of such mutations being transmitted from generation to generation. This is the basis of the evolution of organisms through the process of natural selection.

- At the highest level, sequence similarities and differences allow us to divide all of organismal life into three domains – bacteria, archaea, and eukaryotes.

Studying genome function

- The study of model organisms provides insights into molecular components and processes, helping us to understand how genomes function.

- Viruses exploit the host cellular machinery to ensure their own propagation. Consequently, the study of viruses has helped to illuminate how the host machinery operates.

2 BIOLOGICAL MOLECULES

INTRODUCTION

Despite the extraordinary diversity of Earth's creatures, almost all processes necessary for life depend on a similar set of molecules. The transmission of genetic information, the manufacture of molecules that carry out all the many different tasks of a cell, the synthesis and breakdown of key nutrients, and the very structure of the cell itself, depend on a variety of molecules that are uniquely suited to their tasks.

There are four major classes of biological molecules that play essential roles in all organisms:

- nucleotides
- amino acids
- sugars
- lipids.

All of these molecules contain carbon atoms and are therefore called **organic** compounds. As shown in Figure 2.1, each of them can be found in cells both as individual molecules or combined to form larger molecules known as **polymers** or **macromolecules**. **Nucleic acids** are polymers of nucleotides responsible for carrying genetic information. Proteins, on the other hand, are polymers of **amino acids** that function as workhorses, carrying out most of the chemical reactions in the cell or giving cells their structure and shape. **Fatty acids** are key components of **lipids**, the molecules that form membranes that encapsulate cells, separating the cell contents from the environment and also helping to organize the cell's interior into specialized compartments. Polymers of sugars, called **carbohydrates**, serve diverse functions such as storing energy obtained from food, modulating the function of both proteins and lipids, and providing mechanical strength to cells. All of these molecules sometimes carry out other functions in addition to their principal roles.

The laws of chemistry and physics govern all that goes on in an organism as fully as they govern all other processes in the universe. They determine which molecules can exist, each molecule's shape, and whether or not a chemical reaction can occur and how fast that reaction proceeds. For example, these laws dictate that proteins in the cell, in common with all other macromolecules, will degrade over time even though, for proteins, this process is very slow. Understanding the chemistry and physics in a cell is therefore essential to understanding all of the biological processes that we shall learn about later in this book.

In this chapter, we will review the basic properties of atoms and learn how they are joined together to form molecules. We will then see how the unique properties of the four major classes of biological molecules enable them to carry out distinct functions in the cell. Throughout, we will see how the behavior of all biological molecules is dominated by the properties of water (Figure 2.2). Cells consist of 70%–80% water by mass, with this water influencing every aspect of a cell's biology. Water has also been a driving force throughout the evolution of life: life originated in the primordial seas, and the chemistry of modern-day organisms reflects their origins in that aqueous environment.

We now begin our exploration of biological macromolecules by considering the individual components – the atoms – from which they are formed.

2.1 ATOMS, MOLECULES, AND CHEMICAL BONDS

The structure of atoms dictates their chemical reactivity

All matter in the universe is composed of a limited set of elements that obey the laws of physics and chemistry. The smallest unit of an element that retains its defining chemical properties is known as the atom. Atoms are, in turn, composed of sub-atomic particles: positively charged protons and electrically neutral neutrons, which are clustered in a central nucleus and surrounded by negatively charged electrons. The resultant atom is electrically neutral, the positive protons and negative electrons balancing one another precisely. Different types of atoms (also known as elements) are distinguished by the number of protons they contain; this is known as the atomic number. For example, hydrogen (H), the lightest element, has a single proton and thus an atomic number of 1, whereas calcium (Ca) has 20 protons and thus an atomic number of 20. The atomic mass of an atom, given in units called daltons (Da), is equal to the number of protons plus neutrons in an atom. Carbon, for example, typically has six protons and six neutrons and therefore has an atomic mass of 12 Da. There are altogether 92 naturally occurring elements, of which just a handful predominate in living things (Figure 2.3). Four elements alone – carbon (C), oxygen (O), hydrogen (H), and nitrogen (N) – account for over 96% of the mass of an organism.

Main families of biological molecules	
building blocks	**larger units**
sugars	polysaccharides
fatty acids	lipids, membranes
amino acids	proteins
nucleotides	nucleic acids

miscellaneous metabolites (10%)
nucleic acids (5%)
inorganic ions (3%)
sugars (6%)
lipids (16%)
proteins (60%)

Figure 2.1 The four major classes of biological molecules. The table (top) lists the four major types of biological molecules, along with their building blocks. The pie chart (bottom) shows the relative percentage of the cell mass (excluding water) contributed by each class of molecules.

Figure 2.2 Water influences the structure of cells and biological molecules. Jackson Lake, Grand Teton National Park, Wyoming. Photo courtesy of CW.

Common elements in biology				
element	symbol	atomic number	atomic mass (Daltons)	proportion of total body mass (%)
hydrogen	H	1	1	10
carbon	C	6	12	23
nitrogen	N	7	14	2.6
oxygen	O	8	16	61
sodium	Na	11	23	0.14
magnesium	Mg	12	24	0.027
phosphorus	P	15	31	1.1
sulfur	S	16	32	0.2
chlorine	Cl	17	35	0.12
potassium	K	19	39	0.2
calcium	Ca	20	40	1.4

Figure 2.3 Common elements in biology. The table lists the most common elements in biology, along with the symbol, atomic number (the number of protons), atomic mass of the most common isotope, and the proportion of the total body mass made up by each element.

The arrangement of electrons in each atom determines its interactions with other atoms. Electrons are found in discrete states of varying distance from the nucleus, known as shells. The innermost shell contains up to two electrons, the second and third shells can accommodate eight electrons each, and the fourth shell contains up to 18 electrons. Figure 2.4 shows the electron shells of hydrogen, carbon, nitrogen, and oxygen. An atom is most stable and chemically unreactive when its outermost electron shell is completely filled, that is, when it contains the maximum number of electrons for that shell.

Covalent bonds form when electrons are shared between atoms

An atom whose outer shell is not filled is chemically reactive, meaning that it can participate in chemical reactions with other atoms to form molecules. Why does having an unfilled outer shell make an atom chemically reactive? The answer lies in the way that an atom with an unfilled outer shell can share electrons with another atom. By sharing electrons, the atom can achieve a full outer shell, and becomes more stable as a result. This situation is depicted in Figure 2.5 for hydrogen and oxygen, which can combine to form a water molecule. Hydrogen has just one electron in its outer shell instead of the allowed two, while oxygen has six electrons in its outer shell instead of the allowed eight. When two hydrogen atoms share their electrons with an oxygen atom, each atom now has a complete outer shell and is therefore more stable. Sharing of electrons between atoms is the basis of **covalent bonds,** whereby atoms are joined to one another to form **molecules**. The sharing of one pair of electrons gives rise to a **single bond** between two atoms (Figure 2.5a). Sharing of two pairs of electrons constitutes a **double bond** (Figure 2.5b). Covalent bonds are typically very stable, which means it takes a substantial amount of energy to break them. The precise amount of energy needed to

Figure 2.4 Atomic structure of hydrogen, carbon, nitrogen, and oxygen. Atoms are composed of a central nucleus of protons (red) and neutrons (green) surrounded by shells of electrons (blue) in which the first shell contains up to two electrons and the second up to eight electrons. Hydrogen has only one electron in its outer shell; carbon has four electrons; nitrogen has five electrons; and oxygen has six electrons instead of the allowed eight.

Figure 2.5 Covalent bonds. (a) A water molecule forms when each of two hydrogen atoms shares an electron with an oxygen atom. Each hydrogen atom forms a single covalent bond with the oxygen atom. (b) Oxygen is commonly found as the molecule O_2, in which two oxygen atoms share electrons. Each shared electron pair represents a single covalent bond. The two oxygen atoms therefore form a double covalent bond with one another.

break a covalent bond depends on the types of atoms involved and the number of bonds they form with one another, with double bonds generally being stronger than single bonds. The strength of covalent bonding makes biological molecules stable under the physiological conditions prevailing in cells.

A more complex bonding pattern can occur in cases in which an atom is joined to its neighbors by both single and double bonds. An important example in biology is the **peptide bond** that, as we shall see later, joins amino acids together in a protein. Although we typically draw a double bond between the carbon and oxygen and a single bond between the carbon and nitrogen atom, there is actually a sharing of electrons between the three atoms that gives both the C–O and C–N bonds partial double bond character, as indicated in Figure 2.6. We shall see below how partial bonds constrain the shapes that a molecule containing them can adopt.

Shape and flexibility of a molecule are determined by its electrons

The distribution of the electrons among atoms determines the overall shape of a molecule. When an atom participates in a covalent bond with more than one atom, the covalent bonds will be arrayed in a configuration that is characteristic of the particular atom (Figure 2.7). Carbon atoms, which can share all four of the electrons in their outer shell, make bonds that radiate outward and are equally spaced from one another, as if pointing towards the corners of a tetrahedron. The two bonds that can be formed with an oxygen atom form a V shape, whereas the unshared electron pairs are located on the opposing side of the oxygen atom.

The number of bonds that are formed between two atoms places an additional constraint on molecular geometry. Although atoms can rotate freely about a single bond, double bonds have properties that create a barrier to rotation. Even partial double bond character prevents rotation significantly, which means that electron sharing of the type shown in Figure 2.6 will also constrain the **conformation** of molecules that contain them. Together, bond order and the spatial arrangement of covalent bonds about each atom are important factors in determining the shapes of biological molecules. For example, the peptide bond, which joins adjacent **amino acid residues** in a **polypeptide** chain (Section 2.9), has partial double bond character (Figure 2.6b). As a consequence, free rotation about two of the covalent bonds that comprise the peptide bond 'unit' is not possible, placing restrictions on the three-dimensional conformations that a polypeptide can adopt when **folded**.

The effective size of an atom is determined by its electrons

Atoms that are not joined to one another by covalent bonds behave in many ways like hard spheres with a defined radius. Just as two billiard balls cannot interpenetrate

Figure 2.6 Partial double bonds. (a) Carbon atoms flanked by oxygen and nitrogen atoms are usually written as having one double bond and one single bond. However, there is actually a sharing of electrons as shown in (b), such that both bonds have partial double bond character.

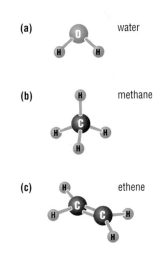

Figure 2.7 Geometry of bonds in oxygen and carbon atoms. (a) The bonds in an oxygen atom are positioned so that two single bonds have a V configuration, as shown for bonds with hydrogen in a water molecule. (b) Carbon can make four single bonds, with a tetrahedral geometry. (c) Where two atoms participate in a double bond, other atoms bound to the same carbon atom are forced to lie in the same plane because the double bond prevents rotation around it.

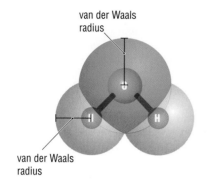

Figure 2.8 van der Waals radii of atoms in a water molecule.

one another, non-bonded atoms can only come within a certain distance of one another before they begin to clash. The radius of the sphere is a characteristic of each type of atom and is called the **van der Waals radius** (Figure 2.8). When atoms come closer than the sum of their van der Waals radii, the interaction energy becomes highly unfavorable and they repel one another. When atoms are very far apart – that is, at distances equal to many times their van der Waals radii – they do not interact at all and the interaction energy is therefore nearly zero. However, atoms that are separated by just around the sum of their van der Waals radii weakly attract one another. Biologists call these attractions **van der Waals interactions**; they are also known as London dispersion forces. We will learn about the physical basis for this attractive force, as well as the role it plays in interactions between biological macromolecules, in the next section. The way in which both attractive and repulsive forces vary with the distance between two atoms is shown in Figure 2.9. Look at this figure, and note how two atoms repulse each other when the nuclei are very close to one another, but then attract each other as the nuclei move slightly further apart. As the distance between the nuclei increases further, however, this attraction decreases until there is no net attraction or repulsion between the two nuclei.

Electrons are not always shared equally by atoms joined by covalent bonds

Electrons are often more closely associated with one of the two atoms in a covalent bond than the other. As a result, the atom with the greater share of the electrons has a partial negative charge and is therefore **electronegative** (denoted δ−), while the atom with the lesser share has a partial positive charge and is therefore **electropositive** (denoted δ+). This separation of charge is known as a **dipole** (Figure 2.10), and such covalent bonds are said to be **polar**. Molecules or chemical groups containing bonds in which electrons are shared more equally are **non-polar**. Whether a chemical group is polar or non-polar has important consequences for its behavior in the aqueous environment of the cell, which consists predominantly of water molecules that are themselves polar. As we shall see, polar chemical groups interact most favorably with polar molecules such as water, while non-polar groups prefer to interact with one another rather than with the polar solvent. Many molecules have both polar and non-polar regions. For example, the lipid molecules that form the double-layered cell membrane have a polar head and non-polar tail. It is this marked difference between the two regions of the molecule that drives the way multiple lipid

Figure 2.9 Interaction energy between two non-bonded atoms. The plot shows the interaction energy between two atoms as a function of the distance between them. The minimum in the plot occurs at a distance equal to the sum of the van der Waals radii, at which two atoms form the most favorable van der Waals interaction. At shorter distances, the atoms start to clash and the energy increases greatly.

molecules interact to form the cell membrane structure – with polar heads associating to form the outer faces of the cell membrane, and the polar tails associating to form the membrane interior. We will revisit lipids and membrane formation later in the chapter, when we see how the tendency of non-polar groups to associate with one another drives much of the behavior of biological macromolecules.

Atoms can become fully charged

As noted above, unequal sharing of electrons in polar bonds can give rise to a partial electrostatic charge on one or more atoms. However, because of their energetic preference to have a completely filled outer shell, atoms sometimes gain or lose electrons entirely and therefore become fully electrostatically charged. Atoms or chemical groups that become charged in this way are said to be **ionized** and are known as **ions**. Elements whose outer shells are just one electron short of being filled can acquire an electron and thereby become negatively charged. Conversely, atoms that have just one or two outer shell electrons can lose these electrons and become positively charged. Atoms with opposite charges interact favorably with one another to form salts. A familiar example is the interaction between the positively charged sodium ions and negatively charged chloride ions in table salt (Figure 2.11). This attraction between fully charged atoms, known as an ionic interaction, can also occur between chemical groups that bear full positive or negative charges. Full charges on chemical groups generally occur when a positively charged hydrogen ion associates with or dissociates from the group, as we shall see later in this chapter. Most biological molecules contain charged groups, and ionic interactions are therefore common in biology.

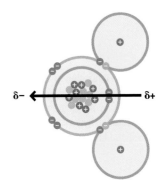

Figure 2.10 Unequal sharing of electrons confers molecular polarity. The oxygen atom (blue) in a water molecule takes a greater share of the shared electrons than each hydrogen atom (red). As a result, the oxygen atom has a partial negative charge (indicated as $\delta-$) and each hydrogen atom has a partial positive charge (indicated as $\delta+$).

2.2 LIFE IN AQUEOUS SOLUTION

The chemistry of life – the biochemical reactions that keep us alive – occurs not in a test tube in the laboratory, but within the cells of our body. Our cells contain the cytoplasm, the water-based (aqueous) environment in which the biochemical reactions of life occur. For such a simple molecule, comprising just one oxygen atom and two hydrogen atoms, water has a number of important properties that make it particularly suited to being the medium of life. These include the ability to form so-called hydrogen bonds (see Section 2.3) and the ability to become ionized. Without these properties, water would be unable to support life in the way that it does; it would be a poor solvent for many biological molecules, and would be a poor medium for biochemical reactions.

Figure 2.11 Ionic interaction between a sodium and a chlorine atom. The sodium atom has only one electron in its outer shell; the chlorine atom has seven. Loss of an outer shell electron from the sodium atom and gain of an electron by the chlorine atom leaves both atoms with filled outer shells. The result is a full positive charge on the sodium atom, which is thus a cation, and a full negative charge on the chlorine atom, which is thus an anion. The oppositely charged ions are strongly attracted to one another and form a salt.

$$H_2O \; + \; H_2O \quad \rightleftharpoons \quad H_3O^+ \; + \; OH^-$$

a proton is transferred from one water molecule to another resulting in the ionization of both molecules

hydronium ion

hydroxide ion

Figure 2.12 The ionization of water. Water molecules can become ionized by gaining or losing a proton. Gain of a proton gives rise to the positively charged hydronium ion; loss of a proton gives rise to the negatively charged hydroxide ion.

Water can become ionized

Any aqueous solution contains some water molecules that have dissociated into positively and negatively charged ions. As shown in Figure 2.12, a water molecule can lose a proton, denoted H^+, from one of its hydrogen atoms, leaving a negatively charged hydroxide ion, OH^-. The proton associates most favorably with the partial negative charge of the oxygen atom of another water molecule, forming the hydronium ion, H_3O^+. In a solution of pure water, only a very small percentage of water molecules will dissociate in this way; the concentration of OH^- and H_3O^+ ions is just 10^{-7}M. Importantly, the concentration of OH^- and H_3O^+ ions in a solution are intrinsically linked: the product of the concentration of these two ions, $[OH^-][H_3O^+]$, is always equal to 10^{-14}. This relationship always holds true, even when the actual concentrations of OH^- and H_3O^+ ions change, as we will see next.

pH denotes the concentration of hydrogen ions in solution

Hydrogen ions drive many reactions in biology, so their concentration is of great interest and importance. The concentration of hydrogen ions can change due to the ionization of molecules present in a given aqueous solution. A variety of compounds can become ionized in aqueous solution to release either OH^- or H^+ ions. The H^+ ions (protons) associate with water molecules to form H_3O^+ ions, whereas the OH^- ions remain 'free' in solution.

When there is an excess of either OH^- or H_3O^+ ions, there will be a decrease in the concentration of the other such that the product of the concentration of OH^- and H_3O^+ ions is preserved, as per the relationship $[OH^-][H_3O^+] = 10^{-14}$ noted above. Molecules that are able to release H^+ when they dissolve in a solution are called **acids**, while molecules that are able to absorb (or 'accept') H^+ to reduce the concentration of H^+ in a solution are called **bases**. Although virtually all free H^+ ions associate with water to form H_3O^+, it is a convention in biology to refer to the excess H^+ as if they were free in solution.

In water, the H^+ concentration can vary greatly. This range in H^+ concentration is conveniently expressed in terms of **pH**, which corresponds to the negative log of the H^+ concentration (Figure 2.13). For example, if the concentration of H^+ in a given solution is 10^{-5} moles/L, the solution has a pH of 5 ($-\log 10^{-5} = 5$). In pure water, the H^+ concentration is equal to 10^{-7} moles/L. Pure water thus has a pH of 7, which we call 'neutral'. Solutions with a pH less than 7 are considered acidic. Conversely, solutions with a pH greater than 7 are called basic or alkaline. In the coming sections, we will see how the pH of a solution affects the ionization state of a variety of biological molecules.

Figure 2.13 The pH scale. The concentration of H^+ in a solution is expressed as the $-\log_{10}$ of the H^+ concentration.

The effect of pH on the properties of a biological molecule is important when we realize that the pH of the aqueous environments to which molecules are exposed in different parts of the human body can vary considerably. For example, the pH of the blood is 7.4 (almost neutral), whereas that of the gastric juices in the stomach is 1.6, which is very acidic. These variations in pH create challenges for pharmaceutical companies in the design of therapeutic drugs: a drug may have certain physical and chemical properties at one pH, but different properties at another pH. Drugs that are meant to be taken orally must be able to pass through the very acidic environment of the stomach without being destroyed, and therefore must be stable even at very low pH. A drug that is formulated for injection into the blood stream, however, simply needs to function at neutral pH.

2.3 NON-COVALENT INTERACTIONS

In Section 2.1, we explored the ways in which atoms come together to form molecules by forming covalent bonds, which hold atoms together as molecules. However, a number of weaker, non-covalent interactions are essential to the way different parts of a single molecule associate to give a molecule its defined three-dimensional shape, and to the way different molecules come together to form larger complexes. In this section, we explore non-covalent interactions, and discover their vital role in biological systems.

Non-covalent interactions allow dynamic interactions between molecules

The binding interactions that occur between biological molecules involve energies that are far lower than those of the covalent bonds, and can therefore be more readily disrupted and re-formed as they mediate dynamic processes in the cell. These weaker interactions, known as **non-covalent interactions**, underlie processes that require rapid or transient interactions, such as proteins binding to DNA to regulate gene expression, muscle proteins associating and then disengaging as the muscle fibers contract, and signaling proteins binding at cell surfaces. Non-covalent interactions also hold together more stable assemblages such as the molecular components of a ribosome or the two strands of a DNA molecule. These complexes have a lifetime of days or even years, but their components must be able to move apart and assemble to allow a protein chain to be synthesized, or the genetic code to be read.

All non-covalent interactions are fundamentally electrostatic in nature, which means that they depend on the attractive forces between opposite charges. There are three principal types of non-covalent interactions (Figure 2.14):

Non-covalent interactions			
type of interaction		distance between interacting atoms	energy of interaction (kJ/mole)
salt bridge	—CH₂—COO⁻·········⁺H₃N—(CH₂)₄—	0.27 nm	12.5–17
hydrogen bond	N—H ········ O=C (donor) (acceptor)	0.27 nm	2–6
van der Waals	—CH₃········H₃C—	0.38 nm	3–4

Figure 2.14 Non-covalent interactions. Three types of non-covalent interactions are shown. The distance over which each interaction operates is measured between the centers of the atoms involved.

- salt bridges, which are ionic interactions
- hydrogen bonds, which are interactions between polar atoms bearing partial charges
- van der Waals interactions, which arise from attractive forces between atoms of chemical groups. As we will see later, biological molecules bind to one another through a combination of these interactions.

A salt bridge is an ionic interaction between charged chemical groups

We saw in the previous section that chemical groups within a molecule can become ionized, and that groups with opposite charges can interact with one another, as in ionic compounds such as salt. We therefore call an ionic interaction between chemical groups a **salt bridge** (Figure 2.14). Since the electrostatic field of a fully charged atom extends at equal strength in all directions, the energy of a salt bridge does not depend on the precise orientation of the two oppositely charged groups and is only a function of the distance between them. This attractive energy between opposite charges falls off with increasing distance, r, between the two charges and is proportional to $1/r$. (By contrast, the force operating between the two charges is proportional to $1/r^2$.)

Hydrogen bonds between two dipoles form a common non-covalent interaction

Electrostatic interactions can also occur between atoms bearing partial charges, such as those that participate in polar covalent bonds. A **hydrogen bond** arises from electrostatic interactions between the positive end of one dipole and the negative end of another (Figure 2.14). The positive end is located on a hydrogen atom that is joined by a polar bond to an electronegative atom, and the negative end is located on a different electronegative atom, usually oxygen or nitrogen.

A simple example of hydrogen bonding occurs between water molecules, with the electropositive hydrogen of one water molecule forming a hydrogen bond with the electronegative oxygen in another (Figure 2.15). We refer to the hydrogen atom as the **hydrogen bond donor** and the electronegative partner – in this case the oxygen atom – as the **hydrogen bond acceptor**. Because hydrogen bonds depend on dipoles, whose electric field varies with angle about the dipole, the energy of a hydrogen bond is greatest when the three atoms involved lie in a straight line, as shown in Figure 2.15. Hydrogen bonds can impose a high degree of specificity on

Figure 2.15 Hydrogen bonds between water molecules. Hydrogen bonds between water molecules with a partial negative charge on the oxygen atoms and a partial positive charge on the hydrogen atoms.

the interactions between two binding partners because they require that hydrogen bond donors and acceptors line up in a precise manner at the binding interface in order to maximize binding energy. As is the case for all types of interactions, there is also an optimal distance at which the energy of hydrogen bonding is greatest.

van der Waals interactions can occur between any two atoms

Any two atoms may attract one another, whether or not they bear a full or partial charge. We saw in the previous section that atoms separated by somewhat more than the sum of their van der Waals radii weakly attract one another. This attraction is the **van der Waals interaction**. The van der Waals interaction arises when the close approach of two atoms causes each atom to induce transient dipoles in the other atom. These dipoles occur because the electron distribution around each atom varies from moment to moment. The distribution is not equal, but varies in an uneven, asymmetrical way. The uneven electron distribution produces a dipole, which in turn can induce a dipole of opposite polarity in a neighboring atom (Figure 2.16). The resulting favorable energy is due to attractive electrostatic forces between the opposing dipoles.

van der Waals interactions can occur between any two atoms, and are particularly important in interactions between non-polar atoms or chemical groups that cannot participate in other types of non-covalent interaction.

The aqueous environment of the cell affects the energy of non-covalent interactions

The behavior of all biological molecules is dominated by the aqueous environment of the cell. The interactions we have described occur in the presence of many water molecules, as well as different types of ions within the cell. This affects the energy of the interactions, and also influences which types of interaction are most favorable. Salt bridges, for example, seldom have the large favorable energy one would expect from the attractive force between full opposite charges because the surrounding water molecules and ions also interact with charged groups, thereby reducing the attractive energy of the salt bridge interaction.

The aqueous environment in the cell influences not only the strength of interactions between molecules, but the types of interactions that occur. Polar molecules and highly charged molecules dissolve readily in water because of the many hydrogen bonding interactions they can make with water. By contrast, highly non-polar molecules do not make energetically favorable interactions with water and prefer, instead, to associate with one another. We say that these non-polar molecules and chemical groups are **hydrophobic**, or 'water-hating', because they separate from water. For example, oils are hydrophobic compounds that separate from water: when an oil and water are mixed the oil molecules associate with one another to form a discrete layer that floats on the surface of the water.

Polar and charged molecules that make favorable interactions with water are termed **hydrophilic**, or water-loving. As we will see throughout this chapter, the distinct behavior of hydrophilic and hydrophobic chemical groups in water dictate the three-dimensional structure of proteins and nucleic acids, the formation of cell membranes, and the way in which biological molecules interact with one another.

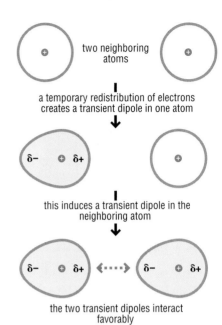

Figure 2.16 van der Waals interactions. Non-polar atoms typically have a uniform distribution of electrons about the nucleus, and hence no dipole moment. From time to time, there is a spontaneous redistribution of electrons about the nucleus that gives rise to a partial negative charge to one side and a partial positive charge at the opposite pole. If another atom is nearby, the transient dipole in the first atom can induce a transient dipole of opposite polarity in the second, such that the two atoms form favorable electrostatic interactions.

Hydrophobic interactions drive the folding of biological macromolecules

The terms 'hydrophilic' and 'hydrophobic' are not only applied to whole molecules. Instead, different portions of a single molecule may show distinct hydrophilic and hydrophobic characters. For example, one region of a polypeptide chain may be hydrophobic in character, whereas another region may be more hydrophilic.

Hydrophobic interactions arise from the way that hydrophobic portions of molecules preferentially cluster together when exposed to an aqueous environment. This 'clustering' usually results in the folding of hydrophobic regions towards the center of a macromolecule, such that these regions are shielded from their aqueous surroundings. By contrast, hydrophilic regions of a molecule can interact favorably with an aqueous environment, and will preferentially lie on the surface of a molecule, where they are exposed to the surrounding aqueous medium.

This partitioning – with hydrophobic regions of a molecule shielded from an aqueous environment, and hydrophilic regions directly exposed to it – is an important driving force behind the way that macromolecules such as nucleic acids and proteins **fold** into their characteristic three-dimensional shapes. For example, the characteristic three-dimensional structure of DNA – the double-helix – sees the hydrophobic bases located at the center of the molecule, and the hydrophilic sugar–phosphate backbone exposed more directly to the aqueous surroundings.

➲ We learn more about the thermodynamic driving forces behind hydrophobic interactions in Chapter 3.

Hydrophobic interactions are often grouped under the general umbrella of 'non-covalent interactions', although they refer more to the partitioning of hydrophobic groups in an aqueous environment.

Non-covalent interactions stabilize the binding of macromolecules at chemically complementary interfaces

Biological processes all depend on specific interactions between molecules in the cellular milieu. Enzymes must bind their substrates preferentially relative to other chemically similar species in the cell, while molecular machines, such as the ribosome, comprise a large number of molecules that interact in a precise way. The fundamental principles underlying all of these interactions are the same. Whether or not molecules bind to one another, and for how long, depends on the strength of the interaction and on the concentration of each component.

Figure 2.17 A chemically complementary interface. Two molecules bind to one another with surfaces that are complementary in shape. The interacting surfaces must also be chemically complementary: hydrogen donors (red) pair with acceptors (blue), hydrophobic patches (green) contact one another, and ionic interactions form between positively and negatively charged side chains.

The strength of the interaction between macromolecules is determined by the shape and chemical nature of the contacting surfaces. Favorable chemical interactions arise when hydrophobic groups on one surface interact with hydrophobic groups on the other, when there is pairing of hydrogen bond donors and acceptors, or when positively and negatively charged side chains interact (Figure 2.17). If the two surfaces have complementary shapes, with the bumps on one surface fitting into depressions on the other, it allows the two surfaces to fit very closely together and helps to maximize the number of favorable non-covalent interactions across the interface (Figure 2.17). In general, the strength of the interaction between two macromolecules increases with increasing size of the interaction interface. Preferential binding between certain molecules relative to others is dictated by relative binding strength, a concept generally referred to as **specificity**.

Specific biological interactions may be weak or strong, their specificity being determined by how well competing species bind relative to the interaction of interest. It is not the strength of a specific interaction, but rather the comparative strength of binding to the correct partner versus the incorrect partner that governs specificity. Two entities that exhibit a high degree of specificity typically experience a stronger binding interaction than two entities that exhibit a lower degree of specificity (and are less likely to interact). This difference underpins the way that different receptors in the cell show specificity for a particular signaling molecule and helps to ensure that the receptors are not inappropriately triggered by an array of different molecules. For example, if a receptor shows specificity for a particular signaling molecule, it is more likely to bind to that molecule than to any other, because it binds to the 'preferred' molecule particularly tightly.

2.4 NUCLEOTIDES AND NUCLEIC ACIDS

The physical laws we have described in the previous sections dictate how atoms are joined to make up the remarkable molecules in living things, as well as dictating the molecules' behavior. In the remainder of this chapter, we will learn about the major types of biological molecules and the properties that allow them to carry out their functions in the cell. We begin by considering the nucleic acids.

Nucleic acids are key molecules in fundamental processes of life

Nucleic acids are responsible for storing and transmitting genetic information and therefore have a central role in biology. Cells contain two types of nucleic acid, deoxyribonucleic acid (DNA) and ribonucleic acid (RNA). The function of DNA is to store genetic information in all cells and in some viruses. RNA can also store genetic information and forms the chromosomes of other viruses, but has many other functions including decoding the genetic information in all forms of life and catalyzing the key chemical reactions during protein synthesis.

DNA and RNA are polymers that are synthesized from building blocks known as **nucleotides**. Other types of nucleotides also participate in a wide range of biochemical processes that include working together with proteins to catalyze chemical reactions and transfer chemical energy, most notably through the activity of adenosine triphosphate (ATP). Nucleotides are also commonly used as carriers of chemical groups that can be transferred to other molecules, to which they are covalently joined. We will see throughout this book several examples of the central role played by nucleotides in the many biochemical processes in cells.

Cells contain nucleotides of two general types

Nucleotides are small molecules composed of an aromatic group known as a **base** (also called a **nucleobase**) a sugar, and one or more phosphate groups (Figure 2.18). RNA and DNA differ in the sugar their nucleotides contain. RNA contains the sugar **ribose**, while DNA contains the sugar **deoxyribose** (Figure 2.19). DNA features two types of bases: the **purine** bases, adenine (A) and guanine (G), and the **pyrimidine** bases, cytosine (C) and thymine (T) (Figure 2.20). RNA also

Figure 2.18 A nucleotide. Shown here is deoxyadenosine 5″-monophosphate (dAMP), which contains a base (adenine, pink), a sugar (deoxyribose, black), and a phosphate group (gold).

Figure 2.19 The sugars deoxyribose and ribose. According to convention, the five carbon atoms on the sugars are indicated by primed numbers. Deoxyribose differs from ribose only by the absence of an oxygen atom at the 2′ carbon (red).

(b) pK_a values for ionizable atoms in bases		
base	atom	pK_a
adenine	N1	3.88
cytosine	N3	4.56
guanine	N7	3.60
guanine	N1	10.0
thymine	N3	10.5
uracil	N3	10.1

Figure 2.20 Purine and pyrimidine bases. (a) Adenine and guanine are purine bases, whereas thymine, cytosine and uracil are pyrimidine bases. The atoms in the rings are numbered according to the conventions for purine and pyrimidine rings. (b) The table shows the pK_a values of ionizable atoms in the bases.

contains adenine, guanine, and cytosine, but with uracil (U) in place of thymine (Figure 2.20). The bases are planar, conjugated rings that are typically uncharged under physiological conditions, but can gain or lose a proton at near physiological pH (Figure 2.20). The **pK_a** values for each base, which are listed in Figure 2.20, correspond to the pH at which this transition generally occurs: a given atom is protonated at pH values below the pK_a and unprotonated at pH values above the pK_a. This property of bases is particularly important for the ability of RNA molecules to participate in chemical reactions, as we will see later in this book.

A base is joined to a sugar by a glycosidic bond, which is a covalent bond between the C1' atom of the sugar and the N9 atom of the purine or the N1 atom of the pyrimidine (the numbering convention is shown in Figures 2.19 and 2.20). A base and sugar covalently linked in this way is called a **nucleoside**. Nucleosides are named for the base and are called adenosine (A), guanosine (G), cytidine (C), thymidine (T), and uridine (U), and have a prefix indicating the linked sugar from which they are formed. Deoxyadenosine, for example, is composed of adenine covalently joined to deoxyribose. As it contains the deoxyribose sugar, deoxyadenosine is found in DNA, and not in RNA. By contrast, adenosine contains the ribose sugar and is therefore found only in RNA.

A nucleotide is formed when one or more phosphate groups are attached by phosphate ester linkages to either the 3' or 5' hydroxyl groups of the nucleoside sugar as illustrated in Figure 2.18, giving rise to **ribonucleotides** and **deoxyribonucleotides**. The term 'phosphate ester' describes the chemical nature of the covalent bond that joins the phosphate group to the nucleoside. The phosphate groups are acidic (hence nucleic acid), which means that they are negatively charged at physiological pH.

DNA and RNA are polymers of nucleotides

DNA is a nucleic acid polymer that is synthesized from deoxyribonucleotide building blocks. The nucleotides in DNA are joined by phosphodiester linkages between the 3' hydroxyl of one nucleotide and the 5' hydroxyl of the next (Figure 2.21). Each nucleotide is added with a 3' to 5' linkage in the same way, so that a DNA strand has defined directionality, with a 5' phosphate group exposed at one end and a 3' OH at the other end. The deoxyribose sugars and phosphodiester linkages comprise a monotonously repeating **sugar–phosphate backbone**, whereas the base attached to each sugar can vary from one nucleotide to the next along the DNA strand. By convention, the sequence of bases along a DNA strand is written starting at the 5' end and proceeding to the 3' end. It is this sequence of bases that we talk of when considering the 'sequence' of a strand of DNA.

Figure 2.21 A single strand of DNA. Nucleotides are connected by phosphodiester linkages, which join the 3′ hydroxyl of one nucleotide to the 5′ oxygen of the next. The sequence of bases (red) is read from the 5′ end of the DNA. In this case, the sequence is ATGC.

A single strand of RNA is similar in most ways to DNA. The key difference is that RNA nucleotides contain ribose in place of deoxyribose (see Figure 2.19) in addition to containing the base, uracil, in place of thymine. We shall see later that the additional 2′ oxygen in the RNA sugar, ribose, makes RNA more chemically reactive than DNA, as well as enabling RNA to adopt a wider variety of three-dimensional structures. This seemingly small chemical difference is thereby responsible for the greater functional versatility of RNA.

Bases can exist as tautomeric isomers

We have shown the arrangement of atoms in the bases in their most commonly found forms. However, these bases are in equilibrium with another form in which a proton has migrated to an alternative position; these forms of the base are known as **tautomeric isomers** or **tautomers** (Figure 2.22). In a typical population of nucleotides or nucleic acids, less than 0.01% of the bases are found in the rare tautomeric form. While seemingly uncommon, this still means that large genomes may contain a significant number of rare tautomers. If one considers the entire human genome of 3 billion base pairs, about 100 000 bases may adopt the rare tautomeric form at any given moment. As we shall see in Chapter 12, this has important consequences for the fidelity of DNA replication during cell division, and provides one mechanism for generating genetic variation.

Nucleotide monomers have diverse important functions in cells

In addition to functioning as the building blocks for DNA and RNA, a large variety of nucleotides play essential roles in the cell. One of these is to store energy derived from the breakdown of food in high-energy covalent bonds within molecules such as ATP. These bonds can then be broken to release energy that can be used to drive chemical reactions and mechanical processes, such as the motion of

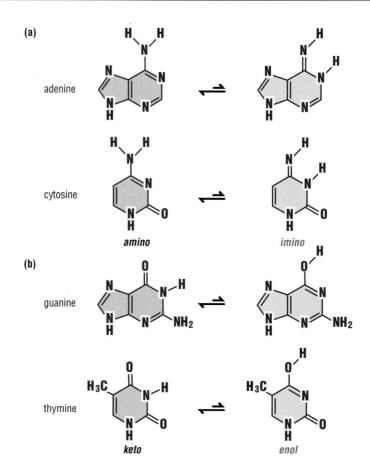

Figure 2.22 Base tautomers. Bases can adopt different arrangements of atoms called tautomers. (a) Adenine and cytosine normally adopt the common amino form (gray), but also occasionally adopt the imino form (blue). (b) Guanine and thymine are normally found in the keto form (gray), but can tautomerize to form much rarer enol forms (blue).

flagella that allows bacteria to 'swim'. We will learn more in the next chapter about how various biological processes can be powered by the input of energy.

Nucleotides also carry chemical groups that can be transferred to other molecules, a process that occurs in a wide variety of biosynthetic reactions as well as in the regulation of biological molecules. Figure 2.23 shows some examples of single nucleotides and the essential roles they play in processes that we shall encounter throughout this book.

2.5 THE STRUCTURE OF DNA

Single strands of DNA associate non-covalently to form a double-helical structure

DNA is unique among biological macromolecules in the simplicity and directness of the relationship between its structure and its biological function. In the cell, DNA molecules are double stranded. The bases on opposing strands form hydrogen bonds that cause them to pair with one another in a precise correspondence that is dictated by their geometry: cytosine on one strand makes three hydrogen bonds with the **complementary** surface of guanine on the other, whereas thymine on one strand makes two hydrogen bonds with the complementary surface of adenine on the other (Figure 2.24). We refer to these as **Watson–Crick base pairs** after the two scientists who correctly deduced the structure of DNA (see Experimental approach 2.1). Since the sequence of bases in one strand dictates the sequence of bases in

Coenzymes		
nucleotide	role	chemical formula
ATP	energy carrier, phosphoryl donor	
CoA	acyl group activation and transfer (e.g. acetyl–CoA, propionyl–CoA)	R: acetyl propionyl butyryl
S-adenosyl methionine	methyl group donor	
NAD⁺/NADH NADP⁺/NADPH	oxidation–reduction	NAD⁺: R = H NADP⁺: R = PO₃⁻
FAD/FADH₂	oxidation–reduction	

Figure 2.23 Table of common nucleotides.

the other, we say that the two strands are complementary to each other. We shall see in Chapter 6 that this complementarity allows genetic information to be copied because each DNA strand is a **template** for synthesizing a copy of the complementary strand. We shall also see in Chapter 8 that the same type of base-pairing interactions are used to allow synthesis of a complementary strand of RNA – the messenger RNA molecule – as the first step in decoding the genetic information.

Figure 2.24 Watson–Crick base pairs. Adenine pairs with thymine while cytosine pairs with guanine. These pairings are due to the formation of hydrogen bonds (red dashed lines) between the bases. The deoxyribose sugars to which the bases are attached are indicated in blue.

Two complementary DNA strands pair readily with one another when the strands are oriented in opposite directions, so that the 5′ end of one strand pairs with the 3′ end of the other. The two strands of DNA are therefore said to be **antiparallel**. The most energetically favorable conformation for double-stranded DNA is one in which the two strands wind around one another, forming a **double helix** (Figure 2.25). The helix shown is right-handed, meaning that the DNA strands spiral counterclockwise as you look down them.

The three-dimensional structure of double-stranded DNA can be explained by the physical properties of the nucleotides that make up the DNA strands. The hydrophilic sugar–phosphate backbone of each DNA strand lies on the outside of the helix where it can interact with water molecules, while the largely hydrophobic base pairs lie on the inside. The base pairs form a spiral stack at the interior of the double helix, with neighboring base pairs along the stack making favorable van der Waals interactions with one another (Figure 2.25a). At the same time, the hydrophobic faces of the bases are sequestered from the surrounding aqueous environment. This arrangement, called **base stacking**, is energetically highly favorable and is a major factor in stabilizing the structure of double-stranded DNA and thus contributing to the durability required for the robust storage of genetic information. Since both A–T and G–C base pairs have similar widths, the helix has a fairly uniform diameter of 20 Å along its length.

DNA in the cell is predominantly B-form

The double helix can adopt several forms, and the differences between the forms have important functional consequences. Double-stranded DNA in cells generally adopts a conformation called **B-DNA**, whose helical structure repeats itself around every 10.5 base pairs as one looks along the double helix (Figure 2.25a). Each successive base pair is separated by 3.4 Å. The two intertwining DNA strands produce a double helix with two grooves of unequal size: the wide

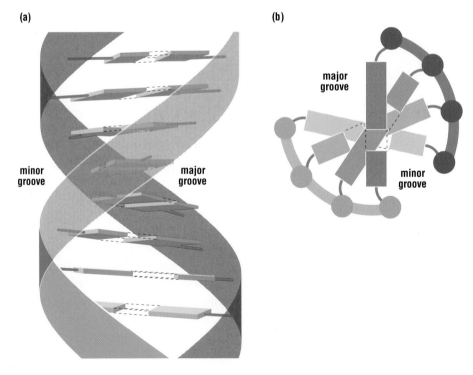

Figure 2.25 Double-helical DNA. (a) A side view of the helix showing that the base pairs (pink) are stacked perpendicular to the axis of the helix. The two DNA strands are antiparallel and wind around one another to form a helix with two different grooves of unequal width: the wider, major groove, and the narrower, minor groove. The helix winds around a full 360° turn about every 10.5 base pairs, with the base pairs about 3.4 Å apart; thus each turn of the helix is approximately 36 Å long. (b) A view looking down the B-DNA helix, showing how the wide major groove and the narrower, shallower minor groove are formed by the asymmetric attachment of the bases to the sugar-phosphate backbone (gray).

major groove (~13 Å) and the narrower, shallower **minor groove** (~9 Å width) (Figure 2.25b). The edges of the base pairs are exposed in these major and minor grooves. We shall see later and throughout this book how the structural features of the DNA helix, and in particular the atoms of the bases that are exposed in the grooves, govern the way in which proteins can interact with DNA in the processes of DNA replication, transcription, recombination, and repair.

DNA can adopt other conformations

Although the B-DNA helix is the dominant form in the cell, DNA can also adopt other helical conformations. One of these is the **A-DNA** helix (Figure 2.26), which has 11 base pairs per turn, with the base pairs tilted relative to the DNA axis as well as being displaced farther from the DNA axis than they are in B-DNA (Figure 2.26). As a result, the A-DNA helix has a narrower, deeper major groove and a shallower, wider minor groove than B-DNA. Although A-DNA differs in several details from B-DNA, they are both right-handed helices.

The A-helix conformation can be induced by some proteins that form a complex with DNA, and it is therefore thought that some segments of chromosomal DNA that are bound to these proteins may adopt a conformation closer to the A-DNA helix. By contrast to its minor role in DNA, we see in Section 2.8 how the A-type helix is particularly important in RNA structure, since this is the only type of double helix that double-stranded RNA can form.

The particular sequence of bases along a DNA strand can influence the structure of the double helix in a number of ways. Although we often draw the DNA helix as following a straight path, DNA can bend somewhat, and certain base sequences are more bendable than others. In general, regions of the DNA double helix containing successive A–T followed by T–A base pairs tend to bend more readily than helices with G–C steps.

An unusual type of DNA helix that spirals in the opposite direction, and is therefore left-handed, can form under special conditions in DNA strands with alternating

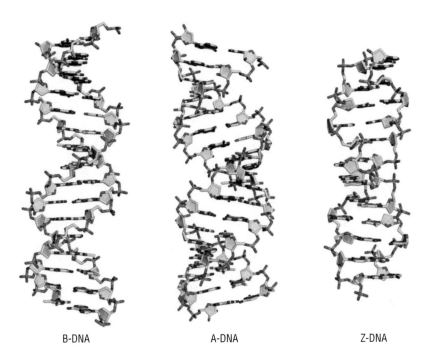

B-DNA   A-DNA   Z-DNA

Figure 2.26 A-, B- and Z-DNA. A comparison of the three principal DNA helix types: B-DNA; A-DNA; and Z-DNA. In contrast with A- and B-DNA, which adopt right-handed helices, the Z-DNA helix spirals in the opposite direction and is therefore left-handed.

2.1 EXPERIMENTAL APPROACH

The discovery of the DNA double helix

The most renowned paper in molecular biology is undoubtedly the 1953 publication by James Watson and Francis Crick of their proposed structure of B-DNA. In just a single page, the authors presented the structure of the DNA double helix and, in doing so, began a new era in biology. This was because the discovery of base-pairing and complementarity between the two DNA strands immediately suggested how genetic information could be copied and transmitted.

How did Watson and Crick arrive at the structure of B-DNA? In 1953, methods for determining the crystal structures of macromolecules had not yet been devised, nor were there the means for isolating or synthesizing a uniformly pure sample of DNA that would have been suitable for single crystal studies. Instead, Watson and Crick relied on a combination of basic chemical information about nucleic acids, model building, some findings about DNA composition in different organisms, and x-ray diffraction data on DNA fibers produced by Rosalind Franklin.

By 1953, the chemical composition of single-stranded DNA was known, and it had been proposed that the bases were most likely to be oriented perpendicular to the ribose sugar. There had been some question about which tautomer the bases adopted, but a chemist named Jerry Donohue informed Watson and Crick that the bases were in the enol form, which proved to be key in determining how base pairs formed. Edwin Chargaff had recently found that adenine and thymine were always found in a 1:1 ratio, as were guanine and cytosine. All of these provided clues to the underlying structure of DNA.

The key information, however, came from x-ray diffraction studies. Long DNA fibers could be isolated from cells that had been disrupted. These fibers, which consisted of many DNA strands, could be lifted out of a solution and oriented in an x-ray beam. The x-rays scattered from the fiber were then recorded on film (Figure 1); this scattering of x-rays is referred to as diffraction. The pattern of streaks and smears that results from x-rays striking the film are produced by x-ray waves that bounce off the DNA (or any other molecule) and combine with one another as they scatter in different directions. The particular pattern of constructive and destructive interference that is observed on the x-ray film depends upon the structure of the DNA fiber. Franklin had discovered that the details of the diffraction pattern changed depending on how hydrated the DNA samples were. The pattern obtained at low humidity was dubbed the 'A' pattern, while that obtained at high humidity was dubbed the 'B' pattern. This is the origin of the names for the A-DNA and B-DNA helix types that produced these patterns.

The unusual pattern of scattered x-rays seen in Figure 1, with the characteristic cross pattern and the presence of individual streaks

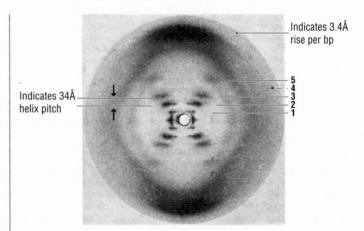

Figure 1 X-ray diffraction from B-DNA fibers. The appearance of the pattern is characteristic of diffraction from a helix, while the location of the smears and spots give some of the helix dimensions. Note that layer line 4 is missing, indicating that the two strands of the double helix are not evenly spaced along the helix axis, but instead are arranged to produce a larger (the major) and smaller (the minor) groove. Reproduced from Franklin & Gosling, 1953.

that fall on regularly spaced lines (known as layer lines), are characteristic of a helical molecule. Importantly, the details of the pattern, including the spacing between the layer lines and the location of the large smear along the vertical line, provided information on the dimensions of the helix and the spacing between each successive repeat. Diffraction from DNA fibers could even reveal that the DNA consisted of two intertwining helices, and that the strands were antiparallel to one another. These and other features could be determined from x-ray fiber diffraction patterns by taking advantage of the mathematical description of diffraction from helices.

Rosalind Franklin's famed diffraction image of B-DNA provided the high-quality data that was needed to deduce the parameters of the helix. From this photograph, it was possible to deduce that the sugar-phosphate backbone lay on the outside and was fairly extended. The location of the large smears along the vertical axis of the photograph (Figure 1) meant that the vertical distance between the repeat unit of the helix was 3.4 Å, while the spacing between the layer lines and the number of layer lines observed (not all are readily apparent in Figure 1) meant that the helix had ten base pairs per turn and that the helix repeated every 34 Å. The absence of a layer line (Figure 1; note missing layer line 4) meant that the two DNA strands were not evenly spaced relative to one another. It was also possible to determine that the helix was about 20 Å in diameter based on other features of the diffraction pattern.

Watson and Crick's leap of insight was to figure out how the base pairs fit together at the core of the helix. For this, they used model building and chemical intuition, along with knowledge of Chargaff's rule about base ratios, to arrive at the Watson–Crick base pairs we now know. The model was immediately appealing because it explained so much about the function of DNA, in addition to revealing its intrinsic beauty.

Definitive proof of the proposed model for B-DNA did not arrive for another 27 years, when the first crystal structure of a B-DNA fragment was determined. In contrast with the model built by Watson and Crick, the crystal structure yielded an experimentally determined atomic model of DNA, based on the x-ray data alone without any assumptions other than basic information about covalent bonds and atomic radii. Dick Dickerson and his colleagues determined the atomic resolution structure of a 12 base pair oligonucleotide, dubbed the 'Dickerson dodecamer', in 1980 (Figure 2). They took advantage of a new method for synthesizing DNA in solution that made it possible to purify large quantities of DNA molecules of a defined length and sequence, which could then be crystallized. By 1980, there were also well-established methods for determining crystal structures of macromolecules – something that had not yet been possible back in 1953. (See Chapter 16 for a description of x-ray crystal structure determination.)

The B-DNA crystal structure confirmed all of the essential features of the Watson and Crick model, with some interesting deviations. The DNA base pairs were not perfectly planar, but, rather, contained various degrees of propeller twist. In addition, the crystal structure had slightly more than ten base pairs per turn. Another key feature was that the helix axis was not perfectly straight, but rather was bent by about 19°. The authors duly noted that this degree of bending, when extended over longer DNA fragments, would be sufficient to allow DNA to wrap around a histone core to form a nucleosome, the fundamental unit of eukaryotic chromatin. Thus began a new era in the understanding of DNA structure and how it relates to its function.

Figure 2 Crystal structure of B-DNA: the Dickerson dodecamer. Note the bend in the DNA axis. Reproduced from Wing *et al.*, 1980.

Franklin RE, Gosling RG. Molecular configuration in sodium thymonucleate. *Nature* 1953;**171**:740–741.

Watson JD, Crick FH. Molecular structure of nucleic acids; a structure for deoxyribose nucleic acid. *Nature* 1953;**171**:737–738.

Wing R, Drew H, Takano T, *et al*. Crystal structure analysis of a complete turn of B-DNA. *Nature* 1980;**287**:755–758.

G and C bases. This helix, shown in Figure 2.26, is known as **Z-DNA**. Formation of Z-DNA is favored by a chemical modification of cytosine, methylation (which we shall learn more about later) and high salt concentrations. It is also promoted by torsional stress on the DNA. While the ability of DNA to adopt the Z-type helix *in vitro* is well established, there is some debate about the role it might play in the cell.

DNA can contain topological variations known as supercoiling

The small amount of flexibility between DNA base pairs allows large fragments of DNA, containing thousands of base pairs, to appear like flexible ropes in electron micrographs. If the DNA is a closed circle, as is the case for many bacterial or viral chromosomes, it typically looks highly intertwined and tangled when it is isolated from a cell. This is because the DNA is **supercoiled**, which places the molecule under tension and causes it to twist on itself (Figure 2.27). In contrast, a **relaxed DNA** circle is open and not under tension.

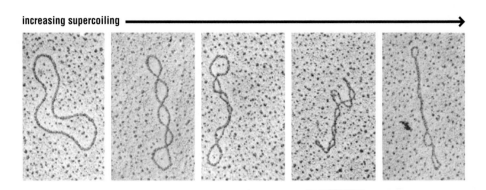

increasing supercoiling

Figure 2.27 Supercoiled and relaxed DNA.
These electron micrographs show a closed,
circular bacterial plasmid as the degree of
supercoiling increases.
From Kornberg and Baker. *DNA Replication*,
2nd ed. New York: WH Freeman.

To understand how supercoiled and relaxed circular DNA differ, imagine doing the following (Figure 2.28): take a covalently closed, circular DNA molecule, cleave both strands in one place, and then rotate one DNA end around several times before covalently closing the circle again. The resulting DNA circle is supercoiled and will wind about itself. This is because rotating the DNA end changes the helical twist of the DNA from its preferred state containing about 10.5 base pairs per turn to one with a fewer number of base pairs per turn. When the DNA is ligated, the untwisted portion of DNA will tend to spring back to adopt its favored state, with 10.5 base pairs per turn. This, however, causes the plasmid to wrap around itself as shown in Figure 2.28. The closed DNA circle is now supercoiled.

There are two 'directions' of supercoiling – positive and negative. An initial unwinding of the DNA – which is in the clockwise direction and is equivalent to separating the DNA strands – leads to accumulation of **negative supercoils** as

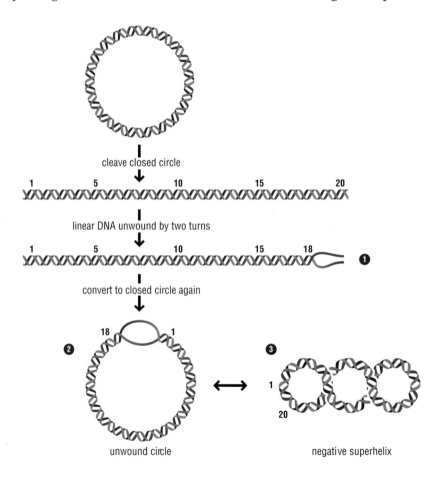

Figure 2.28 Introducing supercoils into DNA. Supercoiling can be introduced into DNA fragments whose ends are not free to rotate. In this example, the DNA in the linear fragment is unwound to remove two complete turns (1), then ligated to form a closed circle (2). The two unwound regions can twist up again to form a right-handed double helix, but this will cause two negative supercoils to form in the DNA (3).

DNA unwound by one helical turn

DNA helix forms one supercoil

Figure 2.29 Supercoils can form in constrained linear DNA. A change in DNA twist can cause supercoils to form in any DNA fragment that is constrained at both ends.

shown in Figure 2.28, whereas an initial twisting of the DNA in the counterclockwise direction leads to formation of **positive supercoils**.

Supercoiling can also occur in linear fragments of DNA in which the ends are immobilized and therefore not free to rotate and release superhelical tension (Figure 2.29). Thus, linear chromosomes or segments in the middle of a chromosome whose boundaries are constrained can also contain supercoils. Supercoiled DNA is under torsional strain (that is, there is tension in the molecule). If a nick or break is introduced into a supercoiled plasmid or chromosome, one end will swivel around the other to relax the supercoiling, and hence alleviate the strain on the DNA molecule.

DNA supercoiling is related to the linking number and twist

The relation between supercoiling and the winding or unwinding of the DNA helix can be described quantitatively. To understand this description, we must first define three quantities: linking number, twist, and writhe (which is another term for supercoiling).

Linking number (Lk) describes the number of times one of the strands of DNA wraps around the other. For each turn of the DNA double helix (roughly every 10.5 base pairs of B-DNA), one single strand wraps around the other once. A closed, circular DNA plasmid of 1050 base pairs that is relaxed therefore has a linking number of 100. An important feature of the linking number is that it cannot change in closed circular DNA – or in a DNA fragment that is constrained at its two ends – without cleaving one or both of the DNA single strands. We say that the linking number is a topological invariant of closed circular or constrained DNA, as it is unaffected by changes in the overall shape of the DNA.

The twist is a property of the double-helical structure of DNA. **Twist (Tw)** represents the number of turns in the DNA helix in a given fragment of DNA; a relaxed plasmid or a straight fragment of DNA that lies in a plane has a twist of +1 for each turn of the double helix. (The positive sign is for a right-handed helix, whereas a negative sign would pertain for a left-handed helix.) For example, a 105 base pair fragment of relaxed B-DNA has a twist of 10 (= 105 bp ÷ 10.5 bp/turn), while a 1050 base pair fragment of relaxed B-DNA has a twist of 100.

Writhe (Wr) describes the supercoiling of closed circular DNA (Figure 2.28) or of DNA constrained at two ends (Figure 2.29). A relaxed plasmid has a writhe of

◆ We see how the toroidal form of negatively supercoiled DNA is similar to the way in which chromosomal DNA winds around histone proteins, which is how DNA is packaged in the nucleus, in Chapter 4.

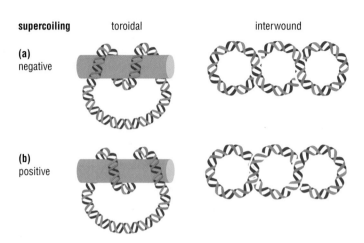

Figure 2.30 The toroidal and interwound forms of supercoiled DNA. Supercoiled DNA can adopt either of two principal forms: toroidal or interwound. (a) Negatively supercoiled DNA can adopt the toroidal form, in which the DNA spirals in a left-handed manner, or an interwound form. (b) In the toroidal form of positively supercoiled DNA, the DNA spirals in a right-handed manner. Note how the interwound forms of positively and negatively supercoiled DNA are twisted in the opposite manner.

zero; positively supercoiled DNA has a value of writhe greater than zero while negatively supercoiled DNA has a value of writhe that is less than zero. Figure 2.30 shows the different forms adopted by positively and negatively supercoiled DNA circles. Note in the figure how supercoiled DNA can adopt either the toroidal form or the interwound form.

Having defined linking number, twist, and writhe, we will now see how changing the twist of a plasmid affects supercoiling. Although the linking number is constant for a closed circular or constrained DNA fragment, twist and writhe are free to vary according to the relation Lk = Tw + Wr. This means that any change in twist must be balanced by an opposite change in the value of writhe (supercoiling).

In the example discussed above and shown in Figure 2.31, the DNA has been unwound by two turns (a decrease in twist and linking number of 2) and the writhe is kept equal to zero before ligating the DNA circle. However, it is not favorable for the DNA to have a region in which two turns have been unwound, and this segment will tend to anneal and twist again to adopt a standard double helix.

If the twist goes up by +2, the writhe must change by –2; in other words, the DNA now contains two negative supercoils. As long as the plasmid is not nicked, such that the linking number cannot change, these two forms – one with no supercoils but a local region of unwinding (Tw = 18, Wr = 0) and the other with two negative supercoils (where the twist increases by 2, and the writhe decreases by 2, leading to Tw = 20, Wr = –2) – can interchange with one another.

Negative supercoiling favors local unwinding of DNA

The balance between DNA twisting and supercoiling has important biological consequences. As shown in Figure 2.31, negative supercoils in DNA can be removed by unwinding one or more turns of DNA, which is equivalent to separating the strands. In this way, the unfavorable energy of strand separation is partially compensated for by relieving superhelical tension due to negative supercoiling. This explains why negatively supercoiled DNA templates are needed for many biological processes that require separation of the DNA strands.

⊘ We look at topoisomerases in more detail in Chapter 4.

There are cellular enzymes called **topoisomerases** that introduce or remove supercoils from DNA in the cell. Topoisomerases change the supercoiling by introducing a transient break in either one or both DNA strands in the double-stranded DNA, depending on the type of enzyme. Since increasing the degree of supercoiling is energetically unfavorable, enzymes that introduce additional

Figure 2.31 The effect of supercoiling on twist and writhe. Building on the example in Figure 2.28, DNA in the linear fragment is unwound to remove two complete turns (1), then ligated to form a closed circle (2) with a linking number (Lk) of 18 and a twist (Tw) of 18. If the two unwound regions twist up again (3) to form a right-handed double helix we see a change in twist of +2; this causes two negative supercoils to form in the DNA, leading to a change in writhe of −2. Note that the linking number remains the same in both cases.

supercoils into DNA – usually negative – must use the energy provided by ATP hydrolysis to drive this process. Some topoisomerases can also separate chromosomes that have become tangled and, as we shall see later, play an important role during DNA replication.

2.6 CHEMICAL PROPERTIES OF RNA

RNA is chemically distinct from the related nucleic acid, DNA

RNA is a nucleic acid that in many ways is similar to DNA, but with key chemical differences that give it far greater functional diversity. Like DNA, an RNA molecule is a polymer of nucleotides that are joined by phosphodiester linkages between the 3′ end of one nucleotide and the 5′ end of the next (Figure 2.32). As already explained above, the principal chemical differences are that RNA nucleotides contain ribose in place of deoxyribose and uracil in place of thymine (see Figures 2.19 and 2.20). These differences do not affect the fundamental property of base-pairing; uracil can form base pairs with adenine just as thymine does. Further, it is the ability of RNA to form base pairs with DNA, as well as with RNA, that allows cells to synthesize RNA copies of genes from a DNA or RNA template. However, it is the presence of the 2′ OH on the ribose and the distinct chemical structure of uracil, as compared with thymine, that allow RNA to adopt diverse structures, which are not readily formed by DNA.

RNA is often chemically modified after it is synthesized

Many RNA molecules require chemical modification of some of their nucleotides in order to become fully functional. These modifications, which alter the chemical structure of the bases, are introduced after the RNA is synthesized. They are generally not found in messenger RNA (mRNA, the RNA that acts as an intermediate in the flow of genetic information from DNA to proteins), but rather in RNA molecules that carry out other functions in the cell. These modifications are typically referred to as **post-transcriptional** modifications. Examples of modified uridine nucleotides are shown in Figure 2.33. Nearly 100 modified nucleosides have been

Figure 2.32 A single strand of RNA. RNA nucleotides are joined by single phosphodiester linkages, just as in DNA. The differences between RNA and DNA lie in the 2′ OH of ribose, rather than the 2′ H of deoxyribose found in DNA, and in the substitution of uracil in place of thymine. (See also Figures 2.19 and 2.20, which highlight these differences.)

⮕ We explore post-transcriptional modifications in Chapter 9.

uridine   dihydrouridine   pseudouridine
                  (D)   (ψ)

Figure 2.33 Examples of modified RNA bases. Left, the standard uridine nucleotide base; middle, dihydrouridine, found in the D-loop of tRNA; right, pseudouridine, found in abundance in eukaryotic rRNAs.

identified to date, and there may well be additional modifications that have not yet been discovered. In contrast with DNA methylation (Section 2.4), which plays a role in regulating whether a gene is transcribed or not and is reversible, chemical modifications of RNA are largely irreversible and do not play a regulatory role.

The best-studied examples of modified nucleotides in RNA are found in the protein synthesis machinery. For example, the transfer RNA (tRNA) adaptor molecule must undergo numerous chemical modifications in order to be able to carry out its function. Around 10% of the roughly 75 nucleotides in tRNA are chemically modified after the RNA is transcribed. The ribosome, which is the large complex of protein and RNA molecules that synthesizes proteins, similarly contains many modified nucleotides in the ribosomal RNA (rRNA). In human rRNAs, approximately 200 nucleotides are modified after transcription, which represents about 3% of the 6500 nucleotides in rRNA.

The position and nature of nucleotide modifications in a particular RNA molecule are often conserved among different species. This conservation suggests that these modifications play important roles in the molecules in which they are found (otherwise they would not have been conserved!). The chemical modifications expand the chemical diversity of RNA that extends its functional repertoire beyond that permitted by the four standard RNA nucleotides, A, U, C, and G. As we will see later, proteins have greater chemical diversity than RNA even without chemical modifications, since they are polymers constructed from 20 chemically distinct building blocks – the amino acids.

The 2′-OH of ribose has important structural and functional consequences

The presence of a 2′ OH group in the RNA ribose, as compared with deoxyribose in DNA, has important consequences for RNA structure and function. The chemical

Figure 2.34 Role of 2′ OH in favoring the A-type helix form in RNA. The 2′ OH in the ribose would clash with other backbone atoms if RNA were to adopt the B-type helix.
From Berg *et al.* Biochemistry, 5th ed. New York: WH Freeman.

stability of RNA polymers is much lower than that of DNA because the oxygen in the 2′-OH of ribose facilitates a reaction that can cleave phosphodiester bonds. The chemical features of this simple reaction will be described in more detail in the next chapter, when we discuss catalysis (see Section 3.4). The instability of single strands of RNA, as compared with the far greater stability of DNA, is very likely the reason that DNA was selected during evolution for the long-term storage of genetic information. The less stable RNA molecule has been retained as the intermediate messenger, since mRNA is typically needed for only a limited period of time before it degrades chemically or is actively broken down by enzymes in the cell when it is no longer needed.

The additional hydroxyl group in the ribose also has important consequences for the three-dimensional structure of RNA molecules. Like DNA, RNA molecules can form double helices with base-pairing interactions between guanine and cytosine, and between adenine and uracil. However, the presence of the 2′ OH on the ribose causes double-stranded RNA to exclusively adopt the A-type helix (one of the helix types that DNA can adopt, as we saw earlier) for two reasons:

- First, the 2′ OH of ribose would clash with the sugar–phosphate backbone if RNA was to form a B-type helix (Figure 2.34).
- Second, the additional hydroxyl group on the 2′ position of ribose causes the sugar ring to buckle in a distinct fashion known as C3′ endo, which favors formation of an A-type helix, instead of the C2′ endo conformation typically found in B-type helices (Figure 2.35). These distinct conformations of the five-membered ribose ring are referred to as the **sugar pucker**.

Figure 2.35 Sugar pucker in RNA and DNA. The sugar in RNA and DNA is non-planar and can adopt a variety of conformations referred to as sugar pucker. The top structure is energetically unfavorable and is shown here to highlight the chemical structure of the ribose sugar. The C3′ endo conformation is adopted in the A-type helix, whereas the C2′ endo prevails in the B-type helix.

Perhaps the most profound consequence of the 2′ OH of RNA derives from its ability to form hydrogen bonds. While, as we saw earlier, DNA can also form hydrogen bonds, the 2′ OH of RNA confers on RNA the ability to form hydrogen bonds more extensively than DNA. This added feature of RNA expands the potential for forming multiple interactions between nucleotides. We shall see in the next section how such hydrogen bonding by the 2′ OH promotes formation of complex structures that are unique to RNA molecules.

2.7 RNA FOLDING AND STRUCTURE

RNA can fold to form compact structures

Although DNA in cells is found overwhelmingly in a helical, double-stranded form, a far greater variety of RNA structures is found in biology. Many of these arise from single-stranded RNA molecules that fold to form compact three-dimensional structures. These RNA molecules do not typically code for proteins, and thus are termed 'non-coding'. Instead, they carry out structural, catalytic, and regulatory roles in many cellular processes including protein synthesis. The precise conformation, or **fold**, of an RNA molecule is determined by its nucleotide base sequence. Together, the fold and nucleotide composition determine the functional properties of the RNA. An ever-increasing number of genes in the human genome are known to be transcribed into non-coding RNA molecules, indicating the widespread involvement of these RNA molecules in many cellular processes.

RNA molecules can be described in terms of three levels of structural organization. The primary structure of RNA refers simply to the sequence of bases along an RNA strand, reading from the 5′ to 3′ direction. Portions of the RNA molecule can form short double-helical stretches, and we refer to these regions as RNA secondary structure, to distinguish them from the primary structure of nucleotides in the RNA strand. The single- and double-stranded portions of the RNA molecule come together to form a precise arrangement, which is its tertiary structure. We will expand upon each of these concepts in the sections that follow.

Short RNA helices are the building blocks of larger RNA structures

The fundamental structural unit of folded RNA molecules is the short, double-stranded helix, generally no longer than six to eight base pairs in length (Figure 2.36). These double-stranded regions of RNA are A-type helices (see Figure 2.26) that form when two antiparallel nucleotide sequences form Watson–Crick base pairs with one another. If these complementary sequences are relatively close together in the linear or primary RNA sequence, the RNA forms what is called a **hairpin** structure (Figure 2.36). The double helical portion of the hairpin is called the stem and the unpaired bases are called the loop. Complementary sequences that are widely separated from one another in the primary structure can also pair to form double-helical stretches, but they lack the hairpin. The collection of these base-paired elements (or helices) within a single RNA molecule is depicted in Figure 2.36a. Both types of helix are seen in this diagram of the secondary structure of a tRNA molecule – the D helix (in green) is formed from neighboring elements in the primary structure whereas the acceptor helix (in red) is formed from more distant elements.

(a)

(b)

Figure 2.36 Double-helical regions of RNA. Single-stranded RNA molecules fold back on themselves to form short regions of double-stranded helices. (a) This RNA, called tRNA^Phe, has four helical regions, three of which are hairpin structures containing a double helical region (blue, green, and orange) and a loop. The helix shown in red brings together the start and the end of the RNA strand to form a long-range pairing interaction. Several post-transcriptionally modified nucleotides are indicated: ψ is pseudouridine, D is dihydrouridine, T is ribothymidine, and H is a highly modified adenosine found proximal to the anticodon. (b) In the three-dimensional structure of this RNA, coaxial stacking of the helical segments to form the overall L-shape can be easily seen.

Figure 2.37 A base pair containing a methylated base. This base pair between a di-methylated guanine and adenine is non-planar because a methyl group in guanine clashes with the adenine ring.

Figure 2.38 Wobble base pair. The wobble base pair U–G, is shown next to the canonical Watson–Crick base pair, C–G.

Unlike DNA helices, RNA helices regularly contain base pairs that do not result from typical Watson–Crick interactions. Termed **non-canonical base pairs**, these base pairs often feature bases that have been chemically modified in some way. A common modification is the addition of a methyl group in a process called methylation, as depicted in Figure 2.37. Non-canonical base pairs are accommodated within double-helical regions, although they can cause distortions in the helix. These distortions are important for function. For example, there is a pairing interaction between a methylated G and an A in certain tRNAs that is critical for formation of a functionally important kink in the structure.

One of the more commonly observed non-canonical base pairs is formed between G and U nucleotides. This base pair is stabilized by two hydrogen bonds, just like a Watson–Crick A–U base pair, and is referred to as a 'wobble' base pair (Figure 2.38). We will discuss wobble interactions again when we learn about tRNA decoding interactions on the ribosome during protein synthesis.

Short helices assemble to form the final folded RNA structure

The full tertiary structure of RNA is formed when double-helical regions interact with one another, as well as with unpaired stretches of the RNA. The strong tendency of base pairs to stack upon one another gives rise to **coaxial stacking**, in which the helices stack with their central axes roughly lined up, as shown in Figure 2.36b. RNA structures typically contain several coaxially stacked helices. As seen in Figure 2.36b, which shows a tRNA molecule, each branch of the L-shaped molecule is composed of a distinct coaxial stack.

A second general feature of RNA structure is typified by interactions between nucleotides in unpaired and base-paired regions of a single RNA molecule. Hydrogen bonding interactions involving the 2′ OH of ribose or the exposed chemical groups on the bases are particularly common. An example is the **base triple** interaction, illustrated in Figure 2.39a, in which three bases interact with

(a)

(b)

Figure 2.39 Tertiary interactions that stabilize RNA structure. (a) CGG base triple in which an isolated G residue interacts with a Watson–Crick G–C base pair in the major groove. (b) A-minor motif, in which an adenosine nucleoside forms hydrogen bonds with the ribose and bases in the minor groove of a helix.

Figure 2.40 Example of an RNA structure. The tertiary structure of the group I intron provides an example of multiple sets of coaxially stacked helices and buried magnesium ions (red) that stabilize the overall fold of the molecule by minimizing charge repulsion of the phosphodiester backbone.

A. laidlawii	CAACGCCCUUUUAAGGCGUGG
A. lipoferum	CAUCUGACUUUUAAUCAGAGG
B. burgdorferi	CAGCGCCCUUUUAAGGCGUUU
B. subtilis	CAUCUGACUUUUAAUCAGAGG
E. coli	CAGUUGACUUUUAAUCAAUUG
H. pylori	CAAUUCCCUUUUAAGGAAUGG
M. genitalium	CAUUUGACUUUUAAUCAAAGG
M. Pg50	CAACUGGCUUUUAACCAGUGG
M. pneumoniae	CAUUUGACUUUUAAUCAAAGG
S. aureus	CAUCUGACUUUUAAUCAGAGG
T. pallidum	CAACGCCCUUUUAAGGCGUGG

covariation

anticodon

Figure 2.41 Phylogenetic analysis in RNA structure. Aligned tRNA^{Lys} sequences are used to identify covarying regions (pink) that are predicted to form paired regions in the final folded RNA structure.

one another, two by Watson–Crick hydrogen bonds and the third, unpaired base by hydrogen bonding with the exposed edge of the base pair. Another common example is the **A-minor motif**, where adenosine from one part of the molecule inserts into the minor groove of a double-helical region found elsewhere in the molecule. The unpaired adenosine hydrogen bonds with one or two 2′ OH groups flanking this groove as well as with the bases themselves (Figure 2.39b). These are just two examples of the many types of intramolecular interactions that stabilize a folded RNA structure.

A potential barrier to the folding of RNA into a compact three-dimensional shape is the high density of negative charge associated with the phosphodiester backbone, which would give rise to electrostatic repulsion when different segments of the backbone come together. RNA molecules overcome this problem by binding a large number of cations, most often magnesium. These positively charged ions balance the negative charges of the oxygen atoms in the sugar–phosphate backbone, allowing the backbones of different regions of the RNA to pack against one another as they are buried in the interior of the molecule. Figure 2.40 shows an example of a folded RNA structure with bound magnesium ions in the interior of the molecule that stabilize structure highlighted in red.

RNA secondary structures can be identified by phylogenetic analysis

The helical or paired regions of an RNA that assemble to form a complex tertiary structure can frequently be identified by searching in the RNA sequence for potential base-pairing interactions. This can be done with a computer program that searches for complementary sequences throughout the RNA molecule. Since this analysis can only identify regions with the potential to form helices, it will invariably pinpoint some short regions of complementarity that occur purely by chance, but that do not actually form helices in the RNA structure.

An approach that helps to identify which of the potential helix-forming regions indeed form helices in the folded RNA structure is known as **phylogenetic analysis**. This approach takes advantage of the fact that functionally equivalent RNA molecules are expected to have very similar secondary and tertiary structures. Related RNA sequences (often from different species) can therefore be analyzed individually for potential double-helical regions, and the location of those regions can then be compared among different species.

For a region of the RNA structure that is indeed double helical, the base-pairing interactions will be conserved even though the actual nucleotide sequence may differ among species (Figure 2.41). This is because there is selective pressure in the course of evolution to preserve the overall structure of the RNA, rather than the base sequence. Differences that conserve interactions are known as **covariations** because sequence changes at interacting bases are coupled in a way that preserves base-pairing. This cross-species comparison is central to the analysis of functional RNA molecules and is similar in concept to comparative approaches used to analyze **protein folds**, as we shall see in Section 2.11.

2.8 THE RNA WORLD AND ITS ROLE IN THE EVOLUTION OF MODERN-DAY LIFE

RNA biology provides insights into the origins of life

Scientists have long been fascinated by the question of how life evolved on Earth. At some point in the distant past, there must have been a transition from a mixture of reactive molecules in the primordial seas – the early environment in which life evolved – to the formation of life. While defining life is a notoriously difficult task, the most fundamental properties of living organisms are that they are able to reproduce themselves and that they evolve as a result of chemical variations that arise as they reproduce. Although we cannot know what the first life forms were like, we can speculate about them based on our knowledge of chemistry and the various forms of life on Earth today.

Any autonomously replicating system that continually grows and divides requires a supply of small molecules and energy. Early life forms must have been chemically far simpler than modern organisms and are presumed to have depended upon a more limited set of biological molecules. As living things evolved to become more complex, they made use of additional classes of molecules and incorporated the necessary chemical reactions into their biological pathways. This led to the evolution of the common ancestor of all modern life (known as LUCA, the last universal common ancestor), which likely depended on RNA, DNA, proteins, carbohydrates and lipids to fulfill the requirements for life. Heritable information in this ancestor was stored as sequences of base pairs in a nucleic acid polymer, while other biochemical functions were, to a large extent, carried out by proteins. There are, as far as we know, no modern-day examples of organisms that resemble an earlier form of life from which all forms of life subsequently evolved, but it is interesting to speculate about the molecules these earlier life forms could have used. In particular, we would like to know how life evolved from simple chemical processes in the primordial world.

There is modern-day evidence for the earlier existence of an RNA world

There is a great deal of circumstantial evidence that nucleic acids or similar molecules were the dominant biological macromolecules in the earliest forms of life, and were responsible for carrying out most biological functions before a sophisticated protein synthesis apparatus evolved. Nucleic acids are replicable by nature (since the base-pairing interactions allow one polynucleotide strand to serve as a template for another), while proteins are not. In addition to being able to store genetic information, RNA molecules in modern organisms can also catalyze chemical reactions, as well as adopt complex three-dimensional structures that carry out a variety of other biological functions. It is therefore possible to imagine an earlier world in which RNA acted both as an information carrier and as the primary functional entity in cells, performing all the tasks needed by a primitive life form.

Some of the most compelling arguments for a putative RNA world derive from the diverse roles that RNA plays in modern-day cells. RNA plays a central role

Figure 2.42 Remnant of the RNA world.
The ribosome contains both RNA and proteins, with the RNA component carrying out most key functions during the synthesis of proteins.

in processes including RNA processing, protein synthesis, and the regulation of gene expression. In these different roles, RNA can be a repository of information, provide the mechanism to decode the stored information, and sometimes be the catalyst to drive chemical reactions. Protein synthesis is the quintessential example of all these facets of RNA: the functional core of the ribosome (Figure 2.42) is composed predominantly of RNA; the molecule encoding a protein is contained in mRNA; and molecules that decode the information in the RNA message and orient amino acids for polypeptide synthesis are provided by RNA. When considering the evolution of protein synthesis, it is difficult to understand how a protein synthesis machine could be made of protein before such machinery exists – the classic chicken or egg first problem. But it is not difficult to imagine a protein synthesis machine composed entirely of RNA that emerged from a world in which RNA was the dominant macromolecule.

Another persuasive argument for life's origins in an RNA world lies in the composition of the many enzyme cofactors used by modern-day organisms. As we will learn in the next chapter, many protein enzymes that catalyze chemical reactions depend on a small molecule that is referred to as a coenzyme. Many of these coenzymes are ribonucleotides that are built from a core structure of the nucleoside, adenosine. For example, the coenzymes involved in oxidation and reduction reactions, including nicotinamide adenosine dinucleotide (NAD), nicotinamide adenosine dinucleotide phosphate (NADP), flavin adenine dinucleotide (FAD), and flavin mononucleotide (FMN), are composed of organic ring structures attached to a core adenosine moiety (Figure 2.43). The adenosine serves as a carrier of the chemically active group but is not, itself, directly involved in the

Figure 2.43 Remnant of the RNA world. Three examples of ribonucleotide cofactors that are all built from adenosine (blue): nicotinamide adenosine dinucleotide (NAD), acetyl-coenzyme A (acetyl-CoA), and S-adenosylmethionine. The chemical groups that can be transferred to other proteins are shown in pink.

facilitated reactions. Similarly, adenosine-based nucleotides are the carriers for different chemical moieties that are transferred to proteins or nucleic acids in a variety of cellular reactions. Because each cofactor is used by many different enzymes to carry out similar functions, it has been suggested that the dependence of all current organisms upon these nucleotide cofactors became fixed at an early stage in evolution. To change a cofactor would compromise the function of all the enzymes using that cofactor, and the organism would struggle to survive such a deficit. Thus, the biological roles of nucleotide cofactors were locked into place during a time when RNA dominated the biochemical scene and cofactors were built from the molecular building blocks that were available.

Other arguments for the existence of an RNA world have rested on the ability of both natural RNA molecules and those derived from *in vitro* experiments to fold into complex three-dimensional structures that catalyze chemical reactions. Some researchers argue that the recently discovered RNA molecules called ribo-switches, which control gene expression throughout the bacterial world, might also be remnants of an RNA world. While these hypotheses are subject to considerable debate, these ideas have helped guide our thinking about the types of molecules that may have been found in an early RNA world whose functions have since been largely taken over by the more chemically diverse proteins.

Prebiotic chemistry and the feasibility of an RNA world

While the molecular evidence makes a strong case for the predominant role of nucleic acid early in the evolution of life, scientists have long struggled to understand how such an RNA world could have emerged from the primordial soup (a term that refers to the early environment in which life evolved mentioned above). Early experiments to simulate the primordial soup, using spark chambers and presumed primitive earth components such as water, methane and ammonia, readily generated reasonably diverse mixtures of amino acids. However, identifying conditions that lead to the robust synthesis of nucleic acids has been much more difficult. First, there are several chemically distinct components that are needed: the nucleotide bases, the sugar moieties, and the phosphate backbone. Although adenine is synthesized efficiently from mixtures of hydrogen cyanide and ammonia, the other bases (G, C, and U) are much less readily synthesized. Prebiotic synthesis of the sugar ring, ribose, presents another significant chemical challenge, as does formation of the glycosidic bond between the bases and the sugars. Despite these and other feasibility issues surrounding the synthesis of nucleic acid-like material on a primitive earth, the concept of an RNA world is securely founded on the historical record provided by modern biology and it is here that we look to understand our origins.

This glimpse of the place of nucleic acids on the primitive earth brings us to the end of our initial exploration of the nucleic acids. What we have seen here is just the foundation for what is to come in later chapters, where we investigate further the pivotal role that nucleic acids play in the biological processes that underpin life. Nucleic acids are only part of the story of genome function, however. Their position center-stage must be shared with another vitally important group of biological macromolecules, the proteins, which we explore in the next three sections.

2.9 FUNDAMENTALS OF PROTEIN STRUCTURE

Proteins are the most versatile macromolecules in biology

Almost every aspect of biology relies on proteins. Proteins can catalyze chemical reactions and form structural components of a cell, they can create channels in a membrane to allow molecules to pass in and out of cells, or they can recognize a vast array of infectious microorganisms. Proteins also form molecular machines that convert energy from one form into another, for example allowing muscles to contract and plants to harness the energy of sunlight. Proteins can carry out such an astonishing variety of functions because they are synthesized from a chemically diverse set of building blocks. This chemical diversity makes it possible for different proteins to adopt highly specialized structures that have been optimized through evolution to perform their particular tasks. We shall see here and in the following sections how the chemical makeup and structure of protein molecules has allowed them to evolve to be the dominant macromolecule in the cell.

Amino acids are the building blocks of proteins

Proteins are polymers composed of **amino acids** that are joined by covalent bonds. This is analogous to the way that nucleic acids are polymers of nucleotides, which are also joined by covalent bonds. Each free amino acid has an amino group, a carboxylate group, and a side chain (Figure 2.44), attached to a central carbon atom (denoted Cα). There are 20 different types of side chain, each with distinct chemical properties (Figure 2.45). There are side chains that are largely hydrophobic in character, and those that contain polar chemical groups. A few side chains contain both hydrophobic and hydrophilic groups, and are as a result termed **amphipathic** side chains. Some side chains have a combination of properties, for example, lysine has a positively charged amino group at the end of a long hydrophobic side chain.

As we saw earlier for nucleic acid bases, some amino acid side chains can gain or lose a proton and therefore carry an electrical charge at neutral pH (Figure 2.46). The basic amino acids, lysine and arginine, bear a full positive charge at neutral pH, while aspartate and glutamate bear a full negative charge. The typical pH at which the transition between the charged and uncharged species occurs is the pK_a.

The unique chemical properties of each side chain govern the interactions of side chains with one another, with other biological molecules, and with the aqueous environment of the cell. **Sulfhydryl** (–SH) groups on cysteine side chains can react with one another to form a covalent –S–S– linkage called a sulfhydryl, which is an important bond for stabilizing protein structures. Importantly, however, disulfide bonds are only found in extracellular proteins (those that are secreted from the cell) because the interior of the cell is a reducing environment – an environment in which the –S–S– linkage is unable to form.

The chemical properties of side chains also allow them to play key roles in enzymatic reactions. The side chain of histidine has a pK_a around 6.5, which means that the transition between the protonated and unprotonated form of histidine occurs near the neutral pH of the cell. This property enables histidine to either donate *or* accept a proton during chemical reactions.

Figure 2.44 An amino acid. Chemical structure of an amino acid (side chains shown in blue), showing a central carbon atom (Cα) to which is attached an amino group (–NH$_3^+$), a carboxylate group (–COO⁻), a hydrogen atom, and a side chain (which is by convention shown as R).

Figure 2.45 The 20 amino acids. Shown for each amino acid is the chemical structure, the full common name, and the three- and one-letter abbreviated name. The pK_a values are shown for ionizable side chains. Here we group the amino acids according to their hydrophobic or hydrophilic character, but there are other ways to classify them.

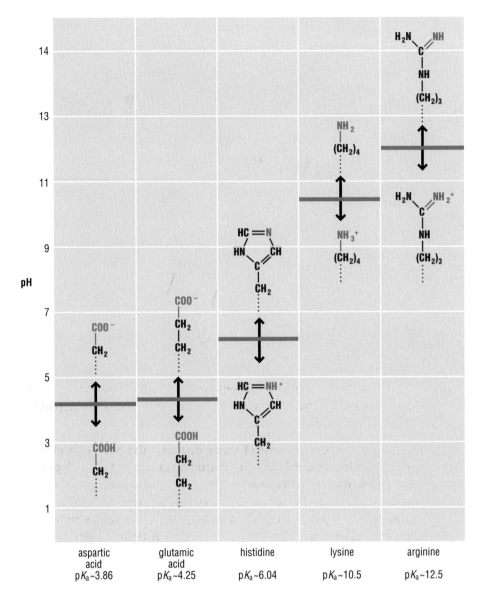

Figure 2.46 Amino acid side chains can become charged. The charge of any amino acid depends upon its pK_a (horizontal gray line) and the pH of the solution. The common charged amino acid side chains are shown, along with their chemical structures at pH values above and below the pK_a. The ionizable group is indicated in red (acidic) or blue (basic).

With the exception of glycine, the amino acids exist as two distinct stereoisomers that differ in the arrangement of atoms about Cα: the L- and D-amino acids. Like all stereoisomers, the L- and D-amino acids contain identical types of atoms and chemical bonds but are mirror images of one another, and are therefore not superimposable in space. Only L-amino acids are used by the main protein-synthesis apparatus of the cell, the ribosome.

Amino acids are joined by peptide bonds to form polypeptides

A protein is composed of amino acids that are joined by characteristic covalent bonds. These linkages, called **peptide bonds**, form through a reaction between the carboxylate group of one amino acid and the amino group of the next, generating one water molecule for each peptide bond that is formed (Figure 2.47). A polymer formed from the joining of amino acids is called a **polypeptide**, while the amino acids that have been incorporated are now called **amino acid residues**. Proteins produced by living organisms range enormously in size; the hormone

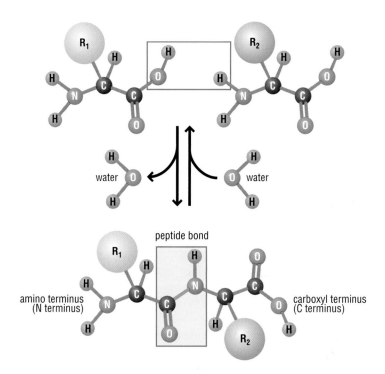

Figure 2.47 The peptide bond. Formation of a peptide bond between two amino acids, with the accompanying release of a water molecule. The addition of a water is required to reverse the reaction and break the peptide bond.

oxytocin, which triggers uterine contractions during childbirth, is just nine amino acids in length, whereas dystrophin, a muscle protein that is defective in patients with Duchenne muscular dystrophy, has 3685 amino acids.

Polypeptides have a repeating series of atoms that form the **peptide backbone**, from which the amino acid side chains protrude. One end of the polypeptide chain has an exposed amino group and is called the **N-terminus**, while the other end has an exposed carboxyl group and is called the **C-terminus**. Since the ribosome synthesizes a polypeptide beginning at the N-terminal end, the amino acid residues in a protein are numbered beginning with the N-terminal residue.

Properties of the peptide backbone govern polypeptide conformation

The peptide bond has properties that limit the conformations that polypeptide chains can adopt. The atoms that make up the peptide bond all lie in a plane (Figure 2.48), due to a phenomenon known as **resonance**, in which electrons are shared among several atoms of the peptide bond unit. This sharing is what we call delocalization. The delocalization of electrons in the peptide bond prevents free rotation of one of the bonds that form the overall peptide bond 'unit', locking the atoms into a planar (flat) conformation (Figure 2.48). Consequently, each amino acid unit in the polypeptide backbone has only two covalent bonds about which free rotation can occur: the N–Cα bond, which is defined by the rotation angle, ϕ, and the Cα–C bond, which is defined by rotation angle ψ (Figure 2.48).

There are many combinations of ϕ and ψ angles that would give rise to collisions between atoms in the polypeptide chain or the side chains. These combinations of ϕ and ψ angles therefore do not occur. A **Ramachandran plot** provides a convenient graphical depiction of the allowable combinations of ϕ and ψ angles (Figure 2.49). While the planarity of the peptide bond restricts the possible conformations of the polypeptide backbone, it still allows proteins to adopt an

(a)

(b)

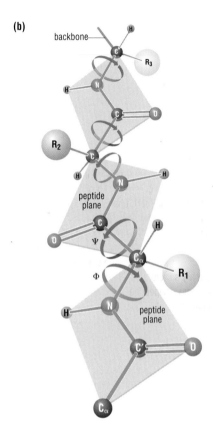

Figure 2.48 Bond rotations in polypeptide backbones. (a) The peptide bond has a partial double bond character due to the sharing of electrons between C–O and C–N atoms. As a result, the O–C–N–H atoms all lie in a plane. (b) A three amino acid polypeptide, showing the ψ and φ bonds that allow the amino acid residues to rotate about each other.

Figure 2.49 Ramachandran plot. Each point on the plot depicts a combination of φ and ψ angles about a single Cα in a polypeptide chain. The black regions show φ,ψ combinations that do not produce clashes between atoms, while the combinations in the gray regions produce minimal clashes. Glycine can adopt many more φ,ψ combinations than depicted here because it lacks a side chain.

enormous number of conformations and contributes to the stability of protein structures, as we shall see.

2.10 PROTEIN FOLDING

Proteins fold to adopt defined three-dimensional shapes

Each protein adopts a characteristic shape that is formed when the polypeptide folds upon itself to yield a much more compact three-dimensional arrangement of atoms. This process of collapse, referred to as **protein folding**, is driven by non-covalent interactions between atoms in the polypeptide chain. The resulting three-dimensional arrangement of the atoms is called the protein's structure, or **conformation**, and is determined by the sequence of amino acids along the polypeptide chain and by the chemical and physical properties of the polypeptide backbone.

There are four levels of organization that describe a protein's conformation (Figure 2.50). The sequence of amino acids along a polypeptide chain is a protein's **primary structure**. Short regions of the polypeptide chain can form regular, repeating regions of structure stabilized by hydrogen bonds called **secondary structure**. When a protein folds, these regular elements come together, along with the other residues in the polypeptide chain, to form a defined shape called the **tertiary structure** of a protein. Many proteins are actually complexes formed by the association of several folded polypeptides, which together constitute the **quaternary structure** of a protein.

The most common forms of secondary structure are alpha helices and beta sheets

Polypeptide chains can form local regions of uniform structure known as **secondary structure**, which are characterized by repeating patterns of φ and ψ angles and of hydrogen bonds between neighboring atoms. There are two common forms of secondary structure, the **alpha helix** and the **beta sheet**. In an **alpha helix**, the polypeptide backbone curves in a right-handed helical path, allowing hydrogen bonds to form between the carbonyl oxygen (C=O) of one residue and the amide nitrogen (N–H) four residues further along the polypeptide chain (Figure 2.51). The resulting structure is cylindrical, with the polypeptide backbone forming a well-packed interior and the side chains protruding on all sides.

(a)   primary

N ┄┄┄┄ T A C E V A E I S Y K K F R Q L I Q V N P
        V K E S T V Q L R R A M Q A S L R M L I D
    G
        N L A F L D V T G R I A Q T L L N L A K Q
        V I Q G I E Q R T I K I Q M G D P H T M A D P
    G
        C S R E T V G R I L K M L E D Q N ┄┄┄┄ C

(b)   secondary

alpha helices   antiparallel beta strands

(c)   tertiary

(d)   quaternary

Figure 2.50 Levels of protein structure. The four levels of protein structure are illustrated for the catabolite activator protein (CAP), which regulates transcription in *E. coli*. (a) The amino acid sequence of a protein is its primary structure. (b) Short regions of regular, repeating structure in the protein are called secondary structure elements. The most common are alpha helices and beta sheets. (c) The three-dimensional structure of a polypeptide is its tertiary structure. (d) The arrangement of two or more polypeptide chains in a protein is its quaternary structure.

Many alpha helices found in proteins have hydrophobic side chains exposed on one face and polar side chains exposed on the opposite face and are hence known as **amphipathic alpha helices**. Helices with this character frequently occur on the surfaces of proteins, where their polar faces are in contact with water while their hydrophobic faces interact with the tightly packed protein core, where water is excluded. The extensive hydrogen bonding in alpha helices also allows the polar atoms of the peptide bond to be buried in the hydrophobic core of a folded protein.

The **beta sheet** is formed when two or more segments of the polypeptide backbone, called **beta strands**, hydrogen bond to one another through their backbone carbonyl and amide groups (Figure 2.52). These aligned strands form a sheet-like structure that is slightly twisted. In contrast with the alpha helix, beta sheets can form between two or more non-contiguous segments of a polypeptide. The strands of the beta sheet can be oriented in the same direction (N- to C-terminus) and form a **parallel beta sheet**, or they can be oriented in opposite directions, forming an **antiparallel beta sheet** (Figure 2.52). Mixed sheets, containing both parallel and antiparallel pairings, also occur. In all cases, the side chains protrude above and below the beta sheet. Like alpha helices, amphipathic beta sheets are also found at protein surfaces.

The ability of a region of polypeptide chain to adopt one of the distinctive secondary structures described above depends on the primary structure of the polypeptide: certain amino acid residues favor the adoption of a given secondary structure more than others. For example, leucine, methionine, glutamine, and glutamic acid are often found in helices, while valine, isoleucine, and phenylalanine are more often found in beta sheets. By contrast, glycine and proline are often found in beta

Figure 2.51 The alpha helix. The alpha helix is a coil in which the carbonyl oxygen of one residue hydrogen bonds with the amide hydrogen four residues further along the chain. Each residue in an alpha helix has an φ angle of about −57° and a ψ angle of about −47°. The helix spirals in a right-handed direction, like a right-handed cork screw.

⮕ We learn more about entropy in Chapter 3.

turns (which link helices or sheets that are adjacent in a polypeptide chain). Glycine shows exceptional conformational flexibility due its lack of a bulky side chain, so it is particularly well suited to forming sharp bends in the polypeptide chain.

Secondary structural elements condense to form the folded protein

The tertiary structure of a protein forms when the alpha helices and beta sheets, which are connected by less regular regions of polypeptide chain, associate in a precise manner to form a defined three-dimensional shape. The polar, aqueous environment of the cell drives protein folding by encouraging hydrophobic groups in the polypeptide chain to interact with one another inside the protein, while favoring the exposure of polar groups on the protein's surface (Figure 2.53). The final, folded protein has a core filled with predominantly non-polar side chains that form favorable van der Waals interactions with one another.

The formation of secondary structural elements helps to bury polar backbone groups in the protein's interior. Recalling that the secondary structure is stabilized by hydrogen bonding of the peptide backbone, it is energetically more favorable to bury hydrogen-bonded polar atoms in the hydrophobic interior of a protein than to bury unpaired hydrogen bond donor or acceptor atoms (that is, hydrogen bond donors or acceptors that are not participating in a hydrogen bond). The atoms in the interior of a protein fit together in a way that leaves very few empty spaces or gaps, which would be energetically unfavorable. On the surface of a protein, there are primarily charged and polar side chains that are exposed to the cell's aqueous environment, along with loops of polypeptide chain that connect the secondary structural elements.

The folded form of a polypeptide chain has more favorable energy in an aqueous environment than the unfolded chain. When hydrophobic side chains are exposed to water, the nearby polar water molecules tend to orient themselves in a way to cancel out one another's electric dipoles. By contrast, when the protein folds, the hydrophobic side chains associate with one another, freeing the water molecules to adopt a larger variety of orientations in their polar surroundings (Figure 2.53). The number of conformations or positions a molecule can adopt is related to a thermodynamic quantity known as entropy – a measure of the disorder, or spread of energy, within a given system. An increase in entropy is energetically favorable. The increase in disorder of water molecules released by protein folding therefore makes the folding process energetically favorable. Even though the polypeptide chain itself has a decrease in disorder (and therefore lower entropy) as it changes from a flexible, freely moving polymer to one with a defined shape, the net change in entropy for the entire system – polypeptide and surrounding water – is positive. We shall see later in this chapter how the formation of cell membranes is driven by a similar process involving the partitioning of hydrophobic chemical groups from the polar environment of the cell.

Amino acid sequence determines protein structure

As we note above, the folded structure of a protein is determined by its amino acid sequence (that is, its primary structure), and most proteins adopt their particular three-dimensional structures spontaneously in an aqueous environment. There are many proteins that can be unfolded by a change in solution conditions – either

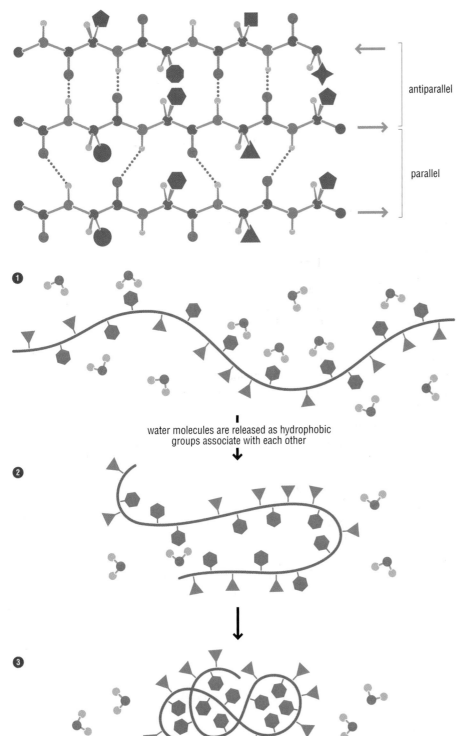

Figure 2.52 The beta sheet. In a beta sheet two or more extended polypeptide chains hydrogen bond with one another through their backbone carbonyl and amide groups. Two strands of a beta sheet can be either parallel (with the same N- to C-terminus orientation) or antiparallel (with opposite orientation). The side chains (blue shapes) project above and below the beta sheet.

water molecules are released as hydrophobic groups associate with each other

Figure 2.53 Protein folding. The folding of a polypeptide chain is driven by the association of hydrophobic side chains (pink) with one another in the interior of the protein, with polar side chains (blue) exposed on the surface. Water molecules are ordered in the vicinity of hydrophobic groups in the unfolded polypeptide (step 1), but are released as the hydrophobic residues cluster together in the folded protein (steps 2 and 3). This release of water is energetically favorable and helps to drive protein folding.

by treatment with chemicals that destabilize the protein or by heating to high temperatures – and then spontaneously refold when the solution is returned to the original conditions. Some polypeptides require the assistance of proteins called chaperones to fold in the cell. The chaperones ensure that the folding process proceeds smoothly but do not determine the final structure. That said, there are some proteins that cannot refold properly; we say that they have been denatured. The white of an egg, which contains the protein albumin, is a classic example:

Figure 2.54 Sequence similarity and protein folding. The Matα2 (blue; PDB code 1APL) and engrailed (green; PDB code 1HDD) homeodomain proteins are only 27% identical in amino acid sequence, but adopt a nearly identical protein fold. Note the three additional amino acids inserted in MATα2 as compared with engrailed.

when the egg white is heated, the albumin denatures, at which point the egg white essentially changes irreversibly from liquid to solid.

As the process of protein folding is so complex and involves an intricate network of non-covalent interactions, we still cannot reliably predict the tertiary structure of a protein solely from its primary amino acid sequence. However, it is possible to infer whether two proteins are likely to adopt a similar structure by comparing their amino acid sequences: two polypeptides that have 50% of their amino acids in common generally have nearly identical structures. Even when two polypeptides have just 25% of their amino acids in common – also referred to as 25% sequence identity – their structures are generally fairly similar. Figure 2.54 shows an example of two proteins that regulate gene expression and are 27% identical in sequence, yet adopt virtually the same structure and carry out similar functions in the cell. Comparisons of amino acid sequence are commonly used to identify whether two proteins are predicted to have the same structure.

While it is not yet possible to predict tertiary structure on the basis of primary structure with any degree of accuracy, it is possible to predict secondary structural elements with some confidence. Indeed, we can now predict the occurrence of secondary structure on the basis of primary structure with an accuracy in excess of 75%.

Amino acid sequence can reveal clues about protein function

Each protein can carry out its particular function in a cell because its structure and the chemical nature of the exposed amino acids are tailored for a particular task. We saw in the example above that amino acid sequence can diverge quite a bit while at the same time conserving overall structure and, generally, function. Since the amino acid sequence of a protein determines its structure and function,

Figure 2.55 Comparing protein sequences can help identify the function of a protein. Flow chart illustrating how the function of an unknown protein can be identified by comparing its amino acid sequence to other proteins of known function. The sequence of a portion of the MATα2 protein (α2) is compared to a database of proteins whose functions are known, using a computer to line the sequence of α2 with each protein and obtain the best match. This identifies the protein engrailed (en), which is identical to α2 at 27% of its amino acids, and adopts a virtually identical structure, as shown in Fig. 2.54.

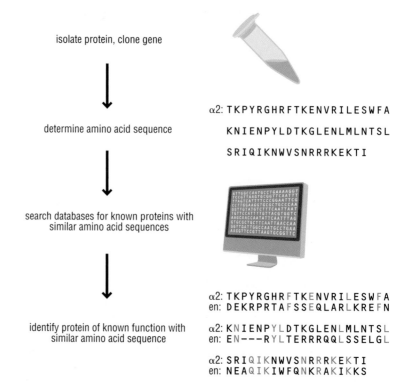

isolate protein, clone gene

determine amino acid sequence

search databases for known proteins with similar amino acid sequences

identify protein of known function with similar amino acid sequence

α2: TKPYRGHRFTKENVRILESWFA
KNIENPYLDTKGLENLMLNTSL
SRIQIKNWVSNRRRKEKTI

α2: TKPYRGHRFTKENVRILESWFA
en: DEKRPRTAFSSEQLARLKREFN

α2: KNIENPYLDTKGLENLMLNTSL
en: EN---RYLTERRRQQLSSELGL

α2: SRIQIKNWVSNRRRKEKTI
en: NEAQIKIWFQNKRAKIKKS

it is sometimes possible to predict the function of a newly identified gene based on the amino acid sequence of the protein it encodes. The sequence of this protein can be compared with the sequences of proteins whose functions are known; if any proteins have at least 25% of their amino acids in common, they likely adopt the same fold (Figure 2.55). Often, although not always, this also means that the proteins perform a similar function.

Further clues to function may come from determining whether certain key amino acids – perhaps those required for catalysis, or binding to another protein – are also conserved in the newly identified protein. Ultimately, the protein's function must be verified with biochemical or genetic experiments. Sequence comparisons must be interpreted with caution, as there are many examples in which a particular protein fold has been adapted to perform a new function. Nevertheless, the ability to search databases of amino acid sequences is a powerful and efficient approach to obtaining important clues about a protein's function.

2.11 PROTEIN FOLDS

Protein folds have been conserved through evolution

Proteins can adopt an astonishing variety of tertiary structures. However, some structures look very much like one another in their overall shape, secondary structure content, and the way in which secondary structural elements pack together and are connected by loops. We refer to the arrangement of secondary structural elements that characterizes a particular protein as a **protein fold**. Proteins of identical structure, meaning that they have the identical arrangement of atoms in space, also have the same fold. However, two proteins that do not appear at first glance to have identical structures may, in fact, have identical folds if they contain essentially the same secondary structural elements that are connected to one another in the same way. As seen in the example in Figure 2.56, two proteins with the same fold can sometimes overlay rather poorly in space because of differences in the precise length of secondary structural elements and the way in which they pack against one another.

The great variety of protein folds is evident in the examples of common protein folds depicted in Figure 2.57. Some proteins, such as the enzyme triose phosphate isomerase (Figure 2.57b), are **globular**, meaning that they are roughly spherical and have a relatively low ratio of surface area to volume. Others are highly elongated, such as the protein, GCN4 (Figure 2.57d), which dimerizes through the association of two alpha helices that form a structure known as a **coiled coil**. The two helices gently coil around one another and interact through a set of hydrophobic side chains at the interface.

Folds also vary greatly in size and secondary structure composition; the homeodomain, which binds DNA, is small and contains only alpha helices (Figure 2.57a), while the much larger immunoglobulins are composed almost entirely of beta sheet (Figure 2.57c). We shall see throughout this book how the structural features of proteins are tailored to carry out specific functions.

While there is tremendous diversity in the types of protein folds that are known, it is also clear that the number of protein folds found in nature is limited. Most arbitrary amino acid sequences would fail to fold into a stable structure. The

Figure 2.56 Two proteins with the same fold yet significant structural differences. Although these two small protein neurotoxins, NTII (blue; PDB code 1NOR) and Fasciculin I (red; PDB code 1FAS), have the same fold, there are many structural differences between them, especially in the length of the beta strands and in the conformation of the loops.

number of potential different sequences for a protein 100 amino acids long is 20^{100} or 1.3×10^{130}! Yet of the >64 000 protein structures that were determined as of 2010, only a thousand or so folds have been observed. A variety of amino acid sequences can be accommodated by the same – or nearly the same – fold. Changes in the amino acid sequence (an amino acid substitution or other mutation) that do not alter the overall fold or properties of a protein may be tolerated – that is, they allow the protein to function as it did prior to the substitution. By contrast, those substitutions (or other mutations) that are more deleterious are not tolerated, and the protein no longer functions as it should. However, as it collects further mutations, or even insertions or deletions in the polypeptide chain, the protein may evolve a new, useful function.

Variants arise from a single ancestral protein through a process of **divergent evolution**, in which proteins with new characteristics, and eventually separate functions, evolve. Proteins whose primary structures may appear to be unrelated, but which have similar folds, are generally considered to have undergone divergent

Figure 2.57 Examples of protein folds. A few examples of protein folds are (a) the homeodomain, which is entirely alpha helical, (b) triose phosphate isomerase, which consists of a TIM barrel, a mixed alpha helix, beta sheet structure, (c) immunoglobulin, which is almost entirely beta sheet, and (d) GCN4, which consists of a pair of long alpha helices that dimerize by forming a coiled coil.

evolution. Occasionally, nature seems to have solved the same problem twice, producing two proteins that carry out a similar function but appear to have evolved independently. This more uncommon process is referred to as **convergent evolution**. An example can be found in the enzymes chymotrypsin and subtilisin, which catalyze peptide bond cleavage using the same triad of amino acids but are otherwise dissimilar in structure (Figure 2.58). It is the continued selective pressure to carry out particular tasks that drives the evolution of proteins, and causes certain efficient solutions to be used again and again in biology.

Proteins can be composed of multiple domains

Many protein polypeptide chains contain more than one structurally distinct region. Each distinctive region is called a **domain**, which is a compact region of protein structure usually made up of a contiguous segment of the polypeptide chain that is capable of folding on its own. While some proteins consist of a single structural domain, most are built up in a modular fashion from several domains fused together in a single polypeptide chain. Each domain generally has a distinct fold and carries out a particular task. For example, proteins that regulate RNA transcription often have a modular composition (Figure 2.59). Such proteins are composed of several different domains, each with a characteristic fold and a distinct function. The particular combination of functional domains gives rise to the unique character of each protein.

Having now seen how proteins adopt an array of different three-dimensional shapes – a diversity that underpins the multitude of different roles proteins play in biological systems – we finish this chapter by considering two other groups of biological macromolecules, the carbohydrates and lipids.

Figure 2.58 Example of convergent evolution. The enzymes chymotrypsin (top) and subtilisin (bottom) do not have the same fold, yet they have the same set of three amino acids known as the catalytic triad, which catalyzes peptide bond cleavage.

2.12 SUGARS AND CARBOHYDRATES

Carbohydrates play a variety of roles in the cell

Carbohydrates comprise a diverse class of molecules that are involved in a wide range of biological processes. We have already encountered a simple type of carbohydrate: the ribose sugar that forms part of the backbone of RNA. Carbohydrates are also a major source of energy for all organisms, and are the form in which many cells story energy for future use. Polymers known as complex carbohydrates can form rigid structures that provide mechanical support, such as the cell walls of bacteria and plants or the rigid exoskeleton of insects. We shall see here how the chemical diversity of carbohydrates enables them to play myriad roles in biology.

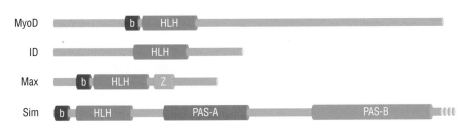

Figure 2.59 The modular nature of HLH proteins. The diagram shows how different proteins that bind DNA and regulate transcription can be built up from discrete domains. All of the proteins shown here contain a helix-loop-helix DNA-binding domain, denoted HLH. These proteins contain one or more other conserved domains: b, basic region, Z, leucine zipper, PAS-A domain, PAS-B domain.

Simple sugars can form linear or ring-like structures

Like the other major biological macromolecules, proteins and nucleic acids, **carbohydrates** are polymers, in this case built up from fundamental building blocks called **monosaccharides**. A monosaccharide is a simple organic molecule that generally has the chemical formula $(CH_2O)_n$, where n is greater than 3, and which cannot be hydrolyzed to form simpler saccharides. The nomenclature for naming monosaccharides indicates the number of carbon atoms in each molecule. For example, a six-carbon monosaccharide has the chemical formula $C_6H_{12}O_6$ and is therefore referred to as a **hexose**. A five-carbon monosaccharide has the formula $C_5H_{10}O_5$ and is called a **pentose**. We refer to these monosaccharides as sugars, but we also use the term sugar to refer to simple carbohydrates containing up to three monosaccharide molecules linked together.

Figure 2.60 shows some hexoses and pentoses that differ only in the location of their carbonyl (C=O) groups. Because the monosaccharides contain at least one asymmetrical carbon, they can exist in two non-interconvertible forms, called isomers (D or L). The D– or L– designation is defined by the arrangement of atoms around the chiral carbon farthest from the aldehyde or ketone group (Figure 2.61). Most naturally occurring pentoses and hexoses are found as the D form.

A structural feature of certain sugars is their ability to form covalently closed, ring-like structures. This occurs when a hydroxyl group reacts with a ketone or aldehyde carbonyl group, yielding a ring-like isomer. For pentose and hexose sugars dissolved in water, the cyclized form is more prevalent than the linear sugar. Examples of the linear and cyclic forms of a sugar are shown in Figure 2.62.

A hexose sugar can cyclize in one of two ways. A **pyranose** ring is formed when the C-1 aldehyde of a sugar such as glucose is attacked by the C-5 hydroxyl, forming a six-member ring (Figure 2.62). A five-member **furanose** ring results when a C-2 keto group reacts with the C-5 hydroxyl. These relatively stable five- and six-member rings predominate in biology. Furanose and pyranose rings can adopt different isomers called **anomers**, which differ in the orientation of the hydroxyl group at the C-1 carbon (illustrated in Figure 2.62). Ribose, when an integral part of nucleotides, is a furanose ring in the β anomeric form. RNA contains β-D-ribose, while DNA contains β-D-deoxyribose.

Monosaccharides can be linked together to form complex carbohydrates

Unlike proteins and nucleic acids, which are linear polymers, carbohydrates can form branched polymers. The building blocks of carbohydrates are cyclic monosaccharides, which can be covalently linked to form larger and more complex polymers. The hydroxyl of one monosaccharide can be joined to another in a neighboring molecule by a condensation reaction that releases a water molecule. The resulting **disaccharide** comprises two sugars joined by a linkage termed a **glycosidic bond** (Figure 2.63a). Disaccharides, in turn, can be linked with one another to form larger carbohydrates termed **oligosaccharides**. The largest carbohydrates, which consist of tens or hundreds of monosaccharides, are called **polysaccharides**.

Figure 2.60 Simple linear sugars. Two five-carbon pentose sugars are shown, ribose and ribulose, along with two six-carbon hexose sugars, glucose and fructose.

Since sugars have many hydroxyl groups and are present in different anomeric forms, there are many different ways in which one sugar can be covalently joined with another. For example, there are 11 chemically distinct ways to join two D-glucose units. In addition, a single monosaccharide can be linked to two others. This allows the assembly of branched polysaccharides such as glycogen (Figure 2.63b), which is the form in which glucose is stored by mammalian cells until it is needed as a source of energy. Because of the many possible ways in which sugars can be linked to form oligo- and polysaccharides, carbohydrates form a much more diverse family of polymers than proteins and nucleic acids, which have only one type of linkage joining one building block with the next. Not surprisingly, this also means that a much larger set of enzymes is needed to assemble or digest different carbohydrates, since a distinct enzyme is needed to catalyze formation or breakage of each type of linkage.

Carbohydrates play structural roles in all organisms

Some polysaccharides form polymers that have great mechanical strength. Plant cell walls contain cellulose, a linear polysaccharide composed of glucose subunits that gives plants their rigid structure. The unbranched glucose polymers that comprise cellulose form long, linear chains that line up and hydrogen bond to one another, thereby strengthening the plant cell wall. Bacteria have cell walls composed of **peptidoglycans**, which are linear polysaccharides that are joined to one another by peptide cross-links. The first specific antibiotic ever discovered, penicillin, acts by disrupting synthesis of the peptidoglycan bacterial cell wall. The hard outer shell of crustaceans and insects is composed of chitin, a linear polysaccharide. All multicellular organisms have carbohydrate polymers in the spaces between cells, which is called the extracellular matrix. The presence of structural carbohydrates in all organisms suggests that they appeared early in evolution.

Carbohydrates can be attached to proteins and lipids

Carbohydrates can be attached to other macromolecules, thereby altering the properties and behavior of the macromolecules in essential ways. Carbohydrates can be covalently attached to proteins to form **glycoproteins**, while lipids with covalently attached carbohydrates are called **glycolipids**. We shall learn more

Figure 2.61 Monosaccharide isomers. Monosaccharides can exist as two non-interconvertible isomers, denoted D and L. This designation is determined by the arrangement of atoms around the carbon furthest from the aldehyde or ketone group. D- and L-glucose are shown here, with the designations determined by the arrangement about the carbon atom highlighted in blue.

Figure 2.62 Cyclized sugars. Glucose (top) can cyclize in one of two ways, producing ring structures that differ only in the orientation of the hydroxyl group attached to the carbon-1 atom. In the linear sugar, the aldehyde group is shown in blue and the hydroxyl is shown in red; the corresponding atoms in the cyclized sugars are colored accordingly. Ribose (bottom) can also cyclize in two ways but only the β form is shown.

Figure 2.63 Polysaccharides. (a) Glucose and fructose can combine to form the disaccharide, sucrose, with the concomitant release of water (red). (b) Glycogen is a complex, highly branched polysaccharide of linked glucose sugars (gray spheres).

about covalent attachment of carbohydrates to proteins and lipids later in this chapter.

Animal cell surfaces are richly decorated with carbohydrates, due to the presence of carbohydrates on cell surface proteins. The attachment of carbohydrates, termed **glycosylation**, stabilizes the structures of cell surface proteins and plays an important role in the interactions of cells with each other and with the free proteins surrounding cells. For example, the A, B, O and Rh blood types are distinguished by different patterns of protein glycosylation on the surface of red blood cells. Carbohydrates linked to proteins and lipids on the cell surface can be recognized by specialized proteins called **lectins**, which bind to particular carbohydrates and mediate interactions between cells. Carbohydrates on the cell surface can also be exploited by some pathogens to gain entry into the cell. Influenza virus binds to sialic acid, a glycosylation found on certain cell surface proteins. *Vibrio cholerae,* which causes cholera, only invades epithelial cells with a particular type of glycolipid, G(M1) ganglioside, which contains a branched pentasaccharide linked to fatty acids. These infectious agents have clearly evolved strategies for infection that exploit the presence of carbohydrates on the surface of the cells they invade.

2.13 LIPIDS

Lipids are hydrophobic molecules

Lipids comprise a diverse class of biological molecules that are all strongly hydrophobic in character. The basic constituents of cell membranes, steroid compounds such as cholesterol, and the fats and oils that are used by cells to store energy are all lipids. Lipid molecules can be attached to proteins in order to tether the proteins to membranes, and lipids can be used to transmit signals within cells. We shall learn here about the basic properties of lipid molecules that are important for their biological functions.

Fatty acids are a key component in most types of lipid

The fundamental building block of most lipids in biology is the **fatty acid**. A fatty acid is a hydrocarbon chain with a carboxylic acid group at one end; several fatty acids are illustrated in Figure 2.64. The hydrocarbon chain is hydrophobic, whereas the negatively charged carboxylic acid is hydrophilic, which means fatty acids are amphipathic. A fatty acid can be **saturated**, meaning that all the carbon atoms are joined by single bonds, or they can be **unsaturated**, meaning that there are one or more double bonds joining the atoms in the hydrocarbon chain.

In biological systems, unsaturated bonds are generally found in the *cis* configuration, which introduces a kink in the hydrocarbon chain, rather than the *trans* configuration, which preserves the straight conformation found in saturated chains. For example, Figure 2.64a shows the structure of the **saturated fatty acid**, palmitic acid (which lacks double bonds in the hydrocarbon chain), and two unsaturated fatty acids, oleic acid and elaidic acid. Notice how the double bond in oleic acid adopts a *cis* conformation, while the double bond in elaidic acid adopts a *trans* conformation. By contrast, the bonds in palmitic acid are all *trans*, which is characteristic of saturated fatty acids. Fatty acids vary in length and in the number and position of double bonds, all factors leading to differences in their chemical properties.

One important consequence of saturation is its effect on how fatty acids associate with one another. In an aqueous environment, the hydrophobic hydrocarbon tails of fatty acids tend to pack against one another rather than with the surrounding water molecules. Saturated hydrocarbon tails are highly flexible molecules that can form extensive interactions with one another. Unsaturated tails, on the other hand, contain kinks caused by their double bonds and therefore these tails do not generally pack together as tightly. (Look again at Figure 2.64a. Note how the hydrocarbon chain of oleic acid, which adopts a conformation, is kinked, whereas the hydrocarbon chain of elaidic acid, whose double bond is in the *trans* conformation, is straight.) As a result, the loosely packed unsaturated fatty acids are more fluid at room temperature than the tightly packed saturated fatty acids (Figure 2.64b). This is why butter, which is high in saturated fatty acids, is solid at room temperature, while vegetable oil, which is high in unsaturated fatty acids, remains liquid. Plant oils can be hydrogenated *in vitro*, which reduces the number of double bonds per fatty acid chain and therefore makes them less fluid at room temperature. This process is used in the production of margarine. These chemically hydrogenated oils have been found to be harmful to human health, because

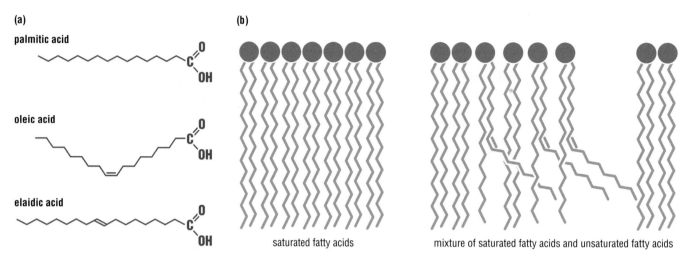

(a)

palmitic acid

oleic acid

elaidic acid

(b)

saturated fatty acids

mixture of saturated fatty acids and unsaturated fatty acids

Figure 2.64 Saturated and unsaturated fatty acids. (a) Palmitic acid is a saturated fatty acid, meaning that there are single bonds connecting all the carbon atoms in the hydrocarbon chain. Oleic acid is an unsaturated fatty acid with a single *cis* double bond, which introduces a kink in the hydrocarbon chain. Elaidic acid is similar to oleic acid but has a *trans* double bond in place of the cis double bond. (b) *Trans* fatty acids can pack closely together and form extensive van der Waals interactions, while mixtures of *cis* and *trans* fatty acids cannot pack as tightly and hence form a less stable – and more fluid – array.

the hydrogenation process yields fatty acids containing *trans* double bonds, which are known as *trans* fats. These *trans* fats do not resemble any of the naturally synthesized fats, and lead to the buildup of arterial plaques for reasons that are as yet poorly understood.

Fatty acid chain length also influences fluidity. Short-chain fatty acids exhibit a higher degree of fluidity than those with longer hydrocarbon chains because their intermolecular packing is less compact. Fatty acid lengths in biological systems can range from 14 to 24 carbons in length, with those 16 and 18 carbons in length being the most common.

Many lipids are composed of fatty acids attached to a head group

Many lipid molecules consist of fatty acids attached to a polar moiety called a **head group** (Figure 2.65). The mixed hydrophobic and hydrophilic nature of lipids, which we refer to as **amphipathic**, is a defining feature of lipids that allows them to form biological membranes, as we shall see below.

There are several different types of lipids common in biology, each distinguished by the type of head group to which the fatty acids are attached. The most common class of lipid found in cell membranes is the **phospholipid**. The head group of the most common phospholipids comprises glycerol, a phosphate group, and an alcohol (Figure 2.65). Two fatty acids are covalently linked directly to the glycerol, forming the hydrophobic portion of the lipid molecule, while the head group is the hydrophilic part. There are many different types of glycerol-derived phospholipids, each distinguished by the type of alcohol attached to the phosphate group (Figure 2.66), and by the length and degree of saturation of the fatty acids. Glycolipids are another class of lipid molecules found in membranes, and are distinguished by the sugar moieties attached to their head groups.

Cholesterol, an abundant constituent of animal cell membranes, is a lipid whose chemical structure differs significantly from the lipids described above. It contains

Figure 2.65 Schematic structure of a phospholipid. Two fatty acids (black) are attached to a polar head group (blue) containing glycerol, a phosphate group, and an alcohol.

Figure 2.66 Examples of glycerol head groups.

several linked hydrocarbon rings, a short hydrophobic tail, and a single polar hydroxyl group (Figure 2.67). Therefore, cholesterol lacks the more clearly defined head group of other lipids. Cholesterol is the metabolic precursor of a broader class of molecules called steroid hormones such as estrogen and progesterone. Its amphiphilic character due to the polar OH group, combined with its relatively rigid fused ring system, results in important consequences for the fluidity of cell membranes, as we shall see in the next section.

Figure 2.67 Cholesterol. This molecule contains the four-ring structure that is typical of steroids, as well as a hydrophobic tail (top right) and a hydrophilic hydroxyl group (bottom left).

Membranes contain a bilayer of oriented phospholipids

The unique properties of phospholipids enable them to form a protective sheath around cells called a **plasma membrane**, which allows the cell to maintain an intracellular environment that is different from its exterior. The amphiphilic nature of phospholipids favors their arrangement in a **lipid bilayer**, with the polar head groups in contact with the surrounding aqueous medium and the hydrophobic fatty acid tails in the interior where they are shielded from the water (Figure 2.68). van der Waals interactions between the fatty acid tails of the lipids help to stabilize the bilayer.

Lipid bilayers can be thought of as two-dimensional fluids because the lipids, along with other molecules in the membrane, can diffuse quite rapidly within each plane of the bilayer. The precise lipid composition of a membrane governs how readily this diffusion occurs. Fatty acid chain length and saturation influence membrane fluidity in the same way that they affect whether fats are solid or liquid at room temperature, as we learned above. Membrane fluidity can also be affected by the presence of cholesterol, which is found in most animal cell membranes. Membranes with significant amounts of cholesterol inserted into the bilayer have reduced fluidity, but also are less likely to become solid at very low temperatures. The ability to fine-tune membrane fluidity is important in regulating the diffusion of the proteins embedded within the membrane, as we will see next.

Proteins within the membrane allow communication between the cytoplasm and the cell's exterior

Cells must have the ability to take in a variety of molecules from the surrounding environment, as well as have a means to communicate with or respond to other cells or organisms. Lipid bilayers, however, form a barrier that is relatively impermeable to most molecules. Cell membranes therefore contain embedded proteins that allow the transfer of molecules and information across the lipid bilayer.

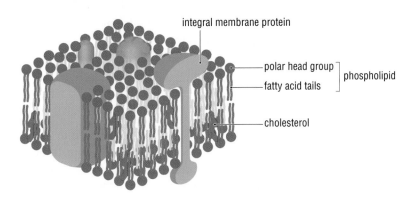

integral membrane protein

polar head group ⎤
fatty acid tails ⎦ phospholipid

cholesterol

Figure 2.68 A cell membrane. The oriented phospholipids form a bilayer 5–10 nm thick, with the fatty acid tails pointing inward and the polar head groups facing the aqueous environment. Cholesterol molecules (red) can insert between the fatty acid tails. Proteins, shown in blue, can lie within the membrane.

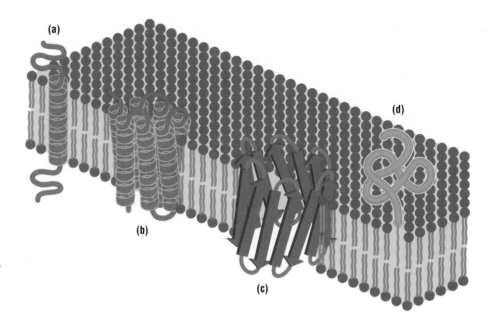

Figure 2.69 Membrane proteins. Integral membrane proteins can span membranes with one or more alpha helices as shown in (a) and (b), or with a beta sheet arrangement (c). Peripheral membrane proteins can be anchored to membranes by the covalent attachment of a fatty acid (d).

The typical plasma membrane contains about 50% protein by mass, but protein content varies with membrane type. It ranges from a low of 18% in the membranes of nerve cells to 75% in the organelle membranes of mitochondria.

Membrane proteins vary a great deal in structure and in the degree to which they are embedded within the lipid bilayer (Figure 2.69). **Integral membrane proteins** are embedded largely within the bilayer, whereas some proteins are anchored to the membrane by inserting just a small portion of the protein into the bilayer. Many membrane proteins have domains on the cell's interior and exterior joined by one or more alpha helices called **transmembrane helices** that span the membrane (Figure 2.69b). There are also integral membrane proteins that span the membrane and are composed primarily of structures rich in alpha helices (Figure 2.69b) or beta sheets (Figure 2.69c). Some of these proteins contain a central pore that allows ions or small molecules to pass across the membrane.

In all cases, the portion of the protein embedded in the membrane must be rich in hydrophobic side chains that interact favorably with the hydrophobic fatty acid chains within the lipid bilayer, with few, if any, polar or charged side chains. This is because of the unfavorable energy that results from burying a polar group within the non-polar lipid bilayer, where there is no water or other polar molecules with which it may form favorable hydrogen bonding interactions. For this reason, peptide backbone that is buried within lipid bilayers is generally found in alpha-helical or beta-sheet form, because in these secondary structures the polar backbone atoms are hydrogen bonded with each other. **Peripheral membrane proteins** do not contain transmembrane regions, and can bind reversibly to the cell membrane.

Covalent attachment of lipids can target proteins to the membrane

Proteins that lack exposed hydrophobic regions can be targeted to membranes by becoming covalently modified with a lipid or fatty acid (Figure 2.69d). Some peripheral proteins, for example, are anchored to membranes by a fatty acid, itself embedded in the lipid bilayer, that is covalently attached to the protein. The particular type of lipid that is attached can determine the intracellular membrane, and hence the organelle, with which the protein associates. Many different types of lipids, ranging from simple fatty acids such as palmitic acid (Figure 2.64) to complex modifications containing carbohydrate and multiple fatty acids, can be attached to proteins. Cholesterol, in addition to being a membrane constituent that regulates the fluidity of a bilayer, can also be covalently attached to proteins and thereby anchor them in membranes. We will learn more about lipid modifications and their biological roles in Chapter 12.

2.14 CHEMICAL MODIFICATION IN BIOLOGICAL REGULATION

We see throughout biology how the structure and function of a biological molecule are intrinsically linked. For example, if the three-dimensional structure of an enzyme's active site changes, it may no longer be able to bind to its correct substrate. If the structure of a membrane-bound receptor changes, it may no longer be able to bind to its complementary signaling molecule, and will therefore fail to pass its signal to the cell interior. Such modification of structure – and its concomitant effect on function – is not always disadvantageous, however, but can often be used to activate or regulate a biological molecule's activity.

Proteins are often chemically modified after translation

After being synthesized, a protein may undergo chemical modifications of its amino acid side chains or of the polypeptide backbone itself in order to be able to carry out its biological function. In some cases, the chemical modification represents the final processing step that produces the fully functional proteins, while other modifications are transient and reversible, and are used to regulate an aspect of the protein's function. We refer to these chemical alterations as **post-translational modifications**, since they occur after the ribosome has translated the information in the messenger RNA.

Addition and subsequent removal of chemical groups is a major mechanism for regulating the activity of a protein. The covalent attachment of inorganic phosphate to specific side chains in a process called **phosphorylation** is a particularly common example of this. **Acetylation** and **methylation** (Figure 2.70), whereby acetyl and methyl groups are attached to amino acid side chains, also often occur in protein regulation. All of these chemical modifications exert their effect in one of two general ways: they can cause the protein to change conformation, perhaps activating or inhibiting the protein, or the chemically modified side chains can bind to other proteins. These events can then trigger changes in any of a wide variety of cellular processes. We shall see examples of such reversible chemical modifications throughout this book.

DNA methylation controls gene expression

Chemical modification of DNA also occurs and plays a variety of regulatory roles. The most common type of modification to DNA, and the only one we will encounter in this book, is methylation. In bacteria, both cytosine and adenine bases can be modified, whereas cytosine methylation occurs in mammals and in

Post-translational modifications of proteins

modification	description	common function of modification	icon
lipid modification	attachment of lipid to protein (examples include palmitylation and myristoylation)	localization to the membrane	
glycosylation	attachment of carbohydrates to side chains; typically Ser, Thr or Asn	protection from degradation, recognition by other proteins	
phosphorylation	attachment of phosphate group to Ser, Thr, His, Tyr, His, Asp	regulation of interactions or activity	
acetylation	attachment of an acetyl group to Lys side chains	regulation of interactions	
methylation	attachment of one or more methyl groups to Lys or Arg side chains	regulation of interactions	
ubiquitination	attachment of small protein, ubiquitin, to lysine side chains (one or more ubiquitins may be attached)	regulation of interactions, targeting of protein for degradation	ubiquitin

Figure 2.70 Reversible modifications that alter the activity of proteins. At the right is a depiction of the protein (blue) along with the icon that represents the modification. Chemical structures are given for phosphate, acetyl, and methyl groups.

plants. Bacteria use DNA methylation to distinguish the parental strand from the newly synthesized strand after DNA replication. DNA methylation is also used by bacteria to distinguish its own genome from that of invading viruses. In mammals and plants, cytosine methylation is used to turn off gene expression. We will learn about DNA methylation in detail in Chapter 4.

✱ SUMMARY

In this chapter, we reviewed how molecules are built up by linking atoms together with covalent bonds, and covered the way in which molecules interact with one another non-covalently in the aqueous environment of the cell. We learned about the four classes of biological macromolecules – nucleic acids (RNA and DNA), proteins, carbohydrates, and lipids – and the way in which they are assembled from a set of chemical building blocks. Each of these macromolecules has a characteristic set of properties that makes it uniquely suited to its biological role. Many biological molecules can be covalently modified in ways that alter their chemical properties and allow their function to be regulated.

Atoms and molecules

- Atoms are joined by covalent bonds to form molecules through the sharing of unpaired electrons.

- Electrons are not shared equally in some covalent bonds, with the result that some atoms have a partial positive or negative charge.

- Molecules with an overall unequal charge distribution are polar, whereas molecules with an overall even distribution of charge are non-polar.

- Water molecules are polar, and this property plays an important role in governing the behavior of molecules in aqueous solution.

- Polar chemical groups and molecules form favorable interactions with polar water molecules; we therefore call them hydrophilic.

- Non-polar molecules and chemical groups prefer to associate with one another, rather than with water; we therefore call them hydrophobic.

- Ions are atoms that become fully charged through the gain or loss of electrons.

- Water becomes ionized to form H^+ and OH^- ions; pH is a measure of the concentration of H^+ ions and is given by $pH = -\log_{10}[H^+]$.

- Molecules interact in the aqueous environment of the cell through three types of non-covalent interactions: salt bridges, hydrogen bonding, and van der Waals interactions.

- Salt bridges are electrostatic interactions between chemical groups bearing opposite charges.

- A hydrogen bond is an electrostatic interaction between a hydrogen atom with a partial positive charge (the hydrogen bond donor) and an atom with a partial negative charge (the hydrogen bond acceptor).

- Van der Waals interactions are attractive forces that arise between two atoms due to formation of transient dipoles that attract one another.

Biological macromolecules

Nucleic acids

- The family of molecules known as nucleic acids includes DNA, RNA, and a variety of nucleotide monomers.

- A nucleotide is composed of a base, a sugar, and one or more phosphate groups.

- In DNA, the sugar is deoxyribose, and the bases are adenine, guanine, cytosine, or thymine.

- A single strand of DNA is a polymer of covalently linked deoxyribonucleotides that are joined by phosphodiester linkages between the 3′OH of one ribose and the 5′OH of the next ribose.

- Two DNA strands associate with one another to form a double helix, with one strand spiraling around the other. The dominant type of DNA helix in cells is known as the B-helix, or B-DNA.

- The two strands of the double helix are antiparallel, with the sugar-phosphate backbone on the outside and the bases on the interior.

- The bases on opposite strands of the double helix hydrogen bond with one another to form Watson–Crick base pairs, with adenine pairing with thymine through two hydrogen bonds and guanine pairing with cytosine through three hydrogen bonds.

- Long double-stranded DNA molecules can be under tension because of topological variations known as supercoiling. Negative supercoiling favors local unwinding of the DNA helix and separation of the two single strands.

- In RNA nucleotides, the sugar is ribose and the base is adenine, guanine, cytosine or uracil. Like DNA, RNA nucleotides are joined to one another by phosphodiester linkages.

- RNA nucleotides are subject to a wide variety of chemical modifications.

- The 2′OH on the ribose confers on RNA additional structural and chemical diversity not found in DNA.

- Double-stranded RNA forms the A-type double helix.

- Some RNA molecules form complex structures containing short stretches of double-stranded regions.

Proteins

- Proteins are polymers of amino acids (also called polypeptides) that are joined to one another by peptide bonds.

- There are 20 amino acids, each consisting of an amino group, a carboxyl group, and a chemically distinct side chain.

- A polypeptide chain condenses in water to form a defined structure, with hydrophobic (non-polar) side chains buried on the inside and hydrophilic (polar) side chains on the outside.

- There are four levels of organization that describe protein structure:

  - primary structure – sequence of amino acids

  - secondary structure – formation of alpha helix and beta sheet, which contain characteristic regular, repeating patterns of hydrogen bonds between residues

  - tertiary structure – overall three-dimensional structure of a single polypeptide

  - quaternary structure – the arrangement of two or more polypeptides that form a complex with one another.

- The amino acid sequence dictates protein structure, which means that proteins that are similar in amino acid sequence adopt similar structures.

Carbohydrates and lipids

- Carbohydrates are polymers of sugar molecules known as monosaccharides.

- A monosaccharide has the general chemical formula $(CH_2O)_n$, where n >3. The five- and six-carbon sugars are the most common in biology.

- Monosaccharides are joined to one another in a condensation reaction that releases a water molecule, yielding disaccharides, oligosaccharides and polysaccharides, depending on the number of monosaccharide units involved.

- Monosaccharides can be linked to one another in a large variety of ways to form branched, as well as linear, polymers.

- Lipids are hydrophobic molecules, often composed of a non-polar fatty acid and a polar head group. This mixed hydrophobic and hydrophilic nature is referred to as amphipathic.

- Fatty acids are long hydrocarbon chains that are hydrophobic.

- The most common type of polar head group consists of glycerol covalently linked to a phosphate group and an alcohol. Lipids containing this type of head group are termed phospholipids.

- In water, phospholipids associate with one another to form a lipid bilayer, with the polar head groups facing the water and the hydrophobic lipids buried in the interior.

Covalent modification of biological molecules

- All types of biological molecules can be modified by the covalent attachment of molecules.

- Carbohydrates can be covalently attached to protein, forming a glycoprotein, or to lipid, yielding a glycolipid.

- Lipids and fatty acids can be covalently attached to proteins.

- Methylation of DNA bases regulates a variety of processes.

- Proteins are subject to a wide variety of chemical modifications that include methylation, acetylation, and phosphorylation.

📖 FURTHER READING

2.1 ATOMS, MOLECULES, AND CHEMICAL BONDS

Pauling LC. *The Nature of the Chemical Bond and the Structure of Molecules and Crystals: An Introduction to Modern Structural Chemistry*, 3rd ed. Ithaca, NY: Cornell University Press, 1960.

2.8 THE RNA WORLD AND ITS ROLE IN THE EVOLUTION OF MODERN-DAY LIFE

Doudna JA, Cech TR. The chemical repertoire of natural ribozymes. *Nature*, 2002:**418**:222–228.

White HB. Coenzymes as fossils of an earlier metabolic state. *Journal of Molecular Evolution* 1976;7:101–104.

2.9 FUNDAMENTALS OF PROTEIN STRUCTURE

Petsko GA, Ringe D. *Protein Structure and Function*. London: New Science Press, 2003.

THE CHEMICAL BASIS OF LIFE

3

INTRODUCTION

The very definition of life requires that the molecules within a living organism be in constant flux. Different molecules are being continuously synthesized, broken down, chemically rearranged, and modified. One macromolecule binds to another to mediate a process and then becomes free again, to be available for yet another interaction in the cell. To understand the different processes that are required to maintain and propagate life, we must review the thermodynamic laws that govern how likely it is that a set of molecules will form a complex with one another, and how changing conditions in the cell affect that likelihood.

Similarly, it is important to know how likely it is that certain biochemical reactions will occur, and how those reactions are harnessed by the cell to carry out a variety of tasks. By studying the thermodynamics and chemistry underlying all of biology, we will be able to understand the inner workings of the cell and how organisms respond to a changing environment. In this chapter, we will learn about the laws of thermodynamics and chemistry that underlie biological processes. We will also see how the cell is able to harness the energy released from one chemical reaction to power a second reaction that would otherwise not occur, much in the same way that a car can be propelled up a hill using the energy released by burning gasoline.

Most of the important chemical reactions in biology do not proceed rapidly enough to be useful to an organism. When, for example, a cell needs a particular protein, it cannot afford to wait for the free amino acids to react with one another and form a protein polymer. Even under very favorable conditions, in which all the free amino acids were available at high concentrations, it would take more than ten years for a 100 amino-acid protein to form spontaneously. Cells therefore need molecules that can accelerate reactions so that they occur on a timescale that is useful to the organism. We shall learn in this chapter about some of the ways in which **enzymes catalyze** chemical reactions, and how enzyme activity is regulated in the cell.

3.1 THERMODYNAMIC RULES IN BIOLOGICAL SYSTEMS

Living things obey the laws of thermodynamics

All cellular functions depend on the synthesis and breakdown of macromolecules and on their intimate interactions. In the previous chapter, we learned how the remarkable physical and chemical properties of biological molecules govern their interactions with one another and allow them to adopt particular shapes. An understanding of these principles is not sufficient, however, to predict macromolecular behavior. For example, proteins and DNA do not assemble spontaneously if their component building blocks are simply added to water; nor can the assembly of specific non-covalent complexes such as ribosomes or muscle fibers be guaranteed simply by mixing their components in an aqueous environment. Instead, to gain a full understanding, we need to consider the biochemical reactions that underpin macromolecular behavior by asking three key questions. First, is a reaction energetically favorable – that is, will it happen at all? Second, how far will the reaction proceed? And, third, how fast will the reaction happen?

Understanding how cellular macromolecules behave in the ordered way that they do, and what the answers to these questions are, requires an understanding of energetics as defined by the basic laws of thermodynamics. These laws provide us with a small number of rules that apply to any molecular process.

The energy associated with a system is indicated by its enthalpy

All biological systems rely on the flow of energy to drive forward the biochemical processes on which they rely: to digest and metabolize food; to power the contraction of muscles; to maintain the body at a constant, stable temperature. However, organisms cannot create energy. Instead, energy is merely converted from one form into another. This concept – the so-called conservation of energy – is known as the **first law of thermodynamics**. Energy can take many different forms: for example, kinetic energy (the energy associated with motion), chemical energy (the energy stored in chemical bonds), and thermal energy.

How much energy does an atom, a molecule, or an organism have? The measure of the internal energy that a given entity possesses is given by its **enthalpy** (represented by the symbol H). If an entity is of high energy, it has a large enthalpy; if it is of low energy, it has a small enthalpy. However, the enthalpy value in isolation is not all that useful in helping us to determine whether a biological process can happen or not. Instead, we are more interested in the change in enthalpy – that is, the difference between the enthalpy value at the start and end of a process. We represent the change in enthalpy by writing ΔH (where the symbol Δ (Greek uppercase delta) means 'change in').

If the final state of a process has a higher enthalpy than the starting state, the change in enthalpy is positive, as illustrated in Figure 3.1. Remembering that enthalpy is a measure of energy, this means that the end-state of the process has more energy than the starting state, so energy must be transferred *to* the system during the course of the process to make up this difference. We say that a process

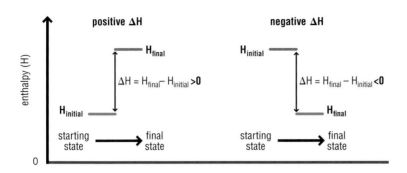

Figure 3.1 Enthalpy changes. If the final state of a process has a higher enthalpy than the starting state, ΔH (the change in enthalpy) is positive. If the enthalpy of the final state is lower than the enthalpy of the initial state, ΔH is negative.

that exhibits a positive enthalpy change (and needs to have energy supplied to it to make it happen) is **endothermic**. By contrast, if the end-state of the process has a lower enthalpy than the starting state, the change in enthalpy is negative (Figure 3.1). Consequently, the end-point of the process has less energy than the starting state, so energy is released from the system during the course of the process. A process with a negative enthalpy change – that releases energy to the surroundings – is **exothermic**.

So why is this significant? Biological systems comprise many interlinked reactions and processes. The enthalpy change for some of these processes will be positive; for others it will be negative. For example, plants carry out a process known as photosynthesis, in which they synthesize the sugar, glucose, from carbon dioxide and water according to the equation:

$$6CO_2 + 6H_2O \rightarrow C_6H_{12}O_6 + 6O_2.$$

This reaction, which also releases oxygen, has an enthalpy change of +2870 kJ/mol (685 kcal/mol) and is therefore endothermic, meaning that it requires an input of energy. That energy comes from sunlight. By contrast, the breakdown of glucose in the reverse reaction

$$C_6H_{12}O_6 + 6O_2 \rightarrow 6CO_2 + 6H_2O$$

has an enthalpy change of –2870 kJ/mol (–685 kcal/mol). This negative number tells us that the breakdown of glucose releases energy; it is highly exothermic.

Entropy gives the measure of disorder in a system

Not all of the energy associated with a change of enthalpy during the course of a reaction or process is free to be channeled into doing something useful (what we describe as being free to do 'work'). Why is this?

As well as being described in terms of its enthalpy, the amount of energy something has can also be described in terms of its **entropy**, which is represented by the symbol S. Entropy is something of an abstract concept, but essentially gives a measure of disorder: something with high entropy is more disordered than something with lower entropy (Figure 3.2a). For example, let us consider a population of water molecules (Figure 3.2b). If these water molecules have very little energy, they will tend to adopt a very organized, low-energy structure – namely, ice. Ice has low entropy. If we transfer energy to the molecules in the form of heat, however, they will start to become more disorganized: the ordered array of molecules in ice will fall apart, and the molecules will become liquid water, which is

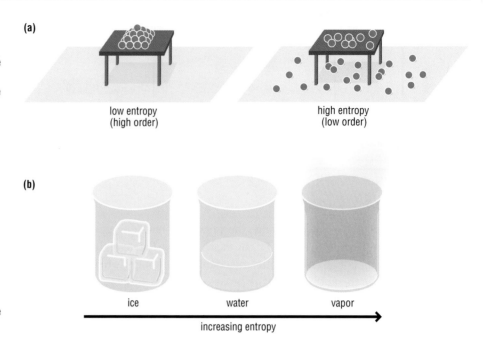

Figure 3.2 Entropy. (a) Entropy, denoted by the quantity *S*, gives a measure of order. The table on the left has low entropy because the balls are stacked neatly in a pile. On the right, the stack has fallen and the balls lie in random positions. This configuration for the balls is of higher entropy than the one on the left. It would take the input of energy to stack the balls up again in a neat pile, whereas a slight jostling of the table will cause them to disperse in a random way. (b) Water is in a low entropy state when it is in the solid state as ice, which contains water molecules aligned and immobilized within a crystal lattice. Liquid water has higher entropy than ice, as the water molecules are free to tumble and diffuse around the beaker. Water vapor has even higher entropy than liquid water, as the water molecules in vapor are free to leave the beaker and diffuse around the room.

of higher entropy than ice. If we add more energy still, the molecules will become even more disorganized, drifting apart from each other to form water vapor, which is of yet higher entropy. The link between disorder and entropy holds true even when we consider biological molecules: a folded protein, which represents a highly organized structure, has lower entropy than an unfolded protein, which is disorganized.

Just as we see a change in enthalpy during the course of a process (written as ΔH), so we also see a change in entropy, written as ΔS. A positive ΔS means that the final state has less order than the initial state, whereas a negative ΔS means that the final state has greater order. Therefore, the freezing of water – from liquid to solid – corresponds to a decrease in entropy, so the entropy change, ΔS, is negative. By contrast, the vaporization of water – from liquid to gas – corresponds to an increase in entropy, so the entropy change is positive.

For a process to happen spontaneously, the overall entropy change must be positive

Why is entropy important? The answer lies in the **second law of thermodynamics** which states, in its simplest form, that disorder in the universe can only increase. Put another way, for a process to happen spontaneously – without outside influence – the change in entropy must be positive.

How, then, do biological systems continually build ordered structures – macro-molecules, cells, complex tissues – out of simple building blocks without defying the second law of thermodynamics? How does the folding of proteins occur if this process – from a disorganized to an organized polypeptide – represents a *decrease* in entropy? This apparent contradiction is resolved if we consider the cell in the context of the universe as a whole. As we shall see, the processes and reactions that generate increased order in biology, such as the synthesis of macromolecules or the assembly of cell membranes, are coupled to other reactions (just as in our discussion of enthalpy above). Together, the net effect is to *increase* the disorder in the surroundings and, hence, the universe.

In the case of protein folding, the decrease in entropy associated with the organized folding of a protein is offset by the *increase* in entropy stemming from an increase in the disorder of water molecules surrounding the protein. When the protein is unfolded, water molecules surround the molecule in a relatively organized way. Upon folding, however, this organized interaction is disrupted, leading to an increase in the entropy of the water molecules.

Let us now come back to our earlier comment that not all of the energy associated with a change in enthalpy is free to do work. The reason for this is that some of the energy is used to drive an increase in the entropy of the system – that is, it is used to satisfy the second law of thermodynamics. The proportion of energy that is used to drive an increase is entropy is therefore not available to do work. The amount of energy that is actually free to do something useful is called the **Gibbs free energy**, *G*.

The free energy measures how favorable a process is

How readily two molecules will bind to one another or whether a chemical reaction will occur spontaneously is governed by the difference in the Gibbs free energy of the initial state and the final state, which is written as Δ*G*. For a process to happen spontaneously, the change in free energy, Δ*G*, must be negative, such that the final state has a lower free energy than the initial state. Such a process gives out energy rather than requiring energy to be supplied. By contrast, a process with a positive Δ*G* requires energy and is not spontaneous (Figure 3.3). These processes can be compared to that of a ball rolling freely down a hill or requiring an input of energy for rolling up the hill. When an ice cube is placed in a glass of warm water, it will rapidly melt because the free energy change associated with the ice melting is negative. The synthesis of sugars by plants has a positive Δ*G* and therefore requires an input of energy, which is supplied by sunlight.

The free energy is related to both the enthalpy and entropy by the equation:

$$\Delta G = \Delta H - T\Delta S$$

where *T* is the temperature at which the reaction occurs, expressed in degrees Kelvin. This equation tells us that negative changes in enthalpy, when heat is released, or positive changes in entropy, when there is an increase in disorder, can both contribute to the value of Δ*G* being negative and, hence, to a reaction being spontaneous.

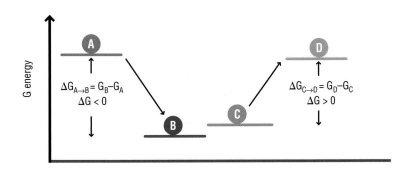

Figure 3.3 Free energy changes. A reaction converting molecule A into B is favorable because the free energy of the final state is lower than the free energy of the initial state. The free energy change, given by $\Delta G = G_B - G_A$, is therefore negative. A reaction such as that converting molecule C to D is unfavorable because the free energy change is positive $(G_D - G_C)$. The situation is analogous to rolling a ball up a hill, which requires energy (C→D), while a ball will roll freely down a hill (A→B) and decrease its energy.

An important consequence of the composite nature of the free energy change is that a process that involves an unfavorable change in either enthalpy or entropy can still occur spontaneously as long as the equation relating the two terms yields a negative overall free energy change. A simple example is the spontaneous dissolving of table salt, sodium chloride (NaCl), in water. The breaking of the ionic lattice holding the Na^+ and Cl^- ions together in the salt crystals requires energy, so this process draws heat from the solution. Consequently, the enthalpy change is positive, and hence unfavorable. However, there is a large favorable change in entropy that arises from freeing into solution the many ions that had been part of an organized crystal lattice, offsetting the amount of energy needed to break apart the lattice. We shall see that in biology, where unfavorable reactions are continually required for the synthesis of macromolecules and the regulation of their behavior, the price of these unfavorable changes is paid for by favorable changes in a second reaction that thereby drives the first. In so doing, we see the second law of thermodynamics in action, with the *overall* entropy – the level of disorder in the system – increasing.

Energy is needed to drive unfavorable reactions

Many reactions that are necessary for life are energetically unfavorable. For example, a DNA strand will not form spontaneously from free nucleoside monophosphates because the free energy of forming phosphodiester linkages is 25 kJ/mol (+6.0 kcal/mol). An unfavorable process can, however, be driven by coupling it with an energetically favorable process. If the sum of the free energy of two coupled reactions is negative, the reactions can proceed together.

The coupling of an unfavorable chemical reaction with a highly favorable one is illustrated by the way in which cells synthesize DNA. Instead of sequentially adding nucleoside monophosphates in a simple condensation reaction, cells use nucleoside triphosphates as precursors in the reaction. These nucleoside triphosphates are referred to as high-energy molecules because hydrolysis of the bonds that join the phosphate groups releases a large amount of energy. As illustrated in Figure 3.4, nucleoside triphosphates are used to synthesize DNA in a reaction that releases the last two phosphate groups, producing **pyrophosphate** (PP), and adds a nucleoside monophosphate to the growing DNA chain. This reaction is still somewhat unfavorable, with a ΔG of 2 kJ/mol (0.5 kcal/mol), but notably much more favorable than the simple condensation of monophosphates. Subsequent hydrolysis of pyrophosphate to organic monophosphate by pyrophosphatase, with a ΔG of –31 kJ/mol (–7.3 kcal/mol), is coupled to the bond forming reaction. Thus the total free energy change of the coupled reactions is –29 kJ/mol (–6.8 kcal/mol), which is quite favorable. There are many reactions in biology in which the energy released by cleaving a high-energy molecule is used to drive an otherwise unfavorable reaction.

The free energy of a reaction is influenced by the concentration of reactants

The free energy change of a reaction depends on both the inherent properties of the participating components and on their concentrations in solution at a given point in time. The concentration of individual proteins, nucleic acids,

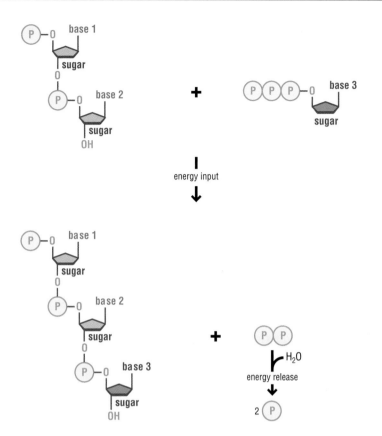

Figure 3.4 Coupling of favorable and unfavorable reactions in DNA synthesis. Synthesis of DNA from nucleoside monophosphates is unfavorable. Using nucleoside triphosphates (NTPs) as the precursors in the reaction allows the coupling of NTP hydrolysis to the unfavorable condensation reaction, thereby making the net reaction of DNA synthesis favorable.

and various small molecules can vary greatly among different types of cells and under different conditions. This means that whether a particular reaction will proceed spontaneously, and ultimately what proportion of the molecules will undergo the reaction, will depend on the particular conditions in the cell, or in the test tube.

To determine how molecules will react with one another under a set of conditions that may be of interest to us, we start by considering the energy change for the reaction in which the reactants and products are in their **standard states**. For compounds in solution, the standard state equates to the compound being present at a concentration of 1M. (By contrast, the standard state for a compound in gaseous form is for it to be at a pressure of 1 atm.) Standard states can be defined at any temperature but, if it is not specified, we usually assume the temperature to be 298 K (25 °C). Measurements made under standard conditions can act as reference values, which we can use to characterize a system when it exists under other, non-standard conditions. The **standard free energy change**, denoted by ΔG°, describes the inherent tendency of a specific reaction to proceed when the reactants and products are in their standard states. This quantity is the result of the particular chemical and physical properties of the reacting molecules and of the products. (The superscript, $^\circ$, tells us that a thermodynamic quantity – such as ΔG – has been measured when the reactants and products are in their standard states.)

In general, however, we are interested in determining the free energy change when the concentration of the reactants differs from the standard state. To illustrate how concentration affects the overall free energy change, we will start by considering a very simple reaction in which X is converted into Y – for example, a *cis* proline amino acid converting to a *trans* proline in a polypeptide chain. This reaction can be written in the following way:

$$X \rightleftharpoons Y.$$

The arrows are shown pointing in both directions to denote that this is a reversible reaction. So how can we determine ΔG for this reaction under non-standard conditions? The free energy change for the formation of complex Y can be calculated from the equation:

$$\Delta G = \Delta G^\circ + RT \ln[Y]/[X]$$

where R is the gas constant (8.31 J K^{-1} mol^{-1} or 1.987×10^{-3} kcal mol^{-1}deg^{-1}), T is the temperature expressed in degrees Kelvin, ln is the natural logarithm, and the square brackets denote concentration in units of mol/L. The first term, ΔG°, provides information about the specific reaction being considered when the reactants and products are in their standard states; we can often look up these values from reference tables rather than having to calculate them ourselves. By contrast, the value of the second term, $RT \ln[Y]/[X]$, is determined by the relative concentrations of reactant (X) and product (Y) in the reaction system that we are interested in.

An important consequence of this equation is that the reaction becomes more spontaneous as the relative concentration of X increases, as this causes the ln[Y]/[X] term to become increasingly negative. This means that increasing the amount of reactants in solution initially can drive the reaction forward towards conversion to Y. If, on the other hand, there is a great excess of the product, Y, the term ln [Y]/[X] will be positive. If the net value of RTln[Y]/[X] is large enough such that adding it to ΔG° gives a net positive value, the reaction will be unfavorable and will not proceed as written. Instead, the reverse reaction will occur, with Y converting to X.

This example illustrates an important point: the initial concentration of both products and reactants determines the direction of the reaction. A sufficiently large excess of reactants will drive the reaction forward, while a large enough excess of product will drive the reaction in the reverse reaction. This is true for all types of reactions – chemical reactions and binding reactions. This fundamental point is central to understanding how biological interactions and reactions are driven throughout the cell.

The ratio of products and reactants at equilibrium is determined by the standard free energy

We have seen how the standard free energy change, ΔG°, and the initial concentration of products and reactants determines whether or not a reaction is initially favorable, and in which direction the reaction will proceed. But what happens after that initial time, as reactants are converted into products (or vice versa) and the balance of reactants and products changes? When hemoglobin in red blood cells carries oxygen to tissues, how much oxygen will it deposit there? For any chemical reaction, how much product will be produced under a particular set of conditions?

To see what happens as a reaction proceeds, we will again consider the very simple reaction in which X is converted into Y (Figure 3.5). This is a reversible reaction, which means that it actually comprises two separate reactions happening in parallel: a forward reaction, X → Y, and a reverse reaction, Y → X.

Imagine starting out with a solution containing only X molecules and a favorable (i.e. negative) ΔG° for the conversion of X to Y. As the reaction proceeds, reactants are consumed and the concentration of product, Y, increases. This change causes the [Y]/[X] term in the equation to increase, which will make the overall free energy less negative, as given by the equation $\Delta G = \Delta G^\circ + RT \ln[Y]/[X]$. In other words, the forward reaction will become less favorable than under the initial conditions as more and more reactant is converted into product.

As the product, Y, accumulates, the rate of the reverse reaction will increase, namely the conversion of Y back to X. As long as the net free energy is negative, this reverse reaction will occur relatively less often than the forward reaction. Overall, Y will continue to accumulate as long as the free energy, ΔG, is negative.

As the reaction proceeds still further, the ratio of product to reactant (Y to X) will reach a point such that ΔG reaches 0. At this point, when $\Delta G = 0$, there is no net change in the system: the rate of formation of product from reactants exactly balances the rate of conversion of product into reactants. We say that the reaction has reached **equilibrium**. The reaction has proceeded as far as is favorable, and the relative amounts of reactant and product will change no further.

At equilibrium:

$$\Delta G = 0 = \Delta G^\circ + RT \ln([Y]/[X])$$

or

$$\Delta G^\circ = -RT \ln([Y]/[X]).$$

We can see from this expression that, under equilibrium conditions, the standard free energy change, ΔG°, can tell us the ratio of product to reactant. In the coming sections, we will see in greater detail how this relationship can be used to tell us how much product and reactant will be present in a chemical reaction, and what proportion of binding partners in a solution will form a complex.

the reaction X ⇌ Y

(a) at time, t = 0

(b) later

(c) at equilibrium

Figure 3.5 The conversion of X to Y. The reaction in which X is converted to Y is a reversible reaction. (a) Initial conditions: all X, no Y. (b) As the reaction proceeds, X is converted into Y and Y begins to accumulate. Some Y is also converted back into X, but the net rate of Y formation exceeds the rate of the reverse reaction. (c) At equilibrium, the relative amount of X and Y remains constant, although some X continues to be converted to Y while an equal amount of Y is converted back to X.

3.2 BINDING EQUILIBRIA AND KINETICS

Virtually all of the biological processes covered in this book depend on the binding of two or more molecules to one another. A protein that regulates whether a gene is on or off must bind to DNA. The ribosome, which synthesizes proteins, must bind to messenger RNA and transfer RNA charged with amino acids before it can form a polypeptide chain. We can use the thermodynamic principles we have discussed thus far to understand binding reactions: how likely it is that a binding reaction will occur, and how much complex will form under a given set of conditions. Since conditions in living things frequently change, it is also important to understand how quickly complexes form and come apart and how long it takes for a binding reaction to reach equilibrium. As we will see, the proportion of molecules that form a complex at equilibrium is directly related to the rates governing binding and dissociation: the binding kinetics.

Association:

k_{on} - association rate constant
k_{off} - dissociation rate constant

Rate of association of AB = k_{on}[A][B]
Rate of dissociation of AB = k_{off}[AB]

At equilibrium:

$$k_{on}[A][B] = k_{off}[A]$$

$$K_{eq} = \frac{k_{on}}{k_{off}} = \frac{[AB]}{[A][B]} = K_a$$

Remember that:

$$\Delta G = \Delta G^0 + RT \ln \frac{[AB]}{[A][B]}$$

And at equilibrium, $\Delta G = 0$

Thus at equilibrium:

$$0 = \Delta G^0 + RT \ln \frac{[AB]}{[A][B]}$$

$$\Delta G^0_{association} = -RT \ln K_{eq}$$

Dissociation:

At equilibrium:

$$K_{eq} = \frac{k_{off}}{k_{on}} = \frac{[A][B]}{[AB]} = K_d = \frac{1}{K_a}$$

$$\Delta G^0_{dissociation} = RT \ln K_d$$

$$\Delta G^0_{association} = RT \ln K_d$$

Figure 3.6 Kinetics of binding and equilibrium constants. Formation of a complex between molecules A and B is governed by the rate constants, k_{on} and k_{off}, which in turn determine the equilibrium binding constant, K_{eq}. Note that the rate constants are denoted by a lower case k. k_{on} is a second-order rate constant with units of M^{-1} s^{-1} while k_{off} is a first-order rate constant with units of per second. Equilibrium binding constants are indicated by an upper case K and, in this example, K_a has units of M^{-1} whereas K_d has units of M.

The ratio of bound and free molecules at equilibrium is determined by the standard free energy

The general equation describing the free energy of any reaction can be adapted to describe a binding interaction as follows: Consider the binding reaction of two molecules, A and B, to form a complex denoted AB. This reaction can be written in the following way:

$$A + B \rightleftharpoons AB.$$

The arrow is pointing in both directions to denote that this is a reversible reaction: just as A associates with B to form a complex, the AB complex can dissociate to release the individual A and B components.

As we saw above, the free energy change for the formation of complex AB is given by the equation:

$$\Delta G = \Delta G^0 + RT \ln([AB]/[A][B]).$$

In this case, the ΔG^0 term provides information about the change in energy when A and B form a complex, relative to their being free in solution. If the two molecules form energetically favorable interactions with one another such that the formation of the complex is favored, the value of ΔG^0 will be negative. At equilibrium (when $\Delta G = 0$):

$$\Delta G^0 = -RT \ln([AB]/[A][B]).$$

The ratio of complex AB to reactants at equilibrium is determined by the standard free energy change, ΔG^0. If ΔG^0 is negative, telling us that complex formation is energetically favorable, the numerator in the above equation (the concentration of AB at equilibrium), must be larger than the denominator (the concentration of free A molecules multiplied by the concentration of free B molecules that are left in solution at equilibrium). In otherwords the complex predominates at equilibrium, and binding is favored.

Reaction rates are influenced by the concentrations of the reactants

Although thermodynamic quantities tell us whether a reaction is spontaneous and how much product will be formed, these constants do not reveal how quickly – or how slowly – the process occurs. For example, once the hormone, estrogen, enters a cell, how long will it take for certain genes to be activated in response? To answer this question, we must know the rate at which the underlying processes occur. The description of chemical reaction rates is known as kinetics.

Let us begin by considering how long it will take for the binding reaction that we have been considering to reach equilibrium. For a simple binding reaction during which molecules A and B associate to form the AB complex, the rate of binding is given by the **rate constant**, k_{on}, multiplied by the concentration of the binding partners: k_{on}[A][B] (Figure 3.6). (Notice how the rate constant carries the subscript 'on'; this indicates that the process represents a binding reaction, or the association of two (or more) species.)

The rate constant for this particular reaction has units of $M^{-1}s^{-1}$, and is termed a bimolecular (second order) rate constant because it describes the rate at which two molecules bind to one another. For a rate constant of 10^7 M^{-1}s^{-1} and initial reactant concentrations of 10^{-3} M for both A and B, the initial rate at which A and B form a complex will be 10 M s^{-1}.

While the rate at which a complex forms can vary quite a bit, the k_{on} cannot be greater than the diffusion limit, which is the maximum rate at which two molecules can move through an aqueous environment and eventually collide with one another. For a typical interaction between proteins, the diffusion limit is on the order of 10^8 M^{-1}s^{-1}. When k_{on} approaches this value, essentially all encounters of the two species result in productive (that is, successful) binding. However, most biological binding reactions exhibit k_{on} values that are considerably lower than this upper limit. The diffusion limit affects all types of chemical reactions and also places an upper limit on the rate of enzyme-catalyzed reactions, as we see below.

So far, we have only considered the association of two species – that is, their binding. Of equal importance in biology is the reverse process, that of dissociation. For example, the process by which oxygen is delivered to our tissues depends on both binding and dissociation: oxygen must participate in a binding reaction with hemoglobin in the lungs, but must then undergo dissociation once hemoglobin has reached the target tissue, such that the oxygen can be taken up by that tissue.

The rate at which the AB complex dissociates is given by a rate constant, k_{off}, multiplied by the concentration of the complex [AB]: k_{off}[AB] (Figure 3.6). Since the rate describes the dissociation of a single species, the AB complex, it is a unimolecular (first order) rate constant and has units of per second (s^{-1}). This value tells us how likely it is that the AB complex will dissociate in a given time frame, and thus permits us to calculate the average lifetime of a biological complex.

The kinetics of a binding process can have important biological consequences. Consider an example of a drug that must bind to a particular protein in the cell in order to be effective. If the k_{on} is too slow, the drug may be excreted from the body before it engages its target. By contrast, a too-slow k_{off} may mean that clearing the drug from the body may take a very long time.

The time it takes for half of all complexes in a given population of identical complexes to come apart is called the half-life, $t_{1/2}$, as illustrated in Figure 3.7. For simple binding equilibria, the half-life of a complex is related to the rate of dissociation of

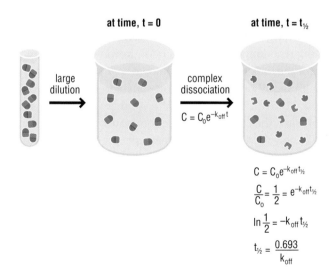

at time, t = 0

large dilution

complex dissociation

$C = C_0 e^{-k_{off} t}$

at time, t = t$_{1/2}$

$C = C_0 e^{-k_{off} t_{1/2}}$

$\dfrac{C}{C_0} = \dfrac{1}{2} = e^{-k_{off} t_{1/2}}$

$\ln \dfrac{1}{2} = -k_{off} t_{1/2}$

$t_{1/2} = \dfrac{0.693}{k_{off}}$

Figure 3.7 The half-life of a complex in solution. The half-life of a complex in solution gives a measure of the off rate, k_{off}. To see the meaning of the half-life, imagine taking a sample of complexes and placing them in a large volume, so that complexes can come apart but cannot re-form. At time 0, the concentration is C_0. How long it will take for half of the complexes to come apart? This is obtained from the equation describing the dissociation of the complexes, $C = C_0 e^{-kt}$, as detailed in the figure.

the complex by the expression $t_{1/2} = 0.693/k_{off}$. So, for example, for a complex with a dissociation rate of 0.1/s, half of all the complexes in the population will come apart in 0.693/0.1 s, or 6.93 seconds.

Unlike the rate of complex formation, the rate at which a complex dissociates is not limited by the speed at which molecules diffuse through liquid, since a single complex can come apart at any time without needing to interact with any of the other molecules in solution. There is therefore no inherent limit on k_{off}.

The equilibrium constant is a ratio of the forward and reverse reaction rates

At equilibrium, the rate at which the complex, AB, dissociates must equal the rate at which A and B associate. In terms of the rates we defined above, it is the point at which

$$k_{on}[A][B] = k_{off}[AB].$$

Rearranging, we obtain

$$k_{on}/k_{off} = [AB]/[A][B].$$

This ratio of forward and reverse rate constants is defined as the **equilibrium constant (K_{eq})**:

$$K_{eq} = k_{on}/k_{off} = [AB]/[A][B].$$

Since the particular reaction we are describing is the association of two molecules with one another to form a complex, we also call this the **equilibrium association constant (K_a)**. (Note that we use the upper case K to denote equilibrium constants, and lower case k to denote a rate constant.)

The equilibrium constant gives a measure of the proportion of the molecules in solution that are bound in a complex at equilibrium. It is not the absolute value of either rate constant that matters; it is the ratio of the on and off rates that determines K_{eq}. The magnitude of the forward rate, k_{on}, will simply determine how long it takes to reach equilibrium once the individual components are mixed in solution, while the k_{off} will determine how long it takes the complex to fall apart (Figure 3.6).

The equilibrium constant, K_{eq}, relates to a general reversible reaction and is always the ratio of the forward and reverse rate constants. When considering the specific case of a binding reaction such as A + B ⇌AB, however, we saw that the equilibrium constant for the forward reaction is known as K_a, the equilibrium association constant. Since the equilibrium association constant is equal to [AB]/[A][B], it has units of M/(M × M), which simplify to M^{-1}. A large value for K_a tells us that the products (in this case, the complex AB) predominate over reactants at equilibrium, whereas a small value for K_a tells us that reactants predominate over products. We can therefore say that two species exhibiting a large value for K_a show a greater degree of binding – that is, they bind more readily – than two species exhibiting a smaller K_a.

By contrast, the equilibrium constant for the reverse reaction, namely the dissociation of complex AB:

$$AB \rightleftharpoons A + B$$

is known as the **equilibrium dissociation constant (K_d)**, and is simply the inverse of the association constant, K_a:

$$K_d = k_{off}/k_{on} = [A][B]/[AB] = 1/K_a.$$

The K_d, which is expressed in units of concentration (M), gives us a measure of the proportion of molecules bound up in a complex at equilibrium. A large value for K_d indicates that the products (in this case the single, dissociated species A and B) predominate at equilibrium, whereas a smaller value suggests that reactants (the complex) predominate (Figure 3.8). The K_d is used nearly universally in biology to describe binding reactions.

The equilibrium constant is related to the free energy of the interaction

The equilibrium constant has a thermodynamic meaning because the relative amounts of free components and of the complex present at equilibrium are directly related to differences in the free energy, ΔG°, of the free and complexed components. Combining the expressions shown above for the free energy change at equilibrium and the equilibrium constant for the binding reaction $A + B \rightleftharpoons AB$ gives the equation:

$$\Delta G^\circ = -RT\ln([AB]/[A][B]) = -RT\ln K_{eq}.$$

This important equation gives the relationship between the equilibrium constant and the free energy of a reaction. In general, if the equilibrium constant is greater than 1, ΔG° is negative, and complex formation is favored. By contrast, if the equilibrium constant is less than 1, ΔG° is positive, and most of the A and B molecules will not be part of a complex at equilibrium. Notice how this mirrors our discussion of the free energy of binding reactions.

The key information one generally needs to know about a given reaction in biology is how favorable it is – that is, what is the ΔG° – and what the ratio of products to reactants will be at equilibrium – that is, how far the reaction proceeds. For a binding reaction, the ratio of products to reactants at equilibrium is given by K_d, which is the equilibrium constant for the dissociation reaction. Combining the information above – that $\Delta G^\circ = -RT\ln K_{eq}$ and $K_a = 1/K_d$ – we see for the binding reaction $A + B \rightleftharpoons AB$ that

$$\Delta G^\circ = -RT\ln K_a = -RT\ln(1/K_d) = RT\ln K_d.$$

Figure 3.9 lists some values for equilibrium dissociation constants along with the corresponding free energy changes for the association reaction. As can be seen from the values, a negative ΔG° corresponds to a K_d that is less than 1, which means that the concentration of the complex at equilibrium exceeds that of the uncomplexed molecules. A biological interaction of reasonable affinity, characterized by a K_d of 10^{-9} M, has a corresponding ΔG° of roughly –51.5 kJ/mol (–12.3 kcal/mol). Keep in mind that we are comparing the energy of the *association* reaction with the equilibrium constant for the *dissociaton* reaction, K_d.

A very important implication of equations relating free energy and the equilibrium constant is that seemingly small differences in binding energy, which contribute to the overall free energy, will result in large differences in K_d. For example, the free energy value of a single hydrogen bond is on the order of 4.2–13 kJ/mol

products dominate at equilibrium
[AB] < [A][B]
K_d is large

reactants dominate at equilibrium
[AB] > [A][B]
K_d is small

Figure 3.8 Relation of K_d value to the equilibrium concentrations of complex and free molecules. A large value for K_d indicates that the products (in this case the single, dissociated species A and B) predominate at equilibrium, whereas a smaller value suggests that reactants (the complex) predominate.

Dissociation Constants (K_d) and Corresponding Free Energy of Association		
$K_d = \left(\dfrac{[A][B]}{[AB]}\right)$	free energy (kJ/mole)	free energy (kcal/mole)
10^3 M	17.1	4.09
1 M	0	0
10^{-3} M (mM)	–17.1	–4.09
10^{-6} M (μM)	–34.2	–8.17
10^{-9} M (nM)	–51.4	–12.3
10^{-12} M (pM)	–68.5	–16.4
10^{-15} M (fM)	–85.6	–20.5

Figure 3.9 Table of equilibrium dissociation constants and free energy change of association. The table lists free energy changes of binding and their corresponding equilibrium dissociation constants, as defined by the equation $\Delta G^0_{association} = RT\ln K_d$.

(~1–3 kcal/mol), which means that altering (adding or subtracting) a hydrogen bond from a binding interface between two proteins has the potential to change the K_d by several orders of magnitude! This fundamental relationship between binding energy and the dissociation constant is the basis for understanding biological interactions in the cell, and how seemingly similar molecular species can have quite distinct binding partners and biological consequences.

3.3 BINDING PROCESSES IN BIOLOGY

Until now, we have been discussing the simple example of binding in which just two molecules bind to one another, such as the binding of a single **substrate** to an enzyme. Much of biology, however, involves many more molecules that come together in a network of interactions. In this section, we explore some of the characteristic features of the binding interactions that are so important in biological systems, and the way they are underpinned by the thermodynamic concepts we have considered so far.

A binding curve shows how complex formation increases with concentration

One of the most central questions about the interaction between any two biological molecules is the proportion that will form a complex at a given concentration of each molecule, A and B. The answer will depend on the concentration of A and B, as well as on the equilibrium dissociation constant, K_d.

The importance of using the K_d to describe a binding reaction is that, under certain conditions, the K_d tells us the concentration at which half of a set of molecules are bound up in a complex while the other half are free in solution. For the example we have been considering, $AB \rightleftharpoons A + B$, we will consider conditions where there is much more B than A in solution (or vice versa). By plotting the amount of complex, AB, formed as a function of the total amount of B added to a reaction system, we obtain a hyperbolic curve that is known as a **binding curve** (Figure 3.10a). The binding curve provides a simple depiction of how much complex is formed at different substrate concentrations. The equilibrium dissociation constant, K_d, corresponds to the concentration of molecule B at which half of the A molecules are bound in a complex and half are free. Therefore, the value of K_d can be readily determined from such a plot, as illustrated in Figure 3.10a.

The reason that K_d gives the concentration of B at which half-maximal binding occurs can be explained as follows: if the concentration of B is very much greater than A, then when some B is 'used up' to form a complex with A, the impact on the concentration of B is negligible. Under such conditions at equilibrium, we can write that $[A] = [AB]$. We learned above that $K_d = [A][B]/[AB]$. So, when $[A] = [AB]$, $[A]/[AB] = 1$ and, therefore, $K_d = [B]$. In other words, when the concentration of B is equal to the K_d, there is as much of A bound to B as there is free A in solution.

Taking this one step further, by comparing the concentration of B to the value of K_d for the complex being examined, we can estimate the number of molecules of A and B that are likely to be bound to one another at a given concentration of B, as depicted in Figure 3.12. If [B] in a reaction mixture is much lower than K_d then we

(a)

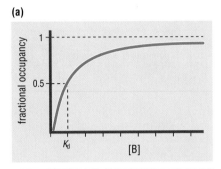

$$K_d = \frac{[A][B]}{[AB]}$$

Solve for [AB]:

$$[AB] = \frac{[A][B]}{K_d}$$

Fractional occupancy $= \dfrac{[AB]}{[A] + [AB]}$

Substitute [AB] in above,

Fractional occupancy $= \dfrac{\left(\dfrac{[A][B]}{K_d}\right)}{[A] + \left(\dfrac{[A][B]}{K_d}\right)}$

$= \dfrac{[A][B]}{K_d[A] + [A][B]} = \dfrac{[B]}{K_d + [B]}$

(b)

concentration of B as a function of K_d (M)	fractional occupancy of A
$0.001\,K_d$	0.001
$0.01\,K_d$	0.01
$0.1\,K_d$	0.091
K_d	0.5
$10\,K_d$	0.91
$100\,K_d$	0.99

Figure 3.10 Fractional binding as a function of concentration. (a) This graph shows the fractional occupancy of A as a function of [B] for the simple binding reaction $A + B \rightleftharpoons AB$ under conditions where B is in excess. The concentration of B where 50% of A is bound (fractional occupancy of A = 0.5) corresponds to the dissociation constant (K_d) for the binding reaction. (b) The table shows specific examples of the fraction of A bound to B as [B] is increased.

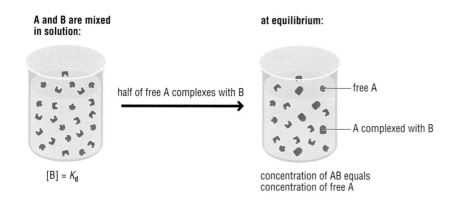

A and B are mixed in solution:

at equilibrium:

half of free A complexes with B

free A

A complexed with B

[B] = K_d

concentration of AB equals concentration of free A

Figure 3.11 K_d and the concentration of complex at equilibrium. Molecules A and B mixed in solution (left) will start to bind to one another. At equilibrium (right), the K_d gives the concentration of [B] at which half of its binding partners, A, will form a complex with B, with the other half free in solution.

infer that only a small proportion of A will have formed a complex with it. (If [B] is less than the K_d then less than half of the available A will have formed a complex.) By contrast, if [B] is much higher than the K_d, then we infer that a large proportion of A will have formed a complex with B.

Comparing the concentration of interacting species to the K_d can help us predict the proportion of the species that are likely to form a complex at a given concentration. The values listed in Figure 3.10b show that when the concentration of B is much higher than the K_d, almost all of A is complexed with B, while at concentrations of B much lower than the K_d, almost all of A is free (uncomplexed). It is also very important to note that it does not matter how *many* molecules of A there are as compared to B; it is the *concentration* of A and B and the K_d for the binding interaction that matter. Even when there are a million molecules of B for every molecule of A, there will only be appreciable complex formation if the concentration of B is roughly equal to or greater than the K_d.

Typical biological interactions have dissociation constants (values of K_d) in the range of 10^{-6} M (micromolar, or μM) to 10^{-9} M (nanomolar, or nM). It is interesting to consider that in the bacterium, *E. coli*, whose cell volume is roughly 10^{-15} L (a femtoliter, fL), a single molecule expressed in the cell will be present at a concentration in the nanomolar range. (By contrast, one molecule within other, larger

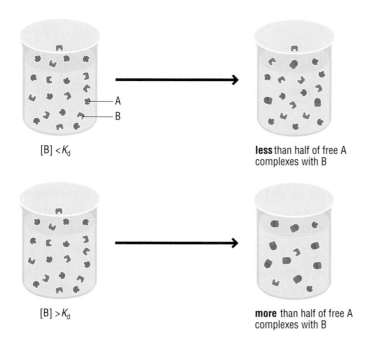

[B] < K_d

A
B

less than half of free A complexes with B

[B] > K_d

more than half of free A complexes with B

Figure 3.12 The ratio of [B] to K_d and the proportion of complex formed at equilibrium. We perform the same experiment as in Figure 3.11, but with different starting concentrations of B. When [B] is less than K_d, less than half of the free A molecules will form a complex with B. When [B] is greater than K_d, more than half of free A molecules will form a complex with B.

cell types would be present at a much lower concentration.) It stands to reason that most cellular components will be present at higher concentrations than this when they are synthesized in an *E. coli* cell (otherwise there would be none present). In general, the K_d corresponding to a particular biological interaction – as dictated by the strength of that interaction – is close to the cellular concentration at which the molecules must bind in the cell to achieve a particular biological outcome.

The chemical details of the bound and free molecules govern the energy of interaction

The net binding energy for two interacting partners represents the energy difference of the entire system before and after the molecules form a complex. The total energy of complex formation will be the net change in enthalpy and entropy before and after complex formation, summed over all the different interactions, according to the equation $\Delta G = \Delta H - T\Delta S$. When two molecules bind to one another, interactions between the surrounding water and the molecules are broken and replaced with hydrogen bonds, salt bridges, and van der Waals interactions between the two molecules at the binding interface. The water molecules are thus released when the two interacting surfaces come together, contributing positively to the entropy of the system.

Sometimes, water molecules are trapped at an intermolecular interface, where they participate in networks of hydrogen bonding interactions (Figure 3.13). Immobilized water molecules have less entropy than free water molecules, and thus contribute unfavorably to the entropy term of the energy equation. Chemical groups located at the binding interface may also become less mobile upon complex formation, which would also contribute unfavorably to the entropic term of the energy equation.

Binding events are most typically associated with the release of energy as heat, and negative changes in enthalpy, resulting in overall favorable contributions to the free energy. As in all aspects of chemistry, it is the *net* energy change in the system, summed over all the changes in entropy and enthalpy in going from the free state to the bound, that determines the overall energy of binding.

Specific binding is of higher energy than non-specific binding

Macromolecules have available to them many potential binding partners in the cell. While a given protein may have one particular macromolecule to which it must bind in order to trigger a particular response, it may also be able to bind to other molecules in the cell without yielding a productive response. The protein is able to select its correct partner from among the other, competing molecules because it binds more tightly to its correct partner than to any other molecule in the cell (see Experimental approach 3.1). Binding to the correct partner is often termed **specific binding**, whereas lower-energy interactions with other molecules are termed **non-specific binding**. The specificity of a given interaction depends on how likely it is that the specific binding partner is selected from among competing ligands.

The specificity of a given binding reaction can be determined by comparing the thermodynamic constants of binding to the correct partner with the constants for binding to other, competing molecules. The free energy of association with the correct partner will be more negative than that associated with binding to other,

water molecule

Figure 3.13 Water molecules immobilized at a binding interface. Solvent molecules can form bridges between two molecules, as in this example of two protein monomers.

3.1 EXPERIMENTAL APPROACH

Insights from kinetics into how DNA-binding proteins find their target sites

The kinetics of a given binding process can give important insights into how the two binding partners locate each other in the cell. A fundamental problem for all proteins that interact with specific DNA sequences is how they locate these relatively small sites (6–30 base pairs) within the vast excess of non-specific genomic DNA. In a classic study by Riggs, Bourgeois, and Cohn of the kinetics of *Escherichia coli* lac repressor binding to its operator sites, the authors found that binding to specific DNA sites occurred very rapidly. Indeed, the observed rate constant for the association of lac repressor with the lac operator on a 50 kb piece of DNA was 7×10^9 M^{-1} s^{-1}, a value considerably higher than the generally accepted figure for the diffusion limit for protein–ligand association reactions, 10^8 M^{-1} s^{-1}. This observation led to the view that lac repressor did not locate its binding sites by simply diffusing randomly through solution before binding to its correct DNA site, but instead bound first to a random site in the chromosome and then slid along the DNA until it reached its correct binding site (Figure 1). This type of binding mechanism was termed facilitated diffusion or one-dimensional diffusion, because the protein was confined to sliding in one dimension along a DNA strand.

Experiments on a number of different proteins appeared to support the view that proteins located their DNA sites by one-dimensional diffusion. For example, it was found that the restriction enzyme, EcoRI, could locate its target sequence more rapidly when the site was located within a long DNA fragment as compared with a short one (Jack *et al.*, 1982). In the experiment depicted in Figure 2, the authors compared binding of EcoRI to a 4361 bp fragment

Figure 1 Model for one-dimensional diffusion along DNA. The protein (1) diffuses through solution and binds to a non-specific site (2). It then randomly diffuses back and forth along the DNA until it locates its specific binding site (3).

and a 34 bp fragment. The enzyme was mixed with both fragments at time 0 and complex formation was monitored. While the protein initially bound overwhelmingly to the long DNA fragment, binding to the two fragments was equalized as the complexes were allowed to equilibrate. This suggested that the equilibrium dissociation constant was the same for the two DNA fragments, but that the on rate (and hence also the off rate) was more rapid for the longer fragment.

Using modern approaches that permit one to manipulate individual DNA molecules, evidence for additional contributions to DNA binding has accumulated. According to this altered view, a protein rapidly 'searches' large amounts of non-specific DNA using a combination of one-dimensional diffusion and three-dimensional 'hops' of the protein between DNA segments (Figure 3a). van den Broek and colleagues came to this conclusion by directly measuring the

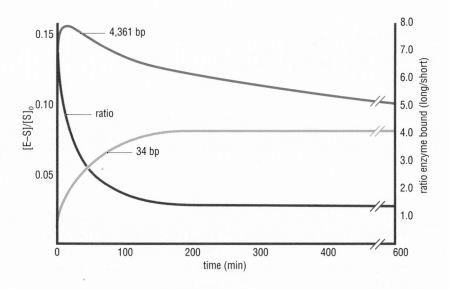

Figure 2 Binding of EcoRI to long and short DNA fragments. At time 0, EcoRI was mixed with two DNA fragments, one long and one short, and the binding to each fragment was then monitored as a function of time. While the protein bound very rapidly to the long DNA fragment at first, with time the binding to the two fragments was nearly equal. Reproduced from Jack *et al.*, *Proceedings of the National Academy of Sciences of the U S A* 1982:**79**:4010–4014.

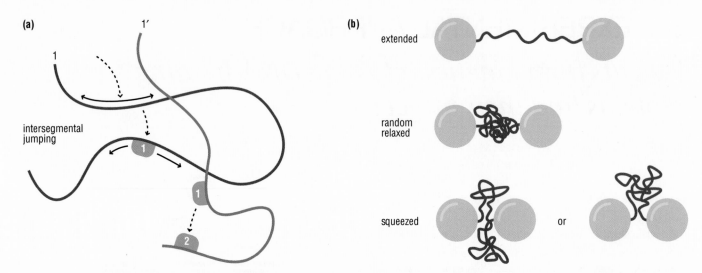

Figure 3 **DNA binding by diffusion and intersegmental hopping.** (a) A protein binds to DNA and can then either diffuse along for short distances or hop from one segment to another. The latter process is shown to be facilitated by DNA compaction in a single molecule analysis. (b) Illustration of the experiment, in which each end of a long DNA fragment is attached to a bead that can be manipulated to condense or extend the DNA. Protein association is facilitated when DNA is condensed (random relaxed), but not too compressed (squeezed), which makes hopping from strand to strand more likely.
Reproduced from van den Broek *et al.*, *Proceedings of the National Academy of Sciences of the U S A* 2008:**105**:15738–15742.

association rate for EcoRV endonuclease binding to DNA, but comparing the rates for the same DNA fragment as a function of how compact it is.

DNA compaction was varied by attaching a bead to each end of the DNA fragment and manipulating the distance between them using a device known as optical tweezers (Figure 3b). This approach is generally referred to as 'single molecule' analysis since it is based on direct observation of specific molecules through a microscope. By studying the binding process as the distance between the two beads was varied, the authors found that the rates of association (the on rates) were somewhat higher when the degree of DNA compaction was higher. In thinking

about their results, the researchers argued that in the crowded intracellular environment where DNA density is extremely high, the contributions of such hopping to the search algorithm of a protein could be quite significant for biochemical function.

Riggs AD, Bourgeois S, Cohn M. The lac repressor-operator interaction. 3. Kinetic studies. *Journal of Molecular Biology* 1970;**53**:401–417.

Jack WE, Terry BJ, Modrich P. Involvement of outside DNA sequences in the major kinetic path by which EcoRI endonuclease locates and leaves its recognition sequence. *Proceedings of the National Academy of Sciences of the U S A* 1982;**79**:4010–4014.

van den Broek B, Lomholt MA, Kalisch SMJ, Metzler R, Wuite GJL. How DNA coiling enhances target localization by proteins. *Proceedings of the National Academy of Sciences of the U S A* 2008;**105**:15738–15742.

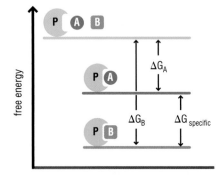

Figure 3.14 Binding specificity. Protein P binds preferentially to protein B, with which it forms a more favorable complex than with protein A. The net free energy change upon binding is greater (i.e. more negative, and therefore more favorable) when P binds to B as compared with binding to A.

random molecules in the cell (Figure 3.14). This will be reflected in the equilibrium dissociation constants, as the K_d for the specific interaction will be lower (reflecting tighter binding) than the K_d for the non-specific interaction.

The ratio of these two constants,

$$K_d^{specific} / K_d^{non\text{-}specific},$$

gives a measure of the specificity of a given interaction. Specificity can vary widely: a ratio of 10 or less typifies an interaction with limited specificity, that is, the two molecules only show limited preference for one another. By contrast, ratios of 1000 or greater mean that the binding interaction is highly specific. The bacterial lac repressor protein, for example, binds to DNA of arbitrary sequence with a K_d of around 10^{-7} M, while it binds to its specific DNA site with a K_d of around 10^{-11} M. This 10 000-fold preference for the correct site over random sites ensures that the lac repressor will regulate the correct set of genes in the bacterial genome. The specificity of different interactions has been fine-tuned in the course of evolution to ensure that they can take place in the crowded cell milieu.

Cooperative binding makes an important contribution to the sensitivity and regulation of biological systems

In many examples involving multiple binding species, the free energy change as each successive subunit is added to the larger complex can depend on the molecules that have previously been incorporated into the complex. We say that the binding is cooperative if the binding of one molecule makes the binding of the next molecule more favorable. This is known as **positive cooperativity**. **Cooperative binding** can also be negative, meaning that the binding of one molecule makes the binding of the next less favorable, but the examples we will encounter in this book will be almost exclusively those of positive cooperativity. Cooperative binding can be exploited by the cell to fine-tune responses to small changes in the concentration of individual binding partners.

A simple example of cooperativity – and one that we shall encounter later in this book – can occur in the binding of a given type of protein to two adjacent sites in a double-helical strand of DNA (Figure 3.15). If there are no interactions between the two proteins when they are each bound to the neighboring DNA sites, the binding affinity for one site is unaffected by whether or not another protein is bound to an adjacent site. However, if proteins bound to neighboring sites can form favorable interactions with one another, the binding of protein to DNA makes the binding of a second protein to the neighboring site more favorable. This additional energy makes the free energy of binding to the second site more negative; in other words, more favorable.

The effect of cooperative binding, and its importance in fine-tuning regulation, can be readily seen in the changes that it produces in a **binding curve**. Instead of the hyperbolic curve that we described above for a simple binding interaction, cooperative binding produces an S-shaped, **sigmoidal curve**, in which complex formation at first rises slowly with increasing protein concentration and then rises steeply over a narrow concentration range (Figure 3.15). The sharper increase occurs because, as one DNA site is filled, the effective affinity for the next site increases and that site will become occupied at a lower protein concentration than it would in the absence of cooperative interactions. Cooperativity has the effect of narrowing the concentration range over which the binding partners make the transition from being largely free to being mostly bound up in a complex.

Cooperative interactions are critical to many biological switches because they enable the cell to respond to smaller changes in the concentration of a particular ligand. If the cellular concentration of a macromolecule or its ligand is near the midpoint of the sigmoidal response curve, where the binding reaction is half-saturated, small fluctuations in concentration can result in large changes in fractional occupancy. This poises a system to rapidly switch processes on and off when ligand concentrations fluctuate in small but biologically significant ways.

A variety of mechanisms can give rise to cooperative binding besides the example described above. In the example above of cooperative binding of protein to DNA, favorable interactions between protein bound to adjacent sites on the DNA are sufficient to make the binding cooperative. In another type of mechanism, the binding of a ligand can induce a conformational change in protein structure, which can then alter the affinity for binding the next ligand. The conformational change can be within a single polypeptide, or in the quaternary arrangements of subunits in a complex. The binding of oxygen to the hemoglobin tetramer in blood cells – while not covered in this book – is a classic example of cooperativity:

(a)

independent binding:
molecules bind to two sites on DNA

cooperative binding:
molecule at site A increases affinity of second molecule for site B

(b)

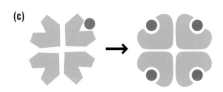

(c)

Figure 3.15 Effect of positive cooperativity on binding curves. (a) Above, two proteins are shown binding independently to two binding sites on DNA; below, two proteins bind cooperatively to adjacent sites, with the binding of one protein to site A increasing the affinity of the second protein for site B. (b) The solid line shows the sigmoidal binding curve characteristic of cooperativity. Note that within a critical range of concentration (shaded area), relatively small differences in concentration B have large effects on binding. Simple binding with no cooperativity gives the hyperbolic curve shown as a blue dotted line. (c) Binding of one molecule of ligand B (red circle) to one subunit of protein A (green) causes a conformational change that both increases the affinity of the bound subunit for the ligand, and induces the same change in the other subunits, increasing their affinity for ligand B.

once a single oxygen molecule binds to one protein subunit, the affinity of oxygen for the remaining three subunits is significantly increased. Protein folding is also cooperative: once the first turn of an alpha helix forms, the remainder folds much more quickly to form the final helical form.

3.4 ENZYME CATALYSIS

An activation energy must be overcome for a reaction to proceed

We note above that the free energy change associated with a reaction determines whether a reaction happens spontaneously or not. However, the spontaneity (or otherwise) of a reaction tells us nothing about how fast a reaction will occur. A reaction may exhibit a favorable (i.e. negative) change in Gibbs free energy, but may still proceed very slowly indeed. For example, the chemical breakdown of a polypeptide chain into individual amino acids through hydrolysis is spontaneous, but happens slowly over months to years. What makes such reactions happen so slowly?

The answer is that every reaction has associated with it an **activation energy** – an energy threshold that must be overcome along the reaction path from reactants to products, as depicted in Figure 3.16. If the activation energy for a reaction is very high at room temperature, as denoted in Figure 3.16, only very few reactant molecules possess the energy required to overcome the activation energy barrier. Consequently, the reaction happens very slowly.

Many important biological reactions have activation energies that are so high that the rates at which they would proceed under 'normal' conditions would be so slow as to prohibit life: few molecules would possess energies higher than the activation energy, and only these few molecules could undergo chemical change. One way to overcome this problem would be to heat the reactant molecules so that more of them possessed energies greater than the activation energy barrier. However, biological systems cannot be exposed to extremes of heat – not least because many essential molecules would become denatured and cease to function. So, how does nature circumvent this problem? The answer lies in the activity of a vital class of biological molecules, the enzymes.

Enzymes accelerate chemical reactions

Cells must continually carry out a remarkable assortment of chemical reactions, which must proceed at a rate that is fast enough to be useful to the cell. All living things therefore depend on molecules called **enzymes** that accelerate the rate of chemical reactions, a process known as **catalysis**. Enzymes catalyze chemical reactions by lowering the activation energy barrier (Figure 3.16). Most enzymes are proteins, although some are composed of RNA and are therefore called **ribozymes**. Enzymes vary in the degree to which they accelerate reaction rates. Some accelerate a chemical reaction by just several orders of magnitude, while some enhance reaction rates by a remarkable factor of 10^{17}.

There are several features that are common to all enzymes. An enzyme binds to one or more molecules known as **substrates**, promotes a chemical reaction in the substrate molecules, and then releases the products of the reaction. The enzyme is not permanently modified in the process, so it is then able to bind another

Figure 3.16 Activation energy. For each chemical reaction, an activation energy barrier must be overcome in the process of converting substrates into products. Even if the net energy change of the reaction is very favorable, few molecules at room temperature may have the energy to overcome the activation energy (blue). An enzyme can increase the rate of product formation by lowering the activation energy for the reaction (red), so a greater proportion of molecules will have the energy needed to overcome the barrier. Note that enzyme catalysis does *not* alter the energy difference between substrates and products.

substrate molecule and perform the reaction again. Enzymes generally catalyze just one type of chemical reaction and are usually highly specific for the types of substrates on which they act. For this reason, cells produce many different types of enzymes, each uniquely designed for a particular chemical task.

A chemical reaction with a negative free energy change should occur spontaneously, given enough time. However, in the course of the chemical reaction, the substrates must undergo transient changes in stereochemistry, charge, or covalent structure before completing the necessary rearrangements to produce the final products. These changes are energetically unfavorable, which means that the first steps in the chemical reaction generally involve a positive free energy change. These are then followed by chemical steps with a negative free energy change of even greater magnitude, resulting in a chemical reaction whose net free energy change, when summed over all the steps, is negative, as indicated in Figure 3.16.

A consequence of this energetic pathway is that there is a free energy barrier between the initial and final states of the chemical reaction. The **transition state**, which is the highest point in free energy on the reaction pathway from substrate to product, marks the top of the free energy barrier (Figure 3.16). This species exists only transiently (for perhaps as little as 10^{-15} seconds) and is generally different in stereochemistry and charge configuration from either the substrate or the product. It is the energy difference between the ground state and the transition state that constitutes the activation energy.

Without the aid of an enzyme, few substrate molecules at any one time typically have sufficient energy to reach the transition state, and, as a result, the uncatalyzed reaction proceeds slowly. Enzymes achieve their catalytic effects by lowering the activation-energy barrier using several different strategies that we will discuss below.

The enzyme active site contains residues that catalyze the chemical reaction

Each enzyme has an **active site**, which is a pocket or region in the enzyme containing precisely positioned chemical groups that catalyze the chemical reaction (Figure 3.17). In addition, there may be other molecules bound in the active site that either assist or actively participate in the chemical reaction. These include metal ions and water, as well as larger molecules such as the nucleotide cofactors we learn about later in this chapter. The great majority of enzymes are proteins, and the key chemical groups in their active sites typically are amino acid side chains, although the peptide backbone itself can occasionally participate. By contrast, the active sites of ribozymes are composed of chemical groups found on the nucleotide bases or on the sugar–phosphate backbone.

The chemical nature and shape of the active sites are uniquely tailored to the reaction that the enzyme catalyzes. Some enzymes also have an additional pocket that binds to more remote regions of the substrate molecule to aid in specific substrate recognition. Evolutionarily divergent enzymes that catalyze the same reaction may vary a great deal in sequence and structure, but are usually very similar in the active site.

Enzymes lower the activation energy in a variety of ways

Enzymes use a number of different mechanisms to lower the activation energy barrier of a reaction. Many enzymes lower the energy barrier by specifically stabilizing the transition state, through favorable free energy of binding between the enzyme and the transition state. If the enzyme binds more tightly to the transition state than to the substrate or the product, the barrier to the reaction will be lowered, making the reaction more likely to proceed (Figure 3.18). Some enzymes are thought to catalyze a reaction by simply bringing two reactive substrates together, thereby pre-organizing them for catalysis. This is often referred to as proximity catalysis and is perhaps the simplest way to stabilize the transition state of the reaction.

Another way of reducing the size of the energy barrier is to destabilize the original substrate, raising its free energy as compared with the molecule free in solution. This is called ground-state destabilization. A substrate bound to the enzyme in a less stable, higher-energy conformation starts out farther up the free-energy hill towards the transition state, and the reaction will therefore proceed at a higher rate.

Figure 3.17 Enzyme active site. The chemical reaction catalyzed by an enzyme takes place in the active site, which contains side chains that bind the substrate molecules, catalyze the chemical reaction and promote the subsequent release of reaction products. The three side chains shown here are found in the active sites of many proteases that cleave polypeptide chains.

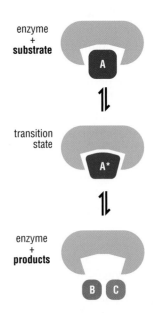

Figure 3.18 Transition state stabilization. Enzymes generally bind the transition state (A*) more tightly than either the substrate (A) or products (B and C). If an enzyme bound most tightly to the substrate, the reaction would be inhibited. If the enzyme bound tightly to the products, the products would not be rapidly released from the enzyme following catalysis.

Finally, an enzyme can lower the barrier to a chemical reaction by causing it to proceed through a different path composed of a series of steps, each with a smaller energy barrier than that of the uncatalyzed reaction. We refer to each of the chemical species that is transiently formed during such a reaction as a **reaction intermediate**. Enzymes commonly employ more than one strategy to catalyze a reaction.

A few catalytic strategies are common among biological enzymes

Having now seen the general methods through which enzymes catalyze reactions, we need to consider another question: what is happening in chemical terms – at the atomic level – when a reaction is catalyzed? Although the human genome encodes many thousands of enzymes, there are a few conserved reaction mechanisms that enzymes adopt to lower the activation barrier for reactions in order to promote catalysis.

The most common type of reaction that we will observe throughout this book is a nucleophilic displacement reaction, in which an atom that is rich in electrons, called a **nucleophile**, donates an electron to an atom that is deficient in electrons, known as an **electrophile**. Another term used for this process is nucleophilic attack. As with any movement of electrons within and between molecules, this reaction – and the transfer of electrons it represents – leads to a change in chemical bonding, as depicted in Figure 3.19. This nucleophilic displacement reaction lies at the heart of DNA, RNA and protein synthesis as well as most reactions involved in nucleic acid and protein transformations.

In each of these polymerization reactions – the joining of subunits to form a larger molecule – the nucleophile is activated to initiate a nucleophilic attack through removal of a proton by a **general base** on the enzyme, as depicted in Figure 3.19. A molecule that is able to donate (or lose) a proton is called an acid, while a molecule that is able to accept (or gain) a proton is called a base. Acids and bases must

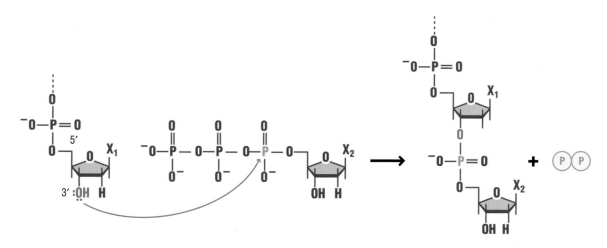

Figure 3.19 Nucleophilic attack in a DNA polymerization reaction. In the synthesis of a DNA strand, the 3′ OH at the end of the growing strand carries out a nucleophilic attack on the innermost phosphate of the incoming nucleoside triphosphate. A general base withdraws a proton from the 3′ OH, making the oxygen a better nucleophile.

Figure 3.20 Pentacovalent phosphate intermediate. During the course of the polymerization reaction catalyzed by DNA and RNA polymerase, a transition state intermediate is formed where the phosphate atom is coordinated by five ligands. This is the pentacovalent intermediate.

act in tandem: for an acid to donate a proton, it must have a base to donate the proton to. Enzymes often facilitate such proton movements by providing moieties that act as acids and bases during the reaction cycle. This type of participation by the enzyme is referred to as general acid–base catalysis. In the reaction shown in Figure 3.19, a base removes a proton from the 3' OH of ribose, allowing the oxygen to carry out a nucleophilic attack on the innermost phosphate of the incoming deoxynucleotide triphosphate. The side chain of histidine is particularly useful in mediating acid–base catalysis; since its pK_a is near 7, it can act as both an acid and a base at neutral pH to both accept and donate protons during the course of an acid–base reaction.

Another common feature of these reactions is that the transition state structure contains an additional bond. For example, during the course of DNA and RNA synthesis a pentacovalent phosphate intermediate is formed (rather than the ground state tetravalent structure) (Figure 3.20). This transition state structure forms favorable interactions with the DNA and RNA polymerase enzymes that help to catalyze the polymerization reaction. Similarly, protein synthesis involves formation of a tetravalent carbon intermediate (rather than the ground state trivalent structure). These transition state structures have charge configurations and covalent bonding arrangements that are quite distinct from those of the ground state. The exquisite specificity of enzymes permits them to recognize these differences in structure – that is, the presence of the additional bonds in the transition state structures – to stabilize specific transition states and thus promote catalysis.

Active sites position chemical moieties to promote catalysis

The cleavage of RNA by the enzyme, RNAse A, provides a well-understood example of nucleophilic catalysis that also highlights some of the ways in which enzymes catalyze chemical reactions. The RNAse A reaction proceeds by acid–base catalysis, which means that the enzyme must provide chemical moieties that function as acids and/or bases to facilitate proton movement. As with all other enzymes, the enzyme contains features that stabilize the transition state structure. We will see in the following how RNAse A accomplishes each of these steps and thereby promotes cleavage of an RNA strand.

The reaction, which is diagrammed in Figure 3.21, begins with the nucleophilic attack on the phosphate backbone of the RNA strand by the 2' OH of ribose. Since this chemical group, on its own, is not sufficiently nucleophilic, a proton must first be removed in order for it to function as a nucleophile, thereby enabling catalysis to occur. RNAse A facilitates this deprotonation with a histidine side chain, which removes a proton from the 2' OH and thereby activates the oxygen atom. This histidine is therefore acting as a general base. The nucleophilic attack results in the formation of a negatively charged pentacovalent phosphate intermediate in the transition state. Positively charged lysine side chains in the enzyme active site surround this group, stabilizing the transition state through favorable ionic interactions between the negatively charged phosphate and the positive charges on the lysines. At this point, another histidine side chain donates a proton to the other part of the cleaved strand, which is the 5' oxygen of the leaving group. The enzyme remains bound to the remaining strand that contains the cyclic phosphate intermediate.

In the second step of the reaction, the cyclic intermediate is hydrolyzed to yield the final product containing a 3′ phosphate group. The two active site histidine side chains, one that gained a proton and the other that lost a proton, now reverse their roles. A water molecule is activated when one histidine removes a proton, and then attacks the cyclic intermediate (Figure 3.21). At the same time, the second histidine donates a proton, leading to formation of the final product. The use of histidine in this enzyme as both a general acid and a general base highlights its special utility in enzymatic reactions.

Transesterification reactions involve the breakage and re-forming of a phosphodiester linkage

There are many reactions in which the phosphodiester linkage is broken and then re-formed with a different nucleic acid fragment in a concerted reaction that maintains the same sort of chemical linkage. This type of reaction is called a **transesterification** reaction, since an ester linkage is broken and then re-formed with a different fragment. An important feature of this type of reaction is that there is no energetic difference between the substrate and the product. Figure 3.22 outlines the key steps of a transesterification reaction that is central to the RNA splicing that occurs when RNA is cleaved to remove an intron. During this reaction, the 3′ end of one exon is joined to the 5′ end of another.

Many enzymes require an additional molecule to catalyze a chemical reaction

Not every biological reaction can be carried out efficiently using the somewhat limited chemical properties of amino acids. Some enzymes require an additional molecule that participates in the reaction, called a **cofactor** or **coenzyme**. An enzyme cofactor may be covalently attached to the enzyme, but is more generally simply tightly bound to the enzyme near the active site. Cofactors typically act as donors or acceptors of chemical groups, ranging from electrons to slightly larger groups such as acetyl groups. Once a cofactor loses or gains atoms during the course of the reaction, it must be regenerated before it can participate again in subsequent rounds of catalysis. Figure 3.23 lists some of the common coenzymes and their chemical function in reactions. All organisms use a universal set of coenzymes, indicating that their use developed very early in evolution. Most of these coenzymes are derived from ribonucleotides, probably because the chemistry of early life forms was dominated by RNA.

Several coenzymes are essential for enzyme-catalyzed **oxidation** and **reduction** reactions. Reduction is the process of gaining electrons, while the loss of electrons is referred to as oxidation. Oxidation and reduction must always occur together, since whenever electrons are lost by one molecule, another must receive them. Reactions in which one species is reduced and another oxidized are therefore called oxidation–reduction, **redox reactions** or electron transfer reactions. This type of reaction, which is one of the most common in biology, typically depends upon high-energy cofactors such as nicotinamide adenine dinucleotide (reduced form; NADH) and nicotinamide adenine dinucleotide phosphate (reduced form; NADPH). Nicotinamide adenine dinucleotide (NAD) is found in the cell in its

Figure 3.21 Mechanism of RNA cleavage by RNAse. His 12 acts as a general base to remove a proton from the 2′ OH to promote a nucleophilic attack by the oxygen on the adjacent phosphate group. His 119, acting as a general acid, donates a proton to the 5′ oxygen of the leaving group, leading to strand cleavage and formation of a 2′,3′ cyclic intermediate. In the second step, a water molecule attacks the cyclic intermediate with His 12 and His 119 reversing their roles, with His 119 acting as a general base and His 12 as a general acid.

Figure 3.22 Transesterification. The schematic diagram shown here illustrates the final steps in exon ligation, in which one RNA fragment is cleaved and its 3' end then joined to the 5' end of a new fragment.

oxidized form, **NAD⁺**, and in a higher-energy reduced form, **NADH**. NAD⁺ is converted to NADH by accepting two electrons and a proton, H⁺, which is equivalent to accepting a hydride ion, H⁻ (Figure 3.24). NADH can then participate in reactions as a donor of H⁻, as illustrated in Figure 3.24.

Nicotinamide adenine dinucleotide phosphate (**NADP⁺**) and **NADPH** play similar roles, but the additional phosphate group generally prevents NADP⁺ and NADPH from binding to enzymes designed to function with NAD⁺ and NADH. This provides the cell with two different pools of electron transfer molecules, which each can be maintained at different levels and which can function in independent pathways. In our cells, these two pools of coenzymes, NAD⁺/NADH and NADP⁺/NADPH, are generally used for catabolic and anabolic reactions, respectively. Cells must maintain adequate and appropriately balanced pools of these and related cofactors in order to drive the many redox reactions needed for life.

Some enzymes require metal ions for catalysis

Many enzymes require a bound metal ion to catalyze chemical reactions. These metal ions are typically tightly bound to the enzyme and may even be covalently attached, such as the iron-carrying heme cofactor. Such enzymes are referred to as **metalloenzymes**. The most common metal ions used by enzymes are iron, copper, zinc and magnesium. Not coincidentally, these metals are among the most abundant in the Earth's crust. Like the coenzymes described above, metal ions can participate in electron transfer reactions through changes in the oxidation state of the metal. In addition, metal ions can assist reactions by providing positive charge density that stabilizes negative charge and, in doing so, can orient substrates and promote catalysis. Finally, metal ions can directly participate in catalysis by lowering the pK_a of bound water, thus providing a ready source of nucleophilic OH⁻ ions. We will see many examples throughout this book of both protein and RNA enzymes (ribozymes) that use metal ion cofactors in catalysis.

Other enzymes bind metal ions more loosely, usually cations of the alkali or alkaline earth metals like sodium, potassium, magnesium or calcium. In these cases, the metal ions more typically play a structural rather than a catalytic role – they stabilize charges to promote proper polypeptide folding and, hence, proper function. Therefore, the corresponding proteins are not referred to as metalloenzymes.

Common Coenzymes		
coenzyme	role	example
NAD⁺/NADH	oxidation-reduction	dehydrogenases (NAD⁺)
NADP⁺/NADPH	oxidation-reduction	reductases (NADPH)
FAD/FADH₂	oxidation-reduction	dehydrogenases (FAD)
CoA	acyl-group activation and transfer	histone acetyltransferases
Biotin	CO₂ transfer	carboxylases

Figure 3.23 Common coenzymes.

Figure 3.24 Oxidation–reduction of nicotinamide adenine dinucleotide (NAD⁺) and nicotinamide adenine dinucleotide phosphate (NADP⁺). Acetaldehyde is reduced to ethanol by alcohol dehydrogenase with the concomitant oxidation of nicotinamide adenine dinucleotide (reduced form; NADH) to NAD⁺. Conversely, glucose-6-phosphate is oxidized by glucose-6-phosphate dehydrogenase with the concomitant reduction of NADP⁺.

3.5 ENZYME KINETICS

We see above how enzymes increase the rate at which a process occurs, most particularly by lowering the activation energy barrier. Understanding the rate at which a given process will happen is often of huge importance: if we administer a drug to a patient we need to know how long the drug will remain in the body. Will it remain long enough for it to elicit a therapeutic effect, or will it be metabolized too quickly? In this section, we learn more about how we can describe the rates of reactions as we explore the field of **enzyme kinetics**.

Enzyme activity can be described by Michaelis–Menten kinetics

The properties of enzymes are well suited to the cellular conditions in which they operate. Enzymes bind to their substrates and catalyze their reactions at rates that allow for the continued life of the cell. **Enzyme kinetics** is the study of how quickly enzymes carry out reactions. Understanding the kinetics of enzymatic reactions can shed a great deal of light on the molecular details of the catalytic mechanism and, potentially, on the conditions in the cell under which the enzyme has evolved to function.

The kinetic behavior of many enzymes can be described according to the Michaelis–Menten model, which can be used to derive useful parameters that describe the activity of an enzyme. Many enzymes carry out a relatively straightforward reaction that begins when the enzyme binds to a single substrate, S, forming an enzyme–substrate complex, ES (Figure 3.25). This first step is analogous to the simple binding reactions that we saw earlier in this chapter, and is therefore associated with an association (k_{on}) and dissociation (k_{off}) rate constant (Figure 3.25). Once the enzyme–substrate complex is formed, the enzyme catalyzes a reaction that converts the substrate into product. The rate constant describing this reaction is denoted k_{cat} and is also referred to as the turnover number of the enzyme. The overall rate at which substrate is converted into product is called the reaction velocity and is expressed in units of $mol^{-1} s^{-1}$.

Many enzymatic reactions proceed according to the following simple scheme:

$$E + S \underset{k_{off}}{\overset{k_{on}}{\rightleftharpoons}} ES \overset{k_{cat}}{\longrightarrow} E + P$$

The initial reaction rate, v_0, can be described by the Michaelis–Menten equation:

$$v_0 = V_{max} [S] / (K_M + [S])$$

where the maximum rate at which the reaction proceeds is given by:

$$V_{max} = k_{cat} [ES]$$

We define the contant K_M as:

$$K_M = \frac{k_{off} + k_{cat}}{k_{on}}$$

Figure 3.25 Michaelis–Menten kinetics. Enzymatic reactions that follow Michaelis–Menten kinetics are characterized by this simple equation, in which E stands for enzyme, S for substrate, ES the enzyme–substrate complex, and P the product. The Michaelis–Menten equation assumes that there is no conversion of product into substrate, which is true if initial reaction rates are measured before the product has accumulated to any significant degree.

According to this model for enzyme behavior, when substrate concentration is high enough that all enzyme is found in the ES form, then the reaction is rate limited by k_{cat}, the rate at which the enzyme-substrate complex is converted into enzyme and product, and is insensitive to higher substrate concentrations. We say that the enzyme is saturated.

Enzymes whose activity can be described in terms of these simple assumptions are generally characterized by the **Michaelis–Menten equation** (Figure 3.25). The Michaelis–Menten equation can be used to describe how the rate of the reaction changes as the concentration of substrate is varied. When studying enzymatic reactions, it is most useful to look at the initial velocity at which the enzyme catalyzes a reaction, before a significant amount of product has accumulated. This is because the rates of some enzymatic reactions are altered – usually decreased – by the presence of large amounts of product, so examining the initial velocity provides a measure of the intrinsic catalytic power of the enzyme.

Measurements of the initial velocity are taken before very much substrate is consumed, which makes valid the assumption that the substrate concentration remains roughly the same over the period of measurement for each data point. Since very little product will have been generated in the initial stages, we can also neglect the rate of the back reaction, namely the conversion of product back into substrate. The hyperbolic plot shown in Figure 3.26 shows how the initial reaction velocity increases as substrate concentration is increased according to the Michaelis–Menten equation. The rate at which the product is formed is equal to k_{cat} multiplied by the concentration of enzyme–substrate complex (Figure 3.26). If the substrate concentration is sufficiently high that all the enzyme has substrate bound to it, then the reaction rate is as high as it can be. This maximal velocity is called V_{max}; it is determined by k_{cat} multiplied by the total enzyme concentration.

The concentration of substrate at which the initial velocity is half that of V_{max} is called the Michaelis constant, K_M, and is defined as $(k_{off} + k_{cat})/k_{on}$. For enzymes where the k_{cat} is slow relative to k_{off}, then the K_M provides a fairly good measure of the equilibrium dissociation constant, $K_d = k_{off}/k_{on}$, for substrate binding to the enzyme.

More generally, the K_M provides a measure of how the reaction rate of an enzyme varies with substrate concentration. A low K_M means that the enzyme reaches its maximal reaction velocity at low substrate concentrations, whereas an enzyme with a higher K_M requires higher concentrations of substrate to approach its maximal reaction velocity. The K_M of a particular enzyme is generally close in value to the physiological concentration of its substrate. This ensures that the enzyme is catalyzing the reaction at a robust rate, while ensuring that a change in the concentration of substrate will result in a significant change in reaction velocity. When the substrate concentration is substantially above the K_M, the reaction velocity changes little in response to modest variations in substrate concentration (Figure 3.26).

Another parameter that derives from the Michaelis–Menten model for enzyme behavior is k_{cat}/K_M, which is often referred to as the specificity factor. This ratio is related to how often the encounter of an enzyme and substrate results in a

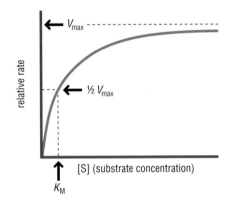

Figure 3.26 Initial velocity of a Michaelis–Menten reaction. The plot shows how substrate concentration affects the initial velocity of a reaction that follows Michaelis–Menten kinetics. At low substrate concentrations, the velocity increases proportionally with substrate concentration. At higher substrate concentrations, the reaction velocity approaches a maximum velocity, V_{max}. V_{max} depends upon the speed at which each enzyme molecule can convert substrate into product and also the concentration of enzyme. The K_M is the substrate concentration at which the velocity is half that of V_{max}; it is a characteristic feature of each enzyme and it does not vary with the enzyme concentration.

productive reaction (to generate product, P) and is therefore used to describe an enzyme's efficiency with a particular substrate.

Enzymes can be inhibited by small molecules

An inhibitor is a molecule that binds to an enzyme and interferes with the enzymatic reaction. Inhibitors can be small molecules or large proteins, and they can be found naturally in the cell where they regulate the activity of an enzyme or can be supplied exogenously. Many drugs are enzyme inhibitors. Drugs used to treat human immunodeficiency virus (HIV) infection, such as azathioprine (brand name Retrovir) and indinavir (brand name Crixivan), inhibit the enzymatic activity of the HIV reverse transcriptase and protease enzymes, respectively. They are both examples of reversible inhibitors, which can bind to the enzyme and then dissociate. Other less commonly seen inhibitors covalently react with the enzyme, causing permanent inactivation, and are commonly referred to as suicide inhibitors.

There are several ways in which a reversible inhibitor can interfere with an enzymatic reaction. A **competitive inhibitor** binds to an enzyme in a way that directly competes with substrate binding. A competitive inhibitor often closely resembles the substrate and binds within the substrate-binding site, but it does not engage in the chemical reaction catalyzed by the enzyme. The effects of a competitive inhibitor, shown in Figure 3.27, can be overcome by simply adding more substrate. A competitive inhibitor therefore does not affect the V_{max} of an enzyme. Sometimes the products of an enzymatic reaction can be competitive inhibitors, since they also bind to the substrate-binding site. This form of competitive inhibition is known as product inhibition, and can occur if the accumulating product reaches a concentration sufficient to favor significant binding to the enzyme.

A **non-competitive inhibitor** reduces the rate of an enzymatic reaction without interfering with the ability of the substrate to bind to the enzyme. These inhibitors typically bind in a location that is removed from the substrate-binding site. An example is the class of anti-HIV drugs known as non-nucleoside reverse transcriptase inhibitors, which bind to the HIV reverse transcriptase enzyme and inhibit synthesis of viral RNA without interfering with substrate binding. Non-competitive inhibitors lower the V_{max} of an enzyme (Figure 3.27) and their effects cannot be overcome by increasing the concentration of substrate. While there are other forms of enzyme inhibition, competitive and non-competitive inhibition are the two principal types that will be encountered in this book.

Figure 3.27 Enzyme inhibition. A competitive inhibitor (pink line) competes with the substrate for binding to the enzyme, increasing the apparent K_M. A non-competitive inhibitor (blue line) binds to the enzyme–substrate complex and prevents catalysis from occurring at its maximal rate, lowering the apparent V_{max}.

✱ SUMMARY

The behavior of molecules in living cells is governed by the laws of chemistry and physics. The thermodynamic and kinetic principles underlying chemical reactions in the cell can be used to describe biological processes and predict how cells will respond to different conditions. We learned in this chapter how to apply these principles to understand and predict binding interactions and enzyme catalysis.

Thermodynamic principles

- The enthalpy, H, is a measure of the internal energy of a system.

- The entropy, S, is a measure of disorder in the system.

- The energy that is available to do useful work is called the Gibbs free energy, G.

- Every process in biology – the binding of two or more molecules, a biochemical reaction – has associated with it a *change* in the Gibbs free energy, ΔG.

- The free energy change is related to enthalpy and entropy by the relation, $\Delta G = \Delta H - T\Delta S$.

- A process must have a negative ΔG in order to be spontaneous.

- An unfavorable reaction (with positive ΔG) can be made favorable by coupling it to a favorable reaction such that the overall free energy change of the coupled reactions is negative.

- Every process has associated with it a standard free energy change, ΔG°, which is the free energy change occurring when the reactants and products are in their standard states.

- The overall free energy change for the reaction $X \rightleftharpoons Y$ is influenced by the concentrations of reactant and product and is given by:

$$\Delta G = \Delta G^\circ + RT \ln[Y]/[X], \text{ where } R = 8.31 \text{ J mol}^{-1} \text{ K}^{-1}$$
$$\text{or } 1.987 \times 10^{-3} \text{ kcal mol}^{-1} \text{ deg}^{-1} \text{ and } T \text{ is the}$$
$$\text{temperature in degrees Kelvin.}$$

- At equilibrium, the rate of the forward reaction equals the rate of the reverse reaction and the overall free energy change is equal to zero. At this point,

$$\Delta G^\circ = -RT \ln[Y]/[X].$$

Binding interactions

- For the binding reaction, $A + B \rightleftharpoons AB$, the ratio of complex to free molecules at equilibrium is given by:

$$\Delta G^\circ = -RT \ln[AB]/[A][B].$$

- The rate of complex formation is given by $k_{on}[A][B]$, where k_{on} is a rate constant.

- The rate of complex dissociation is given by $k_{off}[AB]$, where k_{off} is a rate constant.

- The time it takes for half of all complexes in a given population of identical complexes to come apart is called the half-life:

$$t_{1/2} = 0.693/k_{off}.$$

- The equilibrium constant for the binding reaction, $A + B \rightleftharpoons AB$, is the ratio of k_{on} to k_{off}, and is also related to the ratio of complex to free molecules as follows:

$$K_{eq} = k_{on}/k_{off} = [AB]/[A][B].$$

- This equilibrium constant for this reaction is also written K_a because is describes the association reaction (binding of A and B).

- The free energy change and the equilibrium constant for any reaction are related by the equation $\Delta G^\circ = -RT \ln K_{eq}$.

- In biology we often refer to the equilibrium constant describing the dissociation reaction, $AB \rightleftharpoons A + B$. This equilibrium dissociation constant relates to the reverse of the association reaction, and is given by:

$$K_d = k_{off}/k_{on} = [A][B]/[AB] = 1/K_a.$$

- The equilibrium dissociation constant, K_d (with units of M), gives the concentration at which half of the molecules are bound in a complex while the other half are free.

- The binding curve depicts how much complex is formed at different substrate concentrations.

- Cooperative interactions can occur when two or more ligands bind to the same molecule, and the binding of one molecule makes the binding of the next molecule more (or less) favorable. The resulting binding curve is S-shaped, or sigmoidal.

Enzyme catalysis

- The activation energy is an energy threshold that must be overcome before a reaction can proceed.

- Enzymes accelerate the rate of a chemical reaction without affecting the free energy change between reactants and products.

- Enzymes catalyze reactions by lowering the activation energy barrier.

- Enzymes use a variety of strategies to lower the activation energy barrier, including binding preferentially to the transition state of the chemical reaction and binding simultaneously to two substrates, orienting them favorably for reaction.

- Binding of substrate occurs at the active site, a pocket or region of the enzyme containing precisely positioned chemical groups that effect catalysis by participating in the chemical reaction.

- Many enzymatic reactions obey Michaelis–Menten kinetics, in which the initial rate of product formation is given by:

$$v_o = V_{max}[S]/(K_M + [S]).$$

- The maximum reaction velocity is given by $V_{max} = k_{cat}[E]$, where k_{cat} is the rate at which the enzyme–substrate complex is converted into product and [E] is the enzyme concentration.

- Inhibitors are molecules that bind to enzymes and interfere with the chemical reactions they catalyze. Competitive inhibitors increase the apparent K_M of an enzyme, while non-competitive inhibitors reduce the V_{max}.

 # FURTHER READING

GENERAL

Berg JM, Tymoczko JL, Stryer L. *Biochemistry*, 6th ed. New York, NY: WH Freeman, 2007.

Voet DJ, Voet JG. *Biochemistry*, 3rd ed. Hoboken, NJ: Wiley-Blackwell, 2004.

4 CHROMOSOME STRUCTURE AND FUNCTION

INTRODUCTION

Chromosomes are the vehicles for the transmission of genes

The genetic information for each organism comprises a set of instructions that enables them to live, grow, reproduce, and pass the information along to their progeny. These instructions are in the form of genes, which are organized into very long DNA molecules. Each individual DNA molecule is packaged into what we call a **chromosome**, which acts as the vehicle that ensures stable maintenance of the genetic information and facilitates its transmission to the next generation. Chromosomal DNA is associated with many proteins that help to package the DNA inside the cell; the proteins also help the stored genetic information to be read and used to synthesize the vast array of molecules that are necessary for life.

The central subject of this book is how genetic information is read, transmitted and preserved. In this chapter, we describe how the vehicle for that information—the chromosome—is organized, and give an overview of the changes that take place in chromosomes as cells grow, divide, and respond to changing conditions. We will learn about special features of the DNA and proteins in chromosomes that make it possible for an organism to make faithful duplicates of each chromosome as cells grow and divide. Since key elements of the chromosome are important when cells divide and DNA is replicated, we will briefly review the basic stages of the **cell cycle**, which are the series of events that lead to the replication and segregation of chromosomal DNA during cell division. We will learn about the packaging of eukaryotic DNA into a protein–DNA complex known as **chromatin**, and how modifications of these proteins and of the DNA molecule can give rise to changes in chromatin structure. Then, we will learn about special sequence elements within the DNA that are required for key processes in the cell cycle, and for preserving the integrity of chromosomes. Finally, we will see how chromosomes are organized within the eukaryotic nucleus, and how and why chromosomes change their organization during the cell cycle.

Variations in chromosome structure and chromatin modifications impact all processes that involve chromosomal DNA, including DNA replication, transcription, and DNA repair—all topics that we describe in later chapters. Therefore, while this chapter provides a general introduction to chromosome structure and function, it is not the complete story of how chromosomes influence the molecular processes that determine genome function. Instead, we will gradually build this complete picture during later chapters, when we revisit many of the themes and topics of this chapter in more detail.

➲ Topics introduced in this chapter feature particularly strongly in Chapters 5, 6, 7 and 8.

4.1 ORGANIZATION OF CHROMOSOMES

Chromosomes contain genetic information in double-stranded DNA molecules

A cell's genome – the total DNA content of a cell – is divided among one or more chromosomes that organize, store, and transmit information for producing protein and RNA molecules. Each chromosome contains one double-stranded DNA molecule, along with proteins that help to condense and organize the DNA within the cell. The DNA molecule can be linear or circular: all eukaryotes have linear chromosomes, whereas most bacteria and archaea have circular chromosomes.

A single chromosome contains many genes embedded within a single DNA molecule. Each gene consists of a linear string of DNA bases that contain the information for producing the RNA and protein molecules needed by the cell. Interspersed between the genes are regions of DNA known as **intergenic DNA** that do not, to our knowledge, contain genes. Although once considered by some scientists to be 'junk' DNA with no function, ongoing studies have revealed possible regulatory functions for these stretches of DNA that lack apparent protein-coding sequences. We will return to this intriguing issue later in this book.

There is great variation in the relative amount of gene versus intergenic DNA in chromosomes from different organisms. The genes of less complex, single-celled organisms, such as bacteria or budding yeast, are located fairly close to one another in the chromosome, with little intergenic DNA. More complex organisms have a lower density of genes within their chromosomes and more intergenic DNA. Human chromosome 18, for example, contains one gene for each million base pairs, whereas a budding yeast chromosome contains 500 genes per million base pairs. Even within species, the gene density on different chromosomes is not the same: thus, human chromosome 1, which is the largest chromosome, has twice the gene density of human chromosome 18.

Each species has a characteristic set of chromosomes

Every organism maintains a fixed number of chromosomes in each cell. The number of chromosomes per cell, along with the size of each chromosome, is a defining characteristic of each species. Most bacteria and archaea have a single chromosome in each cell, while eukaryotic cells have multiple chromosomes within a membrane-bound compartment known as the nucleus.

The number of chromosomes per cell varies greatly among different species. Human cells contain 46 chromosomes in the nucleus, half of which are inherited from the mother and half from the father. The fission yeast, *Schizosaccharomyces pombe*, has six chromosomes, while the kingfisher (a bird) has 132. Archaea usually have a single, circular chromosome, as do most bacteria. There are, however, a few examples of bacteria with linear chromosomes or more than one circular chromosome.

Another defining characteristic of each type of cell is how many sets of identical chromosomes it contains, a characteristic described by the term **ploidy**. Human cells, like those of most eukaryotes, are **diploid**, which means that they contain two copies of each chromosome. (In the case of humans, there are two copies each of chromosomes 1 to 22, and either two X chromosomes, or one X and one Y.) The egg and sperm cells that are involved in sexual reproduction are **haploid**, meaning that they contain a single set of chromosomes. When an egg is fertilized by a sperm, the two sets of genetic material come together in one nucleus to produce a diploid, fertilized egg cell that goes on to develop into a diploid individual. The copy of each chromosome inherited from the mother is **homologous** to the copy inherited from the father, meaning that it contains the same genes and gene order, but may contain different versions of each gene.

Exceptions to this simple alternating pattern between haploid and diploid do occur. **Polyploid** cells, which contain multiple sets of chromosomes (more than two), are found in some organisms during certain stages of their lifecycles. **Polyploidy** is particularly common among cultivated plants such as wheat, coffee, and oats, and also occurs in wild maize. However, it is not confined to plants, and is also observed in a small number of specific human cell types. Two examples are hepatocytes, which are cells in the liver that produce proteins, lipids, and bile salts, and bone marrow megakaryocytes, which produce blood platelets.

Other classes of DNA can be found within cells

There is additional DNA found within cells besides the chromosomal DNA contained in the nucleus. Mitochondria, the membrane-enclosed organelles in eukaryotic cells that are the primary source of cellular adenosine triphosphate (ATP), contain a single chromosome that is usually circular. The presence of distinct mitochondrial DNA is one of the reasons that mitochondria are thought to have arisen from independently living bacteria that were ingested by another single-celled organism, within which the bacteria developed a symbiotic relationship. In this scenario, the single-celled organism and ingested bacteria would have had separate, unique genomes, a difference still reflected by mitochondria and the cells they occupy to this day. Chloroplasts, the plant organelles where photosynthesis occurs, also have their own circular chromosome and are thought to have originated from blue-green algae, incorporated into another cell by a similar but independent event during evolution.

⊜ We cover how plasmids are used to introduce novel or altered genes into bacteria in Chapter 16.

In addition to their main genomic DNA, bacterial cells can contain double-stranded, circular DNA molecules known as **plasmids**, which contain a small number of genes. Plasmids are much smaller than bacterial chromosomes and typically encode a few proteins that confer on the bacterium a trait that provides a selective advantage, such as drug resistance. Since plasmids can be readily introduced into some types of bacteria, they have been enormously useful tools in the laboratory for genetic engineering.

Although this chapter focuses on the chromosomes in cells, it is useful to contrast cellular genomes with the genetic information found in viruses. Viruses are infectious agents rather than free-living organisms, and cannot propagate on their own. Their genetic information is stored, replicated, and delivered to progeny in a way that is distinct from mechanisms used by cells. Viruses carry genetic information in the form of either DNA or RNA, and their chromosomes may be either single- or double-stranded. Like cellular chromosomes, viral DNA can be either circular or linear. Viruses propagate by taking over a host cell and using its machinery to produce many new viral genomes and viral proteins, which assemble into active viral particles that can be dispersed. Studies of viruses and the way they harness the host cell machinery have provided key insights, not just into disease, but also into the basic workings of the cell.

4.2 THE CELL CYCLE AND CHROMOSOME DYNAMICS

Chromosomes are duplicated and segregated to daughter cells during the cell cycle

The faithful transmission of genetic information from one generation to the next depends on a cell's ability to make copies of each chromosome and then distribute the complete set of chromosomes to the two daughter cells. The incorrect segregation of chromosomes – for example, where one daughter cell receives an extra chromosome, and the other cell receives none – may have a serious, negative impact on the viability of either or both daughter cells. Chromosomes therefore have many features that ensure the proper copying and distribution of chromosomes during the many rounds of cell division that occur during the lifetime of a single individual. To see how these features function and when they are needed, we will first review the key events that occur during the process of cell division.

Before a cell can divide, it must replicate its DNA and segregate the duplicated chromosomes equally to what will become the two daughter cells. The highly coordinated sequence of events that occurs during the division process is called the cell-division cycle, or **cell cycle**. We focus here on somatic cells, which are the cells that form the body of an organism; these are distinct from germline cells that take part in sexual reproduction, such as those cells that produce the haploid sperm and eggs, and which exhibit a modified form of chromosome duplication and segregation compared to somatic cells. The cell cycle and its regulation will be discussed in detail in Chapter 5 and chromosome segregation will be discussed in Chapter 7. Here we outline the stages of the cell cycle and describe the changes in the chromosomes – the **chromosome cycle** – that occur at each stage.

The eukaryotic cell cycle is divided into four distinct phases

The cell-division cycle of a somatic cell has four distinct phases that happen in sequence: **G1, S, G2,** and **M** (Figure 4.1). In the first phase, **G1** (for gap phase 1), cell growth occurs until cells attain a minimum size that is required to progress to the next phase. During this next phase, **S phase** (for DNA synthesis), the DNA is copied, thereby duplicating all of the chromosomes. The two identical copies of each chromosome, called **sister chromatids**, remain physically associated with one another. S phase is followed by **G2** (for gap phase 2), during which the cell prepares for **mitosis**.

In the final cell-cycle stage the **mitotic (M) phase**, sister chromatids are separated and a complete set of chromosomes is delivered to each pole of the cell. This process is known as chromosome segregation. Mitosis can be readily distinguished

Figure 4.1 The chromosome cycle is correlated with the cell cycle. The cell cycle is divided into four phases: G1 for gap phase 1, S for DNA synthesis, G2 for gap phase 2, and M for mitosis. During S phase, the chromosomes are duplicated, yielding two sister chromatids per chromosome. At the end of G2, the chromosomes condense to prepare for mitosis. During mitosis, the chromosomes are highly condensed to facilitate their distribution to daughter cells. Microtubules (green) emanate from the centrosomes (black), bind to the centromere (yellow) found in each chromosomes, and the sister chromatids are pulled to opposite poles. The cells then divide, producing two daughter cells, each with an identical set of chromosomes. Chromosomes condense during M phase and later decondense during G1 to allow proteins access to the DNA for transcription and repair.

chromosome duplication

S

G1   G2

M

mitosis

mitotic spindle

centrosome

cell division

metaphase

➲ The steps in mitosis are described in greater detail in Chapters 5 and 7.

from **interphase**, when changes in chromosome and nuclear structures during mitosis are clearly evident in the microscope (see below). Mitosis is followed by **cytokinesis**, which completes the division of the mother cell into two daughter cells, each containing the same number of chromosomes. G1, S, and G2 are often collectively referred to as interphase, since cells in these stages are not easily distinguished in the microscope.

During mitosis, the cell must separate all sister chromatid pairs to form two identical chromosome sets for distribution to daughter cells. This process occurs in a series of steps that are illustrated in Figure 4.1. Here we will give a brief overview of the steps.

Before M phase, sister chromatids are held together by protein complexes that bind along their length. As the cell enters M phase, the nuclear membrane is disassembled, and sister chromatids condense into compact structures that remain bound together. Human chromosomes become condensed by a factor of 10, which allows them to segregate without becoming entangled. The **centromeres** of the condensed sister chromatids bind microtubules that emanate from centrosomes, which define each pole of the mitotic spindle (Figure 4.1). Thus, the spindle organizes the sister chromatid pairs at the center of the cell; this portion of M phase is known as metaphase. The ties that bind the sister chromatids together are then removed, and each sister chromatid moves to its pole on the spindle. This ensures that each pole receives one full set of chromosomes. The sets of chromosomes at each pole become encapsulated by a nuclear membrane, and the cell divides to yield two daughter cells.

Chromosomes have a distinctive appearance at mitosis

Condensed mitotic chromosomes have been studied under the microscope for many years. The most familiar picture of human chromosomes comes from a **karyotype**, a display of the chromosome set of an individual, lined up from the largest to smallest, like a collection of butterflies (Figure 4.2a). The chromosomes typically seen in a karyotype are taken from microscope images of metaphase

(a)

1 2 3 4 5 6 7 8

9 10 11 12 13 14 15 16

17 18 19 20 21 22 Y X

(b)

(c)

1 2 3 4 5

6 7 8 9 10 11 12

13 14 15 16 17 18

9 20 21 22 X

Figure 4.2 Karyotype of human chromosomes. (a) Diagram of a human male karyotype. The banding pattern arises when the highly condensed chromosomes are stained with Giemsa dye. Each chromosome has been duplicated and is still attached at the centromere. As dictated by karyotype convention, the chromosomes are arranged and numbered from largest to smallest. (b) Photomicrograph of a spread of Giemsa-stained chromosomes from a diploid cell from a human male at a stage when the chromosomes have become sufficiently condensed to become visible in the light microscope. Each chromosome can be identified by its distinctive banding pattern. (c) Spectral karyotype of human chromosomes. A normal female metaphase was hybridized with specific DNA probes such that each chromosome is labeled with a different ratio of fluorescent colors. The image shown is a computer-generated karyotype in which each chromosome is represented as an arbitrary color. In this karyotype, the pairs of homologous chromosomes are shown beside each other.

chromosomes, when the sister chromatids are maximally condensed but have not yet separated (Figure 4.2b). We will learn in the next section how the binding of special proteins helps to package and condense DNA. When proteins are removed from mitotic chromosomes, the DNA appears as large loops that are apparently anchored near the midline of the chromosome (Figure 4.3). Organisms with smaller genomes do not have such a high level of condensation. *Saccharomyces cerevisiae*, for example, performs only a twofold compaction of the DNA, and metaphase chromosomes cannot be seen in this organism under the microscope.

4.3 PACKAGING CHROMOSOMAL DNA

We noted at the start of this chapter that a chromosome is composed of DNA packaged with specific proteins, which help to condense the DNA into a relatively tiny volume of space. This packaging also helps to protect the genome from mutation and rearrangements. The DNA–protein complex that comprises chromosomes is called chromatin. In this section, we explore how chromatin is formed, and learn about the distinct three-dimensional structures it adopts in order to organize the cell's genome within the nucleus and regulate the access of other proteins to the DNA.

DNA condensation occurs in multiple steps

The length of DNA packed into a cell nucleus is astounding. In a single human cell, over 2 m of DNA must fit inside a nucleus that is 10^{-5} m (10 μm) in diameter. This is like taking 30 miles of string, cutting it into 46 pieces, and then packing the pieces neatly into your backpack. Bacteria have a similar problem: 1500 μm

Figure 4.3 Mitotic chromosomes contain large looped domains of DNA. A mitotic chromosome has been treated to release proteins. Large loops of DNA released from the condensed chromosome can be seen as a halo extending outward.

⊙ Supercoiling is described in Chapter 2.

Figure 4.4 The bacterial DNA-binding protein IHF bends DNA. The mainly alpha-helical IHF protein binds to DNA as a dimer. The two monomers are shown as red and blue ribbons, respectively. IHF binding bends the DNA, which facilitates its packaging and helps to recruit other proteins and enzymes that act on the bent DNA. PDB 1IHF.

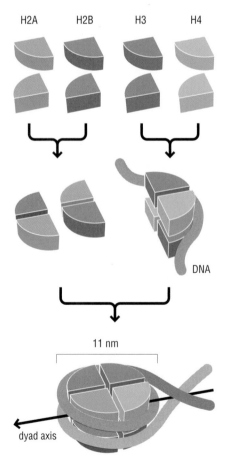

of DNA must be packaged into a cell that is only 2 μm long. How is this packaging achieved? To organize DNA into such a small space, cells use multiple levels of packing.

The first level of DNA packaging in all organisms is the binding of small, basic (positively charged) proteins directly to the DNA along its length. The many positively charged side chains on these proteins help to counterbalance the negatively charged DNA. In bacteria, the binding of small basic proteins bends the DNA, which helps pack it into a more compact structure (Figure 4.4). The compacted bacterial chromosome is called the **nucleoid**. The DNA within this DNA–protein complex becomes supercoiled as a result of its packaging; that is, it is wrapped around itself like a twisted rubber band, and thus is much more compact than non-supercoiled DNA.

The ability to supercoil DNA is essential for bacterial viability, and supercoiled DNA plays a part in many chromosomal functions in eukaryotes as well. Many of the enzymes that act on DNA, such as recombinases, only act on supercoiled DNA, as we shall see in later chapters.

Eukaryotic DNA and histones are packaged into nucleosomes

In eukaryotes, the basic DNA-binding proteins that package DNA into chromatin are called **histones**. There are four core **histone proteins**, H2A, H2B, H3, and H4, which evolved early in the history of eukaryotes. Because core histones are essential and are required to interact with many chromosomal proteins, they represent some of the most highly conserved proteins in eukaryotes. All four of the core histones have a common protein fold known as the histone fold, and are rich in lysine and arginine residues. The side chains of the amino acids lysine and arginine carry positive charges, which help to counterbalance the many negative charges due on phosphate groups in the DNA backbone. This electrostatic interaction – between positive and negative charges – helps to stabilize the overall histone – DNA assembly.

DNA wraps around a complex of histone proteins to form a **nucleosome**, which is the fundamental unit of eukaryotic chromatin. At the center of the nucleosome is a disk-shaped protein complex, the **histone octamer**, which contains two copies of each of the four core histone proteins (Figure 4.5). Two monomers each of histones H3 and H4 come together to form a tetramer, while histones H2A and H2B form heterodimers. The full histone octamer is assembled on the DNA in an active process that is facilitated by a number of different proteins.

Figure 4.5 Histones assemble on DNA to form nucleosomes. Assembly of the histone octamer. Histones H3 and H4 assemble to form a heterotetramer containing two copies of H3 and two copies of H4, while histones H2A and H2B form heterodimers and then into a tetramer. The H3•H4 heterotetramer binds to DNA first (possibly assembling on the DNA from two H3•H4 dimers), followed by two H2A•H2B heterodimers, forming a nucleosome. DNA is wrapped around the histone octamer in a left-handed manner to form a nucleosome.

The H3·H4 tetramer is assembled on the DNA from two H3·H4 dimers; this is followed by the association of two H2A·H2B heterodimers to give the complete histone octamer. The arrangement of histone proteins in the octamer has a twofold axis of symmetry (Figure 4.5), meaning that the octamer looks the same when rotated 180° about the axis shown. The DNA wound around the histone core also conforms to this twofold symmetry (Figure 4.5). This occurs because each of the histone proteins associates with the DNA in a similar manner.

About 146 base pairs of DNA wrap around the histone octamer to form a nucleosome (Figure 4.5). This amounts to somewhat less than two complete windings of the DNA around the histone octamer (1.76 turns, to be more precise). This number has been deduced from two different kinds of experiments: the three-dimensional atomic structure of the nucleosome core particle (shown in Figure 4.6), and from nuclease digestion experiments, which are described in Experimental approach 4.1. The DNA winds around the histone octamer in a left-handed manner (Figures 4.5 and 4.6). The direction of this winding is important: when a histone octamer is stripped away from the DNA, it leaves behind negatively supercoiled DNA. But why is this important? Negative supercoiling makes it easier for double-stranded DNA to be drawn out into separate strands (a requirement for DNA replication and transcription). By contrast, positive supercoiling makes it *more* difficult to separate the two strands.

Each core histone protein has an N-terminal 'tail' that extends away from the nucleosome core between the coils of DNA (Figure 4.6). These tails are stretches of amino acids of up to 25 amino acids in length that do not adopt a defined structure. The histone tails interact with adjacent nucleosomes to condense the chromatin structure further, although their precise role in chromatin compaction remains an area of active study. In addition, as we will see later in the chapter, the histone tails are subject to covalent modifications that alter chromatin structure and function.

> ⊙ The role of chromatin assembly factors and histone chaperones in assembling the nucleosome is covered in Chapter 6.

> ⊙ The role of negative supercoiling in favoring strand separation is discussed in Chapter 2.

Figure 4.6 Crystal structure of the nucleosome core. The histone proteins are shown in colors corresponding to those in Figure 4.5. The DNA wrapped around the histone octamer is shown in gray. The left view is of the top of the disk-like octamer; the right view is from the side. The N terminal tails of the histones project outward from the nucleosome core (see text). PDB 1KX5.

4.1 EXPERIMENTAL APPROACH

Nuclease probes of chromatin organization

nuclease digestion

protected DNA fragments

Figure 1 Protection of nucleosomal DNA from micrococcal nuclease digestion. The DNA that is not in contact with protein is digested into nucleotides, while the nucleosomal DNA is protected from digestion.

In the early 1970s, it was known that chromatin consisted of DNA packaged in some way with histone proteins, but the nature of the packaging was not at all clear. In 1974, Roger Kornberg proposed that chromatin consisted of a regular 'repeating unit of two each of the four main types of histone and about 200 base pairs of DNA'. Empirical evidence for this model of the histone octamer and its role in packaging DNA was reported in the same year by Markus Noll, in the form of nuclease digestion experiments.

The Noll group used nucleases to probe the way eukaryotic DNA is packaged into chromatin. Micrococcal nuclease cleaves accessible double-stranded DNA, and thus can differentiate DNA wrapped around a histone octamer from 'naked' unpackaged DNA. After DNA cleavage the protected regions are ultimately visualized by stripping the histone proteins away from the DNA and analyzing the resulting DNA fragments by gel electrophoresis (see cartoon in Figure 1).

We see in Figure 2 the banding pattern of DNA that results from a nuclease treatment of rat chromosomes. The pattern of DNA fragments is consistent with the model of a regular repeating unit as proposed by Kornberg. Chromatin that was incubated for a short time with micrococcal nuclease yielded a variety of discrete fragments of DNA, corresponding to the protection by one, two, three, or more nucleosomes. The smallest DNA fragments were about 170 bp and 205 bp in length (two discrete bands), while each of the longer fragments was larger than the next by about 200 bp, consistent with the proposed model. It turns out that the 205 bp fragment is the first one typically generated by the nuclease, when it nicks a site more or less halfway between two sequential octamers. Further digestion by the nuclease nibbles closer to the octamer, yielding the smaller 170 bp fragment.

Nuclease digestion of DNA can now be used in conjunction with modern DNA sequencing methods to determine the location of individual nucleosomes on each chromosome. This was done by Mavrich and colleagues in their 2008 study of nucleosome positions in the yeast genome. In this approach, yeast chromosomes were first treated with a chemical cross-linking agent that causes covalent bonds to form between the histone proteins and the DNA in each nucleosome (Figure 3). Following this treatment, the chromosomes were digested with micrococcal nuclease, and the nucleosome particles were purified using antibodies against histones H3 and H4. The chemical links were then removed, and the different DNA fragments in the mixture were sequenced (outlined in Figure 3).

205 ± 15 base pairs

170 ± 10 base pairs

Figure 2 Nuclease digestion pattern of chromatin. The electrophoretic pattern shows the fragments of DNA that were produced from limited digestion of rat liver chromosomes with micrococcal nuclease. Reproduced from Noll, *Nature* 1974:**251**:249–251.

Figure 3 Mapping the positions of nucleosomes. The diagram shows the procedure used to identify the positions of nucleosomes in the yeast chromosome. Once the DNA wrapped around each nucleosome has been sequenced, a computer search of the yeast genome is done to map the sequence to its correct position in the chromosome.

A computer-aided search was used to identify the location of each DNA sequence in the yeast genome, thereby giving the location of each nucleosome in the yeast genome. From such a study comes large amounts of data that must be sifted through to identify trends and insights. For example, Mavrich and colleagues found that the 5′ promoter and 3′ ends of *S. cerevisiae* genes were relatively free of nucleosomes. The authors proposed that such nucleosome-free regions help proteins involved in transcription initiation and termination to gain access to their target locations on the DNA. This study thus provided immediate insights into genome function and gene expression.

Kornberg RD. Chromatin structure: a repeating unit of histones and DNA. *Science* 1974;**184**:868–871.

Mavrich TN, Ioshikhes IP, Venters BJ, *et al*. A barrier nucleosome model for statistical positioning of nucleosomes throughout the yeast genome. *Genome Research* 2008;**18**:1073–1083.

Noll M. Subunit structure of chromatin. *Nature* 1974;**251**:249–251.

Although the four core histones – H2A, H2B, H3, and H4 – are the most common histones in nucleosomes, there are other histone proteins in the cell that can instead be incorporated into a histone octamer. These **histone variants** (Figure 4.7) are highly similar to core histones in amino acid sequence, but are encoded by distinct genes. Histone variant proteins are found in place of the common core histones at different locations in chromatin, where they play special functions. For example, CENP-A (centromere protein-A) is a variant of histone H3 that is incorporated into nucleosomes at centromeres and is essential for the formation and function of the centromere. H2A.Z, a variant of H2A, is more commonly found upstream of genes and has been implicated in transcriptional regulation. Another H2A variant, H2A.X, is phosphorylated at sites of DNA double-strand breaks and is

Variant histones		
histone class	**variant histones**	**proposed function**
H2A	H2A.X	DNA damage signaling
	H2A.Z	transcriptional activation
	MacroH2A1	binds inactive X chromosome
	MacroH2A2	binds inactive X chromosome
	H2A-Bbd	
H2B	H2B.1	
H3	CENP-A	defines centromere chromatin
	H3FA3	
	H3.3B	
	H3FT	
H4		

Figure 4.7 Variant histones and their proposed functions where known.

Figure 4.8 Different types of DNA bends allow the DNA to wrap around the histone octamer. The DNA in the nucleosome is smoothly bent in some regions (a and b) and more sharply bent at some points. (c) The cartoon to the right of each image depicts the bending of the helix. The gray line represents the path of the center of the DNA helix, and the red and yellow bars represent how the bases fit on the bent DNA. In the smooth bend in panel (b), some of the base pairs are shifted from their more typical positions relative to the helix axis. PDB 1KX5.

thought to recruit DNA repair machinery. We will learn later in this book about the role that these and other histone variants play.

Nucleosomes form preferentially at certain DNA sequences

Since all eukaryotic DNA is packaged by histones, the way in which histone octamers bind to DNA is relatively insensitive to the DNA sequence. However, nucleosomes do form preferentially along certain types of DNA sequences. When yeast chromosomes are examined, for example, there are certain regions that contain nucleosomes at particular DNA sequences, whereas nucleosome positions in other regions appear to be more variable.

How do the preferred locations arise? The answer lies in the way in which the DNA wraps around the histone octamer to form a nucleosome. As seen in Figure 4.6, the DNA that is wrapped around the core histone octamer is, for the most part, smoothly bent, but contains sharper bends at some locations (Figure 4.8). The DNA sequence at the sharp bends generally consists of pyrimidine–purine base steps (the sequence thymine/adenine (TA) or cytosine/guanine (CG) as viewed along one strand), which are more highly bendable than other base steps. The minor groove narrowing that must occur in order for the DNA to wrap smoothly around the nucleosome core is energetically more favored in AT-rich sequences. The more smoothly bent regions therefore contain short stretches of more bendable AT-rich sequences, which alternate with stiffer GC-rich sequences, each centered roughly half a helical turn apart from one another. In this way, the base sequence helps to determine the relative location of the DNA around the histone protein core.

Chromatin is organized into 30 nm filaments and higher-order structures

The packaging of chromatin occurs at several successive levels, working up from the small to large scale, leading to condensed forms that can be seen in the electron microscope (Figure 4.9). These levels are visible in the 10 nm fiber, the 30 nm fiber, and loop structures. Let us now consider each of these in turn.

(a)

(b)

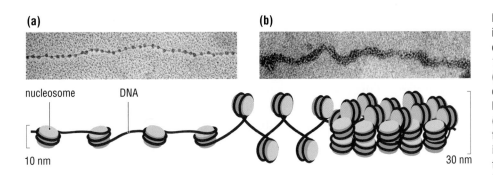

nucleosome   DNA

10 nm

30 nm

Figure 4.9 Nucleosomes become packaged into higher-order structures in eukaryotic chromosomes. (a) Electron micrograph of the 10 nm fiber of chromatin (top) and schematic (bottom), showing the 'beads on a string' organization of nucleosomes along DNA. (b) Electron micrograph (top) and schematic (bottom) of the 30 nm fiber that is formed by further packing. The less compact zigzag model is shown on the left and a more condensed form on the right. Condensation occurs through interactions between nucleosomes that are not yet fully understood. H1 and other proteins are present in the 30 nm fiber, but are not shown here for simplicity.

H1

Figure 4.10 The binding of linker histone H1 to the nucleosome. Histone H1 binds to the linker DNA in the nucleosome, facilitating formation of the 30 nm fiber (shown in Figure 4.9b).

The first stage of packaging of eukaryotic DNA into chromatin is the 10 nm fiber. This level of packaging arises from the way that histone octamers associate with double-stranded DNA to form nucleosomes, which are separated from one another by stretches of DNA to form a fiber that appears in electron micrographs like 'beads on a string' (Figure 4.9a). In this state, which occurs in the test tube at low ionic strength, the individual nucleosomes (the 'beads') are clearly visible and are separated by linker DNA (the 'string').

The 10 nm fiber can be further folded up into a more condensed filament 30 nm in diameter (Figure 4.9b) with the help of a protein, histone H1. In the 30 nm fiber, the nucleosome core particles form a regular, alternating arrangement that brings the nucleosomes in contact with one another. Our current understanding of how chromatin is organized into the 30 nm fiber is shown in Figure 4.9. Histone H1 binds to the linker DNA that connects successive nucleosomes, facilitating the compaction of the chromatin (Figure 4.10). Histone H1 is therefore known as a linker histone.

The tails of the core histone proteins that protrude from each nucleosome are also important in formation of the 30 nm fiber, although the precise way in which they stabilize the fiber remains an open question. One of the salient questions that has occupied researchers has been that of how the linker DNA connects one nucleosome to the next, and how chromatin with linkers of varying length is organized into a 30 nm fiber. As new data emerge from a variety of physical measurements, our understanding of the structure of the 30 nm fiber (and the diagrams you find in different textbooks!) is modified accordingly.

In the cell, the 30 nm fiber is further organized to become part of the compact forms of chromosomes that are seen in electron micrographs (see Figure 4.3).

nucleolus   nucleus

euchromatin   heterochromatin

Figure 4.11 Electron micrograph of a cell showing the heterochromatin localized to the nuclear periphery. The large dark-rimmed circle in the middle of the cell is the nucleus. The white regions within the nucleus represent euchromatin and the dark-staining regions heterochromatin. The heterochromatin is concentrated at the periphery of the nucleus. Many nuclei also contain spherical structures called nucleoli (the nucleus shown has one visible nucleolus). The nucleolus will be discussed further towards the end of this chapter.

➲ DNA repair is covered in Chapter 12 and DNA recombination in Chapter 13.

We know little about the precise nature of how chromatin is further compacted to form these structures. However, it is clear that large loops of chromatin, which has been organized into 30 nm fibers, are anchored to a central scaffold whose precise composition is still not completely understood. Nevertheless, many studies have clarified the role of the histone proteins both in organizing the genetic material and in allowing access to the DNA during transcription, DNA replication, and other processes, as we shall see throughout this book.

4.4 VARIATION IN CHROMATIN STRUCTURE

The chromosomes of a cell undergo a variety of changes in the course of the cell cycle, with marked changes taking place in the packaging and organization of DNA. During interphase in particular, when chromosomes are relatively decondensed and various genes may be actively transcribed, there is great variation in DNA compaction in different regions of the chromosome. We begin by describing some of these variations in chromosome structure, after which we will explore some of the basic mechanisms underlying the variations in chromatin that regulate access of proteins to the DNA at all stages of the cell cycle.

Interphase chromatin is not uniformly packaged

During interphase, chromosomes are in an extended, relatively decondensed form and genes are actively transcribed. However, within the relatively decondensed chromosomes, we find significant variations in structure. Our earliest understanding of these differences came from studies of whole nuclei, such as the electron micrograph shown in Figure 4.11. In these images, some of the chromatin stained lightly with the dye used to visualize DNA while other regions stained more darkly. The light staining regions were interpreted as being relatively decondensed regions, and were named **euchromatin**, and the strongly staining regions were apparently more compacted regions and were named **heterochromatin**. The root 'chroma', from the Greek word for color, derives from these early observations of how chromatin bound to dyes.

Interphase chromosomes contain regions of both heterochromatin and euchromatin. Genes that are actively transcribed typically lie in regions of euchromatin. Euchromatin is relatively decondensed, and it is this open structure that presumably allows ready access to DNA by the enzymes and proteins that must bind to it for transcription, repair and recombination to occur. By contrast, heterochromatin appears to be more condensed and is typically concentrated near the periphery of the nucleus (Figure 4.11). While it was originally thought that genes in heterochromatin were not transcribed, we now know that there is indeed some transcription in these regions. At the same time, there is well-established evidence that the **translocation** of certain genes to the vicinity of heterochromatin actively prevents transcription, as well as other processes, such as DNA repair and recombination.

Additional regions of chromosomes also form heterochromatin. These regions include the DNA at and near the centromeres, which is the segment of the chromosome to which the spindle attaches for segregation, and at **telomeres**, the special DNA sequences at the ends of chromosomes. In addition, those sites in chromosomes that contain highly repetitive DNA sequences are also packaged into heterochromatin. We will learn much more about the nature of centromeres and telomeres later in this chapter.

The differences between heterochromatin and euchromatin are often detected experimentally by the **nuclease sensitivity** of the region. Heterochromatin is more resistant to digestion by DNase I than euchromatin. The relative resistance of heterochromatic DNA to digestion is often interpreted as indicating that heterochromatin is more condensed than euchromatin, although it may equally reflect differences in the ability of nucleases to interact with DNA bound by specific heterochromatin proteins.

➲ The use of nuclease digestion to probe DNA structure and interactions with proteins is covered in Chapter 16.

The structural state of chromatin affects all processes directed by DNA

While most studies of the functional significance of chromatin structure have focused on gene transcription, the way in which DNA is packaged into chromatin affects all processes that involve interactions with chromosomes. The role of chromatin structure in the regulation of transcription will be discussed in Chapter 8. Here we briefly describe some examples of how chromatin structure influences replication, recombination, and chromosome transmission.

The effects of chromatin structure on DNA replication can be demonstrated by experiments in which an origin of replication normally found in euchromatin is placed into heterochromatin. An origin of replication is a location in the chromosome at which DNA replication begins.

➲ We describe replication origins in detail in Chapter 6.

Replication origins located at different positions along a chromosome initiate replication at different times during S phase. Chromosome rearrangements that place an early-replicating origin into heterochromatin result in late replication of that origin. The functional importance of regulating replication timing is not entirely clear; however, observations on the budding yeast *S. cerevisiae*, suggest that late-replicating origins may allow cell division to be delayed if DNA is damaged or other steps in DNA replication are incomplete.

Differences in chromatin structure can also control which parts of the chromosomes are able to undergo recombination, a process by which a strand of DNA is broken and then joined to a different DNA molecule. In yeast, the tandem array of genes encoding ribosomal RNA (rDNA genes) is packaged into heterochromatin and normally has a reduced recombination rate, which helps prevent the loss and rearrangement of genes that can occur between highly repetitive sequences. When heterochromatin is not properly maintained, however, there is increased mitotic recombination in the rDNA gene cluster. The packaging of repetitive DNA into heterochromatin therefore helps stabilize the genome.

➲ We learn in detail about recombination in Chapter 13.

Chromatin structure also plays a role in chromosome segregation and cell division. As described later in this chapter, the ability to form a functional centromere depends on a specific variant histone H3 called CENP-A. In many species, disruption of normal heterochromatin formation causes major defects in chromosome segregation at mitosis, and chromosome pairing at meiosis. Thus in the regulation of chromatin structure is critical for centromere function and chromosome stability.

4.5 COVALENT MODIFICATIONS OF HISTONES

Chromatin is far from being a permanently static entity, in which sections of chromosomal DNA are irreversibly 'frozen' into euchromatin or heterochromatin. Neither, as we note above, is the biological role of chromatin purely structural – merely a convenient means of storing DNA in the cell. Instead, there are dynamic changes in chromatin structure that are mediated by various modifications to the chromatin. These modifications of both the physical location of its nucleosomes, within the chromosome, as well as the biochemical composition of the histone proteins that comprise the nucleosomes, are central to regulation.

Modification of histones affects chromatin structure and function

The histone proteins that package DNA into chromatin are subject to a variety of post-translational modifications. These modifications, which add chemical groups or small proteins to histone side chains, serve as important signals in regulating virtually all processes that involve interactions with chromatin. Acetylation, methylation, phosphorylation, and **ubiquitination**, are covalent modifications that are commonly found in histones.

Post-translational modifications occur both in the histone tails (Figure 4.12) and in the globular core of the histone proteins (Figure 4.13). Once modified, these sites are thought to recruit specific proteins to the chromatin, which then affect chromatin structure and function. The covalent modifications also constitute chemical changes that can alter chromatin structure directly; acetylation, for example, removes the positive charge from lysine side chains and thereby alters interactions with the negatively charged DNA. While **histone modifications** are relatively minor in terms of the changes they introduce at the chemical level, the

➲ Post-translational modification of proteins is covered in detail in Chapter 11. A brief overview can be found in Chapter 2.

(a)

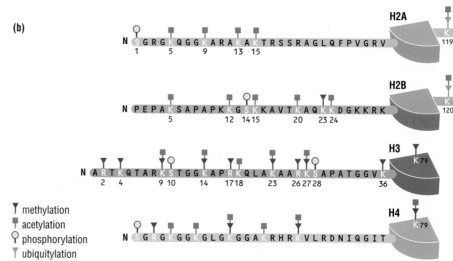

(b)

Figure 4.12 Modification of histone tails. (a) Histone tails are covalently modified on the N-terminal regions that protrude from the nucleosome core region. (b) The tails are methylated, acetylated, phosphorylated, or ubiquitinated at specific amino acids. The combination of specific modifications on a given histone tail can influence the function or accessibility of the associated region of DNA.

biological impact of such changes is far from trivial: histone modifications are an important means of modulating any process that involves accessing the DNA.

An important feature of histone modifications is that they are reversible. There are specialized enzymes that add particular chemical modifications to histone side chains, and other types of enzymes that remove the modification. Figure 4.14 shows some examples of different histone modifications and the enzymes that either add or remove them. The dynamic addition and removal of particular modifications – the biological equivalent of an on/off switch – is used by all eukaryotic cells to regulate a wide array of processes.

Histone acetylation is associated with active chromatin

Virtually all regions of eukaryotic chromosomes contain histones with acetylated lysine residues. However, the relative levels of acetylation vary substantially: actively transcribed regions in euchromatin contain highly acetylated histones, whereas heterochromatin, which contains few transcribed genes, has very low levels of acetylation. The primary targets of acetylation are histones H3 and H4, which contain numerous lysines that can become acetylated (Figure 4.12). Enzymes called **histone acetyltransferases (HATs)** covalently attach acetyl groups to lysine side chains, while **histone deacetylases (HDACs)** remove the acetyl groups (see Figure 4.14). The isolation of the first histone acetyltransferase is described in Experimental approach 4.2.

Histone acetylation generates sites for the binding of proteins containing domains known as **bromodomains**. These domains bind specifically to particular acetyl-lysines (Figure 4.15) – for example, acetylated lysine 14 on histone H3. The bromodomain protein may then recruit other proteins, including nucleosome-remodeling complexes or other enzymes that act on DNA (Figure 4.15). The effect of acetylation may also derive from the resulting change in electrostatic charge, since unmodified lysines are positively charged, while acetyl-lysine is electrostatically neutral.

(a)

(b)

(c)

○ acetylation
● methylation
○ phosphorylation
○ ubiquitination
○ ubiquitination or acetylation

Figure 4.13 Modified residues in the core of the histone octamer. Residues that lie on the surface of the globular core of the histone octamer are also subject to post-translational modifications. The DNA wrapped around the histone octamere is shown in blue. (a) A view down the DNA double helix axis; (b) As in panel a, but rotated 90° around the 'North-South' axis (c) As in panel a, but rotated 90° around the 'East-West' axis.
From Cosgrove *et al.* (2004).

Figure 4.14 Examples of different classes of histone-modifying enzymes in yeast and humans.

Different Classes of Histone-Modification Enzymes		
enzyme family	**species**	**complex**
histone acetyltransferase (HAT) complexes		
Gcn5 family	*S. cerevisiae*	SAGA, ADA
TAF250	human	TFIID
PCAF	human	PCAF
histone deacetylase (HDAC) complexes		
HDAC1, HDAC2 family	*S. cerevisiae*	Sin3
	human	Sin3. NURD
HDAC5	human	N-CoR/SMRT
Sir2	*S. cerevisiae*	Sir
histone methyltransferase (HMT) complexes		
PRMT family	*S. cerevisiae*	RMT1/HMT1
	human	PRMT1
Set domain family	human	Suv39h1
Jumanji family	*S. cerevisiae*	JHDM1
LSD1 family	human	LSD1

4.2 EXPERIMENTAL APPROACH

Discovery of a histone acetyltransferase

Although it was known that histones were acetylated and that histone acetylation was associated with transcriptionally active chromatin, the enzyme responsible for this post-translational modification was not known. David Allis and colleagues were able to crack this problem by developing an assay in which acetyltransferase activity could be detected in proteins that were separated on a gel impregnated with histones.

This special gel was prepared by adding histone proteins to the polyacrylamide-SDS mixture before the gel polymerized. Protein samples could then be electrophoresed on the gel and separated by molecular weight, as on a standard protein gel. The gel was then treated to remove the SDS, allowing the proteins to adopt their native fold, and the gel was then soaked in radiolabeled acetyl-CoA (the acetyl group donor). Active histone acetyltransferases in the gel could then catalyze transfer of the labeled acetyl group onto histone proteins. Since the histone proteins were everywhere in the gel, while the acetyltransferase proteins were only located in particular bands on the gel, an autoradiograph of the gel would show which protein band on the gel contained histone acetyltransferase activity. This approach allowed the researchers to directly analyze still-heterogeneous column fractions and identify which protein in that particular mixture was the active protein.

Taking advantage of the fact that the macronuclei of *Tetrahymena* are an enriched source of the enzyme, Brownell and Allis fractionated extracts using various protein purification techniques and used the histone-impregnated gel to monitor the activity in each fraction. In this way, the researchers observed a 55 kDa protein (which they called p55) that had histone acetyltransferase activity

(Figure 1). Brownell *et al.* then identified the protein by eluting it from the gel and determining the amino acid sequence of several peptide fragments of the purified protein. The partial amino acid sequence made it possible to clone the entire cDNA encoding the protein using PCR methods (see Section 16.9). The resulting DNA sequence of the cDNA matched a known gene called GCN5 in yeast that was required for full transcriptional activity of some genes. This match unambiguously linked histone acetylation and transcriptional activation. (Today, the partial amino acid sequence would lead to direct identification of the gene (as described in Section 16.7) without the intervening step of cDNA cloning, because the complete genome sequence of *Tetrahymena* is available.)

Since the first histone acetyltransferase was identified, numerous enzymes that acetylate and methylate histones at specific residues have been isolated. Since the identity of the modified residue and the location of the modified histone within the chromosome has important functional significance, techniques have been developed to find where in the genome particular modified histone proteins are located. These methods rely upon antibodies that recognize very specific modifications, such as acetylated lysine 5 (K5) or K9 in histone H2A, or acetylated K5, K12, K20, or K120 in histone H2B. A method described in Experimental approach 4.1 can then be used to identify where in the genome histones with particular modifications are located: the histones can be cross-linked to the DNA, after which the DNA sheared, and the histones of interest immunoprecipitated with the appropriate antibody (which binds specifically to the modification of interest). Using this approach, Wang and colleagues mapped histone acetylation and methylation sites across large regions of

Figure 1 Activity gel showing identification of histone acetyltransferase. (a) The gel is stained with Coomassie dye and shows the proteins in the partially purified fractions. (b) Autoradiograph of the same gel as in (a) after the proteins were renatured and the gel was incubated with radiolabeled acetyl-coenzyme A (acetyl-CoA), followed by washing with unlabeled acetyl-CoA to remove the unincorporated radiolabel. The bands indicated a single 55 kDa protein that acetylated histones within the gel matrix. Lane 'L' contains all the proteins in the extract (namely, before fractionation). The arrows point to the protein band that exhibits acetylatransferase activity (in L and fraction 6).

Reproduced from Brownell & Allis, *Proceedings of the National Academy of Sciences of the U S A* 1995;**92**:6364–6368.

Figure 2 Mapping of different histone modifications across a region of human chromosome 20. Each line shows the location of the indicated modified histone. The column marked by red lines indicates the transcription start site of a gene residing in this region. Modifications that were determined experimentally to be statistically significant are labeled with red astrisks on the left.
Reproduced from Wang *et al.*, *Nature Genetics* 2008;**40**:897–903.

the human genome, yielding detailed information that provides insights into the roles of histone modification in dictating biological consequences (Figure 2).

Brownell JE, Allis CD. An activity gel assay detects a single, catalytically active histone acetyltransferase subunit in *Tetrahymena* macronuclei. *Proceedings of the National Academy of Sciences of the U S A* 1995;**92**:6364–6368.

Brownell JE, Zhou J, Ranalli T, *et al*. *Tetrahymena* histone acetyltransferase A: a homolog to yeast Gcn5p linking histone acetylation to gene activation. *Cell* 1996;**84**:843–851.

Wang Z, Zang C, Rosenfeld JA, *et al*. Combinatorial patterns of histone acetylations and methylations in the human genome. *Nature Genetics* 2008;**40**:897–903.

Figure 4.15 Histone tail acetylation allows recruitment of proteins that act on chromatin. (a) Acetylation of histones in chromatin recruits bromodomain-containing proteins that can, in turn, specifically recruit other proteins that may, for example, activate nearby genes. (b) Structure of the bromodomain from GCN5 bound to acetyl-lysine in a histone H4 peptide. PDB code 1E6I.

Figure 4.16 Histone tail methylation recruits proteins that initiate heterochromatin formation. (a) Methylation of lysine 9 on H3 provides a site for recruitment of proteins containing a chromodomain (green). An example of a chromodomain protein is the protein HP1, which helps initiate heterochromatin formation. Some chromodomain proteins interact with other proteins (orange) and recruit them to sites of H3 Lys 9 methylation histone. (b) The structure of the HP1 chromodomain bound to trimethyl-lysine is shown. (PDB code 1KNE).

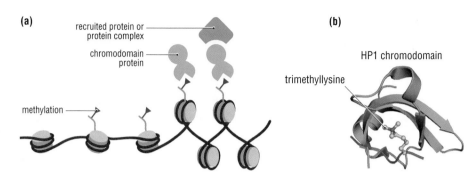

Histone methylation can be associated with either active or inactive chromatin

Histones H3 and H4 can be methylated at both lysine and arginine residues (see Figure 4.12). Up to three methyl groups can be covalently attached to a single lysine, while up to two can be attached to an arginine side chain. The different degrees of methylation – namely mono-, di-, or trimethyl lysine, and mono- or dimethyl arginine – can have different consequences by recruiting different proteins that recognize the different methylated side chains. Histones are methylated by enzymes known as **histone methyltransferases (HMTs)** and are removed by **histone demethylases**. Some examples of these enzymes are listed in Figure 4.14.

Unlike acetylation, which is uniformly associated with active transcription, methylation is associated with either activation *or* repression of transcription. The precise effect of histone methylation depends upon which residue is methylated. For example, methylation of lysine 9 in the tail of histone H3 is associated with transcriptionally silent chromatin – including heterochromatin – while methylation of H3 lysine 4 is associated with active chromatin. It is, therefore, the recognition of specific modifications on histones by other proteins, not the chemical change on the histone itself, that determines the functional consequence of the methylation.

Like acetyl-lysine, methylated lysine and arginine side chains are recognized by proteins that bind specifically to methylated residues. **Chromodomains**, for example, bind to specific methylated lysines (Figure 4.16). Chromodomains are typically found in proteins that mediate transcriptional **silencing**; an example is the protein HP1, which helps initiate heterochromatin formation in mammalian cells (Figure 4.16).

Histone phosphorylation and ubiquitination serve as additional regulatory signals

Histone tails can be phosphorylated at either serine or threonine residues, as depicted in Figure 4.12. The phosphate groups are added by kinases and removed by phosphatases. Histone phosphorylation plays a variety of roles: phosphorylation of histone H3 at serine 10 alone allows the transcription of genes required for cell growth. In contrast, chromosome condensation at mitosis is tightly correlated with the phosphorylation of histone H3 at both serines 10 and 28. Phosphorylated residues can be recognized by specialized protein domains, although the precise mechanism by which phosphorylation of histone exerts an effect remains to be elucidated.

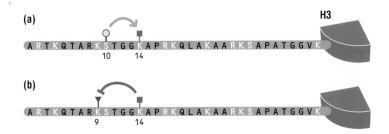

The covalent attachment of ubiquitin to lysine side chains is the largest histone modification, since ubiquitin is a 76-amino acid protein, a notable contrast to the handful of atoms added by each of the previously mentioned histone modifications. Ubiquitin is attached to histones in a series of enzymatic steps, and can be removed by a deubiquitinating enzyme (DUBs). Histones can be similarly modified by the covalent attachment of sumo (small ubiquitin-like modifier), which is similar in structure to ubiquitin. Ubiquitination and sumoylation of histones play roles in regulating different steps of transcription as well as DNA repair.

Different patterns of histone modifications have distinct consequences

Post-translational modification of one histone protein can trigger a cascade of events that leads to further chromatin modifications, either in another histone protein or in the DNA. For example, phosphorylation of serine 10 promotes the acetylation of H3 lysine 14, which, as we saw above, is associated with active chromatin. Acetylation of H3 lysine 14 then inhibits methylation of lysine 9 (Figure 4.17), which, as we saw earlier, is a docking signal for proteins associated with heterochromatin formation.

The concerted interaction of histone modifications, together with the large number of possible combinations of modifications, has led to the suggestion that there is a histone code, in which unique combinations of modifications specify particular functional states of chromatin. According to this model, the particular pattern of histone modifications found in the vicinity of a gene influences whether or not that gene is transcribed. The precise nature of functional information defined by the proposed histone code is an area of active investigation.

4.6 NUCLEOSOME-REMODELING COMPLEXES

ATP-dependent nucleosome remodeling alters the location of nucleosomes on DNA

While the packaging of eukaryotic DNA into chromatin helps to organize the genetic material in the nucleus, it presents a potential barrier to different proteins and enzymes that must access the DNA. For example, the wrapping of DNA around a histone core can block binding of the RNA polymerase enzyme, thereby repressing transcription. To overcome this barrier, the location of nucleosomes along the DNA can be altered in a regulated manner so that transcription

ATP-dependent chromatin remodeling exposes DNA binding sites

DNA-binding protein recognition

Figure 4.18 ATP-dependent chromatin remodeling allows access to specific sites on DNA. The binding sites for specific proteins in DNA (pink) may be inaccessible to DNA binding proteins owing to association of the region with the nucleosome. ATP-dependent chromatin remodeling results in the movement of the nucleosome along the DNA, exposing the binding site. Specific DNA-binding proteins (green) can now bind to their sites.

SWI/SNF family e.g. *S. cerevisiae* - SWI/SNF

HSA short insertion

DExx   HELICc   bromo

ISWI family e.g. humnan - hCHRAC

SANT

SLIDE

CHD family e.g. *Drosophila*- Mi-2

tandem chromo

INO80 family e.g. *S. cerevisiae* - INO80

HSA long insertion

Figure 4.19 The four families of ATP-dependent nucleosome-remodeling complexes. All nucleosome remodeling complexes contain a subunit with a ATPase domain that is split into two parts called DExx and HELICc. These two domains can be separated by either a short or long insertion. Different families are distinguished from one another by the other domains contained within the ATPase subunit. The domains are called HSA (helicase-SANT), bromodomain (binds acetyl-lysine), chromodomain (binds methyl-lysine), and SANT-SLIDE (binds to histone tails).

➲ We discuss the mechanism of mating-type switching in Chapter 8.

can proceed (Figure 4.18). This alteration is mediated by **ATP-dependent nucleosome-remodeling complexes**, which are multisubunit enzyme complexes that use the energy provided by ATP hydrolysis to change the positions of nucleosomes on DNA. Nucleosome-remodeling enzymes can either slide the histone octamer along the DNA or remove the histone octamer entirely and transfer it from one segment of the DNA molecule to another. In addition, nucleosome-remodeling complexes can increase accessibility of DNA by introducing loops into the DNA that is wrapped around a single histone core, thus increasing the accessibility of the DNA.

There are several different classes of nucleosome-remodeling complexes, which are distinguished by their subunit composition (Figure 4.19) as well as by their effect on chromatin. The SWI/SNF complex, for example, disrupts nucleosome positioning, while the ACF complex (the ISWI family) helps to position nucleosomes during chromatin assembly. Despite their diversity, all nucleosome-remodeling enzymes have in common a conserved ATPase domain that hydrolyzes ATP. Some complexes also contain subunits that bind to chemically modified residues in histone proteins. The SWI/SNF complex, for example, contains 11 subunits, including an ATPase domain that catalyzes DNA-dependent ATP hydrolysis and a bromodomain that binds to acetylated lysine residues in histones or in other proteins associated with the DNA.

One of the first chromatin-remodeling complexes to be studied, SWI/SNF, was identified in genetic screens for yeast defective in mating-type switching (SWI) and sucrose fermentation (SNF, sucrose non-fermenter). The mutations found to be responsible for the observed defects, termed *swi/snf*, could be suppressed by mutations in histone genes, suggesting that the wild-type Swi/Snf proteins might mediate function by altering chromatin structure.

A number of different models have been proposed to explain how ATP-dependent nucleosome-remodeling complexes alter the positions of nucleosomes on DNA (Figure 4.20). One proposal is that the enzymes propagate a local distortion in the DNA around the octamer, which gives rise to a nucleosome in a new position on the DNA strand (Figure 4.20a). It is also possible that a global distortion in the DNA, perhaps one that puts the DNA under torsional stress, could trigger repositioning of the nucleosome on the DNA (Figure 4.20b). The precise mechanism by which nucleosome-remodeling complexes alter nucleosome position is an active area of inquiry.

Figure 4.20 Different ways in which nucleosomes might be repositioned along the DNA by the actions of ATP-dependent nucleosome-remodeling enzymes. A number of models have been proposed for how nucleosome remodeling enzymes reposition nucleosomes: (a) propagation of a local DNA distortion around a histone octamer can lead to 'sliding' of the octamer along the DNA; and (b) a global conformational change in the DNA around the histone octamer could trigger repositioning of the nucleosome on the DNA.

(a) propagation of a local DNA distortion

(b) global conformational change in DNA

Nucleosome-remodeling complexes can be targeted to specific sites by interactions with other proteins

Since nucleosome remodeling is typically required in precise locations on the DNA, the cell must have a mechanism for recruiting nucleosome-remodeling complexes to specific sites on the DNA where they are needed (Figure 4.21). The most common way of recruiting these complexes is through association with proteins that bind to specific sequences in the DNA (Figure 4.21a). A wide variety of eukaryotic transcriptional regulators can recruit nucleosome-remodeling complexes through direct, non-covalent binding between proteins. Post-translational modifications in histone proteins can also recruit nucleosome-remodeling enzymes (Figure 4.21b). As mentioned above, some nucleosome-remodeling complexes contain specialized domains that recognize specific histone modifications. In this way, chemical modification of histones can trigger the binding of nucleosome remodeling complexes, which in turn alter transcription or other processes involving access to the DNA.

4.7 DNA METHYLATION

Methyltransferases add methyl groups to cytosine in DNA

Not all chromatin modifications that involve modification of chromosomal proteins; chemical modification can also occur on the DNA itself. In both bacteria and eukaryotes, methyl groups can be added to cytosine residues to generate 5-methyl cytosine (Figure 4.22). This modification, known as **DNA methylation**, is carried out by enzymes known as **DNA methyltransferases** (or **DNA methylases**). The basic mechanism of methylation is similar in bacteria and eukaryotes; however, the biological function of DNA methylation differs in each system, as we shall see below.

DNA methylation carries some risk to the cell, however, of leading to permanent alterations to the DNA. Methylated cytosines are chemically less stable than unmethylated cytosines, and undergo rare deamination. This deamination of methyl cytosine generates a thymine in the DNA (Figure 4.22c), and thus a methylated CpG dinucleotide becomes TpG. This residue change can lead to a permanent change in the DNA base sequence, as sometimes the single nucleotide mismatch is not recognized by the DNA repair machinery prior to replication. In contrast, the less frequent deamination of unmethylated cytosine generates a uracil residue. This is efficiently recognized as damage to the DNA and is correctly repaired to restore the original thymine moiety.

Base flipping is used to access bases to be methylated

The stable structure of DNA makes enzyme-driven base modification reactions, such as cytosine methylation, difficult. The bases are tightly packed inside the double helix, which is stabilized by base-pairing and base-stacking (see Figure 2.25). DNA methyltransferases therefore use an elegant mechanism called base

Figure 4.21 Recruitment of nucleosome-remodeling complexes to chromatin. (a) Proteins (orange) bound directly to a particular DNA sequence can specifically recruit nucleosome-remodeling enzymes (dark blue). (b) Acetylated lysine residues (pink squares) in histone proteins may be specifically recognized by a bromodomain (blue circle) that is part of a nucleosome-remodeling complex.
Adapted from Becker and Hörz (2002).

➲ We describe how DNA lesions are repaired in Chapters 8 and 13.

Figure 4.22 DNA methylation. (a) The chemical structure of unmethylated and methylated cytosine. (b) DNA is preferentially methylated at cytosine in CG pairs. (c) Deamination of methyl cytosine gives thymine, leading to a mutation of C to T.

Figure 4.23 Methylation of DNA requires that the base to be modified is flipped out from the double helix. The structure shown is of HhaI methyltransferase (green) bound to DNA (gray). The single cytosine residue to be modified (red) is flipped entirely out of the DNA helix and placed in the methyltransferase active site. This rather drastic distortion of the DNA is the mechanism used by most DNA methyltransferases studied. PDB code 3MHT.

flipping to gain access to cytosine bases, as was first discovered for the bacterial methyltransferase, HhaI. During base flipping, the cytosine base to be modified is flipped completely out of the DNA helix and an amino acid side chain from the enzyme is inserted in its place to stack temporarily with the adjacent nucleotides (Figure 4.23). In the meantime, the flipped-out cytosine is modified by the methyltransferase before being reinserted into the double helix.

Having now seen how this somewhat modest change in the chemical structure of cytosine is brought about, we now need to ask: what is the biological impact of DNA methylation?

Bacteria use methylation to distinguish newly synthesized DNA and to mark and protect their own DNA

In *Escherichia coli*, cytosine methylation can be used to distinguish the newly replicated DNA strand in repair processes. While both strands of DNA are methylated in the growing cell, just after DNA replication, only the parental strand is methylated. We say that such a half-methylated site on a chromosome is **hemimethylated**. The methylation status of the two strands is 'read' by DNA repair enzymes that are responsible for correcting mismatched bases. These enzymes search for nearby methylated cytosines to identify the parental strand so that they can replace the mismatched, incorrect base in the newly synthesized strand using the correctly identified parental strand as the template.

DNA methylation is also used by bacteria to distinguish their genomic DNA from invading bacteriophage DNA. Bacteria have specific endonucleases called restriction enzymes that cut DNA at sites containing particular recognition sequences; these enzymes are used to destroy foreign DNA that gets into the cell. To avoid destroying their own DNA, bacteria methylate the recognition sites in their own genomes, which prevents cleavage by their own restriction enzymes. For example, all HhaI recognition sites, GCGC (where the first C is methylated), are methylated in the host so that the endogenous HhaI endonuclease will not cleave the DNA at those sites. Foreign DNA from an invading virus will not be methylated, however, and thus will be cleaved at every GCGC site and consequently destroyed.

DNA methylation is associated with gene silencing in plants and mammals

In many eukaryotes, particularly plants and mammals, DNA methylation is used to silence transcription of genes. In other words, the presence of methylated cytosines along a DNA strand is interpreted by the cell to signal that the DNA that is modified in this way should not be transcribed. Any genes that are tagged in this way essentially become switched off.

Cytosine methylation in eukaryotes occurs most commonly at the sequences, **CpG** and **CpXpG**. Up to 60% of CpG base steps (that is, adjacent C and G nucleotides joined by a phosphodiester bond) in the human genome are methylated in somatic cells. An important feature of this form of transcription repression is that the methylated – and hence silenced – state can be passed on to daughter cells, thereby maintaining the silenced state over many generations of cell division. That is, when the DNA is replicated, it is not only the DNA sequence that is preserved, but also the patterning of methylation along that DNA strand. The pattern can be conserved through successive cell generations by fully methylating the hemi-methylated sites created during replication. This type of gene regulation is referred to as **epigenetic silencing**, which means that the information inherited by daughter cells is not due to changes in the DNA nucleotide sequence. (The term 'epigenetic' refers to any type of heritable state or change that is not encoded in the DNA base sequence, but which affects the expression state of that DNA.)

Certain protein complexes bind directly to DNA containing methylated cytosine. DNA methyl-binding domains are specialized proteins that bind to methylated cytosine. These domains are often subunits of larger complexes, which are thus directed to methylated DNA. These complexes may include histone deacetylases or methyltransferases that then modify the histone proteins in the vicinity of the methylated DNA (Figures 4.24a, b). The converse can also occur: there are proteins that bind to modified histones, which in turn recruit DNA methyltransferases that methylate DNA in the vicinity of the modified histones (Figure 4.24c). Nucleosome-remodeling complexes can also be recruited to

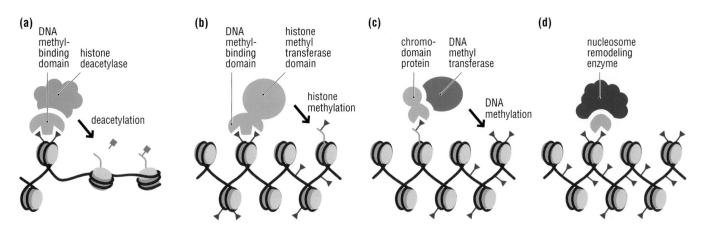

Figure 4.24 DNA methylation can recruit proteins involved in histone modification and vice versa. Methyl groups (blue triangles) can be added to both DNA and histone tails. (a) Methyl groups on DNA recruit methyl-binding proteins (green), which in turn can recruit histone deacetylase (orange) which modifies chromatin by removing acetyl groups (pink squares) from histone tails and thus initiates silencing. (b) Methyl-binding proteins that have histone methyltransferase activity (green) can bind to sites of DNA methylation and methylate histones (blue triangle), which is a signal for chromatin silencing. (c) Methylated histone tails can bind chromodomain-containing proteins (green), which in turn can recruit DNA methyltransferases (aqua) that methylate adjacent DNA. (d) Nucleosome remodeling complexes (dark blue) can bind specifically to regions of methylated DNA, using subunits that specifically recognize methylated cytosine (green).

DNA through specific interactions with methylated DNA (Figure 4.24d). Since some eukaryotes, such as yeast and fruit flies, silence specific genes using histone modification but lack DNA methylation, it is thought that DNA methylation is a special mechanism present in mammals and some other species to lock in a silent chromatin state.

DNA methylation is involved in X-chromosome inactivation and genomic imprinting in mammals

X inactivation is one of the best examples of the interrelation between DNA methylation, formation of heterochromatin and the silencing of chromosome functions. In female mammals, one of the two X chromosomes in each cell is permanently inactivated early in development so that dosage of the gene products encoded on the X is the same as in males, who have only one X chromosome. What do we mean by the 'dosage of gene products'? We can imagine a female carrying two copies of a given gene – one on each X chromosome. By contrast, a male would carry just one copy of the same gene, because they only have one X chromosome. If both X chromosomes in a female were transcriptionally active, the female would express double the amount of gene product relative to the male, because the female has two copies of the gene being transcribed, rather than just the one. X inactivation helps to correct the imbalance by switching off one of the female's X chromosomes such that both females and males have the same number of X chromosomes in an active state. The inactive X bears all the hallmarks of mammalian heterochromatin: the DNA is heavily methylated, the histones are modified as in heterochromatin, and most regions are completely inactive.

DNA methylation is also linked to a form of gene inactivation called **imprinting**, which is the silencing of one copy of a gene on either the maternal or the paternal chromosome, while the homologous genes on the other chromosome are expressed. Imprinting, which occurs in mammals, is thought to control the activity of critical paternal and maternal genes that regulate growth of the embryo. For some genes, the copy inherited from the mother is silenced, while that inherited from the father is active (Figure 4.25). For others, the copy from the father is silenced. The inactive copy of the gene is usually highly methylated. In normal individuals, expression of only one copy of these genes is needed. However, if the active gene

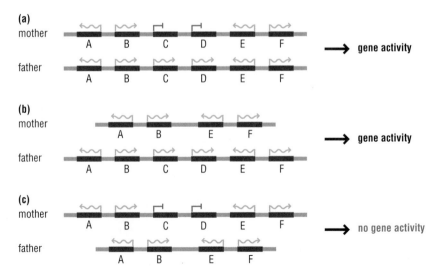

Figure 4.25 The activity of some genes is affected by genomic imprinting. (a) For loci where imprinting occurs, the copy of the gene that was inherited from the mother may be packaged differently and silenced, while the copy inherited from the father is active (green arrows indicate 'active' or expressed, while red bars indicate 'inactive' or silenced). (b) If a deletion removes a regions containing genes C and D from the mother, there is still a functional gene product from the father. (c) If a deletion occurs in the copy from the father, however, no functional gene product for genes C and D will be made. Although this example shows the maternal allele being silenced, which copy is imprinted depends on the particular locus.

is deleted, there is no expression of the gene at all, which usually leads to developmental abnormalities or death of the embryo. An example of an imprinted gene is *IGF2* (insulin-like growth factor 2), which encodes a peptide hormone that plays a role during embryonic development. Normally, only the paternal copy of the gene is expressed because the maternal copy is silenced. In some types of tumor cells, the imprinting of the maternal gene is lost, leading to aberrant expression of both copies of the gene. Imprinted genes are often found in clusters, suggesting that extensive regions of chromatin may be imprinted coordinately.

4.8 BOUNDARY ELEMENTS SEPARATE CHROMATIN DOMAINS

Special DNA sites establish the transition between euchromatin and heterochromatin

The different chemical modifications we have learned about in the preceding sections help to differentiate actively transcribed chromatin – the euchromatin – from transcriptionally silent heterochromatin. Cells appear to have specific boundaries between euchromatin and heterochromatin, implying that there is a mechanism for maintaining the separation between these two distinct forms of chromatin and preventing one from encroaching on the other. These sites on the chromosome are known as **insulators** or **boundary elements**.

Evidence for specific boundaries between euchromatin and heterochromatin came first from classical experiments on **position-effect variegation** in fruit flies. Wild-type fruit flies have red eyes, which comes from the expression of a gene called *white⁺*. (Mutant flies that lack an intact copy of this gene, or which cannot produce a functional protein product, lack the red pigment and therefore have white eyes.) However, an unusual pattern was sometimes observed in which the compound eye of the fruit fly had a mottled appearance, with some facets of the eye colored red while the rest were white (Figure 4.26a). How could the expression of the *white⁺* gene differ from facet to facet in the eye of a single fly? It turns out that there was a rearrangement in the chromosome of the mutant fly that moved the white gene from its normal location in euchromatin to a location immediately adjacent to a heterochromatic region (Figure 4.26b). This rearrangement, known as a chromosomal inversion, also likely removed a barrier element that normally separates the euchromatic and heterochromatic regions, thus allowing the heterochromatin to spread into sequences that are usually euchromatic.

The spreading of the heterochromatin silences genes that are normally expressed and, in this case, causes silencing of the *white⁺* gene in some cells in the compound eye. The degree of heterochromatic spreading varies from cell to cell, such that some cells in which spreading is minimal will continue to express the *white⁺* gene and will be red, while others will not (and will be white). Further, because the new heterochromatin states are heritable, the descendants of a cell still expressing the *white⁺* gene will give rise to a small patch of red cells (Figure 4.26b). By contrast, clones of cells in which the heterochromatin has spread into the *white⁺* gene, however, will not express the gene and will be white. The result is red and white mottled, or variegated, eyes.

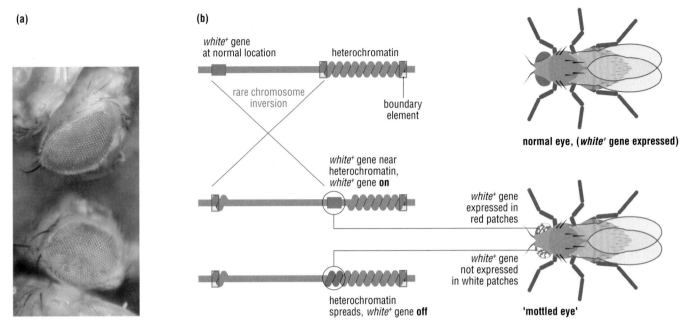

Figure 4.26 Position-effect variegation in *Drosophila*. (a) The normal *Drosophila* eye (top) and a mottled eye due to position-effect variegation (bottom). (b) Wild-type fruit flies that express the *white*⁺ gene have red eyes. The *white*⁺ gene is normally located in euchromatin. There is a nearby region of heterochromatin that is flanked by boundary elements (open boxes). In some mutants, a chromosome inversion occurs that places the *white*⁺ gene adjacent to the heterochromatin and removes the boundary element between them. In some eye cells, the *white*⁺ gene continues to be expressed and the cells are red, but in other cells the heterochromatin encroaches on the *white* locus due to loss of the boundary element. In these cells, *white*⁺ is silenced, giving rise to patches of white cells in the eye. Figure 4.26a reproduced from Lippman and Martienssen (2004).

This phenomenon is called position-effect variegation because it depends on the position of the gene on the chromosome. The inherited alteration in gene expression is an example of an epigenetic effect, since, unlike a mutation, the DNA sequence is no different in the two cell types although the expression state of the gene is.

Boundary elements can affect chromatin modification

Boundary elements that apparently stop the spreading of heterochromatin along the chromosome have been identified in a variety of different settings. A direct connection between the spreading of heterochromatin and post-translational modification of histone proteins has been found in the fission yeast, *S. pombe*. In this organism, two boundary elements border a region of silent heterochromatin around the centromere. The chromatin between the two boundary elements contains a high level of histone H3 methylated at lysine 9 (see Section 4.4), indicating that this region is silenced (Figure 4.27). In contrast, the active chromatin to the

Figure 4.27 Boundary elements mark transitions in different states of histone modification. The *S. pombe* heterochromatic centromere region (dark blue) on chromosome 3 is bounded by two boundary elements. Outside this region, the expressed mating-type locus (*mat1*; in red) and several other expressed genes (red) are located in euchromatin, as indicated by the methylation of lysine 4 on histone H3. Between the boundary elements, shown as open boxes, the DNA is packaged into heterochromatin. The *mat2* and *mat3* loci located in this region are silenced and histone H3 is modified by methylation at lysine 9.

left and right of the silent region has a high level of lysine 4-methylated histone H3. Deletion of the left boundary element results in the leftward spreading of H3 lysine 9 methylation and chromatin silencing, but does not affect the chromatin to the right of the other boundary. Similarly, deletion of the right boundary element results in spreading of the heterochromatin and H3 lysine 9 methylation to the right. Knowing that boundary elements mediate these effects, the obvious next question is: how?

Boundary elements act by binding specific proteins

There are a number of models that can explain heterochromatin spreading and how the propagation of heterochromatin along the chromosome can be delimited by a boundary element. The best-understood boundary elements are specific DNA sequences that bind those proteins that regulate the spreading of histone-modification activity. Heterochromatin can spread by the progressive modification of one nucleosome after the other until a boundary element is reached (Figure 4.28a). Heterochromatin spreading can be initiated by deacetylation of histone H3, followed by methylation of H3 at lysine 9 and subsequent binding of a chromodomain-containing chromatin-silencing protein, such as Swi6 in yeast or HP1 in animals, and/or recruitment of other silencing factors (Figure 4.28a). Boundary elements act as physical barriers to this progressive histone modification, inhibiting modification beyond the boundary element itself.

Another way in which boundary elements could function is to separate large defined looped domains of chromatin, as depicted in Figure 4.28b. By anchoring the domains at the base of the loops, they could delimit distinct, separate regions of the chromosome that may differ in chromatin structure and, possibly, in the relative amount of transcription that occurs in that region. There is also evidence that, in some cases, proteins that specifically bind to boundary elements anchor that region of the chromosome to the nuclear periphery, which is followed by

(a)

(b)

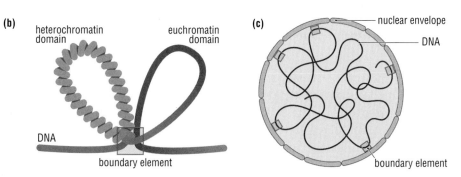

Figure 4.28 Possible ways in which boundary elements might function. (a) Some boundary elements are sites that stop the propagation of progressive histone modification. Histones in euchromatin can become methylated (dark blue triangles), which then allows binding of heterochromatin-inducing proteins such as Swi6 (green balls). Binding of Swi6 facilitates the binding of more Swi6 and thus a region of DNA can become progressively incorporated into heterochromatin. At the boundary element, this propagation stops. The chromatin on the other side can be active euchromatin, with acetylated histones (pink squares). (b) Boundary elements might be sites that organize chromatin into large looped domains that differ in structure. All of the chromatin in one loop will be heterochromatin and in a separate loop it will all be euchromatin. (c) Another suggestion is that boundary elements are sites at which DNA is attached to the nuclear envelope. This is supported by the fact that most heterochromatin is found around the periphery of the nucleus.

(a)

(b)

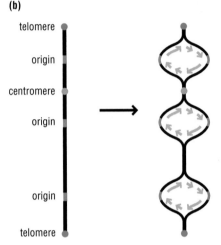

Figure 4.29 Replication origins initiate bidirectional replication. (a) A circular bacterial chromosome has a single origin (red), which initiates replication that proceeds in both directions around the chromosome and terminates at the *ter* region (yellow). (b) The multiple origins on a linear eukaryotic chromosome initiate bidirectional replication independently at many sites along the chromosome. Chromosomes also contain a centrosome (in orange) needed for chromosome segregation, and telomeres at chromosomes end (turquoise circles) that protect the ends from attrition.

➲ Replication origins and the mechanism and regulation of DNA replication are described in more detail in Chapter 6.

➲ The reason the very end of the chromosome cannot be copied is discussed in Chapter 6.

silencing of the genes in that region. Heterochromatin is often found at the nuclear periphery, which suggests that the relocation of a chromosome segment to the nuclear periphery triggers heterochromatin formation (Figure 4.28c).

4.9 ELEMENTS REQUIRED FOR CHROMOSOME FUNCTION

In earlier sections, we saw how chromosomes comprise two key components – DNA and histone proteins. These two components assemble to form the nucleosome, the fundamental 'unit' of chromosome packing. However, there are other structural elements of a chromosome that are central to its correct behavior. In this section, we explore these elements – the origin of replication, the centromere, and the telomere.

All chromosomes contain specialized regions that direct essential chromosomal functions

Chromosomes contain special features that allow them to be duplicated during DNA replication, and to ensure that the resulting copies of the chromosomal DNA are segregated into the two daughter cells at cell division.

Replication in both bacteria and eukaryotes initiates at the **origins of replication** (*ori*). In bacteria, sequences near the origin of replication also governs the distribution of the duplicated chromosomes to the two daughter cells, while a specific **terminus region** (*ter*) specifies termination of replication. In eukaryotes, the **centromere** directs chromosome segregation. In addition, specialized regions called **telomeres** at the ends of eukaryotic chromosomes stabilize the ends and assure the maintenance of the most terminal portions of the DNA. We will briefly introduce replication origins and terminus regions here, and will discuss telomeres and centromeres in detail in the next two sections.

DNA replication initiates at specific chromosomal sites termed origins

For DNA replication to begin, the two strands of the DNA double helix must be unwound and separated to allow the replication machinery access to the DNA. Origins of replication (*ori*) are the sites at which this unwinding occurs. Once initiated, DNA replication is bidirectional, proceeding away from the origin in both directions. In bacteria, DNA replication initiates at a single site. In eukaryotes, whose chromosomes are linear and typically much larger than bacterial chromosomes, DNA replication initiates at multiple sites along the chromosome (Figure 4.29). Origins of replication are typically found in regions that do not contain genes.

In *E. coli* and other bacteria, a distinctive DNA sequence specifies the single origin of replication. Although the exact sequence differs in different bacteria, in all cases it provides a site to which specialized initiator proteins bind (Figure 4.29). Replication origins in eukaryotes are often, but not always, defined by a particular DNA sequence. In budding yeast, replication origins are defined by a relatively short, specific DNA sequence. By contrast, in *Drosophila*, humans, and other multicellular organisms, initiation does not always occur at a unique origin sequence. In these cases, specific histone modifications, rather than a unique DNA sequence, may determine where DNA replication begins.

DNA replication of linear chromosomes poses a special problem

In linear chromosomes, bidirectional replication from internal origins allows copying of all of the sequences in the internal part of the chromosome (Figure 4.29b). However, when the replication fork gets to the very end of the DNA molecule, there is a small region on one strand of DNA that cannot be copied (This end replication problem is discussed in detail in chapter 6).

Thus, when there is no mechanism to compensate for this, the chromosome gets shorter every time the DNA is replicated. Telomeres provide a mechanism of overcoming this end-replication problem, and also protect the chromosomes from degradation and from fusing end-to-end with other chromosomes, as we shall soon see.

4.10 THE CENTROMERE

The centromere is a region of DNA necessary for chromosome segregation

Centromeres are components of all eukaryotic chromosomes that are necessary for correct chromosome segregation at cell division. In most species, each chromosome has a single centromere, which is packaged into heterochromatin and whose structure is stably inherited from one cell generation to the next. In mitotic chromosomes, in which duplication of DNA strands has occurred, but sister chromatids have yet to separate, the centromere appears as the site where paired sister chromatids are constricted (Figure 4.30).

The centromeric DNA binds specific proteins that form a structure called the **kinetochore**. During mitosis, kinetochores of sister chromosomes attach to microtubules emanating from opposite spindle poles. This attachment allows the spindle to pull sister chromatids apart and distribute them to the two daughter cells (see Figure 4.31).

Figure 4.30 Duplicated chromosomes consist of two sister chromatids still joined together. Diagram of two sister chromatids after duplication but before the identical duplicated copies have separated.

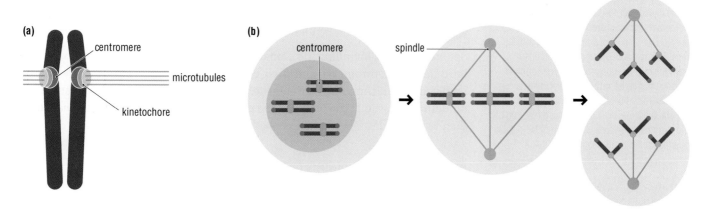

Figure 4.31 Chromosome segregation in eukaryotes. (a) The kinetochore mediates the attachment of spindle fibers to the centromere. (b) At mitosis, the centromere of each duplicated chromosome becomes attached to the mitotic spindle (here represented in highly schematic form in green) and the two sister chromatids are pulled to opposite poles of the cell. Telomeres are shown as turquoise circles.

➡ Kinetochore structure and function is discussed in Chapter 7.

Figure 4.32 Centromeres differ in size and DNA sequence in different species. *S. cerevisiae* centromeres are about 125 bp long and consist of three DNA elements CDE I, II, and III (dark/light blue). *S. pombe* centromeres are 40–100 kb in size and are made of many repeated sequence elements (small purple arrows) that make up larger repeats (large dark blue arrows). Between the two large inverted repeats is a core region of unique sequence, which differs between chromosomes. *Drosophila* centromeres are around 420 kb in size and are made of repetitive sequence; they include many transposable elements (yellow lines). Human centromeres are 100 kb to 4 Mb in size and consist of tandem arrays of alpha-satellite repeats (small light green arrows) grouped into higher-order repeats (large green arrows). *C. elegans* chromosomes are holocentric: a centromere can be established at any point along the chromosome. In all organisms, a homolog of the mammalian histone H3 variant CENP-A (blue quarter-circles) binds centromeric DNA.

➡ The identification of individual chromosomes through the method of fluorescence *in situ* hybridization (FISH) is covered in Chapter 16.

The segregation of chromosomes through the attachment of centromeres to a mitotic spindle is an essential conserved function in all eukaryotes. Yet, remarkably, centromeres of different species have distinct DNA sequences. Instead, it is the protein components of the centromeric chromatin and kinetochores that are conserved. Here we will discuss properties of centromeres.

Centromere sequence and size vary widely between species

Centromeres in different organisms vary not only in sequence but also in size (Figure 4.32). *S. cerevisiae* and some other yeasts have very small, defined DNA sequences at the centromere; these are termed **point centromeres** because of their small size and simple organization. The way in which yeast centromeres were first isolated is described in Experimental approach 4.3. The *S. cerevisiae* centromeric DNA is 125 bp long and is divided into three distinct elements (centromere DNA elements (CDE) I, II, and III). These elements are sufficient to provide centromere function when incorporated into another DNA molecule. CDE I and III each consist of a unique conserved sequence. Mutations in these elements disrupt centromere function. CDE II consists of 80–90 bp of AT-rich DNA but, unlike CDE I and III, the exact sequence is not important to its function. In other yeast species with point centromeres, the DNA sequence differs from that in *S. cerevisiae*.

Regional centromeres, in contrast, consist of hundreds of kilobases (kb) of repetitive DNA sequence. The fission yeast *S. pombe*, the fruit fly *Drosophila*, and mammals all have regional centromeres. Each of the three distinct chromosomes of *S. pombe* has a slightly different centromere DNA sequence, ranging in size from 40 kb to 100 kb. Each centromere has a large inverted repeat sequence, which itself consists of smaller repetitive sequences, surrounding a unique central core sequence of 5–6 kb (Figure 4.32). The core sequence differs on each chromosome, while the repetitive sequence is similar.

The core alone is not sufficient to establish centromere function, indicating that the unique sequence element of a regional centromere is not sufficient for its function, in contrast to point centromeres. *Drosophila* centromeres are about 400 kb in size and also consist mostly of repetitive DNA. Human and mouse centromeric DNA consists of about a megabase (1 million bases) of repetitive sequence termed **alpha (α)-satellite** repeats. It is unrelated to the sequence in *S. pombe* or *Drosophila*, and consists of tandem repeats of 171 bp, which are organized into an array of higher-order repeats (Figure 4.32). Although the sequences of alpha-satellite repeats are all very similar, each chromosome contains its own specific variant and, indeed, can be identified by it.

Some organisms, including the nematode (worm) *Caenorhabditis elegans*, do not have defined centromeres and their chromosomes are said to be **holocentric**. Rather than attaching at a unique site, multiple spindle microtubules attach at numerous sites along the length of the chromosome (Figure 4.32). It appears that there is no specific DNA sequence that defines a *C. elegans* centromere.

Centromeres contain a histone H3 variant called CENP-A

Despite dramatic sequence differences, centromeres of all organisms have a defined region in which nucleosomes contain a histone H3 variant, known as CENP-A (or CENH3, in place of histone H3). CENP-A shows specificity for centromeric regions,

4.3 EXPERIMENTAL APPROACH

The isolation of a yeast centromere

Although centromeres had been seen in cytological studies over a hundred years ago and had long been studied using genetic approaches, their molecular structure was first uncovered by Clarke and Carbon in 1980. These scientists used a combination of genetics and molecular biology to isolate the first centromere sequence from the yeast *S. cerevisiae*. The key to the success of this experiment was their development of an assay for centromere function (Figure 2). Their assay relied on a type of plasmid known as an ARS plasmid (for autonomously replicating sequence) that has a replication origin. When ARS plasmids are introduced into yeast, they can replicate, but do not segregate when the cell divides. As a consequence, the plasmids accumulate in the mother cell, and the daughter cells do not receive the plasmid at cell division. Clarke and Carbon reasoned that if a DNA element that conferred centromere function was added to the plasmid, it would segregate at mitosis so that all of the cell progeny would have the plasmid. This assay turned out to work tremendously well.

To start the hunt for centromere DNA, Clarke and Carbon focused on a region of chromosome III where the approximate location of the centromere was known from genetic studies. They had in their collection DNA clones for two genes that flanked the centromere: LEU2 mapped to the left and CDC10 mapped to the right of the centromere on the genetic map (Figure 1). These experiments were done long before the genome sequence was known, so they had to find a way to isolate DNA fragments between these two known markers. They employed a technique called chromosome walking, in which successive overlapping clones are isolated that span the region of the DNA between the two known markers. Once they had a set of clones that they suspected might contain a centromere, they tested each clone in their plasmid segregation assay (Figure 2). Indeed, three independent clones conferred mitotic segregation on the usually unstable ARS plasmids, these clones were then further characterized.

To identify the sequence responsible for 'centromere' function, they tested smaller and smaller pieces of the initial DNA candidates that they had cloned, finally narrowing in on a 1.6 kb sequence that was sufficient to confer mitotic stability. When this region was deleted from the plasmid, the plasmid no longer segregated to both mother and daughter cells. Moreover, in addition to conferring plasmid segregation at mitosis, this piece of DNA also conferred proper segregation at meiosis. We see how the careful development of a sensitive assay, coupled with the relatively straightforward procedure of cloning, paid off to yield results of longstanding significance.

Clarke L, Carbon J. Isolation of a yeast centromere and construction of functional small circular chromosomes. *Nature* 1980;**287**:504–509.

Figure 1 Genetic and physical map of the centromere region of chromosome 3. The top line represents the genetic map near CEN3. The line below represents the physical map of the region. The lines below the physical map show the DNA fragments that were assayed to test for centromere activity. The orange circle represents the region carrying functional centromere sequences.

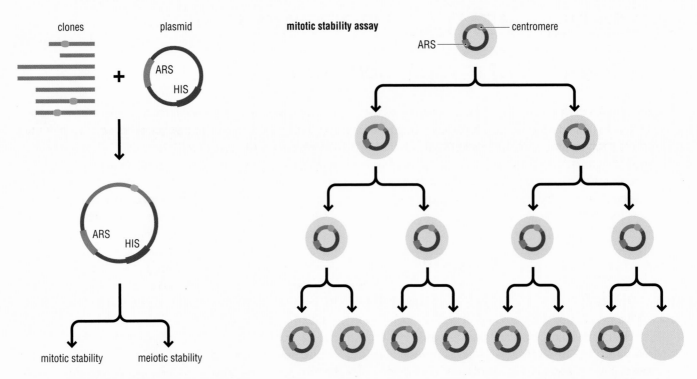

Figure 2 Identification of functional centromeric DNA. Fragments of yeast DNA were cloned into a plasmid containing an ARS (red) and a selectable marker (HIS (blue), which allows certain yeast cells to grow on growth media lacking histidine.). Each clone was tested for its ability to segregate in a mitotic stability assay. For those clones that have a centromere (orange), the plasmid segregates at mitosis and all of the progeny contain the plasmid. In contrast, most of the progeny of cells that do not have the plasmid will lack the selectable marker, HIS, and thus will not grow when plated on media that lack histidine. Occasionally, the plasmid fails to replicate or segregate, resulting in a cells lacking the plasmid (right most cell in the bottom row).

and particularly the AT-rich domain. The region of the centromere demarked by the presence of CENP-A is the domain onto which kinetochore proteins assemble. The presence of CENP-A is crucial for centromere function: cells that lack CENP-A fail to recruit most known kinetochore components to the kinetochore. We do not yet know how this histone variant gets incorporated only at centromeric sites, but it is clear that the levels of CENP-A are carefully regulated, as abnormally high CENP-A protein levels can lead to binding at non-centromeric sites and disrupt kinetochore function. This disruption may result from drawing kinetochore proteins away from their normal sites of assembly, or by directing assembly of an ectopic centromere (meaning a centromere in an abnormal location) that causes chromosome breakage and mis-segregation during mitosis.

Chromatin structure defines centromere function

While the function of many chromosomal elements depends on their DNA sequence, surprisingly it is chromatin structure, not DNA sequence, that defines a functional centromere. This was initially suggested by two observations. First, in human cells, not all large tracts of alpha-satellite sequence form a functional centromere. Human chromosomes that are rearranged to generate two centromeres, and chromosomes into which an extra large tract of alpha-satellite DNA is introduced, often contain just one active centromere, while the second potential centromere is inactivated through changes in a loss of centromeric protein chromatin structure and the proteins it contains. This **centromere inactivation** prevents the chromosome breakage that would occur if two centromeres on the same chromosome were pulled to opposite poles at cell division.

The second observation that chromatin structure defined centromeres was the discovery of new functional centromeres, or **neocentromeres**, that can form on a DNA fragment that has no other active centromere (Figure 4.33). This can occur in a chromosome that has lost its centromere through a deletion of that segment of DNA. Neocentromeres can form at sites that do not contain alpha-satellite DNA, but can nevertheless assemble a functional kinetochore.

An important hallmark of centromeres is the presence of heterochromatin, where processes such as gene expression and recombination are often repressed (see Section 4.4). In the centromeric regions that flank the CENP-A-containing domain, histone H3 and H4 are both hypoacetylated and hypermethylated, which is important for heterochromatin formation and proper centromere function.

The chromatin structure of centromeres is inherited epigenetically, that is, the sites at which centromeres are formed on newly replicated DNA are not solely determined by the DNA sequence of the parent chromosome. Instead, the new centromeres form at sites determined by the location of centrosomes on chromosomes in the parent cell. An epigenetic mechanism can establish and maintain centromeres (Figure 4.34). It has been proposed that repetitive, AT-rich DNA, such as alpha-satellite, has a high probability of binding CENP-A to form a centromere-specific chromatin structure. Thus, alpha-satellite sequences found at all centromeres are most likely to form centromeres. At a low probability, however, CENP-A can bind to a DNA sequence other than alpha-satellite and establish a centromere, thus forming a neocentromere. This newly established functional centromere is self-propagating, irrespective of its DNA sequence, which is why neocentromeres are stably inherited at each cell division. Self-propagation of centromere structure is achieved by epigenetic mechanisms for maintaining heterochromatin, such as histone modification and the recruitment of specific proteins described in Section 4.4. This inheritance of chromatin structure will be discussed in more detail in Chapter 6.

4.11 THE TELOMERE

Having now considered centromeres, the specialized structures that appear within the chromosome, let us move on to consider a specialized element that appears at the end of the chromosome – the telomere.

Telomeres protect the ends of linear chromosomes

Telomeres are specialized structures that define the ends of eukaryotic chromosomes and are essential for their stability. Telomeric DNA consists of simple tandem repeated sequences that have one strand that is G-rich (containing many G residues) and one strand that is C-rich. The **G-rich strand** in all eukaryotes is the DNA strand that extends 5′ to 3′ towards the chromosome end and terminates

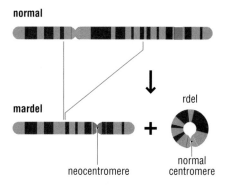

Figure 4.33 Neocentromeres can form away from the usual centromere region. A normal human chromosome 10 is shown at the top with the centromere marked in orange. A deletion removes the centromere, and the region marked in pink becomes a neocentromere; the new chromosome is known as mardel (marker deletion). The region of DNA containing the centromere that was deleted can circularize and form a separate chromosome known as rdel (ring deletion). Neocentromeres appear to be generated very rarely. The formation of neocentromeres is considered an epigenetic phenomenon, as the function of the DNA is altered in a heritable manner without any alteration in its DNA sequence.

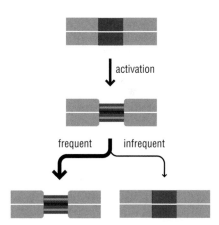

Figure 4.34 Centromeres are formed and lost by epigenetic mechanisms. A region of DNA, usually comprising alpha-satellite sequence, can become activated to form a centromere (striped blue and gray), probably by binding CENP-A. With each cell division there is a high probability that this centromere will self-propagate and a new centromere will form at that location on the daughter chromosomes. Very infrequently, however, the centromere will be inactivated (solid blue box), self-propagation will not occur, and a functional centromere will not be established at that location.

Telomere sequence repeats		
group	**organism**	**repeat sequence**
ciliates	*Tetrahymena*	TTGGGG
flagellates	*Trypanosoma*	TTAGGG
fungi	*S. cerevisiae*	TG$_{1-3}$
	S. pombe	TTAC(A)G$_{2-5}$
nematodes	*C. elegans*	TTAGGC
plants	*Arabidopsis*	TTTAGGG
vertebrates	Human	TTAGGG

Figure 4.35 Telomere repeat sequences in various organisms. The telomeric DNA in most eukaryotes consists of tandem arrays of a simple repeat. Only the G-rich strand of the DNA sequence is shown for simplicity. The G-rich strand always runs 5′ to 3′ toward the end of the chromosome. At the very end of the telomere is a 30–100 bp overhang on the G-rich strand.

in a small single-stranded region (Figure 4.36a). Thus, while most of the telomere repeats are double-stranded, there is a small G-rich overhang that is important for telomere function. Representative repeat sequences from different organisms are listed in Figure 4.35; human telomeres, for example, consist of many repeats of the sequence TTAGGG. The total length of this repetitive DNA region varies from about 50 bp to 30 000 bp, depending on the species.

In some organisms with long telomeres, the G-rich overhang may be processed into a t-loop (Figure 4.36b). This loop is formed by base-pairing of the single-stranded G overhang with the double-stranded region of the telomere DNA. These t-loop structures are thought to protect the very end of the chromosome, though their precise role is not yet understood.

The telomere repeats are binding sites for specific proteins that mark the chromosome ends as natural ends and thereby distinguish them from DNA breaks. This prevents the inappropriate activation of the cell's DNA damage response, which often leads to cell death. If either the telomeric DNA sequences or the proteins that bind them are defective, the cell will either die or accumulate chromosomal abnormalities. Telomeres are also the means of overcoming the end-replication problem, which, as we have already mentioned (Section 4.9), arises from the inability of DNA polymerase to copy the ends of a linear DNA molecule completely. While the bulk of the double-stranded telomere sequence is replicated by conventional DNA replication machinery, the extreme ends are maintained by a specialized enzyme called **telomerase**, which adds new telomere sequences to the ends of chromosomes.

⊘ The telomerase mechanism is described in more detail in Chapter 6.

Telomeres are maintained by the enzyme telomerase

Telomerase is an unusual DNA polymerase that has both a protein component and an essential RNA component. The RNA contains a short template region that specifies the sequence of the telomere repeats that are added onto the chromosome ends. This protein component synthesizes the telomeric DNA, using the RNA component as a template.

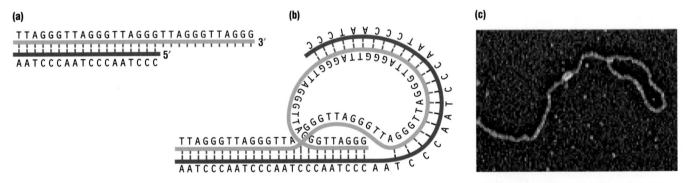

Figure 4.36 Linear and t-loop structure of human telomeres. Telomere DNA may be linear or may form a structure known as a t-loop. (a) In the linear state there is a G-rich overhang that is bound and protected by end-binding proteins (see text). (b) The t-loop forms when the single-stranded G-rich tail is base-paired with the adjacent double-stranded DNA. (c) Electron micrograph showing t-loop.
Figure 4.36c reproduced from the Laboratory for Cell Biology and Genetics, Rockefeller University, New York.

In humans, telomerase is active – and telomere length is maintained – in stem cells that divide frequently to maintain tissues in the body and in germline cells that produce the sperm and eggs for the propagation of the species (Figure 4.37a). Similarly, telomerase is active in unicellular eukaryotes such as yeast, which divide indefinitely.

In many human tissues, there is not enough telomerase to maintain telomere length and cell division, and progressive telomere shortening therefore occurs. The importance of telomere maintenance is highlighted by the fact that mutations in telomerase cause inherited diseases in humans such as bone marrow failure, because short telomeres limit the number of divisions cells can undergo. Inappropriate activation of telomerase, on the other hand, can allow cells to continue to grow when they should not, as occurs in cancer. Indeed, telomerase is active in more than 90% of human tumors, and is required to enable tumor cells to divide indefinitely. Drugs that inhibit telomerase are therefore being tested as treatments for cancer.

Although telomerase activity is the principal mechanism for telomere maintenance, some species use alternative mechanisms. *Drosophila*, for example, has two transposable elements that compensate for sequence loss by repeatedly transposing new elements onto chromosome ends. Some other insects maintain their chromosome ends through DNA recombination, which can copy or transfer sequences between two chromosomes.

In other organisms, recombination mechanisms are used. Such recombination is also used as a backup mechanism when telomerase is absent. For example, when telomerase is artificially inactivated in yeast, the telomeres become shorter at each cell division until the cells stop dividing. A few cells, however, continue to grow by using a specialized form of recombination to elongate the telomeres (Figure 4.37b), and these cells can therefore grow indefinitely despite the absence of telomerase.

Telomere-binding proteins are essential for telomere function

Telomeres contain a variety of proteins that not only physically protect the ends of chromosomes, but also help maintain the telomere at a particular length. There are three classes of telomere-binding proteins. End-binding proteins bind to and protect the G-rich overhang at the extreme terminus, double-stranded DNA-binding proteins bind along the length of the telomere sequence, and there are also proteins that associate with the telomeric DNA-binding proteins (Figure 4.38).

Telomere-binding proteins regulate the length of telomeres; overexpression of the double-stranded DNA-binding telomere proteins results in shorter telomeres than normal, whereas decreasing their binding results in telomere lengthening. These telomere-binding proteins are thought to regulate the access of telomerase to the telomere, although we do not yet fully understand how this occurs.

Although a major role of telomeres is to distinguish natural chromosome ends from DNA breaks, some of the proteins found at telomeres are, surprisingly, recombination proteins normally involved in repairing double-strand breaks.

⊃ We learn about DNA recombination in Chapter 12.

(a) Telomerase

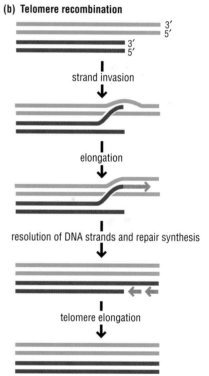

(b) Telomere recombination

Figure 4.37 Both telomerase and recombination can lengthen telomeres. (a) Telomeres contain a single-stranded overhang region on the G-rich 3′ strand. The 3′ end can be elongated by telomerase. The C-rich strand can then be filled in by DNA polymerases. (b) Recombination can lead to the net elongation of telomeres. The sequence at the ends of chromosomes is highly repetitive and is identical on all telomeres in a given species. The 3′ end of one chromosome can invade a second chromosome and use the sequence of the second chromosome as a template for telomere elongation by DNA polymerase. The lagging strand can then be filled in by conventional replication mechanisms.

Figure 4.38 Human telomere-binding proteins. Most organisms have three classes of telomere-associated proteins that carry out telomere function: end-binding proteins, double-stranded DNA-binding proteins, and telomere-associated proteins that do not bind the DNA directly but are associated with other bound proteins. The exact proteins involved vary from organism to organism: the human telomere proteins that comprise the Shelterin complex are shown here. Pot1 (turquoise ovals) is bound to the single-stranded G-rich tail DNA and is also associated with TPP1. TRF1 (dark blue elongated ovals) and TRF2 (orange elongated ovals) are bound along the length of the double-stranded telomeric DNA. Rap1 (orange circle), Tin2 (red ovals). The double-stranded DNA is organized into nucleosomes, not shown here for simplicity.

The exact role of these proteins at telomeres is not yet clear, but they might be involved in processing the DNA at the very end of the telomere to allow access for telomerase or other telomere-binding proteins.

Loss of telomere function leads to a DNA-damage response and chromosomal instability

Telomere function can be disrupted in a variety of ways. Shortening the telomere, for example, leads to loss of binding sites for proteins. The direct inhibition of protein binding will also leave the chromosome ends unprotected. In all cases, the unprotected end can be recognized by the cell as a DNA break, which activates cellular DNA-damage pathways, causing arrest of the cell cycle or cell death. On rare occasions, the exposed end may recombine with DNA on another chromosome, leading to translocation of genetic material and genetic abnormalities; or it may fuse with another chromosome end. Chromosomes fused end-to-end are unstable because they often contain two functional centromeres. If the centromeres are pulled to opposite poles at cell division, the chromosome will break and the daughter cells will inherit broken chromosomes. This will, in turn, lead to further chromosome rearrangements because the broken ends of different chromosomes may become fused to one another.

4.12 CHROMOSOME ARCHITECTURE IN THE NUCLEUS

We end this chapter by exploring the location of chromosomes in the context of the cell as a whole. It is tempting to view the nucleus as a vessel in which chromosomes move around freely to occupy random locations. However, the reality is rather different.

Chromosomes occupy specific territories in the eukaryotic nucleus

During interphase, genes are being actively transcribed in the nucleus to provide the cell with the macromolecules it needs to carry out a myriad of functions. Within the interphase nucleus, each chromosome occupies a characteristic, discrete space termed a **chromosome territory** (Figure 4.39). This organization is remarkable because each interphase chromosome, which is in a relatively extended state, is hundreds of times longer than the diameter of the nucleus. One might imagine that in packing this much DNA into a nucleus, the chromosomes would

Figure 4.39 Chromosomes are localized to defined territories in the interphase nucleus. The chromosomes in a nucleus from a cultured chicken cell are stained with chromosome-specific fluorescently labeled probes, as in the karyotype shown in Figure 4.2c. An optical section through the nucleus shows each chromosome stained with a different color. Each copy of the five chromosomes is localized to a distinct domain within the nucleus (as the cell is diploid, there are two copies of each chromosome).
From Cremer and Cremer, *Nature Reviews Genetics* 2001;**2**:292–301.

be jumbled together like worms in a bag, instead of the organized distribution that is actually found.

The mechanism by which chromosome territories are established and maintained is not yet clear, although their locations may represent where each chromosome came to rest after mitosis and decondensed, with little further movement. Indeed, in some tissues the three-dimensional arrangement of chromosomes during mitosis is quite clearly maintained in the resulting nuclei, with all the centromeres located at one end and the telomeres at the other. This reflects the arrangement the chromosomes had during segregation as they were pulled to a pole of the cell, with the centromeres in front and the chromosome ends trailing behind.

Even though the actual location of a chromosome in the nucleus during interphase appears not to change markedly, the size of the territory can vary. When transcription of a gene on a given chromosome is activated, for example, the volume of the chromosome's territory increases. This probably reflects further unpacking of the chromatin to allow the transcription machinery to access the DNA.

The location of a chromosomal region is correlated with transcriptional activity

It is not just the location of a chromosome, as a complete entity, within the nucleus that follows a characteristic pattern. In addition, particular regions *within a single chromosome* can be found in characteristic locations. Specifically, the nuclear location of a chromosomal region is correlated with its transcriptional activity. Heterochromatic regions, in which genes are silenced, are often associated with the nuclear envelope (see Section 4.7). Electron micrographs first revealed the dense-staining heterochromatin around the periphery of the nucleus (see Figure 4.11), and many subsequent studies have shown that heterochromatin is typically located at the nuclear periphery. In female mammals, for example, the heterochromatic inactive X chromosome is always located at the nuclear periphery. In yeast, the telomeres, which are heterochromatic, cluster together in distinct foci at the nuclear periphery. This peripheral location is associated with the silencing of genes placed at telomeres and with the late firing of origins of replication located in telomeric DNA. In human cells, late-replicating DNA, another hallmark of heterochromatin, is also located at the nuclear periphery.

Figure 4.40 The location of human chromosomes 18 and 19 within the nucleus. Human chromosomes 18 and 19 are of a similar size but chromosome 18 has many fewer genes. Chromosome 18, shown colored in red, is reproducibly located at the periphery of the nucleus, while chromosome 19, shown in green, is located more centrally. From Chubb and Bickmore (2003).

Certain intrinsic chromosomal properties are also correlated with their location. In human cells, for example, the density of genes on a chromosome appears to correlate with its nuclear location. Human chromosomes 18 and 19 are of similar size, but differ greatly in their number of genes. The gene-rich chromosome 19 is reproducibly located in the interior of the nucleus, while the gene-poor chromosome 18 is located near the nuclear periphery (Figure 4.40). The significance of the central location of gene-rich chromosomes is not yet clear.

Chromosomal regions can be silenced if placed adjacent to an area of the nucleus occupied by centromeric heterochromatin

A gene does not need to be physically integrated into heterochromatin for it to be silenced. Simply bringing a gene into the proximity of a region of centromeric heterochromatin, that is, *in trans*, can have the same effect. This has been demonstrated in *Drosophila*, in which, unlike mammals, homologous chromosomes remain paired with one another throughout the mitotic cell cycle.

The *brown* locus in *Drosophila* is at the opposite end of the chromosome from the centromeric heterochromatin. An inactive mutant form of this gene, known as *brown*[D], arises from an inappropriate insertion of a block of heterochromatin in the vicinity of the *brown* locus. In cells that contain one wild-type *brown* gene on one homolog and *brown*[D] on the other, both are inactive. This inactivation of the wild-type copy seems to be due to the relocation of the chromosomal regions containing wild-type and mutant *brown* loci to a site in the nucleus adjacent to the centromeric heterochromatin (Figure 4.41). The heterochromatic region inserted in the mutant gene directs this relocation in some way. Maternally and paternally inherited homologs are generally thought to control gene expression independently of one another in mitotic cells. Thus, *trans* silencing and concerted relocation of both loci reveals an interaction between homologs that is poorly understood.

Human cells show a similar phenomenon. When human B lymphocytes switch from a resting state to active cell division, certain genes move from their typical nuclear locations in resting cells to a location adjacent to centromeric heterochromatin, and are thereby silenced. It is not yet clear whether silencing causes the relocation or the relocation causes the silencing. Conversely, activation of the beta-globin locus in developing human red blood cells is associated with its relocation away from centromeric heterochromatin. We are some way from developing a clear, unambiguous picture of the mechanism behind this silencing.

The nucleolus is a nuclear substructure that forms around the ribosomal RNA genes

Electron micrographs show a prominent feature inside the nucleus called the **nucleolus** (Figure 4.11), which is where actively transcribed rRNA genes are found.

Figure 4.41 Proximity to centromeric heterochromatin in the nucleus can silence genes in *Drosophila*. (1) The paired *brown* gene loci (blue) are usually located far away from the nuclear region occupied by the centromere heterochromatin on the same chromosome (orange). (2) Insertion of a small region of heterochromatin (pink) into one of the two *brown* genes, which inactivates the gene, causes movement of the chromatin containing the paired loci to a new nuclear region adjacent to the centromere heterochromatin. This relocation results in the silencing of the active copy of the *brown* gene.

translocation of heterochromatin to brown locus

paired brown loci relocalize to centric heterochromatin

The ribosome is the large nucleoprotein complex that synthesizes proteins and contains several large RNA subunits termed rRNA, which are encoded by genes called the rDNA. In addition to containing the rDNA along with the proteins, the nucleolus is also rich in other macromolecules required for ribosome assembly. These include enzymes required for processing the RNA molecules after they are transcribed, as well as ribosomal proteins. Although the nucleolus appears in electron micrographs to have distinct boundaries (see Figure 4.11), it lacks a membrane. Instead, it is the proteins and RNA that are specific to the nucleolus, and the way in which these molecules are stained in electron microscopy, that gives the nucleolus its distinctive appearance.

In order to meet the cell's need for new ribosome synthesis, many organisms have multiple copies of each gene encoding an RNA subunit. The total number of repeats – termed rDNA repeats – varies among organisms and ranges from several hundred to 10 000. The rDNA repeats are arranged in clusters in the eukaryotic chromosome; in human chromosomes, there are ten sets of ribosomal gene repeats (five homologous pairs) (Figure 4.42). At the end of M phase, transcription of the rDNA commences and the nucleolus forms as a visible, distinct structure around these highly transcribed sets of genes. All ten sets of genes coalesce in one or more nucleoli, where the rRNA subunits are transcribed and then assembled into ribosomes.

The nucleolus is thus sometimes seen physically associated with the region of the DNA that contains the rRNA genes (Figure 4.42). This region of the chromosome where the nucleolus is associated is termed the nucleolar organizer region (NOR). The nucleolus usually disappears in early M phase and re-forms late in M phase, once the cell has divided, the chromosomes have decondensed, and transcription of various genes commences. In some cell types such as amphibian oocytes, the nucleolus remains visible during the early stages of chromosome condensation.

Despite the high level of transcription in the nucleolus, a subset of the rDNA repeats are transcriptionally silent and have the features of heterochromatin. These heterochromatic regions are found at the periphery of the nucleolus, while the actively transcribed rDNA are found within the nucleolus. In addition to silencing transcription, heterochromatin formation also prevents fragmentation of the rDNA repeats, whose many repeated DNA sequences make them prone to recombination and chromosome instability. Indeed, mutations in proteins that maintain rDNA silencing give rise to instability of the rDNA.

In later chapters, we will see how the activity of a gene can be controlled by sophisticated mechanisms that modulate transcription and translation. However, we saw throughout this chapter how there are numerous ways of controlling this activity at a more fundamental level – the physical structure of the chromosome in which the gene is located. This control is mediated using some elegantly simple cues, from the addition or removal of methyl groups and other chemical entities, to the physical location of the chromosome within the nucleus. These mechanisms show the ingenious ways in which biological systems have evolved to carefully regulate their activity, seemingly with no opportunity for control overlooked.

In this chapter, we learned about the basic features of chromosomes, including how genomic DNA is packaged by proteins into chromatin and how special regions of the chromosomes play a role in ensuring the transmission of genetic information to daughter cells. In subsequent chapters, we will explore in detail how these features impact all processes that involve the chromosomes.

➲ The structure of the ribosome is covered in Chapter 10.

➲ We learn in Chapter 8 about Sir2, a histone deacetylase in yeast that is required for rDNA silencing.

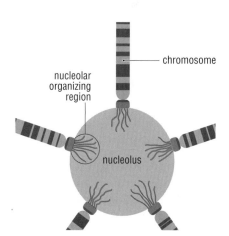

Figure 4.42 rDNA repeats and the nucleolus. In humans, there are rDNA repeats near the ends of five different chromosomes. These regions (called nucleolar organizing regions (NOR)) localize to the nucleolus, where the rDNA genes are actively transcribed. The figure shows five chromosomes: in a diploid cell there are two copies of each, giving a total of ten.

✱ SUMMARY

Chromosome basics

- An organism's genome is divided into one or more chromosomes.

- Each chromosome contains many genes embedded within a single DNA molecule.

- Between the genes lie stretches of intergenic DNA. There is great variation in the relative amount of intergenic DNA found in a given chromosome.

- A chromosome can be linear or circular.

- Each organism contains a characteristic number of chromosomes in each cell.

- Ploidy describes the number of identical sets of chromosomes a cell has. Diploid cells contain two sets of chromosomes in each cell, while haploid cells contain just a single set of chromosomes each.

- Chromosomes are duplicated and distributed to daughter cells in the course of the cell cycle, which is divided into four phases: G1, S, and G2 (known collectively as interphase), and M (the mitotic phase).

- Mitotic chromosomes are highly condensed and have a distinctive appearance, whereas chromosomes have a more diffuse appearance during interphase.

Chromatin

- Chromosomes in all organisms are associated with proteins that help to condense and organize the DNA molecules inside the cell.

- The compacted bacterial chromosome is known as the nucleoid.

- Eukaryotic DNA is packaged by histone proteins into chromatin.

- The basic building block of chromatin is the nucleosome, which consists of around 146 bp of DNA wrapped twice around the histone octamer in a left-handed manner.

- The histone octamer consists of two copies each of the histone proteins, H2A, H2B, H3, and H4. The histone proteins have flexible N-terminal extensions, known as 'tails', that extend between the turns of DNA.

- DNA packaged into nucleosomes adopts a 'beads on a string' appearance known as the 10 nm fiber. This is further compacted into a 30 nm fiber with the assistance of the linker histone, H1.

- Histone proteins are subject to numerous post-translational modifications, including acetylation, methylation, phosphorylation, ubiquitination, and sumoylation, which play key roles in regulating numerous processes that involve interactions with chromatin.

- Nucleosome remodeling complexes can alter the positions of nucleosomes using the energy from ATP hydrolysis.

- DNA can be methylated at cytosine bases.

- Interphase chromosomes contain regions of heterochromatin, which appear compact and stain vividly in the microscope and euchromatin, which appears more diffuse in the microscope and is more easily digested by nucleases.

- While transcribed genes lie primarily in euchromatic regions, it is now clear that there is some transcription in heterochromatic regions, as well.

- Boundary elements separate heterochromatic and euchromatic regions and can limit the spreading of heterochromatin along the chromosome.

Elements required for chromosome function

- DNA replication begins at the origin of replication. Bacterial chromosomes have a single origin of replication, while eukaryotic chromosomes can have multiple origins.

- The centromere is a region in the eukaryotic chromosome that is involved in the pairing of sister chromatids and chromosome segregation.

- The centromere is heterochromatic and contains specific variant histones (for example, CENP-A in place of histone H3).

- Telomeres are special structures at the ends of eukaryotic chromosomes that protect the chromosomes from degradation or end-to-end fusion, and prevent the shortening of chromosomes after many rounds of DNA replication.

- Telomeres consist of many repeats of a short DNA sequence (TTAGGG in human chromosomes). Each telomere is mostly double-stranded but also contains a single-stranded overhanging extension.

- The telomeric repeats are added to the ends of chromosomes by the enzyme telomerase, which has an RNA subunit containing the template for synthesizing the telomeric repeats.

Chromosome architecture in the nucleus

- During interphase, each chromosome occupies a distinct chromosome territory in the nucleus.

- Actively transcribed regions of the chromosome tend to be near the center of the nucleus, whereas heterochromatin is often associated with the nuclear envelope.

- Transcription of the rRNA subunits (encoded in the rDNA) takes place in the nucleolus, a distinct substructure in the nucleus. The nucleolus forms around the actively transcribed rDNA repeats.

- A portion of the repeated rDNA genes are found in heterochromatin, which prevents recombination among the repeats.

 FURTHER READING

4.1 ORGANIZATION OF CHROMOSOMES

Casjens S. The diverse and dynamic structure of bacterial genomes. *Annual Review of Genetics* 1998;**32**:339–377.

Duret L, Bucher P. Searching for regulatory elements in human noncoding sequences. *Current Opinion in Structural Biology* 1997;**7**:399–406.

4.2 THE CELL CYCLE AND CHROMOSOME DYNAMICS

Blow JJ, Tanaka TU. The chromosome cycle: coordinating replication and segregation. Second in the cycles review series. *EMBO Reports* 2005;**6**:1028–1034.

Diffley JF, Labib K. The chromosome replication cycle. *Journal of Cell Science* 2002;**115**:869–872.

Morgan DO. *The Cell Cycle*. London: New Science Press, 2006.

4.3 PACKAGING CHROMOSOMAL DNA

Travers AA. The structural basis of DNA flexibility. *Philosophical Transactions Series A Mathematical Physical and Engineering Sciences* 2004;**362**:1423–1438.

Wu C, Bassett A, Travers A. A variable topology for the 30-nm chromatin fibre. *EMBO Reports* 2007;**8**:1129–1134.

Wyman C, Kanaar R. Chromosome organization: reaching out to embrace new models. *Current Biology* 2002;**12**:R446–R448.

4.4 VARIATION IN CHROMATIN STRUCTURE

Bailis JM, Forsburg SL. It's all in the timing: linking S phase to chromatin structure and chromosome dynamics. *Cell Cycle* 2003;**2**:303–306.

Rusche LN, Kirchmaier AL, Rine J. The establishment, inheritance, and function of silenced chromatin in *Saccharomyces cerevisiae*. *Annual Review of Biochemistry* 2003;**72**:481–516.

4.5 COVALENT MODIFICATIONS OF HISTONES

Berger SL. The complex language of chromatin regulation during transcription. *Nature* 2007;**447**:407–412.

Ruthenburg AJ, Li H, Patel DJ, Allis CD. Multivalent engagement of chromatin modifications by linked binding modules. *Nature Reviews* 2007;**8**:983–994.

Taverna SD, Li H, Ruthenburg AJ, Allis CD, Patel DJ. How chromatin-binding modules interpret histone modifications: lessons from professional pocket pickers. *Nature Structural and Molecular Biology* 2007;**14**:1025–1040.

4.6 NUCLEOSOME-REMODELING COMPLEXES

Becker PB, Hörz W. ATP-dependent nucleosome remodeling. *Annual Review of Biochemistry* 2002;**71**:247–273.

Clapier CR, Cairns BR. The biology of chromatin remodeling complexes. *Annual Review of Biochemistry* 2009;**78**:273–304.

Mellor J. The dynamics of chromatin remodeling at promoters. *Molecular Cell* 2005;**19**:147–157.

4.7 DNA METHYLATION

Feinberg AP. Phenotypic plasticity and the epigenetics of human disease. *Nature* 2007;**447**:433–440.

Ideraabdullah FY, Vigneau S, Bartolomei MS. Genomic imprinting mechanisms in mammals. *Mutation Research* 2008;**647**:77–85.

Suzuki MM, Bird A. DNA methylation landscapes: provocative insights from epigenomics. *Nature Reviews Genetics* 2008;**9**:465–476.

4.8 BOUNDARY ELEMENTS SEPARATE CHROMATIN DOMAINS

Bushey AM, Dorman ER, Corces VG. Chromatin insulators: regulatory mechanisms and epigenetic inheritance. *Molecular Cell* 2008;**32**:1–9.

Wallace JA, Felsenfeld G. We gather together: insulators and genome organization. *Current Opinion in Genetics and Development* 2007;**17**:400–407.

Wei GH, Liu DP, Liang CC. Chromatin domain boundaries: insulators and beyond. *Cell Research* 2005;**15**:292–300.

4.9 ELEMENTS REQUIRED FOR CHROMOSOME FUNCTION

Mechali M. DNA replication origins: from sequence specificity to epigenetics. *Nature Reviews Genetics* 2001;**2**:640–645.

Murray AW, Szostak JW. Artificial chromosomes. *Scientific American* 1987;**257**:62–68.

4.10 THE CENTROMERE

Amor DJ, Choo KH. Neocentromeres: role in human disease, evolution, and centromere study. *American Journal of Human Genetics* 2002;**71**:695–714.

Amor DJ, Kalitsis P, Sumer H, Choo KH. Building the centromere: from foundation proteins to 3D organization. *Trends in Cell Biology* 2004;**14**:359–368.

Bloom K. Centromere dynamics. *Current Opinion in Genetics and Development* 2007;**17**:151–156.

Ekwall K. Epigenetic control of centromere behavior. *Annual Review of Genetics* 2007;**41**:63–81.

Pidoux AL, Allshire RC. The role of heterochromatin in centromere function. *Philosophical Transactions of the Royal Society of London Series B Biological Sciences* 2005;**360**:569–579.

4.11 THE TELOMERE

Armanios M. Syndromes of telomere shortening. *Annual Review of Genomics and Human Genetics* 2009;**10**:45–61.

Smogorzewska A, de Lange T. Regulation of telomerase by telomeric proteins. *Annual Review of Biochemistry* 2004;**73**:177–208.

4.12 CHROMOSOME ARCHITECTURE IN THE NUCLEUS

Cremer T, Cremer M, Dietzel S, *et al.* Chromosome territories–a functional nuclear landscape. *Current Opinion in Cell Biology* 2006;**18**:307–316.

Gilbert N, Gilchrist S, Bickmore WA. Chromatin organization in the mammalian nucleus. *International Review of Cytology* 2005;**242**:283–336.

McKeown PC, Shaw PJ. Chromatin: linking structure and function in the nucleolus. *Chromosoma* 2009;**118**:11–23.

THE CELL CYCLE

INTRODUCTION

Cell division is the process by which one cell divides to produce two daughter cells. Single-celled organisms divide in order to reproduce themselves, while cells of multicellular organisms divide in order to increase in mass, to generate distinct tissues, to form gametes, and to allow tissue renewal. To undergo cell division, cells must first duplicate certain cellular components, such as chromosomes. They then must properly distribute these components to the two daughter cells (Figure 5.1). The series of events that leads to the formation of two daughter cells from one parental cell is called the **cell cycle**, and it encompasses a series of highly coordinated processes that ensure that cell duplication occurs in a precise and timely manner.

The overall principles of cell division apply to all kingdoms of life. However, as we will see in this chapter, the mechanisms of cell cycle regulation differ between bacteria and eukaryotes. Most of the processes that will be discussed apply specifically to eukaryotes, but recent work suggests that eukaryotic proteins and regulatory pathways have parallels in archaea. Our understanding of the cell cycle is often based on experiments that were done in model organisms, such as yeast, flies, and mice. The reason for this is that model organisms are less complex than human cells and are more feasible to handle in laboratory settings. Nonetheless, since the eukaryotic cell cycle is evolutionarily conserved, what we learn from model organisms is often applicable to humans.

In this chapter we will be introduced to the events that occur as cells progress through the cell cycle, and we will learn about the regulatory mechanisms, triggered by internal cues or by signals from outside the cell, that affect cell cycle progression. We will learn more about specific events and processes that make up the cell cycle in Chapters 6 and 7.

5.1 STEPS IN THE EUKARYOTIC CELL CYCLE

Our exploration of the molecular processes and components that underpin the cell cycle first focuses on how the process takes place in eukaryotes. We begin with a general overview of the cell cycle as a whole, before considering the underlying mechanisms and components in more detail.

Figure 5.1 The cell cycle. An essential part of cell division is the duplication of certain cellular components, including the exact duplication of each chromosome in S phase. Shown is a cell with two chromosomes (in red). Chromosomes are divided equally between two daughter cells in M phase (including the chromosome segregation phase (in light green) and cytokinesis (in blue). Other cellular components that are important for this process are the centrosomes (shown in blue), centromeres and their associated kinetochores (shown in orange), and cohesin (shown in black). These will be discussed later in this chapter and in the following chapters.

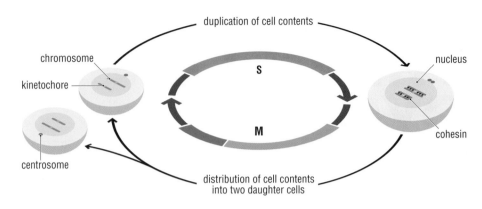

The eukaryotic cell cycle is divided into four distinct phases

As described in Chapter 4, the eukaryotic cell cycle can be divided into four stages, or cell cycle phases, according to the major event that takes place during that time. To recapitulate, the four major cell cycle phases are G1, S, G2, and M (Figure 5.2):

- **G1 phase** is a period of growth before chromosome duplication.

- **S phase** is when DNA synthesis occurs, generating two identical copies of each chromosome, called sister chromatids.

- **G2 phase** is a gap phase that allows both growth and preparation for the separation of the sister chromatids. The mitotic spindle, the apparatus that mediates chromosome segregation, begins to form during this phase.

- **M phase** includes two major events: **mitosis**, when the sister chromatids are segregated to opposite sides of the cell, and **cytokinesis**, which is the process though which a cell divides to create two daughter cells.

Figure 5.2 The events of the eukaryotic chromosome cycle. The central events of cell division are chromosome duplication, which takes place in S phase, followed by chromosome segregation (mitosis) and cell division (cytokinesis), which are collectively called M phase. G1 is the gap phase between M and S phases, immediately after cell division and prior to chromosome duplication. During G1 cells may exit the cell cycle and become quiescent (in a G0 phase), or they may pass the restriction point (designated as R in the figure) and commit to a new cell cycle. G2 is the gap phase between S and M phases, and it is when the mitotic spindle, the apparatus that drives chromosome segregation, is formed (shown in green). The spindle is anchored at both poles by centrosomes (shown as blue dots). Metaphase is the stage in mitosis when sister chromatids are aligned on the mitotic spindle and anaphase is the stage when sister chromatids are separated and pulled to opposite spindle poles. Before anaphase, sister chromatids are held together by cohesin (shown as black dots).

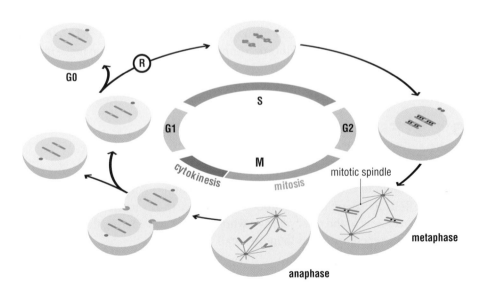

After mitosis the two daughter cells then return to G1 phase. Cells may pause and exit from the cycle in the G1 phase and enter a non-dividing state, called **G0** (G zero). The regulation of each of these specific steps in the cell cycle is necessary for the accuracy of the division process.

Cell growth occurs in the G1 phase

The G1 phase is the stage during which cell growth occurs. The full name of this phase is Gap phase 1, and it is so-called because there is no major observable change in the shape or size of the chromosomes during this stage. This is in contrast to S phase and M phase where noticeable changes in the appearance of the chromosomes do occur. While the term 'gap' may imply that nothing is happening, there are in fact essential processes occurring throughout this stage. During G1 the cell must grow to increase its cellular contents so that there is sufficient material within the cell for it to divide into two daughters later in the cell cycle.

One of the important events in G1 phase is **Start**, or the **restriction point** (Figure 5.2): once cells have passed the restriction point they are committed to go though the whole cell cycle, although they may be paused along the way if certain types of cellular defects arise. Because of this irreversible transition, the decision to embark on a new cell cycle depends, to a large extent, on nutrient availability. In addition, entry into the cell cycle is often tightly linked to cell size: in many cell types a critical size must be reached before the cell can pass the restriction point and commit to another division. Division before this critical size is reached would result in ever-smaller cells at each division. Finally, in animal cells, transition past the restriction point is often regulated by external signals, such as **growth factors**.

Cells may exit the cell cycle and enter G0

During G1, cells will either commit to a new round of division or will enter a resting stage termed G0, also known as quiescence, prior to passing the restriction point. For example, unicellular organisms will exit the cell cycle under conditions of poor nutrient availability, while cells in multicellular organisms become quiescent in response either to negative growth signals from neighboring cells or to lack of positive growth signals. The exit from the cell cycle also occurs during development: initially, during embryogenesis, there is rapid cell division to increase cell mass. Later, cells become specialized, or differentiated, to perform a specific task and they stop dividing.

Some cells in our body, such as adult stem cells, keep dividing to replenish a particular tissue (for example, adult skin stem cells are constantly dividing to provide new skin cells). However, most human cells are naturally arrested in G0. Some can be stimulated to re-enter the cell cycle. Others, however, are permanently arrested in G0. For example, human nerve cells carry out essential functions and yet they are not actively dividing.

It is important to note that quiescent cells are not just in a very long G1 phase, but that cells in G0 have certain markers that distinguish them from G1 cells. For example, quiescent cells have elevated levels of cell cycle inhibitors and they lack certain DNA replication proteins. In contrast, G1 cells have low levels of these inhibitors and DNA replication proteins are present.

DNA is replicated during S phase

Once a cell passes the restriction point, it enters S phase, when DNA replication takes place. The purpose of DNA replication is to produce an accurate copy of all the chromosomes, thereby creating two complete copies of the genome. Each chromosome is duplicated to form two identical sister chromatids, which remain associated with each other until mitosis. The sister chromatids are held together by a protein complex called cohesin.

⊙ The mechanisms of DNA replication and its regulation are discussed in more detail in Chapter 6. The action of cohesin is described in more detail in Chapter 7.

To ensure accurate duplication, there are regulatory mechanisms that stop cell cycle progression until the DNA is fully replicated, and other mechanisms that block replication in a region of the chromosome that has already been duplicated. Once the DNA is completely replicated, the cell can move to the next cell cycle stage, termed the G2 phase.

The G2 phase allows cells to prepare for chromosome segregation

After the completion of DNA synthesis, cells enter the G2 phase, during which they continue to grow and prepare for mitosis. During G2 there is a further increase in cellular content in order to generate the cellular components that will be distributed between the two daughter cells. Other events that occur in G2 include the formation of the **mitotic spindle**, the apparatus that drives chromosome segregation.

⊙ The structure and function of the mitotic spindle is described in detail in Chapter 7.

In addition, G2 phase is a point at which cell cycle progression can be stopped if chromosomes experience DNA damage. For example, if a chromosome has a double-stranded DNA break, there is a regulatory mechanism that will prevent cells from continuing to mitosis until the chromosome is repaired. Progression through mitosis with a broken chromosome could result in the loss of genetic material.

Regulatory mechanism, often called **checkpoint** pathways, monitor the integrity of different cellular structures. If a problem is detected, these checkpoint pathways prevent cell cycle progression, thereby allowing the cell to fix the problem before it results in permanent damage that may be passed on to the daughter cells. (We discuss checkpoint pathways further in Section 5.7.) Thus, although passage through the restriction point commits a cell to complete the cell cycle, a cell may occasionally pause its progression to deal with certain types of damage or error.

Chromosome segregation occurs during M phase

⊙ The process of chromosome segregation and cytokinesis is discussed in greater detail in Chapter 7.

Mitosis is the stage at which chromosomes are segregated into two new daughter cells. The progression through M phase is characterized by specific changes in chromosome organization.

The key event in mitosis is the attachment of the chromosomes to the mitotic spindle, which then pulls the sister chromatids apart to create two identical clusters of chromosomes, each containing the same chromosome composition as that of the parent cell. Errors that occur during this stage can lead to chromosome mis-segregation, and consequently give rise to cells with too few or too many chromosomes. Either of these situations can be deleterious to the cell or the organism.

Just like in S phase and in G2 phase, in mitosis there is also a checkpoint pathway that stops cell cycle progression if, for example, the chromosomes are not attached properly to the spindle. After chromosome segregation is completed, the cell divides in a process called cytokinesis to give rise to two daughter cells, which are in G1 phase.

Some cell types have variations in their cell cycle organization

The general description of the cell cycle given above is accurate for many single-celled organisms and human somatic cells. Note, however, that the total length of the cell cycle and the relative lengths of the different phases can vary greatly between cell types, as illustrated in Figure 5.3. For example, the entire cell cycle in budding yeast takes about 90 minutes, while in many human cells it takes 18 hours or more. Even different yeasts exhibit different types of cell cycle: budding yeast have a long G1 and a short G2, while fission yeast have a short G1 and a long G2.

Not only can the length of the cell cycle phases differ, but in certain cell types some cell cycle phases may be absent altogether. For example, in certain organisms, early embryonic development requires rapid cell divisions. In these early embryos there is no new protein synthesis and the embryo relies instead on proteins that were deposited in the oocyte by the mother. In the cell cycle of these embryos there are no gap phases (G1 and G2) and instead the cells shuttle between S phase and M phase (Figure 5.3).

Nonetheless, in all dividing cells, there is a common engine that drives the cell cycle forward regardless of the length of the various phases or even the types of phases that make up their unique cell cycle. This engine is based on the activity of a family of kinases (which catalyze the transfer of a phosphate group from a donor, such as adenosine triphosphate (ATP), to a target protein) that are regulated by proteins called **cyclins**. In the next section we see how these **cyclin-dependent kinases (Cdks)** move the cell cycle forward, from one cell cycle phase to the next.

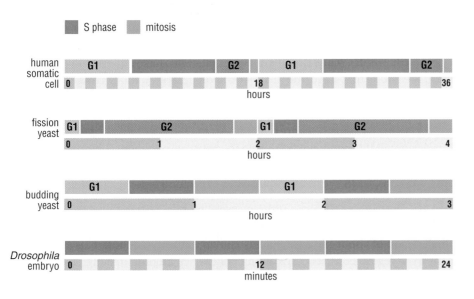

Figure 5.3 Variations can occur in the cell cycle in different cell types. The cell cycle of a human somatic cell follows the outline described in Figure 5.1. However, some cell types, especially embryonic cells, have a variant of this organization. For example, the first 13 *Drosophila* embryonic divisions are much more rapid than those of human somatic cells and they lack gap phases. Cell cycle patterns also differ between different yeast species: fission yeast have a very long G2 phase while the G2 phase of budding yeast is virtually non-existent. In the figure, G1 is colored in light gray, S phase is colored in pink, G2 is colored in dark gray and mitosis in green. Note that the absolute cell cycle length varies greatly between organisms – this figure only shows the relative lengths of each period for a given organism.

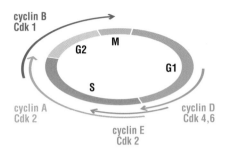

Figure 5.4 Different cyclins and cyclin dependent kinases (Cdks) function in different cell cycle stages. The diagram illustrates the different cyclins and Cdks that act during the cell cycle of a mammalian somatic cell. Different cyclins are present during different phases of the cell cycle and pair with different Cdks. Cyclin D (a G1 cyclin) can activate either Cdk4 or Cdk6, cyclin E (a G1/S cyclin) can activate Cdk2, cyclin A (an S phase cyclin) can also activate Cdk2, and cyclin B (an M phase cyclin) can activate Cdk1.

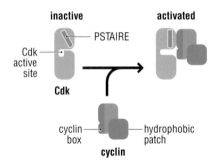

Figure 5.5 Cdk activation. The activation of Cdks at specific cell cycle stages requires binding to a specific cyclin. Cyclins contain a region called the cyclin box and a region called a hydrophobic patch (both shown in dark blue), which are important for cyclin function (see text for further detail). A Cdk can be represented as a bilobed protein with its active site in a cleft between the two lobes. The binding of cyclin to Cdk results in a conformational change in the Cdk, including movement of the PSTAIRE region (shown in red), opening the active site to potential substrates. For simplicity, this diagram does not include changes in Cdk phosphorylation, which also contribute to Cdk activation (see Figure 5.7).

5.2 CYCLINS AND CYCLIN-DEPENDENT KINASES

Cyclin-dependent kinases are key regulators of cell cycle progression

The central components of the machinery that drives and regulates the cell cycle are enzymes called the cyclin-dependent kinases (Cdks). Cdks belong to a family of serine/threonine protein kinases that phosphorylate specific serine and/or threonine residues on their protein substrates. Some organisms, such as mammals, have multiple Cdks, while other organisms, such as yeast, have a single Cdk. Read Experimental approach 5.1 to find out about the discovery of the first human Cdk and the identification of Cdk substrates.

The activity and specificity of Cdks is determined by regulatory proteins called cyclins. Different cyclins are present during different cell cycle phases (Figure 5.4), and in cells with multiple Cdks this leads to the activation of distinct cyclin–Cdk complexes at different cell cycle stages. As the cell progresses through the cycle, specific cyclin–Cdks become active and they phosphorylate specific substrates that carry out particular cell cycle events. This phosphorylation may change the enzymatic activity of a substrate or alter its interaction with other proteins.

Not only do cyclin–Cdks activate the events in their cell cycle phase, they also lead to the activation of the cyclin–Cdk of the next cell cycle phase (for example, by promoting the expression of the next cyclin), thereby pushing the cell cycle forward. Interestingly, despite the presence of multiple Cdks that are active at distinct cell cycle phases, studies in mice and other organisms revealed that there is a significant functional overlap between the Cdks, with one Cdk being able to compensate for another that has been deleted. The exception is Cdk1, which is essential for viability.

Cyclins bind to and regulate Cdk activity

Cyclins are small proteins that bind tightly to Cdks. In doing so, they lead to a conformation change in the Cdk that is necessary for the Cdk to become catalytically active (Figure 5.5), although additional activation steps are still needed (see below). Cyclins are so named because of the cyclical nature of their accumulation in the cell: unlike Cdk levels, which are relatively constant throughout the cell cycle, the levels of cyclins rise in a specific cell cycle stage, and then they abruptly fall at the end of that stage due to their proteolytic degradation. Thus, Cdk activity is regulated by the availability of its partner cyclin. Since cyclins contribute to the substrate specificity of the cyclin–Cdk complex, the presence of specific cyclins will dictate which substrates are phosphorylated: only when an appropriate cyclin is present will its cognate Cdk be able to carry out phosphorylation. Cyclins are often referred to by the phase of the cell cycle that they regulate: accordingly, there are G1 cyclins, G1/S cyclins, S cyclins, and M cyclins.

The overall structure of cyclin proteins is not highly conserved between different organisms. Instead, the sequence similarity between cyclins is concentrated in a small region of about 100 amino acids known as the **cyclin box**. This structurally conserved region consists of five alpha helices, which bind to a highly conserved

5.1 EXPERIMENTAL APPROACH

Cyclin-dependent kinases: of yeast and men

The first Cdks that were identified were those of the budding yeast *Saccharomyces cerevisiae* (a protein called Cdc28) and the fission yeast *Schizosaccharomyces pombe* (called Cdc2). At the time, it was not known that the activity of these kinases was dependent on cyclins, and hence the term Cdk had not yet been coined. It was known, however, that these proteins play key roles in cell cycle progression, and thus it was of interest to identify the human homolog.

To accomplish this, Paul Nurse and colleagues exploited the likely evolutionary conservation between Cdk genes in different species. Although *S. cerevisiae* and *S. pombe* are both yeasts, they are only distantly related from an evolutionary standpoint. Nurse and colleagues recognized a remarkable 60% amino acid identity between the *S. cerevisiae* and *S. pombe* proteins and, based on this, reasoned that the human Cdk is likely to be similar to the yeast Cdks. They further reasoned that, this being the case, they could identify the human Cdk homolog by its ability to complement the growth defect of an *S. pombe cdc2* mutant strain.

The *cdc2* mutant strain used by the Nurse group carried a temperature-sensitive allele of *cdc2* (called *cdc2-33*) that only allowed cell division at a permissive temperature. (For a description of how *cdc* mutants were isolated, see Experimental approach 5.2.) At the non-permissive temperature, the mutant cells could not divide and became elongated. The investigators then introduced a human cDNA library (a large collection of plasmids each carrying a different human cDNA, described in Chapter 16) into the *cdc2-33* mutant strain (Figure 1a). As they had predicted, the investigators found a subset of cells that now could divide at the non-permissive temperature. Several examples are shown in Figure 1b where cells carrying the 'suppressing' plasmid appear normal (pink arrow), while cells that have lost the plasmid or have a different plasmid cannot divide and appear elongated (blue arrow). The investigators isolated the plasmid that allowed for more normal growth at the non-permissive temperature, and identified a gene they named *CDC2*Hs, for the human (*Homo sapiens*) homolog of the fission yeast *cdc2*.

The human Cdc2 protein identified by Nurse and colleagues indeed had a very high degree of similarity to its yeast counterparts (Figure 2). This was a breakthrough finding not only because it identified the first human Cdk (other Cdks would soon follow),

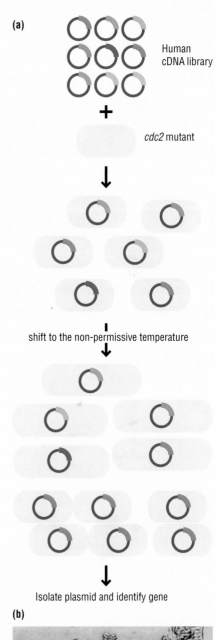

(a)

Human cDNA library

+

cdc2 mutant

shift to the non-permissive temperature

Isolate plasmid and identify gene

(b)

Figure 1 Cloning of human Cdk. (a) The strategy used by Lee and Nurse to clone human CDC2 (from the top): A library of human cDNA clones was transformed into a fission yeast *cdc2* mutant strain to yield a collection of yeast cells, each transformed with a different cDNA clone. These cells were then shifted to the non-permissive temperature. Cells that contained a plasmid that complemented the *cdc2* defect (the plasmid with a pink insert) could divide and form colonies at the non-permissive temperature. All other cells contained non-complementing plasmids and thus could not divide, and instead fo rmed elongated cells. (b) An image of the *cdc2* strain transformed with the human *CDC2* clone. Cells that contained the plasmid divided and had a normal shape (pink arrow). At some frequency the plasmid is lost, and these cells cannot divide and instead have an elongated shape (blue arrow). Figure 1b is taken with permission from Lee and Nurse, *Nature* 1987:**327**:31–35.

2Hs (human)	M E D Y T K I E K I G E G T Y G V V Y K G R H K T T
2Sp (S.pombe)	M E N Y Q K V E K I G E G T Y G V V Y K A R H K L S
28 (S. cerevisiae)	M S G E L A N Y K K L E K V G E G T Y G V V Y K A L D L R P G Q

2Hs (human)	G Q V V A M K K I R L E S E E E G V P S T A I R E I S L L K E
2Sp (S.pombe)	G R I V A M K K I R L E D E S E G V P S T A I R E I S L L K E
28 (S. cerevisiae)	G Q R V V A L K K I R L E S E D E G V P S T A I R E I S L L K E

2Hs (human)	L R H P N I V S L Q D V L M Q D S R L Y L I F E F L S
2Sp (S.pombe)	V N D E N N R S N C V R L L D I L H A E S K L Y L V F E F L D
28 (S. cerevisiae)	L K D D N I V R L Y D I V H S D A H K L Y L V F E F L D

2Hs (human)	M D L K E Y L D S I P P G Q Y M D S S L V K S Y L Y Q I L Q
2Sp (S.pombe)	M D L K K Y M D R I S E T G A T S L D P R L V Q K F T Y Q L V N
28 (S. cerevisiae)	L D L K R Y M E G I P K D Q P L G A D I V K K F M M Q L C K

2Hs (human)	G I V F C H S R R V L H R D L K P Q N L L I D D K G T I K L A D
2Sp (S.pombe)	G V N F C H S R R I I H R D L K P Q N L L I D K E G N L K L A D
28 (S. cerevisiae)	G I A Y C H S H R I L H R D L K P Q N L L I N K D G N L K L G D

2Hs (human)	F G L A R A F G I P I R V Y T H E V V T L W Y R S P E V L L G S
2Sp (S.pombe)	F G L A R S F G V P L R N Y T H E I V T L W Y R A P E V L L G S
28 (S. cerevisiae)	F G L A R A F G V P L R A Y T H E I V T L W T R A P E V L L G G

2Hs (human)	A R Y S T P V D I W S I G T I F A E L A T K K P L F H G D S E I
2Sp (S.pombe)	R H Y S T G V D I W S V G C I F A E M I R R S P L F P G D S E I
28 (S. cerevisiae)	K Q Y S T G V D T V S I G C I F A E M C N R A P I F S G D S E I

2Hs (human)	D Q L F R I F R A L G T P N N E V W P E V E S L Q D Y K N T F P
2Sp (S.pombe)	D E I F K I F Q V L G T P N E E V W P G V T L L Q D Y K S T F P
28 (S. cerevisiae)	D Q I F K I F R V L G T P N E A I W P D I V Y L P D F K P S F P

2Hs (human)	K W K P G S L A S H V K N L D E N G L D L L S K M L I Y D P A K
2Sp (S.pombe)	R W K R M D L H K V V P N G E E D A I E L L S A M L V Y D P A H
28 (S. cerevisiae)	Q W R R K D L S Q V V P S L D P R G I D L L D K L L A Y D P I N

2Hs (human)	R I S G K M A L N H P Y F N D L D N Q I K K M
2Sp (S.pombe)	R I S A K R A L Q Q N Y L R D F H
28 (S. cerevisiae)	R I S A R R A A I H P Y F Q E S

Figure 2 Comparison of the amino acid sequences of human and yeast Cdc2. 2Hs is human *CDC2*, 2Sp is *S. pombe cdc2*, and 28 is the *S. cerevisiae CDC28*. Identities between the human and yeast proteins are shown in pink.
With permission from Lee and Nurse, *Nature* 1987:**327**:31–35.

(a)

(b)

(c)

Figure 3 An analog specific kinase.
(a) Various kinases, including the *S. cerevisiae* Cdc28 (sometimes also called Cdk1), share a similar binding pocket for ATP (shown in pink, binding pocket is shown in white). (b) Kevan Shokat and colleagues devised a Cdc28 mutant protein (Cdc28-as1) with an altered ATP binding site, which can accommodate both ATP and an ATP analog. The ATP analog does not fit into the ATP-binding pocket of other kinases. (c) A representative sample of yeast proteins that were tested for phosphorylation by Cdc28-as1. Strains carrying the indicated genes fused to a tag were grown individually and subjected to affinity purification to enrich for the tagged protein. The extracts were then incubated with purified Cdc28-as1 bound to the Clb2 cyclin and the radioactive ATP analog. At the end of the reaction the extract was run on a gel. The top image is an autoradiograph showing the various radioactive proteins on the gel. Note that some proteins, such as those encoded by *ZRG8*, *GAL3*, and *FKH2* were not phosphorylated, while others, such as those encoded by the ORFs *YOR177C* or *TPR171W*, were heavily phosphorylated. The bottom image is the silver-stained gel from which the autoradiograph was obtained. Note that most lanes contain many proteins, which is typical of this type of affinity purification.
With permission from Ubersax *et al.*, *Nature* 2003; **425**:859–864.

but also because it demonstrated that cell cycle regulation is conserved throughout evolution. This allowed scientists to draw insights from experiments done in simple model organisms, such as yeast, in order to understand pathways of regulation in more complicated organisms, such as humans.

Following the discovery of Cdks and the realization that these kinases regulate cell cycle progression by phosphorylating other proteins, investigators wanted to identify Cdk substrates. This was not a simple task: one could not simply identify all the phosphorylated proteins in the cell because cells have many different kinases and only a fraction of the phosphorylated proteins are Cdk substrates. Instead, there was a need to develop a method to distinguish Cdk substrates from other phosphorylated proteins. This problem was addressed in a clever way by David Morgan, Kevan Shokat, and colleagues who devised a mutant form of the *S. cerevisiae* kinase Cdc28 (also called Cdk1) that had an altered ATP binding pocket (Figure 3). The mutant Cdc28, called Cdc28-as1 (for <u>a</u>nalog <u>s</u>ensitive), could bind an ATP analog that was too bulky to fit into the ATP binding pockets of other kinases, thus allowing its activity to be specifically followed in a sea of other kinases. By

incubating this mutant kinase (Cdc28-as1) and a gamma-labeled radioactive ATP analog in a cellular extract, the scientists could immediately observe that many proteins are targeted by Cdc28-as1 in a phosphorylation reaction.

Specific candidates were next identified using a yeast collection in which each individual open reading frame (ORF) was fused to an affinity tag (GST) that allowed the corresponding protein to be purified from the extract (Figure 3c, lower panel). The 'semi-purified' proteins were then subjected to phosphorylation by Cdc28-as1 again in the presence of the radioactive ATP analog (Figure 3c); of the 700 proteins tested, 360 were found to be Cdc28 substrates, many of them novel targets. Follow up experiments subsequently confirmed many of these proteins as valid *in vivo* Cdk substrates. In Experimental approach 12.2, we will see another approach used to identify the substrates of a protein kinase.

Ubersax JA, Woodbury EL, Quang PN, *et al*. Targets of the cyclin-dependent kinase Cdk1. *Nature* 2003;**425**:859–864.

Lee MG, Nurse P. Complementation used to clone a human homolgoue of the fission yeast cell cycle control gene cdc2. *Nature* 1987;**327**: 31–35.

motif in the Cdk termed **PSTAIRE**, after the single letter code for the motif's amino acid sequence. Upon binding of cyclin to the PSTAIRE region in the Cdk, major structural changes occur that allow the kinase to adopt an active conformation (Figure 5.5), making the active site more accessible to the protein substrate. At least some cyclins can bind directly to specific substrate through a domain in the cyclin called a **hydrophobic** patch (Figure 5.5). Thus, the binding of cyclin to the Cdk not only activates the Cdk but also promotes recruitment of Cdk substrates.

Cyclin–Cdks move the cell cycle forward

The function of different cyclin–Cdk complexes is to promote cell cycle progression. This means that cyclin–Cdks not only promote stage specific processes, but they also promote the transition to the next cell cycle stage. The actual network of interactions that drive cell cycle progression is very complex and it involves dozens of proteins. A simplified example taken from the cell cycle regulatory mechanism in budding yeast is shown in Figure 5.6. Note that there are both positive

Figure 5.6 A regulatory cell cycle circuit acting at the beginning of M phase. Entry into M phase is dependent on the activity of M cyclin–Cdk. However, prior to M phase, M cyclin is unstable due to proteolysis (step 3 in the regulatory circuit). To allow entry into M phase, S cyclin–Cdk, which is needed to promote S phase processes (step 1) inhibits M cyclin proteolysis (step 2). This allows M cyclin to accumulate and activate its Cdk (step 4), thereby promoting entry into M phase (step 5). In addition, M cyclin–Cdk also inhibits the expression of S phase cyclins (step 6), thereby extinguishing processes that occur earlier in the cell cycle. Keep in mind that the actual regulatory circuit is much more complex. Another example of a cell cycle regulatory circuit is shown in Figure 5.10. The relationships between the various players in this scheme are denoted by green arrows, which represent an activation process, or by red bars, which represent an inhibitory process.

interactions, such as those that promote protein activity or gene expression, and negative interactions, which inhibit activity or lead to protein degradation.

When cells are in S phase, S cyclin–Cdk1 activates DNA replication (Figure 5.6). In order to move on to mitosis, however, M cyclin–Cdk1 has to be active. Prior to mitosis – in G1 and early S phase – an M cyclin degradation process causes M cyclins to be unstable. Consequently, the M cyclin–Cdk complex cannot be formed. Part of S cyclin–Cdk1 activity is to promote entry into mitosis, and it does so by inactivating the M phase cyclin proteolysis machinery. This type of regulation – where two negative reactions result in a positive outcome is very common in cell cycle regulation (in this instance, the two 'negative' reactions are the proteolysis of M cyclin and the inhibition of this proteolysis, while the positive outcome is the stabilization of M cyclin).

The cell cycle circuitry is obviously more complex, because not only does the cell cycle machinery have to activate the next cell cycle step, it also has to turn off the previous one. In the example shown in Figure 5.6, the association of M cyclins with the Cdk not only drives progression through mitosis but also inhibits the transcription factors responsible for S cyclin expression, thereby turning off processes that preceded M phase. Such processes, and many others like them, make up the cell cycle engine that drives cell cycle progression.

5.3 REGULATION OF CDK ACTIVITY

Cdk activity is regulated by multiple mechanisms

The fundamental importance of Cdks is underscored by the regulation of their activity at multiple levels. We have seen earlier in this section that Cdk activity is regulated by the presence or absence of a particular cyclin. However, this is not the only way by which Cdk activity is regulated. Cdks are also regulated, both positively and negatively, by phosphorylation. There are specific amino acid residues in the Cdk that must be phosphorylated, and different residues that must be dephosphorylated, for the enzyme to be active. Thus, the careful control of both the kinases that add these phosphate groups, and the phosphatases that remove them, allow the cell to finely tune the activity of the Cdks. In addition, there are small proteins termed cyclin kinase inhibitors that bind to and specifically inhibit Cdk activity. In this section we consider these processes in more detail as we discuss some of the mechanisms that regulate Cdk activity.

Cdks can be activated by phosphorylation and dephosphorylation

The binding of cyclin to a Cdk changes the conformation of the Cdk such that its active site is now more accessible to its substrates. However, even with the cyclin bound, there is still a structural hindrance that prevents cyclin–Cdk from phosphorylating its substrates. For a Cdk to be active, there is a specific threonine residue adjacent to the kinase active site that must be phosphorylated. Phosphorylation at this site is catalyzed by enzymes called **Cdk-activating kinases (CAKs)** (Figure 5.7). CAK activity is usually high throughout the cell cycle. Therefore, although CAK-dependent phosphorylation is necessary for Cdk activation, it can be viewed as a post-translational modification rather than as a

Figure 5.7 Two steps in Cdk activation. Cyclin binding alone causes partial activation of Cdks, but complete activation also requires phosphorylation by Cdk-activating kinase (CAK). In animal cells, CAK phosphorylates the Cdk subunit only after cyclin binding, and so the two steps in Cdk activation are usually ordered as shown here, with cyclin binding occurring first. CAK tends to be in constant excess in the cell, so that cyclin binding is the rate-limiting step in Cdk activation. Cdk is shown in gray and cyclin is shown in blue. Phosphorylation site is in yellow.

Cdk

cyclin

CAK

cell cycle-regulated mode of activation. Thus, both cyclin binding and phosphorylation by CAK are needed to fully activate the cyclin–Cdk complex.

In the case of the M cyclin–Cdk complex, not only does CAK have to activate the Cdk, two specific phosphate groups on Cdk must be *removed* for full Cdk activity. One site of inhibitory phosphorylation is a tyrosine residue (Tyr 15 in human Cdks) that is found in all major Cdks. The second site of inhibitory phosphorylation is found only in animal cell Cdks, and is at the adjacent threonine residue (Thr 14). These two residues, Thr 14 and Tyr 15, are located in the ATP-binding site of Cdk and their phosphorylation interferes with the orientation of ATP in the binding pocket. Therefore, as long as Thr 14 and Tyr 15 are phosphorylated, M cyclin–Cdk cannot phosphorylate its substrates.

The phosphorylation state of Thr 14 and Tyr15 is determined by the opposing activities of a kinase and a phosphatase acting at these sites: Thr 14 and Tyr15 are phosphorylated by the **Wee1** kinase, and dephosphorylated by the **Cdc25** phosphatase, (Figure 5.8a). Wee1 is active during G1, S, and G2 phases while Cdc25 becomes active in G2. Moreover, the activities of both Wee1 and Cdc25 are regulated by their substrate, the M cyclin–Cdk complex: Wee1 is inhibited and Cdc25 is activated by M-phase cyclin Cdk phosphorylation (Figure 5.8b). Thus, M cyclin–Cdk inhibits its inhibitor and activates its activator. This type of regulation is called a **positive feedback loop**; in this instance it leads to a sudden increase in Cdk activity.

So why is this increase so sudden? Imagine that a small fraction of M cyclin–Cdk becomes active. This active cyclin–Cdk inhibits Wee1, making it easier for Cdc25 to activate more cyclin–Cdk. Cdc25 itself becomes more active through phosphorylation by Cdk, further increasing the activity of cyclin–Cdk. This leads to an abrupt rise in Cdk activity. In a system where there is no positive feedback, such that Cdk activity would solely depend on cyclin binding, the increase in cyclin–Cdk activity would be gradual and directly proportional to the levels of cyclin in the cell (Figure 5.9a). However, the presence of a positive feedback loop creates an ON/OFF switch (Figure 5.9b): the levels of cyclin–Cdk complex increase in proportion to the level of cyclin, but the complex remains inactive. The activity of Cdc25, however, leads to an abrupt increase in cyclin–Cdk activity, due to the positive feedback loop, allowing a sharp transition from G2 into mitosis.

Cdc25 is also an important target for regulation of Cdk activity in the presence of DNA damage. Under these conditions, a regulatory pathway called the DNA damage checkpoint pathway leads to the phosphorylation of Cdc25. This inactivates Cdc25 and, as a result, the inhibitory phosphorylations at Thr14 and Tyr15 cannot be removed and the cell cycle is prevented from moving forward.

Cdk inhibitors negatively regulate cyclin–Cdk activity

The binding of specific inhibitory proteins called **cyclin kinase inhibitors (CKIs)** provides yet another level of regulation of cyclin–Cdk complexes. These small inhibitory proteins are not similar at the amino acid sequence level, but they all bind and inactivate cyclin–Cdk complexes.

There are two classes of animal CKIs. The proteins in one class bind to both the cyclin and the Cdk in G1-S or S phase cyclin–Cdk complexes. CKIs of this class bind the hydrophobic patch of the cyclin (see Figure 5.5) and to a large domain

(a)

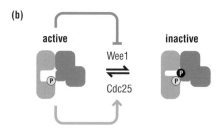

(b)

Figure 5.8 Control of Cdk activity by inhibitory phosphorylation. (a) The fully active cyclin–Cdk complex (phosphorylated by CAK, as indicated by the yellow P) can be inhibited by Wee1 phosphorylation at the active site of the Cdk (inhibitory phosphorylation is indicated by the black P). Dephosphorylation of the inhibitory sites by the phosphatase Cdc25 leads to reactivation of the cyclin-Cdk complex. (b) Wee1 and Cdc25 are regulated by M cyclin–Cdk: phosphorylation of Wee1 by M cyclin-Cdk inhibits the activity of Wee1 (shown as a red bar) while phosphorylation of Cdc25 by M cyclin-Cdk stimulates the activity of Cdc 25 (shown as a green arrow). This leads to a positive feedback loop, where as more active M cyclin–Cdk is formed, Cdc25 becomes more active while Wee1 is further inhibited.

⊙ We discuss the DNA damage response in more detail in Chapter 12.

(a)

(b)

Figure 5.9 The contribution of Wee1/Cdc25 to the switch-like behavior of M cyclin-Cdk. (a) The levels of Cdk in the cell (dashed gray line) are relatively constant and are in excess compared to the levels of M cyclin (in this example cyclin B, shown as a dashed dark blue line). Furthermore, the activity of the Cdk-activating enzyme kinase (CAK) is also constant throughout the cell cycle (not shown) and is not rate limiting. Thus, if the only requirement for Cdk activation was the binding of M cyclin, the activity of M cyclin–Cdk (orange line) would be directly proportional to the levels of cyclin, and it would increase gradually as cyclin is being synthesized. (b) However, cyclin–Cdk is also regulated by the Wee1/Cdc25 positive feedback loop (see Figure 5.8). As a result the activity of cyclin–Cdk is repressed until Cdc25 is activated, when cyclin-Cdk is further stimulated by the positive feedback loop, leading to a burst of cyclin–Cdk activity.

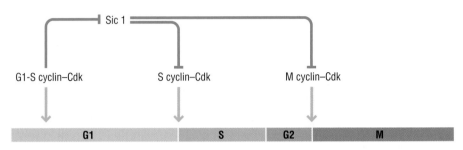

Figure 5.10 A regulatory cell cycle circuit acting at the beginning of S phase. In budding yeast, progression beyond G1 is regulated by the Cdk1 inhibitor Sic1 (budding yeast have only a single Cdk). Sic1 is an inhibitor of S cyclin–Cdk1 and of M cyclin–Cdk1, but not of G1-S cyclin–Cdk1. As G1-S cyclins accumulate, they phosphorylate Sic1 and target it for degradation. This alleviates the inhibition on S cyclin–Cdk1 and allows entry into S phase. Note that at this point M cyclin–Cdk cannot be activated due to the instability of M cyclins (Figure 5.6). Thus, Sic1 and protein degradation are two mechanisms that down-regulate the activity of M cyclin–Cdk during early stages of the cell cycle. Activating processes are shown as green arrows while inhibitory processes are shown as red bars.

on the Cdk, thereby distorting Cdk's structure and directly inhibiting ATP binding. In contrast, inhibitors of the second class bind to G1 Cdk monomers on the face of the protein opposite the active site. This binding induces a global change in the Cdk structure, which reduces cyclin binding and alters the Cdk active site.

An example of a CKI is the budding yeast Sic1 protein. Although Sic1 inhibits cyclin–Cdks, it is not similar in structure to any of the animal Cdks. Sic1 ensures that cells remain in G1 by inhibiting the major S cyclin–Cdk1 and M cyclin–Cdk1 complexes. Sic1, however, does not inhibit G1-S cyclin–Cdk1 complexes, so G1 processes and passage through Start are not inhibited. At the beginning of S phase, Sic1 is phosphorylated by G1-S cyclins and targeted for destruction, thereby allowing S and M phase cyclin–Cdks to become active (Figure 5.10). Thus, CKIs and other regulatory processes play an important role in ensuring the timely execution of the various cell cycle processes.

5.4 CELL CYCLE REGULATION BY CDKS

In the previous section, we discussed how the activity of Cdks is regulated. But what is the outcome of this activity? What effects do Cdks actually have in the cell once active? We consider the answer to these questions in this section.

Cdks have many cellular targets

The activity of cyclin–Cdk regulates the cell cycle through the phosphorylation of specific substrates. (We discuss a way in which Cdk substrates in yeast have been identified in Experimental approach 5.1.) Once phosphorylated, these substrates have an altered activity or altered stability. This, in turn, affects the processes in which these proteins participate. But what are the processes that respond to Cdk-dependent phosphorylation? We know of over 400 proteins in yeast that are specific targets of the cyclin–Cdk kinase, and recent studies have begun to systematically uncover substrates of animal cyclin–Cdks. The identified substrates play roles in many cellular processes. Some proteins have roles in processes that are directly related to cell cycle progression, such as DNA replication, chromosome condensation, mitosis, and cell division. Other protein substrates, such as those involved in protein synthesis, may contribute to cell cycle regulation in a less direct way.

Many Cdk substrates have multiple phosphorylation sites. The method for elucidating how Cdk phosphorylation contributes to protein function often involves the identification of phosphorylated serine or threonine residues, mutating the corresponding DNA sequence so that these sites can no longer be phosphorylated, and then determining the consequences caused by the presence of an 'unphosphorylatable' protein. For some substrates, the role of Cdk phosphorylation has been elucidated, but for many others the role of phosphorylation is not yet known. In the sections that follow we will discuss several examples, out of many, of how Cdk phosphorylation affects cell cycle progression.

Cyclin–Cdks regulate the transcription of specific genes

One mechanism by which cyclin–Cdks regulate cell cycle events is through the regulation of transcription. Indeed, many of the identified cyclin–Cdk targets are transcription factors. In budding yeast, over 1200 genes – nearly 20% of the genome – are transcribed in a cell cycle-regulated manner, and are responsible for carrying out the various cell cycle processes. This cell cycle-regulated pattern of expression is controlled by cyclin–Cdk activity.

Let us consider some examples of how phosphorylation of cyclin–Cdk substrates can regulate transcription. In mammalian cells, entry into G1 is under transcriptional regulation: prior to the restriction point, the transcription factor **E2F** is inhibited through its association with one of the **Rb** family of proteins (Figure 5.11). Phosphorylation of Rb by the G1 cyclin–Cdks, cyclin D/Cdk4 or cyclin D/Cdk6, releases E2F from Rb, allowing E2F to activate transcription of a specific set of genes required for the G1/S transition. Mutations in the Rb gene that abolish its ability to bind E2F lead to inappropriate gene activation, which drives cell division and gives rise to an increased risk of cancer. In fact, the Rb gene gets its name from retinoblastoma, a type of cancer where a mutation in the Rb gene was initially identified. The regulatory pathway controlling Rb function is one of the most frequently altered pathways in cancer, as we will discuss towards the end of this chapter.

Cyclin–Cdks regulate many distinct cellular processes

Many cellular organelles and structures change during cell cycle progression. For example, during mitosis the nuclear envelope disassembles. Many of the details of this process are still to be elucidated, but we know that various proteins associated with the nuclear envelope, including components of the nuclear pore complex (the multiprotein structures that allow selective passage of proteins and other molecules between the cytoplasm and the nucleoplasm) and the nuclear lamina

Figure 5.11 Regulation of gene expression by cyclin–Cdk phosphorylation. In G0 and early G1 of metazoan cells, the transcription activator E2F is bound to Rb. This association prevents E2F from activating transcription. In order to enter the cell cycle, cyclin D associates with either Cdk4 or Cdk6, and the resulting cyclin–Cdk phosphorylates Rb. This causes Rb to dissociate from E2F, freeing E2F to activate the expression of genes necessary for entering the cell cycle. Note that if Rb is inactivated, due to a mutation, for example, E2F may promote unscheduled cell cycle entry, leading to uncontrolled cell division, such as in cancer cells.

are phosphorylated by M cyclin–Cdks, and that this phosphorylation contributes to the disassembly of the nuclear envelope during mitosis.

As cells divide, each daughter cell needs to receive a set of chromosomes; they also need to inherit other cellular structures and organelles. We now know that the inheritance of at least some of these structures is dependent on Cdk activity. For example, in yeast, the movement of a cellular structure called the vacuole during cell division is controlled by Cdk phosphorylation. Likewise, the reorganization during cell division of the Golgi apparatus, an organelle involved in protein sorting, is dependent on Cdk phosphorylation. Finally, in budding yeast, changes in cell shape, which occur during the cell cycle, are directly controlled by Cdk activity. Thus, Cdk phosphorylation controls multiple cellular processes.

5.5 REGULATION OF PROTEOLYSIS BY CDKS

Cyclin–Cdks regulate proteolysis of specific substrates

A way to ensure that the cell cycle moves in a forward direction is to eliminate proteins needed for the previous cell cycle stage, for example by protein degradation. Indeed, cyclin–Cdks drive the cell cycle forward by promoting the proteolytic destruction of key regulatory proteins. To date, the only known mechanism for cell cycle-regulated protein degradation is by the **ubiquitin**-mediated proteolysis machinery. Ubiquitin is a small protein that is highly conserved in evolution. As we saw in Chapter 2, the last step in the attachment of ubiquitin molecules to a protein is catalyzed by the enzyme **ubiquitin ligase**. Ubiquitination can target proteins to different fates. In the case of cell cycle regulation, however, proteins that are decorated by chains of ubiquitin are often targeted for degradation and are destroyed by a large protease complex called the **proteasome**.

There are two major ways in which cyclin–Cdks regulate protein degradation. First, some cell cycle regulators have to be phosphorylated in order to be recognized by certain ubiquitin ligases. Thus, cyclin–Cdk targets these substrates for degradation by phosphorylating them. Second, cyclin–Cdk directly phosphorylates a number of ubiquitin ligases, leading to their activation. In this case, cyclin–Cdk targets proteins for degradation by activating the ubiquitin ligase. To illustrate this, let us discuss two ubiquitin ligases that are central to cell cycle progression.

The SCF ubiquitin ligase complex catalyzes ubiquitination of phosphorylated substrates

The **SCF** complex is a ubiquitin ligase that regulates the transition from G1 into S phase by degrading specific proteins that are phosphorylated by G1-S cyclin–Cdks. SCF is named for three of its central components – **Skp1**, **cullin**, and the **F-box protein** (Figure 5.12). The complex also contains a RING domain protein, often found in ubiquitin ligases. In the SCF complex, the RING domain protein Rbx1 binds to another component of the ubiquitination pathway, the E2 enzyme. The F-box protein is the subunit that binds the protein substrate that will be ubiquitinated. In the cell there are many F-box proteins, each with specificity for a distinct set of substrates.

→ The process of ubiquitination and protein degradation is discussed in greater detail in Chapter 11.

Figure 5.12 SCF is a multisubunit enzyme.
(a) SCF contains several subunits: the subunits that it is named after, which are Skp1, Cul1, and an F-box protein, and a fourth subunit, Rbx1, which contains a RING domain and binds another component of the ubiquitination process, an E2 enzyme. The phosphorylated target protein binds to the F-box subunit, after which ubiquitin is transferred from the E2 to a lysine side chain on the target protein. (b) A composite model structure for human SCF derived from a crystal structure. PDBs 1LDK, 1FBV, 1FQV. Adapted from Zheng *et al.* (2002).

The activity of SCF is constant throughout the cell cycle. However, the substrate must be phosphorylated in order to bind the F-box protein. It is this step that is regulated by cyclin–Cdk phosphorylation. An example of a protein that is targeted for degradation by SCF-dependent ubiquitination is the yeast Cdk inhibitor Sic1, which we encountered earlier in this chapter. As you recall, Sic1 is an S phase cyclin–Cdk inhibitor (Figure 5.10). Sic1 phosphorylation by G1-S cyclin–Cdks leads to its ubiquitination by the SCF and its subsequent degradation. As a result, S phase cyclin–Cdks become active. Other proteins that are degraded by the SCF are the G1-S cyclins themselves, while S phase cyclins are not SCF targets. Thus, through Cdk activity, cyclins that are needed for early cell cycle events are degraded, allowing the cell to move forward in the cell cycle.

The APC ubiquitin ligase, which drives anaphase and mitotic exit, is activated by Cdk phosphorylation

A second mechanism by which Cdk activity regulates proteolysis is by directly activating a ubiquitin ligase known as the **anaphase-promoting complex (APC)** early in mitosis. The APC, as its name implies, is essential for promoting the anaphase stage of mitosis, which is when sister chromatids separate. The importance of the APC was shown genetically in yeast, where the inactivation of the APC blocks the cell's ability to progress through mitosis (Figure 5.13). Although the APC catalyzes the ubiquitination of many proteins, it has three major substrates whose destruction is particularly critical. Two of these substrates must be destroyed at the metaphase-to-anaphase transition to allow proper chromosome separation at anaphase.

The APC also regulates the destruction of the M cyclins. The inactivation of M cyclin–Cdk is needed in order to exit mitosis. Finally, the APC has numerous substrates that it targets for degradation during G1.

⊜ The mitotic substrates of the APC are discussed in detail in Chapter 7.

In most organisms, the APC is a complex of 11–13 subunits, including a cullin subunit and a RING domain subunit, much like SCF. Look at Experimental approach 5.2 to see how some of the genes coding for APC subunits have been identified. Unlike SCF, however, the APC does not have an F-box protein, and instead its substrate specificity is determined by activator subunits. Two activator proteins, **Cdc20** and **Cdh1**, are particularly important. Cdc20 targets proteins for ubiquitination by the APC early in mitosis, while Cdh1 targets proteins for ubiquitination by the APC in late mitosis and G1. Early in mitosis, multiple APC subunits are phosphorylated by mitotic cyclin–Cdk, a step that is necessary for

5.2 EXPERIMENTAL APPROACH

The cell division cycle: from genes in yeast to function in animal cells

A major breakthrough in studies of the cell cycle came in 1970 with the isolation of *S. cerevisiae* mutants that were defective in cell cycle progression – these were referred to as cell division cycle (*cdc*) mutants. Lee Hartwell and colleagues based their genetic screen to find cell cycle mutants on three considerations. First, the group assumed that the cell cycle of budding yeast could be followed by cellular and nuclear morphology. Second, they thought that genes that contribute to cell cycle progression would likely be essential. Thus they would need to isolate temperature-sensitive (*ts*) mutants that are viable at the permissive temperature (23 °C) but are non-viable at some other non-permissive temperature (36 °C in this case). Third, they assumed that cell cycle mutants would arrest with a uniform cell cycle morphology. With these criteria in mind, the Hartwell group screened a yeast *ts* mutant collection, looking for those strains that failed to progress beyond a particular cell cycle stage at the non-permissive temperature (Figure 1a). In real terms, this meant that they looked in the microscope at many different yeast cell cultures and identified those where the population was homogeneous in a manner that seemed relevant to progression through the cell cycle (i.e. all might be 'unbudded' or 'large budded'). Of the 100 initial temperature-sensitive strains that were screened, 15 exhibited a uniform cell cycle arrest. These were called *cdc* genes and were assigned a number according to their order of isolation (*cdc1*, *cdc2*, etc).

To characterize the *cdc* mutants that they had isolated, the investigators took an asynchronous population of cells (containing cells at various cell cycle stages), put them on an agar plate containing the appropriate medium, shifted them to the non-permissive temperature, and took images using a microscope at specific time intervals. While some cells arrested shortly after the temperature shift, within one cell cycle, other cells that were further along in the cell cycle at the time of the temperature shift completed that particular cycle and arrested only in the next cycle. For example, for the *cdc2* mutant strain shown in Figure 1, the unbudded cell marked with the blue arrow is probably in early G1. This cell generates a bud but then arrests as a large budded cell. In contrast, the large budded cell marked with a pink arrow, which is probably in G2 or mitosis, completes the cell cycle and generates two cells, each of which progresses to a large budded cell, and then arrests (Figure 1a).

The interpretation of these observations was that, at the time of the temperature shift, the unbudded cell was at a point in the cell cycle somewhere before the critical step that could not be executed at the non-permissive temperature. Therefore it could not complete the cell cycle. The large budded cell, on the other hand, must have passed that critical point at the time of the temperature shift, and only its progeny encountered the limiting step during the next cell cycle. For both cases, having passed the critical point (or execution point) the cells progressed to the large budded cell stage, which was their terminal point. Specifically, for the *cdc2* mutant, we now know that the Cdc2 protein is involved in DNA replication. Thus, the execution point for Cdc2 function is in S phase, where cells normally have a small bud. By contrast, the terminal point of these *cdc2* mutants is further along the cell cycle, where cells already have a large bud.

For some mutants, the execution point and terminal point are the same. For example, a mutant with a defect in an anaphase-promoting complex (APC) subunit fails to complete anaphase initiation and arrests at that point of the cell cycle. The APC is discussed in Chapter 7.

For other mutants, the execution point and terminal point are not the same. For example, a mutation that causes a defect in the duplication of the yeast centrosome equivalent (the spindle pole body) fails to promote spindle pole duplication in G1-S, but the cells do not arrest until they are in mitosis. These findings underscore an important concept: when analyzing mutants with a particular cell cycle morphology, the defective process that leads to this morphology could have taken place at some earlier stage in the cell cycle.

Hartwell and his colleagues went on to identify numerous additional *cdc* mutants; subsequent work from many labs has demonstrated that these *cdc* genes are involved in a wide range of cell cycle related processes and many are conserved in all eukaryotes, including humans. Indeed, the loss of function of many of these cell cycle genes in human cells is an important contributor to the onset of cancer, as described in section 5.8.

Genetic studies such as those carried out by the Hartwell group could generally pinpoint the cell cycle process that was disrupted by a mutation, but often could not identify the specific molecular function of the proteins encoded by the genes. Twenty years after the initial discovery of the *cdc* genes, Marc Kirschner and colleagues used a biochemical approach to examine the mechanism of cyclin B degradation during mitotic progression. At the time, it was known that cyclin B destruction occurred in a cell cycle dependent manner and that it involved ubiquitination of cyclin B. The role of ubiquitination, the attachment of the small ubiquitin protein, in protein degradation is discussed in Chapter 11. It was also understood that cyclin B destruction was necessary for exit

Figure 1 **The isolation of cdc mutants.** (a) Wild-type control cells (left panel) and mutant cells (right panel) were plated on agar plates containing yeast media, and shifted to 36 °C. Cell growth was then monitored at 1, 3, and 5 hours after the temperature shift. The wild-type cells grew continuously, while the mutant cells arrest as large budded cells. The blue and pink arrows point to cells that at the time of the temperature shift were unbudded or large budded, respectively. (b) Execution point versus terminal point. The diagram follows roughly the fate of the cells indicated by arrows in Figure 1a. Shown is an example where the execution point is at around late G1–S phase, while the terminal point is at G2–M. Cells that are shifted to the non-permissive temperature while in G1 will immediately pass the execution point and arrest as large budded cells, at the terminal point (blue arrow). In contrast, a large budded cell will continue to progress through the cell cycle, and only its progeny will pass through the execution point and arrest at the terminal point (red arrow). Figure 1a adapted from Hartwell *et al.*, *Proceedings of the National Academy of Sciences of the U S A* 1970:**66**:352–359.

from mitosis, though the machinery responsible for cyclin B ubiquitination was not yet known.

To approach this question, the Kirschner group assayed cell extracts from *Xenopus laevis* (frog) oocytes, which readily progress through nearly every cell cycle process in a test tube, including DNA replication, spindle formation, chromosome condensation, and nuclear envelope breakdown and assembly. To identify the machinery involved in cyclin B destruction, Kirschner and colleagues labeled cyclin B with radioactive iodine and searched for an enzymatic activity that resulted in a ladder of higher molecular weight cyclin B, which is typical of ubiquitination (Figure 2). Upon fractionating the *Xenopus* egg extracts, they discovered that at least two of the required components were present in fractions Q1 and Q2 (Figure 2). Size fractionation of Q2 revealed that the activity of interest eluted as a very large 20S complex that contained

an E3 ubiquitin ligase. Previous genetic and biochemical studies in budding yeast suggested that a complex of several proteins, including Cdc16, Cdc23 and Cdc27, is needed for mitotic progression. The Kirschner group next showed that these very same proteins were indeed present in the active complex that they had identified (Figure 3). Based on its activity in cell cycle progression, they called the complex the anaphase-promoting complex (or APC).

Both yeast genetics and the fractionation of frog oocytes were critical to advancing our understanding of the cell cycle. Interestingly, the findings from the genetic screens and biochemical analysis turned out to be somewhat overlapping; proteins identified by mutations that led to defects in the cell cycle in yeast were found to have enzymatic activities required for cell cycle progression in the oocytes. These studies illustrate how

(a)

(b)

Figure 2 Identification of a cell cycle-regulated activity for cyclin B ubiquitination.

Figure 2 Identification of a cell cycle-regulated activity for cyclin B ubiquitination. (a) Fractionation scheme of *Xenopus* egg extract. The investigators used either mitotic egg extract (meaning that the extract was derived from oocytes that were in mitosis), or from interphase egg extracts. After subjecting the extract to centrifugation, which separated the insoluble material (pellet, P), membrane (M) fraction and supernatant (S100), the supernatant was subjected to further fractionation on an anion exchange column, resulting in two fractions, Q1 (the flow through, namely the material that did not bind the column) and Q2, which contained material eluted from the column with a gradient of 0.1–0.6 M KCl, and then combined. (b) Ubiquitination activity was followed by the appearance of a ladder on SDS-polyacrylamide gel electrophoresis (SDS-PAGE) of radiolabeled cyclin B ((^{125}I)-cyclin B). Only the supernatant fraction could promote ubiquitination. Neither fraction Q1 alone, nor Q2 alone could promote cyclin B ubiquitination, but the mixture of both was active. The investigators showed in this study that Q1 contains two different ubiquitin conjugating enzymes, either of which could promote cyclin B ubiquitination in the presence of the Q2 fraction, and that Q2 was active only when prepared from mitotic extracts, but not from interphase extracts (denoted (•) in the figure). The mitotic activity of Q2 was important because it suggested that the cell cycle-regulated activity of oocyte extract was present in this fraction.

Figure 3 The active fraction in Q2 contains the Cdc27 and Cdc16 proteins. Fraction Q2 from Figure 2 was subjected to a sucrose gradient, where proteins and complexes sediment according to their density. The starting material is loaded in the lane labelled L and the sucrose fractions are numbered 1–14. Using antibodies against Cdc27 and Cdc16, the investigators found that the fractions that are active in cyclin B ubiquitination (upper panels) contain Cdc27 and Cdc16 (lower panels).

different experimental approaches can complement each other to provide a more complete understanding of the biological process at hand.

King RW, Peters JM, Tugendreich S, *et al*. A 20S complex containing CDC27 and CDC16 catalyzes the mitosis-specific conjugation of ubiquitin to cyclin B. *Cell* 1995;**81**:279–288.

Hartwell LH, Culotti J, Reid B. Genetic control of the cell division cycle in yeast, I. Detection of mutants. *Proceedings of the National Academy of Sciences of the U S A* 1970;**66**:352–359.

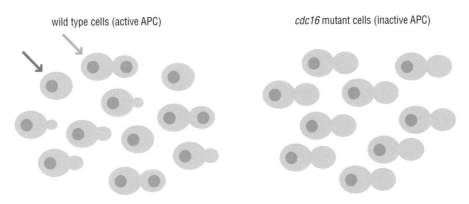

Figure 5.13 The anaphase-promoting complex (APC) is needed for progression to anaphase. The different cell cycle stages of budding yeast can be assessed by the size of the bud (from no bud at all in G1 to a large bud in mitosis) and by the state of the DNA (one nucleus before anaphase; two nuclei – one in the mother and one in the bud – after anaphase). A culture of wild-type cells (left panel) contains a mixture of cells in various cell cycle stages (see for example an unbudded cell indicated by the red arrow, and a cell in anaphase (large bud, two DNA masses), indicated by the green arrow). In contrast, in a culture of cells where one of the APC subunits was inactivated (for example, cells carrying a temperature-sensitive allele of the APC subunit that are grown at the non-permissive temperature), the cell cannot enter anaphase and they accumulate in mid-mitosis, as large budded cells with a single nucleus.

APC activation (Figure 5.14a). This phosphorylation promotes Cdc20 binding to the APC, which, in turn, promotes the ubiquitination, and subsequent degradation, of a protein that inhibits sister chromatid separation. Cdc20 also recruits mitotic cyclins for ubiquitination, thereby initiating the process of mitotic cyclin–Cdk inactivation. Thus, mitotic cyclin–Cdk promotes its own demise by activating the APC.

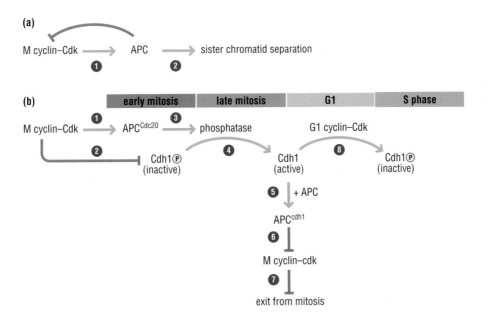

Figure 5.14 The anaphase-promoting complex (APC) and its activator subunits, Cdc20 and Cdh1, regulate progression through, and exit from, mitosis.
(a) The APC is activated through its phosphorylation by M cyclin–Cdk (step 1 in the regulatory circuit). Once activated, the APC promotes progression through mitosis by targeting anaphase inhibitors for degradation, which promotes sister chromatid separation (step 2). The APC also targets M cyclins for degradation (step 3), a process that is necessary for mitotic exit. Thus, by phosphorylating the APC, M cyclins activate the machinery that will eventually lead to their demise. (b) In budding yeast, the activation of the APC early in mitosis by M cyclin-Cdk phosphorylation (step 1) allows the APC to associate with Cdc20. At this point, Cdh1 is still phosphorylated by M cyclin–Cdk, which prevents Cdh1 from associating with the APC. APC^Cdc20 promotes progression through mitosis, leading to the activation of a phosphatase (step 3) that dephosphorylates Cdh1 (step 4). This allows Cdh1 to bind to the APC (step 5) and target M cyclin for degradation (step 6). M cyclin degradation promotes the transition from M to G1 (i.e. exit from mitosis, step 7). Once cells are in G1, the levels of G1 cyclins begin to accumulate, because they are not APC substrates. This results in the activation of the G1 cyclin–Cdk1 that phosphorylates Cdh1 (step 8), thereby inactivating the APC^Cdh1 and allowing S phase cyclins to accumulate (not shown). Activating processes are shown as green arrows while inhibitory processes are shown as red bars.

In yeast, Cdh1 is also phosphorylated by M phase cyclin–Cdk, but unlike the APC itself, this phosphorylation is inhibitory. Cdh1 remains phosphorylated until after anaphase, when a particular phosphatase becomes active (Figure 5.14b). Thus, by phosphorylating Cdh1, the cell cycle machinery ensures that Cdh1 will become active only after anaphase. Once the phosphatase removes the inhibitory phosphorylations from Cdh1, Cdh1 becomes active and, in association with the APC, targets M phase cyclins for degradation. This further reduces the activity of M phase cyclin–Cdk1, thereby making the removal of Cdh1 phosphorylation more efficient and enabling APC-Cdh1 to target more M phase cyclins for degradation.

The activity of M cyclin–Cdk1 blocks exit from mitosis. Therefore, the degradation of M phase cyclins, which leads to the inactivation of M cyclin–Cdk, promotes cell cycle progression from M phase to G1. Since Cdh1 becomes active only in *late* mitosis, this regulatory mechanism ensures the degradation of M phase cyclins, and consequently exit from mitosis, will only occur after the processes of mitosis have been completed.

APC-Cdh1 remains active in G1, thereby preventing the accumulation of M phase cyclin. This ensures that the cell cycle will not go back into mitosis. However, there must come a point where the cell cycle has to move past G1, and if Cdh1 remains active, S phase cyclins will not accumulate. Indeed, G1 cyclins are not targeted for degradation by APC-Cdh1, and when their level of activity rises, they phosphorylate and inactivate Cdh1. This allows the S phase cyclins to accumulate and promote the transition from G1 to S. Therefore, through protein degradation, the Cdk machinery can move the cell cycle forward.

5.6 CHECKPOINTS: INTRINSIC PATHWAYS THAT CAN HALT THE CELL CYCLE

The cell monitors certain cell cycle processes and can arrest at specific transition points

In addition to the mechanisms described earlier in this chapter that drive the cycle forward, the cell also has intrinsic surveillance mechanisms, often called checkpoint pathways, that can stop the cell cycle when certain things go wrong. Thus, checkpoint pathways can be viewed as quality control mechanisms that contribute to the fidelity of the cell division cycle. We discuss the discovery of the first checkpoint gene, and the rationale behind the checkpoint concept, in Experimental approach 5.3.

Different checkpoint pathways monitor various essential cellular processes, and activate a variety of responses when errors are detected. These responses include the activation of repair or correction pathways, and the induction of a delay in the progression of the cell cycle (or even cell cycle arrest). Cell cycle progression is typically halted at the next important cell cycle transition, and it provides the cell time to correct errors. In eukaryotes, cell cycle progression can be arrested at different points in the cell cycle – the G1-S transition, S phase, the G2-M transition and mitosis – depending on the checkpoint pathway that is activated (Figure 5.15).

Figure 5.15 Key cell cycle transitions are blocked by checkpoint pathways. Progression through the cell cycle in the presence of certain types of cellular damage can have dire consequences to the cell. In eukaryotes, different checkpoint pathways can arrest, or delay, cell cycle progression at various key transitions, as described later in this section. The DNA damage checkpoint pathway can arrest cell cycle progression at the restriction point (the G1–S boundary) or in G2/M. The DNA replication checkpoint pathway can inhibit further DNA replication during S phase and can delay entry into mitosis. The spindle assembly checkpoint pathway blocks the initiation of anaphase and the exit from mitosis.

5.3 EXPERIMENTAL APPROACH

Checkpoint pathways: delay and conquer

The presence of DNA damage poses a serious threat to cells because it can affect chromosome integrity. Scientists had noted that cells exposed to DNA damaging agents exhibited delayed cell cycle progression in G2, but until 1988 the mechanism responsible for this delay was not known. To understand the nature of this cell cycle delay, Ted Weinert and Lee Hartwell decided to use budding yeast *S. cerevisiae* to identify genes involved in regulating cell cycle progression in response to DNA damage. They reasoned that mutants in the process would exhibit a greater sensitivity to DNA damaging agents because they would not arrest and thus would continue through the cell cycle even in the presence of DNA damage.

At the time, there already existed a large collection of yeast mutants that were hypersensitive to radiation and thus termed *rad* mutants. Weinert and Hartwell decided to look through this collection to identify mutants in which cell cycle progression failed to arrest after irradiation. The assay they used was simple: cells from

asynchronous populations of haploid wild-type or mutant strains were placed on solid growth media, the cells were then exposed to x-rays, and cell cycle progression was followed in the microscope (by examining how many times each cell divided).

S. cerevisiae cells divide by budding, so Weinert and Hartwell scored the appearance of new buds to mark cells that were dividing; unbudded cells were counted as cells that were not dividing. Haploid cells were used for this assay because, if damaged in G1, they should exhibit the greatest cell cycle delay since they have neither a homologous chromosome nor a sister chromatid that can serve as a template for repair. Indeed, most unbudded cells on the plates exposed to radiation formed a bud (indicative of progression to G2), but then remained arrested as large budded cells for at least 10 hours (Figure 1, top panels). In contrast, the investigators found one mutant strain, *rad9*, in which unbudded cells failed to arrest after irradiation and divided multiple times within the 10-hour

before irradiation   10 hours after irradiation

wild type

rad9

Figure 1 No DNA damage-induced arrest for rad9 mutant cells. Wild-type and *rad9* cells were plated, x-irradiated at a dose of 2 krad, and then allowed to grow for 10 hours at 23 °C. At this temperature, the cell division cycle is about three hours long. Note that although some cells grew in the wild-type irradiated sample, many cells arrested as 'large budded' and in the 'wild type, after irradiation' panel are larger, most likely because these cells continue to accumulate mass but are unable to progress through the cell cycle.

time span (Figure 1, bottom panels). When the results shown in these images were quantified, the investigators found that 80% of unbudded wild-type cells arrested as large budded cells after irradiation, whereas less than 8% of unbudded *rad9* mutant cells arrested as large budded cells under the same conditions.

These findings led Weinert and Hartwell to propose the idea of cell cycle checkpoints, regulatory mechanisms that prevent a cell cycle event from taking place if an earlier event has not been properly completed. Of course, some steps in cell cycle progression cannot take place simply because they directly depend on the product of an earlier event. For example, in yeast, a spindle will not form if the spindle pole body (the yeast centrosome equivalent) did not duplicate, because the spindle pole bodies are essential structural components of the yeast spindle. In other cases, however, a regulatory mechanism prevents a cell cycle event from taking place before an earlier event has been completed. In the case of DNA damage, the DNA damage checkpoint pathway prevents progression through mitosis when

there is, for example, a double-stranded DNA break. This break, by itself, does not prevent the chromosomes from binding to spindle microtubules, nor is the break an impediment for chromosome segregation. Rather, the presence of the DNA break activates a regulatory mechanism, the DNA damage checkpoint pathway, which prevents cells from activating the processes that are necessary for mitosis.

Checkpoint-based control of the cell cycle, first elucidated in *S. cerevisiae*, is critical in all eukaryotes, and components of these checkpoints are conserved throughout evolution. It is interesting to note that mutations that affect the p53 protein, a major regulator of checkpoint function in human cells, are found in over 50% of all human cancers. Why is this? Mutations in p53 result in a failure of cells to arrest in response to DNA damage, and the continued growth of these cells leads to tumor formation.

Weinert TA, Hartwell LH. The *RAD9* gene controls the cell cycle response to DNA damage in *Saccharomyces cerevisiae*. *Science* 1988;**241**:317–322.

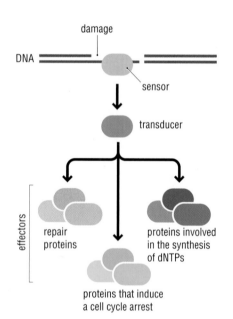

Figure 5.16 The structure of a checkpoint pathway. The example shown is of a checkpoint pathway induced by the presence of DNA damage in the form of a double-stranded DNA break. The sensor protein associated with the damaged site activates a transducer protein. The transducer, which could be a kinase, activates multiple substrates, called effectors. In the case of DNA damage, some effectors are involved in DNA damage repair (orange spheres), other may induce a cell cycle arrest (green spheres) while yet others contribute to the synthesis of deoxyribonucleotide triphosphate (dNTP) that are required for DNA synthesis after the damage is repaired (blue spheres). dNTP is a general term referring to dATP, dCTP, dGTP, and dTTP.

The checkpoint mechanism is not unique to eukaryotes; bacteria have checkpoint mechanisms as well. Both bacteria and eukaryotes have checkpoint pathways that monitor DNA integrity, and eukaryotes also have a checkpoint pathway that monitors the chromosome segregation process. In this section we will briefly touch upon the main properties of different checkpoint pathways in eukaryotes. The bacterial checkpoint, called the SOS response, will be described when we discuss the bacterial cell cycle later in this chapter. We also discuss the eukaryotic checkpoints in greater detail in later chapters in the context of the processes that they monitor.

Checkpoint pathways are composed of sensors, transducers, and effectors

The activation of a checkpoint pathway in response to the presence of a cellular defect requires proteins, referred to as sensors, which sense the cellular defect and activate the checkpoint response. For example, DNA damage is often associated with the formation of long stretches of single-stranded DNA (Figure 5.16). In this case, the activation of the checkpoint pathway is dependent on the binding of sensor proteins to the single-stranded DNA. Different types of cellular damage recruit distinct sensor proteins, and this recruitment is necessary for the appropriate cellular response. Specifically, the involvement of damage-type specific sensors ensures that the correct checkpoint pathway is triggered, and that other checkpoint pathways remain dormant.

Once sensors associate with the damage, they activate a second set of proteins, called transducers, whose function is to launch the checkpoint pathway response. There are different types of transducers, such as protein kinases and proteases. The common function of all transducers is to activate a third group of proteins, called effectors, which actually carry out the checkpoint pathway response (see Figure 5.16). For example, transducers that are activated in response to DNA damage will activate effector proteins that are involved in repair of DNA damage, inhibition of cell cycle progression, and production of nucleotides, the building blocks for DNA synthesis. By activating multiple effectors, transducer proteins can initiate many processes simultaneously, which act together to effectively repair the damage.

Insults to DNA integrity can induce checkpoint pathways that activate repair and stop cell cycle progression

Both bacteria and eukaryotes have mechanisms to respond to DNA damage, although the sensors, transducers, and effectors involved in these processes are quite distinct. There are two main forms of damage (or 'insults') to DNA that are monitored by the cell: stalled DNA replication forks and DNA double strand breaks.

During DNA replication, double-stranded DNA is separated into single strands, with each strand forming the template for a new daughter strand. During this process, the replication fork usually moves along the DNA without perturbation, unwinding the DNA into single strands as it goes. Occasionally, however, replication is impeded – for example, due to a block on the DNA being copied, or due to insufficient levels of nucleotides from which to synthesize the DNA. When this happens, DNA replication stops and single-stranded DNA accumulates. Damaged DNA, such as a DNA double strand break, also results in the formation of single-stranded DNA.

With both stalled replication forks and DNA damage, it is the single-stranded DNA that is recognized by sensor proteins and which signals checkpoint activation. (Note, however, that somewhat different sensor proteins get recruited to single-stranded DNA in the case of replication block versus DNA damage such as a double-stranded DNA break.) These sensor proteins activate transducers, in the form of protein kinases, which in turn activate a variety of effectors that lead to a range of cellular responses, including DNA repair and cell cycle arrest.

The stage at which cell cycle arrest occurs depends on when the damage is detected. For partially replicated DNA, cell cycle progression will stop in S phase, and the checkpoint will inhibit mitotic cyclin–Cdks to ensure that mitosis does not take place with partially replicated DNA. For DNA damage that is detected in S phase or in G2, the checkpoint pathway will block entry into mitosis by inhibiting the activation of mitotic cyclin–Cdk and arresting at G2-M transition (for example, by preventing Cdc25 activation, see Figure 5.8). When the DNA damage is detected in G1, the checkpoint pathway will block passage through the restriction point by inhibiting G1-cyclin–Cdk activity. Indeed, if cell cycle progression continues with broken or partially replicated DNA, the consequences would be disastrous to the daughter cells, which would inherit incomplete genomes.

In mammalian cells, if the perturbation to DNA integrity persists for a long time, checkpoint pathways will induce programmed cell death (apoptosis) to eliminate cells that cannot be properly repaired. Thus, in response to defects in the DNA, checkpoint pathways activate repair mechanisms, block cell cycle progression, and may trigger cell death.

The spindle assembly checkpoint pathway inhibits progression through mitosis

In addition to mechanisms that monitor DNA integrity, eukaryotic cells also monitor the processes required to correctly segregate chromosomes in mitosis. In Chapter 4 we discussed the mitotic spindle that attaches to chromosomes and

⊝ The checkpoint that is activated in response to DNA damage is discussed extensively in Chapter 12.

⊝ The process of DNA replication is discussed in Chapter 6.

moves them to the daughter cells. We will discuss the segregation process, as well as the structure and function of the spindle, in more detail in Chapter 7. A crucial step in the chromosome segregation process is the attachment of the chromosomes to the spindle: lack of attachment or an incorrect attachment can result in the segregation of both copies of a chromosome to one daughter cell and no copies of that chromosome to the other daughter cell.

To ensure proper chromosome segregation, the spindle assembly checkpoint pathway monitors the attachments between chromosomes and the spindle. When a chromosome is not attached to the spindle, for example, the spindle checkpoint pathway is activated, although the exact nature of this pathway's sensor is not yet known. Numerous proteins, some of which are protein kinases, then participate in the transduction of the signal to the main effector of this pathway, the APC. As mentioned earlier, APC activity is required for progression through mitosis. The spindle assembly checkpoint pathway results in APC inactivation, thereby blocking both anaphase initiation and mitotic exit. Thus, as long as there are chromosomes that are not attached to the spindle, the spindle assembly checkpoint pathway will inhibit progression through mitosis, thereby providing more time for the chromosome to become attached to the spindle. Once proper spindle attachments are formed, the checkpoint signal is extinguished and progression though mitosis can resume.

The importance of checkpoint pathways is underscored by the fact that mutations in some checkpoint genes are associated with disease. For example, mutations in the gene encoding for the DNA damage sensor **ATM** are the underlying cause of the disorder ataxia telangiectasia. Ataxia telangiectasia has multiple manifestations, one of which is an increased risk of cancer. ATM is a kinase, and some of its targets are well-known cancer-associated genes, such as p53 and BRCA1.

Thus, checkpoint pathways ensure the integrity of the cell cycle process, and failure in checkpoint pathway activity can result in dire consequences.

⮕ For more on ATM, see Chapter 12.

5.7 EXTRINSIC REGULATORS OF CELL CYCLE PROGRESSION

In the previous sections, we have seen how the monitoring of events within the cell leads to the regulation of cell cycle progression. However, it is not just the internal environment of the cell that influences this progression: factors in the environment surrounding the cell are also important mediators of the cell cycle. We consider some of the key environmental factors – so-called extrinsic regulators – in this section.

Cell growth and division are regulated by extracellular growth factors and mitogens

In addition to internal signals, such as those generated by checkpoint pathways, that regulate cell cycle progression, cells also respond to signals from their environment. These signals include growth factors and **mitogens**, both of which promote the passage through the restriction point (Figure 5.17). Growth factors, as their name implies, are molecules that stimulate cells to increase their mass.

Figure 5.17 External regulation of cell proliferation. External regulation of entry into G1 can be triggered by growth factors and mitogens. Activation at the cell surface leads to a signaling cascade (not shown) that stimulates G1 cyclin–Cdk activity. G1 cyclin–Cdk complexes then promote the activation of gene regulatory proteins that stimulate the expression of numerous G1/S genes, including genes encoding G1/S cyclins, S cyclins and the proteins that carry out the events of the early cell cycle. The resulting G1/S- and S–Cdk activities result in passage through the restruction point and irreversible commitment to cell division.

In some cell types this increase in mass is necessary in order to pass through the restriction point, whereas in others, growth factors stimulate cell growth without cell division. In contrast, mitogens directly stimulate cells to divide and proliferate without stimulating growth, by activating the cell cycle machinery that drives cells past the restriction point.

Growth factors increase growth rate and stimulate division indirectly

In certain cells, where progression through the restriction point requires that cells reach a minimum size threshold before division can occur, growth factors can stimulate cell division, and thus increase proliferation, by promoting an increase in cell mass. Many single-celled eukaryotes monitor the levels of various nutrients in the environment and adjust their rates of growth and metabolism accordingly. If the concentration of important nutrients decreases, cells respond by decreasing their growth rate.

The dependence of cell division on cell growth results from links between the cell's metabolic machinery and the cell cycle control system. In budding yeast, for example, changes in the overall rate of protein synthesis affect the activities of both the G1 cyclin–Cdk1 and G1–S cyclin–Cdk1. There is also a link between growth rate and division in mammalian cells. However, the precise mechanism that links growth and G1 entry is not yet clear.

Growth rate is regulated by a protein kinase called **TOR**, which is central to the response to nutrients and growth factors. Growth factors activate a receptor by binding to it at the cell surface, and this leads to the initiation of a signaling cascade. In this cascade, one regulator stimulates the activity of the next in a series of steps that can be modulated at many levels. A simplified version of the cascade is shown in Figure 5.18, with the TOR kinase being activated at the end of the cascade. Activation of TOR leads to the subsequent activation of many processes, chief among them being an increase in the rate of protein synthesis, which drives an increase in cell mass.

Mitogens control the division of mammalian cells

Cells of mature multicellular organisms usually divide only when the maintenance or repair of tissues is required. Many mammalian cells regulate this kind of cell proliferation by responding to external mitogenic signals. Mitogens are generally soluble peptides or small proteins secreted by neighboring cells, and directly signal cell division by activating cyclin–Cdks. Some are highly specific regulators of division in just one cell type, whereas others have much broader action on many cell types.

An example of a mitogen is **epidermal growth factor (EGF)**, a soluble polypeptide that controls the initiation of division in many different cell types. Though the name that was originally given to this polypeptide was 'growth factor', we now know that it acts as a mitogen by directly stimulating entry to the cell cycle at G1. An example of how a mitogen activates cell proliferation is shown in Figure 5.19.

Mitogens bind to a receptor that initiates a signal transduction cascade involving multiple protein kinases. The presence of multiple steps gives the cell numerous

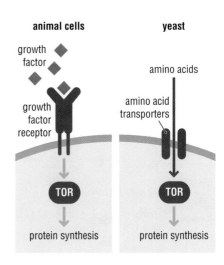

Figure 5.18 Growth factors and nutrients regulate cell division though increasing growth rate. In both unicellular eukaryotes and metazoans, cell growth is sensitive to stimuli from outside the cell. In organisms such as yeast (right panel), one of the environmental cues that is translated into growth signals is nutrient availability, such as the abundance of amino acids. These amino acids are transported into the cell by specific amino acid transporters. In animal cells (left panel), cell growth is often dependent on growth factors, which bind growth factor receptors that have both an extracellular and an intracellular component. Upon growth factor binding to the extracellular part of the receptor, a signaling cascade is triggered by the intracellular part of the receptor. In both animal cells and yeast, signals from nutrients and growth factors converge on the TOR kinase, which controls cell growth in part by regulating protein synthesis.

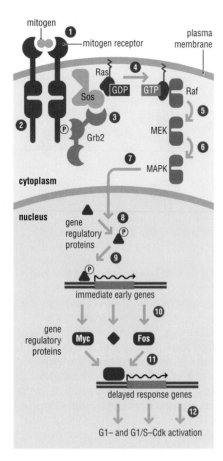

Figure 5.19 Mitogens stimulate cell division through a signaling cascade. The mitogen receptor has an extracellular component, which can bind a mitogen, and an intracellular component, which, upon mitogen binding, can activate a signal transduction cascade. Mitogen binding to its receptor (step 1 in this signaling cascade) results in receptor dimerization (step 2). This triggers the recruitment of several proteins (in the example shown: Sos and Grb2) to the plasma membrane (step 3), which activate the membrane-associated Ras protein (by exchanging Ras-bound GDP with GTP) (step 4). Activated Ras, in turn, activates the first of a series of cytoplasmic kinases, each of which phosphorylates the next kinase in the cascade (Raf phosphorylates mitogen-activated or extracellular signal-regulated protein kinase (MEK, step 5)), which phosphorylates mitogen-activated protein kinase (MAPK, step 6)). The last kinase in this cascade, MAPK, is transported into the nucleus (step 7), where it activates, through phosphorylating a transcription factor (step 8), the expression of a series of genes (step 9), including the genes coding for Myc and Fos (step 10). Myc, Fos, and other proteins then activate the expression of another set of genes (step 11), including G1 and G1–S cyclins (step 12). Thus, through this cascade, the mitogen signal is transmitted from outside the cell, through the plasma membrane, to the cytoplasm and into the nucleus, where it induces the expression of genes required for cell proliferation.

opportunities to regulate this process. Through this cascade, the signal is transduced from outside the cell all the way into the nucleus. This cascade results in the activation of various G1 and S genes, including cyclins. Thus, this cascade links cell surface signaling mediated by the mitogen to entry into the cell cycle. For example, following mitogen stimulation, the G1 cyclin, cyclin D binds and activates Cdk4, which phosphorylates Rb. As we saw earlier in Section 5.5, Rb phosphorylation frees E2F to activate G1 gene expression and this, in turn, promotes cell cycle progression.

In some cases, a single extracellular protein acts both as a growth factor and as a mitogen, resulting in a coordinated increase in the rates of both growth and division. As we will see in the next section, dysregulation of various steps in this pathway can lead to uncontrolled cell division and cancer.

5.8 THE CELL CYCLE AND CANCER

Tumor cells lack growth controls and accumulate mutations

Cancer is a disease that derives primarily from uncontrolled cell division. This inappropriate accumulation of cells can result from either inappropriate cell growth or from insufficient cell death. It is of little surprise, therefore, that the mechanisms that regulate cell cycle progression are frequently altered in cancer. There are three primary changes occurring in cells that fuel cancer growth: dominant activation of extrinsic signaling pathways, loss of regulatory mechanisms (including checkpoint pathways that normally limit cell growth), and genetic alterations that allow the accumulation of multiple mutations.

As we have seen in this chapter, there are multiple overlapping controls to the cell cycle. In order for a cell to become a tumor cell and grow uncontrolled, often more than one event that alters cell growth has to take place. The multiple mutations required for a cell to become cancerous do not occur all at once but rather accumulate over time. For example, one mutation may happen in one cell that allows it to divide more frequently and outgrow the cells that surround it (Figure 5.20). These early cancer cells now represent a significantly larger fraction of the cells and, if additional growth control pathways are mutated, these cells will have a continued advantage. In this way, the cancer cells accumulate mutations and rapidly outgrow the other cell and thus take over the population.

The role of accumulated mutations in the evolution of cancers is highlighted by the increased incidence of cancer as people get older. The accumulation of multiple genetic changes that promote cancerous growth can take years. When cancer appears in relatively young individuals, it is suspected that these individuals have a genetic predisposition for cancer. An example are women with mutations in the BRCA1 gene, who develop breast cancer at a much high frequency than the general population.

While there are many factors that are needed for cancer cells to become a tumor, including changes in the surrounding tissue, here we will focus on the changes in the cancer cell itself that alter cell cycle regulation and lead to increased growth. The genes that are altered in cancer are often classified into two general groups, those that are encoded by **oncogenes**, that promote tumor growth, and those that are encoded by **tumor suppressor genes**, that normally halt the cell cycle but fail to do so in cancer.

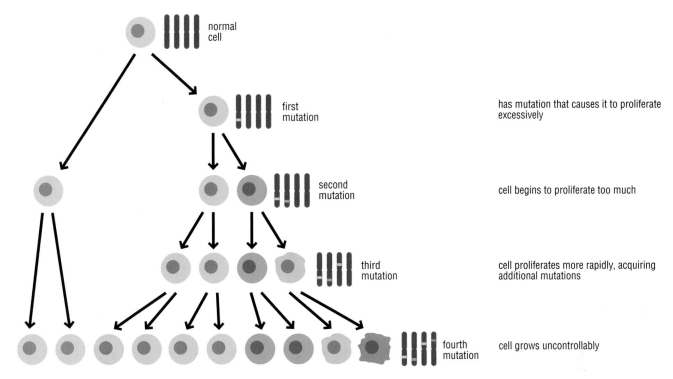

normal cell

first mutation — has mutation that causes it to proliferate excessively

second mutation — cell begins to proliferate too much

third mutation — cell proliferates more rapidly, acquiring additional mutations

fourth mutation — cell grows uncontrollably

Figure 5.20 Evolution of a tumor. Tumor evolution begins with a mutation (shown as a pale red mark on one of the chromosomes) that provides a cell with a competitive advantage – perhaps, as shown here, by enabling the cell to proliferate more rapidly than its neighbors (note that the red/pink cells divide more rapidly than the normal gray cells). Over a period of many years, the descendants of that mutant cell acquire additional mutations (notice the accumulation of red and pink marks on the other chromosomes) that allow them to overcome the various regulatory barriers that restrain cell proliferation.

Oncogenes are often aberrant versions of positive regulators of the cell cycle

The cells of multicellular organisms normally proliferate in response to growth factors and mitogens. As discussed in the previous sections, mitogens and growth factors are essential extrinsic regulators of cell cycle progression, which usually stimulate entry into the cell cycle at G1. The uncontrolled growth of cancer cells is frequently due to defects in the mitogenic signaling pathway (Figure 5.19). Oncogenic proteins may function at any level in this pathway. In some cases, cells aberrantly make and secrete a stimulatory protein that signals its own cell division. This **autocrine loop** stimulates a cell to promote its own division as well as that of its surrounding cells (Figure 5.21).

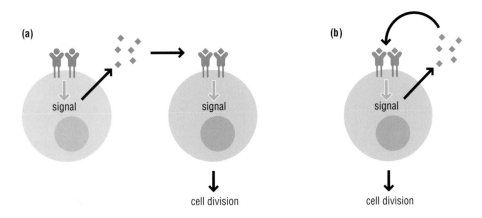

(a)

signal

signal

cell division

(b)

signal

cell division

Figure 5.21 Cell signalling. (a) Under normal conditions, cell proliferation would be stimulated only when appropriate signals are generated, for example, by neighboring cells. Under these conditions, presence of secreted stimulating signal (pink diamonds) from the blue cell will trigger cell division in the gray cell, while absence of a stimulating signal will maintain the cell in a non-dividing state. (b) Under certain conditions, cancer cells generate their own inducing signals, generating an autocrine loop. These signals occur irrespective of the neighboring cells, and as such, independent of whether the tissue requires additional cell divisions. In this case, the cell divides uncontrollably, stimulated by its own signals.

Another way that an oncogenic protein may stimulate cell division is by the constitutive activation of one of the effectors in a mitogenic pathway. A gene that normally would be activated by a signal from the outside may acquire a mutation that makes it active in the absence of that signal. An example of such an oncogene is *ras*, a component of the pathway that induces cell cycle entry following mitogenic stimulation (see Figure 5.19). As we saw in the previous section, the activation of Ras requires the association of a mitogen with its receptor, followed by the recruitment of proteins to the plasma membrane that are necessary for Ras activation. However, the mutant version of the *ras* gene is constitutively active – that is, it can induce the protein kinase cascade in the absence of mitogen. Consequently, the cell responds as if there is mitogen present, even though there is none, and activates the cell division pathway.

Similarly, the machinery of the cell cycle itself may be altered such that the cell enters the cell cycle at G1 constitutively without waiting for the appropriate signal. For example, in some cancers mutations in cyclin D render it insensitive to the cyclin-dependent inhibitors described in Section 5.3, which regulate the activity of cyclin–Cdk kinases.

Another kind of oncogene is one that codes for a protein that normally protects cells against cell death. As we will discuss in detail in Chapter 13, cells often undergo apoptosis or cell death when inappropriate or conflicting signals are present. Overexpression of some genes will block the cell death signals and allow continued growth of cells that would otherwise die. If this occurs in cells that are proliferating, it can lead to considerable expansion of one type of cell, which may, in turn, become cancer.

Tumor suppressor genes are often defective negative regulators of the cell cycle

The uncontrolled growth of tumor cells is often due to their failure to respond to normal signals that inhibit the cell cycle. In such cases, the defect is often in proteins that would normally stop growth. These tumor suppressor mutations can be in checkpoint genes that normally signal for repair and also arrest the cell division when things go amiss in the cell cycle. For example, **p53** is regulated by checkpoint kinases that are activated in response to DNA damage, as described in Section 5.7.

For many cancers, loss of p53 function occurs during the multistep mutational process that results in tumor growth. In fact, greater than 50% of cancers have mutations in p53. If p53 is not functional, cells live, grow, and accumulate more mutations rather than arresting or dying in response to damage or checkpoint signals. Consequently, these cells have an advantage and outgrow their p53 wild-type neighbors, ultimately generating a tumor.

In addition to mutations in checkpoint genes, tumor suppressors can also be normal negative regulators of cell cycle progression. One example is the Rb transcriptional regulator that we described earlier in this chapter. A mutation in Rb that blocks its ability to inhibit the E2F transcription factor (see Figure 5.11), can promote inappropriate entry into the cell cycle in G1. This pathway that regulates G1 entry is so central to cell cycle progression that the vast majority of cancers have mutations in one of the components of this regulatory cascade.

⊙ We see further in Chapter 12 that p53 plays an essential role in the response to DNA damage by triggering either cell cycle arrest or apoptosis.

The link between cell cycle regulation and cancer underscores the importance of understanding the cell cycle machinery. In the next two chapters will we explore various aspects of cell cycle regulation in the context of specific processes: DNA replication (Chapter 6) and chromosome segregation (Chapter 7). As you read through these chapters notice the intricate relationship between the cyclin–Cdk machinery and the proteins that it regulates. It is then easy to understand how mis-regulation can lead to dire consequences for the cell.

5.9 THE BACTERIAL CELL CYCLE

Like single-celled eukaryotes, the rate of bacterial growth is determined by environmental conditions. For example, bacteria accumulate faster in chicken soup, which is nutrient rich, than in tap water, which is nutrient poor. Much like eukaryotes, where passage through the restriction point occurs only when conditions are right, bacteria also have pathways that translate internal and environmental cues, such as nutrient availability, into signals that regulate the cell cycle. Moreover, certain bacteria inhibit cell cycle progression in the presence of DNA damage, again in a similar manner to eukaryotes. Most importantly, however, both eukaryotes and bacteria have cell cycle machineries that are made up of hardwired signaling pathways that drive cell cycle progression.

Unlike eukaryotes, however, in which the cell cycle machinery is conserved across species, different bacteria utilize distinct mechanisms to regulate the cell cycle. This is likely the result of evolutionary diversity caused by the broad range of ecological niches that bacteria inhabit. These may have posed distinct selective pressures that require different cell cycle paradigms. As researchers began to study cell cycle regulation in different bacteria, they uncovered a menagerie of intricate pathways and processes, many of which had no parallels in other bacteria. In the study of the eukaryotic cell cycle, researchers often rely on discoveries made in one organism to decipher similar processes in another. This is not always possible when studying the bacterial cell cycle. As a result, our knowledge of cell cycle regulation in bacteria is less complete than that of eukaryotes, and there is still a wealth of information relating to pathways, proteins and processes remaining to be discovered. In this section we will discuss general features of the bacterial cell cycle and examine specific examples from different bacteria.

The main landmarks of most bacterial cell cycles are DNA replication and cell division

Under slow growth conditions, such as in poor nutrient conditions, the bacterial cell cycle can be divided into three main periods (Figure 5.22a):

- 'B period' – during which the bacterium increases in mass and size
- 'C period' – during which DNA replication and chromosome segregation take place
- 'D period' – the time between the end of DNA replication and the subsequent cell division, during which chromosome segregation and division are completed.

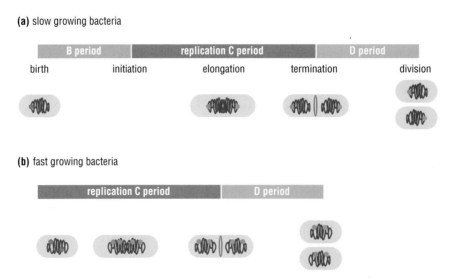

(a) slow growing bacteria

(b) fast growing bacteria

Figure 5.22 The periods of the bacterial cell cycle. (a) Under slow growth conditions, the bacterial cell cycle can be divided into three periods: the 'B period', which is from when the cell is born until DNA replication begins; the 'C period', which encompasses DNA replication and chromosome segregation; and the 'D period', which is the interval from the end of the C period until cell division takes place. Under these conditions, the initiation of DNA replication from the replication origin (red dot) duplicates the origin region (blue and orange dots) and eventually the entire chromosome. This is followed by cell division, which gives rise to two cells, each with a single origin and a completely replicated chromosome. (b) Under fast growth conditions, some bacterial strains can divide before replication is completed. For example, in the cell shown, the cell is born with DNA that has already begun to duplicate in the parent cell, as seen by the presence of separated origins of replication (in red). In this cell, a new round of replication is initiated during the C period: the red origin on the left is duplicated to become the two blue origins, while the red origin on the right is duplicated to become the two orange origins. These origins segregate the same as a single origin would, namely that one segregates to the left cell pole while the other segregates to the right (each cell half contains one blue and one orange origin), giving rise to two cells with origins of replications that have already been duplicated. Under these growth conditions DNA replication is ongoing as the cell is born and there is no B period.

Under conditions that support rapid growth (Figure 5.22b), such as nutrient-rich growth conditions, cells grow in size faster and cell division occurs more frequently, to the point where the length of the cell cycle is shorter than the time it takes complete DNA replication. This is possible because in these cells, a new round of replication begins before the previous one has been completed, a phenomenon known as **multifork replication**. Under these conditions, the 'B period' is eliminated and the newborn cells, which already have partially duplicated chromosomes, continue with DNA replication as soon as they are formed. This situation is different from eukaryotic cells, where the presence of partially replicated DNA will stop cell cycle progression.

⊃ The regulation of DNA replication is described in Chapter 6.

Bacterial cell growth and division are sensitive to nutritional cues

The mechanistic details of bacterial DNA replication, chromosome segregation, and cell division will be discussed in the coming chapters. Here we will focus on the regulation of cell cycle events. For example, what dictates when a bacterial cell divides or when it initiates DNA replication? Bacteria growing in nutrient-rich growth media tend to be larger than those grown in nutrient-poor media, apparently to accommodate the greater amount of DNA they contain due to multifork replication. A larger cell size suggests that the cell division machinery has to be sensitive to nutritional cues. Indeed, in *Bacillus subtilis*, cell division is inhibited by a metabolic pathway that senses the amount of available glucose: under rich growth conditions, where glucose is abundant, an enzyme called UgtP localizes to cell midline and inhibits cell division, thereby forcing the cell to reach a larger size before the cell divides (Figure 5.23).

(a) nutrient-rich conditions

(a) nutrient-rich conditions

(b) nutrient-poor conditions

Figure 5.23 The regulation of cell division by the *Bacillus subtilis* UgtP. Cell division is mediated by the FtsZ ring, shown in red (the nucleoid is shown in blue). Under nutrient-rich conditions (a), cell division is delayed to allow the cell to increase in size. This delay in cell division is mediated by the protein UgtP (in green), which localizes to the cell midzone (the future site of cell division) and binds to FtsZ to inhibit its activity. When the cell reaches the appropriate size, inhibition by UgtP is alleviated by an unknown mechanism and cell division takes place. Under nutrient-poor conditions (b), UgtP is sequestered in foci away from midcell where it cannot inhibit cell division. Hence, under nutrient-poor conditions division occurs at a substantially smaller cell size than during growth in a nutrient-rich medium.

UgtP specifically inhibits the assembly of FtsZ, a protein that forms a ring-like structure and that is needed for cell division in bacteria. Under poor growth conditions, UgtP levels are reduced and the remaining protein is sequestered away from the division machinery in small, punctate foci. As a result, under these conditions division takes place at a smaller cell size. How the inhibition of division by UgtP is relieved when cells reach the appropriate size is not yet known.

Cell size also impacts the timing of initiation of DNA replication. In *Escherichia coli*, the ATP-binding protein DnaA is the primary factor implicated in coordinating cell growth with DNA replication. DnaA is essential for the initiation of DNA replication. About 20 molecules of DnaA have to bind to DNA sites near the origin of replication in order for replication to begin. Thus, processes that affect the association of DnaA with its binding sites will affect when initiation of DNA replication occurs.

⮕ FtsZ is discussed in detail in Chapter 7.

⮕ The process of DNA replication in bacteria, and DnaA action in particular, is discussed in detail in Chapter 6.

Caulobacter crescentus has defined cell cycle stages

The study of cell cycle regulation in any organism or cell type is often facilitated by the existence of visible landmarks, such as the size of the bud in dividing budding yeast, or the organization of chromosomes in animal cells. These landmarks change as the cell progresses through the cell cycle, allowing researchers to easily determine specific cell cycle stages. Determining the cell cycle stage of bacteria is often much harder because many lack easily detectable features. For example, the only morphological cell cycle change in the rod-shaped *E. coli* is the length of the organism. Perhaps for this reason, many advances in our knowledge of cell cycle regulation in bacteria came from *Caulobacter crescentus*, a crescent-shaped bacterium, which has several clearly visible structures (Figure 5.24), including a stalk, a flagellum and several pili. Importantly, the appearance and disappearance of these structures is tightly linked to cell cycle events such as chromosome replication and segregation.

Another reason *C. crescentus* emerged as a powerful model organism for cell cycle studies in bacteria is the way it replicates its DNA. Many types of bacteria, including *E. coli*, can have overlapping replication cycles in a single cell cycle when growth conditions are favorable. This means that the daughter cells have chromosomes that have already begun the duplication process for the next generation when the bacterium divides. This makes it difficult to study the relationship between the

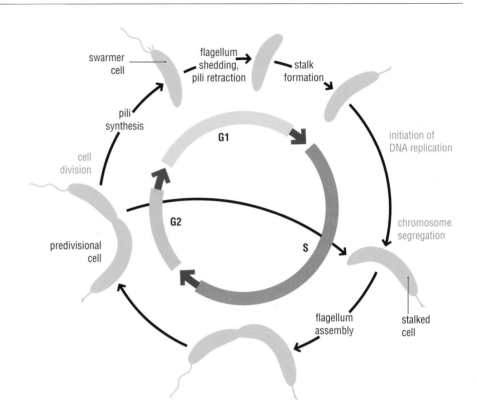

Figure 5.24 The cell cycle phases of *Caulobacter crescentus.* The cell cycle phases (G1, S, G2), key morphological changes, and cell cycle events (pink text). See text for further details.

○➔ The mechanism of chromosome segregation in *C. crescentus* and other bacteria is discussed in Chapter 7.

initiation of DNA replication and other cell cycle events, for example, and limits the experiments to those done in poor growth conditions. In contrast, *C. crescentus* initiates and completes one cycle of DNA replication in the same cell cycle, allowing the demarcation of an S phase, much like in eukaryotes (Figure 5.24).

During the *C. crescentus* S phase, chromosomes are replicated and segregated concomitantly. This S phase is preceded by a gap phase (called G1), during which the flagellum and pili are lost and the stalk begins to form, and it is followed by a second gap phase (called G2), during which cell division takes place (Figure 5.24).

The various *C. crescentus* cell cycle stages are characterized by the expression of distinct sets of genes. Analysis of gene expression revealed that over 550 *C. crescentus* genes, equivalent to 19% of the *C. crescentus* genome, are expressed in a cell cycle-regulated fashion. Three key transcriptional regulators involved in the cell cycle are CtrA, GcrA, and DnaA (Figure 5.25a, b). Phosphorylated CtrA is found in swarmer cells (which are in G1) and predivisional cells (which are cells before cell division), where it inhibits the initiation of DNA replication by binding to the origin of replication. This prevents the binding of DnaA, which, as we discussed earlier, is necessary for the initiation of DNA replication. CtrA also inhibits cell division, and other morphological changes that should occur later in the cell cycle, by inhibiting the expression of the *gcrA* gene. GcrA is needed for the expression of genes involved in late stages of the cell cycle. When the cell enters S phase, corresponding with the transition from a swarmer to a stalked cell, CtrA is degraded and DnaA is stabilized. (Note that, unlike in *E. coli*, DnaA levels in *C. crescentus* do change during the cell cycle; *C. crescentus* DnaA levels are regulated by both proteolysis and at the gene expression level.)

DnaA is not only needed for the initiation of DNA replication, but also controls the transcription of a variety of genes, such as *gcrA* and other genes needed for growth, DNA replication, and cell division. GcrA activates additional genes required for

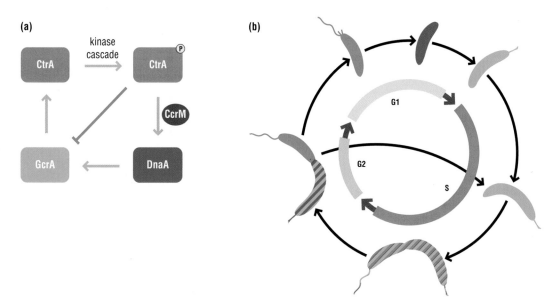

Figure 5.25 The cell cycle regulatory circuit in *C. crescentus*. (a) CtrA is phosphorylated through a kinase cascade that is present in the predivisional cell and swarmer cell. Phosphorylated CtrA inhibits initiation of DNA replication and the expression of genes needed for late stages in the cell cycle, including GcrA (illustrated by red bar). CtrA induces the expression of DnaA indirectly, by inducing the expression of an enzyme, CcrM, that methylates the *dnaA* gene, enabling its expression. Once the cell develops a stalk, CtrA is degraded (not shown) and DnaA induces both DNA replication and the expression of numerous genes including *gcrA*. The GcrA protein, in turn, activates the expression of genes needed for S phase and G2. It also induces the expression of *ctrA*, thus completing the cycle. (b) The presence of CtrA (in red), DnaA (in dark blue) and GcrA (in blue) throughout the cell cycle. The hatched marks indicate the presence of two proteins.

S phase and G2, and it also activates the expression of the gene coding for CtrA. Late in the cell cycle, CtrA is phosphorylated, allowing it to bind the origin of replication and to prevent further cycles of DNA replication until the next cell cycle. This regulatory circuit, and many other additional processes that impinge on this circuit, drive cell cycle progression forward in a manner similar to eukaryotic regulatory loops.

C. crescentus is not only a good model for cell cycle studies, it also provides insights into processes associated with asymmetrical cell division. On cell division, two different cell types are formed: a swarmer cell, which can move thanks to its flagella, and a stalked cell, which is immobile (Figure 5.24). The degradation of CtrA occurs only in cells that have a stalk (Figure 5.25b), thanks to the accumulation of the CtrA degradation machinery near the stalked pole of the cell. Thus, immediately after cell division, the stalked cell contains no CtrA and can initiate DNA replication immediately, while the swarmer cell has stable and active CtrA, and the cell must first differentiate into a stalked cell before CtrA is degraded and the cell can progress into S phase.

In response to DNA damage, the bacterial SOS response induces transcription of DNA repair genes and inhibits cell division

We have seen how DNA replication and cell division are controlled during normal cell growth. In some cases, however, cell cycle progression needs to be halted in order to permit repair of cellular damage, such as DNA damage, and prevent the transmission of damaged genetic material to daughter cells. In eukaryotes, this type of regulation is mediated by checkpoint pathways, such as the DNA damage and DNA replication checkpoint pathways (see Section 5.7). By contrast, in many, but not all, bacteria, the presence of DNA damage induces a cellular response

Figure 5.26 The SOS response. In *E. coli*, LexA (in pink) serves as a repressor of roughly 40 genes by binding as a dimer to the SOS box in their promoter region, thereby blocking SOS access to RNA polymerase. In the presence of single-stranded DNA (which can form as a result of a block to DNA replication or resection of one of the strands from a double-strand DNA break), the RecA protein (in orange) oligomerizes on the single-stranded DNA. This process activates the protease activity of RecA. Activated RecA then cleaves LexA, which allows RNA polymerase to transcribe LexA-controlled genes.

➲ FtsZ and bacterial cell division is discussed further in Chapter 7.

called the SOS response. The molecular aspects of the SOS response will be described in more detail in Chapter 12, and here we will discuss how the SOS response affects the cell cycle.

The SOS response was the first checkpoint pathway to be discovered, although its discovery preceded the coining of the checkpoint term. In the SOS response, the presence of DNA damage leads to the formation of single-stranded DNA. The binding of a protein called RecA to this single-stranded DNA results in the activation of proteolytic activity (Figure 5.26.). The protease activity then cleaves and inactivates a transcription repressor, called LexA, leading to the induction of genes involved in repair and cell cycle arrest. If we compare the SOS response to eukaryotic checkpoint pathways (Section 5.7), then in the SOS response RecA serves as both sensor and transducer, while LexA is the effector protein.

The mechanism of the SOS response is best characterized in *E. coli*. LexA inhibits the expression of roughly 40 genes, including the *sulA* gene, which regulates cell division. Specifically, the SulA protein binds to the FtsZ protein, which is part of a contractile ring necessary for the formation of the septum that separates the two daughter cells. The binding of SulA to FtsZ inhibits the formation of the FtsZ ring, thereby blocking cell division. Taken together, the presence of DNA damage leads to the activation of the SOS response, which, in turn, inhibits cell division, thereby preventing the segregation of broken DNA molecules to the daughter cells.

C. crescentus also has an SOS response that controls over 30 genes, but *C. crescentus* does not contain a homolog of the *E. coli sulA* gene, and thus the mechanism by which the SOS response affects the *C. crescentus* cell cycle machinery is not known. This is perhaps not surprising because in several bacteria, including *B. subtilis*, the SOS response inhibits cell division through proteins that are not homologous to SulA. Thus, while the cellular consequences of SOS response induction are conserved, the proteins that mediate the effect are often not.

✱ SUMMARY

In this chapter we learned about the machinery that drives and regulates cell cycle progression in both eukaryotes and bacteria. We saw how in eukaryotes cyclin–Cdks promote not only the processes that take place during a particular phase, but how cyclin–Cdks activate the next phase and extinguish the previous one. We also learned that superimposed on the cyclin–Cdk machinery are regulatory pathways, called checkpoints, that inhibit cell cycle progression in the presence of damage or certain types of errors. These checkpoint pathways ultimately stop cell cycle progression by affecting cyclin–Cdks. The multiple levels of regulation illustrate the importance in carrying out cell cycle processes with the utmost accuracy. Indeed, we saw that failure to do so can result in cancer. This theme will be further elaborated in the coming chapters.

We also learned that bacteria are more diverse in terms of their cell cycle schemes: some bacteria can carry out multifork replication, whereas others do not. The major regulatory steps in many bacteria are the initiation of DNA replication and cell division. Although the bacterial cell cycle differs substantially for the eukaryotic one, they both have regulatory mechanisms that inhibit cell cycle progression when errors or certain types of damage are detected. In the case of bacteria, the SOS response inhibits cell division in the presence of DNA damage.

The eukaryotic cell cycle

- The cell cycle is divided into four phases: G1, S, G2, and M.

- Cell cycle progression is driven by cyclin–Cdks. Cdks are active only when bound to cyclins.

- Different cyclins associate with Cdks at various cell cycle stages. The specificity of the cyclin–Cdk complex is determined by the cyclin.

- Cyclin–Cdks are regulated at multiple levels, including the expression and degradation of cyclins, activating and inhibitory phosphorylation of the Cdk, and association with Cdk inhibitors.

- Cyclin–Cdks have many targets, and can promote cell cycle progression by activating other kinases, promoting gene expression, and promoting protein degradation.

Checkpoint pathways

- Checkpoint pathways are processes that become activated in response to certain cellular insults. Different checkpoint pathways respond to a different type of cellular damage or defect.

- In general, checkpoint pathways are made of sensors, transducers, and effectors.

- Checkpoint pathways can lead to the induction of repair processes, cell cycle arrest and, in multicellular organisms, apoptosis.

Extrinsic regulators

- Cells respond to extrinsic signals such as growth factors and mitogens.

- Growth factors stimulate the cell to increase its mass.

- Mitogens directly stimulate cells to divide and proliferate without stimulating growth.

Cancer

- The three primary changes that fuel cancer growth are: dominant activation of extrinsic signaling pathways, loss of regulatory mechanisms (including checkpoint pathways that normally limit cell growth), and genetic instability.

- Oncogenic proteins are often positive regulators that are being activated in an aberrant way.

- Tumor suppressors are negative regulators of cell cycle progression and are often defective in cancer cells.

The bacterial cell cycle

- Among bacteria, there are many different cell cycle paradigms, often involving non-conserved proteins.

- Key regulatory points in many bacterial cell cycles are the initiation of DNA replication and cell division.

FURTHER READING

5.1 STEPS IN THE EUKARYOTIC CELL CYCLE

Blomena VA, Boonstraa J. Cell fate determination during G1 phase progression. *Cellular and Molecular Life Sciences* 2007;**64**:3084–3104.

Morgan DO. *The Cell Cycle Principles of Control.* London: New Science Press, 2007.

5.2 CYCLINS AND CYCLIN-DEPENDENT KINASES

Bloom J, Cross FR. Multiple levels of cyclin specificity in cell-cycle control. *Nature Reviews Molecular Cell Biology* 2007;**8**:149–160.

Hochegger H, Takeda S, Hunt T. Cyclin-dependent kinases and cell-cycle transitions: does one fit all? *Nature Reviews Molecular Cell Biology* 2008;**9**:910–916.

Kaldis P, Aleem E. Cell cycle sibling rivalry: Cdc2 vs. Cdk2. *Cell Cycle* 2005;**4**:1491–1494.

5.3 REGULATION OF CDK ACTIVITY

Besson A, Dowdy SF, Roberts JM. CDK inhibitors: cell cycle regulators and beyond. *Developmental Cell* 2008;**14**:159–169.

Kaldis P. The cdk-activating kinase (CAK): from yeast to mammals. *Cellular and Molecular Life Sciences* 1999;**55**:284–296.

5.4 CELL CYCLE REGULATION BY CDKS

van den Heuvel S, Dyson NJ. Conserved functions of the pRB and E2F families. *Nature Reviews Molecular Cell Biology* 2008;**9**:713–724.

5.5 PROTEOLYTIC REGULATION BY CDKS

Harper JW, Burton JL, Solomon MJ. The anaphase-promoting complex: it's not just for mitosis any more. *Genes and Development* 2002;**16**:2179–2206.

Peters JM. The anaphase-promoting complex: proteolysis in mitosis and beyond. *Molecular Cell* 2002;**9**:931–943.

5.6 CHECKPOINTS: INTRINSIC PATHWAYS THAT CAN HALT THE CELL CYCLE

Harper JW, Elledge SJ. The DNA damage response: ten years after. *Molecular Cell* 2007;**28**:739–745.

Lew DJ, Burke DJ. The spindle assembly and spindle position checkpoints. *Annual Review of Genetics* 2003;**37**:251–282.

5.7 EXTRINSIC REGULATORS OF CELL CYCLE PROGRESSION

Hallstrom TC, Nevins JR. Balancing the decision of cell proliferation and cell fate. *Cell Cycle* 2009;**8**:4, 532–535.

5.8 THE CELL CYCLE AND CANCER

Jones RG, Thompson CB. Tumor suppressors and cell metabolism: a recipe for cancer growth. *Genes and Development* 2009;**23**:537–548.

Luo J, Solimini NL, Elledge SJ. Principles of cancer therapy: oncogene and non-oncogene addiction. *Cell* 2009;**136**:823–837.

Malumbres M, Barbacid M. Cell cycle, CDKs and cancer: a changing paradigm. *Nature Reviews Cancer* 2009;**9**:153–166.

5.9 THE BACTERIAL CELL CYCLE

Collier J, Shapiro L. Spatial complexity and control of a bacterial cell cycle. *Current Opinion in Biotechnology* 2007;**18**:333–340.

Erill I, Campoy S, Barbe J. Aeons of distress: an evolutionary perspective on the bacterial SOS response. *FEMS Microbiology Reviews* 2007;**31**:637–656.

Mott ML, Berger JM. DNA replication initiation: mechanisms and regulation in bacteria. *Nature Reviews Microbiology* 2007;**5**:343–354.

DNA REPLICATION

6

INTRODUCTION

Before cells divide, they must make a complete and faithful copy of the DNA in their chromosomes. This process of copying is called DNA replication. During replication, each strand of the DNA double helix is copied to make a new strand, thus producing two new daughter DNA double helices. Replication is a carefully orchestrated process, requiring the activity of several important enzymes, each of whose role in DNA replication has been conserved across evolutionary time. It is also a process for which accuracy is paramount: the viability of successive generations of cells is dependent on the process of DNA replication ensuring that the daughter cells produced from each round of cell division receive an exact copy of their parent cell's DNA. Failure to maintain the DNA sequence of a cell line from generation to generation can have serious consequences both for the cell itself and for the organism of which it forms a part.

In this chapter we explore DNA replication by first considering the various stages that make up the overall process of replication, building a broad picture of how the cell moves from having one copy of a DNA double-stranded helix to having two. We then consider the key components of the molecular machinery that makes replication happen. Finally, we revisit each stage of replication in more detail to discover how the different molecular components come together to ensure that DNA replication happens as it should, and how regulatory mechanisms ensure that DNA is replicated only once in each cell-division cycle.

6.1 OVERVIEW OF DNA REPLICATION

We begin the chapter by looking at DNA replication as an overall process, identifying its key features and key components.

Each strand of the DNA double helix serves as a template for the synthesis of a complementary new strand

DNA replication is a fundamental process carried out by all living organisms. It is therefore not surprising that the basic mechanism of replication had an early evolutionary origin and is the same in bacteria, archaea, and eukaryotes. According to this basic mechanism of replication, both strands of the parental DNA molecule are copied during replication, each strand serving as a template for the synthesis of a new daughter strand (Figure 6.1). We call this process **semi-conservative**: when replication is

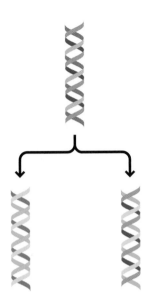

Figure 6.1 Semi-conservative replication. The parental double helix is shown in light and dark gray. After replication, each daughter has one of the two old parental strands (gray) and one newly synthesized strand (green).

complete, two daughter helices have been produced, each containing one old strand (which has been conserved from the parent) and one new strand. The experiments that first showed this to be the case are described in Experimental approach 6.1.

The process of copying the template strand involves the recognition of a base in the template strand and the addition of the complementary base to the new daughter strand (Figure 6.2). This ensures that not only are the daughter molecules identical to the parent DNA, they are also identical to each other. Consequently, when the cell divides and one copy is passed to each daughter cell, both cells receive exactly the same genetic material.

DNA synthesis occurs at replication forks that move outwards from origins of replication

In chromosomes, DNA replication starts at specialized sites called origins of replication (see Section 4.9) and moves away from an origin in both directions, creating a structure known as a **replication bubble** (Figure 6.3). The DNA double helix is opened at the origins of replication and then unwound on both sides of the origin to form structures called **replication forks**. The replication forks are the sites at which single-stranded DNA is exposed and at which DNA synthesis occurs. As replication enters elongation phase, the two replication forks that have originated from a single origin of replication move away from each other, proceeding along the DNA template strand in opposite directions.

DNA replication comprises three distinct phases

The process of DNA replication can be divided into three consecutive phases: initiation, elongation, and termination. The overall process of DNA replication is summarized in Figure 6.4. We will first develop an overall picture of the process of DNA replication, identifying the key components in the molecular machinery that we will examine in more detail later. Towards the end of the chapter, we will further discuss each step in the process.

Initiation of replication occurs at origins

During initiation, the first phase of DNA replication, the helix of the double-stranded parent DNA is opened up in order to give replication enzymes and other proteins access to the single strands that will form the templates for the daughter strands that are to be synthesized (Figure 6.4, step 1). We have seen in Chapter 4 that initiation occurs at specific origins of replication termed *ori* (see Figure 4.29). The origin is recognized by a specific **initiator protein** that opens the double-stranded DNA and recruits a class of enzymes called helicases (Figure 6.4, step 1). **DNA helicases** act to unwind the double-stranded DNA to form the single-stranded templates required for DNA replication (see Section 6.5). As soon as single-stranded DNA is formed, it is coated with single-stranded binding proteins, which are removed as DNA synthesis proceeds. Initiation is under strict control to ensure that DNA replication is initiated only once per cell division.

Figure 6.2 Template-directed DNA synthesis. A parental DNA strand (dark gray) acts as a template for the synthesis of an exactly complementary new strand (green) by the successive base-pairing of incoming nucleotides to the template. The base A pairs with T and C pairs with G. Correctly paired nucleotides are joined to the new strand one at a time by the enzyme DNA polymerase. The new strand is always synthesized in the 5′ to 3′ direction.

6.1 EXPERIMENTAL APPROACH
Discovery of semi-conservative DNA replication

When James Watson and Francis Crick proposed the double-helical structure of DNA (see Experimental approach 2.1), they realized that the structure suggested a potential way to make exact duplicates of the double-stranded DNA during cell division. They said at the end of their paper 'it has not escaped our attention that the specific pairing we have postulated immediately suggests a possible copying mechanism for the genetic material'. From casual examination it would seem that simply unzipping the helix and copying each separate strand would be an obvious solution. Others, however, pointed out that the helical nature of DNA itself posed a problem to simply opening and copying the helix. This led to much debate about whether DNA replication was indeed carried out by such a strand separation and copying mechanism.

Matthew Meselson and Franklin Stahl resolved this theoretical argument by designing an elegant set of experiments showing that the DNA is indeed copied in a semi-conservative manner. That is, one parental strand remains intact while a copy of that complete strand is generated by the synthesis of new DNA. In their experiments, the researchers labeled DNA from *Escherichia coli* by growing the cells in a medium enriched in the heavy isotope of nitrogen, N^{15}, which imparts a different density to DNA than the natural isotope N^{14}. To examine the distribution of heavy and light DNA in their samples, an experimental approach called density gradient centrifugation was developed. As they discovered, when a cesium chloride (CsCl) solution is spun at high speeds in an ultracentrifuge, a density gradient is established in the salt solution. DNA added to this tube will migrate to and form a band at a position in the gradient that is equivalent to its own buoyant density.

Using this method, Meselson and Stahl first showed that DNA isolated from bacteria grown in N^{15}-containing media formed a band at a different position in the gradient than DNA from bacteria grown in the standard 'light' nitrogen source, N^{14}. Having established the method, the researchers then grew a culture of bacteria in N^{15} and shifted it to N^{14}-containing media, and then followed the density of DNA in the cultures closely through four cell divisions. At several time points (reported here as 'generations'), they removed a sample of the culture, isolated DNA from the sample, and 'banded' it in a CsCl density gradient (Figure 1). At the earliest time-point (0 generations, before the addition of the N^{14} media), the entire population of DNA banded as 'heavy' N^{15} DNA, while after one cell division, all of the DNA banded at an intermediate density between heavy and light. After two full cell divisions, there was a band of all light DNA and a band of intermediate density.

The authors interpreted the results shown in Figure 1 with the diagram shown in Figure 2. After the first round of DNA synthesis,

Figure 1 CsCl gradient centrifugation differentiates heavy and light DNA peaks. (a) The image of the banded DNA taken from the ultracentrifuge camera is shown for 12 different time points. (b) The tracing of the bands shown in (a) and the estimated number of generations is shown. The generations were calculated based on the growth rate and doubling time determined for the cells in this experiment. The DNA moved from heavy:heavy to heavy:light to light:light peaks as the cells grow in N^{14} light media. The last two panels show a mixing of 0 and 1.9 generations and 0 and 4.1 generations to show the overlap in the bands.

Reproduced from Meselson and Stahl, *Proceedings of the National Academy of Sciences of the U S A* 1958:**44**:671–682.

all of the DNA was of intermediate density because the old strand was 'heavy' (containing N^{15}) and the new strand was 'light' (containing N^{14}). During the second round of DNA synthesis, the parental light strand was replicated with light DNA giving rise to light:light DNA (which exhibited a light density after centrifugation) and the parent heavy strand was also replicated with light DNA, giving rise to heavy:light DNA, which exhibited the intermediate

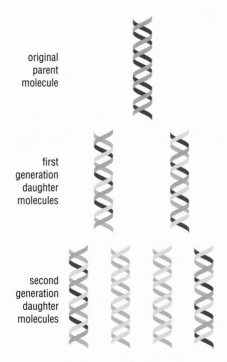

original
parent
molecule

first
generation
daughter
molecules

second
generation
daughter
molecules

Figure 2 Interpretation of the CsCl banding experiment shown in **Figure 1.** The initial helix is made of two strands of heavy DNA. After one cell division both daughter molecules are heavy:light, after two rounds of cell division, there are two light:light molecules and two heavy:light molecules; no heavy:heavy molecules will remain. Reproduced from Meselson and Stahl, *Proceedings of the National Academy of Sciences of the U S A* 1958:**44**:671–682.

density. After two more rounds of synthesis, most of the DNA was light:light with a small amount at the intermediate (light:heavy) density. As a marker of the density for each type, they mixed the 0 time point and the 2 generation time point and saw bands of all three densities.

This elegant set of simple experiments not only clearly showed the semi-conservative nature of DNA replication but also provided a powerful tool for analysis (CsCl density gradient centrifugation) that that became a mainstay for separation and analysis of DNA in molecular biology. Indeed, CsCl-based density gradient centrifugation is still widely used for studying replication. With this method, and knowing the sequence of the entire yeast genome, M. K. Raghuraman and colleagues simultaneously mapped the timing of replication of all 'origins' in yeast. As we discuss in this chapter, some origins fire early in the replication cycle while others fire later. Raghuraman *et al*. again used density labeling, growing yeast first in a 'heavy' media (containing both N^{15} and carbon C^{14}) and then shifting the cells to a 'light' media (with N^{14} and C^{12}). Replication was then followed in a manner similar to the Meselson and Stahl experiment, but in this case the authors added an isolation step wherein they directly purified DNA from the heavy: heavy and heavy:light bands at different time points. This purified DNA was then sheared and hybridized to a microarray chip with the whole genome arrayed (microarrays are discussed in Section 16.9) (Figure 3).

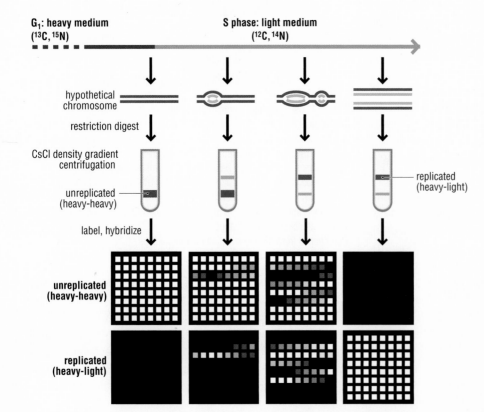

Figure 3 Experimental strategy for genome-wide analysis of DNA replication timing. Top: timeline of experiment: cells are grown in heavy medium for several generations and a sample of DNA is taken at time 0. Samples are then taken at several intervals over the course of one cell cycle. Middle: the replication of a hypothetical region of the chromosome is depicted and the expected banding location in CsCl gradients is shown. Bottom: the heavy-heavy DNA is hybridized to a microarray representing the entire yeast genome and in a separate experiment DNA from the heavy-light region is hybridized to a duplicate microarray. At the beginning of the experiment all of the regions of the genome are represented in the heavy-heavy fraction. At the end of the experiment all of the DNA is in the heavy-light fraction. At the intermediate time point the earliest regions to replicate show up on the array probed with the heavy-light fraction and at the same time disappear from the heavy-heavy fraction. Adapted from Raghuraman *et al*., *Science* 2001:**294**:115–121.

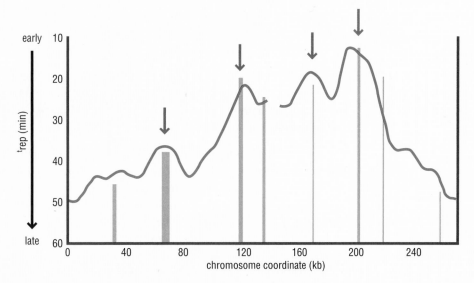

Figure 4 Replication timing across all of yeast chromosome VI. The data from the experiment outlined in Figure 3 is plotted for the length of one chromosome. The Y-axis shows the timing of replication for a given chromosomal region. Dark gray bars correspond to restriction fragments known to contain replication origins. Blue arrows indicate the four origins used in more than 50% of the cell cycles. Reproduced from Raghuraman *et al.*, *Science* 2001:**294**:115–121.

The time at which a sequence disappears from the heavy:heavy fraction and appears in the heavy:light fraction represents the time at which that sequence starts to replicate. Thus, early firing origins, and the sequences near them, are the first to move to the heavy:light fraction, while late firing origins, and the sequences near them, move much later. Using this approach, the timing of replication along each individual chromosome was thus determined (Figure 4).

Meselson M, Stahl FW. The replication of DNA in *Escherichia coli*. *Proceedings of the National Academy of Sciences of the U S A* 1958;**44**:671–682.

Raghuraman MK, Winzeler EA, Collingwood D, *et al*. Replication dynamics of the yeast genome. *Science* 2001;**294**:115–121.

During elongation each base in the parent DNA strand is read by DNA polymerase to direct synthesis of a daughter strand in a 5′ to 3′ direction

After the replication machinery is in place, and the double-stranded DNA has been opened up during initiation, we enter the elongation phase of DNA replication. During this phase, the replication machinery moves along the parent DNA strands, copying the strands into daughter strands as it proceeds. In other words, as the replication machinery moves along the DNA template strands, the daughter strands are gradually elongated.

The synthesis of a new DNA strand is catalyzed by an enzyme called **DNA polymerase**. This enzyme moves along the template, reading the template nucleotides one at a time and adding the complementary nucleotide to the end of the new strand. The DNA polymer has a distinct polarity, as described in Chapter 3, and the chemistry of the nucleotide-addition reaction only allows DNA polymerase to add new nucleotides to the 3′ end of the growing DNA strand. This has the very important consequence that DNA replication can only proceed in a 5′ to 3′ direction.

DNA replication starts with the synthesis of a short stretch of RNA

One unusual feature of DNA polymerase is that it cannot build a new DNA strand from the very start of the parent strand; it can only add nucleotides to the

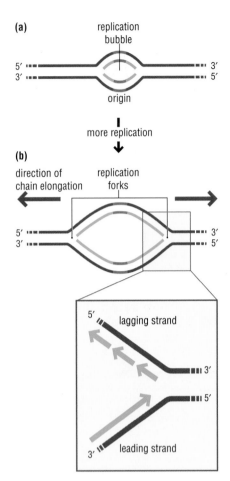

Figure 6.3 Bidirectional DNA replication.
(a) Replication initiates at an origin (red) and proceeds by the copying of both strands of DNA in both directions away from the origin. This creates what is known as a replication bubble. (b) At either end of the bubble are moving replication forks at which the DNA helix is being continuously unwound and DNA is being synthesized. Newly synthesized DNA is shown in green. The yellow box shows a close-up of one replication fork. The strand being synthesized continuously in the 5′ to 3′ direction is the leading strand (solid green arrow on the bottom strand) and the strand being synthesized discontinuously is the lagging strand (short green arrows on the top strand of the fork). These DNA fragments will be eventually joined to form a continuous DNA strand.

3′ end of a nucleotide fragment that already exists. The nucleotide fragment on which the DNA polymerase builds its daughter strand is a short strand of RNA termed a **primer**, which provides the 3′ end the DNA polymerase requires.

So we see that, before elongation can proceed fully, an RNA primer must be generated (Figure 6.4, step 2). The RNA primer is synthesized by a specialized polymerase called **primase**, which – unlike DNA polymerase – can initiate a new RNA strand without the need for a pre-existing 3′ end.

To complete DNA replication, the RNA primer is degraded and replaced with DNA. It is not known why DNA polymerase cannot start a new strand on its own while RNA polymerase can start new strands at will. It may be that the priming of replication by primase and later removal of the RNA allows an extra **proofreading** step to increase the accuracy of the replication process.

DNA polymerase is recruited to the DNA at primer–template junctions

After the primer is synthesized, the stage is now set for the recruitment of DNA polymerase. This recruitment – and the maintained association of DNA polymerase with the template strand – is mediated by two specific protein complexes, the **clamp loader**, and the **sliding clamp**, which we learn about further in Section 6.6. As the name suggests, the clamp loader acts to load the sliding clamp on to the DNA template. The sliding clamp then recruits DNA polymerase to the template, ready for DNA synthesis to commence. Again, as indicated by its name, the sliding clamp remains clamped to the DNA template, sliding along it as elongation occurs, and keeping the DNA polymerase tethered to the template in so doing.

The clamp loader and sliding clamp play an important role in recruiting the DNA polymerase to the appropriate location on the DNA template: they localize specifically at the 3′ primer-template junction, the region where DNA synthesis needs to commence (Figure 6.4, step 3). In eukaryotes the first few nucleotides of the daughter strand are synthesized by DNA polymerase α, which acts in tight association with primase (forming a complex called DNA polymerase α–primase). The replicative polymerase is then recruited to begin processive elongation (Figure 6.4, step 4). In eukaryotes, this involves a switch from DNA polymerase α–primase to the replicative polymerase δ or ε. We learn more about the different types of polymerase in Section 6.4.

Figure 6.4 Steps in DNA replication. DNA replication can be described as a series of discrete steps. Initiation: (1) Helicase loading: helicases (light orange rings) are recruited to the origin (red) by initiator proteins (not shown), the DNA double helix is opened to allow both strands to be copied. This small replication bubble is now ready for DNA replication to start. (2) Priming: primases (light orange) are recruited by the helicases and initiate the synthesis of RNA primers (turquoise) on both strands. The primers allow DNA polymerases to synthesize the DNA strands. Bacterial primase is shown here for simplicity. In eukaryotes, pol α-primase synthesizes an RNA primer followed by a short stretch of DNA. Elongation: (3) Loading of the sliding clamp: the clamp loader (not shown) binds to the 3′ primer–template junction and places the sliding clamp (light orange) around the DNA. (4) Recruitment of the replicative polymerase (purple): interaction with the sliding clamp places the polymerase in the correct position to elongate the 3′ end of the primer. (5) Bidirectional fork movement and strand elongation: the two forks move away from the origin in opposite directions and new DNA synthesis occurs on both strands. DNA synthesis on the lagging strand is discontinuous, producing short segments of DNA. Successive segments are each started at the replication fork and are elongated in the 5′ to 3′ direction. (6) Termination: once replication is complete, DNA ligase (not shown) seals any remaining gaps in the DNA backbone between DNA segments, resulting in two complete daughter molecules.

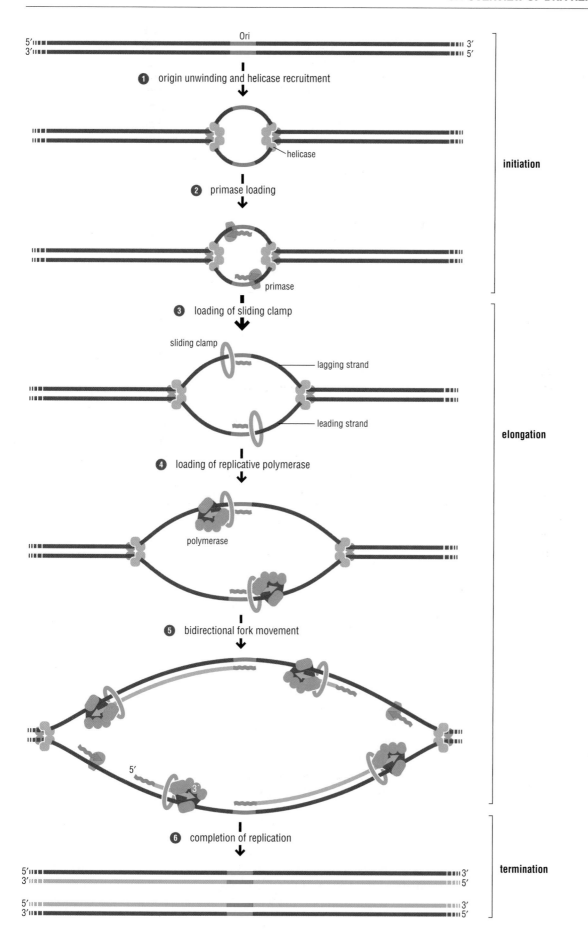

DNA synthesis is continuous on one strand and discontinuous on the other strand

The fact that DNA can only be synthesized in the 5′ to 3′ direction poses a problem at the replication fork: both strands of the DNA molecule must be copied at the same time but yet the two strands have opposite orientations – one strand runs in the 5′ to 3′ direction, while its complementary strand runs in the 3′ to 5′ direction. To solve this problem, DNA synthesis on the two strands is different (Figure 6.3). On one strand, synthesis is continuous: the daughter DNA can be synthesized as a single, uninterrupted strand from 5′ to 3′. This strand is known as the **leading strand**. By contrast, on the other strand, known as the **lagging strand**, synthesis occurs discontinuously: a series of short DNA fragments are synthesized from 5′ to 3′, which are then subsequently joined into a continuous strand. Primer synthesis and polymerase loading occur once on the leading strand, but must occur repeatedly on the lagging strand.

Polymerases, helicases, and primases travel with the replication fork

The helicases that unwind the DNA at the origin travel ahead of the polymerases to unwind the DNA at the fork. Directly behind the helicase, the leading-strand polymerase synthesizes DNA in the 5′ to 3′ direction, while a substantial region of single-stranded DNA is generated on the lagging strand and is bound by single-stranded binding proteins, which bind and protect the single-stranded DNA. Primase at the fork binds this single-stranded region of DNA and synthesizes a new primer. This enables the initiation of further DNA synthesis on the lagging strand to yield a relatively short segment of new DNA (Figure 6.4, step 5).

DNA synthesis on the leading and lagging strands is carried out by separate DNA polymerases, both of which form part of the replication machinery. As we will see in the sections that follow, the two polymerases move together along the DNA. This entails the looping of the lagging strand to allow simultaneous 5′ to 3′ synthesis on both strands. Thus, the illustration in Figure 6.4 is somewhat artificial as it does not show the coupling of the two polymerases. However, it illustrates in a simplified manner the principles and the order of the steps that occur.

Termination of replication occurs either when two forks meet or at the ends of linear chromosomes

Termination of DNA replication occurs when the two forks moving in opposite directions meet and the **replication complexes** are disassembled. In bacteria this occurs at a specific site, *ter* (see Section 4.9), while in eukaryotes no specific termination sequence is required.

After elongation by the replicative polymerases, the RNA primers are removed and the two adjacent newly synthesized DNA strands are connected by DNA ligase (Figure 6.4, step 6). There is presumably a regulated mechanism to disassemble the two large fork complexes that meet between adjacent origins. However, little is known about the details of this final step. Finally, at the very end of a linear chromosome, termination occurs when the fork reaches the end and, presumably, falls off.

DNA REPLICATION: CORE COMPONENTS

Having reviewed the key aspects of DNA replication, we will now examine the proteins that are key players in the replication machinery. After we have learned about the proteins themselves, we will bring both process and machinery together to see in more detail how the cell achieves the remarkable feat of replicating DNA with as few errors as possible.

6.2 DNA POLYMERASES: STRUCTURE AND FUNCTION

DNA polymerases make a complementary copy of one DNA strand

The central enzymes in DNA replication are DNA polymerases, which move along the template strand of DNA, reading the bases and adding the correct complementary nucleotide to the new strand. At each position, DNA polymerase selects the correct nucleotide to be added, catalyzes its linkage to the 3′ OH end of the chain, and then moves to the next template residue, maintaining contact with the template during the process. The accuracy or **fidelity** of DNA replication is essential for correct transmission of the genetic material from one generation to the next. If errors are allowed to creep in, mutations occur that could affect the gene product or regulatory regions, neither of which is desirable.

Much of this accuracy is due to the activity of DNA polymerases: in addition to selecting the correct nucleotide for addition, many polymerases also proofread for errors. If the wrong nucleotide is added, it is recognized and removed by an **exonuclease** activity in the polymerase, and the correct nucleotide is inserted. We will discuss the mechanisms that ensure the accuracy of replication in Section 6.3.

DNA polymerases carry out a number of different functions in the cell

All cells have a number of different DNA polymerases that perform distinct functions. Some polymerases are responsible for replicating the genome, some for the repair of DNA, and some for continuing replication across damaged DNA. Others have even more specialized functions, such as the replication of telomeres.

We discuss telomeres in more detail in Sections 4.11 and 6.11.

Some polymerases comprise just a single protein, which exhibits a range of functions, including nucleotide addition, proofreading and substrate specificity. By contrast, other polymerases comprise multiple subunits, with each subunit having a distinct function, such that full polymerase activity is only achieved when all subunits come together to form a complex. In this section we will discuss the common features of polymerase structure and function focusing on the major replicative polymerases; some of the more specialized polymerases are described in Section 6.4, and in Chapter 12 when DNA repair is discussed.

The major replicative polymerases that are involved with genome duplication are DNA polymerase III in bacteria and DNA polymerases δ and ε in eukaryotes. These enzymes synthesize DNA rapidly and with high fidelity. The high degree of

(a)

(b)

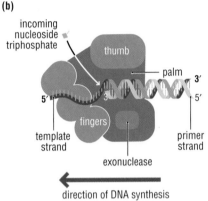

Figure 6.5 DNA polymerase resembles a right hand. (a) Crystal structure of the Klenow fragment of *E. coli* DNA polymerase I, which comprises the polymerase domains and the 3′ to 5′ exonuclease activity, but lacks the 5′ to 3′ exonuclease domain of pol I. The polymerase domain resembles a right hand with the fingers on the side, the palm domain forming a cleft where the DNA will lie and the thumb on the top. The exonuclease domain for this class of polymerases is to one side of the three core polymerase domains (PDB 1KLN). (b) Schematic representation of the polymerase structure in (a). The palm domain is in the interior of the structure. The single-stranded template DNA is fed past the fingers and into the active site in the palm where nucleotide addition is catalyzed. The incoming nucleotide is joined to the free 3′ OH in the polymerase active site in the palm. The newly synthesized double helix exits from the back of the polymerase domains.

conservation of the replicative polymerases in bacteria, archaea, and eukaryotes reflects the early evolutionary origin of their specialized function.

While we note above that some polymerases comprise just a single protein, the major replicative polymerases are composed of multiple protein subunits that help them achieve speed while also enabling them to maintain the fidelity of replication. Bacterial DNA replication proceeds at a rate of around 1000 nucleotides per second, while eukaryotic chromosomes are replicated more slowly, at the rate of around 50 nucleotides per second. To achieve this speed, polymerases have specialized subunits that carry out specific functions. Some polymerase subunits are involved in proofreading – monitoring for errors in the way one would do when proofreading text – while others help attach the polymerase to the DNA template so that it can replicate long stretches of DNA before it dissociates. This ability to remain associated with the template is referred to as the **processivity** of a polymerase.

The structure of a DNA polymerase resembles a right hand

The most highly conserved component of the polymerase complex is the subunit that contains the catalytic site for nucleotide addition. The structure of this core polymerase is exemplified by that of DNA polymerase I (pol I) from *E. coli*. Pol I is a specialized polymerase that is essential for finishing DNA replication and removing the RNA primers that initiated replication. It is composed of a single protein chain that folds into three structural domains that, together, look something like your right hand. Because of this resemblance, the three domains are called the palm, the fingers, and the thumb (Figure 6.5).

The spatial relationship of the three domains to each other and the structure of the finger and thumb regions vary among different polymerases, but the overall structure and catalytic mechanism is conserved in polymerases from all organisms. The most highly conserved domain is the palm, which forms a cleft into which the growing double-stranded DNA fits. Meanwhile, the single-stranded template strand wraps through the fingers.

During the elongation phase of DNA replication, 'free' nucleoside triphosphates are added to the growing daughter strand in a reaction catalyzed by the polymerase. In most polymerases, the finger domain positions the incoming nucleoside triphosphate in relation to the template DNA. The thumb domain helps to hold the elongating duplex DNA and also maintains the contact with the template that permits processive synthesis. In DNA pol I, the 3′ to 5′ exonuclease function that removes incorrect bases is present as an additional domain of the same protein (see Figure 6.5).

The active site of the enzyme, where the reaction catalyzed by the enzyme occurs, lies deep in the cleft of the beta sheet comprising the palm domain. At the active site, carboxylate groups of two aspartate residues coordinate two magnesium ions that participate in catalysis, as described in detail below.

The polymerase active site catalyzes the addition of nucleotides onto the growing DNA chain

The active site of the polymerase catalyzes a phosphoryl transfer reaction that adds a nucleotide to the nascent DNA strand by linking the 5′ phosphoryl group on the nucleotide to the 3′ hydroxyl on the end of the chain to form a phosphodiester

Figure 6.6 Two-metal-ion mechanism of catalysis. The carboxylate groups of two conserved aspartate residues in the polymerase active site coordinate two metal ions (typically magnesium ions [Mg^{2+}] in DNA polymerases) and hold them in the correct orientation to participate in catalysis. One Mg^{2+} interacts with the 3′ OH group of the growing strand, and the other interacts with the incoming nucleoside triphosphate (green and orange). The Mg^{2+} bound to the growing strand facilitates the attack of the 3′ OH on the α-phosphoryl group of the incoming nucleoside triphosphate and a new O–P bond is formed. The two phosphate groups highlighted in orange are released as pyrophosphate and are subsequently hydrolyzed. Both metal ions stabilize the structure of the transition state of the reaction. This two-metal-ion catalyzed phosphoryl transfer is similar to that used in all polymerases, including RNA polymerases.

bond (see Section 2.6). The continuous series of phosphate bonds linking adjacent nucleotides in the DNA molecule comprises the so-called DNA backbone, which lies on the outside of the double helix. Let us now explore the mechanism of this reaction in a little more detail.

Incoming nucleotides are in the form of nucleoside triphosphates. The reaction proceeds through a nucleophilic attack by the nascent chain 3′ OH on the α-phosphate of the incoming nucleoside triphosphate, releasing pyrophosphate (PPi) (Figure 6.6). Critical to this reaction are two Mg^{2+} ions present in the active site that are positioned by carboxylate moieties on two conserved aspartate residues. One of these Mg^{2+} ions functions primarily to activate the 3′ OH of the terminal nucleotide, whereas the second Mg^{2+} interacts with the incoming nucleoside triphosphate and stabilizes a developing negative charge on the leaving oxygen during the reaction. The incoming triphosphate is also stabilized by interactions with side chains on an alpha helix in the finger domain. Subsequent hydrolysis of the released pyrophosphate provides the free energy that drives the reaction forward.

The general chemical features of this polymerization reaction – including the nucleophilic attack, the use of two metal ions in the active site to promote the reaction, and the use of pyrophosphate hydrolysis to drive the reaction forward – are common features in biological systems, which we encounter a number of times in this book.

6.3 DNA POLYMERASES: FIDELITY AND PROCESSIVITY

The fidelity of DNA replication is assured by proofreading mechanisms

The accuracy, or **fidelity**, of DNA replication is of utmost importance to the cell. Errors in DNA synthesis cause changes in the genetic code that will be passed to

purine–pyrimidine mispairs

G–T
G–C

purine–purine mispairs

G–A
T–A

G–G
G–C

◯ incorrect
◯ correct

Figure 6.7 Structure of mismatched base pairs. Incorrectly base-paired nucleotides have a distinctly different molecular shape from that of a correct base pair. The top row shows the shape of a correct G–C base pair outlined in gray. Superimposed in pink is the shape of an incorrect G–T pair. The difference in the two shapes is obvious. The two lower rows illustrate the same point for G–A and G–G mispairings. The correct base pair is outlined in gray in each case; note how similar the shapes of G–C and T–A base pairs are to each other. Correctly base-paired nucleotides fit into the polymerase active site whereas incorrectly base-paired nucleotides do not.
From Johnson and Beese (2004).

all subsequent cells and can alter essential cellular functions. A replicative polymerase typically only makes one uncorrected error for every 100 000 nucleotides synthesized. This remarkable fidelity is achieved by activities that monitor the identity of nucleotides at two distinct stages: during the polymerization reaction itself, and after polymerization has occurred.

At the first level, the polymerase recognizes correct nucleotides because they fit precisely into the active site when base-paired with the template strand. The geometry of an adenine–thymine (A–T) pair and a cytosine–guanine (C–G) pair are similar to each other and will fit neatly in the active site, whereas mismatched nucleotides have a different geometry that does not fit as well (Figure 6.7). In turn, this correct geometry favors the catalytic reaction and the addition of the incoming nucleotide onto the new DNA strand.

Despite these spatial constraints, an incorrect nucleotide is occasionally added to the new DNA strand. This error can be corrected by **proofreading**, which serves as a second level of monitoring. Most polymerases have an exonuclease activity that cleaves the final nucleotide from a DNA strand if it is incorrect before the next one is added. This activity is referred to as 3′ to 5′ **exonuclease** activity as the nucleotide is removed from the 3′ end of the newly synthesized DNA strand.

The polymerase active site and the exonuclease active site are spatially separated on the polymerase such that, when a correct base is added, the polymerase simply moves on to the next template position and proceeds with polymerization. However, when a mispaired nucleotide is incorporated, the rate of incorporation of the next nucleotide by the polymerase is substantially slowed, thus giving the exonuclease time to function. For the excision reaction associated with proofreading, the 3′ end of the newly generated DNA is moved from the polymerase active site into the exonuclease active site, where the terminal nucleotide is removed (Figure 6.8). The 3′ end is then repositioned in the polymerase active site and elongation can proceed. These events are all accomplished while the polymerase enzyme remains associated with the template DNA.

The recognition of correct nucleotides based on their geometry derives from the inherent energetic differences between correct and incorrect nucleotides. As such, proofreading at this level does not need an input of energy for recognition to occur because the enzyme does not actually get as far as adding the wrong nucleotide to the growing chain. By contrast, if an incorrect nucleotide *is* incorporated, the cell does need to expend energy – in this case, to excise the incorrect nucleotide. Such processes that allow an enzyme to achieve greater discrimination in a second step than permitted by the inherent energetic differences in the interactions of the correct and incorrect nucleotide are referred to as kinetic proofreading, and typically are associated with energetic expenditure (the removal of incorrect nucleotide(s) in this case).

The energetic costs associated with kinetic proofreading can be considered a cost to the cell for increasing the fidelity of DNA replication as needed for propagation of the genome, beyond that which would be possible if only the first level of proofreading – the non-energy requiring step – occurred. As we shall see when we discuss the translation of RNA into protein (Chapter 10), two-step mechanisms for ensuring accuracy are a common feature of fundamental molecular processes.

Processive and distributive polymerases have specific roles in DNA replication

After each addition of a nucleotide, there is a chance that the polymerase will dissociate from the template. If this occurred too often, replication would be slow and inefficient. As noted in Section 6.2, the polymerases engaged in general replication have high processivity: that is, they remain associated with the template through many rounds of nucleotide addition. A processive polymerase may copy many thousands of nucleotides before dissociating, thus allowing long regions of a chromosome to be copied efficiently. The main replicative polymerases are intrinsically processive, but, as we shall see later in the chapter, also have several accessory proteins that tether them to the DNA or to other proteins at the fork.

Not all polymerases are inherently processive, however. Some dissociate frequently from the template, polymerizing only ten or 100 nucleotides at a time, depending on the particular polymerase. A new polymerase molecule then binds and continues elongation of the strand. In these cases, repeated rounds of polymerase dissociation and reassociation are required in order for the template strand to be fully copied (Figure 6.9). This is known as a **distributive** mechanism of elongation. Polymerases involved in DNA repair are significantly less processive than replicative polymerases, but these polymerases only need to copy short stretches of DNA (where single-stranded gaps are left by the repair process).

In eukaryotes, immediately after the RNA primer is synthesized, a short stretch of DNA is made by a primase-associated DNA polymerase called polymerase α, which is also not very processive, adding only 30–100 nucleotides before dissociating.

6.4 SPECIALIZED POLYMERASES

Distinct DNA polymerase families have specialized roles in DNA replication

The replicative DNA polymerases – DNA polymerase III in bacteria and DNA polymerases δ and ε in eukaryotes – copy the bulk of the DNA during the replication stage of the cell cycle. In addition to these replicative DNA polymerases, however, there are a number of other important DNA polymerase enzymes in both bacterial and eukaryotic cells. The overall structure and mechanism of these different DNA polymerases is remarkably conserved, but they differ from each other in subtle ways that allow them to carry out distinct functions in the cell.

DNA polymerases can be grouped into functional families according to their evolutionary relatedness

The active site and the catalytic mechanism for nucleotide polymerization are conserved among polymerases. Beyond the active site, however, other regions of the polymerase can vary considerably. DNA polymerases are grouped into six different families – A, B, C, D, X, and Y – on the basis of structural similarity and conserved

Figure 6.8 Steps in proofreading. When an incorrect base pair is incorporated despite the discrimination step described in Figure 6.7, it can be removed from the end by editing. The mispaired base at the 3′ end of the growing strand (shown in pink) is detected and the 3′ end is moved to the 3′ to 5′ exonuclease site on the polymerase. This editing mode involves breaking several base pairs between the nascent strand and the template to relocate the 3′ end. In the editing site the 3′-most nucleotide is removed from the DNA chain and the 3′ end is then allowed to return to the polymerase active site in the palm. Now a new round of nucleotide addition can occur in the polymerase active site.

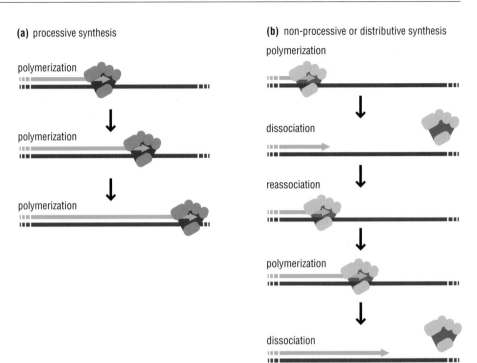

(a) processive synthesis

polymerization

polymerization

polymerization

(b) non-processive or distributive synthesis

polymerization

dissociation

reassociation

polymerization

dissociation

Figure 6.9 Processive and non-processive polymerases. (a) In processive DNA synthesis, the polymerase remains associated with the template DNA strand for many thousands of nucleotides. (b) Polymerases that synthesize only a short region of DNA are usually non-processive or distributive. The polymerase is bound to the template and synthesizes DNA, however with a certain probability it will dissociate from the DNA. A new polymerase molecule can then bind to the primer template DNA and further polymerization will occur.

➲ We learn more about DNA repair and the polymerases involved in Chapter 12.

sequence domains (Figure 6.10). Strikingly, in any given organism the similarities within a family are greater than the similarities between the different polymerases, indicating that these polymerase families diverged early in evolution.

The A family of polymerases includes the DNA polymerase I from *E. coli*, which is involved in finishing DNA replication and removing the RNA **primers** (see

DNA polymerase families			
family	**polymerase**	**source**	**function**
A	pol 1 (Klenow)	*E. coli*	gap repair
	Taq	*Thermus aquaticus*	replication
	T7	phage T7	replication
B	polα	eukaryotes	primase and repair
	polδ	eukaryotes	replication
	polε	eukaryotes	replication
	polζ (Rev3)	eukaryotes	translesion synthesis
	pol II	*E. coli*	
	T4	phage	
	RPB69	phage	
	pol B	archaea	
C	pol III	*E. coli*	replication
D	pol D	archaea	
X	polβ	eukaryotes	gap repair
	polλ	eukaryotes	gap repair
	polμ	eukaryotes	gap repair
	polσ	eukaryotes	gap repair
Y	polη (Rad 30)	eukaryotes	translesion synthesis
	polι	eukaryotes	translesion synthesis
	polκ	eukaryotes	translesion synthesis
	REV 1	eukaryotes	translesion synthesis
	Din B (Pol IV)	*E. coli*	translesion synthesis
	UmuCD (Pol V)	*E. coli*	translesion synthesis
	Dbh	archaea	translesion synthesis
	Dpo4	archaea	translesion synthesis
RT	reverse transcriptase	retrovirus	copy genome
		eukaryotes	copy retrotransposons
	telomerase	eukaryotes	elongate telomeres

Figure 6.10 DNA polymerase families.

Section 6.1). The complete native DNA polymerase I contains a 5′ to 3′ exonuclease domain that allows it to remove DNA or RNA ahead of it; it also features the 3′ to 5′ proofreading exonuclease activity already noted. The B, C, and D families contain the replicative polymerases of eukaryotes, bacteria, and archaea, respectively. These polymerases have high fidelity and possess a 3′ to 5′ proofreading exonuclease. The X family of polymerases is specialized for DNA repair, filling in the gaps generated in DNA during repair. **Y family polymerases** are specialized to replicate past bulky adducts in damaged DNA; other polymerases will stop polymerization when they encounter such lesions.

➡ We discuss the role of the Y family polymerases in DNA repair in Chapter 12.

Reverse transcriptases are DNA polymerases that use RNA templates to make DNA

In addition to the DNA polymerases that are involved in DNA replication and repair, eukaryotic cells and some viruses have DNA polymerases that copy RNA into DNA. These are called **reverse transcriptases**. Unlike all the other DNA polymerases described so far, reverse transcriptases use single-stranded RNA as a template to synthesize a complementary DNA strand. The structure of reverse transcriptase is very much like that of other polymerases: it has thumb and finger domains and a palm domain. The active site of reverse transcriptase is similar to that of other DNA polymerases and uses the same catalytic mechanism for linking nucleotides together.

Reverse transcriptases are encoded by viruses as well as DNA elements called retrotransposons, which are found in eukaryotic genomes. These are a class of the transposable genetic elements, which we will learn more about in Chapter 13. We will discuss telomerase, a specialized form of a reverse transcriptase that is needed to synthesize chromosome ends in Section 6.11.

6.5 DNA HELICASES: UNWINDING OF THE DOUBLE HELIX

DNA helicase unwinds double-stranded DNA for copying

Having now considered the key enzymes involved in the actual synthesis of DNA during replication, let us now turn to an enzyme that has an important function before synthesis can actually occur – the unwinding of the double-stranded helix of the parent DNA.

When base-paired in the double-stranded helical DNA molecule, nucleotides are inaccessible to polymerases, making copying impossible. Both primase and polymerases need access to single-stranded DNA. To achieve this, the internal base-pairing must be broken and the helix unwound.

As we discuss in Section 6.1, the first step in this unwinding process is the initial opening of the helix, a step performed by the initiator protein at the origin of replication. Once opened, the unwinding of the double helix to expose single-stranded DNA for copying can begin. This unwinding process is catalyzed by an enzyme called **DNA helicase**. As will be discussed in Chapter 12, the cell also needs to open the helix for DNA repair and recombination, and there are a unique set of DNA helicases for this particular case of DNA unwinding.

Figure 6.11 Structure of the papillomavirus helicase E1 protein and DNA. The helicase protein is a ring composed of six identical subunits (PDB 2GXA). The structure shows that double-stranded DNA fits through the center on the helical structure. There are five contact points between the phosphates on the single-stranded DNA and specific helices within the helicase that allow the movement of the helicase along DNA.
From Enemark and Joshua-Tor (2006).

Helicases open the double-stranded DNA and then travel with the replication fork, continuously unwinding the DNA to provide a template for the polymerase to copy. The helicases involved in replication in both bacteria and eukaryotes are composed of six subunits that form a ring structure that surrounds one strand of the DNA.

The structure of the replicative DNA helicase from papillomavirus provides insights into the mechanism of helicase action. This viral helicase is a hexamer of one protein, E1. In the co-crystal of protein and DNA, a single strand of DNA fits into the center of the channel formed by the hexamer ring, as depicted in Figure 6.11). The other strand of the DNA is displaced by the helicase and bound by single-stranded binding proteins. Each monomer possesses a series of DNA-binding loops such that, when the six monomers come together to form the hexamer, the DNA binding loops form a spiral staircase on the inside of the channel which binds the DNA, and move along it one nucleotide at a time. Each movement of the DNA binding loop within each monomer requires adenosine triphosphate (ATP) hydrolysis; thus to move six nucleotides requires six ATP molecules, making DNA unwinding an energy costly process (Figure 6.12).

Figure 6.12 Mechanism of helicase action. The five contact points between the helicase subunits (depicted as orange train cars) and the phosphates (balls) of single-stranded DNA allow the helicase to track along the DNA. ATP hydrolysis and subsequent adenosine diphosphate (ADP) release followed by a new round of ATP binding at the subunit interfaces drives the process forward. Each DNA binding site within the protein is in contact with one unique nucleotide. ATP hydrolysis occurring between subunits located near the front of the train leads to the release of the DNA. Upon binding a new ATP molecule, a released subunit moves to the back of the train to pick up the next available single-stranded DNA phosphate, escorting it through the channel, thus allowing the steps to be repeated.
Adapted from Enemark and Joshua-Tor (2008).

In *E. coli*, the helicase associated with DNA replication is known as **DnaB** and is a hexamer of six identical subunits. In eukaryotes and archaea, the replicative helicase is called **MCM** and is a complex of six different proteins, Mcm2–7, which assemble to form a ring. The eukaryotic replicative helicase is conserved throughout evolution and requires accessory factors for helicase activity.

Single-stranded DNA-binding proteins keep DNA strands separated

The helicase exposes a region of single-stranded DNA that must be kept open for copying to proceed. This is achieved by coating the strand with single-stranded binding proteins. In bacteria, a monomeric protein called **Single-Stranded Binding protein** (**SSB protein**) associates to form tetramers around which the DNA is wrapped in a manner that significantly compacts the single-stranded DNA. In eukaryotes, the single-stranded binding protein is a complex of three different subunits called **replication protein A (RPA)**.

The SSB and RPA proteins both stabilize the single-stranded DNA and interact specifically with other proteins needed for replication. Coating of single strands is particularly important on the lagging strand, because long stretches of single-stranded DNA are generated as a result of the discontinuous nature of replication on this strand. (Remember that replication of the lagging strand requires the synthesis of a series of short fragments of DNA, which are later joined to form a continuous strand. The regions of single-stranded DNA are protected by a coat of single-stranded binding proteins before they are copied.)

Topoisomerases assist helicases by removing supercoils from DNA

As helicases and polymerases move along DNA, unwinding and copying it as they go, they generate torsional stress in the double-stranded helical DNA in front of them. Since the ends of the chromosome are not free to rotate, strand separation during DNA replication is accompanied by an overwinding of the DNA ahead of the replication fork (Figure 6.13). To envision how this happens, imagine a long piece of wool (which consists of many thin wool fibers coiled about one another in a helical manner) with one end held immobile. If you tease apart the free end into two strands and start to pull them apart, the wool ahead of the separating strands will become increasingly twisted as more and more of the wool is separated. As more of the wool 'helix' is unwound, the unwinding will be met with increasing resistance, making the process of unwinding more difficult.

As we learned in Chapter 2, changes in the winding of the two DNA strands – defined as twist – lead to changes in DNA supercoiling if the ends of the DNA are prevented from rotating or if the template is a covalently closed circle. The supercoils that build up ahead of the replication fork generate superhelical tension that can make strand separation increasingly more difficult (Figure 6.13). This superhelical tension is the equivalent of the resistance we describe in our example of the teasing apart of strands of wool, above.

One full turn of the double helix is unwound for every ten base pairs of DNA that the replication machinery travels along. With unwinding happening at such a

Figure 6.13 Supercoiling of DNA ahead of the replication fork. As the helicase at the fork unwinds the DNA, the helical twist from the DNA is translated into supercoils ahead of the fork as it moves forward. The circular DNA of a bacterial chromosome or eukaryotic chromosomal DNA packaged into chromatin is not free to swivel, and thus behaves as if it were anchored (bar at right). The supercoils must be removed by either topoisomerase I or II for the fork to continue its movement.

pace, the tension generated ahead of the replication machinery as unwinding occurs can quickly become considerable. The supercoils introduced during replication make it progressively more difficult for the fork to move. So how is this overwinding overcome to enable the cell to generate single-stranded DNA during replication? The answer lies in a family of enzymes called topoisomerases.

Topoisomerases remove supercoils and thereby relieve the torsional stress that is present in the overwound DNA. They achieve this by transiently breaking the DNA backbone, thereby changing the DNA linking number and allowing the supercoils to relax.

The name of this family of enzymes denotes the fact that the substrate and product of a topoisomerase reaction are chemically identical, yet differ in topology.

There are several different ways that topoisomerases couple strand cleavage with a change in template supercoiling, and these enzymes are therefore grouped according to the particular mechanism they use (Figure 6.14). Type IA and IB topoisomerases break one of the two strands and do not require ATP; type II topoisomerases cleave both strands, in a reaction that is powered by ATP hydrolysis. Both types of topoisomerases are present at replication forks in both bacteria and eukaryotes. Because they break the DNA backbone, topoisomerases can also untangle DNA that is inappropriately linked or knotted.

Type I topoisomerases change supercoiling by introducing transient breaks in one strand of DNA

How can an enzyme change the topology of DNA? While each class of topoisomerase carries out a distinct set of steps to achieve this change in the DNA,

> ➲ The relationship between twist, supercoiling, and linking number is discussed in Chapter 2.

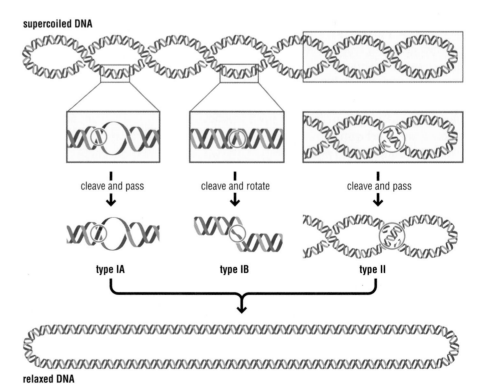

Figure 6.14 Three different mechanisms by which topoisomerases relax supercoils in DNA. (a) Type IA topoisomerases cleave single-stranded DNA and facilitate strand passage of one single strand past the other. (b) Type IB topoisomerases also cleave single-stranded DNA and allow the free end to swivel to release supercoils before religation. (c) Type II topoisomerases cleave the double strand of DNA and pass one strand behind the other. The mechanism by which type II topoisomerases relax supercoils is shown in more detail in Figure 6.15. From Figure 1 of Corbett and Berger (2004).

all topoisomerase reactions share some common features. A central event is an attack on the phosphodiester backbone of DNA by a tyrosine hydroxyl on the topoisomerase enzyme, leading to formation of a phosphoester linkage between the tyrosine and either the 5′ or 3′ end of the cleaved strand. This bond holds one end of a DNA strand in place while the DNA molecule as a whole undergoes a change in topology. The cleavage reaction is then reversed, the two ends of the cleaved strand religated, and the bond with the tyrosine broken. Since the substrate and product in the reaction are chemically identical, no net energy input is needed. As we shall see, the ATP hydrolysis required by some topoisomerases powers conformational changes in the enzyme that are needed to promote the topological change in the DNA.

Type IB enzymes do not use ATP, but rely upon the superhelical tension in the DNA to drive the relaxing of supercoils while the DNA is temporarily cleaved. The type IB enzymes relieve tension ahead of the replication fork by nicking one strand of the DNA and allowing one strand to swivel around the other before the DNA is religated (Fig. 6.14).

Specifically, type IB enzymes cleave one DNA strand and form a transient covalent bond between an active site tyrosine residue and the 3′ end of the cleaved strand. The superhelical tension in the DNA causes the free 5′ end to swivel about the uncleaved strand, with each complete rotation changing the linking number of the DNA in steps of one. After one or more rotations (up to five rotations have been observed), the free 5′ end attacks the bond between the enzyme and the 3′ end and the DNA strand is religated, generating double-stranded DNA with fewer supercoils.

Type IA and type II topoisomerases change supercoiling via strand passage

In contrast to the type IB enzyme described above, the type II and type IA enzymes change the supercoiling of DNA in a different way, by a process known as strand passage. Figure 6.15 illustrates how type II topoisomerases use strand

Figure 6.15 Type II topoisomerases break double-stranded DNA to allow the passage of another double helix. A type II topoisomerase will break both strands of one DNA molecule and allow one double helix to pass through another. This allows two intertwined helices to be resolved and ultimately separated. (1) The topoisomerase binds both double-stranded helices, at this point the blue helix is above the pink one. (2) The double-stranded DNA of the pink helix is broken and the blue helix passed though the broken ends. (3) The pink helix is religated and the blue helix is now below the pink one. Strand passage in this manner allows supercoils to be released.

passage to change the linking of double-stranded DNA. The dimeric enzyme cleaves both strands of the DNA duplex, forming a covalent bond between a tyrosine and the 5′ end of each cleaved strand. Conformational changes in the enzyme separate the two ends of the cleaved strand while allowing another double-stranded segment to pass through the gap in the DNA. The cleavage reaction is then reversed and the DNA is religated. This reaction requires ATP hydrolysis, which presumably helps to power the conformational change in the enzyme. The resulting DNA has fewer supercoils and its linking number is changed by two for each cycle.

While most type II enzymes relieve supercoiling, an important exception is bacterial DNA gyrase, which introduces negative supercoils into DNA. DNA gyrase maintains the circular bacterial chromosome and plasmids in a negatively supercoiled state, strand opening is made energetically more favorable as compared with a relaxed template (one lacking supercoils). This is important in the initiation of both DNA replication and transcription.

Like the type II topoisomerases, type IA enzymes use a strand passage mechanism to remove supercoils from DNA. However, an important difference is that type IA enzymes introduce a transient break in just one of the DNA strands, and linking number changes in steps of one. Instead of passing a double-stranded segment of DNA through a transient break in the DNA duplex, type IA enzymes pass a single strand of DNA through a transient break in another single strand. Like the other topoisomerases, a transient break is introduced into the DNA and the 5′ end of the single DNA strand forms a covalent bond with a tyrosine. This DNA is then religated after the strand passage.

The activities of type IA and type II topoisomerases in allowing the passage of DNA strands past one another facilitates the untangling of daughter chromosomes that are intertwined after replication. This untangling allows the replicated chromosomes to segregate at cell division, the mechanism of which will be discussed in more detail later in the chapter when we discuss termination of replication. The activity of both type I or type II topoisomerases in removing supercoils that accumulate is essential for replication to proceed. For this reason, a variety of chemotherapeutic agents used in the treatment of cancer are topoisomerase inhibitors: since topoisomerase activity is needed by the rapidly dividing cells in a tumor, these inhibitors can kill rapidly dividing cancerous cells.

> Read about the relationship between supercoiling (writhe), twist, and linking number in Section 2.5.

6.6 THE SLIDING CLAMP AND CLAMP LOADER

A sliding clamp and associated clamp loader increase the processivity of DNA polymerase

Let us now move on to consider some other proteins that play important roles during DNA replication.

We noted earlier that the replicative polymerases are highly processive – that is, they remain associated with the DNA template strand for a considerable length of time before dissociating. The high processivity of the replicative polymerases is largely due to an important multisubunit protein called the **sliding clamp** that keeps the polymerase tethered to the DNA. This clamp is assembled by another multisubunit protein known as the **clamp loader**. The structures of these proteins are highly conserved from bacteria to archaea and eukaryotes, indicating that they arose early in evolution and are of fundamental importance to efficient replication.

The clamp loader, as well as other proteins associated with replication that we shall learn about later in the chapter, are members of the **AAA+ family** of ATPases. (AAA+ stands for **ATPases associated with a variety of cellular activities**.) These proteins all share a conserved region of 220 amino acids that includes an ATPase domain. We shall see that many of the actions of AAA+ ATPases associated with replication undergo conformational changes; these changes, which are required for their functions, are driven by ATP hydrolysis.

The ring-shaped sliding clamp keeps DNA polymerase attached to the DNA

The sliding clamp is a donut-shaped protein complex (Figure 6.16) that completely encircles the double-stranded DNA. The clamp is a very stable structure and, once clamped onto the DNA, will remain associated with it (and will ensure that the DNA polymerase is tethered to the DNA template) while thousands of nucleotides are copied. Sliding clamp proteins are present in bacteria, archaea and eukaryotes. In bacteria they are known as the **β protein** and in eukaryotes as **PCNA**. The name PCNA, which stands for Proliferating Cell Nuclear Antigen, reflects the history of discovery of this protein, which was first identified by specific antibodies that recognized a protein present in the nuclei of dividing cells but not resting cells. Because antibody recognition of PCNA changed in the cell cycle, it was thought that the protein levels changed during the cell cycle. However, this protein is actually present at a constant level; it is only antibody accessibility that varies throughout the cell cycle. Thus the protein itself does not undergo periodic destruction as we saw for cell cycle proteins in Chapter 5.

Overall, the sliding clamp has a very similar structure in all organisms, reflecting conservation of the mechanism of DNA replication throughout evolution (Figure 6.16). In *E. coli* the ring is a homodimer of β proteins. Each β protein has three similar domains, so dimerization generates a structure with six symmetrically arranged domains. In eukaryotes, an analogous structure is made by the interaction of three copies of the PCNA protein, each of which is composed of two domains with structural similarity to those of bacterial β protein. Thus, the *E. coli* and eukaryotic sliding clamps have a very similar ring structure consisting of six domains surrounding a central cavity of 35 Å that can accommodate duplex DNA.

During replication, the clamp binds DNA polymerase and slides along the DNA with the polymerase, thus keeping the polymerase tethered to the DNA. PCNA interacts with DNA polymerase δ through a motif of eight amino acids present in each protein subunit, a motif that is conserved in eukaryotic and archaeal sliding clamps. The β protein has a similar short peptide motif that interacts with DNA polymerase III, further indicating the conservation of these proteins. We will see

(a) bacterial sliding clamp

(b) eukaryotic sliding clamp

Figure 6.16 Structures of the bacterial and eukaryotic sliding clamps. (a) The β protein sliding clamp from *E. coli* is composed of two identical β proteins, each of which has three similar domains. Dimerization forms a ring that has six similar domains (PDB 2POL). (b) The eukaryotic sliding clamp is made of three molecules of the PCNA protein. Each PCNA molecule has two similar domains and the trimer of PCNAs forms a ring with six domains (PDB 1AXC). The structures of the two sliding clamps are remarkably similar despite differences in protein sequence and the number of protein subunits. The DNA is threaded through the hole in the center of the ring.
From O'Donnell *et al.* (2001).

Figure 6.17 Structure of the eukaryotic clamp loader. (a) Structure of co-crystal of replication factor C (RFC) clamp loader and the PCNA sliding clamp with DNA modeled into the structure (PDB 1SXJ and 3GLF). PCNA is shown in orange, and each subunit of RFC is shown in a different color. The RFC clamp forms a helix that matches the twist of the DNA. (b) The RFC/PCNA/DNA complex showing each of the 5 RFC subunits and the manner in which the DNA is threaded through the complex. From Bowman *et al.* (2004).

Figure 6.18 Assembly of sliding clamp by the clamp loader onto primer–template DNA.
(1) The binding of ATP to the clamp loader (blue) leads to its association with the sliding clamp (orange) and opening of the sliding clamp.
(2) The opened clamp loader–sliding clamp has high affinity for the 3′ end of a primer–template junction in the DNA. (3) DNA binding triggers ATP hydrolysis leading to clamp closure and the release and recycling of the clamp loader.
(4) The sliding clamp is now loaded on the primer–template junction and can recruit DNA polymerase (purple) to the site to begin elongation.

in Chapter 12 that a similar ring-shaped structure is formed by a complex of proteins that are involved in detecting sites of DNA damage, and is similarly thought to track along the DNA.

The clamp loader assembles the sliding clamp onto DNA

The sliding clamp, which exists as a ring, must somehow be loaded onto the DNA when needed for replication. This requires that the sliding clamp be opened, DNA be inserted into the center, and the clamp be allowed to close again. This is accomplished by the clamp loader, which also has a multiprotein ring structure. The core of the clamp loader in both bacteria and eukaryotes consists of five protein subunits as illustrated in Figure 6.17. In *E. coli* it is composed of three copies of a protein subunit called γ, one δ subunit and one δ′ subunit. In eukaryotes, the clamp loader is known as **Replication Factor C (RFC)**, and consists of five different protein subunits, called Rfc1–5.

Some subunits of both the bacterial and eukaryotic clamp loader are AAA+ ATPases, and the cycle of sliding clamp loading to DNA is driven by conformational changes brought about by the binding of ATP to these ATPases, and subsequent ATP hydrolysis. In the absence of nucleotides, the clamp loader has low affinity for the sliding clamp. Upon binding ATP, however, the clamp loader undergoes a conformational change, which allows it to bind and force open the sliding clamp (Figure 6.18). In bacteria, the δ subunit interacts with and opens the sliding clamp, while, in eukaryotes, the Rfc1 subunit is involved in ring opening.

The clamp loader–sliding clamp complex, which assumes a spiral shape rather than a closed circle, has high affinity for the primer template junction – the region encompassing the 3′ end of the newly synthesized primer as it associates with the template DNA. In eukaryotes, the complex engages with the short sequence of DNA made by polymerase α–primase, which lies at the end of the primer strand. Binding of the clamp loader–sliding clamp complex to this DNA structure stimulates ATPase activity, which leads to another conformational change that closes the sliding clamp and releases the clamp loader from the DNA. The sliding clamp is thus ready to bind DNA polymerase at exactly the right position for the polymerase to begin elongation (Figure 6.18).

On dissociation of the adenosine diphosphate (ADP; the product of ATP hydrolysis), the clamp loader can be recharged with ATP and the cycle repeated. We shall see later that although the clamp loader is released from the sliding clamp and the DNA, it remains associated with the polymerase.

DNA REPLICATION: MECHANISM

The proteins required for DNA replication must be assembled into a replication complex and be loaded onto the DNA in a highly coordinated fashion. After the DNA duplex is opened by initiator proteins, helicases are recruited to the single-stranded DNA. The helicases help recruit primase, then the sliding clamp and DNA polymerases are loaded onto the DNA as soon as a primer is synthesized. Both strands of the DNA are then copied, the leading strand continuously and the lagging strand discontinuously. As replication proceeds, newly synthesized DNA segments are joined together to finally generate two identical daughter DNA molecules. Each step in DNA replication must occur at the proper time and only when the previous steps are completed. This is accomplished by the successive recruitment of proteins to DNA and close regulation of their activity, which also helps guarantee the accuracy of replication.

In this part of the chapter we discuss in more detail the molecular mechanisms adopted to make DNA replication happen, before moving on to consider how regulation is achieved.

6.7 ORIGINS AND INITIATION OF DNA REPLICATION

Replication begins at discrete sites on the chromosome

Replication origins are the sites at which the DNA helix is initially opened (see Section 4.8). Proteins called **initiator proteins** bind to origins to allow loading of the DNA helicases, which drive the unwinding of the helix (Figure 6.19). In addition, the initiator proteins also recruit many of the other proteins required for replication. Although origins in some organisms contain specific DNA sequence elements, it is the ability to bind initiator proteins that marks a site as a DNA replication origin.

The initiator proteins in both bacteria and eukaryotes are members of the AAA+ family of ATP-binding proteins (see Section 6.6). In *E. coli*, the initiator protein is a single-subunit protein called **DnaA**, which self-associates into a multisubunit

Figure 6.19 Initiation of replication occurs at specific sites in DNA. Origins of replication typically have a specific DNA sequence (orange dashes) that binds the initiator protein (green) with high affinity, adjacent to an AT-rich DNA sequence (pink). Binding of the initiator protein to its high-affinity binding site facilitates unwinding of the AT-rich region. Replication proteins, such as DNA helicase (orange), can then bind to the unwound DNA and begin to recruit other proteins for DNA replication.

complex when bound to ATP. In eukaryotes, the initiator is a protein complex called the **Origin Recognition Complex (ORC)**. As described in Experimental approach 6.2, this complex was first identified in the budding yeast *Saccharomyces cerevisiae*, where it has six essential subunits, Orc1–6. Five of these (Orc1–5) are members of the AAA+ family.

The binding and hydrolysis of ATP is used to regulate the initiation process. ATP binding by Orc1, for example, is required for the ORC to bind to DNA, while for DnaA it regulates the ability to form oligomers. There are homologs of ORC proteins in all eukaryotes, and despite the differences in origin structure, it appears that all eukaryotes require the ORC for origin activation.

The initiation of replication requires DNA unwinding and recruitment of helicases

The binding of initiator proteins to DNA specifies where replication begins. In many cases these initiator proteins bind to a well-defined DNA sequence. Typically, an origin of replication has recognizable binding sites for the initiator proteins and an adjacent AT-rich region that facilitates unwinding. A–T base-paired DNA requires less energy to separate than C–G base pairs due to there being fewer hydrogen bonds between the bases, as discussed in Chapter 1. These AT-rich regions are sometimes referred to as DNA unwinding elements. Organisms with well-defined, specific origin sequences include *E. coli*, bacteriophage lambda, *S. cerevisiae*, and mammalian viruses.

The study of *E. coli* has revealed much about the steps by which initiation occurs. The *E. coli* origin, termed *oriC* for 'origin of chromosomal replication', consists of a 245-base sequence. This contains seven 9 bp elements called DnaA boxes, which bind seven or more molecules of the initiator protein DnaA with high affinity. When fully loaded with ATP, the DnaA protein AAA+ domains multimerize into a spiral, or filament (Figure 6.20). This filament interacts with the DnaA boxes along the DNA strand. This interaction distorts the strand, causing a bend in the strand with an angle of up to 40°. The DNA is then further bent by the binding of additional bacterial histone-like proteins. This binding of DnaA, and perhaps also the bending of the DNA, facilitates the local unwinding of DNA in the AT-rich

Figure 6.20 DNA wrapped around initiator protein DnaA. (a) Structure of DnaA oligomers forming a spiral around which the DNA will wrap (PDB 2HCB). (b) Cartoon of the wrapping of the DNA around the ATP-bound DnaA and the subsequent unwinding of the origin sequence. From Erzberger JP *et al.*, *Nature Structural and Molecular Biology* 2006;**13**:676–683.

6.2 EXPERIMENTAL APPROACH

Discovery of the origin recognition complex

While DNA sequences that function as origins of replication had been identified in a number of systems using a variety of approaches, it was not clear what factors were localized to these regions to coordinate the activities of replication initiation. The ORC was discovered by Stephen Bell and Bruce Stillman when they set out to look for proteins that bind specifically to a DNA fragment containing a replication origin from yeast. Bell and Stillman used a DNA footprinting assay to identify cellular components that bound to a well-characterized yeast origin, ARS1.

Briefly, nuclear protein extract from yeast cells was incubated with a ^{32}P-labeled DNA fragment containing ARS1, and a DNase1 protection assay was used to see whether any proteins in the extract bound to the DNA. (See Section 16.11 for a more complete explanation of nuclease protection assays.) As initial footprinting results were promising, the researchers began to fractionate their nuclear extract on different column matrices to identify a fraction enriched in the binding activity of interest. (See Section 16.6 for further discussion of protein purification techniques.) Following each fractionation step, the various column fractions were evaluated for ARS1 binding activity using the footprinting assay. After four column fractionation steps, the active fraction was further subdivided by ultracentrifugation through a glycerol gradient; this approach distinguishes various proteins and protein complexes by size and shape. The proteins residing in the final fractions from the glycerol gradient were resolved on polyacrylamide gels and visualized (Figure 1a). Substantial ARS1 binding activity can be seen in fractions 18–20 (Figure 1b), correlating with the visible proteins present in those two fractions.

It turns out that the proteins found in these fractions represent the six subunits of ORC with sizes 120 kD, 72 kD, 62 kD, 56 kD, 53 kD, and 50 kD. In subsequent publications, Bell and Stillman identified these proteins by cutting them out of a protein gel like the one shown in Figure 1a, and subjecting them to peptide sequencing (described in Section 16.7). The gene encoding each of the subunits was ultimately identified, and cloned, by inferring the codon sequence from the sequence of amino acids in the protein.

The final proof that this complex of six proteins is the ORC that footprints across the ARS1 sequence came from reconstitution experiments. Each of the six cloned yeast proteins was simultaneously expressed in insect cells using a recombinant protein expression system. In this way the entire ORC was able to assemble inside the cells before purification and subsequent activity assays. The researchers found that ARS1 footprint could

Figure 1 Footprinting approach used to identify ORC binding activity.
(a) Protein gel showing the protein complex that was purified after four chromatography steps and glycerol gradient centrifugation. The even fractions from the gradient numbered 6–30 are shown; fraction 18 contains the highest level of a set of proteins that corresponds to origin recognition complex (ORC). (b) Nuclease protection (footprinting) assay showing that fractions 16 through 20 contain a DNA binding activity that protects the ARS1 sequence in a DNA fragment. The proteins are not visible in fraction 16 in the gel above because there is not enough protein present to detect by staining in the protein gel, even though there is enough to detect in the sensitive footprinting assay. Reproduced from Bell and Stillman, *Nature* 1992;**357**:128–134.

be reconstituted when all six proteins were simultaneously expressed. If just one component was omitted, however, no footprint was seen. This reconstitution verifies that the six ORC proteins together make up the active ORC.

Subsequent experiments from Bell and Stillman and a number of other groups showed that the ORC protein complex that was identified in yeast is conserved in many other organisms. The human ORC has been reconstituted and has many of the same properties of the ORC isolated from yeast. Indeed, related ORCs function in the initiation of DNA replication in all eukaryotic species examined.

Bell SP, Stillman B. ATP-dependent recognition of eukaryotic origins of DNA replication by a multiprotein complex. *Nature* 1992;**357**:128–134.

Bell SP, Mitchell J, Leber J, Kobayashi R, Stillman B. The multidomain structure of Orc1p reveals similarity to regulators of DNA replication and transcriptional silencing. *Cell* 1995;**83**:563–568.

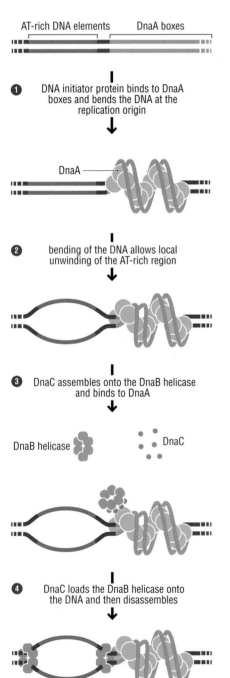

region to yield single-stranded DNA. The DnaA protein may then interact directly with the single-stranded DNA to facilitate strand opening.

As well as generating single-stranded DNA at the origin, DnaA also recruits the bacterial DNA helicase, DnaB, to the origin. Since the DNA helicase forms a ring structure around a single strand of DNA, similar to the loading of the sliding clamp discussed above, DnaB needs to be loaded onto the DNA by another protein called DnaC. To accomplish this, a complex of two hexameric DnaB molecules and the helicase loader protein DnaC interacts with DnaA bound at the origin (see Figure 6.21). These interactions with DnaA and DnaC are essential for bringing the DnaB helicase to the origin. Once the helicase is loaded, primase and polymerase can be recruited.

DNA replication origins in the yeast *S. cerevisiae* have a similar general structure to that of *oriC*. A typical origin in this yeast is about 100 bp long and contains two sites, A1 and B1, at which the ORC binds. Adjacent to these sites are AT-rich regions B2 and B3, which also interact with the ORC and may facilitate unwinding (Figure 6.22). As in *E. coli*, the origin DNA is wrapped around the ORC complex. ORC recruits two helicase-loading proteins Cdc6 and Cdt1, which in turn recruit the helicase Mcm2–7 (see Section 6.5). The complex of all these proteins is called the **prereplicative complex**. Unlike assembly of the initiation complex at the bacterial origin, assembly of this complex is not sufficient to initiate replication. Instead, it is loaded onto DNA well before DNA replication is actually due to begin, and the complex must be specifically activated to initiate replication. This regulation ensures that replication occurs at the correct time in the cell cycle, as we learned in Chapter 5.

Chromatin structure determines replication origins in some organisms

Whereas origins are determined by unique DNA sequences in many unicellular organisms and viruses, chromatin structure appears to play an important role in multicellular organisms. In humans and other animals, replication is reproducibly

Figure 6.21 Initiation of DNA replication at *E. coli oriC*. (1) The initiator protein DnaA binds to three high-affinity DNA elements called DnaA boxes at the *E. coli* origin. The binding of the first DnaA proteins allows the binding of additional DnaA proteins. When complexed with ATP, these proteins oligomerize into a right-handed helical oligomer. (2) The wrapping of the DNA around the protein complex and the DnaA itself facilitate local unwinding of the adjacent AT-rich region. (3) The helicase loader DnaC binds the helicase DnaB and loads it onto the single-stranded DNA. (4) DnaC then dissociates, and the origin is ready for recruitment of primase and other replication proteins.

initiated in specific chromosomal regions, and yet these regions do not contain discrete, highly conserved DNA sequences as found in yeast and bacterial origins, and mammalian viruses. Despite the lack of a specific origin sequence, however, the ORC is still required for the initiation of replication.

How does chromatin structure have its effect? In *Drosophila*, the ORC is recruited to origins where histone tails are hyperacetylated. When the overall level of histone acetylation is artificially increased, there is an increase in origin firing (the initiation of replication from an origin), indicating that chromatin structure can determine origin activity. The regulation of transcription initiation also involves acetylation of histone tails. Thus, proteins such as histone acetyltransferases, which are known to regulate transcription through their effect on chromatin structure, may also play a role in regulating the usage of replication origins.

Other organisms that do not have specific, high-affinity binding sites for ORC may also make use of aspects of chromatin structure, such as histone modification, to specify origin location and usage.

⊙ We learn more about histone acetylation in Section 4.5 and more about the regulation of transcription in Chapter 8.

6.8 LEADING AND LAGGING STRAND SYNTHESIS

DNA primase synthesizes the RNA primers that start all new DNA strands

After an origin of replication has been opened and helicases and primases have been recruited, DNA replication can begin. As discussed in Section 6.1, the first step in DNA replication is the synthesis of primers. Primers are required on both the leading and lagging strands. Because of the discontinuous nature of DNA synthesis, however, priming occurs much more frequently on the lagging strand (see Figure 6.3).

A primer is a short stretch of RNA or RNA + DNA, which is synthesized by a specialized polymerase known as **primase** (Figure 6.23). Primase binds single-stranded DNA that is coated with single-stranded binding proteins. Unlike DNA polymerases, which require a 3′ OH group onto which the DNA polymerase can add nucleotides to effect growth of the new strand, RNA polymerases such as primase can start RNA synthesis *de novo*. On the leading strand, initiation of DNA replication occurs only at the origin. By contrast, during discontinuous DNA replication on the lagging strand, primase must synthesize a primer for each new DNA fragment.

In bacteria, a short RNA primer of around 10–30 bases is synthesized by a primase that then hands off replication directly to DNA polymerase III, the bacterial replicative polymerase. In eukaryotes, however, the primase complex contains four subunits: two subunits that function as a primase, bound in a complex with the DNA polymerase α catalytic subunit and an accessory subunit. The polymerase α–primase complex synthesizes a stretch of 10–30 nucleotides of RNA, which the polymerase α subunit of the complex elongates with a short stretch of DNA. After this primer synthesis, the replicative polymerases δ or ε take over and carry out the rest of the elongation phase of DNA replication. This handover from polymerase α–primase to polymerase δ or ε is known as **polymerase switching** (Figure 6.24). Polymerase switch occurs each time a new DNA strand is started on both the leading and the lagging strands.

Figure 6.22 Budding yeast origin of replication. Origins of replication in *S. cerevisiae* contain two core sequences (A and B1) that together bind the origin recognition complex (ORC). The A element includes an 11 bp region called the ARS consensus, whose sequence is highly conserved among all budding yeast origins. The replication origin also contains a less conserved region (B2) that serves as the major site of DNA unwinding. Some budding yeast origins contain a third region (B3, not shown) that binds an additional protein factor Abf1 (not shown), which helps regulate origin activity, probably through the control of nucleosome boundaries at the ARS.

Figure 6.23 Primases catalyze the synthesis of a short RNA primer for DNA replication. (a) The eukaryotic primase is part of a DNA polymerase α–primase complex. This complex is shown here in simplified form: the native eukaryotic primase (orange) is a two-subunit enzyme, as is DNA polymerase α (pink). (b) Bacterial primases (orange) are single-subunit RNA polymerases. The primase from *E. coli* is also known as DnaG.

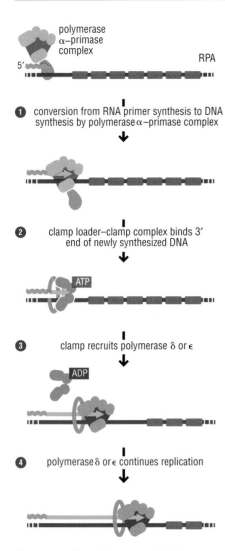

① conversion from RNA primer synthesis to DNA synthesis by polymerase α–primase complex

② clamp loader–clamp complex binds 3′ end of newly synthesized DNA

③ clamp recruits polymerase δ or ε

④ polymerase δ or ε continues replication

Figure 6.24 The polymerase switch. In eukaryotes, the RNA primer is synthesized by the polymerase α–primase complex (pink), which binds to single-stranded DNA coated with the single-stranded binding protein RPA. It synthesizes an initial RNA primer and then changes to DNA synthesis. Polymerase α is soon replaced by one of the more processive polymerases, δ or ε. The polymerase switch occurs as a result of the binding of the clamp loader (blue) onto the junction between the 3′ end of the primer and the single-stranded DNA template. The clamp loader loads the sliding clamp onto DNA (orange), which in turn recruits polymerase δ or ε (purple).

The replicative polymerase is recruited by the sliding clamp, which is loaded onto DNA at the primer–template junction by the clamp loader (see Section 6.6). The sliding clamp is placed around the double-stranded DNA by the clamp loader at the free 3′ end that is to be elongated. The clamp loader is then released and polymerase δ or ε binds to the same face of the clamp where the loader had bound. This sequence of events ensures that the DNA polymerase is loaded onto the DNA only after a primer has been made, and is also in the right place to begin DNA elongation.

The lagging strand is synthesized in short stretches of DNA called Okazaki fragments

As the replication fork progresses along the chromosome, DNA is synthesized on both template strands, a process that requires two polymerases. As the two strands have opposite chemical polarities this creates a problem: for the fork to move in one direction, the polymerases would apparently have to synthesize one strand in the 3′ to 5′ direction and the other in the 5′ to 3′ direction. However, as described in Section 6.1, DNA polymerases cannot synthesize DNA in the 3′ to 5′ direction. Instead, the lagging strand is synthesized in the 5′ to 3′ direction, but in short discontinuous segments. This discontinuous synthesis is coupled to the movement of the replication fork, as we shall discuss later.

The initiation of new DNA synthesis on the lagging strand is carried out in an identical manner to that described above for the leading strand, but occurs much more frequently and rapidly. The short stretches of DNA synthesized on the lagging strand are called **Okazaki fragments** (after their discoverer); once each fragment is completed, it is joined to the previous fragment in a process called **Okazaki fragment maturation** to make a continuous length of DNA.

Okazaki fragment maturation generates a continuous DNA strand

Okazaki fragment maturation takes place in several steps, as illustrated in Figures 6.25 and 6.26. First, the RNA primer is removed and replaced by DNA, and then the two adjacent DNA fragments are ligated – that is, joined together.

The mechanism of Okazaki fragment maturation differs slightly between bacteria and eukaryotes. In bacteria, DNA polymerase III synthesizes DNA up to the beginning of the next RNA primer, then stops and dissociates from the sliding clamp and the DNA. The sliding clamp remains on the DNA and is involved in recruiting the proteins for Okazaki fragment maturation. Primer removal is the task of another polymerase, DNA polymerase I (see Section 6.2), which possesses both 3′ to 5′ (proofreading) and 5′ to 3′ exonuclease activity, and uses the latter to remove the RNA primer of the previous Okazaki fragment. Just after RNA removal, DNA polymerase I uses its polymerase activity to fill in the gap left by the RNA with new DNA. When it reaches the beginning of the next DNA segment fragment, the DNA polymerase cannot seal the final nick in the DNA backbone that remains; this is done by an enzyme called **DNA ligase**, which can catalyze the formation of a covalent bond between adjacent 5′ and 3′ ends. This ligation reaction links the two fragments together into one continuous DNA strand (see Figure 6.25).

Okazaki fragment maturation in eukaryotes is similar in principle, but removal of the RNA primer and the filling-in with new DNA are carried out in a slightly different way. When it reaches a primer, polymerase δ or ε continues synthesizing DNA rather than stopping. This results in the end of the strand that was ahead of the polymerase being lifted like a flap and displaced as the polymerase runs into it. This strand displacement occurs with the assistance of several polymerase accessory proteins. A protein called Fen1 then cleaves the loosened and partly displaced single-stranded DNA, or flap (Figure 6.26). Fen1 is an **endonuclease** that cleaves single-stranded nucleic acids internally. Under some circumstances Fen1 cleavage does not occur right away and a long, displaced single strand is created. Dna2, which has both helicase and endonuclease activity, can then cleave the long flap, which is followed by further processing by Fen1.

Figure 6.25 Okazaki fragment maturation in *E. coli*. DNA polymerase III synthesizes new DNA from the 3' end of a primer until it reaches the beginning of the primer of the previous Okazaki fragment. When polymerase III encounters the primer, it leaves the DNA and polymerase I is recruited. The 5' to 3' exonuclease activity of DNA polymerase I removes the RNA primer and its nucleotide-polymerizing activity fills in the gap with DNA. The gap in the DNA backbone between the end of this fragment and the beginning of the next is joined by DNA ligase, which is initially recruited to the site by binding to the sliding clamp.

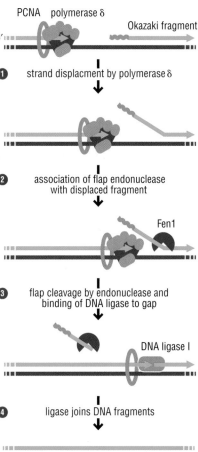

Figure 6.26 Okazaki fragment maturation in eukaryotes. DNA polymerase δ synthesizes DNA up to and past the 5' end of the adjacent Okazaki fragment, displacing a single-stranded 5' tail that contains the RNA primer. The Fen1 endonuclease then cleaves the flap, leaving a simple nick in the DNA backbone that is sealed by DNA ligase. PCNA, proliferative cell nuclear antigen.

After cleavage and consequent removal of the RNA primer, DNA ligase joins the 3′ end of the newly made DNA to the 5′ end of the adjacent DNA. In archaea, a homolog of Fen1 is required for Okazaki fragment maturation, suggesting that this mechanism is conserved between archaea and eukaryotes.

DNA elongation and Okazaki fragment maturation all take place in the context of the replication complex at the replication fork, which is moving along the DNA and synthesizing both the leading strand and lagging strand simultaneously. In the next several sections we will examine how this coordinated synthesis is achieved.

6.9 THE REPLICATION FORK

Leading and lagging strand synthesis is coupled at the replication fork

While leading strand synthesis proceeds continuously from 5′ to 3′, with the polymerase moving in the same direction as the moving fork, lagging strand synthesis is discontinuous, with the polymerase moving in the direction opposite to that of the fork (Figure 6.27a). The movement of the polymerases is nonetheless coordinated, with the lagging strand looping around the fork to allow both polymerases to travel together in the direction of fork movement (Figure 6.27b). The discovery of this looping is described in Experimental approach 6.3. The coordination of leading and lagging strand polymerases allows the replication of both strands to be regulated together. If leading strand replication stalls, lagging strand replication will also be halted. For example, if damaged DNA is encountered on one strand, the entire fork complex will halt and wait for repair.

The coupling of DNA replication on the leading and lagging strands in bacteria is achieved by physically associating the proteins replicating each strand into one large protein complex called the **replication complex** or **replisome**, which associates with the DNA as a whole as the fork moves. Proteins within the complex can engage and disengage from the DNA itself, but remain associated with the fork, ready to act when required.

Leading and lagging strand synthesis in *E. coli* is coordinated by the replisome

The coupling of leading and lagging strand polymerases is best understood for bacterial replication complexes (Figure 6.28). In *E. coli* the multisubunit assembly of DNA polymerases, sliding clamps, and clamp loader is called the DNA **polymerase III holoenzyme** or the **replisome**. It contains two copies of the multisubunit DNA polymerase III together with sliding clamps, and a clamp loader that continually reloads sliding clamps on the lagging strand. One polymerase replicates the leading strand and one replicates the lagging strand as the fork progresses along the double helix.

The polymerases are linked together via two subunits of a protein called tau (τ), which is associated with the clamp loader, and which also links the complex to the helicase. (The helicase unwinds the DNA double strand ahead of the moving

(a)

5′
3′

(b)

5′
3′

Figure 6.27 The replication fork.
(a) Replication bubble showing the direction of DNA synthesis on the leading and lagging strands. (b) Replication bubble showing two replication forks with the lagging strands looped around to allow both polymerases at one fork to travel together in the direction of fork movement.

6.3 EXPERIMENTAL APPROACH

Coupling of polymerases at the replication fork: the trombone model

The architecture of the DNA replication fork is both elegant and complex. Early biochemical studies by Bruce Alberts and colleagues showed that leading and lagging strand polymerase activities are coupled at the fork, raising a topological conundrum: how can polymerization go in two directions at the fork? On the basis of the biochemical data, a model was proposed whereby the template DNA of the lagging strand is looped out, allowing the two polymerases to effectively proceed together in the same physical direction, with the lagging strand polymerase continually reinitiating the extension reaction on an available template. This model was referred to as the trombone model because of the similarity between the looped-out DNA as imagined by this model and the slider on a trombone (Figure 1).

Electron microscopy and a biochemically reconstituted system eventually allowed for direct visualization of this proposed structure. To set up such an experiment, all of the proteins known to be required for DNA synthesis and fork movement were purified and incubated *in vitro* with DNA template in conditions where active replication would occur. The preparation was then fixed and concentrated, and spread on a special carbon grid on which the DNA and protein complexes could lie flat. The grid was then coated with fine tungsten powder, which allowed the contrast in different surfaces to be visualized. (The general technique of electron microscopy is described in Chapter 16.)

In the micrograph in Figure 2a, a circle is being replicated to generate a long double-stranded copy (so-called rolling circle replication). The double helix of DNA of the template circle can be seen as a wavy garden hose winding around on the grid. The polymerase complex, or replisome, sits at the junction between this helix and the circle; there is a looped region that represents the trombone structure adjacent to the circle. This loop is generated by the coupled polymerase as depicted in Figure 2b. Subsequent microscopy work in which specific components were individually labeled and imaged demonstrated unambiguously that each of the expected fork proteins are found in the large visible protein complex.

The power of such techniques that visualize individual molecules has recently been adapted to answer detailed functional questions about replication fork activity. Using powerful optical microscopes it is possible to visualize single DNA molecules in an ongoing biochemical reaction. In one recent study, a DNA molecule was chemically tethered at one end to a glass while the other end was attached to a magnetic bead. Replication fork components were next assembled on the DNA (by adding the required purified

Figure 1 Trombone model for replication forks. Biochemical experiments indicated that both the leading and lagging strand polymerases are coupled together during replication. Such coupling of the leading and lagging strands was predicted in this model to cause looping of the lagging strand DNA to allow both polymerases to travel together.
Reproduced from Alberts, *Trends in Genetics* 1985;**1**:26–30.

(a)

(b)

Figure 2 Visualization of trombone looping. (a) Electron micrograph of rolling circle replication showing the replisome and the looped out lagging strand of DNA. (b) Interpretation of the electron micrograph.
Reproduced from Chastain *et al.*, *Journal of Biological Chemistry* 2003;**278**:21276–21285.

(a)

(b)

Figure 3 Single molecule experiments show dynamics of replication fork progression. (a) A tracing of the position of a single molecule of DNA with an assembled replication fork. The DNA is tethered to a slide at one end and the position of the other end is monitored over time. As the polymerization reaction proceeds initially there is no change in the length of the DNA. As a loop grows, however, the length of the DNA shortens as some of the DNA is incorporated into the loop. However upon release of the looped out strand the DNA abruptly gets longer. (b) Diagram explaining the interpretation of the results in (a). Reproduced from Lee *et al.*, *Nature* 2006;**439**:621–624.

proteins), and buffer was flowed across the DNA–protein solution, effectively stretching the DNA and pushing the bead end away.

Up to 50 individual DNA molecules could be followed simultaneously by precisely measuring the position of each magnetic bead over time. The replication reaction was initiated simultaneously on all molecules at a particular time (by the addition of dNTPs, for example) and the position of the bead was followed over time as each fork synthesized DNA. The change in bead position represents fork movement as synthesis progresses; the *abrupt* changes in bead position represent release of the looped lagging strand

DNA. By following the movement of individual molecules, these investigators obtained a representative picture of how the process is occurring at the molecular level.

The stochastic nature of specific steps such as loop release can be measured as differences in the timing of the release step. One example of such a single molecule tracing is shown in Figure 3a. If all molecules were averaged, as they are in typical 'bulk' biochemical experiments, instead of single molecules being monitored it would not be possible to see the specific steps in the reaction that may vary.

With this system, many parameters about the role of each protein component of the fork can be assayed by altering the reaction conditions. For example, omitting the helicase blocks the

extension of the fork, indicating that unwinding is essential for fork movement. In the experiment shown here, the authors found that lagging strand release from the replisome is a regulated process and that the loop length is regulated by the primase-mediated initiation of lagging strand synthesis.

Alberts BM. Protein machines mediate the basic genetic processes. *Trends in Genetics* 1985;**1**:26–30.

Lee JB, Hite RK, Hamdan SM, *et al*. DNA primase acts as a molecular brake in DNA replication. *Nature* 2006;**439**:621–624.

Nossal NG, Makhov AM, Chastain PD 2nd, Jones CE, Griffith JD. Architecture of the bacteriophage T4 replication complex revealed with nanoscale biopointers. *Journal of Biological Chemistry* 2007; **282**:1098–1108.

replisome.) One of the functions of the replication complex is to keep a lagging-strand polymerase associated with the fork even though it is being released from the DNA at the end of each Okazaki fragment. This association of polymerase, clamp loader, and sliding clamp means that the loading of a new clamp and a polymerase onto DNA at the beginning of each Okazaki fragment can be efficiently coordinated with the progress of the polymerase on the leading strand.

A bacterial replication fork moves at a rate of around 750–1000 bp per second and there is a new Okazaki fragment every 1–2 kb, so every second or two the clamp loader is putting a new clamp onto a newly primed 3′ end and a polymerase is re-engaging with a clamp and DNA.

The eukaryotic replication complex shares many features of the *E. coli* complex

The basic mechanism of replication fork progression in eukaryotes is very similar to that described above for *E. coli*. It is thought that leading and lagging strand

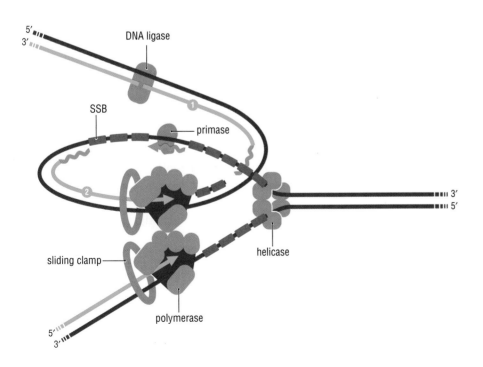

Figure 6.28 Replication fork coupling in *E. coli*. The leading and lagging strands are coupled by looping. The lagging strand is shown looped with a completed Okazaki fragment (1) and one DNA polymerase III molecule almost completing the synthesis of the next Okazaki fragment (2). A new primer is being synthesized by primase on the unwound lagging strand. The leading strand is being synthesized continuously and the whole fork is moving to the right. The replisome that moves with the fork contains two associated DNA polymerase III molecules bound to the sliding clamp and linked by the tau protein (not shown).

synthesis may be coupled in eukaryotes, like they are in *E. coli*. However, unlike the replisome of *E. coli*, the eukaryotic replication proteins do not seem to exist in a single, large complex as evidenced by the way they fail to co-purify as such. Therefore, when we refer to the eukaryotic replication complex, we are really considering a collection of proteins working in concert, but not necessarily as a tightly bound complex.

Both polymerase δ and polymerase ε can be found at the replication fork, and there is good evidence that polymerase ε is primarily responsible for replicating the leading strand and polymerase δ for the lagging strand. Unlike the bacterial replisome, and as we note above, the polymerases in the eukaryotic replication complex do not appear to be physically associated with one another and no protein that links the two polymerases, equivalent to bacterial tau, has been identified. However, as in bacteria, a clamp loader and sliding clamp appear to be part of the replication complex, recycling the lagging strand polymerase after synthesis of each Okazaki fragment. In eukaryotes, there is also a polymerase switch at the beginning of each Okazaki fragment, as described in Section 6.8, and the polymerase α–primase is part of the assembled eukaryotic replication machine.

6.10 TERMINATION OF DNA REPLICATION

We have explored the process by which DNA replication is initiated, and how two daughter strands are synthesized using the original double-stranded parent as a template. We now consider what happens when replication is complete – the process of termination.

The meeting of two replication forks terminates replication

For the circular chromosomes of bacteria, termination of DNA replication occurs when the two replication forks moving away from the single origin meet on the opposite side of the circle (Figure 6.29). As the forks approach each other, the replication complex must disassemble and the two growing strands must be joined. This results in the interlinking of the two new daughter DNAs. Therefore, the two circular DNA molecules must be resolved to allow the duplicated chromosomes to separate. This resolution can be carried out by topoisomerases as described in detail in Section 6.5 above.

The linear chromosomes of eukaryotes have multiple origins of replication, each of which initiates a replication bubble. Consequently, replication terminates at multiple points. As replication proceeds, two replication forks from different bubbles will meet in the same way as the two bacterial replication forks. At that point, replication must be terminated and the two growing strands joined (Figure 6.30). The chromosomes, which are intertwined as a result of replication, must also be unlinked by topoisomerase as described above for bacterial chromosomes.

Termination of replication occurs at a specific site in bacterial chromosomes

The termination of DNA replication occurs when two forks approach each other. In *E. coli*, termination occurs at a DNA region called *ter* (see Section 4.9) on

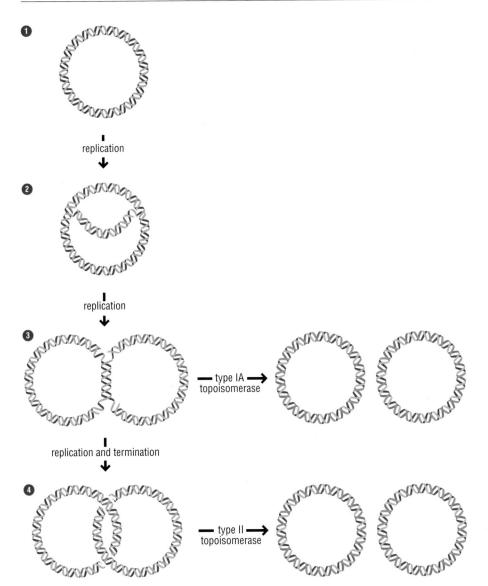

Figure 6.29 Replication of circular chromosomes results in daughter molecules that are still linked to each other. Replication of a circular DNA double helix results in two interlinked or catenated double-stranded daughter molecules. The circular chromosomes must be unlinked or decatenated by type 1A topoisomerases at step 3 or type II topoisomerases at step 4 to allow separation.

the opposite side of the circular chromosome from the replication origin, *oriC*. As the two replication forks that initiate at *oriC* progress at the same pace around the chromosome, they will meet near *ter*, which contains specific protein binding sites. The *ter* element is bound by a replication terminator protein known as Tus in *E. coli*. When the forks approach each other, one fork arrests when it is stopped by the bound Tus protein. When the other fork arrives it too arrests. At this point, disassembly of the forks presumably occurs, leaving a short length of single-stranded DNA uncopied. This gap is assumed to be filled in by a specialized polymerase, DNA polymerase I, which is also used in DNA repair, as discussed in Chapter 12. The molecular details of how the fork is disassembled to allow repair synthesis are not yet clear.

In circular bacterial chromosomes or circular viral genomes, termination of replication results in two daughter molecules that are linked together, or catenated, as described in Chapter 4. These daughters are separated in a process called **decatenation**, which involves strand breakage by either type IA or type II topoisomerases (see Section 6.5). Breakage and strand passage of the single-stranded DNA by type IA topoisomerase before replication is complete will unlink the almost-complete

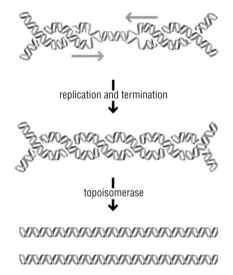

Figure 6.30 Converging forks create interlinked DNA on linear chromosomes. As two forks converge and replicate the region of DNA between them, the double-helical nature of DNA is translated into helical intertwining of the two finished daughter molecules. This intertwining is then resolved by topoisomerase before the daughters are separated.

molecules shown in Figure 6.29, step 3. By contrast, double strand DNA breakage and passage by type II topoisomerases can unlink the fully replicated daughter molecules shown in Figure 6.29, step 4.

Replication does not usually terminate at specific sites in eukaryotic chromosomes

The existence of multiple replication origins in eukaryotes means that there is no one specific site where termination occurs, in contrast to the situation in bacteria. Instead, termination can occur anywhere on the chromosome. We do not yet know how the two converging replication forks interact to allow their disassembly and assure the completion of DNA synthesis between them. There are, however, some regions in eukaryotic chromosomes where replication is blocked in one direction by a specialized structure. In budding yeast, for example, there is a specific termination site termed the **replication fork barrier** in the cluster of repeated genes encoding ribosomal DNA (rDNA). A protein called FOB1 binds to a specific DNA sequence at the fork barrier and blocks progression of replication forks coming from one direction. This ensures that the repeated rRNA genes are always replicated in the same direction; if a fork approaches from the other direction, it is stopped. This mechanism may be specific to sites of repeated genes and may help to prevent errors in copying the repeated DNA sequences.

Due to the double-helical nature of DNA, replication of linear chromosomes is translated into helical intertwining of the two finished daughter molecules as shown in Figure 6.30. As in the case of concatenated circular chromosomes, the intertwining of very long linear eukaryotic chromosomes also requires topoisomerases for resolution. The long bulky nature of the chromosome does not allow for passive untangling from the ends, and topoisomerases are required for the daughter chromosomes to be separated.

In contrast to internal origins, replication termination at the *end* of a linear chromosome can occur by the polymerase simply copying to the end of the molecule. When this happens, however, it poses a problem: how can the polymerase replicate all of the nucleotides at the very end of the DNA sequence? We explore the answer in the next section.

6.11 THE END-REPLICATION PROBLEM AND TELOMERASE

The end of a linear DNA molecule cannot be completely replicated by the replication complex

The mechanism of DNA replication described so far in this chapter allows faithful copying of the vast majority of chromosomal DNA. For organisms with linear chromosomes, however, it results in the non-replication and eventual loss of some sequence from the ends of the DNA molecules. This loss occurs because lagging strand synthesis cannot complete the copying of the very end of a linear DNA molecule. This is known as the **end-replication problem** and, after repeated rounds of replication, leads to considerable loss of sequence from chromosome ends.

Incomplete replication is often described as arising from RNA primer removal. However, in principle, it can occur in several ways. The removal of the last Okazaki fragment synthesized before the replication complex runs off the end of a DNA template strand will leave a single-stranded region of DNA (Figure 6.31). Alternatively, the dissolution of the replication fork when it reaches the end of the leading strand may result in premature termination of DNA synthesis on the looped-out lagging strand and may also lead to incomplete copying of the leading strand (Figure 6.32).

The end-replication problem can be solved by several different mechanisms

Cells and viruses have several different mechanisms for overcoming the end-replication problem (Figure 6.33). In circular chromosomes, such as those of most bacteria, the problem is avoided altogether. Viruses with linear genomes have evolved several different mechanisms to replicate the ends of their DNA. For

Figure 6.31 Inability to fill the gap left after the removal of the last RNA primer is one cause of the end-replication problem. After initiation of replication, leading and lagging strand synthesis proceeds to the end of the chromosomal DNA. During replication, the RNA primers between adjacent Okazaki fragments are removed and replaced with DNA, and the fragments are ligated to eventually produce one long continuous strand. The terminal RNA primer on the lagging strand has, however, no adjacent fragment to ligate to and so once it is removed, the lagging strand is shorter. When this shortened strand is replicated at the next round of DNA synthesis that sequence will be permanently lost.

Figure 6.32 Dissolution of the replication complex may lead to the loss of the terminal Okazaki fragment. A replication fork is shown with the single-stranded DNA bound by SSB. The replication fork machinery disassembles when it reaches the end of a DNA molecule. Because the lagging strand is looped out to allow coordinated leading and lagging strand synthesis, the last Okazaki fragment begun may not be completed before the fork disassembles. This leads to loss of sequence on the lagging strand. A small region on the leading strand may also not be copied after fork collapse.

Mechanisms of overcoming the end-replication problem	
mechanism	**example**
circular chromosome	bacterial chromosome, plasmids
end recombination	mitochondrial genomes, bacteria
transposition to telomeres	*Drosophila* telomeres
genome concatamerization	T4 phage
hairpin ends	vaccinia virus
terminal protein primer	adenovirus, phage ϕ29
telomere repeat addition	eukaryotic chromosomes

Figure 6.33 Mechanisms for overcoming the end-replication problem. The table lists several different mechanisms that are used by a variety of organisms to avoid loss of sequence at the ends of a DNA molecule when it is replicated.

example, bacteriophage T4 strings several copies of its genome together before replication to reduce the number of ends. Another mechanism is found in the vaccinia virus. Its double-stranded DNA genome has the two strands covalently linked to each other at each end of the DNA, creating closed hairpins. At each end, replication proceeds around the hairpin, and the daughter molecules are subsequently processed to generate two complete, new genomes. In bacteriophage ϕ29 and mammalian adenovirus, replication is primed by a terminal protein and not by an RNA primer; by initiating from a protein there is no terminal RNA primer to remove and the full DNA sequence can be maintained.

Eukaryotes, in contrast, overcome the end-replication problem in an entirely different way. As described in Section 4.11, eukaryotic chromosomes have telomeres at their ends, which consist of simple sequence repeats (see Figure 4.35). The length of the whole tract of telomere repeats varies from less than 100 bp to more than 20 000 bp, depending on the species. The bulk of the telomere sequence is replicated by the replicative polymerases as they move along the double-stranded telomere DNA, and, for the reasons described above, the ends of the telomeres grow shorter at each round of replication.

Telomeres can, however, be elongated by the enzyme telomerase independently of the standard DNA replication machinery. This enzyme adds specific telomere DNA sequences *de novo* onto the chromosome ends: when the telomeres shorten, telomerase is recruited to add telomere repeats. This results in net telomere elongation and allows telomeres to be maintained around an equilibrium length, as described in Chapter 4 (see Figure 4.37). In *Drosophila*, telomere length is maintained by yet another mechanism: transposable elements called Het and TART transpose onto the ends of the telomeres to maintain their length.

Telomerase contains a specialized reverse transcriptase and an essential RNA component

Telomerase is a unique DNA polymerase that is made up of a catalytic protein and an integral RNA molecule, which also provides the template for the synthesis of telomere repeats in addition to its functional role in the enzyme activity. In addition to these universal conserved subunits, there are different accessory proteins in different species.

The catalytic protein of telomerase is called telomerase reverse transcriptase (TERT) and is conserved between distantly related eukaryotes. Telomerase has similar catalytic domains to those in other polymerases including reverse transcriptases (see Section 6.4), with the motifs that make up the palm, fingers and thumb domains being conserved. Within these regions, the catalytic residues that coordinate magnesium in other polymerases and allow catalysis (see Figure 6.6) are also conserved in telomerase. Mutation of these residues inactivates the enzyme, indicating that it uses a similar catalytic mechanism to that of reverse transcriptase.

The essential RNA component of telomerase varies in size from 150 nucleotides to over 1300 nucleotides in different species. It also varies in sequence, but its overall secondary structure is conserved. In addition to providing a small stretch of template RNA, specific structures within the telomerase RNA are required for telomerase activity. Mutations in the RNA in sites distant from the template region render the enzyme inactive, indicating that these residues play an as yet undefined role in catalysis.

Telomerase adds telomeric repeats processively

Telomerase binds to single-stranded telomeric DNA that is at the very end of the telomere whereupon the 3′ end of the sequence to be elongated is base-paired with the template sequence of the telomerase RNA (Figure 6.34). The enzyme then uses the RNA as a template to add telomeric sequences to the 3′ end of the

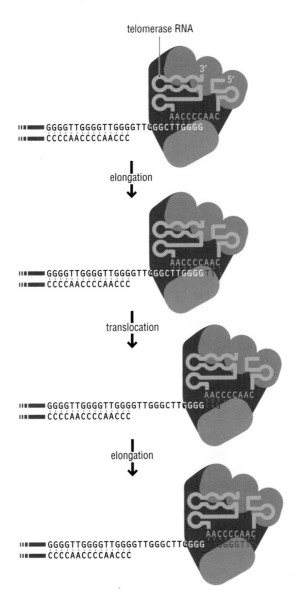

Figure 6.34 Elongation of telomeres by telomerase. Telomerase binds to the single-stranded GT-rich DNA at the end of a telomere. The 3′ OH end of the DNA is base-paired with the RNA template region (the AACCCCAAC sequence in green) in the telomerase. In the elongation step, telomere repeats are copied using the template as a guide and are added to the DNA (shown as red bases). Once the end of the template is reached, the telomerase translocates forward to reposition the new DNA 3′ end with the RNA template for a second round of elongation. Telomerase does not dissociate from the DNA during translocation and can processively add hundreds of base pairs of new telomere sequence.

telomeric DNA. Once the end of the template is reached, the new DNA 3′ end is realigned with the beginning of the RNA template while the enzyme remains associated with the telomere DNA. The telomerase then makes another copy of the template, but this time adding to the 3′ end of the new DNA. As such, the translocation event, and the repeated synthesis from a single template, allows telomerase to processively add hundreds of nucleotides onto a telomere before it dissociates (Figure 6.34).

6.12 CHROMATIN REPLICATION

Chromosome replication requires duplication of both the DNA and the protein components of chromatin

In all organisms, the chromosomes contain not only a DNA molecule but also specific proteins bound to the DNA. In eukaryotes, DNA is packaged into nucleosomes.

Many other proteins are also bound to DNA along the chromosomes, including those proteins essential for DNA repair, DNA recombination, and gene expression. To maintain chromosome function, therefore, not only must the DNA be copied, but also specific proteins must be recruited to both daughter molecules after the replication fork passes, to give faithful copies of both the DNA and the protein components of the parental chromosome. In addition, modifications to the DNA such as methylation (see Section 4.7) must be re-established on both daughter molecules after replication. The re-establishment of the chromatin structure allows the epigenetic inheritance of heterochromatic and euchromatic regions in the chromosomes, as introduced in Chapter 4 and described further in Chapter 8.

Old histones present ahead of the replication fork are recycled onto newly replicated DNA

To replicate chromosomal DNA in eukaryotes, the replication fork must disrupt the nucleosomes on the parental DNA, separate the DNA strands and then reassemble nucleosomes on both daughter DNA molecules behind the fork. Remarkably, the histones that were bound in front of the replication fork rebind behind the fork, maintaining the modification state of the chromosomal region. As the fork proceeds, the modification state of the parental chromatin is thus passed on by a redistribution of the old nucleosomes between the two daughter DNAs in a process called **parental histone segregation** (Figure 6.35).

As we saw in Chapter 4, histones are subject to a number of modifications including methylation, acetylation and phosphorylation, and modifications differ between heterochromatin and euchromatin. During chromatin replication the old H3 and H4 core histones are segregated to each daughter and a reconstituted nucleosome is assembled with H2A and H2B. This segregation of the old, modified histones to both daughter strands is thought to provide a guide to the reconstruction of the parental chromatin state in each daughter chromosome.

> ⊃ We discuss the packaging of DNA into nucleosomes in Section 4.3.

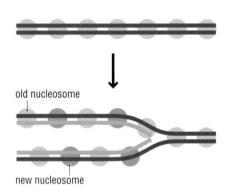

old nucleosome

new nucleosome

Figure 6.35 Histone segregation. Chromosome replication in eukaryotes proceeds through the duplication of chromatin as well as DNA. As the replication fork passes though chromatin, half of the old H3 and H4 core histones (light balls) go to one daughter DNA helix and half to the other and the full nucleosome is then reconstituted behind the fork. Newly synthesized H3 and H4 core histones (dark balls) are then deposited on both daughter molecules to maintain a ratio of one nucleosome approximately every 200 bp. The inheritance of old histones H3 and H4 by each DNA daughter molecule allows for the modification state of the histones in any given region of DNA to be inherited by both daughter chromosomes. The new nucleosomes are presumed to be modified according to the modification state of their neighbors.

The distribution of old histones to both new DNA molecules is carried out by the active transfer of H3 and H4 by a protein complex that contains the histone chaperone Asf1. Asf1 binds a half nucleosome containing a tetramer of two H3 and two H4 histones and guides its addition onto DNA. This segregation of old nucleosomes means that, immediately behind the fork, each of the two daughter DNA strands has only half the number of nucleosomes required to package the DNA. New nucleosomes must be assembled, and these are added directly behind the fork by the **Caf1 (chromatin assembly factor)** together with Asf1. The newly synthesized and deposited histones are surrounded by old, modified histones, which are thought to recruit histone-modification enzymes that rapidly modify the new histones according to the state of the old histone neighbors to which they are adjacent.

Replication-dependent histones are synthesized and modified during replication just before their assembly onto DNA

To generate the large number of histone proteins needed during replication to make new nucleosomes, histone synthesis is tightly linked to DNA synthesis. The cell has a number of different histone variants (see Section 4.3), but only the major histones – H2A, H2B, H3, and H4 – are synthesized during replication to provide the building blocks for new nucleosome assembly. They are therefore known as the **replication-dependent histones** or **S phase histones**.

Histone assembly into nucleosomes occurs stepwise by the formation of subcomplexes, which then assemble into the full nucleosome on the DNA. First, histones H2A and H2B form dimers, and independently two subunits each of histones H3 and H4 assemble into a stable tetrameric complex. The H3–H4 tetramer is first deposited on the DNA, after which two H2A–H2B dimers are added to complete the histone octamer, as described in Chapter 4.

Histones are deposited behind the replication fork by the chromatin assembly factor

Newly synthesized S phase histones H3 and H4 are marked by the acetylation of lysine residues in their N-terminal tails. The specific lysine that is acetylated differs from those acetylated in areas of active chromatin: in humans, *Drosophila* and yeast, lysine 5 and lysine 12 in newly synthesized histone H4 are acetylated.

These acetylated histones are bound by the chaperone proteins Asf1 and Caf1 as described above, and the latter deposits them on DNA (Figure 6.36). Caf1 is a complex of three protein subunits that binds tightly to newly synthesized, acetylated histone H3–H4 tetramers, and interacts with the sliding clamp protein PCNA (see Section 6.6) to target the H3–H4 tetramer to the DNA behind the replication fork and together with Asf1 facilitate assembly of the H3–H4 tetramer into a nucleosome behind the replication fork (Figure 6.37). After loading of H3–H4 tetramers, the addition of newly synthesized histone H2A–H2B dimers into the nucleosome is assisted by additional histone chaperone proteins. Once the histones are deposited on DNA, they are deacetylated, ready to be modified by enzymes thought to be recruited by the adjacent old modified histones.

> ➔ We discuss histone modification enzymes in Section 4.5.

Figure 6.36 Caf1 assembly of nucleosomes. The newly synthesized and acetylated histone H3 and H4 proteins, in association with Asf1, assemble into a half nucleosome. This acetylated half nucleosome binds to the Caf1 protein. Caf1 together with Asf1 then facilitates histone placement on DNA.

Figure 6.37 Caf1-mediated incorporation of H3–H4 tetramers into replicating DNA. To incorporate nucleosomes with newly synthesized H3 and H4 histones (1), Caf1 and Asf1 bound to an acetylated H3–H4 half nucleosome interact with the sliding clamp protein (proliferative cell nuclear antigen (PCNA)) (2), and deposits the H3–H4 tetramer on the DNA. Other chromatin-assembly factors including FACT (not shown) then assemble the H2A–H2B tetramer together with the H3–H4 tetramer to make a full nucleosome (3). Nucleosome spacing is adjusted by chromatin-remodeling complexes, and the histones are subject to other modifications such as deacetylation (4). Caf1 together with Asf1 only loads H3–H4 tetramers onto the newly synthesized DNA behind the replication fork. Only one strand of the fork is shown for simplicity.

DNA REPLICATION: REGULATION

The concluding part of this chapter brings together all that we have seen in the chapter so far to examine DNA replication as a carefully coordinated, tightly regulated sequence of events.

6.13 REGULATION OF INITIATION OF REPLICATION IN *E. COLI*

The initiation of replication is tightly regulated

The initiation of DNA replication commits a cell to divide and so must be tightly regulated and coordinated with cell growth, DNA repair and other processes that ensure accurate cell duplication. Initiation of replication is strictly controlled in both bacteria and eukaryotes. Because eukaryotes contain multiple replication origins along the chromosome whereas bacteria contain a single origin, the mechanisms of regulation used by the two groups differ significantly. Here we will discuss the mechanisms that regulate the firing of the *E. coli* origin, *oriC*, which is highly conserved in all bacteria, and in the following section we will discuss the regulation of eukaryotic origins.

The initiation of replication in *E. coli* is regulated in several ways that not only activate the origin but also ensure that, once the origin has fired, it will be blocked from firing again until sufficient time has passed for cell division to occur. However, as we shall see below, in *E. coli* these latter mechanisms can be bypassed under certain circumstances to allow very rapid reinitiation and faster cell division.

Four principal mechanisms have been implicated in the control of the initiation of DNA replication in bacteria: regulatory inactivation of DnaA (RIDA); binding of SeqA to *oriC* to block both DNA methylation and DnaA binding; the existence of the DatA DNA site that titrates out initiator protein; and a DARS DNA sequence that mediates recycling of DnaA-ADP to DnaA-ATP. We consider each of these in turn, below.

ATP binding and hydrolysis regulates the binding of initiator proteins to the origin of replication

The RIDA mechanism is central to the blocking of replication because it involves the hydrolysis of ATP-bound DnaA to DnaA-ADP and the consequent inactivation of the DnaA initiator activity (see Section 6.7). So why is the hydrolysis of ATP important? As discussed earlier, the *E. coli* initiator protein DnaA is a member of the AAA+ ATPase family (see Section 6.6), and will only multimerize to initiate replication at *oriC* when it is bound with ATP (Figure 6.38). The binding and hydrolysis of ATP helps establish the irreversibility of the initiation process. After binding to *oriC*, the ATP-bound DnaA unwinds the DNA for replication initiation. After DNA unwinding, the β clamp recruits another AAA+ protein, called Hda that stimulates the ATPase activity of DnaA, causing the ATP bound by DnaA to be hydrolyzed to ADP. This ADP bound form of the molecule cannot multimerize and thus cannot initiate origin firing, effectively blocking reinitiation.

SeqA binding to hemi-methylated DNA helps to block the reinitiation of replication

While the RIDA mechanism described above is considered to be the major negative regulatory step that prevents initiation, there are also other mechanisms that prevent reinitiation: one mechanism involves the methylation of the DNA sequences at *oriC*. Bacterial chromosomal DNA is methylated on the A residue at **GATC sites** by the enzyme DNA adenine methylase (Dam methylase). As described in Chapter 4, both strands are usually methylated. Immediately after replication, however, only one strand of the new DNA, the parental strand, will be methylated until Dam methylase modifies the new strand (see Figure 4.23). Hence, the DNA is said to be hemi-methylated.

There are 11 GATC sites in the *oriC* DNA sequence that overlap the DnaA binding sites (Figure 6.39). When hemi-methylated, these sites are specifically bound by a protein known as SeqA, which binds to hemi-methylated, but not fully methylated, GATC sites. Binding of SeqA to the hemi-methylated DNA at the origin blocks the access of methylase to these sites and thus inhibits both the methylation of the new strand, and binding of DnaA. This block only lasts a short time, however, as SeqA dissociates from the hemi-methylated DNA, allowing Dam methylase to regain access and fully methylate the new strand, preventing rebinding of SeqA. The replicated origin is usually fully methylated about ten minutes after initiation, and so SeqA binding can only temporarily block DnaA binding and origin refiring.

DnaA binding to *datA* DNA sequences sequesters the initiator protein after initiation

A third mechanism that prevents reinitiation is the binding of the initiator protein DnaA to a 1 kb region of DNA called *datA* (for <u>DnaA</u> <u>t</u>itrator) to the right of *oriC*. This region has sites that bind DnaA with very high affinity, and can sequester about 25% of the free DnaA molecules in the cell. After replication – but before cell division – there will be two copies of this region in the cell, which together sequester excess DnaA, making the protein unavailable for rebinding to *oriC*. When segregation of chromosomes occurs, however, such that there is just one copy of *datA* per cell again, the levels of DnaA sequestered are not sufficient to affect its activity, promoting initiation once more.

The DARS DNA sequence mediates recycling of DnaA-ADP to DnaA-ATP

Although the three processes just mentioned prevent replication and promote the formation of DnaA-ADP, DnaA-ATP must again become available for the next round of initiation. This is achieved by the new synthesis of DnaA, which is rapidly converted to DnaA-ATP. In addition, the accumulated DnaA-ADP can be recycled to DnaA-ATP. Two DnaA binding sites outside of *oriC*, called DARS (<u>D</u>naA <u>r</u>eactivating <u>s</u>equences), serve as cofactors to stimulate the exchange of ADP bound to DnaA to ATP.

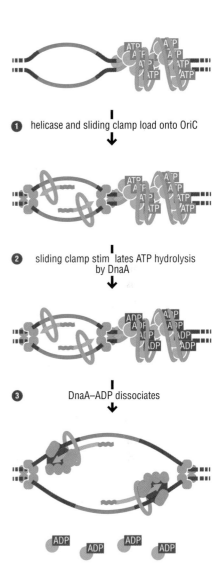

1. helicase and sliding clamp load onto OriC

2. sliding clamp stimulates ATP hydrolysis by DnaA

3. DnaA–ADP dissociates

Figure 6.38 ATP binding of DnaA regulates *oriC* firing. DnaA protein bound to ATP oligomerizes and binds to DnaA sites (orange) in *oriC*. The DNA wraps around the oligomer to initiate unwinding as described in Figure 6.20. The DnaB helicase is loaded by the DnaC protein (not shown) onto the unwound DNA. Primer synthesis occurs and the sliding clamp and DNA polymerase III are recruited. The interaction of the sliding clamp and DNA polymerase III and the Hda protein (not shown) with ATP-bound DnaA stimulates hydrolysis of ATP. The ADP-bound DnaA no longer oligomerizes nor binds DNA and thus dissociates from the DNA. This is one of several levels of regulation that prevent premature refiring of *oriC*.

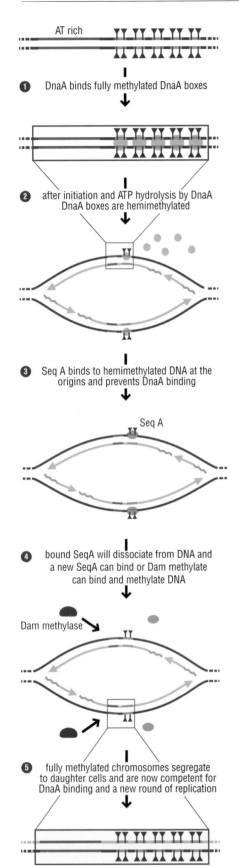

AT rich

1 DnaA binds fully methylated DnaA boxes

2 after initiation and ATP hydrolysis by DnaA DnaA boxes are hemimethylated

3 Seq A binds to hemimethylated DNA at the origins and prevents DnaA binding

Seq A

4 bound SeqA will dissociate from DNA and a new SeqA can bind or Dam methylate can bind and methylate DNA

Dam methylase

5 fully methylated chromosomes segregate to daughter cells and are now competent for DnaA binding and a new round of replication

Figure 6.39 *oriC* **contains GATC sites for methylation by Dam methylase.** Before replication, the GATC sites overlapping DnaA boxes within the *oriC* DNA are fully methylated (blue triangles). ATP-bound DnaA (green balls) binds to these sites (orange) and allows initiation, as described in Figure 6.38. For a short time after initiation, the GATC sites in the newly replicated DNA are in the hemi-methylated state as the newly synthesized DNA strand has not yet been methylated. The SeqA protein binds to these hemi-methylated sites and prevents DnaA rebinding. After some time the SeqA will dissociate and allow full remethylation of the GATC sites by Dam methylase so that replication initiation can occur again in the new daughter cell. SeqA also binds the hemi-methylated promoter of the *dnaA* gene and prevents synthesis of new DnaA (which is rapidly converted to DnaA-ATP) until the *dnaA* promoter is fully remethylated.

Initiation can take place before cell division is complete when *E. coli* cells are growing rapidly

In *E. coli* the timing of initiation is tightly coordinated with growth rate. Under different nutrient conditions, cell division may occur as rapidly as every 20 minutes or as slowly as every 150 minutes. The elongation rate for copying DNA is constant, however, and it takes about 50 minutes to replicate the bacterial chromosome, whatever the growth conditions. To enable a more rapid rate of cell division in good growth conditions, *E. coli* initiates a second round of chromosome replication before the first round is complete. Therefore, when cells are growing slowly, there is one copy of *oriC* before initiation and two copies after initiation; when cells are dividing rapidly, however, there can be four to eight copies of *oriC* in a single cell. Thus the frequency of initiation, which is determined by the growth conditions, determines the frequency of cell division.

Initiation occurs on every origin in the cell synchronously, and so the number of origins is always a power of two. This synchronous initiation is due to the four levels of regulation of initiation described above. Perturbations of any one of these mechanisms results in reinitiation, leading to an odd number of origins per cell. The timing of initiation is proportional to cell size, and it is thought that the ratio of DnaA molecules bound to ATP versus those bound to ADP regulates the ability to reinitiate replication in fast-growing cultures. Thus, with adequate nutrients there will be a higher level of ATP-bound DnaA, which promotes initiation. Although this re-replication in *E. coli* is well regulated, it is in contrast to the situation in eukaryotes where the firing of origins more than once per cell division is tightly blocked, as we will see below.

6.14 REGULATION OF REPLICATION INITIATION IN EUKARYOTES

Eukaryotic chromosomes are duplicated only once per cell cycle

Eukaryotes have multiple origins of replication along the chromosomes and the regulation of initiation is critically important. Each origin must fire once – and only once – in each cell cycle. If unregulated replication occurred, some regions along the chromosome might be copied more than once and others not at all. This would lead to problems in separating the chromosomes at division, and possibly resulting in there being too many copies of some genes in daughter cells. Some of the first experiments to study the regulation of DNA synthesis were carried out with cell fusions as described in Experimental approach 6.4.

6.4 EXPERIMENTAL APPROACH

Cell fusion studies and insights into cell cycle regulation of DNA replication

To understand DNA synthesis, Patu Rao and Robert Johnson set out to study its regulation, asking what might be seen when cells found in different stages of the cell cycle were fused with one another. Using this approach, they could ask how the contents of the nucleus in one cell cycle stage might influence the behavior of another nucleus – perhaps a diffusible factor existed in one nucleus that could affect the state of the other. It was presumed that signals (or factors) might travel from one nucleus to the other, through the intermingled cytoplasm that results from cell fusion.

To study the effects of cell fusion, the researchers first synchronized cells at either the G1, S, or G2 phase. They next fused cells at one stage with cells from another stage and pulsed the cells with H^3-thymidine to follow any DNA replication that takes place in the fused cells. Imagine, for example, that one nucleus of a binucleate cell stimulated the other nucleus to begin S phase (replication) before it normally would. In this case, we would expect an increase in the incorporation of H^3-thymidine in these fusion cells. In these experiments, the individual nuclei could be seen and the level of H^3 incorporation in those loci could be specifically evaluated. (The data were analyzed by visualizing individual nuclei in the population, not by bulk biochemistry.)

Using this general method, Rao and Johnson did three sets of experiments: they carried out S/G1 fusion, S/G2 fusion, and

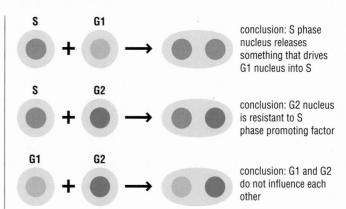

Figure 2 **Summary of results of cell fusion experiments.** When S phase cells are fused with G1 phase cells, the G1 nuclei prematurely enter S phase (shown as gray nucleus becoming red). When S phase cells are fused with G2 phase cells, the G2 nuclei do not begin to enter S phase (gray nucleus remains gray). When G1 phase cells are fused with G2 phase cells, neither cell type prematurely initiates S phase (gray nuclei remain gray).

G1/G2 fusion. Their experimental results showed a rapid increase in the incorporation of H^3-thymidine labeling over time (in the G1 nuclei) *only* in the fusion between the S/G1 cells compared to control samples, where they fused G1 with G1 cells or unfused G1 alone (Figure 1). From these results they concluded that there is a diffusible factor in S phase cells that can act on G1 phase nuclei to stimulate DNA synthesis.

In the other two sets of experiments, when S/G2 cells or G1/G2 cells were fused, they did not see an increase in H^3-thymidine labeling. From these experiments, the authors concluded that the diffusible factor from S phase cells cannot act on G2 phase cells to stimulate DNA synthesis. These results are summarized in Figure 2.

These experiments set the stage for understanding the cell cycle regulation of DNA replication. Today we understand the results of these experiments in terms of ORC binding and pre-replication complex assembly in G1 nuclei. We know that only the G1 nucleus assembles pre-replication complexes; the activities of Cdk1 and Cdk7 trigger the 'firing' of this complex and the initiation of replication. Consistent with the early experiments, pre-replication complexes cannot form in G2 cells, and thus replication cannot be initiated there.

Figure 1 **Induction of DNA synthesis in G1 nuclei.** The incorporation of H3 thymidine into G1 nuclei is monitored after cell fusion. When a G1 nucleus is exposed to the same cytoplasm as an S phase nucleus by cell fusion there is an increase in G1 nuclei that are labeled over time (blue line). In contrast the incorporation of H3 thymidine into G1 nuclei when fused with other G1 cells (pink line) or G1 cells unfused (gray line) happened only after 8 hours when replication is expected to initiate.

Rao PN, Johnson RT. Mammalian cell fusion: studies on the regulation of DNA synthesis and mitosis. *Nature* 1970;**225**:159–164.

⊙ The cell cycle is described in Chapter 5.

The firing of each origin once per cell division is achieved, in part, by dividing the initiation of DNA replication into tightly regulated steps that occur at different times in the cell cycle. First, in the G1 phase, replication origins are selected and prepared for activation. Second, in the S phase of the cell cycle, the preselected origins are activated. Finally, once an origin has fired in S phase, it cannot be reused until it is prepared for activation in the daughter cell in the next G1 phase. Thus, each origin is activated only once in each cell cycle.

The origin recognition complex is the central component of the prereplicative complex

As described for bacteria in Section 6.13, the initiation of replication at an origin depends on the binding of specific initiator proteins to origin DNA. The eukaryotic initiator proteins form a complex of six tightly associated subunits (Orc1–6) called the origin recognition complex (ORC) (see Section 6.7). In *S. cerevisiae*, the ORC binds to a clearly defined origin DNA sequence; in organisms in which origins are not so well defined by sequence, however, it also recognizes elements of chromatin structure that specify replication origins. ORC is the central component of a large assembly of proteins – called the **preReplicative Complex** or **pre-RC** – that forms at the replication origin in G1 (Figure 6.40).

After ORC has bound the origin, it interacts with two more proteins called Cdc6 and Cdt1. Together these components recruit the Mcm helicase complex to the origin. As described in Section 6.5, Mcm is the six-membered (Mcm2–7) DNA helicase that unwinds the DNA at the origin. In the pre-RC, however, the Mcm helicase is not active; only after origin activation in S phase does Mcm become activated to unwind the DNA, and help recruit the DNA polymerases that carry out DNA synthesis (Figure 6.41).

Formation of the prereplicative complex occurs only in G1

One important mechanism that limits origin firing in eukaryotes to once per cell cycle is the fact that a pre-RC can only form in G1. After origin firing, several components of the pre-RC are phosphorylated and either destroyed or exported from nucleus, and thus are not available to initiate a new pre-RC. In yeast, for example, phosphorylation of the pre-RC component Cdc6 triggers the subsequent destruction of Cdc6. These and other mechanisms block the formation of new pre-RCs until the G1 phase of the next cell cycle. In multicellular eukaryotes, pre-RC assembly is also governed by the activity of the anaphase-promoting complex (APC), which was described in Chapter 5. The APC triggers the proteolytic destruction of several pre-RC components until the end of mitosis, when the APC becomes inactive.

Activation of the pre-RC occurs in S phase

Activation of replication origins in S phase is triggered by phosphorylation of the pre-RC, which is carried out by S phase Cdks (described in Chapter 5) with assistance from another protein kinase, Cdc7. By phosphorylating specific proteins at the replication origin, these kinases trigger the transformation of the pre-RC into an active replication complex (Figure 6.41). At this point, the two strands of DNA

Figure 6.41 The prereplicative complex is activated to initiate replication by phosphorylation of Cdc6. The S phase Cdks and the kinase Cdc7 trigger origin activation by phosphorylating Cdc6. Both Cdc6 and Cdt1 are removed from the origin and Cdc6 is then destroyed. Origin activation also results in activation of the Mcm2–7 helicase and unwinding of the DNA helix. RPA proteins (not shown) bind the single-stranded DNA to prevent it from reannealing, and polymerase α-primase then initiates primer synthesis and DNA replication.

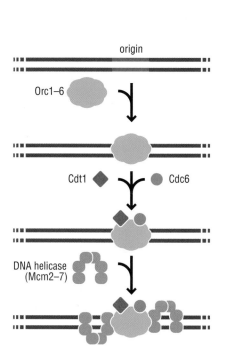

Figure 6.40 Assembly of the prereplicative complex. When bound to an origin, the ORC recruits two additional proteins, Cdc6 and Cdt1, to form a larger complex that is then responsible for loading the six-subunit Mcm2–7 complex onto the DNA. It is not clear how many copies of each component are present in the complex. For simplicity, two copies of the Mcm2–7 complex are shown here, positioned to move in opposite directions along the DNA after origin activation.

at the origin unwind, and the Mcm2–7 helicase complex is activated and recruits DNA polymerases and other proteins to form a pair of replication forks that move out from the origin as described earlier in the chapter.

Replication must be completed before chromosome segregation occurs

It is important that chromosome replication is finished before the cell proceeds to separate the daughter chromosomes. If the cell attempts to segregate partially duplicated chromosomes it could result in broken chromosomes in the daughter cells. In most eukaryotic cells, this order of events is ensured by the DNA damage response system, which monitors progression through S phase. If replication fails, then replication forks will stall. Stalled forks trigger the DNA damage response, and a signal is sent to the cell cycle control system to block progression into mitosis, as described in Chapter 5. Thus, the cell is prevented from attempting to segregate chromosomes that are not fully duplicated. The DNA damage response is discussed further in Chapter 12.

✱ SUMMARY

Principles

- DNA is duplicated via semi-conservative replication, whereby each original strand serves as a template for synthesis of a new complementary daughter strand.

- DNA replication occurs in three distinct phases: initiation, elongation, and termination.

Initiation of replication

- Initiation of replication usually occurs at a unique chromosomal location called an origin, which is recognized by specific proteins.

- DNA strands must be separated by a DNA helicase to allow the replication machinery to gain access to the two parental strands.

- New strands are initiated by RNA or combined RNA and DNA primers, which are synthesized by primase. Primers are then elongated by DNA polymerase.

Elongation during replication

- The position where the parental DNA strands are separated and new DNA is synthesized on both strands is called the replication fork.

- Since DNA polymerase can only add nucleotides to the elongating chain in a 5′ to 3′ direction, the two strands are replicated differently.

- The leading strand is synthesized as one long stretch of DNA while the lagging strand is made as short pieces call Okazaki fragments, which need to be stitched together by DNA ligase.

- DNA polymerases replicating the leading and lagging strand are coupled so that they move together at the replication fork.

Processivity and fidelity of DNA replication

- Sliding clamps facilitate the processivity of the replicative DNA polymerase.

- The assembly of the sliding clamp onto the DNA strand is mediated by the clamp loader.

- Accurate DNA replication is critical and is ensured through two distinct mechanisms mediated by DNA polymerase: the insertion of the nucleotide with the correct geometry and proofreading, the removal of incorrectly incorporated nucleotides.

Termination of replication

- The termination of replication occurs when two replication forks traveling in opposite directions meet. In bacteria this happens at a specific chromosomal position called ter.

- Replicated chromosomes are linked together and must be unlinked or decatenated by topoisomerases.

- DNA polymerases are unable to completely replicate the ends of linear DNA due to the need for RNA primers. This is called the end-replication problem.

- For many eukaryotic chromosomes the end-replication problem is solved by telomerase, which adds telomere sequences onto the ends of chromosomes.

Specialized polymerases

- Specialized DNA polymerases carry out a variety of functions including DNA repair and the bypass of damaged DNA. One specialized polymerase, reverse transcriptase, copies RNA into DNA.

Replication of chromatin

- During the replication of eukaryotic chromosomes, the parental nucleosomes are segregated such that half go to each daughter strand, and new nucleosomes are added to each strand by chromatin assembly factors. In this way, histone modifications are equally segregated to each daughter strand and can direct modification of the new nucleosomes.

Regulation of DNA replication

- The initiation of replication is tightly regulated by multiple mechanisms.

- In bacteria, where there is a single origin, replication initiation is regulated by the availability of the initiation factor DnaA in its ATP-bound state, and the availability of the origin sequence.

- In eukaryotes, where there are multiple origins, replication initiation is tightly linked to the cell cycle.

- In G1 phase, the pre-RC is formed at the origins, but is inactive. After initiation of S phase, the pre-RC is altered allowing initiation.

- Pre-RCs cannot reform until the next G1 due to the inactivation of the components by phosphorylation and degradation in all other phases.

FURTHER READING

6.1 OVERVIEW OF DNA REPLICATION

Davey MJ, Jeruzalmi D, Kuriyan J, O'Donnell M. Motors and switches: AAA+ machines within the replisome. *Nature Reviews Molecular Cell Biology* 2002;**3**:826–835.

Kunkel TA, Burgers PM. Dividing the workload at a eukaryotic replication fork. *Trends in Cell Biology* 2008;**18**:521–527.

Waga S, Stillman B. The DNA replication fork in eukaryotic cells. *Annual Review of Biochemistry* 1998;**67**:721–751.

6.2 DNA POLYMERASES: STRUCTURE AND FUNCTION; AND
6.3 DNA POLYMERASES: FIDELITY AND PROCESSIVITY

Francklyn CS. DNA polymerases and aminoacyl-tRNA synthetases: shared mechanisms for ensuring the fidelity of gene expression. *Biochemistry* 2008;**47**:11695–11703.

Joyce CM, Benkovic SJ. DNA polymerase fidelity: kinetics, structure, and checkpoints. *Biochemistry* 2004;**43**;14317–14324.

Joyce CM, Steitz TA. Function and structure relationships in DNA polymerases. *Annual Review of Biochemistry* 1994;**63**:777–822.

Kunkel TA, Bebenek K. DNA replication fidelity. *Annual Review of Biochemistry* 2000;**69**:497–529.

Patel PH, Suzuki M, Adman E, Shinkai A, Loeb LA. Prokaryotic DNA polymerase I: evolution, structure, and 'base flipping' mechanism for nucleotide selection. *Journal of Molecular Biology* 2001;**308**:823–837.

6.4 SPECIALIZED POLYMERASES

Friedberg EC, Wagner R, Radman M. Specialized DNA polymerases, cellular survival, and the genesis of mutations. *Science* 2002;**296**:1627–1630.

Prakash S, Johnson RE, Prakash L. Eukaryotic translesion synthesis DNA polymerases: specificity of structure and function. *Annual Review of Biochemistry* 2005;**74**:317–353.

Yang W. Damage repair DNA polymerases Y. *Current Opinion in Structural Biology* 2003;**13**:23–30.

6.5 DNA HELICASES: UNWINDING OF THE DOUBLE HELIX

Enemark EJ, Joshua-Tor L. On helicases and other motor proteins. *Current Opinion in Structural Biology* 2008;**18**:243–257.

Forsburg SL. The MCM helicase: linking checkpoints to the replication fork. *Biochemical Society Transactions* 2008;**36**:114–119.

Masai H, You Z, Arai K. Control of DNA replication: regulation and activation of eukaryotic replicative helicase, MCM. *IUBMB Life* 2005;**57**:323–335.

Wang JC. Cellular roles of DNA topoisomerases: a molecular perspective. *Nature Reviews Molecular Cell Biology* 2002;**3**:430–440.

6.6 THE SLIDING CLAMP AND CLAMP LOADER

Bloom LB. Dynamics of loading the Escherichia coli DNA polymerase processivity clamp. *Critical Reviews in Biochemistry and Molecular Biology* 2006;**41**:179–208.

Ellison V, Stillman B. Opening of the clamp: an intimate view of an ATP-driven biological machine. *Cell* 2001;**106**:655–660.

Indiani C, O'Donnell M. The replication clamp-loading machine at work in the three domains of life. *Nature Reviews Molecular Cell Biology* 2006;**7**:751–761.

Johnson A, O'Donnell M. Cellular DNA replicases: components and dynamics at the replication fork. *Annual Review of Biochemistry* 2005;**74**:283–315.

Prives C, Gottifredi V. The p21 and PCNA partnership: a new twist for an old plot. *Cell Cycle* 2008;**7**:3840–3846.

6.7 ORIGINS OF REPLICATION AND THE INITIATION OF DNA REPLICATION

Bell SP, Dutta A. DNA replication in eukaryotic cells. *Annual Review of Biochemistry* 2002;**71**:333–374.

DePamphilis ML. Cell cycle dependent regulation of the origin recognition complex. *Cell Cycle* 2005;**4**:70–79.

Diffley JF. Regulation of early events in chromosome replication. *Current Biology* 2004;**14**:R778–786.

Méchali M. DNA replication origins: from sequence specificity to epigenetics. *Nature Reviews Genetics* 2001;**2**:640–645.

Prasanth SG, Mendez J, Prasanth KV, Stillman B. Dynamics of pre-replication complex proteins during the cell division cycle. *Philosophical Transactions of the Royal Society of London Series B, Biological Sciences* 2004;**359**:7–16.

6.8 LEADING AND LAGGING STRAND SYNTHESIS

Benkovic SJ, Valentine AM, Salinas F. Replisome-mediated DNA replication. *Annual Review of Biochemistry* 2001;**70**:181–208.

Corn JE, Berger JM. Regulation of bacterial priming and daughter strand synthesis through helicase-primase interactions. *Nucleic Acids Research* 2006;**34**:4082–4088.

Kao HI, Bambara RA. The protein components and mechanism of eukaryotic Okazaki fragment maturation. *Critical Reviews in Biochemistry and Molecular Biology* 2003;**38**:433–452.

6.9 THE REPLICATION FORK

Burgers PM. Polymerase dynamics at the eukaryotic DNA replication fork. *Journal of Biological Chemistry* 2009;**284**:4041–4045.

Fang L, Davey MJ, O'Donnell M. Replisome assembly at oriC, the replication origin of E. coli, reveals an explanation for initiation sites outside an origin. *Molecular Cell* 1999;**4**:541–553.

O'Donnell M. Replisome architecture and dynamics in E. coli. *Journal of Biological Chemistry* 2006:**281**;10653–10656.

Waga S, Stillman B. The DNA replication fork in eukaryotic cells. *Annual Review of Biochemistry* 1998:**67**:721–751.

6.10 TERMINATION OF DNA REPLICATION

Mohanty BK, Bastia D. Binding of the replication terminator protein Fob1p to the Ter sites of yeast causes polar fork arrest. *Journal of Biological Chemistry* 2004;**279**:1932–1941.

Neylon C, Kralicek AV, Hill TM, Dixon NE. Replication termination in Escherichia coli: structure and antihelicase activity of the Tus-Ter complex. *Microbiology and Molecular Biology Reviews* 2005;**69**:501–526.

Wang JC. Cellular roles of DNA topoisomerases: a molecular perspective. *Nature Reviews Molecular Cell Biology* 2002;**3**:430–440.

6.11 THE END-REPLICATION PROBLEM AND TELOMERASE

Baird DM. Telomere dynamics in human cells. *Biochimie* 2008;**90**:116–121.

Bianchi A, Shore D. How telomerase reaches its end: mechanism of telomerase regulation by the telomeric complex. *Molecular Cell* 2008;**31**:153–165.

Greider CW. Telomeres and senescence: the history, the experiment, the future. *Current Biology* 1998;**8**:R178–R181.

Lingner J, Cooper JP, Cech TR. Telomerase and DNA end replication: no longer a lagging strand problem? *Science* 1995;**269**:1533–1534.

6.12 CHROMATIN REPLICATION

Akey CW, Luger K. Histone chaperones and nucleosome assembly. *Current Opinion in Structural Biology* 2003;**13**:6–14.

Altaf M, Auger A, Covic M, Cote J. Connection between histone H2A variants and chromatin remodeling complexes. *Biochemistry and Cell Biology* 2009;**87**:35–50.

Bailis JM, Forsburg SL. It's all in the timing: linking S phase to chromatin structure and chromosome dynamics. *Cell Cycle* 2003;**2**:303–306.

Corpet A, Almouzni G. Making copies of chromatin: the challenge of nucleosomal organization and epigenetic information. *Trends in Cell Biology* 2009;**19**:29–41.

Martin C, Zhang Y. Mechanisms of epigenetic inheritance. *Current Opinion in Cell Biology* 2007;**19**:266–272.

Moldovan GL, Pfander B, Jentsch S. PCNA, the maestro of the replication fork. *Cell* 2007;**129**:665–679.

6.13 REGULATION OF INITIATION OF REPLICATION IN *E. COLI*

Kaguni JM. DnaA: controlling the initiation of bacterial DNA replication and more. *Annual Review of Microbiology* 2006;**60**:351–375.

Leonard A, Grimwade JC. Initiating chromosome replication on E. coli: it makes sense to recycle. *Genes and Development* 2009;**23**:1145–1150.

Mott ML, Berger JM. DNA replication initiation: mechanisms and regulation in bacteria. *Nature Reviews Microbiology* 2007;**5**:343–354.

Zakrzewska-Czerwinska J, Jakimowicz D, Zawilak-Pawlik A, Messer W. Regulation of the initiation of chromosomal replication in bacteria. *FEMS Microbiology Reviews* 2007;**31**:378–387.

6.14 REGULATION OF REPLICATION INITIATION IN EUKARYOTES

Bell SP. The origin recognition complex: from simple origins to complex functions. *Genes and Development* 2002;**16**:659–672.

Diffley JF. Regulation of early events in chromosome replication. *Current Biology* 2004;**14**:R778–R786.

MacAlpine DM, Bell SP. A genomic view of eukaryotic DNA replication. *Chromosome Research* 2005;**13**:309–326.

Sclafani RA, Holzen TM. Cell cycle regulation of DNA replication. *Annual Review of Genetics* 2007;**41**:237–280.

CHROMOSOME SEGREGATION

INTRODUCTION

Cell proliferation is the process by which one cell gives rise to two daughter cells. This process is critical for the survival of single-celled organisms such as bacteria and fungi. It is also essential for the development of multicellular organisms, during the replenishments of renewable tissue such as blood and skin, and for wound healing. The key challenge in this process is to ensure that each daughter cell receives one, and *only* one, copy of each chromosome. Indeed, failure to meet this goal can have dire consequences. For example, Down syndrome, in which chromosome 21 appears in three copies instead of just two, is due to a **chromosome segregation** error that took place during the formation of an egg or sperm. Cancer cells also tend to exhibit an abnormal number of chromosomes.

In many instances, inheriting too few or too many copies of a chromosome can lead to abnormal cell behavior, such as in the case of cancer, or even cell death. Thus, successful cell proliferation depends on a flawless transmission of chromosomes from one generation to the next. This process, known as chromosome segregation, involves many structural and regulatory components that will be discussed in this chapter.

Most of this chapter will be dedicated to **mitosis**, the cell-division program of somatic cells (all the eukaryotic cells other than those involved in sexual reproduction). We will also discuss chromosome segregation in **meiosis**, the specialized cell-division program required to generate the **gametes** that are involved in sexual reproduction. Finally, we will discuss the process of chromosome segregation in bacteria, whose principles are similar to those that govern mitosis, but with different structural and regulatory machineries.

Before considering the specific cellular and molecular processes that underpin mitosis and meiosis, we begin with a broad overview of the stages of mitosis, to give us a general framework to which we can add more detail later.

7.1 THE STAGES OF MITOSIS

Processes that take place early in the cell cycle set the stage for mitosis

To understand the mechanism and regulation of chromosome segregation in eukaryotes, we must first become familiar with cellular events that lead up to and take place during mitosis. These events are part of the eukaryotic cell cycle,

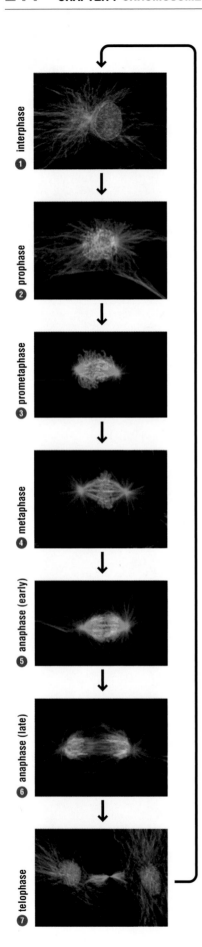

which was discussed in Chapter 5. Dividing cells spend most of their lifecycle in **interphase**, which encompasses the G1, S, and G2 phases of the cell cycle (Figure 7.1, step 1). Two events critical for chromosome segregation take place during interphase – DNA replication in S phase, and the initial stages of spindle formation in late G1 (see Section 5.1). As we learned in Chapter 6, DNA replication doubles the chromosome content of the cell, producing two identical sister chromatids from each chromosome. After DNA replication, sister chromatids remain associated with each other by the presence of a protein complex called **cohesin,** which links the two sister chromatids to each other. This association, or 'cohesion', is important for later stages of mitosis in order to direct the segregation of sister chromatids to different daughter cells (see Section 7.2).

During mitosis of animal cells, the movement of chromosomes to the daughter cells occurs through the attachment of the chromosomes to a structure called the **mitotic spindle** (or 'spindle', for short), which is made of filaments called **microtubules** (see Section 7.3). The first steps in spindle formation take place during interphase, in the cytoplasm. This means that during interphase the chromosomes, which are in the nucleus, cannot bind to the spindle. Therefore, at some point during mitosis, the nuclear envelope has to disassemble, as indeed is the case.

Early mitosis is marked by changes in chromosome structure and nuclear integrity

The first visible signs of mitosis occur in **prophase** (Figure 7.1, step 2), when the duplicated chromosomes begin to condense and individual pairs of sister chromatids become apparent. Prophase is also the time when the spindle grows in length, forming a structure that has two poles, with microtubules aligned in parallel between them. Immediately after prophase, in a stage called **prometaphase** (Figure 7.1, step 3), the nuclear envelope breaks down and the spindle microtubules make contact with the chromosomes.

The attachment of chromosomes to microtubules leads to chromosome alignment at the spindle mid-zone

The spindle microtubules contribute to chromosome segregation by binding to pairs of sister chromatids and pulling them apart, such that each sister chromatid moves to a different spindle pole. This means that each sister chromatid in a given pair has to bind spindle microtubules that are connected to a different spindle pole. Following nuclear envelope breakdown, the attachments between the chromosomes and spindle microtubules are initially random, meaning that both sister chromatids may bind microtubules that are connected to the same spindle pole. Eventually, however, with the help of regulatory proteins, sister chromatids bind stably to microtubules that are associated with opposite spindle poles, a state called **bi-orientation**. At this point, the spindle microtubules exert pulling forces on the sister chromatids,

Figure 7.1 The stages of mitosis. The cell cycle can be divided into (1) interphase, which spans the phases G1, S, and G2 (see Section 5.1), and (2–7) mitosis, which is followed by cytokinesis (not shown). The stages of mitosis are (2) prophase, (3) prometaphase, (4) metaphase, (5, 6) anaphase, and (7) telophase. The chromosomes are shown in pink and microtubules are shown in green. Note the changes in microtubule organization and chromosome condensation between interphase and mitosis. The nuclear envelope cannot be seen, but it is present during interphase, prophase, and telophase. Images courtesy of Tamara A. Potapova and Gary J. Gorbsky, Oklahoma Medical Research Foundation.

but the chromatids do not separate because they are still connected to each other by cohesin complexes. However, because the paired sister chromatids are subjected to forces from microtubules attached to two *different* spindle poles, the chromatids move to the spindle mid-zone equidistant between the two poles, much like in a tug-of-wag game between equally strong teams. The stage at which all chromosomes are bi-oriented and aligned on the spindle mid-zone is called **metaphase** (Figure 7.1, step 4) – a stage that we consider in more detail in Section 7.4.

Separation of sister chromatids at anaphase is a point of no return

For chromosome segregation to take place, cohesion of sister chromatids has to be dissolved. This occurs only when *all* chromosomes have bi-oriented. Cohesion dissolution is accomplished by the cleavage of a subunit of the cohesin complex, thereby inactivating the complex. This occurs during the **anaphase** stage of mitosis and it allows the sister chromatids to be pulled to opposite spindle poles by forces acting through the spindle microtubules (Figure 7.1, steps 5 and 6). Anaphase is a critical point in mitosis because it is irreversible: once sister chromatids separate, it is no longer possible to correct erroneous chromosome–spindle attachments that could lead to inaccurate chromosome segregation. Therefore, multiple regulatory processes control the onset of anaphase, and in particular the timing at which sister chromatid cohesion is dissolved (see Section 7.5).

Events in late mitosis set the stage for the return to interphase

Once anaphase is completed, the cells begin the journey back to interphase. During **telophase**, which is the last stage of mitosis (Figure 7.1, step 7), the spindle disassembles, the chromosomes decondense and the nuclear envelope reassembles to create two daughter nuclei. All that is left to complete cell division is the physical separation of the two daughter cells, in a process called **cytokinesis**. The plane of cell division is determined in anaphase, and it corresponds to the location of the spindle mid-zone. This ensures that the plane of cell division is situated between the two daughter nuclei, resulting in two cells, each with a full complement of chromosomes. Upon completion of cytokinesis the cells are back in interphase.

Having discussed mitosis in general terms, and having identified the key stages that make up the process as a whole, we can now move on to examine in more detail the molecular and cellular events that characterize each stage. We begin by considering cohesion and **condensation** of chromosomes.

7.2 CHROMOSOME CONDENSATION AND COHESION

Mitosis is associated with changes in chromosome structure

Chromosome segregation is accompanied by major changes in chromosome organization. Before entering mitosis, chromosomes must compact, because in interphase, the chromosomes are in an extended state that is unlikely to yield orderly chromosome movement. Compaction also protects the chromosomes from getting their trailing arms caught in the plane of cell division during cytokinesis. This process of compaction is called **chromosome condensation** and it requires a protein complex called **condensin**. We learn more about this condensation process later in this section.

Another structural feature of chromosomes in dividing cells is **sister chromatid cohesion**, which links the two sister chromatids to each other prior to segregation. Sister chromatid cohesion ensures that the sister chromatids remain attached to each other so that they can bi-orient and make the appropriate attachments to the spindle microtubules. Sister chromatid cohesion is mediated by the **cohesin** complex, which we introduced in Section 4.2. Cohesin is related in structure and protein composition to condensin.

Condensin and cohesin form ring-like structures

Condensin and cohesin belong to the **structural maintenance of chromatin (SMC)** protein complex family. In eukaryotes, SMC protein complexes, such as condensin and cohesin, are composed of a heterodimer of SMC proteins, each containing an ATPase domain, and several associated subunits (Figure 7.2). In bacteria, the SMC protein complexes identified so far contain a homodimer of SMC proteins. Adenosine triphosphate (ATP) binding and its subsequent hydrolysis, together with the actions of the associated subunits, is believed to result in conformational changes in the SMC heterodimer that lead to changes in chromosome structure. For example, ATP hydrolysis may open and close the SMC coiled-coil arms, allowing the complex to associate with or dissociate from chromosomes. We do not yet know how these complexes contribute to chromosome cohesion and condensation, but it is thought that the SMC arms are long enough to encircle one, and possibly two, chromosomes.

Of particular note is the kleisin subunit (from the Greek word for 'closure'), which bridges the ATPase domains of the SMC subunits (Figure 7.2). The kleisin

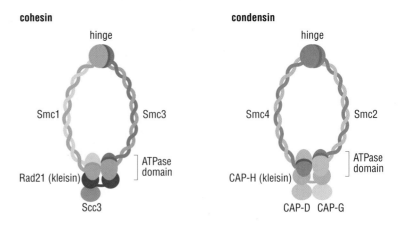

Figure 7.2 Cohesin and condensin belong to the SMC protein complex family. Cohesin and condensin are composed of two SMC proteins: cohesin is made of Smc1 and Smc3 while condensin is made of Smc2 and Smc4. Both cohesin and condensin contain additional subunits, including a subunit that is referred to as kleisin, which are thought to bind near the ATPase domain of the SMC subunits. In many organisms the kleisin subunits of cohesin and condensin are called Rad21 and CAP-H, respectively. In both complexes the kleisin subunit bridges between the ATPase domains of the two Smc molecules. Electron microscopy of purified complexes showed that the cohesin complex forms a round, ring-like structure whereas the SMC arms of condensin form a V-shaped structure. The structure of these complexes when associated with chromosomes is not known.

subunit of cohesin is cleaved at the onset of anaphase (as discussed further in Section 7.5). Whether the kleisin subunit of condensin regulates condensin function is not known.

Chromosome condensation requires the condensin complex

Despite the striking visible changes that accompany chromosome condensation, which are illustrated in Figure 7.3, we do not know much about the molecular mechanisms that drive chromosome condensation. Although the condensin complex appears to be needed for stable chromosome condensation, depletion of condensin in a variety of cellular systems results in a condensation defect, particularly in later stages of mitosis (that is, anaphase and telophase), but not in a complete loss of chromosome compaction. Thus, is it possible that other factors also contribute to the condensation process, and that condensins are needed to stabilize the condensed chromosome configuration. It is thought that condensins associate with widely separated chromosomal regions, thereby bringing them closer together and compacting the chromosome. Chromosome condensation requires functional ATPase domains in the SMC proteins, but how ATP binding or hydrolysis contribute to condensation is not clear.

Cohesin is essential for accurate chromosome segregation

By contrast with our rather sketchy understanding of chromosome condensation, we know more about the action of the cohesin complex. Sister chromatid cohesion enables chromosomes to segregate accurately by ensuring that, at metaphase, the two sisters are attached to opposite poles of the spindle. We learn how sister chromatid separation has been studied experimentally in Experimental approach 7.1. In the absence of sister chromatid cohesion, there would be no way of coordinating the binding of the two sister chromatids to different spindle poles. The importance of cohesion in the segregation process is illustrated by the observation that, if cohesin is inactivated, neither of the daughter cells will receive a full complement of chromosomes. As a result, chromosomes grossly missegregate and the cell dies.

Cohesin complexes assemble on chromosomes during DNA replication, so that the sister chromatids are held together from the time they are formed in S phase. Establishing cohesion during DNA replication circumvents the difficulty that would arise if sister chromatids were allowed to separate after replication and then had to be identified out of a mixture containing all the other chromosomes.

Cohesins cannot associate with chromosomes on their own, but they are loaded onto chromosomes by a specific loading complex. Although cohesins show a preference for localizing to AT-rich regions on DNA, known as cohesin-associated regions (CARs), they do not bind to a specific DNA sequence. It is possible that cohesions (or their loading factors) favor particular types of chromatin modification that are abundant at CARs. Alternatively, CARs often coincide with interagenic regions, and it is therefore possible that cohesions are excluded from active sites of transcription. Note that cohesin binding to chromosomes is not sufficient to generate cohesion. Once cohesins are loaded onto the chromatids, the action of yet another protein, called Eco1, is required for the establishment of cohesion,

Figure 7.3 Chromosome condensation. During interphase, chromosomes are decondensed, and it is difficult to distinguish between individual chromosomes. In prophase, chromosomes are condensed: they become thicker and appear as discrete units. Shown are light microscopy images of chromosomes from a Chinese hamster ovary cell in (a) interphase and (b) late prophase/early prometaphase. Adapted from Kireeva *et al.* (2004).

7.1 EXPERIMENTAL APPROACH

The advantage of being able to see: following chromosome movement in single-celled organisms

Often significant leaps in our understanding of a cellular process have come from a new development in microscopy, which gives us a newfound ability to see structures or detect specific proteins or DNA or RNA sequences in the cell in a way that was not previously possible. Such leaps forward have been instrumental for our understanding of chromosome movement. It was known from the observation of large animal cells by light microscopy that chromosomes are dynamic structures, changing conformation and position throughout the cell cycle. The monitoring of individual chromosomes by light microscopy even allowed the detection of some chromosomal abnormalities. However, small changes in chromosome structure cannot be detected by light microscopy and it is not possible to follow individual chromosomes in eukaryotic microorganisms such as yeast.

A powerful development in both research and diagnostics was fluorescence *in situ* hybridization, or FISH (see Section 16.8), wherein DNA is visualized through hybridization to a fluorescently labeled DNA probe. In this method, cells are immobilized on a slide and the DNA is denatured. The fluorescent DNA probe is then added to the treated cells where it anneals only to the complementary sequence located within the chromosome. With this technique one can monitor any locus in the cell, in both metaphase and interphase cells, and one can detect even small deletions and chromosomal rearrangements if the probe is designed carefully.

Doug Koshland and colleagues modified the FISH method to directly observe yeast chromosomes in strains that were suspected to have defects in chromosome segregation. Unlike animal cells, individual yeast chromosomes cannot be distinguished by light microscopy. The investigators knew from genetic analysis that when *pds1-1* mutant cells were stalled in mitosis with the use of drugs, they lost viability when subsequently allowed to continue through the cell cycle. (By contrast, under the same conditions, wild-type cells remained viable.) To determine the cause of this phenotype, they used FISH to examine sister chromatid cohesion in wild-type cells and in *pds1-1* mutant cells that were held in prometaphase. They found that in haploid wild-type cells in prometaphase, a FISH

Figure 1 Fluorescence *in situ* hybridization (FISH). Wild type and *pds1-1* mutant cells were treated with the drug nocodazole, which depolymerizes microtubules. This results in the activation of the spindle checkpoint pathway and leads to a cell cycle arrest in prometaphase. The cells were then exposed to a FISH probe against a region of chromosome 16. In wild-type cells, sister chromatids maintain their cohesion, and consequently these cells mostly have one FISH signal (Figure 1a (top panel), b) *pds1-1* mutant cells, on the other hand, fail to maintain cohesion and a large fraction of cells exhibit two FISH signals (Figure 1a (bottom panel), b). Figure 1b is a quantification of 200 nuclei such as those shown in Figure 1a.
From Yamamoto *et al.*, *Journal of Cell Biology* 1996;**133**:99–110.

Figure 2 **Following a specific chromosomal locus in live cells.** (a) A schematic drawing of the *lacO* array/lacl-GFP system. (b) A haploid budding yeast cell expressing lacl-GFP (in green) and carrying a *lacO* array (256 repeats) near the centromere of chromosome III. The first image on the left shows a live cell prior to sister chromatid separation. The following images (from left to right) show sister chromatid segregation during anaphase. Note that each cell received one copy of chromosome III.
Figure 2b is from Thrower *et al.*, *Molecular Biology of the Cell* 2001;**12**:2800–2812.

probe to a particular region on one of the chromosome arms gave only a single fluorescent signal because the sister chromatids were held very close together (Figure 1a (top panel), 1b). However, in *pds1-1* mutant strains, two fluorescent signals were observed, indicating that the sisters drifted apart (Figure 1a (bottom panel), b). Based on this and other results they concluded that Pds1 (which stands for precocious dissociation of sisters) is required to maintain sister chromatid cohesion in metaphase.

Although the FISH method can be used to follow almost any chromosomal locus in any cell type, it cannot be used to follow chromosome movement in live cells. Live cell imaging is of particular importance when studying the dynamics of chromosome segregation. To follow a particular chromosomal region in live cells, Andrew Murray, Andy Belmont, and colleagues established a method where *in vivo* expressed fluorescent fusion proteins are targeted to specific loci in the genome through a DNA binding motif that recognizes an array of binding sites in the chromosome. This group took advantage of the tight binding between the *Escherichia coli* Lacl transcriptional repressor and its DNA binding site, *lacO*. As shown in Figure 2a, an array containing many tandem *lacO* binding sites is integrated at a particular chromosomal locus such that when a Lacl-GFP (green fluorescent protein) fusion protein is expressed in these cells, multiple Lacl-GFP proteins bind to the *lacO* array. We note that in such an experiment, the signal from the many bound Lacl-GFP proteins coalesces to a single fluorescent spot; a single Lacl-GFP generates a signal that is too weak to

detect by conventional fluorescence microscopy. This approach provides a powerful tool that enables researchers to follow chromosome movement in live cells.

In one nice example, Kerry Bloom and colleagues used this approach to follow a haploid budding yeast cell undergoing mitosis (Figure 2b). The yeast in this case were engineered to have a *lacO* array near the centromere of chromosome III. As we see in Figure 2b, the yeast have undergone DNA replication, but because the sister chromatids remain next to each other until anaphase, only a single GFP spot is observed (left-most image). When anaphase begins, the sister chromatids separate, resulting in the appearance of two GFP spots, each corresponding to the centromeric region of the duplicated chromosome III. As anaphase continues, the spindle (which cannot be seen) elongates and the centromeres of the sister chromatids move away from one other, such that each cell (the mother and the bud) receives one copy of chromosome III. This approach is useful for many applications, but it requires the insertion of the *lacO* array into the desired place in the genome.

Straight AF, Belmont AS, Robinett CC, Murray AW. GFP tagging of budding yeast chromosomes reveals that protein–protein interactions can mediate sister chromatid cohesion. *Current Biology* 1996;**6**:1599–1608.

Yamamoto A, Guacci V, Koshland D. Pds1p, an inhibitor of anaphase in budding yeast, plays a critical role in the APC and checkpoint pathway(s). *Journal of Cell Biology* 1996;**133**:99–110.

(a)

(b)

Figure 7.4 Two ways in which sister chromatids may be held together by cohesin. (a) Cohesin complexes may mediate sister chromatid cohesion by encircling two sister chromatids simultaneously (chromatids are represented by the gray rods). (b) An alternative proposal is that cohesin encircles a single chromatid and that chromatids are held together by interactions between cohesin complexes on different sister chromatids. In this representation we show the two cohesin complexes interacting through their kleisin subunits, but other types of cohesin–cohesin interactions are possible.

namely the physical linking of the two sister chromatids. Thus, in the absence of Eco1, chromosomes are decorated with cohesin complexes but the two sister chromatids are not associated with each other. Eco1 has acetyltransferase activity (that is, it transfers an acetyl group from one compound to another) and it acetylates one of cohesin's SMC subunits. This activity is essential for cohesion, but the exact mechanism is unknown. We also do not know whether all of the bound cohesin is engaged in cohesion or whether only a fraction of chromosome-bound cohesin is actually engaged in holding the sister chromatids together.

The mechanism by which cohesin complexes hold the sister chromatids together is unknown, but given the size of the SMC protein arms, it is possible that a single cohesin complex traps both sister chromatids (Figure 7.4a). Another possibility is that each cohesin complex encircles a single sister chromatid, and cohesion is mediated by the interactions between cohesin complexes on different sister chromatids (Figure 7.4b). In either case, it is thought that ATP binding and hydrolysis by the ATPase domain of the SMC heterodimer is responsible for conformational changes in the cohesin complex that allow it to open, encircle one or both chromatids, and then close again. Because cohesion is established as chromosomes duplicate, the physical proximity of the newly formed sister chromatids dictates that cohesin will link sister chromatids rather than unrelated chromosomes.

The exception to the S phase cohesin loading is when DNA breakage occurs after DNA replication. In this case, cohesin is loaded during late G2/early mitosis in the vicinity of the break, as well as at CARs genome-wide, and cohesion is established at these sites site. This process is thought to facilitate the repair of the break by tethering the broken sister chromatid to its intact twin, which can serve as a template for repair. Indeed, a role for cohesin in repair of DNA damage is evident from the name of one of cohesin's subunits, Rad21, which stems from its initial identification as a gene which when mutated can lead to radiation sensitivity.

The separation of sister chromatids during anaphase is induced by the proteolytic cleavage of the kleisin subunit of cohesion (see Section 7.4). This event inactivates the cohesin complex and destroys cohesion, allowing the sister chromatids to be pulled apart by the spindle microtubules (Figure 7.5). Thus, the next step in understanding chromosome segregation is learning how the mitotic spindle functions, described in the following section.

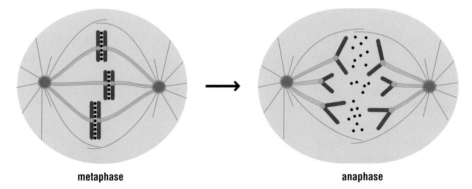

metaphase   anaphase

Figure 7.5 Sister chromatid cohesion is necessary for accurate chromosome segregation. Shown is a hypothetical cell with three different chromosomes after DNA replication, so that each chromosome is composed of two sister chromatids. During metaphase, the sister chromatids are held together by the cohesin complex (black dots). The spindle microtubules, shown in green, are pulling on the chromosomes but cohesion prevents the sister chromatids from separating. At anaphase, the cohesin complex dissociates from the chromosomes, allowing the sister chromatids to separate and be pulled away from each other by microtubules of the mitotic spindle. Had cohesin been absent, the sister chromatids would have separated immediately following DNA replication, and would have attached at random to the spindle microtubules.

7.3 THE MITOTIC SPINDLE

Chromosome segregation is mediated by the mitotic spindle

A key feature of chromosome segregation is chromosome movement: first as the paired sister chromatids align during prometaphase and metaphase, and then as sister chromatids separate in anaphase (Figure 7.1). Chromosome movement is mediated through the attachment of chromosomes to the mitotic spindle. Therefore, to understand the process of chromosome movement we need to know how the spindle exerts force on the chromosomes. We also need to know how chromosomes attach to the spindle such that they can be moved to the right place at the right time. In this section we will learn about the components of the spindle. In the coming sections we will learn how chromosomes attach to the spindle and how the movement of chromosomes is regulated.

The spindle of most animal cells has two main structural components: protein filaments called microtubules, and two large structures called **centrosomes**, which are located one at each pole of the spindle (Figure 7.6). Spindle microtubules are anchored at one of their ends to a centrosome. At their other end, microtubules can interact with different targets: a chromosome, a microtubule emanating from the opposite spindle pole, or the plasma membrane at the cell periphery (Figure 7.6).

Microtubules are classified according to the structure with which they associate: microtubules that interact with chromosomes and facilitate their movement are called **kinetochore microtubules**, named after the chromosome-associated structure with which these microtubules interact (Section 7.4). Microtubules that interact with other microtubules emanating from the opposite spindle pole are called **polar microtubules**, and they serve to elongate the spindle during

Figure 7.6 The mitotic spindle. (a) An image of a spindle from a cell in metaphase. The DNA is shown in pink, microtubules in green and γ-tubulin, a component of a structure that is located at the spindle pole, called the centrosome, is shown in blue. (b) A schematic diagram of a mitotic spindle. Microtubules (in green) are anchored at one of the spindle poles, to the centrosome (in blue). Microtubules that bind to chromosomes (in pink) are called kinetochore microtubules because of the structure on the chromosomes with which microtubules interact (see Section 7.5). Microtubules interacting with other microtubules emanating from the opposite spindle pole are polar microtubules. Microtubules emanating from the spindle poles towards the cell cortex are called astral microtubules.
Figure 7.5a adapted from Wong and Stearns, *BMC Cell Biology* 2005;**6**:6.

anaphase (Section 7.5). Finally, microtubules that extend toward the cell periphery are called **astral microtubules**, and they play a role in positioning the spindle within the cell throughout mitosis.

Microtubules are dynamic structures, undergoing periods of shortening and lengthening throughout their lifetime. Chromosome movement, spindle elongation and spindle positioning are made possible by this dynamic behavior of microtubules and by the presence of specialized proteins, called **motor proteins**, which associate with microtubules. Motor proteins, as the name implies, have the ability to move proteins and other structures along microtubules, which serve as racetracks. Motor proteins utilize energy, in the form of ATP hydrolysis, to mediate this movement. Different motor proteins have different functions: some motor proteins facilitate the movement of chromosomes along kinetochore microtubules, some help slide polar microtubules relative to each other, some affect microtubule dynamics (namely lengthening and shortening), and some move astral microtubules along the cell cortex. In this section we describe the special properties of microtubules that promote their dynamic nature, and we will see how microtubules, centrosomes, and motor proteins create a functional spindle that can promote chromosome segregation.

Microtubules are made of tubulin dimers

Microtubules are hollow protein filaments (Figure 7.7). The building blocks of microtubules are **tubulin heterodimers** made of one α-tubulin and one β-tubulin subunit. Tubulin heterodimers join head to tail to form linear **protofilaments** (Figure 7.8); 13 protofilaments are aligned side by side to form a microtubule. Because the tubulin heterodimer is asymmetrical, microtubules have two distinct ends: the face of one end (called the **microtubule plus end**) is made of β-tubulin subunits whereas the face at the other end (called the **microtubule minus end**) is made of α-tubulin subunits (Figure 7.8). In the spindle, the minus ends of microtubules are usually anchored at the centrosome whereas the plus ends extend toward chromosomes, the cell periphery, or microtubules from the opposite spindle pole.

In addition to their role in the spindle, microtubules are involved in other various cellular processes, such as transport, cell motility, and cell shape. For example, in animal cells during interphase (Figure 7.1, part 1), microtubules form an extensive network that serves as a highway system along which motor proteins move macromolecular assemblies and vesicles within the cell. The interphase microtubule network disassembles before mitosis, and the tubulin subunits become dedicated to spindle assembly.

Microtubules are dynamic structures

Microtubules grow and shrink rapidly in the early stages of mitosis. This property is referred to as **dynamic instability**. Since microtubules find their targets by chance encounters, the dynamic nature of microtubules allows them to probe a large volume of the cell, increasing the likelihood that they will encounter their target. The changes in microtubule length are due to the addition or removal of tubulin heterodimers at microtubule ends, with tubulin exchanging faster in the plus end than the minus end. It is important to note that tubulin heterodimers can

Figure 7.7 Microtubules are hollow. A three-dimensional reconstruction of a section of a microtubule based on data from cryoelectron microscopy, showing the 13 side-by-side protofilaments that form the hollow microtubule. Adapted from Li *et al.* (2002).

Figure 7.8 Microtubules are made of tubulin dimers. Tubulin heterodimers are made of α-tubulin (light-green spheres) and β-tubulin (dark-green spheres). The β-tubulin subunit can bind guanosine triphosphate (GTP; pink square) and hydrolyze it to guanosine diphosphate (GDP; gray square). Tubulin heterodimers bind head to tail (namely the α-tubulin subunit of one dimer binds to the β-tubulin subunit of another) to form long protofilaments. The protofilaments line up next to each other to form a cylindrical, hollow microtubule. The protofilaments are slightly off-register, creating a helical pitch. The end of the microtubule that has the α-tubulin subunits exposed is designated the minus end, and the end of the microtubule that has the β-tubulin subunits exposed is designated the plus end. Tubulin heterodimers can only be added or removed from the ends of the microtubule.

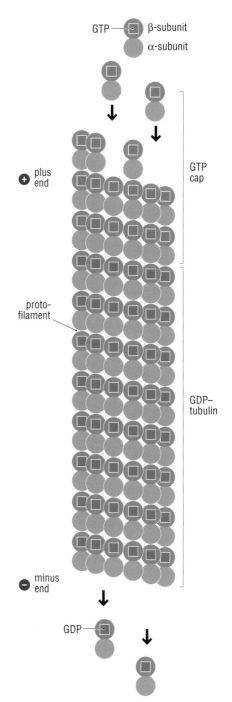

only be added or removed from microtubule ends. The process of rapid microtubule shortening is called **catastrophe** and the reverse process, during which microtubules grow again, is called **rescue** (Figure 7.9).

The β-tubulin subunit of the tubulin heterodimer is able to bind and hydrolyze guanosine triphosphate (GTP), and the transition between catastrophe and rescue is governed by whether the tubulin heterodimers (and specifically their β subunits) at the microtubule end are bound to GTP or guanosine diphosphate (GDP). Free tubulin heterodimers that are added to the microtubule end contain β-tubulin in a GTP-bound state. Once bound, this GTP is gradually hydrolyzed to GDP. Whether the tubulin heterodimers at the microtubule ends are bound to GTP or GDP will determine the behavior of the microtubule. GTP-bound tubulin heterodimers have a high affinity toward each other; at the microtubule end they tend to associate and recruit additional GTP-bound tubulin, causing the microtubule to grow. On the other hand, GDP-bound tubulin heterodimers have a low affinity toward each other and, when present at the microtubule end, they tend to dissociate, causing the microtubule to shorten.

When tubulin heterodimers are added faster then the rate of GTP hydrolysis, the microtubule will have a cap of GTPβ-tubulin, which protects the microtubule from disassembly and allows further heterodimers to be added (Figure 7.9). However, if the rate of tubulin heterodimer addition is slow, such that GTP is hydrolyzed to GDP faster than new GTP-tubulin heterodimers are added, then heterodimers at the microtubule end will be GDP bound, causing these heterodimers to dissociate from the microtubule end and the microtubule will shorten. (Remember that tubulin heterodimers can only be added or removed from the microtubule end.) When the rate of GTP-tubulin addition increases again, the process of microtubule shortening will be rescued and the microtubule will once again grow. These cycles of catastrophe and rescue are responsible for the dynamic nature of microtubules and play an important role in chromosome movement, as we will see in the next section.

The centrosome is a microtubule-nucleating structure that duplicates once per cell cycle

The centrosome is a cytoplasmic structure in animal cells that provides a platform on which microtubules form, or 'nucleate'. Cells normally inherit a single centrosome from their parent cell. During interphase, this centrosome serves as a cytoplasmic microtubule-organizing center for the cell-wide microtubule network (Figure 7.1, part 1). Toward the end of G1, the action of G1 and S phase cyclin-dependent kinases triggers centrosome duplication, marking the beginning of spindle formation.

➔ We discuss cyclin-dependent kinases in Section 5.2.

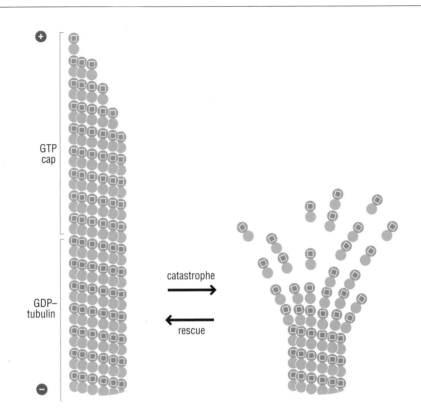

Figure 7.9 Dynamic instability. Tubulin heterodimers bound to guanosine triphosphate (GTP; pink squares) form stable interactions within the microtubule, whereas tubulin heterodimers bound to guanosine diphosphate (GDP; gray squares) form weak interactions, causing the microtubule to depolymerize. Microtubule depolymerization involves the dissociation of tubulin heterodimers as well as the splaying apart of the protofilaments. When the rate of GTP-bound tubulin addition is slower than that of GTP hydrolysis, the GTP cap is lost and the microtubule begins to depolymerize (catastrophe). When the GTP cap re-forms, growth can resume (rescue). Repeated episodes of catastrophe and rescue contribute to the dynamic instability of microtubules.

Upon entry into prophase, the interphase network of microtubules disassembles and the two centrosomes separate and begin to assemble a mitotic spindle (Figure 7.1). Normally, at this point, a cell will have just two centrosomes. However, if a cell ends up with an aberrant number of centrosomes, this could cause the spindle to adopt an abnormal shape: if there is a single centrosome, a situation that can arise from a failure in centrosome duplication, a spindle will form with only one pole. Conversely, if there are more than two centrosomes, a situation that is often seen in cancer cells, this could lead to the formation of a spindle with multiple poles (Figure 7.10). Monopolar and multipolar spindles are incapable of supporting accurate chromosome segregation; only a bipolar spindle is capable

Figure 7.10 Centrosomes dictate the structure of the spindle. Errors in cell division, drug treatments and defects in centrosome duplication can result in cells with an abnormal number of centrosomes. Since centrosomes nucleate microtubules, this, in turn, can lead to the formation of an abnormal spindle. (a) A normal bipolar spindle with two centrosomes and (b) a multipolar spindle with four centrosomes. The DNA is in blue, microtubules are in green and the centrosomes are marked by a centrosomal protein, called pericentrin, in red. Images courtesy of Neil Ganem and David Pellman.

(a) bipolar spindle

(b) multipolar spindle

of supporting chromosome segregation such that at the end of mitosis two identical chromosome clusters will form. Thus, inheriting a single centrosome and duplicating the centrosome only once per cell cycle are crucial for accurate chromosome segregation.

Each centrosome contains a pair of **centrioles**, which are short cylindrical structures made of microtubules that are positioned at right angles to each other (Figure 7.11). Centrioles perform two functions: first, they are required for centrosome duplication. Centrosome duplication begins with the separation of the centrioles, followed by centriole duplication and the recruitment of pericentriolar material to form two centrosomes. The mechanisms that promote centriole duplication and restrict centrosome duplication to once per cell cycle are not yet understood, although they clearly are a crucial step in the formation of a proper spindle.

Centrioles are also required to recruit proteins that form the pericentriolar matrix, a protein-dense region that surrounds the centrioles. One of the components of the pericentriolar matrix is a type of tubulin called γ-tubulin, which, together with several other proteins, makes up γ-tubulin ring complexes (γ-TuRCs). These complexes provide templates onto which the α/β tubulin heterodimers assemble to form the cylindrical microtubule – in other words, γ-TuRCs act as the foundations on which the microtubules are constructed. γ-TuRCs can also nucleate microtubule formation outside centrosomes. This is especially true for plant cells, which lack centrosomes but contain γ-TuRCs.

It is important to note that spindles can also form independently of centrosomes. For example, in female meiosis (the chromosome segregation process that generates oocytes), spindles are formed entirely without centrosomes. In this case, spindle self-assembly is attributed to the ability of chromosome-associated proteins to nucleate microtubules and to the activity motor proteins (see below) that cross-link microtubules to form a stable bipolar array of microtubules. This process is subjected to the cell cycle machinery such that spindle self-assembly does not happen before nuclear envelope breakdown. It is thought that during mitosis both centrosome-dependent and centrosome-independent microtubule nucleation events contribute to spindle assembly.

The spindle is organized by the action of motor proteins

In addition to centrosome duplication, the formation of a bipolar spindle also requires the activity of microtubule-associated proteins, commonly referred to as MAPs, including motor proteins. The motor proteins that are associated with the spindle have two functional domains: a motor domain that promotes the movement of the motor along the microtubule, and a non-motor domain, called a cargo-binding domain, which binds to a particular cellular structure (Figure 7.12a).

During interphase, the cargo domain of certain motor proteins may bind proteins that are to be transported along microtubules. In mitosis, motor proteins contribute to chromosome segregation in a variety of ways: some motor proteins bind to chromosome-associated proteins and facilitate chromosome movement along microtubules. The cargo-binding domain of other motor proteins

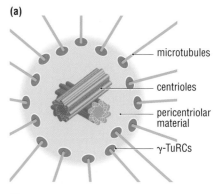

(a)

- microtubules
- centrioles
- pericentriolar material
- γ-TuRCs

(b)

- cross-section of a centriole
- longitudinal section of a centriole
- microubule

Figure 7.11 The structure of the centrosome.
(a) Each centrosome contains a pair of centrioles (shown in green) that are positioned at right angles to each other. The centrioles are surrounded by pericentriolar material (shown in gray). Within the pericentriolar material are multiple γ-tubulin ring complexes (γ-TuRCs) (shown in pink), which nucleate microtubules (green). γ-Tubulin serves as a template for microtubule formation and caps the microtubule at the minus end. (b) Electron micrograph of centrioles within a centrosome. Because of the orientation of the centrioles relative to each other, one centriole is seen in cross-section whereas the other is seen in a longitudinal section. Each centriole is composed of a cylindrical array of nine groups of three microtubules, as can be seen in the cross-section.
Micrograph in Figure 7.11b from McGill *et al.* (1976).

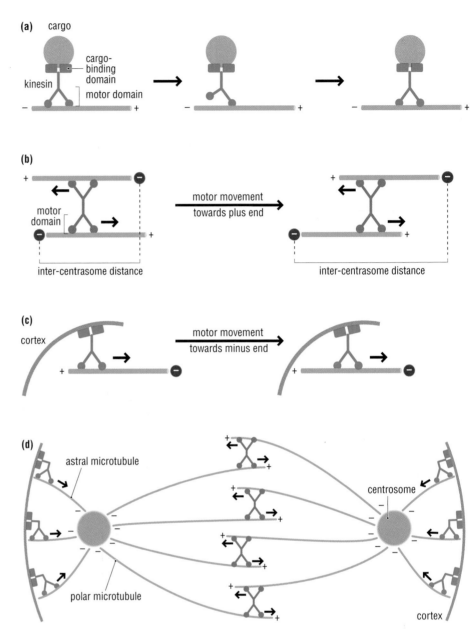

Figure 7.12 Motor proteins. (a) The motor protein shown (turquoise) is a homodimeric plus-end-directed motor containing two motor domains that are associated with a microtubule (green), and two cargo-binding domains associated with a cargo (orange). Movement of the motor protein is mediated by the ATPase activity of its motor domains, which leads a conformation change in the motor domain, causing the motor to 'walk' along the microtubule toward the plus end. As the motor moves along the microtubule, so does its cargo. (b) The motor protein shown (in blue) has plus-end-directed motor domains at both its ends. Therefore, it can cross-link polar microtubules. These microtubules are attached to a centrosome at their minus end (in gray). When the motor walks toward the microtubule plus end, its activity moves the microtubules relative to each other, thereby increasing the distance between the two centrosomes. (c) The cargo-binding domain of the motor protein shown (in turquoise) is bound to the cell cortex. The motor domain has minus-end-directed motor activity. Because the motor protein is anchored to the cell cortex and is therefore stationary, and as this motor protein moves towards the microtubules minus end, it reels in the microtubule, along with its associated centrosome. As a result the centrosome gets closer to the cortex. Motors that have this kind of activity include the dynein family of motor proteins. (d) The actions of different motor proteins in the spindle (the chromosomes are not shown). A plus-end-directed motor protein with motor domains at both ends cross-links the polar microtubules and pushes the spindle poles apart. Cytoplasmic dynein anchors the plus ends of microtubules at the cell cortex and reels in the centrosome towards the cortex. Not shown are motor proteins associated with chromosomes that help move chromosomes along microtubules, as well as motor proteins that affect microtubule depolymerization.

can bind proteins associated with the plasma membrane, while their motor domain binds a microtubule. In this configuration, motor proteins can reel in microtubules toward the cell periphery, helping to position the spindle within the cell. In some cases, motor proteins do not have a cargo-binding domain but rather have motor domains at both ends (Figure 7.12b). Such motor proteins can bind microtubules at both ends and they can help slide microtubules relative to each other.

The motor domains of particular motor proteins can move in one direction only – some motor proteins move toward the microtubule plus end whereas others move toward the minus end. One way to determine the direction of movement of a particular motor protein is to mark it with a fluorescent probe and to follow its movement on purified microtubules, under the microscope. A combination of different motor proteins, some that move to the plus end and others that move to the minus end, is needed for the variety of microtubule movements to take place. The two main types of motor protein involved in spindle assembly and function are the kinesin-like motor proteins and dynein.

Kinesin-like motor proteins are typically plus-end-directed motors, while dynein is a minus-end-directed motor. Kinesin motor proteins that have motor domains on both ends can bind simultaneously to two microtubules from opposite spindle poles. Because these microtubules are in opposite orientations, the movement of each motor domain toward the plus ends pushes the spindle poles apart (Figure 7.12b). This action is important both in separating the centrosomes after their duplication and in elongating the spindle during anaphase. When dynein is localized to the cell cortex, its minus end motor activity reels in the microtubule toward the cell cortex, pulling the whole spindle with it. This action helps position the spindle in the cell (Figure 7.12c). The actions of motor proteins in shaping and positioning the spindle are summarized in Figure 7.12d. Many other non-motor proteins participate in spindle assembly, including some that provide structural support by bundling microtubules, and some that act in a regulatory capacity to affect microtubule dynamics.

The initial steps of spindle assembly occur in the cytoplasm, before nuclear envelope breakdown (see Section 7.1). The completion of spindle assembly occurs when the spindle microtubules engage with the chromosomes once the nuclear envelope disassembles, in prometaphase, as we describe next.

7.4 PROMETAPHASE AND METAPHASE

The kinetochore links chromosomes to spindle microtubules

At the end of prophase, the nuclear envelope breaks down, allowing spindle microtubules to gain access to the chromosomes. This marks the beginning of prometaphase, during which chromosomes become associated with microtubules of the mitotic spindle. Prometaphase ends when all the chromosomes are properly attached to spindle microtubules and the chromosomes have moved to the spindle mid-zone. At this point, the cell is said to be in metaphase.

Microtubules interact with chromosomes through a complex called the **kinetochore** (Figure 7.13), which assembles at a specified chromosomal locus called the centromere, a region of the chromosome that specifies the assembly of the

(a)

(b)

Figure 7.13 Structure of the kinetochore.
(a) Microtubules (green) attach to chromosomes (gray rods) through a multisubunit protein structure called the kinetochore (orange). Kinetochores of animal cells can bind more than 20 microtubules. In metaphase, the pulling forces exerted by the microtubules are opposed by the presence of cohesin complexes (in black) that link the two sister chromatids.
(b) The inner and outer layers of an animal cell kinetochore can be distinguished in the electron micrograph, which shows a kinetochore bound to microtubules.
Figure 7.13b courtesy of Jeremy Pickett-Heaps.

⮕ We consider centromeres in more detail in Section 4.10.

kinetochore. In most cell types each chromosome has one centromere, but in some cell types, such as somatic cells of the nematode *Caenorhabditis elegans*, chromosomes have multiple centromeres distributed along the entire length of the chromosomes.

The kinetochore has multiple functions. In addition to interacting with microtubules, it associates with motor proteins that help move chromosomes along microtubules and it also ensures that chromosomes are oriented properly with respect to the spindle: because sister chromatids need to move away from each other during chromosome segregation, the binding of microtubules to chromosomes must be coordinated so that sister chromatids bind microtubules that originate from opposite spindle poles. Proteins associated with the kinetochore have the ability to distinguish between correct and incorrect kinetochore–microtubule attachments and detach improperly attached kinetochores from microtubules. In addition, certain kinetochore proteins have the ability to induce a cell cycle arrest if kinetochores are not attached to microtubules, as we discuss further in Section 7.5. In this section we will discuss the properties of the kinetochore and how correct kinetochore–microtubule attachments are formed.

Kinetochores vary in size but have conserved proteins and functions

Like centromeres, which vary in size from around 100 base pairs in budding yeast to hundreds of kilobase pairs in animal cells (see Chapter 4), kinetochores also vary greatly in size. The budding yeast kinetochore binds a single microtubule, whereas kinetochores of animal cells are much larger and can bind more than 20 microtubules. Nonetheless, kinetochore proteins are highly conserved in species from yeast to humans, suggesting that the basic mechanism for microtubule binding by kinetochores is evolutionarily conserved. Identifying the exact protein composition of a kinetochore has been a very challenging task because even the simplest known kinetochore, that of budding yeast, is over 5 MDa in size.

Many kinetochore proteins have been identified in simple model organisms through genetic screens for mutations that disrupt chromosome segregation. With the advent of protein identification techniques such as mass spectrometry, some kinetochore proteins were identified biochemically, for example, by purifying known kinetochore proteins and identifying other proteins with which they associate. In all cases, the designation of a protein as a kinetochore protein is done only after confirming that it associates with the centromeric region of the chromosome. To date, more than 80 proteins that participate in building a kinetochore of animal cells have been identified; some have known structural, motor, or regulatory functions, but the function of most is unknown. Read Experimental approach 7.2 to learn how kinetochore proteins have been identified.

Many kinetochore-associated proteins are called CENP-X, where CENP stands for centromere protein, and X represents the order in which the proteins were discovered: the list of CENP proteins currently runs from A through U. Kinetochore proteins can be roughly assigned to two groups based on their physical proximity to the centromere: inner kinetochore proteins, which are physically associated with the DNA or located very close to it, and outer kinetochore proteins, which

7.2 EXPERIMENTAL APPROACH
The hunt for kinetochore proteins

Kinetochore structures in metazoans have long been observed by microscopy, but the molecular composition of these structures was not known. To identify kinetochore components, Phil Hieter and colleagues conducted a genetic screen in budding yeast. As we saw in the screen for cell cycle mutants in Experimental approach 5.1, key in the design of any genetic screen is (1) the prediction of the phenotypes that might arise from a disruption (mutation!) in the process of interest and (2) the means to detect the predicted phenotype. When screening for mutations in genes involved in kinetochore function, the Hieter group predicted that such mutations would lead to increased rates of chromosome loss. As a strategy for detecting this predicted loss, the investigators followed the presence (or loss) of a non-essential truncated short chromosome (called a chromosome fragment) containing a marker gene that affects the color of the colony. When the chromosome fragment, and with it the marker gene is lost, red sectors appear in otherwise white yeast colonies (Figure 1).

The use of the chromosome fragment that contained a functional centromere was a critical aspect of this screen. Since kinetochores are essential for chromosome segregation, mutations that severely diminish kinetochore function would likely result in massive chromosome loss, and thus in cell death. However, as a chromosome fragment is less stable than a normal chromosome, the group was able to isolate phenotypically mild mutations that had a more pronounced effect on the chromosome fragment compared to an endogenous chromosome, thus having a minimal effect on cell viability. We might refer to this strain carrying the chromosome fragment marker as a 'sensitized' background for conducting the screen.

To conduct the screen, Hieter and colleagues treated yeast cells with a chemical that induced point mutations, and then isolated cells that showed a high rate of chromosome fragment loss, as evidenced by the appearance of red sectors within the white colonies. Through this screen, the researchers isolated over 130 independent mutants that corresponded to mutations in 11 different genes. After their initial identification, evidence that these proteins were *bona fide* kinetochore proteins was still needed. Subsequent studies have shown that while some of the genes identified in this screen code for kinetochore proteins that are conserved through the vertebrate lineage, other are involved in processes such as sister chromatid cohesion. This result was not surprising given that defects in sister chromatid cohesion would also be expected to increase chromosome loss rate.

Kinetochore proteins were also identified through biochemical approaches. Investigators suspected the kinetochore to be a large complex, but the identity of only a few kinetochore proteins, such as CENP-A, was known. To identify additional kinetochore components, investigators turned to biochemistry to isolate proteins that co-purified with a known kinetochore protein. Don Cleveland and colleagues were interested in identifying proteins that interact with the

Figure 1 A genetic screen for mutations in kinetochore genes. The yeast strain used in this screen carried a mutation in a gene involved in adenine biosynthesis (the *ADE2* gene). Yeast cells that are proficient in adenine biosynthesis give rise to white colonies, while yeast carrying the *ade2-101* allele give rise to red colonies. In this assay, cells also carried a non-essential chromosome fragment that had a centromere and a gene for a tRNA suppressor, which allowed the synthesis of the complete Ade2 protein from the *ade2-101* allele. Therefore, when the tRNA suppressor was present, the Ade2 protein was made and the colonies were white. When the suppressor was missing, for example due to the loss of the chromosome fragment, the Ade2 protein could not be made and the colonies were red. Loss of chromosome fragment during the growth of the colony resulted in the appearance of a red sector. Wild-type cells (a) give rise to mostly white colonies, unless the chromosome fragment was lost before the cells were plated. Mutations that affected kinetochore function (b – a mild defect; c – a severe defect) increased the loss rate of the chromosome fragment, resulting in the appearance of red sectors. (The red sector appears as black in this black and white photograph.) Adapted from Spencer *et al.*, *Genetics* 1990;**124**:237–249.

kinetochore protein CENP-A and so they designed a CENP-A gene fused to a protein tag and expressed it in mammalian cells (Figure 2). Immunoprecipitation was then used to purify the tagged CENP-A and its associated proteins from a protein extract prepared from these cells. The purified proteins were separated on a gel and identified by mass spectrometry. The techniques of immunoprecipitation and protein identification by mass spectrometry are discussed in Chapter 16.

In this analysis, the Cleveland group found a number of proteins that were associated with CENP-A. The group then pursued two approaches to determine if these proteins were indeed kinetochore proteins. First, the researchers generated gene fusions of the newly identified proteins (designated CENP-K through T) with GFP and showed that these proteins localized to the same place as previously identified kinetochore proteins (an example for CENP-N is shown in Figure 3a). Second, the group hypothesized if these proteins were involved in kinetochore function, inactivation of these genes would lead to chromosome segregation defects. To examine this, the researchers down-regulated the levels of each of the identified proteins using small inhibitory RNAs (siRNA, see Sections 8.15 and 16.4), and confirmed the participation of a number of the identified proteins in kinetochore function. For example, in the

(a)

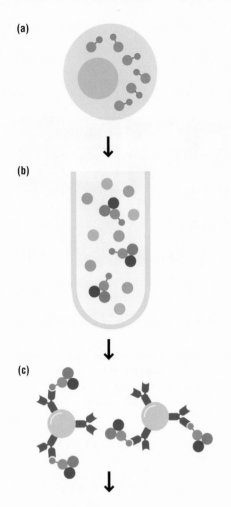

(b)

(c)

(d) **identify proteins by mass spectrometry**

Figure 2 Identifying kinetochore proteins through a biochemical approach. Protein purification takes advantage of a unique property of the protein of interest. Here, cells synthesized CENP-A (shown in blue) that was fused to a protein tag (in pink) for which antibodies are available (a). Under the right conditions, the only protein that these antibodies will recognize is the tagged CENP-A. To purify tagged CENP-A, a protein extract is prepared (b) and mixed with beads that are bound to the antibodies (c). When the beads are removed from the protein mixture they will bring with them the tagged CENP-A, and any proteins to which CENP-A is bound (in gray), leaving behind the rest of the proteins (in green and orange). To identify the proteins to which CENP-A is bound, the beads are stripped of the proteins, which are then separated according to their size on a polyacrylamide gel (d). The identity of each protein band, as indicated on the left side of the stained gel, is determined by mass spectrometry. The protein names shown in bold are those that were not previously known to be associated with CENP-A. CENP-A-ST is the tagged version of CENP-A.
Figure 2d adapted from Foltz *et al.*, *Nature Cell Biology* 2008;**8**:458–469.

case of CENP-N (Figure 3b), down-regulation resulted in defects in chromosome movement towards the metaphase plate, as would be expected from a defect in kinetochore–microtubule binding.

These examples nicely demonstrate how genetic and biochemical approaches often yield complementary information. Neither approach typically yields a complete picture, but through a combination of approaches, scientists can get increasingly detailed pictures of biological processes.

Foltz DR, Jansen LE, Black BE, *et al.* The human CENP-A centromeric nucleosome associated complex. *Nature Cell Biology* 2006;**8**:458–469.

Spencer F, Gerring SL, Connelly C, Hieter P. Mitotic chromosome transmission fidelity mutants in *Saccharomyces cerevisiae*. *Genetics* 1990;**124**: 237–249.

Figure 3 CENP-N is a kinetochore protein. To determine if the proteins that were identified by mass spectrometry were indeed kinetochore proteins, Foltz *et al.* made protein fusions with the green fluorescent protein (GFP), expressed them in cells, and determined where they localize (panel a). Shown is an example of the co-localization of GFP:CENP-N and the target of the anti-centromere antibodies (ACA), which was previously shown to identify kinetochore proteins. The merged images on the far right show the co-localization of the two proteins. To confirm that the proteins identified by mass spectrometry have a role in kinetochore function, Foltz *et al.* examined what happens when the levels of these proteins is reduced by the use of siRNA (see Section 16.4). For example, whereas in the control-treated cell the chromosomes are all aligned on the metaphase plate (Figure 3b, left image), down-regulation of CENP-N leads to the appearance of lagging chromosomes that have not properly attached to the spindle. (Figure 3a right image).
Adapted from Foltz *et al.*, *Nature Cell Biology* 2008;**8**:458–469.

are further away from the centromere. Several of the proteins defined as inner kinetochore proteins are constitutively associated with the centromere, including the histone H3 variant CENP-A (see Section 4.10). Others, including all of the outer kinetochore proteins, only join the kinetochore during mitosis.

Microtubules are captured by proteins of the outer kinetochore

Kinetochores interact with microtubule plus ends. The association between the kinetochore and microtubules is mediated by the conserved rod-shaped outer-kinetochore subcomplex, called the Ndc80 complex (Figure 7.14). However, since microtubules can grow and shrink at the plus end, this raises the question of how kinetochores can maintain an attachment to dynamically unstable microtubules. In fact, one of the driving forces of sister chromatid separation in anaphase is the depolymerization of the microtubule plus end (namely the loss of tubulin dimers from the microtubule end), which shortens the microtubule and brings the chromosome closer to the spindle pole. The chromosome, however, must remain attached to the microtubule even when this shedding of tubulin molecules occurs, and therefore the kinetochore must somehow accommodate attachment to a dynamic microtubule end. The mechanism by which this occurs lies, at least in part, in the way by which the Ndc80 complex to bind microtubules. The Ndc80 complex can track along microtubules, such that when the microtubule end depolymerizes, the complex will move to a microtubule region that is still intact, thus maintaining an association between the kinetochore and the microtubule.

Correct kinetochore–microtubule attachment creates tension between sister kinetochores

Capture of chromosomes by spindle microtubules depends on chance encounter of the plus end of a microtubule with a kinetochore. To ensure proper chromosome segregation, these chance encounters must ultimately result in the attachment of kinetochores of sister chromatids to microtubules emanating from opposite spindle poles. The attachment of sister chromatids to opposite spindle poles is called bi-orientation or **amphitelic attachment** (Figure 7.15a).

When kinetochores are bi-oriented, the pulling forces exerted by the microtubules are opposed by the cohesion between the two sister chromatids, and the resulting tension between the sister kinetochores is thought to stabilize kinetochore-microtubule attachments and promote the binding of more microtubules. Because microtubules encounter their target by chance, many of the initial kinetochore-microtubule attachments are incorrect or incomplete. For example, sister kinetochores could both attach to microtubules that emanate from the same spindle pole (Figure 7.15b), or one of the sister kinetochores could be without associated microtubules (Figure 7.15c). The difference between these interactions and the correct attachment is that tension is generated only when sister kinetochores are bi-oriented.

Incorrect attachments, where no tension is generated, are destabilized by the activity of a protein kinase called aurora protein kinase, which detects the lack of tension by a yet unknown mechanism. In the absence of tension, aurora protein kinase phosphorylates several kinetochore proteins, including the region of the

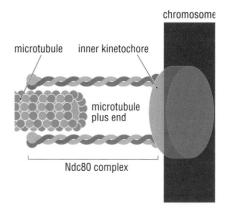

Figure 7.14 Proposed organization of the protein complexes that attaches kinetochores to microtubules. The Ndc80 complex (in blue and red), which is part of the outer kinetochore, binds directly to microtubules (in green). Note that the end of the microtubule is available for the addition of tubulin. When microtubules depolymerize, kinetochore proteins slide further toward the minus end, thus maintaining an association between the kinetochore and the microtubule.

Figure 7.15 Correct and incorrect kinetochore–microtubule attachments. (a) In the correct attachment of microtubules (green) to kinetochores (orange), sister chromatids (gray rods) are bound to microtubules emanating from opposite spindle poles. This attachment is called an amphitelic attachment and the sister chromatids are bi-oriented. (b) Examples of incorrect attachments include the association of kinetochores with microtubules emanating from the same spindle pole (1), or when one of the kinetochores is unattached (2).

Ndc80 complex that interacts with microtubules. Aurora phosphorylation reduces the binding affinity between the kinetochore and microtubules, causing the kinetochore to release the associated microtubules and creating another opportunity for microtubule binding. It is possible that in the presence of tension, the aurora kinase substrates are pulled away from the kinase and their phosphorylation is thus prevented, thereby preserving kinetochore-microtubule attachments. We learn more about the investigation of kinetochore tension in Experimental approach 7.3.

This process of attachment and release mediated by aurora protein kinase is repeated until the sister chromatids bi-orient and tension is produced. Since the persistence of incorrect kinetochore–microtubule interactions leads to chromosome missegregation, scientists have targeted this process for drug therapy to disrupt cell division in cancerous cells. Indeed, inhibition of aurora protein kinase leads to massive chromosome missegregation, which is accompanied by cell death (Figure 7.16).

Incorrect kinetochore–microtubule attachments also lead to the activation of a regulatory pathway, called the **spindle checkpoint pathway** (also known as the spindle assembly checkpoint pathway), which inhibits further progression through mitosis as long as incorrect attachments exist. The mitotic delay induced by this checkpoint pathway will be discussed in the next section.

The length of prometaphase depends on how quickly chromosomes are able to bi-orient; even a single unattached kinetochore will activate the spindle checkpoint pathway and delay further progression through mitosis. The formation of proper kinetochore–microtubule attachments is crucial for cell survival, and since incorrect attachments can only be corrected during prometaphase, before sister chromatids separate, these cycles of microtubule attachments and detachments continue until all chromosomes have bi-oriented. At the end of prometaphase, the bi-oriented chromosomes are held aligned under tension along the spindle mid-zone. When this occurs, the cell is in metaphase, and the chromosomes are poised for segregation, as we will discuss next.

Figure 7.16 Inactivation of aurora protein kinase leads to the persistence of incorrect kinetochore–microtubule attachments. (a) When aurora protein kinase is active, incorrect kinetochore–microtubule attachments do not persist and all chromosomes eventually bi-orient. This type of binding leads to a balance of forces acting on the chromosomes (blue), such that all the chromosomes align at the spindle midline. (b) When aurora protein kinase is inactivated by drug treatment, the microtubules are still arranged in a bipolar spindle, but incorrect microtubule–kinetochore attachments cannot be corrected. As a result, pulling forces on the chromosomes are greater toward one pole of the spindle than the other, resulting in a failure to align at the spindle midline.
Adapted from Lampson *et al.* (2004).

7.3 EXPERIMENTAL APPROACH

Chromosome bi-orientation: how to generate and detect tension

One of the most important steps in the cell division cycle is the segregation of one copy of each sister chromatid to each of the daughter cells. In preparing for chromosome segregation, the two sister chromatids initially bind microtubules at random, but eventually they bi-orient. The spindle checkpoint pathway prevents progression to anaphase until *all* paired sister chromatids have bi-oriented. A similar situation occurs in meiosis, where homologous chromosomes (rather than sister chromatids) also need to bi-orient (see Section 7.7).

What is the mechanism that signals that bi-orientation has been accomplished? Richard McIntosh was the first to propose that the phenomenon distinguishing bi-orientation in mitosis from any other kind of chromosome-microtubule attachment is the presence of tension between the two sister chromatids due to pulling forces, in opposite directions, exerted by the microtubules.

To test this hypothesis, Xiatong Li and Bruce Nicklas took advantage of the ability to directly micro-manipulate chromosomes in praying mantis spermatocytes. These spermatocytes are unusual in that they normally contain three sex chromosomes: one Y and two X chromosomes, designated X_1 and X_2. In order to bi-orient in meiosis, homologous chromosomes normally pair and form 'bivalents', but these sex chromosomes have to form a trivalent complex, with the Y chromosome facing one pole, and the X_1 and X_2 facing the other (X and Y chromosomes can pair

thanks to a short region of homology). In other words, when these chromosomes bi-orient, the X_1 and X_2 chromosomes are pulled by spindle microtubules in one direction, while the Y chromosome is pulled in the other direction (Figure 1a). Sometimes one of the X chromosomes fails to associate with the X–Y pair, resulting in a checkpoint-dependent cell cycle arrest (Figure 1b).

Li and Nicklas argued that if the delay was due to a lack of tension, because this lone X chromosome was not paired with any other chromosome and therefore could not bi-orient, artificially creating tension by pulling on this lone X chromosome with a micro-manipulator should alleviate the checkpoint-induced delay. Indeed, when they pulled on the lone X chromosome in the appropriate direction (Figure 2), the cell cycle delay was considerably shortened. This was an elegant demonstration that tension plays a direct role in signaling bi-orientation.

Once the link between tension and bi-orientation was established, researchers began to search for the actual source of the tension and a mechanism by which that tension regulates anaphase. One possibility was that tension in mitosis was created between sister chromatids when their kinetochores were pulled apart (referred to as inter-kinetochore tension, Figure 3b). Alternatively, the source of tension could be within the kinetochore itself (intra-kinetochore tension, Figure 3c). To distinguish between these models, Thomas Maresca and Ted Salmon constructed a

Figure 1 Tension between sex chromosomes in grasshopper spermatocytes. Grasshopper spermatocytes contain three sex chromosomes: two X chromosomes, designated X_1 and X_2, and one Y chromosome. During the first meiotic division, the three chromosomes need to form a trivalent species, with the X chromosomes being pulled by microtubules from one spindle pole and the Y chromosome being pulled by microtubules from the opposite pole (Figure 1a, microtubules are not shown, the arrows point to the direction at which the chromosomes are pulled. Orange circles are kinetochores). This configuration generates tension and the cells can proceed through meiosis ('go ahead' situation). When one of the X chromosomes is not attached to the other X and Y chromosomes, the lone chromosome is being pulled by microtubules in only one direction and therefore is not experiencing tension, resulting in checkpoint activation and inhibition of cell cycle progression (Figure 1b, 'wait' situation). If, however, the lone X chromosome is pulled by a micro-needle in the opposite direction to that of the microtubules, this creates tension within the chromosome and its kinetochore, thereby satisfying the tension requirement and allowing the cell cycle to proceed. Adapted from Li and Nicklas, *Nature* 1995;**373**:630–632.

Figure 2 Creating tension using a micro-needle. To test the tension idea, Li and Nicklas identified spermatocytes in which one of the X chromosomes was not attached to the Y and other X chromosome. In the first panel (time 0 seconds), the X$_1$ and Y chromosomes are attached and are being pulled in opposite directions. (Notice the stretched configuration of the X–Y pair, as well as the other chromosomes in this image.) The chromosomes are aligned on a metaphase plate, but anaphase does not take place because of the presence of an unattached X chromosome (next panel). The same cell contains another X chromosome, X$_2$ (time 0.1 seconds), which can be seen on a different focal plane (that is, it is either above or below the other chromosomes). This chromosome is not under tension – it is being pulled toward one pole and it is not stretched. To create tension, the investigators inserted a micro-needle into the arm of chromosome X$_2$ (time 0.1 second, the needle can be seen as a white triangle and the arrow is pointing toward its tip) and pulled the chromosome towards the opposite pole (time 1 second, note how the chromosome now appears stretched). Anaphase can be seen after 55 seconds, where homologous chromosomes separate to opposite poles, and separation is completed at 69 seconds. Adapted from Li and Nicklas, *Nature* 1995;**373**:630–632.

(a) no tension   **(b) inter-kinetochore tension**   **(c) intra-kinetochore tension**

Figure 3 Inter- and intra-kinetochore tension. In mitosis, chromosomes bi-orient when sister kinetochores (in orange) bind microtubules (in green) from opposite poles. Cohesion between the sister chromatids (illustrated as black dots) opposes the pulling forces of the microtubules and can create two types of tension (black arrows): between sister kinetochores (b) or within each kinetochore (c). Note that both types of tension can be present concomitantly (such as in (c)). There is no tension when microtubules are not attached (a).

Figure 4 Creating a tension monitor. (a) To measure both inter- and intra-kinetochore tension in live cells, Maresca and Salmon constructed a fly cell line where a centromere-associated protein (CID) and outer kinetochore protein (Ndc80) were labeled with mCherry and GFP, respectively (illustrated in red and green in Figure 4a). The inter-kinetochore distance was measured between the two mCherry signals (ΔCID). The intra-kinetochore distance, designated as ∂, was determined by measuring the distance between the two Ndc80-GFP signals (ΔNdc80), subtracting the ΔCID value, and dividing the result by two, as there are two kinetochores between the two Ndc80-GFP signals. The measurements were done from the center of each fluorescent signal. (b) Top panel: In prophase the kinetochores have not yet bound microtubules, and therefore there is no tension. Consequently, the signals from CID-mCherry and Ndc80-GFP overlap almost entirely. In metaphase (bottom panel), in contrast, the pulling forces of the microtubules cause the kinetochores to extend, resulting in a partial displacement of the Ndc80-GFP signal from the CID-mCherry signal.

Drosophila melanogaster (fruit fly) cell line in which they could follow both inter- and intra-kinetochore tension. In this cell line, the inner part of the kinetochore was labeled with a fly centromere-binding protein, CID, fused to a red fluorescent protein (mCherry), whereas the outer kinetochore was labeled with an outer kineto-chore protein, Ndc80, fused to GFP (Figure 4a). The use of these color markers enabled the investigators to directly measure the separation between the two proteins by microscopy. When there was no tension, such as in prophase, the signals from the inner and outer kinetochore proteins nearly overlapped (Figure 4b, top panels). However, during metaphase when the tension is known to be high, the two Ndc80-GFP signals were clearly separated from the two CID-mCherry signals (Figure 4b, bottom panels). These data argued that the kinetochore itself was being stretched, and thus that intra-kinetochore tension could be critical for dictating the checkpoint control.

Having developed an assay to detect tension, the research-ers then tested a variety of conditions that altered microtubule behavior (and thus changed the intra-kinetochore distance, inter-kinetochore distance, or both) and examined when the checkpoint was activated. These studies showed that when the kinetochore was stretched (hence increasing the intra-kinetochore distance), checkpoint activation was decreased. By contrast, inter-kinetochore distances did not correlate with checkpoint activation. Thus, tension could be sensed specifically by intra-kinetochore stretching.

The studies described here illustrate how investigators often need to adopt creative approaches to address the questions they are seeking to answer. How can one generate and measure ten-sion in a single cell and follow it in a microscope? Here, praying mantis spermatocytes, with their large chromosomes, and the ability to mark and monitor the distance between two fluorescent protein spots in fly cells provided important insights. It is strik-ing how often in biology an organism with unusual properties has allowed the field to propel itself forward.

Li X, Nicklas RB. Mitotic forces control a cell cycle checkpoint. *Nature* 1995;**373**:630–632.

Maresca TJ, Salmon ED. Intrakinetochore stretch is associated with changes in kinetochore phosphorylation and spindle assembly checkpoint activity. *Journal of Cell Biology* 2009;**184**:383–381.

7.5 ANAPHASE: AN IRREVERSIBLE STEP IN CHROMOSOME SEGREGATION

At anaphase sister chromatids separate and move to opposite spindle poles

Anaphase is the stage in mitosis when sister chromatids separate and are pulled to opposite poles of the spindle by kinetochore-associated microtubules. As we learned in Section 7.2, sister chromatids are associated with each other from the time of DNA replication through metaphase, by the presence of cohesin complexes. At the onset of anaphase, however, these cohesin complexes must be removed to allow the spindle microtubules to pull the sister chromatids apart (Figure 7.17). In the first part of this section we will discuss how cohesin is removed.

The metaphase-to-anaphase transition marks a point of no return in mitosis: mistakes in chromosome–microtubule attachments must be corrected before this stage or the daughter cells will inherit an abnormal number of chromosomes. Thus, the dissolution of sister chromatid cohesion must be delayed until proper kinetochore–microtubule attachments are formed. This is the function of the spindle checkpoint pathway that was introduced in Chapter 5 and will be dis-cussed further in the second part of this section.

Once cohesin is removed, chromosome movement occurs in two steps: during **anaphase A** sister chromatids move towards their respective spindle poles, and during **anaphase B** the spindle poles move away from each other, further increas-ing the distance between the sister chromatids (Figure 7.17). This microtubule-dependent movement will be discussed in the last part of this section.

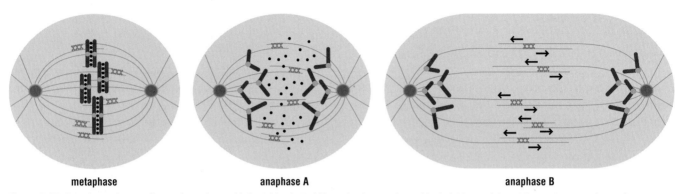

metaphase   anaphase A   anaphase B

Figure 7.17 Anaphase. At metaphase, sister chromatids have bi-oriented. The cohesin complexes (black dots) remaining at the centromeres (orange) oppose the pulling forces of the spindle microtubules (green). At this point the microtubule pulling forces acting on each of the bi-oriented sister kinetochores are roughly the same, causing the chromosomes to align on the spindle mid-zone. At the metaphase-to-anaphase transition, cohesin is destroyed, allowing the spindle microtubules to pull the sister chromatids apart. At first, the chromosomes move toward the poles (in gray), in what is known as anaphase A. Subsequently, motor proteins associated with polar microtubules (shown as blue crosses) cause the spindle to elongate. This, along with motor proteins associated with the cortex, which pull on astral microtubules (not shown), cause the spindle poles to move apart, in what is known as anaphase B.

Dissolution of cohesin at the beginning of anaphase depends on separase and is regulated by the APC

Sister chromatid separation at the onset of anaphase depends on the complete removal of the cohesin complexes that hold sister chromatids together. In animal cells, most cohesin complexes that associate with chromosome arms are removed during prophase, while cohesin complexes located around centromeres persist until the beginning of anaphase. Centromere-associated cohesin is removed by the activity of a protease called **separase**, which cleaves the kleisin subunit of the cohesin complex (see Section 7.2). As a result, the integrity of the cohesin ring is disrupted, leading to the dissociation of cohesin from the chromosomes.

The proper timing of kleisin cleavage by separase is crucial, because if sister chromatid cohesion is dissolved before chromosomes bi-orient, there will be no mechanism to guide the proper attachment of sister chromatids to microtubules from the opposite spindle pole. Instead, chromosomes will attach randomly to the spindle, leading to aberrant chromosome segregation. Thus, separase activity must be under tight regulation. Indeed, until anaphase, separase is kept inactive by multiple regulatory mechanisms, thereby ensuring that cohesin does not dissociate prematurely. Three proteins participate in regulating separase activity: cyclin B, **securin**, and **shugoshin** (a Japanese word that means guardian spirit; also called Sgo1) (Figure 7.18).

Figure 7.18 Sister chromatid separation is regulated by multiple proteins. (a) Separase (shown in dark blue) is inhibited by securin (in light blue), which binds to the N and C termini of separase and inhibits the protease activity. Separase is also inhibited by cyclin–Cdk phosphorylation (yellow P in black circle). (b) Cohesin (black circles) at centromeres (marked in orange) is protected from the cleavage by the presence of shugoshin (Sgo1; light blue), which blocks the access of separase to centromere-bound cohesin.

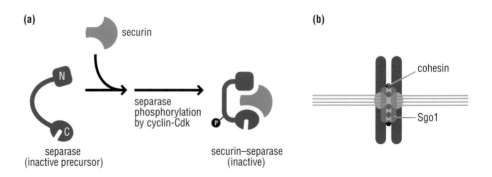

(a)

securin

N

separase
phosphorylation
by cyclin-Cdk

P

C

separase
(inactive precursor)

securin–separase
(inactive)

(b)

cohesin

Sgo1

Prior to anaphase, cyclin B–Cdk phosphorylates separase, thereby inhibiting separase's catalytic activity (Figure 7.18a). In addition, separase binds to a protein called securin, which prevents separase's catalytic site from gaining access to cohesin's kleisin subunit (Figure 7.18a). Finally, cohesin complexes located at centromeres are protected from separase's proteolytic action by the kinetochore-associated protein shugoshin (Figure 7.18b). These three proteins, cyclin B, securin, and shugoshin, are degraded at the onset of anaphase, thereby lifting the inhibition of separase activity and allowing it to promote sister chromatid separation.

The degradation of cyclin B, securin, and shugoshin occurs through a ubiquitin-mediated protein degradation process in which the proteins are targeted for degradation by the addition of polyubiquitin chains to lysine side chains on the proteins in question. Ubiquitination of these proteins is carried out by the anaphase-promoting complex (APC), a ubiquitin ligase that was described in Chapter 5. At the metaphase-to-anaphase transition, the APC acts in conjunction with the substrate-recognition subunit **Cdc20**, which recognizes numerous cell cycle proteins, including cyclin B, securin, and shugoshin. Thus, the chain of events that results in anaphase initiation activation begins with the APC^{Cdc20}-dependent ubiquitination and subsequent degradation of cyclin B, securin, and shugoshin, resulting in the activation of separase, which cleaves cohesin's kleisin and leads to the dissolution of sister chromatid cohesion (Figure 7.19). Cdc20 is therefore a key player in promoting anaphase and, as well shall discuss next, it is one of the main targets of the spindle checkpoint pathway.

The spindle checkpoint inhibits anaphase initiation by inactivating APC^{Cdc20}

The spindle checkpoint is activated by kinetochores that are not associated with microtubules. Unattached kinetochores are present in early mitosis, before nuclear envelope breakdown, when microtubules have yet to gain access to chromosomes. Unattached kinetochores also result from the activity of aurora protein kinase, which disassembles incorrect kinetochore–microtubule associations, as we discussed in Section 7.4. The presence of an unattached kinetochore causes several spindle checkpoint proteins to bind and inhibit Cdc20, which inactivates the APC^{Cdc20} complex (Figure 7.20). Therefore, as long as there are unattached kinetochores in the cell, the spindle checkpoint inhibits the APC^{Cdc20}, and anaphase is delayed. Once all the chromosomes successfully bi-orient, the signal for checkpoint activation is extinguished, the APC^{Cdc20} becomes active, and anaphase can begin. Thus, the spindle checkpoint pathway links the state of kinetochore

⊘ We discuss ubiquitination further in Chapter 11.

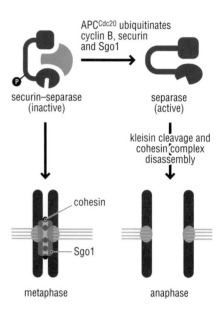

Figure 7.19 The anaphase-promoting complex (APC) promotes the degradation of proteins that inhibit separase. At the onset of anaphase, the APC, in conjunction with Cdc20, promotes the ubiquitination and subsequent degradation of shugoshin (Sgo1), cyclin B, and securin (light blue). Cyclin B degradation facilitates the removal of the inhibitory phosphorylation from separase (dark blue). Securin degradation leads to further activation of separase, which can then cleave the kleisin subunit of cohesin, thereby dissolving cohesion and allowing the sister chromatids to separate at anaphase.

Figure 7.20 Unattached kinetochores inhibit anaphase by activating the spindle checkpoint pathway. An unattached kinetochore generates a signal that activates the spindle checkpoint pathway. Proteins of this pathway bind and inactivate Cdc20. Since Cdc20 is necessary for APC-mediated ubiquitination of shugoshin (Sgo1), cyclin B, and securin, the presence of an unattached kinetochore effectively inhibits the initiation of anaphase.

attachment to cell cycle progression, and ensures that anaphase initiation does not occur prematurely.

The spindle checkpoint pathway is crucial for ensuring proper chromosome segregation. Indeed, components of this pathway are essential for viability; mice in which spindle checkpoint genes are deleted die *in utero*. The importance of the spindle checkpoint pathway in the mouse model system is further underscored by the association between mutations in spindle checkpoint genes and the predisposition to cancer. One of the hallmarks of cancer cells is an abnormal number of chromosomes, referred to as **aneuploidy**. It is thought that when the spindle checkpoint pathway is inactive, the cell cycle proceeds aberrantly despite the presence of unattached kinetochores, and this leads to chromosome missegregation and aneuploidy. Changes in chromosome copy number cause many cells to die, but in a small fraction of cells the combination of chromosomes will be such that it could allow selection for cancer cells that have lost tumor suppressor genes or gained oncogenes as described in Chapter 5. Thus, the spindle checkpoint pathway acts as a quality control mechanism that inhibits APC^{Cdc20} activation, and hence prevents progression into anaphase, until all chromosomes are properly attached.

Chromosome segregation in anaphase is driven by microtubule flux, microtubule depolymerization, and spindle-associated motor proteins

Once separase is active and cohesion has been dissolved, spindle microtubules pull the sister chromatids apart. As we mentioned in the beginning of this section, sister chromatids first move towards their respective spindle poles (known as anaphase A), and then the spindle poles begin to move away from each other, further increasing the distance between the sister chromatids (known as anaphase B). The movement of the chromosomes to the spindle poles at anaphase A is driven by a combination of **microtubule flux** and depolymerization of the kinetochore microtubules at their plus ends.

Microtubule flux is the gradual movement of tubulin heterodimers from the plus end of a microtubule, where tubulin is being added, toward the minus end at the spindle pole, where tubulin is being removed, as illustrated in Figure 7.21. Microtubule flux exerts a pulling force on sister chromatids towards the poles to

Figure 7.21 Microtubule flux. Under conditions where tubulin heterodimers are added at the microtubule plus end and removed from the microtubule minus end, the incorporated tubulin heterodimers gradually move closer to the microtubule's minus end. If a protein were to be tethered to a newly incorporated tubulin heterodimer, it would gradually more towards the minus end (imagine a person standing on an escalator step at one end of the escalator, and is carried with it as the step moves toward the escalator's other end).

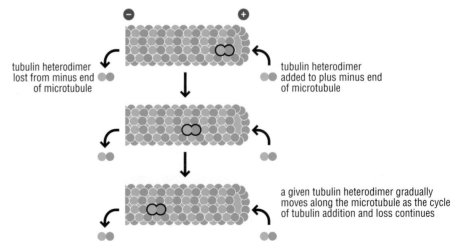

tubulin heterodimer lost from minus end of microtubule

tubulin heterodimer added to plus minus end of microtubule

a given tubulin heterodimer gradually moves along the microtubule as the cycle of tubulin addition and loss continues

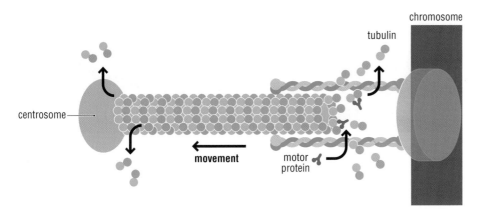

Microtubule flux results in the movement of tubulin within the microtubule from the plus end towards the minus end. During anaphase, there is a loss of tubulin dimers from the microtubule minus end, leading to microtubule flux. At the same time, depolymerization of the microtubule plus end results in splaying of the protofilaments and microtubule shortening. The Ndc80 complex (red-blue) maintains contact with the microtubule as the microtubule shortens from its plus end, causing the chromosome to be pulled toward the spindle pole.

which they are connected. Imagine the kinetochore as being associated with a particular site on a microtubule: as tubulin subunits are moved from the minus end, this site gets closer to the spindle pole, and so does the chromosome. Kinetochore-associated proteins, such as motor proteins and the Ndc80 complex, also contribute to the poleward movement of chromosomes towards the microtubule minus end: minus-end–directed motor proteins may actively move their chromosome cargo toward the minus end, other motor proteins induce microtubule depolymerization at the microtubule plus end, and kinetochore proteins, such as the Ndc80 complex, ensure that the kinetochore remains bound to the part of the microtubule that has yet to disassemble (Figure 7.22).

There are two driving forces for anaphase B: the activity of plus-end-directed bipolar motors, which bind simultaneously to two polar microtubules from opposite poles of the spindle and push the poles apart (Figure 7.12b), and pulling forces exerted by minus-end-directed motor proteins that bind to the cell cortex and pull on the astral microtubules (Figure 7.12c). By the end of anaphase B, the two chromosome clusters are at the poles of the elongated spindle. If there were no errors in chromosome attachments to the spindle, these two clusters will have an identical chromosome composition. The next step will be to physically separate these DNA masses by cell division, creating two cells that are genetically identical to their parent.

7.6 THE COMPLETION OF MITOSIS AND CYTOKINESIS

Mitosis ends with spindle disassembly and nuclear envelope reassembly

The final phase of mitosis is telophase. Once chromosome segregation is completed, the spindle disassembles (Figure 7.1). The chromosomes, which have been condensed throughout mitosis, become decondensed, and the nuclear envelope re-forms around the two chromosome clusters (Figure 7.23). These processes can take place only when mitotic Cdk has been inactivated by the APC-dependent ubiquitination and subsequent degradation of mitotic cyclins. To illustrate this point, in yeast, expression of a non-degradable cyclin B (namely, a mutated cyclin B that cannot be ubiquitinated by the APC) prevents exit from mitosis. Mitotic

Figure 7.23 Nuclear envelope formation at the end of mitosis. The images show a cell from a human cell line synthesizing histone H2B fused to a red fluorescent protein (shown in red), which marks the chromosomes, and a nuclear envelope protein (Sun1) fused to a green fluorescent protein (in green). During metaphase, nuclear envelope proteins localize to the cytoplasm or the endoplasmic reticulum. These proteins are recruited to the chromosomes as cells exit mitosis (telophase and cytokinesis). The bar depicts 10 μM in length.
Adapted from Anderson and Hetzer (2008).

cyclin degradation begins at the onset of anaphase and is completed by telophase. The inactivation of Cdk allows for the dephosphorylation of proteins that were phosphorylated earlier in the cell cycle by Cdk; this, in turn, triggers mitotic exit and cytokinesis.

Cytokinesis in animal cells, but not in plant cells, is driven by the contraction of an actomyosin ring

Cytokinesis, which happens after telophase, is the process by which a parent cell divides to give rise to two daughter cells. Cell division requires the placement of a membrane between the two newly formed nuclei. Animal cells and plant cells do this in different ways. Animal cells pull the plasma membrane inward, toward the cell center, whereas plant cells form a membrane at the cell center and expand it towards the cell cortex (Figure 7.24).

(a) animal cell

(b) plant cell

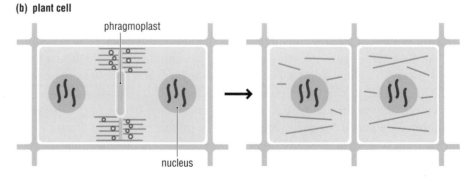

Figure 7.24 Cytokinesis in animal cells occurs from the cell surface inward whereas cytokinesis in plant cells occurs from the cell center outward. (a) During cytokinesis in animal cells, the contraction of an actomyosin ring associated with the cell membrane leads to the invagination of the membrane. At the end of this process the cytoplasm of the two daughter cells is separated by a membrane. (b) During cytokinesis in plant cells, a phragmoplast forms in the middle of the cell and recruits, via microtubules (green), membrane vesicles (blue) that form a membranous structure extending toward the cell cortex. Once the membrane is fully extended, the cytoplasm of the parent cell has been divided into two daughter cells.

In animal cells, cytokinesis is mediated by a **contractile ring**, called the **actomyosin ring**, which is made of actin filaments bound to motor proteins called myosin. This ring binds to the plasma membrane at a site called the **cleavage furrow**, the site where cell division will take place. (The question of how the site of the cleavage furrow is determined is an interesting one, which we will discuss later in this section.) Ring contraction is facilitated by the motor activity of myosin, which pulls the actin filaments towards each other, thereby pulling the plasma membrane inward (Figure 7.24a). Many proteins besides actin and myosin are involved in this process, and some act by controlling the motor activity of myosin and the polymerization or depolymerization of the actin filaments. The contraction of the actomyosin ring, and the invagination of the plasma membrane with it, proceeds until a small hole is left. Cytokinesis is completed when this hole is filled. This occurs by the fusion of membrane vesicles that are targeted to the site of cytokinesis, forming an intact membrane that separates the cytoplasms of the two daughter cells.

In plants, the membrane that separates the two daughter cells is formed independently of the plasma membrane. In this case, a cytoplasmic structure called the **phragmoplast** is responsible for delivering membrane vesicles to the plane of cell division by targeted secretion. Successive rounds of vesicle fusion generate a membrane that extends from the cell center towards the cell cortex to separate the daughter cells (Figure 7.24b). In plants, cytokinesis does not involve a contractile ring.

Cytokinesis is regulated both spatially and temporally

For the successful completion of chromosome segregation, the timing of cytokinesis must be regulated so that the cleavage furrow forms only when the two sets of chromosomes are fully separated. In addition, the position of the cleavage furrow must be such that it is situated between the two daughter nuclei. In animal cells, the position of the cleavage furrow corresponds to the spindle mid-zone, and is thought to be determined by a signal from the spindle, although the nature of the signal is not known. Nonetheless, placing the cleavage furrow at the spindle mid-zone ensures that cytokinesis takes place between the two daughter nuclei.

The temporal regulation of cytokinesis is also poorly understood. However, what we do know is that cytokinesis depends on proteins associated with the mid-zone of the elongated anaphase spindle, which comprises bundled polar microtubules. This structure forms only *after* chromosome segregation is completed. Linking cytokinesis to the spindle mid-zone therefore ensures that cell division does not take place before the end of anaphase. This combination of spatial and temporal cues ensures that cell division results in the formation of two daughter cells, each containing a full set of chromosomes.

7.7 MEIOSIS: GENERATING HAPLOID GAMETES FROM DIPLOID CELLS

In the previous sections of this chapter, we focused on the strategy adopted by the cell to produce two daughter cells that are faithful copies of the parent, down to the level of possessing complete copies of the parental cell genome. However, the division of a parent cell to produce two daughter cells that contain only *half* of the genetic material of their parent, is also a central process in biology. We describe this process, meiosis, in this section.

Meiosis generates cells that are involved in sexual reproduction

During mitotic proliferation of somatic cells, a dividing cell gives rise to two progeny cells that are genetically identical to their parent (Figure 7.25a). By contrast, during sexual reproduction, a gamete from one parent, for example the female egg, fuses with a gamete from the other parent, for example the male sperm, to create an offspring whose genetic composition differs from that of its parents (Figure 7.25b). If gametes had the same DNA content as somatic cells, the total DNA content, or ploidy, of the progeny would double with each successive generation. To circumvent this problem, the ploidy of gametes is half that of somatic cells and, therefore, the fusion of two gametes results in a cell with the same number of chromosomes as the cells of the parent organism. In most animals, gametes are haploid (having one copy of each chromosome), and the fusion of male and female gametes gives rise to a diploid offspring (Figure 7.25b). The specialized cell division process that creates gametes is called meiosis. It is also referred to as a reductional division because it leads to a reduction in ploidy from parent cell to gamete.

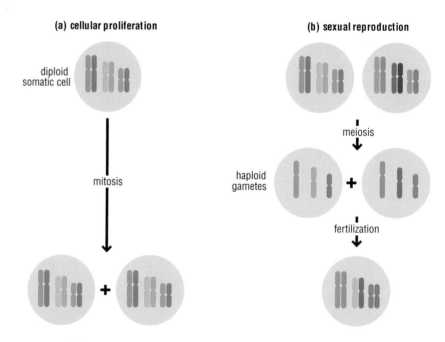

Figure 7.25 Somatic cell proliferation versus sexual proliferation. (a) Somatic cells proliferate by undergoing mitosis, during which cell division results in two daughter cells that are identical to their parent. (b) During sexual reproduction, two parents (whose genotypes are represented as the two diploid cells on the top), form haploid gametes by a meiotic division (in humans these would be the egg and sperm). Gametes are fused to give rise to a progeny cell that differs in chromosome composition from both its parents. The haploid gametes shown for each parent are just one of many possible combinations. Not shown are the exchanges between homologous chromosomes that occur during meiosis. Those will be explained later in this section. Note that some organisms have a higher ploidy (for example, tetraploid organisms have four homologs of each chromosome). Nonetheless, while in mitosis the number of chromosomes is the same in the parent and daughter cells, during sexual reproduction the number of chromosomes in the gametes is always half that of the parent.

Meiosis involves two successive rounds of chromosome segregation

As in mitosis, meiosis begins with a single round of DNA replication, so that each chromosome is made of two identical sister chromatids that are associated with each other by cohesin. Unlike mitosis, however, DNA replication in meiosis is followed by *two* successive rounds of chromosome segregation (Figure 7.26): in the first round, called meiosis I, **homologous chromosomes** first pair, exchange pieces of DNA, and then segregate away from each other. This is unlike mitosis, where the homologous chromosomes do not interact. In the second round of chromosome segregation, called meiosis II, sister chromatids segregate away from each other, much like in mitosis.

The meiotic segregation process has several requirements: homologous chromosomes need to associate with each other before chromosome segregation begins, sister chromatids have to remain attached during the first meiotic division

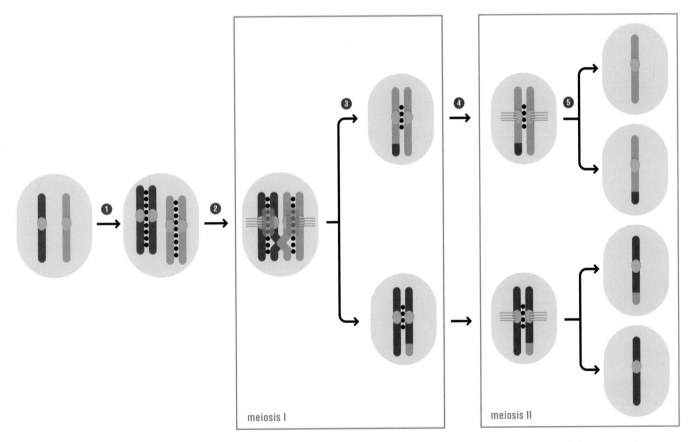

Figure 7.26 Chromosome segregation during meiosis. Animal cells that undergo meiosis are typically diploid, containing one set of chromosomes from the mother and one from the father (for simplicity, only one homologous pair is shown). During DNA replication (1) sister chromatids become attached to each other by cohesin complexes (shown as black circles). Prior to meiosis I, homologous chromosomes align by the action of homologous recombination, creating recombination crossovers that result in exchange of DNA between homologous sister chromatids (2). The homologous chromosomes, which are attached to each through the crossover, bi-orient on the meiosis I spindle (green). Note that at this stage kinetochores of sister chromatids co-orient (namely, attached to microtubules from the same spindle pole). During meiosis I, cohesin is removed from chromosome arms but not around centromeres due to the presence of shugoshin (light blue). This allows homologous chromosomes to separate while sister chromatids remain associated (3). During meiosis II, kinetochores of sister chromatids bi-orient on the meiosis II spindle (4). This is followed by chromosome segregation, resulting in four gametes, each containing a single copy of each chromosome (5).

(a)

homologous
chromosomes

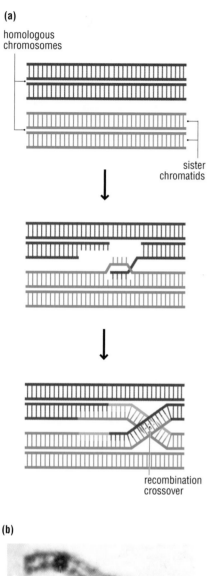

sister
chromatids

recombination
crossover

(b)

homologous
chromosomes

chiasmata

centromeres

meiosis I

but separate during the second meiotic division, and the kinetochores of sister chromatids have to co-orient during meiosis I but bi-orient during meiosis II. The details of these processes are described below.

Homologous chromosomes become physically attached before meiosis I through homologous recombination

A prerequisite for chromosome segregation in both meiosis and mitosis is the physical association between the chromosomes that need to be separated from each other. (Imagine sorting a basket full of unpaired socks of different colors into two identical piles by first matching each sock with its partner.) Earlier in this chapter we learned that in mitosis, sister chromatids associate with each through cohesin complexes before being separated by spindle microtubules. In meiosis I, homologous chromosomes need to associate with each other before segregation. However, unlike sister chromatids, which are bound to each other by cohesin complexes as they are formed, homologous chromosomes must find one another in a nucleus full of other, non-homologous chromosomes.

The feature that distinguishes two homologous chromosomes from all other chromosomes is their high degree of sequence similarity, and this is the basis for their selective pairing. This sequence similarity allows the formation of complementary base-pairing between DNA strands of homologous chromosomes. It is not entirely clear how DNA strands from homologous chromosomes initially interact: one idea is that DNA breaks are necessary for strands from one chromosome to probe other chromosomes for complementary sequences. It is also possible that the initial search for complementary sequences may take place without the formation of DNA breaks, and that an intact chromosomal region interacts with its homologous sequence on the homologous chromosome by a yet unknown mechanism. Either way, homologous chromosomes rely on the presence of highly similar (though not identical) sequences in order to pair.

In meiosis, homologous recombination is the basis for the physical association between homologous chromosomes. In order to undergo recombination, a specialized nuclease creates multiple double-strand breaks along the chromosomes, allowing a strand of DNA from a chromatid of one chromosome to stably interact with complementary sequences in a chromatid of the homologous chromosome (Figure 7.27a). Many such interactions take place along any

Figure 7.27 Homologous chromosomes interact through homologous recombination. (a) The diagram shows two homologous chromosomes (gray and turquoise) each composed of two sister chromatids. A DNA double-strand break in one of the chromatids of the gray chromosome allows a DNA segment near the break to anneal to its homologous sequence on a chromatid of the homologous (turquoise) chromosome. As cells progress through meiosis, some of these recombination intermediates are converted to covalent attachments between chromatids of homologous chromosomes, such that a turquoise chromatid is attached to the end of the gray chromatid and vice versa. This type of exchange is called recombination crossover. 1PV4. (b) A recombination crossover as visualized by light microscopy. Shown is a homologous pair of grasshopper chromosomes in prophase of meiosis I. The X-shaped structure that is the crossover is called a chiasmata. Crossovers are stabilized by the presence of cohesin (black circles) between the crossover and the chromosome end. During meiosis I cohesin complexes are removed from chromosome arms, allowing spindle microtubules (not shown) to interact with kinetochores that assemble at each chromosome's centromere and pull the homologous chromosomes apart. Cohesin remains associated with the centromere region, thereby maintaining an association between the two sister chromatids until meiosis II.
Electron micrograph in Figure 7.27b from B. John, 1990.

two homologous chromosomes, thereby aligning the two homologous chromosomes side by side. These breaks are then repaired. In the process, a strand from one chromatid can become permanently joined to a DNA strand from a chromatid of its homologous chromosome, creating a continuous **recombinant DNA** molecule (Figure 7.27a). This process, which is known as **homologous recombination**, also plays an important role in the accurate repair of broken chromosomes in somatic cells.

➔ We discuss the process of homologous recombination further in Chapter 13.

These physical connections between chromatids of homologous chromosomes are called **recombination crossovers**, and their cytological manifestations, as viewed by microscopy, are called **chiasmata** (Figure 7.27b). Since the sister chromatids are themselves held together by cohesin, recombination crossovers create a stable connection between the two homologous chromosomes. Note that despite the proximity and sequence identity between sister chromatids, meiotic homologous recombination takes place preferentially between complementary sequences of homologous chromosomes, not between sister chromatids. The mechanism that dictates this preference is not fully understood. Recombination crossovers in meiosis occur at least once per homologous chromosome pair, ensuring that all chromosomes are stably associated with their homolog. Keep these recombination crossovers in mind: they are also responsible for genetic variation, which we will discuss at the end of this section.

The synaptonemal complex provides an additional link between homologous chromosomes

Homologous recombination is essential for the initial interaction between homologous chromosomes. However, the creation of stable interactions also requires a protein structure called the **synaptonemal complex**, which forms along the entire length of the paired homologous chromosomes. A subset of the proteins that make up the synaptonemal complex form two structures, called **lateral elements**, along each of the paired sister chromatids, which organize the sister chromatids within each homolog into discrete loops. The lateral elements are connected by other proteins of the synaptonemal complex that form two rows of a coiled-coil protein known as the **transverse filament** (Figure 7.28).

There is a tight link between synaptonemal complex formation and the recombination process: mutants that fail to undergo recombination are also defective in synaptonemal complex formation, and the synaptonemal complex helps to stabilize recombination intermediates. This massive structure is dissolved prior to chromosome segregation in meiosis I, leaving behind the DNA crossovers and sister chromatid cohesion to maintain the physical attachment between homologous chromosomes. It is possible that another role for the synaptonemal complex is to direct recombination events between homologous chromosomes rather than between sister chromatids.

After the synaptonemal complex disassembles, the first of the two chromosome segregation events (namely meiosis I) is ready to begin. During sperm formation, chromosome segregation follows immediately after synaptonemal complex disassembly. Mammalian oocytes, on the other hand, remain arrested prior to meiosis I, awaiting the appropriate hormonal trigger. For women, this arrest can last over 40 years. In fact, loss of cohesion over time that may lead to subsequent missegregation (for example, having two homologous chromosomes segregate to

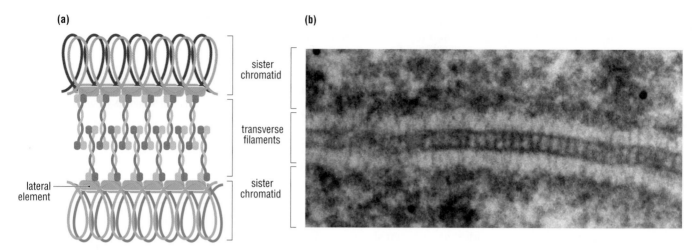

(a)

sister
chromatid

transverse
filaments

sister
chromatid

lateral
element

(b)

Figure 7.28 Components of the synaptonemal complex. (a) During early meiosis the paired sister chromatids (one shown in gray and the other in turquoise) compact to create a series of loops held together at their base by a protein structure called the lateral element (green). Lateral elements of homologous chromosomes are linked via the transverse filament. Within the transverse element the proteins (in pink) form homodimers that interact with each other via their N-terminal domain and with the lateral element via the C-terminal domain. (b) An electron micrograph showing the synaptonemal complex. Figure 7.28b from Schmekel and Daneholt (1998).

the same spindle pole during meiosis I) is thought to be one of the contributing factors to the increase in the incidence of trisomy in offspring as women get older, such as in Down syndrome, where there are three copies of chromosome 21.

In meiosis I, kinetochores of sister chromatids co-orient and cohesion around centromeres remains intact

We just learned that in meiosis I, sister chromatids do not separate. This means that unlike in mitosis, where kinetochores of sister chromatids bi-orient, in meiosis I kinetochores of sister chromatids must face the same spindle pole, or co-orient. The mechanism that allows co-orientation of sister kinetochores during meiosis I has been deciphered in budding yeast, where it was found that co-orientation is achieved by the binding of a protein complex, called monopolin, to sister kinetochores in early meiosis I. How monopolin affects co-orientation is not known. However, it is assumed that either it inactivates one of the sister kinetochores, or it fuses them so that they effectively work as a single kinetochore. Further, monopolin is not active during meiosis II when sister kinetochores need to bi-orient.

The existence of both crossovers and sister chromatid cohesion poses a problem when separating homologous chromosomes in meiosis I, because these two links effectively lock the two homologs together and would resist chromosome segregation if spindle microtubules were to try to pull the homologs apart. In theory, this could be resolved by dissolving all sister chromatid cohesion in meiosis I, but that would pose a problem for meiosis II, in which sister chromatid cohesion is needed to ensure sister chromatids faithfully segregate from one another. To allow homologous chromosomes to separate in meiosis I, while still holding sister chromatids together until meiosis II, cohesin complexes are removed in two steps. First, during meiosis I, cohesin complexes on chromosome arms are removed while cohesin complexes near centromeres remain in place. Then, during meiosis II, all the remaining cohesin is removed.

In both meiosis I and meiosis II, cohesin removal is mediated by separase, the protease that is also responsible for cohesin removal in mitosis as described earlier in the chapter. How is it, then, that cohesin close to the centromere is spared from the action of separase during meiosis I? At least two proteins participate in this process: the shugoshin protein, whose role in mitosis has already been described, and a meiosis-specific kleisin subunit of cohesin, called Rec8.

We learned earlier that in mitosis, shugoshin prevents separase from cleaving centromeric cohesin prior to anaphase. In meiosis I, as in mitosis, shugoshin protects centromeric cohesin from removal by separase (Figure 7.26, steps 2 and 3). However, while in mitosis shugoshin is removed at anaphase onset, in meiosis I shugoshin is not removed, thereby protecting centromere-associated cohesin from separase and allowing sister chromatid cohesion to be maintained until meiosis II. In meiosis II, as in mitosis, shugoshin is degraded prior to sister chromatid separation, allowing sister chromatids to separate. This ability of shugoshin to protect centromeric cohesion in meiosis I depends in an unknown way on Rec8, because if Rec8 is experimentally removed and its mitotic counterpart is expressed in its place, centromeric cohesion is lost during meiosis I.

The kinetochore also plays a role in determining whether cohesin will be removed or spared from degradation (Figure 7.29): in mitosis and meiosis II the tension between kinetochores of sister chromatids could displace shugoshin from cohesin, allowing separase to infiltrate and cleave cohesin's kleisin subunit. On the other hand, during meiosis I, where there is no tension between kinetochores of sister chromatids, shugoshin remains in close proximity to cohesin and protects it from cleavage by separase.

Recombination crossovers and sister chromatid cohesion are needed for homologous chromosome bi-orientation

In previous sections we saw that during mitotic chromosome segregation, bi-orientation of sister kinetochores leads to tension, and that the absence of tension is indicative of incomplete or incorrect kinetochore–microtubule attachments

meiosis I   mitosis and meiosis II

Figure 7.29 Shugoshin protects centromeric cohesion during meiosis I. To explain how shugoshin protects centromeric cohesion during meiosis I but not during meiosis II or mitosis, it was proposed that during meiosis I shugoshin co-localizes with cohesin, thereby protecting it from separase-dependent cleavage, while in meiosis II shugoshin and cohesin do not co-localize, and thus cohesin can be cleaved by separase and removed from chromosomes. Specifically, during meiosis I, when there is no tension between the co-oriented kinetochores (shown in orange) of sister chromatids, shugoshin (shown in light blue) is present near cohesin (black circles), and this presumably prevents separase from accessing cohesin. However, in mitosis and meiosis II, when sister kinetochores bi-orient, pulling by spindle microtubules (in green) in opposite directions displaces shugoshin from cohesin, thereby exposing cohesin to separase's action.

(see Sections 7.4 and 7.5). An analogous situation exists in meiosis I, where tension is created as kinetochores of homologous chromosomes bi-orient. In this case, it is the recombination crossovers that counteract the pulling forces of the spindle microtubules (Figure 7.26, meiosis I). This reveals another role of homologous recombination: not only is it involved in identifying homologous chromosomes and holding them together, but it is also necessary for the tension which signals that the homologous chromosomes have bi-oriented. Note that crossovers alone are not sufficient to counteract the pulling forces of the spindle, and sister chromatid cohesion between the crossover and the chromosome ends must be in place to prevent the exchanged chromosomes from being pulled apart. As in mitosis, if kinetochores of homologous chromosomes attach to microtubules from the same spindle pole so that no tension is created, regulatory proteins reverse these attachments. This generates unattached kinetochores that activate the spindle checkpoint pathway (see Section 7.5), which in turn inhibits further progression through meiosis.

Chromosome segregation in meiosis II is similar to chromosome segregation in mitosis

At the end of meiosis I the homologous chromosomes, each still comprising a pair of joined sister chromatids, have separated from each other. The cell now moves on to meiosis II, the ultimate goal of which is to separate the sister chromatids to produce haploid gametes.

Once chromosome segregation in meiosis I is complete, the meiosis I spindle disassembles and meiosis II can begin. Meiosis II is very much like mitosis, except that meiosis II is not preceded by DNA replication. Initiation of DNA replication, as we saw in Section 6.7, can occur only after Cdk activity has declined at the end of mitosis to levels low enough to allow the components at the replication origin to assemble. In meiosis, during meiosis I, Cdk activity declines to a level low enough to allow completion of meiosis I and dissolution of the meiosis I spindle, but it is maintained at a level too high for re-initiation of DNA replication. In this way, replication between meiosis I and meiosis II is prevented.

At the onset of meiosis II, kinetochores of sister chromatids bi-orient on the meiosis II spindle (Figure 7.26, step 4). At anaphase of meiosis II the remaining cohesin is removed by separase and sister chromatids segregate away from each other (Figure 7.26, step 5). Thus, following two rounds of segregation without an intervening replication step, each gamete has half the number of chromosomes compared with somatic cells.

Meiosis generates diversity

Mitotic divisions generate progeny that are identical to their predecessors. In meiosis, however, the resulting gametes are all different. Two aspects of meiosis contribute to the distinct genetic makeup of gametes. First, consider that prior to meiosis the diploid parental cell has two set of chromosomes, each contributed by one of the organism's two parents (for the purpose of this explanation we will designate them maternal and paternal sets). During meiosis I, when homologous chromosomes separate, some maternally contributed chromosomes will go to one pole while the rest will go to the other. The same is true for the paternally contributed chromosomes. Therefore, at the end of meiosis I, not only does each spindle pole contain a distinct combination of maternal and paternal chromosomes, but two identical cells undergoing meiosis are unlikely to give rise to genetically identical gametes (Figure 7.30).

Second, homologous recombination creates unique chromosome combinations because of the exchange between maternal and paternal chromosomes during homologous recombination. Different cells have recombination crossovers at different chromosomal loci so that the chromosomes resulting from meiosis are likely to possess allele combinations that are different from the 'parent' chromosomes. This explains why gametes of an individual are likely to have a different genetic makeup to one another. Thus, the combination of a random assortment of maternal and paternal chromosomes, combined with homologous recombination, is the genetic reason for why siblings look alike but are not identical.

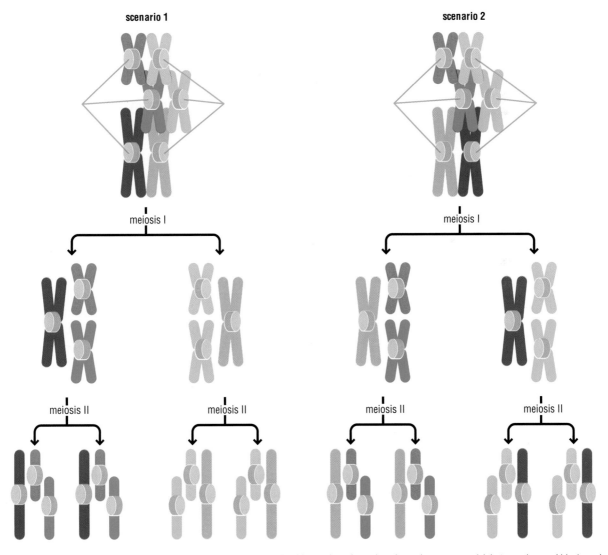

Figure 7.30 Meiosis as a way of generating diversity. The hypothetical diploid organism shown has three chromosomes (pink, turquoise, and blue), each represented by two homologous chromosomes (dark and light). As the cells of this organism undergo meiosis, different gametes are generated, depending on the orientation of the homologous chromosomes on the meiosis I spindle. In scenario I, all the dark chromosomes are facing one pole while the light chromosomes face the other. In scenario II, the dark pink, dark turquoise, and light blue chromosomes face one pole, whereas the light pink, dark turquoise, and dark blue chromosomes face the other pole. As we follow these chromosomes through meiosis we can see that the gametes generated in scenario I are different from the gametes generated in scenario II. Not shown in this illustration are the consequences of homologous recombination, which further add to genetic diversity (see main text).

7.8 CHROMOSOME SEGREGATION IN BACTERIA

The need for cellular machinery that drives chromosome segregation is not unique to eukaryotes: organisms from all kingdoms rely on regulated chromosome segregation to ensure the proliferation of the species. While the nature of the segregation process differs between prokaryotes and eukaryotes, all cells must deliver one copy of each chromosome to the future daughter cells. So far we have discussed the nature of the eukaryotic segregation machinery. Less is known about the process of chromosome segregation in bacteria. Only recently, thanks in part to technical advances in cell biology methodologies that allow observations in cells as small as bacteria, scientists have begun to understand the bacterial chromosome segregation process at a molecular level. We now know that many distinct chromosome segregation processes have evolved among bacteria. Here we will discuss the general principles of bacterial chromosome segregation and provide examples from specific organisms.

Our understanding of chromosome segregation machineries in bacteria stems from the study of a small number of strains

We should begin our consideration of bacterial chromosome segregation by stressing something that should be borne in mind when reading this section: the vast majority of bacteria have never been studied in a laboratory setting and our current knowledge is derived from only a small number of strains. Most of these have a single circular chromosome with one origin of replication and a defined replication terminus (see Sections 6.7 and 6.10), although some studied bacteria, such as *Borrelia* and *Streptomyces*, have linear chromosomes, and others, such as *Vibrio cholera*, have more than one circular chromosome. As in eukaryotes, the segregation of the bacterial chromosome appears to be driven by an active segregation machinery. This is evident from the rate of chromosome movement following chromosome duplication, which exceeds the rate of bacterial cell growth: if chromosomes were simply tethered to a site in the cells, for example the cell membrane, and segregation was driven solely by cell growth, the rate of chromosome movement would equal that of cell growth, rather than exceed it. During chromosome segregation of most bacteria, the origin of replication is at the leading edge of the moving chromosome, indicating that the site of chromosome attachment to the segregation machinery, namely the bacterial equivalent of a centromere, is located near the replication origin.

There are several features that distinguish bacterial chromosome segregation from eukaryotic segregation. Unlike in eukaryotes, where DNA replication and chromosome segregation occur at distinct cell cycle phases, in bacteria the two processes are coupled. Namely, chromosome segregation in bacteria occurs as the DNA is being replicated. Also, under favorable growth conditions, certain types of bacteria are capable of supporting mass doubling and undergoing cell division in less time than is required for chromosome replication. This is possible because bacteria can undergo **multifork replication** (see Section 5.9). Indeed, we should be careful not to confuse bacterial multifork replication, where a single origin of replication fires before replication from the previous origin is completed, with eukaryotic DNA replication, where each chromosome has multiple origins and hence multiple forks, but replication must be completed before cell division can occur. Thus, while eukaryotes have checkpoint mechanisms that inhibit cell cycle progression until DNA replication is completed, no such mechanism exists in bacteria. Finally, while in eukaryotes chromosome segregation relies on motor proteins associated with a microtubule-based spindle, no such motor proteins have been implicated in chromosome segregation in bacteria.

⊖ We learn more about chromosome duplication in Chapter 6.

Chromosome segregation involves the binding of the ParB protein to the *parS* site and the recruitment of SMC complex proteins

The first components of the bacterial segregation machinery were identified through genetic studies on plasmid partitioning, and later shown to be also involved in chromosome segregation. These genes and chromosomal loci were given the name *ParX*, where X is a different letter for each gene/protein (such as **ParA**, **ParB**, etc). (Homologous proteins in different bacteria may have different names. To simplify the discussion, here we will use the ParX nomenclature.) Many, but not all, bacterial chromosomes contain a DNA sequence, called ***parS***, which functions as a centromere (Figure 7.31). Low copy plasmids also utilize *parS* sequences to promote their segregation. Although *parS* sites from different bacteria vary in sequence, they are almost always located near the origin of replication and they are often present in multiple copies.

In the bacterium *E. coli*, *parS* recruits a protein called **ParB**, which assembles into a nucleoprotein filament along the DNA flanking *parS* (Figure 7.31; note that other bacteria have a protein similar to ParB in function, but not in sequence). ParB, in turn, recruits to the *parS* site a second protein, ParA. There are different types of **ParA** proteins in different bacteria, all of which share the ability to bind and hydrolyze ATP. In *Caulobacter crescentus*, the *parS*, ParA, and ParB equivalents are all essential for viability, and their inactivation results in chromosome segregation defects. In *Bacillus subtilis*, on the other hand, inactivation of the ParA homolog has few consequences, but inactivation of the ParB homolog results in chromosome disorganization. In the absence of ParB, regions that are normally located near the cell poles, for example, can be found anywhere in the cell, indicating loss of chromosome organization.

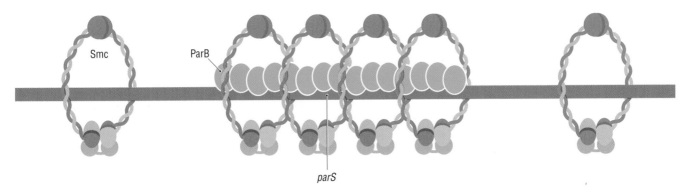

Figure 7.31 The bacterial centromere. The *parS* site (blue), present in many plasmids and bacterial chromosomes, is situated close to the origin of replication (not shown). *parS* recruits ParB (green), which assembles into a nucleoprotein filament. ParB, in turn, recruits ParA (not shown) and the SMC complex (red/pink). Although purified SMC complex forms ring structures, the structure of the SMC complex when it is bound to DNA is not known, nor is it known how the binding of the SMC complex affects DNA structure. Although the SMC complex concentrates around the *parS*/ParB site, it can also be found throughout the bacterial chromosomes.

How do ParA and ParB promote chromosome segregation? Several hypotheses have been proposed for the involvement of these proteins in plasmid segregation. Since ParA can form filaments, it may provide tracks on which plasmids can move, or ParA filaments may push chromosomes into their final destination as the filaments polymerize. The pushing mechanism is reminiscent of a plasmid partitioning mechanism, involving a protein **ParM**. ParM, which is needed for the partitioning of the low copy R1 plasmids, can form a long polymer that associates with two replicated plasmids at its ends (Figure 7.32). As the polymer grows, it pushes the plasmids away from each other towards the cell poles, thereby increasing the likelihood that when the bacterium divides the two plasmids will end up in two different cells. Whether this type of mechanism applies to chromosome segregation remains to be seen.

The ability of ParB to organize the chromosome for segregation may be attributed, at least in *B. subtilis*, to its role in recruiting to the *parS* region a complex that belongs to the structural maintenance of chromosomes (**SMC** family of proteins; Figure 7.31). Bacteria have only one type of SMC complex, which is similar in structure and function to the eukaryotic condensin (see Section 7.2). Loss of bacterial SMC function leads to chromosome decondensation and defects in chromosome organization and segregation.

The SMC complex is bound throughout the bacterial chromosome but localizes at high concentrations around *parS*. Even bacteria that lack *parS* sites, such as *E. coli*, still have SMC-like complexes that bind near the origin of replication. How the SMC complex promotes chromosome segregation is currently not known, but it has been suggested that SMC complexes at the two newly replicated origins compact the DNA and by that move the two sister chromatids away from each other. It is interesting that although the eukaryotic and prokaryotic segregation

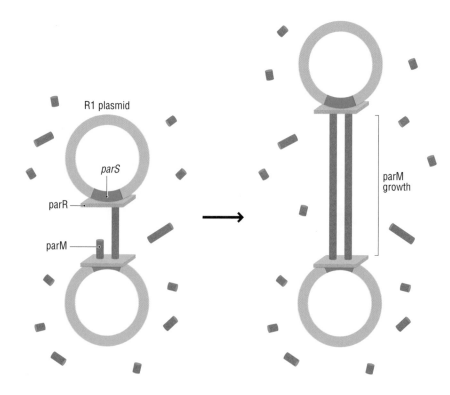

Figure 7.32 The segregation machinery of the R1 plasmid. The R1 plasmid (green) requires two proteins for its segregation: ParR (orange), which binds to the *parS* centromere site on the plasmid (blue), and ParM (red), which can form filaments and is associated with the R1 plasmid through its binding to ParR. The ParM filaments grow by incorporating free ParM near the attachment site to ParR. The elongation of the ParM filament pushes the R1 plasmids away from each other.

machineries are quite distinct, they both involve proteins of the SMC family. This also highlights the importance of chromosome compaction in the segregation process.

Different bacteria adopt different chromosome segregation mechanisms

The geometry of the segregation process differs among bacteria. For example, in *E. coli,* the replication machinery is positioned near the cell mid-zone, and the duplicated chromosomes move in a symmetrical fashion away from the mid-zone, towards the cell poles (Figure 7.33, left). In contrast, in *C. crescentus,* DNA replication occurs at one end of the cell, and one of the replicated chromosomes migrates to the opposite cell end (Figure 7.33, right). In either case the origin of replication is at the leading edge of the moving chromosome(s), and in the case of *C. crescentus* the *parS* site and its associated ParA and ParB proteins are at the leading edge as well. However, as mentioned above, the mechanism that directs chromosome movement to the poles in any bacteria remains to be discovered.

In Chapter 5 we discussed the asymmetrical cell-division process in *C. crescentus.* The *B. subtilis* system provides another example of asymmetrical cell division that poses an interesting problem during chromosome segregation. During sporulation in *B. subtilis*, cells divide asymmetrically, generating a large mother cell and a smaller forespore (Figure 7.34). However, as the replicated chromosomes move away from each other, they adopt an elongated shape, with the origins of replications at the two poles and the replication termini near the cell center. Therefore, at this point, most of the DNA destined to be in the forespore still resides within the mother cell. This DNA is actively pumped into the forespore by a DNA transporter

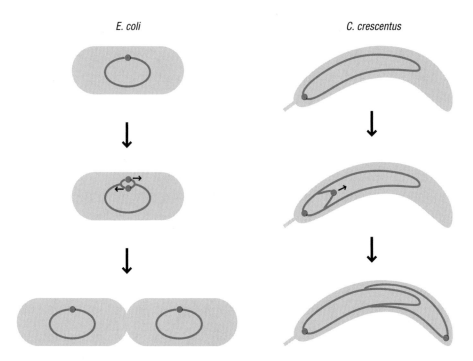

Figure 7.33 **Chromosome segregation in** **E. coli and C. crescentus.** The *E. coli* origin of replication (red) is situated near the cell center. As the chromosome duplicates, the origins move apart and end up near the centers of the two new daughter cells. In contrast, the *C. crescentus* origin of replication is located near one of the poles (the stalked pole). When DNA replication occurs, one origin stays at that pole (the old pole), while the other moves to the opposite pole (the new pole).

⊖ FtsK is discussed further in Chapter 13.

called **FtsK** (or SpoIIIE in *B. subtilis*). This is the same transporter that moves the DNA away from the plane of cell division during a regular cell cycle in *B. subtilis* and other bacteria (see below). FtsK-type transporters can move along DNA at a rate of 4–7 kb per second, but since FtsK/SpoIIIE is fixed in the *B. subtilis* forespore membrane, it is the DNA that is actively moved through the pump.

During chromosome segregation, centromeres of certain bacteria become tethered to the cell's pole

Although the mechanism that drives chromosome movement during segregation remains to be discovered, we now know that in certain bacteria, the origin of replication that is moved to the cell pole becomes anchored there by specialized machinery. As we discussed earlier, in *C. crescentus*, one copy of the chromosome moves from one cell pole to the other, led by *parS* and its associated ParA and ParB (Figures 7.33 and 7.35). When it reaches the new pole, ParB becomes associated with a protein network made of a proline-rich protein called PopZ. Prior to DNA replication, PopZ is present at the old pole, where it is needed to anchor the centromere. When DNA replication begins, PopZ appears in the new pole. In the absence of PopZ, the origin regions become detached from both cell poles, and chromosome segregation is severely disrupted. A similar type of anchoring mechanism, although with other types of proteins, also exists in *B. subtilis* during sporulation, and thus it is probably a common feature of bacteria that segregate their *parS* sites to the cell poles. Anchoring of the chromosome to the poles may be important for setting up the proper cell-division site, as we will discuss next.

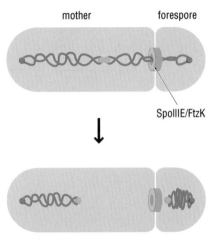

Figure 7.34 Chromosome segregation during *B. subtilis* sporulation. During sporulation, *B. subtilis* cells divide asymmetrically, forming a large mother cell (on the left) and a small forespore (on the right). The DNA segregates such that the origins (red) are each near a cell pole, and the rest of the chromosome stretches between the origins, with the chromosome termini (in green) situated near the cell's center. As a result, most of the DNA that should be in the forespore is in the mother. FtsK (orange) is a DNA translocase that pumps the chromosome into the forespore.

Bacterial cells divide by the contraction of an FtsZ ring

In both bacteria and eukaryotes (but not plants), cell division is mediated by a contractile ring. However, the nature of this ring differs greatly between the three kingdoms. In eukaryotes the ring is made of actin and myosin, whereas in bacteria and some archaea, it is made of **FtsZ**, a homolog of tubulin, and numerous associated proteins. The bacterial contractile ring, called the **Z ring**, is tethered to the inner face of the membrane at the cell mid-zone. As the ring contracts, it pulls the membrane with it, forming a septum that separates the two daughter cells.

A potential problem during cell division is that the septum, which is formed after the Z ring contracts, could bisect lagging chromosomes. To avoid this problem, in both eukaryotes and bacteria the DNA is condensed into a compact structure by a conserved complex containing SMC proteins. Eukaryotes have a further regulatory mechanism that is linked to mitotic cyclin–CDK levels and inhibits cell division until telophase, when the two DNA masses are well apart. Bacterial chromosomes, on the other hand, often remain catenated (that is, interlinked) after DNA replication (see Section 6.10) and, as a result, linked chromosomes could get stuck in the cell's mid-zone.

Figure 7.35 PopZ anchors the origin-proximal centromere to the poles in *C. crescentus*. In *C. crescentus*, as the DNA is being replicated, one origin of replication (red), with its associated *parS* site (not shown) and ParB protein (green), moves from the old pole (with the stalk) to the new pole. The origin becomes tethered to the new pole through the interaction of ParB with PopZ (orange). PopZ is also responsible for tethering of the origin to the old pole.

To avoid these linked chromosomes getting bisected by the FtsZ contractile ring, bacteria actively decatenate and remove lagging DNA from the plane of cell division by an ATP-dependent DNA translocase, called FtsK, that associates with the FtsZ ring. We encountered FtsK (or SpoEIII in *B. subtilis*) earlier in this section when we discussed the segregation of the chromosome into the *B. subtilis* forespore. As in the forespore case, since FtsK is tethered to the FtsZ ring, ATP hydrolysis does not move the protein along the DNA but rather moves the DNA through the protein, thereby pumping the ends of the chromosomes away from the contractile ring.

It is important that the FtsK translocates the DNA at the terminus of the chromosome toward the pole and not the DNA at the origin end back toward the midline, because reeling in the origin will result in one daughter cell with two chromosomes and one daughter cell with none. The directionality of FtsK action is determined by DNA sequences called the FtsK-orienting polar sequences. These sequences are dispersed throughout the chromosome and serve as DNA arrows that point from the origin of replication towards the terminus. FtsK recognizes these sequences and reels in the DNA in the appropriate direction. Thus, the action of FtsK serves as a mechanism that couples cell division to the completion of chromosome segregation, as we will discuss further in Chapter 13.

Two mechanisms control the placement of the division site: nucleoid occlusion and the MinCDE system

Once the sister chromosomes have segregated, the bacterial cell is ready to divide. A universal challenge faced by all dividing cells is how to position the division site correctly, so as to bisect the cell with the segregated chromosomes on either side. We have seen that in eukaryotes the spindle mid-zone guides the placement of the cleavage furrow (Section 7.6). Two of the best-understood mechanisms that cooperate to specify the cell-division site in bacteria are called **nucleoid occlusion** and the **MinCDE system**, and these operate in *E. coli* and other bacteria.

The bacterial chromosome is organized into a protein-containing structure called a nucleoid, which was introduced in Section 4.1. It is thought that nucleoid-associated proteins create a negative signal of unknown nature in the vicinity of the nucleoid. Because of this signal, the FtsZ ring can assemble either between the two nucleoids or between one nucleoid and the poles of the bacterium (Figure 7.36a). The MinCDE system also inhibits the assembly of the cell-division apparatus. MinCDE proteins appear to form a gradient of inhibitory signal that is highest at the cell poles and lowest at the cell mid-zone. This allows the FtsZ ring to assemble in a wide region around the cell's center (Figure 7.36b). By combining the two mechanisms, the FtsZ ring can only assemble in a narrow region in the center of the cell, between the segregated chromosomes (Figure 7.36c).

In *C. crescentus*, which lacks the MinCDE system, the assembly of the FtsZ ring is inhibited by a protein called MipZ. MipZ interacts with ParB, and therefore the accumulation of MipZ at the cell poles and away from the cell mid-zone is dependent on PopZ, the protein that forms a filamentous network at the cell pole. Prior to chromosome segregation, PopZ is absent from the new pole and MipZ can inhibit FtsZ throughout the cell. Once chromosome segregation begins and PopZ accumulates in the new pole, MipZ concentrates at the two cell poles as well, clearing the mid-zone for FtsZ assembly. Thus, the association of MipZ and ParB with PopZ couples the assembly of FtsZ to the chromosome partitioning processes, thereby ensuring that cell division will take place only when chromosome segregation is completed.

(a) nucleoid occlusion

(b) MinCDE system

(c) both mechanisms

Figure 7.36 Determining the site of cell division. Two mechanisms act in concert to define the site at which cell division will take place: (a) the nucleoid occlusion pathway and (b) the MinCDE system. In the nucleoid occlusion pathway, the nucleoid (blue oval shapes) emits a negative signal (yellow) that blocks the assembly of the FtsZ ring (green) in the vicinity of the nucleoid. When only this pathway is working, the FtsZ ring can assemble either between the nucleoids or near the cell poles. The MinCDE system generates a gradient of negative signal (shown in red), that is highest at the cell poles and lowest at the cell mid-zone, allowing FtsZ ring assembly in a wide area between the two nucleoids. (c) The combination of the two signals directs the assembly of the FtsZ ring to a narrow region between the two nucleoids.

✳ SUMMARY

In this chapter we learned about the mechanism and regulation of chromosome segregation in mitosis and in meiosis as well as in bacteria. In all cases chromosomes attach to a segregation apparatus, which separates the chromosomes to form two identical copies of the genome. The one overarching principle of chromosome segregation is that it must be accurate: cells must receive one, and only one, copy of each chromosome (remember that in meiosis, each gamete must receive one, and only one, copy of either of the two homologous chromosomes). Deviation from this rule will have dire consequences: an abnormal number of chromosomes can lead to severe cellular dysfunction or even cell death.

Mitotic chromosome segregation

- Mitosis includes five main steps: prophase, prometaphase, metaphase, anaphase, and telophase. Mitosis concludes at cytokinesis, when the parent cell divides.

- Chromosomes are segregated by the spindle apparatus, a bipolar structure made of microtubules. The attachment between the chromosomes and the microtubules is mediated by the kinetochore, which assembles at the chromosome's centromere.

- Sister chromatid cohesion plays an essential role in chromosome segregation: it allows the cell to recognize the two identical chromosomes as sister chromatids and it enables the generation of tension when proper microtubule attachment (bi-orientation) is achieved.

- Sister chromatid cohesion is dissolved at the onset of anaphase by the protease separase. Separase is negatively regulated in multiple ways to ensure that sister chromatids do not separate precociously.

- The spindle assembly checkpoint pathway inhibits progression into anaphase until all chromosomes bi-orient on the mitotic spindle.

Meiotic chromosome segregation

- Meiosis is the chromosome segregation process that generates gametes. The products of meiosis have half the ploidy of the parent cell.

- Meiosis I involves the separation of homologous chromosomes while meiosis II involves the separation of sister chromatids.

- Recombination crossovers, one of the products of homologous recombination, are needed in order to align homologous chromosomes and facilitate the generation of tension when homologous chromosomes bi-orient during meiosis I.

- In meiosis, cohesin is removed in two steps: cohesin complexes associated with the chromosome arms are removed during meiosis I while cohesin complexes associated with centromeres are removed during meiosis II.

- Kinetochores of sister chromatids co-orient during meiosis I and bi-orient during meiosis II.

- Homologous recombination and random assortment of maternal and paternal chromosomes during meiosis I create genetic variation.

Bacterial chromosome segregation

- Different bacteria utilize different segregation machineries.

- Some bacterial chromosomes are segregated by a machinery that involved a sequence that serves as a centromere (*parS*) and centromere binding proteins called ParA and ParB. ParB recruits the SMC protein complex that promotes chromosome compaction.

- Certain plasmids (e.g. R1) are segregated by ParM filaments, which bind plasmids at the filament ends and push the plasmids apart as the filament elongates.

- In most bacteria studied to date, the origins of replication are at the leading edge of the segregating chromosomes. In some bacteria, the parS locus, which is near the origin, becomes tethered to the cell pole. In *C. crescentus*, this tethering is mediated by the PopZ protein.

- The FtsK protein (SpoEIII in *B. subtilis*) acts as a DNA pump that moves the DNA away from the plane of cell division. The directionality of this movement is determined by FtsK orienting polar sequences.

- Bacterial cell division is mediated by a contractile ring made of the FtsZ protein, which is similar to the eukaryotic tubulin.

- In certain bacteria, the site of cell division is determined by two mechanisms: nucleoid occlusion, whereby proteins associated with the nucleoid inhibit the assembly of the FtsZ ring, and the MinCDE system, which forms a FtsZ inhibitory gradient that is highest at the cell poles and lowest at the cell mid-zone. The combination of these two mechanisms confines the assembly of the FtsZ ring to a narrow region at the cell's mid-zone.

 # FURTHER READING

7.2 CHROMOSOME CONDENSATION AND COHESION

Hudson DF, Marshall KM, Earnshaw WC. Condensin: architect of mitotic chromosomes. *Chromosome Research* 2009;**17**:131–144.

Onn I, Heidinger-Pauli JM, Guacci V, Unal E, Koshland DE. Sister chromatid cohesion: a simple complex with a complex reality. *Annual Review of Cell and Developmental Biology* 2008;**24**:105–129.

7.3 THE MITOTIC SPINDLE

Walczak CE, Heald R. Mechanisms of mitotic spindle assembly and function. *International Review of Cytology* 2008;**265**:111–158.

7.4 PROMETAPHASE AND METAPHASE

Cheeseman IM, Desai A. Molecular architecture of the kinetochore-microtubule interface. *Nature Reviews Molecular Cell Biology* 2008;**9**:33–46.

7.5 ANAPHASE: AN IRREVERSIBLE STEP IN CHROMOSOME SEGREGATION

Kops GJ, Weaver BA, Cleveland DW. The road to cancer: aneuploidy and the mitotic checkpoint. *Nature Reviews Cancer* 2005;**5**:773–785.

Pines J. Mitosis, a matter of getting rid of the right protein at the right time. *Trends in Cell Biology* 2006;**16**:55–63.

Pinsky BA, Biggins S. The spindle checkpoint: tension versus attachment. *Trends in Cell Biology* 2005;**15**:486–493.

7.6 THE COMPLETION OF MITOSIS AND CYTOKINESIS

Burgess DR, Chang F. Site selection for the cleavage furrow at cytokinesis. *Trends in Cell Biology* 2005;**15**:156–162.

Jurgens G. Plant cytokinesis: fission by fusion. *Trends in Cell Biology* 2005;**15**:277–283.

Kutay U, Hetzer MW. Reorganization of the nuclear envelope during open mitosis. *Current Opinion in Cell Biology* 2008;**20**:669–677.

7.7 MEIOSIS: GENERATING HAPLOID GAMETES FROM DIPLOID CELLS

Gerton JL, Hawly RS. Homologous chromosome interaction in meiosis: diversity amidst conservation. *Nature Reviews Genetics* 2005;**6**:477–487.

Jones KT. Meiosis in oocytes: predisposition to aneuploidy and its increased incidence with age. *Human Reproduction Update* 2008;**14**:143–158.

Marston AL, Amon A. Meiosis: cell cycle controls shuffle and deal. *Nature Reviews Molecular Cell Biology* 2004;**5**:983–997.

7.8 CHROMOSOME SEGREGATION IN BACTERIA

Errington J, Murray H, Wu LJ. Diversity and redundancy in bacterial chromosome segregation mechanisms. *Philosophical Transactions of the Royal Society of London Series B, Biological Sciences* 2005;**360**:497–505.

Graumann PL. Cytoskeletal elements in bacteria. *Annual Review of Microbiology* 2007;**61**:589–618.

Margolin W. FtsZ and the division of prokaryotic cells and organelles nature reviews. *Molecular Cell Biology* 2005;**6**:862–871.

Needleman DJ. Plasmid segregation: is a total understanding within reach? *Current Biology* 2008;**18**:R212–R214.

Ramamurthi KS, Losick R. Grasping at origins. *Cell* 2008;**134**:916–918.

Thanbichler M. Closing the ring: a new twist to bacterial chromosome condensation. *Cell* 2009;**137**:598–599.

8 TRANSCRIPTION

INTRODUCTION

The information stored in chromosomes represents instructions for synthesizing all of the macromolecules needed by an organism to live and reproduce. This information is encoded in the sequence of bases in the chromosomal DNA. We covered in the previous chapters how genetic information is copied and distributed to newly divided cells. In this chapter we consider how the information in a linear sequence of adenine (A), thymine (T), guanine (G), and cytosine (C) bases in DNA is used to produce a functional protein or RNA molecule in the cell.

The flow of genetic information from DNA to RNA to protein (Figure 8.1) is a universal feature of all cells. Although important differences and adaptations have evolved among different organisms, we shall see that the basic mechanisms underlying gene expression are remarkably conserved from bacteria to humans, suggesting that these essential processes evolved well before the divergence of the three kingdoms of life.

In this chapter, we will describe the individual steps that must occur to produce an RNA copy of a gene, and will explore the range of mechanisms – from the simple to the sophisticated – that organisms utilize to regulate this essential process. We shall see that, despite the many differences between bacteria, archaea, and eukaryotes, there are common principles underlying RNA transcription and the way in which this process is regulated in all cells.

We begin this chapter by identifying the key features of transcription in general terms, and describe the important molecular components that are involved in this process.

8.1 OVERVIEW OF TRANSCRIPTION

Transcription produces a copy of the coding strand of DNA using ribonucleotides

⊖ The process of translation is described in detail in Chapter 10.

To produce a functional RNA molecule, the cell must make an RNA copy of a DNA sequence. This process is known as **transcription**, because a copy of one strand of the DNA known as the **coding strand** (Figure 8.1) is transcribed into a single strand of ribonucleotides. To produce a protein, the sequence of ribonucleotides in the RNA copy of a gene must then be read by the ribosome, which translates the information encoded by the bases into a polypeptide chain (Figure 8.1).

The RNA is therefore named **messenger RNA** (**mRNA**) for its role in carrying a copy of the genetic information to the ribosome.

DNA is transcribed into RNA by the **RNA polymerase** enzyme, which makes a copy of the coding strand by using the complementary strand of DNA, known as the **template strand** (or **non-coding strand**). RNA polymerase must first separate the two strands of DNA to expose the template DNA strand. This makes it possible for the precursors of RNA – the **ribonucleoside** triphosphates – to form Watson–Crick base pairs with the non-coding strand, a critical step in ensuring the faithful transfer of information from DNA to RNA.

The bases in the ribonucleoside triphosphates are the same as those in DNA, with one exception: RNA contains uracil (U) in place of thymine (T). RNA also has the sugar, ribose, whereas DNA contains the sugar, deoxyribose. The RNA polymerase enzyme catalyzes formation of phosphodiester linkages as successive ribonucleotides are added to the 3′ end of the growing RNA chain. The chemical steps required to synthesize an RNA molecule are thus quite similar to those in DNA replication. The final product of transcription, however, is a single-stranded RNA molecule called the **transcript** – whose base sequence is a copy of that of the coding strand.

Transcription proceeds through three key stages

The process of transcription, outlined in Figure 8.2, can be divided into three main stages: **initiation**, **elongation**, and **termination**. Transcription begins when RNA polymerase locates and binds to a DNA sequence immediately preceding the gene. This is called the **promoter** and signals where transcription should start. The transcription start site is the first base of the coding strand that is transcribed and is denoted +1. Since transcription proceeds in the 5′ to 3′ direction, sites located 3′ to this or any given site are said to be located **downstream**, whereas those located to the 5′ side are said to be **upstream**.

In the **initiation** phase, the enzyme separates the two strands of DNA, creating a short unpaired region known as a **transcription bubble** (Figure 8.2). With the template strand available for base-pairing with incoming ribonucleoside triphosphates, RNA polymerase can now catalyze formation of the RNA transcript. RNA polymerase begins by synthesizing the first few ribonucleotides of the RNA transcript while remaining at the promoter. When RNA polymerase succeeds in synthesizing an RNA product of sufficient size, the enzyme moves past the promoter and changes conformation, becoming more stably associated with the DNA.

After this transition, the polymerase enzyme continues to move along the template strand and elongate the RNA transcript. In this **elongation** phase, the enzyme unwinds the DNA ahead of it to expose the single-stranded template DNA. Behind the enzyme, the template strand pairs again with the coding strand to re-form the DNA duplex. In this way, RNA polymerase maintains the transcription bubble as it proceeds along the chromosome. Within this open region, a short stretch of DNA is base-paired with the most recently synthesized RNA, while the rest of the growing RNA strand is extruded through RNA polymerase (Figure 8.2).

Elongation of the RNA transcript continues until RNA polymerase encounters a DNA sequence known as a **terminator** that signals where RNA synthesis should stop (the **termination** phase) (Figure 8.2). The polymerase enzyme can then dissociate from the DNA template and is free to initiate a fresh round of transcription.

→ We introduce the building blocks of RNA in Chapter 2.

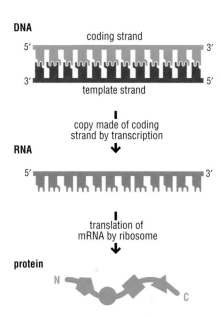

DNA
coding strand
5′   3′
3′   5′
template strand

copy made of coding strand by transcription

RNA
5′   3′

translation of mRNA by ribosome

protein
N   C

Figure 8.1 Information flow in the cell: DNA → RNA → protein. Genetic information in all cellular organisms is stored in the base sequence of double-stranded DNA. The cell harnesses this information by making an RNA copy of the coding strand of the DNA, using the complementary non-coding strand as a template. For protein-coding genes, this RNA is called messenger RNA (mRNA). In the process of translation, the ribosome synthesizes a protein whose amino acid sequence is specified by the sequence of bases in the mRNA, with each three consecutive bases specifying a single amino acid.

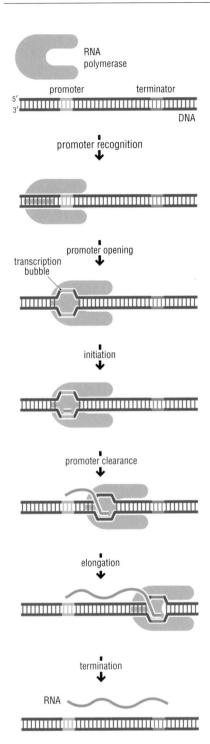

Figure 8.2 Transcription consists of a number of distinct stages. RNA polymerase recognizes and binds to the promoter (green) The enzyme separates the two DNA strands and initiates polymerization of the first ribonucleotides of the RNA transcript. Once it has moved past the promoter, the conformation of RNA polymerase changes. The enzyme then elongates the transcript until it encounters a terminator sequence (red) where it is released from the DNA template along with the RNA transcript.

Transcription is regulated in all cells

Although the genome encodes every protein and RNA molecule needed by a cell, not all of these molecules are needed at any given moment. Instead, the needs of a cell vary significantly with time, developmental stage, and changes in the environment. Sometimes the cell requires many copies of a given protein or RNA molecule, and at other times none may be needed, or just a few copies would suffice. The amount of each protein and RNA molecule in the cell can be adjusted by controlling the number of RNA transcripts made of each gene. It is for this reason that an estimated 6–10% of human genes are devoted to regulating transcription. We explore the regulation of transcription in Sections 8.7–8.14.

The packaging of eukaryotic DNA into chromatin poses a barrier to transcription

The packaging of eukaryotic DNA into chromatin affects all aspects of transcription from initiation through elongation. We saw in Chapter 4 that DNA in eukaryotic cells is packaged by histones and other proteins, with the nucleosome as the fundamental building block of chromatin. Recall that each nucleosome consists of 147 bp of DNA wrapped around a histone octamer (two each of histone H2A, H2B, H3, and H4). This wrapping can present a barrier to transcription because it can prevent the appropriate molecular components from binding to DNA.

Eukaryotic cells overcome the chromatin barrier by using three general classes of enzyme that must also be recruited to the transcribed gene in addition to RNA polymerase. Nucleosome-remodeling enzymes, which reposition histones and histone octamers from one location on the DNA to another, help remove the barrier to transcription imposed by the structure of chromatin and allow binding of the necessary proteins to the promoter. (These enzymes can also be marshaled to block transcription by relocating nucleosomes to positions that interfere with transcription.) Histone chaperones, which disassemble and reassemble the histone octamer, also help to remove the chromatin barrier from promoters and from the path of the elongating RNA polymerase enzyme. They are also needed to insert variant histones at particular locations in a gene. Finally, the enzyme-catalyzed reversible modification of histone proteins has a central role in modulating chromatin structure and in governing the recruitment of proteins that bind to particular histone modifications. Acetylation, methylation, ubiquitination, sumoylation, phosphorylation, and adenosine diphosphate (ADP) ribosylation are all post-translational modifications found in histone proteins. These modifications are important in all processes that involve DNA and chromatin, including replication and DNA repair, as well as transcription. The structure of the nucleosome, ATP-dependent nucleosome remodeling, and post-translational modification of histones are discussed in Chapter 4.

We will see throughout this chapter how enzymes that target chromatin play a central role in both facilitating and regulating transcription in eukaryotes.

We will examine each stage of transcription in more detail during the course of this chapter. Before doing so, however, we need to consider a key molecular component without which transcription would be unable to occur. This key component is the macromolecular enzyme that produces an RNA transcript from a DNA template: RNA polymerase.

TRANSCRIPTION: CORE COMPONENTS

8.2 RNA POLYMERASE CORE ENZYME

RNA polymerase is a multisubunit enzyme with a conserved core

The RNA polymerases that transcribe cellular DNA are large, multisubunit enzymes. Bacteria and archaea each possess a single type of RNA polymerase, while eukaryotes typically have three RNA polymerase enzymes that are each responsible for transcribing a different class of genes. RNA polymerase I directs transcription of large ribosomal RNA (rRNA) genes, RNA polymerase II directs transcription of all genes encoding mRNAs, and RNA polymerase III directs transcription of a variety of RNA molecules including transfer RNA (tRNA) and 5S rRNA. Plants are distinct in having a fourth RNA polymerase that transcribes regulatory RNA molecules. Although RNA polymerase II has been studied most extensively because of its role in transcribing protein-coding genes and will be covered in greatest detail in this chapter, RNA polymerase I and III together are responsible for more than 80% of the total RNA synthesized in growing eukaryotic cells.

At the heart of all polymerase enzymes is a set of subunits called the **core enzyme**, which catalyzes the synthesis of RNA using a DNA template. The core enzyme cannot operate in isolation, however, it depends on additional proteins that target it to promoter sequences. The actual number of subunits in the core enzyme and the sequences of the individual protein subunits vary among the three domains of life: bacteria have the smallest enzymes, whereas eukaryotes and archaea have larger, more complex RNA polymerases. Despite these variations, the RNA polymerase core enzymes from bacteria, archaea, and eukaryotes share the same basic architecture, reflecting a common mechanism for synthesizing RNA using a DNA template. This conservation of the enzyme core indicates that all cellular RNA polymerases evolved from a common ancestor.

The bacterial core enzyme illustrates key features of RNA polymerases

Bacterial RNA polymerase is the smallest and simplest of the cellular polymerases. The bacterial core enzyme is a complex of roughly 400 kD, comprising five polypeptides: two copies of the α subunit and one each of β, β', and ω (Figure 8.3a). In some bacteria, the large β and β' subunits are produced as a single polypeptide. The α subunit is composed of an N-terminal domain (known as the αNTD) and a C-terminal domain (known as the αCTD) joined by a flexible linker. The five core subunits assemble to form a large complex with two jaw-like lobes separated by a central cleft (Figure 8.3b).

The upper and lower jaw (as viewed in Figure 8.3b) are formed by the β and β' subunits. At the base, where the β and β' subunits meet, are two α subunit N-terminal domains. The ω subunit completes the outer base of the core polymerase structure. The C-terminal domain of the α subunit does not form part of the globular core enzyme, but rather extends outward, where it is available for interactions with the DNA template and with transcriptional regulators.

Figure 8.3 Common features of bacterial, archaeal and eukaryotic RNA polymerases. (a) The enzymes from all three domains of life contain a conserved core corresponding to the five subunits of bacterial RNA polymerase, colored in pruple (β′), light pruple (β), red (α), yellow (second α subunit), and blue (ω). The corresponding core subunits in archaeal and eukaryotic RNA polymerases are shown in the same colors. The eukaryotic RNA polymerase subunits are denote Rpb1, Rpb2, and so on. Archaeal and eukaryotic polymerases contain subunits in addition to the conserved core. Structures of (b) *Thermus aquaticus* (bacterial, PDB 1HQM), (c) *Sulfolobus solfataricus* (archaeal, PDB 2PMZ) and (d) yeast (eukaryotic, PDB 1Y1W) RNA polymerase with subunits color-coded as in (a). The Mg^{2+} ion at the active site is shown as a pink sphere, and bound structural Zn^{2+} ions are shown as green spheres.

The active site of the polymerase enzyme lies at the base of the long cleft separating the two jaws (Figure 8.3b). The active site is also accessible to the outside of the jaw through an opening in the β′ subunit. These two points of access to the active site are key structural elements that position the components of the chemical reaction – the ribonucleotides that are joined together during the synthesis of an RNA molecule – in an optimum way for the reaction to occur.

Eukaryotic and archaeal RNA polymerases share a common central architecture with the bacterial enzyme

Eukaryotic and archaeal RNA polymerases (Figure 8.3) contain a central core that is remarkably similar to that of bacterial RNA polymerase (Figure 8.3b), even though the eukaryotic and archaeal enzymes are larger and contain more subunits (Figure 8.4). RNA polymerase enzymes from all three domains of life contain five central subunits that are similar in amino acid sequence (Figure 8.4) and in three-dimensional structure (Figure 8.3b). The similarities are highest in the region containing the active site, where the amino acid sequence is nearly identical in bacterial, eukaryotic, and archaeal enzymes. This invariance of key catalytic residues means that all cellular RNA polymerases use an identical mechanism for synthesizing RNA.

RNA poymerase subunits					
RNA polymerase	**Pol I**	**Pol II**	**Pol III**	**bacterial**	**archaeal**
ten-subunit core					
	A190	Rpb1	C160	β′	A′/A″
	A135	Rpb2	C128	β	B
	AC40	Rpb3	AC40	α	D
	AC19	Rpb11	AC19	α	L
	AC12.2	Rpb9	C11		X
	Rpb5 (ABC27)	Rpb5	Rpb5		H
	Rpb6 (ABC23)	Rpb6	Rpb6	ω	K
	Rpb8 (ABC14.5)	Rpb8	Rpb8		
	Rpb10 (ABC10α)	Rpb10	Rpb10		N
	Rpb12 (ABC10b)	Rpb12	Rpb12		P
Rpb4/7 subcomplex					
	A14	Rpb4	C17		F
	A43	Rpb7	C25		E
Pol III-specific subcomplex					
	-	-	C82		
	-	-	C34		
	-	-	C31		

Figure 8.4 RNA polymerase subunits. The table shows the components of eukaryotic RNA polymerase II (PolII) that can collectively transcribe DNA, and their similarities to the eukaryotic polymerase I and III enzymes as well as the archaeal and bacterial RNA polymerases.

The correspondence between the core subunits of the bacterial enzyme and their counterparts in the eukaryotic and archaeal enzymes can be seen in Figures 8.3 and 8.4. The eukaryotic RNA polymerase II subunits, Rpb1 and Rpb2, correspond to the large bacterial β′ and β subunits, Rpb6 corresponds to ω, and Rpb3 and Rpb11 correspond to the two α subunits.

The additional subunits found in eukaryotic and archaeal RNA polymerases are arranged around the five central subunits (Figure 8.3a). Eukaryotic RNA polymerase II, for example, contains seven additional subunits, giving a total of 12 subunits for the enzyme. Most of the additional subunits are conserved in eukaryotes and archaea (Figure 8.4).

Eukaryotic RNA polymerases share certain common subunits

The three types of eukaryotic RNA polymerase share a structurally conserved core of ten subunits (Figure 8.4). Each of the polymerases then contains additional subunits that are particular to RNA polymerase I, II, or III. RNA polymerase I has a total of 14 subunits and RNA polymerase III has 17 subunits, in comparison with the 12 subunits of RNA polymerase II. The subunits that are unique to a particular eukaryotic polymerase are presumably responsible for functions specific to each type of polymerase, including the regulation of transcription.

The RNA polymerase II enzyme, alone of the three eukaryotic polymerases, carries out an additional function: it couples transcription to processing of the RNA transcript to produce the mature mRNA. This coupling depends on an important feature unique to RNA polymerase II called the C-terminal domain (**CTD**). This additional domain comprises the C-terminus of the Rpb1 subunit, which forms part of the conserved enzymatic core (Figure 8.4). The CTD consists of many repeats of seven amino acids (a heptad) with the consensus sequence Tyr-Ser-Pro-Thr-Ser-Pro-Ser; this heptad is repeated 26 or 27 times in the yeast enzyme and 52 times in the human enzyme. The CTD has a central role in the transition from initiation to elongation and in processing of the RNA transcript, as we shall see in later sections.

TRANSCRIPTION: MECHANISM

Having examined transcription in general terms, and learned the basic features of the enzyme that is central to this process, we can now begin to explore each of the three stages of transcription in more detail.

8.3 PROMOTER RECOGNITION IN BACTERIA AND EUKARYOTES

The RNA polymerase core enzyme can synthesize an RNA copy of any DNA template, but is unable to locate the transcription start site on its own. Additional proteins are needed to guide the core enzyme complex to the promoter, which is the DNA sequence adjacent to the transcription start site where RNA polymerase first binds. Although the RNA polymerase core enzyme that we learned about above has been well conserved through evolution across all three domains of life, bacteria and eukaryotes have evolved distinct solutions to the problem of locating the promoter sequence. We begin by describing how the bacterial enzyme is brought to the transcription start site.

Sigma factors direct bacterial RNA polymerase to promoters

Bacterial RNA polymerase core enzymes can transcribe RNA on their own, but they do not recognize promoter sequences. Instead, the bacterial core enzyme requires a specialized subunit called a **sigma factor** (σ), which makes direct contacts with the promoter sequence (Figure 8.5) and thereby ensures that transcription begins at the proper position in the DNA. The complex formed by the sigma subunit and the polymerase core enzyme constitutes the bacterial **holoenzyme**. Bacteria contain a variety of sigma factors that specifically recognize different promoter sequences. It is therefore the particular sigma factor utilized by a given RNA polymerase that determines which genes will be transcribed by that polymerase holoenzyme.

All cells have a **primary sigma factor**, which directs transcription from the promoters of genes encoding essential proteins that are needed by growing cells. These genes are therefore often referred to as housekeeping genes. In addition, bacteria have a variety of **alternative sigma factors** whose levels or activities are increased in response to specific signals or stress conditions. The alternative sigma factors direct transcription of genes that are only required under certain conditions, or whose expression must be increased or decreased in response to changing conditions. For example, *Escherichia coli* responds to heat shock (a sudden increase in temperature) by synthesizing an alternative sigma factor (called σ^H or σ^{32}) that it needs to transcribe genes encoding proteins that help the cells respond to heat shock.

Figure 8.5 Promoter recognition by the bacterial RNA polymerase holoenzyme. The core RNA polymerase associates with a sigma factor that binds to a specific promoter sequence (green) in the DNA. The αC-terminal domains contact the UP element (orange), which is located upstream of the −35 region of some promoters.

Bacterial species differ significantly in the number of sigma factors they encode, reflecting the different environmental conditions to which they must respond. *E. coli*, a bacterium that leads a relatively sheltered life in the gut of other organisms (including humans), has only seven sigma factors. In contrast, the genome of the soil bacterium, *Bacillus subtilis*, encodes 17 sigma factors, while *Streptomyces coelicolor*, which is also free living in the soil, has 65. The large number of sigma factors enable these organisms to initiate transcription from different promoters at different times (depending on the specific sigma factors produced at a given time), hence changing the molecular 'machinery' that is expressed – and is in operation – within the organism. This enables these microorganisms to respond to changing conditions in their environment and undergo major morphological changes when threatened with starvation.

Sigma factors bind to specific sequences that define bacterial promoters

The sigma subunit guides the bacterial RNA polymerase holoenzyme to the promoter by binding to specific sequences upstream of the start site of transcription, whose position along the DNA sequence is denoted as +1. Bacterial promoters generally contain two key elements that are recognized by a sigma factor (Figure 8.6a): a conserved sequence most commonly centered at 35 nucleotides upstream of the start of transcription and therefore termed the **–35** (minus 35) element, and a second conserved sequence centered at 10 nucleotides upstream of the start site known as the **–10** (minus 10) element. Each of these promoter elements is contacted by the sigma factor protein, as we shall see shortly.

Each sigma factor has a preferred promoter sequence to which it binds. A given sigma factor will favor binding to promoters with particular DNA base sequences in the –35 and –10 regions. In addition, there is a preferred spacing between the –35 and –10 promoter elements that favors avid binding. Some examples of different sigma factors and the optimal promoter elements that they recognize are shown in Figure 8.6b. Variations in both promoter element spacing and sequence allow for differential regulation of genes, since these DNA sequence variations will have distinct effects on the relative binding of different classes of sigma factors. The more closely the promoter sequence matches the consensus for a given sigma factor, the more tightly sigma will bind. Tighter binding to a particular promoter will lead to higher relative rates of transcription, which in turn will lead

Figure 8.6 Structure of bacterial consensus promoters and regions of sigma subunits that contact DNA. (a) The four major regions of conservation are indicated on the diagram of a sigma factor. Regions of contact between the sigma factor and the promoter are indicated by arrows. (b) The consensus sequences for promoters recognized by the major *E. coli* and *B. subtilis* sigma factors and the *E. coli* σ^H, σ^N, and σ^F alternative sigma factors. N, any nucleotide; W, A or T. Sigma factors are referred to either by a letter designation or by a number indicating the molecular weight in kilodaltons.

to higher levels of protein expressed, whereas weak binding to another promoter will result in lower rates of transcript production. Promoter sequence variations also influence the rates of promoter opening and transcription initiation, as we explore further below.

Some bacterial promoters contain sequences in addition to the –35 and –10 regions that can further modulate RNA polymerase binding and initiation of transcription. At some promoters that lack an optimal –35 sequence, transcription initiation depends on the recognition of an **extended –10 region** consisting of an additional two bases located on the upstream side of the –10 region (Figure 8.6a). These bases provide additional sites for favorable contacts with the sigma subunit, partly making up for the absence of an optimal –35 region. Furthermore, some bacterial genes that are transcribed at a high rate contain an additional AT-rich sequence called the **UP element** to the upstream side of the –35 region (Figure 8.5). The UP element serves as a binding site for the C-terminal domain of the α subunit of RNA polymerase, thereby strengthening the binding of RNA polymerase to the DNA. This tight binding helps to promote high rates of transcription of housekeeping genes such as those encoding rRNA.

The different promoter elements are recognized by distinct domains of the sigma subunit. The sigma subunit contains a poorly structured region at the N-terminus, followed by three structural domains (domains 2–4) (Figure 8.6a). When sigma binds to the RNA polymerase core enzyme to form the RNA polymerase holoenzyme, the three structural domains of sigma are positioned to recognize specific elements of the promoter sequence (Figure 8.7). Domain 2 binds to the –10 region and is implicated in separating the DNA duplex, a process that is known as promoter melting. The immediately adjacent sigma domain 3 recognizes the two bases comprising the extended –10 region. Domain 4 recognizes the –35 region and is bound to a flexible flap of the polymerase β subunit that may allow movement of domain 4 to accommodate variably spaced –35 and –10 elements. An interesting feature that can be found between domains 3 and 4 is a loop called region 3.2 (Figure 8.7), which extends into the active site. Residues in this loop help to position incoming nucleotides in the active site.

An N-terminal region of sigma called region 1.1, which is found in a subset of sigma factors (for example, σ^{70}), prevents DNA binding by a free sigma subunit (that is, sigma that is not part of the holoenzyme); it also accelerates the rate of DNA strand separation at certain promoters. Region 1.1 is unstructured when it is part of the holoenzyme and thus is not depicted in Figure 8.7.

Sigma factor activity can be regulated in many ways

While the amounts of the core subunits of RNA polymerase remain relatively constant under different growth conditions, the levels and activities of many sigma factors can vary significantly in response to environmental or developmental signals. This enables the cell to alter its pattern of gene expression to meet changing needs. The relative amount of sigma factor in the cell can be regulated by modulating the transcription or translation of the sigma factor, or by modulating the stability of the sigma factor protein or of its mRNA transcript.

Figure 8.7 Structure of a bacterial RNA polymerase holoenzyme bound to a promoter. The structure of the *T. aquaticus* holoenzyme bound to promoter DNA shows how the three structural domains of the sigma subunit bind to the core enzyme in a position to recognize the promoter elements. The core component of the holoenzyme is shown as a molecular surface, with the αI, αII, and β' subunits labelled accordingly. The sigma factor is shown as a Cα cartoon depiction. The DNA is numbered relative to the transcription start site at +1. The $\sigma2$ domain (orange) recognizes the −10 region (green), while the $\sigma3$ domain (blue) binds to the flanking base pairs of the extended −10 region. The $\sigma4$ domain (orange) binds to the −35 element (green) (hybrid of PDB 1L9Z and 1IW7).

The incorporation of a sigma factor into a holoenzyme can also be regulated. Some sigma factors are synthesized as pro-sigma factors, meaning that they contain inhibitory domains that must be cleaved off before the sigma factor can associate with the core RNA polymerase. There are also regulatory proteins called **anti-sigma factors** that bind to certain sigma factors and inhibit their function, for example by preventing the interaction of the sigma factor with the RNA polymerase core enzyme. The activities of anti-sigma factors, in turn, can be regulated by controlling their expression in the cell or by sequestering them through interactions with other cellular proteins (called anti-anti-sigma factors).

In the example of sigma factor regulation illustrated in Figure 8.8, the anti-sigma factor FlgM forms a complex with the sigma factor, σ^F (also called σ^{28}), which directs transcription of genes required for completing the assembly of flagella in the bacterium *Salmonella typhimurium*. (Flagella are tail-like structures that allow the cell to propel itself.) While the proteins that form the base of the flagella are being synthesized, FlgM is bound to σ^F, preventing incorporation of the sigma subunit into the holoenzyme and thereby preventing expression of the genes whose transcription depends upon σ^F. During the late stages of flagellar synthesis, however, FlgM is exported from the cell through the incomplete flagellar apparatus, releasing σ^F to **activate** transcription of the genes necessary for final assembly of the flagella.

Eukaryotic and archaeal RNA polymerases require general transcription factors for promoter recognition

Having explored the initiation of transcription in bacteria, let us now consider how the process occurs in eukaryotes and archaea. Like the bacterial polymerase described in Section 8.2, the RNA polymerase core enzymes in eukaryotes and

Figure 8.8 Regulation of *S. typhimurium* sigma factor activity during flagellum biosynthesis. σ^F is needed for the expression of genes late in the assembly of the flagellar motility motor. The genes required for the synthesis and assembly of the initial hook–basal body complex of the flagellum (gray), as well as the genes for σ^F and the anti-sigma factor, FlgM (pink triangles), that inhibits σ^F, are transcribed by an RNA polymerase containing the housekeeping sigma factor σ^70 (blue). FlgM binds σ^F and prevents it from interacting with the RNA polymerase core enzyme. Once the intermediate hook and basal body structure of the flagellum has been completed, however, FlgM, which carries signals that cause its export at this stage, releases σ^F and is secreted from the cell. The released σ^F then can interact with the RNA polymerase core enzyme and directs transcription of genes required for the completion of the flagellum.

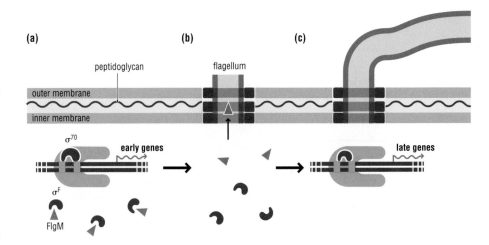

archaea also depend on additional proteins to target them to specific sites in DNA where transcription is initiated. These proteins are called **general transcription factors** and come together to form a large complex that binds RNA polymerase and helps it initiate transcription at specific promoters. But while the bacterial sigma factors target the polymerase to the promoter and mediate promoter melting, the eukaryotic and archaeal general transcription factors also catalyze additional steps in the initiation of transcription, as we discuss further below.

The general transcription factors that are required to initiate transcription by eukaryotic RNA polymerase II have been the most extensively studied. These factors, which are either single proteins or protein complexes, are named for the distinct, biochemically purified species that were required to reconstitute promoter-specific transcription initiation *in vitro*. Denoted TFIIA (Transcription Factor for RNA polymerase II, species A), TFIIB, TFIID, TFIIE, TFIIF, and TFIIH, these proteins (or protein complexes) assemble at promoters together with the 12-subunit polymerase core enzyme to form the **pre-initiation complex**. Figure 8.9 presents an overview of their functions.

The eukaryotic RNA polymerase I and III enzymes require distinct sets of proteins to form a transcription initiation complex. However, the one protein required for transcription initiation by all eukaryotic polymerases is called **TATA binding protein (TBP)**, which is also a subunit of the TFIID complex mentioned above. TBP is also conserved in archaea, where it is also required for transcription initiation by the single archaeal RNA polymerase.

Eukaryotic and archaeal promoters are often defined by a TATA box and other core promoter elements

The promoters of eukaryotic genes contain a variety of sequences called **core promoter elements** that direct the binding of the transcription initiation complex, thereby determining the start site and direction of transcription (Figure 8.10). Promoters for each of the three eukaryotic RNA polymerases typically contain a characteristic set of promoter elements, although these vary significantly in sequence and conservation among individual promoters.

General transcription factors associated with RNA polymerase II			
factor	number of subunits in human	key subunits	functions
TFIIA	3		stabilizes binding of TBP and TFIIB
TFIIB	1		promoter recognition stabilize early transcribing complex
TFIID	15	TBP, TAFs	promoter recognition, DNA bending interacts with regulatory factors
TFIIE	2		recruits TFIIH
TFIIF	2		suppresses non-specific DNA binding captures non-template strand upon melting
TFIIH	9	p62, p52 XPD, XPB MAT1, Cdk7 Cyclin H p34, p44	unwinds promoter DNA, phosphorylates CTD

Figure 8.9 Basal transcription factors required for transcription initiation by eukaryotic RNA polymerase II.

The promoters of genes transcribed by RNA polymerase II commonly contain a **TATA box**, named for its TATAAA consensus sequence, which is centered approximately 25–30 bp to the upstream side of the transcription start. Additional promoter elements, shown in Figure 8.10, that may also be present include the TFIIB recognition element (BRE), the initiator element (INR), and the downstream promoter element (DPE), the latter of which occurs downstream of the start site of transcription. RNA polymerase II promoters can contain any combination of these elements. However, many promoters have been identified that do not contain any of these elements and may possess as yet undefined characteristics that direct the correct start of transcription. Promoters from higher eukaryotes deviate most significantly from the consensus described above. Understanding how transcription is initiated from promoters that lack a TATA box and other promoter elements is an ongoing area of research.

TFIID nucleates assembly of the RNA polymerase II pre-initiation complex

The first step in the assembly of the transcription initiation complex occurs when TFIID binds to the TATA box. TFIID plays the same role as the bacterial sigma factor in locating the promoter. TFIID was first identified as a purified biochemical fraction that was needed for transcription *in vitro*, but turned out to consist of a complex of proteins. The subunit of the TFIID complex that binds to DNA and recognizes the TATA box sequence is **TBP** (TATA Binding Protein). TBP is a saddle-shaped protein that binds to the minor groove of the DNA with its concave face, inducing profound bending and partial unwinding of the DNA double helix (Figure 8.11). The remaining components of TFIID, called **TBP-associated factors** (**TAFs**), include at least a dozen proteins with various functions. Some of the TAFs mediate recognition of other core promoter elements, including the INR and DPE sequences (see Figure 8.10). Other TAFs have enzymatic activities and are required for activation of transcription by gene-specific transcriptional regulators.

Once TFIID is bound to the promoter region, assembly of the pre-initiation complex proceeds with the recruitment of TFIIB, a single polypeptide that recognizes

Figure 8.10 Structure of RNA polymerase II promoters and the regions contacted by TFIIB and TBP. Conserved DNA sequences called core promoter elements (green) are found in many eukaryotic promoter regions. The TATA box is the binding site for TBP (blue) which promotes binding of TFIIB (red) to the adjacent BRE element. These help nucleate assembly of the rest of the transcription initiation complex.

Figure 8.11 Structure of TBP binding to the minor groove of a TATA promoter element. TBP (blue) binds the TATA box and induces a profound distortion in the DNA, while the TFIIB (red) and TFIIA (green) bind to either side. Hybrid of PDB 1NVP and 1D3U.

the adjacent BRE promoter element (Figure 8.10). The asymmetrical binding of TFIIB helps to establish the correct direction of transcription, since TFIIB forms direct contacts with the core enzyme.

Some features of TFIIB are similar to those of the bacterial sigma factor. When TFIIB binds to the polymerase core enzyme, an extended N-terminal loop called the beta finger domain extends deep into the enzyme, with its tip near the active site. This loop, which is reminiscent of the bacterial sigma loop 3.2 (Figure 8.7), is thought to enhance formation of the early transcribing complex.

The additional binding of TFIIA further stabilizes the TBP–DNA interaction (Figure 8.11). Assembly of the complex proceeds with the recruitment of the core polymerase enzyme and TFIIF, and binding of TFIIE and TFIIH.

The TFIIH complex catalyzes the ATP-powered unwinding of DNA to expose the template strand. This is aided by a subunit of TFIIH called XPB, an enzyme that promotes DNA unwinding (also known as a helicase). In addition to facilitating the initiation of transcription, TFIIH also plays a role in the repair of damaged DNA; mutations in XPB, as well as in another TFIIH subunit, XPD, give rise to human diseases such as xeroderma pigmentosum and Cockayne syndrome, which are characterized by defects in DNA repair. TFIIH also contains enzymatic subunits that catalyze phosphorylation of the polymerase RPB1 CTD, which occurs later in the transcription cycle during the transition to the elongating complex.

Figure 8.12 Proteins required for regulated RNA polymerase II transcription. The RNA polymerase II holoenzyme consists of three sets of proteins: the multisubunit core polymerase enzyme (light pruple), which catalyzes synthesis of the RNA transcript, the general transcription factors (blue, green and red), which help RNA polymerase recognize specific promoters, and the Mediator complex (purple), which is required for regulated transcription at most promoters. This large complex of proteins assembles at the promoter region of a gene before transcription is initiated. Histone modifying enzymes and nucleosome remodeling enzymes that are recruited to particular promotor regions further stimulate transcription initiation and elongation.

Additional protein complexes are required for transcription by RNA polymerase II

The pre-initiation complex, consisting of the RNA polymerase II core enzyme and the general transcription factors, is sufficient to initiate transcription at specific promoters *in vitro*. Additional protein complexes, however, are required for transcription of most genes *in vivo* (Figure 8.12). These include a variety of enzymes that act on chromatin, as well as a complex known as **Mediator**.

Changes in histone modifications and chromatin structure accompany transcription initiation. These changes clear the way for the assembly of the pre-initiation complex, and facilitate movement of nucleosomes to allow RNA polymerase to transcribe through chromatin. Histone-modification enzymes, which covalently modify histones, and nucleosome-remodeling enzymes, which reposition histones and histone octamers, help remove the barrier to transcription imposed by the structure of chromatin. Many of these enzymes are recruited to promoter regions through the action of proteins that regulate transcription, as we will see below.

Another large protein complex, called **Mediator** is required to activate many genes transcribed by RNA polymerase II (Figure 8.12). The mediator complex contains more than 20 individual subunits, many of which are conserved between yeast and humans. This complex was isolated after the general transcription factors had been identified, when it was discovered that there appeared to be a missing component that was needed for transcriptional regulation. The experiments that led to the discovery of Mediator are described in Experimental approach 8.1. While the requirement for Mediator in eukaryotic transcription is clear, much remains to be learned about the function of individual subunits. Mediator, along with chromatin modifying enzymes, does not form part of the pre-initiation complex, but instead is recruited by regulatory proteins.

RNA polymerases I and III use TBP to recognize their characteristic promoters

Eukaryotic RNA polymerases I and III rely on a distinct set of proteins to initiate transcription. Both enzymes are very similar to RNA polymerase II, with several of the core enzyme subunits identical in all three eukaryotic RNA polymerases (Figure 8.4). RNA polymerase I and III, however, recognize quite distinct promoter sequences and have unique general transcription factors (Figure 8.13). The feature in common among all three RNA polymerase enzymes is their dependence upon TBP to initiate transcription. In each case, TBP is a subunit of a larger complex required for transcription initiation: SL1, TFIID, and TFIIIB, which are required by RNA polymerases I, II, and III, respectively.

RNA polymerase I, which transcribes rRNA genes, binds to a promoter containing a core promoter element and an upstream control element (UCE). TBP, which is part of a larger complex called SL1, helps RNA polymerase I to recognize the core promoter (Figure 8.13a).

Genes transcribed by RNA polymerase III, which include a variety of RNA genes such as tRNA and the 5S RNA subunit of the ribosome, have an unusual promoter structure: some of the key promoter elements are located *downstream* of the transcription start site. Figure 8.13 shows the different pre-initiation complexes formed by RNA polymerase III upstream of tRNA genes and the 5S RNA gene. Promoter recognition by RNA polymerase III is mediated by TBP – which in this case is a subunit of TFIIIB – mirroring promoter recognition by polymerases I and II.

(a) rRNA gene

(b) tRNA gene

(c) 5S RNA gene

Figure 8.13 Pre-initiation complexes formed by RNA polymerase I and III. (a) RNA polymerase I promoters contain an upstream control element (UCE) and a core region, which are contacted by the general transcription factors, UBF and SL1, in human cells. The latter contains a TBP subunit. (b) RNA polymerase III pre-initiation complex bound to the promoter of a yeast tRNA gene. (c) The promoter region of the 5S RNA gene transcribed by RNA polymerase III has a different sequence from tRNA genes and also requires an additional protein, TFIIIA. Note that, in both (b) and (c), some of the general transcription factors bind downstream of the transcription start site.

⮕ We learn about tRNAs and the ribosome in Chapter 10.

8.4 INITIATION OF TRANSCRIPTION AND TRANSITION TO AN ELONGATING COMPLEX

In the preceding section, we discussed how bacterial, archaeal, and eukaryotic RNA polymerases bind to the promoter of a gene in the correct position and orientation to start transcription. At this stage, the complex of RNA polymerase and

8.1 EXPERIMENTAL APPROACH

Identification of Mediator

The existence of Mediator was first proposed on the basis of indirect evidence suggesting that there was an unidentified component in the yeast nucleus that was needed for transcription activation. Experiments had shown that when excess amounts of the activator protein, GAL4-VP16 (a genetically engineered hybrid protein containing the GAL4 DNA-binding domain fused to the VP16 activation domain), were added to yeast nuclear extracts, transcription decreased from fully activated levels. This phenomenon was dubbed 'squelching'. The hypothesis was that the decrease was the result of the binding of GAL4-VP16 to factors in the extract that were needed for full activation of transcription. The addition of excess RNA polymerase II or general transcription factors (such as TFIID, TFIIA, etc.) did not overcome squelching, suggesting that those proteins were not the ones being sequestered by the excess GAL4-VP16.

If RNA polymerase and the general transcription factors were not the target of squelching, could there be other undiscovered proteins required for transcriptional activation? To address this question, Roger Kornberg and colleagues fractionated yeast nuclear extracts by ion exchange chromatography, and added those fractions (eluted from the column at different salt concentrations, see Section 16.6) to an *in vitro* transcription assay. They found that addition of one of the fractions stimulated transcription and overcame the squelching effect in a dose-dependent manner (see Figure 1). Without knowing what was in the fraction, they dubbed the unknown factor 'Mediator'. Subsequent studies have shown Mediator to be a complex containing around 20 proteins with a mass of about 1 MDa. The individual components of Mediator eventually were identified through a combination of biochemistry and genetics.

Mediator complexes from other organisms can now be analyzed rapidly using modern mass spectrometry methods that make it possible to identify the individual proteins found in a single enriched sample (see Section 16.7). An example can be found in a report by Joan and Ron Conaway and their co-workers, who identified the subunits in human Mediator complexes isolated from HeLa cells. Several groups had independently purified Mediator from mammalian cells, but each reported a different subset of proteins in their complexes. This suggested that not all Mediator proteins had been identified, or that there were several different Mediator complexes. The Conaways took advantage of a highly sensitive mass spectrometry method known as MudPIT (Multidimensional Protein Identification Technology) to resolve this confusion. To attack the problem, the researchers transfected HeLa cells with a DNA vector expressing one

Figure 1 **Reversal of activator squelching by yeast fraction isolated from an ion exchange column.** The fraction containing the putative Mediator (pink) stimulates transcription by overcoming the squelching effect, while a control fraction (blue) fails to do so.
Adapted from Kelleher *et al.*, *Cell* 1990;**61**:1209–1215.

of the putative Mediator subunits fused to an N-terminal FLAG epitope (a type of tag used for affinity purification; see Section 16.10). Antibodies that recognized the FLAG epitope were used to immunoprecipitate the Mediator complex, and the individual subunits were then directly identified by mass spectrometry. To ensure that the search was not biased by the particular subunit that was tagged, or by the position of the tag, the experiment was repeated with several different putative Mediator subunits. Figure 2 shows the polypeptides that had been reported previously (Figure 2a: red boxes) along with the subunits identified in the Conaway study (Figure 2b: red boxes).

By using this very sensitive detection method, a consistent set of human Mediator subunits could be identified. Importantly, all the subunits that had previously been found in one or another experiment were present in the complexes isolated in this study. Differences between the sets of proteins identified in this and previous studies may reflect differences in the sensitivity of the approach, or in the purity or homogeneity of the particular preparation characterized.

Kelleher RJ 3rd, Flanagan PM, Kornberg RD. A novel Mediator between activator proteins and the RNA polymerase II transcription apparatus. *Cell* 1990;**61**:1209–1215.

Sato S, Tomomori-Sato C, Parmely TJ, *et al.* A set of consensus mammalian Mediator subunits identified by multidimensional protein identification technology. *Molecular Cell* 2004;**14**:685–691.

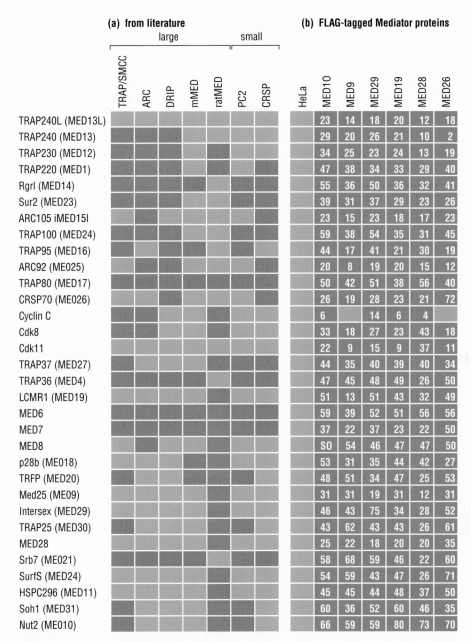

Figure 2 Identification of human Mediator subunits by MudPIT. On the left is a list of Mediator subunits found in one or more independent isolates. (a) The names of complexes purified by various groups are listed across the top; a red square indicates that the subunit is present. (b) Mediator complexes isolated by immunoprecipitation from the indicated FLAG-tagged Mediator protein were characterized by MudPIT. Protein subunits that were identified are indicated with a red square (with the number corresponding to the percent amino acid sequence coverage by the MudPIT approach). The column labeled 'HeLa' corresponds to control cells that did not express a FLAG-tagged protein.
Reproduced from Sato *et al.*, *Molecular Cell* 2004;**14**:685–691.

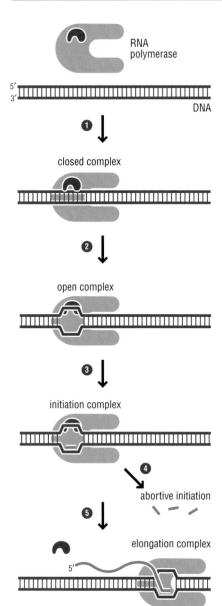

Figure 8.14 Steps in promoter recognition, promoter opening, transcription initiation, and promoter clearance by bacterial RNA polymerase. (1) The RNA polymerase holoenzyme binds to the promoter region, forming the closed complex. (2) The enzyme then separates the template and non-template strand, forming the open complex. (3) Synthesis of an RNA strand is initiated, using base-pairing interactions with the template strands. (4) RNA polymerase frequently synthesizes only a few base pairs at first and then releases the product, in a process known as abortive initiation. (5) When an RNA product of around 12–15 nucleotides is successfully synthesized, RNA polymerase undergoes a conformational change and loosens its grip on the sigma subunit to become an elongating complex that continues to synthesize RNA as it translocates along the DNA template.

the promoter is called a **closed complex**. The enzyme is now ready to begin the cascade of events required to initiate transcription.

RNA polymerase must separate the two strands of DNA

The first step in transcribing a gene is to separate the two strands of DNA in the region of the transcription start site to expose the template strand (see Figure 8.14). RNA polymerase opens up a short region of around 14 bases known as the **transcription bubble**. The unpaired template strand binds to the polymerase near the active site and is kept apart from the non-template strand by regions of the RNA polymerase called the 'rudder', 'lid', and 'zipper' (Figure 8.15). The complex of RNA polymerase that is bound to an opened region of DNA and poised to initiate transcription is called the **open complex**.

The transition to the open complex can either occur spontaneously or may require the input of energy, depending upon the class of polymerase enzyme. Eukaryotic RNA polymerase II depends on the helicase subunits of the general transcription factor, TFIIH, to unwind and separate the two DNA strands in a process that is fueled by hydrolysis of ATP. The requirement of ATP-derived energy to separate the DNA strands is unique to RNA polymerase II; it is not required by RNA polymerases I or III, or by the archaeal enzyme, for whom the change from closed to open complex occurs spontaneously. For bacterial RNA polymerase, the change from the closed to the open complex is usually also spontaneous and does not require additional energy. A special exception to this is the transcription of genes that depend on σ^{54}, which requires energy from ATP hydrolysis.

RNA polymerase initiates synthesis of the nascent RNA chain

Once the template strand is exposed and positioned near the active site of the RNA polymerase, the enzyme can begin synthesis of RNA. Free ribonucleoside

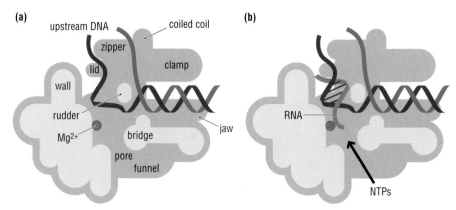

Figure 8.15 The path of the DNA and RNA in RNA polymerase. This is a cutaway view of the interior of the RNA polymerase enzyme. (a) The open complex consists of RNA polymerase bound to the promoter DNA (gray), with the transcription bubble opened and the template strand positioned near the Mg^{2+} ion (pink) in the enzyme's active site. The rudder, lid, and zipper help keep the two strands of DNA apart. The downstream DNA is held in the cleft between the two large subunits. (b) Transcription is initiated, and incoming ribonucleoside triphosphates (NTPs) enter the active site and are added to the growing RNA chain (blue). A conformational change that closes the clamp head around the upstream DNA helps prevent dissociation of the enzyme–DNA–RNA complex, thereby helping to ensure processivity.
Adapted from Cramer *et al.* (2002).

triphosphates enter through the funnel region of RNA polymerase (Figure 8.15) and form complementary base pairs with bases on the exposed template strand. The first base in mRNA is typically a purine (A or G). Once base-paired with the template strand, successive nucleotides are added to the 3′ end of the growing RNA chain, as outlined in Figure 8.16. The 3′ OH of the last base in the growing RNA chain initiates a nucleophilic attack on the innermost phosphate of the incoming ribonucleoside triphosphate. This is an energetically favorable reaction that leads to formation of a phosphodiester bond with the 5′ carbon of the newly added nucleotide and release of pyrophosphate. The 5′ end of the growing RNA chain exits through a special exit channel (Figure 8.15b) while a loop within the lid helps to separate the RNA from the DNA template.

The catalytic mechanism for ribonucleotide addition is remarkably similar to that in DNA replication. See Section 6.2 (in particular, Figure 6.6) to review the mechanism of DNA synthesis.

Just like DNA polymerase, RNA polymerase has two Mg^{2+} ions in the active site (Figure 8.17) that play key roles in activating the 3′-OH for nucleophilic attack and stabilizing a developing negative charge on the leaving oxygen. One Mg^{2+} ion is stably bound in the RNA polymerase active site (Figure 8.3), whereas the second is brought in by the incoming ribonucleoside triphosphate. A feature in the active site known as the trigger loop helps to position the incoming ribonucleoside

Figure 8.16 The biochemical reaction catalyzed by RNA polymerase. An incoming ribonucleoside triphosphate is selected for its ability to base pair with the exposed DNA strand. The innermost phosphate of this ribonucleotide (the α phosphate) is then subjected to nucleophilic attack by the 3′-OH group of the last nucleotide of the growing RNA chain. The release and further cleavage of the resulting pyrophosphate generates energy for the reaction.

(a)

RNA

template
DNA

R1020

trigger loop

R766

H1085

K752

■ template DNA
■ RNA
■ STP in A site
■ trigger loop, Rpb1
■ Rpb2

(b)

Figure 8.17 Two metal ions and the trigger loop in the RNA polymerase II active site. (a) The active site contains two bound Mg^{2+} ions (pink spheres): one that is always bound to the enzyme, and a second that is brought in by the incoming ribonucleoside triphosphate. The trigger loop (purple) positions the ribonucleoside triphosphate and helps to discriminate whether the correct base has been inserted. (b) RNA polymerase II elongation complex, with the template DNA (gray) and RNA (blue). PDB 1R9T.
With permission from R Kornberg.

> ⮕ We learn more about the nature of nucleophilic attack as a reaction mechanism in Chapter 2.

triphosphate and ensure that it is a correct match to the template strand (Figure 8.17a). Once the new phosphodiester bond is formed, pyrophosphate is released and then further broken down by hydrolysis, releasing additional energy and thereby favoring the polymerization reaction.

Surprisingly often, RNA polymerase fails to achieve productive synthesis of the full-length RNA on the first attempt and enters into a series of abortive cycles of synthesis (Figure 8.14). These cycles, referred to as **abortive initiation**, result in the production and release of short RNA products (2–9 nucleotides in length), which are subsequently degraded. The amount of abortive initiation varies from one promoter to another and can be regulated by sequence-specific DNA-binding proteins or by the availability of particular nucleotides. If insufficient cellular components are available to support transcription, the process will be aborted. See Experimental approach 8.2 to see how the products of abortive initiation were first detected *in vitro*, and how they have more recently been visualized *in vivo*.

Abortive initiation is controlled by the proteins that direct the core polymerase to the promoter. Both the bacterial sigma protein and the eukaryotic TFIIB basal transcription factor play similar roles in abortive initiation. But how do they exert their effect? We learned in Section 8.3 that loop 3.2 of sigma (Figure 8.7) and the beta finger loop of TFIIB extend into the active site of the core enzyme. Each of these loops is in a position to block the elongating transcript after a short initial transcript has been synthesized. In order for transcription to continue unimpeded, the loop must somehow move out of the way so that the transcript can continue to grow. If the loop does not move, nucleotide addition ceases and the short transcript is released. If the loop is displaced, nucleotide addition can continue. Displacement of the loop (or beta finger) is thought to help the polymerase enzyme break away from the promoter by disrupting contacts with proteins that helped to recruit the core polymerase, setting the stage for the next step in the transcription cycle: **promoter clearance.**

8.2 EXPERIMENTAL APPROACH

Abortive initiation by E. coli RNA polymerase

The phenomenon of abortive initiation was discovered by examining the transcripts produced by *E. coli* RNA polymerase at the *lacUV5* promoter. Earlier experiments had hinted that the progression from transcription initiation to elongation was not necessarily smooth. Some of those experiments, however, were conducted under conditions in which one or more nucleotide precursors was limiting. In addition, the detection methods that had previously been used – such as paper chromatography – could not resolve many different short nucleotide products. Agamemnon Carpousis and Jay Gralla (1980) decided to use the relatively new method of high-resolution gel electrophoresis to see what nucleotide products were produced *in vitro* when all four ribonucleoside triphosphates were present.

In the experiment shown in Figure 1, RNA polymerase holoenzyme was incubated with DNA containing the *lacUV5* promoter, along with radiolabeled nucleotide precursors, ATP, GTP, CTP, and UTP. The products were then separated by gel electrophoresis (see Section 16.6) and visualized by autoradiography on film.

6-MER

4-MER

3-MER

2-MER

a b c d e f g

Figure 1 Separation of abortive initiation products by gel electrophoresis. Lane a shows a control in which DNA was omitted from the reaction. The remaining lanes show the products of abortive initiation in the presence of different nucleotide concentrations.
Reproduced from Carpousis and Gralla, *Biochemistry* 1980;**19**:3245–3253.

The authors found that many short RNA transcripts, in the range of two to six nucleotides in length, were produced at each concentration of nucleotide tested (Figure 1). Near the top of the gel are the much longer transcripts that resulted from productive elongation. The relative absence of transcripts of an intermediate length supported the idea that cycles of abortive initiation were followed by highly processive transcription once RNA polymerase entered into the elongation phase.

Despite longstanding observations of abortive initiation by both bacterial and eukaryotic RNA polymerase enzymes, the phenomenon had until recently only been observed in a test tube. While it was speculated that abortive initiation was a universal feature of transcription initiation that also took place *in vivo*, there had been no direct evidence that abortive transcripts were generated *in vivo*, much less that these transcripts accumulated in the cell. Was abortive initiation an *in vitro* artifact? Could there be factors in cells that suppressed abortive initiation? Even if abortive initiation occurred in cells, would the products not be rapidly degraded?

To answer these questions, Seth Goldman, Richard Ebright, and Bryce Nickels (2009) searched for abortive initiation products in *E. coli* using methods that had been recently devised to detect small RNA molecules in the cell. Instead of studying the *lacUV5* promoter, they chose a bacteriophage promoter (the phage T5 N25anti-promoter) that was known to exhibit particularly high rates of abortive initiation *in vitro*. Since the cell could express abortive initiation products from many different promoters, the researchers used oligonucleotide hybridization (described in Section 16.9) to detect the specific products of this promoter. The probe used was an oligonucleotide complementary to the first 15 nucleotides of the putative abortive transcripts, but made with special nucleotides (to yield 'locked nucleic acid' or 'LNA') that hybridize with exceptionally high affinity and specificity.

Experiments were first performed *in vitro* to compare detection by oligonucleotide hybridization with the radiolabeling method used by Carpousis and Gralla (Figure 2a, right and left, respectively). The hybridization method with the LNA probe successfully identified abortive transcripts of 11–15 bp in length. The same experiments were then carried out on lysates prepared from cells containing a plasmid with the T5 N25anti-promoter and the gene complementary to the LNA probe; Figure 2b shows that the same set of short transcripts is detected *in vivo*.

These experiments confirmed that abortive initiation takes place *in vivo*, and that its products accumulate – at least in *E. coli*. An important question that remains is whether the abortive products play regulatory roles *in vivo*. Among the many possibilities is

Figure 2 Abortive initiation occurs *in vivo.* (a) Abortive initiation *in vitro.* Left panel: Gel electrophoresis followed by autoradiography shows transcripts generated in 'hot' (radioactive) reactions. Right panel: Detection of abortive initiation products by hybridization of locked nucleic acids (LNA) (non-radioactive) reactions. (b) Abortive initiation *in vivo.* Detection was performed by hybridization of LNAs to transcripts generated *in vivo.* Reproduced from Goldman *et al., Science* 2009;**324**:927–928.

that these small nucleic acid products could serve as sequence-specific primers for RNA synthesis or lagging-strand DNA synthesis, or perhaps could stabilize other RNA molecules through base-pairing interactions.

Carpousis AJ, Gralla JD. Cycling of ribonucleic acid polymerase to produce oligonucleotides during initiation in vitro at the lac UV5 promoter. *Biochemistry* 1980;**19**:3245–3253.

Goldman SR, Ebright RH, Nickels BE. Direct detection of abortive RNA transcripts in vivo. *Science* 2009;**324**:927–928.

RNA polymerase undergoes a conformational change as it leaves the promoter

When RNA polymerase succeeds in synthesizing an RNA about 9–11 nucleotides long, the enzyme breaks free from the promoter and becomes a productive 'elongating complex.' During this process, termed **promoter clearance** (or **promoter escape**), the polymerase enzyme undergoes a conformational change that closes the clamp region around the DNA and helps make the association of the enzyme with the DNA very stable (Figure 8.18, step 3). At the same time, RNA polymerase

❶ abortive initiation

❷ end of abortive initiation

❸ promoter clearance

❹ transcription elongation complex

Figure 8.18 Structural transitions from abortive initiation through elongation for bacterial RNA polymerase. Shown are cutaway views of the RNA polymerase holoenzyme (with σ, blue; rest of RNA polymerase gray; catalytic Mg^{2+}, pink sphere), promoter DNA (template strand, dark gray; non-template strand, light gray) and the RNA transcript (blue). (1) Abortive initiation, with sigma factor loop 3.2 blocking path of RNA. (2) End of abortive initiation, with loop 3.2 swung out of the way. (3) Promoter clearance. (4) Transcription elongation complex. Adapted from Murakami and Darst (2003).

loosens its grip on initiation factors – the bacterial sigma subunit or the eukaryotic general transcription factors – leading to their eventual release. The enzyme is now free to translocate along the DNA, allowing the growing RNA chain to exit through a special channel (see Figure 8.18, step 4).

Eukaryotic RNA polymerase II becomes phosphorylated as it converts from the initiating complex to the elongating complex. Specifically, the general transcription factor, TFIIH, phosphorylates the CTD of the large subunit, RPB1 (see Section 8.1), at the serine that is the fifth residue of the heptad repeat. This is the first in a series of CTD phosphorylation events that are needed to recruit proteins required for processing the mRNA transcript, as we shall learn next.

8.5 TRANSCRIPTION ELONGATION

Transcription proceeds in a highly processive manner

Once RNA polymerase has cleared the promoter region and made the transition to an elongating complex, transcription proceeds in a processive manner, such that one nucleotide after another is added in quick succession by a single RNA polymerase enzyme for hundreds or even thousands of bases. This processivity indicates that the elongating polymerase enzyme is very stably bound to the DNA; the stability results from the way that the two large subunits of RNA polymerase close around the DNA like a set of jaws, preventing the elongation complex from falling off the DNA. The energy released by cleavage of the incoming nucleoside

Figure 8.19 Transcription bubble. The transcription bubble consists of 14 base pairs of unpaired DNA, with the nine most recently added RNA nucleotides base-paired with the template strand.

triphosphates and subsequent pyrophosphate hydrolysis also favors continuation of the polymerization process, such that the polymerase enzyme can synthesize the growing transcript at a rate of about 20–50 nucleotides per second.

A transcription bubble containing around 12–14 unpaired bases is maintained by the enzyme as it moves along the DNA, as illustrated in Figure 8.19. RNA polymerase separates the downstream DNA into unpaired strands, while base pairs re-form behind the translocating enzyme. Within the transcription bubble are the nine most recently added ribonucleotides of the RNA transcript, which remain base-paired to the template strand (Figure 8.19). These characteristics of the transcription bubble persist as successive ribonucleotides are added to the 3′ end of the RNA transcript (see Figure 8.15). The association of the polymerase with the RNA–DNA hybrid within the transcription bubble contributes to the stability of the elongation complex.

Elongation factors suppress the pausing of transcription and rescue stalled polymerase enzymes

Despite the processivity of the reaction, chain elongation does not proceed at a uniform rate. Instead, the polymerase pauses periodically during RNA synthesis, most often due to physical obstructions that prevent the smooth passage of the polymerase along the DNA strand. This phenomenon, called **transcriptional pausing**, can be caused by physical features that arise due to the way in which the template DNA sequence and complementary RNA transcript interact.

One type of sequence-dependent pausing occurs when the RNA transcript contains short complementary stretches that form a transient hairpin, which then interacts with the polymerase and temporarily halts RNA synthesis. Transcriptional pausing can also be brought about by the presence of a weak DNA–RNA hybrid in the transcription bubble, the result of an AU-rich sequence or a misincorporated base. The polymerase can eventually resume RNA synthesis on its own, although pausing can be either relieved or enhanced by proteins called **elongation factors**. **Transcriptional arrest** occurs when the enzyme is unable to resume RNA synthesis without the assistance of elongation factors. As we shall see later in this chapter, transcriptional pausing and arrest can be regulated, providing another point at which cells can modulate gene expression.

Eukaryotic cells rely on proteins called elongation factors to promote transcription elongation, although their precise mechanisms of action remain obscure. The critical importance of these proteins is clear from the many human diseases that arise from defects in one or more elongation factors (Figure 8.20). For example,

Functions of eukaryotic elongation factors and their roles in human disease		
factor	function	disease/note
TFIIS	transcript cleavage, suppresses pausing	
ELL	disrupts interaction with TBP and TFIIB	acute myeloid leukemia
CSB	suppresses pausing	Cockayne syndrome (associated with a defect in nucletoide excision repair)
elongins ABC	suppress pausing, associate with VHL protein	von Hippel-Lindau disease (familial cancer predisposition syndrome)
p-TEFb	phosphorylates CTD of Rpb1 subunit of RNA PolII (functions in concert with HIV Tat)	HIV infection
NELF, DSIF	negative elongation factor promotes pausing	implicated in Wolf-Hirschhorn syndrome
elongator complex	aids in elongation (Elp1-6) through chromatin; contains a histone acetyltransferase subunit	

Figure 8.20 Examples of eukaryotic elongation factors, their functions and their roles in human disease.

the ELL protein is one of a class of elongation factors that suppress pausing of RNA polymerase and thereby stimulate the overall rate of elongation. Disruption of the ELL protein can give rise to acute myeloid leukemia.

In many genes transcribed by RNA polymerase II, the enzyme stalls after transcribing around 35–50 base pairs and only resumes efficient transcription with the assistance of other proteins. Termed **promoter proximal pausing**, this phenomenon was originally thought to be restricted to a handful of genes. We will learn about promoter proximal pausing in the regulation of heat shock genes in Section 8.14. More recent findings, however, indicate that many eukaryotic genes contain a paused RNA polymerase enzyme just a few dozen base pairs from the start site of transcription, awaiting the action of proteins that can restart transcription. These observations suggest that regulation of promoter proximal pausing may be more important to eukaryotic gene regulation than had been previously thought.

Pausing and arrest of transcription can be alleviated by cleavage of the RNA strand

When there is a pause in RNA synthesis, the elongation complex may **backtrack**, reversing direction on the DNA (Figure 8.21). This causes the most recently synthesized portion of the RNA to separate from the template strand and protrude through the funnel region of RNA polymerase as the transcription bubble moves backward (Figure 8.21). Transcription can be restarted with the assistance of **transcript cleavage factors**, which cause cleavage of the protruding 3′ end of the RNA transcript and allow transcription to resume. Transcript cleavage has been postulated to have a role in error correction: if mismatched bases are incorporated into the growing mRNA, they may destabilize the RNA–DNA hybrid and cause distortions in the duplex region, which causes the RNA polymerase to stall. By cleaving this mismatched region, the RNA polymerase can then correct this error by resynthesizing the portion that has been cleaved.

Transcript cleavage factors do not cleave RNA on their own, but instead promote the **intrinsic** endonuclease activity of the RNA polymerase enzyme. We saw in Chapter 3 how chemical reactions most often comprise both a forward and a

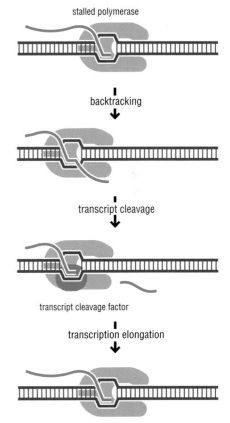

Figure 8.21 Transcript cleavage factors rescue a backtracked polymerase. A stalled RNA polymerase can backtrack, meaning that the transcription bubble moves backwards and the 3′ end of the RNA protrudes through the enzyme. By promoting cleavage of the exposed 3′ end of the RNA, transcript cleavage factors can allow RNA synthesis to resume.

Figure 8.22 Eukaryotic and bacterial transcript cleavage factors bound to RNA polymerase. (a) GreB, a bacterial transcription cleavage factor, extends into the active site of bacterial RNA polymerase and positions a second metal ion (green). (b) TFIIS, the eukaryotic transcript cleavage factor, similarly extends into the active site of RNA polymerase II, where it positions a second metal ion (green). From Cramer (2004).

reverse reaction. Just as all enzymes can catalyze both forward and reverse reactions, RNA polymerase can catalyze transcript cleavage as well as nucleotide addition. In the absence of transcript cleavage factors, the cleavage reaction proceeds at a negligible rate, but is greatly accelerated when transcript cleavage factors are present. GreA and GreB are the transcript cleavage factors in *E. coli*, while TFIIS is the transcript cleavage factor for the eukaryotic RNA polymerase II. These proteins bind in the funnel region of RNA polymerase, where the 3′ end of the mRNA protrudes. Figure 8.22 shows how GreB and TFIIS bind to RNA polymerase and extends into the funnel, where it is thought to position a metal ion in the active site. The metal activates a water molecule for hydrolysis of the phosphodiester bond, thereby promoting cleavage of the last two to ten nucleotides of the transcript. This exposes a new 3′-OH at the end of the RNA transcript that is correctly positioned in the active site for a fresh round of nucleotide addition.

Phosphorylation of the RNA polymerase II CTD couples elongation with pre-mRNA processing

The mRNA synthesized by RNA polymerase II is rarely in its mature, fully functional form. Instead, it is termed a pre-mRNA, and must be subjected to various processing events that yield the final, mature mRNA. An additional set of chemical events unique to RNA polymerase II couple transcription elongation to this processing of mRNA in eukaryotes. For example, the first stage in mRNA processing is the addition of a guanosine cap to the 5′ end of the nascent transcript. The phosphorylated CTD of the RNA polymerase II Rpb1 subunit provides a scaffold for linking transcription elongation with this, as well as subsequent, mRNA processing events.

How are transcription elongation and mRNA processing coupled in practice? We have already seen that phosphorylation at the fifth serine in the CTD heptad repeats takes place when RNA polymerase II clears the promoter region. As outlined in Figure 8.23, phosphorylation of the CTD triggers recruitment of RNA-processing enzymes that add a guanosine cap to the 5′ end of the nascent transcript. In order to allow sufficient time for recruitment of the processing enzymes and addition of the cap, transcription elongation is temporarily arrested by negative elongation factors, the proteins NELF and DSIF (Figure 8.20). Capping of the mRNA triggers recruitment of a kinase, p-TEFb (Figure 8.23), which phosphorylates the CTD at the second serine in each heptad repeat

We cover RNA processing in depth in Chapter 9.

and causes polymerase to resume elongation. The more heavily phosphorylated CTD tail can now recruit additional RNA-processing enzymes that are required for later processing of the complete transcript.

Nucleosomes must be removed from the path of the elongating polymerase for transcription to proceed

Just as nucleosomes can interfere with the assembly of the transcription pre-initiation complex on DNA, they also pose a barrier to the elongating RNA polymerase enzyme. The DNA must be unwrapped from the histone core in order to open a transcription bubble, yet this has an energetic cost that could cause the elongating polymerase to stop transcribing. Eukaryotic cells therefore have proteins that actively remove nucleosomes from the path of the transcribing polymerase and reassemble them in its wake (Figure 8.24). This is accomplished with the help of histone chaperones, which aid in the temporary disassembly and reassembly of histone octamers. Histone chaperones such as FACT (Facilitates Chromatin Transcription), Asf1, and Spt6 bind to one or more of the proteins belonging to a partially disassembled histone octamer; for example, FACT binds to H2A–H2B dimers.

There is some evidence that the eviction of histones is either preceded by or coupled to a wave of histone modifications, such as acetylation and H2B ubiquitination. Precisely how all these events are coupled during the process of transcription remains an exciting area of research.

The need to dismantle and then reassemble nucleosomes during transcription has been established primarily for RNA polymerase II. However, there is evidence that similar changes in chromatin are also required for transcription elongation by the other eukaryotic RNA polymerases, I and III.

Transcription elongation produces supercoiling that must be relieved by topoisomerases

As the transcription bubble moves through a double-stranded DNA template, RNA polymerase is continually unwinding the DNA template at the downstream end of the enzyme (Figure 8.25). At the same time, the unpaired strands at the upstream end must twist around one another to re-form the DNA double helix. As we learned in Chapter 2, changes in the winding of the two DNA strands – defined as twist – lead to changes in DNA supercoiling if the ends of the DNA are prevented from rotating or if the template is a covalently closed circle. As transcription proceeds, the DNA flanking the transcribed region is not free to rotate even in linear chromosomes due to the packaging of the DNA into chromatin. Unwinding the downstream DNA therefore reduces the twist of the DNA, which is compensated for by an increase in positive supercoiling ahead of the transcribing polymerase. Similarly, the reannealing DNA behind the polymerase enzyme increases in twist as the double helix re-forms, leading to an accumulation of negative supercoils in the upstream region.

The build-up of superhelical tension could cause the elongating complex to stall and therefore must be relieved. This is achieved through the action of topoisomerase enzymes. In *E. coli*, DNA gyrase removes the positive supercoils ahead

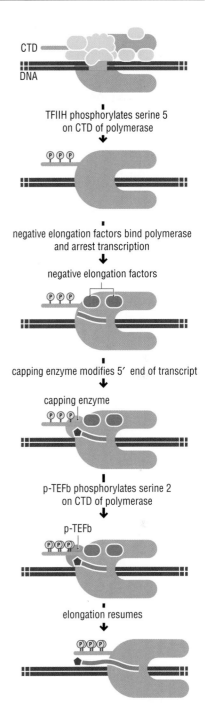

CTD

DNA

TFIIH phosphorylates serine 5 on CTD of polymerase

negative elongation factors bind polymerase and arrest transcription

negative elongation factors

capping enzyme modifies 5′ end of transcript

capping enzyme

p-TEFb phosphorylates serine 2 on CTD of polymerase

p-TEFb

elongation resumes

Figure 8.23 Phosphorylation of the RPB1 C-terminal domain (CTD) couples pre-mRNA processing with transcription elongation by RNA polymerase II. Phosphorylation of the fifth serine residue in the heptad repeats in the CTD of the RNA polymerase II large subunit leads to binding of several proteins (pink) that cause transcriptional arrest. This then allows recruitment of enzymes (green) that catalyze formation of a 5′ cap on the nascent RNA. The capping of the mRNA is followed by further phosphorylation of the CTD at serine 2 residues, which leads to resumption of transcription elongation.

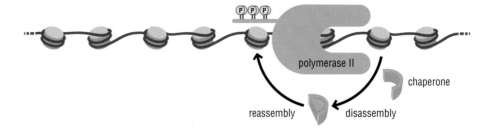

Figure 8.24 Nucleosome disassembly and reassembly during elongation. Nucleosomes are disassembled ahead of the elongating polymerase and reassembled in its wake with the help of histone chaperones. From Krebs and Tora, *Molecular Cell* 2009;**34**:397–399.

 We learn about histone chaperones and topoisomerases in Chapter 6.

of the transcribing polymerase and DNA topoisomerase I removes the negative supercoils behind the enzyme.

8.6 TRANSCRIPTION TERMINATION

Transcription is terminated by a variety of mechanisms

Once RNA polymerase has succeeded in transcribing a complete gene or set of genes, the enzyme must stop transcribing at the proper position and release both the completed transcript and the DNA. This prevents inappropriate read-through transcription of genes further downstream and frees the polymerase to initiate transcription again. As described earlier, RNA polymerase stops transcription at defined DNA sequences called terminators. In essence, the effect of terminators on transcription is opposite to that of promoters: a promoter is a defined sequence that promotes transcription, whereas a terminator is a defined sequence that stops it. Various mechanisms are used to terminate transcription, depending on the type of polymerase. In common with all other steps in transcription, termination provides another opportunity for regulating the process of gene expression.

Intrinsic terminators promote termination in the absence of additional factors

At certain terminator sites, RNA polymerase terminates transcription in the absence of any other factors. These DNA sites are called **intrinsic** or **simple terminators**. In bacteria, intrinsic terminators have two characteristic features. One is an inverted repeat sequence that specifies a secondary structural element in the RNA in which the RNA strand folds back on itself and base-pairs to form a

Figure 8.25 DNA supercoiling results from transcription elongation. RNA polymerase moving along the template. DNA gives rise to the accumulation of negative supercoiling behind the enzyme and positive supercoiling ahead of it.

hairpin-like structure called a **stem-loop**, as illustrated in Figure 8.26a. The second is a series of around eight to ten A residues (**polyadenosine** or poly(A)) immediately downstream that encodes a stretch of poly(U) in the RNA (Figure 8.26a).

Once RNA polymerase has transcribed the terminator sequence, the complementary halves of the inverted repeat in the RNA transcript base-pair with one another and form the stem–loop structure (Figure 8.26a). At the same time, the poly(U)–poly(A) RNA–DNA hybrid that results from transcribing the poly(A) region has an unusually weak base-paired structure and is hence unstable. The opening up of this region within the transcription bubble is thought to arrest transcription while the RNA hairpin part of the terminator sequence interacts with the polymerase and somehow helps to pull the RNA out of the active site, causing the elongation complex to dissociate.

In eukaryotes, specific terminator sequences are also recognized by RNA polymerase III. These terminators are defined by a stretch of U residues in the RNA transcript, which again may destabilize the RNA–DNA hybrid.

Termination of some transcripts requires additional protein factors

In many cases, the DNA sequence alone is not sufficient to terminate transcription. At these sites, denoted **enzymatic terminators**, additional protein factors are required for RNA polymerase to terminate transcription and be released from the DNA template.

Certain *E. coli* genes depend on a protein called Rho to terminate their transcription. Terminators that require Rho for transcription termination are called **Rho-dependent termination** sites, while those that do not are called **Rho-independent terminators**. The terminators of Rho-dependent genes do not contain the characteristic hairpin and run of U residues that is found in intrinsic terminators (Figure 8.26).

Rho is a hexameric ATPase that binds to C-rich sites on the nascent RNA chain termed *rut* sites (for *rho ut*ilization). The Rho hexamer can exist as an open ring (Figure 8.27), which facilitates its loading onto the RNA. Once RNA is bound inside the Rho hexamer, it is thought that the ring closes, and concerted cycles of ATP hydrolysis by Rho drive structural changes that pull the RNA through the center of the ring and away from the RNA polymerase (Figures 8.26b and 8.27b).

Transcription termination by eukaryotic RNA polymerase I also requires additional proteins. The terminators recognized by RNA polymerase I contain U-rich sequences just as intrinsic terminators do, but termination requires the action of a DNA-binding protein (Reb1p in yeast and TTF1 in mouse) that binds to sequences downstream of the **transcription unit**. While the instability of the RNA–DNA hybrid is sufficient to lead to transcript release and termination in yeast once the DNA-binding protein has triggered termination, an additional release factor is required in animals.

Termination by RNA polymerase II is coupled to RNA processing

Transcription termination by RNA polymerase II is coupled to RNA processing. Just as the early stages of transcription are coupled to processing the 5′ end of the nascent transcript, termination of transcription is coupled to processing of the

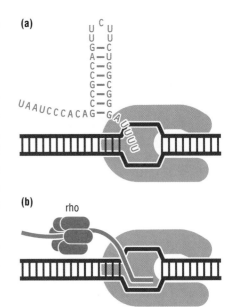

Figure 8.26 Terminators in bacteria.
(a) Intrinsic, Rho-independent terminators encode an RNA stem–loop hairpin and a stretch of eight to ten U residues. The poly(U)–poly(A) RNA–DNA hybrid that is generated on transcribing this region has an unusually weak base-paired structure and the RNA hairpin helps to pull the RNA out of the active site.
(b) At enzymatic, Rho-dependent terminators, the Rho hexamer (pink) uses the energy from ATP hydrolysis to pull the transcript away from RNA polymerase.

(a)

(b)

5′

3′

Figure 8.27 Rho termination factor. (a) The six subunits of the hexameric Rho helicase form a ring that is open at one end. PDB 1PV4. (b) Proposed binding of RNA to Rho. The red ovals are the RNA-binding domains that are not present in the crystal structure shown in (a). Upon binding of the mRNA, Rho forms a closed ring, and sequential rounds of ATP hydrolysis by Rho drive structural changes that pull the RNA along the channel of the ring and away from RNA polymerase.
Figure 8.27b is adapted from Skordalakes and Berger (2003).

3′ end. Most mRNAs transcribed by RNA polymerase II have a poly(A) tail that is added to the 3′ end of the transcript, whereas those RNAs that are not polyadenylated (for example, small non-coding RNAs or the mRNA encoding histones) are processed at their 3′ ends nonetheless. The information that signals what type of processing should occur is found in the template DNA, which is in turn transcribed into RNA.

Polyadenylation of mRNA is linked to transcription termination. RNA polymerase II continues to transcribe beyond the nucleotide sequence that signals polyadenylation. As the nascent transcript exits the polymerase enzyme, the mRNA is cleaved at the poly(A) signal and the 3′ end of the mRNA is processed, yielding a polyadenylated transcript. At the same time, however, RNA polymerase II continues to transcribe. What then signals polymerase to stop and dissociate from the template?

Two models for how transcription is terminated have been proposed (Figure 8.28). Recall that the RNA-processing proteins associate with both the processing 'signals' in the RNA and with the phosphorylated CTD of the RNA polymerase II large subunit. According to the 'allosteric' model (Figure 8.28a), recognition of the poly(A) or 3′-end processing signals, along with RNA cleavage at the processing sites, could lead to a conformational change in the polymerase enzyme that causes it to dissociate from the DNA template and release the remaining 3′ RNA fragment. ('Allostery' is a general term used to describe the way that the activity of a protein may be modulated through a change in its conformation, whereby the conformational change is stimulated by some kind of 'signal' – often the binding of a signaling molecule to the protein in question.)

A second model for how transcription is terminated, however, has emerged from the recent finding that termination is linked to degradation of the second RNA fragment by a 5′ to 3′ ribonuclease. This is known as the 'torpedo' model and is shown in Figure 8.28b. After the mRNA is cleaved at the polyA signal, a ribonuclease called Rat1 begins to degrade the nascent transcript emerging from the still-transcribing RNA polymerase II. The ribonuclease tracks along the RNA, degrading it in the 5′ to 3′ direction until it runs into the polymerase like a torpedo, disrupting polymerization and causing the enzyme to dissociate from the DNA template. It is quite possible that a combination of the torpedo and allosteric models may be responsible for transcription termination.

TRANSCRIPTION: REGULATION

We have now considered the mechanisms by which each of the three key stages in transcription occur. Central to the correct development and ongoing health of an organism, however, is the regulation of transcription: a series of concerted mechanisms that ensure genes are transcribed in the right cells, at the right time. The elegant and intricate patterns of cellular development, growth and maintenance depend on the precise orchestration of transcription. In the remaining sections of this chapter, we explore the principal ways in which transcription is regulated in bacteria, eukaryotes and archaea, identifying both the differences between the three branches of life, and the common themes and strategies that unite them.

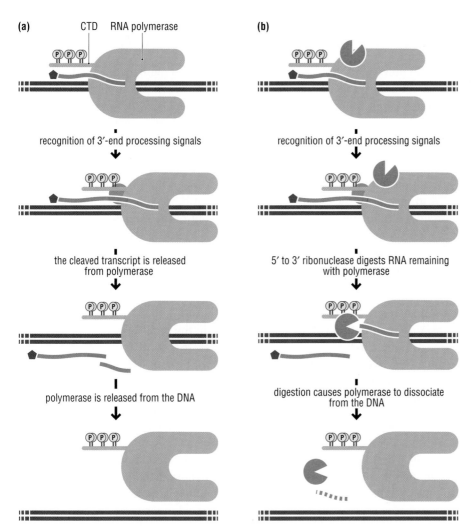

(a)

CTD   RNA polymerase

↓ recognition of 3′-end processing signals

↓ the cleaved transcript is released from polymerase

↓ polymerase is released from the DNA

(b)

↓ recognition of 3′-end processing signals

↓ 5′ to 3′ ribonuclease digests RNA remaining with polymerase

↓ digestion causes polymerase to dissociate from the DNA

Figure 8.28 Transcription termination by eukaryotic RNA polymerase II is coupled to RNA processing. Two proposed models for transcription termination are illustrated. (a) Allosteric model: RNA polymerase II transcribes through the polyadenylation and 3′-end processing signals (pink), and RNA-processing complexes (green) associate with both the processing signals and the phosphorylated C-terminal domain (CTD) of the large subunit. Recognition of the processing signals and/or cleavage at the processing sites causes termination. The cleaved RNA is released and RNA polymerase dissociates from the DNA template. (b) Torpedo model: RNA downstream of the poly(A) cleavage site (blue line) is digested by a 5′ to 3′ ribonuclease (the 'Torpedo', shown in red), which remains with RNA polymerase II throughout the length of the gene. After poly(A) site cleavage, the exonuclease torpedo causes polymerase to dissociate from the DNA template.

8.7 PRINCIPLES OF TRANSCRIPTION REGULATION

While every gene in a genome plays a role in the life cycle of the organism, the need for a particular gene product – whether it is a protein or an RNA molecule – can vary greatly with time, stage of development, environment, and cell type. The reason a skin cell is different from a neuron is that a different subset of genes expressed in each cell type gives each tissue type its unique characteristics. Regulation of gene expression therefore lies at the heart of cell differentiation and development. The importance of transcriptional regulation in determining cell identity is highlighted by the remarkable finding that just four transcription factors are needed to 'reprogram' B lymphocytes (which play a role in the immune response) into pluripotent stem cells (which are capable of differentiating into any tissue type). Organisms have various mechanisms for regulating the expression of individual genes, thereby ensuring that appropriate levels of protein and RNA molecules are available in the cell.

In this section, we review the key principles that underpin the process of transcriptional regulation, and the key players that participate in this process. We then go on to explore specific mechanisms by which transcription is regulated, in both bacteria and eukaryotes, later in the chapter.

Any step in the transcription cycle can be regulated

The RNA polymerase holoenzyme can transcribe any gene that has an intact promoter. Yet at any given time, the relative rate at which a complete transcript is generated varies tremendously among different genes. While some of the variations are due to intrinsic differences in promoter strength – and hence how avidly polymerase binds to DNA and changes conformation to give the elongating complex – most of the differences are due to targeted gene regulation.

Any step in the transcription cycle, from binding of the polymerase enzyme to DNA to release of the complete transcript, can be regulated. Cells of all types contain a remarkable array of proteins that control the relative amount of transcription from each promoter, increasing or decreasing the rate at which RNA transcripts of selected genes are produced. Regulatory proteins that decrease the rate of transcription initiation are termed **repressors**, while proteins that increase transcription are denoted **activators**.

Most gene regulation occurs at the level of transcription initiation. Any aspect of transcription initiation can be modulated, from the binding of RNA polymerase to the promoter, to the ability of bound polymerase to form an open complex and begin productive RNA synthesis. However, we will also see examples in which later steps in the transcription cycle, such as elongation and termination, are regulated to either increase or decrease synthesis of a complete transcript.

The transcribed RNA itself can also play a role in regulating transcription. Some RNA transcripts fold to adopt structured regions that can inhibit or promote transcription. These structures can be influenced by the binding of proteins or small metabolites that exert control over transcription through their effect on RNA structure. We will see several examples of this form of regulation later in this chapter.

There are also RNA transcripts whose sole function is to regulate gene expression rather than to encode a functional protein or RNA molecule. Although most of these target translation, some regulatory RNA molecules control transcription, as we will also learn about in this chapter.

Proteins that bind to DNA target specific genes for regulation

Most genes are regulated by proteins that bind to the chromosome in the vicinity of the transcribed gene. Specific regions in the chromosome called **regulatory sequences** are the sites in the DNA to which regulatory proteins bind. Regulatory sequences are most commonly located upstream of the promoter, and can be immediately adjacent to the promoter (or even overlapping) or up to many kilobases away. A gene might be regulated by a single protein, or by several different proteins that bind to distinct sites within the regulatory sequence.

In bacteria, the DNA sequences recognized by transcriptional regulators are termed **operator** sites. Bacterial operators are located upstream of the transcription start site, typically close to or overlapping the promoter (Figure 8.29a,b). This arrangement allows the regulatory proteins to interact with the polymerase enzyme at the promoter site. However, some bacterial operators are located several hundred or more base pairs away from the promoter. In this case the intervening DNA must loop out to allow the regulatory protein to interact with the polymerase enzyme, as illustrated in Figure 8.29c. Proteins called **architectural**

(a) activator   promoter   ON

regulatory
sequence

(b) repressor   OFF

(c) ON

architectural
DNA-binding protein

Figure 8.29 The position of different regulatory sequences relative to bacterial promoters. (a) A regulatory sequence located just upstream (5′) of the promoter region is recognized by a DNA-binding protein that activates transcription from the promoter immediately downstream by helping the polymerase to bind. (b) The regulatory sequence overlaps the promoter. In this case, binding of the regulatory protein to its DNA site can block binding of RNA polymerase to the promoter, thereby repressing transcription. (c) Regulatory proteins that bind several hundred or more base pairs upstream influence transcription of the downstream promoter as the result of DNA looping, which brings the regulatory proteins in contact with the downstream polymerase enzyme an architectural DNA-binding protein can promote looping by bending the DNA.

DNA-binding proteins sometimes assist in the process of looping out DNA by promoting bending of the intervening sequence (Figures 8.29c).

Complex regulatory sequences located far from the gene are common in higher eukaryotes

Eukaryotic genes are commonly controlled by regulatory sequences located farther from the promoter than in bacteria, with a greater number of proteins regulating each gene. In multicellular organisms, regulatory sequences can encompass binding sites for a half dozen or more individual proteins and may be located at great distances from the promoter region – sometimes several thousand base pairs or more away (Figure 8.30). These sequences, called **enhancers**, bind proteins that activate transcription. An enhancer can be located either upstream or downstream of the target gene.

Enhancers are able to regulate transcription at distant promoters because the DNA forms a loop that allows proteins bound to the enhancer region to contact proteins bound to the promoter region (Figure 8.30c). In higher eukaryotes, there are even more complex regulatory regions known as **locus control regions**, which contain a combination of enhancer and insulator elements (Figure 8.31). For example, expression of the globin genes, which encode different subunits of hemoglobin during the course of development, is under control of a locus control region (Figure 8.31).

Proteins that regulate transcription are modular in nature

Proteins that regulate transcription must be able to bind to the correct set of regulatory sequences, which must be distinguished from the very large number of

Figure 8.30 Eukaryotic enhancers can be located far from the promoter. Enhancers can be located several thousand base pairs (a) upstream or (b) downstream of the promoter and coding regions. (c) Looping of the DNA that separates the enhancer region from the promoter enables the enhancer-bound regulatory proteins to contact the transcription initiation complex. This looping can be promoted by architectural DNA-binding proteins that induce sharp bends in the DNA.

Figure 8.31 Locus control region (LCR) of the beta-globin genes. Model for activation of beta-globin genes by proteins bound to the locus control region. Several different proteins bind to the locus control region, leading to recruitment of additional proteins and activating transcription of the downstream gene. One model is that DNA looping brings the LCR near the promoter region, where the bound proteins can promote assembly of a transcription initiation complex.
Adapted from Spector and Gasser (2003).

⊙ We learn about insulators in Chapter 4.

co-activator

ON

co-repressor

OFF

Figure 8.32 Recruitment of co-activator and co-repressor proteins to the DNA. Some sequence-specific DNA-binding proteins can recruit co-activator proteins to the DNA, which activate transcription of downstream genes, whereas other DNA-binding proteins recruit co-repressors, which repress transcription of downstream genes.

potential binding sites in the genome. To achieve this specific binding, each regulator contains a **DNA-binding domain** that recognizes specific DNA sequences (Section 8.8). In addition to the DNA-binding domain, transcriptional regulators usually contain additional domains that may be responsible for oligomerization, the activation or repression of transcription, and interactions with other transcriptional regulators. Indeed, eukaryotic regulatory proteins are typically modular in nature, with separable DNA-binding and activation (or repression) domains. However, some regulatory proteins – primarily in bacteria – consist of a single domain that carries out all the necessary functions.

DNA-binding proteins can recruit additional proteins that regulate transcription

In eukaryotic cells, DNA-binding proteins typically do not control transcription directly, but rather exert their function by recruiting additional proteins that either activate or repress transcription. These proteins, denoted **co-activators** or **co-repressors**, are not able to bind DNA on their own and must therefore be brought to the regulatory region by DNA-binding proteins (Figure 8.32). Most DNA-binding proteins selectively recruit particular co-activators or co-repressors. These intermediary proteins or protein complexes can interact directly with the transcription machinery or harbor enzymatic activities that modify chromatin.

The activity of regulatory proteins can be modified

The transcription of many genes is altered in response to a variety of changing conditions. One way in which gene expression can be altered in response to an external signal is through a direct change in the activity of the DNA-binding protein. The transcription of some genes can be altered by small molecules that bind directly to regulatory proteins and act as **allosteric effectors**. Allosteric effectors trigger a conformational change in the proteins to which they bind, altering the protein's DNA-binding properties or interactions with other proteins. The nucleotide, cyclic adenosine monophosphate (cAMP), and the hormone, estrogen, are two examples of allosteric effectors that bind to transcriptional regulators, altering the way in which they modulate transcription.

Covalent modification of transcriptional regulators, such as phosphorylation or acetylation, can also be used to alter the DNA-binding properties of a protein or its effect on transcription. These modifications may be the end result of a signaling cascade, which is a series of biochemical events triggered by an extracellular signal that we describe in more detail in Section 8.13. The subcellular localization of a transcriptional regulator also can be altered by regulated nuclear import or export.

Modification events such as those mentioned above rarely happen in isolation. Instead, multiple strategies are often used to fine-tune regulation of a single gene.

Chromatin modifications play a central role in the regulation of eukaryotic transcription

All of the steps in transcription that we have described are affected by the fact that, in eukaryotes, DNA is packaged into chromatin. We have already seen that the packaging of eukaryotic DNA into chromatin affects transcription initiation and elongation, and that chromatin structure must therefore be altered through the combined action of ATP-dependent nucleosome remodelers, enzymes that mediate post-translational modification of histones, and histone chaperones. The activities of all of these enzymes must be coordinated in order to transcribe a gene.

An important mechanism by which genes are activated or repressed in eukaryotes is through the recruitment of enzymes that modify chromatin. Turn to Chapter 4 for a comprehensive discussion of chromatin structure, histone modifications, and nucleosome remodeling. Here we summarize the two ways by which post-translational modification of histone proteins regulates transcription: by altering the charges on the modified side chains (with the exception of methylation) and by creating binding sites for protein domains that recognize specific modifications.

We learned previously (Section 4.4) that regions of chromatin that are **hyperacetylated** (that is, are heavily acetylated) are sites of active transcription, whereas low rates of transcription are found in regions of **hypoacetylated** chromatin (those sites that are acetylated at low levels). The activation of many eukaryotic genes is linked to recruitment of histone acetyltransferase enzymes, also known as HATs, which acetylate lysine residues within specific histone proteins (as well as lysine residues in other proteins that are involved in transcriptional regulation). Histone acetylation, in turn can help to recruit proteins that contain bromodomains, which bind specifically to acetyl-lysine. Thus, for example, the SWI/SNF nucleosome remodeling enzyme, which contains a bromodomain, is recruited to acetylated regions of chromatin. Similarly, genes can be repressed by recruitment of histone deacetylase enzymes, known as HDACs.

Post-translational modifications other than acetylation also play important roles in transcription regulation. Such modifications include methylation, phosphorylation, ubiquitination, and sumoylation. Like acetylation, these histone modifications both alter the chemical properties of side chains and create binding sites for specialized proteins that bind specifically to the modified side chains. However, there is no simple correlation between any of these modifications and transcriptional activity. We learned in Section 4.5 that methylation of histone tails, for example, is correlated with both repressed *and* activated transcription depending on the lysine residue that is modified.

Although there is much yet to be discovered about the precise chromatin modifications that must take place in order to transcribe any given gene, a picture has been emerging thanks to modern methods that make it possible to screen whole genomes for the positioning of nucleosomes and the presence of particular histone modifications, variant histones, and DNA methylation. Figure 8.33 shows the histone modifications found across a typical yeast gene that is actively transcribed.

Figure 8.33 The pattern of histone modifications across a eukaryotic gene. The pattern of histone modifications found across the promoter and transcribed region of a yeast gene is shown, along with the enzymes responsible for the modifications. Adapted from Saunders *et al.* (2006).

(a)

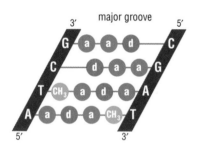

(b)

Figure 8.34 The chemical groups on the base pairs that are exposed in the major and minor grooves. The exposed hydrogen bond donors on Watson–Crick base pairs are in red and the acceptors are in blue. The hydrophobic methyl group of thymine is in green.

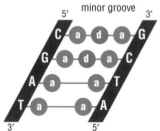

Figure 8.35 Pattern of chemical groups exposed in the major and minor groove. This schematic diagram shows the pattern of hydrogen bond donors and acceptors that are exposed in the major and minor grooves of DNA for the sequence, 5′–CGAT–3′. Note that each base pair has a unique chemical signature in the major groove, but not in the minor groove.

The observed connection between particular histone modifications and subsequent outcomes has led to the proposal that there is a 'histone code' consisting of particular combinations of covalent modifications that specify precise outcomes. Indeed, characteristic patterns of histone modifications have been observed in actively transcribed genes (Figure 8.33). Whether or not there is truly a precise histone code remains an active area of investigation.

8.8 DNA-BINDING DOMAINS IN TRANSCRIPTIONAL REGULATORS

We note above how proteins that regulate transcription possess specific DNA-binding domains, which recognize specific DNA sequences. The specificity of this interaction – between DNA-binding protein and target DNA sequence – is central to the successful regulation of gene expression. It is therefore worth pausing to consider the range of DNA-binding domains that are found in transcriptional regulators, and the commonalities and differences that exist between them.

Gene regulation depends on proteins that recognize specific DNA sequences

A protein that regulates transcription of a particular gene must be able to bind to the correct site within a regulatory sequence, selecting it from among countless other sites within the genome. Proteins that recognize specific DNA sequences must be able to form a set of non-covalent interactions with DNA that favor binding to the 'correct' DNA base sequence over binding to other locations in the chromosome. Many different protein folds that recognize specific DNA sequences have arisen in the course of evolution. Despite great variation in their overall structure, there are some common themes and principles that underlie the way in which such proteins recognize specific DNA sequences.

Each DNA sequence has a distinct chemical signature that can be recognized

As we learned in Chapter 2, the DNA double helix is fairly uniform in structure, with a negatively charged sugar–phosphate backbone facing outward and the edges of the base pairs exposed in the major and minor grooves. Each base pair has a characteristic set of chemical groups that is exposed in the major and minor grooves: hydrogen bond donors and acceptors and, in the case of thymine, a methyl group (Figure 8.34). A given DNA sequence is therefore characterized by a distinct array of hydrogen bond donors and acceptors and methyl groups that are exposed in the grooves (Figure 8.35).

The pattern of functional groups exposed in the major groove is unique to each DNA sequence, while there is less variability in the chemical surface exposed in the minor groove. This limited variability in the minor groove occurs because the pattern of hydrogen bond donors and acceptors is the same for A-T and T-A base pairs, and for G-C and C-G base pairs, as seen in Figure 8.35. These therefore cannot be distinguished from one another through hydrogen bond interactions in

Figure 8.36 Examples of contacts between side chains and bases. (a) An asparagine side chain can form two hydrogen bonds with an adenine base. (b) An alpha helix in the major groove with side chains contacting bases. Some side chains (N51 and R54) form direct hydrogen bonds with the base pairs, while others (S50) form water-mediated hydrogen bonds. PDB 1YRN.

the minor groove. Not surprisingly, most DNA-binding proteins that recognize a particular sequence of base pairs do so primarily through major groove contacts.

A protein recognizes a specific base pair sequence by forming a chemically complementary set of non-covalent interactions with the DNA. The surface of the protein that recognizes the DNA sequence contains a set of side chain and main chain atoms that can form favorable interactions with the particular array of DNA functional groups that are exposed. Thus, a hydrogen bond donor on the protein is paired with a hydrogen bond acceptor in the DNA, while a hydrophobic side chain might be in van der Waals contact with the methyl group of thymine. Figure 8.36 shows examples of side chain interactions with DNA bases. As in other examples of macromolecular interactions, there are often water molecules at the protein–DNA interface that mediate interactions with both the protein and the DNA via hydrogen bonds.

Shape and charge complementarity are important in protein–DNA interactions

A protein must form a sufficient number of non-covalent interactions with the DNA to ensure binding to the correct DNA sequence. (What constitutes a 'sufficient number' depends on many different factors, however, and will vary with both protein, DNA sequence, and the details of the regulation.) If the shape of the protein is complementary to that of the DNA, the chance is increased that side chains on the surface of the protein will approach functional groups on the DNA closely enough to form favorable non-covalent interactions. Many proteins that contact bases in the major groove of the DNA therefore do so with an alpha helix, whose shape and dimensions allow it to fit in the major groove of B-DNA (Figure 8.36). A two-stranded beta sheet can also fit readily into the major groove of B-DNA.

While the minor groove of B-DNA is normally too narrow to accommodate either a helix or a beta sheet, distortions can be induced by protein binding that open up the groove, as we saw in the case of TBP (Figure 8.11). While there is an energetic penalty in distorting the DNA, it can be repaid by sufficient favorable interactions with the protein.

Any protein that interacts with DNA – whether or not it binds to a specific sequence – must contain chemical features that favor binding to the highly negatively charged sugar–phosphate backbone of DNA. For this reason, protein domains that interact with DNA generally contain many lysine and arginine side

Figure 8.37 Electronegative DNA and electropositive protein surface. Surface rendering of DNA and protein is colored according to electrostatic potential. Electropositive surfaces are colored blue, electronegative surfaces red, and neutral light gray. The region of the protein surface that contacts the DNA is electropositive and therefore interacts favorably with the negatively charged DNA backbone.

chains, which are positively charged and therefore interact favorably with the negatively charged phosphate groups (Figure 8.37). Hydrogen bond donors such as serines or tyrosines, which bear a partial positive charge, can also form favorable electrostatic interactions with phosphate groups.

Sequence-specific DNA-binding proteins can be classified broadly by different DNA-binding motifs

A remarkable variety of domains that recognize specific DNA sequences are found among transcriptional regulators. Called **DNA-binding motifs**, these protein folds are tailored to accommodate the structural features characteristic of B-DNA. Some of these motifs are found universally in all organisms, while others appeared later in evolution and may be found just in eukaryotes, or only in metazoans (multicellular organisms). Here we touch upon some of the most common DNA-binding motifs and how they interact with DNA.

The helix-turn-helix is a common DNA-binding motif found in all organisms

The **helix-turn-helix** was the first DNA-binding motif to be identified. It consists of two alpha helices related by a fixed angle and connected by a tight bend (the 'turn') (Figure 8.38). The second, the C-terminal helix, is the **recognition helix** that fits in the major groove of the DNA and forms contacts with the base pairs that 'read' the DNA sequence. As can be seen in Figure 8.38, the helix-turn-helix can be found in a variety of protein folds that otherwise have little in common. Many helix-turn-helix proteins are dimers that position the recognition helices of the two monomers from which they are formed 34 Å apart, which matches the spacing of successive major grooves in B-DNA. The DNA sequence that is recognized has a symmetry that roughly matches that of the dimer (Figure 8.38).

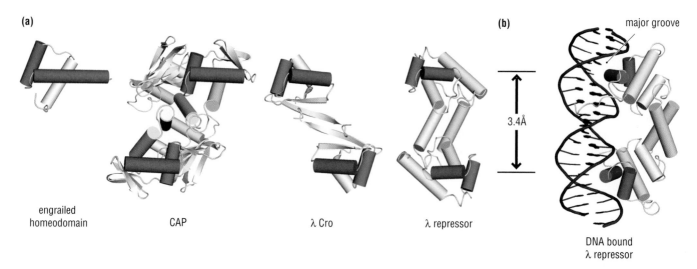

Figure 8.38 Examples of helix-turn-helix proteins. (a) The helix-turn-helix in each protein is shown in blue (first helix) and red (second, recognition helix). The *E. coli* CAP protein (also called CRP), phage λ repressor, and λ are dimers in which the respective recognition helices are separated by 34 Å (3.4 nm). The engrailed homeodomain protein from *Drosophila*, which also contains a helix-turn-helix, can bind DNA as a monomer. (b) The structure of λ repressor bound to DNA. In addition to the helix-turn-helix, each monomer has an N-terminal 'arm' that binds in the major groove on the opposite face of the DNA. PDB codes 1HDD, 1CGP, 6CRO, 1LMB.

In this way, each subunit forms very similar side chain contacts with each half of the binding site in DNA.

The **homeodomain**, one of the most common types of DNA-binding domains in eukaryotes, is a monomeric helix-turn-helix protein whose recognition helix is somewhat longer than its prokaryotic counterparts (Figure 8.39). An N-terminal arm that inserts into the adjacent minor groove allows homeodomain proteins to form additional contacts with DNA bases. We will see later that some homeodomain proteins actually bind DNA cooperatively with other DNA-binding proteins, thereby forming a complex that binds DNA very tightly and with high specificity for particular DNA sequences.

The zinc finger is the most common DNA-binding domain in the human genome

A number of different DNA-binding folds are organized around one or more zinc ions whose role is to stabilize the overall structure of the protein. Most prominent of these is the **zinc finger**, a small domain of about 30 amino acids whose single alpha helix and two antiparallel beta strands are held together by a central zinc ion (Figure 8.40a). The zinc is coordinated by two cysteine and two histidine residues that are universally conserved in this family of proteins; these domains are therefore known as Cys_2His_2 zinc fingers.

A single protein will typically contain several zinc fingers arranged in tandem, with each domain recognizing two to three base pairs (Figure 8.40b). Each zinc finger inserts its alpha helix end-on into the major groove, where side chains form base contacts. Variation in the number of zinc fingers per protein, as well as the amino acid sequence of each finger, allows different proteins to recognize DNA sequences of differing length and sequence. It is perhaps for this reason that the zinc finger is the most common DNA-binding domain encoded in the human genome.

Basic region-leucine zipper and helix-loop-helix motifs mediate DNA binding and dimerization of eukaryotic regulators

Coiled coils, which are pairs of parallel alpha helices that gently wind around each other, form an integral part of some DNA-binding domains. Basic region-leucine zipper (bZIP) proteins consist of long, uninterrupted alpha helices of about 60

Figure 8.39 The MATα2 homeodomain bound to DNA. Helix 3 (red) of MATα2 homeodomain lies in the major groove of the DNA, where side chains form contacts with individual base pairs. An N-terminal arm (green) forms additional contacts in the minor groove of the DNA (PDB 1APL).

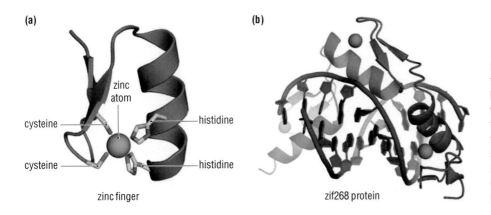

(a)

zinc atom

cysteine — histidine

cysteine — histidine

zinc finger

(b)

zif268 protein

Figure 8.40 The zinc finger domain. (a) A zinc finger contains a central zinc atom, coordinated by two histidine and two cysteine side chains, which stabilizes the entire structure. (b) Zinc finger domains typically occur in multiple copies, with each domain recognizing a short DNA sequence. In zif268, it is the end of each alpha helix that inserts into the major groove, with each zinc finger contacting two to three base pairs (PDB code 1ZAA).

(a)   **(b)**

bZIP protein   bHLH protein

Figure 8.41 Coiled coils in leucine zipper and helix-loop-helix proteins. Both these classes of DNA-binding domains have in common a coiled coil, which consists of two alpha helices that gently coil around one another and interact through a series of hydrophobic residues. (a) In bZIP proteins, each monomer is a single helix that extends into the major groove of the DNA. (b) In bHLH proteins, each monomer consists of two helices joined by a loop. One portion of the dimerization interface consists of a four-helix bundle, in addition to the coiled coil interface. PDBs 1YSA, 1HLO.

residues. These proteins associate via their C-terminal halves, forming a parallel coiled coil with leucine residues at the hydrophobic dimer interface (Figure 8.41). (The leucine residues lie along one face of each alpha helix and 'zip' together when the helices associate, stabilizing the coiled coil structure.) The N-terminal portions of the dimerized helices splay out and insert into the major groove on either side of the DNA.

A striking feature of these proteins is that the helical structure of the DNA-binding region is coupled to DNA binding: these residues are unstructured in the absence of DNA and only form the organized DNA-binding motif in the presence of DNA.

Basic region-helix-loop-helix (bHLH) proteins also contain a coiled coil that helps to mediate dimerization, and an alpha helix that lies in the major groove of DNA (Figure 8.41b). Each bHLH monomer consists of two helices joined by a loop. Portions of the four helices – two from each monomer – meet to form what is known as a four-helix bundle, which forms part of the dimer interface. The coiled coil that extends past the four-helix bundle mediates additional dimer contacts. The basic region-leucine zipper and helix-loop-helix protein families have many members that form both homodimers and heterodimers.

DNA recognition can also be mediated by beta sheets or loops

All the structures described so far recognize DNA through alpha-helical elements. As mentioned above, the major groove of B-DNA can also accommodate a pair of beta strands. The *E. coli* MetJ repressor protein binds DNA using two antiparallel beta strands, which fit into the major groove of B-DNA (Figure 8.42). The two proteins, p50 and p65, which comprise the mammalian transcriptional regulator, nuclear factor (NF)-κB, are both composed mostly of beta sheet. The two proteins belong to what is denoted the Rel homology domain family of DNA-binding proteins, which all resemble an immunoglobulin protein. Immunoglobulin proteins are composed entirely of beta sheet and connecting loops, and play a variety of roles in the immune response. Rel homology domains use the loops connecting successive beta strands to contact the DNA.

Figure 8.42 Beta sheets in DNA recognition. The MetJ repressor binds DNA by inserting two antiparallel β strands in the major groove. PDB 1CMA.

We have now reviewed the key structural features that mediate protein–DNA interactions, which underpins the regulation of transcription. But how do regulatory proteins effect regulation in practice? What mechanisms are used by different organisms to ensure that transcriptional regulation occurs as it should? We discover the answers to these questions in the next sections of this chapter.

8.9 MECHANISMS FOR REGULATING TRANSCRIPTION IN BACTERIA

We have seen the various ways in which different transcriptional regulators locate their target sequences in the genome. In bacteria, these regulatory sequences generally lie close to the promoter. Once a protein that regulates gene expression binds to its target sequence, it is in a position to influence transcription of the downstream gene. A variety of mechanisms are used to either repress or activate transcription initiation in bacteria. Several of these are illustrated in the examples that follow.

Trp repressor blocks RNA polymerase binding

One of the simplest ways in which a protein can repress transcription is by preventing RNA polymerase from binding to the promoter. This type of gene regulation is illustrated by the *E. coli* Trp repressor, a helix-turn-helix protein whose DNA-binding activity is controlled by the level of tryptophan in the cell (Figure 8.43). The genes encoding the enzymes needed to synthesize the amino acid tryptophan, are clustered together in the *E. coli* chromosome and are under the control of a single promoter and operator. The genes are transcribed as a single mRNA transcript. Such an arrangement of genes, whereby they are regulated and transcribed as a unit, is known as an **operon.**

When cellular levels of the amino acid tryptophan are low, genes that encode enzymes required for the synthesis of tryptophan are expressed. When levels of tryptophan are sufficiently high, transcription from the *trp* operon is repressed. This is accomplished by the Trp repressor, which binds to a site directly overlapping the promoter and thereby blocks access by RNA polymerase.

The Trp repressor is able to respond directly to the concentration of tryptophan in the cell because tryptophan is an allosteric effector of Trp repressor's DNA-binding activity. When sufficient amounts of tryptophan are present, one molecule of tryptophan binds to each monomer in the Trp repressor dimer and orients the helix-turn-helix motif of the repressor so that it is presented properly to the DNA major groove. The Trp repressor then binds DNA and prevents RNA polymerase from binding to the promoter. In the absence of tryptophan binding, which occurs when intracellular concentrations of tryptophan are low, the helix-turn-helix unit collapses inward. With the recognition helices now incorrectly aligned to fit in successive major grooves of the DNA, the protein cannot bind to the operator or repress transcription and the genes in the Trp operon are transcribed (Figure 8.43).

(a)

(b)

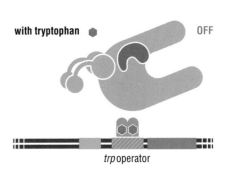

Figure 8.43 Regulation of transcription by the Trp repressor. (a) The *trp*, *trpE, D, C, B, A* operon. The genes, *trpE, D, C, B* and *A* are transcribed as a single mRNA. (b) In the absence of tryptophan, Trp repressor does not bind DNA and the genes of the *trp* operon are transcribed. When tryptophan is present, it binds to the Trp repressor and renders the protein capable of binding DNA. The Trp repressor binds to regulatory sites that overlap the promoter region, thereby preventing RNA polymerase binding and repressing transcription of the downstream genes.

Figure 8.44 Mechanism of catabolite activator protein (CAP) activation of class I and class II promoters. CAP activates transcription from a variety of promoters that fall into two distinct classes. Class I promoters (top) have a CAP-binding site just upstream of the promoter, allowing CAP to contact the C-terminal domain (CTD) of the α subunit. At class II promoters (bottom), the CAP operator overlaps the –35 region and allows CAP to interact simultaneously with the CTD and N-terminal domain (NTD) of the α subunit of RNA polymerase. The positioning of the –35 region and the CAP-binding site allow polymerase and CAP to bind simultaneously to the DNA.

The CAP activator protein recruits RNA polymerase by contacting the CTD of the RNA polymerase α subunit

Just as blocking the binding of RNA polymerase to the promoter represses transcription, enhancing the binding of polymerase to the promoter can activate transcription. Most bacterial regulators activate transcription by helping to recruit RNA polymerase to a promoter. One example is the Catabolite Activator Protein (CAP; also known as the cAMP receptor protein (CRP)). CAP activates transcription of more than 100 promoters in *E. coli* under conditions in which the levels of carbon sources such as glucose are too low to allow rapid metabolism. Glucose depletion in bacteria leads to increased synthesis of cAMP, which binds to the dimeric CAP and increases its affinity for DNA. The CAP dimer binds to specific DNA sites in or near target promoters and enhances the ability of the RNA polymerase holoenzyme to bind to its promoter and initiate transcription (Figure 8.44).

CAP can activate transcription in several ways, and promoters can be grouped according to the mechanism by which CAP stimulates transcription at that particular promoter. At class I promoters, CAP binds DNA upstream of the promoter and recruits RNA polymerase by binding directly to the CTD of the RNA polymerase α subunit (Figure 8.44). By helping the RNA polymerase holoenzyme to bind DNA, CAP promotes formation of the closed complex. At class II promoters, the CAP-binding site overlaps the binding site for RNA polymerase (Figure 8.44). At these promoters, interactions between CAP and the α subunit CTD also facilitate binding of RNA polymerase to promoter DNA. In addition, interactions between CAP and the α subunit N-terminal domain promote isomerization of the closed complex to the open complex.

CAP and LacI repressor regulate the *lac* operon

Many genes are regulated in response to multiple signals. The regulation of the *lac* operon is a classic example of how transcription of a set of genes can be modulated in response to two distinct signals: the relative availability of glucose and lactose. The simple sugar, glucose, is a preferred carbon source for *E. coli*, but the bacterium can also metabolize lactose, a disaccharide. Lactose utilization is a more complex process and therefore costs the cell more energy. Consequently, when both sugars are available, the bacterium will preferentially metabolize glucose over lactose, so the genes required for lactose utilization are only expressed at a low level. Only when glucose levels are low and lactose levels are high will the genes required for lactose utilization be switched on.

The *lac* operon encodes the three genes required for lactose uptake and utilization (Figure 8.45). The key to regulating the *lac* operon in response to two separate signals is the use of two different regulatory proteins: CAP, which indirectly senses glucose levels, and the *lac* repressor (or LacI), which senses lactose levels. These two proteins, which are both members of the helix-turn-helix family, have opposite effects on transcription and also respond differently to the binding of an allosteric regulator. As we learned above, CAP binds DNA and *activates* transcription only when it is bound to cAMP, which is found in abundance when glucose levels are low. The *lac* repressor, by contrast, binds DNA and *represses* transcription only in the *absence* of its allosteric regulator, allolactose. The binding of allolactose, a metabolite of lactose that is only present when lactose is being actively metabolized by the bacterium, inactivates the *lac* repressor.

Figure 8.45 Regulation of transcription by the Lac repressor and CAP activator. (a) The *lac* operon. (b) Diagram of how transcription of the *lac* operon is regulated by CAP and the Lac repressor (LacI) in response to different levels of glucose and lactose. CAP binding activates transcription of the *lac* genes, while LacI binding blocks polymerase binding and represses transcription. The maximum level of expression occurs when LacI does not bind DNA (high lactose concentration) and glucose levels are low (CAP binds DNA).

We can now see how the *lac* repressor and CAP regulate transcription in response to the availability of glucose and lactose (Figure 8.45). The *lacZ, lacY,* and *lacA* genes are transcribed as a single transcript (just as we saw in the case of the *trp* operon). The *lacI* gene, which encodes the *lac* repressor (Figure 8.45), is constitutively expressed (meaning that it is always transcribed), so the *lac* repressor is always present in the cell. In the absence of lactose, the *lac* repressor binds to a DNA site (*lacO*, the *lac* operator) that overlaps the promoter and represses transcription. What happens when lactose is present? In this case, the *lac* repressor is inactivated by the binding of allolactose, which means it no longer represses transcription, causing it to dissociate from DNA.

The relative *level* of transcription, however, will depend on whether glucose is also present. If glucose levels are low, CAP will bind upstream of the operon and activate transcription of the *lac* operon, since cAMP binding enables CAP to bind DNA. The proteins encoded by the *lac* operon will therefore be synthesized at high levels and the bacterium will be able to make efficient use of lactose. If glucose levels are high, however, CAP does not bind DNA (because cAMP levels are low) and the *lac* genes are transcribed at a much lower level. The bacterium will make less efficient use of lactose since there is adequate glucose to serve the cell's metabolic needs.

MerR-family activators alter promoter structure

The variable spacing between –10 and –35 sequences provides another opportunity for transcriptional regulation in bacteria. If the spacing between these two sites is much larger than is optimal, RNA polymerase binding to the promoter is significantly weakened. The MerR family, named for a protein that activates expression of genes required for mercury detoxification, forms a unique group

Figure 8.46 Mechanism of promoter opening by MerR-family members. The structure of BmrR explains how MerR-family members facilitate the opening of promoters in which the −10 (green) and −35 (pink) sequences are spaced 19 bp apart. (a) Promoter with a 19 bp spacer. (b) BmrR binding leads to a distortion of the DNA such that the −10 and −35 elements now have the same spacing and orientation as the promoter element with −35 and −10 spaces, 17 bp apart, shown in (c). (d) Structure of BmrR bound to DNA (PDB 1EXJ).
Figure 8.46a–c from Heldwein and Brennan (2001).

of transcriptional regulators that can change the way the promoter elements are organized in space relative to each other. The −35 and −10 sequences of promoters regulated by MerR are spaced an unusually long 19 bp apart, as compared with the ideal spacing of 17 bp (Figure 8.46). The two additional base pairs not only place the −35 and −10 promoter elements farther apart, but also affect the relative orientation of the two promoter elements about the DNA axis. This is because each additional base-pair step is rotated roughly 34° about the helix axis and translated a distance of about 0.34 nm along the helix.

MerR activates transcription by altering the relation between the −10 and −35 regions to more closely resemble an ideal promoter (Figure 8.46). The binding site for the MerR protein overlaps the promoter region. As seen in the structure of a related protein, BmrR, proteins in this family are able to change the distance and angular spacing between the −10 and −35 regions by distorting the DNA structure between the two promoter elements (Figure 8.46). With the alignment of the two promoter regions now closer to that found in optimally spaced promoters, RNA polymerase binds more readily. Notice how BmrR binds on the opposite face of the DNA and therefore does not interfere with polymerase binding.

Activation by NtrC involves DNA looping and ATP hydrolysis

⊙ The initial characterization of NtrC is described in Experimental approach 11.2.

Promoters that are transcribed by RNA polymerase holoenzyme containing the σ^{54} (also known as σ^N) subunit must be activated by protein complexes that use ATP hydrolysis to drive promoter opening. A well-characterized example of this class of activators is the *E. coli* NtrC protein, which activates genes that are important for nitrogen metabolism (Figure 8.47). Genes regulated by NtrC generally contain two or more binding sites for NtrC well upstream of the promoter. NtrC can bind to these sites but it does not activate transcription until a regulatory domain within the protein is phosphorylated. This phosphorylation event triggers oligomerization of NtrC, which has two effects on the protein's activity. First, oligomerization of NtrC promotes interaction of the NtrC protein with the polymerase enzyme that is bound to the promoter, causing the intervening DNA to loop out (Figure 8.47). At some promoters, the architectural DNA-binding protein integration host factor (IHF) helps promote the looping by binding to the intervening sequences and bending the DNA.

Oligomerization of NtrC also stimulates its rate of ATP hydrolysis, which in turn helps to promote open-complex formation at σ^{54}-dependent promoters. Somehow, the σ^{54} holoenzyme bound to a promoter is unable to isomerize to the open complex in the absence of ATP hydrolysis. It is thought that NtrC uses the energy released by ATP hydrolysis to help the holoenzyme overcome this block.

8.10 COMPETITION BETWEEN cI AND CRO AND THE FATE OF BACTERIOPHAGE LAMBDA

It is frequently necessary for an organism to respond to a complex set of signals adjusting transcription of multiple genes. This type of regulation is often the result of the concerted action of a network of proteins, which may exert antagonistic effects on one another. The outcome – whether transcription happens or not – then depends on the balance of power between regulators that have opposing effects on transcription. In this section we learn about a prime example of how transcription can be regulated by a network of proteins – regulation of bacteriophage lambda genes.

Bacteriophage lambda can undergo lytic growth or lysogeny

Bacteriophage lambda presents a well-studied example of how the competition between several different sequence-specific DNA-binding proteins can lead to very different physiological outcomes. This bacterial virus infects *E. coli* by injecting its short chromosome into the bacterium's cytoplasm, where the **phage** can have one of two fates (Figure 8.48). Phage lambda can undergo **lytic growth**, reproducing many virus particles within the cell, leading to its eventual lysis (bursting of the host cell). Alternatively, the phage can undergo **lysogeny**, whereby the phage DNA integrates into the bacterial chromosome and becomes dormant.

The integrated phage is denoted a **prophage** and a bacterial cell whose chromosome contains an integrated copy of the bacteriophage chromosome is called a **lysogen**. In the lysogen, the viral DNA is replicated along with the bacterial chromosome. Conditions that threaten the viability of the bacterial host cell, such as DNA damage, can trigger the prophage to excise itself from the chromosome, undergo lytic growth, and produce viral progeny that burst out of the damaged host cell.

The choice between lysis and lysogeny is controlled by the levels of the cI, cII, and Cro proteins

The choice between lysis and lysogeny is determined by the relative levels of three DNA-binding proteins encoded by the phage: cI (also called the lambda repressor), cII and Cro. These three regulators, which are the first proteins that are

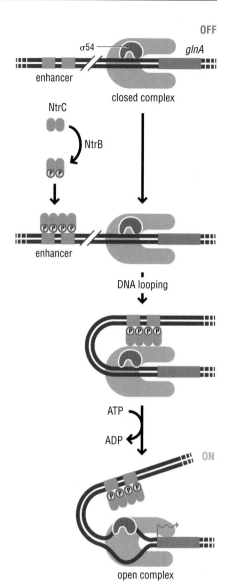

Figure 8.47 Mechanism of action of NtrC. The NtrB kinase phosphorylates NtrC, causing NtrC to oligomerize and bind to its regulatory region in the DNA. The intervening DNA loops out, allowing NtrC to contact σ^{54} RNA polymerase. Oligomerization also triggers the hydrolysis of ATP to ADP by NtrC, which helps promote open-complex formation by the σ^{54} RNA polymerase.

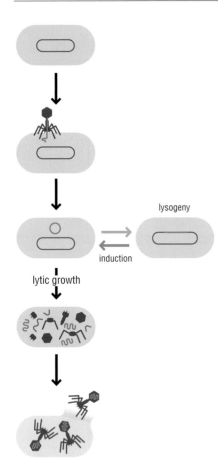

Figure 8.48 The lifecycle of bacteriophage lambda. Bacteriophage lambda infects a cell by injecting its linear chromosome (pink) into a host *E. coli* cell. The lambda chromosome circularizes and has one of two fates: it can direct lytic growth of the phage, directing synthesis of new phage particles that are released by bursting the host cell, or it can incorporate itself into the host cell's chromosome and become a prophage, remaining in this quiescent state indefinitely. Damage to the host cell's chromosome, such as that caused by irradiation with ultraviolet light, can cause the integrated prophage to excise itself from the host cell's chromosome, circularize, and commence lytic growth.

synthesized after phage lambda infects an *E. coli* cell, bind to specific sites in the phage chromosome and control transcription from the four promoters illustrated in Figure 8.49: P_R (rightward promoter), P_L (leftward promoter), P_{RE} (promoter for repressor establishment) and P_{RM} (promoter for repressor maintenance). As we shall see next, whether the phage grows lytically or establishes a lysogen depends upon the rate at which these proteins build up in the infected cell and occupy their DNA binding sites.

After a lambda phage injects its genetic material into a bacterial cell, the linear phage chromosome circularizes (Figure 8.48), and the host cell RNA polymerase initiates transcription from P_L and P_R. The first genes in the P_R transcript encode the Cro and cII regulators (Figure 8.49). If sufficient cII protein accumulates, it binds DNA and stimulates transcription from the P_{RE} promoter, producing an mRNA transcript encoding the cI regulator. The newly synthesized cI protein, in turn, stimulates transcription from the weak P_{RM} promoter, yielding another transcript encoding cI. As cI builds up in the cell, the protein binds to sites in the P_L and P_R promoter regions and represses transcription of the lytic genes encoded by the P_L and P_R transcripts. cI also activates transcription of the gene that encodes integrase, which mediates lambda DNA integration into the bacterial chromosome (see Experimental approach 8.3). Thus, lysogeny is established and maintained by high levels of cI.

By contrast, if cII does not accumulate in sufficient amounts and too little cI protein is synthesized, the Cro protein will bind the P_{RM} promoter and repress transcription of the cI gene. By preventing cI synthesis, Cro ensures expression of the lytic genes from P_L and P_R.

Ultimately, whether or not the bacteriophage undergoes lysis or lysogeny after infecting a bacterial cell hinges on the activity of cII. cII is sensitive to the host cell proteases, making it unstable in the cell. If the environmental factors are such that cII accumulates and directs expression of enough cI, a lysogenic state is established. If cII is degraded, cI is not synthesized and all other bacteriophage genes are expressed, allowing autonomous viral replication, packaging, and, ultimately cell lysis.

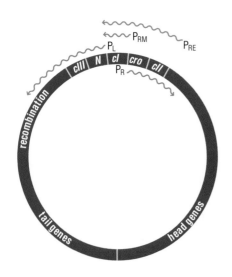

Figure 8.49 A map of the lambda chromosome. Map of the 48.5 kb bacteriophage lambda chromosome showing some of the genes and the promoters that are important in the early stages of the phage lifecycle. The P_L and P_R promoters direct transcription of genes required for lytic growth, whereas the P_{RM} transcript encodes cI, which is required for establishment and maintenance of the lysogenic state. Following infection of a host cell, transcription from P_R leads to synthesis of the Cro and cII proteins. The cII protein stimulates transcription from P_{RE}, which leads to synthesis of cI. The cI and Cro proteins compete for regulation of the lytic genes: cI represses their transcription whereas Cro turns off synthesis of cI and ensures that the lytic genes will be expressed properly.

The DNA-binding affinities of cI and Cro are a key component in the choice between lysis and lysogeny

A closer look at the P_L and P_R promoters reveals that a key component of the molecular switch controlled by cI and Cro is the relative affinities of these proteins for six DNA binding sites. The dimeric cI protein is composed of two discrete domains: an N-terminal DNA-binding domain with a helix-turn-helix motif and a C-terminal dimerization domain (Figure 8.50a). cI dimers further oligomerize to form tetramers, which in turn can associate to form an octamer. These higher-order interactions are required for cooperative cI binding to DNA. In contrast, the Cro protein consists of a single, small domain (also with a helix-turn-helix) that binds DNA as a dimer.

The operators, O_L and O_R, at the P_L and P_R promoters, respectively, each contain three sites to which either cI or Cro can bind (Figure 8.50b). The three sites in O_L – O_L1, O_L2 and O_L3 – overlap the P_L promoter, while O_R1, O_R2 and O_R3 overlap the divergent P_R and P_{RM} promoters. At both O_L and O_R, cI binds first to its highest-affinity sites, O_L1 and O_R1, and recruits a second cI dimer to the weaker O_L2 and O_R2 binding sites through cooperative interactions. The pairs of cI dimers bound to O_R and O_L interact with one another to form an octamer, looping out the intervening DNA (Figure 8.50c). This arrangement of cI proteins on the DNA prevents RNA polymerase binding to

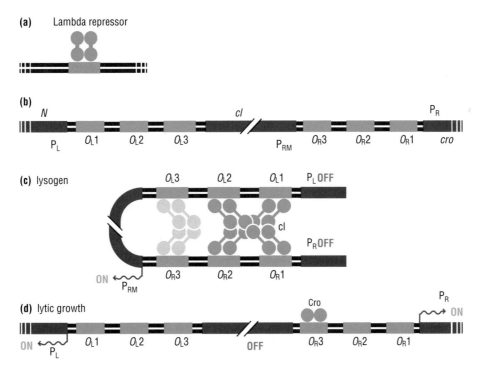

Figure 8.50 Binding of cI and Cro to the rightward and leftward operators. (a) Lambda repressor dimer. Each monomer of cI contains an N-terminal DNA-binding domain and a C-terminal dimerization domain. (b) The arrangement of DNA binding sites and promoters in the leftward (O_L) and rightward (O_R) operators. Each operator region contains three DNA sequences that can be recognized by both cI and Cro: O_L1, O_L2, and O_L3, and O_R1, O_R2, and O_R3, respectively. (c) In a lysogen, cI dimers bind cooperatively to O_L1 and O_L2 in the leftward operator and to O_R1 and O_R2 in the rightward operator. The four cI dimers bound to these sites interact via their C-terminal domains, looping out the intervening DNA. This arrangement of cI proteins on the DNA blocks RNA polymerase binding to the P_R and P_L promoters, while stimulating transcription from P_{RM}. The DNA loop promotes cooperative interactions between cI dimers bound to O_R3 and O_L3. Binding of cI to O_R3 turns off transcription of cI. (d) The Cro protein binds to the same DNA sites as cI, but with different relative preferences. If the cI protein is inactivated as a result of UV-induced damage to the cell, or if insufficient amounts of cI are present following infection of the host cell, the Cro protein binds first to O_R3 and represses transcription of P_{RM}, turning off synthesis of cI. As lytic growth proceeds and concentrations of Cro build up in the cell it binds to the remaining sites in O_L and O_R, turning down expression of the early viral genes.

8.3 EXPERIMENTAL APPROACH

Cooperative binding and DNA looping in gene regulation by the lambda cI repressor

One of the key components involved in regulating the switch between lysis and lysogeny in bacteriophage lambda depends on cooperative binding of the cl protein to DNA. This property of cl was discovered in 1979 by Alexander Johnson, Barbara Meyer, and Mark Ptashne, who measured the binding affinity of cl to the three sites within O_R: O_R1, O_R2, and O_R3. It had already been established that cl dimers bound to these three sites, as could an N-terminal fragment containing just the DNA-binding domain. Without the C-terminal domain, the DNA-binding domains could not dimerize efficiently and therefore bound DNA more weakly. Ptashne and colleagues uncovered a second property of the C-terminal domain: the ability to mediate cooperative binding between cl dimers.

The way in which the bacteriophage lambda cl repressor fills successive binding sites in O_R was determined from nuclease protection experiments (see Section 16.11). As seen in Figure 1, these experiments show vividly where in the operator DNA cl binds. Lane 1 shows the pattern of DNA fragments produced when a radiolabeled DNA fragment containing O_R is subjected to limited digestions by DNase I and the products are then separated by gel electrophoresis. As cl repressor is added in successively higher

Figure 1 DNase I protection experiment showing the binding of the bacteriophage lambda cl repressor to O_R1, O_R2, and O_R3. The concentration of cl increases in each successive lane. Reproduced from Johnson *et al.*, *Proceedings of the National Academy of Sciences of the USA* 1979;**76**:5061–5065.

concentrations (lanes 2–8), the protein dimers bind to each of the individual DNA sites – O_R1, O_R2, and O_R3 – protecting those base pairs from digestion. This causes the bands in that region to fade and eventually disappear, as more and more DNA fragments are protected from DNase I digestion by the bound protein.

By noting how the sites on the DNA filled as a function of cl concentration, the equilibrium dissociation constants (K_d) could be determined for the three sites (O_R1, O_R2, and O_R3) within the rightward operator. Recall that the K_d is the protein concentration at which half-maximal binding occurs. (Turn to Section 3.2 to review the principles of equilibrium binding.) As shown in Figures 1 and 2a, cl fills O_R1 and O_R2 at roughly the same concentration in the titration; this results from the fact that cl dimers bind cooperatively to these two sites (see Section 3.3 for discussion of cooperativity). By contrast, the O_R3 site was only filled when the protein concentration was tenfold higher (a tenfold higher K_d). However, if a mutation was introduced into O_R1 that abrogated DNA binding (O_R1^-), cl bound to O_R2 and O_R3 with the *same* affinity (Figure 2a). In addition, the affinity of cl for O_R3 became higher (~ fivefold) such that it matched the affinity for O_R2. This observation suggested that, in the absence of cl binding to O_R1, there were now cooperative interactions between dimers bound to O_R2 and O_R3. When the same experiments were performed with the N-terminal domain of cl (missing the C-terminal domain), cooperative binding was no longer observed; indeed, this fragment bound more tightly to O_R1 than to O_R2 (Figure 2b). In addition, mutations in O_R1 had no effect on binding of the N-terminal fragment to the other two sites (Figure 2b).

The results of these experiments could be explained by the model shown in Figure 3a, dubbed 'alternate pairwise cooperativity.' The C-terminal domain of cl was proposed to mediate cooperative interactions between adjacently bound dimers. The nature of the interaction was such that one dimer could interact with only one other dimer on the DNA, giving the pairwise interactions shown in Figure 3a. This meant that, in the wild-type operator, a cl dimer bound to O_R3 would not bind DNA cooperatively because the dimer bound to the adjacent O_R2 site was already interacting with a dimer at O_R1 and was therefore not available for additional interactions. O_R3 would only fill at much higher concentrations of cl (Figure 2a), thereby regulating its own expression by shutting off transcription from P_{RM} (Figure 3b).

The model shown in Figure 2b for how cl regulated transcription in a lysogen stood for over 20 years until a relatively recent study revealed a new aspect to the regulatory switch that had gone undetected. One of the long-standing puzzles about lambda had to do with the roles of O_R3 and O_L3 in autoregulation of cl expression. The

	relative repressor concentration					
	intact repressor			N-terminal domain		
DNA	O_R1	O_R2	O_R3	O_R1	O_R2	O_R3
O_R^+	25	2	1	25	25	1
O_R1^- (vs326)	5	5	-	25	25	-
O_R2^- (virC23)	25	-	2	25	-	1
O_R1^- O_R2^- (Δ265 virC23)	25	-	-			
O_R1^- O_R3^- (Δvc1 vc3)	-	25	-	-	25	-

Figure 2 **Relative concentration of cl repressor required for half-maximal binding to wild-type and mutant O_R.** The concentration of lambda repressor and the N-terminal domain is expressed as a multiple of the concentration of each protein needed to fill O_R3. (a) Intact cl. (b) N-terminal domain of cl. Johnson *et al.*, 1979.

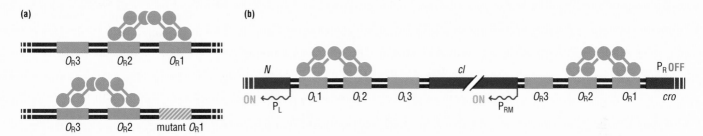

Figure 3 **Alternate pairwise cooperativity model.** (a) Figure illustrating that just two dimers at a time could interact with one another. (b) Model for how cl bound cooperatively to the O_R and O_L operators. O_R3 was predicted to fill at much higher cl concentrations, due to the absence of cooperativity at that site in the wild-type operator. The DNA looping shown in Figure 8.43 was discovered over 20 years later. Adapted from Hochschild, *Current Biology* 2002;**12**:R87–R89.

affinity of cl for O_R3 seemed too weak for cl to shut off its own transcription, at least at concentrations of cl known to exist in a lysogen; moreover, there was no known role for O_L3 in a lysogen. In 2001, Ian Dodd and colleagues made the surprising discovery that binding of cl to O_R3 was aided by cooperative interactions with O_L3. They replaced the *cl* gene with the gene expressing β-galactosidase, whose levels can be readily measured in solution, while the cl protein itself was encoded on a separate plasmid whose synthesis could be controlled independently. This system made it possible to measure transcription from P_{RM} as a function of increasing cl concentration. As shown in Figure 4, repression of P_{RM} by cl requires an intact O_L operator (compare the level of transcription in the absence and presence of O_L as cl concentration increases).

How could binding to O_L affect binding to a site in O_R several thousand base pairs away? The model that explains these observations is that the intervening DNA loops out as depicted in Figure 8.50c, allowing cl dimers bound to O_L to directly contact dimers bound to O_R. This idea did not come out of thin air – it had been recently discovered that cl bound to distant sites (>3.5 kb) could induce a loop in the DNA, although it was not clear what the physiological significance of the looping could be. Independently, biochemical experiments had shown that cl could form octamers at high concentrations. The model in Figure 8.50c brings together all of these observations: a pair of cl dimers bound to O_R1 and O_R2 interacts with a pair of cl dimers bound to O_L1 and O_L2, forming an octamer, which is mediated by the C-terminal domains and looping out of the intervening DNA. The looping then facilitates cooperative binding between dimer bound to O_R3 and O_L3, as shown in Figure 8.50.

Figure 4 **Transcription from P_{RM} in the presence and absence of O_L.** In the absence of O_L, as the concentration of cl increases, there is little repression of transcription from P_{RM} with a wild-type O_R3. In the presence of O_L, high concentrations of cl repress transcription from P_{RM}. Figure kindly supplied by Ian Dodd.

Cooperativity between these two distant sites ensures that P_{RM} is actually repressed 50% of the time in a lysogen, resulting in a cl concentration that perfectly poises the switch for induction.

Dodd IB, Perkins AJ, Tsemitsidis D, Egan JB. Octamerization of lambda CI repressor is needed for effective repression of P_{RM} and efficient switching from lysogeny. *Genes and Development* 2001;**15**:3013–3022.

Johnson AD, Meyer BJ, Ptashne M. Interactions between DNA-bound repressors govern regulation by the lambda phage repressor. *Proceedings of the National Academy of Sciences of the U S A* 1979;**76**:5061–5065.

P_R and P_L and stimulates transcription from P_{RM} as a result of interactions between RNA polymerase and the cI dimer bound to O_R2. The increased transcription from P_{RM} leads to increased cI levels.

As cI builds up, the protein can bind cooperatively to the weakest binding sites, O_R3 and O_L3, which now are close in space due to the DNA looping (Figure 8.50c). Occupancy of O_R3 occludes P_{RM} and suppresses synthesis of cI. This **autoregulation**, whereby cI regulates its own synthesis, ensures there is sufficient cI in the cell to maintain the lysogenic state, but not so much that prophage induction is prevented.

A lysogen can persist for many generations, with the phage stably incorporated in the bacterial genome. With each cell division, the daughter cells receive a copy of the circular chromosome containing the integrated prophage, along with enough cI protein to stimulate fresh cI synthesis, maintaining the lysogenic state. This inheritance of the lysogenic state is a form of epigenetic inheritance, since the information that the prophage remain in the 'off' state is transmitted by the cI protein and not by differences in the DNA sequence.

What if, early in the stage of phage infection, there is insufficient cI protein to establish lysogeny and ensure sufficient transcription of the cI gene? In this case, the Cro protein will build up and bind to the same operator sites as cI (Figure 8.50d). A key difference is that, in contrast to cI, Cro binds with greatest affinity to O_R3, where it blocks RNA polymerase binding to P_{RM} and represses cI synthesis. As the concentration of Cro builds up in the cell later during lytic growth, it binds to the remaining operator sites in O_L and O_R (Figure 8.50d). This leads to the autorepression of *cro* as well as the repression of all other early lytic genes.

The prophage is induced on cI cleavage

The prophage is stably integrated into the *E. coli* chromosome and can be maintained through many rounds of DNA replication and cell division. However, under conditions that threaten survival of the host cell and hence of the integrated phage DNA, the prophage can excise itself from the host chromosome and undergo lytic growth, thereby producing new phage and escaping the compromised host. This switch from lysogeny to lytic growth is brought about by a response to DNA damage that disables the cI repressor.

Damage of the host cell DNA – for example by chemical agents or by ultraviolet light – triggers the bacterial **SOS response**, which leads to activation of the RecA protein. RecA promotes autocatalytic cleavage of the C-terminal domain of cI, separating it from the N-terminal DNA-binding domain (Figure 8.50a). Without the C-terminal domain, cI can no longer dimerize efficiently or form highly cooperative interactions between adjacent cI dimers. On its own, the N-terminal domain has relatively low affinity for the operator sites and therefore dissociates from the DNA. Now transcription of the early lytic genes from P_R and P_L can occur; there is little further synthesis of cI because P_{RM} is a relatively weaker promoter than P_L. One of the first lytic genes to be transcribed from the P_L promoter is *cro* (Figure 8.49), which represses transcription of the *cI* gene from P_{RM} and prevents re-establishment of the lysogenic state (Figure 8.50c) allowing lytic growth to occur. The resulting phage particles lyse the damaged host cell and are free to infect fresh cells.

⊙ We introduce the concept of epigenetic inheritance in Chapter 4.

⊙ We learn about the response to DNA damage in Chapter 12.

The types of transcription feedback loops we see in the bacteriophage lambda underlie complicated environmental sensing and developmental decisions from systems as 'simple' as a bacteriophage to systems as 'complicated' as humans.

8.11 MECHANISMS FOR MODULATING EUKARYOTIC TRANSCRIPTION

In the preceding sections we described several ways in which sequence-specific DNA-binding proteins can modulate bacterial RNA polymerase activity. In most of these examples, the DNA-binding protein itself modulates transcription, either by directly contacting the polymerase holoenzyme and stimulating transcription, or by physically preventing polymerase from binding to the promoter. In all cases, only a very limited number of proteins are required to regulate transcription of a given bacterial promoter. In eukaryotes, by contrast, transcription is regulated by more elaborate mechanisms that depend on much larger sets of proteins. Eukaryotic genes are typically regulated by proteins that bind DNA and then recruit co-activator or co-repressor protein complexes that activate or repress transcription, respectively. The examples that follow illustrate some of the diverse mechanisms by which transcription is controlled in eukaryotes.

ELK1 recruits the Mediator complex and activates transcription

We learned earlier that the multisubunit Mediator complex is generally required for the activation of transcription by RNA polymerase II. Mediator serves as a kind of molecular bridge between activators and the polymerase enzyme and is thought to stimulate transcription either by promoting assembly of the transcription pre-initiation complex, or by stimulating an increase in the rate of transcription initiation.

One way in which transcriptional activators can function is by recruiting the Mediator complex directly. The mammalian ELK1 protein activates transcription in response to extracellular signals known as mitogens, which are molecules that cause a cell to enter into mitosis. In the absence of a mitogen, ELK1 binds DNA together with another protein (called serum response factor (SRF), shown in Figure 8.51), but does not activate transcription. However, mitogen binding to a cell surface receptor activates protein kinases in the cytoplasm, which subsequently phosphorylate the activation domain of ELK1. The phosphorylated activation domain recruits the Mediator complex by binding to a specific subunit in the complex (called Med23). In addition, a histone acetyltransferase complex is also recruited (Figure 8.51). This increases the rate of transcription initiation, thereby stimulating an increase in the overall level of transcription.

Gal4 recruits RNA polymerase through a co-activator complex

The regulation of transcription by the yeast Gal4 protein provides an example of a more complex interplay between several proteins that regulate transcription in response to the presence of a particular signal. Gal4 regulates genes that

Figure 8.51 Recruitment of Mediator by the activation domain of ELK1. The ELK1 protein binds to DNA along with a second protein called serum response factor (SRF). By itself, this protein complex does not activate transcription. When mitogens bind to receptors on the cell surface, this leads to activation of an enzyme, mitogen-activated protein (MAP) kinase, which phosphorylates the activation domain of ELK1. The phosphorylated ELK1 recruits the Mediator complex by interacting with a subunit of mediator, Med23. A histone acetyltransferase (HAT) complex is also recruited. As a result, the rate of transcription initiation is increased.

Figure 8.52 Mechanism of Gal4 recruitment of RNA polymerase II. In the absence of galactose (top), Gal4 is bound to an upstream activating sequence (UAS), but Gal80 binds to Gal4 and blocks its ability to activate transcription. In the presence of galactose (bottom), Gal3 binds to Gal80 and transports it to the cytoplasm, relieving the inhibitory effect on Gal4. Gal4 is now able to recruit the SAGA co-activator complex, which in turn promotes assembly of the transcription initiation complex.

Figure 8.53 Transcriptional repression through recruitment of a histone deacetylase. The Ume6 protein binds DNA and recruits the Sin3/Rpd3 co-repressor complex, which deacetylates histone tails and helps cause transcriptional repression (top). Starvation triggers phosphorylation of Ume6, which causes Sin3/Rpd3 to dissociate and be replaced by a co-activator, Ime1 (bottom).

encode proteins required for galactose metabolism. When galactose levels are elevated, Gal4 activates transcription by binding to a target sequence known as the UAS_G (Upstream Activating Sequence for Gal) (Figure 8.52, step 1). As is typical for most eukaryotic activators, Gal4 has separate DNA-binding and activation domains. The activation domain is rich in acidic amino acids, a property common to many, though not all, activation domains. The separation of the DNA-binding and activation functions of Gal4 into discrete domains has been exploited to produce chimeric proteins that are useful tools to molecular biologists, as will be discussed in Section 16.11.

The way in which Gal4 regulates transcription in response to elevated galactose levels in yeast is depicted in Figure 8.52. Gal4 forms a dimer that binds to a specific DNA sequence located upstream of genes whose expression is required in the presence of galactose. Unlike the bacterial Trp repressor and CAP activator, Gal4 does not bind to the allosteric effector, galactose, directly. Instead, Gal4 activity is controlled by two proteins, Gal80 and Gal3 (see Figure 8.52).

In the absence of galactose, Gal80 binds to the activation domain of Gal4 and prevents it from activating transcription (Figure 8.52). However, an increase in cellular galactose levels somehow activates the Gal3 protein and causes it to bind to Gal80 instead of to Gal4. The Gal80–Gal3 complex translocates to the cytoplasm, and the Gal4 activating region is free to recruit a multiprotein co-activator complex called SAGA, as well as Mediator. The SAGA and Mediator co-activator complexes help promote assembly of the transcription initiation complex by contacting the basal transcription factors, which in turn help to recruit RNA polymerase II.

Ume6 represses transcription by recruiting a co-repressor complex with histone deacetylase activity

The great majority of transcriptional regulators in eukaryotes activate transcription, but some proteins actively *repress* transcription of their target promoters. One protein that can both activate and repress transcription is the yeast Ume6 protein. Ume6 ensures that cells respond appropriately to their nutritional environment, binding upstream of genes whose expression is modulated in response to changes in the levels of metabolites such as glucose, nitrogen, and inositol.

When there is adequate nitrogen and a source of carbon, Ume6 represses transcription by binding DNA and recruiting co-repressor complexes. One of these complexes contains a protein called Sin3 and a histone deacetylase called Rpd3 (Figure 8.53a). In addition, Ume6 recruits a nucleosome remodeling enzyme, Isw2. The deacetylase targets the N-terminal tails of adjacently bound histones, removing acetyl groups from the lysine residues. Deacetylation of residues of the tail of histone H4, in particular, allows the formation of more compact chromatin states, and consequently represses transcription. At the same time, the nucleosome-remodeling enzyme helps to establish an altered chromatin structure that further contributes to repression of transcription.

In the absence of nitrogen and glucose, however, Ume6 is phosphorylated (Figure 8.53b), which causes Sin3 and Rpd3 to dissociate. A transcriptional co-activator, Ime1, is recruited instead by Ume6, which now activates transcription.

8.12 COMBINATORIAL REGULATION OF EUKARYOTIC TRANSCRIPTION

We saw in Section 8.10 how bacteriophage lambda uses a combination of regulatory proteins to modulate transcription. In this section we see how a similar network of regulators is often employed in eukaryotic systems.

The complexity of higher organisms depends on elaborate regulatory networks that can respond differently to a variety of signals. A response to multiple signals can be accomplished by regulating a single gene with several different proteins, each of which is controlled by a different parameter. Whether or not the target gene is transcribed depends upon the particular combination of proteins that bind to the regulatory sequences. This type of regulation, termed **combinatorial control**, is a hallmark of eukaryotic gene regulation. Regulating a gene with multiple proteins also makes it possible to activate (or repress) a gene only when a certain set of conditions are met. Regulating transcription with more than one protein also provides a simple way to regulate genes that specify cell type, as we shall see next.

Yeast cell type is determined by different combinations of four transcriptional regulators

A remarkably elegant combinatorial circuit controls the genes that determine cell type in the budding yeast, *Saccharomyces cerevisiae*. These single-celled eukaryotes can exist as one of three cell types: **a**, α, or **a**/α cells (Figure 8.54). The **a** and α cell types are haploid, meaning that they each have just a single complement of chromosomes. The **a**/α cell type is diploid, and forms when an **a** and an α cell fuse with one another, in a process that is known as **mating** (Figure 8.54). Only haploid cells of opposite **mating type** can mate with one another – that is, **a** and α cells can mate, but not two **a** cells, or two α cells. The diploid **a**/α cell that results from mating of an **a** and an α cell is incapable of mating but, under conditions of starvation, can undergo meiosis to form haploid spores.

The different phenotypes of **a**, α and **a**/α cells result from distinct patterns of gene expression patterns in each cell type (Figure 8.55). These patterns are specified by four proteins, a1, α1, α2 and MCM1, that bind DNA in different combinations (Figure 8.56). These regulatory proteins are encoded at the *MAT* locus, which is located on yeast chromosome III. The DNA sequence of this segment of chromosome III differs with cell type: **a** cells encode the a1 protein at the *MAT* locus, while α cells encode the α1 and α2 proteins.

In **a** cells, the genes whose transcription confers the **a** phenotype – the **a**-specific genes – are activated by a cellular protein called Mcm1 (Figure 8.56, top panel). The α-specific genes, which require an activator to be expressed, are not transcribed in **a** cells. The a1 protein, although present, does not bind DNA in **a** cells but plays an important role in diploid cells, as explained below.

The pattern of activation and repression is reversed in α cells (Figure 8.56, middle panel): the α1 protein activates transcription of α-specific genes while α2 represses transcription of **a**-specific genes by recruiting a co-repressor complex

Figure 8.54 The three yeast cell types. Budding yeast can exist as one of three cell types: the a or α haploid cell types, or the a/α diploid. Each haploid cell type produces a distinct type of signaling molecule known as a pheromone that is secreted into the surrounding medium. **a** cells and α cells also express pheromone receptors on their surfaces that detect the presence of an opposite cell type. **a** and α cells mate by fusing with one another, forming diploid **a**/α cells, which are incapable of mating. When deprived of adequate nutrients, a diploid cell will undergo meiosis to produce four haploid cells.

	a-specific genes	α-specific genes	haploid-specific genes
a-cell (haploid)	ON	OFF	ON
α-cell (haploid)	OFF	ON	ON
a-cell (diploid)	OFF	OFF	OFF

Figure 8.55 Expression patterns in the three yeast cell types. Three classes of genes are important in specifying yeast cell type.

called Tup1/Ssn6. Interestingly, both α1 and α2 exert their effect in concert with the cellular Mcm1 protein, which activates transcription when recruited by α1 but helps to repress transcription when in complex with α2.

The fusion of an **a** and an α cell during mating gives rise to a new regulatory scenario in the resulting **a**/α diploid cell (Figure 8.56, bottom panel). The a1 protein and α2 proteins combine to form a heterodimer that binds to a different set of DNA sites than either α2 or a1 alone. The a1/α2 heterodimer represses transcription of a variety of genes that are needed only in haploid cells. One of the repressed genes encodes α1; without this protein, the α-specific genes are not activated. As in α cells, the α2 protein also represses transcription of **a**-specific genes. With no expression of either **a** or α genes, the diploid cells are incapable of mating.

The a1/α2 heterodimer also represses transcription of genes that encode proteins required by both types of haploid cell. One of these haploid-specific genes is *RME1*, which encodes a DNA-binding protein that represses genes required for meiosis. When a1/α2 repress transcription of *RME1*, the Rme1 repressor protein is no longer synthesized and the genes required for meiosis can be expressed. Diploid cells are thus able to undergo meiosis to produce haploid spores when the yeast cells are starved. Interestingly, one of the genes that Rme1 represses is *IME1*, which encodes the transcriptional activator recruited by Ume6 that we learned about in Section 8.11.

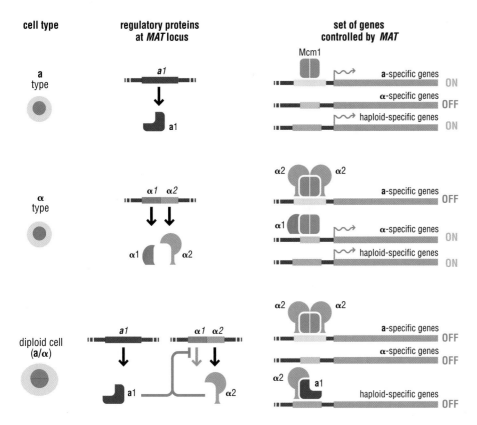

cell type

regulatory proteins
at *MAT* locus

set of genes
controlled by *MAT*

Figure 8.56 Control of cell type in budding yeast. The budding yeast *S. cerevisiae* can exist as one of three cell types: the **a** or α haploid cell types, or the **a**/α diploid. Each cell type is established by regulating three groups of genes: the **a**-specific genes, α-specific genes and haploid-specific genes. These genes are regulated by proteins expressed by the MAT locus.

The regulation of cell type in yeast provides an example of how four transcriptional regulators present at different levels can cooperate to set up different patterns of gene expression that lead to cell differentiation.

Regulation of the human interferon-β enhancer integrates signals from several pathways

The regulation of the human interferon-β gene provides an example of how synergy between several regulatory proteins makes it possible to regulate transcription of a single gene in response to multiple signals. The interferon protein encoded by the interferon-β gene helps to defend cells from viral infection by inhibiting synthesis of the viral genome and stimulating the host immune response. When viruses initially infect human cells, genes that encode interferon proteins are induced, allowing the cells to mount an antiviral response before the immune system has had time to respond.

Induction of the interferon-β gene depends on the binding of a number of different proteins to the interferon-β enhancer. As shown in Figure 8.57, these proteins bind to precisely positioned sites within the enhancer to form a complex of DNA-bound transcriptional regulators called the **enhanceosome**. Cooperative interactions between neighboring proteins are required for stable complex assembly. These interactions are facilitated by binding of an architectural DNA-binding protein called HMG-I(Y) (Figure 8.57a), which helps to nucleate assembly of the

Figure 8.57 Proteins bound to the interferon-β enhanceosome. (a) Multiple proteins are required to bind cooperatively to the interferon-β enhancer to trigger activation of the interferon-β gene. The architectural DNA-binding protein HMG-I(Y) (dark blue) introduces pronounced bends into the DNA, which facilitates the binding of the other transcriptional regulators. (b) Structures of the regulatory proteins bound to the interferon-β enhanceosome. C-Jun and ATF2 are bZIP DNA-binding domains, p65 and p50 belong to the Rel homology domain family, and IRFeA, 3C, 7B, and 7D all belong to the IRF family. PDB hybrid of 2O6G, 2O6I, 1T2K.
From Panne *et al.* (2007).

complex of DNA-bound transcriptional regulators, although it is not present in the final assembled complex (Figure 8.57b). When HMG-I(Y) binds DNA in the first step in enhanceosome assembly, it introduces bending of the DNA, which facilitates binding of the other transcriptional regulators (see Figure 8.57). In the final complex, however (Figure 8.57b), the DNA is not bent.

Binding of all of the proteins to DNA proceeds in a highly cooperative manner, with the binding of certain proteins early in complex formation facilitating the binding of further components of the enhanceosome. Once the enhanceosome is complete, the interferon-β gene is activated. The interferon-β gene is not transcribed in cells that synthesize only a subset of the activator proteins. In general, the requirement for using a large number of proteins to activate a single gene ensures that the gene is activated only under a very precise set of conditions.

8.13 SIGNALING CASCADES AND REGULATION OF TRANSCRIPTION

Cells are continually altering their patterns of gene expression in order to respond to changes in their surroundings. Shifting hormonal levels within an organism, alterations in the environment, and responses to infections are all examples of changes that trigger alterations in gene expression. We have already seen several examples in which changes in the level of sugars or amino acids within the cellular environment can lead to increased or decreased transcription of particular genes. However, extracellular signals can also trigger a variety of biochemical events whose end result is a change in gene expression. These signaling cascades can range from the simple to the elaborate and can change the activities of gene-specific regulators through a variety of mechanisms. In this section, we describe several examples that illustrate how cells modulate transcription patterns in response to signals from the environment.

Nuclear receptor proteins respond to effector molecules that diffuse through the membrane

One of the simplest ways in which an extracellular signal can be transmitted to the nucleus is through steroid hormones that serve as allosteric effectors of regulatory proteins. These signaling molecules include the sex hormones, estradiol (commonly known as estrogen) and testosterone, as well as the thyroid hormone, which plays an essential part in controlling metabolism. These hormones are lipid soluble and can diffuse through membranes into the cell, where they bind directly to **nuclear receptor proteins**. At least 50 members of the nuclear receptor superfamily have been identified in the human genome, each of which is thought to bind specifically to a particular signaling molecule.

Nuclear receptor proteins contain a DNA-binding domain and a ligand-binding domain, which binds the hormone or other small ligand. The binding of ligand induces a conformational change in the ligand-binding domain of the receptor (Figure 8.58). Rather than affecting the binding of the receptor to DNA, as in the case of the bacterial Trp repressor, the ligand-induced conformational change instead affects the ability of the nuclear hormone receptor to recruit different co-repressor and co-activator complexes to the DNA (Figure 8.58a).

While most nuclear receptor proteins bind DNA as homodimers, some, such as the thyroid hormone receptor–retinoid X receptor, can bind DNA as heterodimers. The ability to heterodimerize makes more complex regulatory networks possible. In the absence of thyroid hormone, for example, the thyroid hormone receptor–retinoid X receptor heterodimer recruits two complexes, N-CoR and SMRT,

Figure 8.58 Co-activator recruitment and nuclear translocation of steroid hormone receptors upon hormone binding. (a) Some nuclear receptor proteins bind to DNA in the absence of ligand. These receptors recruit co-repressor complexes in the absence of ligand, and to co-activator complexes in the presence of ligand. (b) Some nuclear receptor proteins are found in the cytoplasm, sequestered by the heat shock protein HSP90. Ligand binding leads to dissociation of HSP90, allowing the receptor to enter the nucleus, bind to DNA, and recruit a co-activator complex.

which are co-repressors that contain histone deacetylase subunits. In contrast, the thyroid-hormone–bound heterodimer recruits two other proteins, CREB-binding protein (CBP) and P300/CBP-associated factor (PCAF), which are activators with histone acetyltransferase activity.

In the case of other nuclear receptor proteins such as the glucocorticoid receptor, ligand binding also triggers translocation of the protein to the nucleus (Figure 8.58b). In the absence of hormone, glucocorticoid receptor is in a complex with an inhibitory protein, HSP90, which causes the glucocorticoid receptor to remain in the cytoplasm. The binding of hormone causes HSP90 to dissociate and allows the receptor to enter the nucleus.

The very direct effect of ligand binding on transcription makes it possible to design drugs that have potent effects on nuclear receptor signaling pathways. Tamoxifen, for example, is highly successful in treating some types of breast cancer in which the tumor cells require estrogen to proliferate. Tamoxifen competes with estradiol for binding to the estrogen receptor protein but, although it can also bind to the estrogen receptor, its different molecular structure prevents the ligand-binding domain from adopting its normal estradiol-bound conformation. The altered protein conformation in the presence of tamoxifen prevents recruitment of the co-activator complex, thereby blocking activation of estrogen-responsive genes and preventing cell proliferation.

Two-component signal transduction pathways in bacteria, fungi and plants are composed of histidine kinases and response regulators

There are many examples in which the first response to an extracellular signal is the phosphorylation of a cytoplasmic protein. The phosphorylated protein, in turn, triggers additional biochemical events that eventually lead to a change in transcription. A relatively simple example of this type of signaling cascade is the **two-component signal transduction pathways** found in bacteria, fungi and plants. Two-component systems have a **sensor kinase**, which detects the relative amounts of specific molecules in the cell's environment, and a **response regulator**, which is activated when it is phosphorylated by the sensor protein (Figure 8.59).

The sensor protein is a histidine kinase, typically membrane-bound, that becomes autophosphorylated on a histidine residue when it detects a certain signal. This phosphate is then transferred to an aspartate residue on the response regulator, which often contains a DNA-binding domain. So the environmental signal leads to sensor protein autophosphorylation, which leads to response regulator phosphorylation and a downstream effect.

There are many variations of this type of signaling. There are sensor kinases that are cytoplasmic, response regulators that have functions other than transcriptional regulation, and more complex networks that involve more than two proteins. The common feature of these signaling systems is the transfer of a phosphate from the histidine kinase sensor to a downstream effector protein.

The response of *E. coli* to changing levels of inorganic phosphate is a simple example of two-component signal transduction. As shown in Figure 8.59, the PhoR sensor kinase is a transmembrane protein that binds a free phosphate in

Figure 8.59 Two-component signaling mechanisms in response to extracellular stimuli. The basic two-component phosphotransfer system, found in bacteria, consists of a dimeric transmembrane receptor histidine kinase and a cytoplasmic response regulator. Information flows between the two proteins in the form of a phosphoryl group that is transferred from the histidine kinase to the response regulator. Histidine kinases catalyze ATP-dependent autophosphorylation of a specific conserved His residue (H). The activities of histidine kinases are modulated by environmental signals such as nutrients or osmotic stress. The phosphoryl group (P) is then transferred to a specific aspartic acid residue (D) located within the conserved regulatory domain of a response regulator. Phosphorylation of the regulator typically activates an associated (or downstream) effector domain, which ultimately elicits a specific cellular response.

the periplasmic space of the bacterial cell. At low levels of phosphate, the phosphate dissociates from the periplasmic sensor domain triggering a change in the cytoplasmic domain. This change activates the sensor kinase, causing it to catalyze transfer of the γ-phosphate of ATP to a histidine in the sensor kinase domain. The PhoR sensor kinase then transfers the phosphate from its histidine residue to an aspartate residue in PhoB, the response regulator. Phosphorylated PhoB then activates transcription of a number of genes required by the cell when phosphate concentrations are low.

A cascade of events involving protein phosphorylation, proteolysis and nuclear translocation regulates NF-κB activity

Many signaling cascades involve multiple steps and may take advantage of a number of different mechanisms for transmitting a signal from one molecule to the next. This is particularly common in metazoans, which require more complex regulatory networks than unicellular organisms. NF-κB, a DNA-binding dimer that plays a key role in mammalian inflammatory and immune responses, provides a well-characterized example of a multistep cascade of events that leads to transcriptional activation (Figure 8.60).

NF-κB is a heterodimer of the p50 and p65 proteins. As mentioned earlier, these DNA-binding domains belong to the Rel homology domain family. In unstimulated cells, the NF-κB heterodimers are held in the cytoplasm by an inhibitory protein called I-κB, which binds to a specialized region of NF-κB (the nuclear localization signal) that is responsible for targeting NF-κB to the nucleus. However, extracellular signals that are induced by infection trigger a cascade of events that lead to the activation of the I-κB kinase (IKK). Activated IKK phosphorylates two serine residues in the regulatory domain of I-κB. The phosphorylated I-κB protein is now recognized by an enzyme called an E3 ubiquitin ligase, which attaches a chain of ubiquitin proteins to I-κB. This ubiquitin chain targets IκB for degradation by a large enzymatic complex called the **proteasome**. Thus the ubiquitin-tagged I-κB is degraded, exposing the nuclear localization signal of NF-κB and freeing the protein to translocate to the nucleus, where it binds DNA and activates gene expression.

Figure 8.60 Change in subcellular localization of nuclear factor (NF)-κB in response to a diverse set of extracellular signals. In response to extracellular signals, a cascade of events (arrows) leads to the phosphorylation and thus the activation of the inhibitory (I)-κB kinase (IKK) (blue). Activated IKK then phosphorylates the regulatory domain of I-κB (red). Phosphorylated I-κB is targeted for degradation, releasing NF-κB dimers (orange) and exposing the nuclear localization signal (green). NF-kappaB is now free to translocate to the nucleus and activate gene expression.

⊖ Ubiquitin and the proteasome are described in detail in Chapter 11.

8.14 REGULATION OF ELONGATION AND TERMINATION BY RNA AND PROTEINS

Most of the mechanisms for regulating transcription that we have discussed up to this point have the effect of controlling whether or not transcription of a given gene is initiated. Yet, even after RNA polymerase has initiated transcription and broken free of the promoter, the subsequent steps in elongation and termination provide additional opportunities for the cell to regulate whether the full-length transcript is made. Here we discuss further mechanisms for regulating whether a full-length transcript is synthesized by the RNA polymerase. We will see that, in several cases, the transcribed RNA folds to form a defined structure that plays a central role in the regulation of elongation and termination: that is, the RNA performs a function, rather than merely serving as a template for protein synthesis.

(a)

(b)

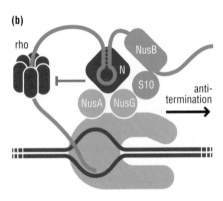

Figure 8.61 Anti-termination by the phage lambda N protein. (a) In the lytic phase, transcription is initiated at both the P_R and P_L promoters, but transcription initially terminates at the t_R and t_L terminators, respectively. This leads to synthesis of the Cro and N proteins encoded by these two transcripts. The N protein prevents termination at the two terminator sites, allowing production of longer transcripts. (b) A number of proteins are required along with N to relieve termination. The N protein binds to a stem–loop structure formed by the *nut* site (shown in (a)). This RNA–protein complex interacts with several other cellular proteins, including NusA, NusG, NusB and S10, as well as RNA polymerase. In some way that is not yet understood, this relieves termination by the cellular rho protein and allows polymerase to transcribe through this region.

The bacteriophage lambda N and Q proteins prevent stalling and termination by altering the properties of bacterial RNA polymerase

Some genes in the bacteriophage lambda chromosome are transcribed only when premature termination is prevented. This type of regulation is therefore denoted **anti-termination**. As we saw in Section 8.10, the transcription of the phage lambda lytic genes begins at the divergent P_L and P_R promoters (Figure 8.61a). Initially, the transcripts produced are short because RNA polymerase encounters sequences in the RNA, t_L and t_R, that lead to termination of transcription (Figure 8.61a). Thus, only the Cro and N proteins are synthesized at the early stages of lytic growth. We previously described the role of Cro in binding DNA and preventing re-establishment of the lysogenic state (Section 8.10); the other important lytic gene product is the N protein, which also begins to build up in the cell. At sufficient concentrations, the N protein alleviates termination at the t_L and t_R sites (Figure 8.61a).

The N protein causes anti-termination by interacting with special features in the transcribed RNA termed *nut* sites (for *N ut*ilization) (Figure 8.61a). Within each *nut* site are two RNA sequences called Box A and Box B. Box B forms an RNA stem–loop to which N binds, while Box A is a binding site for the *E. coli* protein NusB (Figure 8.61b). The binding of N protein, together with NusB and other cellular proteins (NusA, NusE, and NusG; see Figure 8.61b), alters the properties of RNA polymerase such that it gains the ability to transcribe efficiently through the transcription termination site. The precise mechanism by which these proteins act in concert to relieve termination is not yet understood.

With termination suppressed, RNA polymerase can now continue to transcribe the downstream genes, which encode several proteins including those needed for phage chromosome replication. Another of the downstream genes encodes the lambda Q protein, which overrides transcription termination at the $P_{R'}$ promoter and allows transcription of the late lytic genes (Figure 8.62).

In contrast to the N protein, the Q protein prevents termination by binding directly to DNA. The binding site for the Q protein, called QBE, is located between the –35 and –10 regions of the $P_{R'}$ promoter, which is transcribed by RNA polymerase holoenzyme containing the σ^{70} subunit. In the absence of Q, transcription terminates just downstream of the –10 region. When sufficient Q protein is present, it binds to the QBE site and directly contacts the σ^{70} subunit, causing elongation to resume (Figure 8.62). The Q protein remains associated with the elongation complex as the downstream sequences are transcribed.

The human immunodeficiency virus-1 anti-termination protein Tat is essential for transcription of the retroviral genome

Transcription of the Human Immunodeficiency Virus (HIV)-1 genome depends upon a viral anti-termination factor whose function is similar to that of the phage lambda N protein. The HIV-1 genome contains a sequence called *tar* that, much like the phage lambda *nut* site, causes formation of an RNA stem–loop in the transcribed RNA called the TAR element. Formation of the stem–loop will lead to premature termination of transcription without the action of Tat, a small viral

Figure 8.62 Anti-termination by the phage lambda Q protein. (a) The late lytic genes are under control of the $P_{R'}$ promoter, which is transcribed by RNA polymerase holoenzyme containing σ^{70}. In the absence of Q, transcription pauses upstream of the lytic genes. (b) Q protein relieves termination by binding directly to a DNA sequence located between the −10 and −35 sequences in the promoter. Q interactions with σ^{70} relieve the pause, and the holoenzyme complex with Q resumes elongation.

(a)

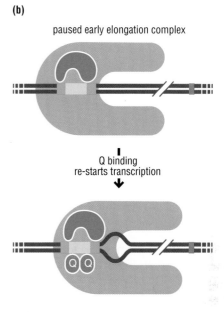

(b)

protein that binds to the *tar* sequence. Tat binds cooperatively to *tar* together with the cellular kinase, P-TEFb, which phosphorylates the CTD of RNA polymerase II as well as the two regulators of elongation, DSIF and NELF (Figure 8.63). As we learned in Section 8.5, hyperphosphorylation of the RNA polymerase CTD relieves pausing, thereby preventing premature termination of the viral transcript.

Except for the involvement of polymerase phosphorylation, anti-termination by the HIV-1 Tat protein is remarkably reminiscent of the way in which lambda N protein overcomes transcription termination in bacteria (Figure 8.61). This extends not only to the use of a viral (Tat or N) protein and a cellular protein (Cdk9 or NusB) to bind the RNA cooperatively, but also to sequence similarity between the bacterial NusG protein and Spt5, which is one of the subunits of DSIF.

Attenuation is a mechanism for controlling the effect of terminator sequences

Another form of regulation in bacteria takes advantage of the ability of certain mRNA sequences to form alternative RNA secondary structures – one that leads to transcription termination and another to productive gene expression. Called **attenuation**, this mechanism is used to control transcription of genes encoding enzymes required for amino acid biosynthesis. Consequently, under certain conditions, expression of the full length transcript is promoted, whereas, under other conditions, it is the termination of transcription that is favored.

An example of attenuation is found in the regulation of the *E. coli trp* operon (Figure 8.64). Regulation of the *trp* operon is determined by the levels of tryptophan present in the cellular environment. Different levels of tryptophan cause the transcript to adopt different secondary structures that directly influence how transcription proceeds. So how does this regulation occur in practice?

We learned earlier (Section 8.9) that the genes of the *trp* operon are transcribed as a single **polycistronic message**. This mRNA contains a leader sequence near its 5′ end that contains an intrinsic terminator (also called an **attenuator**) along with two different additional RNA sequences capable of forming stem–loop structures.

The leader sequence, which encodes a small open reading frame that is rich in codons for the amino acid, tryptophan, contains four key blocks of ribonucleotides – regions 1, 2, 3, and 4 – which can form alternative pairing arrangements. Regions 1 and 2 are complementary and can form a stem–loop structure, while regions 3 and 4 can also base-pair and form a Rho-independent terminator (Section 8.6). Regions 2 and 3 can form an alternative stem–loop structure, which then leaves regions 1 and 4 unpaired. Figure 8.64 shows how these different possible pairings are exploited to regulate transcription.

Figure 8.63 Anti-termination by the human immunodeficiency virus (HIV)-1 Tat protein. The Tat protein from the retrovirus HIV-1 causes anti-termination in a way that is reminiscent of the λ N protein. Tat binds to an RNA sequence in the HIV-1 genome called *tar*, causing premature termination. The association of the Tat–*tar* complex with several cellular proteins allows RNA polymerase to overcome the block and continue transcribing. CTD, C-terminal domain.

Figure 8.64 Transcription attenuation regulates amino acid biosynthesis in bacteria. The *trp* operon leader transcript can fold into two alternative RNA secondary structures. At low levels of tryptophan, the structure with the attenuator stem–loop 2:3 is formed and transcription proceeds through the entire operon. At high levels of tryptophan, the structure with a stem–loop formed between RNA sequences 1:2 and a terminator stem–loop between sequences 3:4 is formed, leading to termination.

➲ The process of tRNA charging is described in Chapter 10.

Once RNA polymerase transcribes the beginning of the leader sequence, the ribosome binds to the message and begins to translate the leader peptide while RNA polymerase continues to transcribe the downstream sequences (Figure 8.64). When levels of tryptophan are low, however, the ribosome stalls within the short open reading frame while waiting for a tRNA to be charged with tryptophan. This occludes region 1 in the leader sequence of the RNA, which then cannot base-pair with region 2 (Figure 8.64). If region 4 has not yet been transcribed, regions 2 and 3 form an alternative stem-loop known as an anti-terminator. This prevents the terminator (the 3-4 pair) from forming and thereby ensures that transcription elongation will continue. At high levels of tryptophan, however, there is enough charged tRNA to allow the ribosome to move rapidly to the end of the small open reading frame. This allows the terminator stem-loop to form, leading to transcription termination.

Direct binding of metabolites to RNA riboswitches can control transcription termination

The formation of alternative RNA structures can be regulated even more directly by the binding of small metabolites, as is the case for riboswitches. A **riboswitch** is a portion of an RNA transcript that can bind directly to a small molecule that controls the secondary structure that the RNA transcript adopts, which in turn determines whether a gene is expressed. Riboswitches can regulate either transcription or translation; we focus here on their role in regulating transcription. Riboswitches are found primarily in bacteria; indeed, 2% of the genes in *B. subtilis* are regulated by riboswitches. Riboswitches have also been identified in plants and fungi and are thought to exist in archaea.

Riboswitches regulate transcription by controlling whether or not a terminator or anti-terminator stem–loop forms in the mRNA, which determines whether transcription is completed appropriately. All riboswitches (including those regulating translation, which we will learn about in Chapter 10) contain two key regions (Figure 8.65a): the aptamer, a sometimes complex RNA structure that binds to the metabolite being sensed, and an expression platform, which consists of segments of the RNA that control transcription (or translation). The binding of metabolite to the aptamer alters the structure of the riboswitch and thereby regulates whether or not the downstream gene is expressed. Examples of riboswitches include those regulated by purine nucleotides, thiamine pyrophosphate, *S*-adenosyl methionine, glycine, and lysine.

Purine riboswitches provide a simple illustration of how metabolite binding can regulate transcription. The adenine riboswitch of *B. subtilis* regulates transcription of genes required for adenine synthesis and transport. Expression of these genes depends upon whether a terminator or an anti-terminator forms in the 5′ end of the mRNA transcript (Figure 8.65b); the formation of a terminator or anti-terminator is determined, in turn, by the binding of adenine to the aptamer. As in the case of the *trp* operon, the RNA contains four regions that can pair in alternative ways (Figure 8.65b). In the absence of adenine, the downstream genes are transcribed because the riboswitch forms the RNA secondary structure shown in Figure. 8.65b, with regions 2 and 3 forming an anti-terminator. By contrast, when the adenine concentration is high, adenine binds directly to the aptamer and triggers formation of an alternative structure in which regions 1 and 2 pair (Figure 8.65c). This allows regions 3 and 4 to form a terminator, thereby suppressing transcription of the downstream genes. Figure 8.65d shows how

(a)

(b) low adenine

(c) high adenine

(d)

aptamer

adenine

1 2 3 4

expression platform

1

2 3

4

UUUUU

AUG

transcription is ON

1 2 3 4

UUUUU

transcription is terminated

Figure 8.65 Riboswitch regulation of transcription termination. (a) All riboswitches contain an aptamer, which binds to the metabolite, and an expression platform. In the example shown, regions 3 and 4 of the RNA form a terminator. (b) The adenine riboswitch. When concentrations of adenine are low, the RNA forms the secondary structure shown and the downstream gene is transcribed. The pairing of regions 2 and 3 form an anti-terminator, which prevents regions 3 and 4 from forming a terminator. (c) When the concentration of adenine is high, adenine binds directly to the aptamer and triggers formation of an alternative RNA structure. Regions 3 and 4 pair to form a terminator. (d) The structure of the adenine riboswitch (PDB Y26).

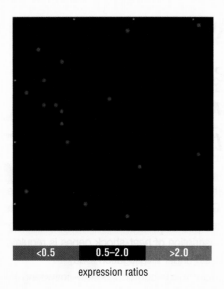

<0.5	0.5–2.0	>2.0

expression ratios

Figure 3 Identification of new heat shock genes by microarray analysis. Fluorescein-labeled probes from Jurkat cells subjected to heat shock were compared with Cy5-labeled probes from untreated cells. Visible spots represent genes where differences in expression are readily detected. Reproduced with permission from Schena *et al.*, *Proceedings of the National Academy of Sciences of the U S A* 1996;**93**:10614–10619.

The very strong and clear transcriptional induction of the specific genes on heat shock, as is shown in Figure 2, has been exploited in a wide range of experiments including the development of microarrays to examine the transcript levels of thousands of genes in parallel. When Patrick Brown, Ronald Davis, and colleagues first set out to develop DNA arrays by attaching (printing) 1046 human cDNAs to glass slides, heat shock was one of the first conditions they tested. Even though only a limited number (by today's standards) of cDNAs were printed on the slides, they discovered new genes whose expression was induced when human T (Jurkat) cells were subjected to heat shock (Figure 3). In the experiment shown, the cDNAs prepared from samples grown under two conditions (37°C and 43°C) were labeled through the incorporation of two different fluorescent nucleotides, fluorescein-dCTP or Cy5-dCTP. When the resulting two cDNA populations were hybridized to the array, differences in the ratio of the two dyes indicated whether a given gene was induced or repressed upon the heat shock treatment.

Pelham HR. A regulatory upstream promoter element in the *Drosophila hsp 70* heat-shock gene. *Cell* 1982;**30**:517–528.

Ritossa F. A new puffing pattern induced by temperature shock and DNP in *Drosophila*. *Experientia* 1962;**18**:517–573.

Schena M, Shalon D, Heller R, *et al*. Parallel human genome analysis: microarray-based expression monitoring of 1000 genes. *Proceedings of the National Academy of Sciences of the U S A* 1996;**93**:10614–10619.

8.15 TRANSCRIPTIONAL SILENCING

So far, the regulatory mechanisms we have discussed are those controlling synthesis of an individual mRNA transcript. We now turn to a different type of regulatory mechanism found in eukaryotes, whereby transcription in large regions of the chromosome containing multiple promoters may be suppressed for long periods, often over the course of many cell divisions. This mechanism is referred to as silencing, to distinguish it from other modes of repression. Although transcriptional repression is generally a transient and reversible state, the silenced state is stable and can be inherited through many cell generations. This mode of inheritance, as we have mentioned before (see Section 4.7), is known as epigenetic inheritance.

Transcriptional silencing is brought about by changes in chromatin structure

Transcriptional silencing is brought about by an alteration in chromatin structure that can often be detected by assays that measure differences in nuclease accessibility and staining. In Chapter 4, we saw that expressed genes usually lie in euchromatic regions of the chromosome, while heterochromatic regions are condensed in a way that generally leads to gene silencing. Another feature that distinguishes transcriptional silencing from simple repression is that silenced regions can spread along a chromosome. Spreading can be blocked by boundary elements, as we saw for example in Section 4.8.

or translation; we focus here on their role in regulating transcription. Riboswitches are found primarily in bacteria; indeed, 2% of the genes in *B. subtilis* are regulated by riboswitches. Riboswitches have also been identified in plants and fungi and are thought to exist in archaea.

Riboswitches regulate transcription by controlling whether or not a terminator or anti-terminator stem–loop forms in the mRNA, which determines whether transcription is completed appropriately. All riboswitches (including those regulating translation, which we will learn about in Chapter 10) contain two key regions (Figure 8.65a): the aptamer, a sometimes complex RNA structure that binds to the metabolite being sensed, and an expression platform, which consists of segments of the RNA that control transcription (or translation). The binding of metabolite to the aptamer alters the structure of the riboswitch and thereby regulates whether or not the downstream gene is expressed. Examples of riboswitches include those regulated by purine nucleotides, thiamine pyrophosphate, *S*-adenosyl methionine, glycine, and lysine.

Purine riboswitches provide a simple illustration of how metabolite binding can regulate transcription. The adenine riboswitch of *B. subtilis* regulates transcription of genes required for adenine synthesis and transport. Expression of these genes depends upon whether a terminator or an anti-terminator forms in the 5′ end of the mRNA transcript (Figure 8.65b); the formation of a terminator or anti-terminator is determined, in turn, by the binding of adenine to the aptamer. As in the case of the *trp* operon, the RNA contains four regions that can pair in alternative ways (Figure 8.65b). In the absence of adenine, the downstream genes are transcribed because the riboswitch forms the RNA secondary structure shown in Figure. 8.65b, with regions 2 and 3 forming an anti-terminator. By contrast, when the adenine concentration is high, adenine binds directly to the aptamer and triggers formation of an alternative structure in which regions 1 and 2 pair (Figure 8.65c). This allows regions 3 and 4 to form a terminator, thereby suppressing transcription of the downstream genes. Figure 8.65d shows how

Figure 8.65 Riboswitch regulation of transcription termination. (a) All riboswitches contain an aptamer, which binds to the metabolite, and an expression platform. In the example shown, regions 3 and 4 of the RNA form a terminator. (b) The adenine riboswitch. When concentrations of adenine are low, the RNA forms the secondary structure shown and the downstream gene is transcribed. The pairing of regions 2 and 3 form an anti-terminator, which prevents regions 3 and 4 from forming a terminator. (c) When the concentration of adenine is high, adenine binds directly to the aptamer and triggers formation of an alternative RNA structure. Regions 3 and 4 pair to form a terminator. (d) The structure of the adenine riboswitch (PDB Y26).

⮑ We learn about molecular chaperones in Chapter 11.

the aptamer folds about the bound adenine nucleotide to form a globular three-dimensional structure.

Heat-shock factor converts RNA polymerase into a productive elongating complex by recruiting the Mediator–co-activator complex

Regulation of transcription elongation can also be brought about entirely by the recruitment of additional proteins, without the involvement of RNA. We learned earlier (Section 8.5) that RNA polymerase sometimes stalls after transcribing a few dozen nucleotides, a phenomenon known as promoter proximal pausing. An example can be found in the transcription of the *Drosophila hsp70* gene. The Hsp70 protein is a molecular chaperone that plays a variety of protective roles when cells are subject to abnormally high temperatures, a condition known as heat shock.

One of the functions of Hsp70 is to prevent aggregation of partially synthesized or unfolded proteins, which can accumulate under conditions of heat shock. In the regulation of the *hsp70* gene, regulation of promoter proximal pausing is used to poise genes for a rapid transcription in response to heat shock. The discovery of the transcriptional response to heat shock in *Drosophila* is described in Experimental approach box 8.4.

The regulation of the *hsp70* gene under normal and heat-shock conditions is illustrated in Figure 8.66. Upstream of the *hsp70* TATA elements, there are multiple binding sites for two sequence-specific DNA-binding proteins, GAGA factor and Heat-Shock Factor (Hsf). Under normal conditions, GAGA factor binds to the *hsp70* promoter and recruits the ATP-dependent NUcleosome-Remodeling Factor (NURF). NURF helps to maintain the promoter in a nucleosome-free open configuration that allows binding of the RNA polymerase II basal transcription machinery. The polymerase enzyme can initiate transcription and synthesizes transcripts of 21–35 nucleotides in length. However, at normal temperatures of around 25°C, the polymerase enzyme stalls (Figure 8.66a) because it fails to undergo sufficient phosphorylation to be converted into a productive elongating complex (see Section 8.5).

The stalled polymerase can be restarted by a sudden rise in temperature, leading to expression of the heat-shock gene and synthesis of Hsp70. When subjected to heat shock, in which the temperature is shifted to 37°C, the previously monomeric Hsf protein forms trimers that are able to bind to DNA sequences called heat-shock elements (HSEs) in the *hsp70* promoter. The conversion of Hsf from the monomeric to trimeric form is thought to involve a conformational change, caused by the heat itself or by a heat-induced redirection of chaperone proteins, that leads to the formation of the active trimer. Once bound to DNA, Hsf contacts the Mediator complex, either recruiting it to the promoter or by simply interacting with Mediator that is already bound to the stalled polymerase (Figure 8.66b). This complex in turn interacts with other factors, including a kinase (p-TEFb) that phosphorylates the CTD of the Rpd1 subunit of RNA polymerase II, thereby converting the stalled enzyme into a productive elongating complex. The *hsp70* gene is transcribed and the encoded protein synthesized, allowing the cell to mount a protective response.

(a) without heat shock

(b) with heat shock

Figure 8.66 Mechanism of heat-shock factor (Hsf) conversion of RNA polymerase II into a productive elongating complex. The *Drosophila hsp70* gene contains upstream binding sites for the GAGA and Hsf regulatory proteins. (a) Under normal conditions, Hsf monomers are present in the cell but cannot bind DNA. GAGA binds DNA and recruits the nucleosome-remodeling factor (NURF). RNA polymerase II synthesizes a short stretch of RNA and then stalls. (b) Under conditions of heat shock, the Hsf monomers trimerize to form an active DNA-binding complex. The DNA-bound Hsf trimer contacts the mediator complex, which helps rescue the stalled polymerase and convert it into a productive elongating complex.

8.4 EXPERIMENTAL APPROACH

Serendipitous discovery of the heat-shock response opened the door for numerous studies of transcriptional regulation

Sometimes new directions in research are brought about by serendipity. This is the case for studies of the heat-shock response, which have yielded many important insights into transcriptional regulation in both bacteria and eukaryotes because the response is rapid and easy to induce. That an increase in temperature led to a rapid change in transcription was discovered by Ferruccio Ritossa in 1962 when someone in the laboratory changed the temperature of the incubators in which he was maintaining his stocks of *Drosophila*. At the time, Ritossa was examining the polytene chromosomes from the salivary glands of *Drosophila* larvae (which are easy to see under the microscope because they are polyploid). He and others were finding that different regions of the chromosomes 'puffed up' during different developmental stages and that these 'puffs' corresponded to regions where radioactive uridine was incorporated, indicating that RNA was being made. As shown in Figure 1, Ritossa reported that if the larvae were shifted from 25°C to 30°C for 30 minutes, the puffing pattern changed markedly at a defined number of chromosomal positions.

This exciting discovery, brought about because of a change in incubator temperature, spawned many important research directions that led to the discovery of protein chaperones (discussed in Chapter 11), and produced many insights into transcriptional regulation. A key question that arose immediately was how the transcription of heat-shock genes is regulated, aspects of which are still under investigation today.

Hugh Pelham made contributions to answering this question by defining the promoter regions of the *Drosophila hsp70* promoter required for the heat-shock induction. He cloned portions of the *hsp70* gene into an SV40 vector (which contains the replication origin from the SV40 virus) and transfected this vector into COS cells (monkey cells) (see Section 16.3 for a discussion of cloning technology). By following the end of the *hsp70* transcript using 5′ S1 mapping (described in Section 16.9), he showed that transcription of the *hsp70* promoter could be induced in a heterologous system (Figure 2). By deleting different regions of the hsp70 promoter, he further showed that induction was lost when sequences downstream of –66 were deleted. These experiments indicated sequences between –10 and –66 were sufficient for the heat-shock induction.

Figure 1 Increased transcription of heat shock puffs. The 2L 14 and 15 regions of the salivary gland chromosome of *Drosophila buschii* larvae reared at 25 °C (a) and 30 °C (b). Reproduced with permission from Ritossa, *Experientia* 1962;**18**:517–573.

Figure 2 Promoter regions required for heat shock induction. S1-mapping of *hsp70* transcripts (indicated by the arrow) synthesized in COS cells for different deletion mutants of the *hsp70* promoter without (-) and with heat (+). Reproduced with permission from Pelham, *Cell* 1982;**30**:517–528.

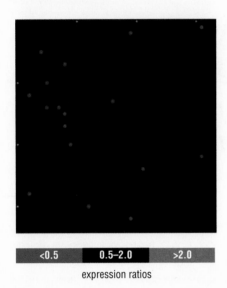

expression ratios

Figure 3 **Identification of new heat shock genes by microarray analysis.** Fluorescein-labeled probes from Jurkat cells subjected to heat shock were compared with Cy5-labeled probes from untreated cells. Visible spots represent genes where differences in expression are readily detected. Reproduced with permission from Schena *et al.*, *Proceedings of the National Academy of Sciences of the U S A* 1996;**93**:10614–10619.

The very strong and clear transcriptional induction of the specific genes on heat shock, as is shown in Figure 2, has been exploited in a wide range of experiments including the development of microarrays to examine the transcript levels of thousands of genes in parallel. When Patrick Brown, Ronald Davis, and colleagues first set out to develop DNA arrays by attaching (printing) 1046 human cDNAs to glass slides, heat shock was one of the first conditions they tested. Even though only a limited number (by today's standards) of cDNAs were printed on the slides, they discovered new genes whose expression was induced when human T (Jurkat) cells were subjected to heat shock (Figure 3). In the experiment shown, the cDNAs prepared from samples grown under two conditions (37°C and 43°C) were labeled through the incorporation of two different fluorescent nucleotides, fluorescein-dCTP or Cy5-dCTP. When the resulting two cDNA populations were hybridized to the array, differences in the ratio of the two dyes indicated whether a given gene was induced or repressed upon the heat shock treatment.

Pelham HR. A regulatory upstream promoter element in the *Drosophila hsp 70* heat-shock gene. *Cell* 1982;**30**:517–528.

Ritossa F. A new puffing pattern induced by temperature shock and DNP in *Drosophila*. *Experientia* 1962;**18**:517–573.

Schena M, Shalon D, Heller R, *et al.* Parallel human genome analysis: microarray-based expression monitoring of 1000 genes. *Proceedings of the National Academy of Sciences of the U S A* 1996;**93**:10614–10619.

8.15 TRANSCRIPTIONAL SILENCING

So far, the regulatory mechanisms we have discussed are those controlling synthesis of an individual mRNA transcript. We now turn to a different type of regulatory mechanism found in eukaryotes, whereby transcription in large regions of the chromosome containing multiple promoters may be suppressed for long periods, often over the course of many cell divisions. This mechanism is referred to as silencing, to distinguish it from other modes of repression. Although transcriptional repression is generally a transient and reversible state, the silenced state is stable and can be inherited through many cell generations. This mode of inheritance, as we have mentioned before (see Section 4.7), is known as epigenetic inheritance.

Transcriptional silencing is brought about by changes in chromatin structure

Transcriptional silencing is brought about by an alteration in chromatin structure that can often be detected by assays that measure differences in nuclease accessibility and staining. In Chapter 4, we saw that expressed genes usually lie in euchromatic regions of the chromosome, while heterochromatic regions are condensed in a way that generally leads to gene silencing. Another feature that distinguishes transcriptional silencing from simple repression is that silenced regions can spread along a chromosome. Spreading can be blocked by boundary elements, as we saw for example in Section 4.8.

In this section, we describe several different mechanisms by which silencing is established. We shall see that each of these examples involves chemical modifications of either histone proteins or DNA, but that these modifications are established and maintained in rather different ways. Notably, the final example we will cover shows a heterochromatic state that can be established by transcription within the heterochromatic region itself, through a mechanism known as RNA interference, or RNAi.

For our first example we turn again to the budding yeast mating-type loci. This organism is one of the most amenable experimental systems for exploring transcriptional regulatory mechanisms, yet the same basic mechanisms also appear to function in animal cells. In all cases, the silent state must first be established at a specific chromosomal location from which the silent state then often spreads. Subsequently, the silent state must be maintained over an extended period of time.

Figure 8.67 Silencing of the mating-type cassettes in budding yeast. Yeast chromosome III contains a silent copy of the a cassette at the HML locus and a silent copy of the a cassette at the HMR locus. The *MAT* locus contains an actively transcribed copy of either the a or the α cassette.

The budding yeast mating-type loci and telomeric regions are silenced by establishing a heterochromatic state

We saw in Section 8.12 how cell type in budding yeast is determined by the DNA-binding proteins that are expressed in a region of chromosome III called the *MAT* locus. Yeast **a** and α cells differ only in the genes encoded by the *MAT* locus: **a** cells contain the a1 regulatory gene at the *MAT* locus and α cells have in its place the α regulatory genes, α1 and α2 (see Figure 8.67). In addition to the actively transcribed sequence at the *MAT* locus, chromosome III contains an additional copy of the **a** and of the α gene sequences that are, however, *not* expressed. As shown in Figure 8.67, to the right of the *MAT* locus is a silenced copy of the **a** sequence called *HMRa* and to the left is a silenced copy of the α sequences called *HMLα*. These silent copies are present in all yeast cells, irrespective of which genes are expressed at the *MAT* locus.

Silencing of the *HMRa* and *HMLα* **mating-type cassettes** (also called the *HM* loci) depends upon the Silencing Information Regulator (Sir) proteins, which establish and maintain a region of heterochromatin at *HML* and *HMR*. Abf1, along with two other proteins called Rap1 and Orc1, binds to regions within *HMR* and *HML* (Figure 8.68). These proteins help establish silencing by recruiting the other three SIR proteins, Sir2, Sir3, and Sir4. These proteins spread along the chromatin fiber through a network of interactions between each other and with the N-terminal tails of histones H3 and H4. The silencing activity depends upon the enzymatic activity of Sir2, a nicotinamide adenosine dinucleotide (NAD⁺)-dependent deacetylase. Sir2 maintains the chromatin in a hypoacetylated state, targeting Lys16 in histone H4. Sir2 is unique among lysine deacetylases in its dependence on NAD⁺, which makes silencing in yeast sensitive to intracellular levels of NAD⁺.

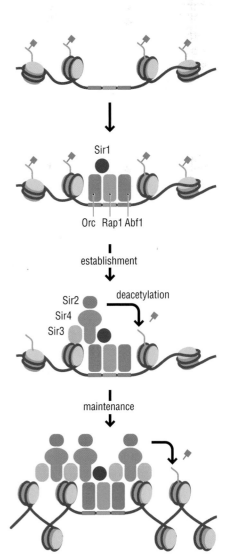

Figure 8.68 Establishment of silencing of the *HML* and *HMR* loci in budding yeast. Silencing in *S. cerevisiae* is initiated by the recruitment of SIR proteins to specific regions of the chromosome by the DNA-binding proteins Rap1, Orc1, and Abf1. The Sir2, Sir3, and Sir4 proteins then spread along the chromatin and the Sir2 histone deacetylase maintains the chromatin in a hypoacetylated state. Sir1 plays a role in establishing silencing, although its precise function is not well understood.

⊙ We learn about the rDNA repeats and the role of heterochromatin in preventing recombination among them in Sections 4.4 and 4.11.

Genes that are near the yeast telomeres are also silenced, in a mechanism that is very similar to the silencing of the HM loci. The Rap1 protein, which helps to nucleate silencing at the HM loci, also binds to the telomeric repeat sequences. Rap1 recruits the Sir2/Sir3/Sir4 complex, which spreads along the chromatin and establishes a heterochromatic state that silences gene expression. When localized to the genes near the telomere, the silencing is referred to as telomeric silencing.

Sir2 is also involved in a third form of silencing, in which it represses RNA polymerase II transcripts in the intergenic regions between rDNA genes. This helps to establish a heterochromatic state that prevents recombination between the multiple copies of the rDNA genes. This form of silencing relies on a different set of partner proteins that act along with Sir2 to inhibit transcription.

Telomeric silencing in a pathogenic yeast is linked to niacin levels in the urinary tract

Telomeric silencing is exploited by *Candida glabrata*, a pathogenic yeast that causes urinary tract infections, to sense the host environment. An important step in infection occurs when this organism adheres to the host urinary tract, through the action of adhesion proteins. Expression of one of the genes encoding an adhesion protein coding genes is controlled by telomeric silencing and is therefore switched off when there is adequate NAD+ for Sir2 deacetylase activity. Importantly, while other yeast species can synthesize NAD+ from a variety of precursor molecules, *C. glabrata* requires niacin (nicotinic acid) as a precursor, which it obtains from its surroundings. In the urinary tract, however, there is typically inadequate niacin. *C. glabrata* therefore cannot make enough NAD+, which reduces Sir2 deacetylase activity and relieves silencing of the adhesion gene. The pathogenic yeast can now adhere to the urinary tract tissues and infect the host. Note how the level of NAD+ indirectly controls adhesion of the yeast by determining whether the Sir2 deacetylase is active and, hence, whether the required adhesion genes are expressed. More generally, the dependence of the entire Sir2 family of enzymes on NAD+ allows the activity of these enzymes to be coupled to cellular metabolism, in which NAD+ plays a critical role.

DNA methylation of an enhancer-blocking insulator regulates silencing of an imprinted gene

Transcriptional silencing can be mediated by modifications to DNA, as well as to histone proteins. As we learned in Chapter 4, mammalian chromosomes are subject to methylation at cytosine residues. One way such DNA methylation can mediate transcriptional silencing is by blocking the binding of proteins that bind preferentially to unmethylated DNA. This mechanism underlies the regulation of the imprinted genes, *IGF2* and *H19*.

⊙ We introduce DNA methylation, imprinting and insulator elements in Chapter 4.

The *IGF2* gene encodes an insulin-like growth factor hormone that is required by the developing embryo, while *H19* encodes a non-protein coding RNA transcript that is thought to play a role in cancer. These genes are located near one another on human chromosome 11 and can be activated by the same enhancer element (Figure 8.69a). However, only one of these genes is expressed in each chromosome: the *IGF2* gene is expressed in the maternal chromosome, while the *H19* gene is expressed in the paternal chromosome. What gives rise to this pattern of expression?

The difference between maternal and paternal gene expression arises because of different patterns of DNA methylation in the insulator control region (ICR), which lies between the *IGF2* and *H19* genes. The ICR is the binding site for a protein called CCCTC-binding factor (CTCF), which contacts the ICR with a set of zinc finger domains. CTCF can only bind to this DNA sequence when it is *not* methylated. In the paternal chromosome, cytosine bases in the ICR are methylated and CTCF cannot bind. By contrast, this DNA sequence is not methylated in the maternal chromosome and CTCF is able to bind (Figure 8.69a).

It is the binding of CTCF to the ICR that helps to control which gene the enhancer acts upon. So how does this happen? In the maternal chromosome (Figure 8.69a), proteins bound to the enhancer activate transcription of the *H19* gene. However, CTCF bound to the ICR blocks the ability of the enhancer to activate the *IGF2* gene, which remains inactive.

In paternal chromosomes, however, both the ICR and the promoter of the adjacent *H19* gene are methylated. Methylation of the ICR means that CTCF can no longer bind to it. Consequently, the effect of the enhancer on the *IGF2* gene is not blocked and transcription of *IGF2* is activated. Methylation of the promoter region of the *H19* gene, however, compromises the binding of the transcriptional machinery and the *H19* gene remains inactive. As shown in Figure 8.69b, CTCF is thought to control formation of separate chromosome loops, which in turn influence whether transcription of one or the other gene is activated by proteins bound to the enhancer.

The importance of imprinting at this locus is seen in the developmental disorder called Beckwith–Wiedemann syndrome. Children with this syndrome are much larger than normal at birth and are prone to a variety of cancers in childhood. The syndrome arises when the *IGF2* gene is expressed in both copies of chromosome 11, while the *H19* gene is inactive. This can arise due to defects

Figure 8.69 Imprinting and silencing. (a) The maternal chromosome (top) is not methylated, so CTCF can bind DNA. The enhancer activates transcription of the *H19* gene but CTCF blocks activation of *IGF2*. In the paternal chromosome (bottom), the CTCF binding site and *H19* are methylated. CTCF does not bind to DNA and the enhancer activates transcription of *IGF2*. (b) Model of proposed looping that accounts for the patterns of expression in part (a). It has been proposed that CTCF mediates formation of a loop that represses transcription of *IGF2* and allows transcription of *H19*.

Figure 8.70 MeCP2 binds to methylated DNA and recruits a co-repressor complex. MeCP2 binds specifically to methylated CpG and recruits a complex containing a histone deacetylase (HDAC) that represses transcription of the target genes.
From Bienvenu *et al.* (2006).

➲ We discuss the discovery of RNA interference and the mechanisms by which the small regulatory RNAs are generated and how they function in a variety of contexts in Chapter 9.

in methylation of the paternal chromosomes, or chromosomal rearrangements that disrupt the ICR.

Methylated DNA can recruit enzymes that modify chromatin and repress transcription

DNA methylation also can have an opposite effect on regulatory protein binding than is the case for CTCF: instead of blocking the binding of proteins to DNA, methylation can lead to the recruitment of proteins that bind specifically to methylated DNA. Methyl-binding domains are protein domains that bind specifically to methyl cytosine. Some proteins that contain a methyl-binding domain recruit complexes that contain nucleosome-remodeling or histone-modifying enzymes, which repress transcription. The DNA methylation is the stable epigenetic signal that is maintained through cell division, while the methyl-binding domain helps to 'interpret' the signal to repress transcription.

The protein MeCP2 is a methyl-binding domain that binds to methylated DNA and recruits the Sin3A transcriptional co-repressor, which contains a histone deacetylase (Figure 8.70). MeCP2 represses a variety of genes in humans, including genes in neuronal cells. Mutations in the gene encoding MeCP2 can lead to a devastating disease known as Rett syndrome, which affects 1 in 10 000 girls. Girls with this disorder appear normal at birth but soon develop autism-like behaviors, repetitive hand movements, and mental retardation, along with physical symptoms such as scoliosis and seizures. The syndrome is found primarily in girls because the gene for MeCP2 is located in the X chromosome. As in the case of other X-linked traits, this means that affected girls will have one mutant and one wild-type copy of MeCP2, whereas an affected boy has only a single mutant copy. Without a normal copy of the gene, most male fetuses do not survive to term, and the few that are born die within the first two years of life.

The synthesis of small RNA molecules triggers silencing of the fission yeast mating-type loci

Like the budding yeast, *S. cerevisiae*, the fission yeast, *Schizosaccharomyces pombe*, also silences its mating-type cassettes by establishing a heterochromatic state. However, the mechanism of silencing used in *S. pombe* is quite different from that of its evolutionarily distant cousin, the budding yeast. Establishment of the silenced chromatin in *S. pombe* exploits a more recently discovered mechanism that depends upon small regulatory RNA molecules; this mechanism is known as **RNA interference**.

As we saw in Section 4.8, the silent mating-type cassettes in *S. pombe* lie in a region of heterochromatin that flanks the centromere. The establishment of this heterochromatic region, as outlined in Figure 8.71, begins with transcription of the centromere, which consists of a series of repeated nucleotide sequences. Although transcription of silenced regions seems counterintuitive, *some* transcription is required at many loci where production of small RNAs leads to silencing of *other* genes. Both strands of the centromeric DNA are transcribed, which gives rise to complementary RNA strands that can base-pair with one another. These double-stranded RNA molecules are then cleaved into 21 bp fragments that are

Figure 8.71 Establishment of heterochromatic silencing in the fission yeast. Silencing of the mating-type loci in *S. pombe* requires transcription of the coding and non-coding strands of a set of repeated sequences. These two transcripts form double-stranded RNA (dsRNA, shown in blue) that is processed into short fragments by the RNAi machinery. The resulting short siRNAs lead, through unknown mechanisms, to formation of a repressive chromatin structure that includes the Clr3 histone deacetylase, Clr4 histone H3 lysine 9 methyltransferase and Swi6, which binds to H3 tails methylated at lysine 9.

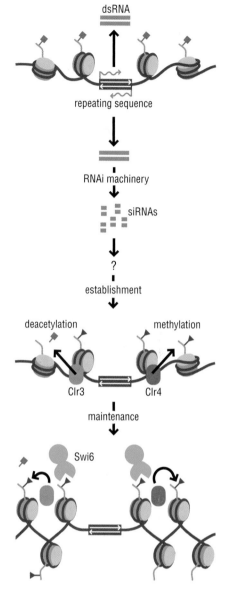

incorporated into a protein complex (called RITS, which stands for RNA-induced initiation of transcriptional silencing), which then associates with the actively transcribed centromeres (Figure 8.71). The RITS complex, in turn, recruits a histone deacetylase and a histone methylase that specifically target lysine 9 of histone H3. The histone-modifying enzymes ensure that the histones in this region have the chemical hallmarks of heterochromatin in *S. pombe*, namely the hypoacetylation of histones, and methylation at lysine 9 of histone H3.

The methylated lysine 9 in the N-terminal tail of histone H3 recruits additional histone-modifying enzymes to the chromatin template and facilitates the spreading of heterochromatin. The protein, Swi6, binds to the methylated histone tail via its chromodomain, a specialized protein domain that binds to methylated lysine (which we learned about in Section 4.8). Swi6, in turn, recruits additional histone methylase and deacetylase enzymes, which target the neighboring histones. As additional histones are modified, additional Swi6 proteins bind and the cycle continues, spreading the heterochromatin further down the chromatin fiber. As we learned in Chapter 4 (Section 4.8) the spreading of heterochromatin continues through the mating-type genes, thereby silencing them. The spreading of heterochromatin only ceases when a boundary element is reached.

RNA interference has been found to similarly silence the transcription of many other eukaryotic genes, particularly repeated genes, although small regulatory RNA molecules also target steps subsequent to transcription. The important role played by RNA in regulating gene expression was not appreciated until recently, when several observations pointed to a role for small non-coding RNA molecules in regulating protein levels. The discovery of RNA interference revealed that there was a common mechanism underlying the different forms of regulation by small RNA molecules. We learn more about this exciting aspect of biology in the next two chapters.

✱ SUMMARY

In order for the information in the genome to be expressed, the cell must synthesize an RNA copy of the DNA – a process called transcription. Since this is the key step in gene expression, transcription is highly regulated in all cells. The RNA polymerase enzyme catalyzes the synthesis of the RNA from ribonucleotide triphosphate precursors, using the DNA template strand to assemble a copy of the coding strand. RNA is synthesized from the 5′ to the 3′ end, just as in DNA replication, in a defined series of events known as the transcription cycle: initiation, elongation, and termination. Cells can regulate the transcription of individual genes by targeting any of the steps in the transcription cycle. In the next chapter, we will learn how the transcribed RNA is processed to produce the final functional RNA molecule in the cell.

RNA polymerases

- RNA polymerase enzymes in all three domains of life share a structurally similar core and synthesize RNA using a common mechanism.

- Bacteria and archaea have a single RNA polymerase enzyme, while eukaryotes have three:

  - RNA polymerase I – transcribes genes encoding rRNAs.

  - RNA polymerase II – transcribes all protein-coding genes as well as some non-coding regulatory RNA molecules.

  - RNA polymerase III – transcribes tRNA and other RNA molecules needed for RNA splicing.

- Plants are an exception among eukaryotes in possessing a fourth RNA polymerase (IV) that transcribes regulatory RNA molecules.

- Transcription by RNA polymerase II is closely coupled to RNA processing. RNA polymerase II has a C-terminal domain that recruits proteins required for processing the mRNA. This recruitment is triggered by changes in the phosphorylation of residues in the CTD.

- Bacterial RNA polymerase is targeted to its promoter by the sigma subunit, which recognizes the –35 and –10 regions of the promoter.

- Eukaryotic and archaeal RNA polymerases require a set of proteins called general transcription factors that recognize a set of promoter elements, and promote some of the later events in the transcription cycle. All of these polymerases depend on a subunit called TBP for promoter recognition.

- The complex of RNA polymerase with the double-stranded DNA is called the closed complex.

Steps in transcription

- Transcription begins with isomerization of RNA polymerase to form the open complex, in which the DNA strands are separated and synthesis of the RNA begins, using the template strand to position the incoming ribonucleotide triphosphates.

- RNA polymerase may first go through several cycles of abortive initiation, producing very short transcripts, before breaking free of the promoter and entering into the elongation phase.

- The elongating polymerase maintains an open transcription bubble consisting of a melted region of the DNA (around 12–15 base pairs) and the 8–9 most recently added nucleotides of the RNA transcript base-paired with the template DNA.

- During transcription elongation, RNA polymerase occasionally stalls to correct an error due to a misincorporated base, or because of features of the DNA or RNA that cause the polymerization to pause.

- If the stalled RNA polymerase backtracks, transcript cleavage factors promote cleavage of the most recently added ribonucleotides and thereby help to restart transcription.

- Transcription termination in bacteria can occur by two different mechanisms. In rho-independent termination, the DNA contains an intrinsic terminator sequence that gives rise to a stem–loop in the transcribed RNA that causes transcription to terminate. Rho-dependent termination requires the action of the ATP-dependent enzyme, rho.

- Termination of transcription by eukaryotic RNA polymerase II is coupled to the final steps in mRNA processing.

- Since eukaryotic DNA is packaged into chromatin, chromatin-modifying and nucleosome-remodeling enzymes are required for efficient transcription initiation and elongation, as well as regulation of transcription in eukaryotes.

Regulation of transcription

- Any step of the transcription cycle can be regulated to control the rate at which RNA is transcribed.

- Proteins that decrease the level of transcription are called repressors, whereas those that increase the level of transcription are called activators.

- Genes are typically regulated by one or more proteins that bind to a regulatory site in the DNA. A regulatory site may lie near the promoter (as is typical in bacteria) or many thousands of base pairs upstream or downstream of the gene (as is more common in eukaryotes).

- Many genes in eukaryotes are regulated by multiple proteins that bind to one or more regulatory regions.

- The binding of an allosteric effector molecule can alter the ability of a protein to bind DNA or regulate transcription. In this way, proteins such as the Trp repressor in bacteria and the glucocorticoid receptor protein in humans can regulate transcription in direct response to the level of a ligand present in the cell.

- A variety of mechanisms are used to repress or activate transcription initiation in bacteria. Transcription can be repressed by blocking the binding of RNA polymerase (Trp repressor, cI). Transcription can be activated by recruiting the polymerase holoenzyme (CAP), stimulating open complex formation (CAP), or by altering the structure of promoter DNA (MerR).

- Eukaryotic genes are regulated by co-activator and co-repressor complexes, which are recruited to the DNA by sequence-specific DNA binding proteins. These complexes typically contain enzymes that reposition nucleosomes, or that add or remove post-translational modifications from histones and other transcription factors. An additional complex called Mediator is required for appropriate gene regulation by RNA polymerase II.

- Histone acetyltransferases are used to activate transcription in eukaryotes, while histone deacetylases repress transcription. Other enzymes, such as histone methyltransferases and ATP-dependent nucleosome-remodeling complexes, can be used either to activate or repress transcription.

- The expression of some bacterial and eukaryotic genes is regulated by controlling transcription elongation or termination. The bacterial N and Q proteins, and the HIV Tat protein, regulate transcription in this way.

- Small molecule metabolites can regulate elongation and termination by binding to a region of the mRNA called a riboswitch, which adopts different secondary structural elements depending upon whether or not the metabolite is bound. Many bacterial genes and some eukaryotic and archaeal genes are regulated in this way.

- Transcription in large regions of eukaryotic chromosomes can be repressed or silenced by a variety of mechanisms that alter chromatin structure and sometimes depends on small regulatory RNAs (RNA interference).

 FURTHER READING

8.1 OVERVIEW OF TRANSCRIPTION

Lee TI, Young RA. Transcription of eukaryotic protein-coding genes. *Annual Review of Biochemistry* 2000;**34**:77–137.

Orphanides G, Reinberg D. A unified theory of gene expression. *Cell* 2002;**108**:439–451.

8.2 RNA POLYMERASE CORE ENZYME

Cramer P, Armache KJ, Baumli S, *et al*. Structure of eukaryotic RNA polymerases. *Annual Review of Biophysics* 2008;**37**:337–352.

Murakami KS, Darst SA. Bacterial RNA polymerases: the whole story. *Current Opinion in Structural Biology* 2003;**13**:31–39.

Paule MR, White RJ. Survey and summary: transcription by RNA polymerases I and III. *Nucleic Acids Research* 2000;**28**:1283–1298.

8.3 PROMOTER RECOGNITION IN BACTERIA AND EUKARYOTES

Gruber TM, Gross CA. Multiple sigma subunits and the partitioning of bacterial transcription space. *Annual Review of Microbiology* 2003;**57**:441–466.

Hughes KT, Mathee K. The anti-sigma factors. *Annual Review of Microbiology* 1998;**52**:231–286.

Juven-Gershon T, Hsu JY, Theisen JW, Kadonaga JT. The RNA polymerase II core promoter – the gateway to transcription. *Current Opinion in Cell Biology* 2008;**20**:253–259.

Kornberg RD. Mediator and the mechanism of transcriptional activation. *Trends in Biochemical Sciences* 2005;**30**:235–239.

Woychik NA, Hampsey M. The RNA polymerase II machinery: structure illuminates function. *Cell* 2002;**108**:453–463.

8.4 INITIATION OF TRANSCRIPTION AND TRANSITION TO AN ELONGATING COMPLEX

Saunders A, Core LJ, Lis JT. Breaking barriers to transcription elongation. *Nature Reviews Molecular Cell Biology* 2006;**7**:557–567.

Shilatifard A, Conaway JW, Conaway RC. The RNA polymerase II elongation complex. *Annual Review of Biochemistry* 2003;**72**:693–715.

8.5 TRANSCRIPTION ELONGATION

Borukhov S, Lee J, Laptenko O. Bacterial transcription elongation factors: new insights into molecular mechanism of action. *Molecular Microbiology* 2005;**55**:1315–1324.

Gilmour DS. Promoter proximal pausing on genes in metazoans. *Chromosoma* 2009;**118**: 1–10.

Pandit S, Wang D, Fu XD. Functional integration of transcriptional and RNA processing machineries. *Current Opinion in Cell Biology* 2008;**20**:260–265.

Workman JL. Nucleosome displacement in transcription. *Genes and Development* 2006;**20**:2009–2017.

8.6 TRANSCRIPTION TERMINATION

Proudfoot NJ, Furger A, Dye MJ. Integrating mRNA processing with transcription. *Cell* 2002;**108**:501–512.

8.7 PRINCIPLES OF TRANSCRIPTION REGULATION

Li B, Carey M, Workman JL. The role of chromatin during transcription. *Cell* 2007;**128**:707–719.

8.8 DNA-BINDING DOMAINS IN TRANSCRIPTIONAL REGULATORS

Garvie C, Wolberger C. Recognition of specific DNA sequences. *Molecular Cell* 2001;**8**:937–946.

8.9 MECHANISMS FOR REGULATING TRANSCRIPTION IN BACTERIA

Lawson CL, Swigon D, Murakami KS, *et al*. Catabolite activator protein: DNA binding and transcription activation. *Current Opinion in Structural Biology* 2004;**14**:10–20.

Schumacher MA, Brennan RG. Structural mechanisms of multidrug recognition and regulation by bacterial multidrug transcription factors. *Molecular Microbiology* 2002;**45**:885–893.

Xu H, Hoover TR. Transcriptional regulation at a distance in bacteria. *Current Opinion in Microbiology* 2001;**4**:138–144.

8.10 COMPETITION BETWEEN CI AND CRO AND THE FATE OF BACTERIOPHAGE LAMBDA

Hochschild A. The lambda switch: cI closes the gap in autoregulation. *Current Biology* 2002;**12**:R87–R89.

Ptashne M. *A Genetic Switch*, 3rd edn. Cold Spring Harbor, New York: Cold Spring Harbor Laboratory Press, 2004.

8.11 MECHANISMS FOR MODULATING EUKARYOTIC TRANSCRIPTION

Buchwalter G, Gross C, Wasylyk B. Ets ternary complex transcription factors. *Gene* 2004;**324**:1–14.

Traven A, Jelicic B, Sopta M. Yeast Gal4: a transcriptional paradigm revisited. *EMBO Reports* 2006;**7**:496–499.

Vershon AK, Pierce M. Transcriptional regulation of meiosis in yeast. *Current Opinion in Cell Biology* 2000;**12**:334–339.

8.12 COMBINATORIAL REGULATION OF EUKARYOTIC TRANSCRIPTION

Herskowitz I. A regulatory hierarchy for cell specialization in yeast. *Nature* 1989;**42**:749–757.

Kassir Y, Adir N, Boger-Nadjar E, *et al.* Transcriptional regulation of meiosis in budding yeast. *International Review of Cytology* 2003;**224**: 111–117.

Panne D. The enhanceosome. *Current Opinion in Structural Biology* 2008;**18**:236–242.

8.13 SIGNALING CASCADES AND REGULATION OF TRANSCRIPTION

Bain DL, Heneghan AF, Connaghan-Jones KD, Miura MT. Nuclear receptor structure: implications for function. *Annual Review of Physiology* 2007;**69**:201–220.

Stock AM, Robinson VL, Goudreau PN. Two-component signal transduction. *Annual Review of Biochemistry* 2000;**69**:183–215.

Vallabhapurapu S, Karin M. Regulation and function of NF-kappaB transcription factors in the immune system. *Annual Review of Immunology* 2009;**27**:693–733.

Zhang J, Lazar MA. The mechanism of action of thyroid hormones. *Annual Review of Physiology* 2000;**62**:439–466.

8.14 REGULATION OF ELONGATION AND TERMINATION BY RNA AND PROTEINS

Henkin TM, Yanofsky C. Regulation by transcription attenuation in bacteria: how RNA provides instructions for transcription termination/antitermination decisions. *BioEssays* 2002;**24**:700–707.

Roberts JW, Shankar S, Filter JJ. RNA polymerase elongation factors. *Annual Review of Microbiology* 2008;**62**:211–233.

Roth A, Breaker RR. The structural and functional diversity of metabolite-binding riboswitches. *Annual Review of Biochemistry* 2009;**78**:305–334.

8.15 TRANSCRIPTIONAL SILENCING

Grewal SIS, Jia S. Heterochromatin revisited. *Nature Reviews Genetics* 2007;**8**:35–46.

Moazed D. Small RNAs in transcriptional gene silencing and genome defence. *Nature* 2009;**457**:413–420.

Rusche LN, Kirchmaier AL, Rine J. The establishment, inheritance, and function of silenced chromatin in Saccharomyces cerevisiae. *Annual Review of Biochemistry* 2003;**72**:481–516.

RNA PROCESSING

9

INTRODUCTION

RNAs have diverse roles in the flow of information from DNA to protein. Messenger RNAs (mRNAs) are informational molecules, which specify the amino acid sequence of proteins: they carry the information encoded by genes and direct ribosomes to synthesize proteins from this information. Other RNAs function to regulate gene expression or act as parts of the machinery for translating the mRNA into protein. For example, ribosomal RNAs (rRNAs) provide structural and catalytic components of the protein synthesis machinery (as we see in Chapter 10), and transfer RNAs (tRNAs) function as adaptors between the nucleotide sequence of an mRNA and the amino acid sequence of the encoded protein.

We saw in Chapter 8 how RNAs are synthesized from their DNA template by RNA polymerases. However, many RNA molecules that emerge from RNA polymerase are not fully functional. Instead, these **precursor RNAs (pre-RNAs)** first need to be modified in one or more ways to generate the functional or **mature RNA**. In this chapter we consider the great variety of different alterations, collectively known as **RNA processing**, to which an RNA molecule may be subjected, and the machinery that executes them. We also discover how these modifications play an important part in generating biological diversity, enabling single RNA molecules to encode multiple end-products.

We begin this chapter by briefly reviewing the key forms of modification that occur. We will then revisit each of these approaches in more detail throughout the rest of the chapter.

9.1 OVERVIEW OF RNA PROCESSING

RNA is processed in many different ways

Figure 9.1 summarizes the many ways in which RNAs can be processed. In the simplest case, the ends of the transcript are removed from the larger precursor RNA by RNA-cleaving enzymes called **ribonucleases**.

The coding sequence of many genes, particularly eukaryotic mRNAs, is not continuous but is interrupted by non-coding sequences known as **introns**. Despite their seemingly disruptive nature, introns are included in the precursor RNA produced by RNA polymerase along with the coding segments, or **exons**. However, introns must be removed to generate a functional rRNA, tRNA, or mRNA that will

Figure 9.1 Types of RNA processing. The mature RNA is indicated by the blue lines. Some RNAs are cleaved from longer precursors; for others, introns (gray lines) are removed and exons are spliced together. The 3′ ends of RNAs can also be modified by polyadenylation, and eukaryotic mRNAs undergo 5′-end capping. Finally, some transcripts are edited by the insertion, deletion, or modification of a base. m⁷G, 7-methyl guanosine.

⊃ We describe the termination of transcription in Section 8.6.

be translated into protein. The removal of the introns and rejoining of the exons that remain is known as **RNA splicing**.

Eukaryotic pre-mRNAs also undergo a modification known as **5′ capping** whereby the 5′ end becomes modified with a 7-methyl guanosine; they also undergo **polyadenylation** in which a long stretch of adenosine nucleotides is added to the 3′ end as part of the transcription termination process.

RNA processing can also involve the insertion or deletion of bases or chemical modification of bases: these changes are collectively known as **RNA editing**. Finally, during the last stage of their cellular life, RNA molecules are degraded by nucleases into individual nucleotides that can be recycled by the cell.

RNA processing events are points for regulation and quality control, and are sources of diversity

Fundamentally, RNA processing is required to produce functional RNAs, but processing also brings with it three associated benefits. First, processing events contribute to the regulation of gene expression: the efficiency with which a particular RNA is spliced, capped, polyadenylated, and edited governs how quickly the RNA can function or be translated. In addition, the concentration of any RNA in a cell is regulated by the balance between the transcription of the RNA (the rate at which a given RNA is synthesized) and the degradation of the RNA by nucleases. For example, if the transcription of a gene is induced by an environmental signal such as a steroid hormone, the synthesis of the protein can be switched off by the degradation of the RNA when the hormone signal is no longer present.

Second, as well as providing a means for regulating gene expression levels, RNA processing can generate many different RNAs from just one gene. Many genes contain multiple introns and exons. Exons can be spliced together in a variety of combinations to produce different proteins (Figure 9.2). An extreme example of the multiplicity of products that can be generated from a single gene is provided

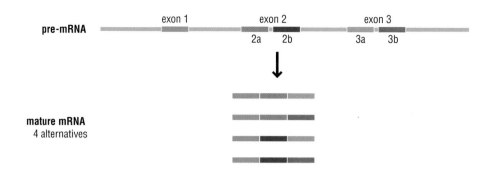

Figure 9.2 Protein diversity brought about by alternative splicing. The putative pre-mRNA shown has one exon 1, two possible exon 2s (2a and 2b) and two possible exon 3s (3a and 3b). Alternative splicing of this example could lead to the production of four different RNA products (exons 1+2a+3a, 1+2a+3b, 1+2b+3a and 1+2b+3b).

by the *Drosophila melanogaster* gene *dscam*. Proteins produced from this gene help to guide nerve cells in the central nervous system. The *dscam* gene has 115 exons; of these, 95 exons are in four clusters that are alternatively spliced. Thus there is the potential to produce 38 016 different proteins. This diversity contributes to the exquisite specificity observed for neuronal connections.

Alternative splicing does not change the sequence of nucleotides in an RNA but determines whether a given sequence is present or not in the synthesized product: each exon encodes an amino acid sequence that may or may not be included in a protein synthesized from the precursor RNA once splicing has occurred. RNA editing, on the other hand, involves removing, inserting, or chemically modifying individual nucleotides, all events that can change the amino acid sequence of the encoded protein. We will see an example where up to half the nucleotides in a fully processed RNA are the result of editing (see Section 9.7).

Third in our list of associated benefits, RNA processing also serves as a quality control step. mRNAs that are defective in some way (for example, as a result of errors during transcription or splicing) are detected and degraded by cellular machinery. In addition to these systems that monitor endogenous RNA quality, eukaryotic cells possess a system referred to as **RNA interference** (**RNAi**) to detect and degrade foreign RNAs produced by parasites such as viruses. The RNAi machinery shares enzymes with the small RNA-based systems involved in transcriptional silencing, which we discussed in Chapter 8. All of the quality control steps prevent the accumulation of RNAs (or eventually proteins) that could be detrimental to the cell.

Many RNA processing machines have RNA components

Most RNA processing steps are performed by multicomponent macromolecular machines. In many cases these machines interact with and influence one another. In addition, many complexes are localized to specific compartments in eukaryotic cells: some RNA processing complexes are found in the nucleus, either distributed throughout or localized to specific regions such as the nucleolus, whereas others are found in the cytoplasm. Specific interactions and localization allow the processes to occur in the appropriate sequential manner and provide checks to ensure the proper processing has occurred. Thus, for example, mRNAs are not exported from the nucleus to the cytoplasm until they are appropriately modified.

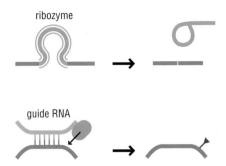

Figure 9.3 Functions of RNAs in RNA processing machines. Some RNAs, called ribozymes, have intrinsic enzymatic activity. Although these RNAs can often catalyze reactions in the absence of proteins, proteins generally increase the activity of ribozymes. Other RNAs, referred to as guide RNAs (green), direct proteins (orange) to carry out reactions at specific sequences by base-pairing with the target mRNAs.

Many RNA processing machines contain both protein and RNA components and are therefore referred to as **ribonucleoproteins** (**RNPs**). While a subset of the RNAs function only as structural components of the RNP, many RNAs play active parts in the functions of an RNP. For example, some RNAs have catalytic activity and are termed ribozymes (Figure 9.3). Others are called **guide RNAs** because they base-pair with the precursor RNAs and guide the catalytic components of an RNP to where the processing events should take place. In the course of this chapter, and in the following chapter on protein synthesis, we shall see many examples of structural, catalytic, or guide functions of RNAs in diverse RNP complexes.

9.2 tRNA AND rRNA PROCESSING

Having reviewed the main modifications that occur to RNA molecules, and some of the key components involved in the modification processes, we will now consider a range of specific processing events in more detail. We begin with the processing associated with the synthesis of mature tRNAs and rRNAs.

tRNAs and rRNAs are synthesized as longer precursors

The transcripts for most tRNAs and rRNAs in all organisms are synthesized as long precursor molecules that must be processed to produce functionally active RNAs. The precursor might simply need one or both ends trimmed; alternatively, more extensive processing could be required. Often, a precursor RNA includes two or more functional RNAs that are cut, or excised, from the precursor by enzymatic cleavage. Two examples of this are shown in Figure 9.4. An *Escherichia coli* rRNA precursor encodes three rRNAs as well as several tRNAs, whereas the *Saccharomyces cerevisiae* precursor encodes three rRNAs. Synthesis of a single long precursor ensures that similar amounts of the functional RNAs, such as all three rRNAs, are produced, simply because the synthesis of one is intrinsically linked to another as they come from the same precursor molecule.

Ribonucleases, the enzymes that cleave the RNAs into smaller parts, can operate in one of two ways (Figure 9.5). Some ribonucleases, denoted exonucleases, digest the RNA strand by successively removing nucleotides from the end. Most exonucleases recognize the 3′ end and digest in the 3′ to 5′ direction, although some bacterial and eukaryotic exonucleases begin digestion at the 5′ end and proceed in the 3′ direction. Generally these enzymes act on single-stranded ends and do not exhibit sequence specificity. Other ribonucleases, termed endonucleases, cleave the RNA internally (that is, within a strand, and not at an exposed end). Endonucleases differ in their specificities for RNA targets; some endonucleases cleave double-stranded RNA while others cleave single-stranded RNA. Features of two representative and well-characterized endonucleases, RNase III and RNase P, are discussed below.

E. coli
16S rRNA   tRNA   23S rRNA   5S rRNA

S. cerevisiae
18S rRNA   5.8S rRNA   25S rRNA

Figure 9.4 *E. coli* and *S. cerevisiae* precursors encoding tRNAs and rRNAs. The mature, functional tRNAs and rRNAs in bacteria and eukaryotes are usually cleaved from longer precursor RNAs.

The processing of tRNAs and rRNAs can involve many steps and can involve more than one pathway, as illustrated for the processing of the rRNA precursor from *S. cerevisiae* (Figure 9.6). The advantages of having multiple processing pathways are not known. In general, the various processing steps do not occur in isolation, but instead are coordinated with each other and with other cellular processes. For example, in eukaryotes, rRNA transcription and early processing steps both occur in the nucleolus and are tightly coupled. Further rRNA processing is tied to the early steps of ribosomal protein assembly on the rRNA as well as, in some cases, export to the cytoplasm.

The endonuclease RNase III cleaves and trims rRNA precursors

Excision of the bacterial rRNAs from longer precursors and the trimming of some rRNAs and tRNAs is carried out by the endonuclease RNase III, which recognizes double-stranded RNA. This class of ribonuclease binds stem structures in precursor RNAs and then cleaves the double-stranded RNA on one or both sides of the stem.

Endonucleases similar to RNase III are involved in many processes in the cell. In subsequent sections we will see how these enzymes in eukaryotes are important in generating both microRNA (**miRNA**) regulators and the small interfering RNAs (siRNAs) that inhibit the expression of detrimental genes.

The RNA-containing endonuclease RNase P cleaves the 5′ ends of tRNAs

The precise trimming of the 5′ end of tRNAs is accomplished by the endonuclease ribonuclease P, which cleaves single-stranded RNA. Unlike RNase III, the bacterial, archaeal, eukaryotic, and mitochondrial RNase P enzymes all contain an RNA component in addition to one or more protein components. The RNA component of bacterial RNase P alone can act as an enzyme to cut the RNA, and was one of the

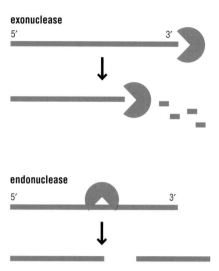

Figure 9.5 The action of exonucleases and endonucleases. Exonucleases can sequentially digest RNA from the 3′ end in a 5′ direction or from the 5′ end in a 3′ direction, while endonucleases cleave within the RNA.

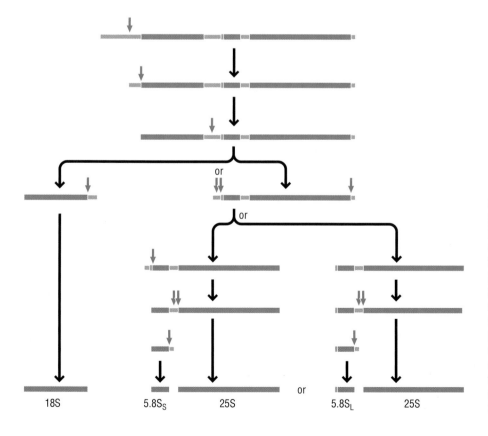

18S   5.8S$_S$ 25S   5.8S$_L$ 25S

Figure 9.6 Cleavage steps during the processing of mature *S. cerevisiae* rRNAs from the precursor RNA. Multiple cleavage steps (indicated with pink arrows) are required for the processing of mature rRNAs from the precursor RNA. The products of this processing – two large rRNAs (labeled 18S and 25S) and the short rRNA (5.8S) – are discussed in Chapter 10. At one stage during the processing of the precursor, one of two different steps is carried out, leading to the production of either the short (5.8SS) or long form (5.8SL) of the 5.8S rRNA. This diagram does not show the additional processing events in which nucleotide bases are chemically modified.

first RNAs shown to act as a ribozyme (see Experimental approach 9.1). The single protein subunit of the bacterial enzyme enhances performance relative to that achieved by the RNA component alone, and may broaden the range of substrates the enzyme can cleave. By contrast, the isolated RNA component of archaeal, eukaryotic, and mitochondrial RNase P does not act as a ribozyme, although the RNA is essential for catalysis, and is probably the functional core of these enzymes.

Several tRNAs and some rRNAs have introns that require splicing

The precursor RNAs for several tRNAs and some rRNAs are interrupted by introns, which must be removed to produce the functional RNA, providing an additional step for regulation and quality control. tRNA precursors undergo splicing that is catalyzed by protein factors. In contrast, some introns found in rRNA precursors catalyze their own removal and are thus **self-splicing**. In these cases, the precursor rRNAs themselves act as ribozymes that catalyze the removal of introns and joining of the exons that remain (referred to as intron excision and exon ligation, respectively). The details of the different splicing reactions will be discussed in subsequent sections.

CCA is added to tRNAs in the absence of a nucleic acid template

The 3′ termini of tRNAs contain a universally conserved CCA sequence, which is the site of attachment for the amino acid that the tRNA carries to the ribosome during protein synthesis.

⊙ We describe the attachment of amino acids to tRNAs in more detail in Chapter 10.

For some bacterial tRNAs, this CCA sequence is encoded by the tRNA gene, but for most tRNAs this sequence is added post-transcriptionally by a CCA-adding enzyme. Indeed, even in organisms in which the CCA sequence is encoded by the tRNA gene, the CCA-adding enzyme is required for the repair and maintenance of the CCA sequence because the single-stranded 3′ termini of these tRNAs are particularly susceptible to digestion by exonucleases.

Although the CCA-adding enzyme shares sequence similarity with a DNA repair polymerase, the enzyme does not use a nucleic acid template to direct the addition of two C residues followed by a single A residue to the tRNA. Instead, the enzyme contains a binding pocket that can accommodate either a cytidine triphosphate (CTP) or an adenosine triphosphate (ATP). The size and shape of this binding pocket change depend on which form of the tRNA is bound – the tRNA without the CCA tail, or the tRNA with a single C added, and so on. These changes to the binding pocket determine which nucleotide is added to the 3′ end of the tRNA at each step, and hence drive the synthesis of the correct CCA tail.

9.3 tRNA AND rRNA NUCLEOTIDE MODIFICATIONS

Nucleotides in tRNAs and rRNAs are modified in numerous ways

Many of the nucleotides in tRNAs and rRNAs are chemically modified after transcription. The modifications can be found on the nucleotide base or the ribose sugar ring, and can be relatively small: they may comprise the addition of hydrogen atoms, the methylation of nitrogen or oxygen, or the addition of selenium (Figure 9.7a). The modifications also can be quite large, such as the incorporation of the amino acid threonine. Some nucleotides even carry multiple independent modifications. Typically, particular positions in different tRNA and rRNA species are specifically modified and, for those modifications that have been studied in detail, they often appear to be required for the optimal survival and growth of organisms. Some modifications do not have any obvious functional role, yet are well conserved during evolution, suggesting that they do nevertheless confer a selective advantage to the organism.

More than 80 different modifications have been seen in tRNAs. Some of these modifications seem to contribute to the overall structural stability of tRNAs and to their ability to interact with other molecules. It has been argued that the chemical diversity of the modifications adds to the limited repertoire of the four standard ribonucleotides, thus allowing for higher order function. For example, we will see in Chapter 10 that some modifications allow tRNA to interact properly with mRNA during protein synthesis to ensure that this process occurs with high accuracy.

The most common rRNA modifications are **ribose 2′-O-methylation** (to produce methylribose) and **pseudouridylation** (the conversion of uridine to pseudouridine) (Figure 9.7b). Many of the rRNA modifications are found in regions of the ribosome that are known to be important for ribosome function. It is thought that the extra hydrogen bonding capacity of pseudouridine relative to uridine could allow this modified nucleotide to help stabilize the structure of the ribosome.

Mutations in the protein dyskerin, the enzymatic component of a molecular complex involved in pseudouridylation, lead to a human disease of skin and bone marrow failure known as dyskeratosis congenita, underlining the importance of the modified nucleotide pseudouridine in biological function.

Specific enzymes catalyze individual tRNA and some rRNA modifications

In bacteria, the modifications found on tRNAs and rRNAs are added by site-specific enzymes that catalyze a given modification at a specific site in the RNA. Some of the more complex modifications require the actions of several enzymes to synthesize the final modified nucleotide. In a few cases, such as that of the pseudouridine synthases (which catalyze the conversion of uridine to pseudouridine) and several of the methylases (which add methyl groups to a target molecule), we have a detailed understanding of the catalytic mechanisms of the modifying enzymes. For most of the modifications, however, we know little about the enzymes involved and how specific sequences are recognized. Given the large number and diversity of modifications found and the large number of proteins that are committed to their synthesis, it is clear that these modifications must play a biologically significant role.

Site-specific enzymes also catalyze some of the tRNA and rRNA modifications in archaea and eukaryotes, though as we shall learn next, the majority of the modifications in these organisms are carried out by specialized RNA-guided machinery.

Figure 9.7 Examples of nucleotide modifications found in tRNA and rRNA. Modifications can occur to the ribose sugar or the nucleotide base, and they can consist of the addition of a single methyl group or be more extensive. Shown here are (a) two tRNA modifications, the methylation of guanosine and the addition of a methyl amino methyl group and selenium to uridine, as well as (b) the two most common rRNA modifications, the conversion of uridine to pseudouridine and the methylation of ribose to produce 2′-O methylribose.

snoRNAs guide methylation and pseudouridylation of eukaryotic and archaeal rRNAs and some tRNAs

We mentioned above that two particularly common modifications to rRNA are ribose methylation and the conversion of uridine to pseudouridine. Although the enzymes that carry out these modifications in all organisms are similar, the mechanism by which the sites are recognized in archaea and eukaryotes is very different from the mechanism found in bacteria. In particular, the archaeal and eukaryotic machineries make use of short RNA molecules called **small nucleolar RNAs (snoRNAs)** to guide the enzymes to the site in the RNA that is to be modified. The snoRNAs associate with specific sets of proteins, which include the enzyme that carries out the modification reaction, to form an RNP known as a **snoRNP**.

SnoRNAs range in length between 60 and 300 nucleotides and are found in the nucleolus, where the majority of the rRNA processing occurs. Most vertebrate snoRNAs are processed from the introns of precursor mRNAs, illustrating the fact that introns can themselves encode functional RNAs. The snoRNAs fall into two major classes that possess distinct conserved sequence elements and direct different nucleotide modifications: the RNAs that direct ribose methylation are referred to as the box C/D snoRNAs, and the RNAs that direct pseudouridylation are referred to as the H/ACA snoRNAs.

How does the snoRNA exert its effect? The snoRNA within the ribonucleoprotein complex base-pairs with the target RNA and thus directs the modification enzymes to particular positions (Figure 9.8). While the function of most of the snoRNAs is to guide modifications of the eukaryotic rRNAs and tRNAs, a subset have been found to direct modifications of specific mRNAs. Similar RNA-guided reactions will be seen again in Section 9.5 where we describe how RNPs containing guide RNAs direct the precise removal of introns from the mRNA.

Figure 9.8 Ribose methylation and pseudouridylation guided by box C/D and box H/ACA snoRNAs. (a) snoRNAs containing conserved C/D box motifs (green) base-pair with a 10–21-nucleotide region of a target RNA (blue) and direct the methylation of specific bases by a fibrillarin-containing protein complex (pale orange). (b) snoRNAs containing conserved H/ACA box motifs (green) base-pair with two 3–10-nucleotide regions of a target RNA (blue) and direct the conversion of uridines to pseudouridines by a dyskerin-containing protein complex (pale yellow).

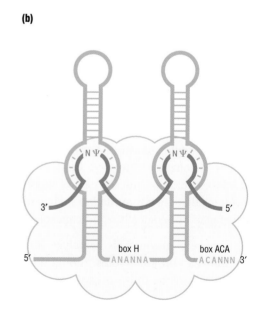

9.4 mRNA CAPPING AND POLYADENYLATION

We now move on from tRNAs and rRNAs to consider the modification of mRNAs. Although mRNAs are typically not functional molecules in their own right, they play a vital part in the way biological information is transferred from genome to protein, and their processing both during and after transcription is key to their correct activity.

Eukaryotic mRNAs are capped at the 5′ end and polyadenylated at the 3′ end

Both the 5′ and 3′ ends of eukaryotic mRNAs are modified during transcription by RNA polymerase II. These modifications protect the newly synthesized mRNAs from degradation by nucleases, and allow the mRNAs to interact with many of the proteins they encounter during their cellular existence. The 5′ ends of the mRNAs are 'capped' with a special 7-methylguanine nucleotide that is attached to the end of the message via a unique 5′–5′ triphosphate linkage (Figure 9.9). This structure, called the **5′ cap**, is essential for the efficient elongation and termination of transcription as well as the subsequent processing of the mRNA. The cap structure also functions as a binding site for proteins that export the mRNA from the nucleus to the cytoplasm, and for directing the initiation of protein synthesis from the mRNA, as we will see in the next chapter. In addition, the cap protects the 5′ end of the mRNA from degradation by 5′ to 3′ exonucleases.

The 3′ ends of all eukaryotic mRNAs, except those encoding metazoan histones, are modified by the addition of a tail of approximately 200 adenosine residues, called a **polyadenosine**, or **poly(A) tail**. Like the 5′ cap, the 3′ poly(A)

Figure 9.9 Structure of the 5′ cap of eukaryotic mRNAs. A guanine nucleotide (pink) is added to the 5′ end of an mRNA (blue) via a 5′–5′ linkage. The guanine is subsequently methylated at the N7 position (pink arrow). In higher eukaryotes, the 2′-O position of the second base and sometimes the third base is also methylated (green arrows).

circulating B lymphocyte

antibody-secreting B lymphocyte

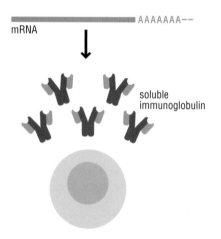

Figure 9.10 Alternative polyadenylation leads to switch between membrane-bound and secreted immunoglobulin. In circulating B lymphocytes, polyadenylation at one site leads to the synthesis of a longer transmembrane domain-containing version of the immunoglobulin protein, while in antibody-secreting B lymphocytes, polyadenylation at a second site leads to the synthesis of a shorter, secreted version. Such diversity in polyadenylation site selection allows for multiple protein products to be specified by a single gene in the genome.

tail protects the mRNA from degradation by exonucleases and plays a role in the initiation of protein synthesis. Some mRNAs have more than one site, called the **polyadenylation site**, at which the precursor mRNA is cleaved and the poly(A) tail added. These multiple polyadenylation sites contribute to the regulation of protein synthesis and expand the range of protein products made from a single mRNA species. For example, specific cells of the immune system called **B lymphocytes** make antibodies, also known as **immunoglobulins**, in both membrane-bound and secreted forms (Figure 9.10). Polyadenylation at one site in the immunoglobulin mRNA leads to the synthesis of a longer protein containing a transmembrane domain. This membrane-bound form of the immunoglobulin acts as a cell surface receptor, which, upon binding microorganisms, activates the B lymphocyte to proliferate, providing immune cells needed to defend the body against the invading pathogen. Polyadenylation at a different site produces a shorter, secreted form of the immunoglobulin, which circulates in the blood and targets microorganisms to which it binds for destruction.

Bacteria also contain poly(A) polymerases, and bacterial RNAs have been shown to be polyadenylated. However, the fraction of RNA that is polyadenylated is lower than in eukaryotic cells and polyadenylation in bacteria is generally associated with the destabilization of the mRNA.

Capping and polyadenylation in eukaryotes are multistep processes

The 5′ cap is added to eukaryotic mRNA in three successive steps that occur soon after the growing RNA emerges from RNA polymerase II, when only 20–30 nucleotides have been synthesized (Figure 9.11). First, an RNA 5′ triphosphatase catalyzes the removal of one phosphate from the triphosphate at the 5′ end of the mRNA. A guanyl transferase then attaches a molecule of guanosine monophosphate (GMP; in a reaction utilizing guanosine triphosphate (GTP)) to this end through a 5′–5′ triphosphate linkage. Finally, the transferred guanine is methylated by a guanine-7-methyltransferase.

In yeast, these three reaction steps are performed by three separate enzymes, while the first two reactions in *Caenorhabditis elegans* and mammals are catalyzed by a single bifunctional enzyme. In higher eukaryotes, the 2′ O positions of ribose in the second and sometimes the third nucleotides of the mRNA are methylated, which seems to allow more efficient translation of these mRNAs.

Polyadenylation of the 3′ end of eukaryotic mRNAs begins with an initial cleavage of the mRNA (Figure 9.12). For most transcripts, the cleavage occurs after a CA nucleotide pair that lies somewhere between a conserved AAUAAA hexamer sequence and a U- or GU-rich region further downstream. Following cleavage, a tail of approximately 200 adenosines is added by poly(A) polymerase. In general, a larger complex of proteins is required for this sequence of reactions than is required for the 5′ capping reactions: it is speculated that the larger complex is required because the polyadenylation enzymes must be able to perform the difficult job of recognizing and processing the different polyadenylation sites in different mRNAs.

Figure 9.11 The 5′ cap is added in three successive steps. The 5′ capping process begins as soon as 20–30 nucleotides have been synthesized by RNA polymerase II. First, one phosphate is removed from the triphosphate at the 5′ end of the mRNA. A molecule of GMP is then attached to the end by a guanyl transferase. Finally, the transferred guanine is methylated by a guanine-7-methyltransferase.

Figure 9.12 Polyadenylation occurs between a conserved AAUAAA sequence and a downstream U- or GU-rich region. The precursor mRNA is cleaved after a CA dinucleotide, which is found somewhere between the AAUAAA hexanucleotide sequence and a downstream U- or GU-rich region. Following cleavage, up to 200 adenine nucleotides are added to the new 3′ end of the mRNA by poly(A) polymerase.

Capping and polyadenylation are intimately coupled to each other and to transcription by RNA polymerase II

There is an intimate association between 5′ capping and 3′ polyadenylation, and also with other processes that occur during transcription via RNA polymerase II. As we saw in Chapter 8, capping is required during transcription to allow RNA polymerase II to continue elongation of the mRNA (see Figure 8.23), and polyadenylation is required for the efficient termination of transcription (see Figure 8.28). In between capping and polyadenylation, RNA polymerase II recruits proteins that remove introns from the mRNA, as we will see in the next few sections on RNA splicing.

The region of RNA polymerase II that is responsible for orchestrating all these mRNA processing events is the C-terminal domain (CTD) of RPB1, the largest subunit of the polymerase. Figure 9.13 shows how there is a sequential interaction

Figure 9.13 The C-terminal domain of RNA polymerase II coordinates many steps in the processing of mRNAs. The C-terminal domain of RNA polymerase II (CTD) plays a critical role in assembling and coordinating the activities of mRNA processing complexes. On the initiation of transcription, repeated amino acid sequences in the RNA polymerase CTD become partially phosphorylated (indicated by yellow balls). This form of the CTD recruits the enzymes (green) that add the 5′-cap (pink pentagon) to the growing RNA. After the RNA is capped, RNA polymerase can continue elongation, which is associated with further phosphorylation of the CTD. This additional phosphorylation in turn allows the recruitment of RNA splicing machinery (orange), which removes introns (gray lines) from the RNA. The additional phosphorylation also leads to the recruitment of the complex (blue) that cleaves and polyadenylates the 3′ end of the mRNA. Once RNA polymerase passes the polyadenylation site (pink), the 3′ end processing complex is offloaded from RNA polymerase onto the nascent mRNA.

1 5′ capping enzyme complex is recruited by partially phosphorylated CTD

2 additional phosphorylation of the CTD upon transcription elongation allows recruitment of splicing machinery

3 transcription past 3′ end processing signals leads to recruitment of 3′ end processing complex

4 spliced mRNA is cleaved and polyadenylated

between the CTD and the different processing complexes. RNA polymerases I and III do not have a domain equivalent to the RPB1CTD (of RNA polymerase II), and RNAs that are synthesized by these polymerases are not capped and polyadenylated.

Transcription and polyadenylation are also coupled to appropriate mRNA localization

Although the transcription and processing of eukaryotic mRNAs occurs in the nucleus, the translation of mRNAs occurs in the cytoplasm. Thus, mRNAs must be exported out of the nucleus before translation can occur. Here too, downstream consequences are tightly linked to both transcription and processing of the mRNA; the protein factors required for nuclear export are loaded onto the precursor mRNA during transcription, but the export-competent mRNA–protein complexes are not released from the transcription complex until they receive signals from polyadenylation enzymes.

Beyond merely being transported out of the nucleus, some mRNAs are localized to very specific regions of the cytoplasm. For example, a number of neuronal mRNAs are only translated in the distant synapses of the neuron. This specific localization also requires complete processing of the mRNA as well as the presence of 'localization elements', generally found at the 3′ end of the transcript. These localization sequences are bound by specific RNA binding proteins that often have dual functions – affecting both localization and regulating translation. For example, in the developing *Drosophila* oocyte, the bicoid mRNA is localized to the anterior pole through binding to Staufen, an RNA-binding protein that interacts with sequence elements in the 3′ **untranslated region** (**UTR**) of the bicoid mRNA as well as with components of the cell cytoskeleton.

The biochemical modifications made to mRNA molecules, which we have explored throughout this section, effect somewhat subtle changes to the primary structure of the molecules involved. However, we now go on to discover how many mRNAs are also subjected to a much more invasive form of processing, called splicing, in which entire sections of primary structure are simply removed from the transcribed molecule.

9.5 RNA SPLICING

Introns must be removed from precursor RNAs

As we note in Section 9.1, the mature transcript for many genes is encoded in a discontinuous manner in a series of discrete exons, which are separated from each other along the DNA strand by non-coding introns. mRNAs, rRNAs, and tRNAs can all contain introns that must be removed from precursor RNAs to produce functional molecules. There are a number of different classes of intron, with their relative distribution and abundance depending on the kingdom. Some are excised by proteins, some by RNPs, and some excise themselves, yet all use some variation of a series of transesterification reactions that we describe below. Most introns do not themselves contain genes and are simply removed from the precursor RNAs and degraded (Figure 9.14). Exceptions to this rule

Figure 9.14 RNA splicing. RNA splicing involves cutting an intron (gray) from RNA and joining together (splicing) the two neighboring exons (blue). The spliced exons form the functional RNA and the intron is usually degraded.

are the snoRNAs (see Section 9.2) and certain miRNAs in invertebrates and mammals (which will be discussed in Section 9.9) as they can be present in the introns of mRNAs.

While introns are widespread, they are far more common in eukaryotes than bacteria. Indeed, the existence of introns has allowed for a process referred to as **exon shuffling** wherein the DNA encoding exons can be exchanged and reordered through genetic recombination between intronic DNA. This process thus allows for the emergence of novel genes and is presumed to have played a major part in the evolution of eukaryotic genomes. As we will see in Section 9.7, the differential removal of introns from an RNA transcript can result in the production of different spliced RNAs from a single gene, thereby increasing the number of gene products encoded in an organism's genome.

RNA splicing occurs by two transesterification reactions

There are several different pathways by which introns are removed from genes, but most of them involve the same underlying chemical reactions. The two exons that surround an intron are not directly joined to one another in a single step. Instead, a two-step reaction occurs in which the intron is first detached from exon 1, freeing this exon to react with exon 2. Both of these reaction steps are **transesterifications** in which a single phosphodiester bond between two nucleotides is broken and replaced with an energetically equivalent phosphodiester bond (Figure 9.15).

There are two implications for replacing one chemical bond with a similar chemical bond: first, the reaction can occur without the net input of energy from an energy source such as ATP; and, second, the reaction can be readily reversible. In fact, reverse splicing reactions by certain classes of intron, which allow them to insert into duplex DNA, are thought to have played a significant role in dispersing introns and thus in shaping the evolution of the genome.

Most introns in eukaryotes are removed by a complex ribonucleoprotein machine called the spliceosome through a mechanism that we will discuss in the next section. However, a few introns in eukaryotes, bacteria, and organelles are able to catalyze their own removal from RNA and are therefore called **self-splicing introns**. The discovery of self-splicing introns is described in Experimental approach 9.1. There are two groups of self-splicing introns, which illustrate some general features of RNA splicing that we will look at, before discussing the more common but complex case of splicing by spliceosomes.

Group I introns can be self-splicing

Group I introns are a class of introns found in a number of genes in bacteria, viruses, microeukaryotes and plants. Many of these introns are autocatalytic RNAs, or ribozymes, which can catalyze their own excision from a primary transcript in a splicing reaction involving two sequential transesterifications (Figure 9.16a). The first transesterification reaction has the unusual feature that a free guanosine nucleotide (rather than a specific nucleotide within the RNA molecule) is the attacking species that detaches the 5′ end of the intron from its neighboring exon 1. In the second transesterification reaction, the released terminus of exon 1 attacks

Figure 9.15 The transesterification reaction of RNA splicing. The 3′ OH of an incoming nucleotide (or RNA) (pink) attacks and breaks the 3′–5′ phosphodiester bond between two nucleotides in the target RNA (orange). As a result, a new 3′–5′ phosphodiester bond is formed between the incoming nucleotide (or RNA) and the target RNA. Alternatively, the 2′ OH of the incoming RNA can attack the target RNA, in which case the incoming and target RNA become joined though a 2′–5′ phosphodiester bond (not shown).

9.1 EXPERIMENTAL APPROACH

Catalytic RNA: discovery and tertiary structure elucidation

While discussions in the 1960s by Francis Crick and Leslie Orgel fueled ideas of an RNA world wherein RNA played a primary role both as information carrier and catalyst in our earliest ancestors, broad support for this notion arrived with the discovery in the mid 1980s of catalytic RNAs in modern-day organisms. In a scientific twist of fate, two different examples of such catalytic RNA were independently discovered around the same time in the laboratories of Thomas Cech, Sidney Altman, and Norman Pace, forever changing our views on the plausibility of an RNA world. Here we will focus on the discovery by Thomas Cech and colleagues of a self-splicing intron in the rRNA gene of *Tetrahymena thermophila*.

The Cech laboratory was interested in studying the basic mechanistic features of intron removal (or splicing); the group was working on the model organism *T. thermophila* in which individual nuclei could be manipulated and where the intron of interest was found in the very abundant rRNA transcript. Initial characterization of intron removal suggested that the process was similar to that of known introns in the mRNAs of other eukaryotic genes, in that it happened in the nucleus by a cleavage-ligation process. However, the 'nature of the activity responsible for splicing remained perplexing' and ultimately led the scientists to the surprising discovery of catalytic RNAs.

In their efforts to isolate the factors responsible for splicing, the authors found the splicing activity was able to withstand procedures intended to remove proteins from the splicing reaction, including phenol extraction, treatment with the detergent sodium dodecyl sulfate (SDS), proteinase digestion, and heat denaturation. These observations led the group to propose that the catalytic entity was the RNA itself, for which they coined the term 'ribozyme' (an RNA enzyme).

The key experiment that supported this hypothesis was the observation of splicing activity with a transcript of the *T. thermophila* rRNA gene precursor synthesized in a test tube by a heterologous RNA polymerase enzyme from *E. coli* in the presence of radioactive GTP. The original figure from the paper is shown in Figure 1, where we focus on specific lanes. Lanes 1 and 3 were loaded with the transcription reactions for two different intron-containing constructs. The two bands that migrate at the sizes of the linear form of the excised intron (L IVS) and the circular form of the excised intron (C IVS) provide evidence that a splicing reaction has taken place. Lanes 4 and 6 were loaded with similar transcription reactions that were further incubated with free GTP, Mg^{+2}, and salt – all

Figure 1 Splicing activity observed in *in vitro* transcription reactions with purified *E. coli* RNA polymerase. Various plasmid DNA templates were transcribed in reactions containing α-[^{32}P]-GTP and the products loaded directly on a 4% polyacrylamide, 8 M urea gel. C IVS is the circular form of the excised intervening sequence RNA. L IVS is the linear form of the excised intervening sequence RNA.
Reproduced with permission from Kruger *et al.*, *Cell* 1982;**31**:147–157.

components found to stimulate the splicing reaction, as seen here by the diminished amount of full length transcript.

As there was no possibility that *T. thermophila* proteins were present in any of these reactions, it became clear that the catalytic activity must derive from the RNA itself. While many subsequent experiments more elegantly defined how GTP stimulates the splicing reaction, what the reaction products are, and how they are made, this experiment was simple and the results sufficiently clear to change the mindset of a field to believe that RNA on its own can function as an enzyme.

A year later, Sidney Altman, Norman Pace, and colleagues reported that the RNA component of the ribonuclease P enzyme responsible for processing the 5′ end of the precursor tRNA transcript is also a ribozyme. Since that time, a number of other natural ribozymes have been identified, and there is now considerable support for the idea that key RNPs in modern-day biology (in particular the spliceosome and the ribosome) are at their functional core driven by RNA. Together, these discoveries forever changed the way biologists view evolution, and have lent support to our understanding of the ever-expanding roles of RNA throughout biology that we explore in this book.

Figure 2 Overall secondary and tertiary structure of the group I intron. (a) Secondary structure of the group I intron from the pre-tRNA[Ile] from the purple bacterium *Azoarcus sp. BH72*. The principal features of secondary structure, and many tertiary interactions, were determined through phylogenetic analysis. Paired regions are sequentially numbered (P1 through P10) as are joining segments (e.g. J4/5 connect P4 and P5). (b) Tertiary structure of the same intron with coloring corresponding to the secondary structure in (a). The gray U1A feature represents an RNA-binding protein used to aid crystallization.

As biochemical studies established the catalytic prowess of the rRNA intron in *T. thermophila*, phylogenetic analyses by Francois Michel and Bernard Dujon identified a whole class of introns in mitochondrial, chloroplast, and nuclear-encoded genes in various fungi that appeared to share characteristic sequence features with this intron. These introns were referred to as 'group I' and could easily be distinguished from the more standard eukaryotic pre-mRNA introns as they lacked the canonical GU…AG splice site recognition elements. Such comparative sequence analysis led to the secondary structural model predictions for the group I intron class on which the field relied for many years to guide biochemical studies (see Figure 2a).

More recently, Scott Strobel and colleagues obtained crystals of the group I intron, with both exons attached, which diffracted to a resolution of 3.1 Å. The structure solved for these crystals provides a view of the global three-dimensional structure of this catalytic RNA, with several co-axially stacked helices (seen in green and blue) converging to provide a binding site for the substrate helix (yellow) involved in the splicing reaction (Figure 2b). Even in the absence of a hydrophobic core, the RNA is able to form a secure binding site where specific catalysis can be promoted.

The structure also revealed key features of the intron that allow for efficient catalysis – the 3′ oxygen of the 5′ exon is beautifully poised for inline nucleophilic attack on a conformationally constrained phosphate at the intron–3′ exon junction. In this structure, metal ions were visible for the first time in the active site. (In fact, there were two of them.) This finding corroborated much biochemistry over the past several decades arguing for the existence of several Mg^{+2} ions important for catalysis. This structure of an intact group I intron precursor RNA has provided both an unprecedented view of a catalytic RNA, and a wealth of information on the structural motifs that underpin the complex three-dimensional structures of RNA throughout biology.

Adams PL, Stahley MR, Kosek AB, Wang J, Strobel SA. Crystal structure of a self-splicing group I intron with both exons. *Nature* 2004;**430**:45–50.

Guerrier-Takada C, Gardiner K, Marsh T, Pace N, Altman S. The RNA moiety of ribonuclease P is the catalytic subunit of the enzyme. *Cell* 1983;**35**:849–857.

Kruger K, Grabowski PJ, Zaug AJ, *et al.* Self-splicing RNA: Autoexcision and autocyclization of the ribosomal RNA intervening sequence of *Tetrahymena. Cell* 1982;**31**:147–157.

→ We discuss wobble pairs in Section 2.7.

at the intron–exon 2 junction, thus joining the two exons together and releasing the free intron.

The splice sites are precisely defined by the three-dimensional structure of the intron and by the recognition of a conserved G–U wobble pairing interaction found within a paired helical region.

The activity of other group I introns is dependent on the assistance of cellular and intron-encoded proteins, and so these are not referred to as self-splicing.

Group II introns splice via the same chemical mechanism as spliceosomal introns

Group II introns comprise another class of introns that are found in bacteria and within genes found in the organelles of plants, fungi, and yeast. Group II introns are somewhat larger in size than their group I cousins (group I introns range from ~120 to 450 nucleotides in size and group II introns range from ~400 to 1000 nucleotides in length, excluding internal open reading frames) and they perform a slightly different set of transesterification reactions (Figure 9.16b). In this case, the 2′ OH of a specific adenosine within the intron functions as the initial attacking nucleophile in a two-step series of transesterification reactions that link exon 1 with exon 2. As we have seen for group I introns, some group II introns can function without the assistance of any cellular proteins, whereas others do depend on protein cofactors. As was true for the group I intron, the splice sites of the group II intron are defined by the three-dimensional structure of the intron itself and by specific base-pairings and tertiary interactions between intron and exon segments that position the critical sites within the catalytic center of the ribozyme.

→ We discuss base-pairing and tertiary interactions in RNA in Section 2.7.

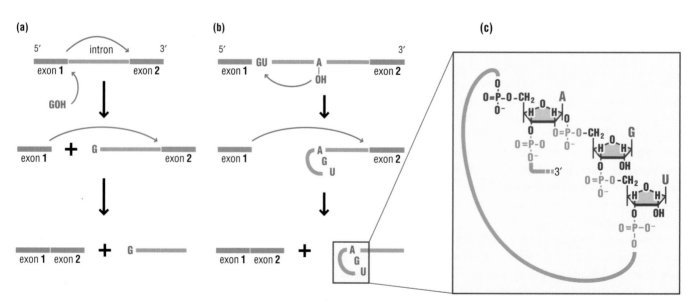

Figure 9.16 Splicing schemes for group I and group II introns. (a) The group I intron first uses the 3′ OH of an exogenous guanosine nucleotide (pink) to attack the boundary between exon 1 and the intron, releasing the intron from the exon. Next, the 3′ end of exon 1 attacks the boundary between the intron and exon 2, detaching the intron and joining the two exons together. (b) The group II splicing reaction uses an internal adenosine 2′ OH (pink) to attack the exon 1–intron boundary. The second splicing reaction occurs as it does in group I introns, with exon 1 attacking the intron–exon 2 boundary to join the two exons and release the intron. The resulting branched structure of the intron (pink box) is called a lariat and is shown in more chemical detail in (c).

The group II introns are of particular interest in furthering our understanding of the evolution of splicing: the utilization of an intron-internal adenosine (the branch point adenosine) in the chemistry of the reaction results in the formation of a branched **lariat** intermediate (Figures 9.16b, c), as is observed for the more common eukaryotic spliceosomal mechanism (see Section 9.6). These parallels speak to likely common evolutionary origins and inform our mechanistic understanding of both processes.

Group II introns are also interesting in that they can act as mobile genetic elements that attack and insert themselves into duplex DNA by a reversal of the splicing mechanism. Also of interest is the fact that group II introns often contain large open reading frames, which encode 'maturase' proteins that assist structurally in splicing and directly in the catalysis of intron mobility reactions.

◉ We discuss transposition by group II introns further in Chapter 14.

9.6 EUKARYOTIC mRNA SPLICING BY THE SPLICEOSOME

Splicing of eukaryotic mRNAs is carried out by a large RNP machine called the spliceosome

Genes in higher eukaryotes usually contain several introns. These introns are often of considerable length, sometimes extending to many thousands of bases. As such, they can account for 90% of the length of a typical precursor mRNA (Figure 9.17). By comparison, lower eukaryotes such as yeast have fewer introns, and they are generally short (<300 nucleotides). The vast majority of eukaryotic introns are not self-splicing, and the formidable task of identifying and splicing together exons among all the intronic RNA is performed by a large ribonucleoprotein machine, the spliceosome. The spliceosome is composed of several individual **small nuclear ribonucleoproteins** (**snRNP**, pronounced 'snurps', which comprise both RNA and protein components) and many more additional proteins that come and go during the splicing reaction.

The splicing reaction catalyzed by the spliceosome is similar to that of the group II introns we discussed in the preceding section (see Figure 9.16b). In both cases, the 2' OH of an adenosine residue located within the intron itself attacks the exon 1–intron boundary, detaching the intron from the exon and producing a branched intron structure (the lariat). Next, the terminal 3' OH of the newly released exon attacks the intron–exon 2 junction, splicing together the two exons and releasing the lariat intron.

The high similarity in mechanism between the transesterification reactions catalyzed by the group II intron and the spliceosome has led to the idea that the simpler self-splicing intron is the evolutionary predecessor of the more complicated multicomponent eukaryotic machinery.

| 1 | 2 | 5 | 10 | 15 | 20 | 30 | 40 | | 45 | 50 | 55 | 60 | | 70 | 79 |

Figure 9.17 Exon–intron structure of the human dystrophin gene. The dystrophin gene contains 78 introns (gray), which comprise most of the length of the pre-mRNA. As a result, exons (blue) can be many thousands of bases apart. From Roberts (1995).

Figure 9.18 Sequences in precursor mRNAs recognized by spliceosomes. Spliceosomes recognize a 5′ splice site motif (at the exon 1–intron border), a 3′ splice site motif (at the intron–exon 2 border), the branch-point sequence (with the branch-point adenosine in pink) and a polypyrimidine (Y) tract upstream of the 3′ exon. This diagram shows the consensus sequences of mammals; other eukaryotes have their own functionally equivalent sequences.

Splice site recognition and spliceosome assembly are dominated by RNA–RNA interactions

Before splicing can occur, the spliceosome must identify the splice sites between introns and exons – the sites at which exons are separated from their neighboring introns, and at which two exons are subsequently attached. In contrast to the group I and II introns, where the splice sites are defined by the three-dimensional structures of the introns themselves, the spliceosome identifies splice sites by recognizing short sequence motifs found in each pre-mRNA. Key sequences are the **5′ and 3′ splice sites**, a **branch-point nucleotide** within the intron (equivalent to the branch-point adenosine we saw in the preceding section), and a **polypyrimidine tract** before the 3′ splice site (Figure 9.18).

In addition to these essential sequence elements, there are other sequences within the exons and in nearby intronic regions that also help to define the exon–intron junctions. These additional sequence determinants allow cells to produce different mature RNAs from a precursor RNA depending on which introns and exons are recognized by spliceosomes, a process we explore further in Section 9.7.

The spliceosome is composed of five snRNPs (U1, U2, U4, U5, and U6) each containing an RNA molecule called an **snRNA** that is usually 100–300 nucleotides long, plus additional protein factors that recognize specific sequences in the mRNA or promote conformational rearrangements in the spliceosome required for the splicing reaction to progress. These snRNAs perform many of the spliceosome's mRNA recognition events by forming Watson–Crick base pairs with the precursor mRNA and one another. Figure 9.19 shows a basic outline of the steps that are thought to occur during splicing. First, the 5′ splice site is recognized by the U1 snRNP (Figure 9.19, step 1). Other splice site consensus sequences are recognized by non-snRNP factors; the branch-point sequence is recognized by the branch-point-binding protein (BBP), and the polypyrimidine tract and 3′ splice site are bound by two specific protein components of a splicing complex referred to as U2AF (U2 auxiliary factor), U2AF65 and U2AF35, respectively (Figure 9.19, step 1).

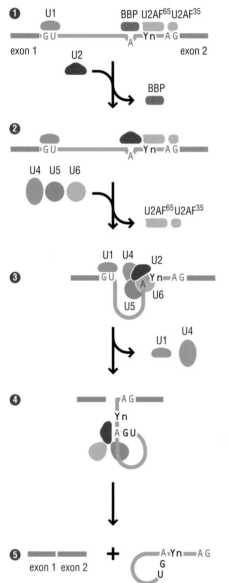

Figure 9.19 Spliceosome assembly and the splicing reaction. This schematic diagram shows the assembly of the spliceosome and the subsequent splicing reactions on an mRNA transcript. (1) The 5′ splice site (at the exon 1–intron border) is bound by the U1 snRNP, the branch-point sequence is bound by the branch-point-binding protein (BBP), and the polypyrimidine tract (Y(n)) and the 3′ splice site (at the intron–exon 2 border) are bound by a complex of the U2AF proteins, U2AF65, and U2AF35, respectively. (2) The U2 snRNP displaces BBP from the branch point, and a complex of the U4–U5–U6 snRNPs displaces the U2AF complex. (3) A rearrangement of the snRNPs releases the U1 and U4 snRNPs and (4) the initial transesterification reaction occurs to form the lariat intermediate. (5) Further rearrangements bring the two splice sites together, and the two exons are spliced together and the lariat intron is released.

Following the binding of these initial components, the remainder of the splicing apparatus assembles around them, in some cases displacing some of the previously bound components (Figure 9.19, steps 2 and 3). The precursor mRNA then undergoes a structural rearrangement, the first transesterification step takes place and the lariat forms (Figure 9.19, step 4). Other rearrangements in the spliceosome next bring together the newly freed exon 1 and the intron–exon 2 junction – the two RNA pieces that will be joined to create the final mRNA product – and the second transesterification reaction completes the splicing (Figure 9.19, step 5).

While the molecular details of the individual steps are not fully understood, it is clear that the spliceosome is a dynamic ribonucleoprotein machine, and that conformational changes in the RNA components themselves are central to the spliceosome's function. Indeed, the known ATP-dependence of the splicing reactions is thought to reflect the requirements of the many helicases that promote and regulate these RNA-based conformational rearrangements. Based on similarities in mechanism to the group II intron, it is also generally believed that the snRNAs themselves (the U2 and U6 snRNAs being the primary candidates) must play a central functional role in catalysis.

Protein factors contribute in multiple ways to pre-mRNA splicing

In addition to the snRNPs and proteins we have already mentioned, there are hundreds of other proteins that are associated with the splicing machinery and play important roles in the process. For example, PRP8 is an essential, large protein found near the active site of the assembled spliceosome, and which has long been thought to be well positioned (and seemingly ubiquitously present in biochemical experiments) to play a critical role in catalysis. A number of ATPases probably facilitate structural rearrangements of the snRNAs during the sequential splicing steps as well as promoting mRNA and intron lariat release after the reaction is complete. It also has been proposed that these same ATPases are important for 'proofreading' mechanisms that promote fidelity in splice site selection. Missplicing in the cell can have dire consequences as the desired product is not produced, and often the wrong products can be toxic for the cell. Such an expenditure of energy for the purpose of promoting fidelity is a common feature of biological reactions.

Other proteins appear to regulate the splicing reactions. For example, the SR proteins (named after the serine(S)/arginine(R) dipeptide repeats that they possess) bind to the pre-mRNA where they are thought to help recruit other components of the spliceosome to the 5′ and 3′ splice sites. Some SR proteins play an important constitutive role in splicing reactions, whereas others may be produced in a cell-type specific manner where they are involved in controlling alternative exon splicing reactions (as we will discuss in Section 9.7). SR proteins can be heavily modified by phosphorylation on their abundant serine residues and the extent of this phosphorylation may be altered throughout the splicing cycle to modulate splicing events.

Extensive study of the human spliceosome has shown that some of its integral components are also involved in transcription, polyadenylation, nuclear mRNA export, and translation broadly highlighting the intimate coordination of biochemical events in mRNA biogenesis (see Figure 9.13). In one well-understood example of such coordination, the process of splicing leaves behind a set of proteins, the **exon junction complex** (**EJC**), at splice junctions, thus 'marking' the transcript as processed and ready for export and translation. These EJCs are eventually recognized by the translational machinery in a process that evaluates whether the mRNA represents a full-length product (a form of quality control).

A minority of genes in animals and plants are processed by an alternative set of splicing snRNPs (the AT–AC spliceosome)

While the majority of pre-mRNA introns have GU and AG dinucleotides at their termini and are recognized by the 'major' spliceosome described above, some introns in multicellular organisms are processed by a spliceosome that includes the U5 component of the spliceosome as above, but with four different snRNPs (U11, U12, U4atac, and U6atac) functionally replacing the U1, U2, U4, and U6 snRNPs. This machinery is referred to as the minor or U12-dependent spliceosome and is far less abundant in cells than the major spliceosome.

Certain introns spliced by the U12-dependent spliceosome have unusual splice site sequences with AU and AC dinucleotides at the intron termini. Consequently, the spliceosome was initially referred to as the AT–AC spliceosome. In general, the splicing pathway for this minor spliceosome appears to be similar to that followed by the major spliceosome. It is not clear why this secondary splicing pathway was retained during the evolution of metazoans.

Trans splicing joins exons from two separate RNA transcripts

The different splicing processes described above are all examples of *cis* **splicing** – that is, the splicing events occur between parts of the *same* molecule. Another unusual form of splicing is *trans* **splicing**, which results in exons from two separate RNA molecules becoming joined together. *Trans* splicing has mostly been explored in nematode worms and trypanosomes (the single-celled animals that cause African sleeping sickness) where these events are reasonably common (trypanosomes splice all of their mRNAs in this way). Similar *trans* splicing has also been observed in *Drosophila*, in which it occurs for a few genes.

In *trans* splicing, the 5′ exon is a short RNA called the spliced leader RNA (**SL RNA**) that is generally 20–50 nucleotides long. During the *trans* splicing process, this SL RNA is joined to a variety of different mRNA transcripts. Following the *trans* splicing reaction, the SL RNA represents the first exon in the mature mRNA and is located just upstream of the AUG initiation site.

During the *trans* splicing reaction, the SL RNA is bound to proteins to form the SL snRNP, effectively replacing the U1 snRNP bound to the 5′ splice site during the more standard splicing reactions. This complex then interacts with the introns and exons within the mRNA and the U2, U4, U5, and U6 snRNPs to carry out a splicing reaction essentially analogous to the *cis* splicing reactions described above. The sequence and cap structure of the SL RNA are thought to allow the highly efficient recognition of this mRNA by the ribosome and critical translation factors, thus resulting in overall robust translation.

9.7 EXON DEFINITION AND ALTERNATIVE SPLICING

Diversity in gene expression in eukaryotes derives from alternative splicing

In a gene comprising multiple introns and exons, the process of splicing may yield more than just a single mature mRNA in which the exons are joined in only one way. Instead, at least 75% of genes in higher eukaryotes undergo alternative splicing, in which different combinations of exons in a gene are spliced together to produce different mature mRNAs. There are many ways in which the splicing pattern of a multiexon mRNA can be altered. Although most exons are constitutive, meaning that they are always included in the final mRNA, regulated exons are also found and these can be included, or not.

In addition to the potential exclusion or inclusion of a particular exon, the splicing pattern of an mRNA can be disrupted by the utilization of alternative 5′ or 3′ splice sites, or even through the utilization of alternative transcriptional start or termination (polyadenylation) sites. Some of the possible outcomes are shown schematically in Figure 9.20. These alterations all depend on a variety of factors such as the cell type, the developmental stage of the cell or organism, or cues from the cellular environment, and are typically mediated through interactions with a large group of auxiliary proteins involved in the splicing reactions that we will describe below.

Alternative exons represent an immense source of potential genetic diversity. As mentioned in Section 9.1, an extreme example is the *Drosophila dscam* pre-mRNA that can in principle be processed to generate more than 38 000 different mature transcripts. Indeed, some of the most extensive regulation through alternative splicing has been observed in the mammalian nervous system, which is not surprising given the complexity of neuronal interactions.

In Section 9.6 we learned about the sequence motifs – the 5′ and 3′ splice sites, the branch-point consensus and the polypyrimidine tract – which are the primary determinants used by the spliceosome to identify splice sites in precursor RNAs. We now look in more detail at how spliceosomes distinguish and discriminate between potential splice sites in an mRNA, and how cells can regulate alternative splicing.

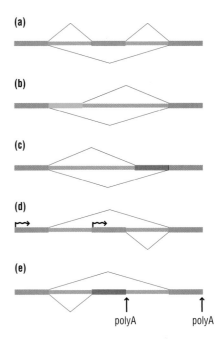

Figure 9.20 Possible outcomes of alternative splicing. A particular exon may be included or excluded (a) or alternative 5′ (b) or 3′ (c) splice sites may be used. Additional variation can come from the utilization of alternative transcriptional start (d) or termination (polyadenylation) (e) sites.

Exon and intron definition models help to explain how the ordered recognition of splice sites can occur

The fundamental task of the spliceosome is to recognize splice junctions and accurately join together the appropriate exons. The sequences that define the exon–intron junctions are, as we have seen, relatively simple. Consequently, similar sequences may occur in introns purely by chance: such accidental splicing signals are known as **cryptic splice sites**. The spliceosome must therefore not only reliably detect true splice sites, but must also reliably ignore cryptic splice sites.

9.2 EXPERIMENTAL APPROACH

Determining the roles of exon and intron definition in eukaryotic splicing though mutational and computational analyses

Early studies of eukaryotic pre-mRNA splicing identified short consensus splice site recognition sequences that appeared to be rather ubiquitously used to define complex splicing patterns. These rules were relatively simple: GU at the 5′ splice site and AG at the 3′ splice site, with several adjacent positions that are biased toward particular bases and determine the intrinsic strength of the site. However, the low information content of these sites immediately suggested that there must be many other determinants that are involved in defining the appropriate and sequential splice sites with high precision (and indeed many of these have been defined).

As discussed in Section 9.7, there have been two dominant models for how splice site recognition occurs, through exon definition or through intron definition, wherein interactions between the 5′ and 3′ splice sites first occur spanning over an exon or an intron. The rules appear to be different in different organisms, and these differences appear to correlate with critical features in the pre-mRNAs themselves. While insights into these models and where they apply have come from numerous approaches,

certainly some of the early biochemical studies by Susan Berget and colleagues were instrumental in guiding our thinking about these problems. Berget realized that exon and intron definition models could be distinguished by following the efficiency of spliceosome assembly and actual splicing in *in vitro* reconstituted mammalian splicing extracts with a carefully chosen set of pre-mRNA reporters carrying different combinations of complete and partial exons and introns.

If we first look at Figure 1, we see the authors compare the formation of a spliceosomal assembly intermediate (α) on a native gel using two sets of pre-mRNA constructs (radioactively labeled). The constructs have either a complete exon (with both 5′ and 3′ splice sites) or a partial exon (with only the 5′ splice site present) (a vs. b) followed by variable length 'intronic' sequences. We see in Figure 1a (the E2 series) that, when a complete exon sequence is present, an intermediate (α) forms in each case, though with diminishing efficiency as the intronic sequence length is increased. By contrast, we see in Figure 1b (the E2Δ5 series) that with an incomplete exon, formation of the assembly

Figure 1 Enhancement of complex assembly for mRNA substrates containing a full exon. Complex assembly in mammalian splicing extract was followed on 3.5% native acrylamide-agarose gels following incubation with [³²P]-radiolabeled mRNA substrates. Cartoons of pre-mRNA constructs are shown below the gel with complete (a) or incomplete (b) blue exons and gray introns (of variable length). Equal amounts of substrate were incubated in the extracts for the indicated times before gel electrophoresis. I represents the uncomplexed mRNA, whereas α represents a known splicing intermediate observed in the absence of a full intron.

Reproduced with permission from Robberson *et al., Molecular and Cellular Biology* 1990;**10**:84–94.

Figure 2 **Enhancement of complex assembly on two-exon precursor mRNAs with a 5′ splice site in exon 2.** Splicing assembly reactions were followed as in Figure 1. Cartoons of pre-mRNA constructs are shown below the gel with complete (a) or incomplete (b) blue exons and gray introns (of variable length) A, A′, and B represent productive splicing complexes for two-exon containing pre-mRNAs. Reproduced from Robberson *et al., Molecular and Cellular Biology* 1990;**10**:84–94.

intermediate is substantially less efficient, and the stability of the complex over time is compromised. These data immediately suggest that the presence of both consensus splice sites (and the exon between them) are important determinants of spliceosome assembly. However, these experiments do not directly compare the efficiency of spliceosomal assembly on a complete exon with that on a complete intron (since there is no complete intron in these pre-mRNAs).

In the next experiment (Figure 2), the researchers again compared spliceosomal assembly efficiency on a set of pre-mRNAs containing a full exon (E1) and intron, as well as portions of another exon (E2). In Figure 2a (the E1–E2 series), complex assembly was quite efficient (now forming three distinct complexes, A, A′, and B) on all constructs carrying two complete exons surrounding an intron. Splicing is also efficient with each of these constructs (not shown). In Figure 2b (the E1-E2Δ5 series), complex assembly was severely compromised when the second exon was incomplete (missing the 5′ splice site recognition sequence). Since there is a complete intron present in these constructs in both parts of Figure 2, and yet the efficiency of assembly is very different, the immediate conclusion was that recognition of splice sites across the exon (but not the intron) is critical to the assembly process. While the experiments described here (Figures 1 and 2) followed the assembly of the splicing machinery, the results correlate very well with overall splicing activity observed in the same *in vitro* extracts.

From these experiments, the exon definition model emerged as a robust framework for thinking about splicing in mammals.

The exon definition model (and its role in mammals) has been supported by a recent bioinformatic analysis by Christopher Burge and colleagues using the vast wealth of information available in the annotated genomes. These studies modeled the coevolution of splicing elements in orthologous exon pairs in several different eukaryotic taxa using network formalisms, called Principal Variation-based Subset Analysis (PVSA), more commonly applied to the analysis of protein interaction networks.

The basic idea behind the analysis was to use comparative genomics to ask whether covariations (or signs of co-dependence) were apparent between the 5′ and 3′ splice sites located across exons from one another or across introns. In essence, the method asked whether large decreases in the strength of the 5′ splice sites of exons in one organism relative to another were associated with compensating increases in the strength of the upstream (cross-intron) or downstream (cross-exon) 3′ splice site. This analysis revealed major trends in splicing element coevolution that are different between mammals and other eukaryotic taxa.

In terms of the exon versus intron definition models, these computational studies observed clear cross-exon (but not cross-intron) compensatory interactions between the 5′ and 3′ splice sites in human and mouse genes, whereas cross-intron compensatory interactions were more commonly observed in plants, fungi, and invertebrates (indicated with green arrows in Figure 3a). These observations are strikingly consistent with the length variability of introns and exons across these diverse taxa (Figure 3b) and with the relative extents of exon exclusion versus intron retention observed experimentally in these systems (see columns in Figure 3a). These data provide experimental support (in the terms of evolutionary history) for distinct modes of splice site definition across the eukaryotic lineage, and highlight the power of phylogenetic analysis in identifying functionally relevant biological interactions.

Figure 3 **Coevolution of splice sites in diverse eukaryotic taxa.** (a) Using PVSA, the coevolution of five sequence elements within the pre-mRNA were modeled (the 3′ and 5′ splice sites of two consecutive exons and intron length). An arrow from one node to another (the ovals) indicates significant covariation seen in the analysis, with compensatory relationships shown in green, and relationships with unresolved directionality in black. The percentages of exon skipping and intron retention events among observed alternative splicing events are also listed. (b) Semilog plot of exon and intron length distributions in five representative genome classes.
Reproduced from Xiao *et al.*, *Proceedings of the National Academy of Sciences of the U S A* 2007;**104**:18583–18588.

Robberson BL, Cote GJ, Berget SM. Exon definition may facilitate splice site selection in RNAs with multiple exons. *Molecular and Cellular Biology* 1990;**10**:84–94.

Xiao X, Wang Z, Jang M, Burge CB. Coevolutionary networks of splicing *cis*-regulatory elements. *Proceedings of the National Academy of Sciences of the U S A* 2007;**104**:18583–18588.

There are two models for how the spliceosome correctly identifies splice sites. These models are called **exon definition** and **intron definition**. Experiments that have explored the alternative models are discussed in Experimental approach 9.2. The relative contributions of each model to splice site identification are thought to be somewhat species dependent.

In mammals, the likely general explanation for how the spliceosome correctly identifies splice sites is through **exon definition**, in which the exon is 'defined' for the spliceosome to recognize (Figure 9.21a). According to this model, the 5′ and 3′ ends of an exon are brought together by interactions between the U1 snRNP and the U2 complex (composed of the U2 and U2AF snRNP). The exon is typically marked by SR proteins that are needed to 'define' the exon and assemble the activated splicing complex, while the introns are typically bound by a heterogeneous group of proteins (referred to as hnRNPs), which can help in masking the numerous potential cryptic splice sites (Figure 9.22). This hypothesis nicely

(a) exon definition - predicted consequence is exon exclusion

(b) intron definition - predicted consequence is intron inclusion

Figure 9.21 Two possibilities for how correct splice sites are identified. (a) In the exon definition hypothesis, splice sites are brought together across the short exon sequences by direct interactions between the U2–U2AF and the U1 snRNP complexes (or between these components and the cap binding complex at the 5′ end of the mRNA and the poly(A)-binding proteins at the 3′ end of the mRNA). This idea predicts exon exclusion when mutation of the 5′ splice site occurs. (b) Alternatively, interactions between the U2–U2AF and the U1 snRNP complexes might occur across the relatively longer intron sequences. This idea predicts intron inclusion when mutation of the 5′ splice site occurs.

explains several general observations from mammalian splicing systems: that exons tend to be short relative to introns and thus simpler to define; and that mutation of 5′ splice sites tends to result in exon exclusion rather than intron inclusion.

The alternative to the exon definition model is that introns instead are defined through interactions between sequential 5′ and 3′ splice site bound factors, denoted **intron definition** (Figure 9.21b). The prediction of this model is that intron inclusion will result from the mutation of a 5′ splice site, as has been observed in certain systems.

While both models help us to understand how sequential splice sites are identified to promote ordered and accurate splicing, it is clear that the final active splicing complex must be one in which 5′ and 3′ splice site protein complexes interact with one another across the intron (to facilitate its removal). In the cases where exons are 'defined', this requires reorganization of the splicing intermediates, whereas in the cases where introns are 'defined', no such reorganization is required. How such reorganization might take place in the case of exon definition remains an open question in the field.

An important feature of both models (exonic vs. intronic) for splice site definition is that the 5′ and 3′ splice sites are marked co-transcriptionally (that is, while transcription is still in progress). These events are facilitated by specific interactions between the unusual C-terminal tail of RNA polymerase (Section 9.4) and splicing components including the U1 snRNP. Co-transcriptional

Figure 9.22 Definition of exon and intron sequences. Binding of the SR proteins to exons and hnRNP complexes to introns helps the spliceosome to differentiate between exon and intron sequences.

splicing allows sequential splice sites to 'see' one another as they emerge from the polymerase, thus minimizing the chance that distant exons are mistakenly joined together. Moreover, after the splicing reactions are complete, a residual protein 'mark' referred to as the exon junction complex is left behind, providing important information about the splicing status for the cytoplasmic machinery that will eventually encounter this mRNA transcript (see Chapter 10). This mark effectively tells the translational machinery that the splicing machinery has completed its task.

Alternative splicing is regulated by a complex set of pre-mRNA sequences in exons and introns that are recognized by diverse cellular factors

Exon definition and the co-transcriptional assembly of the splicing machinery explain in general terms how exons can be identified by the spliceosome and joined to the next exon in the pre-mRNA transcript. However, the splice site consensus sequences do not provide sufficient information to determine whether a site will be identified as functional by the spliceosome. Instead, other information and interactions are required to activate their use. Moreover, the selection of particular splice sites is often tightly regulated in order to include or omit specific exons, or portions of them, from the mature mRNA (as described in Figure 9.20).

In addition to the exon or intron 'defining' interactions described above, there are many non-splice site regulatory sequences that strongly affect spliceosomal function. RNA elements that positively affect splicing are referred to as enhancers. Such **intronic** and **exonic** splicing **enhancer sequences** (ISE and ESE, respectively) typically bind proteins that increase the ability of spliceosomes to identify nearby splice sites in a pre-mRNA. The SR family of proteins appears to be broadly involved in binding ESE sequences (and recruiting other components) and can be thought of as important determinants of exon definition.

Intronic and **exonic** splicing **silencer sequences** (ISS and ESS, respectively) have the opposite effect, often masking splice sites from the spliceosome and causing exons to be omitted from the mature mRNA. A well-characterized example of a silencer protein is the polypyrimidine tract binding protein (PTB), that binds to intronic elements and acts to silence neighboring weak exons. In some cases, PTB competes directly with U2AF^{65} for binding to the polypyrimidine tract sequence of the intron, whereas in other cases PTB binds to other pyrimidine-rich sites and interferes with spliceosome assembly by other mechanisms.

One particularly well understood example of alternative splicing is found for the c-*src* mRNA, which encodes the SRC tyrosine kinase. Here a combination of positive and negative inputs regulates the inclusion of the neural-specific N1 exon in the final mRNA product. In non-neuronal cells, inclusion of the N1 exon is repressed by the PTB protein bound to intronic silencer elements on both sides of the exon (Figure 9.23). In neuronal cells, PTB is replaced by a related nPTB protein that binds to the same elements as PTB but does not form a repressive complex. Instead, a nearby enhancer sequence is activated that stimulates N1 splicing

non-neuronal cells

exon 3   N1 exon   exon 4

exon 3   exon 4

neuronal cells

exon 3   N1 exon   exon 4

Figure 9.23 Alternative splicing of the c-*src* mRNA in neuronal cells. In non-neuronal cells, inclusion of the N1 exon is repressed by the PTB protein bound to intronic silencer sequences. In neuronal cells, the PTB protein is replaced by a related nPTB protein such that nearby sequences can function as intronic enhancer sequences.

through binding to the regulatory proteins FOX-1 and FOX-2. There are additional regulatory proteins that also bind to the region of the N1 exon to affect its splicing. This combination of positive and negative control affecting the same exon is seen in most examples of tissue specific splicing.

We see from this example how the overall regulation of alternative splicing involves a complex interplay of pre-mRNA sequences with a variety of protein factors that may be expressed at different levels in different cell types. Further, mutations that emerge in the genes themselves can also affect the recognition of splice sites, leading to the activation of cryptic splice sites and the creation of new exons, generating even more genetic diversity.

9.8 RNA EDITING

The splicing pathways explored in the previous sections provide an important means of generating diversity from a specific pool of genes. In this section we see how the pool of gene products yielded from a single gene can be broadened still further through a series of additional modifications, which are collectively termed RNA editing.

mRNAs can be edited by the modification, insertion, and deletion of specific nucleotides

In addition to splicing, capping, and polyadenylation, some mRNAs also undergo RNA editing, whereby specific nucleotides are modified, inserted, or deleted. This editing can affect only one or two nucleotides (akin to the correction of a few typographical errors) or it can be surprisingly extensive (akin to the complete

```
UGAUACAAAAAAACAUGACUACAUGAUAAGUA
UCAUUUUAUGUUAUUUUUUGGUAGUUUUUUUAC
AUUUGUAUCGUUUUACAUUUG*GUCCACAGCA
UCCCG***CAGCACAUG**GUGUUUUAUGUUG
UUUAUUUGUAUUUUUGUGGUGA*AUUUAUUGUU
UA**UAUUGAUUGUAUUAUA***GGUUAUUUG
CAUCGUGGUACAGAAAAGUUAUGUGAAUAUAA
AAGUGUAGAACAAUGUCUUCCGUAUUUCGACA
GGUUAGAUUAUGUUA*GUGUUUGUUGUAAUGA
GCAUUUGUUGUCUUUA***UGUUUUGAGUAUA
UGUUGCGAUGUUGUUUGUCGUUACGUUGUGCA
UUUAUUUAAUUG***GAAUUUAU***CCGUAG
UUUUAAUGGUUUGUUGUGUAUAUCAUGUAUGG
UUUUUGG*AUUUAGGUUGUUUGUCUCCGUUG*U
UAUGAUCAUUUGAGGAA***CG*UGACAAAUU
GAUGACAUUUUUUGAUUUAUG**UUGUGGUUG
UCGUAUGCAUUUGGCUUUCAUGGUUUUAUUA*
GGUAUUCUUGAUGAUUUUGUUUUUGGUUUUGU
UGAUUUUUUGUUGUUUGA***UAAUAUCAU
GUUUGUUUGUUAUGGAUUGUUAUGAUUGUUA
UUUGUGGGUAAUCGUUUAUUUUAUUUUAUUUG
CGUUUGC***GUGGUUUGUCAUUUUUUGAUUU
AUAUGUUUA**GUUUUUA**A**UAGUUUAA
GUGGUGUUUUGUCUCGUUCGUUAGGUAUGGUG
UGAGAUUGUCGUUUAUUUAGUUGUUA****UG
A****GUUGUAUUUUAUGUUUGUUUUUAUGAUUA
UUGUUUUUGUUUUUAUAGGUGAUGCAUUUGA*U
CGUUUAUUUUUACGUUUGUUUGAUAUUUG***
**GUUUUGUUUGUUUGUUUG**AUUAUUUAUA
UUGUGAUUAUUACCAUUG***AGACCAUUAUU
AUGUUAUUUUAUAGUUUGUGGGUGUUGUUUU
GCCGGGUAUA*UCAUUUGC*UUGUGUUGAACA
CCCCAAAGGUGA***GUAUUGUUUGUAACAU
***UGUUUUUUGUGUUGGUUUAUGUUCUCGUUU
GCGUUGUGCGGAUUUUUUUGCA*UAUUUGUUUA
UUGGAUGUUUGUUUGCGUGGGUUUUUUUAUUGCA
UGAUUUAGAUGUC***C*GUUUUAGGUAAUAUU
GAUGUUGUUUUUGGAUCCGUAGAUCGUUA*GU
UUUAUAUGUG**A******GGUUAUUGUAGGA
UUGUUUAAAAUUGAAUAAAAA-poly(A)
```

Figure 9.24 An example of extensive RNA editing. The sequence of the NADH dehydrogenase 7 gene of *Trypanosoma brucei* after RNA editing. Black nucleotides are encoded by mitochondrial DNA, blue asterisks indicate uridines that are deleted and red nucleotides indicate uridines that are inserted.

Figure 9.25 Deamination of adenosine and cytidine. The deamination of adenosine and cytidine produces inosine and uridine, respectively.

rewriting of a text), involving half of the nucleotides in the mRNA as the example in Figure 9.24 shows. Editing is another way for a cell to vary the amino acid sequence of a protein that is encoded by a given mRNA, and makes it extremely difficult for us to predict the protein sequence encoded by a gene whose precursor RNA is processed in this way. In some cases, RNA editing provides cells with a mechanism to produce more than one form of a protein from a single gene. In this way, RNA editing can function like the alternative splicing mechanism that we discussed in Section 9.7 in that it increases the coding capacity of the genome.

Two widespread forms of editing are the deamination of adenosine to form inosine and the deamination of cytidine to form uridine (Figure 9.25), the effects of which we discuss below. These nucleotide conversions have been seen in RNA viruses, the organelles of a variety of eukaryotes, and the nuclei of metazoan organisms. By contrast, the insertion or deletion of nucleotides is far less widespread. Uridine insertion or deletion has been found almost exclusively in the mitochondrial RNAs of a few single-celled eukaryotes such as trypanosomes (the parasites which cause Chagas disease and African sleeping sickness). Cytidine insertion has thus far only been seen in the mitochondrial RNAs of slime molds, such as *Physarum polycephalum*. The origins of these unusual mechanisms for gene control are not well understood.

Adenosine deaminases that act on RNA convert adenosine to inosine

The deamination of adenosine to inosine is the most widespread form of mRNA editing in higher eukaryotes and can occur in coding sequences, introns, and the 5′ and 3′ UTR of transcripts. The replacement of adenosine with inosine has the same effect on coding as a change to guanosine – that is, inosine is interpreted as guanosine. Consequently, changes of this type within coding regions generally result in changes in the sequence of the protein translated from the mRNA. Indeed, there are several well-studied examples in the nervous system where adenosine deamination allows more than one form of a protein to be expressed from the same gene.

The enzymes that catalyze adenosine to inosine editing are referred to as **ADARs** (adenosine deaminases that act on RNA). These enzymes contain one or more double-stranded **RNA-binding motifs** (which will be described in Section 9.11) and are known to target double-stranded regions of RNA. Insights into how such double-stranded sites in the mRNA are targeted for editing can be seen in Experimental approach 9.3. Interestingly, ADARs deaminate as many as half of the adenosines in some RNAs while in others they target a specific adenosine among hundreds. These differences in specificity are not well understood, but undoubtedly derive from differences in the structures of the RNA substrates.

Sequence elements in mRNA direct cytidine to uridine deamination

Numerous plant mitochondrial and chloroplast mRNAs, as well as the mRNA encoding mammalian apolipoprotein, are edited by the deamination of cytidine to uridine. Whether this type of editing is more widespread in mammalian

9.3 EXPERIMENTAL APPROACH

Identification of RNA editing sites through serendipity and comparative genomics

The discovery of RNA editing emerged initially from routine comparison of the sequences of cDNAs of various genes with their genomic equivalent. cDNA samples are typically prepared from poly(A) selected mRNAs (mRNAs that are fully processed and are found in the cytoplasm) that have been reverse transcribed and prepared for sequence analysis. In the case of certain ionotropic glutamate channel genes in mammalian cells, these approaches identified single nucleotide discrepancies in the sequence of the cDNA and genomic DNA copies of the gene (an apparent A to G change) that would result in the exchange of a glutamine (Q) residue for an arginine (R) in the protein.

Other studies had shown that these amino acid changes are found in a transmembrane segment of the protein and profoundly alter the properties of ion flow in membranes in which these proteins are found. These sequence changes are thus important for overall function. Of course, to be certain that such changes in the mRNA sequence result from editing, the potential existence of multiple copies of the sequence within the genome needed to be eliminated as a possibility through other forms of analysis.

In the study by Peter Seeburg and colleagues, the authors demonstrated that, for several glutamate channels (GluR-B, GluR-5, and GluR-6) where an arginine codon had been observed in at least some fraction of the cDNA clones, the genomic copy clearly specifies glutamine (Q) and there is only one copy of the exon that encodes this transmembrane segment. Further analysis of a number of cDNA clones of these three genes (GluR-B, GluR-5, and GluR-6), prepared using the polymerase chain reaction, as compared with close, apparently unedited, homologs (GluR-C and GluR-D) provided clear evidence for the specificity of the editing phenomenon (Figure 1). For example, for the GluR-B gene, all 61 sequenced clones carry the edited version (R) of the gene; for GluR-C, and GluR-D, all sequenced clones carry the non-edited version (Q); and GluR-5 and GluR-6 appear to be partially edited.

These observations provided compelling support for the idea of RNA editing and motivated experiments that ultimately identified the enzymes responsible for what turned out to be A to I editing activity (the adenosine deaminases acting on RNA or ADARs) and the particular sequence features that these enzymes depend on for the selection of editing sites.

Compilation of glutamine and arginine codons in transmembrane II of glutamate receptor subunit cDNAs					
	GluR-B	GluR-C	GluR-D	GluR-5	GluR-6
Q	0	150	426	122	89
R	61	0	0	78	260

Figure 1 **Analysis of independent cDNA clones of the indicated glutamate receptor genes provide evidence for highly specific RNA editing.** cDNA constructed from rat brain RNA was used as a template in PCRs to amplify the coding region for the relevant transmembrane segment in these proteins. Sequence identity of the various clones was determined through 'plaque' hybridization (each phage plaque on a plate derives from a single cDNA sequence) to Q and R-specific radioactive probes.

Biochemical studies during subsequent years conducted on the ADAR enzymes found that the pre-mRNA editing substrate is typically an imperfect RNA duplex formed between the exon containing the adenosine that is to be modified and an intronic noncoding element called the editing site complementary sequence (ECS). Because the ECS can be rather distantly located relative to the edited site (from hundreds to thousands of nucleotides away), and because the pairing interactions are not precise, it was notoriously difficult to identify RNA editing sites by simple inspection of sequences.

In an effort to identify further editing sites, Robert Reenan and colleagues took a phylogenetic approach and analyzed the characteristics of the editing site of the *paralytic* gene (a Na+ channel) in *Drosophila*. The researchers hypothesized that if RNA editing of this particular site were conserved between species, then conservation might be seen in the intronic sequences (ECSs) that are critical for editing. To their surprise, an alignment of the *paralytic* gene of 18 distinct *Drosophila* species revealed a striking level of conservation in the exonic sequence surrounding the edited site (Figure 2). In the alignment shown, we see that even third position degeneracy, typically found within highly conserved protein coding sequences, is completely absent. In Figure 3, we see how exonic conservation emerges from such analysis; in the case of the *DSC1* gene, exon 15 is seen to be more highly conserved than any others, a clue that ultimately led to its identification as an authentic editing site. Indeed, similar comparative analysis of exonic conservation in 914 genes (annotated as potentially relevant for neurological behavior)

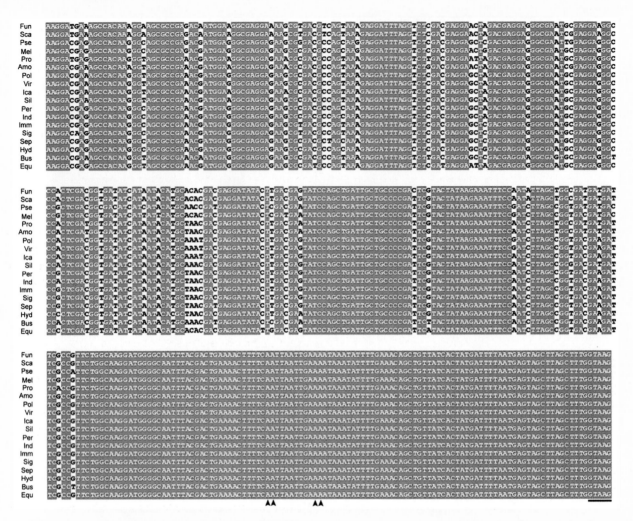

Figure 2 Conservation of coding sequence in the vicinity of an RNA editing site. Alignment of sequences of the para Na+ channel T/M site from 18 *Drosophila* species spanning 60–80 million years divergence time. Sequence of the entire exon containing the editing site is shown, with the 5′ splice site underlined. Nucleotides shaded in red are conserved among all 18 species. Arrowheads indicate the position of the adenosine residues that are modified by dADAR (d for *Drosophila*). Reproduced with permission from Hoopengardner *et al.*, *Science* 2003;**301**:832–836.

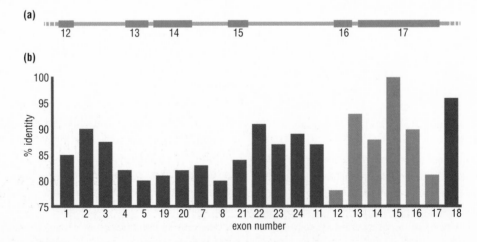

Figure 3 Experimental identification of novel edited site through analysis of exonic conservation. Bar graph showing percent sequence identify for the exons of *DSC1* transcript in *Drosophila*. Red bar indicates the edited exon, blue bars correspond to unedited neighboring exons shown in (a), and black bars represent the unedited more distant exons.
Reproduced with permission from Hoopengardner *et al.*, *Science* 2003;**301**:832–836.

identified 16 previously unknown ADAR target genes that could be experimentally confirmed.

The success of this particular study highlights the power of comparative genomics to identify biologically essential events that might escape identification through other more targeted, organism-specific approaches.

Hoopengardner B, Bhalla T, Staber C, Reenan R. Nervous system targets of RNA editing identified by comparative genomics. *Science* 2003;**301**:832–836.

Sommer G, Köhler M, Sprengel R, Seeburg PH. RNA editing in brain controls a determinant of ion flow in glutamate-gated channels. *Cell* 1991;**67**:11–19.

systems is a subject of current study. Apolipoprotein B is involved in the binding and transport of lipids throughout the human body. The deamination of a single cytidine in the apolipoprotein B mRNA introduces a stop codon into the transcript causing premature termination of translation, and results in the formation of a short form of apolipoprotein B called APOB48 (Figure 9.26). In humans, the editing occurs in the small intestine but not in the liver. Thus, APOB48 is produced in the small intestine, where it is required for the absorption of lipids from food, while the long form of the protein (APOB100) is synthesized in the liver where it acts to transport cholesterol from the liver to the rest of the body.

RNA editing allows these two specialized forms of the protein to be produced from a single gene. The site of cytidine deamination in the apolipoprotein B mRNA is determined by specific 'mooring' sequences within the mRNA: several sequence elements surrounding the cytidine base pair with one another and are thought to form a double-stranded element which positions the cytidine such that it is recognized and modified by a cytidine deaminase enzyme.

Uridine addition and deletion is directed by guide RNAs

One of the most unusual forms of editing is that of uridine addition and deletion in trypanosomes. Figure 9.24 shows an example of the numerous uridine nucleotides that are inserted into or deleted from a single mRNA encoded by trypanosome mitochondria. It is a mystery why trypanosomes edit their mRNAs so extensively, but the editing mechanism itself has been well characterized (Figure 9.27). The editing involves guide RNAs approximately 20–50 nucleotides long that bind to the mRNA and define where the insertions or deletions of uridines occur. The guide RNAs are partially complementary to the unedited RNA but are more extensively complementary to the fully edited mRNA.

The process of editing begins when the guide RNAs first base-pair with part of the mRNA. The editing is then guided by the base-pairing interaction. During each editing step, an endonuclease cuts the mRNA at a mismatch between the guide RNA and mRNA, and the guide RNA then serves as a template to direct the addition or deletion of uridines. The addition of uridines, which base-pair with adenosines or guanosines in the guide RNA, is catalyzed by a 3′ terminal uridylyl transferase, as illustrated in Figure 9.27a. Alternatively, a 3′–5′ exonuclease *removes* uridines that are unable to base-pair with the guide RNA (Figure 9.27b). The final step in both the addition and deletion steps of uridine editing is the ligation of the resulting edited fragments of the mRNA.

Figure 9.26 Cytidine deamination changes the protein expressed by the Apolipoprotein B mRNA. In the liver, the APOB mRNA is not edited and produces a long form of the protein. In the small intestine, the deamination of a single cytidine to uridine introduces a translation stop codon resulting in the synthesis of a shorter protein.

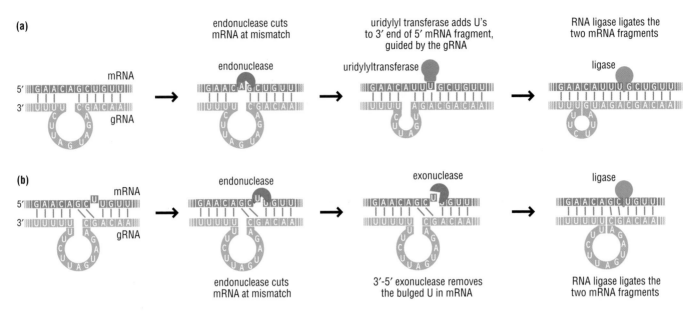

Figure 9.27 The insertion and deletion of uridine nucleotides is directed by guide RNAs. During both the insertion (a) and deletion (b) of uridine nucleotides, guide RNA molecules (gRNA, green) base-pair with the target mRNA. RNA editing occurs where there are mismatches between the guide RNA and the mRNA. The guide RNA acts as a template for the insertion or deletion of uridines, as directed by the base-pairing.

9.9 DEGRADATION OF NORMAL RNAs

In the preceding sections, we have considered ways in which the properties of an RNA molecule can be modified through the implementation of chemical modifications to the molecule. However, a much starker way of modulating the activity of an RNA is to remove it from a biological system completely, rather than merely modify what is there. In this section we explore such controlled removal of RNA – the process of degradation.

RNA degradation can influence gene expression

All cells have mechanisms to degrade RNA, digesting the RNA back to individual nucleotides. This degradation allows cells to remove RNAs that are no longer required or which might be present at too high a level for the current needs of the cell. The precious nucleotides in the RNA can then be recycled in new RNA transcripts. Cells have methods to degrade 'normal' RNAs – those RNAs that perform useful functions to cells – and different methods to efficiently destroy foreign or defective RNAs, which we will discuss in the next section.

Some RNAs, such as rRNAs, are needed for long periods of time and are comparatively stable. In contrast, other RNAs, such as some mRNAs, are only needed briefly and are rapidly degraded. The stability of a given RNA species in the cell is described by the **RNA half-life**, the time in which the amount of the RNA is reduced by one half. RNA half-life varies widely: *E. coli* transcripts have a half-life of less than one minute to one hour, and the half-life of vertebrate mRNAs range from 20 minutes to more than 24 hours.

A variety of determinants impact RNA stability

Various factors can affect the stability of an RNA. Among these factors are the sequences and structures at both the 5′ and 3′ ends of RNAs. As we mentioned above, the 5′ cap structure present on eukaryotic mRNAs protects the 5′ ends from degradation by 5′ to 3′ exonucleases. Similarly, bacterial RNAs with a 5′ triphosphate are more stable than mRNAs with a 5′ monophosphate. Further protection is provided by stem–loop structures present at the very 5′ end of some bacterial RNAs.

Stem–loop structures found at the 3′ end of bacterial RNAs, including the stem-loops corresponding to the Rho-independent terminators described in Section 8.6, also generally stabilize the transcript by protecting it from 3′ to 5′ exonucleases. On the other hand, the presence of a poly(A) tail in bacteria is associated with *decreased* stability. By contrast, in eukaryotes, the poly(A) tail generally lends *increased* stability to RNA transcripts whereas other elements, including the AU-rich elements (AREs) found in the 3′ UTRs of some mRNAs, cause instability by directing poly(A) tail removal.

A dramatic example of the importance of AREs is illustrated by *c-fos*, a short-lived mRNA with an ARE, which encodes a transcriptional regulator involved in promoting the growth of vertebrate cells. Some tumor-causing viruses contain a mutated form of *c-fos*, called *v-fos*, which lacks an ARE. When these viruses infect a host cell, the absence of an ARE prevents the *v-fos* mRNA from being rapidly degraded, leading to high levels of the *v-fos* protein, which promote the uncontrolled growth of infected cells.

In all organisms, processes that act on RNAs, such as RNA splicing, transport of the RNA through cellular compartments, or the translation of an mRNA into protein, also impact the half-life of RNAs. In addition, specific proteins and non-coding RNAs can bind to RNA transcripts to regulate their stabilities, as we shall see in the examples below. In all of these cases, the RNA processing event can either block or open up a site for digestion by the degrading enzymes.

RNA degradation in eukaryotes and bacteria involves multiple steps

Although the degradation processes for bacterial and eukaryotic RNAs were originally thought to be distinct, the discovery of an increasing number of common degradation machinery components suggests an unexpected degree of similarity between the two pathways. For many *E. coli* RNAs, degradation is initiated by an endonuclease (Figure 9.28). The products of the initial endonuclease digestion are then degraded by 3′ to 5′ exonucleases. Most often, the initiating endonuclease is RNase E, which is part of a protein complex called the **degradosome**. The degradosome contains a number of proteins likely to facilitate degradation, including an RNA helicase that uncoils double-stranded regions of the RNA and a 3′ to 5′ exonuclease denoted polynucleotide phosphorylase (PNPase). Other bacterial species that lack RNase E instead use an enzyme (RNase J) with both 5′ to 3′ exonuclease activity and endonuclease activity to degrade RNA.

The presence of a triphosphate at the 5′ ends of RNAs inhibits RNase E activity and protects against 5′ to 3′ exonuclease degradation. Thus, the conversion of a

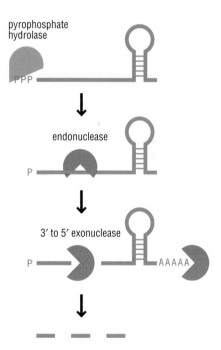

Figure 9.28 Pathways for degradation of bacterial mRNAs. The degradation of most bacterial mRNAs is initiated by an endonucleolytic cleavage. The RNA fragments are then degraded to individual nucleotides by 3′ to 5′ exonucleases. The removal of 5′ triphosphates stimulates the endonucleolytic cleavage, and the polyadenylation of the 3′ ends facilitates 3′ to 5′ exonuclease degradation, particularly of those RNAs with 3′ structures (such as stem–loops) that are resistant to exonuclease cleavage.

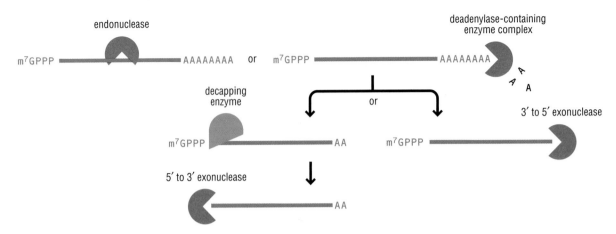

Figure 9.29 Pathways for the degradation of eukaryotic mRNAs. While the degradation of some eukaryotic mRNAs can be initiated by endonuclease cleavage, the first step in the degradation of most eukaryotic mRNAs is the shortening of the poly(A) tail, catalyzed by a deadenylase. Subsequently, degradation can proceed by further digestion by 3′ to 5′ exonucleases, or by the removal of the 5′ cap structure, which then allows digestion by 5′ to 3′ exonucleases. m⁷G, 7-methyl guanosine.

Figure 9.30 Some bacterial mRNAs are degraded in response to low cellular iron levels. When intracellular iron levels are high, expression of the RyhB RNA (green) is repressed and its target mRNAs are stable. When intracellular iron levels are low, expression of the RyhB RNA is induced, and RyhB RNA base-pairing with specific mRNAs blocks ribosome binding and targets the mRNAs for degradation.

5′ triphosphate to a 5′ monophosphate by the RppH pyrophosphate hydrolase enzyme can stimulate degradation. All 3′ ends can eventually be digested by 3′ to 5′ exonucleases. However, RNA stem–loop structures can block the accessibility of the 3′ end, making RNAs resistant to exonuclease action. The digestion of such RNAs is facilitated by polyadenylation of the 3′ ends, which provides an accessible, unstructured end for the action of exonucleases.

In eukaryotes, RNA degradation also can be initiated by the action of endonucleases. However, poly(A) tails generally prevent rather than facilitate degradation. Thus the first step in the degradation of many normal eukaryotic mRNAs is the shortening of poly(A) tails, after which degradation proceeds in one of two ways (Figure 9.29). Either decapping enzymes, which are related to the bacterial RppH enzyme, remove the 5′ cap structure and the transcript is digested by a 5′ to 3′ exonuclease, or degradation proceeds from the 3′ end and is catalyzed by the **exosome**, a 3′ to 5′ exonuclease related to bacterial PNPase.

mRNA degradation of certain RNAs is regulated in response to iron levels

Some mRNAs are degraded in response to specific stimuli. For example, in both bacteria and mammalian cells, the stability of certain mRNAs is regulated in response to cellular iron levels. In *E. coli*, several transcripts that encode proteins that bind to iron are degraded when iron levels are low, allowing the limiting iron pools to be available to proteins that are most important for the cell. This degradation is mediated by a 90 nucleotide RNA (RyhB) that is synthesized under conditions of low iron. The small RNA base pairs with the mRNA targets, blocking ribosome binding and promoting degradation by RNase E (Figure 9.30).

In mammalian cells, the intracellular concentration of iron is regulated in part by the transferrin receptor, a protein that imports transferrin-bound iron into the cell

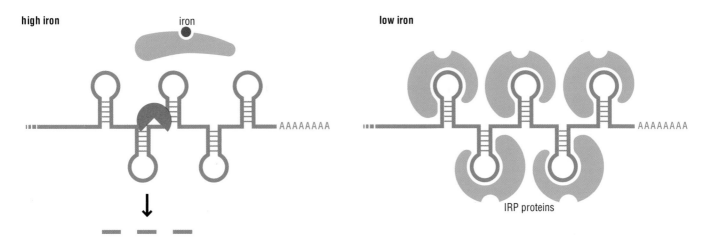

Figure 9.31 Degradation of mammalian transferrin receptor mRNA is inhibited by low cellular iron levels. When intracellular iron levels are high, the transferrin receptor mRNA is susceptible to cleavage by endonucleases (pink). When intracellular iron levels are low, the mRNA is protected by iron regulatory proteins (green, representing either IRP1 or IRP2), which block nuclease access.

(transferrin itself is a glycoprotein that binds iron very tightly). The mRNA encoding the transferrin receptor is rapidly degraded by nucleases when iron levels in a cell are high and the cell has no need to import more iron. However, when iron levels are low, the transferrin receptor mRNA is protected from degradation. This protection is provided by two Iron Regulatory Proteins, IRP1 and IRP2, that bind to stem–loop structures called Iron Response Elements (IREs) found in the 3′ UTR of the transferrin receptor mRNA (Figure 9.31). When iron is plentiful in the cell, the IRPs are no longer able to bind RNA, thus allowing the transferrin receptor mRNA to be degraded.

By regulating the stability of the mRNA in this way, mammalian cells are able to promote or prevent the synthesis of transferrin receptor in response to cellular iron levels. As we shall see in the next chapter, IRP proteins also directly regulate the synthesis of proteins from other mRNAs by binding to IRE stem–loop structures located in 5′ UTRs.

The RyhB RNA and the IRP proteins provide just two examples of how mRNA stability can be regulated in response to a specific signal. More and more proteins and a plethora of small, regulatory RNAs, including the miRNAs described below, are being found to modulate mRNA stability in all organisms, in response to a wide range of environmental inputs or in specific cell types.

9.10 DEGRADATION OF FOREIGN AND DEFECTIVE RNAs

Foreign and defective RNAs can be harmful to cells

Several types of RNA molecules can have harmful effects on a cell. 'Foreign' RNAs, derived from foreign DNA such as viruses or transposons, can be detrimental to the cell, for example, if the RNAs are translated into proteins that inhibit critical cell functions or promote transposition. In addition, an RNA that is defective in some way, due to errors during transcription or RNA processing or RNA folding, has the potential to disrupt normal cellular processes by encoding a product with altered function. Just as for the degradation of normal RNA transcripts, cells have various means to identify and destroy these potentially harmful RNAs.

Foreign RNAs are removed by an RNAi mechanism

The RNA molecules derived from the transcription of many viruses that enter the cell or from the transcription of transposons or other repetitive sequences in the chromosome are typically double stranded, and eukaryotic cells possess defense mechanisms that detect and destroy these RNAs. This phenomenon has been observed in a number of different organisms and was first called **RNA interference** in *C. elegans* and **post-transcriptional gene silencing** (**PTGS**) in plants. The discovery of RNA interference is described in Experimental approach 9.4.

Our current understanding of the mechanism of RNAi is illustrated in Figure 9.32. The process is initiated by the cleavage of the RNA into small double-stranded segments. Across species, these small RNAs range from 20 to 30 nucleotides in length, though the small RNAs generated by a specific complex in a specific organism are precise in their length. Thus for example, in flies, the RNAs are nearly always 21 nucleotides long. The cleavage is carried out by a protein complex containing a nuclease called **Dicer**, which is similar to the RNAse III nuclease we discussed in Section 9.1. The double-stranded RNAs are loaded into a second protein complex termed the RNA-induced silencing complex (**RISC**), on which the two strands of the duplex are separated and one strand is released. The retained RNA strand, also denoted the guide strand, facilitates interactions with a complementary full-length RNA target. (Note that the guide strand is distinct from the guide RNA involved in RNA editing, discussed in Section 9.7.) The RISC-mediated binding interaction ultimately results in cleavage of perfectly complementary target RNA at a specific position within the duplex. It follows that this cleavage event disrupts target RNA function. The cleavage event is catalyzed by the **Argonaute** protein, a core protein component of RISC. Argonaute has a domain with sequence similarities to the RNase H family of endonucleases, which cleave the RNA strand of DNA–RNA duplexes.

Mechanisms exist both to spread RNAi throughout an organism, and to defend against it

In some organisms like plants and nematodes, amplification mechanisms exist for spreading the RNAi response from cell to cell throughout an organism. Although the details of this process are unclear, it apparently depends on

1 double-stranded RNA is recognized by and cleaved by the Dicer complex

2 double-stranded RNA fragments are loaded onto the RISC complex and one strand is released

3 RISC complex bound siRNA is recruited to target RNA

4 RNA is targeted for degradation

Figure 9.32 RNA degradation by RNA interference (RNAi). (1) Double-stranded RNA is recognized by the Dicer endonuclease and digested into 21–25 nucleotide siRNA fragments that have two or three nucleotide 5′ overhangs. (2) The RNA fragments are loaded on to the RISC protein complex and one strand is released. (3) The bound siRNA guides the RISC complex to homologous single–stranded target RNA molecules. (4) The homologous RNA is then cleaved by the Argonaute protein within the region of base pairing with the siRNA.

9.4 EXPERIMENTAL APPROACH

The discovery of RNA interference and its broad application to forward genetics

While in the 1990s it was known that the introduction of RNA into cells could sometimes mediate 'interference' with the function of endogenous genes, a mechanistic understanding of such interference was lacking. The consensus at the time was that such interfering RNAs (typically designed as antisense to the gene of interest) hybridize directly with their target to somehow block subsequent gene expression. Of particular interest to Andrew Fire, Craig Mello, and colleagues was the surprising observation that both sense and antisense RNA preparations could cause 'gene silencing' in the nematode *C. elegans*, and that the observed interference could persist into the next generation of organisms. A major breakthrough in understanding the process of RNA interference (or RNAi) came with the insightful experiments of Fire and Mello in 1998 where they established that the most potent trigger of gene silencing was neither a sense nor an antisense RNA transcript, but instead a double-stranded RNA species. These scientists reasoned that such duplex RNA would be naturally generated by phage RNA polymerases during the transcription reactions used to prepare the samples, and might explain the surprising potency of the sense transcripts in their silencing experiments.

To address this hypothesis, the authors injected 'deliberately prepared' double-stranded RNA (as well as purified single-stranded RNA) targeting a non-essential gene, the myofilament protein encoding *unc-22*. Reductions in *unc-22* gene expression are easily measured in the progeny organism since loss of function *unc-22* worms exhibit a rather obvious 'twitching' phenotype. As shown in Figure 1, targeting of several different exons (covering exons 21-22 or 27) in *unc-22* with double-stranded RNA (sense + antisense, S + AS) resulted in dramatic loss of function phenotypes (100% strong twitchers), whereas injection of the single-stranded RNAs (either S or AS) did not.

Several other genes, *fem-1* and *hlh-1*, were also targeted by these approaches as their elimination from *C. elegans* also conferred an easily measured, but distinct, phenotype. Again, the injection of double-stranded RNA targeting exons of the genes of interest resulted in the expected loss of function phenotypes but the injection of the equivalent single-stranded RNAs did not. Importantly, in no case did non-specific RNAs produce the very particular phenotypes being assayed. Finally, in a visually appealing experiment, green fluorescent protein (GFP)-synthesizing worms were targeted with double-stranded RNAs to GFP and the amount of GFP protein expressed in these organisms was dramatically reduced (Figure 2).

The mechanism of action of the observed RNA interference was more fully explored when the authors compared the efficacy of

Effects of sense, antisense and mixed RNAs on progeny of infected animal				
gene fragment	segment	size (kb)	injected RNA	F1 phenotype
unc22A	exon 21-22	742	S	wild type
			AS	wild type
			S + AS	strong twitchers (100%)
unc22B	exon 27	1033	S	wild type
			AS	wild type
			S + AS	strong twitchers (100%)
fem1A	exon 10	531	S	hermaphrodite (98%)
			AS	hermaphrodite (>98%)
			S + AS	female (72%)
fem1B	intron 8	556	S + AS	hermaphrodite (98%)
hlh1A	exons 1-6	1033	S	wild type (<2% lpy-dpy)
			AS	wild type (<2% lpy-dpy)
			S + AS	lpy-dpy larvae (>90%)
hllh1D	intron 1	697	S + AS	wild type (<2% lpy-dpy)

Figure 1 Effects of sense, antisense, and mixed RNAs on progeny of infected animals. The indicated RNA was injected into six to ten adult hermaphrodite worms and their eggs collected after an appropriate time. Progeny phenotypes were scored upon hatching and at appropriate intervals.

Figure 2 Direct visualization of RNA interference effects in the whole *C. elegans* animal. Fluorescence micrographs show progeny of injected animals from GFP-reporter strain PD4251. a, b Progeny of animals injected with a control RNA (ds-*unc22A*). c, d Progeny of animals injected with ds-*gfpG*. a, c Young larva and b, d adult.
Reproduced from Fire *et al.*, *Nature* 1998;**391**:806–811.

double-stranded RNAs targeting either the introns or the promoters of the genes, and observed no phenotypic changes (Figure 2). These observations suggested that the mechanism of action of gene silencing involved the targeting of an mRNA species that has already been transcribed and processed, and thus was most likely to take place in the cytoplasm following export of the mRNA from the nucleus. Moreover, the fact that the knock-down phenotypes could be discerned in progeny organisms, where the initial input RNA concentrations would be very dilute, indicated that the mechanism of action included features that allowed for an amplification of the input signal.

Many of the mechanistic features uncovered in this landmark publication have turned out to be broadly accurate in a wide range of organisms and these same features have allowed for a revolution in forward genetic studies. By simply delivering double-stranded RNAs to a given organism, loss of function variants can be directly evaluated without the necessity of directly manipulating the genome through more complicated approaches, thus greatly expediting genetic analysis. In the ten years that have passed since these earliest observations, RNA interference has arguably become the single most commonly used tool for probing gene function in eukaryotic systems.

Even more importantly, these initial discoveries foreshadowed and dovetailed with the discovery of multiple classes of small RNAs (including miRNAs, piRNAs, and siRNAs) now known to play ubiquitous and central roles in modulating gene expression throughout the eukaryotic lineage. The implications of these earliest insightful experiments will be far-reaching for the foreseeable future.

Fire A, Xu S, Montgomery MK, *et al.* Potent and specific genetic interference by double-stranded RNA in *Caenorhabditis elegans*. *Nature* 1998;**391**:806–811.

both a transcription step by an RNA-dependent RNA polymerase (RdRp) to produce additional RNA and on membrane features that allow the RNAs to be distributed.

We note above how many of the foreign RNAs that stimulate RNAi are double stranded. Double-stranded molecules can also be derived from host RNA; some cellular RNAs have the potential to base-pair with each other and form long double-stranded duplexes. The cell must have mechanisms in place to distinguish between these two classes of dsRNA, as one class should be recognized as self, and the other not. It turns out that both foreign and host double-stranded RNAs are processed into **siRNAs**, but there appears to be discrimination at the next step wherein only certain single-stranded siRNAs are converted into double-stranded RNAs by RdRp. How RdRps distinguish between the different single-stranded siRNAs is an open question.

Interestingly, some viruses have evolved mechanisms to suppress RNAi. For example, a plant virus called the carnation Italian ringspot virus produces a caliper-like protein, which specifically binds double-stranded siRNAs and prevents their incorporation into RISC.

Dicer and RISC complexes are also involved in other forms of gene regulation

In the preceding chapter, we discussed how short RNAs are involved in transcriptional gene silencing by directing chromatin modifications. In the next chapter we will learn how short RNAs also regulate translation, and in Chapter 14 we will encounter other short RNAs that silence transposons. Although these short RNAs direct very different forms of regulation, they are all associated with RISC-family complexes, and most are generated by complexes that contain Dicer or Dicer-like enzymes. While these short RNA-mediated regulation mechanisms are collectively referred to as RNA silencing, the short RNAs can be subdivided into categories including: small interfering RNAs (siRNAs), microRNAs (miRNAs), and Piwi-interacting RNAs (**piRNAs**).

siRNAs, which direct RNA interference and transcriptional gene silencing (as discussed above and in Chapter 8), do not derive from specific genes but are generated through excision along the length of a double-stranded RNA of either external (such as viruses) or endogenous (such as repeat sequences) origin. Thus, multiple siRNAs can be produced from a given duplex RNA, generally phased with a periodicity about the length of an siRNA.

miRNAs, which modulate translation and RNA degradation, are different in that they are encoded by distinct genes within the genome. As shown in Figure 9.33, these specific RNA products are excised from characteristic hairpin structures in the nucleus by a Dicer-like enzyme called Drosha in animals and Dcl1 in plants, and are further processed by Dicer-like enzymes or Dcl1 in the cytoplasm (along with the related siRNAs). As such, miRNAs are specific gene products that target specific gene products (mRNAs) in the cells where they are produced.

The final group of small RNAs are the **piRNAs**, which primarily act to silence transposons in the germlines of animals (and will be discussed further in Chapter 14). These small RNAs appear to derive from single-stranded RNA precursors and are ultimately loaded onto the Piwi subfamily of Argonaute proteins to bring about downstream cellular consequences.

Figure 9.33 Precursor for the miR-35 through miR-41 microRNAs in *C. elegans*. The six microRNAs (indicated in green) are cleaved out of the longer precursor RNA by the Drosha enzyme. Note how all are cleaved from the right side of a characteristic stem–loop structure.
Based on Figure 1D of Lau *et al.* (2001).

The existence of these diverse and widely distributed classes of small RNAs eluded identification for many years. In the light of the genomic era and high throughput approaches to decipher genome function, we can anticipate that the known roles of RNA in biology will continue to expand.

Defective mRNAs are recognized and removed by specific decay mechanisms

In the next chapter we will discuss how a molecular machine called the ribosome synthesizes a protein molecule by reading the sequence of nucleotides in an mRNA. Normally, when the protein has been fully synthesized the ribosome recognizes a particular sequence of three nucleotides in the mRNA called a *stop codon* (or *nonsense codon*), which causes the ribosome to terminate protein synthesis. Some defective mRNAs, however, contain a premature translation termination codon that causes the ribosome to stop synthesis before the natural protein is completed, others completely lack a termination codon that prevents the termination reaction from occurring, while still others have problematic coding sequences that can promote stalling during translation.

At the very least, the production of a truncated or aberrant protein is a waste of a cell's energy and chemical resources; at worst, it results in the generation of a protein product that disrupts cellular functions. Eukaryotic cells have evolved means to specifically eliminate such defective mRNAs, degrading them in processes termed nonsense-mediated mRNA decay (NMD), non-stop decay (NSD), and no-go decay (NGD) respectively. As we will discuss in more detail in the next chapter, the decay of such defective mRNAs depends on molecular signals. In the case of NMD, the signals are well understood, and include the EJC protein complex deposited at splice junctions during pre-mRNA splicing (see Section 9.6).

Throughout this chapter we have considered a range of processes, many of which depend on the interaction of an RNA molecule with one or more proteins. Often, these interactions are specific, and are mediated by characteristic features of both RNA and protein, which facilitate specific interactions. We end this chapter by exploring structural features of some of the RNA binding proteins essential to the processes described.

9.11 RNA-BINDING DOMAINS IN PROTEINS

RNA structure can be specifically recognized by a variety of RNA-binding domains

Most of the processes described in this chapter depend on precise interactions between the RNAs being processed and the various protein machines involved in the processing reactions. For these interactions to be specific, a protein must only recognize the RNA to be modified. Just as proteins that bind to specific DNA sequences possess a variety of different DNA-binding motifs, the proteins that specifically recognize RNA also possess one or more of a limited number of different **RNA-binding motifs**.

There are, however, fundamental differences in DNA and RNA structure that lead to differences in the way that proteins bind to these two types of nucleic acid. DNA is generally found in a regular B-form helical structure, and proteins most often interact with nucleotide bases in the wider major groove (see Section 8.9). By contrast, double-stranded RNA is generally found in an A-form helical structure in which the deep major groove is relatively less accessible compared with the shallower minor groove. Moreover, RNAs rarely assume an extended helical structure and instead are composed of a series of shorter interrupted helices that pack together to form more complicated three-dimensional structures (see Figure 2.40 and Experimental approach 9.1).

RNA bases in single-stranded regions of an RNA molecule are also accessible to interact via extensive stacking interactions with the aromatic side chains of proteins. While some RNA-binding motifs in proteins recognize some single strands by their sequence (and constellation of hydrogen-bonding interactions, for example), others recognize specific loops or bulges associated with a unique backbone configuration (that typically differs from standard A-form double-stranded helix). In general, RNA-binding motifs exhibit a greater structural variety than DNA-binding motifs, reflecting the greater structural versatility of RNA. Here we discuss the most prevalent of these motifs.

Single-stranded RNA is bound by RRM domains in a variety of configurations

The most abundant **RNA-binding motif** is the **RNA-recognition motif** (**RRM**), also called the **RNA-binding domain** (**RBD**) or **ribonucleoprotein** (**RNP**) **domain**. The RRM domain was first identified in the poly(A) binding protein (PABP) and in one of the hnRNP proteins (discussed in Section 9.6); it is now known to occur in multiple splicing factors.

The motif provides a versatile platform for binding RNA that is used to direct RNA processing factors to many different RNAs in the cell. In these domains, alpha helices and beta strand elements come together in a sandwich structure such that a four-stranded antiparallel beta sheet is packed against two alpha helices, forming a platform that predominantly binds single-stranded RNA. A feature of this platform is the adaptability of its interactions with RNA, allowing it to bind to RNAs that can be very different in structure (Figure 9.34).

There are also two other relatively common single-stranded RNA binding motifs: the KH (for Hn RNP K homology) and the PAZ (for PIWI/Argonaute/Zwille) domains. In KH domains, such as those also found in hnRNP proteins (described in Section 9.6), an RNA binding groove is formed by two alpha helices and one beta strand. The PAZ domain is found in a number of proteins involved in the RNAi processing pathway including Dicer, the key enzyme in siRNA generation, and Argonaute, the key catalytic component in initiating RNAi-mediated RNA degradation (discussed in Section 9.10). In these proteins, a pocket formed by beta strands and an alpha helix of the PAZ domain specifically interacts with the single-stranded overhangs of siRNAs generated by Dicer cleavage (Figure 9.35).

(a)

(b)

Figure 9.34 Examples of two different RRM platforms binding to single-stranded RNA. The RRM domains in hamster nucleolin (top) and human PABP protein (bottom) in complex with RNA illustrate the flexibility with which the RRM platform of four beta-strands supported by two alpha helices can interact with RNA targets. PDBs 1FJE, 1CVJ.

Figure 9.35 The PAZ domain forms a pocket for single-stranded RNA ends. In the PAZ domain of human Argonaute 2, four beta strands and an alpha helix form a pocket for the single-stranded overhangs of siRNAs. PDB 1SI3.

Double-stranded RNA-binding domains distinguish between double-stranded RNA, DNA, and RNA–DNA hybrids

The RNase III family enzymes that are central to both rRNA processing and RNAi have **double-stranded RNA-binding domains (dsRBD)**. Examples of proteins carrying this RNA binding motif include Dicer and the ADAR protein involved in adenosine deamination in eukaryotic cells (discussed in Section 9.8). These proteins have an alpha helix and two loop regions that contact 2′ OH and phosphate groups in three adjacent minor, major, and minor grooves of an RNA helix (Figure 9.36). These interactions allow this class of proteins to recognize double-helical RNA and to discriminate against DNA or DNA–RNA hybrids.

Some RNAs are recognized by proteins with multiple, repeated motifs

Several RNA binding proteins contain more than one RNA-binding motif in tandem. For example, some proteins contain two or more RRM domains that form more extensive contacts with their target RNAs and thus allow for increased binding affinity (or specificity). A subset of RNA binding proteins even have multiple, repeated copies of an individual binding motif. These motifs can be present in a single protein or in oligomers, where each subunit contributes a binding motif. Striking examples are provided by members of the eukaryotic PUF family of proteins (named after the founding *Drosophila* Pumilio and *C. elegans* FBF proteins), of which one is shown in Figure 9.37.

Figure 9.37 An example of RNA binding by a protein with a repeated domain. The human Pumilio 1 homology domain contains multiple repeats of a two-helix motif, each of which interacts with a single nucleotide in the RNA and contributes a histidine, tyrosine, or arginine residue that stacks with the nucleotide. The side chains for these stacking amino acids are shown. PDB 1M8Y.

minor groove

major groove

minor groove

Figure 9.36 dsRBM domains make multiple contacts with adjacent grooves in double-stranded RNA. Helix 1, loop 4, and loop 2 of the dsRBM of *Saccharomyces cerevisiae* RNase III make contacts with sequential minor, major, and minor grooves, respectively. PDB 1T4L.

These proteins contain eight imperfect repeats of 36 amino acids and bind to single-stranded sequences in the 3′ ends of mRNAs to repress translation. Three conserved amino acids in each of the repeats in the protein contact an individual RNA nucleotide base, with one residue in particular making extensive stacking interactions with the base. The eukaryotic Sm proteins that form the core of eukaryotic splicing complexes (and related proteins found in archaea and bacteria) present another example. These proteins form oligomeric rings comprised of six or seven protein subunits, where each subunit can provide stacking interactions with uridine residues found in a variety of single-stranded RNAs.

RNA recognition can also be mediated by small basic motifs or zinc finger domains

For a subset of proteins, RNA binding requires the protein to possess small, basic stretches that are rich in arginine and lysine residues. While these regions, which can fold into beta hairpins or alpha helices, are required for RNA recognition, they do not bind RNA in isolation and often do not fold into secondary structures in the absence of RNA. Two proteins encoded by the human immunodeficiency virus (HIV), Tat and Rev, which bind the TAR and RRE sequence elements of the HIV RNA, respectively, are well studied examples of proteins with such small basic regions.

Other proteins containing the zinc finger DNA-binding domain described in Section 8.9 also bind to RNA. Unlike the binding of a zinc finger to DNA, where the finger inserts its alpha helix end-on into the major groove and side chains form base contacts, the interactions between zinc fingers and RNA are much more varied and include contacts with both single- and double-stranded regions of RNA.

In the next chapter we will learn how the processed tRNAs and rRNAs act to translate the processed mRNAs.

 SUMMARY

Overview

- RNA processing events are points for regulation and quality control, and are sources of diversity.

- Many RNA processing reactions are directed by RNA components.

tRNA and rRNA processing

- tRNAs and rRNAs are processed out of longer precursor transcripts and the nucleotides are post-transcriptionally modified.

mRNA processing

- The maturation of eukaryotic mRNAs requires the addition of a 5′ cap and poly(A) tail, processes that are closely tied to transcription, splicing, transport out of the nucleus, and ultimately to translation.

RNA splicing

- RNA splicing allows the generation of great diversity in RNA products and can be catalyzed by the RNA itself (self-splicing) or by a large protein and RNA-containing complex called the spliceosome.

- Splice site recognition depends on many protein factors that help in defining which exons are brought together. These choices can be regulated to generate alternatively spliced products.

RNA editing

- The sequence of mRNAs can be edited (nucleotides deleted, inserted, or changed) such that it might not be possible to predict the protein sequence from the DNA sequence of a gene.

- The sites of editing are specified by diverse mechanisms involving the direct recognition of RNA sequences or structures or interactions with guide RNAs.

RNA degradation

- The degradation of RNAs is another mechanism by which gene expression can be regulated and can be initiated by endonuclease cleavage or by removal of 5′ or 3′ end RNA features that lead to further exonucleolytic degradation.

siRNAs, miRNAs, and piRNAs

- Cells have mechanisms to sense and destroy foreign and defective RNA.

- Eukaryotic small RNAs, which are involved in a range of regulatory processes, are generated by Dicer family enzymes and act in association with Argonaute-containing complexes. These small RNAs include:

  - siRNAs, which direct RNA interference and transcriptional gene silencing, do not derive from specific genes but are generated through excision along the length of a double-stranded RNA of either external (such as viruses) or endogenous (such as repeat sequences) origin.

  - miRNAs, which modulate translation and hasten RNA degradation, are encoded by distinct genes within the genome.

  - piRNAs, which primarily act to silence transposons in the germlines of animals appear to derive from single-stranded RNA precursors and are ultimately loaded onto the Piwi subfamily of Argonaute proteins.

RNA-binding domains

- The structures of RNA binding domains, which include the RNA recognition motif (RRM), PAZ domain, and double-stranded RNA-binding domain (dsRBD), exhibit more variety than DNA-binding domains.

📖 FURTHER READING

9.1 OVERVIEW OF RNA PROCESSING

Hattori D, Millard SS, Wojtowicz WM, Zipursky SL. Dscam-mediated cell recognition regulates neural circuit formation. *Annual Review of Cell and Developmental Biology* 2008;**24**:597–620.

9.2 tRNA AND rRNA PROCESSING

Hopper AK, Phizicky EM. tRNA transfers to the limelight. *Genes and Development* 2003;**17**:162–180.

Weiner AM. tRNA maturation: RNA polymerization without a nucleic acid template. *Current Biology* 2004;**14**:R883–R885.

9.3 tRNA AND rRNA NUCLEOTIDE MODIFICATIONS

Decatur WA, Fournier MJ. RNA-guided nucleotide modification of ribosomal and other RNAs. *Journal of Biological Chemistry* 2003;**278**:695–698.

Kiss T. Small nucleolar RNAs: an abundant group of noncoding RNAs with diverse cellular functions. *Cell* 2002;**109**:145–148.

Ofengand J. Ribosomal RNA pseudouridines and pseudouridine synthases. *FEBS Letters* 2002;**514**:17–25.

9.4 mRNA CAPPING AND POLYADENYLATION

Jensen TH, Dower K, Libri D, Rosbash M. Early formation of mRNP: license for export or quality control? *Molecular Cell* 2003;**11**:1129–1138.

Moore MJ, Proudfoot NJ. Pre-mRNA processing reaches back to transcription and ahead to translation. *Cell* 2009;**369**:688–700.

9.5 RNA SPLICING

Doudna JA, Lorsch JR. Ribozyme catalysis: not different, just worse. *Nature Structural and Molecular Biology* 2005;**12**:395–402.

Qin PZ, Pyle AM. The architectural organization and mechanistic function of group II intron structural elements. *Current Opinion in Structural Biology* 1998;**8**:301–308.

9.6 EUKARYOTIC mRNA SPLICING BY THE SPLICEOSOME

Jurica MS, Moore MJ. Pre-mRNA splicing: awash in a sea of proteins. *Molecular Cell* 2003;**12**:5–14.

Liang XH, Haritan A, Uliel S, Michaeli S. trans and cis splicing in trypanosomatids: mechanism, factors, and regulation. *Eukaryotic Cell* 2003;**2**:830–840.

Staley JP, Guthrie C. Mechanical devices of the spliceosome: motors, clocks, springs, and things. *Cell* 1998;**92**:315–326.

9.7 EXON DEFINITION AND ALTERNATIVE SPLICING

Berget SM. Exon recognition in vertebrate splicing. *Journal of Biological Chemistry* 1995;**270**:2411–2414.

Black DL. Mechanism of alternative pre-messenger RNA splicing. *Annual Review of Biochemistry* 2003;**72**:291–336.

Matlin AJ, Clark F, Smith CW. Understanding alternative splicing: towards a cellular code. *Nature Reviews Molecular Cell Biology* 2005;**6**:386–398.

9.8 RNA EDITING

Bass BL. RNA editing by adenosine deaminases that act on RNA. *Annual Review of Biochemistry* 2002;**71**:817–846.

Horton TL, Landweber LF. Rewriting the information in DNA: RNA editing in kinetoplastids and myxomycetes. *Current Opinion in Microbiology* 2002;**5**:620–626.

Keegan LP, Gallo A, O'Connell MA. The many roles of an RNA editor. *Nature Reviews Genetics* 2001;**2**:869–878.

Simpson L, Sbicego S, Aphasizhev R. Uridine insertion/deletion RNA editing in trypanosome mitochondria: a complex business. *RNA* 2003;**9**:265–276.

9.9 DEGRADATION OF NORMAL RNAs

Aiba H. Mechanism of RNA silencing by Hfq-binding small RNAs. *Current Opinion in Microbiology* 2007;**10**:134–139.

Houseley J, Tollervey D. The many pathways of RNA degradation. *Cell* 2009;**136**:763–776.

9.10 DEGRADATION OF FOREIGN AND DEFECTIVE RNAs

Carthew RW, Sontheimer EJ. Origins and mechanisms of miRNAs and siRNAs. *Cell* 2009;**136**:642–655.

Isken O, Maquat LE. Quality control of eukaryotic mRNA: safeguarding cells from abnormal mRNA function. *Genes and Development* 2007;**21**:1833–1856.

Malone CD, Hannon GJ. Small RNAs as guardians of the genome. *Cell* 2009;**136**:656–668.

Siomi H, Siomi MC. On the road to reading the RNA-interference code. *Nature* 2009;**457**:396–404.

9.11 RNA-BINDING DOMAINS IN PROTEINS

Chen Y, Varani G. Protein families and RNA recognition. *FEBS Journal* 2005;**272**:2088–2097.

Maris C, Dominguez C, Allain FH. The RNA recognition motif, a plastic RNA-binding platform to regulate post-transcriptional gene expression. *FEBS Journal* 2005;**272**:2118–2131.

10 TRANSLATION

INTRODUCTION

In the preceding two chapters we have examined how the information stored in a specific region of a genome is extracted in the form of messenger RNA, and how this mRNA is processed to reveal the information required to direct the synthesis of the protein products needed by the cell. How does the cell decode this information? How can a sequence of nucleotides be used to determine the composition of a biochemically distinct molecule, a polypeptide?

In this chapter we conclude our journey from gene to protein by exploring the process of **translation**, the series of events through which the cell uses the information in an mRNA molecule to direct the synthesis of a protein. We explore the key molecular components that play vital roles both in the process of translation itself and in its regulation; we describe what happens in each individual step of this overall process; and see how – in parallel with the various other processes we encounter throughout this book – the process of translation is regulated, to ensure that it happens at the right time, in the right amount, with an end result that meets the required standards.

While we consider individual aspects of translation in some detail later in the chapter, we begin by painting an overall picture of the process, to give a framework on which the subsequent detail can be placed.

10.1 OVERVIEW OF TRANSLATION

Translation is the final step in decoding the genetic code

Translation is the process by which a polypeptide chain is synthesized according to the instructions in an mRNA template molecule; it is thus the final step in the flow of genetic information from DNA to protein. Translation transforms genotype into phenotype. The earlier steps of this information flow – DNA replication and transcription – are essentially two similar processes in which a new DNA or RNA molecule is synthesized based on the formation of complementary nucleotide base-pairings with a pre-existing DNA template, and the joining together of these bases to form a continuous complementary strand. The synthesis of protein from mRNA is an altogether more complex task: the language of proteins has an alphabet of 20 amino acids, as opposed to the four-nucleotide alphabet of mRNA, and there are no simple amino acid–nucleotide **interactions** that substitute for the direct Watson–Crick base-pairings between nucleotides.

Transfer RNA molecules provide the physical link that decodes the mRNA into the encoded polypeptide

How can a polypeptide chain, whose amino acid sequence is determined by the sequence of nucleotides in an mRNA molecule, be synthesized? The solution to the problem of coupling nucleotide and amino acid sequence is provided by a class of RNA called **transfer RNA** (**tRNA**). tRNAs are bifunctional: they decode the mRNA through direct base-pairing interactions, and participate in the synthesis of the polypeptide as they carry a chemically reactive amino acid at their 3′ end. These amino acids are the raw ingredients of the polypeptide that is synthesized during translation. The enzymes responsible for attaching the appropriate amino acids to the various tRNAs are the **aminoacyl-tRNA synthetases**; this attachment event has a key role in deciphering the **genetic code**.

Each tRNA molecule recognizes a sequence of three nucleotides (collectively called a **codon**) in the mRNA through direct pairing interactions with a region within the tRNA called the **anticodon**, and each tRNA molecule carries a particular amino acid that corresponds to the codon that it recognizes and to which it binds (Figure 10.1). By having these two functions – binding to mRNA and carrying an amino acid – tRNAs can decode the information stored in mRNA, and ensure that the correct amino acid encoded by this information is inserted into the growing polypeptide.

Translation is carried out by a macromolecular RNA–protein complex called the ribosome and is facilitated by a number of protein factors

The complex task of protein synthesis is carried out by a large molecular machine called the **ribosome**. These machines are composed of RNA and protein, and in all organisms have two subunits, the 'small' and the 'large', which are responsible for the two principal events of translation. The small subunit is responsible for deciphering the mRNA, whereas the large subunit mediates the formation of chemical bonds (Figure 10.2). The mass of the large subunit is about twice that of the small subunit, and the overall mass of the ribosome is roughly two-thirds RNA relative to one-third protein. The overall diameter of an intact bacterial ribosome with both large and small subunits together is about 220 Å, about twice as large as that of the *Escherichia coli* RNA polymerase holoenzyme.

The ribosome moves processively (5′ to 3′) along an mRNA molecule, forging peptide bonds between amino acids along the way to synthesize a polypeptide chain.

In bacterial cells, a ribosome synthesizes protein at a rate of about 15 amino acids per second, and amino acids are incorporated with an error rate of around 10^{-3} to 10^{-4} per residue. To put this another way, roughly nine out of ten proteins that are 1000 amino acids long will be synthesized accurately (i.e. without any mistakes) if fidelity is 10^{-4} (0.9999^{1000}), whereas one in three will be accurately synthesized if fidelity is 10^{-3} (0.999^{1000}). While neither the rate nor the fidelity of protein synthesis come close to those reported for DNA replication (which proceeds at a rate of almost 1000 nucleotides per second with an astonishing accuracy of 10^{-10}), they are similar to those reported for transcription (which proceeds at a rate of 50–100 nucleotides per second and an accuracy of around 10^{-4}).

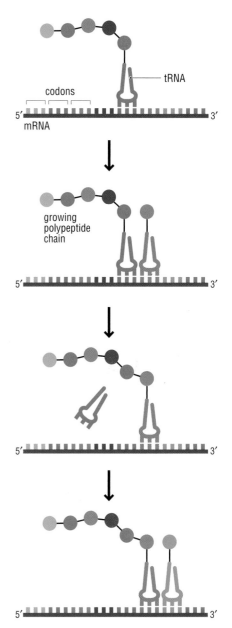

Figure 10.1 tRNA links the mRNA sequence with the amino acid sequence. Distinct tRNA molecules bind to different three-nucleotide long sequences (called codons) in mRNA. Each tRNA also carries a particular amino acid (balls) that are incorporated into a polypeptide chain. Thus tRNAs are able to specify the sequence of amino acids that are 'read' from an mRNA sequence during protein synthesis. In the figure, several tRNAs are shown aligned with the mRNA to emphasize how tRNAs link nucleotide and amino acid sequences, but inside cells, the tRNAs only bind to mRNA found inside the ribosome, two or three at a time.

⊘ We learn about the chemical structure of peptide bonds in Chapter 2.

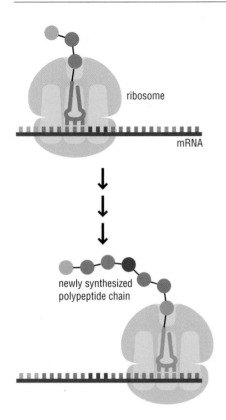

Figure 10.2 Ribosomes are protein synthesis machines. A ribosome moves along an mRNA molecule synthesizing a protein as it goes. The sequence of amino acids in the polypeptide chain is determined by the sequence of nucleotides in the mRNA.

As we have seen in other systems and will see again for translation, fidelity is energetically costly (and can slow down the overall process) and speed is often important for cellular robustness. Ribosomes have evolved to provide a suitable compromise for the cell between high speed and accuracy during protein synthesis.

Many of the proteins that contribute to translation are not permanent components of the ribosome. These proteins, referred to as **translation factors**, temporarily associate with the ribosome and leave when their task is complete. Protein factors help to coordinate the orderly progression of events during translation, ensuring that the right events happen at the right time. Many of these translation factors are GTPases (or G proteins) that couple the energy of guanosine triphosphate (GTP) hydrolysis to events on the ribosome (movement or dissociation, for example). Translation factors are also important for regulating protein synthesis: cells can modify the activity of certain factors to increase or decrease the amount of protein the cells synthesize to suit their needs.

The translation cycle is composed of four basic stages

The translation cycle is composed of four basic stages: **initiation**, **elongation**, **termination** and **ribosome recycling**, which we explore in detail in later sections. At each of these stages, we will see that RNA elements in the ribosome itself play critical functional roles in the process; we will also see that, with the exception of peptide bond formation, all steps in protein synthesis depend on the actions of auxiliary protein factors. In the final sections of the chapter, we will discuss examples of how translation is controlled by cells to finely tune the amount of gene product yielded at a given time, and how these delicately poised systems can be exploited by pathogens.

TRANSLATION: CORE COMPONENTS
10.2 TRANSFER RNA AND THE GENETIC CODE

With a broad picture of translation now in our minds, we need to consider a question that is central to what happens during translation: how is a sequence of nucleotides in a nucleic acid converted into a sequence of altogether different molecules, the amino acids that comprise a polypeptide? The answer lies in a remarkable molecule that has a foot in the camps of both nucleic acids and proteins: tRNA.

tRNA is a bifunctional molecule that links gene sequence to protein sequence

Two processes are essential to translation: the deciphering of triplet **codons** in mRNA, and the incorporation of amino acids encoded by the triplets into a growing polypeptide chain. **tRNAs** link these two processes by virtue of their structure.

We will first look at some general features of tRNA structure and then see how these features allow tRNAs to function during translation.

tRNA molecules are small RNAs ranging from 75 to 94 nucleotides in length. There are many different tRNAs in an organism, each specific to a particular amino acid; the specificity is determined by the enzymes that attach the amino acids to the tRNA. (This topic will be discussed in the next section.) Many of the nucleotides in the tRNAs form base pairs that are best visualized by drawing a tRNA in a cloverleaf pattern in two dimensions (Figure 10.3a). In this representation, tRNA can be seen to consist of four base-paired regions referred to as stems or arms that are interspersed with single-stranded regions called loops.

The 5′ and 3′ ends of the tRNA meet in a double-stranded region called the **acceptor stem**. The acceptor stem ends with a short, universally conserved sequence of bases called the **3′ CCA tail**, which is the attachment point for amino acids that will later be incorporated into the polypeptide chain. At the other end of the cloverleaf, in a loop called the **anticodon loop**, are three nucleotides collectively called the **anticodon** that base-pair in an antiparallel fashion with mRNA codons. (That is, the 5′ end of the mRNA codon base-pairs with the 3′ end of the anticodon, and vice versa.) In this way, the tRNA provides a physical link between the amino acid and the triplet codon that specifies it.

The nucleobases of the anticodon are typically stacked on one another in a structure that follows a well-defined turn in the RNA backbone known as a U-turn. This anticodon loop conformation positions the nucleotides of the anticodon to effectively base-pair with the mRNA. A hypermodified purine residue occurs just after the three nucleotides of the anticodon (typically at position 37). For example, the wybutosine base (Y) found in certain eukaryotic tRNAs is shown in Figure 10.4. The site of modification on the Watson–Crick face of this base prevents it from base-pairing with the mRNA and thus helps in aligning the codon–anticodon base-pairing interactions. Modification at this site is known to be critical for ensuring the high fidelity of **decoding**.

Two other regions of tRNA, the **DHU (D) loop** and **TpsiC (T) loop**, are named after the modified bases they contain: dihydrouridine (D) in the D loop; ribothymidine (T) and pseudouridine (ψ) in the T loop. Figure 10.4 shows the chemical structures of several post-transcriptionally modified bases that are found in the D-, T-, and anticodon loops of various tRNAs.

Although about 10% of the nucleotides in a given tRNA are post-transcriptionally modified, the group of such modified nucleotides is specific to a given tRNA species. The many different post-transcriptional modifications are proposed to promote tRNA function in a number of ways including by aiding folding, stabilizing overall structure, and increasing the specificity of interactions with various components of the translation system (including the codons themselves).

The overall structure of a tRNA in three-dimensional space is an L (Figure 10.3b), where the elbow region (the 'bend' in the L shape) is stabilized by a number of unusual base–base interactions between nucleotides in the D and T loops of the molecule. For example, there are several places where three nucleotides interact together (base triple interactions) and a conserved methylated G–A base pair is critical for kinking the D-stem of the tRNA into a particular conformation (Chapter 2, Figure 2.37). The L shape of the tRNA appropriately spaces the

(a)

(b)

Figure 10.3 The structure of tRNA.
(a) A cloverleaf depiction of the pattern of base-pairing in tRNA. The structures of some of the modified nucleotides indicated as D (dihydrouridine), H (hypermodified purine) and ψ (pseudouridine) are provided in Figure 10.4. (b) The L-shaped three-dimensional structure of a folded tRNA. PDB 1TRA.

➲ We learn about CCA tail addition and post-transcriptional modification in Chapter 9.

Figure 10.4 Examples of post-transcriptionally modified nucleotides found in tRNA. Dihydrouridine and pseudouridine are modifications of uridine and are found in conserved positions in the D and T loops of tRNA, respectively. Inosine and base Y (an example of a hypermodified purine) are modifications of purines (A or G) and are found in the anticodon loop where they play critical roles in decoding mRNA. The Watson–Crick face of each nucleotide is indicated.

anticodon and acceptor stems with respect to one another to interact both with the aminoacyl-tRNA synthetases (that attach the amino acids to the tRNA) and with the two subunits of the ribosome where protein synthesis takes place.

Triplet nucleotide codons specify the genetic code

We have discussed how two regions of tRNA – the 3′ CCA tail to which amino acids are attached and the anticodon that base-pairs with mRNA – enable tRNA to link the **genetic code** to protein sequences. The code that defines the relationship between mRNA sequences and protein sequences is explained in Chapter 1 and is summarized in Figure 10.5. Each triplet mRNA codon specifies a single amino acid (a **sense codon**) or no amino acid (a nonsense codon). A triplet code is the simplest code that can specify as many as 20 amino acids with only four nucleotides. (By contrast, a singlet code – whereby one base = one amino acid – could only specify 4 amino acids, 4^1; a doublet code 16, 4^2; and the triplet code 64, 4^3.) **Stop codons**, often referred to as **nonsense codons**, signal the end of the protein-coding region of an mRNA molecule.

Some amino acids, for example tryptophan and methionine, are specified by a single codon whereas others are specified by multiple codons. (Arginine, for example, typically is specified by six.) If we examine the codon table (Figure 10.5) we see that, in many cases, the first two nucleotides of a codon are sufficient to specify an amino acid while the third nucleotide is irrelevant. For example, the four codons GCA, GCC, GCG and GCU (or GCN where N is any nucleotide) encode

	AGA AGG						GGA			UUA UUG				AGC AGU				GUA	
GCA	CGA				GGA				CUA				CCA	UCA	ACA			GUA		
GCC	CGC				GGC		AUA	CUC				CCC	UCC	ACC			GUC	UAA	
GCG	CGG	GAC	AAC	UGC	GAA	CAA	GGG	CAC	AUC	CUG	AAA		UUC	CCG	UCG	ACG		UAC	GUG	UAG
GCU	CGU	GAU	AAU	UGU	GAG	CAG	GGU	CAU	AUU	CUU	AAG	AUG	UUU	CCU	UCU	ACU	UGG	UAU	GUU	UGA
Ala	Arg	Asp	Asn	Cys	Glu	Gln	Gly	His	Ile	Leu	Lys	Met	Phe	Pro	Ser	Thr	Trp	Tyr	Val	stop
A	R	D	N	C	E	Q	G	H	I	L	K	M	F	P	S	T	W	Y	V	

Figure 10.5 Codon table. The amino acids that are decoded from each trinucleotide codon in the mRNA are indicated. At the bottom are listed their standard three-letter and single-letter abbreviations.

alanine, while the GGN family encodes glycine. These observations do not mean that a single tRNA can decode all four of the GCN alanine codons. Indeed, the multiple codons in such a four-member group are typically decoded by more than one species of tRNA; the set of tRNAs that are loaded with the same amino acid are referred to as **isoacceptors**.

How then does recognition of the codon take place? The first two positions of the codon (reading from the 5′ to 3′ direction in the mRNA) are read by strict Watson–Crick pairing interactions with the third and second positions of the anticodon (typically positions 36 and 35, respectively, numbering sequentially from the 5′ end of the tRNA); the mRNA and tRNA anticodon effectively form an antiparallel mini RNA helix (Figure 10.6a). Complexities in deciphering the code are seen at the third position of the codon where certain deviations from Watson–Crick pairing are permitted. This promiscuity in reading of the third codon position by the first position of the anticodon (the 5′ proximal nucleotide, position 34) is called **wobble pairing**. Wobble pairing allows certain relatively specific non-Watson–Crick pairing interactions to be accepted by the translational machinery (Figure 10.6b). Wobble pairing most often involves G–U base pairs, with the G found either in the codon or in the anticodon. More seldom, the unusual post-transcriptionally modified purine nucleotide inosine is found in the anticodon (at position 34) where it effectively pairs with U, C, or A in the mRNA.

As a result of this relaxation in the stringency of the pairing rules, tRNAs can decode several different codons by forming non-Watson–Crick base-pairing interactions with mRNA at the third position. Because of wobble pairing, the number of tRNAs needed to decode all mRNA codons is less than 61 (the number of sense codons), yet still more than 20 (the number of amino acids) – generally, there are about 40 different tRNAs in an organism that decode all 61 codons. It is interesting to note that not all tRNAs are equally abundant and that some codons are used less frequently than others. (We call these rare codons.) Not surprisingly, the rare codons tend to be decoded by the less abundant tRNA species.

The genetic code is nearly universal across all life forms. AUG, for example, codes for methionine in all organisms whereas CCC codes for proline. In addition, there are typically three stop codons – UAA, UAG, and UGA. Generally speaking, the genetic code has evolved to minimize the deleterious effect of genetic mutations: single nucleotide changes tend to result in the insertion of chemically similar amino acids into a protein. For example, mutation of the first nucleotide of the phenylalanine codon (UUU) will result in the replacement of phenylalanine with other hydrophobic amino acids such as leucine (CUU), isoleucine (AUU), or valine (GUU).

Codon reassignment does happen occasionally

There are exceptions to the universality of the genetic code. Mitochondria, which have much of their own protein translation machinery, decode AUA as methionine rather than as isoleucine thus providing two distinct codons for specifying methionine, AUG and AUA. In *Mycoplasma* species, the stop codon UAG is decoded as tryptophan, in addition to the standard tryptophan codon UGG. In some organisms, stop codons are programmed to incorporate non-standard amino acids, thus expanding the coding potential of the DNA. For example, we shall see later in this chapter (Section 10.11) that selenocysteine is incorporated in a number of

(a)

wobble pairing

(b)

1st base of anticodon	3rd base of codon
C	G
A	U
U	A or G
G	U or C
I	U, C or A

(c)

Figure 10.6 Wobble pairing. (a) The first two nucleotides of the mRNA codon (grey) must interact in a strict Watson–Crick pairing interaction with the anticodon (blue), while the third position (highlighted in pink) may interact through certain non-canonical pairing interactions. (b) The rules for allowed wobble pairings seen in bacteria. For example, a tRNA with inosine in the wobble position can pair with three different mRNA codons that have U, C, or A at the third position. (c) A specific example is shown indicating how tRNA^Phe can decode both UUC and UUU codons by taking advantage of 'wobble' pairing.

systems (bacterial, archaeal, and eukaryotic) through recognition of a UGA codon found in a specialized mRNA context.

Although exceptions to the basic rules of the genetic code are of considerable biological interest, the universal features of the code indicate that a sophisticated protein translation apparatus (and the basic code) evolved before the divergence of the three kingdoms of life.

10.3 AMINOACYL-tRNA SYNTHETASES

Transfer RNAs are not the only vital components of the molecular machinery that make translation happen. Vital, too, are the enzymes that couple tRNAs with the appropriate amino acids for addition to the growing polypeptide chain. Only if the right amino acid is coupled to the right tRNA will the reading of a codon by the translational machinery result in the insertion of the correct amino acid. In this section we discuss the enzymes that are responsible for this essential coupling step, the aminoacyl-tRNA synthetases.

Aminoacyl-tRNA synthetases are responsible for the first step in decoding

Attachment of the appropriate amino acids to the specified tRNAs is the key step in the decoding of genetic information. This attachment, termed **aminoacylation**, is carried out by enzymes called **aminoacyl-tRNA synthetases** (we often refer to them as aaRS, for example, GlyRS refers to the glycyl-tRNA-synthetase) that are specific both for the amino acid and the tRNA species to which the amino acid is attached. As such, this class of enzymes establishes the rules of the genetic code. How these enzymes recognize particular tRNAs and amino acids is central to translation of the nucleotide code into protein.

The direct attachment of an amino acid to a tRNA is not energetically favorable so aminoacyl-tRNA synthetases use a two-step process that is driven by the hydrolysis of adenosine triphosphate (ATP) (Figure 10.7). The amino acid is first activated by the attachment of adenosine monophosphate (AMP) with release of pyrophosphate; this chemical reaction produces an aminoacyl-adenylate that remains bound to the synthetase. As in many biological reactions, hydrolysis of the pyrophosphate moiety to free phosphate by pyrophosphatases in the cell provides critical energy that drives the overall process.

The activated amino acid is then directly transferred to the 2′ or 3′ OH of the ribose of the terminal adenosine of tRNA (the nucleophile in the reaction) with the release of AMP. The resulting **aminoacyl-tRNA** is composed of a chemically reactive amino acid attached to the 3′ CCA tail of the tRNA. The amino acid is chemically reactive because of the high energy bond that joins it to the tRNA. The chemical lability of the bond between the amino acid and the tRNA is a useful property since the amino acid will next be transferred to a growing peptide chain. The resulting aminoacyl-tRNA is protected from spontaneous hydrolysis by immediately binding to an abundant cellular protein known as EFTu in bacteria (and eEF1A in eukaryotes and archaea), which plays a key role in shepherding aminoacyl-tRNAs through the translation process.

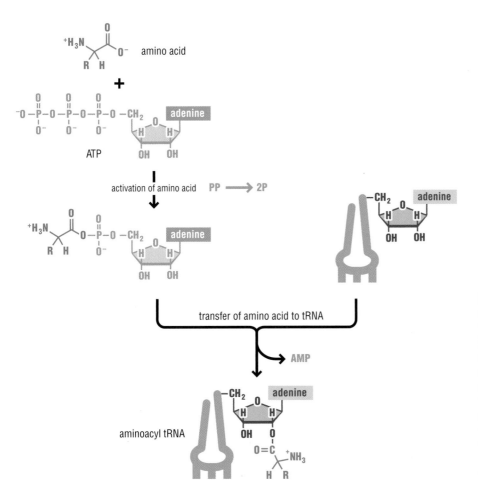

Figure 10.7 Aminoacyl-tRNA synthetases catalyze the aminoacylation of tRNA in a two-step process. Amino acids are activated by the attachment of AMP (yellow) to form an aminoacyl adenylate in which the carboxyl group of the amino acid is linked to the phosphoryl group of AMP. The high-energy bond between the amino acid and AMP can be easily broken, allowing transfer of the aminoacyl group to the 2′ or 3′ hydroxyl group of the ribose of the terminal adenosine of the tRNA (gray). The hydrolysis of pyrophosphate (PP), producing two phosphates (2P), drives the reaction forward.

Aminoacyl-tRNA synthetases recognize specific tRNA species

Each tRNA species must be loaded with the amino acid that corresponds to the mRNA codon to which the tRNA binds – this amino acid is referred to as **cognate**. For example, the codon AUG codes for methionine, so the tRNA recognizing this codon must be loaded with methionine and not a different amino acid. The specific amino acid with which a given tRNA is loaded is indicated by a superscript three-letter shorthand for the amino acid – tRNA[Met], tRNA[Gly], and so on. This loading reaction is performed by the aminoacyl-tRNA synthetases with a remarkable accuracy of less than one error per 10^4 aminoacylation events. There is generally one aminoacyl-tRNA synthetase for each of the standard 20 amino acids in a cell.

How do the synthetases select their tRNAs and amino acids? Aminoacyl-tRNA synthetases recognize tRNAs on the basis of specific sequence and structural features (termed **identity elements**) found primarily in the anticodon loop and acceptor stems. A notable example of a well-defined identity element is the G3:U70 base pair in the acceptor stem of the tRNA[Ala]. The G3:U70 base pair serves as the principal determinant for recognition of alanine tRNAs in all kingdoms, and can confer alanine identity on most tRNAs when the element is simply transplanted to the acceptor stem. (That is, when a tRNA molecule is engineered to carry the G3:U70 base pair, it will become charged with alanine.) Thus, while the overall three-dimensional structure of all tRNAs is highly similar, there are sufficient differences among them to allow synthetases to specifically recognize their cognate tRNAs.

Aminoacyl-tRNA synthetases select the correct amino acid in a two-step process

Aminoacyl-tRNA synthetases use various chemical features, such as charge, hydrophobicity, size, and shape, to discriminate between the different amino acids. It is easy to understand how synthetases can discriminate against amino acids larger than their cognate amino acid. For example, the synthetase for glycine tRNA (GlyRS) would be unlikely to bind phenylalanine or other large amino acids in its active site simply because they are so much larger than glycine. Consequently, these substrates are readily excluded. It is more difficult for synthetases to discriminate against amino acids that are similar to or slightly smaller than the cognate amino acid, as these amino acids are likely able to enter the active site. Indeed, certain **non-cognate** amino acids are activated by the synthetase and in some cases are even transferred to the tRNA.

At this point, a quality control step plays an important role in increasing the fidelity of aminoacylation. Most synthetases have a secondary site (the **editing site**) that accommodates either the aminoacyl-adenylate species (the activated amino acid prior to joining to its cognate tRNA) or the aminoacyl-end of the tRNA, and promotes a hydrolytic reaction (Figure 10.8a). If the hydrolytic reaction takes place prior to transfer of the activated amino acid to the tRNA, the aminoacyl-adenylate itself is hydrolyzed (pre-transfer); if it occurs after transfer to the tRNA the aminoacyl linkage between the amino acid and the tRNA is hydrolyzed (post-transfer). These quality control steps are referred to as editing, and both pre- and post-transfer versions are outlined in Figure 10.8b (steps 2 and 3). Typically, the cognate aminoacyl-adenylates or aminoacyl-tRNAs cannot enter the editing site and so remain intact (Figure 10.8b, step 4).

Let us consider one well-documented example: valine is smaller, but chemically similar to the isoleucine side chain, and as such is readily activated and transferred onto tRNA^{Ile} by the isoleucyl-tRNA synthetase (IleRS). This non-cognate aminoacyl-tRNA species is recognized as mismatched and the aminoacyl linkage is rapidly hydrolyzed in the editing site of the synthetase (post-transfer, as in Figure 10.8b, step 3), thus increasing the overall fidelity of the aminoacylation process.

There are two classes of aminoacyl-tRNA synthetases

Synthetases can be divided into two classes of approximately ten members each, termed class I and II. These two classes have entirely different structures and so they evolved from distinct ancestors. Each binds opposite faces of tRNA as illustrated in Figure 10.9, allowing them to explore somewhat different molecular features of tRNA – class I synthetases typically recognize the minor groove of the acceptor stem whereas the class II synthetases recognize the major groove. In addition, each aminoacyl-tRNA synthetase class has its own regio-specificity such that class I synthetases initially attach amino acids to the 2′ OH of the terminal ribose of the tRNA while class II synthetases attach them to the 3′ OH. However, since migration of the acyl group between the two positions of the ribose sugar

Figure 10.8 Editing by aminoacyl-tRNA synthetases. (a) A tRNA synthetase bound to a tRNA, showing the aminoacylation and editing sites of the synthetase. PDB 1QF6. (b) The editing site increases the fidelity of the aminoacylation reaction: (1) Large non-cognate amino acids are excluded from the aminoacylation site (top); (2) smaller non-cognate amino acids are occasionally activated (to form the aminoacyl-adenylate), but are recognized by the editing site and are hydrolyzed (pre-transfer); or (3) are transferred to the tRNA in the aminoacylation site, and then enter the editing site and are hydrolyzed from the tRNA (post-transfer); (4) the cognate amino acid for the tRNA cannot enter the editing site and therefore remains attached to the tRNA.

class I

class II

Figure 10.9 The evolutionarily distinct class I and class II synthetases recognize different faces of tRNA. The synthetases are shown in blue, the tRNA in gray and the CCA end of the tRNA in red. The protein structures of these classes of synthetase are completely different, and recognize distinct faces of the tRNA. PDBs 1EXD, 1ASY.

Figure 10.10 Mischarging of tRNA^Gln with glutamate is recovered by a transamidase enzyme. In certain organisms missing the gene encoding GlnRS, GluRS has dual specificity for both tRNA^Glu and tRNA^Gln. The product of the reaction, Glu-tRNA^Gln is then transformed into Gln-tRNA^Gln through the actions of a transamidase enzyme.

is extremely rapid, there is no real difference in the attachment chemistry of the aminoacyl-tRNA products released from the two synthetase classes.

It is not clear why two synthetase classes have evolved. Since we saw in Section 10.2 that cells typically produce more than 40 different tRNA species, it is possible that this large number of tRNAs can be more easily distinguished if the two synthetase classes recognize different faces of the tRNA.

Throughout the three kingdoms of life, class I and class II synthetases typically recognize the same groups of amino acids (e.g. the arginine synthetase belongs to class I whereas the alanine synthetase belongs to class II, in all species), although there are noted exceptions.

Some organisms do not have a complete set of 20 aminoacyl-tRNA synthetases

While most eukaryotic organisms have 20 synthetases, one for each standard amino acid, bacterial and archaeal organisms often have fewer than 20. Most typically in these cases, the asparaginyl and the glutaminyl synthetases are missing. A more complicated example can be found for *Methanococcus jannaschii*, where genome sequencing at first revealed only 16 of the 20 known aminoacyl-tRNA synthetase homologs. The lysyl synthetase, normally a class II enzyme, was ultimately identified as a class I synthetase (representing the first example of class switching among the synthetases) – truly missing were the asparaginyl, glutaminyl, and cysteinyl synthetases.

In a variety of organisms missing the genes encoding the asparagine and glutamine tRNA synthetase (AsnRS and GlnRS), the aspartate and glutamate tRNA synthetases (AspRS and GluRS) enzymes have dual specificity for tRNA^Asp and tRNA^Asn, or tRNA^Glu and tRNA^Gln, respectively. And, following misacylation of tRNA^Asn with aspartate or tRNA^Gln with glutamate, a **transamidase** enzyme chemically transforms the side chain of the bound amino acid from an acid to an amide, thus producing the desired asparagine- and glutamine-bound tRNA species (Asn-tRNA^Asn and Gln-tRNA^Gln) (Figure 10.10).

What about the missing cysteine tRNA synthetase (CysRS) in *M. jannaschii*? It is now known that in *M. jannaschii*, a phosphorylated serine is first loaded onto the tRNA^Cys by a specialized class II synthetase (SepRS) and then a second enzyme, a **cysteine desulfurase**, directly converts the phosphorylated serine (on the tRNA) to a cysteine. This mode of formation of Cys-tRNA^Cys appears to be relatively common in archaea as many are missing the CysRS. These exceptions to the rule of at least one synthetase per amino acid reveal interesting clues about the early evolution of tRNAs and the genetic code.

10.4 STRUCTURE OF THE RIBOSOME

In the previous sections, we have seen how the link between nucleic acid and amino acid is forged by tRNA molecules, which act as mediators between the mRNA template and the growing polypeptide chain. However, despite their key role in translation – and in transducing the flow of biological information from a gene to its protein product – tRNA molecules depend for their correct function on a remarkable molecular assembly that orchestrates the overall process. In the following sections we explore this assembly: the ribosome. We begin by describing key molecular and structural features of the ribosome, before considering how it achieves the remarkable feat of coordinating the individual steps of translation.

Ribosomes are composed of two conserved subunits

The ribosome is the macromolecular machine that facilitates the interaction between the tRNAs and the mRNA and forms peptide bonds between amino acids carried by the tRNAs to yield the end product of translation: a polypeptide. The ribosome is thus responsible for protein synthesis. It is a large machine, ranging in mass from 2.5 mDa in bacteria to more than 4 mDa in eukaryotic cells, which reflects the complexity of the tasks it must carry out. In both eukaryotes and bacteria, approximately two-thirds of the ribosome's mass is composed of RNA, termed ribosomal RNA (**rRNA**), and about one-third is composed of a collection of proteins referred to as **ribosomal proteins** (**r-proteins**). We will see that the rRNAs are central to all of the activities of the ribosome. Given the ribosome's fundamental role in synthesizing protein, it is not surprising that ribosomes are highly conserved across all life forms both in structure and in function.

All ribosome particles are composed of a large subunit and a small subunit, each of which contains both rRNA and protein (detailed in Figure 10.11). The small subunit mediates the interactions between mRNA and tRNA and the large subunit catalyzes peptide bond formation. These distinct events are integrated at the interface between the two subunits, where gross movements of the interface can shift the position of the tRNA–mRNA complex as amino acids are sequentially added to the growing polypeptide chain. These movements are coordinated by specific bridge elements between the two subunits, which are composed of RNA and proteins.

Another important structural feature of the ribosome is the exit tunnel in the large subunit through which the polypeptide is extruded as protein synthesis proceeds. We will see in Section 10.12 how many **antibiotics** function by blocking the exit tunnel and in Chapter 11 how external factors involved in protein secretion and folding assemble at the site on the exterior of the large subunit where the tunnel opens to the cytosol.

Protein and rRNA are differentially distributed on the exterior and interface surfaces of the ribosome

The interface between the small and large subunits of the ribosome, where the tRNA substrates bind and act, is rich in rRNA elements and relatively poor in proteins, while ribosomal proteins are more evenly distributed on the exterior. Figure 10.12 shows the distribution of protein and rRNA in the small and large subunits. A number of the ribosomal proteins have unusual structures, consisting of globular domains embedded in the exterior face of the subunit particle with long extended arms that snake their way into the core of the rRNA structure (Figure 10.13). These arms are generally enriched in basic amino acids and are thought to act as mortar in packing the negatively charged rRNA phosphate backbone into a compact tertiary structure.

bacterial ribosome
70S

50S (large)

30S (small)

5S rRNA (120 nts)
23S rRNA (2900 nts)
~34 proteins (L1-L34)

16S rRNA (1540 nts)
~21 proteins (S1-S21)

eukaryotic ribosome
80S

60S (large)

40S (small)

5S rRNA (120 nts)
5.8S rRNA (160 nts)
28S rRNA (4700 nts)
~49 proteins

18S rRNA (1900 nts)
~33 proteins

Figure 10.11 Composition of the ribosome. Both bacterial and eukaryotic ribosomes are composed of one large and one small subunit. The RNA and approximate number of protein components are listed beneath each subunit. The subunits and their components are named historically according to how fast they sediment during centrifugation: the large bacterial 50S subunit sediments faster than the small 30S subunit (where S is the Svedberg unit of sedimentation speed).

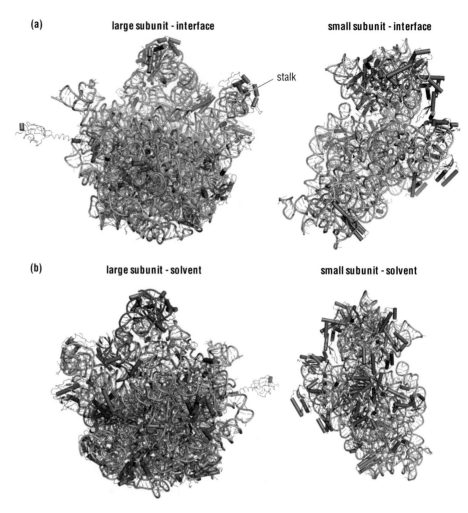

(a) large subunit - interface   small subunit - interface

stalk

(b) large subunit - solvent   small subunit - solvent

Figure 10.12 Distribution of rRNA and protein in the small and large ribosomal subunits. The interface (a) of the small and large subunits, which interact with one another and the tRNAs, is composed mainly of rRNA, while the solvent face (b) of both subunits has a greater concentration of protein. RNA is shown in gray and proteins in turquoise. PDBs 1GIX, 1GIY.

protein L4

protein L15

Figure 10.13 Unusual structure of ribosomal proteins. Several large subunit proteins show unusual structures with globular domains attached to extended positively charged arms. PDB 1FIN.

The large rRNAs from the two ribosomal subunits can be subdivided into distinct regions, or domains, based on their secondary structures. The l6S (or 18S in eukaryotes) rRNA is divided into three major and one minor domain whereas the 23S (or 28S) rRNA is divided into six different domains (Figure 10.14). The organization of the rRNA domains in the two ribosomal subunits is strikingly different. In the small subunit, each rRNA domain is largely limited to one area of the subunit, whereas in the large subunit the rRNA elements are intricately interwoven. It has been suggested that this organization reflects the functional differences of the subunits: interdomain flexibility is suited to large-scale movements performed by the small subunit whereas stability provided by the interwoven RNA domains is suited to protection of the active site for protein synthesis by the large subunit. In addition to the large rRNAs in both subunits (16S/18S and 23S/28S), the large subunit in bacteria and eukaryotes has a smaller RNA known as 5S RNA. In eukaryotes, there is a second small RNA, the 5.8S rRNA; this species is really the 5′ end of the large subunit large RNA that has undergone additional processing in the cell (see Section 9.2).

The crucial roles played by rRNAs in the function of the ribosome are reflected in their remarkably conserved secondary and tertiary structure and even in the conservation of the primary structure. Certain nucleotide stretches in the rRNA

Figure 10.14 Organization of rRNA in the large and small subunits. Secondary structure of (a) the small bacterial subunit rRNA (16S) with domains labeled 5′ for 5′ domain, C for centre, 3′M for 3′ major and 3′m for 3′ minor, (b) the large bacterial subunit rRNA (23S and 5S) with six domains labeled (5′ to 3′) I–IV, and (c) their three-dimensional organization in the two subunits, respectively. PDBs 1GIX, 1GIY. The rRNA elements have corresponding colors in the secondary structure and three-dimensional representations. The domains of the 23S (and 5S) rRNA are more intricately interwoven than the domains of the 16S rRNA.

are conserved across all species: for example 16 of 17 contiguous nucleotides in the large rRNA of the large subunit are more than 95% conserved across the three kingdoms of life. A great many ribosomal proteins also are highly conserved, and these probably are those most critical to ribosome function; a number of other ribosomal proteins are not even present in all species.

rRNAs are central to each of the ribosomal functions

At first glance, the observation that ribosomes are made of RNA and protein appears to raise an intriguing 'chicken-and-egg' problem: how could a machine that synthesizes proteins be made of proteins before proteins actually existed? The answer to this apparent conundrum lies in the composition of the ribosome's functional center.

Importantly, the primary molecular component in the functional centers of the ribosome is RNA. In the small subunit, the so-called 'decoding' region of 16S rRNA directly mediates the interaction between tRNA and mRNA. In the large subunit, the 23S rRNA in the catalytic or **peptidyl transferase** center (**PTC**) directly interacts with the aminoacyl end of the tRNAs. As there are no protein elements within 18 Å of the region where peptide bond formation occurs, it seems likely that rRNA, and not protein, catalyzes this reaction. In Chapter 9 we saw that catalytic RNAs are involved in the processing of introns in mRNA, providing clear evidence for the potential of RNA catalysis in an RNA world early in evolution. The RNA-driven ribosome arguably provides the strongest evidence for this RNA world, explaining how proteins could be synthesized before they had themselves evolved into the components of a protein synthesis machinery.

It is believed that the protein components of ribosomes were late additions to the pre-existing protein synthesis machinery, allowing protein synthesis to become increasingly efficient and more finely regulated. Ribosomes certainly evolved early in the evolution of life because they are remarkably similar in each of the three kingdoms of life – bacteria, archaea, and eukaryotes.

TRANSLATION: MECHANISM

We have now encountered all of the key components of the cellular machinery that mediate translation, and have seen in general terms how this machinery – with the ribosome as its centerpiece – deciphers the information in an mRNA molecule to synthesize a polypeptide chain. In the following sections, we explore each step in the translation cycle at a higher resolution to reveal the molecular mechanisms that underpin the processes we have described in only quite general terms so far.

10.5 THE TRANSLATION CYCLE: THE RIBOSOME IN ACTION

Ribosomes have three binding sites for tRNAs

We saw earlier in this chapter how the small ribosomal subunit is responsible for binding the mRNA and facilitating interactions between the mRNA codons and the anticodons of aminoacyl-tRNAs, and the large subunit of the ribosome is responsible for catalyzing peptide bond formation between amino acids. A polypeptide chain is synthesized as the ribosome moves from codon to codon along an mRNA molecule. The ribosome binds an aminoacyl-tRNA at each step along the way and transfers its attached amino acid to the end of a growing polypeptide chain. Messenger RNA is threaded through the ribosome in the 5′ to 3′ direction and the polypeptide chain is synthesized from the N-terminus to the C-terminus. The subunit interface region of each ribosomal subunit has three binding sites for tRNA, shown nicely through cryoelectron microscopy in Figure 10.15.

Once translation is underway, each aminoacyl-tRNA first binds in the **aminoacyl (A) site** before moving to the **peptidyl (P) site** after a peptide bond is formed between its amino acid and the growing polypeptide chain. A third site, the **exit (E) site**, is occupied by each tRNA as it leaves the ribosome. In the A and P sites, but possibly not in the E site, there is a requirement for appropriate base-pairing between the tRNA and mRNA for active and accurate translation.

Figure 10.15 Ribosomes have three binding sites for tRNA. The tRNAs are bound to ribosomes in the RNA-rich interface between the large (light gray) and small (dark gray) subunits. This molecular model derived from cryoelectron microscopy experiments shows three tRNAs (orange, red and green) bound in the Exit (E), Peptidyl (P), and Aminoacyl (A), sites of the ribosome. PDBs 1GIX, 1GIY.

Figure 10.16 The translation cycle. The basic steps of translation are initiation, elongation (1–3), termination (4) and recycling (5), which are described in the main text. Briefly, the first amino acid of a polypeptide chain is generally methionine, and it is found attached to a tRNA in the P site of the ribosome. This methionine is then attached to the subsequent amino acid at step 2, extending the polypeptide chain. The ribosome moves one codon further along the mRNA, and the elongation cycle then repeats (pink arrow). When a stop codon is reached in the mRNA, termination takes place (4) wherein the completed polypeptide chain is released from the ribosome. Translation ends when the ribosomal subunits are dissociated during ribosome recycling (5), and are thus ready to begin the synthesis of a new protein.

With a picture of the structure of a ribosome – its two subunits and three tRNA binding sites – in mind, we can now look at the basic steps of translation from the beginning to the end of protein synthesis, a process called the translation cycle.

Translation of a message involves initiation, elongation, termination, and ribosome recycling

Figure 10.16 shows the four basic steps of translation: initiation, elongation, termination, and ribosome recycling. These steps are briefly outlined here and will be considered in more detail in Sections 10.7–10.10. Each step in translation is facilitated by protein factors that help to ensure high accuracy and efficiency – translation can proceed in the absence of these factors on certain simple messages but at greatly reduced speed and accuracy.

During the first step, **initiation**, the AUG codon that signals the start of an open reading frame (the protein coding region of mRNA) is identified by a process that involves the interaction of the ribosome, a specialized initiator methionine tRNA, and proteins called **initiation factors** (**IFs**). Initiation effectively decodes the AUG start codon and generates a ribosome with a methionine-loaded tRNA^{Met} bound in the P site; the ribosome is now poised to move along the open reading frame (Figure 10.16, top cartoon). While the early steps of initiation involve only the small subunit of the ribosome, the final step in the process involves the joining of the large ribosomal subunit. At this point elongation proceeds, freeing up the AUG **initiation codon** for the next round of initiation. Indeed, initiation does typically occur again and again, thus piling up ribosomes on a single mRNA, which become dispersed along the open reading frame. These large complexes are referred to as polysomes and are easily visualized by electron microscopy (Figure 10.17). The details of initiation vary substantially between bacteria and eukaryotes. For example, only three initiation factors appear to be critical in bacteria while as many as 28 polypeptides are critical in a mammalian eukaryotic system.

Following initiation, the **elongation** cycle can begin wherein amino acids are sequentially added to a growing polypeptide chain as the ribosome moves directionally (5′ to 3′) along the mRNA. The elongation cycle consists of three basic steps. First, the codon of the mRNA located in the A site of the ribosome specifies which aminoacyl-tRNA species is to be loaded next into the ribosome (Figure 10.16, step 1). In Section 10.3 we mentioned that free aminoacyl-tRNAs in cells are always complexed with specific proteins, which protect the linkage between the amino acid and tRNA from hydrolysis. These **elongation factors**, called EFTu in bacteria and eEF1A in eukaryotes, play an essential role in loading the aminoacyl-tRNAs into the ribosome as they use the energy of GTP hydrolysis to 'evaluate' the interaction between the codon and anticodon, and deposit the appropriate aminoacyl-tRNA in the A site.

Figure 10.17 Polysomes form when multiple ribosomes initiate and elongate the same mRNA. The polysomes here are visualized by electron microscopy: the thin strand is the DNA and the blobs emerging from it are ribosomes bound to newly synthesized mRNA transcripts.

The second step of elongation is the catalysis of peptide bond formation to join the amino acid brought into the A site by its cognate tRNA with the growing polypeptide chain. Before formation of the peptide bond, the polypeptide chain being synthesized is found attached to the tRNA located in the P site. This polypeptide–tRNA species is called a peptidyl-tRNA. During peptide bond formation, the growing polypeptide chain found on the peptidyl-tRNA in the P site of the ribosome is transferred to the amino acid carried by the aminoacyl-tRNA bound in the A site. This peptide bond forming reaction is catalyzed by the peptidyl transfer center (or active site) of the large ribosomal subunit (Figure 10.16, step 2). The energetic driving force for peptide bond formation derives originally from the ATP that was used in the activation of the amino acids by the aminoacyl-tRNA synthetases.

Finally, another GTP-dependent elongation factor (EFG in bacteria and eEF2 in eukaryotes) promotes the three-nucleotide movement of the mRNA–tRNA complex through the ribosome (Figure 10.16, step 3), bringing the newly minted peptidyl-tRNA into the P site and the next codon into the A site. This process, termed **translocation**, effectively opens the A site for the next incoming aminoacyl-tRNA. The elongation cycle then repeats.

At the end of the open reading frame, the ribosome reaches a stop codon (UAG, UAA, or UGA) in the mRNA. This signals the next step in the translation cycle, **termination**, when the fully synthesized polypeptide chain is released from the ribosome. Stop codons are not recognized by specific tRNAs but by protein factors known as **class 1 release factors** (Figure 10.16, step 4). In bacteria, there are two class 1 release factors; RF1 recognizes the codons UAA and UAG, and RF2 recognizes the codons UAA and UGA. In eukaryotes, there is a single class 1 release factor, eRF1, which recognizes all three stop codons. The interaction between the release factor and the ribosome promotes release of the completed peptide chain from the final peptidyl-tRNA, allowing the peptide chain to diffuse away from the ribosome. There is an additional GTPase involved in termination both in bacteria and eukaryotes, RF3 and eRF3, respectively. These GTPases are both referred to as **class 2 release factors** though their mechanisms of action appear to be rather distinctive. (We explore these mechanisms of action in more detail in Section 10.10.)

In the final step of translation, **ribosome recycling**, the large and small ribosomal subunits dissociate from each other, releasing the remaining tRNA and mRNA (Figure 10.16, step 5). In bacteria, a protein known as ribosome recycling factor (RRF) works together with EFG to promote ribosome recycling; there is no obvious RRF homolog in eukaryotes. After subunit dissociation, the liberated ribosomal subunits can then bind again to initiation factors in preparation for new rounds of protein synthesis.

The hybrid states model explains how tRNAs move through the ribosome

How do tRNAs move through the three tRNA binding sites on the ribosome during the translation cycle? One idea for this is the **hybrid states model**, in which tRNA molecules ratchet their way through the subunit interface region of the ribosome, always maintaining contact with one subunit while moving with respect to the other. In this model, the tRNAs move independently with respect to the large subunit directly following peptide bond formation and then, promoted by EFG, the anticodon ends of the tRNA associated with the mRNA move with respect to the small subunit of the ribosome (Figure 10.16, steps 2 and 3). After these two independent movements have taken place, the A site is fully accessible for the next round of elongation to continue. As the tRNAs move, the ribosomal subunits themselves also move with respect to each other. In this view, the ordered process of translation is maintained by the oscillating movements of the ribosomal subunits and the two ends of the tRNAs, during which the tRNAs are never fully released from contact with the ribosome. Initial support for the hybrid state of tRNA binding came from chemical modification experiments as outlined in Experimental approach 10.1. These results have been further supported by a number of independent approaches including cryo-electron microscopy to visualize the state and detailed kinetic studies using both bulk and single molecule approaches to characterize it.

10.6 PROTEIN FACTORS CRITICAL TO THE TRANSLATION CYCLE

The translation cycle is facilitated by the actions of protein factors

In the brief overview of the translation cycle in the preceding section, we have seen that each step in the cycle, with the exception of peptide bond formation, is facilitated by the action of protein factors. Broadly speaking, there are two ways that translation factors act:

First, many factors are GTPases (or G proteins), which catalyze the hydrolysis of GTP. The conformational changes that they undergo during GTP hydrolysis are intimately linked with the orderly, stepwise progression of the ribosome through the translation cycle. EFTu, for example, depends on GTP hydrolysis (and the associated conformational rearrangements) in its role as guardian and deliverer of the aminoacyl-tRNAs during elongation. Such G proteins probably represent the evolutionary origins of important players in cellular signal transduction and cancer, the Ras protein perhaps being the best-known example.

Second, some translation factors simply bind to the ribosome and prevent inappropriate interactions with tRNAs or other components of the translational machinery. For example, in bacteria, the initiation factors IF1 and IF3 are believed to function in this manner by preventing binding of the **initiator tRNA** to the A site and by preventing the large and small subunits of the ribosome from associating before the initiator tRNA has been loaded into its binding site.

An elongation factor that we introduced in Section 10.5, EFG (or eEF2 in eukaryotes), triggers the movement of the mRNA–tRNA complex through the ribosome during the elongation phase of translation. EFG both mimics the structure of tRNA and is a GTPase. In later sections we look in more detail at how translation factors act during the translation cycle, but first we will outline further the

10.1 EXPERIMENTAL APPROACH

Chemical modification analysis to study RNA structure and function without high-resolution structures

Chemical modification analysis has been extensively used to define molecular features of a large number of biologically interesting RNAs and ribonucleoproteins (RNPs). These approaches were especially useful for early analysis of the ribosome structure and function because more high-resolution approaches, such as x-ray crystallography, were not yet technically feasible.

The basic idea behind chemical modification analysis is to bombard an RNA species of interest with chemical reagents that attack and chemically modify RNA elements such that the sites of modification can be easily identified. Accessible RNA sites will be modified, whereas those found within the core of the macromolecule, or those that are protected by various ligands, will not. The sites of modification are typically identified as pauses or stops on a sequencing gel that follows the progress of a reverse transcription reaction as it extends primers along the chosen (rRNA) template. The strength of a given band in the primer extension gel is dependent on whether that site has been modified, and how extensively. If a given RNA molecule escapes modification, reverse transcriptase should fully extend the primer, yielding a full-length complementary DNA strand. If the RNA is modified, reverse transcription will be halted at the modified site, yielding a visibly shorter band on the gel. This general approach is described more thoroughly in Section 16.9.

How do these methods allow the researcher to identify positions of functional importance? Well, when different ligands are bound to the ribosome, for example, specific regions are protected from chemical modification, thus identifying potential ligand interaction sites. In an informative study by Harry Noller and colleagues in 1989, structural probing was performed on a series of ribosomal complexes containing various tRNA substrates stalled at different points in the translational cycle. These experiments provided a structural view of tRNA movements within the ribosome, and surprisingly indicated that tRNAs can move independently with respect to the large and small subunits of the ribosome. The intermediate tRNA binding state (where tRNA has moved with respect to the large, but not the small, subunit) was referred to as the 'hybrid' state of binding (as discussed in Section 10.4).

Representative data from these experiments are shown in Figure 1. In the primer extension sequencing gels, we see that several different nucleotide modification patterns can be followed within lanes documenting the modification patterns of the 16S and 23S rRNAs found in the small and large ribosomal subunits,

respectively. Lanes 1 through 5 contain a set of ribosome complexes with tRNAs bound during different stages in the elongation cycle, and we see that the patterns of accessibility change in response to distinctly bound ligands. When empty ribosomes – with nothing bound to them – are bombarded with chemical reagents (lane 1) we see a number of potential ligand interaction sites being modified. In lanes 2–5, ribosomes with tRNA ligands bound show different modification patterns – for example, nucleotide 926 (Figure 1b) is protected from modification when any tRNA substrate is bound to the ribosome (lanes 2–5), while the modification patterns for U2585 and A1492/93 (Figure 1a and c, respectively) show variability in modification that is dependent on whether peptide bond formation and/or translocation have

Figure 1 Chemical modification experiments of 23S and 16S rRNA with DMS (A and C specific), kethoxal (G specific) and CMCT (U specific). Panel (a) examines a region in 23S rRNA while (b) and (c) examine two regions in the 16S rRNA. Lane 1, 70S ribosomes and polyU; lane 2, 70S, polyU, N-Ac-Phe-tRNA^Phe^; lane 3, 70S, polyU, deacylated tRNA^Phe^; lane 4, 70S, polyU, deacylated tRNA^Phe^, N-Ac-Phe-tRNA^Phe^; lane 5, 70S, polyU, deacylated tRNA^Phe^, N-Ac-Phe-tRNA^Phe^ plus EFG:GTP. C, U, A, and G refer to sequencing lanes, K to unmodified control.
Reproduced from Moazed and Noller, *Nature* 1989;**342**:142–148.

occurred. These different modification patterns ultimately provided Noller and colleagues with experimental support for the hybrid states model of tRNA movement.

This study provides an excellent example of the power of chemical modification analysis to assess properties of functional RNAs at the nucleotide level. For the ribosome, such approaches successfully identified most of the functionally critical rRNA nucleotides long before high-resolution structures clarified their three-dimensional location and function. Even now, when high-resolution structures of the ribosome provide increasingly detailed atomic insight into core molecular features, chemical modification approaches

remain useful as relatively complicated biological questions can be addressed in the absence of a perfectly behaved homogeneous (and crystallize-able) sample. Recent studies have further expanded the application of these approaches by combining structural probing with kinetic analysis to watch in real time the assembly of large macromolecular complexes. Such studies are particularly useful still as they allow for rapid analysis and can address questions where high resolution structures remain elusive.

Moazed D, Noller HF. Intermediate states in the movement of transfer RNA in the ribosome. *Nature* 1989;**342**:142–148.

general principles by which GTPase activities and molecular mimicry are thought to facilitate different steps in the translation cycle.

The evolutionarily related GTPases interact with the protein-rich flexible stalk region of the ribosome

GTPases act at several steps during translation. During initiation, eukaryotic eIF2 is required for loading Met-tRNA$_i^{Met}$ into the AUG-coded P site of the ribosome while bacterial IF2 (or eIF5B in eukaryotes) is required for subunit joining. During elongation, EFTu (eEF1A) is critical for depositing the cognate aminoacyl-tRNAs into the A site, EFG (eEF2) promotes the directional movement of the mRNA–tRNA complex through the ribosome and is critical for ribosome recycling, while RF3 (eRF3) is involved in the termination of protein synthesis.

These GTPases are all related and contain a **P loop motif** (Gly-X-X-X-X-Gly-Lys) common to many nucleotide-binding proteins including Ras. This motif specifically interacts with the phosphate groups of GTP and undergoes substantial conformational changes on GTP hydrolysis (Figure 10.18).

As with other G proteins, the identity of the bound nucleotide (GTP or GDP) dictates the conformation of the protein, and thus the ligands with which it interacts. In translation, the ligands that the GTPases recognize are the distinct states of the ribosome during the translation cycle. So, for example, EFG in its GTP bound (active) state recognizes pre-translocation state ribosomes (that is, ribosomes ready to move to the next mRNA codon) whereas eIF2 in its GTP-bound state recognizes small ribosomal subunits ready to accept a P-site tRNA. The GDP forms of translation factors typically dissociate from the ribosome after they have performed their functions.

As with related GTPases, both the background rate of GTP hydrolysis and the rate of exchange of GDP for GTP can be slow. In situations where GTP hydrolysis or exchange is slow, the reactions are typically promoted by **GTPase-activating proteins (GAPs)** and **guanine-nucleotide exchange factors (GEFs)** (Figure 10.19). A number of additional translation factors play essential roles in translation by acting as GAPs or GEFs. For example, bacterial EFTs (related to eukaryotic eEF1-Balpha) is the GEF for EFTu (or eEF1A) while eukaryotic eIF5 is the GAP for eIF2 (that loads tRNA into the P site during initiation in eukaryotes). In some cases,

(a)

EF-Tu GTP tRNA

(b)

EF-Tu GDP

Figure 10.18 The P loop of GTPases undergoes a large structural rearrangement on GTP hydrolysis. The GTPase in this example is EFTu (blue), shown with bound GTP (a) and GDP (b). An aminoacyl-tRNA (gray) that is bound to EFTu (in its GTP bound form) interacts with the P loop (green). The shift in position of the P loop following GTP hydrolysis causes EFTu to dissociate from the aminoacyl-tRNA. PDBs 1D2E, 1TTT.

Figure 10.19 GTPase-activating proteins (GAPs) and guanine-nucleotide exchange factors (GEFs). Various protein factors and the ribosome itself are thought to function as the (a) GAPs and (b) GEFs for various GTPases involved in translation.

the ribosome itself functions as a GAP (or potentially even as a GEF). The best-documented example of this is with EFTu: once a cognate aminoacyl-tRNA is base-paired with the codon in the ribosome's A site, the ribosome promotes the hydrolysis of GTP by EFTu, thus promoting full acceptance of the aminoacyl-tRNA into the A site and release of EFTu-GDP.

The GTPases involved in translation appear to interact extensively with the protein-rich flexible stalk region of the large ribosomal subunit. The rRNA portion of the stalk region is seen as the prominent extension on the right in Figure 10.12 (though the full extent of the protein-rich stalk is not seen since the protein components are not well resolved in the various atomic resolution structures determined to date). As a group, the GTPases probably communicate with the interior of the ribosome via extended protein domains, associated protein factors, or tRNA substrates, and ultimately bring about rather distinct functional consequences (e.g. translocation, tRNA loading, class 1 release factor dissociation).

Protein factors are not absolutely required for very minimal translation *in vitro*; such reactions can proceed slowly in their absence on certain unstructured mRNAs such as polyuridine (encoding poly-phenylalanine). As a whole, the factors are highly conserved and it is thought that they evolved to contribute speed, precision, and processivity to a crude but already functioning translation apparatus that initially depended only on the coordinated interactions of tRNAs and the ribosome itself. Once a single GTPase became integrated into the translation cycle, related GTPases could be derived from this original template molecule allowing for the evolution of increasingly complex regulation of the process. We will see examples of such regulation later in this chapter.

Molecular mimicry of tRNA may be used by some translation factors to access the interior of the ribosome

One striking feature of several translation factors, including some GTPases, is their apparent mimicry of the shape and size of tRNA substrates of translation. Indeed, looking like a tRNA seems to be an efficient way for translation factors

EFTu-tRNA EF-G

Figure 10.20 Molecular mimicry of tRNAs by translation factors. EFG shares overall structural features with the aminoacyl-tRNA–EFTu complex. In the view shown, domain IV of EFG is colored in gray to highlight its similarity to the tRNA bound to EFTu (also in gray). This structural similarity is consistent with the fact that both molecules bind in the A site of the ribosome. PDBs 1TTT, 1DAR.

to access the functionally critical interface region of the ribosome. Perhaps the most compelling example of tRNA mimicry is seen with the elongation factor (EFG), which binds in the A site of the ribosome to promote the translocation of the mRNA–tRNA complex. Figure 10.20 shows how there is marked similarity in the shape of EFG and the shape of the tRNA–EFTu–GTP complex, both of which are known to function bound within the A site of the ribosome. In functional and structural terms, the class 1 release factors also mimic the A site tRNA in their role in termination; we will explore these similarities more thoroughly in Section 10.10.

Having now seen how the translation cycle is facilitated by various protein factors, let us now consider the molecular mechanisms that are employed by the cell during each stage in this cycle. We start by considering the first stage, that of initiation.

10.7 TRANSLATION INITIATION (MOSTLY IN BACTERIA)

During initiation, a methionyl-tRNA and the AUG start site of the mRNA engage in the P site of the small ribosomal subunit.

The initiation step of translation is the process by which a ribosome identifies the translation start site in an mRNA and traps it in the P site with an initiating methionyl-tRNA in preparation for the beginning of the elongation cycle (see Figure 10.16, top cartoon). The consequence of these steps is that the first codon of the open reading frame is decoded as methionine and polypeptide synthesis can begin when the next aminoacyl-tRNA is loaded into the A site of the ribosome. There are three main tasks that take place during translation initiation. First, the small ribosomal subunit must identify the first codon of an open reading frame, the **initiation codon**, within the mRNA. Second, a methionyl-tRNA must be loaded into the P site of the ribosome where it base-pairs with the initiation codon. Finally, the large ribosomal subunit must join the small subunit that now has the methionyl-tRNA positioned in the P site. Both bacteria and eukaryotes must achieve these three tasks, but bacteria and eukaryotes have differently structured genes and the details of how initiation occurs in these organisms differ accordingly.

Figure 10.21 Features of the initiator tRNAs in bacteria and eukaryotes. (a) Bacterial and eukaryotic initiator tRNAs are different in their chemical composition, as the bacterial Met-tRNA species is formylated on the amine functionality while the eukaryotic Met-tRNA is not (highlighted boxes). (b) Both bacterial and eukaryotic initiator tRNAs have three consecutive GC base pairs in the anticodon stem, and unusual (for tRNAs) base pairs at the top of the acceptor stem (CA and AU, respectively) that prevent binding to EFTu (eEF1A).

Initiator tRNAs have distinctive features

In both eukaryotes and bacteria the initiation codon is typically AUG (which specifies methionine) and is decoded by a specialized tRNA species called the **initiator tRNA**. These initiator molecules, tRNA^{fMet} in bacteria and tRNA$_i^{Met}$ in eukaryotes, are distinct from other tRNAs in several ways that prevent them from associating with EFTu (or eukaryotic eEF1A), which escorts aminoacyl-tRNAs to the A site during elongation. The lower case 'f' on the bacterial initiator tRNA refers to the 'formyl' moiety found on the amine function of the methionine of the aminoacylated tRNA species (formyl-Met-tRNA^{fMet}) (Figure 10.21a). The formyl group is added to the methionine by a transformylase enzyme that recognizes Met-tRNA^{fMet} following the aminoacylation reaction. This chemical moiety functions to stabilize the ester linkage between the amino acid and the tRNA body. In so doing, the stabilizing formyl group found in the bacterial system may help to compensate for protection usually afforded to aminoacylated tRNAs by EFTu/eEF1A (for elongator tRNAs) or by eIF2 (for eukaryotic initiator tRNA).

In bacteria, the identity elements of the initiator tRNA include a C–A wobble mismatch at the start of the acceptor stem and a stretch of three consecutive G–C base pairs in the anticodon stem. Eukaryotic initiator tRNA identity elements include the same consecutive G–C base pairs in the anticodon stem, and an A–U base pair at the start of the acceptor stem (Figure 10.21b). In each system, these identity elements are important anti-determinants for binding of the initiator tRNAs to EFTu (eEF1A), and so these tRNAs are not loaded into the A site by the normal pathway of elongator tRNAs.

How then are initiator tRNAs loaded directly into the P site in these two systems? In eukaryotes, the Met-tRNA$_i^{Met}$ is bound by a GTPase, eIF2, that plays a direct role in start codon recognition and loading of the P site. In bacteria there is no known homolog of this protein, though a different GTPase, IF2, clearly plays an important role in directing binding of f-Met-tRNA^{fMet} to the AUG-programmed P site.

In the remainder of this section, we shall describe what is known of the features of the initiation process in bacteria and in Section 10.8, we shall see how in eukaryotes initiation proceeds in a similar yet distinct way. In later sections (10.13–10.16), we will focus on the many layers of regulation that target translation initiation in bacteria and eukaryotes.

Locating the AUG start site in bacteria is relatively simple

Figure 10.22 shows the basic organization of a typical bacterial mRNA. Bacterial mRNA molecules are often **polycistronic**, meaning that they contain multiple

Bacterial mRNA structure

- non-coding
- ribosome binding sequence (Shine–Dalgarno)

Figure 10.22 General features of bacterial mRNAs. Bacterial transcripts often encode more than one open reading frame (that is, they are polycistronic). Each open reading frame begins with an AUG initiation codon (green) and ends with a stop codon (red). The initiation codons are typically preceded by a Shine–Dalgarno sequence that is recognized by the 16S rRNA (anti-Shine–Dalgarno sequence) of the small ribosomal subunit during initiation.

open reading frames, each encoding a different protein. Each open reading frame in the mRNA has an initiation and a termination codon and we will see in Section 10.10 how ribosomes are able to process each reading frame either independently, or sometimes in a coupled fashion. For now, we will simply consider the means by which bacterial ribosomes identify the initiation codons within these multicistronic mRNAs.

Initiation codons (AUG or sometimes GUG) in bacteria are most typically accompanied by a sequence element in the mRNA called a **Shine–Dalgarno sequence** (or **ribosome-binding site**), which is a polypurine tract located six to eight nucleotides upstream of the AUG initiation codon. The Shine–Dalgarno sequence has the consensus sequence AGGAGGU and it base-pairs with a polypyrimidine tract in the 3′ end of the bacterial 16S rRNA (the **anti-Shine–Dalgarno sequence**) of the small ribosomal subunit (Figure 10.23). Deviations from the consensus sequence result in weaker interactions with the ribosome and are important in dictating how efficiently a given mRNA is translated in the cell. Experimental approach 10.2 describes some of the early observations of Shine and Dalgarno and how they led to the identification of this crucial interaction involved in bacterial translation initiation.

With this spacing of six to eight nucleotides relative to the start codon, base-pairing between the Shine–Dalgarno sequence and the anti-Shine–Dalgarno sequence on the small ribosomal subunit guides the AUG initiation codon into the region of the P site of the ribosome where it is poised to engage with the initiator f-Met-tRNA^{fMet}. There is also an additional class of mRNAs in bacteria (and archaea) that has no leader sequence upstream of the AUG start site; these mRNAs appear to depend on somewhat different mechanisms to initiate translation (so called 'leaderless initiation'), although the mechanism is not yet clear.

Initiator tRNA binding is guided by a set of protein factors

The initiator f-Met-tRNA^{fMet} is guided to the P site of the ribosome with the help of three initiation factors, IF1, IF2, and IF3. In the absence of mRNA or f-Met-tRNA^{fMet}, IF1 and IF3 independently associate with the small ribosomal subunit in the regions where the A and E site tRNAs usually bind, leaving only the P site unfilled (Figure 10.24a). These factors also appear to block sites on the small subunit that are involved in intersubunit interactions. As such, these two proteins effectively direct the initiator tRNA to the P site and prevent the premature association of the two ribosomal subunits. The third factor, the GTPase IF2, also seems to play an important role in this early stage of initiation, helping to direct f-Met-tRNA^{fMet} to the AUG in the P site.

```
        U U C C U C C                          anti-Shine–Dalgarno
T A C T A A G G A G G  T T G T  – –  A T G   lambda cro  ⌉
A C C T G A A G A T T A A A C  – –  A T G   rpsA
  G T G G A G G A C T A A G A A  –  A T G   rplA
    T T A G A G G G A C A A T C G A T G   tufB
      G G G A G T A T G A A A A G T A T G   araC        Shine–Dalgarno (in blue)
T A A C A G G T A G T G A A T  – –  A T G   galS
    G T A A G G A A A T C C A T T A T G   lacY
  A C G A G G G G A A A T C T G  –  A T G   trpA
A A A T G A G G A G A G G T A  – –  A T G   alkB  ⌋
```

Figure 10.23 The purine-rich Shine–Dalgarno sequence specifically interacts with a pyrimidine-rich region of the 16S rRNA. Shown here are a number of bacterial genes with varying levels of complementarity to the anti-Shine–Dalgarno region of the 16S rRNA.

(a)

(b)

f-Met

mRNA

AUG

Shine–Dalgarno

anti-Shine–Dalgarno

AUG

IF3 IF1

GTP

IF2

GTP

IF3 IF1

GDP

IF1

IF3

Figure 10.24 Initiation factors orchestrate bacterial initiation complex assembly.
(a) IF1 and IF3 gray and dark gray, respectively block regions of the small ribosomal subunit where the A and E site tRNAs usually bind, in addition to blocking sites that are involved in intersubunit contacts in the 70S ribosome. These initiation factors thus help in guiding the initiator tRNA (green) into the P site. PDB 1GIX, 1HR0. (b) Diagram showing the ordered steps of assembly of bacterial initiation. IF1 and IF3 initially bind to the small subunit A and E sites, guiding binding of the initiator fMet-tRNA^{fMet} to the P site (with the help of IF2-GTP). The Shine–Dalgarno sequence of the mRNA directly interacts with the anti-Shine–Dalgarno sequence of the 16S rRNA to position the AUG in the P site. Once the initiator tRNA finds the AUG, IF2 facilitates subunit joining (using the energy of GTP hydrolysis), and the three IFs are released from the complex.

The large ribosomal subunit joins the small subunit in the final step in bacterial translation initiation

Once the initiator tRNA has been deposited in the P site, the large subunit needs to join the small subunit so that elongation can begin. The GTPase IF2 couples the energy of GTP hydrolysis to this subunit joining step, ultimately resulting in the displacement of all three initiation factors (IF1, IF2, and IF3) and formation of the initiation complex, which comprises both ribosomal subunits, a bound mRNA and an initiator f-Met-tRNA^{fMet} poised in the P site. At this stage initiation is complete and translation elongation can proceed. These steps are outlined in Figure 10.24b.

10.8 TRANSLATION INITIATION IN EUKARYOTES

Translation initiation in eukaryotes is more complex

Eukaryotic initiation of translation resembles bacterial initiation in that it occurs at an AUG codon that is decoded by a special initiator Met-tRNA$_i^{Met}$ (albeit distinct from the formyl-Met-tRNA^{fMet} of bacterial initiation, see Figure 10.21). However, there are radical differences between the two systems; in fact, no other aspect of translation is so extremely different. The first difference is that eukaryotic 40S small ribosomal subunits do not bind directly to the mRNA, which reflects the fact that eukaryotic mRNA initiation sites do not have an equivalent of the oligopurine Shine–Dalgarno motif that is the key to start site recognition in bacteria. (The few exceptions are discussed in Section 10.16.) Consistent with this fact is the observation that, despite the highly homologous nature of the 16S and 18S rRNAs, the anti-Shine–Dalgarno polypyrimidine tract is wholly missing in eukaryotic 18S rRNA.

Without this direct base-pairing interaction to guide the ribosome to the appropriate AUG start site, eukaryotic ribosomes depend on a distinctive mechanism for identification of the AUG start site. The greater complexity of eukaryotic initiation is evident from the fact that, whereas initiation in bacterial systems require just three distinct initiation factor proteins (IF1, IF2, and IF3), each of which is a single polypeptide, eukaryotic mammalian initiation requires no fewer than 11 distinct factors: eIF1, 1A, 2, 2B, 3, 4A, 4B, 4E, 4G, 5, and 5B. Some of these factors are composed of several polypeptide chains: eIF2 has three polypeptide chains, eIF2B has five, and eIF3 no fewer than 12 (with a mass about half the size of the small ribosomal subunit). All told, the minimal number of polypeptides involved in the mammalian system is 28.

Eukaryotic mRNAs have unique features that lead to the formation of a closed loop initiation complex

Mature eukaryotic mRNAs are generally monocistronic (that is, they encode a single protein product), and initiation usually occurs at the first AUG triplet from the 5′ end of the mRNA. However, for some mRNAs, initiation at this first AUG is inefficient, and there is more initiation at the second (or even third) AUG from the 5′ end. These observations suggest a mechanism whereby the initiation site is

10.2 EXPERIMENTAL APPROACH

Phylogenetic covariation analysis and its utility

Phylogenetic covariation analysis has been used extensively in the study of functional RNAs to identify potential interacting regions, allowing for the elucidation of secondary structures and even longer-range tertiary interactions. In general terms, this approach uses comparative sequence analysis – the analysis of related sequences, often from different organisms – to identify patterns of variation within the RNA that are non-random and instead reflect a co-dependence during evolution. For example, if two sequences interact closely through base-pairing, and one of the two sequences undergoes change, their interaction can only be sustained if the second sequence undergoes a compensatory change. If such concerted changes are observed, the two positions in the RNA are said to covary. The process of evolution will select for such concerted changes if the interaction is important for the function of the RNA.

Covariation analyses have been used with great success to determine the secondary structure of numerous RNA species beginning with the cloverleaf structure of tRNA in the 1960s (Figure 10.3), the 16S rRNA in the 1970s, and the growing body of non-coding RNAs. Such analyses even permit the identification of long-range interactions within these RNAs that stabilize their three-dimensional structures. For example, Michael Levitt saw covariations within the tRNA sequences that led him in the 1960s to propose the existence of the known base triple interactions. And, in perhaps the most stunning of such analyses, Francois Michel and Eric Westhof in 1990 used phylogenetic analysis to identify a number of long-range interactions in the group I intron. Their predictions were remarkably accurate; interactions that they identified in these studies represent some of the most common and important core structural motifs in large RNAs. Indeed, these motifs have since then been observed repeatedly in atomic resolution structures of the group I and II introns and the ribosome.

In an example pertinent to this chapter on translation, two researchers, John Shine and Lynn Dalgarno, were exploring sequence variations in the pyrimidine-rich 3′ end of the 16S rRNA of the small subunit of the ribosome (Figure 1) when they noticed that variation in this sequence (in ribosomes from different organisms)

anti-Shine–Dalgarno sequence

C A C C U C C U U A	*E. coli*
U C C U U U C U A	*B. stearothermophilus*
C U C U C C U U A	*P. aeruginosa*
U C C U U U C U	*C. crescentus*

Figure 1 Secondary structure of 16S rRNA with the anti-Shine–Dalgarno region at the 3′ end highlighted in red. Listed on the right are several of the sequences from different bacteria determined by Shine and Dalgarno, exhibiting variability that they noted correlated with the varying levels of protein expression for certain phage transcripts.

Figure 2 Specificity swap Shine–Dalgarno experiment. Human growth hormone (hGH) accumulation in various cells measured at various times. Red line, wild-type ribosomes and an hGH mRNA with a standard *E. coli* upstream sequence; green and purple lines, mRNAs carrying altered upstream sequences but no specialized ribosomes; pink and blue lines, have mRNAs with altered upstream sequences and specialized ribosomes that match. Reproduced from Hui and de Boer, *Proceedings of the National Academy of Sciences of the U S A* 1987;**84**:4762–4766.

correlated with the level of expression of certain mRNA cistrons (or genes). Shine and Dalgarno further noted that there was a purine-rich region just upstream of the AUG initiation codon in these genes, and that there was sequence variation in this purine-rich region that correlated with the variation in the pyrimidine-rich sequences in the 16S rRNAs. Based on these patterns, Shine and Dalgarno proposed that the pyrimidine-rich 3′ end sequence of the 16S rRNA paired directly with the purine-rich upstream leader sequence of mRNA cistrons, and that the more extensive the pairing (and, hence, the more direct complementarity between the two sequences) the more effectively the mRNA would be translated.

The role of these base-pairing interactions was tested directly when Herman de Boer and colleagues constructed a 'specialized' ribosome system. In these experiments, the human growth hormone (hGH) gene was cloned in a plasmid where the mRNA, carrying one of several highly mutated Shine–Dalgarno sequences, was constitutively expressed in a number of different *Escherichia coli* strains carrying either wild-type or mutant 16S rRNA genes (in addition to wild-type chromosomally expressed ribosomes). The basic idea behind these experiments was to switch the sequences in the Shine–Dalgarno and anti-Shine–Dalgarno motifs, and ask whether expression of the hGH protein depended on the presence of ribosomes with a complementary region in the 16S rRNA. The results were straightforward: the hGH protein was efficiently expressed only in cells where ribosomes with a complementary anti-SD sequence are found (Figure 2). We broadly refer to this as a 'specificity swap' experiment – an approach that works especially well for RNA because of the simplicity of the Watson–Crick pairing interaction and the fact that all base pairs are more or less isosteric with one another.

We now know that the so-called Shine–Dalgarno sequence found upstream of the start site in a majority of bacterial genes plays a key role in the initiation of translation, directly guiding the ribosome to the appropriate AUG. Moreover, the Shine–Dalgarno sequence upstream of the AUG start site remains an important predictor for identifying start sites in gene finder programs in bacteria (and archaea) (Figure 3).

Phylogenetic approaches continue to play a critical role in defining the secondary structure (and occasionally tertiary interactions) of RNA species of interest. New computational tools and computing power have increased the sophistication of phylogenetic analysis, allowing models to be evaluated in more systematic and robust ways. The relatively recent discovery of riboswitches has provided the field with a new set of challenges to solve.

Hui A, de Boer HA. Specialized ribosome system: Preferential translation of a single mRNA species by a subpopulation of mutated ribosomes in *Escherichia coli*. *Proceedings of the National Academy of Sciences of the U S A* 1987;**84**:4762–4766.

Shine J, Dalgarno L. Determinant of cistron specificity in bacterial ribosomes. *Nature* 1975;**254**:34–38.

Figure 3 Nucleotide conservation plot for *E. coli* genes in the regions just upstream of the AUG start site. The conserved purine-rich Shine–Dalgarno sequence located six to eight nucleotides upstream of the start site is readily apparent. Adapted from Tom Schneider 'A Gallery of Sequence Logos'.

selected by a process of ribosome 'scanning' starting from the 5′ end. The reason why recognition of the first AUG may not be 100% efficient is that recognition is sensitive to the sequence context of the AUG. The preferred sequence context in mammals is (A/G)XX**A**U**G**G and is commonly referred to as the **Kozak sequence**. The extent to which the context of a given AUG differs from the consensus in part dictates how efficiently it is recognized by the scanning ribosome.

Other distinctive features of eukaryotic mRNA that help guide initiation are the 7-methyl guanosine cap (at the 5′ end) and the poly(A) tail (at the 3′ end) (Figure 10.25). As we saw in Chapter 9, both of these features are added to the mRNA after transcription. The 7-methyl guanosine cap and poly(A) tail allow for several preparatory steps, which depend on the eIF4 family of factors, to prime the mRNA for the subsequent scanning stage. The 5′ cap structure is specifically bound by a protein factor eIF4E (known as the cap binding protein) and the 3′ poly(A) tail is bound by a protein factor called the poly(A) binding protein (PABP). These factors interact with eIF4A/4B (an ATPase complex that assists in the unwinding of mRNA structure) and eIF4G (Figure 10.26, step 1).

The consequence of the engagement of these multiple binding interactions is the formation of a **closed loop complex** on the mRNA that seems to be important for the most efficient translation. This particular form of the complex appears to serve as a useful 'quality control' feature, allowing the system to immediately distinguish between competent and incompetent mRNA substrates (for example, those complexes lacking a poly(A) tail are not efficiently translated).

Figure 10.25 General features of eukaryotic mRNAs. Eukaryotic mRNAs are usually monocistronic, where the open reading frame begins with an AUG initiation codon (green) and ends with a stop codon (red). At the 5′ end is the 7-methyl guanosine cap and at the 3′ end is the poly(A) tail.

The initiation codon in a eukaryotic mRNA is identified by a ribosome scanning mechanism that begins at the 5′ cap structure

If formation of the closed loop complex is a first step required for the identification of the AUG start site, how then is the actual start site found? This process depends

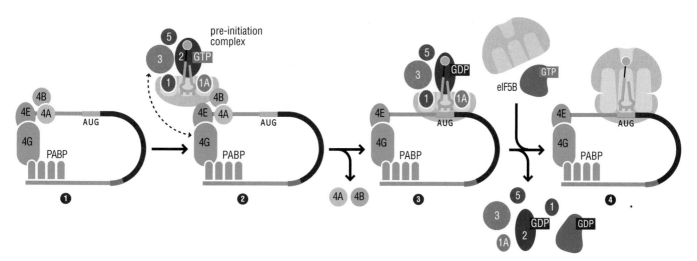

Figure 10.26 The scanning model for eukaryotic translation initiation. (1) The closed loop of mRNA results from the interactions of eIF4G with eIF4E (which binds the mRNA 5′ cap) and with the poly(A) binding protein (PABP, which binds the mRNA poly(A) tail). eIF4A/4B are also present and are important for unwinding the structured mRNA. (2) After formation of the mRNA loop the pre-initiation complex associates with the mRNA complex. (3) This initiation complex scans along the mRNA in search of the AUG initiation codon (green). Once the start codon is found, eIF5 acts as a GTPase-activating protein (GAP) to stimulate GTP hydrolysis on eIF2. (4) Finally, eIF5B-GTP promotes the final step in initiation, subunit joining, and triggers the departure from the complex of the bound initiation factors.

on a second set of preparatory steps during which most other initiation factors (including eIF1, eIF1A, eIF2, eIF3, and eIF5) prime the 40S ribosomal subunit for scanning. eIF1A, the eukaryotic homolog of IF1, similarly binds in the area of the A site of the 40S subunit and blocks premature subunit joining, while eIF1, clearly related to IF3, positions itself around the E site of the 40S subunit. These two proteins thus guide the initiator tRNA to the P site, as was the case for bacterial initiation. An important distinction is that the initiator tRNA, Met-tRNA$_i^{Met}$, is bound to a protein, eIF2, in a ternary complex with GTP, that assists in the loading of the P site. The primed complex is generally referred to as '48S' based on its slightly faster migration in a sucrose gradient (relative to a bare 40S subunit) – a more general term would be the 'pre-initiation complex' (Figure 10.26).

At this stage, the complex is ready for scanning and is brought to the eIF4 factor-bound mRNA through direct interactions between eIF3 and the 'scaffold protein' eIF4G (Figure 10.26, step 2). Scanning now proceeds, with eIF4A (the ATPase) and its associated factor (eIF4B) unwinding secondary structural blocks in the mRNA, and with the initiator tRNA evaluating triplet codons as they go by to see whether they are complementary to the anticodon. Once the scanning ribosome complex identifies an appealing AUG codon (subject to the context features described above and with the specific help of eIF1 in evaluating the pairing interaction), eIF5 functions as a GAP and triggers GTP hydrolysis on eIF2. This GTP hydrolysis associated step serves as a fidelity checkpoint for translation, only rapidly firing on recognition by the complex of an authentic start site (Figure 10.26, step 3).

With start site recognition now complete, initiation factors must dissociate (though the precise order of these departures is unknown) and subunit joining must take place. In this final step, the eukaryotic homolog of IF2, the GTPase eIF5B, promotes large subunit (60S) joining in a GTP hydrolysis-dependent step (Figure 10.26, step 4). Once the large subunit has been recruited to join the smaller subunit, the completed initiation complex (now 80S) is prepared to begin translation elongation.

10.9 TRANSLATION ELONGATION: DECODING, PEPTIDE BOND FORMATION, AND TRANSLOCATION

The initiation step of translation, described in the previous two sections, essentially primes the ribosome for peptide synthesis: it locates the site on the mRNA from which the sequential decoding of codons will occur (in the 3′ direction), and brings both ribosomal subunits together, ready for active synthesis to commence. In this section, we describe the core events underpinning peptide synthesis itself.

The aminoacyl-tRNA–EFTu(eEF1A)–GTP ternary complex interacts with the ribosome to ensure high accuracy during translation

Following initiation, the ribosome sits poised on the mRNA with an initiator methionyl-tRNA in the P site and an empty A site ready to begin the elongation cycle. The elongation cycle comprises three basic steps in all organisms: decoding, peptide bond formation, and translocation.

During the decoding step, the ribosome selects an aminoacyl-tRNA with an anti-codon that is complementary to the mRNA codon in the A site. This aminoacyl-tRNA is bound to elongation factor EFTu-GTP (or eEF1A-GTP in eukaryotes), which interacts with the ribosome to ensure high fidelity in the selection process. In its role as an escort, EFTu acts as another guardian of high accuracy in protein synthesis, the first guardian being the aminoacyl-tRNA synthetases that load tRNAs with only their cognate amino acids (see Section 10.3). The role of EFTu is thus to increase the accuracy of decoding beyond the levels brought about by the simple differences between the binding energy of **cognate**, **near-cognate** (single mismatches) and **non-cognate** (more than one mismatch) **interactions** between the codon and the anticodon. We recall that the rules of the genetic code demand Watson–Crick pairing interactions between the codon and anticodon at the first and second codon positions, but certain non-standard pairing interactions are permitted at the third codon position. The overall fidelity of this EFTu-facilitated selection process is thought to be of the order of 10^{-3}, meaning that an incorrect aminoacyl-tRNA is accepted by the ribosome only once in a thousand times.

How are cognate tRNAs recognized and selectively incorporated by the ribosome and EFTu ternary complex? First, cognate aminoacyl-tRNAs associate more tightly with the ribosome complex than near (or non)-cognate ones, favoring their ultimate acceptance into the ribosome. Indeed, these binding differences can be utilized twice by the ribosome, before and after GTP hydrolysis on EFTu. This provides the ribosome with two opportunities to discriminate between cognate and near-cognate tRNAs and, hence, increase the fidelity of selection. These two different phases of the selection process are referred to as '**initial selection**' and '**proofreading**'. We think of these two cognate-discriminated binding steps as thermodynamic contributors to fidelity (red arrows in Figure 10.27).

initial selection   proofreading

Figure 10.27 Decoding (tRNA selection) by the ribosome and EFTu. The selection of a cognate aminoacyl-tRNA in the A site of the ribosome involves two distinct steps separated by the hydrolysis of GTP by EFTu. In the 'initial selection' phase, an aminoacyl-tRNA complexed with EFTu can be rejected from the ribosome before EFTu hydrolyzes GTP. Following GTP hydrolysis and the departure of EFTu, ribosomes still can reject near-cognate tRNA in a 'proofreading' phase. Both thermodynamic rejection steps are shown with red arrows. Additionally, cognate tRNAs promote more rapid GTPase activation and accommodation (shown in green arrows), allowing for kinetic discrimination mechanisms to contribute to the overall fidelity of tRNA selection.

⊜ We learn about the difference between the major and minor grooves of a double-stranded nucleic acid in Chapter 2.

In addition to the thermodynamic contributions to fidelity, the geometry of the small helix of RNA formed by the codon and anticodon is carefully evaluated in its minor groove by specific ribosomal elements located in the 'decoding center', the region of the small subunit surrounding the codon–anticodon complex.

Recognition of a cognate helix promotes conformational changes in proximal regions of the subunit (see below), and these structural changes are in turn propagated to more distant regions of the ribosome. These transmitted signals ultimately tell the ribosome to act as a GAP for EFTu to increase the rate of GTPase activation and to promote more rapid '**accommodation**' of the aminoacyl-tRNA fully into the A site (there is a rather large movement of the aminoacyl-tRNA following dissociation from EFTu). We think of these two cognate-stimulated forward steps as kinetic contributors to fidelity (green arrows in Figure 10.27).

As a result, cognate tRNAs move more rapidly than near-cognate tRNAs through the initial selection process – first because their ternary complexes are less likely to fall off the ribosome complex than near-cognate tRNAs and, second, because the GTPase activation step is more rapid. These cognate tRNAs are also more efficiently accepted by the ribosome during the second 'proofreading' phase – first, because they fall off less often than near-cognate tRNAs and, second, because the accommodation step is more rapid. So, while near-cognate tRNAs do bind to the ribosome, they are readily discriminated against by this two-step selection process. As we shall discuss in the next section, there are other mechanisms in place that monitor the fidelity of protein synthesis following the misincorporation of an incorrect amino acid into the polypeptide chain. These mechanisms depend on premature termination as a mechanism to increase the overall fidelity of protein synthesis.

Conformational changes in the rRNA near the codon–anticodon pair are critical for tRNA selection (decoding)

We now know in detail some of the conformational changes that take place in the ribosome when a cognate tRNA is recognized. In the regions surrounding the anticodon–codon complex, the decoding center, there are several universally conserved nucleotides in the 16S rRNA that respond in a dramatic way to cognate tRNA recognition. The nucleotides A1492, A1493, and G530 undergo substantial conformational rearrangements on binding a cognate tRNA; in this rearranged (or induced) state they stably interact with the minor groove of the cognate codon–anticodon helix (Figure 10.28). These structural changes, which do not occur efficiently with near- and non-cognate tRNAs, are ultimately responsible for inducing more remote structural changes in the ribosome that stimulate GTPase activation on EFTu and accommodation of the cognate aminoacyl-tRNA into the A site of the ribosome, serving to ensure the fidelity of the process.

Having now considered the decoding step of elongation, let us now go on to consider the second step – peptide bond formation.

Peptide bond formation is catalyzed in an active site containing only RNA

After a cognate aminoacyl-tRNA has been accommodated in the A site of the ribosome, the aminoacyl and peptidyl-tRNAs are ready for the catalysis of peptide

Figure 10.28 Structural changes in the ribosome induced by the binding of a cognate tRNA. Three universally conserved nucleotides in the 16S rRNA in the A site of the ribosome (G530, A1492, and A1493) change conformations on recognition of a cognate codon (blue):anticodon (red) helix. The structure of the nucleotides in the absence of the mRNA–tRNA interaction is shown on the left. In their rearranged state, these nucleotides directly interact with the minor groove of the decoding helix. These structural changes are thought to be critical for triggering GTPase activation by EFTu and for accommodation of cognate tRNA in the A site. PDBs 2WDG, 1J5E.

bond formation wherein the growing polypeptide chain is transferred to the aminoacyl-tRNA (Figure 10.29a). As we discussed earlier, the peptidyl transferase active site is tightly packed with universally conserved rRNA elements that surround the two tRNA substrates and position them for catalysis. (There are no nearby ribosomal proteins.)

How is catalysis promoted in this active site? The chemistry of the reaction is relatively simple and consists of a nucleophilic attack of the amino group of the aminoacyl-tRNA on the electron deficient carbonyl of the peptidyl-tRNA substrate (Figure 10.29b).

Prior to accommodation of the aminoacyl-tRNA, the peptidyl-tRNA is bound in an unreactive configuration for nucleophilic attack. However, when an aminoacyl-tRNA is accommodated in the A site through the tRNA selection process, conformational rearrangements occur in the large subunit catalytic center, which orient the peptidyl-tRNA substrate in a position appropriate for nucleophilic attack. At this stage, the two substrates are poised for catalysis with the conserved 3′ CCA tails forming specific Watson–Crick base pairs with universally conserved

Figure 10.29 Peptide bond formation. (a) Schematic showing the tRNA substrates on the ribosome before and after peptidyl transfer. As a result of catalysis, the growing polypeptide chain is transferred from the P site to the A site tRNA. At some point following peptidyl transfer, the tRNAs are thought to assume a hybrid state of tRNA binding in preparation for translocation. (b) A more detailed chemical view of the two substrates prior to peptidyl transfer. Watson–Crick pairing interactions between the ribosome (the P and A loops) are indicated. The 2′ OH of the peptidyl-tRNA substrate is shown to be facilitating proton transfer to promote the nucleophilic displacement reaction (dashed lines). R refers to the growing polypeptide chain. Adapted from Parnell and Strobel (2003).

guanosine residues in characteristic loops of the 23S rRNA (or the 28S rRNA in eukaryotes) (the A and P loops in Figure 10.29b). A set of universally conserved nucleotides within the large subunit rRNA are positioned nearby, though a role for them in catalysis *per se* has not been identified.

In addition to these required positioning elements of the ribosome (the A and P loops and highly conserved active site nucleotides), the 2′ OH of the peptidyl-tRNA substrate, which is critically positioned in the active site (Figure 10.29b), appears to play an essential role in catalysis. This essential chemical group has been proposed to be involved in facilitating the movement of protons during the reaction. Such a tRNA-dependent mechanism is consistent with the idea that this chemical reaction emerged very early in the evolution of protein synthesis, before the arrival of a highly sophisticated ribosome.

Translocation of the mRNA–tRNA complex is promoted by the GTPase EFG

In the final step of elongation – translocation – the complex of mRNA bound to tRNAs in the A and P sites moves through the ribosome, opening up the A site to receive the next incoming aminoacyl-tRNA (Figure 10.30a). Translocation entails large-scale movements of the tRNAs (and associated mRNA) and conformational changes in the ribosome wherein numerous tertiary contacts must be disrupted and reformed. We have seen how each aminoacyl-tRNA first binds in the A site before moving to the P site (after a peptide bond is formed), and we have learned that each tRNA occupies the exit (E) site as it leaves the ribosome (Section 10.5). We also discussed the hybrid states of tRNA binding as a model to explain how tRNA movement seems to occur independently with respect to the large and small ribosomal subunits. In Section 10.6 we discussed how the GTPase EFG (eEF2 in eukaryotes) mimics the shape and structure of tRNA bound to EFTu and apparently uses this mimicry to promote the structural rearrangements involved in the translocation process. Here we see evidence for this tRNA mimicry in a cryoEM view of EFG bound to the A site of the ribosome (Figure 10.30b). Structural studies indicate that the ribosome itself can assume several different states that are probably critical to the translocation process, one state favoring 'hybrid' pre-translocation-bound tRNAs and another favoring non-hybrid ('classical') post-translocation-bound tRNAs.

Despite these many insights, we still have much to learn about how EFG directly couples the energy of GTP hydrolysis to discrete three-nucleotide movements of the mRNA–tRNA complex in the ribosome.

10.10 TRANSLATION TERMINATION AND REINITIATION

As the three-step cycle of elongation described in the previous section continues, the ribosome glides along its mRNA template, continually extending the polypeptide chain as it moves. Though it might seem plausible that elongation should continue until the ribosome runs out of mRNA to decipher, the reality is different. The end-point of the process of translation is signalled by three specific nonsense

> ⊙ We learn more about the nature of nucleophilic attack in Chapter 2.

(a)

(b)

Figure 10.30 Translocation of the mRNA–tRNA complex. (a) Schematic showing the movements of the mRNA–tRNA complex with respect to the ribosome that occur during the process of translocation. EFG interacts with pre-translocation state ribosomes, with tRNAs probably bound in the 'hybrid' state. A domain of EFG mimics the anticodon of a tRNA, and interacts with the small subunit decoding center, biasing forward movement. (b) Cryo-electron microscopy image of EFG stalled on the ribosome (orange). The image clearly shows how domain IV directly engages the normal A site tRNA binding site on the small subunit. PDBs 2WRK, 2WRL. Taken from Valle *et al.* (2003).

codons that are recognized by specific protein factors. This process of termination, like other steps in translation, is subject to various levels of regulation and control, as we shall discover in this section.

Class 1 and class 2 release factors are critical to termination in bacterial and eukaryotic systems

In most organisms, the end of a protein-coding region of mRNA is signaled by one of three **stop codons**: UAA, UAG and UGA. There are usually no tRNAs that recognize these stop codons; instead, they are recognized by proteins known as **class 1 release factors** (RFs). In bacteria there are two different class 1 release factors, RF1 and RF2, with overlapping specificity – both recognize UAA, while RF1 recognizes UAG and RF2 recognizes UGA. In eukaryotes, only one class 1 release factor, eRF1, recognizes all three stop codons.

In addition to recognizing stop codons in the A site of the ribosome, release factors promote a hydrolytic reaction in the peptidyl transferase center that releases the completed peptide chain from the peptidyl-tRNA in the P site. Thus the class 1 release factors are bifunctional molecules, like tRNAs, linking an interpretation of the genetic code in the small subunit decoding center with a catalytic event in the large subunit catalytic center. In this case, however, a protein factor instead of an RNA facilitates specific recognition of codons and stimulates the catalytic event that results in the release of the peptide chain.

Figure 10.31a shows how the structure of a tRNA and a class 1 release factor (from bacteria), both in their ribosome-bound state, superimpose quite nicely on one another. Notice how there are protein domains of the class 1 RF that correspond to the functional ends of the tRNA, the anticodon loop and the acceptor CCA end. Moreover, the RF bound to the ribosome is very clearly positioned in the A site, spanning the decoding center in the small subunit and the catalytic center in the large subunit (10.31b). It is interesting to note that the class 1 RFs in bacteria and eukaryotes are unrelated to one another (Figure 10.32), though both have evolved to perform the same core functions, and even have the same universally conserved GGQ motif that is essential for catalysis.

In addition to the class 1 release factors, a set of GTPases (known as the **class 2 release factors**) is critical to termination in both the bacterial and eukaryotic

(a)

RF1 aligned to tRNA

(b)

Figure 10.31 Class 1 release factors are protein mimics of functional tRNAs.
(a) Bacterial RF1 structure (blue) superimposed on tRNA^{Phe} structure (gray), both in their ribosome-bound conformation. (b) Atomic resolution structure of RF1 bound to the ribosome, showing how it occupies the A site of the small and large subunits as it deciphers the codon and facilitates peptide release. PDBs 1PNS, 3D5C, 3D5D.
Figure 10.31a from Youngman *et al.* (2008).

Figure 10.32 Bacterial and eukaryotic class 1 release factors are evolutionarily distinct. Atomic resolution structures of class 1 RFs from bacteria (a) and eukaryotes (b) indicate that the structures are distinct, though both rely on the conserved GGQ motif for efficient peptide release. PDBs 3D5C, 1DT9.

systems. However these factors appear to play quite distinct roles in these two systems. In bacteria, the GTPase RF3 appears to function after peptide release to promote the dissociation of the class 1 RF, coupling GTP hydrolysis to this event. In eukaryotes, the GTPase eRF3 appears to interact with eRF1 prior to associating with the ribosomal termination complex, directly coupling GTP hydrolysis to peptide release (Figure 10.33). These substantial differences are not surprising

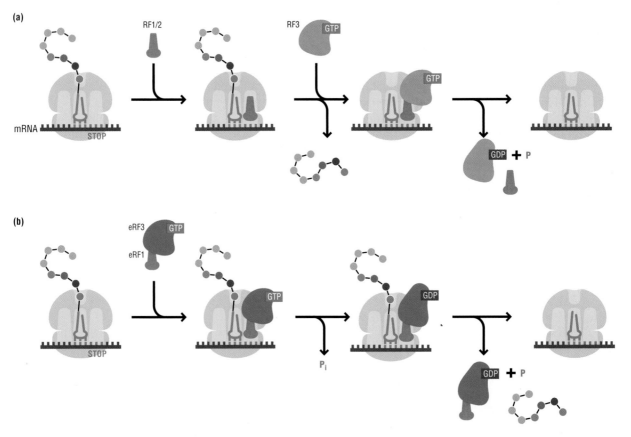

Figure 10.33 Roles of class 1 and 2 release factors in bacteria and eukaryotes. (a) In bacteria, a class I release factor (RF1 or RF2) recognizes the stop codon and triggers the release of the synthesized polypeptide chain from the protein. RF3-GTP then associates with the termination complex, GTP hydrolysis occurs and the class 1 RF and RF3-GDP dissociate. (b) In eukaryotes, eRF1 and eRF3 are associated with one another prior to binding to the ribosome. Moreover, peptide release is faster in the presence of both eRF1 and eRF3, as well as GTP. Following peptide release, the eRF1–eRF3 complex dissociates.

given that the class 1 RFs in these two kingdoms evolved independently, and thus their interaction with a class 2 GTPase also had to evolve separately.

Stop codon recognition by class 1 release factors is highly accurate

Release factor selection in the A site takes place with extremely high fidelity, mirroring tRNA selection. Studies in bacteria suggest that premature peptide release on non-stop (or sense) codons occurs only once in every 100 000 events. Indeed, class 1 RFs appear to have exquisite mechanisms in place to distinguish between stop and sense codons, depending on strategies of induced conformational rearrangements in the small and large subunits to promote catalysis only when true stop codons are recognized.

As for tRNA selection, specific conserved nucleotides in the decoding center (including nucleotides G530 and A1492) undergo distinct conformational rearrangements on recognition of the stop codon by the class 1 RF. And, as for peptide bond formation, the 2′ OH of the peptidyl-tRNA substrate appears to be a key component in the peptide-releasing hydrolytic reaction, working in concert with the conserved GGQ motif on the class 1 RF. Such coupling of recognition of stop codons in the decoding center to conformational rearrangements in the somewhat remote catalytic center (~ 80 Å away in the large subunit) allows for an extremely low rate of premature termination of protein synthesis.

Abortive termination on the ribosome increases the overall fidelity of protein synthesis

Although the processes of tRNA and RF selection have evolved to take place with extremely high accuracy, mistakes nevertheless do happen and the natural consequence is either the incorporation of another wrong amino acid in the growing polypeptide chain, or a premature termination event that takes place on a sense codon, as we describe here. Such decoding errors are increasingly likely during times of stress or starvation when a particular aminoacyl-tRNA might be depleted.

How are such miscoding events detected? Misincorporation events on the ribosome appear to be recognized because the codon:anticodon helix, positioned in the P site following translocation, carries a mismatch. When this perturbation of the P-site decoding helix is detected by the bacterial ribosome, the fidelities of tRNA and RF selection on the next codon in the A site appear to be substantially decreased. This decrease in fidelity initially leads to an iteration of errors, which eventually leads to premature termination of the error-containing polypeptide (Figure 10.34). The prematurely truncated peptide is likely degraded by cellular peptidases and proteases, while the ribosome and mRNA are effectively liberated for additional rounds of translation. This quality control system on the bacterial ribosome can be compared with the proofreading or editing functions of both DNA polymerases and aminoacyl-tRNA synthetases discussed earlier in the book.

➲ We discuss the proofreading/editing functions of DNA polymerase and aminoacyl-tRNA synthetases in Sections 6.3 and 10.3, respectively.

Ribosome recycling is the final step in protein synthesis

Following the release of the completed polypeptide chain and release factors from the ribosome, the ribosome complex must be disassembled to allow the

Figure 10.34 Quality control on the ribosome following a misincorporation event. When a misincorporation takes place due to an error during tRNA selection, a mismatched decoding helix is found in the P site following translocation (notice that tRNA colour does not match codon). This mismatch triggers structural changes in the ribosome complex that substantially decrease the fidelity of tRNA and RF selection in the A site, leading most immediately to an iteration of errors, and ultimately to premature termination at sense codons.

components to take part in another round of protein synthesis. The exact order of events that follows peptide release is not well described in either bacteria or eukaryotes, though the substrate of recycling is in both cases a complex consisting of the two subunits of the ribosome, the mRNA and a deacylated tRNA that once carried the growing peptide chain. In bacteria, a protein factor known as ribosome recycling factor (RRF) appears to act in concert with EFG (and GTP hydrolysis) to promote complex disassembly, and IF3 stabilizes the dissociated ribosomal subunits (Figure 10.35). Less is known of how ribosome recycling takes place in the eukaryotic system, in which no RRF homolog has been identified.

Reinitiation of translation on downstream open reading frames in polycistronic mRNAs is routine in bacteria

In bacteria, genes are typically transcribed into mRNAs that carry more than one open reading frame; such transcripts are called polycistronic mRNAs. In these mRNAs, open reading frames are typically preceded by a Shine–Dalgarno sequence that can independently guide the small subunit to the appropriate AUG for loading of the initiator f-Met-tRNA^{fMet} (as we described in Section 10.6). Following initiation, translation elongation, termination, and ribosome recycling can then take place such that ribosomal subunits, tRNA and mRNA are fully released from the actively translating complex. In some cases, however, there

Figure 10.35 Ribosomal recycling in bacteria. The combined actions of ribosome release factor (RRF) and EFG-GTP cause dissociation of the two subunits of the ribosome. The tRNA and mRNA remain bound to the small subunit until IF3 promotes their departure. The small ribosome subunit–IF3 complex is then free to begin the initiation process with another mRNA.

appears to be coupling between closely spaced open reading frames in a **polycistronic message** such that ribosome recycling is somewhat inefficient. In this situation, the small ribosomal subunit is thought to remain associated with the mRNA and scan in both the forward (3′) and backward (5′) direction in search of another AUG to initiate the next round of protein synthesis. The rate at which the small subunit dissociates from the mRNA is a crucial factor in determining how much 'scanning' can occur before the subunit completely departs from the mRNA.

While a vast majority of eukaryotic genes are monocistronic, a few are polycistronic in the sense that there are multiple sites within the transcript where translation initiation and termination take place, though the 'cistrons' are typically extremely small (with just a few codons) and the product that they encode is not thought to have any cellular function. In these cases, the **upstream open reading frames** (**uORFs**) are commonly used to regulate gene expression. We mention these examples here since reinitiation of translation is the key to how these uORFs can play a role in gene regulation. In brief, some form of incomplete ribosomal recycling must occur at the end of an uORF, which allows the still engaged ribosome to search for the next AUG for reinitiation. These examples can be distinguished from those in bacteria as the scanning is unidirectional (5′ to 3′). A classic example of this form of translational regulation is used to control expression of the transcriptional activator protein Gcn4, which we will examine more closely in Section 10.14.

Deleterious effects from truncated bacterial mRNAs are minimized by the tmRNA protein tagging system

In eukaryotes, only full-length mRNAs are generally translated into protein because formation of the initiation complex depends so strongly on the presence of both the 5′ cap and the 3′ poly(A) tail structures. In bacteria, the translation initiation process does not appear to distinguish between full-length and truncated mRNA species and so these organisms have evolved another system to ensure that only full-length mRNA species are translated into stable protein products.

When bacterial ribosomes encounter the physical end of an mRNA (i.e. before a termination codon is reached), the complex is effectively stalled, taking a valuable ribosome out of circulation, and at best generating a truncated polypeptide. This situation is addressed by an unusual RNA species known as **tmRNA**. The 'tm' stands for the dual features of tmRNA – tRNA-like and mRNA-like; tmRNA acts in multiple steps to recover ribosomes and stimulate degradation of the truncated mRNA and the incomplete polypeptide chain. The tmRNA first acts as a tRNA, adding an alanine residue to the growing peptide chain. It then acts as an mRNA, which is composed of a short open reading frame and a termination codon (Figure 10.36). This mRNA sequence is then translated, resulting in 11 amino acids

Figure 10.36 tmRNA directs a quality control step on stalled ribosomes in bacteria. Bacterial ribosomes occasionally stall on truncated mRNAs that lack a stop codon. This stalled complex is recovered by tmRNA, which acts first as an aminoacyl-tRNA (with a tRNA^{Ala}-like moiety) and second as an mRNA encoding an 11-residue peptide terminated with a stop codon. Following the tmRNA-mediated events, free ribosomes are liberated for further protein synthesis while the truncated proteins carry an 11-residue peptide that targets them for degradation (to avoid likely dominant effects on the cell).

→ We will explore the degradation of proteins in Chapter 11.

(encoded by the tmRNA species) being appended to the truncated protein. This peptide sequence acts as a tag that targets the protein to a set of cellular proteases for degradation. While the basic details of this process have been elucidated, there is little mechanistic understanding of how this unusual RNA interacts with the ribosome to perform these diverse tasks.

TRANSLATION: REGULATION

The importance of regulation in ensuring the correct and appropriate execution of the processes on which life depends is a recurring theme throughout this book. The process of translation is no exception. In the remaining sections of this chapter we explore the key strategies that are employed by both bacteria and eukaryotes to ensure that the translation machinery operates as it should, and that translation as a whole yields the correct end-products. We will consider strategies for regulating translation both at the level of specific mRNAs and more globally in the cell. In Experimental approach 10.3, several experiments are described that provide a sense of how translational regulation has been studied in the past, and more recently. As seen in this panel, our molecular understanding of how regulation might occur often starts with analysis of translation in cells using the tool of polysome profiling.

(a) nonsense suppression

(b) frameshifting

Figure 10.37 Nonsense suppression and frameshifting are relatively common recoding events. (a) Nonsense suppression happens when a nonsense codon is misread by a near-cognate tRNA species, thus reading past the stop codon to extend the polypeptide sequence. (b) Frameshifting happens when the mRNA slips in register, such that protein synthesis continues in a different reading frame.

10.11 RECODING: PROGRAMMED STOP CODON READ-THROUGH AND FRAMESHIFTING

In the preceding sections, we discussed the molecular details of the translation cycle, from initiation to ribosome recycling, and saw the general principles governing this process, including the nature of the genetic code from which successive three-letter nucleotide sequences are transformed into the primary sequence of a protein. However, there are few hard and fast rules in biology; the genetic code is just one more example of a rule that fails to hold absolutely. In this section we discover how biological systems can often flex the rules of the genetic code, introducing more diversity to the way the nucleic acid sequence of a messenger RNA is interpreted.

In contrast with codon reassignment described in Section 10.2, the rules of the genetic code are superseded in a different way in **recoding**, wherein readout is reprogrammed in an mRNA-specific fashion. In such instances, the meaning of a codon is not globally reassigned such that all translation events in the organism or organelle interpret the codon in this different way. Instead, the codon is reinterpreted in the context of a specific mRNA, *in competition with* a standard reading of the mRNA. Such recoding can lead to the production of multiple distinct proteins from a single gene and can be used for regulatory purposes. We will first look at several well characterized examples of the recoding of stop codons (Figure 10.37a) and then at some more complex examples of recoding wherein the frame in which an mRNA is read is altered during the process of translation (Figure 10.37b).

Figure 10.38 Programmed nonsense suppression. In the Murine Leukemia Virus A, a pseudoknot structure in the RNA downstream of the stop codon (UAG) triggers misreading by Gln-tRNA^{Gln}, and thus the production of the GagPol viral gene product. This miscoding must compete effectively with the normal termination factors (eRF1-eRF3), though eRF1 is apparently sequestered through interactions with another viral gene product, RT.

Recoding in response to specific mRNA signals can lead to nonsense suppression

When any of the three normal stop codons are misread by aminoacyl-tRNA species such that the termination of translation fails to occur, we refer to the process as nonsense suppression. Although such read-through normally happens with very low frequency, the likelihood of it occurring can be significantly affected by specific sequence elements in the mRNA. These phenomena have been studied most deliberately in viral mRNAs where a relatively simple hexanucleotide motif (CARYYA) found 3′ to the stop codon triggers increased read-through by near cognate aminoacyl-tRNAs. UAG codons are typically misread by tRNA^{Gln} while UGA codons are typically misread by tRNA^{Trp}.

In one well-studied example in the Murine Leukemia Virus, the GagPol precursor polypeptide required for the viral lifecycle is synthesized through a nonsense suppression mechanism where misreading of a UAG codon is specifically promoted by a 3′ proximal structured **pseudoknot** element in the mRNA through an unknown mechanism. An additional contributor to nonsense suppression in this case is the partial sequestration of eRF1 by a virally encoded gene product, reverse transcriptase (RT). This mechanism is outlined in Figure 10.38. Note that the nonsense suppression must compete with normal stop codon recognition by release factors and so is not 100% efficient. The extent of misreading in this system plays a key role in regulating the levels of production of several critical viral gene products.

Stop codons can be recognized by systems that incorporate non-standard amino acids into proteins

Stop codons can also be recoded to allow a non-standard amino acid (not one of the 20 common amino acids) to be inserted into a protein, often at positions of functional importance for the activity of the protein. The best-characterized example of this mechanism is the insertion of selenocysteine, an amino acid similar to cysteine that carries a selenium atom instead of a sulfur atom in its side chain (Figure 10.39a). This so-called 'twenty-first amino acid' is incorporated into

(a)

(b)

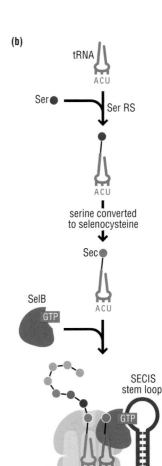

Figure 10.39 The recoding of UGA stop codons for selenocysteine insertion.
(a) The chemical structure of the amino acid selenocysteine. (b) Schematic showing how a specialized tRNA in bacteria is first loaded with a serine that is then enzymatically converted to selenocysteine. The resulting Sec-tRNA^{Sec} is delivered to the ribosome by an EFTu homolog, SelB, that simultaneously recognizes the selenocysteine insertion sequence (SECIS) RNA element downstream of the UGA codon. This mechanism allows for the insertion of selenocysteine only at certain UGA codons in the cell.

10.3 EXPERIMENTAL APPROACH
Translational profiling

The translational activity of ribosomes on a given mRNA has traditionally been followed using a technique referred to as polysome analysis. In a cell, an mRNA that is translationally active tends to be loaded up with ribosomes all along its length, with initiation occurring again (and again) once the previous ribosome has elongated beyond the initiating AUG codon (see Figure 10.17). Such 'ribosome-loaded' mRNAs migrate as large particles in a sucrose (or glycerol) gradient. The absorbance of the sample at around 260 nm is proportional to the amount of nucleic acid present. Therefore this measure directly reports on the predominant RNA in the cell, the rRNA found within the ribosomes. With this readout, the various peaks in the sample can be easily identified – the first peak (starting from the top of the gradient) corresponds to small ribosomal subunits (30S or 40S), the second peak to large ribosomal subunits (50S or 60S), the next to full ribosomes (70S or 80S, or monosomes), then disomes (two ribosomes attached to a single mRNA), trisomes (three attached ribosomes) and so on (Figure 1). mRNAs are sometimes associated with isolated small subunits (presumably in a pre-initiation phase), but primarily are found associated with one or more intact ribosomes (monosomes and higher).

When the various fractions collected from the sucrose gradient are analyzed for their mRNA content (for example, by northern analysis), then clues can be obtained about the translational activity of an mRNA of interest in the cell, and even the type of regulation that might be occurring. Genes that are actively expressed tend to be found in the heavy polysomal fractions where mRNAs are loaded with multiple ribosomes, while genes that are translationally repressed at the level of initiation tend to be excluded from the polysomal fractions, migrating instead as smaller particles closer to the top of the sucrose gradient. Genes that are translationally repressed, but at a stage following initiation (for example, during

Figure 2 **Primer extension analysis of ribosome pausing during translation of the preprolactin mRNA in a wheat-germ translation extract.** Ribosome-protected fragments of mRNA were prepared from an *in vitro* translation reaction either in the presence or absence of signal recognition particle (SRP) or 7-methyl guanosine cap (m⁷G) (to compete with cap dependent translation). Four specific stalls can be identified (a–d) that correspond to distinct sites in the mRNA. Band a represents the initiating population of ribosomes at the AUG start site while band d corresponds to the terminating population at the UAA stop codon. Bands b and c are both centered on rare GGC codons, but the intensity of band b is increased in the presence of SRP. m⁷G competes with translation of the mRNA, and so none of the pauses are seen. Sequencing lanes on the right are labeled A, C, G, T. Three different concentrations of ribosome-protected fragment were used in the primer extension reactions (1.0, 0.3 and 0.1 μl input, top of gel). Reproduced from Wolin and Walter, *EMBO Journal* 1988;**7**:3559–3569.

Figure 1 **Polysome profile using A_{260} to follow the fractionation of ribosomes in a sucrose gradient following centrifugation.** Ribosomal subunits (30S and 50S), 70S, disomes, trisomes and larger are clearly resolved.

elongation or termination) are predicted to be found within the polysomal fractions, but in a state that is stalled or slow. In such a situation, we might expect to observe even more ribosomes than normal loaded on a repressed mRNA. We see from this discussion that polysomal analysis can be used to explore how the translation of a given mRNA might be regulated.

In an insightful early study, Sandra Wolin and Peter Walter developed an experimental approach to map the specific sites of ribosome occupancy on an mRNA (found in polysomes) using what was essentially a ribosome protection assay. The basic idea was that mRNA bound to a ribosome would be protected from exogenously added nuclease, and could be isolated and identified following such a treatment. The real challenge of the experiment was to develop a means of identifying small amounts of a gene product of interest.

In the study by Wolin and Walter, the authors chose to hybridize the isolated ribosome protected fragments to a single-stranded template carrying the anti-sense strand of the gene of interest. The template complex was then analyzed by reverse transcription (see Section 16.9) using a downstream primer to yield a cDNA strand, and the products of the reaction were resolved on a sequencing gel. In the analysis by Wolin and Walter, the authors identified a pause site within the bovine preprolactin mRNA where ribosome occupancy was increased in the presence of exogenously supplied signal recognition particle (SRP) (Figure 2, band b).

As we have come to understand, the ribosome occupancy site in this case corresponds to pausing by the ribosome as the signal peptide emerges from the exit channel of the ribosome (on the solvent side of the large subunit) and engages machinery involved in downstream steps in the protein translocation process (described in Section 11.2). In addition to this primary pause site, the authors identified other ribosome protected fragments that appeared to be centered on rare glycine codons (GGC) in the mRNA template; the authors argued that these sites resulted from the ribosome spending a bit more time paused at these positions during the decoding step of elongation. Both examples show how detailed mapping of ribosome positioning on the mRNA template can provide information on translational events within the cell and their timing. Moreover, these results foreshadow more recent high-throughput approaches as described briefly below.

In recent years, two different high throughput approaches have tackled this same general problem. In the first example, Daniel Herschlag, Patrick Brown, and colleagues coupled microarray approaches with polysomal profiling to globally assess the translational status of mRNAs in the cell. To do this, polysomal fractions were isolated, intact mRNA was extracted from them, and the distribution of each mRNA across the gradient assayed through hybridization to a microarray. The approach yielded a number of insights into broad features of translational activity in the cell. For example, most mRNAs were seen to be somewhat under-populated with

Figure 3 Ribosome profiling explores translation activity at the GCN4 locus in yeast under normal (top, red) and starved (bottom, yellow) conditions. The number of reads of a particular region in the gene (the X-axis) are tabulated on the Y-axis both for a total yeast fragmented mRNA population (green) and for ribosome-protected fragments (blue). As expected (see 10.14), GCN4 expression is induced (more 'reads' within the GCN4 open reading frame (ORF)) under conditions of amino acid starvation. Boxed in orange is another upstream location (containing no AUG codons) where there is clear increased ribosome occupancy during starvation; within this region there are several proposed non-canonical initiation sites.

ribosomes, relative to their anticipated capacity based on length, suggesting that initiation might be the rate-limiting step during translation. (If elongation or termination were rate-limiting, the ribosomes should be quite densely populated on the mRNAs.) Despite the broad success of this approach, its resolution was limited because whole mRNAs were followed rather than the individual 'footprints' of a ribosome on the mRNA.

More recently, ribosome mRNA occupancy has been more precisely assessed by Jonathan Weissman and colleagues, now using high-throughput sequencing to take a very comprehensive look at mRNA status in the cell. (They call the approach 'ribosome profiling'.) In these experiments, ribosome footprints ~28 nts in length were isolated from yeast ribosomes using an RNAse protection approach (similar to that described by Wolin and Walter) and these footprints were directly sequenced, with tens of millions of individual sequence reads being obtained for a given set of conditions. New insights into translation are rapidly emerging from these studies. For example, non-canonical initiation sites (UUG or GUG, instead of AUG) appear to be more heavily utilized during times of cellular stress, as indicated by increased ribosome occupancy on these sites during amino acid starvation (Figure 3). We can already see that the level of resolution that comes from such a deep sequencing approach will yield an unprecedented view of translational complexity both at the level of the single gene, and of the genome.

Arava Y, Wang Y, Storey JD, *et al*. Genome-wide analysis of mRNA translation profiles in *Saccharomyces cerevisiae*. *Proceedings of the National Academy of Sciences of the U S A* 2003;**100**:3889–3894.

Ingolia NT, Ghaemmaghami S, Newman JR, Weissman JS. Genome-wide analysis in vivo of translation with nucleotide resolution using ribosome profiling. *Science* 2009;**324**:218–223.

Wolin SL, Walter P. Ribosome pausing and stacking during translation of a eukaryotic mRNA. *EMBO Journal* 1988;**7**:3559–3569.

several different enzymes where it appears to play a role as a powerful reducing agent in the active sites of these proteins.

In organisms ranging from *E. coli* to humans, there are structural elements in the mRNA that are critical for specifying the insertion of the selenocysteine residue. In *E. coli*, an RNA stem–loop known as a **SelenoCysteine Insertion Sequence** (**SECIS**) is found immediately 3′ to the UGA codon that will be recoded. In mammals the RNA structure is more complex and seems to function when found anywhere within the 3′ untranslated region (UTR) of the mRNA (much as transcriptional enhancers can act by binding to DNA some way downstream of the gene's promoter). In addition to these RNA elements, a number of other factors are required for selenocysteine incorporation. First, there is a system for generating the selenocysteine-tRNA. This system includes a specialized tRNA (tRNA^Sec) that is charged with serine (by SerRS) and has an anticodon that matches the UGA stop codon; additionally, there are enzymes for converting the serine attached to this specialized tRNA to selenocysteine. Second, there is a system that determines which UGA codons will be recognized. For this, a bacterial protein, SelB, that is homologous to EFTu is key. SelB in *E. coli* appears to be bifunctional: in addition to delivering the selenocysteinyl-tRNA to the ribosome (as EFTu does) it has an RNA binding domain that allows it to interact with the SECIS stem–loop, thus directing selenocysteine to be incorporated in the polypeptide chain only at specific stop codons (Figure 10.39b).

In eukaryotes, two different proteins are involved in the actual selenocysteine decoding event, one responsible for binding the SECIS element in the RNA (SBP2) and the other an EFTu like protein with a C-terminal extension (eEFsec) that binds both tRNA^Sec and SBP2.

A twenty-second amino acid called pyrrolysine – a modified lysine residue (Figure 10.40a) – has been found in the active sites of methyltransferase enzymes in certain methane-producing organisms. As in the case of selenocysteine, the presence of this unusual amino acid in the active site seems to be critical for the catalytic activity of the enzyme. Pyrrolysine is incorporated into these enzymes at UAG stop codons in somewhat the same way as selenocysteine. A specialized tRNA (tRNA^{Pyl}) and aminoacyl-tRNA synthetase (PylRS) are required (Figure 10.40b); it is not yet known whether RNA elements that function like the SECIS element are also required to ensure that pyrrolysine is inserted into proteins only at certain UAG codons.

Programmed frameshifting is triggered by local mRNA elements

During translation, ribosomes move sequentially from one mRNA codon to the next by advancing three nucleotides along the mRNA. If the ribosome moves some different number of nucleotides (not comprising a multiple of 3), then this wholly changes the protein sequence decoded from the mRNA, and the reading frame of the mRNA is said to be shifted (Figure 10.41). By contrast, when the mRNA:tRNA complex moves in steps comprising integers of three (so steps of 6, 9, 12, etc.) the frame is maintained, though some number of amino acids encoded by the gene will be omitted (or duplicated).

The change of the reading frame, called **frameshifting**, has most commonly been seen to occur as a single nucleotide shift in the +1 or –1 direction. Such events can take place in very specific mRNA contexts (leading to the term 'programmed frameshift') and can control gene expression in bacterial, mammalian, and viral systems.

An illustrative example of frameshifting is in the expression of the bacterial translation termination factor, RF2. In 70% of bacterial species, the complete RF2 protein is encoded in two open reading frames, a short one immediately followed by a longer one in the +1 frame. Almost without exception, the site of frameshifting within the open reading frame of the mRNA is demarked by the sequence CUU UGA C with the stop codon (shown in red in Figure 10.42) being recognized by RF2. When there is sufficient RF2 in the cell, the UGA codon is efficiently recognized by RF2 that is already present and translation is terminated. When the level of RF2 in the cell is low, however, this termination event is slow and +1 frameshifting occurs, leading to the production of the complete RF2 protein. A +1 frameshift is stimulated because Leu-tRNA^{Leu} in the P site of the ribosome can pair with either of two different codons – most effectively with the CUU codon in the initial frame and less effectively with the near-cognate UUU codon in the second frame that is shifted in the +1 direction (Figure 10.42).

Following the frameshift, translation proceeds normally and RF2 is synthesized. Thus the establishment of competition between somewhat slow termination (on UGA) and the sampling of a new frame (and a different tRNA selection event on GAC) results in the regulated expression of the RF2 gene. Other documented examples of +1 frameshifting are typically stimulated by kinetically sluggish decoding, for example at codons decoded by rare tRNAs.

Another well-known example of frameshifting is the –1 event that is necessary for the production of the polymerase gene product (Pol) in many **retroviruses** and

Figure 10.40 The recoding of UAG stop codons for pyrrolysine insertion. (a) The chemical structure of the non-standard amino acid pyrrolysine. (b) Schematic showing how a specialized tRNA is directly aminoacylated with pyrrolysine by a pyrrolysine specific aminoacyl-tRNA synthetase (PylRS). EFTu is thought to load this tRNA like a standard elongator tRNA, albeit in competition with the release factor present in the cell.

Figure 10.41 Frameshifting disrupts the reading frame of protein synthesis. The initial reading frame of mRNA can be shifted either in the forward or backward direction, here by +1 (blue) or –1 (pink) nucleotides, which changes the sequence of the protein (shown in single letter code) synthesized from the mRNA.

Figure 10.42 Programmed +1 frameshifting controls the amount of RF2 produced in bacteria. The initial reading frame (frame 0) of the mRNA encoding RF2 contains a stop codon (red) that prematurely terminates synthesis. When RF2 levels in the cell are low, termination is inefficient. Given sufficient time, the aminoacyl-tRNA in the P site of the ribosome can pair with the mRNA codon in the +1 reading frame, thus allowing sampling of the A site by a cognate Asp-tRNA^{Asp}, resetting the frame to generate full-length RF2 gene product.

retrotransposons. During expression of the viral genome, 90% of ribosomes terminate translation at the end of the *gag* gene to produce the main structural protein of the viral capsid. However, a small percentage of ribosomes undergo a –1 frameshift that results in the synthesis of the Gag protein fused to the *pol* gene product (Gag-Pol). In the case of MuLV described above (which uses nonsense suppression to regulate read-through between the *gag* and *pol* genes), this –1 frameshifting provides a mechanism for the virus to modulate the relative amounts of the same two proteins. In this case, the –1 frameshifting is thought to occur by slippage of the mRNA in the ribosome, facilitated by a downstream RNA pseudoknot element that seems to perturb interactions between the ribosome and the mRNA (Figure 10.43). The mRNA sequence where recoding occurs is said to be 'slippery' because the tRNA in the P site of the ribosome can bind reasonably well to both the initial and the –1 reading frames, thus allowing the reading frame to slip backwards one nucleotide along the mRNA during the recoding event.

Figure 10.43 Programmed –1 frameshifting controls the relative amounts of Gag and Gag-Pol gene products for retroviruses. The viral Gag protein of Rous sarcoma virus is produced from reading frame 0 of the mRNA. A pseudoknot element in the mRNA downstream of the programmed frameshift site stimulates ~10% of ribosomes to undergo a –1 frameshift, resetting the frame to generate sufficient full-length Gag-Pol fusion protein.

10.12 ANTIBIOTICS THAT TARGET THE RIBOSOME

In the previous section we saw how biological systems can flex the genetic code, and the way in which mRNAs can contribute to this flexing. However, other more invasive ways of interrupting the 'standard' process of translation exist. Primary among these is the use of antibiotics to disrupt translation – or even prevent it altogether – as we discover in this section.

Many antibiotics target the ribosome or translation factors

Antibiotics are small molecules that kill bacteria or fungi or stop these organisms from growing. Antibiotics can be extremely effective drugs if they disrupt a critical cellular process in the bacteria or fungi without disrupting the equivalent process in mammals, if there is one. Mammalian cells, for example, do not have a cell wall like that of bacteria, and the antibiotic penicillin stops bacterial growth by blocking the synthesis of peptidoglycans required to build this cell wall. The antibiotic rifampicin selectively inhibits bacterial transcription by targeting a distinctive pocket on the bacterial RNA polymerase. Similarly, the seemingly subtle structural differences between the active sites of ribosomes from eukaryotes and bacteria can also be discriminated between by small molecules, and indeed antibiotics that disrupt many steps in protein synthesis have been identified (Figure 10.44).

Some antibiotics bind directly to the ribosome while others interact with translation factors to inhibit their function. Since antibiotics typically have a molecular weight of around 10^3 Da, 1000 times smaller than the size of a ribosome (2.5×10^6 Da) (see examples in Figure 10.45), antibiotics must target functionally critical regions of the ribosome to be effective. Erythromycin (and others in this class, such as telithromycin) sits neatly in the exit tunnel of the large subunit, blocking the growing peptide chain from entering the tunnel, and thus preventing translation from proceeding (Figure 10.46). Chloramphenicol binds more centrally in the active site of the large ribosomal subunit, blocking access to critical ribosomal nucleotide residues that are thought to be important for binding tRNAs and for the peptide bond forming and release reactions catalyzed by this active site. A number of antibiotics appear to inhibit more than one step in the translation cycle (e.g. the aminoglycoside paromomycin affects decoding and recycling), suggesting that functionally critical regions of the ribosome may be close together or that movement common to multiple steps might be targeted.

Antibiotic resistance mutations can reveal how antibiotics and the ribosome function

Bacteria can become resistant to antibiotics; consistent with this genetic mutations conferring resistance to antibiotics that disrupt translation are often found in components of the ribosome. By identifying these mutations, we can gain insights into how an antibiotic acts and also about how the ribosome functions more generally.

For example, streptomycin is an antibiotic that acts as a **miscoding agent** – it causes ribosomes to misread mRNA with the result that synthesized proteins have

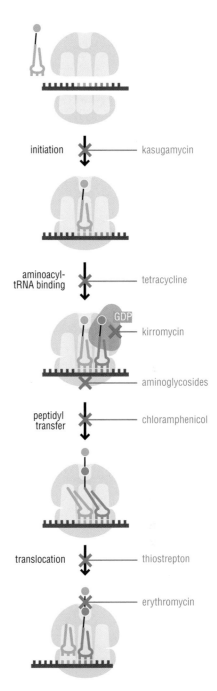

Figure 10.44 Distinct antibiotics inhibit the various steps in translation. Kasugamycin inhibits formation of the initiation complex. During elongation, tetracycline destabilizes binding of aminoacyl-tRNA to the A site and kirromycin prevents the dissociation of EFTu from the ribosome. Aminoglycosides (such as streptomycin and kanamycin) are miscoding agents, which stimulate the ribosome to accept near-cognate tRNAs. Chloramphenicol inhibits the peptidyl transferase reaction and thiostrepton blocks translocation of tRNAs. Erythromycin blocks the exit channel through which synthesized peptide chains leave the ribosome.

(a)

kanamycin

(b)

erythromycin

Figure 10.45 The chemical structures of antibiotics from two important classes. Two naturally occurring classes of antibiotics are the (a) aminoglycosides such as kanamycin and the (b) macrolides such as erythromycin.

Figure 10.46 Many antibiotics function by sterically obstructing the peptide exit tunnel of the ribosome. This cross-section view of the ribosome (cut through the interior of the large subunit) shows how the exit tunnel is blocked by tylosin (an erythromycin derivative). PDB 1K9M.

errors in their amino acid sequence. Streptomycin exerts it effect by binding to the small ribosomal subunit and, in the presence of near-cognate tRNAs, inducing conformational changes that are normally seen only in the presence of cognate tRNA. These rearrangements trick the ribosome into incorporating an incorrect amino acid. Streptomycin-resistant bacteria are routinely found to have mutations in the small subunit protein S12 that is located near the streptomycin binding pocket proximal to the decoding center on the interface of the subunit. In the absence of streptomycin, these resistant bacteria show **restrictive** protein synthesis, meaning they have unusually accurate and slow protein synthesis. Thus, the mutations in the S12 protein appear to counteract the effects of streptomycin by increasing the inherent fidelity of the ribosome. In the absence of streptomycin the resistant bacteria grow more slowly than their wild-type relatives.

Additional mutations can be isolated that increase the growth rates of these streptomycin-resistant bacteria (carrying S12 mutations). These additional mutations are typically found in two small subunit proteins, S4 and S5, that interact with one another on the back side of the small subunit. These mutations by themselves normally cause bacteria to have highly error-prone translation (referred to as a **ram** phenotype, for **ribosomal ambiguity**). The analysis of this series of variant ribosomes has over many years yielded some of the most important insights into how the ribosome balances the demands of high speed and fidelity during protein synthesis.

A general observation that has emerged from the analysis of antibiotic resistance is that mutations tend to be found in the ribosomal proteins rather than the rRNA. Given that the ribosome is fundamentally an RNA machine, why might this be the case? The answer seems to be that although bacteria have a single copy of the gene for each ribosomal protein, they have multiple copies of the rRNA genes. Therefore resistance is only observed when a reasonable fraction of the rRNA genes acquire the requisite mutation. An interesting exception to this general rule is seen in bacterial resistance to the macrolide antibiotics such as erythromycin. In this case, the most common mechanism for resistance is the acquisition by the bacteria of a methylase gene that specifically modifies the binding site for this class of antibiotic. Methylation of a particular adenosine (A2058) in the active site of the large subunit prevents binding of macrolide antibiotics and thus confers resistance. While resistance to erythromycin can also be achieved by direct mutation of the same nucleotide (A2058) in multiple 23S rRNA genes, the acquisition of the methylase gene through horizontal transfer from other bacteria is apparently an easier means of acquiring resistance.

New classes of antibiotics continue to be discovered

The most prolific antibiotic producers are organisms from the genus *Streptomyces* (Gram-positive bacteria) that use these secondary metabolites to challenge microbial competition in their environment. Over time, the targeted bacteria have evolved a variety of mechanisms to combat the antibiotics and these mutations determine the clinical efficacy of the antibiotic. Indeed, there is currently an acute need for the development of new classes of antibiotics for which resistance has not yet emerged. A recent success has been the identification of a new synthetic antibiotic class, the oxazolidinones, whose founding member, linezolid, was recently approved for use in humans for last line of defense anti-bacterial therapy. This antibiotic is structurally unrelated to naturally occurring antibiotics and so resistance has been relatively slow to develop (Figure 10.47). Mechanistically, it appears to target peptide bond formation by binding directly to the large subunit catalytic center much like chloramphenicol.

Ribotoxins and ribosome-inactivating proteins target a highly conserved loop in eukaryotic 28S rRNA and inactivate interactions with the elongation factors

Ribosome-inactivating proteins (RIPs) and ribotoxins are different cytotoxic enzymes that inhibit translation in eukaryotes by modifying a highly conserved and functionally important loop in the 28S rRNA. Ribosome-inactivating proteins are *N*-glycosidases most often produced by plants; the best known is the highly poisonous ricin from *Ricinus communis*. Ribotoxins are ribonucleases secreted by certain fungi with the most noted, alpha-sarcin, produced by *Aspergillus giganteus*. Both cytotoxins target a surface feature of the large subunit rRNA aptly named the sarcin/ricin loop. Ribotoxins cleave a specific phosphodiester bond and ribosome-inactivating proteins depurinate a specific nucleotide, inactivating interactions with the elongation factor eEF1A or eEF2 and leading to a complete shut-down of protein synthesis. Studies of these toxins and their interactions with the ribosome have led to important insights into the role of this highly conserved rRNA element in overall ribosome function.

Figure 10.47 The chemical structure of a new class of synthetic antibiotics (the oxazolidinones). Linezolid is a member of a new class of approved synthetic antibiotics that do not resemble any naturally occurring ones.

10.13 GLOBAL REGULATION OF INITIATION IN BACTERIA AND EUKARYOTES

Translation is globally regulated in response to amino acid starvation

Protein synthesis in the cell needs to be globally regulated in response to external conditions and stimuli. Protein synthesis rates should be high when nutrients are plentiful, to allow maximal cell growth, and low in times of nutrient deprivation. Both bacteria and eukaryotes respond generally to the levels of precursors in the cell; for translation, the precursors are the amino acids. Under normal circumstances, tRNAs are immediately aminoacylated once released from a round of elongation on the ribosome, and are bound in the cell by EFTu (or eEF1A) once they are loaded with the appropriate amino acid. Uncharged tRNA is an indicator of low levels of amino acid and a trigger for global shut-down of protein synthesis in both bacterial and eukaryotic cells, although their mechanisms of response are completely different.

In bacteria, the uncharged tRNA induces a stress response when it begins to compete with decreasing concentrations of ternary complex (EFTu-GTP-aa-tRNA) and binds to the ribosomal A site. This event immediately blocks overall translation as the A site of the cellular population of ribosomes is now occupied with unproductive pseudo-substrate. More importantly, however, this binding event on the ribosome recruits a protein called relA to the complex. RelA then synthesizes high (mM) concentrations of (p)ppGpp (a pentaphosphate guanine nucleotide known as 'magic spot') from GTP/GDP and ATP precursors. (p)ppGpp itself brings about profound effects on transcription, which include a decrease in quantity of stable RNAs (such as rRNA and tRNA) and the induction of stress response factors that go about the task of replenishing the cell's declining resources (for example, inducing the expression of genes involved in amino acid biosynthesis and transport). This process is broadly referred to as the 'stringent response' and is the best understood example in bacteria of a global translational response (Figure 10.48a).

The response in eukaryotes is quite distinct. In these organisms, the uncharged tRNA binds to a protein Gcn2 (in yeast), and more specifically to a domain in this protein related to the histidyl-tRNA synthetase (HisRS). Gcn2 appears to bind uncharged tRNAs with higher affinity than the corresponding charged tRNAs. Notably, this protein also has a kinase domain that is activated (on binding uncharged tRNA) to phosphorylate the translational initiation factor eIF2. We remember from Section 10.8 that eIF2 is a GTPase that is responsible for binding initiator $tRNA_i^{Met}$ and guiding it into the P site during the initiation process. Phosphorylation of eIF2 increases its affinity for GEF, eIF2B, to the point where active eIF2 is effectively depleted from the cell (Figure 10.48b).

The result of this phosphorylation event is an overall shut-down of protein synthesis. However, as we saw in bacteria, the cells also respond to this stress by inducing the expression of certain genes that are important for restoring cellular resources. While some of this induction takes place at the level of transcription, we will see in Section 10.14 how translation of the protein Gcn4 (a transcriptional activator of many amino acid biosynthetic genes) is translationally up-regulated during this global translational shut-down.

Much global translational control is achieved through modulation of eIF2 and its GEF partner eIF2B

The example above shows how a general translation factor (eIF2) can be directly modified to globally affect the efficiency of cellular translation. This type of mechanism is broadly utilized in eukaryotic cells, with proteins involved in multiple different steps in the initiation process serving as targets of regulation.

Figure 10.48 Amino acid starvation triggers distinct responses in bacteria and eukaryotes for global translational shutdown. (a) In bacteria, when amino acids are depleted, uncharged tRNAs become increasingly abundant and eventually compete for binding to the ribosomal A site. RelA protein recognizes this ribosomal complex, in turn stimulating the production of (p)ppGpp by RelA, and activation of the stringent response pathway. (b) In eukaryotes, amino acid depletion also leads to increased amounts of uncharged tRNA. In this case, these tRNAs are bound by a protein factor, Gcn2, thus activating a kinase domain on Gcn2 which targets eIF2. Once phosphorylated, eIF2 binds with high affinity to its guanine-nucleotide exchange factor (GEF), eIF2B, thus effectively depleting cellular eIF2 and shutting down overall translation.

Phosphorylation of eIF2 is a common means of affecting the equilibrium between free eIF2 and eIF2B:eIF2 as the specific phosphorylation of serine 51 on the alpha subunit of eIF2 (eIF2alpha) results in the constitutive binding of eIF2–GDP to eIF2B. This binding interaction sequesters eIF2 in its GDP-bound state, thus preventing the protein from participating in initiation (Figure 10.49). In this role, phosphorylated eIF2 acts as a competitive inhibitor of the GEF function of eIF2B.

At least three different kinases that phosphorylate serine 51 of eIF2alpha in response to stress have been identified in a variety of systems. For example, PKR encodes a mammalian double-stranded RNA-activated kinase that is important in the cellular response to viral invasion; GCN2, as mentioned above, is a member of a kinase family that is responsive to amino acid and glucose starvation, and high salinity; and the PERK kinase family responds to unfolded proteins in the endoplasmic reticulum. Each of these kinases, activated by diverse stresses, phosphorylates serine 51 of eIF2alpha and affects global translation. GCN2 homologs have been identified in yeast, flies, worms, and mammals, suggesting that this protein may be the founding member of this family of related kinases.

The other partner in the functionally important equilibrium with eIF2 is the GEF protein eIF2B. This protein can also be phosphorylated under certain stress conditions, a modification that typically prevents it from interacting with eIF2 and thus from promoting GDP/GTP exchange. The effect of this modification is again global translational down-regulation.

mRNA-associated initiation factors are also commonly targeted to mediate translational control

Generally, mRNAs are at a higher concentration than the components of the translational machinery and therefore mRNAs must compete for the available initiation factors if they are to be translated. We learned in Section 10.8 that formation of the initiation complex in eukaryotic cells begins with a closed loop mRNA complex where eIF4E is bound to the cap, and PABP to the poly(A) tail; these two proteins are linked by interactions with eIF4G (see Figure 10.26). Each of the components in this complex offers a potential target for the regulation of initiation.

How do cells typically regulate this group of factors? First, the transcription of eIF4E is regulated in response to external stimuli. Second, eIF4E is phosphorylated in response to external stimuli. This phosphorylation increases the affinity of eIF4E for the 5′ cap of mRNA, leading to globally increased levels of initiation. Finally, eIF4E can be sequestered by binding proteins known as **4E-BPs** (eIF4E-Binding Proteins), which effectively sequester eIF4E from the active population (decreasing overall translation) (Figure 10.50). Not surprisingly, the binding of eIF4E to

(a)

(b)

Figure 10.49 Phosphorylation status of eIF2 determines the amount that is free in solution. Under normal circumstances, eIF2B acts as a GEF for eIF2, stimulating eIF2 to exchange GDP for GTP (a). Phosphorylated eIF2 binds more tightly to eIF2B, thus preventing GTP/GDP exchange and the release of eIF2 from the complex (b).

Figure 10.50 eIF4E availability is regulated by 4E-BPs. eIF4E can bind to both 4E-BPs (left) and eIF4G (right). The phosphorylation of either eIF4E or the 4E-BPs (center) prevents them from interacting with one another as effectively, thus increasing the overall effective concentration of eIF4E. In addition, phosphorylated eIF4E has higher affinity for the 5′ cap structure.

4E-BPs can be modulated by post-translational modifications: phosphorylation of the 4E-BPs can occur in response to different stimuli, decreasing affinity of eIF4E for the 4E-BP and leading to a global increase in translational initiation.

A number of distinct 4E-BPs have been identified in mammalian cells, suggesting that this is a diverse family of proteins that enables cells to increase or decrease their protein synthesis rates in response to a number of different stimuli.

10.14 REGULATION OF INITIATION VIA THE 5′ UTR IN BACTERIA AND EUKARYOTES

In the previous section, we considered how organisms regulate translation at a general, global level. We now explore some examples of mRNA-specific regulation conferred by particular sequences in the mRNA.

Gene-specific translational regulation depends on sequence features in the mRNA

The rate at which any particular protein in the cell is synthesized is regulated according to the needs of a cell. The rate of synthesis of a protein depends on many factors – the rate of transcription of the gene, the stability of the mRNA and the rate of its translation, to name the most obvious. Arguably, the most rapid changes in the rates of protein synthesis can be achieved by regulating the rate of translation of an mRNA population that is already present in the cell; in this case, there is no requirement to produce new mRNA to increase the downstream step of protein synthesis. Indeed, cells often respond rapidly to environmental stimuli by controlling the rates of translation initiation and, more occasionally, elongation.

As we see below, the alternative structural conformations that mRNAs can assume under differing conditions can be critical for the regulation of gene expression. In bacteria, the best-known examples of the regulation of initiation involve direct obstruction of the Shine–Dalgarno sequence in the 5′ UTR of an mRNA, preventing formation of competent initiation complex. Initiation in eukaryotes involves both the 5′ and 3′ UTRs of the mRNA, thus allowing for more diverse mechanisms of control.

Shine–Dalgarno sequestration is a common mechanism for the regulation of translation in bacteria

We saw in Section 10.7 that a critical component in the initiation of translation in bacteria is the presence and composition of the Shine–Dalgarno sequence upstream of the AUG start site in an mRNA. If the Shine–Dalgarno sequence is not accessible to form interactions with the small subunit of the ribosome then initiation cannot take place. Indeed, bacteria commonly sequester the Shine–Dalgarno sequence through various strategies including the binding of small regulatory RNAs, RNA binding proteins and small molecules. We explore each of these strategies below.

Likely the simplest mechanism for sequestering the Shine–Dalgarno sequence depends on the mRNA 5′ leader sequence itself and its three-dimensional structure. One such example is seen in *Listeria monocytogenes*, a bacterium that infects warm-blooded animals. The expression of PrfA, a key transcriptional activator of genes required for infection by this organism, seems to be regulated by temperature-dependent changes of the structure of the 5′ UTR. At low temperatures outside host animals, the Shine–Dalgarno sequence of the *prfA* mRNA is locked in an RNA structure that renders it inaccessible to the ribosome, thus blocking translation. Once a bacterium enters a host, temperature-dependent structural reorganization of this region of the mRNA makes the Shine–Dalgarno sequence accessible to the ribosome, and translation of the *pfrA* mRNA can occur.

There are a growing number of examples where the expression of biosynthetic genes is subject to autoregulation, that is, where the end product of the pathway directly regulates its own synthesis. For such examples in bacteria, the small molecule product often binds to the mRNA sequence upstream of the open reading frame and changes the structure of the mRNA. Such elements are generally referred to as **riboswitches**. As we discussed in Chapter 8 the binding of these small regulatory molecules to some transcripts can lead to transcription termination. For other mRNAs, the binding of the small molecule can directly block the initiation of translation.

For example, the conserved 5′ regions of the mRNAs of the *thiM* and *thiC* thiamine biosynthetic genes in *E. coli* fold into a structure that binds thiamine pyrophosphate (TPP), an intermediate in the synthesis of thiamine. When thiamine pyrophosphate is bound to the mRNA, the Shine–Dalgarno region is sequestered in the tertiary structure of the mRNA and translation is inhibited. By contrast, when thiamine pyrophosphate levels are low, the free mRNA adopts a structure in which the Shine–Dalgarno sequence is no longer sequestered and expression of the biosynthetic genes is induced (Figure 10.51). It is plausible that metabolite-sensing systems based simply on structural changes in RNA are remnants of ancient regulatory systems that evolved in the RNA world, before the evolution of regulatory proteins.

A class of small, regulatory RNAs (generally 100–150 nts in length) in bacteria also have been shown to either promote or limit translation by directly base-pairing with the 5′ UTR such that the Shine–Dalgarno sequence is either sequestered or revealed. The establishment of these base-pairing interactions is often dependent on the RNA binding protein Hfq, which is synthesized in high levels in many of these organisms. It is also believed that base-pairing interactions between small RNAs and the initial codons of the gene (up to five codons) can act to inhibit translation initiation.

(a)

thiamine pyrophosphate

SD

5′

3′

(b)

5′

SD

3′

Figure 10.51 Translation of thiamine biosynthetic genes is regulated by the binding of thiamine pyrophosphate (TPP) to the 5′ UTR of the mRNA. (a) In the presence of TPP, the Shine–Dalgarno sequence (SD) in the mRNA of thiamine biosynthetic genes is sequestered through interactions with portions of the 5′ UTR, thus preventing ribosome binding and translation. (b) In the absence of TPP, the SD is revealed through conformational rearrangements in the 5′ UTR of the mRNA, thus increasing translation of the downstream gene.

ThrRS mRNA

tRNA^Thr

Figure 10.52 The threonine-tRNA synthetase (ThrRS) mRNA and tRNA^Thr share identity mRNA. These similarities allow ThrRS to bind either species, depending on the amount of deacylated tRNA^Thr available in the cell. Binding to the operator region by ThrRS blocks access to the Shine–Dalgarno sequence, and thus translation.

Another example of regulated Shine–Dalgarno sequestration is seen in the control of the expression of the gene for threonine-tRNA synthetase (*thrS*) in *E. coli*. In this case, sequestration is mediated by a protein rather than by a small molecule or by temperature-dependent changes of RNA structure. Threonine-tRNA synthetase is required to aminoacylate tRNA^Thr, and the demand for the synthetase is particularly high when protein synthesis rates are high. As well as binding unacylated tRNA^Thr, the synthetase can also bind to its own mRNA (through related structural features), sterically blocking the Shine–Dalgarno sequence and thus preventing the translation of more synthetase. Thus *thrS* mRNA and unacylated tRNA^Thr directly compete with one another for binding to threonine-tRNA synthetase (Figure 10.52), depending on the overall rates of protein synthesis.

Initiation in eukaryotes can be regulated by blocking access to the ribosome-binding site

While the sequestration of 5′ UTRs seems to be the rule in bacterial translational control, this mode of regulation has been less commonly observed in eukaryotes. That said, there are examples in eukaryotes where protein binding in the 5′ UTR does regulate translation.

Iron is an essential nutrient whose levels in eukaryotic cells must be tightly controlled as high levels can be toxic. Inside cells, iron is bound to a storage protein known as ferritin; the higher the concentration of iron in a cell, the more ferritin the cell requires. In the 5′ UTR of ferritin mRNA there are several stem–loop structures called Iron Response Elements (IREs), which bind to proteins known as the Iron-Regulatory Proteins. When iron is scarce in the cell, the IRPs binds to the IRE and prevent the initiation of translation by preventing ribosome access to the AUG start site, probably by blocking scanning (Figure 10.53). When iron is plentiful, the IRPs are prevented from binding to the IREs, which in turn allows for efficient initiation of ferritin translation. We recall that the same IRP proteins also regulate the expression of a different gene, the tranferrin receptor, modulating mRNA stability through IREs located in the 3′ UTR (see section 9.9). This example again illustrates how feedback control can be used to exquisitely tune the amount of a protein produced under a specific set of conditions.

Upstream open reading frames regulate the synthesis of Gcn4 protein in yeast

One of the best-characterized examples of translational control in eukaryotes is the sophisticated regulated expression of the Gcn4 protein in yeast, a transcriptional

Figure 10.53 Initiation of translation of ferritin mRNA is regulated through the 5′ UTR. (a) When iron is scarce in the cell, ferritin translation initiation is prevented by the binding of IRP½ to iron-response elements (IREs) in the 5′ UTR of the mRNA. (b) When iron is plentiful, the IRPs (IRP½) no longer bind the IREs and ferritin translation initiation proceeds.

(a)   **(b)**

activator of many amino acid biosynthetic genes. As we discussed in Section 10.13, when amino acid levels in eukaryotic cells are low, the yeast eIF2 kinase Gcn2 is activated, eIF2alpha is phosphorylated, and overall protein synthesis in the cell is reduced. However, the rate of translation of the gene *GCN4* is specifically increased under these conditions.

This upregulation occurs by an unusual mechanism that depends on the presence of multiple uORFs (uORF1–4) found in the transcript upstream of the actual Gcn4 coding region. Under all conditions – whether the cell is starved of amino acids or not – uORF1 is effectively translated. What happens beyond uORF1, however, differs depending on the cellular levels of eIF2 available. Central to all possible outcomes is that the uORF1 stop codon and the sequence context in which it is located favors the resumption of scanning following termination, rather than full ribosome drop-off and recycling – that is, the 40S subunit remains associated with the mRNA and travels further along the mRNA to encounter downstream sequence elements.

Under conditions of amino acid abundance (non-starved), eIF2alpha is largely unphosphorylated and the ternary complex (eIF2 : Met-tRNA$^{:Met}$:GTP) is maximally available. Hence, virtually all 40S subunits scanning downstream from uORF1 rebind initiator tRNA before reaching uORF 2, 3, or 4, and reinitiate translation at one of their start codons. Unlike uORF1, however, the sequence context of these uORFs does not favor the reinitiation of scanning, and ribosome dissociation happens efficiently. As a result, the 40S subunit rarely makes it to the actual *GCN4* open reading frame, and overall expression of the Gcn4 protein is low (Figure 10.54).

By contrast, under conditions of amino acid depletion (starved), the concentration of the eIF2-GTP-Met-tRNA$_i^{Met}$ ternary complex is low. As a consequence, a reasonably large fraction of the 40S ribosomes scanning downstream from uORF1 fails to rebind the ternary complex until reaching the interval between uORF4 and the *GCN4* open reading frame, and as a result these ribosomes only reinitiate translation after this point. As a result, Gcn4 protein is translated at relatively high levels. This model for GCN4 regulation is shown in Figure 10.54. In Experimental approach 10.3, we see how polysome profiling has revealed new insights into this complex form of translational control.

We see that globally decreased levels of protein synthesis can lead to increased expression of certain genes found in an appropriate mRNA context. In this case, 'mRNA context' includes the series of uORFs in the 5′ UTR as well as differing stop codon sequence contexts that affect the balance between ribosome dissociation (recycling) and reinitiation. Similar examples of translational control that depend on uORFs, although a relatively rare form of regulation, have been observed in mammals, maize, and certain fungi.

Figure 10.54 Gcn4 synthesis is regulated by upstream open reading frames and the availability of eIF2. Translation of the polycistronic GCN4 mRNA begins with the translation of uORF1 (left). When amino acids are abundant in the cell (top pathway), eIF2 is largely unphosphorylated and as a result eIF2 activity is high. Efficient reinitiation following the translation of uORF1 leads to the translation of uORF4 (and possibly also uORF2 and uORF3) (top pathway) and subsequent full recycling (ribosome drop-off). When amino acids are not abundant (bottom pathway), eIF2 is largely phosphorylated and is therefore less available to participate in initiation. In this case, reinitiation at uORF4 is inefficient and reinitiation at the *GCN4* open reading frame becomes more likely.

10.15 REGULATION OF TRANSLATION VIA THE 3' UTR IN EUKARYOTES

In Section 10.14, we discussed ways in which translation of specific genes in bacteria and eukaryotes can be regulated by structural changes or by interactions with the 5' UTR of an mRNA that disrupt translation initiation. While direct obstruction of the Shine–Dalgarno sequence is probably the most likely means of regulating translation in bacteria, the 5' UTRs of eukaryotic genes are typically shorter than the 3' UTR and are less commonly involved in such regulatory processes. Instead, the 3' UTRs of eukaryotic genes can be very large and are the driving force for the regulation of a number of developmentally important genes (Figure 10.55).

The basic understanding for such regulation is that elements within the 3' UTR recruit different cellular factors that impact downstream events, for example disrupting formation of the closed loop mRNA initiation complex and thus inhibiting translation initiation. Not surprisingly, these mechanisms target many of the same steps as we saw for global translational repression in Section 10.13; in these cases, however, the effects are gene-specific as they are mediated through factors specifically recruited to the 3' UTR.

Polyadenylation levels can be regulated by 3' UTR binding proteins

One mechanistically well understood example of 3' UTR-mediated translational control is seen during development in the *Xenopus* oocyte. The oocyte contains maternally derived mRNAs that are not immediately translated into protein – these mRNAs are said to be dormant or translationally repressed. Dormant mRNAs have short poly(A) tails, which only during maturation are extended to allow the mRNAs to be translated into protein.

Figure 10.55 3' UTRs can be the driving force in regulating gene expression in certain developmental programs.
(a) Gene fusions carrying a strong constitutive promoter, the gene encoding the green fluorescent protein (GFP) and the 3' UTR of the indicated *C. elegans* genes. (b) When inserted into the worm genome, these fusions show diverse expression patterns in the reproductive tract of the organism. Because the 3' UTR is the only portion of the reporter that differs, the different expression patterns are believed to reflect the inherent properties of this element in regulating translation.
Taken from C. Merritt, G. Seydoux.

How does this repression and subsequent activation of translation occur? The 3' UTRs of the dormant mRNAs have sequences referred to as Cytoplasmic Polyadenylation Elements (CPEs), which play a key role in controlling their expression. When translation is repressed, the CPEs are bound by a protein known as CPEB. CPEB interacts with a protein called Maskin, which interacts in turn with eIF4E (Figure 10.56, step 1). The interaction of eIF4E with Maskin directly precludes the binding of eIF4G to eIF4E, thus 'masking' the expression of an mRNA by preventing the formation of the closed circle initiation complex.

The region of Maskin that directly interacts with eIF4E contains a conserved sequence element YXXXLP (where Y corresponds to tyrosine, L to leucine, P to proline, and X to any amino acid), which mimics a related element in eIF4G. Many cellular factors that regulate translation initiation use this same trick to compete with natural interactions in the cell between eIF4E and eIF4G; these proteins have been described in 10.13 and are broadly referred to as 4E-BPs. Maskin is a type of 4E-BP with additional specificity for the protein CPEB that allows for gene-specific regulation through 3' UTR binding.

The activation of the dormant mRNAs requires maskin to be displaced from eIF4E. The first step in this process is the phosphorylation of CPEB by a kinase, Eg2 (Figure 10.56, step 2), which increases the affinity of CPEB for a protein known as CPSF (cytoplasmic polyadenylation specificity factor) (Figure 10.56, step 3). Once recruited by CPEB, CPSF binds to the standard AAUAAA polyadenylation signal and recruits poly(A) polymerase to the mRNA. The poly(A) polymerase then extends the poly(A) tail of the mRNA. Once extended (Figure 10.56, step 4), PABP (the poly(A) binding protein) can bind to the 3' tail of the mRNA (Figure 10.56, step 5). Having associated with the poly(A) tail, PABP recruits eIF4G, which then displaces Maskin from eIF4E. Once eIF4E and eIF4G interact, the initiation complex is formed and translation of the activated mRNAs can proceed (Figure 10.56, step 6).

This rather complicated sequence of events allows mRNAs that are held dormant by the CPE sequence element to be activated by a single event – the activation of the kinase Eg2. This method of translation control is probably more generally utilized; related proteins appear to be important in specifying critical functions in developing neurons.

RNAs or proteins bound to the 3' UTR can regulate translation at multiple steps following polyadenylation

We see above how the cytoplasmic polyadenylation element in the 3' UTR of a dormant mRNA in *Xenopus* oocytes regulates translation of the mRNA before the mRNA is fully polyadenylated. The expression of many genes in eukaryotic systems can be controlled by other 3' UTR regulatory sequences, and often the regulation is applied to different stages in the initiation (or perhaps elongation) process. The

Figure 10.56 Translational regulation of *Xenopus* oocyte dormant mRNAs through RNA elements found in the 3' UTR.
(1) Dormant mRNAs contain short poly(A) tails and cytoplasmic polyadenylation element sequences (CPEs). Translation of these mRNAs does not proceed because eIF4E is blocked from binding to eIF4G by Maskin. (2) Following the phosphorylation of CPEB by the kinase Eg2, (3) CPEB recruits CPSF to the hexanucleotide sequence AAUAAA in the mRNA. (4) CPSF in turn recruits poly(A) polymerase (PAP) to the mR NA and the poly(A) tail of the mRNA is extended. (5) When PABP (the poly(A) binding protein) binds to the newly extended poly(A) tail, eIF4G can bind and (6) is ultimately able to displace maskin from eIF4E such that translation can proceed.

basic theme is the same as for the CPEB story described above. Sequence elements in the 3′ UTR recruit a variety of different proteins that impinge on different steps in the translational cycle.

In one example, expression of the *lox* mRNA that encodes a lipoxygenase in erythroid precursor cells is controlled via specific sequences (the differentiation control element or DICE) in the 3′ UTR that bind the hnRNP K and E1/E2 proteins. These specific interactions in the 3′ UTR prevent the large ribosomal subunit from joining the small subunit during the initiation process (Figure 10.57a). How such interactions block this step, which is normally catalyzed by the GTPase eIF5B, is not well understood.

In another example, the expression of a protein called Nanos (which is involved in the development of *Drosophila* embryos) is regulated by a 3′ UTR element known as a TCE. In the case of Nanos the TCE is bound by a protein known as Smaug, which in turn interacts with Cup, an eIF4E-binding protein (4E-BP). This complex of proteins represses translation by competing with the normal eIF4G interaction (Figure 10.57b). This example represents a gene-specific version of what we described in Section 10.13 as a mechanism for global translational shut-down.

Finally, the microRNA (miRNA) class of small RNAs in eukaryotes is proposed to regulate translation (as well as mRNA stability). These small RNAs are known to act through binding interactions with partially complementary sequences in the 3′ UTRs of genes. The pairing interactions between the miRNAs and their targets have been well characterized and typically consist of perfectly complementary pairing between positions 2–7 or 2–8 of the miRNA (the 'seed') and the 3′ UTR, with more variability in the extent of pairing with the 3′ portion of the miRNA (Figure 10.58). Importantly, miRNAs typically do not pair with nucleotides in the region directly downstream of the seed from positions 9–11 or so. This is important because perfect complementarity in this region

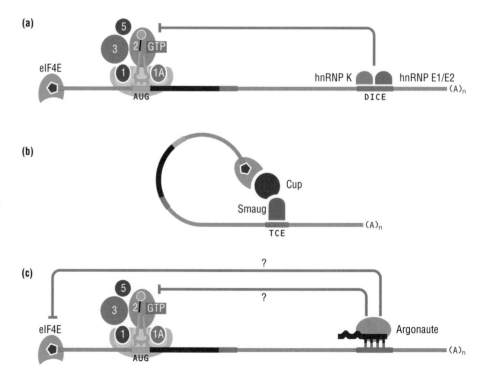

Figure 10.57 Translation can be regulated by the binding of proteins and small RNAs to the 3′ UTR of genes. (a) Expression of the *lox* mRNA in erythroid precursor cells is blocked through a binding element in the 3′ UTR (DICE) that recruits hnRNP K and E1/E2 to repress the subunit joining step of initiation.
(b) Expression of the Nanos protein in *Drosophila* embryos is blocked by the binding of Smaug to a translational control element (TCE) in the 3′ UTR of the mRNA. TCE-bound Smaug then interacts with Cup, a 4E binding protein (4E-BP), which directly binds eIF4E and prevents productive interactions with eIF4G. (c) Expression of *lin14* and *lin28* mRNA during *C. elegans* development is blocked during certain larval stages through interactions with the miRNA lin-4. The binding of the lin-4-miRISC complex to the 3′ UTR of the gene somehow blocks translation, through a mechanism that has not yet been defined.

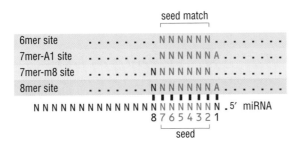

Figure 10.58 Schematic showing proposed interactions between miRNAs and their mRNA target. Base-pairing interactions are thought to be critical between nucleotides 1–8 of the miRNA (the 'seed') and the mRNA, while interactions with the 3' end of the miRNA are more variable.

typically causes the Argonaute protein in the effector complex (miRISC) to cleave the mRNA target in an siRNA-like response. Experimental approach 10.4 describes the initial discovery of miRNAs through genetic analysis in *C. elegans*.

The miRNAs are thought to repress gene expression either by directly blocking translation initiation or elongation (shown in Figure 10.57c), or perhaps by triggering RNA or protein degradation.

Although a number of other factors in the cell appear to be important for this type of regulation, there is no clear mechanism yet to describe how these 3' UTR tethered complexes interfere with a specific molecular event.

> → We discuss the generation of miRNAs and their potential role in mRNA degradation in Section 9.9.

mRNA context downstream of the stop codon (in the 3' UTR) is important in determining stop codon authenticity

As we learned in Section 10.10, normal stop codons are specifically recognized by release factors in bacteria and eukaryotes, which release the growing polypeptide chain and allow for subsequent ribosomal recycling. In certain situations, **premature termination codons** (**PTC**) are found within an mRNA transcript such that a termination codon is encountered by the translation machinery 5' to the normal stop codon; these PTCs may be the result of an error in transcription or splicing, or the product of normal cellular events such as programmed DNA rearrangements in the immune system (where reading frame is lost in two of three events). Most of these PTCs are recognized in eukaryotic cells and the mRNAs themselves are targeted for degradation by a process referred to as **nonsense mediated decay** (**NMD**).

The logic behind this regulation is that prematurely truncated proteins often have dominant negative effects on the cell, and so it is worth targeting mRNAs coding for such proteins for removal from the system. As might be anticipated, nonsense stop codons must be recognized by the ribosome, but be distinguished from authentic stop codons in some way. So how can this recognition occur? In a beautiful example of the integration of nuclear and cytoplasmic events, the splicing machinery in the nucleus deposits a protein complex or 'mark' at exon–exon junctions on the completion of splicing. This mark is then exported with the processed mRNA transcript into the cytoplasm. The complex is referred to as the exon junction complex (EJC), and contains a set of proteins that are largely conserved in the eukaryotic kingdom.

How is this mark recognized? It turns out that stop codons in eukaryotic genes are typically found in the final exon, and so EJCs are typically not found downstream (3') of authentic stop codons. By contrast, PTCs will very often be found in upstream exons, and thus EJCs will be bound to the mRNA downstream of them. NMD is thought to be triggered by elongating ribosomes that encounter a stop codon upstream of a bound EJC; recognition of the EJC is likely mediated through

10.4 EXPERIMENTAL APPROACH

Analysis of the lin4–lin14 *genetic interaction and the development of current miRNA target prediction approaches*

miRNAs are ubiquitously involved in the regulation of gene expression in plants and animals. The first miRNA was identified through genetic analyses in the experimentally tractable worm *C. elegans*. Genetic screens identified a number of worms with a class of mutation that resulted in a particular type of recognizable developmental defect; with these so called lineage (lin) mutations, specific cell fates occur either earlier or later than they normally would. Several of the mutations that caused the developmental defect were mapped in the genome to two distinct genes, *lin–4* and *lin–14*, where other mutations resulted in similar defects.

Further analysis of the *lin–4* and *lin–14* genes led to the discovery of miRNAs and the first ideas about how they might work. Understanding the *lin–14* mutations proved to be relatively straightforward; the *lin–14* gene encodes a protein whose synthesis varies during development with high levels seen in larval stage 1 (L1) and lower levels throughout subsequent stages (L2/L3). The mutations causing lineage defects yielded unregulated expression patterns, with inappropriately high levels of protein detected in the larvae at post-L1 stages. Gary Ruvkun's group found that most of the interesting mutations in *lin–14* were located in the 3′ UTR, immediately suggesting the presence of regulatory elements that are responsible for down-regulating *lin–14* expression under normal circumstances.

Interpretation of the *lin–4* mutations was less straightforward but more exciting. Like the *lin–14* variant worms, the *lin–4* variant worms exhibited a loss of regulation of LIN-14 protein levels, failing to down-regulate expression in the post-L1 stages of development. The observed phenotype argued for the existence of a *lin–4* gene product, assumed to be a protein acting as a repressor of LIN-14 protein synthesis. However, when Victor Ambros' group identified the *lin–4* gene they discovered that it did not encode a protein. Instead, they showed that the gene specifies the production of an RNA transcript that is ultimately processed into a small RNA (around 22 nucleotides in length) now classified as a miRNA (Figure 1).

Through what mechanism then does the LIN-4 RNA product regulate the expression of LIN-14 protein? The levels of the *lin–14* mRNA were shown to be roughly equivalent in larval stages L1 and L2, whereas the amount of LIN-14 protein present during these same stages was very different (Figure 2). Thus, consistent with the genetics, an increase in *lin–4* RNA correlates with the down-regulation of LIN-14 protein levels in larval stage L2,

Figure 1 Identification of two small *lin-4* transcripts. Northern blot of total RNA from a *C. elegans* wild-type strain, a *lin-4* mutant strain, and from a strain with a rescuing plasmid carrying the *lin-4* wild-type gene. Two RNA products are seen, a less abundant longer product (Lin-4L, 61 nucleotides in length) and a more abundant shorter product (lin-4S, ~22 nucleotides long); we now know these correspond to pre-miRNA and fully processed miRNA, respectively. Reproduced from Lee *et al.*, *Cell* 1993;**75**:843–854.

Figure 2 Analysis of mRNA and protein levels in different larval stages during *C. elegans* development. (a) Western analysis of Lin-14 levels (and control myosin) in synchronized preparations of wild-type worms. Lin-14 is down-regulated in wild-type embryos at late stages, but this regulation is lost in the mutant worms. (b) RNAase protection analysis of *lin-14* mRNA levels (and control myosin). In this analysis, the *lin-14* mRNA levels are essentially the same in the early and late extracts. L1 refers to early stage larva and L2/L3 to late stage.
Reproduced from Wightman *et al.*, *Cell* 1993;**75**:855–862.

(a)

(b)

Figure 3 **Conserved sequences in the** *lin-14* **3′ UTR are complementary to the** *lin-4* **RNA.** (a) Representation of regions in the *C. elegans* 3′ UTR that are conserved in between *C. elegans* and *C. briggsae* (blue). Shown in pink (overlaid on gray) are conserved regions that are complementary to the *lin-4* RNA. The mutant breakpoints for several lineage mutants and polyadenylation sites are indicated. (b) Predicted pairings between *lin-4* RNA and the proposed *lin-4* target sites in lin14 3′ UTR. Reproduced from Wightman *et al.*, *Cell* 1993;**75**:855–862.

suggesting that *lin-4* regulates (and, in fact, represses) LIN-14 synthesis. Together, these observations led to the proposal that the mechanism of regulation by miRNAs is post-transcriptional, potentially affecting mRNA stability, some step in translation or protein stability.

Perhaps the most informative insight coming from the early analyses of these genes (and noted by both the Ambros and Ruvkun groups) was that the *lin-4* RNA product could form direct pairing interactions with as many as seven potential complementary regions within the 3′ UTR of the *lin-14* mRNA (Figure 3). These targeted regions in the *lin–14* mRNA are evolutionarily conserved, and fall within the region that is deleted in the *lin–14* mutants. Collectively, these observations suggested that the *lin–4* RNA exerts its effect by specifically binding to complementary regions of the *lin–14* 3′ UTR. These initial predictions about how targets are identified by the miRNAs turn out to be broadly accurate: miRNAs indeed recognize their mRNA targets through direct base-pairing interactions.

These early genetic discoveries opened a new and exciting area of biology that has been well served by the existence of fully sequenced genomes. When a second small RNA (*let–7*) with similar genetic and biochemical properties was discovered, and was seen to be conserved from *C.elegans* to humans, researchers began to wonder if these small regulatory RNAs were not more widespread throughout nature. The flood-gates then opened as the sequencing of small RNAs isolated from worms, flies, and human cells revealed the existence of hundreds of miRNA genes. High-throughput sequencing approaches were used to explore the numbers and diversity of this class of biological macromolecule in multiple organisms – current numbers confidently identify more

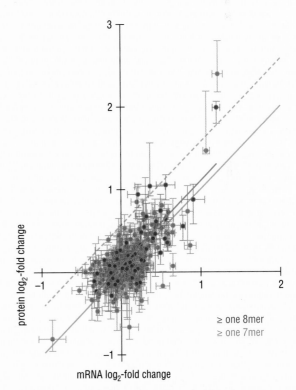

Figure 4 **Comparison of protein and mRNA changes (for a subset of genes) accompanying the loss of a specific miRNA (miR-223) from mouse neutrophils.** Protein and mRNA changes (on the Y and X axis, respectively) for quantitated proteins where messages carrying at least one 8-base long site complementary to miR-223 in the 3′-UTR site are shown in blue, and those with at least one 7-base long site are shown in orange. The least-squares best fit to the 8-base long data is shown as a gray reference line; both lines have a slope of 1.0. Vertical and horizontal error bars indicate variability in the protein and RNA measurements.

Reproduced from Baek *et al.*, *Nature* 2008;**455**:64–71.

than 100 miRNA genes in *C.elegans*, 140 in D. melanogaster, and 400 in humans.

The next task has been to identify the biological targets of these many small RNAs. Two basic approaches have been taken to identify these targets. First, bioinformatic approaches have been used to identify potential mRNA pairing partners for miRNAs, with particular emphasis on those potential targets that are conserved between organisms. Second, such data sets have been compared with the output from high-throughput assays of RNA and protein levels in the cell – microarray and proteomic analyses – to evaluate the impact of miRNAs on these levels. These approaches have exhibited a marked overlap in potential mRNA target sets – RNA and protein levels are both affected by a given miRNA – (Figure 4), and together have been used to develop target prediction software that can identify many of the significant targets of a given miRNA. As each conserved miRNA has evolutionarily conserved interactions with hundreds of mRNAs in the cell, it is clear that miRNAs play an important and broad role in metazoan

biology. While there are significant gaps in the molecular understanding of how miRNAs regulate gene expression, the target predictions place the field in a position to identify the most interesting biology to pursue.

Baek D, Villén J, Shin C, *et al*. The impact of microRNAs on protein output. 2008;**455**:64–71.

Lee RC, Feinbaum RL, Ambros V. The *C. elegans* heterochronic gene encodes small RNAs with antisense complementarity to *lin–14*. *Cell* 1993;**75**:843–854.

Lewis BP, Shih IH, Jones-Rhoades MW, Bartel DP, Burge CB. Prediction of mammalian microRNA targets. *Cell* 2003;**115**:787–798.

Lim LP, Lau NC, Garrett-Engele P, *et al*. Microarray analysis shows that some microRNAs downregulate large numbers of target mRNAs. *Nature* 2005;**433**:769–773.

Selbach M, Schwanhäusser B, Thierfelder N, *et al*. Widespread changes in protein synthesis induced by microRNAs. 2008;**455**:58–63.

Wightman B, Ha I, Ruvkun G. Posttranscriptional regulation of the heterochronic gene *lin–14* by *lin–4* mediates temporal pattern formation in C. elegans. *Cell* 1993;**75**:855–862.

contacts between the termination factor eRF3 and proteins associated with NMD referred to as **UPFs** (a name deriving from their ability to increase or 'up' frameshifting). This recognition step between the UPF proteins and the EJC recruits cellular machinery that ultimately results in the degradation of PTC-containing mRNA transcripts (Figure 10.59). This elegant quality control mechanism nicely highlights the complexities of eukaryotic gene regulation.

Figure 10.59 Nonsense mediated decay (NMD) is triggered by the recognition of exon–exon junction complexes downstream of the premature termination codon (PTC).
(1) Unspliced transcript in the nucleus showing normal start (green) and stop (red) codons as well as premature termination codon (PTC).
(2) Spliced transcript carries exon junction complexes (EJCs) at each exon–exon junction, marking the sites that have undergone splicing.
(3) Ribosomes translating the transcript displace EJCs until a stop codon is encountered. Upon stop codon recognition by eRF1-eRF3, interactions between the termination complex, UPFs and the EJC (4) recruit factors responsible for degradation of the faulty mRNA transcript in the cell.

10.16 VIRAL CORRUPTION OF THE TRANSLATIONAL MACHINERY

The previous sections have illustrated the often sophisticated mechanisms used by the cell to ensure that translation proceeds when it should and with appropriate accuracy, either through the implementation of general mechanisms that have a broad effect on translation, or through more specific mechanisms that target individual mRNAs. The machinery that is so central to the survival of a cell is not immune to exploitation from intruders, however. We end this chapter by describing how the translation machinery is hijacked quite spectacularly by viruses, which depend on the machinery for survival every bit as much as the host cell.

Viruses utilize the host cellular protein synthesis machinery in order to propagate

The goal of a virus is to enter a host cell and take advantage of the host's cellular machinery to promote its own replication and propagation.

As viral genomes do not encode the basic components of the translation machinery, they rely on their host to provide these resources. As such, the viral mRNAs must directly compete with the host mRNAs for access to the translational machinery, and many of the battles between host and virus over protein synthesis machinery are fought in the initiation step of translation. Some viruses have evolved specific mechanisms for selectively turning off host protein synthesis while others have simply developed mechanisms for out-competing the host mRNAs. Exploring the mechanisms by which viruses co-opt the host translation machinery has taught us much about how the normal process of translation can be controlled.

⊝ We discuss the viral lifecycle briefly in Chapter 1.

Viruses can disrupt cap-dependent translation initiation

As described in Section 10.7, mRNAs that are efficiently translated in eukaryotes form a closed loop complex in which a 5′ cap structure and a 3′ poly(A) tail are brought together by a set of protein factors including eIF4E, eIF4G, and PABP. Translation of these mRNAs is said to be **cap-dependent** because formation of the closed loop initiation complex depends on recognition of the 5′ mRNA cap. A number of viruses disrupt cap-dependent initiation in order to redirect the host's translation machinery to their own mRNAs.

For example, picornaviruses (such as poliovirus) produce proteases that target eIF4G and PABP for cleavage, disrupting the formation of the closed loop initiation complex. Influenza viruses disrupt the formation of closed loop complexes by producing an endonuclease that cleaves the host mRNA 5′ caps (but not the viral 5′ caps). In this case, cap-dependent translation can still be utilized by these cleavage-resistant viral mRNAs. These different points of inhibition are outlined in Figure 10.60a. Yet another picornavirus, encephalomyocarditis (EMCV), synthesizes viral proteins that dephosphorylate the 4E-BPs, which results in the sequestration of eIF4E (Figure 10.60b). Again, global cellular cap-dependent translation is inhibited.

Figure 10.60 **Schematics indicating various targets of translational control executed by different viruses.** (a) Closed circle initiation complex indicating targets for viral proteases and RNAases. (b) Viral phosphatases can dephosphorylate 4E-BPs (eIF4E-binding proteins), thus favoring sequestration of eIF4E by the 4E-BPs.

Viral mRNAs must be effectively translated despite disruptions to the host's translational machinery

Whatever mechanism a virus utilizes to disrupt translation of a host's own mRNAs, the virus must also have mechanisms in place that allow it to translate its own mRNAs. For example, the picornaviruses, which inhibit cap-dependent translation by cleavage of eIF4G and PABP, have evolved a cap-independent method of initiating translation of their mRNAs; this method involves RNA elements known as **Internal Ribosome Entry Sites** (**IRES**) found in the 5′ UTR of their mRNAs. These sometimes large and complex IRES RNA structures (for example, Figure 10.61a) allow the viral mRNAs to directly recruit small ribosomal subunits, typically using a reduced set of initiation factors.

Cryo-electron microscopic structures of IRES RNAs bound to the ribosome (Figure 10.61b) show that they bind just where the Shine–Dalgarno interaction would be located in bacterial ribosomes. Indeed, they function in much the same way, positioning the mRNA appropriately, without scanning, to identify the appropriate AUG start site. For example, picornavirus IRES-dependent initiation does not depend on the eIF4 factors that it has destroyed, but instead utilizes distinct factors referred to as **IRES-Transacting Factors** (**ITAFs**). It is interesting to note that the discovery of viral IRESs eventually led researchers to identify IRESs in eukaryotic genes that might, for example, function during times of cellular stress. These molecular species have been less well characterized, but probably use strategies similar to those of the sophisticated viral IRESs.

In other examples, the viral RNA appears simply to compete for limiting translation factors. The rotaviruses, for example, express mRNAs that carry a 5′ cap structure but lack a 3′ poly(A) tail. Instead, the 3′ end of the viral RNA contains a conserved tetranucleotide motif that binds specifically to a virally encoded protein, NSP3,

Figure 10.61 **IRES elements in the 5′ UTR allow for cap-independent translation initiation.** (a) A typical complex picornavirus internal ribosome entry sites (IRES) with complex RNA structure. (b) Cryo-electron microscopic view of the hepatitis C virus internal ribosome entry sites (IRESs) directly interacting with the small ribosomal subunit in the region where mRNA is known to bind. PDBs 2AGN, 2ZKQ. From Spahn, *et al.* (2001).

that itself binds to eIF4G with higher affinity than PABP. In this manner, the viral RNA effectively competes with cellular mRNAs for available eIF4G.

One viral RNA obviates the need for the cellular factors that promote closed loop complex formation by simply utilizing direct RNA–RNA interactions between the 5′ and 3′ UTRs, effectively circularizing the RNA to promote efficient translation.

The hepatitis C virus (HCV) and the cricket paralysis virus (CrPV) have been of particular interest since they both depend on a minimal set of cellular initiation factors to initiate – HCV has dispensed with all but eIF3 and the eIF2 ternary complex, whereas CrPV has even figured out how to initiate translation independent of the eIF2 ternary complex as we discuss below.

eIF2 dependent Met-tRNAi^Met loading is also a critical control point

As we discussed in Section 10.13, the availability of eIF2 for loading initiator met-tRNA in the ribosome is a critical control point for protein synthesis. As we recall, various environmental stimuli can affect the phosphorylation state of eIF2alpha subunit, and thus the availability of eIF2. Of particular interest in this respect is PKR, a cellular protein that phosphorylates eIF2alpha when activated by the presence of double-stranded viral RNA. This phosphorylation promotes an overly stable interaction between eIF2 and eIF2B, slowing down global protein synthesis.

As might be expected, viruses have evolved specific responses to this global attenuation of host cell protein synthesis by acquiring mechanisms for initiation that are independent of eIF2-mediated loading of Met-tRNA$_i$^Met into the ribosome. In two remarkable examples, the viral RNAs themselves appear to have RNA elements either in the 5′ or 3′ UTR that mimic the function and/or shape of a tRNA, thus allowing initiation to take place at non-standard initiation codons (non-AUG).

In the turnip yellow mosaic virus (TYMV), an RNA element in the 3′ UTR of the viral transcript mimics a tRNA^Val and is directly aminoacylated by cellular ValRS. This aminoacylated RNA element appears to function to initiate translation on an AUG codon (albeit with a non-canonical first codon position pairing mismatch). The subsequent alanine codon (GCC) is then normally decoded and the elongation cycle continues (Figure 10.62a). In the cricket paralysis virus, the initiation of translation depends on base-pairing interactions between three nucleotides in the 5′ UTR of the viral transcript and three nucleotides directly upstream of an alanine codon. These interactions appear to mimic a codon–anticodon interaction in the P site of the ribosome. The true initiation of protein synthesis (where the first amino acid is incorporated) then begins at the alanine codon (GCU) positioned in the A site of the ribosome (Figure 10.62b).

Thus, for both viruses, translation initiation is cap- and eIF2-independent (no Met-tRNA$_i$^Met is involved). As eIF2alpha is heavily phosphorylated during infection by these viruses, these mechanisms of initiation allow them to bypass the global cellular translational shut-down triggered by viral RNA in the cell.

Although there are numerous other examples of elegant mechanisms through which viruses take advantage of cellular machinery to propagate, the examples described here capture the key points.

(a) Turnip yellow mosaic virus

(b) Cricket paralysis virus initiation

Figure 10.62 Unusual strategies used by viral RNA elements to promote eIF2-independent translation initiation. (a) Turnip yellow mosaic virus initiates translation in an eIF2-independent manner by utilizing a tRNA^Val mimic found in its 3′ UTR to directly fill the P site. (b) Cricket paralysis virus initiates translation in an eIF2-independent manner by utilizing sequence in the 5′ UTR to create an artificial 'decoding helix' that somehow bypasses normal initiation requirements in the P site.

Ultimately, translation in the cell is a competitive process: different mRNAs are in competition with each other to recruit the molecular components required for translation; the translation assemblies (the ribosome and associated protein factors) on different mRNAs must compete with each other for the building blocks of translation – the tRNAs and amino acids – if translation is to proceed. Beyond this, a cell must compete with external scavengers, viruses, to secure the resources it needs. Any changes in the relative balance of reagents (initiation factors, capped mRNAs, tRNAs etc) will affect the outcome of these battles for cellular resources; translation, like all the other processes that mediate genome function, is a sensitive, finely balanced operation.

✴ SUMMARY

- Translation is the process by which the genetic information, stored as polymers of four nucleotides, gets transformed into proteins composed of amino acids.

- The genetic code specifies how each three nucleotide codon corresponds to a specific amino acid, or stop signal, in the mRNA transcript.

Molecular components

- tRNAs are bifunctional molecules that decipher the genetic information through Watson–Crick pairing interactions with the mRNA, bringing the activated amino acid along for peptide bond formation.

- Aminoacyl-tRNA synthetases are responsible for the first, and primary, event in decoding as they determine which of the 40 or so tRNAs in the cell get attached to which amino acids.

- The ribosome is the macromolecular ribonucleoprotein machine that coordinates the process of translation through events in two distinct subunits, one primarily responsible for decoding and the other for making peptide bonds.

- Many extra-ribosomal protein factors are GTPases that facilitate ribosome-based events through GTP hydrolysis-dependent conformational rearrangements. Other translation factors act by simply binding and occluding specific sites on the ribosome sometimes by mimicking tRNAs.

The mechanism of translation

- Translation can be divided into multiple stages: initiation, elongation, termination and ribosome recycling.

- Each stage of protein synthesis depends on the action of the ribosome, tRNAs, mRNA, and a variety of extra-ribosomal protein factors.

- Bacteria identify the AUG start site through direct interactions between the small subunit rRNA and the Shine–Dalgarno region of the mRNA.

- Eukaryotes identify the AUG start site through a more complex 5′ to 3′ scanning process.

- Elongation is composed of three basic steps: tRNA selection (or decoding), peptide bond formation, and translocation of the mRNA–tRNA complex.

- Termination involves recognition of stop codons by specific protein factors that promote peptide release.

- Ribosome recycling involves dissociation of the mRNA and tRNAs, and dissociation of the two ribosomal subunits; the features of recycling are better defined in bacterial systems.

The regulation of translation

- The genetic code can be altered in a global or mRNA-specific manner, in some cases allowing the introduction of non-standard amino acids.

- Translation initiation can be globally affected through the modification of core translational factors in eukaryotes.

- Both bacterial and eukaryotic organisms respond to amino acid limitation and shut down overall protein synthesis, although the mechanisms differ.

- Initiation of specific bacterial mRNAs can be regulated through obstruction of the Shine–Dalgarno sequence by intrinsic RNA structure, metabolites, small RNAs or proteins.

- Eukaryotic initiation is most often regulated through interactions of various factors (proteins and RNAs) with the 3′ UTR of the transcript.

- Viral systems have evolved ingenious mechanisms for slowing down host protein synthesis, so that their own proteins can be more efficiently synthesized.

 FURTHER READING

10.1 OVERVIEW OF TRANSLATION

Ramakrishnan V. Ribosome structure and the mechanism of translation. *Cell* 2002;**108**:557–572.

10.2 TRANSFER RNA AND THE GENETIC CODE

Chapeville F, Lipmann F, Von Ehrenstein G, *et al.* On the role of soluble ribonucleic acid in coding for amino acids. *Proceedings of the National Academy of Sciences of the U S A* 1962;**15**:1086–1092.

Nirenberg MW, Matthaei JH. The dependence of cell-free protein synthesis in E. coli upon naturally occurring or synthetic polyribonucleotides. *Proceedings of the National Academy of Sciences of the U S A* 1961;**47**:1588–1602.

10.3 AMINOACYL-TRNA SYNTHETASES

Fersht AR. Sieves in sequence. *Science* 1998;**280**:541.

Hou YM, Schimmel P. A simple structural feature is a major determinant of the identity of a transfer RNA. *Nature* 1988;**333**:140–145.

10.4 STRUCTURE OF THE RIBOSOME

Ban N, Nissen P, Hansen J, *et al.* The complete atomic structure of the large ribosomal subunit at 2.4Å resolution. *Science* 2000;**289**:905–920.

Wimberly BT, Brodersen DE, Clemons WM Jr, *et al.* Structure of the 30S ribosomal subunit. *Nature* 2000;**407**:327–339.

Yusupov MM, Yusuova GZ, Baucom A, *et al.* Crystal structure of the ribosome at 5.5Å resolution. *Science* 2001;**292**:883–896.

10.5 THE TRANSLATION CYCLE: THE RIBOSOME IN ACTION

Moazed D, Noller HF. Intermediate states in the movement of transfer RNA in the ribosome. *Nature* 1989;**342**:142–148.

Ramakrishnan V. Ribosome structure and the mechanism of translation. *Cell* 2002;**108**:557–572.

10.6 PROTEIN FACTORS CRITICAL TO THE TRANSLATION CYCLE

Brodersen DE, Ramakrishnan V. Shape can be seductive. *Nature Structural Biology* 2003;**10**:78–80.

Rodnina MV, Stark H, Savelsbergh A, *et al.* GTPases mechanisms and functions of translation factors on the ribosome. *Biological Chemistry* 2000;**381**:377–387.

10.7 TRANSLATION INITIATION (MOSTLY IN BACTERIA)

Mayer C, Stortchevoi A, Köhrer C, Varshney U, RajBhandary UL. Initiator tRNA and its role in initiation of protein synthesis. *Cold Spring Harbor Symposia on Quantitative Biology* 2001;**66**:195–206.

Shine J, Dalgarno L. Determinant of cistron specificity in bacterial ribosomes. *Nature* 1975;**254**:34–38.

Simonetti A, Marzi S, Jenner L, *et al.* A structural view of translation initiation in bacteria. *Cellular and Molecular Life Sciences* 2009;**66**:423–436.

10.8 TRANSLATION INITIATION IN EUKARYOTES

Kapp LD, Lorsch J. The molecular mechanics of eukaryotic translation. *Annual Review of Biochemistry* 2004;**73**:657–704.

Sonenberg N, Hinnebusch AG. Regulation of translation initiation in eukaryotes: mechanisms and biological targets. *Cell* 2009;**136**:731–745.

10.9 TRANSLATION ELONGATION: DECODING, PEPTIDE BOND FORMATION AND TRANSLOCATION

Beringer M, Rodnina MV. The ribosomal peptidyl transferase. *Molecular Cell* 2007;**26**:311–321.

Daviter T, Gromadski KB, Rodnina MV. The ribosome's response to codon-anticodon mismatches. *Biochimie* 2006;**88**:1001–1011.

Moore PB, Steitz TA. The structural basis of large subunit function. *Annual Review of Biochemistry* 2003;**72**:813–850.

Shoji S, Walker SE, Fredrick K. Ribosomal translocation: One step closer to the molecular mechanism. *ACS Chemical Biology* 2009; in press.

10.10 TRANSLATION TERMINATION AND REINITIATION

Karzai AW, Roche ED, Sauer RT. The ssrA-smpB system for protein tagging, directed degradation and ribosome rescue. *Nature Structural Biology* 2000;**7**:449–455.

Youngman E, McDonald ME, Green R. Peptide release on the ribosome: mechanisms and implications for translational control. *Annual Review of Microbiology* 2008;**62**:353–373.

Zaher HS, Green R. Fidelity at the molecular level: Lessons from protein synthesis. *Cell* 2009;**136**:746–762.

10.11 RECODING: PROGRAMMED STOP CODON READ-THROUGH AND FRAMESHIFTING

Craigen WJ, Caskey CT. Expression of peptide chain release factor 2 requires high-efficiency frameshift. *Nature* 1986;**322**:273–275.

Farabaugh PJ. Programmed translational frameshifting. *Annual Review of Genetics* 1996;**30**:507–528.

Gesteland RF, Atkins JF. Recoding: dynamic reprogramming of translation. *Annual Review of Biochemistry* 1996;**65**:741–768.

Sheppard K, Yuan J, Hohn MJ. From one amino acid to another: tRNA-dependent amino acid biosynthesis. *Nucleic Acids Research* 2008;**36**:1813–1825.

10.12 ANTIBIOTICS THAT TARGET THE RIBOSOME

Carter AP, Clemons WM, Brodersen DE, *et al.* Functional insights from the structure of the 30S ribosomal subunit and its interactions with antibiotics. *Nature* 2000;**407**:340–348.

Elseviers D, Gorini L. Direct selection of mutants restricting efficiency of suppression and misreading levels in E. coli B. *Molecular and General Genetics* 1975;**137**:277–287.

Hansen JL, Ippolito JA, Ban N, *et al.* The structures of four macrolide antibiotics bound to the large ribosomal subunit. *Molecular Cell* 2002;**10**:117–128.

Schlunzen F, Zarivach R, Harms J, *et al.* Structural basis for the interactions of antibiotics with the peptidyl transferase centre in eubacteria. *Nature* 2001;**413**:814–821.

Vicens Q, Westhof E. RNA as a drug target: the case of aminoglycosides. *Chembiochem* 2003;**4**:1018–1023.

10.13 GLOBAL REGULATION OF INITIATION IN BACTERIA AND EUKARYOTES

Potrykus K. (p)ppGpp: still magical? *Annual Review of Microbiology* 2008;**62**:35–51.

Sonenberg N, Hinnebusch AG. Regulation of translation initiation in eukaryotes: mechanisms and biological targets. *Cell* 2009;**136**: 731–745.

10.14 REGULATION OF INITIATION VIA THE 5′ UTR IN BACTERIA AND EUKARYOTES

Hinnebusch AG. Translational regulation of yeast GCN4. A window on factors that control initiator tRNA binding to the ribosome. *Journal of Biological Chemistry* 1997;**272**:21661–21664.

Johansson J, Cossart P. RNA-mediated control of virulence gene expression in bacterial pathogens. *Trends in Microbiology* 2003;**11**: 280–285.

Pantopoulos K. Iron metabolism and the IRE/IRP regulatory system: an update. *Annals of the New York Academy of Sciences* 2004;**1012**:1–13.

Storz G, Opdyke JA, Zhang A. Controlling mRNA stability and translation with small, noncoding RNAs. *Current Opinion in Microbiology* 2004;**7**:140–144.

Vitreschak AG, Rodionov DA, Mironov AA, Gelfand MS. Riboswitches: the oldest mechanism for the regulation of gene expression? *Trends in Genetics* 2004;**20**:44–50.

10.15 REGULATION OF TRANSLATION VIA THE 3′ UTR IN EUKARYOTES

Dever TE. Gene-specific regulation by general translation factors. *Cell* 2004;**108**:545–556.

Lasko P, Cho P, Poulin F, Sonenberg N. Contrasting mechanisms of regulating translation of specific Drosophila germline mRNAs at the level of 5′-cap structure binding. *Biochemical Society Transactions* 2005;**33**:1544–1546.

Richter JD. CPEB: a life in translation. *Trends in Biochemical Sciences* 2007;**32**:279–285.

10.16 VIRAL CORRUPTION OF THE TRANSLATIONAL MACHINERY

Jackson RJ. Alternative mechanisms of initiating translation of mammalian mRNAs. *Biochemical Society Transactions* 2005;**33**: 1231–1241.

Kieft JS. Viral IRES RNA structures and ribosome interactions. *Trends in Biochemical Sciences* 2008;**33**:274–283.

Martinez-Salas E, Pacheco A, Serrano P, Fernandez N. New insights into internal ribosome entry site elements relevant for viral gene expression. *Journal of General Virology* 2008;**89**:611–626.

PROTEIN MODIFICATION AND TARGETING

11

INTRODUCTION

In Chapter 10 we learned how proteins are synthesized and how this process is regulated to determine when a given gene product is produced. However, many polypeptide chains are not fully functional upon their release from the ribosome. The proper three-dimensional folding of numerous polypeptides is only achieved with the help of accessory proteins. In addition, a protein may only be needed in a specific location in the cell and thus must be transported to this site after synthesis. The correct localization of proteins depends on specific amino acid motifs that are recognized by the sorting machinery of the cell. The covalent addition of lipids may also be required for a protein to be associated with a membrane. Further, many proteins need to be cleaved or modified by the addition of sugars to become active.

As we have already seen in a number of examples in this book, the activity of a protein can also be modulated by the addition and subsequent removal of one or more small molecule adducts such as phosphate. The covalent additions of lipids, sugars, and small molecule adducts are termed **post-translational modifications**. Because so many chemical changes can occur in a polypeptide chain after it has been synthesized, simply knowing the complete genome sequence of an organism, and hence the amino acid sequence of all translated proteins, is not enough to provide a complete picture of the true diversity of proteins in a cell.

The modifications to which proteins are exposed post-translationally do not always enhance their activity. Proteins can be subject to detrimental covalent modifications as well as misfolding, which in some cases can lead to disease. Degradation of proteins to their component amino acids is also utilized to maintain the proper concentration of a particular protein in the cell and is generally regulated. In this chapter we explore the range of fates to which different proteins may be destined, and the mechanisms by which they occur.

11.1 CHAPERONE-ASSISTED PROTEIN FOLDING

Chaperone proteins assist the proper folding of many proteins

The activity of a protein is dependent on it possessing the correct three-dimensional structure. While the primary sequence of a protein contains all of the information required for the protein to adopt its correct three-dimensional structure, the sequence alone usually is not enough to direct proper folding in an appropriate time span within the context of a cell. Instead, protein folding in the cell requires the assistance of a class of proteins known as **chaperones**, which like the adult Victorian counterpart, can prevent illicit interactions. Chaperone proteins have multiple functions: they prevent a polypeptide from folding too early, aid the correct folding at the proper time, promote the correct folding of a misfolded polypeptide, and assist the unfolding of certain proteins.

Protein misfolding is associated with several neurodegenerative diseases

Several features of protein synthesis and the cellular environment necessitate chaperone-assisted protein folding. The final fold of some proteins contains interactions between amino acid residues that are widely separated in the polypeptide chain. However, the directional nature of protein synthesis – from the N- to the C-terminus – means that folding of the polypeptide chain may begin as the N-terminus of the protein is extruded from the ribosome, long before the protein has been completely synthesized. Thus, if improper associations occur between neighboring regions of the polypeptide chain before the protein is fully synthesized, the protein will be trapped in a misfolded state.

Another factor impacting protein folding is the remarkably crowded nature of the intracellular environment, with protein concentrations of approximately 300 mg/mL. These crowded conditions increase the likelihood of inappropriate interactions between partially folded polypeptides. For example, we learned in Section 2.3 that the driving force for protein folding is the sequestration of hydrophobic residues into the core of the folded protein, where these residues form favorable interactions with one another and are protected from water. However, such hydrophobic regions can readily interact with one another in inappropriate ways early in the folding process – that is, before they have folded to form the intended hydrophobic core – resulting in the formation of undesirable, insoluble protein aggregates.

Insoluble protein aggregates are observed in a number of neurodegenerative diseases, such as Alzheimer's disease and Parkinson's disease. Another example is the spongiform encephalopathies such as mad cow disease, which are caused by the prion protein. In these diseases, a misfolded, abnormal form of the prion protein induces the conversion of normal forms of the protein into the abnormal state, again leading to aggregates. Given the extremely detrimental consequences of protein aggregation, how is misfolding prevented?

Co-translational folding is facilitated by ribosome-associated chaperones

Nascent polypeptide chains begin to emerge from the exit tunnel of the ribosome after approximately 30 amino acids have been covalently joined to one another. The ribosomal exit tunnel is composed of both ribosomal RNA (rRNA) and ribosomal proteins. It is thought that this composition makes the tunnel chemically diverse, and thus inert enough to prevent most peptide sequences from associating with it too strongly. The tunnel has a width of about 15 Å and can potentially accommodate an alpha helix, but not more globular protein structures. So, while it is thought that some local structures, such as alpha helices, begin to assemble within the exit tunnel, proteins generally emerge from the ribosome in an extended conformation. As the peptides emerge, they immediately encounter a number of chaperone proteins, which in some cases are physically associated with ribosomes. Binding of the emerging nascent chain by these chaperones prevents immediate (and potentially inappropriate) folding (Figure 11.1).

⦿ We describe the ribosomal exit tunnel in Section 10.4.

In bacteria, the nascent chain makes contact with the ribosome-associated chaperone known as trigger factor and either concordantly or subsequently makes contact with an Hsp70 chaperone called DnaK (described in more detail below). In eukaryotes, the first encounter is thought to be with the nascent chain-associated complex (NAC) and subsequently with another Hsp70 protein. Both trigger factor and the Hsp70 proteins facilitate proper folding of the emerging polypeptide chain by using a hydrophobic region on their surface to interact with hydrophobic sequences in the

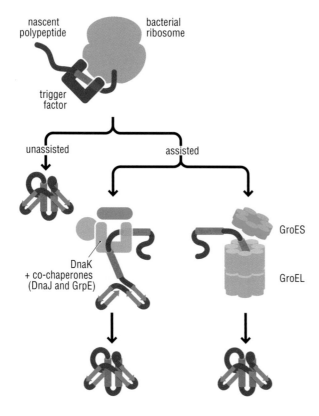

Figure 11.1 Types of protein chaperones. The trigger factor chaperone interacts with and cradles the nascent polypeptide as it emerges from the ribosome. Although some proteins fold autonomously after release from the ribosome, most require the actions of the ATP-dependent Hsp70 or Hsp60 chaperones. The Hsp70 proteins (DnaK) act like a clamp to hold the hydrophobic part of the protein whereas the Hsp60 proteins (GroES + GroEL) form a barrel that provides a protected environment for folding. Examples are shown for bacteria, but homologs or analogs of all proteins are present in eukaryotic cells.
Adapted from Deuerling and Bukau (2004).

nascent polypeptide, thus providing an environment that shields these regions from inappropriate interactions with hydrophobic regions in other parts of the peptide or in other proteins.

Post-translational folding is assisted by Hsp70 and Hsp60 chaperones

Proteins usually have to pass through several distinct stages of folding to be able to adopt their final three-dimensional shape. Consequently, after the early engagement with ribosome-associated chaperones, many proteins still require the assistance of other chaperones—which promote productive over unproductive folding steps—to reach their final folded state (Figure 11.1). A number of conserved classes of chaperone are found in all organisms. The chaperones are collectively known as heat-shock proteins, because many of the proteins were first identified due to their strong induction when cells are exposed to elevated temperature. This heat-shock induction suggests that chaperones are needed to prevent proteins from becoming unfolded or denatured by heat and to refold the proteins if they do become denatured. The best-known classes are Hsp70 and Hsp60, where the number attributed to each heat-shock protein corresponds loosely to its molecular weight.

In general terms, the workings of the Hsp70 and Hsp60 classes of chaperone are quite similar: the chaperones recognize and bind to hydrophobic regions on the non-native polypeptide in a transient fashion with adenosine triphosphate (ATP) driving the conformational changes required for the chaperone to bind and then release the polypeptide. Members of the Hsp70 class of proteins, which include the well-characterized bacterial DnaK protein, are monomeric and act like a clamp that firmly holds an extended hydrophobic segment of polypeptide chain. On ATP binding, the chaperone releases the polypeptide chain, giving it a chance to fold properly.

The Hsp60 class of chaperones, called chaperonins, comprise a structurally distinct class of proteins that form oligomeric barrel-like structures with detachable 'lids' consisting of co-chaperones (depicted in Figure 11.1). The cavity of the barrel is lined with hydrophobic residues, which capture the non-native protein through the exposed hydrophobic surfaces. The binding of ATP and the 'lid' change the character of the cavity wall from hydrophobic to hydrophilic. The protein captured in the closed cavity is released from the wall and can now fold as a monomeric protein, protected from aggregating with other cellular components. On ATP hydrolysis, the lid opens and the folded protein is released. Examples of Hsp60 chaperones include the bacterial GroEL-ES proteins and TRiC or CCT chaperones in eukaryotes.

Once the protein reaches its stable, folded state, the hydrophobic surfaces are largely buried and thus the protein is no longer recognized by the chaperones or by the degradation machinery of the cell. Deciphering the complex patterns that specify the interactions between chaperones and nascent or unfolded peptides still remains a significant challenge.

Disulfide bond formation depends on oxidative reactions

A critical step in the folding of some proteins is the formation of one or more **disulfide bonds** that covalently link pairs of sulfur-containing cysteine residues. Cysteine residues can exist stably either in a reduced form known as a thiol (–SH)

Figure 11.2 Disulfide bond formation. The thiol groups of two cysteine residues are oxidized to form a covalent bond between the two sulfur atoms.

cysteine

or in an oxidized form in which one cysteine residue forms a covalent S-S bond, a disulfide bond, with another cysteine (Figure 11.2). Since the cytosol of the cell tends to be a reducing environment, disulfide bonds are rarely found in cytosolic proteins. However, the bacterial periplasm and the eukaryotic endoplasmic reticulum are oxidizing compartments that favor the formation of disulfide bonds. Thus disulfide bonds are very common in secreted proteins and the correct formation of these bonds is part of proper folding.

Not surprisingly, cells contain enzymes and organic molecules that catalyze the formation or rearrangement of disulfide linkages to favor the final folded native state of the protein (Figure 11.3); functionally similar catalysts have been discovered in both bacteria and eukaryotes. Specifically, two different types of oxidative reaction are carried out: sulfhydryl oxidation results in the *de novo* formation of disulfide bonds, and the isomerization of disulfide bonds allows for the swapping of disulfide bonds until the correct folded state is achieved. In bacteria, the protein DsbA is responsible for sulfhydryl oxidation to facilitate *de novo* disulfide bond formation, while a second protein, DsbC, is responsible for the isomerization of disulfides. In eukaryotic cells, Protein Disulfide Isomerases (PDIs) have been shown to carry out both reactions.

In order to carry out the cysteine oxidation reactions, the thiol-disulfide oxidoreductases DsbA and PDI must themselves be reoxidized after each reaction.

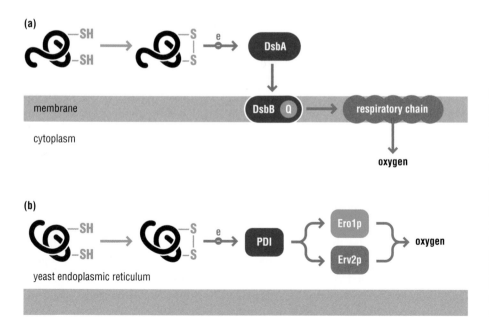

Figure 11.3 Proteins required for protein disulfide bond formation. (a) In bacteria, the oxidoreductase DsbA catalyzes disulfide bond formation in the periplasm. DsbA is reoxidized by DsbB which transfers the reducing equivalents to coenzyme Q from which the reducing equivalents are transferred to the respiratory chain and ultimately to oxygen during aerobic growth. (b) In eukaryotes, sulfhydryl oxidoreductases denoted protein disulfide isomerases (PDIs) catalyze disulfide bond formation in the endoplasmic reticulum. In yeast, the PDI is reduced by FAD-dependent oxidases Ero1p and Erv2p. Again the reducing equivalents are ultimately transferred to oxygen. Adapted from Thorpe and Coppock (2007).

This is achieved by the transfer of reducing equivalents (electrons) from the oxidoreductase to other oxidases such as DsbB in bacteria and Ero1p and Erv2p in yeast. The reducing equivalents are ultimately transferred from these proteins to the electron transport chain.

11.2 TARGETING OF PROTEINS THROUGHOUT THE CELL

In Chapter 1, we learned that the cell has many different compartments. In this section, we discuss how the subcellular localization of a protein can affect its function, and how targeting to the proper location is achieved.

Many proteins need to be targeted to specific regions of the cell

Subcellular targeting by the cell allows for proteins to be involved in distinct processes in different areas of the cell. Localization of proteins also increases their effective concentration by confining many copies of a single protein to a defined part of the cell, thus helping to drive by mass action the reactions and interactions that the protein is involved in. Distinct isoforms of a protein can be differentially localized, thus increasing the number of cellular functions for a gene, without the burden of evolving novel genes.

In bacterial cells, certain proteins must be targeted to the membrane of the cell or to the extracellular environment to perform their function. In the more complicated eukaryotic cells, proteins are targeted to a number of membrane-bound organelles. These include, for example, the nucleus, the endoplasmic reticulum, and the mitochondria. In addition to the membrane-bound structures, eukaryotes contain cytologically distinguishable substructures, such as nuclear speckles and cytoplasmic P bodies, which contain distinct subsets of cellular proteins. So, given the array of possible destinations, how are proteins sorted to their correct location?

The imaging of proteins in different subcellular compartments is described in Section 16.12.

Sorting motifs direct proteins to the proper subcellular compartments

Protein localization in the cell depends on recognition of a diverse set of amino acid sequences within proteins that are recognized by the various sorting and targeting machineries. **Sorting motifs**, also denoted sorting signals or signal sequences in some cases, are sequence identifiers within proteins that target the proteins to the appropriate localization machinery (Figure 11.4). These sequence motifs have often been compared to postal codes within the cell: each location has a unique sorting motif (a unique amino acid code) which directs proteins to that location just as each street in a city has its own postal code to direct mail.

Sorting motifs may be present at either the N- or C-terminus end of a protein (or both) and can be specific amino acids, or a patch of amino acids with specific features. For example, the related secretion machineries in bacteria and eukaryotes recognize a series of hydrophobic amino acids. Often the N-terminal signal sequences are cleaved from proteins during their translocation through the membrane, so that the sorting motif is not present in the mature protein. In addition,

Examples of sorting motifs		
bacteria		
inner membrane	amino-terminus	positively charged/hydrophobic/polar
eukaryotes		
endoplasmic reticulum	amino-terminus	positively charged/hydrophobic/polar
nuclear import	amino-terminus	positively charged (PKKKRKV)
nuclear export	amino-terminus	leucine-rich (LQLPPLERLTL)
mitochondria	amino-terminus	positively charged amphiphilic α helixes
peroxisomes	carboxy-terminus	SKL
ER retention	carboxy-terminus	NPXY or YXXØL; [DE]XXXXL[LI] or DXXLL
lysosomes + endosomes	carboxy-terminus	KDEL

Figure 11.4 Examples of sorting motifs. Example sequences are given in parenthesis. The motifs that direct proteins to the endoplasmic reticulum in eukaryotes and the inner membrane in bacteria are often referred to as signal sequences.

some proteins have multiple sorting motifs. Thus a protein may have both a sequence that targets the protein for insertion into the endoplasmic reticulum as well as a second sequence that signals retention in the endoplasmic reticulum upon its arrival there.

The importance of the cellular sorting process can be illustrated by looking at the consequences of mutations in the sorting motifs themselves. For example, in one form of familial hypercholesterolemia, an inherited condition in which an accumulation of abnormally high levels of low-density lipoprotein (LDL) in the bloodstream leads to premature atherosclerosis, there is a substitution of cysteine for tyrosine in the sorting motif of the LDL receptor. This single mutation prevents uptake of the LDL receptor into the endoplasmic reticulum, thus blocking the clearing of the cholesterol-containing lipoprotein particles. The resulting deposits that accumulate on artery walls can lead to premature heart attacks.

Protein targeting is determined by multiple distinct machineries in the cell

The trafficking of proteins to, and translocation into, the correct subcellular compartment requires two processes to occur. First, the sorting motif must be recognized and the protein must be transported to the proper membrane. Then, the protein must be moved into or across the membrane. The composition of the transport and translocation complexes varies from one compartment to another, but all carry out the same basic steps as illustrated in Figure 11.5.

The initiation of targeting can take place either as the protein is being synthesized (co-translationally) or following synthesis (post-translationally). After the protein with the sorting motif is recognized and transported to the proper membrane, it is transferred to the proteins carrying out the translocation step. In many cases, this transfer step is associated with a conformational change that results in opening of the channel or pore through which the protein is threaded. Finally, the protein must cross or be inserted into the membrane. This can be accomplished by 'pushing' the protein across the membrane with the energy provided by the extrusion of the nascent polypeptide from the ribosome. Alternatively, the protein may be 'pulled' across the membrane with the energy provided by the chaperone proteins that refold the protein as it is extruded on the other side of the membrane.

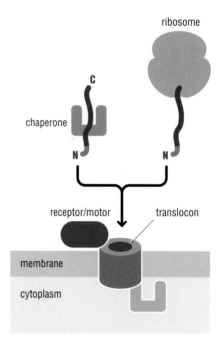

Figure 11.5 General mechanism of protein translocation across a membrane. The sorting motif (red) of the protein to be translocated can be recognized by the receptor as the protein is being translated or after synthesis is complete. If the protein is folded it may need to be unfolded by a chaperone protein. After the signal sequence is recognized and the protein is brought to the membrane, it is threaded through the translocon. The protein can be pushed across the membrane as it is being extruded from the ribosome or it may be pulled across the membrane by the refolding promoted by chaperones on the other side of the membrane.

Protein targeting in bacteria is relatively simple and principally involves the secretion apparatus, which is responsible for directing most membrane-localized and secreted proteins to their final destinations. It is estimated that up to 40% of proteins synthesized in the bacterial cell are targeted using this basic machinery. By contrast, eukaryotic protein targeting mechanisms are considerably more diverse, requiring trafficking of proteins from their site of synthesis in the cytoplasm to a much wider range of membrane-enclosed compartments. Examples of the strategies used to transport proteins into these different destinations are described below.

Secreted proteins are recognized and targeted by related machineries in bacteria and eukaryotes

The machinery involved in targeting proteins to the endoplasmic reticulum and beyond in eukaryotes is evolutionarily related to the secretion machinery in bacteria. Both systems depend on co-translational recognition of a hydrophobic signal sequence by signal recognition particles (SRPs) that greet the polypeptide chain as it emerges from the exit tunnel of the ribosome. These SRPs have a methionine-rich hydrophobic pocket; the unbranched, flexible nature of the methionine residues are thought to make this binding pocket especially suited to broad recognition of hydrophobic amino acids. The SRP in both systems is composed of RNA and protein and depends on guanosine triphosphate (GTP) hydrolysis for downstream steps in the translocation pathway.

Following signal sequence recognition, the SRP interacts with the SRP receptor located either on the extracellular membrane of bacteria or on the surface of the endoplasmic reticulum where the nascent polypeptide is handed over to the ATP-driven translocase – SecA in bacteria and Sec61 in eukaryotes. This translocase then guides the polypeptide through a channel into the periplasm of the bacteria or the lumen of the endoplasmic reticulum.

Other membrane-embedded translocases operate post-translationally to transport unfolded proteins

Unlike the secretion machinery described above, many cellular targeting machineries recognize and target proteins upon release from the ribosome. For example, the mitochondrial targeting machinery recognizes an amphipathic helix, with positively charged amino acids on one face and hydrophobic residues on the other, in a fully synthesized protein. The targeting machinery then engages the protein with the energy-dependent translocation machinery. A key feature of such membrane-embedded translocases is that they typically transport *unfolded* proteins. The chaperone proteins that we described in the previous section (the Hsp70 family) seem to play an important role both in unfolding the proteins prior to the engagement with the translocation machinery and in refolding the proteins within the mitochondria.

Movement into and out of the nucleus depends on a gated transport process

Transport of proteins into and out of the nucleus is highly regulated in both directions. In the case of ribosome assembly, ribosomal proteins synthesized in the cytoplasm must be imported into the nucleus where they assemble with the rRNA in a nuclear sub-region known as the nucleolus. The assembled ribosome must then be exported to the cytoplasm where translation takes place. It is worth pausing for a moment to reflect on the level of movement here: the protein-based ribosome components are synthesized in the cytoplasm, after which they must be transported to the nucleus to assemble with their RNA counterparts, before being transported *back* to the cytoplasm where the complete ribosome has its activity. Such shuttling back and forth between cellular compartments demonstrates just how extensive a process the trafficking of proteins (and other biological molecules) is.

Unlike the translocases discussed above, transport into and out of the nucleus takes place through elaborate structures referred to as nuclear pore complexes (NPCs). While this pore is relatively permeable to small molecules it is quite selective for large molecules. Protein import depends strictly on the recognition of nuclear localization sequences, which are rich in positively charged amino acids. The nuclear localization sequence is recognized by soluble nuclear import receptors that associate with the nucleoporin proteins that extend into the cytoplasm from the nuclear pore complex (Figure 11.6). Once bound by the receptors, the imported proteins pass through a gated aqueous pore, without unfolding. Nuclear export appears to work in a similar fashion, although it depends on a different set of signals and receptors. The pore can be as large as 26 nm in diameter, a dimension that can readily accommodate the ribosomal particles that must pass through in their assembled state.

Since transport through the nuclear pore goes in both directions, how is directionality established? This is the function of the GTPase Ran, a molecular switch, which is found predominately in the GTP-bound form in the nucleus and in the

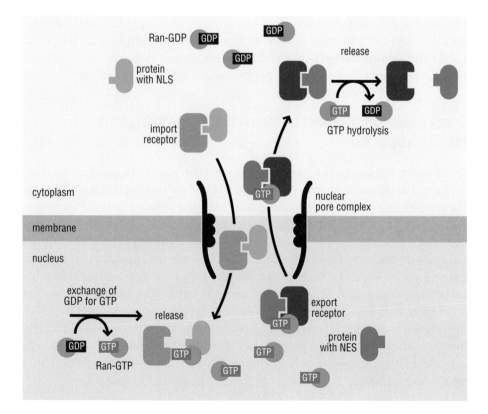

Figure 11.6 Transport in and out of the eukaryotic cell nucleus. For proteins trafficking into the nucleus, the nuclear localization sequence (NLS) is recognized by soluble nuclear import receptors that also bind to components of the nuclear pore complex. After passing through the aqueous pore, the GTP-bound Ran protein stimulates release of the imported protein from the receptor. For proteins trafficking out of the nucleus, binding of an export receptor to a protein with a nuclear export signal (NES) is promoted by Ran-GTP. This tri-partite complex passes through the pore, whereupon release of the exported protein from the receptor is stimulated by hydrolysis of the GTP bound to Ran.

guanosine diphosphate (GDP)-bound form in the cytoplasm. In the nucleus, Ran-GTP promotes the release of imported proteins from their receptors and the binding of proteins to be exported to their receptors, as illustrated in Figure 11.6. That is, Ran-GTP facilitates the drop-off of imported proteins once they reach their intended destination, and the pick-up of proteins for export, preparing them to begin their journey out of the nucleus. In the cytoplasm, hydrolysis of the Ran-bound GTP promotes release of the exported protein from the export receptor, resulting in a predominance of Ran-GDP in the cytoplasm. Another protein called nuclear transport factor 2 (NTF2) transports Ran-GDP to the nucleus, where the GDP on Ran is exchanged for GTP, thus allowing the cycle to repeat itself.

Diverse machineries for cellular protein targeting are critical for high specificity

Though the molecular machineries involved in these different protein localization events are different, each system depends on intrinsic signals encoded within the primary sequence of the protein itself. These distinct signals allow the cellular machineries to accurately select and target the wide range of proteins that must reach specialized cellular locations. In Sections 11.4 and 11.5, we will learn about two other strategies utilized by the cell to localize proteins to different subcellular localizations, lipid modification, and **glycosylation**.

11.3 POST-TRANSLATIONAL CLEAVAGE OF THE POLYPEPTIDE CHAIN

Correct folding and transport to the right cellular location are only some of the processes a polypeptide might have to undergo following translation in order to attain their full biological activity. In this section we learn how an immature polypeptide may undergo cleavage following translation by the ribosome, and see how this cleavage is linked to a polypeptide attaining its proper activity.

Some polypeptides must be cleaved in order to become biologically active

Proteins synthesized by the ribosome naturally carry an N-terminal methionine (or formyl-methionine) residue as specified by the translation initiation process. For many proteins in bacteria and eukaryotes, the removal of this N-terminal methionine by a methionine aminopeptidase is a critical step in their maturation. Once this step is completed, most proteins are biologically active once the polypeptide chain is completed and the folded. However, some cellular systems depend on the initial synthesis of an inactive precursor polypeptide that must be proteolytically cleaved in order to become active. The insertion of such a processing step into the lifetime of a protein provides another point at which the cell can regulate gene expression. The following examples illustrate why it is beneficial to produce certain proteins in an inactive state, and to specifically activate them only when they reach an appropriate location or when their activity is specifically required.

Proteolytic processing is widely used for the timely activation of polypeptides

There are many examples of proteases that cleave polypeptide chains and that require proteolytic processing for their own activation. Chymotrypsin serves as a digestive enzyme in the gut, but must be prevented from digesting the pancreatic cells that produce it. To achieve this selective activity, the enzyme is translated as an inactive precursor called chymotrypsinogen in pancreatic cells. Chymotrypsinogen is then transported to the gut where a protease cleaves the inactive precursor at specific amino acid positions to generate the active chymotrypsin enzyme (Figure 11.7). The precursor lacks enzymatic activity because the uncleaved portions of the enzyme prevent the precursor from folding into a catalytically active configuration.

The hormone insulin provides another example of how proteolytic processing is required for activation. Insulin triggers glucose uptake from the blood. An inactive preproinsulin precursor contains a signal sequence, which directs its co-translational import into the endoplasmic reticulum, where the signal sequence is cleaved off and three disulfide bonds are formed (Figure 11.8). Subsequent cleavage by three additional proteases gives rise to active, secreted insulin. The many steps required to generate the hormone provide multiple opportunities for regulation. In patients with type 1 (juvenile) diabetes, insulin is no longer produced and processed because of immune-mediated destruction of insulin-secreting pancreatic beta cells.

A third example of proteolytic processing is found in the retrovirus family that includes the human immunodeficiency virus (HIV), where a number of essential viral proteins are synthesized as a single precursor **polyprotein**. In Chapter 10, we learned how the translation of the polyprotein is regulated by a programmed frameshift. For this discussion, it is important to know that the precursor polypeptide contains domains corresponding to a number of distinct proteins including a reverse transcriptase that copies viral RNA to give a DNA copy, and an integrase that integrates the viral genome into that of the host cell. After the polyprotein is synthesized, the individual domains are separated from one another by proteolytic cleavage catalyzed by the retroviral-encoded protease (Figure 11.9). This process is particularly relevant to molecular pharmacology since a number of successful anti-retroviral therapies depend on inhibition of the protease that catalyzes these proteolytic events.

Some proteins contain self-excising domains called inteins

Several proteins found in bacteria, archaea, and single-celled eukaryotes contain an internal domain referred to as an **intein**, which must be removed from the final expressed protein product for the protein to become active. Like intron removal by self-splicing RNA, the intein catalyzes its own excision from the

inactive

active

Figure 11.7 Chymotrypsin generation by specific cleavage. The chymotrypsin enzyme is translated as an inactive precursor called chymotrypsinogen in pancreatic cells. Chymotrypsinogen is then transported to the gut, where a protease cleaves the inactive precursor at specific amino acid positions to generate the active enzyme. The three fragments generated by the proteolysis remained connected by disulfide bonds.

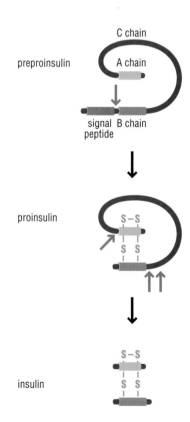

preproinsulin

C chain

A chain

signal B chain
peptide

proinsulin

insulin

Figure 11.8 Insulin maturation by specific cleavage and disulfide bond formation. The inactive preproinsulin precursor contains a signal sequence (red), which directs the co-translational import of the protein into the endoplasmic reticulum. There the signal sequence is cleaved off. Three disulfide bonds are formed. Subsequently, cleavage by three additional proteases gives rise to active insulin.

Figure 11.9 Proteolytic processing of the HIV polyprotein. Five viral proteins (MA, matrix; CA, capsid; PR, protease; RT, reverse transcriptase; and IN, integrase) synthesized as part of a single precursor polyprotein are released by specific proteolytic cleavage.

(a)

(b)

Figure 11.10 Autoprocessing reactions catalyzed by proteins. (a) Self-excision catalyzed by inteins. Key nucleophilic attacks are catalyzed by conserved serine, cysteine or threonine residues. (b) Autoprocessing reaction catalyzed by the Hedgehog protein. In this case, the key nucleophilic attack is catalyzed by an exogenous cholesterol molecule.

translated polypeptide while ligating the flanking polypeptides to form a second protein product (Figure 11.10a). The flanking protein sequences are referred to as **exteins** (with analogy to exons). In this way, two proteins, the ligated exteins and the excised intein polypeptide, can be derived from a single polypeptide chain (Figure 11.10a).

The intein is not always junk that is discarded following excision. It often encodes an endonuclease domain that specifically excises the DNA sequence corresponding to its own amino acid sequence, and mediates the insertion of this DNA fragment into genes that lack the intein sequence. This feature thus allows the intein to propagate throughout genomes as a mobile genetic element. In some cases the endonuclease domain has been replaced with a short peptide linker, thus resulting in an element with no potential mobility. While the joined exteins can in principle be any protein, approximately 70% of inteins reside in proteins involved in DNA replication and repair. The reason for this is not clear.

As for self-splicing introns, conserved residues located within the intein itself are essential for catalyzing the proteolytic autoprocessing reactions through which the intein is excised. The processing reaction is initiated by the residue at the N-terminus of the *intein* (a serine, cysteine, or threonine) attacking the carbonyl group of the preceding peptide bond (Figure 11.11). The same carbonyl group is then subject to a nucleophilic attack by the first residue (again a serine, cysteine, or threonine) in the C-terminal extein. Finally, the last residue of the intein internally cyclizes. This set of reactions results in the formation of a branched intermediate that leads to release of the intein and the formation of a genuine peptide bond between the N- and C-terminal extein moieties. Although the reactions involved in intein excision are chemically somewhat more sophisticated than those involved in RNA splicing, the ultimate outcomes of the two processes are quite comparable.

Autoprocessing of the metazoan Hedgehog protein is chemically similar to intein splicing

The chemical reaction carried out by inteins is remarkably similar to that performed by the eukaryotic Hedgehog protein, which plays a key role in developmental patterning in vertebrates and invertebrates. Central to the mechanism of action of the Hedgehog protein is its ability to catalyze a series of self-processing reactions. As with intein-containing proteins, Hedgehog is expressed as a precursor protein that undergoes autocatalytic cleavage to yield two protein fragments, in this case, C-terminal and N-terminal domain products.

Both Hedgehog autoprocessing and the intein reactions begin with chemical rearrangement of the peptide backbone. However, in the case of the Hedgehog protein, this rearrangement is followed by a nucleophilic attack from an exogenous cholesterol molecule. As a result, cleavage of the peptide bond between the two domains of Hedgehog occurs with concomitant formation of a covalent bond between the C-terminus of the N-terminal domain and a cholesterol molecule (Figure 11.10b). Cleavage of the precursor protein allows this

Figure 11.11 **Intein cleavage reaction.** The intein processing reaction is initiated by the residue (a serine, cysteine, or threonine) at the N-terminus of the intein (red) attacking the carbonyl group of the preceding peptide bond linking the N-terminal extein (blue) with the intein. The same carbonyl group is then subject to a nucleophilic attack by the first residue (again a serine, cysteine, or threonine) in the C-terminal extein (green). Finally, the last residue of the intein internally cyclizes leading to the release of the intein and the formation of a genuine peptide bond between the N- and C-terminal extein moieties.

N-terminal signaling domain to diffuse to adjacent cells where it can mediate its patterning effect. Although inteins are only found in single-celled organisms and Hedgehog proteins are only found in metazoans, the similarities in the sequences of their self-splicing domains suggest a common evolutionary origin (Figure 11.12).

Figure 11.12 Alignment of Hedgehog and intein proteins. Although the overall sequences of Hedgehog and intein proteins are very different, a few key residues (red) are found in all proteins, suggesting a common processing reaction and a common evolutionary origin. Adapted from Hall (1997).

11.4 LIPID MODIFICATION OF PROTEINS

We have explored a number of ways in which the primary sequence of a protein may need to be altered before the protein can adopt its final, fully functional form. However, changes to many proteins do not stop once this final length is reached. A number of covalent additions or modifications occur to modulate localization and function. In this section, we will discuss lipid modifications and their properties.

Covalent attachment of lipids targets proteins to membranes

Lipid modifications generally target proteins to the cell membrane, allowing them to insert themselves and reside in the hydrophobic lipid bilayer. Different types of lipid modifications are thought to target eukaryotic proteins to compositionally distinct cellular membranes, such as the plasma membrane, the Golgi, or lysosomes, as well as to particular regions within a given type of membrane. Lipid modifications are not specific to eukaryotic cells and are used for similar purposes in bacterial cells. More than one type of lipid can be covalently attached to a protein, thereby allowing the cell to use a combination of lipids to fine-tune the localization of a protein in a membrane, just as a postal address features a combination of items – street name, city name, country name – to direct an item of mail to a very specific location. In some cases, modification with a lipid is a reversible process, thus allowing for dynamic changes in the targeting of proteins to membranes. Lipid modification is therefore a strategy used by cells both to dispatch proteins to the appropriate intracellular membrane, and to regulate cellular processes that are mediated by these membrane-associated proteins.

Lipids are covalently attached to specific amino acid residues, typically at or near either the N-terminus or the C-terminus of the protein. In this section, we shall learn about three broad types of lipid modification that have somewhat distinct functional properties, and that are classified according to the type of attached lipid (Figure 11.13):

- **acylation** by fatty acyl groups such as myristoyl or palmitoyl
- **prenylation** by isoprenoid groups such as farnesyl or geranylgeranyl
- addition of glycoinositol phospholipids (referred to as **glycosylphosphatidylinositol (GPI) anchors**).

Acylation processes target the N- and C-termini of proteins

Two different fatty acid modifications, myristoyl and palmitoyl groups, typically have different attachment points. The relatively rare 14-carbon saturated fatty acid myristoyl is attached to proteins at the N-terminus via a stable amide linkage to an N-terminal glycine residue, yielding an *N*-myristoylated protein (*N* refers to the resulting linkage to nitrogen, Figure 11.13a). The myristoyl moiety is donated by the carrier myristoyl CoA, and is attached to the protein co-translationally before folding of the polypeptide chain occurs. The 14-carbon myristoyl chain allows the modified protein to associate with cell membranes and is generally a stable

(a)

myristoyl group
attached to
N terminal glycine

palmitoyl group
attached to cysteine
near C terminus

(b)

farnesyl group
attached to cysteine
near C terminus

(c)

phosphatidylinositol

inositol

N-acetylglucosamine

mannose

glycan

mannose

COO⁻

mannose

GPI anchor attached
to C terminal acyl group

Figure 11.13 **Major classes of lipid modifications.** Proteins can be (a) acetylated by the addition of a fatty acid such as a myristoyl or a palmitoyl group or (b) prenylated by the addition of an isoprenoid group such as farnesyl. The third type of lipid modification (c) involves the addition of a GPI anchor, which is composed of both carbohydrates and lipids.

modification throughout the life of the protein. Proteins that are *N*-myristoylated include certain alpha subunits of heterotrimeric G proteins and the Src family of tyrosine kinases, both of which are involved in mediating the transmission of signals across cell membranes.

The more common 16-carbon saturated palmitoyl is generally attached post-translationally to C-terminal proximal cysteines through their nucleophilic sulfurs (Figure 11.13a). The thioester linkage that attaches the palmitoyl moiety can be broken with relative ease (and can thus be considered to be chemically labile). This ease of modification appears to be used by the cell to reversibly modulate the localization of target proteins and, in some cases, to affect protein stability. For example, lymphoma proprotein convertase, an enzyme that activates proproteins in the secretory pathway, is found in both palmitoylated and unmodified forms. Interestingly, the palmitoylated form of the protein is degraded much faster than the unmodified one, allowing the lifetime of this enzyme to be regulated by lipidation. Other fatty acid chains, such as the 18-carbon stearoyl, can sometimes substitute for the palmitoyl group, leading to a more general term for this modification known as **S-acylation** (*S* refers to the resulting linkage to sulfur).

Some proteins are both *N*-myristoylated and *S*-palmitoylated. This dual modification appears to have the important consequence of targeting the dually modified species to distinct membrane subdomains termed lipid rafts: the presence of both the myristoyl and palmitoyl groups targets the modified protein to a location distinct from that to which a protein carrying only a myristoyl or palmitoyl group alone might be directed.

Prenylation is a stable modification of C-terminal cysteine residues

Prenylation is the modification of a protein through the addition of a group belonging to the isoprenoid family – generally either a C_{15} farnesyl or C_{20} geranylgeranyl group – via a thioether linkage to the sulfur of a cysteine residue (Figure 11.13b). The prenylation machinery recognizes several different cysteine-containing motifs at the C-terminus of the protein to be modified. One of these motifs is CaaX, where 'a' stands for any aliphatic amino acid residue and where 'X' determines which prenyl group will be added. When X is glutamine, methionine, or serine, a farnesyl group is attached to the protein and when X is leucine, a geranylgeranyl group is attached.

Following modification with the prenyl group, the C-terminus is proteolytically processed, removing the three terminal amino acids to yield a protein terminating in the lipidated cysteine residue. Finally, the newly exposed C-terminal carboxyl group is modified with a methyl group that neutralizes its negative charge. Other prenylation motifs include C-terminal CXC or CC amino acid residues, which can be modified with either one or two prenyl groups. The distinctions between these modifications may seem subtle, but probably have important functional consequences that have not yet been completely deciphered. In all examples of prenylation, the modification is stable and so is not believed to act in a regulatory manner.

Almost all prenylated proteins are small GTP-binding proteins. A number of *S*-prenylated proteins appear to also be subject to *S*-acylation at a nearby cysteine residue. Again, this dual tagging has been observed to drive the variable targeting of proteins. For example, such dual lipid modification is used to determine the subcellular localization of the signaling protein Ras, a GTPase that, when mutated, gives rise to a variety of human cancers. All Ras proteins are farnesylated, and can in addition be acylated with palmitoyl groups. Ras proteins bearing both lipid modifications associate with the plasma membrane, whereas removal of the palmitoyl

Figure 11.14 Subcellular localization of Ras is regulated by farnesylation and palmitoylation. Ras proteins bearing both farnesyl and palmitoyl groups are localized to the plasma membrane whereas Ras proteins with only a farnesyl group are localized to the Golgi membrane. PDB 1Q21.

group to leave just the farnesyl group causes Ras to be targeted to the Golgi membrane instead. By cycling between palmitoylated and depalmitoylated states, the cell can carry out Ras signaling in different cellular compartments (Figure 11.14).

GPI anchors are comprised of both carbohydrates and lipids

The modifications we have just been considering involve the addition of lipid groups to a protein. The attachment of a C-terminal GPI anchor to a protein is distinct from the lipid modifications described above, both in chemical composition and in the manner of its attachment. The GPI anchor is composed of a carbohydrate moiety that is tethered to the membrane by two fatty acid chains (Figure 11.13c). The identity of the fatty acids and of the core tetrasaccharide can vary lending diversity to this modification with likely functional consequences.

The GPI anchor is assembled in the endoplasmic reticulum through a complex series of enzymatic steps. The final step involves attachment of the GPI anchor to the C-terminus of the protein with simultaneous cleavage of the protein's transmembrane segment so that the protein is only tethered to the membrane via the GPI anchor. The presence of the carbohydrate moiety directs GPI-anchored proteins to the cell surface via transport through the secretory pathway.

Since the proteins are only anchored to the membrane via the GPI anchor, the protein is readily released from the membrane by the action of phospholipases, which cleave the lipid. GPI-anchored proteins participate in a wide variety of processes in the cell including nutrient uptake, cell adhesion and membrane signaling. A number of human parasites have GPI-anchored enzymes on their cell surfaces that are inactive while tethered to the membrane but which become activated when the parasite encounters host phospholipases.

Multiple lipid modifications on the Hedgehog protein are important for specifying its subcellular localization

There are still other lipid modifications that play critical roles in targeting proteins to membranes. The Hedgehog protein, which plays a central role in regulating embryonic development, must be modified with cholesterol in order to carry out its biological function (Figure 11.15). We learned earlier in this chapter that Hedgehog undergoes an autocatalytic cleavage that separates the protein into two fragments, and the cleavage is accompanied by the attachment of a covalent cholesterol adduct to the newly exposed C-terminus of the N-terminal domain. The discovery of this modification is described in Experimental approach 11.1. Of particular interest here is that the N-terminus of the N-terminal domain is also modified with a palmitoyl residue, suggesting that multiple lipid modifications may aid in specifying the localization of Hedgehog to particular membranes within the cell.

11.5 GLYCOSYLATION OF PROTEINS

Having seen how a polypeptide may be modified through the attachment of a lipid molecule, we now consider a second widespread form of covalent modification, that of glycosylation.

Figure 11.15 **Lipid modifications of the Hedgehog protein.** The Hedgehog signaling protein undergoes autocatalytic cleavage that is accompanied by the attachment of a cholesterol molecule to the exposed C-terminus of the N-terminal domain. A second lipid modification, the palmitoylation of the N-terminal cysteine residue completes the processing of this signaling domain.

Glycosylation can change the solubility of a protein and provide protein recognition sites

The most complex and diverse post-translational modifications to proteins are carbohydrate chains, referred to as **glycans**, which are attached through a process called **glycosylation**. As carbohydrates are hydrophilic in nature these modifications are usually found on the outside of the protein where they have substantial effects on the solubility of the protein. Almost all secreted and membrane-associated proteins in eukaryotic cells are modified with glycans. Collectively, the glycosylated proteins are referred to as **glycoproteins**.

Protein glycosylation serves a variety of functions. Inside the cell, the attachment of glycans generally increases the solubility of nascent glycoproteins and can prevent their aggregation. Small monosaccharides attached to proteins within the cell also are thought to have roles in intracellular signaling. Once the protein is secreted, glycosylation can provide protection from proteases and from non-specific protein–protein interactions. Glycosylated proteins on the cell surface can also be bound by other proteins surrounding the cell that recognize specific glycans; this type of interaction commonly plays a role in the recognition of one

11.1 EXPERIMENTAL APPROACH

The discovery of cholesterol modification for membrane tethering during signaling

The Hedgehog (hh) family of secreted proteins is required for developmental patterning of a wide variety of embryonic structures in diverse organisms. Work by Phil Beachy's group in the early 1990s established much about how this protein is produced and what its *in vivo* roles might be. The group found that the Hedgehog protein (Hh) undergoes several proteolytic processing steps to yield the final active species. First, the signal sequence at the N-terminus of the protein is removed by standard processing. In addition, it became clear that the 45 kD precursor underwent an autocatalytic cleavage event to yield two products – a 20 kD N-terminal domain (Hh-N_p) and a 25 kD CTD (Hh-C). The required molecular components for this autocleavage event were principally contained within the CTD (a particular histidine found there is critical), while the N-terminal domain constituted the extracellular signal responsible for dictating essential tissue-patterning effects during development. The mechanisms for Hh and intein cleavage ultimately turned out to be related, consistent with sequence similarities throughout these proteins (Figure 11.12), and in particular the essential cysteine at the cleavage site.

The group noticed that the internal thioester intermediate of the reaction (akin to that of intein cleavage) yielded a covalent modification of Hh-N_p, which increased its 'hydrophobic character' and then set out to uncover the molecular explanation. It turns out that this autoprocessing-derived hydrophobic modification is at the heart of Hedgehog function, specifying where the signaling molecule Hh-N_p goes and thus where it acts.

The study referred to here focused on identifying the covalent modification and its chemical linkage, and thus how the movement of this protein through tissues is regulated. Since previous studies had determined that the intermediate of the reaction was a thioester carbonyl, it was already clear that high concentrations of thiol reagents (or strong nucleophiles) could attack the intermediate, resulting in release of Hh-C, and the formation of an adduct between the nucleophile and Hh-N_p. Given the hydrophobic nature of Hh-N_p, it was proposed that a lipid might function as the nucleophile *in vivo*, thus producing the more hydrophobic adduct. Beachy and colleagues took this idea as a starting point in their quest to identify this adduct. Bulk lipids isolated from *Drosophila* S2 cultured cells clearly stimulated the processing reaction (much like DTT), yielding the visible 25 kD Hh-C fragment (Figure 1; the construct used in these studies had a truncated N-terminus yielding only a 5 kD Hh-N_p product not visible in this gel system).

The authors next fractionated the lipid extract by thin-layer chromatography (Figure 2a), and found that one specific excised band from the TLC (spot B) stimulated the *in vitro* processing reaction (Figure 2b). Spot B was ultimately identified as cholesterol, first through comparisons to lipid standards, and ultimately through *in vivo* labeling experiments. The chemical linkage to the cholesterol was also determined. Given the potential intein type reaction, it was proposed that the reactive nucleophile on cholesterol would be the 3β hydroxyl group; the base-labile nature of the predicted ester adduct of this reaction was readily confirmed.

The identification of cholesterol as the covalent adduct of Hedgehog protein, and its relatives in other organisms (Sonic hedgehog in mice, for example), had a huge impact on the field. Understanding of this molecular mechanism provided immediate perspective on defects associated with a number of mutant

Figure 1 Lipid stimulation of Hh autoprocessing *in vitro*. Coomassie blue-stained SDS-polyacrylamide gel showing *in vitro* autocleavage reactions of the Hh precursor protein (~29 kD) incubated for three hours at 30 °C with no additions (lane 1), 50 mM DTT (lane 2), 1 mM DTT (lane 3), or 1 mM DTT plus bulk S2 cell lipid extract (lane 4). The Hh-C product of the reaction migrates as an ~25 kD species (lanes 2 and 4), and the ~5 kD NH$_2$-terminal product is not seen in this gel.
Reproduced from Porter *et al.*, *Science* 1996;**274**:255–259.

organisms, as well as on the effects of cholesterol levels on development and growth. The similarities between the Hedgehog autoprocessing domain and the inteins provide insights into the assembly of the hedgehog signaling protein from evolutionarily ancient domains. Indeed, more extensive comparative genomic approaches argue that the intein-related Hedgehog processing domain first came together with the signaling domain into a single gene as the first patterned multicellular animals evolved.

Porter JA, Young KE, Beachy PA. Cholesterol modification of hedgehog signaling proteins in animal development. *Science* 1996;**274**:255–259.

Figure 2 Identification of cholesterol as the stimulatory lipid in the Hh autoprocessing reaction. (a) TLC plate coated with silica gel G (Merck) showing the fractionation of bulk S2 lipids with a heptane:ether:formic acid solvent (80:20:2). Six major spots are visualized by acid charring and are labeled A through F. (b) Coomassie blue-stained SDS-polyacrylamide gel showing *in vitro* autocleavage reactions of the Hh precursor protein incubated with 1 mM DTT plus either unfractionated S2 cell lipids (lane 1) or lipids extracted from spots A through F (lanes 2 to 7, respectively). Addition of lipids from spot B, but not from any other, resulted in processing of the Hh precursor protein. Reproduced from Porter *et al.*, *Science* 1996;**274**:255–259.

cell by another. Important examples of this include the targeting of leukocytes to sites of inflammation and host cell recognition by viruses and bacteria. For example, the influenza virus binds to the carbohydrate sialic acid that is attached to a glycoprotein on the mammalian host cell's surface. This binding results in internalization of the virus by the target cell, enabling the virus to enter and infect the host cell.

Glycans vary greatly in complexity: they range in size from a simple monosaccharide or just a few linked monosaccharides to many hundreds of linked saccharides. The nature of the glycan attached to any given protein varies from species to species and often from one tissue to the next. Protein glycosylation is almost exclusively a eukaryotic property and the complexity of modifications increases as one proceeds up the evolutionary tree. Lower eukaryotes, like yeast, attach only a simple set of sugars to their proteins whereas mammals modify their proteins with highly branched oligosaccharides composed of a wide range of carbohydrates. The size of the attached carbohydrate can be significant, sometimes approaching 90% of the total mass of the glycoprotein. The hydrophilic nature of the glycan group prevents it from **condensing** into a compact structure and so the volume of a glycoprotein relative to its mass is generally large (Figure 11.16).

Figure 11.16 Volume added by carbohydrate addition. The *N*-linked glycan (light yellow, 3 kDa), *O*-linked glycan (dark yellow, 1 kDa) and glycosylphosphatidylinositol (GPI) anchor (orange, 1.5 kDa) only add a fraction of the mass to the CD59 protein, an immune defense protein (blue, 20 kDa). However, the carbohydrate additions which spread out in space, rather than folding inward as is typical for amino acid chains, more than double the volume of the molecule. PDB 1CDR.

Carbohydrates are attached to proteins via nitrogen or oxygen

The diverse glycans that are attached to proteins are most often polymers built from a variety of hexose building blocks (Figure 11.17) and can be either linear or branched. Unlike many of the polymerization reactions that we have discussed throughout this book, the sequential attachment of hexoses depends on distinct enzymes to perform each addition and is specified without a guiding nucleic acid template. Carbohydrates are covalently attached to proteins in two different ways: through *N*-linked or *O*-linked glycosidic bonds. **N-linked carbohydrates** are covalently bonded to the amino nitrogen in asparagine side chains, whereas **O-linked carbohydrates** are covalently bonded to the oxygen of serine or threonine side chains (Figure 11.18). The *N*-linked oligosaccharides are the most common, found on 90% of glycoproteins.

N-linked carbohydrates are usually more complex in nature than *O*-linked carbohydrates and their addition can require 23 or more enzymatic steps. *N*-linked glycosylation begins in the endoplasmic reticulum (Figure 11.19) with the co-translational attachment of the precursor oligosaccharide to an asparagine residue found in a glycosylation signal sequence, Asn-X-Ser, Asn-X-Thr, or Asn-X-Cys, where X can be any amino acid except proline. The precursor oligosaccharide is composed of 14 monosaccharides consisting of glucose, mannose and *N*-acetylglucosamine in a particular branched configuration.

To produce the mature oligosaccharide, the *N*-glycosylation precursor is first trimmed by glucosidases and mannosidases in the endoplasmic reticulum. Additional sugars are then added when the protein is trafficked to the Golgi complex, giving rise to diverse complex sugars. The glycosyl transferase enzymes in this process all depend on sugars activated by attachment to the intracellular nucleotide uridine diphosphate (UDP). As we will see in several more examples, most modifications that are added to proteins are coupled to nucleotide carriers. The breaking of the bond between the carrier and the modification group provides the energy to drive the modification reactions. Many of the processing steps are not dependent on one another, that is, enzymes do not necessarily need to act in a specific order. Thus there can be significant heterogeneity in the final product due to the limited availability of individual enzymes or the varying efficiency of enzymatic steps, for example.

O-glycosylation is similar to *N*-glycosylation, in that it generally begins with the covalent attachment of a precursor sugar, *N*-acetylgalactosamine, which is subsequently derivatized with a variety of other saccharides to yield a number of final

D-glucose (Glc)   D-galactose (Gal)   D-mannose (Man)   N-acetyl-D-glucosamine (GlcNAc)   N-acetyl-D-galactosamine (GalNAc)

Figure 11.17 Chemical structure of some hexoses which are common building blocks for many oligosaccharides that are attached to proteins. The common abbreviation is indicated in parentheses.

structures. Although these carbohydrate appendages are generally not as extravagant as their *N*-linked cousins, they are still diverse in structure.

The importance of these glycosylation steps in human development is underscored by the findings that mutations in any of the roughly 15 genes encoding proteins involved in *N*-glycosylation result in a variety of developmental disorders, including mental retardation and motor deficits.

Reversible *O*-GlcNAc modification may serve a role in signaling

One form of *O*-glycosylation is different from the rather large, diverse set of carbohydrate modifications described above. The attachment of a single *N*-acetylglucosamine residue to a serine or threonine residue, termed *O*-GlcNAc, is a reversible modification that is introduced by *O*-GlcNAc transferase (OGT) and removed by *O*-acetylglucosaminidase (*O*-GlcNAcase). In contrast with other forms of protein glycosylation, which occur largely on membrane or secreted proteins, monoglycosylation by *O*-GlcNAc occurs on cytoplasmic and nuclear proteins.

It appears that the addition and removal of *O*-GlcNAc plays a role akin to that of other small, reversible modifications such as **phosphorylation** and acetylation, which we will discuss in Section 11.6. The sites of monoglycosylation are often identical to the sites of phosphorylation, suggesting that *O*-GlcNAc modification can be used to block phosphorylation and vice versa. The transcriptional regulator, c-Myc, contains a threonine residue that is either glycosylated or phosphorylated, depending on serum conditions. We learned earlier in Chapter 8 that the C-terminal domain (CTD) of RNA polymerase II becomes phosphorylated at serine 2 and serine 5 of the heptapeptide repeats in the course of the transcription cycle. The CTD can also be monoglycosylated with GlcNAc at position

Figure 11.18 *N*-linked and *O*-linked glycosylation. Glycans are attached to (a) asparagine to give *N*-linked carbohydrates or (b) serine or threonine to give *O*-linked carbohydrates.

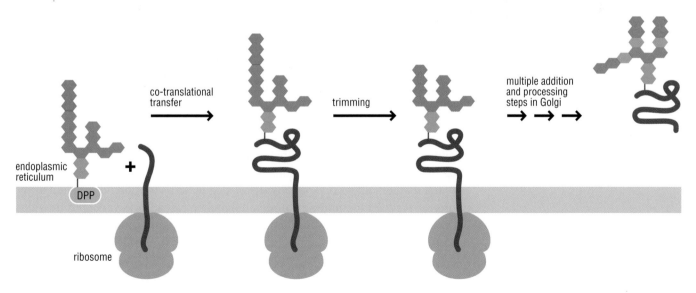

Figure 11.19 Protein glycosylation in the endoplasmic reticulum and Golgi. The oligosaccharide is first transferred from its membrane-bound dolichol pyrophosphate (DPP) carrier to a polypeptide chain while the latter is still being synthesized on the ribosome. As the polypeptide chain grows within the endoplasmic reticulum, monosaccharides may be cleaved from the ends of the oligosaccharide. After completion of the synthesis of the polypeptide chain, the immature glycoprotein is transported to the Golgi apparatus where further modification occurs. New monosaccharides may be added and others removed in a multi-step process involving many different enzymes. The completed glycoprotein is finally transported to its ultimate destination. Yellow hexagons represent *N*-acetyl-glucosamine, blue hexagons represent mannose and red hexagons represent glucose.

4 (threonine) of the heptapeptide. While the precise functional role of *O*-GlcNAc modification, other than its ability to block phosphorylation, is not well understood, there is evidence that monoglycosylation may play a role in nuclear localization and in modulating protein–protein interactions.

The levels of *O*-GlcNAc modification are influenced by the availability of the GlcNAc donor, UDP-GlcNAc, whose levels are in turn controlled by the availability of sugars and other nutrients in the cell. *O*-GlcNAc modification thereby provides a way for the cell to respond to changes in nutrition. It is interesting to note that the carrier molecule for *O*-GlcNAc is also UDP. Other nucleotides are carriers for the reversible modifications that play a role in signaling, such as phosphorylation (ATP), acetylation (acetyl-CoA), and **methylation** (*S*-adenosyl methionine), which we will discuss in the next section.

11.6 PROTEIN PHOSPHORYLATION, ACETYLATION, AND METHYLATION

Small reversible covalent modifications play central roles throughout biology

We have now explored a range of post-translational modifications, which mostly involve the addition of quite large entities to a protein. However, the attached entity does not need to be large for it to have a significant impact on the properties of the molecule to which it is attached. There are a number of post-translational protein modifications utilized in all kingdoms of life that involve the addition and removal of small molecules. The most prevalent of these, which we have already encountered in a number of instances in this book, are phosphorylation, acetylation, and methylation. These three modifications share the property of being reversible and are used very broadly to modulate protein function in the cell.

As with all protein modifications, phosphorylation, acetylation, and methylation can alter protein conformation (and hence protein function) or localization (and hence the site of function). The phosphate, acetyl, and methyl groups provide new binding surfaces and thus new potential binding partners. Many proteins possess specialized domains that can recognize side chains that have undergone phosphorylation, acetylation, or methylation, as illustrated in Figure 11.20. The reversible nature of these modifications enables them to serve as a type of chemical switch that can affect diverse processes such as transcription, chromatin structure, intracellular membrane traffic, and signaling cascades. Reversibility provides a means for the cell to respond rapidly to stimuli without the need to synthesize new proteins. In addition, since the same amino acid can sometimes be the target of more than one modification enzyme, the addition of one modification may serve to block the addition of another modification. So, for example, acetylation of a particular residue might prevent methylation at the same site, hence inhibiting the event that would have been triggered had the methylation occurred.

A common feature of phosphorylation, acetylation, and methylation is that the physical modification itself (a phosphate group, an acetyl group, or a methyl group) is transferred by an enzyme from a nucleotide donor molecule to a target

SH2 domain   bromodomain   chromodomain

Tyr   Lys   Lys

Figure 11.20 Specific modifications are recognized by specialized protein domains. Thus phosphorylated tyrosines are bound by SH2 domains, acetylated lysines are bound by bromodomains and methylated lysines are bound by chromodomains.

protein. The identity of the nucleotide donor, and the amino acids targeted for modification, differ for each type of modification, as we shall discuss below. In each case, a distinct set of enzymes reverses the modification.

Phosphorylation can alter protein function

Surely the best studied and perhaps the most common post-translational modification in cells is the attachment of a phosphate group. The source of the phosphate group that is added to a protein is the nucleoside triphosphate ATP. ATP has three phosphate groups: the third ('terminal') group is transferred from ATP to the target protein through the action of enzymes known as **protein kinases** in a process referred to as **phosphorylation** (Figure 11.21). Reversal of the modification is catalyzed by a different class of enzyme called **protein phosphatases**. In eukaryotes, the amino acid side chains that are usually phosphorylated are serine, threonine, and tyrosine, all amino acids containing a hydroxyl group to which the phosphate group is attached. Eukaryotic kinases are commonly named according to the amino acid they modify, such as serine/threonine kinase. In bacteria, phosphorylation on histidine and aspartate residues predominates (although some eukaryotic-like serine/threonine kinases and tyrosine kinases have been found in bacteria and some histidine kinases are present in yeast and plants).

The importance of phosphorylation in biology is underscored by the significant proportions of both eukaryotic and bacterial genomes that are devoted to phosphorylation; about 2% of all human genes encode kinases that phosphorylate serine, threonine, or tyrosine residues (representing the third most common protein domain) and 1.5% of all genes in *Escherichia coli* are histidine and aspartate kinases.

Phosphorylation, like other post-translational modifications, can affect the target protein in several ways. One possible effect is the creation of a new recognition site to which another protein can bind. For example, as we saw in Chapter 8, phosphorylation of RNA polymerase II triggers recruitment of proteins required for transcriptional elongation, which bind to the phosphorylated CTD portion of the polymerase enzyme but not to the unphosphorylated version. There are specialized protein domains that bind specifically to particular types of phosphorylated side chains. For example, SH2 domains, which are found in some metazoans, bind to phosphotyrosine (a phosphorylated tyrosine residue).

Figure 11.21 Reversible phosphorylation of serine side chains. Kinase enzymes catalyze the nucleophilic attack by the OH group of serine on the γ-phosphate of ATP, yielding phosphoserine. The phosphate group is removed by phosphatases.

Phosphorylation can also change the activity of the target protein itself, either considerably or subtly. This change in activity may come about from the added bulk and the negative charge of the phosphoryl group, or from the hydrogen-bonding potential of the phosphate oxygens. These physical properties of phosphorylated residues can sometimes trigger large conformational changes in the protein that can in turn have consequences for binding and catalysis. For example, in Chapter 10 we saw how phosphorylation of the translation initiation factor eIF2 modifies its interactions with its GTP exchange factor (eIF2B), thus down-regulating its activity in the translation initiation pathway.

Protein phosphorylation is a key feature in the Jak–STAT signaling pathway

The Jak–STAT signaling pathway (Jak stands for Janus Kinase and STAT for Signal Transducers and Activators of Transcription) in eukaryotes provides an illuminating example of how binding of a signaling molecule to an extracellular receptor triggers a chain of phosphorylation events in the cell, which leads to activation of certain genes, thus allowing the transmission of information from the cell surface to the nucleus. The signaling molecule, α-interferon, is a protein secreted by cells in response to viral infection. The α-interferon protein binds to the surface of target cells, triggering a series of biochemical events that results in the transcription of genes encoding proteins that help the cell respond to the viral infection.

As shown in Figure 11.22, the cytokine receptor for α-interferon is a transmembrane protein containing an extracellular domain that binds α-interferon, and a cytoplasmic domain to which an intracellular tyrosine kinase is bound. The α-interferon ligand associates with two extracellular domains, drawing them together in the membrane to form a receptor domain with a single α-interferon protein bound. This dimerization brings together the cytoplasmic domains of the two monomers, along with the tyrosine kinases to which they are bound. When brought together in this way, the tyrosine kinases phosphorylate one another.

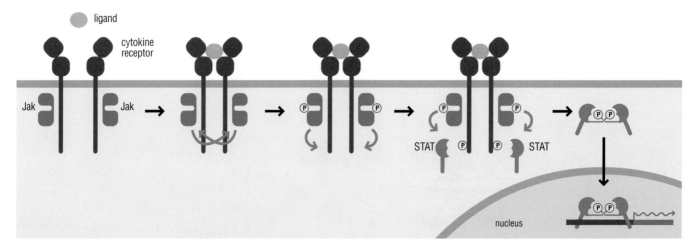

Figure 11.22 The Jak–STAT signaling pathway. Ligand binding induces cytokine receptor chain oligomerization. Weakly active Jaks associated with the receptor cross-activate one another, probably by phosphorylation of the activating site in their kinase domains. The resulting activated Jaks then phosphorylate key tyrosines in the cytoplasmic tail of the cytokine receptor. These sites attract signaling molecules, including STATs and other signaling proteins containing SH2 domains (only STAT is shown). Bound STATs become phosphorylated, after which they dissociate from the receptor and dimerize via their phosphorylation sites and the SH2 domains. The STAT dimers rapidly enter the nucleus and activate transcription of specific target genes.

11.2 EXPERIMENTAL APPROACH

Two-component systems in bacteria: how to interact with your partner

Two-component signal transduction systems are ubiquitous in bacteria where they help the cells respond to their environment. These systems are typically composed of a sensor kinase (usually a membrane protein), which is activated by an environmental signal, and in turn phosphorylates a response regulator (usually a transcription factor). These kinases and their substrates were first identified in genetic screens that searched for mutants defective in responding to environmental factors. For example, bacterial variants defective in responding to nitrogen starvation led to the identification of three genetic loci important for directing transcription from the nitrogen-regulated glnA P2 promoter: NtrC (also called NRI and now understood to be the response regulator), NtrB (also call NRII and now understood to be the sensor kinase) and a specific alternate sigma factor σ^{54} (σ^N).

In a landmark study that followed, Alexander Ninfa and Boris Magasanik carried out *in vitro* transcription assays with purified NRI and NRII2302 (a mutant form of NRII that does not depend on nitrogen) to examine the requirements for glnA P2 promoter activation. In the course of these experiments, the investigators found that pretreatment of NRI with NRII2302, in the presence of mixtures of nucleotides, resulted in increased levels of transcription. By mixing NRI and NRII2302 with different combinations of nucleotides (where A+C+G, A+C, A+G and A alone worked but C+G did not) (Figure 1), the investigators determined that a rate-limiting step in glnA activation depends on the direct interaction between NRI and NRII and the presence of ATP.

Using radiolabeled ATP, the investigators demonstrated that NRII2302 catalyzes the transfer of the γ phosphate of ATP to NRI (Figure 2), and that this modification is required for the increased ability of NRI to activate transcription. These data established the underpinnings of our understanding of the relationship between sensor kinases and response regulators, and their roles in activating transcriptional responses.

As the genes for additional cellular responses to environmental stimuli were characterized (e.g. for chemotaxis, cell envelope stress, and sporulation), it became apparent that cellular responses were quite often dictated by two proteins whose genes were adjacent to one another in the genome. Although it is hard to appreciate today how limited the sequence databases were in 1986, sequence comparisons revealed that these 'two-component' regulators, important for a variety of responses in a wide range of bacterial species, were related and thus must have evolved from a common ancestor.

Figure 1 Effect of nucleotides on NRII2302 and NRI activation of transcription from the glnA P2 promoter. In each case NRI was pretreated with NRII2302 for 20 minutes at 37°C in the presence of the indicated nucleotides. These mixtures were then incubated with RNA polymerase and σ^{54} for the times listed above the gel before UTP and heparin were added to start the transcription reaction. The arrow indicates the transcription product from the glnA P2 promoter.
Reproduced from Ninfa and Magasanik, *Proceedings of the National Academy of Sciences of the U S A* 1986;**83**:5909–5913.

As more and more of these homologous sensor kinase and response regulator proteins were identified, with some organisms containing as many as 200–300 pairs, the question of how specificity is maintained within the cell arose. Using batteries of purified kinase and response regulator proteins, Michael Laub and colleagues recently showed that most of the kinases are in fact quite specific, and only phosphorylate their own cognate response regulator. They surmised that this specificity must result from very specific interactions between the cognate proteins, and that the amino acids involved in this interaction (on the two surfaces) would co-vary. Taking advantage of 1300 sequences that were available in 2008, the investigators identified two clusters of co-varying residues (Figure 3a). By substituting specific residues in the EnvZ kinase with those found in the RstB kinase, they were able to convert EnvZ from a protein that transfers phosphate to OmpR to one that transfers phosphate to the RstA response regulator (Figure 3b), thus effectively 'rewiring the specificity' of the system.

We previously discussed such a specificity swap experiment for an RNA (as applied to the identification of the Shine–Dalgarno

(a)

Figure 2 **Covalent modification of NRI by NRII2302.** NRI was incubated with [γ-³²P]-ATP for various times in the absence (lanes 6–10) or presence (lanes 1–5) of NRII2302. (a) Proteins separated by SDS-PAGE stained with Coomassie blue and (b) an autoradiograph of the gel in (a). Reproduced from Ninfa and Magasanik, *Proceedings of the National Academy of Sciences of the U S A* 1986;**83**:5909–5913.

interaction in Experimental approach 10.2), and we see here how very general and powerful the approach is for defining interactions between a range of biologically interesting molecules.

Ninfa AJ, Magasanik B. Covalent modification of the *glnG* product, NR₁, by the glnL product, NR₁₁, regulates the transcription of the *glnALG* operon in *Escherichia coli*. *Proceedings of the National Academy of Sciences of the U S A* 1986;**83**:5909–5913.

Nixon BT, Ronson CW, Ausubel FM. Two-component regulatory systems responsive to environmental stimuli share strongly conserved domains with the nitrogen assimilation regulatory genes *ntrB* and *ntrC*. *Proceedings of the National Academy of Sciences of the U S A* 1986;**83**:7850–7854.

Skerker JM, Perchuk BS, Siryaporn A, *et al*. Rewiring the specificity of two-component signal transduction systems. *Cell* 2008;**133**:1043–1054.

Figure 3 **Primary sequence alignment of the histidine kinases EnvZ, RstB and CpxA.** (a) Residues showing the strongest covariation with residues in the cognate response regulator are highlighted in green and pink. Sequences of point mutants constructed in which amino acids with high inter-protein mutual information scores were mutated in EnvZ to match the corresponding amino acids in RstB. (b) Effects of mutations on ability of EnvZ histidine kinase (HK~P) to transfer phosphate to its cognate response regulator (RR~P) OmpR or to the RstB cognate RstA or to the CpxA cognate CpxR. In these assays the purified wild-type or mutant kinase protein was autophosphorylated and then incubated alone or examined for phosphotransfer to purified RstA, OmpR, or CpxR. The proteins were separated by SDS-PAGE and phosphorylated proteins were detected by autoradiography.

This phosphorylation activates the kinases, causing them to phosphorylate tyrosine residues in the cytoplasmic domain of the receptor. The phosphorylated tyrosine of the receptor can then recruit STAT, a DNA-binding protein bearing an SH2 domain, which recognizes the phosphotyrosine residue. The STAT protein itself is then phosphorylated at a tyrosine residue, which allows it to dissociate from the receptor and form a dimer that is stabilized by the binding of the SH2 domain of one STAT monomer to the phosphotyrosine residue on the opposing monomer. The STAT dimer can now translocate to the nucleus, where it binds DNA and activates transcription. Thus a signal from outside the cell is transformed into a signal that activates transcription in the nucleus, all based on a series of inter-connected post-translational modification events.

In bacteria, signals from the environment are also converted into changes in gene expression through the activation of kinases at the cell membrane (termed sensor kinases), ultimately leading to the phosphorylation of transcription factors (termed response regulators). As we already learned in Section 8.13 and saw in Figure 8.59, in these cases, the sensor kinases autophosphorylate in the presence of the inducing signal. Subsequently, the phosphate groups are transferred to another protein; either directly to the response regulator, which often regulates transcription, or to intermediary proteins, from which the phosphate groups are then transferred to the response regulators. The discovery of this class of proteins and the interacting amino acids by comparative genomics is described in Experimental approach 11.2.

Proteins can be acetylated on side chains and at their amino termini

Another small, reversible post-translational modification is the acetyl group, which plays an important role in transcriptional regulation. The acetyl group derives from acetyl-CoA, a high-energy nucleotide species, and is transferred to target proteins in a reaction called acetylation, which is catalyzed by proteins known as **acetyl transferases** (Figure 11.23). Two general types of protein acetylation occur in the cell, both targeting amine functional groups: acetylation at either the epsilon position of lysine side chains (frequently on histone proteins) or at the N-terminus of proteins. In the latter case, acetylation can occur on the amino group of N-terminal methionines, or can occur at other amino acids exposed at

Figure 11.23 Reversible acetylation of lysine side chains. Acetyl transferase enzymes catalyzed the transfer of an acetyl group from acetylCoA to the amino group of lysine (for histones, acetyl transferases are denoted HATs). The acetyl group can be removed in two different ways by deacetylases (for histones, deacetylases are denoted HDACs) or by NAD-dependent sirtuins.

◉ We discuss the role of acetylation in transcriptional regulation in Chapter 8.

the N-terminus after the initiator methionine is removed by a methionine amino-peptidase. It has been estimated that more than one-third of all yeast proteins may be acetylated at the N-terminus, and upwards of 80% of proteins in higher eukaryotes. This modification is, however, rare in bacteria, where just three ribosomal proteins are known to be acetylated in this way.

As with phosphorylation, the acetyl group can be readily removed from lysine side chains by **deacetylase** enzymes, which accomplish this by one of two different chemical pathways (Figure 11.23). One pathway is catalyzed by hydrolases that use simple hydrolysis to remove the acetyl group, releasing acetate and restoring the lysine side chain to its unmodified state. A second class, known as sirtuins (named for their similarity to the yeast Sir2 enzyme), use NAD, nicotinamide adenosine dinucleotide (NAD^+) as a co-substrate in the deacetylation reaction. Sirtuins catalyze the cleavage of NAD^+ to release nicotinamide and transfer the acetyl group from the lysine side chain to the remaining portion of the NAD^+ molecule, yielding nicotinamide 2'-*O*-acetyl-ADP-ribose. The biological relevance of this surprising use of NAD^+ as a cofactor by this class of enzymes remains to be understood. In contrast with the acetylation of lysine side chains, *N*-acetylation of the N-terminus is usually not reversed.

What is the consequence of acetylation for the target protein? As with phosphorylation, acetylation affects the overall charge of the protein, in this case effectively neutralizing the positive charge of the lysine side chain. This neutralization of positive charge on lysine could, for example, reduce the affinity of proteins for negatively charged nucleic acids. Lysine side chain modifications on histone proteins have been shown to be critical for their ability to regulate access to chromosomal DNA.

The presence of the acetyl group also provides new chemical functionality for forming different interactions than the unmodified side chain. Just as phospho-tyrosine is recognized specifically by an SH2 domain, acetylated side chains are recognized by specialized protein domains, known as bromodomains, which we learned about in Chapter 4. N-terminal *N*-acetylation blocks the action of amino-peptidases that degrade proteins, thus altering the lifetime of proteins in the cell.

A variety of side chains can be methylated

Methylation involves the addition of a methyl group, CH_3, to a target protein. A variety of proteins in both bacteria and eukaryotes have been found to be methylated. It is a common modification of histone proteins, which, like acetylation, functions to regulate chromatin structure and transcription in the nuclei of eukaryotic cells. Methylation of eukaryotic proteins is performed by a variety of **methyl transferases**, which use *S*-adenosylmethionine (SAM) as the methyl donor. Both lysine and arginine can be methylated but, unlike acetylation, more than one methyl group can be added to a given side chain (Figures 11.24 and 11.25). In addition, only some forms of methylation alter the overall electrostatic charge of the side chain.

Methylated lysine and arginine groups are chemically very stable, and it is only recently that several enzymes that remove the methyl group have been identified. The enzyme LSD1 is an amine oxidase that demethylates mono- or di-methyl lysine, releasing formaldehyde. Methylarginine can be converted to citrulline,

Figure 11.24 Reversible methylation of lysine residues. Lysine may be mono-, di-, or tri-methylated by methyl transferases. Mono- and di-methylated lysines can be demethylated by amino oxidases. The methyl donor is *S*-adenosylmethionine (SAM) resulting in the production of *S*-adenosylhomocysteine (SAH).

Figure 11.25 Reversible acetylation of arginine residues. Unmethylated and mono-methylated arginine can be converted to citrulline or methylated citrulline by the peptidyl arginine deaminase PADI4.

a non-canonical amino acid, by PADI4 (Figure 11.26). While this is not formally a reversal of arginine methylation, it does result in a chemical change that is proposed to play a role in transcriptional regulation.

A number of other side chains can also be methylated. Methyl groups can be covalently attached to oxygen atoms in aspartic and glutamic acid yielding a methyl ester form of these amino acids. This process is referred to as *O*-methylation. This modification is readily reversed because of the chemical nature of an ester linkage. During bacterial chemotaxis, the cytoplasmic domain of a key transmembrane receptor becomes *O*-methylated at several glutamate residues by the methyl transferase, CheR. The modification is fully reversible by CheB, a glutamate ester demethylase. Other less common side chain methylations include *N*-methylation of histidine, glutamine, and asparagine and *S*-methylation of cysteine.

There are several different types of protein domains that recognize methylated lysine or arginine residues. As we saw in previous chapters, methyl-lysine found on histones can be recognized by chromodomains. Another well-characterized protein domain, the tudor domain, recognizes di-methylated lysines, whereas proteins with the WD40 domain recognize di- and tri-methylated lysine. Proteins that recognize methylated side chains often selectively bind to methylated side chains within a particular protein, suggesting that they recognize the modified side chain within a specific context of surrounding amino acid residues.

Figure 11.26 Examples of amino acid oxidation. Formation of (a) cysteine sulfinic acid, cysteine sulfenic acid, cysteine sulfonic acid, and (b) methionine sulfoxide.

Other small reversible modifications also modulate protein function

We have focused on the reversible modifications phosphorylation, acetylation, and methylation, and the enzymes that catalyze the addition and removal of these small adducts. It should be mentioned however, that there are still other small reversible modifications, for example the addition of ADP-ribose to arginine, glutamic acid, or aspartic acid residues, which can regulate proteins in unique ways and also compete for the amino acids that are phosphorylated, acetylated, and methylated.

11.7 PROTEIN OXIDATION, NITROSYLATION, AND NITRATION

Direct chemical modification of some amino acids can have both regulatory and detrimental effects

In addition to modifications that are catalyzed by specific enzymes as discussed in the previous section, there are modifications that occur through direct chemical modification of specific amino acids without the action of an enzyme. These covalent changes may or may not be reversible. In some cases, the modifications are used to regulate protein activity whereas, in others, the covalent changes are detrimental and constitute a form of protein damage. Here we will discuss two of these chemical modifications – oxidation, involving the addition of oxygen, and nitration, involving the addition of nitrogen. We will also explore some consequences of these modifications.

Sulfur-containing amino acids are particularly sensitive to various forms of oxidation

A property of oxygen, which contributes to its chemical reactivity, is that this molecule can lose electrons to give rise to reactive oxygen species such as superoxide (O_2^-), the hydroxyl radical ($\bullet OH$), and peroxides (ROOH). (It is something of a paradox that we depend on oxygen for our survival, yet its chemical reactivity makes it toxic to the human body.) These reactive oxygen species can react with amino acids to covalently modify specific side chains. Cells possess a number of enzymes capable of eliminating reactive oxygen species to protect the cell from oxidative damage. However, a variety of conditions, such as the body's response to infection, can give rise to oxidative stress whereby higher than normal levels of reactive oxygen species are produced in tissues, leading to oxidative damage of protein and other biological molecules. The amount of oxidative damage generally increases in aging cells and is thought to contribute to the progression of age-related diseases.

Several examples of oxidatively modified amino acids are given in Figure 11.26. The chemical properties of the cysteine residue make it particularly well suited to facilitating chemical reactions, as well as binding to transition metals, as we learned in Chapter 2. These same properties render the thiol side chain of cysteine vulnerable to oxidation. As we saw in Section 11.1, the formation of disulfide bonds between cysteine side chains is a positive consequence of oxidation: the formation of disulfide bonds between cysteine residues is critical to the proper folding of some proteins. Cysteine residues also can be directly oxidized by the covalent addition of one, two, or three oxygen atoms, producing sulfenic, sulfinic, and sulfonic acid, respectively (Figure 11.26a). The tripeptide glutathione (γ-glutamylcysteinylglycine), which helps to maintain the reducing environment inside many cells, can also form adducts with single cysteine side chains through a disulfide bond, in a process called glutathionylation. All these modifications can alter the activity of the protein.

Glutathionylation and sulfenic acid formation can be reversed by thiol reductase enzymes or by other thiol-containing molecules (with reduced sulfur groups) in the cell. The enzyme, sulfinic reductase, specifically reverses the two-oxygen modification, while oxidation to a sulfonic acid is thought to be irreversible.

Two examples involving cysteine oxidation to sulfenic acid illustrate the varied effects of this modification. On the one hand, the oxidation of a unique cysteine in a bacterial transcriptional regulator is required for the protein to bind to DNA and activate transcription of a gene encoding an oxidative stress defense gene. On the other hand, oxidation of an active site cysteine to sulfenic acid completely inactivates certain protein tyrosine phosphatases in eukaryotes. Thus when cells are exposed to elevated levels of oxidants, inactivation of the enzyme disrupts pathways that depend upon the enzyme to remove phosphate groups from target proteins. Inactivation of the phosphatase enzymes by this oxidation event can be reversed by reduction of the active site cysteine, restoring the original thiol group.

The other sulfur-containing amino acid, methionine, can also be oxidized by oxygen derivatives, giving rise to methionine sulfoxide (Figure 11.26b). Enzymes that reverse methionine oxidation, called methionine sulfoxide reductases, are widespread in biology, indicating a broad need for these enzymes. As a specific example, part of the pathology of emphysema can be explained by the deleterious consequences of the oxidation of a critical methionine residue on a proteinase inhibitor. The methionine sulfoxide modification greatly lowers the affinity of the proteinase inhibitor for target proteases, such as elastase, which are found in lung tissue. The resulting increase in elastase activity results in deterioration of the lung.

Proteins can be modified by reaction with nitric oxide

Covalent modification of side chains can also be produced by reactive nitrogen species in the cell, resulting in *S*-**nitrosylation** and *O*-nitration. Reactive nitrogen

Figure 11.27 Examples of amino acid nitration. Formation of *S*-nitrosylcysteine and *O*-nitrated tyrosine.

species arise from the synthesis of the reactive gas, nitric oxide (NO), which is produced by specialized enzymes in eukaryotes, where it acts as a second messenger. Nitric oxide and other reactive nitrogen species can also be produced as a side product of nitrogen metabolism in bacteria.

Reactive nitrogen species have chemical properties similar to reactive oxygen species and can therefore also react with proteins, giving rise to a form of covalent modifications. Some of the same amino acids that are oxidized can also be subject to modification by reactive nitrogen species. As has been found for oxidation, these modifications can have a range of consequences. The reactive cysteines of many proteins have been shown to be modified directly by NO, thereby becoming *S*-nitrosylated. Tyrosine can also react with NO-derived species becoming *O*-nitrated (Figure 11.27). *S*-nitrosylation is chemically reversible by free thiols, although the mechanism by which this modification is removed in cells is not known in most cases.

As we have seen for other side chain modifications, protein activity can be regulated by the adducts derived from reactive nitrogen species. An example can be found in zinc-containing metalloproteases, whose active site is generally blocked by a zinc atom coordinated by three histidine side chains and one cysteine side chain. If the cysteine becomes *S*-nitrosylated, the zinc–cysteine coordination is disrupted, and the protease becomes activated.

It is easy to see the similarities between the post-translational modifications that we describe here and in earlier sections of this chapter. The distinguishing feature of the oxidation and nitration reactions is that these modifications are connected to the redox state of the cell, that is, the balance of reducing and oxidizing equivalents, allowing another key feature of cellular homeostasis to be monitored.

11.8 UBIQUITINATION AND SUMOYLATION OF PROTEINS

In the previous two sections, we discussed relatively small post-translational modifications that are broadly used by the cell to modulate protein function. As we saw in our discussion of protein glycosylation, however, post-translational modifications can also be quite large and elaborate. Here we discuss another set of post-translational modifications that involves the addition of an entire small protein to a polypeptide chain.

Figure 11.28 Ubiquitination. (a) Isopeptide bond between ubiquitin and a lysine side chain. (b) Three-dimensional structure of ubiquitin. PDB 2BF8.

Covalent modifications by ubiquitin and related proteins are required for a variety of cellular functions

A distinct set of post-translational modifications involves the addition of a small protein to specific lysines on a target protein. The most common and best-characterized of these modifications is the addition of ubiquitin, a compact, 76 amino acid protein found in all eukaryotes. Ubiquitin is covalently attached to lysine side chains in a process known as **ubiquitination** (also denoted ubiquity-lation) (Figure 11.28). This modification adds about 8500 Da for each ubiquitin attached, which is more than 100 times the molecular weight of a single phosphate group. A protein can be modified by the attachment of a single ubiquitin protein or by a chain of ubiquitin proteins assembled by the ubiquitination of ubiquitin itself. There are several types of ubiquitin polymers as we shall see below.

Ubiquitin was initially identified as the modification that marked proteins for proteolytic degradation in the cell by a large multiprotein complex known as the **proteasome** (see Experimental approach 11.3). However, like other post-trans-lational modifications that we have learned about, ubiquitination is reversible; in this capacity, ubiquitin plays a non-degradative role in a variety of processes including the DNA damage response and the subcellular localization of biological molecules. The precise chemical nature of the ubiquitin modification (position on the protein, length and type of ubiquitin polymer) determines the functional consequences and the fate of the modified protein.

Ubiquitin attachment and chain elongation occur through a series of enzymatic steps

Ubiquitination is carried out in a series of reactions – ubiquitin activation, ubiq-uitin conjugation and ubiquitin–protein ligation – by a corresponding series of enzymes known generically as E1, E2, and E3 enzymes (Figure 11.29). In the first

11.3 EXPERIMENTAL APPROACH

Discovery of ubiquitin-dependent proteolysis

Aaron Ciechanover, Avram Hershko, Irwin Rose, and colleagues set out to address the basic question of why protein degradation in mammalian cells requires energy. This path of exploration led them to discover that degradation depended on the covalent attachment of a 76 amino acid long protein known now as ubiquitin, and guided them to the elucidation of a pathway of central importance in eukaryotic cell biology. These scientists took a biochemical approach and simply examined proteolysis in cellular lysates (extracts) made from reticulocyte cells (immature red blood cells). They identified two important fractions in this lysate – a low-molecular weight fraction (fraction I), which contained a small heat-stable protein (initially named APF-1, but later found to be ubiquitin), and a second higher-molecular weight fraction (fraction II), which in retrospect probably contained the proteasome and enzymes required for ubiquitination – that when combined yielded ATP-dependent proteolysis *in vitro*.

Ciechanover *et al.* next asked whether APF-1 directly associated with components in the system by incubating [125I]-labeled APF-1 with fraction II. Their experiments showed that APF-1 did in fact associate with proteins in the high-molecular weight fraction, in an ATP-dependent manner, and the association was strikingly resistant to high pH (suggesting that the interaction was covalent). In a subsequent paper, Hershko *et al.* reported that denatured substrates, such as lysozyme, were covalently labeled with multiple APF-1 monomers to generate a discernible ladder (Figure 1, bands C_1–C_5). Since the levels of the APF-1-labeled form of the protein decreased over time, Hershko *et al.* proposed that the ATP-dependent linkage of APF-1 to the substrate is followed by proteolytic breakdown. So, in answer to the original question, ATP is required both for the activation of ubiquitin (so that it can be transferred to targeted proteins), and for the proteasome that uses the energy to catalyze proteolysis.

Since the initial discovery that ubiquitin became covalently attached to lysozyme, hundreds of other proteins that are modified by ubiquitin (or other ubiquitin-like proteins) have been identified. Moreover, it has become clear that ubiquitin can be attached via any one of its seven lysine residues, and that the specific residue that is utilized impacts the fate of the ubiquitinated protein. Many questions arise from this diversity, including what determines the specificity for ligation and what are the functions of the different conjugations.

In one recent study, Mark Hochstrasser, Daniel Finley, Junmin Peng, and colleagues set out to identify lysine 11 (K11)-specific linked proteins in *Saccharomyces cerevisiae*, knowing these linkages to be abundant *in vivo*, but poorly characterized. The investigators

Figure 1 Formation of covalent compounds between APF-1 and lysozyme in an ATP-dependent reaction. [125I]-APF-1 was incubated together with fraction II without ATP (lane 1), with ATP (lane 2), with ATP and 5 µg (lane 3), 10 µg (lane 4) and 25 µg (lane 5) of unlabeled lysozyme. In addition [125I]-lysozyme was incubated with unlabeled APF-1 without ATP (lane 6) and with ATP (lane 7). The same five bands C1–C5 are observed when either protein is labeled and incubated with the other unlabeled protein in the presence of ATP. Reproduced from Hershko *et al.*, *Proceedings of the National Academy of Sciences of the U S A* 1980;**77**:1783–1786.

took a high throughput approach and profiled the total yeast proteins and ubiquitinated proteins in a wild-type strain and in a mutant strain in which K11 of ubiquitin was mutated to arginine (Ub-R11). The authors used a quantitative mass spectrometric approach termed SILAC (stable isotope labeling with amino acid in cell culture, discussed in Section 16.10) wherein wild-type (Ub-K11) and mutant (Ub-R11) cells were labeled with heavy and light stable isotopes, respectively, and then mixed. Total proteins from cell lysates, and those purified on the basis of being conjugated to ubiquitin, were separated on SDS gels. The separated proteins were then excised from the gel, digested with trypsin and analyzed by mass spectrometry (LC/MS/MS) (Figure 2). The peptide fragment masses detected by mass spectrometry allowed the different proteins to be identified, while the relative levels of the various fragments (and thus intact protein amounts) were assessed by comparison of the two isotopes.

(a)

R11 strain (light media)   wild-type (heavy media)

combine at 1:1 ratio

extract total cell lysate (TCL)

purify Ub conjuagtes (UC) by nickel column

run TCL and UC on a 1D SDS gel

excise gel bands and digest proteins

identify and quantify proteins by LC-MS/MS

verify Ub conjugates by MW shift

identify candidates with K11 linkages

(b)

Figure 2 Use of SILAC for proteomic analysis of wild-type and R11 mutant yeast strains. (a) Outline of the SILAC method for comparing total cell lysate (TCL) and purified ubiquitin conjugates (UC) from wild-type and R11 mutant strains. (b) Comparison of the TCL and UC by SDS-PAGE showing the great number of proteins analyzed. The gel was excised into ~50 gel bands. Reproduced from Xu *et al.*, *Cell* 2009;**137**:133–145.

Xu *et al.* anticipated that potential K11-ubiquitin-linked proteins would be present at higher levels in the R11 mutant strain as a result of reduced rates of degradation in the absence of K11-specific ubiquitin conjugation. A number of candidate ubiquitinated proteins were identified (347 to be exact), and a few of these were subsequently confirmed as K11-ubiquitin modified proteins through independent assays. Although a fairly broad set of proteins appears to be targeted by this modification, one of the most intriguing candidates identified was Ubc6 (Figure 3), itself an E2 ubiquitin-conjugating enzyme with known roles in the endoplasmic reticulum-associated degradation (ERAD) pathway. This lead was supported by subsequent experiments showing hypersensitivity of the Ub-R11 strain to reagents that induce stress in the endoplasmic reticulum. Like more traditional genetic approaches, the excitement for high throughput approaches comes from their potential to identify new connections in our quest to understand biology.

(a)

a peptide of Cdc48
(AVATEVSANFISVK)

(b)

a peptide of Ubc6
(VISLEEILDPEDR)

Figure 3 Isotope-labeled peptide pairs of Cdc48 (a) and Ubc6 (b) in the total cell lysate. The light- and heavy-labeled peptides were distinguished by their different mass-to-charge (m/z) ratios. While the intensity of the peak for Cdc48 is roughly equivalent in the two strains, the intensity of the Ubc6 peptide is markedly different, consistent with the idea that ubiquitin-dependent degradation is reduced in the R11 mutant strain.
From Xu *et al.*, *Cell* 2009;**137**:133–145.

Ciechanover A, Heller H, Elias S, Haas AL, Hershko A. ATP-dependent conjugation of reticulocyte proteins with the polypeptide required for protein degradation. *Proceedings of the National Academy of Sciences of the U S A* 1980;**77**:1365–1368.

Hershko A, Ciechanover A, Heller H, Haas AL, Rose IA. Proposed role of ATP in protein breakdown: conjugation of protein with multiple chains of the polypeptide of ATP-dependent proteolysis. *Proceedings of the National Academy of Sciences of the U S A* 1980;**77**:1783–1786.

Xu P, Duong DM, Seyfried NT, *et al*. Quantitative proteomics reveals the function of unconventional ubiquitin chains in proteasomal degradation. *Cell* 2009;**137**:133–145.

Figure 11.29 Steps in protein ubiquitination. Protein ubiquitination begins with the ATP-dependent covalent attachment of the C-terminus of the 76-residue ubiquitin protein to the ubiquitin-activating enzyme (E1), which then transfers the ubiquitin to one of a family of ubiquitin-conjugating enzymes (E2). The E2 then collaborates with a target-specific ubiquitin-protein ligase (E3) to catalyze the formation of a peptide bond between the activated C-terminal glycine of ubiquitin and a lysine side chain on the target protein. Additional ubiquitins can then be added to other lysines on the target or on ubiquitin itself, sometimes resulting in multiple ubiquitin polymers on the protein surface.

step, ubiquitin activation, ubiquitin peptide is covalently attached through its C-terminus to the sulfhydryl group of a cysteine in the active site of the ubiquitin-activating enzyme (E1). This reaction is powered by ATP hydrolysis using a two-step mechanism similar to that described for tRNA synthetases in Chapter 10. The E1–ubiquitin conjugate then interacts with a ubiquitin-conjugating enzyme (E2) that catalyzes the transfer of the ubiquitin to an active-site cysteine in E2. The final step is the transfer of the ubiquitin to the target protein. This depends on a third enzyme, the ubiquitin–protein ligase (E3), which helps catalyze transfer of ubiquitin from the sulfhydryl linked E2–ubiquitin conjugate to the amino group of a lysine side chain in the target protein (see Figure 11.29).

As mentioned above, ubiquitin ligases sometimes catalyze the attachment of ubiquitin to lysine residues within ubiquitin itself, giving rise to polyubiquitin chains. There are seven lysine residues in the highly conserved ubiquitin protein, suggesting the potential for tremendous product diversity. However, polyubiquitin chains usually have a uniform linkage type, meaning that each ubiquitin is joined to the same lysine residue in the next ubiquitin. For example, K48-linked polyubiquitin chains are formed by linking the C-terminus of one ubiquitin to lysine 48 in the next, while K63-linked chains contain bonds between the C-terminal residue and lysine 63. Although the functions of K48- and K63-linked polyubiquitin chains are best understood, ongoing studies are finding roles for K11-, K29- and K6-linked polyubiquitin chains, and perhaps others as well.

Although most eukaryotic cells contain a single type of E1 enzyme, they typically have a number of different E2 enzymes and hundreds of E3 enzymes. It is not surprising that only a single E1 enzyme is needed for the uniform task of chemically activating the ubiquitin protein in an ATP-dependent reaction that is independent of the diversity of downstream targets. The great variety of E3 enzymes reflects the need of the cell to very specifically target particular protein substrates for ubiquitination, as well as its need to interact with a variety of E2 enzymes. E2 enzymes are generally well conserved in overall structure and function consistent with the fact that they all must interact with the same E1. By contrast, the variability in the E2 enzymes likely reflects their need to interact with a wide range of E3 enzymes.

As mentioned above, ubiquitination is reversible. Deubiquitinating enzymes are typically cysteine proteases that break the very stable isopeptide bond between ubiquitin and lysine. Other enzymes are involved in breaking down the ubiquitin chains into monomers, thus allowing them to be recycled within the cell. These two systems allow ubiquitin to serve as a reversible signal, much in the way that covalent modifications such as phosphorylation and acetylation play a role in the cell. The human genome encodes over 500 deubiquitinating enzymes, suggesting that these enzymes play diverse roles in very specific pathways, with different pathways having their own deubiquitinating enzymes.

The number of eukaryotic genes devoted to ubiquitination and deubiquitination is on the same order as the number of genes devoted to phosphorylation and dephosphorylation, emphasizing the comparable importance of both processes. Homologs of ubiquitin-modification genes have been found in some bacteria, but the functions of these genes is generally not known.

Different types of ubiquitin modifications have distinct functional consequences

Different types of ubiquitin modification play distinct roles in the cell (see Figure 11.30). Perhaps the best understood is the targeting of proteins for degradation by covalent attachment of a K48-linked polyubiquitin chain. The K48-linked chain is recognized by the proteasome, which proteolyses the protein into short peptides. We shall learn more about the mechanism of protein degradation by the proteasome in the next section.

In contrast to modification by a K48-linked polyubiquitin chain, modification by K63-linked polyubiquitin chains does not lead to degradation of the modified protein. Instead, the K63-linked chains modulate the activities of proteins with critical roles in the DNA damage response described in Section 12.9 as well as in the activation of kinases that lead to activation of the NF-κB transcription factor described in Section 8.13. In the case of the DNA damage response, the addition of different ubiquitin moieties to the DNA processivity factor Proliferative Cell Nuclear Antigen (PCNA) has significantly different consequences. PCNA modified by K63-linked polyubiquitin is associated with error-free DNA replication and repair. In contrast, the attachment of a single ubiquitin at the same position on PCNA results in an error-prone DNA replication, which permits the bypass of bulky DNA lesions that have not been repaired. Each type of modification must somehow recruit a distinct set of proteins to the DNA.

As another example, we learned in Chapter 4 that histone monoubiquitination plays an important role in nucleosome structure and transcriptional regulation. Again, ubiquitin exhibits a non-degradative role in this process.

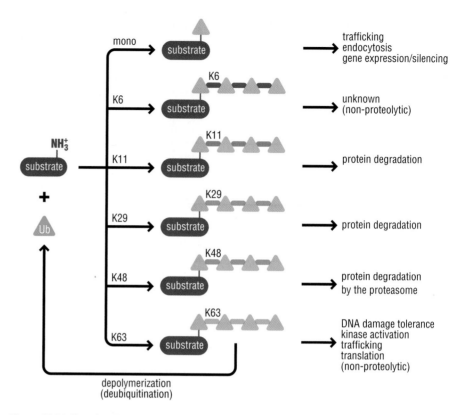

Figure 11.30 Functional consequences of different ubiquitin modifications. Each type of polyubiquitin chain is classified according to the lysine side chain in ubiquitin to which the C-terminus of the next ubiquitin is covalently attached. Each type of linkage is denoted by a different color, as indicated atop each bar.
Adapted from Pickart and Fushman (2004).

Since ubiquitin acts as a signal, there must be protein domains that specifically recognize the different types of ubiquitin modifications. Several domains that interact with mono- or diubiquitin have been identified. These include the UIM (Ub interacting motif), CUE (coupling of ubiquitination to endoplasmic reticulum degradation), and UBA (Ub-associated) domains. In some cases, proteins with more than one ubiquitin recognition domain have been found to bind to longer polyubiquitin chains. It seems clear that recognition of specific ubiquitinated proteins is a relatively complex process that depends on recognition of the structure of the protein surrounding the modification, the length and type of ubiquitin polymer, and perhaps other post-translational modifications found on the protein.

Modification by distinct ubiquitin-like proteins regulates protein activity

There are other small proteins that are similar to ubiquitin in their structure and in the enzymatic steps used to attach them to substrates. Called ubiquitin-like proteins, they serve signaling functions that are distinct from ubiquitin. SUMO (which stands for Small Ubiquitin-related Modifier) is one example of a ubiquitin-like protein; it is covalently attached to lysine side chains in a process known as **sumoylation**. The attachment of SUMO to a substrate is carried out by a specialized set of enzymes in a series of chemical steps that are similar to those used for ubiquitination. The enzymes involved are related to the ubiquitin specific E1, E2, and E3 enzymes.

Sumoylation plays an important role in gene expression and in targeting proteins to the nuclear pore complex. SUMO may also control protein degradation by competing with ubiquitin for common lysine residues. For example, attachment of SUMO to a lysine side chain can prevent attachment of a polyubiquitin chain and hence protect the protein from degradation through the ubiquitin–proteasome pathway. Although polymeric SUMO has been detected, there is currently no known role for SUMO chains in the cell.

11.9 PROTEIN DEGRADATION

Cells must balance requirements for protein synthesis and degradation

Throughout this chapter, we have discussed the post-translational events that influence a protein's life following its release from the ribosome. The folding pathways and the post-translational modifications described in the previous sections illustrate the complexity of the processes that dictate the localization and activity of a protein molecule. The last step in the cellular life of a protein is its degradation. The cell must balance the requirements for the production of fully folded functional proteins with the need to eliminate protein activity when their function is no longer needed. Here we discuss the machineries in the cell that are dedicated to the destruction of proteins. Not surprisingly, these systems are tightly regulated to meet the ever-changing needs of the cell and to prevent unwarranted destruction of proteins that are still required. In contrast to many of the other processes that we have discussed in this chapter (primarily post-translational modifications), the process of degradation is not reversible; such finality requires this process to be strenuously controlled.

The bacterial and eukaryotic machineries for energy-dependent protein destruction are highly related

Related molecular machines power the energy-dependent unfolding and degradation of most proteins in bacteria and eukaryotes. These proteolytic machines are composed of two basic functionalities: an ATPase and a peptidase, which can reside either on the same or on different polypeptide chains. The polypeptide chains oligomerize to form a barrel-like structure with ATPase activities and small pores at both ends, and a protected internal cavity that contains the active sites of the degradative protease (Figure 11.31). The best-known eukaryotic degradation machine, referred to as the proteasome, is composed of about 30 different polypeptide chains. The bacterial versions of this machine tend to be somewhat less complex. For example, the FtsH and Lon proteases are composed of just a single type of monomer (containing both the ATPase and proteolytic functionalities). While there are at least five ATP-dependent proteases in bacteria, there are probably many more in eukaryotes. Here we will focus our discussion on the well-characterized proteasome.

Figure 11.31 Proteasome structure. Proteins targeted for destruction (gray) are fed into a multiprotein complex where they are degraded in the tunnel-like enzymatic core. Bacterial ATP-dependent proteases have a similar structure although they lack the cap (dark blue) at either or both ends. PDB 1PMA.
Adapted from Petsko and Ringe (2004).

The structure of the protease has obvious functional consequences

Even the smallest folded proteins have dimensions that make them unable to enter the cavity of the proteases, thus providing a clear answer to how most proteins in the cell escape the actions of roving proteolytic machines. The ATPase domains surround the entry portal, an ideal location for allowing them to select desirable proteins and to then mechanically denature and thread them into the protected central cavity. The amount of ATP that is needed to denature a range of substrates (of varying stability) is quite variable and, for the ClpXP protease, can range from 30 to 80 ATPs hydrolyzed per about 100 amino acids of protein sequence.

Once substrates are translocated into the central cavity, the effective concentration of protease active site in the small chamber is extremely high (on the order of hundreds of mM) thus resulting in rapid degradation of the protein into peptide fragments ranging from five to 20 amino acids in length. Smaller fragments are thought to readily diffuse out of the chamber whereas larger fragments may depend on further ATP-dependent processes to be released.

Ubiquitin, molecular recognition sequences, and adaptor proteins collaborate to identify proteins that need to be degraded

The finality of protein degradation demands that this cellular event be highly selective. How then are the proper proteins recognized? In eukaryotes, selectivity is in part provided by the polyubiquitin chains (Figure 11.32a), which we discussed in the preceding section. For example, proteins tagged with K48-linked polyubiquitin chains are targeted by the proteasome.

In bacteria, substrate recognition by the proteases appears to be mediated by a diverse set of unstructured peptide signals (Figure 11.32b). Such diversity in the recognition sequences helps to explain how these machines can target disparate classes of proteins. Not surprisingly, the primary recognition sequences tend to be located at both the N- and C-termini of proteins where they are less likely to be sequestered in the core of the molecule. In general, there appears to be one sequence that acts as a primary degradation tag, whereas other sequence elements contribute to the binding in more subtle ways. Other important players in targeting the appropriate proteins for degradation are adaptor proteins (Figure 11.32c), which bind to both the protease and the substrate protein to increase the interaction between these molecules.

Figure 11.32 Protein recognition by ATP-dependent proteolytic complexes. Proteins to be degraded are recognized (a) by the ubiquitin tag in eukaryotes or (b) by an exposed N- or C-terminus in bacteria. (c) Bacterial substrates can also be brought to the ATP-dependent proteases by adaptor proteins.

Stress response programs maintain the balance between protein synthesis and degradation

A number of stresses to which cells can be exposed (including elevated temperature as well as exposure to various chemical reagents) ultimately result in the increased presence of aberrantly folded proteins. These unfolded proteins are detrimental to cell function and are dealt with by the cell in a number of general ways – by a decrease in overall protein production, or by an increase in chaperone or degradative activity. Both bacterial and eukaryotic cells have evolved systems that very specifically respond to such stresses by turning on transcriptional (or sometimes translational) programs that up- or down-regulate the requisite genes, which, themselves, act to rebalance the protein status of the cell.

In bacteria, a pathway regulated by σ^E, an alternative sigma factor discussed in Section 8.3, is responsible for regulating a transcriptional program that is turned on in response to unfolded proteins in the envelope of cells (Figure 11.33a). In this system, a membrane-spanning protein, RseA, has a periplasmic domain, which is important for the recognition of unfolded proteins in the membrane and periplasm, and a cytoplasmic domain that is bound by σ^E. Recognition of unfolded proteins initiates a proteolytic cascade that ultimately leads to the release of σ^E from RseA so that it can bind to RNA polymerase and stimulate transcription of the genes in response to the insult. The genes expressed include those encoding proteases, which degrade the unfolded proteins, and non-coding RNAs, which repress the translation of abundant outer membrane proteins.

A eukaryotic system that is roughly equivalent to the RseA-σ^E system is known as the **unfolded protein response** (UPR) (Figure 11.33b). In this response,

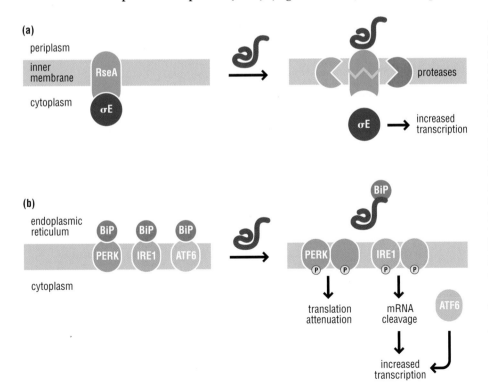

Figure 11.33 Responses to unfolded proteins. (a) Bacterial membrane stress response. The membrane-spanning protein, RseA (orange), has a periplasmic domain, which is important for the recognition of unfolded proteins in the membrane and periplasm, and a cytoplasmic domain that is bound by σ^E (blue). Recognition of unfolded proteins initiates a proteolytic cascade that ultimately leads to the release of σ^E from RseA so that it can bind to RNA polymerase and stimulate transcription of the genes that help to reduce the number of unfolded proteins. (b) Eukaryotic unfolded protein response (UPR). In the absence of stress, three membrane-spanning proteins (IRE1, PERK, and ATF6) interact with the chaperone protein BiP (red) located in the endoplasmic reticulum. Upon stress, BiP preferentially binds unfolded proteins rather than IRE1, PERK, or ATF6. Consequently, IRE1, PERK, and ATF6 are released from sequestration by BiP, resulting in their activation, which includes the dimerization and phosphorylation of IRE1 and PERK. The activation of IRE1, PERK, and ATF6 leads to numerous protective downstream responses.

cells sense unfolded proteins in the endoplasmic reticulum (the approximate equivalent of the periplasm in bacteria) that have been generated under various stress conditions. The UPR pathway in mammalian cells features three membrane-spanning proteins (IRE1, PERK, and ATF6) that respond to stress in the endoplasmic reticulum and communicate to other parts of the cell via their cytoplasmic domains. In the absence of stress, each of these proteins interacts with the chaperone protein BiP located in the endoplasmic reticulum. Upon stress, BiP preferentially binds unfolded proteins rather than IRE1, PERK, or ATF6. Consequently, IRE1, PERK, and ATF6 are released from sequestration by BiP, resulting in their activation, which includes the dimerization and phosphorylation of IRE1 and PERK. The activation of IRE1, PERK and ATF6 leads to numerous downstream responses, including mRNA cleavage to remove an intron of the transcript encoding a key transcription activator (activated IRE1 is an endoribonuclease), translational shut-down of most other genes (activated PERK directly regulates the activity of the translation initiation factor eIF2a), and the transcriptional activation of multiple genes (activated ATF6 is a transcription activator).

The intricate regulation described above, in addition to other regulatory responses to unfolded or misfolded proteins not described here, is not surprising given the burden imposed on the cell by these inactive proteins. Not only do non-functional proteins represent a waste of resources, but these proteins may also block the action of other proteins or have detrimental activities such as the formation of the protein aggregates associated with many neurodegenerative diseases.

In subsequent chapters where we examine how cells cope with changes to the genome, we will see even more examples of how the protein modifications impact all aspects of cell metabolism.

✹ SUMMARY

- Most polypeptide chains are not functional upon release from the ribosome.

Protein folding

- Protein chaperones assist the proper folding and unfolding of proteins by interacting with hydrophobic regions of the protein.

- Some proteins require the formation of disulfide bonds for their correct folding.

Protein targeting

- Proteins are targeted to specific subcellular locations by amino acid sorting motifs as well as by the attachment of lipids such as fatty acyl groups, isoprenoid groups or GPI anchors.

Post-translational cleavage

- Some proteins need to be cleaved by site-specific proteases before they are active.

- Some proteins contain self-excising inteins, which must be removed from the protein product for it to become active.

Covalent modifications of proteins

- Lipid modifications target proteins to membranes.

- The attachment of diverse carbohydrate chains increases protein solubility and can serve as recognition motifs or help protect proteins from degradation.

- Phosphorylation, acetylation, and methylation, and other small molecule additions reversibly regulate protein activity.

- Some non-enzymatic modifications such as oxidation and nitrosylation can be regulatory or detrimental depending on the adduct and the amino acid modified.

- The addition of monomeric or polymeric ubiquitin protein moieties by a series of reactions catalyzed by E1, E2, and E3 enzymes can have a variety of consequences.

- Many modifications added to proteins including sugar, methyl, acetyl, and ubiquitin groups, are coupled to nucleotide carriers which activate the group for transfer to the protein.

- Protein degradation

- Barrel-like ATP-dependent proteolytic complexes, termed the proteasome in eukaryotes, degrade most proteins.

- The substrate protein is threaded into the inside of the barrel, where the protease active sites are located, in a process powered by ATP hydrolysis.

 # FURTHER READING

11.1 CHAPERONE-ASSISTED PROTEIN FOLDING

Bukau B, Weissman J, Horwich A. Molecular chaperones and protein quality control. *Cell* 2006;**125**:443–451.

Chiti F, Dobson CM. Protein misfolding, functional amyloid, and human disease. *Annual Review of Biochemistry* 2006;**75**:333–366.

Deuerling E, Bukau B. Chaperone-assisted folding of newly synthesized proteins in the cytosol. *Critical Reviews in Biochemistry and Molecular Biology* 2004;**39**:261–277.

Mamathambika BS, Bardwell JC. Disulfide-linked protein folding pathways. *Annual Review of Cell and Developmental Biology* 2008;**24**:211–235.

11.2 TARGETING OF PROTEINS THROUGHOUT THE CELL

Bonifacino JM, Traub LM. Signals for sorting of transmembrane proteins to endosomes and lysosomes. *Annual Review of Biochemistry* 2003;**72**:395–447.

Stewart M. Molecular mechanism of the nuclear protein import cycle. *Nature Reviews Molecular Cell Biology* 2007;**8**:195–208.

Terry LJ, Shows EB, Wente SR. Crossing the nuclear envelope: hierarchical regulation of nucleocytoplasmic transport. *Science* 2007;**318**:1412–1416.

Wickner W, Schekman R. Protein translocation across biological membranes. *Science* 2005;**310**:1452–1456.

11.3 POST-TRANSLATIONAL CLEAVAGE OF THE POLYPEPTIDE CHAIN

Jiang J, Hui CC. Hedgehog signaling in development and cancer. *Developmental Cell* 2008;**15**:801–812.

Saleh L, Perler FB. Protein splicing in *cis* and in *trans*. *Chemical Record* 2006;**6**:183–193.

Wlodawer A, Vondrasek J. Inhibitors of HIV-1 protease: a major success of structure-assisted drug design. *Annual Review of Biophysics and Biomolecular Structure* 1998;**27**:249–284.

11.4 LIPID MODIFICATION OF PROTEINS

Mann RK, Beachy PA. Novel lipid modifications of secreted protein signals. *Annual Review of Biochemistry* 2004;**73**:891–923.

Paulick MG, Bertozzi CR. The glycosylphosphatidylinositol anchor: a complex membrane-anchoring structure for proteins. *Biochemistry* 2008;**47**:6991–7000.

Resh MD. Trafficking and signaling by fatty-acylated and prenylated proteins. *Nature Chemical Biology* 2006;**2**:584–590.

Rocks O, Peyker A, Bastiaens PI. Spatio-temporal segregation of Ras signals: one ship, three anchors, many harbors. *Current Opinion in Cell Biology* 2006;**18**:351–357.

11.5 GLYCOSYLATION OF PROTEINS

Hart GW, Housley MP, Slawson C. Cycling of *O*-linked beta-*N*-acetylglucosamine on nucleocytoplasmic proteins. *Nature* 2007;**446**:1017–1022.

Imperiali B, O'Connor SE. Effect of *N*-linked glycosylation on glycopeptide and glycoprotein structure. *Current Opinion in Chemical Biology* 1999;**3**:643–649.

Wells L, Vosseller K, Hart GW. Glycosylation of nucleoplasmic proteins: signal transduction and *O*-GlcNac. *Science* 2001;**291**:2376–2378.

11.6 PROTEIN PHOSPHORYLATION, ACETYLATION, AND METHYLATION

Aaronson DS, Horvath CM. A road map for those who don't know JAK-STAT. *Science* 2002;**296**:1653–1655.

Hunter T. The age of crosstalk: phosphorylation, ubiquitination, and beyond. *Molecular Cell* 2007;**28**:730–738.

Kouzarides T. Chromatin modifications and their function. *Cell* 2007;**128**:693–705.

Laub MT, Goulian M. Specificity in two-component signal transduction pathways. *Annual Review of Genetics* 2007;**41**:121–145.

11.7 PROTEIN OXIDATION AND NITRATION

Hancock JT. The role of redox in signal transduction. *Methods in Molecular Biology* 2008;**476**:1–9.

Hess DT, Matsumoto A, Kim SO, Marshall HE, Stamler JS. Protein S-nitrosylation: purview and parameters. *Nature Reviews Molecular Cell Biology* 2005;**6**:150–166.

Poole LB, Karplus PA, Claiborne A. Protein sulfenic acids in redox signaling. *Annual Review of Pharmacology and Toxicology* 2004;**44**:325–347.

Stadtman ER. Protein oxidation and aging. *Free Radical Research* 2006;**40**:1250–1258.

11.8 UBIQUITINATION AND SUMOYLATION OF PROTEINS

Elsasser S, Finley D. Delivery of ubiquitinated substrates to protein-unfolding machines. *Nature Cell Biology* 2005;**7**:742–749.

Pickart CM, Fushman D. Polyubiquitin chains: polymeric protein signals. *Current Opinion in Chemical Biology* 2004;**8**:610–616.

Ravid T, Hochstrasser M. Diversity of degradation signals in the ubiquitin-proteasome system. *Nature Reviews Molecular Cell Biology* 2008;**9**:679–690.

11.9 PROTEIN DEGRADATION

Ahang E, Kaufman RJ. Signaling the unfolded protein response from the endoplasmic reticulum. *Journal of Biological Chemistry* 2004;**279**:25935–25938.

Alba BM, Gross CA. Regulation of the *Escherichia coli* sigma-dependent envelope stress response. *Molecular Microbiology* 2004;**52**:613–619.

Baker TA, Sauer RT. ATP-dependent proteases of bacteria: recognition logic and operating principles. *Trends in Biochemical Sciences* 2006;**31**:647–653.

CELLULAR RESPONSES TO DNA DAMAGE

INTRODUCTION

The preservation of genetic information from one generation to the next requires the DNA sequence of the cell's genome to be maintained without alteration. Yet the integrity of the DNA sequence is under constant threat of damage from errors in normal cellular processes such as DNA replication and the production of reactive metabolites and environmental agents. DNA damage from such errors can result in changes in base sequence or even in chromosome structure. Despite the sophisticated mechanisms that have evolved to ensure accuracy of DNA replication, errors do occur and are one of the causes of sequence change.

If errors remain uncorrected, changes in DNA are propagated as mutations. If damage is encountered by the replication machinery in the course of DNA replication, chromosomes may be broken or replicated incompletely and fail to segregate properly during cell division.

The effects of damage from replication errors and other processes on DNA are greatly reduced by specialized processes of **DNA repair**. There are many pathways for the detection and repair of damaged DNA. As a result, most damaged DNA is repaired and only a small fraction of DNA damage – likely less than 1 in 1000 incidents – results in a permanent change in DNA sequence or a failure of chromosome replication and segregation.

In this chapter and the next we shall discuss the major mechanisms by which damage and change are introduced, and the mechanisms that have evolved to repair damage and limit change. Many strategies for repair depend on the double-stranded structure of DNA to provide the intact sequence information on which correction of the damaged region can be based: these are the repair pathways we describe in detail in this chapter. Where both strands of the DNA molecule are broken and there is no intact strand to serve as a template for repair, different pathways are required, and these are the subject of Chapter 13.

For some types of DNA damage, repair proteins are always present at levels sufficient for rapid recognition and repair. In other cases, however, especially where DNA damage has arrested DNA replication, synthesis of repair proteins is increased in a global response to DNA damage in both bacteria and eukaryotes.

➔ We learn about the process of DNA replication in Chapter 6.

The inducers of such global responses are single-stranded DNA, the formation of which is the result of many types of DNA damage. Single-stranded DNA is always present when damage has blocked DNA replication. Double-strand breaks also induce such a global damage response. We shall describe the mechanism of such global responses to both single-stranded and double-stranded DNA breaks later in the chapter.

Before we embark on our exploration of DNA repair, a word on the history of repair protein nomenclature may help to make the terminology less confusing. As with many fundamental biological processes that were first elucidated through genetic analysis, the nomenclature of the proteins contributing to DNA repair pathways often reflects the way they were first discovered rather than their biochemical functions or structural relationships. For example, many repair proteins were discovered through the analysis of bacterial or yeast mutants showing abnormal sensitivity to DNA-damaging agents, such as x-rays or ultraviolet light, and are consequently called Rad proteins or Uvr proteins. Human homologs of some of these proteins, discovered later, have sometimes inherited this nomenclature, and we shall encounter human Rad proteins in the course of this chapter. Many bacterial repair proteins play a part in recombination reactions and are consequently known as Rec proteins.

In humans, defects in repair proteins often lead to the development of diseases that are characterized by increased incidence of cancer, and repair proteins identified in this way are named after the diseases they cause when they are defective. For example, the XP proteins we shall meet in this chapter were identified in studies on the human disease xeroderma pigmentosum, in which there is increased sensitivity to DNA damage.

We begin the chapter by describing the types of DNA damage that can occur.

12.1 TYPES OF DNA DAMAGE

Misincorporation of nucleotides occurs during DNA synthesis

The integrity of genetic information in cells depends on the fidelity of DNA replication and the stability of the DNA molecule itself. The accuracy of DNA replication depends critically on the selection of the correct nucleotide to add to the growing DNA strand. Such accurate selection depends on correct base-pairing between the template strand of the DNA and the incoming nucleotide. The fidelity of DNA polymerization, as we have seen in Section 6.3, is increased by the proofreading activities of exonucleases associated with the replisome. The exonucleases excise inappropriate nucleotides and allow resynthesis of DNA with a high probability of incorporating the correct nucleotide. Despite this, however, DNA polymerization is not error-free: mispaired bases do occur in newly synthesized DNA every 10^5 to 10^6 base pairs.

One reason for such errors is the occasional transient formation of tautomeric isomers, or tautomers, of the four DNA nucleotides, in which some of the hydrogen atoms that mediate base-pairing have changed position.

⬁ We discuss tautomers in more detail in Section 2.4.

Such tautomers can form non-Watson–Crick base pairs (Figure 12.1) so that an incorrect nucleotide pairs with the template and is incorporated into the new strand. After the next round of replication, the presence of mispaired bases results in a permanent change in sequence at the position of the mismatch in one of the daughter duplexes, that is, a mutation (Figure 12.2).

If a purine base, adenine or guanine, is replaced by the other purine base or a pyrimidine base, thymidine or cytosine, is replaced by the other pyrimidine, the mutation is called a **transition mutation**. If a purine base is replaced by a pyrimidine base or a pyrimidine base is replaced by a purine base, the mutation is called a **transversion mutation**.

DNA is subject to spontaneous and induced damage that can lead to mispairing and mutation

Alterations to DNA that can arise from both intracellular and extracellular agents disrupt the accurate base-pairing essential to accurate DNA replication, and hence can cause mutations. Bulky alterations can also block DNA replication. In this section, we describe the different sources of such damage and their mechanisms of disrupting DNA. In the next sections, we shall explain how this damaged DNA is repaired.

Spontaneous damage that occurs under physiological conditions includes hydrolytic **depurination**, resulting in loss of purine bases (Figure 12.3a); hydrolytic **depyrimidation**, resulting in the loss of pyrimidine bases; and **deamination**, resulting in changes in base structure (Figure 12.3b). In mammalian cells, for example, about 18 000 bases per cell are lost by depurination each day, while deamination affects around 500 bases per cell per day. Such changes can also occur to free nucleotides, thus poisoning the pool of DNA precursors.

A second important cause of DNA damage is the production of reactive metabolites in the course of normal intracellular metabolism. For example, several cellular processes, including energy production via electron transport chains, produce reactive oxygen species. These species include superoxides, peroxides, and other oxygen ions that cause oxidative damage. Although cells have enzymes such as superoxide dismutase and catalase that specifically target and destroy such species, they are not completely destroyed and remain at sufficient levels to cause significant DNA damage.

Bases, sugars, and the phosphodiester backbone are all subject to oxidation. In fact, oxidative damage to the DNA backbone is the most dangerous type of DNA damage. Why is this? Because if the damage is not repaired, it can result in potentially lethal double-strand breaks that can lead to genome instability as a consequence of subsequent errors in chromosome replication and segregation.

Other natural products of metabolism can also damage DNA: among the most important of these are **alkylating agents**, which are potent modifiers of the structure of DNA bases, adding methyl groups (Figure 12.3c) or ethyl groups, which disrupt normal base-pairing interactions. Common reactions of this type include the formation of O^6-methylguanine, which pairs with thymine instead of cytosine, and 7-ethylguanine, which pairs with thymine instead of cytosine. Alkylating agents can also cause interstrand crosslinks that block strand separation.

Both natural products and human-made chemicals in the environment can interact with and damage DNA. Indeed, the vast majority of substances that are carcinogens work by interacting with DNA and causing mutations. For example,

(a) normal pairing

(b) adenine tautomer pairing

Figure 12.1 **Mispairing due to tautomeric forms of the DNA bases.** (a) The normal base-pairing of thymine and adenine. (b) In contrast, an imino tautomer of adenine (adenine*) mispairs with cytosine.

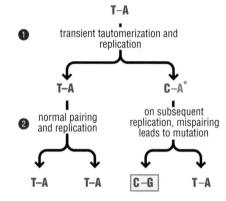

Figure 12.2 **Misincorporation of bases due to tautomerization.** A normal T–A base pair is shown at the top. (1) Tautomerization of adenine (indicated by an asterisk) leads to mispairing with cytosine in the first round of replication. (2) In the second round of replication, the mispairing gives rise to a mutant C–G pair in one of the two copies of the DNA instead of the original T–A pair.

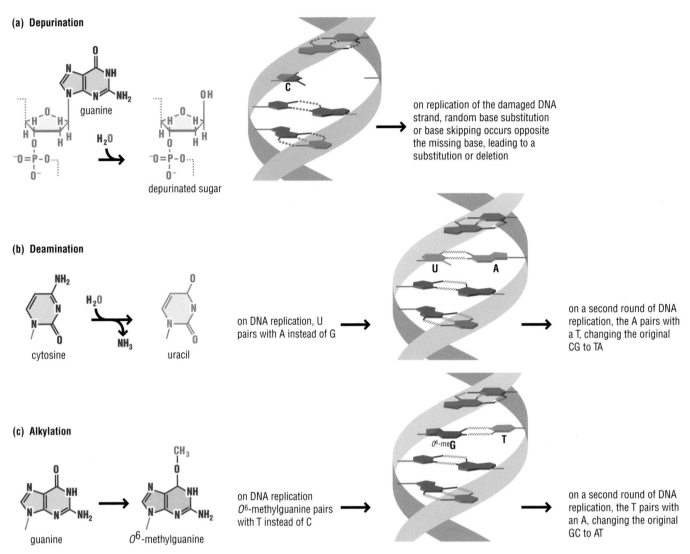

Figure 12.3 Some of the most common causes of base damage. (a) Possibly the most frequent of all types of DNA damage is the hydrolytic breakage of the glycosidic bond between a purine and the sugar–phosphate backbone of DNA, leaving a gap in the structure that causes substitutions or deletions when the DNA is replicated. (b) Another common form of damage is the hydrolytic deamination of cytosine or adenine, which substitutes an oxygen for the amino group. In the case of cytosine this generates uracil, which pairs with adenine, leading to the substitution of an A–T base pair for C–G; deamination of adenine (not shown) generates hypoxanthine, which pairs with cytosine, leading to the substitution of a G–C base pair for A–T. (c) Alkylating agents add methyl or ethyl groups to bases: O^6-methylguanine mispairs with thymine, leading to the substitution of an A–T base pair for C–G. Altered bases are shown in blue; mutant bases with which they mispair are shown in red.

nitrosamines, which are present in some foods, can act as alkylating agents. Degradation products of nitrosamines can also oxidize adenine and cytosine.

Most of the lesions we have discussed so far affect single nucleotides and, if unrepaired, cause single-nucleotide changes, or **point mutations**, in the DNA sequence. Many DNA-damaging agents have more widespread effects, however. Among these are the **intercalating agents**, which are aromatic or flat hydrophobic molecules that become inserted between the stacked bases in the DNA double helix. This insertion pushes the bases apart and distorts the helical structure, which becomes partly unwound and extended (Figure 12.4a). When such DNA is replicated, base pair insertions or deletions are generated in the newly synthesized DNA. In protein-coding regions, DNA insertions or deletions result in **frameshifting** during protein synthesis, such that essential proteins are not expressed and

(a) Intercalation

ethidium bromide

intercalating agents extend and partly unwind the DNA helix and this leads upon replication to insertions and deletions

(b) Pyrimidine dimers

pyrimidine dimers perturb the stacking interactions between DNA bases, distorting the phosphodiester backbone and preventing DNA replication

Figure 12.4 Agents that distort DNA. (a) Intercalating agents such as ethidium bromide insert themselves between adjacent bases, extending and partly unwinding the DNA helix and leading to insertions and deletions on DNA replication. (b) Irradiation by ultraviolet light generates pyrimidine dimers, which cannot fit into the active site of the replicative polymerase, preventing replication of the DNA.

dividing cells exposed to intercalating agents often die. For this reason, intercalating agents have sometimes been used in cancer therapy, to target dividing cells.

DNA is also subject to damage from physical sources. One of these is ultraviolet light, which promotes the formation of intrastrand **pyrimidine dimers** (Figure 12.4b): the coupled nucleotides cannot be accommodated in the active site of highly accurate replicative polymerases, and insertions and deletions are generated in the newly synthesized DNA. Another source of DNA damage is ionizing radiation, such as x-rays, which can introduce single- and double-strand breaks into DNA.

The cell has at its disposal a number of different strategies for repairing damaged DNA, with the precise strategy adopted depending on the nature of the damage. In the next five sections we will consider several repair strategies in turn, before examining how the cell detects the damage in the first place, prior to eliciting an appropriate response. We begin by considering the strategy of post-replication **mismatch repair**.

12.2 POST-REPLICATION MISMATCH REPAIR

Post-replication mismatch repair restores mismatched base pairs to their parental state

The *in vivo* mutation rate of about 10^{-9}–10^{-11} per base pair is considerably lower than the error rate of DNA polymerase of about 10^{-5} to 10^{-6} per base pair. This low *in vivo* mutation rate reflects the activity of the polymerase proofreading

➔ We discuss the sliding beta clamp in more detail in Section 6.6.

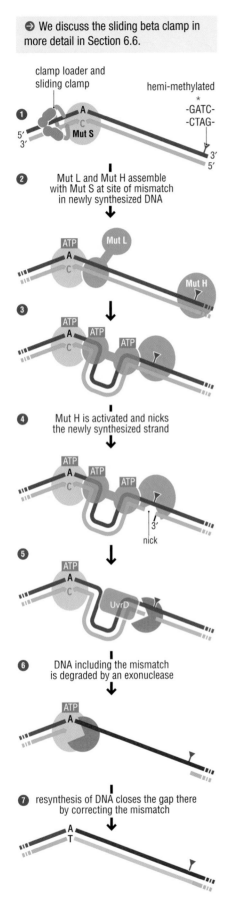

exonuclease and also a post-replication system that can detect and correct remaining mismatches immediately after they occur, a process known as **mismatch repair**.

Mismatch repair, like other repair systems described in this chapter, takes advantage of the fact that DNA is a duplex and contains a complete copy of its genetic information on both strands. Thus, if a base on one strand of a duplex is incorrect or damaged, the other strand can serve as a template to restore the original sequence.

Two types of protein have a central role in mismatch repair in all organisms that have been studied. Inactivation of these proteins in organisms from *Escherichia coli* to humans results in a high mutation rate. Therefore, these proteins are called MutS and MutL proteins, where Mut stands for mutator. In *E. coli*, there is only one MutS and one MutL, each acting as a homodimer; in eukaryotes there are often multiple versions of each protein that interact to form heterodimers.

An effective strategy for mismatch repair demands, first, that mismatches be specifically recognized, and second, that the newly synthesized strand be distinguished from the original strand so that the new, incorrect, nucleotide and not the original, correct, one is replaced. The first of these functions is performed by MutS, which specifically recognizes and binds mismatched base pairs, apparently sensing the way that such mismatches alter the bending properties of DNA. MutS can recognize virtually all mismatched base pairs. MutS protein also interacts with the sliding clamp components of the replisome, which thus serves to focus the repair machinery on newly synthesized regions of DNA where other components of the mismatch repair system assemble with MutS (Figure 12.5).

In *E. coli*, these components include both MutL and a third protein, MutH, which is an endonuclease that can specifically nick the newly synthesized daughter strand, targeting repair to this strand. MutH recognizes newly replicated DNA which, as we saw in Chapter 6, contains hemi-methylated GATC sites in which only the adenine residue of the parental strand is methylated. MutH specifically binds to GATC sites (Figure 12.5, step 2), sometimes hundreds of base pairs away from the site of damage.

If the DNA is hemi-methylated and a mismatch MutS–MutL complex is present, a MutSLH complex forms (step 3) and MutH introduces a nick into the newly synthesized unmethylated DNA strand (step 4). MutL then recruits helicase II

Figure 12.5 The MutSLH repair pathway of *E. coli*. (1) A newly replicated DNA duplex with a parental strand (gray) and newly synthesized strand (green) that contains an A–C mismatch base pair formed during DNA synthesis is shown. The mismatch base pair is bound by MutS, which is recruited to the site of replication errors both by the mismatch base pair and by the sliding clamp (blue ring), which is an essential component of the replisome. Because there is a delay in methylation at GATC sites of newly synthesized DNA, only the parental DNA strand in the region of DNA closest to the replication fork is methylated (blue T) and the new strand (green) can thus be distinguished when mispairing occurs. (2) One lobe of MutL interacts with MutS. (3) The other lobe of MutL interacts with MutH and binds to the hemi-methylated GATC site. (4) The MutSLH complex is assembled, activating MutH. (4) MutH nicks the newly synthesized, non-methylated DNA strand. (5) MutL and MutH dissociate from the complex and UvrD (helicase II) is recruited. (6) UvrD unwinds the nicked strand from the duplex and an exonuclease, shown here as a 5′–3′ exonuclease, degrades the newly synthesized strand from the site of the nick up to the site where MutS is bound to the mismatch. Degradation by a 3′–5′ nuclease occurs if the MutH nick is on the other side of the mismatch. (7) The DNA is then resynthesized with the correct base pair and the new DNA religated to the adjacent undamaged DNA.

(MutU/UvrD) that displaces the nicked strand from the duplex (step 5); exonucleases then degrade the newly synthesized DNA between the nick and the mismatch (step 6). Depending on the position of the nick relative to the mismatch, degradation occurs via a 5′–3′ exonuclease (as shown) or a different 3′–5′ exonuclease. New DNA synthesis fills the gap, replacing the mismatch with the correct base (step 7).

As in the *E. coli* system, the eukaryotic MutS proteins recognize mismatched base pairs. Eukaryotic MutLα does have a latent endonuclease activity that can be activated by eukaryotic MutS bound to a mismatch but it is not yet clear how the eukaryotic mismatch repair system distinguishes template from newly synthesized strands. Although DNA methylation occurs in many eukaryotes, it is not confined to newly replicated DNA, and there is no known eukaryotic homolog of MutH.

The mismatch repair system can also correct insertions and deletions

Mismatched base pairs are not the only errors that can occur during DNA synthesis. Repetitive DNA can form temporary hairpin structures while the DNA is in single-stranded form during DNA replication. DNA polymerase can skip past these hairpins, so that the new strand will contain a deletion of one or more repeats (Figure 12.6). Hairpin formation can also occur in the new DNA strand as it is being synthesized, leading to repeated replication of the same template region, which results in the insertion of one or more extra repeats in the newly synthesized DNA. Taken together, these processes are often called **slipped-strand misreplication**.

The hairpins are detected by mismatch repair systems and their effects are eliminated by degradation of the newly synthesized strand and subsequent resynthesis. In the case of a hairpin in the template strand, for example, degradation of the newly synthesized strand leads to unfolding of the hairpin and accurate resynthesis of the new strand, thus restoring the original sequence. In eukaryotes, different MutS variants mediate correction of base-pair mismatches and repair of hairpins due to slipped-strand misreplication.

Slipped-replication provides an important mechanism for promoting the variable expression of bacterial cell surface proteins in a process called phase variation, allowing evasion of immune responses. The genes whose expression varies contain short runs of the same base pair at the beginning of their open reading frames. Slipped-strand misreplication results in frameshifting, variably turning off and on the expression of different genes. Slipped-strand replication in repeat regions (for example, triplet nucleotide repeats present in some human genes that encode proteins important for brain function) underlies some neurodegenerative diseases.

Figure 12.6 Slipped-strand misreplication. A region of DNA consisting of AT repeats is shown. (1) When replication occurs, hairpin formation in one of the template strands (gray) may occur when the strands are separated at the replication fork. This leads to the newly synthesized complementary strand (green) having a deletion of several repeats. (2) If this strand is not corrected by the mismatch repair system, the deletion will become permanent at the next round of replication (blue).

Defects in mismatch repair result in increased mutation rates

The existence of mismatch repair was first suggested by the discovery of bacterial mutants that had an increased rate of spontaneous mutation – that is, the **mutator phenotype** – due to defects in the mismatch repair process.

Defects in mismatch repair in humans also lead to an increased rate of mutagenesis and are known to cause an increased predisposition to various types of cancer, arising because of the increased probability of mutation in genes that control cell growth and division. For example, mutation of human MUT genes is the cause of many cases of hereditary non-polyposis colon cancer, one of the most common hereditary diseases.

The post-replication mismatch repair described in this section is really a means of correcting errors in newly synthesized DNA, rather than repairing damage that has been inflicted by some environmental factor – more akin to correcting typographical errors on the page of a book than to repairing the damaged binding of that book. In the next sections, however, we consider strategies that are adopted by the cell to repair physical damage.

12.3 REPAIR OF DNA DAMAGE BY DIRECT REVERSAL

Damaged regions of DNA are repaired either by reversal or by excision and resynthesis

Lesions to bases in DNA that result from the reactions described in Section 12.1 are repaired in two main ways, direct reversal of particular types of damage or by recognition of DNA damage, followed by damage excision (removal of a damaged site) and DNA resynthesis. We will consider several types of repair by resynthesis in the following sections. In this section, however, we focus on the strategy of direct reversal.

In the case of some modifications, specialized enzymes recognize particular modified bases and reverse the modification, thus restoring the original state without cutting the DNA. For example, in organisms other than placental mammals (which seem to have lost this pathway when they diverged from marsupials), pyrimidine dimers can be recognized and repaired by enzymes known as **photolyases**, which use energy from near-ultraviolet light to split the dimer and thus reverse the damage, restoring the DNA to its original condition. Bulky lesions such as pyrimidine dimers can also be repaired by the nucleotide excision pathway described in Section 12.5, below.

Base methylation can be reversed by specialized methyltransferases

Alkylating agents can add methyl groups to bases (see Figure 12.3c) or to the DNA backbone. The modification to bases frequently occurs on guanine, which is particularly susceptible to alkylating agents, and is thus distinct from the methylations that occur through the actions of cellular methylases, which always methylate either adenine or cytosine.

⊘ We learn more about the methylation of chromosomal DNA in Section 4.7.

Methylguanine, and methyl groups on the backbone, are recognized by specialized repair alkyltransferases that reverse the damage by removing the methyl group from the base or the backbone and binding it covalently to a cysteine residue in the alkyltransferase active site. This is suicidal for the repair protein because the methylation is irreversible and inactivates the alkyltransferase activity.

In *E. coli*, with remarkable evolutionary ingenuity, the alkyltransferase, Ada, which has acquired a methyl group from the DNA backbone, is a sequence-specific DNA binding protein, which activates its own expression, thereby ensuring its replacement (Figure 12.7). Methylated Ada also activates the expression of a glycosylase enzyme (AlkA), which specifically removes methylated guanine bases from DNA by cleaving the linkage between the base and the sugar. This is the first step in the **base excision repair** pathway, as described below, which thus provides an alternative repair pathway for alkylated bases in *E. coli*. Mice in which the gene encoding the repair alkyltransferase has been deleted are more susceptible to developing tumors after treatment with alkylating agents, revealing the importance of this enzyme in preventing the tumors that may arise as a result of this type of DNA damage.

Specialized proteins sanitize the free nucleotide pool by removing damaged nucleotides

Like DNA itself, the nucleotide precursors of DNA are also subject to damage and may be another source of errors in DNA if allowed to remain in the precursor pool. This is prevented by hydrolases that recognize damaged nucleotides and convert them to monophosphates, which cannot be incorporated into DNA. For example, the *E. coli* hydrolase MutT promotes the destruction of nucleotides containing the oxidized purine 8-oxo-guanine. Inactivation of the human homolog of mouse MutT results in an increased incidence of cancer.

12.4 REPAIR OF DNA DAMAGE BY BASE EXCISION REPAIR

Most DNA repair occurs by excision of the damaged DNA and resynthesis

While some DNA damage is repaired by the direct reversal processes described above, in *most* cases it occurs through the excision of the damaged region and its replacement through the use of the complementary DNA strand as a template. This process is called **excision repair**. Excision repair follows the same principles as in mismatch repair, but is catalyzed by different proteins: specialized enzymes detect and remove the damaged DNA, and specialized DNA repair polymerases replace the excised region.

There are two types of excision repair: **base excision repair** in which a single damaged nucleotide requires replacement; and **nucleotide excision repair**, acting on more bulky lesions, requiring a short stretch of nucleotides to be excised and replaced. In both these repair pathways, the complementary undamaged strand serves as a template for resynthesis of the correct sequence on the damaged strand. In this section, we will describe base excision repair. Nucleotide excision repair will be discussed in the next section.

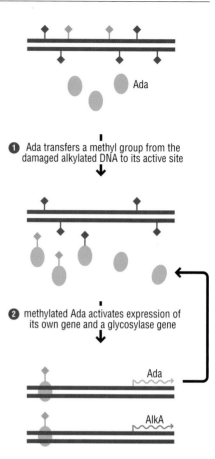

① Ada transfers a methyl group from the damaged alkylated DNA to its active site

② methylated Ada activates expression of its own gene and a glycosylase gene

Figure 12.7 Damage reversal by the O^6-alkylguanine-DNA alkyltransferase (Ada). A DNA duplex that has been alkylated on the O^6 position of guanine (light blue squares) and the DNA backbone (dark blue squares) is shown. A low level of the alkyltransferase O^6-alkylguanine-DNA alkyltransferase (Ada) (green ovals) is constitutively present. (1) Ada removes the methyl group from DNA by covalently attaching it to a cysteine residue in the active site of the protein, thus rendering it inactive for further repair. (2) The alkylated enzyme, however, now acts as a transcription factor, activating transcription of both its own gene and that for AlkA, a glycosylase specific for methylated bases and involved in base excision repair.

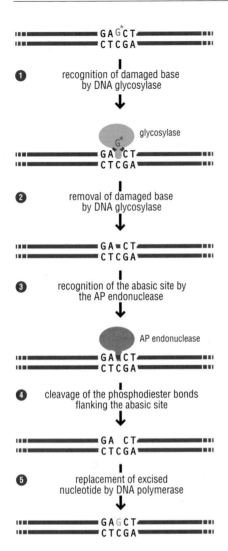

Figure 12.8 The base excision repair pathway. A DNA duplex containing a damaged base (pink) is shown. (1) The damaged base is recognized by a glycosylase (green). (2) The gylcosylase cleaves the glycosidic bond attaching it to the sugar–phosphate backbone of the DNA and releases the base, leaving an abasic site. (3) The abasic site is recognized by the AP endonuclease (pink). (4) The AP endonuclease cleaves the phosphodiester bonds in the backbone on either side of the site, creating a gap. (5) The gap is repaired by a repair DNA polymerase that introduces an intact nucleotide (blue) and sealed by DNA ligase.

Figure 12.9 Schematic diagram of a repair glycosylase bound to DNA. A DNA duplex that contains a damaged base (blue triangle) that disrupts base-pairing is shown. The repair enzyme (green) has an alpha helix that extends into the groove of the DNA helix and flips out the damaged base (pink) that is to be cleaved.

Damaged DNA bases and abasic sites are repaired by base excision and DNA resynthesis

One of the simplest ways in which DNA can be damaged is through breakage of the glycosidic bond linking a base to the sugar–phosphate backbone of the DNA molecule, thus creating an **abasic site** at which the base is missing and there is a gap in the DNA (an example of depurination is shown in Figure 12.3a). Such lesions occur frequently and their repair is initiated by an enzyme, the **apurinic/apyrimidinic endonuclease** (**AP endonuclease**). This enzyme recognizes the abasic site and cleaves the phosphodiester backbone of the DNA strand on either side of the abasic site (Figure 12.8, step 4). The gap is then filled in by repair DNA polymerases discussed in Section 6.2, such as Pol I in *E. coli* and Pol β in mammals, using the undamaged strand as a template.

The same pathway is used in the repair of damaged bases. Many of the mutagenic processes described in Section 12.1 result in the alteration of bases, with consequent mispairing at DNA replication (see Figure 12.3b, c) that may result in point mutations. These lesions are repaired by base excision repair.

Cells contain a number of **DNA glycosylases**, enzymes that recognize and remove a particular damaged base. Some of these enzymes flip the base out of the DNA as illustrated in Figure 12.9, after which it can be excised by hydrolysis of the glycosidic bond (Figure 12.8). The resulting abasic site is then recognized by the AP endonuclease, which cleaves the phosphodiester backbone as described above, excising the nucleotide and sometimes a few additional flanking nucleotides. The gap is filled in by a repair DNA polymerase.

12.5 NUCLEOTIDE EXCISION REPAIR OF BULKY LESIONS

DNA damage that results in helix distortion is repaired by nucleotide excision repair

Some DNA-damaging agents, including ultraviolet irradiation and some chemicals, produce bulky lesions that distort the DNA double helix (see Section 12.1). Although pyrimidine dimers are bulky lesions that can be specifically reversed by photolyases, such lesions are usually repaired by the **nucleotide excision repair** pathway, in which, as with base excision repair, a region of the damaged strand is excised to allow accurate resynthesis of the original DNA sequence using the

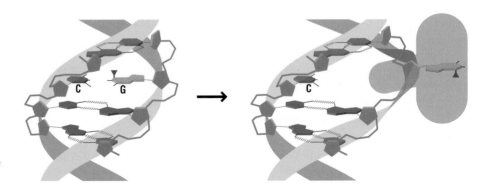

undamaged strand as a template. In this case, however, an oligonucleotide of 10–30 nucleotides including the damaged site is excised, rather than just a single nucleotide.

As in base excision repair, the nucleotide excision repair machinery includes proteins that search for and identify damaged DNA, cleave phosphodiester bonds on either side of the site of damage, and cut away the damaged DNA strand, forming a single-strand gap. This gap is refilled by DNA synthesized by a DNA repair polymerase.

Although the proteins that mediate base excision repair are largely homologous between bacteria and eukaryotes, those that mediate nucleotide excision repair are not evolutionarily related, although they have the same functions and the pathways are similar. Nucleotide excision repair proteins in different eukaryotes are, however, homologous. As the bacterial pathway is much better understood, we will use it to illustrate the process.

Nucleotide excision repair enzymes excise and remove short DNA regions containing bulky lesions

Nucleotide excision repair in bacteria is mediated by the UvrA, UvrB, UvrC, and UvrD proteins. These proteins were initially identified as the products of genes necessary for resistance to damage by ultraviolet light, which causes the formation of pyrimidine dimers: the initial Uvr stands for 'ultraviolet resistant'. The UvrA and UvrB proteins together scan DNA and recognize most types of bulky lesions in DNA, resulting in the loading of UvrB at the position of the damage. UvrB then recruits UvrC, which nicks the DNA backbone upstream and downstream of the lesion, releasing a damaged residue-containing oligonucleotide. DNA that is complementary to the undamaged DNA strand is then resynthesized.

The multi-step ATP-dependent nucleotide excision repair process is schematically illustrated in Figure 12.10. UvrA and UvrB scan DNA for DNA damage in an ATP-dependent manner, likely as a complex containing two UvrAs and two UvrBs (UvrA$_2$B$_2$) whose formation requires ATP hydrolysis. A likely DNA binding surface on UvrA has been determined but we have yet to characterize how damaged DNA is preferentially bound by UvrA. An attractive idea is that UvrA, like MutS, senses variations in DNA structure by detecting variation in DNA deformability (Figure 12.10, step 1).

The DNA strands at the site of damage are separated by the helicase activity of UvrB, and UvrB then stably associates with the DNA, while UvrA dissociates (Figure 12.10, step 2). UvrC, which has endonuclease activity, is then recruited to the UvrB–DNA complex, and cuts the phosphodiester backbone at either side of the lesion, first introducing a nick on the 3′ side of the lesion and subsequently another nick on the 5′ side of the lesion (Figure 12.10, step 3). The combined action of UvrD (DNA helicase II) and DNA polymerase I releases the oligonucleotide, UvrC, and UvrB (Figure 12.10, step 4). Then repair synthesis refills the gap (Figure 12.10, step 5).

The proteins that execute these steps in yeast and humans are listed alongside the functionally analogous bacterial proteins in Figure 12.11. The XP proteins were identified from studying the human disease xeroderma pigmentosum, in which there is increased sensitivity to DNA damage (below).

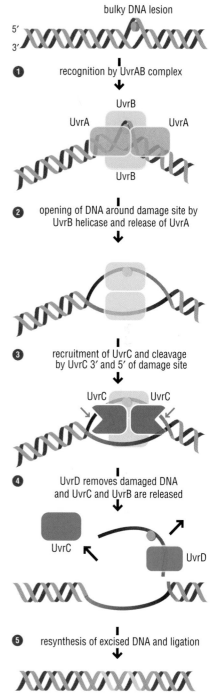

Figure 12.10 The *E. coli* nucleotide excision repair pathway. A DNA duplex containing a bulky lesion (red) is shown. (1) UvrA and UvrB recognize a damaged strand through conformational changes in the DNA helix at the site of the lesion. (2) The UvrB helicase opens the DNA at the site of the lesion and UvrA is released. (3) The UvrB–lesion complex recruits and activates the UvrC endonuclease that will nick the DNA five or ten bases to either side of the lesion (red arrows). (4) UvrC cleavage releases an oligonucleotide containing the lesion that is displaced by UvrD helicase. UvrB and UvrC are also released. (5) The gap is filled by new DNA synthesis, and ligation reseals the DNA.

Some components of the nucleotide excision repair systems in bacteria and eukaryotes			
steps	*E.coli*	*S.cerevisiae*	humans
damage recognition (general)	UvrA, UvrB	Rad14, RPA1–3	XPC-RAD23B, XPA, RPA1–3
transcription-coupled repair	Mfd	Rad26, Rad28	CSA, CSB
DNA unwinding	UvrAB complex	Rad3, Ssl1–2, Tfb1–4	XPB, XPD
5′ excision of damaged site	UvrC	Rad1, Rad10	XPF, ERCC1
3′ excision of damaged site	UvrC	Rad2	XPG
displacement of damaged oligonucleotide	UvrD		
repair DNA polymerase	Pol I	Pol β	Pol δ, Polε

Figure 12.11 Some key components of the nucleotide excision repair machinery. Proteins involved in nucleotide excision repair from different organisms. Eukaryotic proteins are structurally related from yeast to humans. XP proteins were identified from humans with xeroderma pigmentosum. CSA and CSB were identified from humans with Cockayne's syndrome. ERCC1 (excision repair cross-complementing) was identified from a mouse mutant with a defect in excision repair. Although they are functionally similar there is no sequence homology between the bacterial and eukaryotic nucleotide excision repair proteins.

Transcription-coupled nucleotide excision repair targets DNA repair to the template strand of genes that are being transcribed

There are two branches of the nucleotide excision pathway, which differ only in the proteins that recognize DNA damage, converging for the subsequent repair steps. In both bacteria and eukaryotes, nucleotide excision repair operates anywhere in the genome and on several types of damage; in this mode, it is often termed **global genomic repair** (GGR). If the damage occurs in genes that are being transcribed, however, repair is preferentially targeted to these genes to remove stalled RNA polymerases that could disrupt gene expression. This process, **transcription-coupled repair** (TCR), involves the additional recognition proteins listed in Figure 12.11.

Transcription-coupled repair appears to be initiated by recognition of RNA polymerase molecules that have stalled as a result of encountering DNA damage. In *E. coli*, the DNA-binding protein, Mfd, interacts directly with stalled RNA polymerase complexes to dissociate them and remove them from the DNA. Mfd also recruits the nucleotide excision repair components to the site of damage. The analogous Rad26 and CSB proteins in eukaryotes belong to the SWI/SNF family of nucleosome-remodeling enzymes, which change the accessibility of DNA to proteins by altering chromatin structure. They may promote local transcription-coupled repair by facilitating the recruitment of the repair proteins to the damaged DNA.

⊙ We learn about nucleosome remodeling in Section 4.6.

Repair is then specifically targeted to the template DNA strand, that is, the DNA strand that is being transcribed into RNA. The stalled polymerase is removed along with the damaged DNA and the strand restored to its original state by the excision and DNA resynthesis steps outlined above. Transcription-coupled repair ensures that the vital coding sequences in the genome are maintained in good repair, and also enables transcription of a damaged gene to proceed.

Defects in nucleotide excision repair result in mutator phenotypes and disease

As in DNA mismatch repair, inactivation of the nucleotide excision repair proteins in bacteria leads to a **mutator phenotype** (see Section 12.2). In humans also, rare inherited defects in components of the nucleotide excision repair system cause diseases in which the underlying defect is an increased mutation rate. One of these diseases is xeroderma pigmentosum, in which individuals cannot repair pyrimidine dimers, and perhaps other adducts, which are formed in DNA on exposure to ultraviolet light. The result is a greatly increased incidence of skin cancer, caused by the greater likelihood of mutations in genes that control cell growth and division in skin cells, which are the cells most exposed to sunlight.

Xeroderma pigmentosum can be caused by defects in any one of the XP proteins, the eukaryotic proteins functionally equivalent to UvrA, UvrB, and UvrC (see Figure 12.11). A rare inherited disease linked to defective transcription-coupled repair is Cockayne's syndrome, which is characterized by sensitivity to sunlight and the appearance of premature aging, among other symptoms, but not by an increased incidence of cancer. Patients with this disease have lost the function of either CSA or CSB, which are important for transcription-coupled repair (see Figure 12.11), and thus have lost the capacity to repair transcribed DNA preferentially after UV-induced damage.

12.6 TRANSLESION DNA SYNTHESIS

Error-prone polymerases can promote translesion DNA synthesis

The DNA repair pathways we have discussed above promote the restoration of damaged DNA to its pre-damage sequence. However, DNA lesions may not be repaired before they are encountered by a replication fork, especially in cells in which DNA has been heavily damaged, leading to stalled or even broken replication forks.

One possible response to a stalled replication fork is the recruitment of specialized 'translesion' polymerases (TLS), many of which are members of the Y family polymerases, a structurally distinct class of polymerase (Figure 12.12).

➔ We discuss polymerase families in Section 6.2.

In contrast to the highly accurate replicative polymerases, most TLS polymerases are much more error prone and thus the cost of continued replication at an unrepaired lesion is the increased risk of mutagenesis (Figure 12.13). Translesion DNA

Key subunits of translesion polymerases							
polymerase family	polymerase	domain	*E. coli*	*Sulfolobus solfataricus*	*S. cerevisiae*	mouse	human
Y	Pol IV	bacteria	DinB	-	-	-	-
Y	Pol V	bacteria	UmuC	-	-	-	-
Y	Pol IV	archaea	-	Dpo4	-	-	-
Y	Pol IV	archaea	-	Dbh	-	-	-
Y	REV1	eukaryotes	-	-	Rev1	Rev1	REV1
B	Pol zeta (ζ)	eukaryotes	-	-	Rev3	Rev3L/Polζ	REV3L/POLZ
Y	Pol eta (η)	eukaryotes	-	-	Rad30	Polh	RAD30A/XPV/POLH
Y	Pol iota (ι)	eukaryotes	-	-	-	Poli	RAD30B/POLI
Y	Pol kappa (κ)	eukaryotes	-	-	-	Polk	DINB1/POLK

Figure 12.12 Translesion DNA polymerases.

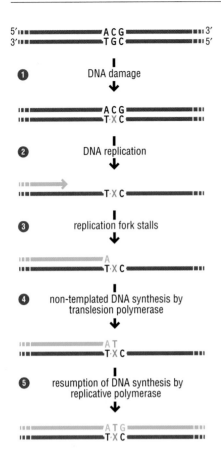

1 DNA damage

2 DNA replication

3 replication fork stalls

4 non-templated DNA synthesis by translesion polymerase

5 resumption of DNA synthesis by replicative polymerase

Figure 12.13 Translesion DNA synthesis. Undamaged DNA (gray) is shown. (1) The bottom strand G (pink X) is damaged and not repaired. (2) A replisome containing a replicative polymerase synthesizes new DNA (green) and approaches the lesion. (3) The replicative polymerase stalls at the unrepaired lesion. (4) A translesion polymerase joins the replisome and incorporates an untemplated T (blue) at the position of the lesion, resulting in a mutation from C to T. (5) Replication by the replicative polymerase resumes.

synthesis may be more accurately considered as a 'damage tolerance' mechanism than a 'DNA repair' mechanism.

TLS polymerases display low fidelity of synthesis, poor processivity, and lack 3'–5' proofreading activity. They are able to bypass bulky lesions because their active sites are much more open and flexible than those of the highly accurate replicative polymerases such as *E. coli* pol III and the eukaryotic pol δ and pol ε (Figure 12.14). Thus, altered base structures not compatible with the active sites of replicative polymerases can be accommodated in the active sites of TLS polymerases, in which tight binding of the substrate base pair does not occur.

The error rates of these TLS polymerases however are 10^{-2} to 10^{-4} compared with error rates of 10^{-5} to 10^{-6} for replicative polymerases, and the cost of their flexibility is thus an increased rate of mutation. Indeed, it is the action of TLS polymerases that actually underlies most mutagenic events in response to DNA damage. Error-prone replication is, however, better than no replication at all.

Interactions between TLS polymerases and processivity clamps are central to translesion synthesis

To preserve the generally high fidelity of DNA replication, access of TLS polymerases to replication forks must be limited. Recruitment of TLS polymerases to

(a) Dpo4 polymerase

(b) A- and B-family polymerases   Dpo4

minor groove

Figure 12.14 Structure of a translesion DNA polymerase. (a) The structure of the Y family Dop4 polymerase with normal DNA is shown. Although Y family polymerases have little sequence homology with replicative polymerases, they have similar palm (red), finger (blue) and thumb (green) domains but contain an additional little finger (pink) domain that is involved in substrate specificity and processivity. The template and primer strand are shown as gray and turquoise the incoming nucleotide is gold. PDB 1UNN. (b) Comparison of the interfaces between a replicating base pair (gray and yellow) and the protein surfaces (blue) of highly accurate replicative polymerases (A- and B-family) and the Dpo4 Y-family is shown. These interfaces are highly complementary in the case of A- and B-family polymerases but much more flexible in the Y-family polymerase. From Yang (2005).

replication forks can occur by interaction with processivity clamps, such as the *E. coli* sliding clamp and eukaryotic proliferative cell nuclear antigen (PCNA) (Figure 12.15). All polymerases including TLS polymerases can interact with the sliding clamp at the same site although it remains to be determined how polymerase switching (from one to the other) is actually modulated. As we shall see in the next section, expression of the *E. coli* TLS polymerases DinB and UmuCD is greatly increased in response to DNA damage. We shall see in Section 12.9 how the modification of PCNA by ubiquitination controls a eukaryotic TLS polymerase as part of a **DNA damage response**.

We have considered a wide variety of mechanisms for the recognition, repair, and tolerance of DNA damage. We explore in Experimental approach 12.1 some of the methods used to identify genes involved in DNA repair.

Figure 12.15 Structure of a translesion DNA polymerase and the beta clamp. The structure of a complex between the 'little finger' domain of the UmuCD Y-family polymerase (pink) is shown with the *E. coli* beta clamp (brown and yellow).
From Bunting *et al.* (2003).

12.7 THE DNA DAMAGE RESPONSE

Extensive damage to DNA induces mechanisms that preserve the viability of the organism

The repair mechanisms we have discussed so far depend on recognition of a specific type of DNA damage by a component of the repair pathway that recruits the proteins appropriate for its repair. Where damage is extensive, however – after exposure to ultraviolet or ionizing radiation, for example – or where it interferes with DNA replication, a variety of cell-wide responses are invoked to amplify the recruitment of repair proteins and to halt the cell cycle while repair is completed. These responses are collectively known as the **DNA damage response**.

Although these responses have been studied largely in cells that have been experimentally subjected to large doses of radiation, it is likely that, especially in cells with large genomes, they are required continuously in dividing cells, to prevent permanent damage that would otherwise result from unrepaired damage in the path of the polymerase during DNA replication.

⊙ We discuss the control of the cell cycle in more detail in Chapter 5.

Single-stranded DNA or double-strand breaks induce the damage response

There are two possible consequences if a polymerase encounters unrepaired DNA damage at a replication fork. Distorting lesions, such as unrepaired oxidative damage, pyrimidine dimers, or DNA adducts cause the polymerase to stall, and delay or prevent the completion of DNA synthesis (Figure 12.16a). (As discussed in the previous section, translesion DNA synthesis by specialized polymerases can occur across a distorting lesion, but such synthesis is generally mutagenic.) Stalling of the replication fork can lead to regions of single-stranded DNA.

Equally as hazardous as distorting lesions are nicks in the DNA, which can be caused by ionizing radiation but also occur through incomplete mismatch, base excision or nucleotide excision repair, all of which involve nicking of the damaged strand by nucleases, as we have seen in the preceding sections. Single-strand nicks in replicating DNA not only cause the polymerase to stall but result in double-strand breaks and the collapse of the replication fork (Figure 12.16b).

12.1 EXPERIMENTAL APPROACH

Finding genes involved in the repair of damaged DNA

Insights into many cellular processes have come from the isolation of mutants with defects in the process. Indeed, this approach has been used with great success for the identification of proteins involved in the repair of DNA damage. When using such genetic approaches, it is generally easier to obtain or select for mutants that are *resistant* to a particular cellular insult, such as a DNA damaging agent. In such an experiment, cells are exposed to a condition that kills the wild-type cells, and survivors are directly isolated. However, some activities or genes are only identified by the isolation of mutants that lead to increased *sensitivity* to a particular condition. How can such mutants, for example, mutants that are sensitive to DNA damaging agents such as ultraviolet (UV) light or x-rays, be isolated?

One commonly used method has been to mutagenize bacterial or yeast cells, spread them on agar plates such that a population of single colonies is present, then replica-plate those same colonies to new agar plates where their growth can be monitored in the presence and absence of the DNA damaging agent. Colonies that do not carry mutations in DNA repair genes grow normally while colonies with mutations grow poorly. The challenge of this 'screening' approach is that the experimenter is looking for the rare single colony that will not grow among thousands of colonies that do. This is quite different from the direct isolation of mutants under a generally repressive set of growth conditions – scientists broadly refer to this approach as a genetic 'selection.'

Paul Howard-Flanders and Lee Theriot exploited a genetic selection approach in which only *E. coli* cells defective in DNA repair would grow. These researchers took advantage of the fact that damage to the DNA of a lytic phage, caused by irradiation with UV light, for example, must be repaired for the phage to grow and kill the cell. Thus, after infection with a damaged lytic phage, those cells that repair the damaged phage DNA die and those that cannot repair the DNA damage live. Howard-Flanders and Theriot plated a population of mutagenized *E. coli* and exposed the cells to lytic phage that had been damaged by UV light. After infection with the phage, they identified colonies that were not killed by the damaged phage, that is, those that did not repair damaged phage DNA. Figure 1 shows that the 14 mutants isolated in this way were much more sensitive than wild type to killing by UV irradiation. These 14 mutations mapped to *uvrA*, *uvrB*, and *uvrC*, three genes that we now know mediate nucleotide excision repair (NER).

A powerful new genetic method exploits a collection of mutant strains in which each non-essential gene of the organism has been disrupted or 'knocked out', that is, where the coding region of the gene has been replaced by another marker, such as a drug resistance marker. In the process of constructing the disruption, one or

Figure 1 UV sensitivity of *E. coli* mutants defective in DNA repair. The ability of wild-type and mutant *E. coli* strains to grow in the presence of increasing doses of UV irradiation was measured. Cell survival was determined by the number of cells that can grow when plated on agar plates. Reproduced from Howard-Flanders and Theriot, *Genetics* 1962;**47**: 1219–1224.

two unique oligonucleotide sequences, called tags or barcodes, are introduced into each strain flanking the marker. These unique tags allow each mutant to be followed within the population of mutants by sequence analysis or hybridization to a microarray. We discuss these collections of disruption strains in Experimental approach 14.2 and Section 16.4.

Such 'knockout collections' have been constructed in several bacteria and the yeasts *Saccharomyces cerevisiae* and *Schizosaccharomyces pombe*. Large collections of knockout mice are also under construction.

Guri Giaver and co-workers used the *S. cerevisiae* knockout collection to identify DNA repair genes. To do this they compared a population of cells from the knockout collection grown with and without DNA damaging agents. Those genes that conferred sensitivity to DNA damage when knocked out were less abundant in the population after growth in the presence of DNA damaging agents. Figure 2a shows the results of their genome-wide screen, using a cluster analysis in which the growth of each knockout strain in the presence of each DNA damaging agent is represented. Blue indicates knockout strains that are most sensitive (grow least well in the presence of DNA damaging agents), and yellow indicates knockout strains that are least affected by these agents. It is notable that mutations in particular groups of genes result in the same (or a very similar) pattern of sensitivity to different DNA damaging

Figure 2 Relative growth of *S. cerevisiae* knockout strains in the presence of DNA damaging agents. (a) The yeast 'knockout' collection was grown in the presence of various DNA damaging agents (horizontal axis). The growth of individual mutants (vertical axis) was evaluated by isolating DNA from the population of cells, amplifying all of the tags (see section 16) and hybridizing the amplified tags to a microarray. The amount of hybridization observed for a particular tag is considered to be proportional to the abundance of the corresponding mutant. The relative growth of different knockout strains is indicated by the colored bar where yellow indicates the least effect on growth and blue indicates the most inhibition of growth. The knockout strains with similar phenotypes are clustered on the vertical axis. Colored bars on the right represent gene clusters of known function, including nucleotide excision repair (NER, blue); error-prone translesion DNA synthesis (red); post-replication repair (yellow); homologous recombination (green); cell-cycle checkpoint control (orange); and a cluster of genes of unknown function shown in (b) (magenta). (b) Zoom view showing one cluster containing the class I NER genes and a second cluster containing several uncharacterized DNA-repair genes. Four of these five genes *(SHU1, SHU2, CSM2, and PSY3)* are known to encode proteins that physically interact. Reproduced from Lee *et al.*, *PLoS Genetics* 2005;**1**:e24.

agents. Those with the most similar patterns are clustered together in the hierarchical cluster analysis shown in the figure, allowing groups of genes with likely related functions to be identified.

This screen reidentified some DNA repair genes known from previous studies, which served to validate the method. Importantly, however, it also identified new genes involved in the process. Genes were implicated in repair if they had a similar pattern of sensitivity to known repair genes. For example, a cluster of several previously uncharacterized genes had a pattern that was very similar to a cluster of genes involved in nucleotide excision repair (Figure 2b). Four of these five uncharacterized genes, *SHU1*, *SHU2*, *CSM2*, and *PSY3*, were known from other studies to encode proteins that physically interact with each other, suggesting that they are found in a protein complex.

While the actual function of these genes remains to be determined, this genome-wide functional analysis suggests that they are involved in repair, and this can now be tested directly for each of these genes. Knockout collections, which are being screened for phenotypes under hundreds of conditions, are providing important clues to the functions of many genes with previously unknown functions, but as in the case of *SHU1*, *SHU2*, *CSM2*, and *PSY3*, these clues often are most helpful when related clues come independently from other lines of experimentation.

Howard-Flanders P, Theriot L. A method for selecting radiation-sensitive mutants of *Escherichia coli*. *Genetics* 1962;**47**:1219–1224.

Lee W, St Onge RP, Proctor M, *et al*. Genome-wide requirements for resistance to functionally distinct DNA-damaging agents. *PLoS Genetics* 2005;**1**:e24.

Double-strand breaks in DNA, which can also be directly caused by ionizing radiation, are a special hazard because in this case there is no intact complementary strand for repair polymerases to copy, and they always elicit a global DNA damage response. The specialized repair mechanisms required to mend double-strand breaks are discussed in Chapter 13; here we are concerned with the DNA damage-response pathways through which the cell cycle is arrested and repair proteins are recruited to the site of double-strand breaks or stalled replication forks.

The first step in the induction of the DNA damage response in both bacteria and eukaryotes is the recognition of damaged DNA by a damage-sensor protein. The critical signal in bacterial cells and the most important signal for eukaryotic cells is the presence of single-stranded DNA arising either from stalled DNA replication or from the actions of exonucleases on damaged DNA. In eukaryotic cells, damage sensors also directly recognize double-strand breaks.

Figure 12.16 Effects of DNA damage at a replication fork. (a) Polymerase stalling: A replication fork with parental (gray) and newly synthesized DNA (green) containing an unrepaired lesion (red) in the parental DNA is shown. The replication fork stalls at the lesion, resulting in single-stranded DNA ahead of the lesion on one strand because the replisome helicase moves ahead of the polymerase. (b) Replication fork collapse: A replication fork with parental (gray) and newly synthesized DNA (green) containing a nick in the parental DNA is shown. The fork collapses when the DNA with the nick becomes unwound, effectively causing a double-strand break.

DNA damage can result in programmed cell death

Single-celled organisms eventually resume division even when the damage cannot be repaired. In multicellular organisms, however, unrepaired DNA damage can ultimately trigger the programmed death of the damaged cell, preventing the propagation of mutations that may result in unregulated growth and thus to cancer, which can be lethal to the organism. The failure of these DNA damage-response mechanisms underlies a variety of human diseases related by the common feature that they result in high sensitivity to DNA damaging agents and increased susceptibility to cancer.

12.8 THE DNA DAMAGE RESPONSE IN BACTERIA

The first identified and best understood example of a cell-wide response to DNA damage is the bacterial **SOS response**, and we discuss this response in the next section, before turning later to the more complex responses of eukaryotic cells to DNA damage.

The bacterial SOS response depends on recognition of single-stranded DNA by the RecA protein

Many of the proteins involved in DNA repair in bacteria are expressed constitutively, but the presence of extensive damage to DNA leads to increased expression of repair genes to cope with the increased demand for repair. The SOS response is a multifaceted cellular response in which more than 40 genes, many of which are directly involved in repair, are induced when DNA is damaged. Central to the activation of the bacterial SOS response is the interaction of the **RecA** protein with single-stranded DNA exposed by DNA damage (Figure 12.17).

RecA forms a nucleoprotein filament on a single strand of DNA which can act as a **co-protease**, and stimulates the self-cleavage of several proteins that contribute to or inhibit the SOS response, and thereby activate or inactivate them. RecA also plays a central part in the repair of double-stranded DNA breaks and in genetic recombination because the RecA–single-stranded DNA filament can also promote the specific pairing of its single strand with a complementary strand in a DNA duplex. We discuss the role of this RecA strand-pairing activity in the repair of double-strand breaks and homologous recombination in more detail in Chapter 13.

One of the proteins whose cleavage is promoted by RecA is LexA, a specific DNA-binding protein that acts as a repressor to block transcription of many of the SOS genes. In the presence of the RecA-single stranded DNA filament, LexA cleaves itself, inactivating its repressor function. This inactivation results in the induction of the genes it usually represses (Figure 12.17b). Many of these genes encode proteins that participate directly in DNA repair. These include some of the Uvr proteins, which act early in the SOS response to promote nucleotide excision repair (see Section 12.5). Indeed, the *recA* gene itself is repressed by LexA although a low level of RecA is synthesized even in the presence of LexA. De-repression of the

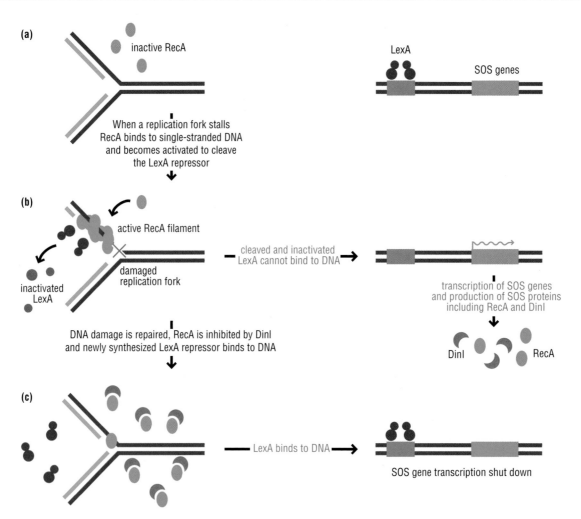

Figure 12.17 Activation and switching off of the SOS response. (a) The replication fork with parental (gray) and newly synthesized (green) DNA proceeds without impairment on undamaged DNA. No active RecA (orange) nucleofilament is present because there is little single-stranded DNA at the replication fork. LexA (dark blue) represses the SOS genes by binding to the operators (turquoise) of. (b) When DNA replication is stalled because of a damaged (red) replication fork, RecA protein binds to the exposed single-stranded DNA and forms a nucleoprotein filament. This RecA nucleoprotein filament promotes the cleavage of the LexA repressor. Cleaved LexA cannot bind to DNA, thus releasing the SOS genes from repression. SOS proteins including RecA itself, other SIS proteins that promote DNA repair and the regulator protein (red) are produced. (c) The SOS response results in DNA repair, which reduces the amount of RecA filament present. In turn, this reduces the amount of LexA cleavage and uncleaved LexA can again bind to regulatory sequences preceding SOS genes, resulting in their repression. Also, DinI binds to RecA, blocking its interaction with single-stranded DNA.

recA gene upon LexA inactivation results in increased production of RecA protein to promote double-strand break repair.

As discussed below, the synthesis of two translesion polymerases is also part of the SOS response. Other SOS-induced proteins include RuvA and RuvB, which we shall encounter in Chapter 13 when we describe the repair of double-strand breaks.

It is interesting to note that a number of the genes induced by the SOS response do not encode obvious DNA repair functions; moreover, DNA damage also results in the induction of many genes that are not regulated by LexA. The mechanisms of their contributions to cell survival and of their induction remain to be determined.

Another SOS response that contributes to cell survival in the presence of DNA damage is a LexA-mediated inhibition of cell division that increases the time window for DNA repair. This inhibition of part of the bacterial cell cycle is similar conceptually to the much more elaborate cell-cycle checkpoints of eukaryotes (see Chapter 5). LexA inactivation leads to the increased expression of SulA, which inhibits bacterial cell division by interacting with FtsZ, a key protein in the formation of the septum that divides a bacterium into two cells (see Section 7.8).

The SOS response is self-regulating. As DNA repair proceeds, the amount of single-stranded DNA decreases and less RecA filament assembles, reducing LexA cleavage. This, in turn, leads to re-repression of the SOS genes. Another SOS protein is DinI (damage inducible I) protein, which interacts with the RecA filament to inhibit the co-protease activity. The negatively charged surface of DinI helix probably acts as a DNA mimic, decreasing the amount of the RecA–single-stranded DNA filament (Figure 12.18). As a result, LexA is no longer degraded, and SOS genes are again repressed (Figure 12.17c).

The SOS response includes increased expression of translesion polymerases and induced mutagenesis

A notable feature of the SOS response to DNA damaging agents is an increased frequency of mutagenesis that is dependent on RecA and LexA. Thus mutagenesis is not a passive consequence of DNA damage. Instead, the activity of SOS-induced proteins is required. The principal source of SOS-induced mutagenesis is a Y-family translesion polymerase composed of UmuC and UmuD, both of which are regulated by LexA. However, the active polymerase also requires proteolysis of UmuD, promoted by the co-protease activity of RecA bound to single-stranded DNA, to generate the UmuD′ fragment that associates with UmuC to form the active UmuD′$_2$C complex. The polymerase active site lies in the UmuC protein. Another Y-family translesion polymerase, DinB (Pol V), is also induced as part of the SOS response but its major role may be in mutagenesis that is induced by other stress conditions such as starvation stress.

Figure 12.18 DinI is a DNA mimic. The structure of DinI reveals a highly negatively charged alpha helix that can act as a mimic of DNA that can interact with RecA, blocking the interaction of single-stranded DNA with RecA that is required for the RecA co-protease activity. PDB 1GHH.
From Voloshin *et al.* (2001).

12.9 THE DNA DAMAGE RESPONSE IN EUKARYOTES

The eukaryotic DNA damage response requires sensor, regulator, mediator, and effector proteins

Two key cellular events are perceived by the cell to be indicative of DNA damage: the sustained existence of single-stranded DNA, and the existence of double-stranded DNA breaks. The eukaryotic DNA damage response, like the bacterial SOS response, is initiated by the binding of **sensor proteins** at the site of such damage (Figure 12.19). But instead of acting directly on the transcriptional regulation of repair proteins (as RecA does in bacterial cells) the sensor proteins of eukaryotic cells recruit specialized protein kinases that signal the presence of damage and coordinate a more complex response. The kinases are known as the **regulator kinases** of the eukaryotic damage response.

Regulator proteins phosphorylate and thereby activate **mediator** proteins that, when phosphorylated, provide binding sites for the recruitment of the **effector proteins** that repair DNA damage. They also recruit checkpoint protein kinases that halt the cell cycle to provide a window for DNA repair. Most DNA repair

1

exogenous and endogenous   replication
DNA damaging agents   stress

2

DNA double-stranded break
single-stranded DNA is
detected by sensor

3   recruitment of ATM or ATR kinases

4   phosphorylation of mediators
and effectors

recruitment of further effector molecules

apoptosis   transcription   cell-cycle
          regulation   control

global cellular changes

+

DNA repair   DNA repair
by homology-   or   by non-homologous
directed repair   end-joining

Figure 12.19 The eukaryotic DNA damage response. (1) DNA is subject to damage from exogenous and endogenous agents that result in DNA double strand breaks and from replication stress that leads to the exposure of single-stranded DNA, for example at stalled replication forks. (2) DNA damage is detected by a sensor protein (such as RPA or MRN in Figure 12.21). (3) A regulator kinase, ATM or ATR (see Figure 12.21) interacts with the sensor protein at the site of DNA damage. (4) Phosphorylation of downstream mediator molecules (such as MDC1 and H2AX in Figure 12.25) that serve as a platform for the recruitment of effector proteins that control the cell cycle and (CHK2 in Figure 12.26), control apoptosis (p53 as in Figure 12.27) or promote DNA repair by homology-directed repair or non-homologous end joining (see Chapter 13).

➲ We discuss the SMC complex family in more detail in Chapter 4.

proteins are constitutively expressed although the transcription of some genes encoding repair proteins is increased (as in the bacterial damage response). However, the mechanism of the eukaryotic transcriptional response is much less well understood. In the next two sections, we focus on the regulation of the DNA damage response and will discuss the mechanisms of the repair of double-strand breaks in Chapter 13.

Many of the proteins and the processes of the eukaryotic DNA damage response are conserved from yeast to humans (Figure 12.20). A notable exception is the programmed cell death response that is elicited in multicellular eukaryotes when damage is not repaired, as we have already mentioned in Section 12.7. The specialized machinery that condemns such damaged cells to self-destruction is described in the last section of this chapter.

In addition to the phosphorylation cascade initiated by the regulator kinases, other post-translational protein modifications such as ubiquitination, acetylation, and others also play important roles in the DNA damage response. Such protein modifications likely provide rapid and reversible protein activation in response to DNA damage.

Let us now consider the mechanism of the eukaryotic DNA damage response in more detail. We begin by describing the eukaryotic damage sensors and the orchestration of the damage response by the regulator kinases.

Two major sensors initiate the DNA damage response in eukaryotic cells

The two major sensors of damaged DNA in human cells are the single-stranded binding protein, replication protein A (**RPA**), which binds DNA ahead of the polymerase at the replication fork (see Section 6.5), and **MRN**, which recognizes DNA double-strand breaks (Figure 12.21) and is discussed below in more detail.

When RPA binds to single-stranded DNA ahead of DNA polymerase at a replication fork, it is normally rapidly removed as the replication machinery progresses. If the polymerase is stalled by DNA damage, however, the replicative helicase can disengage and continue to separate the two strands ahead of the rest of the replication machinery allowing RPA to accumulate and persist over extended regions of DNA (Figure 12.21a, step 1). This continued presence of RPA–single-stranded DNA acts as a signal to the cell of DNA damage, and results in the recruitment of a regulator kinase and the initiation of the damage response.

The other major sensor of DNA damage is MRN, which belongs to the SMC (Structural Maintenance of Chromosomes) family of proteins.

SMC proteins include the cohesins that are required for chromosome condensation and segregation during mitosis (see Section 6.2). MRN is so called because it contains MRE11 (hence M) protein, and the SMC protein RAD50 (hence R). These proteins are highly conserved in all domains of life. Mutations in either protein lead to genome instability as a result of radiation hypersensitivity, for example.

Mammalian MRN includes NBS1, a protein named for Nijmegen breakage syndrome, a human genetic disease that occurs when the protein is defective, and is characterized by radiation sensitivity, chromosomal instability and predisposition to cancer. NBS1 binds to MRE11 and, as we shall see shortly, provides the binding site for one of the two major regulator kinases, as well as other proteins involved in DNA double-strand break repair.

The analogous protein in yeast is called Xrs2, so named for the x-ray sensitive phenotype of cells in which it is defective; Xrs2 has weak sequence similarity to NBS1. We shall see in the next chapter that the mutant phenotypes of these proteins reflect the importance of MR complexes not only in activating the global DNA damage response, but also locally in the actual repair of individual double-strand breaks.

At a double-strand break, the MRE11 subunits interact directly with DNA on each side of the break and may also dimerize, forming a bridge between the two sides of the double-strand break (Figure 12.21b). RAD50, like all eukaryotic SMC proteins, contains two half-site ATPase domains whose dimerization promotes interaction between the N-terminal and C-terminal domains of the protein.

RPA and MRN recruit two distinct regulator kinases

The global regulator kinases that are recruited to the DNA damage sensors RPA and MRN, and which subsequently initiate the DNA damage response in mammals, are **ATR** and **ATM**. Both kinases are highly conserved. ATR is recruited to persistent single-stranded DNA by RPA; by contrast, ATM is recruited to double-strand breaks by MRN (Figure 12.21b, step 2). ATR stands for 'ataxia telangiectasia-related' while ATM more simply stands for 'ataxia telangiectasia-mutated'.

Ataxia telangiectasia is a hereditary disease of humans that is characterized by, among other things, a greatly reduced ability to repair radiation-induced

Alternative names for DNA damage response components		
S. cerevisiae	*S. pombe*	vertebrates
sensor regulator		
Mec1	Rad3	ATR
Tel1	Tel1	ATM
ATR regulatory subunit		
Ddc2/Lcd1	Rad26	ATRIP
effector kinases		
Chk1	Chk1/Rad27	CHK1
Rad53	Cds1	CHK2
MRN complex		
Mre11	Rad32	MRE11
Rad50	Rad50	RAD50
Xrs2	Nbs1	NBS1
9-1-1 (PCNA-like) complex		
Ddc1	Rad9	RAD9
Mec3	Hus1	HUS1
Rad17	Rad1	RAD1
Rad17–RFC complex		
Rad24	Rad17	RAD17
Rfc2-5	Rfc2-5	RFC2-5
adaptors and mediators		
Rad9	Crb2/Rhp1	BRCA1 53BP1 MDC1/NFBD1
Mrc1	Mrc1	CLASPIN

Figure 12.20 Some key components of the DNA damage response in different organisms. From Morgan (2007).

Figure 12.21 Recruitment of ATR and ATM to damaged DNA. (a) Recruitment of ATR. Single-stranded DNA adjacent to duplex DNA can occur, for example, at a stalled or collapsed replication fork (see Figure 12.16). (1) RPA binds to single-stranded DNA. (2) RPA recruits ATRIP, which recruits the kinase ATR. (b) Recruitment of ATM. DNA to a double-strand break is shown. (1) The MRN complex binds to the broken DNA ends. The MRN complex is an SMC protein complex containing MRE11 (green), which can bind and, in some cases, cleave DNA, and RAD50 (dark blue and purple), which is an SMC protein that may also bind broken DNA ends and contains half ATPase domains (A,B) at its N-terminus and C-terminus that can dimerize and are linked by a coiled coil region. The MRN complex also contains NBS1 (orange) (in mammals) or Xrs2 (in yeast). (2) The MRN complex recruits ATM via the NBS1 subunit (Xrs2 in yeast). From Wyman et al.

Figure 12.22 Recruitment of ATR to a single-stranded DNA. (1) RPA binding to single-strand gap resulting from DNA damage. (2) ATR is recruited to RPA bound to single-stranded DNA via its adaptor ATRIP. At the junction of single-stranded and double-stranded DNA, the damage-specific sliding clamp 9-1-1 is loaded by the damage-specific Rad17-containing clamp loader. (3) Interaction of ATR with TOPBP1 interacting with 9-1-1 leads to the activation of the ATR kinase that phosphorylates downstream target proteins.
From Cimprich and Cortez (2008).

discuss the sliding clamp-clamp
...lexes and their roles in DNA
...apter 6.

double-strand breaks and an increased risk of developing cancer: similar deficiencies, as we have seen, result from mutations in components of the MRN complex, reflecting the participation of these proteins in the same DNA damage response pathway. The single known mutation of human ATR causes a splicing variant, resulting in Seckel syndrome, which is characterized by growth retardation, small head size, and mental retardation; ATR-deficient mice are non-viable and both human and mouse cells lacking ATR are highly sensitive to DNA damaging agents.

Stalled replication forks, are far more common than double-strand breaks. The essential role of ATR likely reflects its role in the rapid repair of damaged forks, such that embryos lacking ATR do not survive.

ATM and ATR are both serine-threonine phosphoinositide 3-kinase-related protein kinases (PIKKs). Each kinase initially responds to a different damage signal, and they are functionally non-redundant, although there is some overlap in their downstream phosphorylation targets.

Recognition of single-stranded DNA bound by RPA activates ATR

We saw earlier how RPA binds to single-stranded DNA, and the persistent accumulation of this protein–DNA complex acts as a signal of DNA damage. RPA at the DNA damage site does not directly recruit ATR, however. Instead, it recruits a protein called ATR-interacting protein (**ATRIP**) which acts as an adaptor, binding both to the single-strand-binding protein RPA and to ATR (Figure 12.21a step 2). The presence of RPA–single stranded DNA also recruits another complex to the junction of single stranded DNA and double-stranded DNA; this complex is a specialized damage repair-specific DNA sliding clamp-clamp loader complex (Figure 12.22).

The damage-specific sliding clamp is the trimeric protein 9-1-1 (RAD9-RAD1-HUS1); the clamp loader is the RAD17–RFC2-5 complex. The function of 9-1-1 at sites of DNA damage is to recruit a third protein, TOPBP1. Interaction of TOPBP1 with ATR and 9-1-1 then leads to the activation of the ATR kinase. The initial targets of the ATR kinase are RAD17, 9-1-1, and TOPBP1 themselves at the break site: this increases their ability to activate ATR and amplifies the damage response (Figure 12.22).

Two important consequences of ATR activation are: to control replication forks by phosphorylation of replication proteins; and to control cell cycle transitions by phosphorylation of CHK1 (Figure 12.23). Phosphorylation of replication proteins slows the movement of replication forks, likely allowing more time for repair, and delays the firing of new origins of replication. We will discuss cell cycle control by CHK1 in more detail below when we discuss how ATM activation also controls cell cycle transitions.

Another response provoked by stressed replication forks is the recruitment of TLS polymerases, which facilitate the resumption of replication by allowing untemplated DNA synthesis at a fork-blocking lesion. In addition to recruitment of the regulatory kinase ATR, RPA bound to single-stranded DNA also recruits the ubiquitination complex RAD6–RAD18, which results in the mono-ubiquitination of PCNA by RAD18. TLS polymerases are subsequently attracted to ubiquinated PCNA by their PCNA- and ubiquitin-binding domains and allow the resumption of DNA synthesis (Figure 12.24). Although, as we saw earlier (Section 12.6), such

Figure 12.23 ATR activation modulates DNA replication. Activated ATR can modulate multiple aspects of DNA metabolism by phosphorylation of proteins that control the cell cycle, that act at replication forks and that control the firing of replication origins.

TLS DNA synthesis is error prone, it does at least allow progression of the replication fork. The molecular mechanisms by which the actual polymerase switch on the template strand occurs are not yet understood, however.

Ubiquitination of the yeast 9-1-1 clamp by Rad6-Rad18 also likely facilitates TLS polymerase activity. Intriguingly, recent studies in yeast suggest that ubiquitinated 9-1-1 may play an important role in the regulation of transcriptional induction genes encoding DNA repair functions, although the mechanism of induction remains to be determined.

ATM is activated by recruitment to double-strand breaks by MRN

In mammalian cells, several proteins essential to the DNA damage response interact directly with MRN. One is the regulatory kinase ATM, which is recruited to double-strand break sites though interaction with the NBS1 subunit of MRN (Figure 12.21b). In the absence of DNA damage, ATM exists as an inactive dimer. The interaction of the ATM dimer with the MRN-DNA double-strand break complex results in activation of the ATM kinase activity through dissociation of the dimer and autophosphorylation of ATM. ATM-mediated phosphorylation of MRN also occurs. We discuss other proteins including MDC1 below.

Recruitment of repair proteins and chromatin modification are key in the DNA damage response

So what is the downstream effect of ATM? A key target of ATM phosphorylation is the histone variant H2AX, which as we saw in Chapter 4, makes up some 10–15% of the histone H2A to form a modified chromatin domain on each side of the DSB (Figure 12.25, step 1). Phosphorylated H2AX serves as a binding site for MDC1 (mediator of the DNA-damage checkpoint 1), which is also a target of ATM phosphorylation (Figure 12.25, step 2). Phosphorylated MDC1, in turn, recruits

Figure 12.24 Ubiquitination of PCNA recruits translesion polymerases. In addition to activating ATR, the formation of RPA-single stranded DNA when DNA damage occurs also facilitates the activity of translesion polymerases that allow synthesis, albeit error prone, past DNA lesions. The Rad6–Rad18 complex interacts with RPA–single-stranded DNA and ubiquitinates Lys124 of PCNA. TLS polymerases are recruited to PCNA by specific binding to ubiquitin and PCNA itself.

Figure 12.25 ATM activation results in chromatin modification at the site of a DNA double-strand break. (1) A DNA double-strand break (DSB) is shown in DNA wrapped around nucleosomes (light purple) (2) MRN (MRE11 (green); RAD50 (dark blue and purple); NBS1 (orange)) interacts with DNA at the site of the break and recruits ATM to the break via its interaction with the NBS1 subunit. The resulting activation of the ATM kinase results in phosphorylation of H2AX flanking the break. (3) MDC1 (green) interacts with phosphorylated H2AX nucleosomes, resulting in phosphorylation of MDC1. More MRN complexes that interact with MDC1 are recruited, leading to more ATM at the breaks. Thus MDC1 amplifies the ATM signal. The resulting modified chromatin domain can extend for megabases into the DNA flanking the break. (4) The ubiquitin ligase RNF8 (pink) and its ubiquitin conjugating enzyme partner UBC13 (not shown) interact with phosphorylated MDC1. RNF8 ubiquitinates (orange triangles) H2AX, providing binding sites for other proteins (not shown) that recruit BRCA1 (blue) and 53BP1 (orange).

additional MRN and ATM to the site of the DSB. MDC1 thus plays a critical role in *amplifying* the activation of ATM.

The formation of extended domains of phosphorylated H2AX and MDC1 provides a platform for the recruitment of large complexes of repair proteins to the sites of double-strand breaks to form subnuclear foci called ionizing-radiation–induced foci (IRIF). In Experimental approach 12.2, we examine the initial identification of ATM and how these ATM-dependent foci have been characterized. Indeed this protein recruitment to the sites of double-strand breaks can be considerable, extending to include a megabase of DNA around the damage.

MDC1 and H2AX both serve as mediators that recruit effector proteins (Figure 12.25 step 3). A key protein that interacts with MDC1 is the ubiquitin ligase RFN8 and its partner ubiquitin conjugating enzyme UBC13. Histone ubiquitination by RNF8 then recruits other repair proteins such as the 53BP1, which provides a binding site for p53, and which we will discuss in detail below. Ubiquitination also recruits the tumor suppressor BRCA1, a breast cancer susceptibility gene that is mutated in some families with hereditary predisposition to breast and ovarian cancers, which is also a target for ATM and ATR phosphorylation.

The repair process requires the recruitment of more proteins, including repair proteins and cohesins, which can link the broken DNA to a sister chromatid where the damaged DNA is undergoing replication; we shall see in the next chapter that one of the two pathways for repair of double-strand breaks relies on sister chromatids as a template for repair. A third class of proteins that are recruited to DNA by MDC1 and H2AX are those that control the cell cycle. We will consider in Chapter 13 how these and other proteins recruited to DNA double-strand breaks promote the repair of these breaks.

Although it is a reasonable generalization that ATM is recruited in response to double-stranded DNA breaks and ATR is recruited in response to stalled or collapsed replication forks (signaled by the presence of single-stranded DNA), this distinction is not absolute. As we shall see in the next chapter, DNA at a double-strand break that is not repaired will be resected, that is, one DNA strand will be preferentially degraded to generate single-stranded tails, which will be bound by RPA and will eventually recruit ATR. Conversely, stalled replication forks that are not rapidly repaired will result in double-strand breaks, with recruitment of ATM.

Cell cycle arrest is a short-term response to DNA damage

It is clearly disadvantageous for any cell to divide with damaged DNA. Therefore it is of little surprise that, as part of the DNA damage response, mechanisms to arrest

12.2 EXPERIMENTAL APPROACH
Discovering ATM and its functional characterization

One important goal of research in molecular biology is to identify the mutations that cause human disease and to understand why these mutations lead to the disorder. The mapping and characterization of the *ATM* (ataxia telangiectasia mutated) gene that is defective in ataxia telangiectasia illustrates how scientists can approach this problem. Ataxia telangiectasia is an autosomal recessive disease characterized by a variety of symptoms. It first manifests as a childhood neurological disorder with delayed motor skills, poor balance, and slurred speech, but individuals with ataxia telangiectasia also show a marked increase in radiation sensitivity and increase in cancer susceptibility. One percent of the population is heterozygous for ataxia telangiectasia, carrying a defect in one of the two *ATM* alleles.

By the mid 1990s, a variety of genetic studies, which included the mapping of the ataxia telangiectasia disease region in families through linkage analysis, had narrowed the location of the ataxia telangiectasia locus to a region of about 3 Mb of DNA on the q arm of chromosome 11. Linkage analysis is described in Sections 15.8 and 16.15. As this mapping was performed well before the first draft sequence of the human genome was complete, how then could the scientists determine the actual identity of the *ATM* gene?

In an example of broad collaborative effort, a consortium of research groups each focused on a defined segment of the ~ 3 MB 11q region for further characterization. Using positional cloning, which involves the systematic isolation and characterization of overlapping segments of the genome (similar to the cloning described in Experimental approach 4.3), the Shiloh group obtained additional markers that could be used in further linkage analysis. They initiated this analysis by isolating several large DNA segments (about 500 kb in size) (Figure 1a) that were subsequently cloned into yeast artificial chromosomes (Figure 1b). Using this set of constructs, it was then

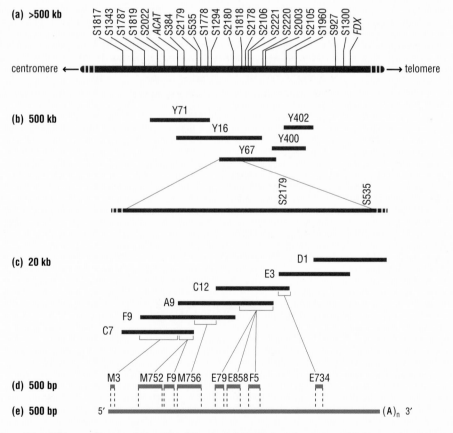

Figure 1 Positional cloning of the *ATM* gene. The general location of the ataxia telangiectasia locus mapped to human chromosome 11q. To identify the defect, the region needed to be narrowed down and the gene itself identified. Beginning with a set of markers for the 11q region (a), the Shiloh group cloned large >500 kb chromosome segments into yeast artificial chromosomes (b). These large fragments were used to identify smaller 20 kb fragments which were used to identify transcripts from the region (c). The fragments C7, F9, A9 and C12 were later all shown to contain part of the gene. The boxes in (d) show restriction fragments where mRNA transcripts (blue) were found. The COOH terminus of ATM is encoded by the cDNA constructed from the transcribed exons (e). Reproduced from Savitsky *et al.*, *Science* 1995;**268**:1749–1753.

possible to identify a polymorphism (a difference between individuals) on these segments, which could be followed to identify the DNA sequence that was most frequently associated with disease when family members who had the disease were compared with those who did not. We discuss the use of polymorphisms to carry out such linkage analysis in Sections 15.8 and 16.15.

Eventually, the 500 kb fragments were subdivided into smaller 20 kb segments (Figure 1c), and the entire set of genes transcribed in this region was identified by isolating cDNAs. The transcribed regions were mapped (Figure 1d), and a 5.9 kb cDNA clone containing the 3′ end of the AT gene was isolated. (With the availability of the complete human genome sequence today the isolation of overlapping clones and construction of transcription maps would no longer be necessary as this information is available online in the annotated sequence. When a new genome is sequenced, however, the construction of transcription maps is still an important phase of genome analysis.)

The 3′ end of the *ATM* gene was encoded by the 5.9 kb cDNA (Figure 1e) and had a predicted open reading frame of 1708 amino acids; the entire *ATM* gene was later shown to encode a protein of 3056 amino acids. Sequencing of the exons of this gene in a number of ataxia telangiectasia patients revealed deletions and frameshift mutations, immediately indicating why the gene might not be functional in individuals with the disease (Figure 2). This sequence analysis also revealed that the protein contained a region related to protein kinases, suggesting that ATM is involved in signaling mediated by protein phosphorylation. Within a year of the identification of the human *ATM* gene, several groups made knockouts of the homologous genes in mice and observed radiation sensitivity similar to that which had been documented in human patients.

Although the path to identifying the *ATM* gene was laborious (largely because the complete genome sequence was not yet in hand), the same basic strategy of associating DNA changes with disease, or some other phenotype, is still required to establish that a gene is involved in a particular process.

The studies described above identified *ATM* as a gene that, when defective, results in radiation sensitivity and other phenotypes.

What then does the gene product do? Many studies have shown that phosphorylation by the ATM kinase plays an important role in DNA damage signaling. In addition, a homologous protein that is also a kinase, termed ATR (ATM related) is also a key player in DNA damage response. Knowing a kinase is involved in regulation is an important step forward, however, knowing what substrates are phosphorylated gives even deeper insight into the mechanism of ATM function. A recent large-scale proteomic analysis of protein phosphorylation in irradiated cells by Elledge and colleagues set out to probe this question.

The basic strategy of the Elledge group was to identify proteins that were phosphorylated by ATM or ATR in response to irradiation. As these kinases share substrate specificity for Ser-Gln (SQ) and Thr-Gln (TQ) motifs, they used antibodies that recognize these phosphorylated motifs to immunoprecipitate proteins of interest (in irradiated and untreated cells). These isolated proteins were then analyzed by SILAC (stable isotope labeling with amino acids in cell culture) following trypsin digestion. This mass spectrophotometric approach facilitates quantitative peptide analysis allowing for direct comparisons of two related samples (as discussed in more detail in Experimental approach 11.3). In this manner, very many phosphorylated proteins were identified; indeed, more than 700 potential protein substrates were identified that displayed more than a fourfold increase in phosphorylation at ATM and ATR consensus phosphorylation sites. We learned about a related but different experimental strategy to identify kinase substrates in Experimental approach 5.1.

The Elledge group then went on to show through direct biochemical methods that phosphorylation of a subset of the identified proteins did increase in response to irradiation. They experimentally tested whether 37 of the newly identified targets of phosphorylation were involved in the DNA damage response by using siRNAs to knockdown expression of the genes encoding the phosphorylated proteins (described in Section 16.4) and then measuring cellular responses to DNA damage. In one particular experiment (Figure 3), the group directly demonstrated increases in the levels of γ H2AX in response to γ irradiation in wild-type

Figure 2 The *ATM* gene carries mutations in ataxia telangiectasia patients. The coding region of the entire *ATM* gene is shown at the top of the figure with the locations of the mutations in 66 different individuals shown as hatched lines. Mutations in *ATM* are most frequently null mutations that result in complete loss of full-length protein. Deletions, insertions, nonsense mutations, and splice site mutations are also commonly found. Each type of mutation is illustrated below the diagram of the gene and the location of the mutation is mapped onto where it occurs in the coding region. The numbers at the right indicated the total number of times the particular kind of mutation was found in this set of patients. For example, frameshift deletions were found in 14 of the 66 individuals analyzed. Reproduced from Sandoval *et al.*, *Hum Mol Genet* 1999;**8**:69–79.

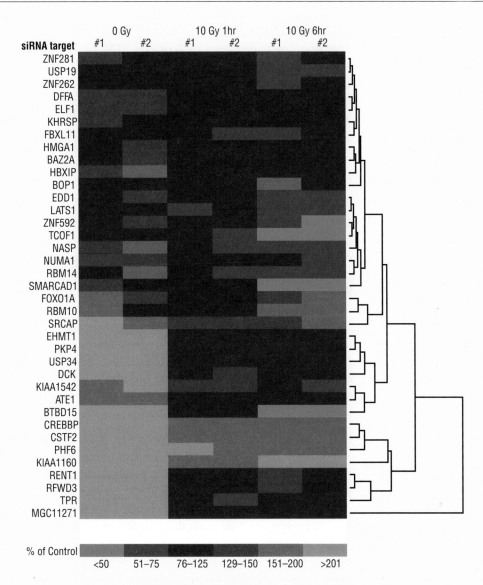

Figure 3 Analysis of the level of γ H2AX in cells treated with siRNA against specific genes. The levels of γ H2AX were measured by western blot assays using an antibody specific to γ H2AX. Cells were treated with the designated siRNA and then either not irradiated (0) or irradiated with 10 Gy and then protein levels measured one hour (#1) or six hours (#2) later. Each cell line treated with a specific siRNA had a characteristic response level at one hour and recovery at six hours. The colors indicate the percent change in level of γ H2AX compared with cells treated with a control unrelated siRNA. The color key is indicated at the bottom. The results are displayed as a hierarchical clustering in which cell lines giving similar responses are grouped next to one another. This clustering analysis helps to identify proteins that might be most functionally similar.
Reproduced from Matsuoaka *et al.*, *Science* 2007;**316**:1160–1166.

cells, a prediction that arose from the observation that γ H2AX levels appeared to be altered when genes of interest from the screen were knocked down. This experiment nicely confirmed the hypothesis that the newly identified target proteins are important in the DNA damage response.

Of the 700 target proteins identified, 31 had been previously shown to be involved in DNA damage; this study thus revealed a much wider range of proteins and cellular processes targeted by ATM than previously anticipated. ATM is now known to be a central regulator of the DNA damage response and thus has been extensively studied. While each of the many studies have provided information about the function of ATM, much still remains to be learned about the substrates and regulation of this important protein. This overview of the analysis of ATM thus illustrates the challenge of understanding a human disease, even after the defective gene or genes have been identified.

Matsuoka S, Ballif BA, Smogorzewska A, *et al*. ATM and ATR substrate analysis reveals extensive protein networks responsive to DNA damage. *Science* 2007;**316**:1160–1166.

Sandoval N, Platzer M, Rosenthal A, *et al*. Characterization of *ATM* gene mutations in 66 ataxia telangiectasia families. *Hum Mol Genet* 1999;**8**:69–79.

Savitsky K, Bar-Shira A, Gilad S, *et al*. A single ataxia telangiectasia gene with a product similar to PI-3 kinase. *Science* 1995;**268**:1749–1753.

cell cycle progression become activated shortly after the DNA damage is detected. This response is mediated by ATR and ATM, both of which phosphorylate and activate the specialized checkpoint kinases, CHK1 and CHK2, which arrest the cell cycle to allow time for DNA repair. The activation of ATR results in phosphorylation and activation of the checkpoint kinase CHK1 (Figure 12.26). Similarly, activated ATM phosphorylates and activates the checkpoint kinase CHK2.

Once phosphorylated, these activated checkpoint kinases halt the cell cycle by inactivating cyclin-dependent kinases (CDKs) that promote transitions in the cell cycle.

➔ We discuss CDKs further in Section 5.2.

This inactivation is indirect: the checkpoint kinases directly phosphorylate the CDC25 phosphatases, blocking their ability to subsequently remove inhibitory phosphates from the CDKs and thereby preventing cell cycle transitions. Detection of damaged DNA during S and G2 in mammalian cells, for example, leads to phosphorylation of CHK1. CHK1 then phosphorylates the CDC25C phosphatase. Without dephosphorylation of the CDK1-CyclinB by the CDC25C phosphatase, the G2→M transition does not occur, blocking the entry into mitosis, preventing the segregation of damaged DNA into daughter cells. The strategy for cell cycle arrest is slightly different in budding yeast cells: they can enter mitosis in the presence of DNA damage but mitosis is then blocked in metaphase. Both strategies, blocking mitotic entry or blocking chromosome segregation, allow increased opportunity for repair and decrease the possibility of permanent changes in DNA sequence.

This cell cycle arrest is readily reversible in single-celled organisms, where each cell must divide in order for the organism to reproduce. In multicellular organisms, by contrast, cell cycle arrest is generally irreversible because severe DNA damage in one cell can be a threat to the organism as a whole through the increased likelihood of cancerous changes in the progeny of the damaged cell. Multicellular organisms meet this threat with a specialized system for inducing the programmed death of cells with damaged DNA – the process of **apoptosis**. The key effector protein in this pathway is p53, a gene regulatory protein that is recruited to sites of damage by the ubiquitinated H2AX histone proteins that occur at DNA double-strand breaks. We will discuss in the next section the programmed cell death response to DNA damage that is mediated by p53.

DNA-PK signals DNA damage and is involved directly in DNA repair

Another **PIKK** in mammalian cells that is activated by double-stranded breaks is DNA-protein kinase (DNA-PK), kinase catalytic subunit called DNA-PKcs and a DSB binding subunit Ku70/80 heterodimer that binds avidly to DNA ends. Interactions between the protein-bound ends of a broken DNA strand can keep the ends together. The ends are then rejoined through the **non-homologous end joining (NHEJ)** pathway, which we describe in Chapter 13.

Activation of the kinase by interaction with Ku bound to DNA is important in the actual repair step, possibly for autophosphorylation to promote conformational changes within the synaptic complex. However DNA-PK phosphorylation of downstream targets is also important for cell survival via pathways involving

(a)

(b)

(c)

Figure 12.26 ATR and ATM control cell cycle checkpoints. Activation of the DNA damage senor kinases ATM and ATR control transitions between multiple cell cycle stages. (a) Control of the G1 to S transition. ATM is activated by double strand breaks and ATR is activated by single stranded DNA that results from replication fork stalling. Their activation results in the phosphorylation and activation of the cell cycle regulators CHK2 and CHK1. These kinases then phosphorylate the CDC phosphatases, CDC25A such that there is no active CDC25A phosphatase to dephosphorylate cycCDK1 and promote the G1 to S transition. (b) Additional G1 and S controls. Activated CHK2 and CHK1 phosphorylate and block the action of CDC7 and DBF4 that are required at the initiation of S-phase and within S-phase. (c) Controlling entry into M-phase. The active CDK2 and CHK1 kinases phosphorylate the phosphotase CDC25C, blocking its ability to dephosphorylate CDK2 and promote the G2 to S transition.

cell cycle arrests and cell death. Yeast lacks DNA-PK but does carry out NHEJ that involves the Ku70/80 heterodimer.

12.10 DNA DAMAGE AND CELL DEATH IN MAMMALIAN CELLS

In the final section of this chapter, we consider the ways in which multicellular organisms cope with the wider impacts of DNA damage – damage that may require the organism to sacrifice some of its component cells to ensure its well-being. We focus in particular on the role of one protein, p53, which plays a major role in a multicellular organism's response to DNA damage.

p53 can block the proliferation of cells containing damaged DNA

Although the persistence of DNA damage in a single-cell organism can result in DNA change or death of that cell, the remaining cell population is not affected by the fate of the mutated cell. The consequences of DNA damage for a multicellular organism are quite different: if cells with damaged DNA can replicate within a multicellular organism there is an increased possibility that some of those damaged cells will acquire genetic changes that lead to unregulated growth of that cell. Such unregulated growth underlies the development of cancers that can be fatal to the whole organism. Thus in multicellular organisms, there are special regulatory responses to unrepaired DNA damage: damaged cells are either permanently arrested in G1; or they are removed by **apoptosis**, a specialized process of programmed cell death. This response is orchestrated by the gene regulatory protein p53.

Importantly, p53 is the single most frequently mutated protein in human cancers, being mutated in at least half of all cases. In the absence of functional p53, the DNA damage response is deficient, mutations accumulate, and the chance of cancer is greatly increased.

Multicellular organisms activate p53 for long-term inhibition of cell proliferation

p53 is a specific DNA-binding protein that directly binds to the promoters of target genes and alters the rate at which they are transcribed. In most cases, expression of target genes is increased, resulting in the increased production of proteins that inhibit cell cycle progression or stimulate apoptosis. p53 can also repress the transcription of some target genes, thereby reducing the levels of inhibitors of apoptosis. The result of p53 action is either cell cycle arrest or cell death.

> ➲ We discuss the acetylation of histones in Section 4.5.

The level of p53 in the cell is controlled in a number of ways

Transmission of a DNA damage signal to p53 is mediated by the adaptor protein 53BP1 (p53 binding protein 1), which binds to modified chromatin, that is, ubiquitinated H2AX, at the site of DNA damage (Figure 12.25). Because of the central

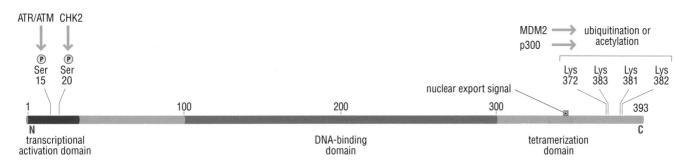

Figure 12.27 Domain structure and regulatory features of the human p53 protein. p53 contains several domains, including a large central region that interacts directly with the DNA target sequence and an N-terminal region that interacts with the transcriptional machinery to stimulate gene expression. When activated, p53 forms a tetramer, and a small region near the C-terminus is required for tetramerization. This region also contains a nuclear export signal that is blocked in the p53 tetramer, enhancing the nuclear localization of the activated protein. Additional regulatory regions are clustered near the termini of the protein. The key regulators MDM2 and p300 both interact with the N-terminal region and modify a group of lysines at the C-terminus, presumably because the termini are more closely apposed in the folded protein. The N-terminal region also contains several phosphorylation sites, including two serines that are phosphorylated during the damage response.

importance of p53 in the response to DNA damage and other cellular stresses, and because its activation can cause the death of the cell, this protein is subject to an unusually large array of regulatory modifications, which, together, ensure that p53 is present and active only when necessary. Most of these modifications act to increase its concentration or its intrinsic gene regulatory activity, or both, when DNA damage occurs.

The major regulators of p53 include MDM2 and p300

The location of functional domains of p53 are illustrated schematically in Figure 12.27. Like many transcriptional regulators, p53 has a DNA-binding domain and a separate activation region that interacts with the transcriptional machinery. This activation region, which is at the N-terminal end of the protein, along with a region at the C-terminal end, are the targets of an array of regulatory molecules that catalyze the ubiquitination, phosphorylation, and acetylation of specific amino acids – modifications that, collectively, control the activity of p53.

The major regulator of p53 is MDM2, an E3 ubiquitin ligase that ubiquitinates several lysine residues near the C-terminus of p53 (Figure 12.28a), thereby targeting it for destruction (see Section 11.8). Because it binds to the N-terminal region of p53, which contains the activation domain that is important in the stimulation of target-gene transcription, MDM2 is thought also to inhibit the gene regulatory activity of p53 until p53 can be destroyed. Mice lacking MDM2 die early in embryonic development, apparently because excessive amounts of p53 block the proliferation of many cell types. Normally, in the absence of DNA damage, MDM2 associates with p53 and keeps its activity at a minimum. When DNA damage occurs, however, the activity of MDM2 is suppressed and p53 is stabilized, leading to expression of p53 target genes.

A second regulator, p300, which promotes the activity of p53, is a histone acetyltransferase. p300 associates with p53 during the DNA damage response and helps promote local gene expression by acetylating histones.

p300 also acetylates p53 itself, at the same lysines that are ubiquitinated by MDM2 in the absence of damage (Figure 12.27). Acetylation of these lysines therefore

(a) undamaged DNA

MDM2

ubiquitin

p53

ubiquitination and
p53 destruction

**(b) damaged DNA and activation of
ATM, ATR and CHK2**

① MDM2–p53 association is disrupted
by phosphorylation

② tetramerization of p53 blocks
nuclear export

③ phosphorylated p53 interacts with
histone acetylase p300

p300

gene expression

cell cycle arrest   cell death

➲ We describe the structure and
function of telomeres in Section 4.11.

Figure 12.28 Activation of p53-dependent gene expression by DNA damage. (a) Metabolism of p53 in the absence of DNA damage. MDM2 (turquoise) binds to and ubiquitinates (orange triangles) p53 (green). Ubiquitinated p53 is then destroyed by the proteasome. The small amount of p53 in undamaged cells is further restrained by its export from the nucleus. (b) Metabolism of p53 in the presence of DNA damage and activation of ATM, ATR, and CHK2. In the absence of DNA damage, MDM2 binds to p53. (1) Following DNA damage, phosphorylation of MDM2 and p53 by activated ATR and ATM and also CHK2 disrupts MDM2 and p53 association, resulting in stabilization and activation of p53. (2) Tetramerization of active p53 blocks nuclear export. The concentration of phosphorylated p53 thus increases in the nucleus. (3) Phosphorylation of p53 also enhances its interactions with transcriptional proteins, including the histone acetyltransferase p300. p300 associates with p53 and promotes the acetylation (pink) of histones and of p53 itself, both of which act to increase p53-dependent gene expression, resulting in cell cycle arrest and cell death.

blocks their ubiquitination, further ensuring the stabilization of p53 during the damage response.

p53 is activated in response to DNA damage

Early in the DNA damage response, MDM2 is phosphorylated, probably by ATM and perhaps also by ATR, inhibiting its association with p53 (Figure 12.28b). The same kinases also phosphorylate p53 itself within its activation domain. This both inhibits MDM2 binding and increases p300 binding and acetylation, thereby increasing the stability and gene regulatory activity of p53. Finally, the effector kinase CHK2 (and probably CHK1 as well) phosphorylates p53 at serine 20, which also reduces MDM2 binding and helps stabilize p53.

Nuclear retention of p53, an indirect effect of stabilizing the protein, ensures its sustained activity during the damage response. In the absence of DNA damage, p53 is tagged for export from the nucleus by a nuclear export signal near the C-terminus (Figure 12.27), and is thereby prevented from associating with target genes. When p53 is stabilized and activated after DNA damage, however, it forms a tetramer in which the nuclear export signal is blocked, ensuring that the active tetramer is retained in the nucleus (Figure 12.28).

So we see how the DNA damage response signal employs multiple overlapping mechanisms, including ubiquitination, phosphorylation, acetylation, and binding to inhibitor and activator proteins, to ensure the rapid and robust activation of p53-dependent gene expression.

Telomere degeneration promotes cell cycle arrest

Another important pathway to p53-dependent cell cycle arrest is initiated at the telomeres that cap the ends of chromosomes.

Telomerase, the enzyme primarily responsible for maintaining telomere length, is not expressed in many differentiated cells; in these cells, telomeres gradually decrease in length over many cell generations. This telomere shortening is associated with the loss of telomere function that is probably due to the deterioration of the protein cap. Since these chromosome ends now appear to resemble a double-break, ATM is activated, leading to p53-dependent cell cycle arrest or apoptosis like that seen in response to double-strand breaks elsewhere in the genome. In other cases, in the absence of telomere shortening, loss of the proteins that bind to telomeres can allow access of nucleases that generate single-stranded DNA tracts, which can also activate the damage response pathway.

✱ SUMMARY

DNA damage

- DNA damage threatens the integrity of the genome and can result in changes in DNA sequence (i.e. mutation).

- Errors in DNA replication result in mismatched base pairs.

- DNA can be damaged by both intracellular and extracellular agents.

- Many different kinds of lesions can occur.

- All components of DNA – the phosphodiester backbone, the deoxyribose sugar and the bases – are subject to damage.

DNA repair pathways

- There are multiple pathways for DNA repair in the cell.

- Because of DNA repair, only a small fraction of DNA damage events result in mutation.

- Defects in DNA damage repair pathways result in an increased rate of mutagenesis, and in sensitivity to DNA damaging agents.

- Many genes encoding DNA repair proteins have been identified by isolation and study of mutants with increased sensitivity to DNA damaging agents.

- Defects in DNA repair pathways underlie many heritable human diseases, which result in an increased incidence of cancer.

- Cancer can result from DNA repair failure in response to increased DNA damage or from a heritable decreased capacity for repair, both of which may lead to the mutation of growth control genes.

- Some repair pathways can recognize specific lesions and reverse them.

- Base methylation can be reversed by specialized methyltransferases.

- Specialized proteins sanitize the free nucleotide pool by removing damaged nucleotides.

- Multiple DNA repair pathways recognize and excise damaged DNA, and use the information on the undamaged strand to direct the resynthesis following damage excision, exploiting the fact that DNA is a duplex.

Postreplication mismatch repair

- Postreplication mismatch repair corrects DNA mismatches resulting from nucleotide misincorporation during DNA synthesis.

- Although DNA replication is highly accurate because of proofreading by exonucleases, misincorporation does occur.

- The mismatch repair system recognizes mismatched base pairs, specifically removes the newly synthesized strand, and resynthesizes DNA to replace the removed base pair.

- Mismatch repair also corrects expansions and contractions that can occur at repeated sequences due to misalignment during DNA synthesis.

- Defects in human mismatch repair genes are associated with hereditary non-polyposis colon cancer.

Base excision repair

- Base excision repair recognizes positions of base loss or damage, removes the damaged nucleotide or a few surrounding nucleotides, and resynthesizes DNA to replace the removed DNA.

- Specialized glycosylases release the damaged base from its sugar.

- The AP (apurinic/apyrimidinic endonuclease) nicks the phosphodiester backbone at positions of base loss.

Nucleotide excision repair

- Nucleotide excision repair recognizes bulky lesions, removes an oligonucleotide containing the damage, and resynthesizes DNA to replace the removed oligonucleotide.

- Although the basic steps of excision repair are the same in bacteria and eukaryotes, there is no homology between the proteins involved.

- Nucleotide excision repair can be targeted to actively transcribed regions by interaction between the repair machinery and transcription machinery.

Translesion DNA synthesis

- Lesions in DNA can result in stalling of replication of the replication fork.

- Specialized translesion DNA polymerases allow error-prone DNA synthesis through sites of DNA damage where other polymerases would be unable to function.

- Although mutagenic, error-prone DNA synthesis allows a chance of cell survival whereas failure to replicate is lethal.

- The active site regions of translesion polymerases are looser, allowing use of an imperfect template.

- Translesion polymerases can be induced in response to DNA damage.

- Nicks in DNA can lead to replication fork collapse.

The DNA damage response

- The presence of unrepaired DNA damage provokes cellular responses that lead to an increased chance for cell survival by inhibition of the cell cycle, and increased activity of DNA repair pathways.

- Single-stranded DNA and DNA double-strand breaks induce the DNA damage responses.

- In multicellular organisms, the continued presence of unrepaired DNA damage can lead to programmed cell death to avoid the multiplication of cells with decreased genome integrity.

The bacterial SOS response

- The bacterial SOS response to DNA damage includes transcriptional induction of DNA repair genes and an increase in activity of translesion polymerases, which results in an increase in the rate of mutagenesis.

- Transcriptional derepression of the SOS genes results from the activation of a co-protease activity of a RecA protein–single-stranded DNA filament that cleaves and inactivates the LexA protein, a repressor of DNA repair genes.

- The RecA co-protease activates a translesion DNA polymerase proteolytic processing of an inactive precursor.

The eukaryotic DNA damage response

- The DNA damage response in eukaryotes acts through phosphorylation signaling cascades and other protein modifications.

- ATR is the principal regulator kinase that responds to stalled replication forks and other kinds of damage that result in the accumulation of single-stranded DNA.

- The slowing of the DNA replication machinery to allow repair and an increase in the activity of error-prone translesion polymerases are responses to increased ATR activity.

- ATM is the principal regulator kinase that responds to double-strand breaks in DNA.

- Large complexes of repair proteins are recruited to the sites of double-strand breaks in DNA.

- Cell cycle regulators are major phosphorylation targets of ATR and ATM, such that the cell cycle stops in the presence of DNA damage.

- Components of the DNA damage response pathway are defective in many human diseases.

DNA damage and cell death

- DNA damage in multicellular organisms can lead to programmed death of damaged cells, avoiding the propagation of cells in which genome integrity has been compromised.

- The programmed cell death response is mediated by p53, a transcriptional regulator.

- The level of p53 is highly controlled, often by post-translation modification.

- Mutations in p53 occurs in at least half of all human cancers.

📖 FURTHER READING

12 INTRODUCTION

Friedberg E, Walker G, Siede W, *et al.*, eds. *DNA Repair and Mutagenesis.* Washington DC: ASM Press, 2006.

Hakem R. DNA-damage repair; the good, the bad, and the ugly. *EMBO Journal* 2008;**27**:589–605.

12.1 TYPES OF DNA DAMAGE

Feuerhahn S, Egly JM. Tools to study DNA repair: what's in the box? *Trends in Genetics* 2008;**24**:467–474.

Garinis GA, van der Horst GT, Vijg J, Hoeijmakers JH. DNA damage and ageing: new-age ideas for an age-old problem. *Nature Cell Biology* 2008;**10**:1241–1247.

Yang W. Structure and mechanism for DNA lesion recognition. *Cell Research* 2008;**18**:184–197.

12.2 POST-REPLICATION MISMATCH REPAIR

Jiricny J. The multifaceted mismatch-repair system. *Nature Reviews Molecular Cell Biology* 2006;**7**:335–346.

Li GM. Mechanisms and functions of DNA mismatch repair. *Cell Research* 2008;**18**:85–98.

McMurray CT. Hijacking of the mismatch repair system to cause CAG expansion and cell death in neurodegenerative disease. *DNA Repair* 2008;**7**:1121–1134.

Modrich P. Mechanisms in eukaryotic mismatch repair. *Journal of Biological Chemistry* 2006;**281**:30305–30309.

12.3 REPAIR OF DNA DAMAGE BY DIRECT REVERSAL

Eker AP, Quayle C, Chaves I, van der Horst GT. DNA repair in mammalian cells: Direct DNA damage reversal: elegant solutions for nasty problems. *Cellular and Molecular Life Sciences* 2009;**66**:968–980.

Sabharwal A, Middleton MR. Exploiting the role of O6-methylguanine-DNA-methyltransferase (MGMT) in cancer therapy. *Current Opinion in Pharmacology* 2006;**6**:355–363.

Sedgwick B, Bates PA, Paik J, Jacobs SC, Lindahl T. Repair of alkylated DNA: recent advances. *DNA Repair* 2007;**6**:429–442.

Tubbs JL, Pegg AE, Tainer JA. DNA binding, nucleotide flipping, and the helix-turn-helix motif in base repair by O6-alkylguanine-DNA alkyltransferase and its implications for cancer chemotherapy. *DNA Repair* 2007;**6**:1100–1115.

12.4 REPAIR OF DNA DAMAGE BY BASE EXCISION REPAIR

Fishel ML, Vasko MR, Kelley MR. DNA repair in neurons: so if they don't divide what's to repair? *Mutation Research* 2007;**614**:24–36.

Hitomi K, Iwai S, Tainer JA. The intricate structural chemistry of base excision repair machinery: implications for DNA damage recognition, removal, and repair. *DNA Repair* 2007;**6**:410–428.

Robertson AB, Klungland A, Rognes T, Leiros I. DNA repair in mammalian cells: Base excision repair: the long and short of it. *Cellular and Molecular Life Sciences* 2009;**66**:981–993.

Wilson DM 3rd, Bohr VA. The mechanics of base excision repair, and its relationship to aging and disease. *DNA Repair* 2007;**6**:544–559.

12.5 NUCLEOTIDE EXCISION REPAIR OF BULKY LESIONS

Bergoglio V, Magnaldo T. Nucleotide excision repair and related human diseases. *Genome Dynamics* 2006;**1**:35–52.

Goosen N, Moolenaar GF. Repair of UV damage in bacteria. *DNA Repair* 2008;**7**:353–379.

Schumacher B, Garinis GA, Hoeijmakers JH. Age to survive: DNA damage and aging. *Trends in Genetics* 2008;**24**:77–85.

Tornaletti S. DNA repair in mammalian cells: Transcription-coupled DNA repair: directing your effort where it's most needed. *Cellular and Molecular Life Sciences* 2009;**66**:1010–1020.

Truglio JJ, Croteau DL, Van Houten B, Kisker C. Prokaryotic nucleotide excision repair: the UvrABC system. *Chemical Reviews* 2006;**106**:233–252.

12.6 TRANSLESION DNA SYNTHESIS

Broyde S, Wang L, Rechkoblit O, Geacintov NE, Patel DJ. Lesion processing: high-fidelity versus lesion-bypass DNA polymerases. *Trends in Biochemical Sciences* 2008;**33**:209–219.

Lehmann AR. New functions for Y family polymerases. *Molecular Cell* 2006;**24**:493–495.

Waters LS, Minesinger BK, Wiltrout ME, *et al.* Eukaryotic translesion polymerases and their roles and regulation in DNA damage tolerance. *Microbiology and Molecular Biology Reviews* 2009;**73**:134–154.

12.7 THE DNA DAMAGE RESPONSE

de Bruin RA, Wittenberg C. All eukaryotes: before turning off G1-S transcription, please check your DNA. *Cell Cycle* 2009;**8**:214–217.

Harper JW, Elledge SJ. The DNA damage response: ten years after. *Molecular Cell* 2007;**28**:739–745.

Matic I, Taddei F, Radman M. Survival versus maintenance of genetic stability: a conflict of priorities during stress. *Research in Microbiology* 2004;**155**:337–341.

12.8 THE DNA DAMAGE RESPONSE IN BACTERIA

Butala M, Zgur-Bertok D, Busby SJ. The bacterial LexA transcriptional repressor. *Cellular and Molecular Life Sciences* 2009;**66**:82–93.

Erill I, Campoy S, Barbé J. Aeons of distress: an evolutionary perspective on the bacterial SOS response. *FEMS Microbiology Reviews* 2007;**31**:637–656.

Galhardo RS, Hastings PJ, Rosenberg SM. Mutation as a stress response and the regulation of evolvability. *Critical Reviews in Biochemistry and Molecular Biology* 2007;**42**:399–435.

Kelley WL. Lex marks the spot: the virulent side of SOS and a closer look at the LexA regulon. *Molecular Microbiology* 2006;**62**:1228–1238.

12.9 THE DNA DAMAGE RESPONSE IN EUKARYOTES

Bergink S, Jentsch S. Principles of ubiquitin and SUMO modifications in DNA repair. *Nature* 2009;**458**:461–467.

Cimprich KA, Cortez D. ATR: an essential regulator of genome integrity. *Nature Reviews Molecular Cell Biology* 2008;**9**:616–627.

Cohn MA, D'Andrea AD. Chromatin recruitment of DNA repair proteins: lessons from the Fanconi anemia and double-strand break repair pathways. *Molecular Cell* 2008;**32**:306–312.

Hurley PJ, Bunz F. ATM and ATR: components of an integrated circuit. *Cell Cycle* 2007;**6**:414–417.

Lavin MF. Ataxia-telangiectasia: from a rare disorder to a paradigm for cell signalling and cancer. *Nature Reviews Molecular Cell Biology* 2008;**9**:759–769.

van Attikum H, Gasser SM. Crosstalk between histone modifications during the DNA damage response. *Trends in Cell Biology* 2009;**19**:207–217.

Yang XH, Zou L. Dual functions of DNA replication forks in checkpoint signaling and PCNA ubiquitination. *Cell Cycle* 2009;**8**:191–194.

12.10 DNA DAMAGE AND CELL DEATH IN MAMMALIAN CELLS

Halazonetis TD, Gorgoulis VG, Bartek J. An oncogene-induced DNA damage model for cancer development. *Science* 2008;**319**:1352–1355.

Kruse JP, Gu W. Modes of p53 regulation. *Cell* 2009;**137**:609–622.

Vazquez A, Bond EE, Levine AJ, Bond GL. The genetics of the p53 pathway, apoptosis and cancer therapy. *Nature Reviews Drug Discovery* 2008;**7**:979–987.

Wang W, El-Deiry WS. Restoration of p53 to limit tumor growth. *Current Opinion in Oncology* 2008;**20**:90–96.

13

REPAIR OF DNA DOUBLE-STRAND BREAKS AND HOMOLOGOUS RECOMBINATION

INTRODUCTION

In Chapter 12, we explored the strategies used by the cell as it detects and responds to damage that may be incurred to its genome. Without such repair mechanisms, a cell would accumulate potentially deleterious mutations, with dire consequences for the cell or the organism it may be part of. The repair strategies outlined in Chapter 12 are those available to the cell when one of the two strands of a DNA duplex remains intact, and can act as a template for the repair of its damaged complementary strand. But what happens if both strands of a duplex are damaged? How does the cell respond at times such as these?

In this chapter we build on Chapter 12 to consider the repair of double-strand DNA breaks to complete our picture of the range of strategies adopted by the cell to maintain an intact genome. However, double-strand DNA breaks are not always due to unwanted damage; such breaks are also an essential part of the function of a genome, and underpin the process of **homologous recombination** through which the exchange of genetic material – and ultimately the generation of biological diversity – occurs.

We begin with a general overview of DNA double-strand break repair and homologous recombination to set out the key concepts that we will develop further during the course of the chapter. We will also learn more about homologous recombination toward the end of this chapter.

13.1 AN OVERVIEW OF DNA DOUBLE-STRAND BREAK REPAIR AND HOMOLOGOUS RECOMBINATION

Double-strand breaks are repaired by two different mechanisms

Double-strand breaks in DNA, unlike other types of DNA damage, must be repaired in the absence of an undamaged template strand that could otherwise have been used to direct the repair process. There are two major mechanisms for repairing double-strand breaks that have distinct requirements and different consequences (Figure 13.1). In the first of these, non-homologous end joining (NHEJ) (Figure 13.1a), the broken ends are simply rejoined. The second mechanism of repair, **homology-directed repair** (Figure 13.1b), involves (as its name implies) pairing between the damaged DNA and homologous sequences in another chromosome, which serves as the template for repair. Both mechanisms are deployed in eukaryotes and in most bacteria. Because *Escherichia coli* lacks an NHEJ system, it was thought for many years that bacteria lacked NHEJ; however, the sequencing of other bacterial genomes has revealed NHEJ systems in many bacteria.

Although NHEJ can occur at any time during the cell cycle, it occurs predominantly in non-dividing (or 'quiescent') cells. It is particularly important for bacteria and haploid yeast in stationary phase. Although NHEJ has the advantage that it does not require a template and can therefore occur during G1 of the cell cycle when no sister chromatid is available, it has the disadvantage that it can cause mutations. Nucleotides at the breakage site are often lost through degradation by cellular nucleases before repair is initiated, and further loss is frequently incurred because of trimming that occurs during the repair process itself. Such mutagenesis is less hazardous in plants and animals, whose genomes contain a high proportion of non-coding DNA and are therefore much less likely to suffer loss of coding sequences as a result of inexact repair.

Homology-directed repair is the major mechanism for repairing double-strand breaks that occur during DNA replication. In eukaryotes, the majority of homology-directed repair occurs during late S phase or G2 phase of the cell cycle only, when closely apposed sister chromatids are available to serve as a template. *E. coli* often contains multiple copies of its genome because segregation can take place before the initiation of a new round of replication and thus double-strand breaks are repaired by homology-directed repair.

(a) non-homologous end joining (NHEJ)

imprecise join

(b) homology-directed repair using homologous chromosome

Figure 13.1 The major pathways of DNA double-strand break repair and their consequences.
(a) Non-homologous end joining. DNA broken by a double-strand break is shown. Non-homologous end joining can occur at any phase of the cell cycle but is most likely during G1 when a sister chromatid is not present to provide a template for repair. Repair by this pathway can often be inexact (jagged line). (b) Homology-directed repair using a homologous chromosome. A chromosome broken by a double-strand break and its homolog (pink) are shown. Repair results in the conversion of information in the region of the break (gray) to information from the homolog (pink). Thus, the information copied from the homologous chromosome may not be identical to that of the damaged sequence if the template region contains different versions of the same gene. Like NHEJ, homology-directed repair can also occur in G1 using information from a homolog (as shown here), but this is infrequent. Homology-directed repair in which an identical sister chromatid provides the information for accurate repair, is the preferred pathway in late S phase. (c) Homologous recombination. Homologous recombination can exchange large segments of DNA, so that AB and ab markers on the parental (gray and pink) duplexes become Ab and aB on the daughter duplexes, after the recombination reaction.

(c) homologous recombination

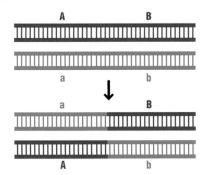

Homology-directed repair generally results in repair without mutation or loss of sequences because it involves copying intact information from an undamaged identical sister chromatid. More rarely, a homologous chromosome may provide the template instead of a sister chromatid; this allows homology-directed repair to also occur in G1, but may not restore the original sequence to the damaged chromosome, because the sequences of the two homologous chromosomes may not be identical. In this case, the damaged region will acquire the sequence of the intact homologous chromosome, a phenomenon known as **loss of heterozygosity**. We shall see later that this loss of heterozygosity can have serious consequences if the sequence on the undamaged homolog contains a defective gene.

So what determines which repair strategy is adopted in cells in which both NHEJ or homology-directed repair are possible? The initial recognition of a double-strand break is key not only to the cellular response to double-strand breaks, as we saw in Chapter 12, but also to the choice of repair pathway. In eukaryotes, double-strand breaks resulting from damage to fully replicated DNA are quickly bound by either an MRN complex (see Section 12.9), or a heterodimer composed of two proteins called Ku70 and Ku80. As well as tethering broken ends, each of these complexes recruits repair proteins to the site of damage to determine the specific pathway adopted: we shall see in a later section how the MRN complex contributes to the cell-cycle restriction of the homology-directed repair pathway; the Ku70/Ku80 heterodimer contributes specifically to the NHEJ pathway.

In this chapter, we shall describe the steps that are required for each of these pathways, and some of what is known of the proteins that mediate them – many of which also play a role in DNA replication. We shall focus mainly on those proteins that have been identified through mutations that increase the susceptibility of single cells to radiation-induced damage, or confer genetic predisposition to cancer in humans and are therefore implicated in specialized roles in DNA repair.

⊃ We learn about proteins involved in DNA replication in Chapter 6.

The double-strand break repair machineries also underlie mechanisms to generate diversity in the immune system and mediate homologous recombination

The mechanistic principles that underpin double-strand break repair are also exploited in other biological contexts. Both the NHEJ and homology-directed repair systems have been recruited in the course of evolution to specialized pathways of regulated DNA exchange or rearrangements that are fundamentally important in the adaptability and survival of multicellular eukaryotes. The NHEJ pathway provides the means for the DNA joining steps in the programmed rearrangement of the DNA of specialized cells of the immune system. This rearrangement generates the highly variable proteins – of which antibodies are the most familiar – that protect us from infection. Defects in the NHEJ pathway can lead to immune deficiency as well as sensitivity to DNA damage that may result in cancer.

⊃ We describe the programmed rearrangement of the DNA of specialized cells of the immune system in detail in Chapter 14.

The proteins of the homology-directed repair pathway have also been adapted to mediate the pairing interactions between homologous DNA molecules that underlie **homologous recombination** (Figure 13.1c). Homologous recombination is the reciprocal exchange of information between chromosomes that is intimately involved in the production of gametes during meiosis (see Chapter 7) and occurs occasionally in mitosis in eukaryotes. Homologous recombination also occurs in bacteria, often during **conjugation**, when specific plasmid-encoded proteins mediate the programmed transfer of DNA from one cell to another.

Other paths of entry for DNA into bacteria include phage infection, in which phage particles can carry cellular DNA, and transformation by naked DNA. Incorporation of these exogenous DNAs into the genome by homologous recombination is relatively common and makes a significant contribution to the horizontal gene transfer that accounts for much bacterial diversification in the course of evolution. These foreign DNAs can also be the 'launch pads' for the discrete self-mobilizable elements that can insert into and diversify recipient genomes. Thus, homologous recombination is universal, and provides a major mechanism for increasing the diversity of individuals in a population. Indeed, homologous recombination is an essential requirement for the generation of germ cells, and defects in this process can lead to infertility.

➲ We discuss mobile elements in Chapter 14.

Homologous recombination in meiosis is a highly regulated process involving closely apposed chromosomes where exchange of DNA can occur without loss or gain of sequences. In somatic cells, by contrast, homologous recombination can occasionally occur between two stretches of homologous DNA located anywhere in the genome, leading to **deletions** or **translocations**, both of which can have grave consequences for the cell. Such events are particularly likely in the genomes of higher vertebrates, which, as we have seen in Chapter 1, have a very high proportion of repetitive DNA, providing large regions of homology between non-homologous chromosomes, and between different regions of the same chromosome. We discuss homologous recombination in somatic cells in later sections of this chapter.

The conserved molecular components of the homology-directed repair and homologous recombination pathways highlight the close relationship between the mechanisms they adopt. We learn more about these commonalities during the course of this chapter.

13.2 DOUBLE-STRAND BREAK REPAIR BY NON-HOMOLOGOUS END JOINING

Having now reviewed the key principles that underpin double-strand break repair we begin our exploration of how this repair occurs in reality by considering the first of the two mechanisms for repair mentioned in Section 13.1, namely non-homologous end joining.

The NHEJ pathway of double-strand break repair does not require a template

The process of NHEJ occurs simply by the rejoining of broken DNA ends and does not require another template duplex such as a sister chromatid, available only in replicating DNA. A homologous chromosome can also serve as a template for repair, but appropriate pairing is not easily be achieved by chance. NHEJ is therefore the only readily available repair pathway for cells in the G1 phase of the cell cycle. (Despite being the only repair pathway available in G1, it can actually occur in any cell cycle phase.)

Since most mammalian cells spend the majority of the time in G1, NHEJ is the predominant mode of double-strand break repair in mammalian cells. In the budding yeast *Saccharomyces cerevisiae*, and the majority of its double-strand break repair occurs by homology-directed repair. NHEJ in yeast occurs mostly in G1 in haploid cells, where there is no homologous chromosome or sister chromatid as an alternative, for example in stationary phase or during sporulation.

The NHEJ pathway is generally mutagenic

In some cases, especially in yeast, repair by the NHEJ pathway may occur through simple rejoining of the broken ends by a DNA ligase, with little or no sequence loss (Figure 13.2i). In most cases, however, the ends are rejoined only after considerable processing of the ends at the break (Figure 13.2ii).

NHEJ events usually occur within about 10 bp of the break but can be up to 100 bp from the break. In one pathway (Figure 13.2i), the ends at the break undergo only limited processing with perhaps the loss of a few base pairs, and two blunt ends are ligated together. In another pathway (Figure 13.2ii), resection reveals short complementary regions that may occur by chance anywhere in the tails. These tails may then base pair, after which the single-stranded tails are trimmed back to the base-paired DNA, the gaps are filled in by repair polymerases, and the ends are ligated. This process can involve considerable sequence loss before the trimmed ends are ligated to repair the break. NHEJ is therefore often mutagenic and may be deleterious if, for example, it occurs in a region of functional DNA sequence such that important coding information is lost. However, such mutation is minor compared with the consequence of no repair happening at all, which could result in the loss of a chromosome arm, for example.

Figure 13.2 Pathways of non-homologous end joining. A DNA broken by a double-strand break is shown, flanked by short regions of homology – ABCD on the top strand and their complements on the bottom. NHEJ can occur through one of two pathways. Pathway (i): (1) nucleases and/or polymerases remove a few base pairs at either side of the break to generate blunt ends; (2) blunt ends are aligned and joined by ligation. In the process, some sequence information is usually lost. Pathway (ii): (1) resection from the site of the break occurs to generate single-stranded tails; (2) base-pairing occurs between stretches of microhomology a few nucleotides long; (3) the tails are trimmed back and gaps are filled by DNA synthesis (green), and some sequence information is lost.

(a) Table of proteins that mediate NHEJ in different organisms

functional component	bacteria	eukaryotes	
		Saccharomyces cerevisiae	multicellular eukaryotes
DNA end binding	Ku (30–40 kDa)	Ku 70/80	Ku 70/80
polymerase	POL domain of LigD	Pol4	Pol μ and λ
nuclease	?	Rad50:Mre11:Xrs2	Artemis: DNA-PKcs
kinase/phosphatase	phosphodiesterase domain of LigD	Tpp1 and others	polynucleotide kinase/ phosphatase and others
ligase	LIG domain of LigD	Nej1:Lif1:Dnl4	Ligase IV-XRCC4 XLF/Cernunnos

Figure 13.3 Proteins and steps in non-homologous end joining (NHEJ). (a) Proteins that mediate NHEJ in different organisms (b) NHEJ in mammals. (1) The C region of a DNA is broken by a double-strand break. (2) The ends are resected by a nuclease. (3) Ku heterodimers bind to the DNA ends at a double-strand break and are thought to help tether the ends to one another. The protein kinase DNA-PKcs is recruited by Ku and activated by binding DNA. (4) DNA-PKcs recruits and phosphorylates the nuclease Artemis leading to its activation. (5) Artemis trims the single-stranded tails. (6) Ligase IV and its cofactors XRCC4 and XlF-Cernunnos assemble on the ends and ligate the ends together. (7) Although now intact, the DNA was imprecisely repaired because resection and trimming of the ends leads to sequence loss.

The NHEJ pathway depends on conserved end-binding and ligating proteins

The NHEJ process in bacteria and eukaryotes requires a protein, Ku, which binds to broken DNA ends and recruits the nucleases, polymerases, kinases, phosphatases and ligases that mediate the joining reaction (Figure 13.3a). The principal known proteins in the mammalian non-homologous end joining pathway, which is the best understood, are illustrated in Figure 13.3b. What are the main steps in this pathway?

First, the Ku complexes slide onto each end of the double-strand DNA break and act as a scaffold for the recruitment of DNA repair enzymes, in particular a protein kinase, DNA-PKcs. Second, DNA-PKcs is activated by binding to Ku, resulting in autophosphorylation, and also phosphorylates and controls other important repair proteins. These include the overhang trimming nuclease Artemis, which is activated on phosphorylation by DNA-PKcs. Third, other DNA end-processing enzymes, including DNA polymerases and/or exonucleases and DNA kinases and/or phosphatases, further process the ends before ligation.

Finally, a specialized DNA ligase, DNA ligase IV, and its accessory proteins, XRCC4 and XLF/Cernunnos, form a complex that ligates the broken DNA ends (XRCC stands for x-ray repair cross-complementing – a reference to the experimental analysis whereby the proteins were discovered).

Ku is conserved from bacteria to mammals. By contrast, DNA-PKcs seems to be a specialization found mainly in vertebrates and may play a part in the relatively high frequency of repair by NHEJ in mammals. DNA-PKcs can also activate cell cycle arrest and programmed cell death in mammals, like ATM (ataxia telangiectasia-mutated) and ATR (ataxia telangiectasia-related) (see Section 12.9).

(b)

1 double-strand break and degradation

A B C D E

2 ends are partially resected by nuclease

A B C D E

3 Ku heterodimers bind broken ends and recruit DNA-PKcs

Ku

DNA-PKcs

4 DNA-PKcs recruits and phosphorylates the nuclease Artemis

Artemis

5 Artemis trims single-stranded tails

6 DNA ends are ligated by ligase IV in complex with XRCC4 and XLF/Cerunnos

XLF/Cernunnos — ligase IV

XRCC4

7 ligation results in deletion

A B D E

imprecise join

The MRX/MRN complex is involved in NHEJ

As we have seen in Section 12.9, MRN complexes play a central part in eukaryotic DNA damage repair and may play a key role in holding broken DNA ends. In budding yeast, the MRN complex, denoted MRX after its components Mre11, Rad50, and Xrs2, is known to also be involved in double-strand break repair by NHEJ, where the complex is thought to collaborate with Ku in holding the DNA ends together.

Rad50 belongs to the SMC (structural maintenance of chromosomes) family of proteins, which are implicated in several processes in which DNA molecules must be juxtaposed. Mre11 has nuclease activity and also plays a role in generating single-stranded tails.

The mammalian MRE11 has been shown to have a role in NHEJ but it remains to be seen if the MRN complex containing MRE11, RAD50, and NBS1 (a functional homolog of Xrs2), also has a primary role in the NHEJ pathway. MRN does have a critical role in initiating the cellular response to double-strand breaks. Indeed, the severe effects of rare mutations in components of the human MRN complex underscore its importance in this initiation process.

As we shall see, the MRN and MRX complexes also participate in homology-directed repair of double-strand breaks. We learn more about their role in this pathway below.

➲ We first encounter Rad50 in Section 12.9.

13.3 HOMOLOGY-DIRECTED REPAIR OF DOUBLE-STRAND BREAKS

The NHEJ pathway allows the repair of double-strand breaks in the absence of a template, but at the cost of mutagenesis. By contrast, homology-directed repair exploits the availability of homologous sequences elsewhere, most often in a sister chromatid, to provide a template for error-free DNA repair. In this section, and the two subsequent sections, we consider how this repair mechanism operates.

Accurate repair of double-strand DNA breaks involves pairing of homologous duplexes

Although the details of homology-directed repair vary between different organisms, the fundamental process is the same in all cases. In contrast to NHEJ, homology-directed repair is not accompanied by the loss of base pairs. Instead, when recombination occurs between sister chromatids, the original sequence is restored.

The reactions that underlie homology-directed double-strand break repair occur within four main phases:

- generation of single-stranded DNA at the site of a double-strand break
- pairing by one of the single-stranded ends with the intact duplex by invasion of the duplex to form a **heteroduplex**
- repair of the damaged duplex by DNA synthesis using strands from the undamaged duplex as a template
- the separation of the two duplexes.

The formation of the heteroduplex is known as **synapsis**, and the steps that precede or follow it as **presynaptic** and **postsynaptic**. In this section we will follow in detail what happens to the DNA during homology-directed repair. In subsequent sections, we discuss the specialized proteins that make these reactions possible, and how they can sometimes cause changes in genomic DNA.

Homology-directed repair requires long single-stranded regions

The first step in homology-directed repair is the generation of single-stranded 3′ ends at the double-strand break by specialized helicases and exonucleases (Figure 13.4, step 1). Whereas in NHEJ only short regions of homology are required to enable the single-stranded DNA at the break to base pair, homology-directed repair requires much longer regions of homology, and the single-stranded tails at the break are usually at least 50 nucleotides long. These long tails are needed to ensure accurate pairing with the template DNA in another chromosome, and thus to ensure error-free repair. In both bacteria and eukaryotes, specialized helicases and nucleases generate the long single-stranded DNAs that are the substrates for homology-dependent repair.

Formation of a D-loop containing heteroduplex DNA provides a template for DNA repair synthesis

In the second step in homology-directed repair, **strand invasion**, the single-stranded DNA from one end of the damaged duplex locates and pairs with its complementary strand in the related unbroken duplex to form DNA **heteroduplex** – a region in which a single strand from one DNA molecule is base-paired with a single strand from a different DNA molecule. This pairing displaces the DNA strand of the same polarity from the undamaged duplex with the formation of a three-stranded structure called a Displacement loop (D loop) (Figure 13.4, step 2). The degree of DNA sequence similarity required for efficient base-pairing and heteroduplex formation in these reactions is usually 95–100% identity extending over at least 20–30 base pairs. Thus there is an extremely small likelihood of pairing between unrelated DNA molecules.

Homology-directed repair can proceed via synthesis-dependent strand annealing

Once specific heteroduplex pairing has occurred between the invading strand and the undamaged duplex to form the D-loop, homology-directed repair can proceed in any of several ways. The most frequent pathway is known as **synthesis-dependent strand annealing (SDSA)** (Figure 13.5).

Figure 13.4 The initiation of homology-directed repair of a DNA double-strand break occurs by formation of a D-loop. A DNA broken by a double-strand break is shown. (1) Resection of the 5′ ends at the break by nucleases generates 3′ single-stranded DNA tails. (2) A single-stranded tail invades a region of undamaged homologous duplex (pink) that may be a homolog or sister chromatid to form a heteroduplex. This displaces the other strand to form a three-stranded D-loop.

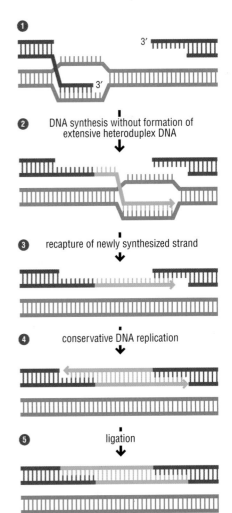

Figure 13.5 Synthesis-dependent strand annealing. In this pathway, both strands of the repaired duplex are formed by new DNA synthesis. (1) A D-loop is formed between a 3′ single-stranded tail from a broken DNA duplex (gray) and intact template duplex (pink) as described in Figure 13.4. (2) New DNA synthesis (green) is initiated at the 3′-OH end of the invading strand of the D-loop using the undamaged DNA as a template. As DNA synthesis progresses, the newly synthesized strand is displaced from the template duplex. Note that the newly synthesized strand and the template strand do not remain paired as in semi-conservative DNA replication. (3) DNA synthesis continues across the pink template strand until the newly synthesized (green) strand can pair with the gray strand from the other side of the gap in the gray duplex. (4) The single-strand gaps in the gray and green DNA are repaired using the gray and green 3′ ends as primers. Both strands of the repaired gap are newly synthesized, i.e. DNA replication is conservative. (5) Ligation regenerates an intact duplex, the gap in the gray duplex being replaced with newly synthesized (green) copies of the template (pink) duplex.

Repair by this pathway begins with DNA synthesis that is initiated in the D-loop using the 3′-OH end of the invading strand as a primer and the undamaged strand as a template (Figure 13.5, step 2). As synthesis proceeds, the newly synthesized strand is released from the template strand. This is sharply different from the semi-conservative DNA synthesis seen during DNA replication, during which the newly synthesized strand remains paired to the template strand.

When it is long enough, the newly synthesized strand pairs with the complementary strand of the damaged duplex on the other side of the break from where D-loop formation began (Figure 13.5, step 3). Notably, the DNA that now bridges the double-strand gap is newly synthesized DNA, generated from the undamaged template duplex; there is no permanent formation of heteroduplex DNA between the damaged duplex and the template duplex. The damaged duplex now contains a region of single-stranded DNA rather than a double-strand gap. This gap is filled in using the newly synthesized strand as a template (Figure 13.5, step 4). Ligation regenerates intact duplex DNA (Figure 13.5, step 5).

Thus, in homology-dependent repair by synthesis-dependent strand annealing, no DNA strand from the undamaged duplex (pink) is joined to or incorporated into the repaired duplex (gray), and the DNA synthesis is conservative: that is, both strands of the repaired duplex result from new DNA synthesis.

Having reviewed the overall events that are characteristic of homology-directed repair, we are now ready to consider the mechanism of this process in more detail. In the following sections, we describe the molecular machinery that operates to generate the single-stranded tails that invade the homologous duplex to generate the D-loop and to align the invading strand with the complementary strand of the intact homolog.

13.4 GENERATION OF SINGLE-STRANDED DNA BY HELICASES AND NUCLEASES

The RecBCD and MRN complexes generate the single-stranded DNA needed for strand invasion

The process of generation of the long single-stranded tails that are required for homology-directed repair is called the presynaptic step. This step depends on both DNA helicases, which separate the two strands of DNA by unwinding (see Section 6.5), and on nucleases. These nucleases include both exonucleases and endonucleases that expose single strands by degradation of one strand in a duplex.

In *E. coli*, in which this step is best understood, there are two distinct presynaptic pathways: the **RecBCD** and the **RecF** pathways. The RecBCD pathway is the predominant presynaptic pathway and is mediated by a protein complex called by the same name: RecBCD (Figure 13.6). This complex, which has combined helicase and nuclease activities, attaches at a double-strand break (Figure 13.6, step 1) and translocates along the DNA, its helicase activity separating the two strands and the nuclease activity cleaving the unwound strands into small fragments (Figure 13.6, step 2). The activity of this enzyme is modulated by a specific DNA sequence, known as a **Chi sequence**, that occurs at frequent intervals

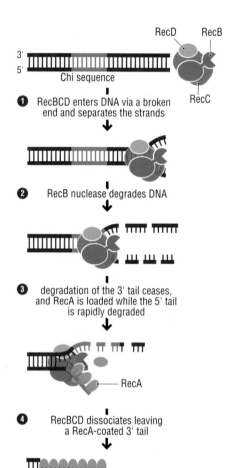

Figure 13.6 Schematic diagram of the generation of single-stranded tails at double-strand breaks by the RecBCD complex. The RecBCD protein complex has dual helicase and nuclease functions. (1) RecBCD loads onto DNA at a double-strand break. (2) RecBCD translocates along the DNA, its helicase activity separating the two strands to generate a fork. (3) RecB nuclease activity cleaves the unwound strands into small fragments. (4) When the RecBCD complex arrives at a specific DNA sequence motif known as a Chi sequence (pink), the nuclease activity is reduced and now preferentially degrades the 5′-ending strand. This results in a duplex DNA with a single-stranded 3′ tail, containing the Chi sequence, to which RecA binds in a step facilitated by RecBCD. (5) RecBCD dissociates leaving a nucleoprotein filament containing RecA bound to a 3′ single-stranded DNA end.

in the *E. coli* genome. When the RecBCD complex arrives at a Chi sequence the nuclease activity is reduced (Figure 13.6, step 3) and the enzyme now preferentially degrades the 5'-ending strand, leaving a duplex DNA with a single-stranded 3' tail containing the Chi sequence (Figure 13.6, step 4). In Experimental approach 13.1 we will explore how the study of single-molecule complexes of RecBCD and Chi-containing DNA defined how Chi modulates RecBCD activity.

RecBCD also facilitates the loading of **RecA** onto the 3' single-stranded tail. RecA protein promotes homologous pairing between single-stranded DNA and a DNA duplex, preferentially recognizing a DNA 3' end of this sort for strand invasion to form a D-loop. Thus, Chi sequences activate homology-directed repair as they generate 3' ended single-stranded DNA. Cells defective in RecBCD are very sensitive to DNA-damaging agents because they cannot repair DNA double-strand breaks.

The other presynaptic system in *E. coli* is known as the RecF pathway, which includes a helicase called RecQ and a nuclease called RecJ that, together, generate single-stranded DNA. This pathway appears to be used mainly in the homology-directed repair of replication forks that have stalled as a result of single-strand lesions in the DNA, which we discuss in Section 13.7. Humans also have homologs of the bacterial RecQ helicases, and defects in these genes are known to contribute to the hereditary diseases Bloom syndrome, Werner syndrome, and Rothmund-Thomson syndrome, which are associated with an increased risk of cancer as a result of increased genome instability.

In eukaryotes, the generation of the single-stranded DNA tails that are the substrates for strand invasion is carried out by several proteins (Figure 13.7). The activation of the specialized helicases and exonucleases required to produce these long single-stranded DNA tails is the decisive factor determining whether a DNA break will be repaired by the homology-dependent repair pathway or by NHEJ.

The repair process is initiated by MRN which, as we have seen, arrives early at the damage site and may also function to hold the DNA ends together at a break site. The MRN complex in yeast, MRX, also has nuclease activity that is key for the initiation of end resection although its exact nature remains to be determined. Following MRN action in the S and G2 phases of the cell cycle, phosphorylation of the CtIP protein by the cyclin-dependent kinase complex occurs and phosphorylated CtIP associates with the MRN complex. The cell cycle dependence of CtIP thus restricts homology-directed repair to S and G2. Two different proteins, EXOI, a 5'–3' exonuclease, and BLM, a 3'–5' RecQ family helicase and exonuclease DNA2, then resect the 5' ends at the break to yield 3' single-stranded tails.

13.5 THE MECHANISM OF DNA STRAND PAIRING AND EXCHANGE

A central process in homology-directed repair is the sequence-specific recognition that occurs between single-stranded DNA and DNA duplexes, and the exchange of the single strand with one of the strands of the duplex to form heteroduplex DNA. This step, as we have seen, is called synapsis. In this section we describe the molecular components that are the key players in mediating synapsis.

◉ We encounter the RecA protein briefly in Section 12.8.

Figure 13.7 Schematic diagram of the generation of 3' single-stranded tails in eukaryotes. An intact DNA is shown. (1) The DNA duplex is broken by a double-strand break. (2) MRN complex binds to the broken ends of the DNA. (3) In S/G2, CtIP is phosphoryled by CDK and associates with MRN. (4) Initial degradation of the 5' ends at the double-strand break is then carried out by CtIP and MRN. (5) The 5' ends are further degraded to expose longer 3' ends by an exonuclease (EXO1) or by a complex containing BLM, a RecQ helicase, and the nuclease DNA2. From Bernstein and Rothstein (2009).

13.1 EXPERIMENTAL APPROACH

DNA sequences that stimulate homologous recombination: Chi sites and RecBCD

The *E. coli* RecBCD complex has been known to play a role in recombination since *recB* and *recC* mutants that were defective in the process were isolated. However, the precise role of these proteins and how they might interact with DNA was not understood. Models for their mechanisms of action have been evolving and are still evolving from both careful observations of phage lambda and *E. coli* genetics and biochemistry.

A key finding in understanding the role of RecBCD was the recognition that the complex interacts directly with Chi sites in DNA. The initial biochemical studies of RecBC (the presence of the RecD subunit was a later discovery) identified it only as a DNA exonuclease. It was subsequently found that RecBC also had helicase activity that could interact with a duplex and expose single-stranded DNA. The fact that RecBC recombination activity was influenced by an 8 bp DNA sequence called Chi was discovered serendipitously in experiments looking at the growth of bacteriophage lambda as discussed below.

When the bacteriophage lambda grows lytically, it produces long concatenated linear DNA molecules that need to be packaged into the phage head. (See Section 8.10 for a review of the lambda lifecycle.) However, if certain lambda genes (*red* and *gam*) are inactive, lambda replication produces only monomeric circles that are poorly packaged, and thus these lambda mutants do not grow well. If the circles are able to recombine, however, they will generate long circular concatamers that can again be packaged, and the phage will grow well.

Several spontaneous lambda mutants were identified that resulted in higher recombination frequencies as reflected by the production of more lytic phage. The production of more phage could be readily visualized by the presence of a larger plaque size (indicating a larger zone of bacterial killing), when the propagation of individual viruses was analyzed on a confluent lawn of bacteria (Figure 1). In mapping these lambda mutations, Gerald Smith and colleagues found that the mutations did not affect proteins required for recombination, but were rather in DNA sites that stimulated recombination. These spontaneous mutations created 8 bp Chi sequences from natural sequences in the phage genome that were already almost consensus Chi sites (i.e. they already had seven of eight Chi nucleotides). Further work demonstrated explicitly that a Chi sequence was indeed a critical DNA site that stimulated *recBC*-dependent recombination. Although Chi was discovered because of its role in the phage lambda lifecycle, we now know that the *E. coli* chromosome has about one Chi site per 5 kb, and that these sites play a key role in stimulating homologous recombination in wild-type cells.

Biochemical studies of RecBC subsequently revealed that there was actually a third polypeptide present in complexes containing RecBC: this turned out to be the RecD protein. The *recD* gene was not identified in screens for recombination-defective mutants because *recD* mutants are hyper-recombinogenic, due to constitutive, unregulated activation of *recBC*-mediated recombination. Indeed, when researchers looked for revertants of the

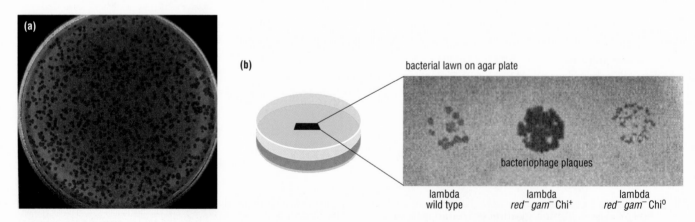

Figure 1 Growth of bacteriophage lambda on *E. coli*. (a) The ability of a lytic phage to grow can be monitored by the size of the plaque, or zone of bacterial killing, produced by the propagation of individual viruses on a confluent lawn of bacteria. (b) The effect of Chi on the growth of bacteriophage lambda: wild-type lambda lacks Chi sites and yields plaques of moderate size; *red⁻ gam⁻* lambda with Chi sites (Chi⁺) yields large plaques; and *red⁻ gam⁻* lambda without Chi sites (Chi⁰) yields small plaques.
Image courtesy of SK Amundsen.

(a)

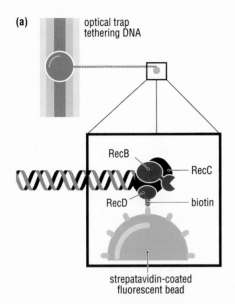

optical trap
tethering DNA

RecB

RecC

RecD

biotin

strepatavidin-coated
fluorescent bead

recBC recombination defects, many of those revertants that were hyper-recombinogenic turned out to be mutations in the *recD* gene. It was also observed that the presence of a Chi sequence did not increase recombination in the hyper-recombingenic *recD⁻* mutants. This lack of Chi stimulation in the *recD⁻* mutants suggested a direct link between the Chi sequence and modulation of RecBCD activity, and eventually led to the explicit suggestion that Chi activates recombination by inactivating RecD. As such, hyper-recombinogenic *recD⁻* variants represent loss of function mutations in the RecD protein; in these variants, recombination is relieved of the inhibitory presence of RecD protein.

How does Chi modulate the activity of RecBCD to stimulate recombination? As mentioned above, biochemical experiments had revealed that RecBCD has helicase and nuclease activity. This protein complex enters DNA at a double-strand break and moves through the duplex as a helicase/nuclease, degrading DNA as it proceeds. A key finding was that when RecBCD encountered a Chi site, the nuclease activity was decreased such that a 3′ single-stranded segment of DNA persisted. This single-stranded DNA is the substrate that RecA uses for recombination. Thus Chi stimulates homologous recombination by switching the activity of RecBCD from a DNA-degrading enzyme to one that makes 3′ single-stranded DNA.

What is the molecular mechanism by which Chi switches the activity of RecBCD? A popular (and entirely reasonable) hypothesis was that when RecBCD encountered Chi as it moved along DNA, RecD was ejected from the complex, giving rise to a RecBC machine that had decreased degradation activity (while still having the helicase activity), thus resulting in the generation of 3′ single-stranded tails. Recent single molecule experiments by Stephen Kowalczykowski and colleagues have shown that this is not the case, however, since RecD remains associated with the RecBCD complex even after encountering Chi.

To follow RecBCD movement along single DNA molecules, Kowalczykowski's group used an 'optical trap' to hold a single DNA molecule in a fixed position such that the movement of proteins (the RecBCD complex) along the immobilized DNA could be analyzed. Long linear ~ 50 kb DNA molecules that either contained Chi sites (here called χ⁺) or did not contain Chi sites (here called χ⁰) were covalently linked to large beads, which were then positioned by 'optical tweezers' in a flow cell such that the DNA was extended by the flow to the right (Figure 2a); the Chi sites (in the χ⁺ DNA) are

(b) χ⁰ DNA   χ⁺ DNA

trap

Figure 2 Video images of the translocation of a single RecBCD molecule along a DNA. (a) Schematic of the experiment. Either χ⁰ or χ⁺ DNA is attached to a large bead (turquoise) that can be stably positioned in the optical trap (red), and the DNA is extended by flow to the right. The RecD subunit of RecBCD is labeled with biotin and thus associates with a small fluorescent nanoparticle covered with streptavidin (green). The nanoparticle tracks the movement of RecBCD along DNA towards the large optically 'trapped' sphere (b) Time lapse photography of the movement of the RecBCD bead along the DNA. The nanoparticle is at the tip of the arrow. The rate of RecBCD movement is plotted in Figure 3.

Figure 3 Plot of the translocation of a single RecBCD complex along DNA. (right) The size of the DNA positioned in the optical trap is shown with the positions of the Chi sites. (left) The position of the RecBCD molecules associated with the nanoparticle from the optically trapped bead followed in Figure 2 is plotted as a function of time. RecBCD initially moves at the same rate on χ^0 (blue line) and χ^+ (pink line) DNA, but the rate of movement of RecBCD slows at the χ^+ site (downward arrow), reflecting changes in the RecBCD nuclease activity. Most notably, RecD remains associated with RecBC and continues to move along DNA, albeit at a slower rate.

found at sites proximal to the open right end of the DNA (Figure 3). The movement of RecD can be specifically followed as it is tagged with a fluorescent nanoparticle (bead). The fluorescent complex enters the DNA at the exposed right duplex end and travels to the left (Figure 2b), towards the χ sites and the large 'optically trapped' bead. (The RecBCD complex is fluorescent because RecD forms part of it.) The positions of the beads were followed over time and those trajectories plotted as in Figure 3.

The initial rate of movement of RecBCD on both the χ^+ and χ^0 DNA was virtually the same (Figure 3). Indeed, the speed of RecBCD movement along the χ^0 DNA lacking the Chi sites did not change over the entire length of the DNA. By contrast, RecBCD abruptly changed speed at the Chi sites as it moved along the χ^+ DNA. Most importantly, however, the fluorescently labeled RecD protein remained associated with the DNA (and thus RecBC) even after the abrupt change in speed, thus directly demonstrating that

RecD is not ejected from the RecBCD complex on encountering Chi. The researchers propose that this change in speed (and the associated production of free single-stranded DNA) might simply reflect conformational rearrangements in the complex that result from recognition of Chi by RecD. This example nicely demonstrates how single-molecule experiments allow direct visualization of molecular events and can be used to test very specific biochemical predictions. New experiments can now be designed to further probe the mechanism by which Chi site recognition alters RecBCD function.

Handa N, Bianco PR, Baskin RJ, Kowalczykowski SC. Direct visualization of RecBCD movement reveals cotranslocation of the RecD motor after χ recognition. *Molecular Cell* 2005;**17**:745–750.

Sprague KU, Faulds DH, Smith GR. A single base-pair change creates a Chi recombinational hotspot in bacteriophage lambda. *Proceedings of the National Academy of Sciences of the U S A* 1978;**75**:6182–6186.

DNA strand pairing is promoted by specialized recombinases

At the heart of the reactions underlying homology-directed repair, as well as the homologous recombination reactions that we discuss later, are the **strand-exchange recombinases**. The best-understood of these are the **RecA** proteins of bacteria and the homologous **Rad51** protein in eukaryotes; homologous proteins also exist in archaea. These proteins catalyze the ATP-dependent sequence-specific recognition events between single-stranded DNA and DNA duplexes, and the exchange of the single strand with one of the strands of the duplex to form heteroduplex DNA. Together, these two events comprise synapsis.

(a) RecA   **(b)** Rad51

Figure 13.8 Structures of RecA and Rad51 filaments. These models of the surfaces (pale gray) of *E. coli* RecA (left) and *S. cerevisiae* Rad51 (right) filaments bound to DNA were generated by three-dimensional reconstruction of many electron micrograph images of these proteins bound to DNA. Structures of RecA and Rad51 determined by crystallography are shown within each model.

Bacterial cells can live without RecA, and yeast cells can live without Rad51, but they grow poorly and are highly sensitive to DNA-damaging agents. Mammals without RAD51 die *in utero*. These growth defects reflect the fact that homology-dependent repair reactions are required to reassemble broken replication forks resulting from DNA damage which we will discuss in Section 13.7.

A second specialized eukaryotic recombinase, Dmc1, collaborates with Rad51 to promote strand exchange during meiotic recombination. The RecA, Rad51, and Dmc1 proteins are closely related structurally (Figure 13.8) and the mechanisms by which they promote strand exchange are the same. However, these mechanisms have been most extensively studied for the bacterial RecA strand-exchange recombinase, so we shall focus our discussion on RecA. Unlike RecA, Rad51 and Dmc1 do not act as co-proteases that, as we see in Section 12.8, promote the cleavage of particular repressors including LexA and thereby control the expression of DNA repair genes during the bacterial SOS response to DNA damage.

Strand exchange begins with the loading of the RecA recombinase monomers onto single-stranded DNA to form the **presynaptic filament** (Figure 13.9, step 1). This process requires ATP, which binds at the interface between the monomers of RecA and thereby drives a conformational change in the RecA molecules that is essential for DNA binding by the assembling RecA filament. DNA binding in turn promotes the assembly of further RecA monomers; therefore, the assembly of the presynaptic filament occurs through cooperative binding between RecA, ATP, and DNA.

Binding of the RecA nucleoprotein filament to the DNA in the invaded intact duplex distorts the DNA molecule so that stretches of extended and untwisted DNA alternate with three-nucleotide segments that assume the conformation of normal B-DNA. Consequently, the whole molecule is extended by about 50%

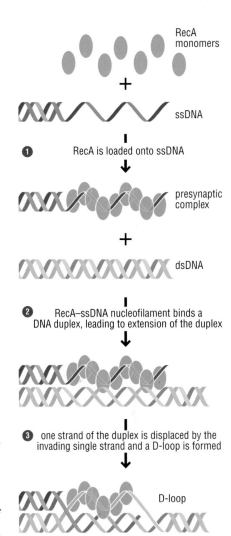

RecA monomers

+

ssDNA

1 RecA is loaded onto ssDNA

presynaptic complex

+

dsDNA

2 RecA–ssDNA nucleofilament binds a DNA duplex, leading to extension of the duplex

3 one strand of the duplex is displaced by the invading single strand and a D-loop is formed

D-loop

Figure 13.9 The recombinase nucleoprotein filament. A broken DNA end with a 3′ single-stranded tail (gray) and RecA monomers (orange circles) are shown. (1) RecA binds helically around the single-stranded DNA (ssDNA) to form a nucleoprotein filament called the presynaptic filament that can invade an intact duplex DNA (dsDNA; pink). (2) Binding of the RecA–single-strand complex to the duplex extends the duplex such that the bases in the single strand and base pairs in the duplex have the same internucleotide spacing. (3) The invading strand (gray) pairs to its complementary strand in the duplex (pink), displacing the other pink strand. The bottom structure shown here corresponds to the start of a D-loop such as the one shown in Figure 13.4.

compared with standard B-form DNA (Figure 13.9, step 2). This arrangement provides the basis for accurate pairing with the recipient DNA duplex during strand exchange.

Untwisting of the extended duplex DNA is thought to enable the interrogation of the duplex for DNA sequence homology by the invading single strand, while the three-nucleotide segments of B-DNA promote correct base pairing between the invading and the recipient strands, since incorrect pairing would disrupt the stacking interactions of the DNA.

The RecA protein plays very little part in the sequence-specific binding interactions between the presynaptic filament and the recipient duplex. Instead, these interactions are largely dependent on the base-pairing interactions between the two DNA molecules. Thus binding is strongly disfavored by mismatched bases, and this is thought to account for the requirement for very high sequence similarity between the interacting DNA molecules, and hence the accuracy of repair.

Once homologous pairing is established between the single strand of the nucleofilament and the complementary strand of the duplex DNA, strand exchange follows and heteroduplex DNA is formed (Figure 13.9, step 3). This is accompanied by hydrolysis of ATP by the recombinase, the disassembly of the nucleoprotein filament, and the displacement of the unpaired strand of the original duplex DNA to form the D-loop, in which the 3′ end of the invading single strand can prime DNA synthesis.

Accessory proteins assist the strand-exchange recombinases

Although RecA or Rad51 recombinases can execute homologous pairing and strand exchange in the absence of other proteins *in vitro*, many other proteins are involved in recombination *in vivo*. For example, single-stranded DNA binding proteins prevent the formation of secondary DNA structures that would obstruct interaction of the single-stranded DNA with the strand exchange recombinase; these proteins, however, must be removed to allow the recombinase protein to bind to form the presynaptic filament. When single-stranded DNA persists, it recruits accessory proteins that bind to the junction of single-stranded and double-stranded DNA. These proteins then catalyze the loading of the recombinase and the displacement of the single-strand binding protein.

Among the accessory proteins in mammals are the BRCA proteins and 'partner and localizer of BRCA2' (PALB2), a protein that mediates interaction between BRCA1 and BRCA2. Strikingly, BRCA2 and BRCA1 were first identified as gene products that, when defective, predispose individuals to breast and ovarian cancers.

We now know that BRCA1, BRCA2, and PALB are all essential for homology-directed repair. Why mutations in these genes specifically cause predisposition to breast and ovarian cancer, rather than other kinds of cancer, is not known. However, it is not unusual for a protein with a very general role in the maintenance of genome stability to give rise to a surprisingly specific range of tumor types in families in which it is defective.

As we saw in Section 12.9, BRCA1 is one of the early proteins recruited to a double-strand break via an ATM response, thus recruiting PALB2 and BRCA2. Moreover,

BRCA1 is a ubiquitin ligase and is thought to be important in the regulated recruitment of repair proteins to sites of DNA damage. This recruitment is driven by the addition of ubiquitin tags to molecules at the damage site that are recognized by components of repair pathways.

Several other ubiquitin ligases are also known to assemble at DNA double-strand breaks, although the exact role of each is not yet clear. In Experimental approach 13.2, we explore how single-molecule studies demonstrate that BRCA2 stimulates RAD51 action by promoting the binding of RAD51 to single-stranded DNA and inhibiting its binding to double-stranded DNA. Although yeast lack homologues of the BRCA and PALB2 proteins, the Rad52 and Rad55–Rad57 complex promote Rad51 function.

> ➔ We see in Chapter 12 how the ubiquitination of the PCNA sliding clamp in response to DNA damage facilitates the action of translesion polymerases at stalled replication forks.

13.6 GENE CONVERSION THROUGH HOMOLOGY-DIRECTED REPAIR

Homology-directed repair at a double-strand break can change gene sequence by replacing one version of a gene with another

We have just seen that, in order to be aligned by recombinases, the participating DNA molecules must be highly similar in sequence; but they need not be identical. Thus although the template for homology-directed repair of a double-strand break is *usually* the sister chromatid of the damaged DNA molecule, this is not always the case. For breaks that occur in G1 of the cell cycle, when sister chromatids are not available, a homologous chromosome may serve as the template.

Sister chromatids are exact replicas of each other (see Section 4.2), and if one serves as a template for repair of the other, the damaged chromatid will in effect be restored to its pre-damage sequence. Homologous chromosomes, on the other hand, generally do not have the same DNA sequence throughout: most diploid organisms are likely to be heterozygous at many genetic loci. Thus, when DNA of a homologous chromosome serves as the repair template, the repaired chromosome will acquire a new sequence – that of the homolog. This process is called **gene conversion**: a sequence on one homologous chromosome is converted into the sequence on the other, which itself remains unchanged (Figure 13.10).

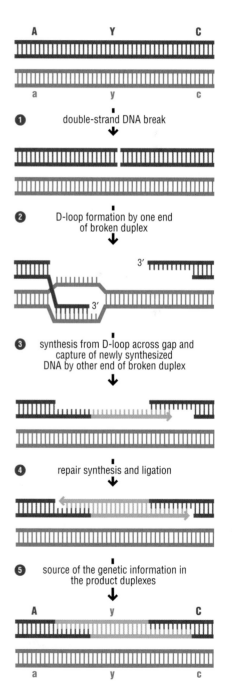

① double-strand DNA break

② D-loop formation by one end of broken duplex

③ synthesis from D-loop across gap and capture of newly synthesized DNA by other end of broken duplex

④ repair synthesis and ligation

⑤ source of the genetic information in the product duplexes

Figure 13.10 Homology-directed repair results in gene conversion. Two homologous parental DNA duplexes are shown. (1) A double-strand break occurs in the Y region of the gray duplex. (2) As also shown in Figures 13.8 and 13.9, the 5′ ends of the broken duplex are resected, and the 3′ end (gray) of the broken DNA on the left invades the homologous pink duplex to form a D-loop. The 3′ end of the invading single strand serves as a primer for DNA synthesis (green) using the intact pink duplex as a template. (3) DNA synthesis extends across the gap until the newly synthesized DNA (green) can pair with the other side of the gray broken duplex. (4) Single-strand gaps are filled by DNA synthesis (green) and ligation occurs to regenerate intact duplex DNA. (5) The initially broken gray helix is now 'healed' by newly synthesized DNA (green), one strand of which results from using the intact duplex (pink) as a template and the other strand from using this newly synthesized (green) strand as a template. Thus the repaired duplex (gray) acquires the sequence of the template duplex (pink) and the gray Y parental information is converted to the pink template y information.

13.2 EXPERIMENTAL APPROACH
Modulating the activities of RecA and RAD51

The function of a gene product is rarely understood from a single assay. Rather an ever-clearer picture emerges from a mosaic of insights that come from a variety of assays and approaches. This has been particularly true for the recombinases RecA and RAD51 (in bacteria and eukaryotes, respectively). While these proteins have been shown to have an astounding number of measurable activities, it has been an ongoing challenge to obtain a comprehensive picture of how these activities interface in the cell.

The *recA* gene was the first recombination gene to be identified. To search for genes involved in recombination, Alvin Clark and Ann Dee Margulies mutagenized cells and looked for variants in which recombination between the chromosome and a DNA fragment introduced into the recipient cell by conjugation was defective. Using this approach, they isolated several mutants, all of which were within the *recA* gene. These mutants also proved to be sensitive to a variety of DNA-damaging agents, establishing an early link between recombination and DNA repair. Subsequent work showed that the RecA protein also has functions in DNA strand exchange, signaling in the bacterial SOS response through proteolysis, and in mutagenic lesion bypass.

How can a single protein carry out so many different functions? Part of the answer comes from an understanding of its interactions with other proteins. Even though RecA is crucially involved in the DNA strand exchange event that plays a pivotal role in recombination, it does not act alone. Indeed, the type of genetic screens that led to the discovery of *recA* also resulted in the identification of other *rec* mutants that showed similar recombination defects. One of these genes encodes the RecJ protein, which was characterized by Susan Lovett and colleagues. Although RecJ protein has 5′ to 3′ exonuclease activity – which can generate the single-stranded DNA that is a substrate in recombination – Lovett considered the possibility that RecJ might also stimulate the strand-exchange process itself.

To test this idea, the group examined RecA-mediated strand exchange using purified components *in vitro*. In these assays, linear duplex DNA was mixed with complementary single-stranded circular DNA, and the mixture then incubated with RecA in the presence or absence of RecJ (Figure 1). The extent of strand exchange was determined by measuring the amount of nicked circular duplex DNA (NC) formed during the incubation (resulting from the repartitioning of a strand of DNA from the linear duplex to the circular DNA). In the presence of RecJ, the band corresponding to the NC duplex was detected within 15 minutes of starting the reaction, while in the absence of RecJ this band was not seen until 60 minutes later. The interpretation of this result is that RecJ

stimulates strand exchange by degrading the strand that would otherwise compete with the exchange reaction by reannealing with the displaced strand.

The mammalian homolog of RecA is RAD51. Like RecA, the monomer RAD51 assembles to form filaments along the DNA and promotes strand exchange between a single-stranded DNA and its complementary strand in a DNA duplex. As is true in bacteria, other proteins in eukaryotes are also required for homologous recombination, including BRCA2 in mammalian cells. BRCA2 (Breast Cancer 2) was so named because it was first identified as a 'disease' gene associated with a predisposition to breast cancer. The finding that BRCA2 accumulated at double-strand DNA breaks *in vivo* suggested that it also plays a role in recombination.

What role does BRCA2 play in homologous recombination? The protein has two notable features: a DNA-binding domain and two regions that interact with RAD51 via short domains called BRC repeats. We now know that the BRC repeats are important because mutations in these repeats block the interaction of BRCA2 with RAD51 in families with a predisposition to breast cancer.

The large size of BRCA2 (3418 amino acids) made it difficult to study the protein activity *in vitro*, as it proved difficult to express and purify. Stephen Kowalczykowski and colleagues gained significant insight into the role of BRCA2 in recombination by studying RAD51-mediated reactions *in vitro* in the presence of a fragment of BRCA2 containing a BRC domain, BRC4. Kowalczykowski adapted the classic assay used to examine RecA-mediated strand exchange (Figure 1) to measure the effects of BRCA2-BRC4 on RAD51-mediated strand exchange (Figure 2). As before, a labeled linear duplex DNA (asterisks) and complementary single-strand circular DNA were incubated with RAD51 and ATP and the formation of joint molecules (JM, formed as strand exchange begins) and a nicked circular duplex DNA (NC, the final reannealed exchange product) were measured. The addition of the BRC4 protein domain stimulated RAD51-mediated strand exchange in these *in vitro* reactions, whereas no stimulation was seen when a mutant version of BRC4 (Δ1524-30) was used in the assay.

Further analysis showed that BRC4 stimulates recombination through two distinct mechanisms. First it stabilizes the interaction of RAD51 with single-stranded DNA as evaluated in an electrophoretic gel mobility shift assay. (Mobility shift assays are described in Chapter 16.)

As we see, the addition of increasing amounts of BRC4 increases the extent of binding of RAD51 to the single-stranded DNA (Figure 3). A second role BRC4 plays is to inhibit the nonproductive interaction of RAD51 with double-stranded DNA.

Figure 1 RecJ stimulates RecA-mediated strand exchange. (a) Schematic diagram of intermediates in the strand exchange assay in presence and absence of RecJ. Single stranded circular DNA (black circle) was mixed with double stranded DNA (black and pink) in the presence of RecA (orange), with (bottom pathway) or without (top pathway) RecJ (blue). RecA faciliates the transfer of the pink strand from the double stranded DNA to the circular DNA resulting in a nicked circle (black and pink). (b) *In vitro* assay for RecA strand exchange activity. A single-stranded circular DNA (gray) is incubated with a linear double-stranded DNA molecule, with one labeled strand (pink) and the other (gray) unlabeled. In the presence of RecA and with or without RecJ as described in panel (a). RecA facilitates strand exchange, measured as the accumulation of the nicked circle product, though the reaction proceeds more quickly in the presence of RecJ protein.
Reproduced from Corrette-Bennett and Lovett, *Journal of Biological Chemistry* 1995;**270**:6881–6885.

Figure 2 The BRC4 domain of BRCA2 stimulates RAD51-mediated strand exchange. (a) Schematic diagram of intermediates in the strand exchange assay. Incubation of an end-labeled double-stranded duplex (dsDNA, pink and black) with a complementary single-stranded circle (ssDNA) initially results in formation of joint molecule (JM) intermediates and culminates in the formation of labeled nicked circle (NC) and labeled displaced single-stranded linear DNA. (b) *In vitro* strand exchange assay wherein all lanes contain Rad51 and different concentrations of the mutant BRC4 domain T1526A or different concentration of the wild-type BRC4 domain. Impressive increases in the production of JM and NC, indicative of strand exchange are observed when the wild-type BRC4 domain is added.
Reproduced from Carreira *et al.*, *Cell* 2009;**136**:1032–1043.

Using a single-molecule approach, Kowalczykowski and colleagues directly examined the binding of RAD51 to double-stranded DNA with and without BRC4 (Figure 4). Fluorescently labeled RAD51 protein was added to a long (50 kb) double-stranded DNA attached to an optically trapped bead. As we saw in Experimental approach 13.1, the position of the bead can be manipulated by optical tweezers in a flow cell. In this experimental set-up, the optical bead and attached DNA was moved using the

Figure 3 BRC4 promotes the binding of RAD51 to single-stranded DNA. Gel showing RAD51 incubated with either GST-BRC4 or GST-BRC4 Δ1524-30 prior to incubation with [^{32}P]-labeled single stranded poly-deoxythimidine DNA (dT40 ssDNA) for one hour in the presence of ATP, Mg^{2+}, and Ca^{2+}. In the presence of GST-BRC4, the majority of the dT40 single-stranded DNA is shifted to the top of the gel, while in the mutant (Δ1524–30) less complex is formed. Reproduced from Carreira *et al.*, *Cell* 2009;**136**:1032–1043.

Figure 4 BRC4 inhibits the binding of RAD51 to double-stranded DNA. (a) Schematic of experimental design to examine binding of fluorescently labeled RAD51 to double-stranded DNA. In the capture channel (bottom row), a 50 kb double-stranded DNA, is tethered to a streptavidin-coated bead (green) (that can be manipulated with optical tweezers (red). (1) One single-labeled DNA molecule is positioned in the observation channel of a flow cell (middle row); pictures are taken over time. (2) The DNA strand is moved into a protein channel containing fluorescently labeled RAD51 protein (orange) (top row). (3) The DNA is moved back to the observation channel and a strong RAD51 specific fluorescent signal will be seen on the DNA if the RAD51 was able to bind the DNA. (b) Single molecule analysis of the effect of BRC4 on RAD51 binding to double-stranded DNA. (1) RAD51 binding to double-stranded DNA is imaged above over a time course from 20 to 120 minutes. (2) When BRC4 is added to the reaction containing RAD51 (as above), no RAD51 fluorescence coating of the DNA is observed. (3) When the mutant BRC4 fragment BRC4 Δ1524–30 is added, RAD51 coats the double-stranded DNA effectively. Reproduced from Carreira *et al.*, *Cell* 2009;**136**:1032–1043.

optical tweezers into a flow cell containing buffer (step 1). The DNA was then moved into a channel that contains fluorescent RAD51 (with a distinct fluorescent signature) in the presence or absence of the BRC4 domain (step 2). After incubation with RAD51, the DNA strand was then moved back into the observation channel, to see whether the RAD51 had bound to the DNA (step 3). RAD51 binding to double-stranded DNA was observed in the absence of BRC, but not when the BRC4 domain was included. As a control, the mutant protein, BRC4 Δ1524-30, fails to inhibit the binding of RAD51 to double-stranded DNA.

Together these experiments demonstrated that the BRC4 domain affects RAD51 activity in at least two ways. These experiments also highlight the challenges of any one experimental approach in yielding the 'mechanism' of action of a particular protein. Molecular insight comes most typically from the gradual accretion of clues that assemble bit by bit into an ever more detailed model.

Carreira A, Hilario J, Amitani I, *et al*. The BRC repeats of BRCA2 modulate the DNA-binding selectivity of RAD51. *Cell* 2009;**136**:1032–1043.

Clark AJ, Margulies AD. Isolation and characterization of recombination-deficient mutants of Escherichia coli K12. *Proceedings of the National Academy of Sciences of the U S A* 1965;**53**:451–459.

Corrette-Bennett SE, Lovett ST. Enhancement of RecA strand-transfer activity by the RecJ exonuclease of Escherichia coli. *Journal of Biological Chemistry* 1995;**270**:6881–6885.

Note that this exchange of sequence is specifically targeted to a particular region, that is at the site of the double-strand break that initiates homology-directed repair. The gray 'Y' marker on the parental gray duplex in which the double-strand break occurred is converted to the pink 'y' marker as a consequence of repair (Figure 13.10, step 5). Thus while one parental duplex contains the Y marker and the other parental duplex the y marker, the two progeny duplexes each contain one y marker; thus the Y marker is converted to y and there are two y markers in the progeny duplexes. This conversion of markers is a signature of interaction between the two parental duplexes. However, note that the relationship and identity of the DNA segments that *flank* the double-strand break do not change: the A and B versions of these genes are still on one duplex and the a and b versions on the other.

This directed repair mechanism has been exploited in some organisms to carry out specific changes in gene structure and expression. Let us now consider several of these.

Yeast mating-type switching is the best understood example of directed gene conversion

Haploid cells of the yeast *S. cerevisiae* exist in two mating types, **a** and α, determined by the expression of either **a** or α mating-type information at a site called MAT on chromosome III (Figure 13.11). In some strains of *S. cerevisiae*, the haploid cells can switch mating type by a process involving **directed gene conversion** by homology-directed repair.

To the left and right of the MAT expression site on chromosome III are silent, non-expressed copies of MATα and MAT**a**, which are packaged into chromatin that is transcriptionally silent (see Section 8.16). Mating-type switching between α and **a** cell types occurs by the transfer of information from a silent MAT locus to the MAT expression site.

The transfer of DNA sequence information from the silent MAT sites to the MAT expression site occurs by directed gene conversion which is mediated by the homology-directed double-strand break repair pathway described in detail in

➜ We learn more about the two mating types of *S. cerevisiae* in Section 8.12.

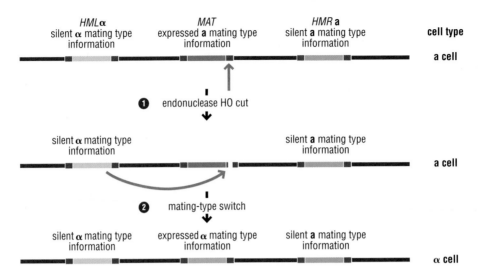

Figure 13.11 Yeast mating-type switching occurs via directional gene conversion. Three different loci in *S. cerevisiae* encode the transcriptional regulators that determine whether a yeast cell is of mating type α or a. The center MAT site is the active site and actually expresses the mating type transcriptional regulators. The flanking sites, *HML* (light green) and *HMR* (light blue) encode α mating regulators and a mating regulators, respectively. All the sites have regions of homology (dark blue). An a cell has the information for the a transcriptional regulators (blue) at the MAT expression site, as well as at the silent *HMR* site (on the right). (1) A switch in mating type is initiated when the HO endonuclease cleaves the DNA at the MAT expression site (pink arrow up), generating a double-strand break. (2) The broken DNA at the expressed site invades homologous sequences (dark blue) bounding the mating-type information at the silent *HMLα* site (on the left). Homology-directed repair copies α (green) information from the silent *HML* site into the MAT expression site in a gene conversion reaction. The source of information transfer into the expression site alternates between the silent α and the silent a donor sites at each mating-type switch.

Section 13.3. The pathway is activated by the introduction of a double-strand break in DNA at the MAT expression locus by a specialized endonuclease called HO (Figure 13.11, step 1). Using the DNA at either the silent *HML* α or silent *HMR* a site as a template, homology-directed repair copies sequence information from one of these sites into the MAT expression site (Figure 13.11, step 2). The *HMLα* and *HMRa* silent loci serve alternately as donors, so that the expression of a and α cell-type information, and thus the mating type of the cell, alternates in a cell lineage.

Gene conversion is one cause of antigenic variation

Several pathogenic microorganisms, notably *Neisseria gonorrhoeae*, the bacterium that causes gonorrhea, *Borrelia burgdorferi*, the bacterial agent of Lyme disease, and the protozoan trypanosomes that cause sleeping sickness, can change the cell surface antigens they express, enabling them to evade the host's immune response. These organisms contain an active gene-expression site from which the cell-surface protein is synthesized and also contain many silent genes encoding variants of this protein. During the course of an infection the gene in the expression site can be replaced by gene conversion with a copy from one of the silent sites, resulting in expression of a different surface protein (Figure 13.12). The detailed mechanisms of these reactions remain to be established but likely resemble yeast mating-type switching.

Figure 13.12 A schematic diagram of antigenic variation due to gene conversion. In some pathogens, the changes that occur in their surface protein antigens are due to directed gene conversion. (a) There is a single site from which the surface protein is expressed and a number of silent storage sites containing different versions of the surface protein genes. (b) Variation results from gene conversion of the sequence in the expression site using the sequence of one of the storage sites as template, changing the surface proteins of the organism.

13.7 HOMOLOGOUS RECOMBINATION

Having now explored how the cell responds to repair damage to its genome in the form of double-strand breaks, we can now go on to consider how the cell uses the same double-strand breaks to its advantage – the process of homologous recombination.

Large segments of DNA can be exchanged between two DNA molecules

Homology-dependent repair does not change the relative positions of the DNA segments flanking the initiating double-strand break although local gene conversion can occur (Figure 13.13a). However, the cellular machinery that carries out homology-directed repair has evolved into a major mechanism for generating new combinations of genetic information. This process is called **homologous recombination** and can result in the reciprocal exchange of large segments of DNA between homologous DNA duplexes (Figure 13.13b). Following recombination, the linkage of the markers that were on the same parental duplex in our hypothetical example in Figure 13.13b (A and B on one parent (gray) and a and b on the other (pink)), is changed in the progeny duplexes to give A and b on one duplex and a and B on the other. Many enzymes involved in homology-directed DNA repair also play a role in homologous recombination.

So how is homologous recombination exploited by the cell? Homologous recombination in meiosis plays a key role in the production of haploid gametes and also provides another pathway for DNA repair and underlies most genetic variation in humans and most other eukaryotes. Homologous recombination also contributes to the diversification of bacterial genomes by mediating the incorporation of exogenous DNA into bacterial genomes during conjugation, transduction, and transformation. This is one way bacteria can acquire new metabolic functions or resistance to an antibiotic.

➲ We discuss the production of haploid gametes in Section 7.7.

We have seen in Chapter 7 that recombination between homologous chromosomes in meiosis plays a key role in the identification of homologs that will be paired and then disjoined during chromosome segregation. Establishment of a connection between duplexes, which identifies them as homologous, can result in the exchange or re-assortment of genetic information between homologous chromosomes during the formation of the gametes, and is observed during meiotic recombination.

Homologous recombination can occur between any DNA molecules with extensive regions of identical sequence, (as with two sister chromatids), or very similar sequence (as with two homologous chromosomes); no specialized recombination sequences are required. Such exchanges generate diversity by reshuffling segments of the genome to make new combinations of existing characteristics, rather

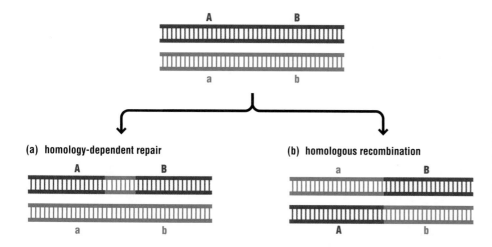

(a) homology-dependent repair

(b) homologous recombination

Figure 13.13 Homology-directed repair and homologous recombination differ in their effects of DNA. Parental homologous DNA duplexes contain the markers A and B (gray duplex) and a and b (pink duplex). (a) Homology-directed repair at a double-strand break between A and B results in local gene conversion (gray to pink) but does not change the relationship of the flanking AB and ab markers. (b) Homologous recombination results in the reciprocal exchange of markers on the parental duplexes to yield progeny duplexes containing a segment from each parent; A and b now lie on the same duplex and a and B are on the other duplex.

than through the introduction of new random mutations or by the translocation of discrete DNA segments, as will be discussed in the next chapter. Homologous recombination is an occasional accidental event in virtually all cells including the somatic cells of multicellular eukaryotes; however, it is a defined, non-accidental feature of meiosis.

The formation and resolution of Holliday junctions is the basis for homologous recombination

Central to the reciprocal exchange of DNA segments between two molecules that constitutes homologous recombination is the formation and **resolution** of **Holliday junctions** (named for Robin Holliday, see Experimental approach 13.3). A Holliday junction is formed by the reciprocal exchange of DNA strands between two duplexes to form regions of heteroduplex DNA (Figure 13.14, step 1). As we shall see below, Holliday junctions can be formed via RecA-like recombinase mediated formation of D-loops.

After formation, the two duplexes that have joined to form the Holliday junction are then separated, a step known as **resolution**. Resolution requires cleavage of the DNA strands at the Holliday junction at the point where the molecules are joined, and is promoted by enzymes called **resolvases**.

Cleavage of Holliday junctions can result in permanent exchange of DNA between two duplexes

There are two ways in which Holliday junctions can be resolved, and only one of these results in the permanent exchange of DNA – that is, recombination – between the participating DNA molecules. Although drawn between two extended helices in the top panel of Figure 13.14, Holliday junctions actually adopt a planar, unfolded structure when they interact with recombination proteins (Figure 13.14, step 2). The Holliday junction has two axes of symmetry, one depicted here as horizontal and the other as vertical (Figure 13.14, step 2). Whether the separation of the two duplexes results in recombination of the participating chromosomes depends on the direction of cleavage across these axes at resolution, that is, where the Holliday junction is cut to separate the two molecules.

If cleavage occurs in the horizontal direction (Figure 13.14, step 3i), the duplexes are separated without exchange of the DNA outside of the Holliday junction. These products are called **non-recombinant** (Figure 13.14, step 4i). The parental segments – A and B, and a and b – are still linked to each other and the only trace of interaction between the parental duplexes is the region of heteroduplex DNA formed by the exchange of strands to form the junction.

In contrast, if resolution occurs in the vertical direction (Figure 13.14, step 3ii), the DNA outside of the Holliday junction is exchanged or crossed over between the two parental duplexes. A and b are now on the same duplex and a and B are on the other. These DNA products are called **recombinant** (Figure 13.14, step 4ii): extensive genetic information has been exchanged between the parental duplexes.

Resolvases are nucleases that can cleave DNA at symmetrical positions in Holliday junctions and thereby separate the duplexes. The *E. coli* resolvase is called **RuvC**. *E. coli* also contains another resolvase, RusA. In Experimental approach 13.3, we

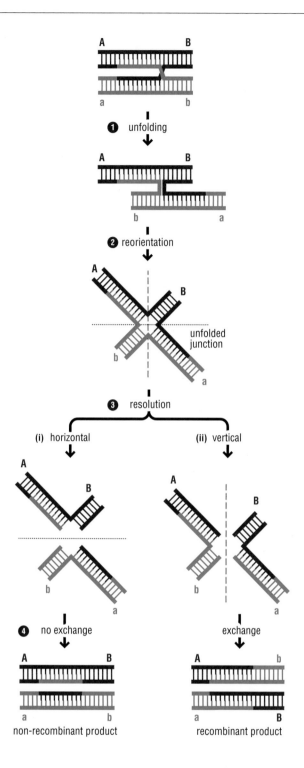

Figure 13.14 Cleavage of a Holliday junction can yield either of two products. Shown is a Holliday junction containing a region of heteroduplex DNA in which there has been a reciprocal exchange of DNA strands from each parental duplex. (1) Rotation of the bottom double helix provides an alternative view of the Holliday junction. Note that both structures contain the same homoduplex and heteroduplex regions and that the connection between the duplexes involves four DNA strands. (2) Reorientation of the junction emphasizes two axes of symmetry (lines) in the junction. The junction can be cleaved by the specialized endonucleases called resolvases in either of two planes, indicated by the dotted horizontal and dashed vertical lines, respectively. (3) Resolution: (i) Horizontal resolution: cleavage along the dotted horizontal plane, also called resolution of the crossing strands, cuts the two DNA strands that were originally broken and exchanged to form the junction. The two parental DNA molecules are regenerated almost exactly (4i): A and B, and a and b, are still on the same duplexes. Such duplexes are called 'non-recombinant'. The progeny duplexes contain regions of heteroduplex DNA flanking the initiation of strand exchange (left) and the point of junction resolution (right). (ii) Vertical resolution: cleavage along the dashed vertical plane cuts the two DNA strands that were not originally cleaved. (4ii) A physical exchange of duplex DNA between the two molecules and the shuffling of genetic information occurs: A and b are now on one duplex, a and B on the other. The duplexes are called 'recombinant'. The progeny duplexes contain regions of heteroduplex DNA flanking the initiation of strand exchange (left) and the point of junction resolution (right).

describe the recent identification of several new resolvases in the yeast *S. cerevisiae* (Yen1) and in humans (GEN1) that cleave Holliday junctions in the same way that RuvC and RusA do. These eukaryotic resolvases are, however, structurally unrelated to RuvC; they belong to the same family as the XP nucleases that participate in nucleotide excision repair. Another new type of resolvase, the eukaryotic SXL1/SLX4, is related to the UvrC endonucleases involved in bacterial nucleotide excision repair.

⊜ We describe the process of nucleotide excision repair in Section 12.5.

13.3 EXPERIMENTAL APPROACH

The difficulties in identifying Holliday junction resolvases

The idea that homologous recombination occurs through the formation and resolution of a DNA intermediate involving the reciprocal exchange of DNA strands was proposed entirely based on genetic observations. This intermediate, the so-called 'Holliday junction', was named after Robin Holliday who first proposed the model to explain certains products of fungal meiosis.

It only became possible to biochemically analyze Holliday junctions after the bacterial RecA protein was isolated, and Stephen West and colleagues showed using homologous plasmids that the protein promoted the formation of paired molecules. Electron microscopy of such paired molecules confirmed that they had the anticipated structure of a Holliday junction. The West group also asked whether the paired molecules formed *in vitro* could give rise to recombinants when transformed into *E. coli*: could the Holliday junctions be resolved (cleaved)? Indeed, both dimeric and monomeric plasmids were recovered from *E. coli* following the transformation, indicating that resolution could occur in the two different orientations as first suggested by Holliday (Figure 1a).

The RecA protein promoted the formation of the Holliday junction in bacteria, but what protein was responsible for resolving the intermediate? Using *in vitro* generated Holliday intermediates made with RecA (Figure 1a) or by annealing of small oligonucleotide

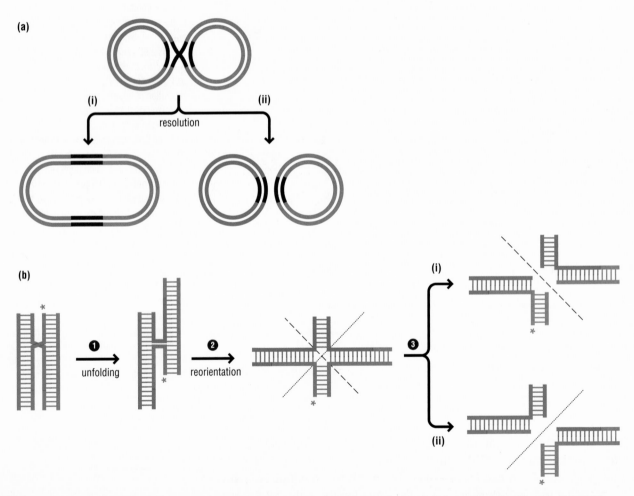

Figure 1 Holliday junction resolution. (a) Holliday junction resolution *in vivo*. (1) A joined plasmid molecule is formed by purified RecA *in vitro*. (2) When transformed into *E. coli*, the plasmid molecule is resolved to either dimeric or monomeric form, indicating that resolution of Holliday junctions can occur *in vivo*. (b) Resolution of a synthetic Holliday junction *in vitro*. A Holliday junction can be made *in vitro* by annealing four oligonucleotides. (See Figure 2b below). (1) The junction is unfolded by rotation of one duplex. (2) The junction is reoriented to emphasize two axes of symmetry. This open "four way" junction is the form of Holliday junctions (*in vivo*). (3) Resolution of the junction can occur by the cleavage of two strands across either axis of symmetry.

Figure 2 **Resolution of a synthetic Holliday junction by RuvC resolvase**. (a) Mapping of cleavage positions on a synthetic Holliday junction. A synthetic Holliday junction was constructed using oligonucleotides; four different forms were made in which a single oligonucleotide (1, 2, 3, 4) was labeled at its 5′ end. The Holliday junctions were incubated with purified RuvC (FV) and the products of the reaction displayed on a denaturing gel. GA and TC lanes serve as markers. Cleavage of oligonucleotides 2 and 4 is observed as indicated by the arrows. (b) Positions of RuvC cleavage on the synthetic Holliday junction. (The position of cleavage and the cleavage axis observed on such synthetic junctions are highly dependent on the sequence of the junction; symmetrical cleavage on both axes has been observed with other junctions.)
Reproduced from Connolly *et al.*, *Proceedings of the National Academy of Sciences USA* 1991;**88**:6063–6067.

substrates (Figure 1b), West and colleagues were able to purify a resolvase activity from *E. coli* and establish that it was the product of the *ruvC* gene (Figure 2a). The *ruvC* gene had originally been identified genetically because of the way mutations in the gene rendered the bacteria extremely sensitive to killing by UV irradiation, a phenotype not inconsistent with resolvase activity. Analysis of RuvC resolution of a Holliday junction made from small oligonucleotides allowed the West group to further show that cleavage was symmetric about the junction (Figure 2b).

Despite many subsequent genetic and biochemical studies in eukaryotes, it would be over 25 years before nuclear Holliday junction resolvases such as RuvC were isolated from these organisms. As we now know, the search for these enzymes was hampered by the little sequence homology between the resolvases identified in bacteria and eukaryotes. Moreover, no single mutations in yeast appear to give rise to cells that have the properties expected of cells defective in resolvase function. Identification of human and yeast resolvases happened recently, emerging from biochemical

assays that exploited synthetic Holliday junctions as substrates, as well as from sequence comparisons between several organisms (and the increasing sophistication of search algorithms).

The first human resolvase activity was isolated by West and colleagues in 2008, some 17 years after they identified the activity in bacteria. In a brute force biochemical approach, the group fractionated large quantities of human tissue culture cells (200 L) and followed Holliday junction resolution activity through multiple protein purification steps (Figure 3). Even after the many purification steps (Figure 3a), the final fraction had multiple polypeptides (Figure 3b), although final fractionation on a gel allowed the researchers to determine that a single gel slice had the resolvase activity (Figure 3c). Mass spectrometric analysis of the proteins located in that gel slice allowed them to identify 20 different proteins (from the defined polypeptide fragments). Of the identified target proteins, the most interesting candidate had homology to a known nuclease in human cells. In follow-up experiments, this protein, designated GEN1, was shown to cleave a synthetic Holliday

Figure 3 Isolation of the human Holliday junction resolvase GEN1. (a) Fractionation scheme for the purification of resolvase activity by multiple column chromatography steps. The presence of resolvase activity in the various fractions was monitored the ability to resolve a synthetic Holliday junction, as outlined at the bottom of the panel: the products of the reaction were displayed by gel electrophoresis. (b) Protein gel displaying polypeptides from fraction 29 of the last column in the purification, the MonoS column, which has the most enriched resolution activity. (c) Assay of gel slices from the gel lane containing proteins from MonoS fraction 29. Slice e contains the resolution activity. Mass spectrometry analysis of the polypeptides in this slice identified 20 polypeptides. One polypeptide, now called GEN1, has a nuclease motif related to the Rad2/XPG family.
Reproduced from Ip *et al.*, *Nature* 2008;**456**:357–361.

junction symmetrically (the hallmark signature of the bacterial RuvC protein). A related protein, Yen1, with similar activity, was subsequently isolated from yeast.

Recently four different groups reported the identification of another Holliday junction resolvase in eukaryotes. In contrast to the extract fractionation strategy used by the West group, these groups used a combination of sequence alignments, protein–protein interaction studies and biochemical assays. A complex of two yeast proteins, Slx1 and Slx4, was previously known to have some endonuclease activity. (An Slx1 homolog was known in humans but an Slx4 homolog had not been identified.) Using

sequence alignments between human, yeast, and *Drosophila* proteins known to be required for recombination, a candidate Slx4 homolog was identified based on similarities in sequence motifs in the C-terminal regions (Figure 4). Biochemical purification and *in vitro* assays using both the yeast and human Slx1/Slx4 complexes showed this complex to have robust cleavage activity on Holliday junctions. Furthermore, inactivation of these genes in yeast and human cells resulted in a DNA damage response, consistent with the conclusion that these proteins encode a resolvase activity and that failure to properly resolve Holliday junctions will lead to DNA breaks.

```
Hsap BTB012 1561  TPMPQYSIMETPVLKKELORFGVRPLP-KRQMVLKLKEIF...(152)...
Drer BTB012 1487  TPPPGFSDMETPFLKNRLNRFGVRPLP-KKQMVLKLKEIH...(101)...
Tcas MUS312  696  TPPANYDEMNTPQVCKELDKFGLKPLK-RSKGAKLLKYIY....(55)...
Dmel MUS312  924  TPKPQFATLPESEILQQLYKYGIKPLK-RKQAVKMLEFIY...(100)...
Cneo BSPI    608  RGPPQYOSWOVKALRLLIAOYGYRPIKOQSPLVQVAAECW....(64)...
Scer S1×4    620  KFCEIMMSQSMKELRQSLKTVGLKPHRTKVEIIQSLQTAS....(24)...

Hsap BTB012       TDEALRCYIRSKPAVYQKVVLYQPFELRELQAELRQNG----LRVSSRRLLDFLDTHCITF 1808
Drer BTB012       KLLAVRQFILSDPELYSRVLQYQPLPLAELRASDRAAG----IBLAAAKLLDFLDSQCITF 1682
Tcas MUS312       LHIAWHNLVMSNPKIREDILLYEPLQLENLHSMLKEQG----FRYNIQDLLTFLDKKCITI  846
Dmel MUS312       LHIAWHNLICANPQLHESVLTYEPIDLQAVYLHLKHMG----HRYDPKDLKTFFDRRCIIF 1119
Cneo BSPI         LHGGFQGMLTSDHDLYLRILRYEPIAFDELVSKAIASG--MTRRGWKKELKNYLDLQCVTY  769
Scer S1×4         IFDHLTELIEAFPDFLERIYTFEPIPLNELIEKLF-SAEPFVSQIDEMTIREWADVQSICL  743
```

Figure 4   Identification of SLX4 homologs through sequence alignment. Alignments of C-termini from proteins that were identified as homologs of yeast SLX4 derived from a BLAST search using the *D. melanogaster* MUS312 protein as the starting sequence. In the alignment shown, sequences from widely divergent lineage are compared. Hsap is *H. sapian* (human), Drer is *Danio rerio* (zebra fish), Tcas is *Tribolium castaneum* (red flour beetle), Dm is *Drosophila melanogaster*, Cneo is *Cryptococcus neoformans* (fungus), and Scer is *S. cerevisiae* (baker's yeast). The blue letters indicate positions conserved in at least four of the six species whereas the pink letters indicate positions conserved in all. The conserved predicted alpha helical structures are indicated with the bar underneath the sequence.
Reproduced from Andersen *et al.*, *Molecular Cell* 2009;**35**:128–135.

Why have two different resolvases been identified in eukaryotes? It is possible that eukaryotic cells possess multiple resolvase activities that may function at different times or places, mirroring the way that *E. coli* has multiple resolvase activities. Further, it is not uncommon for different groups to isolate different proteins with the same activity or for scientists to come to different conclusions, driving the need for further experimentation. It is also not uncommon for several groups to publish similar results at the same time, especially given that research groups all strive to answer the next important biological question. Further studies will be required to understand these multiple eukaryotic resolvases.

Andersen SL, Bergstralh DT, Kohl KP, *et al*. *Drosophila* MUS312 and the vertebrate ortholog BTBD12 interact with DNA structure-specific endonucleases in DNA repair and recombination. *Molecular Cell* 2009;**35**:128–135.

Connolly B, Parsons CA, Benson FE, *et al*. Resolution of Holliday junctions *in vitro* requires the *Escherichia coli* ruvC gene product. *Proceedings of the National Academy of Sciences U S A* 1991;**88**:6063–6067.

Dunderdale HJ, Benson FE, Parsons CA, *et al*. Formation and resolution of recombination intermediates by *E. coli* RecA and RuvC proteins. *Nature* 1991;**354**:506–510.

Fekairi S, Scaglione S, Chahwan C, *et al*. Human SLX4 is a Holliday junction resolvase subunit that binds multiple DNA repair/recombination endonucleases. *Cell* 2009;**138**:78–89.

Ip SC, Rass U, Blanco MG, *et al*. Identification of Holliday junction resolvases from humans and yeast. *Nature* 2008;**456**:357–361.

Munoz IM, Hain K, Declais AC, *et al*. Coordination of structure-specific nucleases by human SLX4/BTBD12 is required for DNA repair. *Molecular Cell* 2009;**35**:116–127.

Svendsen JM, Smogorzewska A, Sowa ME, *et al*. Mammalian BTBD12/SLX4 assembles a Holliday junction resolvase and is required for DNA repair. *Cell* 2009;**138**:63–77.

Branch migration changes the position of Holliday junctions

When two identical duplexes are joined at a Holiday junction, (where sister chromatids become joined in the course of homologous recombination, for example), the junction can move along the DNA without any net change in base-pairing. Such movement is known as **branch migration** (Figure 13.15 left). Branch migration can occur spontaneously in either direction over short distances. This can either increase or decrease the region of heteroduplex DNA between the linked duplexes. In the case of sister chromatids, where the duplexes are identical, branch migration makes no difference to the sequence of the two participating DNA molecules when they are separated.

However, branch migration can also occur between two homologous but non-identical DNA molecules, as in the case of two homologous chromosomes joined by Holliday junctions during meiosis. In this instance, migration requires

Figure 13.15 Schematic diagram of formation and branch migration of a Holliday junction by the RuvABC complex. A Holliday junction formed by a reciprocal exchange of DNA strands between homologous duplexes is shown on the top left. The unfolded form is shown on the right. A and a represent different alleles of a gene on one side of a Holliday junction; B and b represent a gene on the other side. The position of the Holliday junction can move by a process called branch migration (left bottom). This reaction can occur spontaneously because the same number of paired bases are always present, the only change being which DNA strands are paired to each other (right). Branch migration driven by RuvAB. (1) All four arms of the Holliday junction are bound by a tetramer of RuvA, which converts it to the planar conformation. RuvB, which is a helicase, binds to two opposite arms. The RuvAB complex drives migration of the junction along the joined DNAs in the direction indicated by the arrows (pink and gray). (2) Branch migration has occurred to decrease the region of heteroduplex DNA. Imagine heteroduplex pink + gray base pairing. being broken on either side of the center of the junction (top left and bottom right), and reforming with the opposite strands (top right and bottom left) to extend the regions of homoduplex. This moves the Holiday junction along the duplexes with no net change in the number of base pairs. (3) The Holliday junction is shown without RuvAB after branch migration. This open planar form is equivalent to that on the left. The RuvAB complex recruits the resolvase RuvC to the junction and it cleaves the DNA as shown in Figure 13.14.

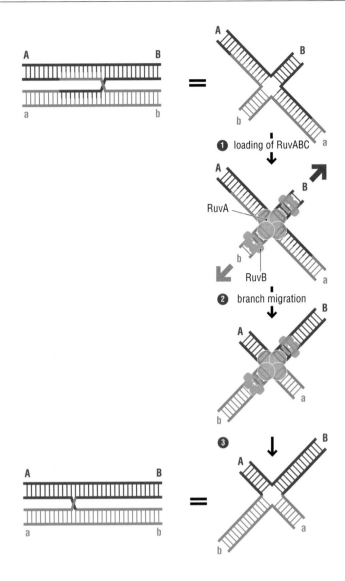

specialized helicases and has important consequences for the participating DNA molecules, in terms of their composition following resolution.

Specific enzymes drive branch migration

Although branch migration is a universal process, occurring both in eukaryotes and bacteria, the enzymatic machinery that drives it has been isolated only in bacteria, and most of what is known to date has been learned from an enzyme complex called **RuvAB** in *E. coli*. As illustrated in Figure 13.15, step 2, RuvA is a tetramer that binds at the Holliday junction, keeping it in the unfolded and open conformation. RuvA also recruits RuvB to the junction. RuvB is a helicase that uses the energy of ATP hydrolysis to move the Holliday junction along the DNA, breaking and re-forming base pairs in one direction, and increasing or decreasing the region of heteroduplex in its wake (Figure 13.15, step 3), depending on which way the junction is moved. Thus, by changing the positions of the Holliday junction, the helicases can determine the point at which the junction is resolved, and thus the point at which exchange between the participating chromosomes can occur. The *E. coli* resolvase RuvC is actually recruited to Holliday junctions by interaction with the RuvAB branch migration machine.

Holliday junction formation can lead to a change in the primary structure of DNA following mismatch repair

Formation of a Holliday junction involves the pairing of strands from different DNA helices to form heteroduplex DNA; branch migration thus influences the extent and location of this heteroduplex DNA. If heteroduplex DNA forms between strands from duplexes that are slightly different in sequence (for example, as may occur between homologs) the heteroduplex DNA will contain mismatches. As we have seen, the process of mismatch correction can operate on such heteroduplexes to remedy the mismatches. However, this process can also change the primary structure of the DNA.

In post-replication mismatch repair (see Section 12.2; Figure 12.5), correction is guided by the parental DNA strand, that is, the newly synthesized strand is identified and the mismatched base specifically removed from the newly synthesized strand. In the case of mismatch correction of recombination intermediates, however, correction is not biased towards one strand or the other, and either DNA strand can serve as the template for repair (Figure 13.16). Thus, while one parental duplex contained the gray marker Y and the other the pink y marker, gene conversion by mismatch correction can result in two progeny duplexes with the Y marker (step 2, i) or two progeny duplexes with the y marker (step 2, ii) without any change in the relationship of the markers, A and B, and a and b, that flank the region of gene conversion. Other possible patterns of correction are described in Figure 13.16.

So, we see how gene conversion of the Y and y markers reveals that strand exchange that formed DNA heteroduplex DNA did occur between the parental duplexes although exchange of the flanking segments did not occur.

Although the mismatch correction described above *can* occur, it does not happen universally: DNA replication may happen before correction can take place; in this

Figure 13.16 Mismatch correction of heteroduplex DNA can lead to gene conversion. Two parental homologous duplexes with different alleles at several positions are shown. (1) Formation of heteroduplex DNA can occur during formation of a Holliday junction. (2) Mismatch correction of the heteroduplex (blue box) can have different outcomes, depending on which strand is used as the template for correction. Note that depending how correction occurs, the Y (gray) and y (pink) markers can become associated with different flanking markers, for example, formation of a AyB duplex. (i) Use of the Y strands in both heteroduplexes as the templates for correction results in 2 Y-containing duplexes (AYB, aYb). Note that the Y (gray) marker has become associated with different flanking markers in aYb. (ii) Use of the y strands in both heteroduplexes as the templates for correction results in two y-containing duplexes. (iii) Use of the y strand in the upper heteroduplex and the Y strand in the bottom heteroduplex as the templates for correction results in one y-containing and one Y-containing duplex. (iv) Use of the Y strand in the upper duplex and the y strand in the bottom heteroduplex as the templates for correction results in one Y-containing and one y-containing duplex. In this case Y and y markers remain associated with the flanking markers of the parental duplex (AYB and ayb).

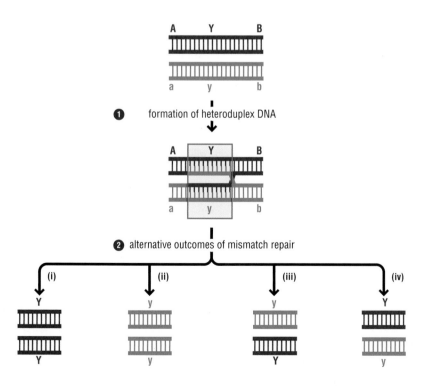

case, no gene conversion will occur. Mismatch correction can also occur within heteroduplex regions formed during homology-dependent repair.

Overall, we see that formation of a Holliday junction and branch migration can lead to significant changes in the primary structure of a DNA molecule: the formation of heteroduplex DNA may lead to sequence changes in the parental duplexes by gene conversion through mismatch correction, and can also lead to the exchange of the DNA segments that flank the junction, depending on how the Holliday junction is resolved.

In the above discussion, we have focused on the molecular processes associated with the resolution of a single Holliday junction. During homologous recombination, however, *two* Holliday junctions are formed and resolved. Let us now go on to consider the mechanism of homologous recombination in more detail, during which two Holliday junctions are formed and resolved. We begin by considering the initiation of the process of homologous recombination.

Homologous recombination is initiated by a double-strand break

In all organisms, from bacteria to yeast to humans, the initiation of homologous recombination occurs at a double-strand break: in meiotic recombination, this break is made by the programmed action of specialized nucleases. The break then undergoes reactions closely related to those underlying homology-directed repair as outlined in Figure 13.17.

The broken DNA ends are resected to generate single-stranded 3′ tails (Figure 13.17, step 1). These ends are bound by strand exchange recombinase molecules, RecA or Rad51, or the specialized meiotic recombinase, Dmc1, exactly as in the synthesis-dependent strand annealing pathway (SDSA) we have already encountered in Section 13.3. The resulting nucleoprotein filament then invades the intact duplex (step 2). The next step, again as in SDSA, is DNA synthesis, with the invading 3′ serving as a primer and its complementary strand in the participating homologous duplex as a template (step 3). Subsequent events depart from those of SDSA, however: instead of unwinding from the template strand, the newly synthesized strand remains paired to the template in the homologous duplex, and the D-loop is extended by continued DNA synthesis.

This continued DNA synthesis displaces more of the non-template strand in the recipient duplex, and the displaced strand pairs with the complementary strand

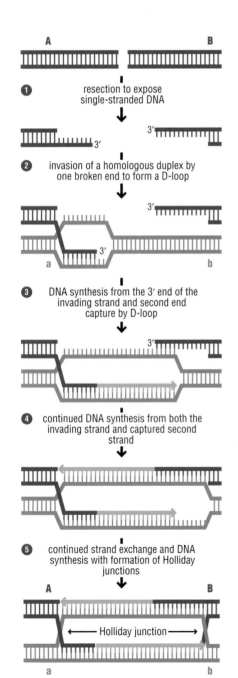

① resection to expose single-stranded DNA

② invasion of a homologous duplex by one broken end to form a D-loop

③ DNA synthesis from the 3′ end of the invading strand and second end capture by D-loop

④ continued DNA synthesis from both the invading strand and captured second strand

⑤ continued strand exchange and DNA synthesis with formation of Holliday junctions

Holliday junction

Figure 13.17 Homologous recombination initiates with a double-strand break. One parental duplex with markers A and B (gray) containing a double-strand break is shown. (1) As previously discussed, the 5′ ends of the broken DNA are resected to give 3′ tails. (2) One 3′ tail from the broken gray duplex invades the homologous pink duplex with the a and b markers to form a region of heteroduplex, displacing the other pink strand to make a three-stranded D-loop, the same reaction that initiates homology-dependent repair, synthesis-dependent strand annealing (SDSA) (Figure 13.5). (3) New DNA synthesis (green) initiates at the 3′ OH end of the invading strand using the homologous pink DNA as a template. In contrast to SDSA, extensive heteroduplex between the newly synthesized green DNA and the template duplex is formed, and DNA synthesis continues such that the pink displaced strand of the D-loop can pair with the gray DNA from the other parental duplex on the other side of the break. This structure could also form by direct formation of D-loops on each side of the double-strand break. (4) DNA synthesis continues from both ends of the broken DNA using the intact pink duplex as a template to regenerate the region of the double-strand break. Notice that in the region of gap repair, each duplex contains one strand of the template (pink) duplex and one newly synthesized strand (green); this DNA replication is thus semi-conservative. (5) Continued DNA synthesis and strand exchange results in the formation of two Holliday junctions linking the damaged and undamaged duplexes. The resolution of these linked duplexes is considered in Figure 13.19.

Figure 13.18 Homology-directed repair and homologous recombination differ in the processing of one of the double-strand break ends. Both homology-directed repair by synthesis-dependent strand annealing, and homologous recombination, initiate by formation of a D-loop between an invading 3′ single strand (gray) and a homologous duplex (pink). The 3′ end of the invading strand serves as a primer for DNA synthesis (green) across the template (pink) duplex. (a) In homology-directed repair by synthesis-dependent strand annealing, the newly synthesized strand (green) does not continue to pair with the template (pink) duplex and pairs instead with complementary sequences from the other broken end (gray). (b) In homologous recombination, the displaced strand from the template duplex (pink) captures the other broken end (gray). This interaction results in formation of Holliday junctions that can be resolved to reciprocally exchange DNA between the two parental duplexes.

from the cut duplex (Figures 13.17, step 4, and 13.18b). In this way, the remaining cut strand on the other side of the break also has an intact template for repair, and the gaps in the cut strands can be closed by repair polymerases using the free 3′-OH end of the cut strands as a primer (Figure 13.17, step 4). Note that DNA replication in this instance is semi-conservative: that is, the regions of DNA synthesis contain one newly synthesized and one old template strand.

The capture of DNA on the other side of the break is key to homologous recombination: without capture, the broken chromosome will be permanently fragmented. This failure to capture the other side of the broken DNA is also a hazard during homology-directed repair. We will see below that the uncaptured chromosome fragment can be replaced by continued DNA synthesis from the invading strand of the D-loop through the process of **break-induced replication** (**BIR**).

The end result of the reactions described above is the formation of two intact DNA helices linked by two Holliday junctions (Figure 13.17, step 5). These junctions flank the position of D-loop formation by one end of the broken DNA, drawn on the left of Figure 13.17, step 2, and the position of capture of the other broken end by the displaced strand of the D-loop, drawn on the right of the same figure. Regions of repair synthesis, which will result in gene conversion (Figure 13.10) and heteroduplex DNA, where gene conversion may occur by mismatch correction (Figure 13.16) lie between the two Holliday junctions. Once formed, however, the positions of the Holliday junctions and thus the distance between them may change by branch migration.

The nature of resolution of the double Holliday junction intermediate determines whether recombination occurs

The processes of strand invasion and templated repair result in the formation of two Holliday junctions joining the two parental duplexes (Figure 13.17). As described above, the separation of the parental duplexes and the completion of

Figure 13.19 Resolution of a homologous recombination intermediate containing two Holliday junctions can yield multiple products. Two parental homologous DNA duplexes are shown. (1) DNA duplexes joined by two Holliday junctions result from the initiation of homologous recombination by a double strand break (Figure 13.18). Separation of the duplexes results from resolution of the Holliday junctions which can occur in alternative ways (Figure 13.15). In both pathways shown, the junction on the left is resolved along the horizontal axis, but resolution in the vertical axis may occur instead Pathway (i): (2) Cleavage can occur along the horizontal plane of both Holliday junctions (blue arrows) are cleaved in the horizontal plane. The two parental DNA molecules are re-formed almost exactly, such that A and B, and a and b, are still on the same duplexes, and are non-recombinant. Regions of heteroduplex (pink and gray strands) and where newly synthesized DNA (green) is paired with its template DNA (pink), lie between the positions of the Holliday junctions. Pathway (ii): (2) Resolution of the double Holliday junction can occur on the horizontal axis of one junction (left) and the vertical axis of the other (right). There is a physical exchange of duplex DNA between the two duplexes and the shuffling of genetic information: A and b are now on one duplex (top), and a and B are on the other (bottom) and are thus recombinant. Regions of heteroduplex (pink and gray strands) and where newly synthesized DNA (green) is paired with its template DNA (pink) lie between the positions of the Holliday junctions.

recombination occurs by the cleavage of the Holliday junctions by resolvases. Recall that, at each junction, resolution can at occur in either of two ways, which differ according to which pair of strands is cut at each junction (Figure 13.15).

How does the resolution of two Holliday junctions affect the relationship of the duplexes involved? Two of the four possible outcomes of the resolution of double Holliday junction are shown in Figure 13.19. In each case, the same pair of strands is cut on the left junction but different pairs are cut at the right junction. The consequence of the different cleavages on the right hand junction is a change in the relationship of the DNA segments outside of the junctions. Resolution in one direction (Figure 13.19, pathway i, step 2) results in non-recombinant progeny duplexes, that is, A and B on one duplex and a and b on the other, as in the parental duplexes. Note that even when resolution occurs without exchange of the flanking A,a and B,b markers, the DNA sequence between the junctions may change because of gene conversion, leaving a permanent reflection of the connection between the parental duplexes that did exist. By contrast, resolution in the other direction (Figure 13.19, pathway ii, step 2), results in recombinant progeny duplexes, with one duplex containing A and b and the other a and B.

Thus recombination, a change in the spatial relationship of the DNA segments that flank a double-strand break, results from Holliday formation and resolution. By contrast, homology-directed repair does not involve Holliday junction formation so no exchange of flanking DNAs is possible.

13.8 REPAIR OF DAMAGED REPLICATION FORKS BY HOMOLOGY-DIRECTED REPAIR

DNA damage can collapse replication forks

We have already seen in Section 12.7 that, if a DNA replication fork encounters a nick in the DNA, the arm of the fork containing the nick breaks, generating a double-strand break that arrests replication. Abasic sites, base lesions or cross-links, or defects in the replication complex itself can all block progression of the replication fork. In *E. coli*, it is estimated that 20–50% of the forks that initiate at OriC are blocked by such events. All these causes of the blocking of replication generate regions of single-stranded DNA at the fork that are targeted by repair proteins. Here, we describe how the homology-directed repair mechanisms operate at collapsed or blocked replication forks.

Homology-directed repair can restore a collapsed replication fork

As we saw in Section 12.7, most collapsed replication forks occur during DNA replication as a consequence of nicks introduced into the phosphodiester backbone by reactive oxygen species that are a result of normal cellular metabolism. In such instances, interactions between the undamaged and damaged arms of the fork mediated by homology-directed repair can restore the fork (Figure 13.20).

So how does such repair unfold? The broken arm of a replication fork (Figure 13.20, step 1) interacts with the undamaged arm through the invasion of a single DNA strand from the damaged arm into the undamaged arm to form a D-loop (Figure 13.20, step 2), thus reestablishing a connection between the broken arm and the rest of the fork. This invading strand then serves as a primer for DNA synthesis (Figure 13.20, step 3). Subsequently the other single strand of the damaged duplex also specifically interacts with the undamaged arm by binding to the displaced strand of the D-loop (Figure 13.20, step 4). Thus single DNA strands from each parental duplex are reciprocally exchanged between the parental duplexes. This structure is then resolved to restore the replication fork (Figure 13.20, step 5). The reciprocal exchange of DNA strands results in the formation of a Holliday junction that can be resolved to separate the arms of the restored fork (Figure 13.20, step 6).

Figure 13.20 DNA replication fork that has broken because of DNA damage in a template strand can be rescued by homology-directed repair. A replication fork containing parental (gray) and newly synthesized (green) DNA strands is shown. The fork is broken by a nick in the gray parental strand that serves as the template for lagging strand synthesis. (1) The broken fork is redrawn to show the double-stand break resulting from the nick. (2) The 3′ end of the parental (gray) strand of the broken lagging strand template pairs with its complementary parental strand, creating a D-loop. (3) The D-loop is extended beyond the position of the original nick that led to fork collapse by DNA synthesis (green) using the invading 3′ OH as a primer. An intact DNA strand is generated at the position of the nick. (4) The newly synthesized lagging strand (green) pairs with the newly synthesized leading strand that was displaced by D-loop formation, resulting in the reciprocal exchange of DNA strands between these two homologous duplexes to form a Holliday junction. (5) Lagging strand synthesis is restarted. (6) The leading and lagging arms of the restarted replication fork are separated by cleavage of the exchanged DNA strands along the horizontal axis in the Holliday junction (blue arrows), and the replication fork is restored.

The resumption of replication also requires reassembly of the replisome. **Replication restart** may need to occur at many different sites (wherever fork collapse has occurred). Therefore, no specific template DNA sequence that binds a specific initiator protein is involved in replication restart, in contrast to replisome assembly at *OriC*. Instead, in bacteria, a specialized protein called PriA can recognize a D-loop and promote loading of the replicative helicase and other replication components onto the D-loop, and thus promote initiation from a DNA intermediate assembled by homology-dependent interactions.

Replication fork reversal can repair a blocked replication fork

➲ We discuss the blocking of replication in more detail in Section 12.7.

As we discussed in Section 12.6, translesion polymerases can promote untemplated error-prone DNA synthesis at positions of DNA damage at stalled replication forks. However, there are other repair pathways related to the replication fork restart including one related to the homology-directed repair pathway described above, which can also facilitate the resumption of DNA synthesis.

The restoration of a damaged replication fork containing unrepaired lesions, such as an abasic site or pyrimidine dimer (Figure 13.21, step 1), can be achieved by a pathway that involves the pairing of the two newly synthesized strands of DNA with each other instead of with the template strands. In this case the fork reverses so that the damaged strand re-pairs with its complementary strand. This provides another opportunity for lesion repair by systems such as base and nucleotide excision repair, which use the complementary DNA strand as a template (Figure 13.21, step 2).

Once repair has occurred, the replication fork may be restarted in several ways. First, the regressed fork may be cleaved to yield a collapsed fork (Figure 13.21, step 3, i) which may be repaired by homology-directed repair. Such repair allows reassembly of the replisome, as described above. The structure in Figure 13.21 (step 3, i) contains a Holliday junction, the site of reciprocal exchange of DNA strands between duplexes, that can be specifically recognized and cleaved by resolvases, with the subsequent reinitiation of replication. Second, the newly synthesized annealed strands may be degraded by a nuclease (Figure 13.21, step 3, ii) or third, the regression of the fork may be reversed (Figure 13.21, step 3, iii). In all cases, the replisome must again be assembled as above to provide replication restart.

13.9 ABERRANT REPAIR AND RECOMBINATION AND CHROMOSOME REARRANGEMENTS

Misdirected repair and recombination reactions can be damaging

The NHEJ reactions, and the recombination reactions that underlie homology-directed repair and homologous recombination at meiosis, are essential to life; but when they are misdirected they can cause genetic changes that can be lethal to the cell, or can considerably increase the risk of cancer. Recombination in

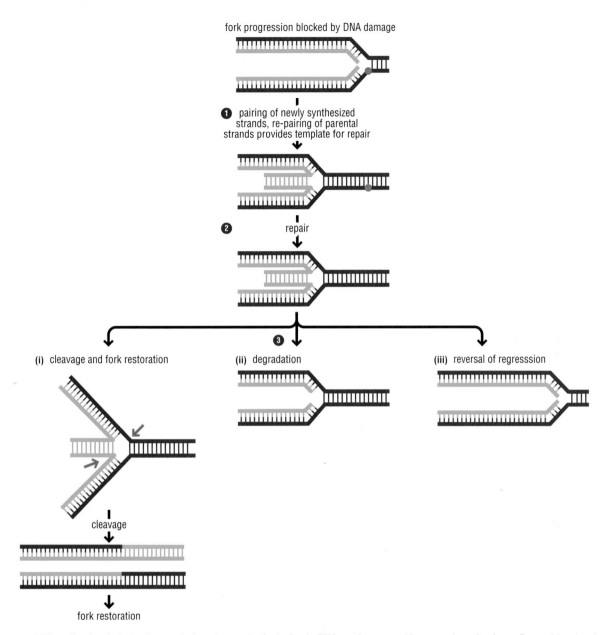

Figure 13.21 DNA replication fork that has stalled as the result of a lesion in DNA can be rescued by several mechanisms. Parental (gray) and newly synthesized (green) DNA strands at a replication fork are shown after the fork has stalled because of a lesion (red ball). (1) Regression of the fork leads to pairing of the newly synthesized strands and re-pairing of the template strands such that the region containing the lesion is again paired to an intact strand. (2) The parental strand is used as a template for repair. (3) The replication fork can be restarted in several ways: (i) the regressed form, actually a Holliday junction, is cleaved by a Holliday junction resolving enzyme such as RuvABC, collapsing the fork which is then restored by recombination, as in Figure 13.20; (ii) the paired newly synthesized strands are degraded by a nuclease, for example RecBCD, or (iii) fork regression is reversed. In cases (ii) and (iii) normal replication is resumed.

somatic cells, for example, can occur either in the course of double-strand break repair or through juxtaposition of chromatids from different chromosomes after DNA replication, with consequences that can be damaging. NHEJ reactions can join unrelated chromosomes, also with damaging consequences. And meiotic recombination is subject to mispairing interactions that cause deletions and **duplications** of DNA on the participating chromosomes. In the closing section of this chapter, we consider some of these aberrant processes, and explore the consequences to the cell of such processes occurring.

Homologous recombination and homology-dependent repair in somatic cells can lead to loss of heterozygosity

One possible consequence of homologous recombination in somatic cells is a phenomenon known as **loss of heterozygosity**, to which we briefly alluded at the beginning of this chapter. Diploid organisms are heterozygous at many of their genes; that is, there are likely to be different alleles of any given gene on any two homologous chromosomes. In most cases, both alleles are functional, and loss of heterozygosity will have no adverse effects on the organism. However, when one of the alleles is defective, the heterozygous cell depends on the normal allele to grow and behave normally. Loss of the functional copy of the gene – that is, loss of heterozygosity – can then have grave consequences. A particular hazard for multicellular animals is loss of the functional copy of a gene that normally suppresses cell proliferation and prevents the cell from giving rise to a cancer.

One mechanism by which loss of heterozygosity can occur as a result of homologous recombination is illustrated in Figure 13.22. After chromosomes have replicated (Figure 13.22, step 1), recombination may occur between two chromatids from different homologs, resulting in the reciprocal exchange of information between the two chromatids so that the defective gene (for example, the b marker) is now present in one chromatid of each homolog on the same template strand (Figure 13.22, step 2). After the chromatids separate and segregate at mitosis, one daughter cell could receive two copies of the normal gene while the other cell receives two copies of the defective gene (Figure 13.22, step 3).

Loss of heterozygosity can also occur through homology-directed repair that initiates within the chromosome and extends to the end of a chromosome, in a process called **break-induced replication (BIR)** (Figure 13.23). In BIR, a DNA end from one side of a double-strand break initiates repair by formation of a D-loop (Figure 13.23, step 2) and, as in homology-directed repair by synthesis-dependent strand annealing, a new strand using a homolog is synthesized (Figure 13.23, step 3). In BIR, however, the newly synthesized strand fails to pair with the other side of the double-strand break and synthesis continues on the same template strand (Figure 13.23, step 4). New duplex DNA attached to only one side of the double-strand break is generated by the assembly of a replisome and initiation of lagging strand synthesis (Figure 13.23, step 5) that extends back to the side of the double-strand break that initiated D-loop formation (Figure 13.23, step 6). Notice that the newly synthesized chromosome segment is derived entirely from copying the intact duplex (Figure 13.23, step 7). Any markers that may have been different on the parental duplexes, for example B and b, are now the same on the progeny duplexes, b and b. Thus BIR results in loss of heterozygosity.

To avoid such loss of heterozygosity, an important step in homology-directed repair by synthesis-dependent strand annealing and homologous recombination is the capture of the other side of the double-strand break.

When loss of heterozygosity involves genes that regulate cell growth, such as the tumor suppressor genes described in Chapter 1, the consequences can be severe. For example, loss of heterozygosity underlies the greatly increased susceptibility to a retinal cancer called retinoblastoma in children who have inherited one defective gene for a protein known as Rb. The Rb protein suppresses the progression of cells from G1 to the S phase of the cell cycle (see Chapter 5). Growth control

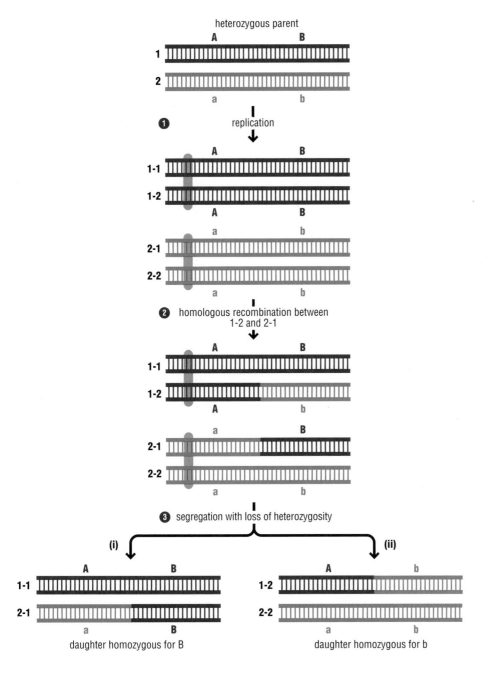

Figure 13.22 Loss of heterozygosity can occur by homologous recombination. Two homologs (1 = gray, 2 = pink) in a diploid that are heterologous at two loci, A/a and B/b, are shown. (1) DNA replication generates two sister chromatids (1-1 and 1-2 for the gray chromosome; 2-1 and 2-2 for the pink chromosome). The sister chromatids are associated with one another at their centromeres (light gray oval). (2) Prior to chromosome segregation, recombination takes place between a gray chromatid (1-2) and a pink chromatid (2-1). As a result, chromatid 1-2 is no longer identical to 1-1 at the B/b locus, and the same is true for 2-1 and 2-2. (3) During chromosome segregation, sister chromatids separate from each other. Thus, chromatid 1-1 could end up with either chromatid 2-1 or 2-2. If 1-1 and 2-1 are in the same daughter cell (i), that cell will have two copies of the B allele and no copies of the b allele. In this case, the other chromatids, 1-2 and 2-2, will be in the second daughter cell, which will have two copies of allele b, and no copies of allele B. This segregation pattern resulted in loss of heterozygocity at the B/b locus. Note that if chromatid 1-1 co-segregates with chromatid 2-2, and chromatid 1-2 co-segregates with chromatid 2-1 (not shown), heterozygocity is maintained despite the recombination event.

is normal in somatic cells heterozygous for Rb (Rb^+/Rb^-), because the undamaged gene is sufficient to maintain the essential functions of Rb. However, homologous recombination can result in a cell homozygous for the defective version of Rb (Rb^-/Rb^-). Such a cell does not properly observe the cell cycle checkpoint that normally prevents DNA replication and cell division, and can proliferate to form a tumor.

Homologous recombination or homology-dependent repair are not the only ways in which loss of heterozygosity can occur. An entire member of a chromosome pair can be lost as a result of a malfunction of the cellular machinery that distributes copies of chromosomes evenly to the two daughter cells at mitosis (see Section 7.1).

A   **B**

① resection to expose
single-stranded DNA

② invasion of a homologous duplex by
one broken end to form a D-loop

③ DNA synthesis without formation of
extensive heteroduplex DNA

④ failure to capture other end

⑤ reassemble fork and initiate lagging strand
synthesis

⑥ cleavage

⑦ ligation

A   **b**

a   **b**

⊖ We learn more about Alu in Chapter 14.

Figure 13.23 Loss of heterozygosity can occur by break-induced replication. When a double-strand break in DNA occurs, sometimes only the DNA at one side of the double-strand break forms a D-loop, resulting in loss of the DNA on the other side of the break. A duplex with markers A and B (gray) containing a double-strand break between the two markers is shown. The cell also contains an intact homologous chromosome with a and b markers (see below) (1) As previously discussed for homology-directed repair via synthesis-dependent strand annealing (Figure 13.5), the 5′ ends of the broken DNA are resected. (2) One of the 3′ tails from the broken gray duplex invades the homologous a,b duplex (pink) to form a region of heteroduplex, displacing the other strand to make a three-stranded D-loop. (3) New DNA synthesis (green) initiates at the 3′ OH end of the invading strand using the undamaged DNA as a template. (4) In contrast to synthesis-dependent strand annealing and homologous recombination, neither the newly synthesized strand (green) nor the complement of the template strand (pink) captures the gray DNA on the other side of the break. Copying of the template strand continues to the end of the template chromosome (note that for illustration purposes, the top and bottom strands of the gray chromosomes were flipped (i.e. the 5′ end on the side of the break is now shown on the bottom). (5) Replication of the lagging strand initiates to form DNA duplex. (6) DNA synthesis continues to the side of the broken duplex (left) where D-loop formation initiated, and the strands in the junction are cleaved. (7) Ligation completes synthesis of the chromosome arm. Note that loss of heterozygosity has occurred. The parental duplexes were heterozygous, that is, Aa and Bb, but that the product of break-induced replication has become homozygous (Aa and bb) because of copying of the template (pink) duplex.

Break-induced replication can restore telomeres

Telomerase mediates telomere maintenance in mammalian germline cells, as we discuss in Section 6.11. However, telomerase is not expressed in many differentiated cells and, when telomeres become sufficiently shortened such that shelterin no longer protects them, they can provoke a DNA damage response that results in apoptosis, that is, cell death. Thus the replicative potential of many differentiated cells is limited, a phenomenon termed 'replicative senescence'. In a small fraction of cells, however, telomere length can be maintained by '**alternative lengthening of telomeres**' (**ALT**). In yeast, the ALT pathway depends on BIR (Figure 13.23) in which the shortened ends can invade another chromosome end at similar sequences, thus priming DNA replication that extends to the end of the template chromosome and adding a new telomere to the invading chromosome end.

A notable feature of tumor cells is that they can continue to divide and avoid the cell death response that usually results from the decreasing length of telomeres. In many of these cells, the expression of telomerase is reactivated, thus avoiding telomere shortening. Alternatively, induction of ALT is frequently observed in some types of tumor, which allows them to avoid the shortening of telomeres and resulting cell death.

Recombination between repeated sequences can result in chromosome rearrangements

Homologous recombination between repeated sequences is a particular hazard in multicellular eukaryotes, where such sequences are common. For example, the human genome contains about a million copies of a sequence called Alu, which is on average about 300 bp long.

The DNA rearrangements that result from recombination between repeated sequences depend on the relative position of the repeats: that is, whether they occur through intrachromosomal recombination, or through inappropriate inter-chromosomal recombination between repeats on sister chromatids, homologous chromosomes or non-homologous chromosomes. Such rearrangements can include **deletions**, **duplications**, and **translocations**.

Figure 13.24 Homologous recombination between repeated sequences can lead to chromosome rearrangements. (a) Intra-chromosomal deletion. Recombination between directly repeated sequences on the same chromosome results in deletion of the sequences between the repeats. A DNA duplex with repeat sequences (blue) is shown. (1) The repeat sequences are aligned. (2) Homologous recombination between the repeats results in the loss (deletion) of the material between the repeats from the chromosome arm and formation of a DNA circle containing a repeat. (b) Unequal crossing over between sister chromatids. This is called unequal crossing over because the DNA segments are unequal, that is, not perfectly aligned. Recombination between repeated sequences (blue) at different positions on sister chromatids (gray) results in deletion and expansion. This type of recombination is also called unequal crossing over because the chromosomes are not perfectly aligned, i.e. are unequal. Sister chromatids (gray) with repeat sequences (blue) are shown. (1) Repeats at different positions in the sister chromatids (blue) align. (2) Recombination between the aligned repeats results in the exchange of DNA segments between the sister chromatids. This example yields a reciprocal translocation, one chromosome losing sequences that are incorporated in the other. (c) Unequal crossing over between homologs. Recombination between repeated sequences on homologs results in deletion and expansion. Homologs (gray and light gray) with repeat sequences (blue) are shown. (1) Repeats at different positions in the homologs align. (2) Recombination between the aligned repeats results in the exchange of DNA segments between the homologs. This example yields a reciprocal translocation, one homolog losing sequences that are incorporated in the other. (d) Translocation between non-homologous chromosomes. Recombination between repeated sequences (blue) on non-homologous chromosomes (gray, pink) results in a reciprocal translocation. A sequence common to two different chromosomes is shown in blue. (1) The common sequences align. (2) Homologous recombination between the common sequences results in a reciprocal translocation.

Recombination between repeats such as Alu on the same chromosome results in deletion of all the DNA between the repeats (see Figure 13.24a). When this DNA includes some essential gene, its loss can have severe consequences. Recombination between misaligned repeats on sister chromatids or on homologous chromosomes, often called **unequal crossing over**, can also occur, resulting in deletions and duplications on the different partners (Figure 13.24b and c). In a translocation, part of one chromosome is moved to another a non-homologous chromosome (Figure 13.24d).

Translocations and other chromosomal rearrangements are frequently found in cancer cells; some are specifically associated with particular cancers and are used to aid diagnosis and in defining cancer subtypes. Translocations that fuse two genes together, disrupting their function or regulation, are known to be directly responsible for several types of cancer, including chronic myelogenous leukemia and Burkitt lymphoma, both cancers of white blood cells.

As we will see in Section 16, Alu-mediated recombination occurring in germline cells at meiosis is the cause of about 0.3% of all cases of human genetic disease, including cases of insulin-resistant diabetes, immune deficiency due to the loss of a serum protein critical to resistance of bacterial infection, and familial hypercholesterolemia leading to heart disease.

Proteins that mediate homology-directed repair and homologous recombination in various organisms

step	E. coli	activity		S. cerevisiae	mammalian	activity
presynapsis: resection at a DSB to form single-stranded DNA with 3'OH ends	RecBCD or	helicase-nuclease	initial processing:	**MRX**	**MRN**	DNA binding and nuclease
	RecJ +	nuclease +		Sae2	CtIP	nuclease
	RecQ	helicase	subsequent processing:	Exo1	EXO1	nuclease
				Dna2 Sgs1-	**DNA2** BLM-	nuclease helicase
removal of RPA from ssDNA and formation of the pres-synaptic filament: loading of the strand exchange recombinase onto single-stranded DNA	RecFOR			Rad52 Rad55/57	BRCA2	
synapsis: detection of homology & strand exchange	RecA			Rad51 Dmc1 (meiosis)	RAD51	strand transfer recombinase
resolution:	RuvABC					
branch migration		RuvAB	branch migration	RecQ?	RecQ?	
cleavage of Holliday junctions		RuvC	cleavage	Yen1 Slx1/Slx4	GEN1 SLX1/SLX4	

Figure 13.25 Proteins that mediate synthesis-dependent strand annealing. Proteins from different organisms that mediate particular steps on synthesis-dependent strand annealing are shown. Proteins shown in the same colour are structurally related.

Aberrant NHEJ can lead to chromosome rearrangements

Misdirected homologous recombination is not the only mechanism of chromosome rearrangement. If multiple double-strand breaks occur, for example through environmental agents such as radiation, or as a result of DNA damage by oxidation, replication errors, or inappropriate cleavage by nucleases – the broken ends of the chromosomes may be incorrectly rejoined by NHEJ (see Section 13.1), so that the repair process joins a broken end to the wrong partner. While such chromosomes can be faithfully transmitted during mitosis, they cause missegregation during meiosis because of errors in the homologous pairing necessary for accurate recognition and segregation of homologs and are often lethal.

✷ SUMMARY

Double-strand breaks

- DNA double-strand breaks are particularly dangerous lesions.

- Failure to repair them can lead to chromosome fragmentation and cell death.

- Cells defective in their ability to repair double-strand breaks are very sensitive to DNA damage.

- There are two major strategies for double-stranded break repair: NHEJ and homology-directed repair.

NHEJ

- NHEJ rejoins the ends across a double-stranded break.

- It is often mutagenic because of nucleolytic processing of the ends prior to joining.

Homology-directed repair

- Homology-directed DNA synthesis can repair double-strand breaks by synthesizing new DNA across the break.

- The repair mediated by HDR is templated by a sister chromatid or homolog.

- HDR-mediated repair changes the sequence of the broken duplex to that of the template duplex in the region of the break, resulting in gene conversion

- Homology-directed repair occurs without changing the relationship of the DNA segments flanking the site of the double-strand break, that is, recombination or 'crossing over' does not occur.

- Identification of a template duplex occurs through the specific base-pairing of a single DNA strand from the broken double strand with its complement in a DNA duplex.

The mechanism of homology-directed repair

- The base-pairing of single-stranded DNA from the broken DNA to a template strand forms a D-loop containing heteroduplex DNA.

- The single strand from the broken DNA duplex serves as a primer for DNA synthesis to replace the DNA lost at the break.

- The newly synthesized strand serves as the template to generate duplex DNA across the region of the break. This DNA synthesis is non-conservative.

- The ability to pair homologous DNA strands provides an important mechanism to reassemble replication forks that have broken or collapsed because of lesions such as nicks in DNA.

- Specialized proteins including helicases and nucleases generate the single-stranded DNA necessary for heteroduplex DNA formation.

- Strand exchange recombinases, which include the bacterial RecA protein and eukaryotic Rad51 proteins, mediate the pairing reactions necessary for heteroduplex D-loop formation.

- The proteins that mediate homology-dependent repair have evolved to mediate homologous recombination.

Homologous recombination

- Homologous recombination is the reciprocal exchange of large segments of DNA between homologous duplexes

- Homologous recombination occurs during meiosis to generate gametes, and occurs between the bacterial chromosome and exogenous DNAs that enter the cell by conjugation, transformation, or in viruses.

- Homologous recombination can occur between any two duplexes that are nearly identical in sequence; it does not require a special site in either participating DNA.

- Homologous recombination is initiated by a double-strand break in one duplex.

- As in homology-dependent repair, a single-stranded DNA from the broken duplex forms a region of heteroduplex DNA in a homologous duplex via formation of a D-loop.

- The single strand from the broken duplex in the D-loop serves as a primer for DNA synthesis using the homologous chromosome as a template.

- The template strand that is displaced from the D-loop by semi-conservative DNA synthesis captures the other side of the broken DNA.

- This capture of the broken end by the template duplex is the key difference between homology-dependent repair and homologous recombination.

Holliday junctions

- Holliday junctions are key recombination intermediates.

- In Holliday junctions, two single DNA strands are reciprocally exchanged between homologous DNA duplexes to form a branched four-way.

- The DNA duplexes that are linked by D-loop formation on one side of an initiating double-strand break, and end capture on the other side of the break, are joined by Holliday junctions, which result from the reciprocal exchange of single strands between two homologous duplexes.

- The Holliday junctions bound regions of heteroduplex DNA and a region in which template-directed DNA synthesis has occurred.

- Gene conversion can result from mismatch repair of the heteroduplex regions and from template-directed DNA synthesis.

- A Holliday junction connecting two duplexes can be resolved (cleaved) in several ways in a process mediated by specialized enzymes called resolvases.

- Whether or not recombination occurs, i.e., whether a change in the relationship of the DNA segments flanking Holliday junctions occurs, is determined by the Holliday junction resolution pathway.

Inappropriate repair and recombination

- Inappropriate homology-directed repair or homologous recombination between heterozygous duplexes can result in progeny that contain information from only one parental duplex.

- Such 'loss of heterozygosity' can be hazardous, most notably leading to cancer when a cell contains only mutant versions of growth control genes.

- Inappropriate homologous recombination between repeated sequences can lead to chromosome rearrangements such as deletions and translocations.

 FURTHER READING

13.1 AN OVERVIEW OF DNA DOUBLE-STRAND BREAK REPAIR AND HOMOLOGOUS RECOMBINATION

Helleday T, Lo J, van Gent DC, Engelward BP. DNA double-strand break repair: from mechanistic understanding to cancer treatment. *DNA Repair* 2007;**6**:923–935.

Pardo B, Gómez-González B, Aguilera A. DNA repair in mammalian cells: DNA double-strand break repair: how to fix a broken relationship. *Cellular and Molecular Life Sciences* 2009;**66**:1039–1056.

Wyman C, Kanaar R. DNA double-strand break repair: all's well that ends well. *Annual Review of Genetics* 2006;**40**:363–383.

13.2 DOUBLE-STRAND BREAK REPAIR BY NON-HOMOLOGOUS END JOINING

Haber JE. Alternative endings. *Proceedings of the National Academy of Sciences of the U S A* 2008;**105**:405–406.

Lieber MR. The mechanism of human nonhomologous DNA end joining. *Journal of Biological Chemistry* 2008;**283**:1–5.

Shuman S, Glickman MS. Bacterial DNA repair by non-homologous end joining. *Nature Reviews Microbiology* 2007;**5**:852–861.

Weterings E, Chen DJ. The endless tale of non-homologous end-joining. *Cell Research* 2008;**18**:114–124.

13.3 HOMOLOGY-DIRECTED REPAIR OF DOUBLE-STRAND BREAKS

Haber JE, Ira G, Malkova A, Sugawara N. Repairing a double-strand chromosome break by homologous recombination: revisiting Robin Holliday's model. *Philosophical Transactions of the Royal Society of London Series B, Biological Sciences* 2004;**359**:79–86.

Hiom K. DNA repair: common approaches to fixing double-strand breaks. *Current Biology* 2009;**19**:R523–525.

13.4 GENERATION OF SINGLE-STRANDED DNA BY HELICASES AND NUCLEASES

Bernstein KA, Rothstein R. At loose ends: resecting a double-strand break. *Cell* 2009;**137**:807–810.

Dillingham MS, Kowalczykowski SC. RecBCD enzyme and the repair of double-stranded DNA breaks. *Microbiology and Molecular Biology Reviews* 2008;**72**:642–671.

Kanaar R, Wyman C. DNA repair by the MRN complex: break it to make it. *Cell* 2008;**135**:14–16.

Mimitou EP, Symington LS. DNA end resection: Many nucleases make light work. *DNA Repair* 2009;**8**:983–995.

Wigley DB. RecBCD: the supercar of DNA repair. *Cell* 2007;**131**:651–653.

13.5 THE MECHANISM OF DNA STRAND PAIRING AND EXCHANGE

Antony E, Tomko EJ, Xiao Q, *et al.* Srs2 disassembles Rad51 filaments by a protein-protein interaction triggering ATP turnover and dissociation of Rad51 from DNA. *Molecular Cell* 2009;**35**:105–115.

Cox MM. Motoring along with the bacterial RecA protein. *Nature Reviews Molecular Cell Biology* 2007;**8**:127–138.

Finkelstein IJ, Greene EC. Single molecule studies of homologous recombination. *Molecular BioSystems* 2008;**4**:1094–1104.

Kowalczykowski SC. Structural biology: snapshots of DNA repair. *Nature* 2008;**453**:463–466.

Thorslund T, West SC. BRCA2: a universal recombinase regulator. *Oncogene* 2007;**26**:7720–7730.

13.6 GENE CONVERSION THROUGH HOMOLOGY-DIRECTED REPAIR

Chen JM, Cooper DN, Chuzhanova N, Férec C, Patrinos GP. Gene conversion: mechanisms, evolution and human disease. *Nature Reviews Genetics* 2007;**8**:762–775.

Deitsch KW, Lukehart SA, Stringer JR. Common strategies for antigenic variation by bacterial, fungal and protozoan pathogens. *Nature Reviews Microbiology* 2009;**7**:493–503.

Palmer GH, Brayton KA. Gene conversion is a convergent strategy for pathogen antigenic variation. *Trends in Parasitology* 2007;**23**:408–413.

13.7 HOMOLOGOUS RECOMBINATION

Déclais AC, Lilley DM. New insight into the recognition of branched DNA structure by junction-resolving enzymes. *Current Opinion in Structural Biology* 2008;**18**:86–95.

Li X, Heyer WD. Homologous recombination in DNA repair and DNA damage tolerance. *Cell Research* 2008;**18**:99–113.

Nimonkar AV, Kowalczykowski SC. Second-end DNA capture in double-strand break repair: how to catch a DNA by its tail. *Cell Cycle* 2009;**8**:1816–1817.

San Filippo J, Sung P, Klein H. Mechanism of eukaryotic homologous recombination. *Annual Review of Biochemistry* 2008;**77**:229–257.

Walsh T, King MC. Ten genes for inherited breast cancer. *Cancer Cell* 2007;**11**:103–105.

13.8 REPAIR OF DAMAGED REPLICATION FORKS BY HOMOLOGY-DIRECTED REPAIR

Heller RC, Marians KJ. Replisome assembly and the direct restart of stalled replication forks. *Nature Reviews Molecular Cell Biology* 2006;**7**:932–943.

Michel B, Boubakri H, Baharoglu Z, LeMasson M, Lestini R. Recombination proteins and rescue of arrested replication forks. *DNA Repair* 2007;**6**:967–980.

Moldovan GL, Pfander B, Jentsch S. PCNA, the maestro of the replication fork. *Cell* 2007;**129**:665–679.

Rudolph CJ, Dhillon P, Moore T, Lloyd RG. Avoiding and resolving conflicts between DNA replication and transcription. *DNA Repair* 2007;**6**:981–993.

13.9 ABERRANT REPAIR AND RECOMBINATION AND CHROMOSOME REARRANGEMENTS

Haber JE. Transpositions and translocations induced by site-specific double-strand breaks in budding yeast. *DNA Repair* 2006;**5**:998–1009.

Jeggo PA. Genomic instability in cancer development. *Advances in Experimental Medicine and Biology* 2005;**570**:175–197.

Llorente B, Smith CE, Symington LS. Break-induced replication: what is it and what is it for? *Cell Cycle* 2008;**7**:859–864.

Weinstock DM, Richardson CA, Elliott B, Jasin M. Modeling oncogenic translocations: distinct roles for double-strand break repair pathways in translocation formation in mammalian cells. *DNA Repair* 2006;**5**:1065–1074.

MOBILE DNA

INTRODUCTION

In the preceding two chapters we learned how the sequence of DNA can be changed by mutations, and by homologous recombination between large regions of DNA. In this chapter we will learn about two ways by which discrete segments of DNA can move from one region of DNA to another: by **transposition**, whereby **transposable elements** can insert in many different sites around the genome; and by **conservative site-specific recombination (CSSR)**, sometimes called site-specific recombination, whereby recombination occurs between pairs of specific related sites to rearrange sequences bounded by the specific sites.

These systems are widespread and have profound impact on chromosome structure and function. Virtually all organisms contain transposable elements, and their mobility is a significant contributor to genetic variation. CSSR systems are found mostly in bacteria where they play important roles in bacterial chromosome segregation and the interactions of viruses and plasmids with their hosts. The mechanisms of DNA breakage and joining associated with transposition and CSSR, and the recombination enzymes that promote them, are different. They involve, however, recombination enzymes that act only at specific sequences, and form oligomeric complexes whose correct assembly with multiple DNA segments is required to coordinate and regulate the DNA breakage and joining reactions involved.

The strategy of controlling a biological transaction by ordered assembly of nucleo-protein complexes through protein binding to *specific* sites in nucleic acids is common to DNA replication and transcription, and RNA processing and translation. By contrast, homologous recombination can occur between any two DNA segments that are highly similar to each other *without* a requirement for particular sequences.

We begin the chapter with an overview of transposition before considering in more detail the mechanisms of action employed by different elements, and the strategies used to regulate these processes. We will then explore CSSR, including some specific well-characterized examples.

14.1 TRANSPOSABLE ELEMENTS: OVERVIEW

Transposable elements are DNA sequences that can move to different sites in the genome

Transposable elements, or **transposons**, are discrete mobile segments of DNA that can move around the genome and insert into different DNA **target sites**

using a specialized type of recombination called transposition. Transposable elements are found in all living organisms; indeed, they are often the largest component of the genomes of multicellular organisms. For example, about 45% of the human genome and 85% of the maize genome derives from transposable element sequences. Transposable elements typically encode the proteins necessary for their mobility (that is, they encode proteins that catalyze the molecular events underpinning transposition). Elements that encode their mobility proteins are called '**autonomous elements**'; '**non-autonomous elements**' do not encode their own mobility proteins but can move when the necessary proteins are supplied from an autonomous element within the same cell. In addition, transposable elements may also encode gene regulatory segments such as enhancers and promoters that effect the expression of adjacent host genes. In bacteria, transposable elements may also encode genes such as drug resistance genes, virulence genes, and genes for a variety of metabolic processes that provide the host with new phenotypes.

Thus, the information encoded by transposable elements can have substantial impact on the host. Indeed, there are multiple examples of the host domestication of transposons and transposable element proteins, whereby transposons have evolved to perform useful functions within a host cell or organism. For example, transposons are major functional constituents of heterochromatin, and centromeres. Further, the reverse transcriptase telomerase is likely to have evolved from a protein involved in nucleic acid mobility within the cell.

In short, transposons are certainly not 'junk DNA'. Although in one view, they can be considered to be genome parasites, with the selfish function of simply propagating themselves, they also have many effects on genome structure and function, and they contribute to the genetic variation that is the substrate for evolution.

Transposable elements are powerful agents for genetic change

Because of their ability to move transposable elements have great potential to generate changes in the genetic constitution of their host cells. For example, when a transposon inserts into the coding region of a gene, it acts to interrupt that coding region. As a result, the gene is usually inactivated because of the truncation of the coding region. Further, many transposable elements carry regulatory sequences such as promoters and enhancers. Therefore, when a transposable element inserts into a gene regulatory region, it may influence expression of nearby genes.

Transposons may also influence gene expression by modifying chromatin structure. For example, heterochromatin is transposon-rich and gene-poor and euchromatin is the reverse. Moreover, transposons are often present in high copy numbers within a genome; the extensive regions of sequence homology resulting from their ubiquitous presence means that transposons can often provide substrates for homologous recombination. Such recombination may lead to intramolecular deletions and an increase in unequal sister chromatid exchange, ultimately causing a change in genome structure and function (see Section 13.9).

The frequency at which transposable elements move and thus contribute to changes in host genotype (and, potentially, phenotype) varies considerably between elements and organisms. Active mobile elements have been found in all model organisms, but transposition is generally rare, except when it is used as a mechanism for viral replication (Figure 14.1). Experimental approach 14.1 describes the experiments that led to the identification of the first human mobile

	fraction of genome derived from transposable elements	fraction of *de novo* mutations resulting from transposable element insertion
human	45%	0.3%
mouse	39%	10%
Drosophila	5.5%	>50%

Figure 14.1 Frequency of *de novo* mutations resulting from transposable element insertion in different species.
From Belancio *et al.* (2008).

element, discovered because a *de novo* insertion resulted in hemophilia. There are about 65 examples of human disease resulting from transposon insertion (Figure 14.2). Also described in Experimental approach 14.1 is the first analysis of mobile elements in a genome of a single human, allowing new information about the variability of transposable element populations between individuals to be

Insertions and disease			
insertion	locus	chromosome	disease
Alu	F8	X	Hemophila B
	F9	X	Hemophila A
	CLCN5	X	Dent's disease
	BTK	X	X-linked agammaglobulinemia
	IL2RG	X	X-linked severe combined immunodeficiency
	GK	X	Glycerol kinase deficiency
	ABCD1	X	Adrenoleukodystrophy
	ATP7A	X	Menkes disease
	CD40LG	X	Hyper-immunoglobulin M syndrome
	CRB1	1	Retinal blinding
	SERPINC1	1	Type 1 antithrombin deficiency
	ZEB2	2	Muckle-Wells syndrome
	MSH	2	Hereditary non-polyposis colorectal cancer
	CASR	3	Hypocalciuric hypercalcemia and hyperparathyroidism
	BCHE	3	Cholinesterase deficiency
	HESX1	3	Aplasia anterior pituitary
	MLV12	5	Associated with leukemia
	APC	5	Hereditary desmoid disease
	NT5C3	7	Chronic hemolytic anemia
	CFTR	7	Cystic fibrosis
	EYA1	8	Branchio-oto-renal syndrome
	LPL	8	Lipoprotein lipase deficiency
	CHD7	8	CHARGE syndrome
	FAS	10	Autoimmune lymphoproliferative syndrome
	FGFR2	10	Apert syndrome
	SERPING1	11	Complement deficiency
	HMBS	11	Acute intermittus porphyria
	MLV12	12	Mucolipidosis type II
	BRCA2	13	Breast cancer
	BRCA1	17	Breast cancer
	NF1	17	Neurofibromatosis
L1	CHM	X	Choroideremia
	CYBB	X	Chronic granulomatous disease
	DMD	X	X-linked Duchenne muscular dystrophy
	F8	X	Hemophilia A
	F9	X	Hemophilia B
	RP2	X	X-linked retinitus pigmentosa
	RPS6KA3	X	Coffin-Lowry syndrome
	APC	5	Colon cancer
	FKTN	9	Fukuyama-type congenital muscular dystrophy
	HBB	11	Beta-thalassemia
	PDHX	11	Pyruvate dehydrogenase complex deficiency

Figure 14.2 **Human diseases cause by transposable element insertion.** Mutations have resulted from insertion of two types of non-LTR elements, Alu and L1.
From Belancio *et al.* (2008).

revealed. It is estimated that in about 1 in every 200 human births, the genome of the newborn has a new transposition event compared to its parents' genomes.

Transposons can insert into many different locations in the genome, resulting in target-site duplications

A characteristic feature of transposable elements is that they can insert into many different sites in the genome. This is because transposition does not require any sequence homology between the transposon DNA and the target site, unlike homologous recombination, which was discussed in Chapter 13, and the CSSR reactions that will be discussed later in this chapter.

Another hallmark of transposition is that element insertion results in **target-site duplications**, that is, short direct repeats that flank the newly inserted transposon (Figure 14.3). Target-site duplications flanking newly inserted elements result from repair by the host cell's DNA repair proteins of single-strand gaps at each transposon end, which are formed upon element insertion. The length of the duplication is characteristic of the particular element. With some elements, DNA synthesis is limited to target-site duplications; with other elements, the entire element is also replicated.

Most transposable elements are confined to moving within the DNA of a single cell. Exceptions are viruses such as the bacteriophage Mu (discussed in Section 14.3) and the retroviruses (discussed in Section 14.5), which use multiple cycles of **replicative transposition** to replicate their genomes during the viral lifecycle.

Transposition can occur via element replication or element excision

Transposable elements move, or transpose, in two main ways (Figure 14.3): they are either excised from one site and reinserted into a new site (pathway a) or they replicate (with the duplicate copy being integrated at a new site in the genome, pathway b). Most elements use only one of these pathways.

The mechanism of movement of some transposons requires only DNA, and does not require an RNA intermediate. These so-called DNA-only elements are also termed Class II elements. Class II elements move through the two steps of excision, whereby they simply cut themselves out from the donor site, and integration, during which they insert themselves into a new target site (Figure 14.3a).

Figure 14.3 Transposons can move by excision and integration or by replication. Transposons move from a donor DNA (gray) to a target DNA (light gray) by either of two general pathways: (a) excision and integration in which the element itself moves from one site to another or (b) replication in which an element copy moves to the insertion site. There is no change to the donor site in the replication pathway, whereas excision and integration requires repair of the gap in the donor backbone. In both pathways, transposon insertion is accompanied by duplication of a short sequence (blue arrow) at the target site.

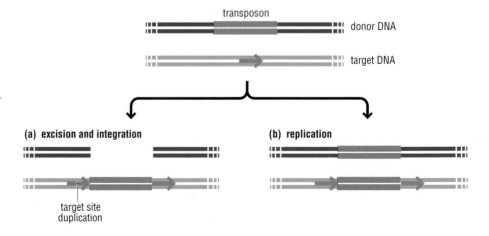

14.1 EXPERIMENTAL APPROACH

Mobile DNA in humans

Transposable elements were first discovered by Barbara McClintock on the basis of the discernible phenotypes that they brought about upon their insertion into DNA. McClintock identified the DNA element Ac (for 'activator') in the late 1940s because it altered the pigment color of corn kernels. It was not until much later, in 1988, that Haig Kazazian and colleagues discovered a human disease, a form of hemophilia, similarly caused by a transposable element.

Hemophilia is a group of hereditary diseases resulting from an inability to control blood clotting. Hemophilia A is a common form of the disease in which a protein required for clotting, called factor VIII, is missing. As the gene for this factor is on the X chromosome, hemophiliacs are usually male; in females, two X chromosomes are present, and having one wild-type copy of the gene is sufficient for effective clotting. Such heterozygous females are, however, carriers of the disorder and can transmit the defective gene to their progeny. Homozygous mutant females are rare, however, because relatively few males survive to reproductive age. Hemophilia A occurs in about 5000–10 000 male births; about 30% of these cases result from new spontaneous mutations.

To examine genomic changes associated with the disease, Kazazian and colleagues analyzed the DNA of 240 unrelated males with hemophilia A by restriction digestion and Southern blot analysis. The factor VIII gene is large, consisting of 186 kb, where 26 exons contain about 9 kb of the total coding sequence. To analyze the locus, they digested genomic DNA with a restriction enzyme, fractionated the fragments on a gel and examined the size of the

factor VIII gene fragments by hybridization. Although the genes of 15 patients carried detectable deletions, the researchers were most interested to see that two patients yielded a *larger* fragment than expected for exon 14. In neither of these two cases were the DNA changes present in the DNA of the asymptomatic parents, suggesting that these insertions were new mutations that resulted in disease. The exon 14 DNA fragments from the two patients were isolated and sequenced, revealing the presence of truncated forms of a human L1 retrotransposon (Figure 1).

Since this initial identification of mutations caused by transposition in humans, sequencing has provided a very powerful tool for in-depth analysis of transposable elements and their presence throughout the genome. An unanticipated finding from the initial sequencing of the human genome by the International Human Genome Sequencing Consortium (Eric Lander and colleagues, 2001) and Celera Genomics (Craig Venter and colleagues, 2001) was that transposable elements make up a very high fraction of the genome. The astounding prevalence of transposons in the genome, most significantly LTR and non-LTR retrotransposons, is discussed in Section 14.1.

While most of these transposable elements are no longer active due to the accumulation of mutations within them, there are important exceptions. Some elements do move, and both the recombination between dead elements and the movement of active transposable elements continue to contribute to structural variations between individuals. And, though most of the structural variations resulting from such events are selectively neutral, some

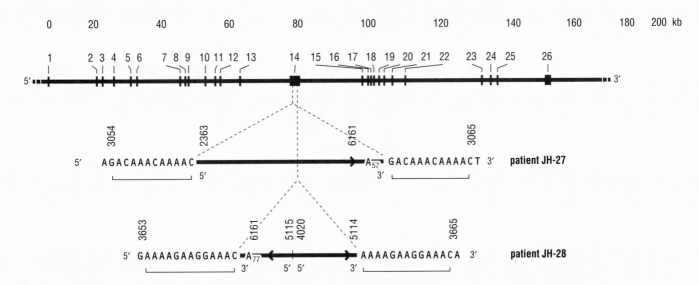

Figure 1 **Diagram of L1 insertions in exon 14 of the factor VIII gene in two patients.** The factor VIII gene spans 186 kb and contains 26 exons. The positions of insertion of segments of the human non-LTR element L1, a LINE element, into intron 14 are shown. The target site duplications are indicated by brackets.

insertions do have phenotypic consequences, as first described by Kazazian and colleagues (1988).

The complete sequence obtained by the Human Genome Project in 2002 represented an assembly of DNA sequences that were derived from a few different people. To understand more fully variations *between* people it is necessary to compare the sequences of individuals, rather than use a composite sequence. Venter and colleagues (Levy *et al.* 2007) determined the complete DNA sequence of a single person, employing a new computer algorithm to correctly assign sequence information to each of the two homologous chromosomes. This complete diploid sequence of an individual is called the HuRef sequence. Having this sequence information from individuals allows for an understanding of the degree of heterozygosity between individuals and for fine structure mapping of the human genome, as discussed in Chapter 15.

The HuRef sequence also has proved to be very useful for understanding structural variations that arise from both the movement of 'live' transposable elements and recombination between inactive transposable elements, as alluded to above.

To specifically examine the role of transposable elements in genomic change, Xing *et al.* (2009) took advantage of the complete HuRef sequence and compared it to the sequence from the Human Genome Project. This comparison identified a total of 643 992 sites at which either insertions or deletions (so called indels) occurred between the two sequences. A small fraction of these indels (8500 or 0.013%) were larger than 50 bp, and a subset of these larger indels was associated with transposable element sequences. In this manner, 846 'mobile element associated structural variants' (MASVs) were identified in the HuRef genome. Based on more detailed analysis of the sequences of the MASVs, Lynn Jorde and colleagues (Xing *et al.*) described four general mechanisms by which they were generated: (a) retrotransposon insertions, (b) non-classical insertions, (c) non-allelic homologous recombination and (d) non-homologous end

Figure 2 Four types of common mobile element associated structural variants (MASVs) in the HuRef genome and how they are thought to arise.
(a) Insertion of non-LTR L1 (red) at a TTAAAA insertion site. (b) Non-classical insertion where a DNA fragment (red) is ligated into a double-strand break.
(c) Non-allelic homologous recombination-mediated insertion/deletion. Interchromosomal non-allelic recombination between homologous sequences resulting in a deletion and insertion; Intrachromosomal recombination between homologous sequences resulting in a deletion. (d) Non-homologous end-joining (NHEJ) mediated deletion. Rejoining at breaks in different chromosomal locations (dark blue and blue) resulting in deletion.

joining (NHEJ)-mediated deletions (Figure 2). The latter three of these classes are mediated by recombination of repetitive DNA. Notably, only one of these classes, retrotransposition, requires active transposition, principally involving L1 and Alu retrotransposons. The role of transposons in genome evolution is discussed in Chapter 15.

At the end of this comparative analysis of HuRef and HGP sequences, retrotransposon insertions were the most frequent structural changes found (small *indels* are not considered in this calculation). These studies estimate an Alu transposition rate of about one in 21 births, and an L1 transposition rate of about one per 212 births. To date 56 disease-causing mobile element insertions have been identified. Comparing this number to the total number of 'disease' mutations known suggests that 0.14% of disease-causing mutations are caused by mobile element insertions. Thus, although Haig Kazazian's initial finding of a retrotransposon causing hemophilia A was a rare find, mobile element associated structural variation has a significant role in disease and an even more significant role in genome evolution.

Kazazian HH Jr, Wong C, Youssoufian H, *et al*. Haemophilia A resulting from de novo insertion of L1 sequences represents a novel mechanism for mutation in man. *Nature* 1988;**332**:164–166.

Lander ES, Linton LM, Birren B, *et al*. Initial sequencing and analysis of the human genome. *Nature* 2001;**409**:860–921.

Levy S, Sutton G, Ng PC, *et al*. The diploid genome sequence of an individual human. *PLoS Biology* 2007;**5**:e254.

Venter JC, Adams MD, Myers EW, *et al*. The sequence of the human genome. *Science* 2001;**291**:1304–1351.

Xing J, Zhang Y, Han K, *et al*. Mobile elements create structural variation: Analysis of a complete human genome. *Genome Research* 2009; **19**:1516–1526.

Many transposable elements do not physically move from their original site, the donor site, but instead generate copies of themselves in another site, the target site. That is, they adopt a replicative mechanism of movement (Figure 14.3b). A few DNA-only elements move by a replicative mechanism. Most replicative elements, however, move via an RNA intermediate, which is then turned into DNA by reverse transcriptase. Elements that move via reverse transcription dominate many eukaryotic genomes but are present in all domains of life and are called retrotransposons or Class I elements.

Beyond being classified as either Class I or Class II according to the presence or absence of an RNA intermediate in the transposition process, most of the different transposons that exist can be divided according to different criteria into three other groups: the **DNA-only transposons**, the **long terminal repeat (LTR) elements**, and the **non-LTR** elements. These classes are distinguished by element structure (Figure 14.4), the type of recombination proteins they encode, and the mechanism of transposition (Figure 14.5). The numbers and types of transposable elements

Figure 14.4 There are three major types of transposable elements. (a) DNA-only cut-and-paste transposons. These elements are bounded by yellow terminal inverted repeat sequences (TIRs), which are recognized by the element-encoded transposase (orange). Autonomous elements encode their own transposase whereas non-autonomous transposons lack a transposase gene but do have TIRs. Non-autonomous transposons can be mobilized by a transposase from an autonomous copy of the element elsewhere in the genome. (b) Long terminal repeat (LTR) elements. These elements are retroviruses or retroviral-like transposons, both of which transpose via an RNA intermediate. LTR elements are flanked by directly repeated 'long terminal repeats' (yellow arrows). LTR elements encode multiple proteins: GAG, which interacts with the element's nucleic acid; protease (PRT), which is important for protein processing; reverse transcriptase (RT) and integrase (IN), which is the transposase. Retroviral elements also have ENV, a coat protein important for cell exit and entry. (c) Non-LTR elements. LINE elements are one type of autonomous non-LTR element that encodes: ORF1, an RNA chaperone protein important for making an RNA-protein complex (RNP); ORF2, which has reverse transcriptase activity (RT) and in some cases endonuclease (EN) activity; and a poly(A) tail. SINE elements are non-autonomous elements that can be mobilized by LINE proteins and have A- and B-box signals for pol III polymerase.

(a) DNA-only cut and paste element

autonomous   TIR   transposase   TIR

non-autonomous

(b) LTR element

LTR   LTR

GAG PRT RT/IN ENV

(c) non-LTR element

En/RT

LINE autonomous   ORF1   ORF2 polyA

SINE non-autonomous   A B polyA

(a) transposition via DNA-only intermediates

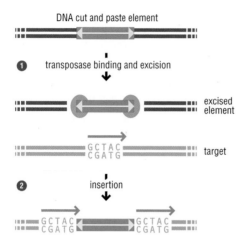

DNA cut and paste element

① transposase binding and excision

excised element

target

② insertion

(b) transposition via an RNA intermediate

(i) transposition of an LTR element

LTR element

① transcription

RNA

② reverse transcription

cDNA

target

③ insertion

(ii) target-primed reverse transcription

non-LTR transposon

donor DNA

① transcription

non-LTR transposon RNA

target DNA

② nick by endonuclease

③ reverse transcription of transposon RNA

④

new non-LTR transposon

Figure 14.5 Major transposition pathways. (a) Cut-and-paste transposition of a DNA-only transposon. The transposon (red) is a discrete DNA segment whose ends are terminal inverted repeat sequences (yellow triangles). These sequences are the binding sites for the transposase (orange). (1) The transposase makes double-strand breaks at these termini, which separate the element from the flanking host DNA (gray). (2) The transposon then inserts into the target DNA (light gray) in a reaction called integration. Insertion is accompanied by a duplication of the sequence at the target site (blue arrow). (b) Transposition via an RNA intermediate. These elements transpose via an RNA copy of the element and do not excise from their donor site. (i) Transposition of a Long Terminal Repeat (LTR) element. The LTR element in its 'provirus' form in the donor DNA (gray) is shown. (1) The element is transcribed into RNA (blue) by the host's RNA polymerase. (2) The element RNA is converted by an element-encoded reverse transcriptase into double-stranded DNA (green). This DNA is equivalent to the excised DNA intermediate of the DNA-only transposon in (a). (3) The integrase (transposase, in orange) then inserts the LTR element DNA into the target DNA by the same mechanism as the cut-and-paste transposons, creating a target-site duplication. (ii) Transposition of a non-LTR element by target-primed-reverse transcription. The non-LTR element in the donor site is shown. (1) The non-LTR element is first transcribed into RNA (blue) by the host RNA polymerase. (2) An element-encoded endonuclease nicks the target DNA. (3) The polyA tail of the non-LTR element anneals to the T-rich target strand resulting from endonuclease cleavage. The 3′ OH end exposed by target DNA nicking is used as the primer for reverse transcription. This process uses the element RNA (blue) as a template and generates an element DNA strand (green) covalently linked to the target DNA. (4) The mechanism of the synthesis of the second strand of DNA is not yet clear but insertion results in a new DNA copy of the element in the target DNA flanked by a target-site duplication.

present in the genome vary considerably between organisms (Figure 14.6). We consider each of these classes in the following sections.

One major class of transposable elements are DNA-only transposons

As implied by the name, all steps in the movement of **DNA-only transposons** involve DNA. Most of these elements move by simple excision and integration, as described below, and are known as DNA-only **cut-and-paste** elements (Figure 14.5a). As we will discuss below, some DNA-only elements use a related nick-and-paste mechanism for replicative transposition.

DNA-only cut-and-paste elements encode a specialized recombinase called a **transposase**, which recognizes particular sequences at the transposon ends and executes the DNA breakage and joining reactions that underlie transposition. These recognition sequences are called terminal inverted repeats (TIRs). The binding specificity of a transposase – such that it recognizes only specific TIRs – ensures that it acts only on a transposable element with cognate binding sites. As such, we see how different transposases catalyze the transposition of only certain transposons.

The TIRs at the ends of each transposon are composed of the same short DNA sequence, but in opposite (that is, inverted) orientations to each other. Binding to these TIRs ensures that the transposase breaks DNA at the termini of the element on both transposon ends.

DNA-only transposons are widespread in bacteria and eukaryotes. In bacteria, almost all transposable elements are autonomous elements (Figure 14.4): that is, they encode their own transposition proteins. Eukaryotic genomes, by contrast, often contain large numbers of non-autonomous elements.

Two other major classes of transposable elements are LTR and non-LTR retrotransposons

The other two major classes of transposable elements, the **LTR elements** and **non-LTR** elements, move by replicative mechanisms rather than through excision, as

Transposable elements in various genomes						
organism	common name	genome size (Mb)	number of ORFs	derived from TEs	% DNA TEs	% RNA TEs
bacteria						
Escherichia coli K-12 MG1655		4.6	4,300	1.0	100	0
archaea						
Mathanocaldococcus jannaschii		1.7	1,700	0.7	100	0
eukaryota						
Saccharomyces cerevisiae	budding yeast	12.1	6,300	3.1	0	100
Schizosaccharomyces pombe	fission yeast	14.0	4,800	1.1	0	100
Arabidopsis thaliana	(plant)	120.0	26,000	10	40	60
Oryza sativa	rice	420.0	23,300	30	85	15
Zea mays	maize	2300.0	32,500	85	11	89
Caenorhabditis elegans	worm	100.0	19,000	12	85	15
Drosophila melanogaster	fly	165.0	13,600	5.5	15	85
Mus musculus	mouse	2500.0	24,000	39	2	98
Homo sapiens	human	3280.0	25,000	45	7	93

Figure 14.6 Genome components in a variety of organisms.

described below. With these elements, an RNA copy of the element is first generated. The RNA copy is then converted into DNA by an element-encoded reverse transcriptase. Because a reverse transcription step is involved (in which DNA is synthesized from RNA), these elements are also known as **retrotransposons** (Figure 14.5b).

The LTR elements, which include retroviruses and retroviral-like elements, are characterized by long terminal repeats of hundreds of base pairs at each end of the element (Figure 14.5b, i). These LTRs are required for reverse transcription by the element-encoded reverse transcriptase, which converts the RNA form of the element to the double-stranded DNA form. This DNA is then inserted into the target site by an element-encoded **integrase**; the mechanism of insertion is identical to that of the DNA-only cut-and-paste transposons and thus retroviral integrases are also transposases. To date, LTR elements have been identified only in eukaryotes.

Non-LTR elements (Figure 14.5b, ii) also move via an RNA intermediate but lack long terminal repeats. These elements use a different transposition mechanism called **target-primed-reverse transcription**. In eukaryotes, one autonomous family of non-LTR elements called **long interspersed elements (LINEs)**, which are prominent in the human genome, encode proteins that mediate their own transposition. These proteins include ORF1, an RNA chaperone protein whose function is not yet known in detail, and ORF2, a reverse transcriptase and often an endonuclease. Non-autonomous non-LTR elements that do not encode transposition proteins (and so rely on 'non-self' proteins for their movement) are called **short interspersed elements (SINEs)** and can be mobilized by LINE-encoded proteins.

An RNA copy of a non-LTR element associates with the new target site through the action of an element-encoded endonuclease that makes a single-strand nick in the target DNA. This nick provides the primer for an element-encoded reverse transcriptase, which makes a DNA version of the element at the target site using the RNA copy of the element as a template. **Non-LTR transposons** are present in eukaryotic nuclei, mitochondria and chloroplasts, and in bacteria.

Host proteins play important roles in transposition

While most transposons encode the proteins required for their movement, host-encoded proteins are also involved in transposition reactions. Therefore, transposons cannot operate as completely independent entities: they rely on the cell whose genome they occupy for their ongoing propagation and movement. The target-site duplications flanking newly inserted elements (Figure 14.5) result from repair by the host cell's DNA repair proteins of single-strand gaps at each transposon end formed upon element insertion. In some cases, host DNA-bending proteins may promote the formation of the protein–DNA complexes that mediate DNA cut-and-paste transposition. Host proteins also play an important role in target-site selection for many elements and in the regulation of transposition. Much remains to be learned about the interesting interplay between the host and transposons during transposition. We learn more about the regulation of transposition in Section 14.8.

⊙ We learn about the synthesis of DNA from an RNA template in Section 6.11.

Having now considered transposition in general terms, let us now examine in more detail the mechanism by which DNA-only 'cut-and-paste' transposons move, and also elements that move by a '**nick-and-paste**' mechanism that involves replication.

14.2 DNA-ONLY TRANSPOSONS

DNA-only cut-and-paste elements move by DNA breakage and joining

As described in the previous section, the **DNA-only cut-and-paste transposable elements** move by DNA breakage reactions that separate the transposon ends from the donor DNA during excision, followed by DNA joining reactions that link the element to a new insertion site. Thus, all the substrates and intermediates in these reactions are DNA. In most cases, the element is completely excised from the donor site by double-stranded breaks before joining to the target site (Figure 14.7a). A very limited amount of DNA replication is required to repair the joints between the transposon ends and the target site. The product of this excision and integration reaction is called a simple insertion.

Unlike cut-and-paste elements, **nick-and-paste** elements move by a variation of the excision and integration mechanism, which results in copying of the DNA-only element. This mechanism is called replicative transposition, and results in a DNA structure called a **cointegrate**. A cointegrate contains two copies of the transposon linked by the donor and target DNA. To date, elements that move by this type of replicative transposition have only been found in bacteria. We will discuss the mechanism of this variation on cut-and-paste transposition below.

DNA cut-and-paste elements are widespread in both bacterial and eukaryotic genomes

DNA cut-and-paste elements have been found in virtually all organisms that have been sequenced (Figure 14.6). Although different terms are often used to describe elements in bacteria and eukaryotes, as detailed in Figure 14.8, their fundamental features, that is, an element bounded by short inverted repeats that are recognized by an element-encoded transposase, are the same.

One important difference between the DNA cut-and-paste elements found in eukaryotes and bacteria is that many bacterial transposons carry other genes in addition to those encoding their transposition functions. These include antibiotic-resistance genes, determinants of bacterial pathogenicity such as toxin genes, or genes for metabolic functions that enable the bacterium to utilize unusual substrates. These additional genes are of great importance to both bacteria and humans. Indeed the rapid dissemination of antibiotic resistance among bacterial populations represents the ability of antibiotic resistance-encoding transposons to 'hitchhike' onto other DNAs, such as plasmids and viruses, using these DNAs as vehicles to move between cells. By contrast, in eukaryotes, other elements encoding non-transposition functions have not yet been identified.

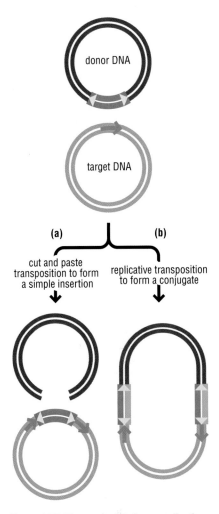

Figure 14.7 The products of non-replicative cut-and-paste transposition and replicative nick-and-paste transposition by DNA-only transposons. The donor DNA (gray) containing the transposon (red) and the target DNA (light gray) containing the target site (blue) are shown. (a) In non-replicative cut-and-paste transposition, the element excises from the donor DNA (gray) by double-strand breaks at the element termini and inserts into a target, forming a simple insertion. The newly inserted element is flanked by target site duplication (blue). (b) In replicative nick-and-paste transposition, the transposon remains attached to the donor DNA and joins to the target DNA, followed by replication of the element and formation of a cointegrate. The cointegrate contains the donor and target DNA linked by two copies of the transposon and target site duplications flank the transposon ends joined to target DNA. Each transposon copy contains one original strand (red) and one newly replicated strand (green). This structure is eventually resolved to form two separate DNA molecules, each with a transposon (see section 14.11).

Figure 14.8 A selection of DNA-only transposable elements from bacteria and eukaryotes. The yellow triangles are the terminal inverted repeat sequences to which the transposase binds, positioning it to act on the ends of the element. The genes for transposase are in orange, and those for other transposition proteins are in shades of red. Non-transposition genes including antibiotic resistance genes in bacterial elements are indicated in gray. (a) Bacteria. *IS911*, *IS10* and *IS50* are small (~1 kb) transposable elements called insertion sequences (IS elements). They encode only their mobility functions, that is, terminal inverted repeats and a transposase. In *IS911*, the transposase OrfAB is formed by translational frameshifting. IS elements can exist independently or as part of compound elements such as the transposons Tn5 and Tn10 in which IS50 and IS10, respectively, flank antibiotic resistance genes. *Tn7* is a non-compound transposon with multiple transposition genes. Mu is a bacteriophage that inserts by transposition into the bacterial genome to form a lysogen and then replicates its DNA by replicative transposition during the lytic phase. (b) Eukaryotes. Eukaryotic elements encode only terminal inverted repeats and the transposase. While some transposases are encoded by a single exon, for example *mariner* of the *Tc1/mariner* superfamily, other transposase genes contain exons and introns. *Tol2* and *piggyBac* are other well-studied elements. Additional elements are described later in this chapter and in Chapter 16.

For historical reasons, bacterial DNA cut-and-paste elements that encode only their own transposition functions are called **insertion sequences (IS elements)**. In some cases, a pair of IS elements flanks another gene (or genes) and the whole segment of DNA can act as a single transposable element; in such instances, the genes between the IS elements are mobilized along with the elements themselves (Figure 14.9). Such combinations of unrelated genes with flanking IS elements are sometimes called **transposons or compound transposons**. Transposon *Tn10*, for example, contains a tetracycline-resistance cassette bounded by two *IS10* elements. Coordinated breakage and joining at the outside ends of the two IS elements results in the movement of the tetracycline-resistance cassette along with the flanking *IS10*s.

14.3 MECHANISM OF DNA-ONLY CUT-AND-PASTE TRANSPOSITION REACTIONS

We have seen above how DNA-only cut-and-paste elements move around the genome first by severing links to the host DNA on either side of the transposon, and then by forming new links to DNA at a target site elsewhere in the genome. But what is happening at the molecular level during this transposition process, and what molecular components are required to make it happen?

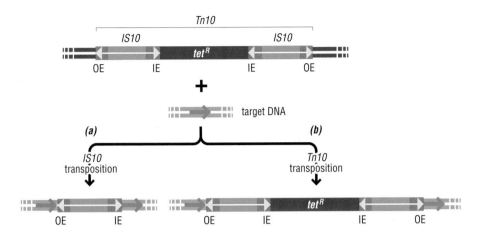

Figure 14.9 Comparison of *IS10* and *Tn10* transposition. *Tn10* is composed of two *IS10* elements flanking a tetracycline resistance cassette. (a) Transposition using the outside end (OE) and inside end (IE) of one *IS10* results in transposition of *IS10*. (b) Transposition using the two outside ends results in *Tn10* transposition.

DNA breakage and joining reactions are coordinated within protein-DNA structures called transpososomes

The movement of all the DNA-only cut-and-paste elements that have been studied in bacteria and eukaryotes adopt the same fundamental mechanisms. Indeed, only slight variations on this mechanism mediate the replicative transposition of nick-and-paste elements such as bacteriophage Mu and the integration of the DNA versions of LTR transposons such as retroviruses.

Three DNA segments are involved in DNA cut-and-paste transposition: the two transposon ends and the target DNA. The key steps in moving the transposon from the donor site to the insertion site are:

- recognition of the transposon ends by the transposases
- the pairing of the transposon ends
- the excision of the transposon from the donor site
- interaction of the excised transposon with the target DNA
- integration of the transposon into the target DNA.

In transposition, these steps are all mediated by protein–DNA and protein–protein interactions: there is no requirement for the type of base-pairing reactions that, as we saw in Chapter 13, underlie homologous recombination and are involved in some other transposition and CSSR reactions (as we will see later in the chapter).

Transposition occurs within sophisticated protein-DNA structures called **transpososomes**, which contain the transposon DNA, the transposase, and host proteins that may be required for transposition. The correct assembly of the transpososome coordinates the interaction of the transposon ends and target DNA and thereby controls the DNA breakage and joining reactions that underpin the transposition process. We first will consider the transpososomes that actually mediate transposition and then consider the chemistry of the DNA transactions in more detail.

Transpososomes mediate and control the progression of the DNA breakage and joining reactions

An overview of the way the transpososome mediates the DNA breakage and joining reactions that underpin DNA-only cut-and-paste transposition is shown in

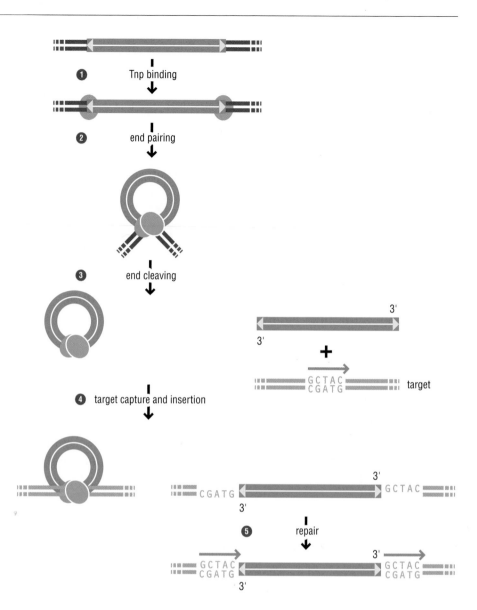

Figure 14.10 Transposition of a DNA cut-and-paste element. A transposon in a donor site (gray) is shown. (1) Transposition begins by the specific binding of the transposase (Tnp, in orange) to the terminal inverted repeats (in yellow). (2) Oligomerization of the transposases results in the pairing of the ends. (3) The DNA breaking activity of the transposase is activated by transposase pairing and transposon is excised from the donor backbone, exposing the 3' OH ends of the transposon. From this point on the panels on the left show protein–DNA complexes, whereas panels on the right show only the DNA. (4) The excised transposon–transposase complex binds to a target DNA (light gray) and the 3' OH transposon ends join to staggered positions on the target DNA, resulting in single-strand gaps flanking the 5' ends of the transposon. (5) The single-strand gaps are filled in by host repair enzymes, resulting in the target-site duplications (blue arrows).

Figure 14.10. The initiation of transposition is marked by the recognition of the transposition substrate DNA by the transposase (Figure 14.10, step 1). This recognition involves the specific binding of the transposase, which is usually encoded by the mobile element itself, to the terminal inverted repeats at the transposon ends.

In some cases, the formation of the complexes, in which transposase binding to the transposon ends occurs, is facilitated by host DNA-bending proteins such as the bacterial integration host factor (IHF), which bends DNA at specific sequences, or HU, a non–sequence-specific DNA-bending protein.

A key step in the formation of an active transpososome is the bringing together, or synapsis, of the transposase molecules bound at the transposon ends through their oligomerization (Figure 14.10, step 2). Such synapsis results in the pairing of the transposon ends. In many instances, such oligomerization leads to a more stable complex, which activates the DNA breakage and joining activity of the transposase. It is important to assure that no uncoupled DNA breakage occurs at a single transposon end that could lead to potentially lethal isolated double-strand breaks.

A common strategy to assure the coupling of end synapsis and DNA cleavage is cleavage *in trans* – literally, where the cleavage happens at the other transposon end from where the transposase is specifically bound. How does this happen? The active form of the transposase includes two monomers, which each bind specifically through their binding domains (yellow in Figure 14.11) at opposite ends of the transposon. However, the active site domain (green) of each monomer catalyzes cleavage at the opposite end of the transposon from where it is specifically bound. Thus, only when transposase binds to both ends and oligomerizes is there an active site domain for DNA breakage and joining positioned at *both* ends of the transposon.

Diverse DNA-only elements use the same chemical reactions for DNA breakage and rejoining

Following the interaction or synapsis of the transposon ends, the transposon is excised from the donor site by transposase-mediated double-strand breaks at both transposon ends (Figure 14.10, step 3). These double-strand breaks can be generated in a variety of ways as illustrated in Figure 14.12. For some elements, transposase promotes a direct double-strand break by hydrolysis of phosphodiester bonds on both strands of the element (Figure 14.12b); alternatively, the transposase may have two subunits, each promoting cleavage on one strand at each end (Figure 14.12c). Hydrolysis results in the loss of the high energy of the phosphodiester linkage, such that a ligation step is necessary to regenerate intact duplex DNA.

For other elements, the hydrolysis of one strand (nicking) exposes a 3′ end, which can then attack the complementary strand by a direct transesterification

Figure 14.11 **X-ray crystal structure of a** *Tn5* **transposase dimer bound to cleaved** *Tn5* **ends.** The cleaved *Tn5* ends bind specifically to the N-terminal regions of the transposase (yellow). Dimerization of the transposase occurs by interaction between C-terminal domains (blue). DNA breakage and joining is mediated by the catalytic domains (green). Note that the cleaved terminus of the *Tn5* end specifically bound to protomer 1 occupies the catalytic domain of protomer 2, reflecting the cleavage *in trans* strategy that couples end cleavage to end pairing. PDB 1MUS.
From Steiniger-White *et al.* (2004).

Figure 14.12 Strategies for transposon excision. The double-strand breaks that excise DNA-only elements from the donor DNA backbone (gray) can occur by a variety of combinations of nicking (hydrolysis) reactions, and hairpin formation reactions via attack of a 3′ OH exposed by nicking on a complementary DNA strand. The key result in all cases is the exposure of 3′ OHs at the transposon termini. (a) *hAT* elements excise by nicking one nucleotide beyond the 5′ end of the transposon (P-N-transposon), followed by hairpin formation on the flanking donor DNA by the attack of the donor 3′OH exposed by nicking on its complementary strand. This hairpin is subsequently opened by a nuclease to allow repair of the donor backbone. (b) *mariner/Tc1* elements excise by hydrolysis at the transposon 3′ and 5′ ends. (c) The *Tn7* transposase has two subunits, TnsA and TnsB. TnsA promotes hydrolysis at the 5′ end and TnsB at the 3′ end. (d) *Tn5*, *Tn10*, and *piggyBac* elements excise by hydrolysis at the 3′ transposon end, followed by hairpin formation via the attack of the transposon 3′OH end on its complementary strand. The transposase then opens the hairpin on the transposon end by hydrolysis to expose the transposon 3′ OH end. In the case of *piggyBac*, the transposon always inserts at TTAA and excision occurs via formation of a TTAA hairpin. With *Tn5* and *Tn10*, the 3′ transposon end hairpins directly to the 5′ end of the transposon.

➲ We discuss nucleophilic attack in general terms in Section 3.4.

reaction, forming a DNA hairpin in which the top and bottom strands of a DNA duplex are covalently linked and excising the transposon (Figure 14.12a, d). In contrast to hydrolysis of a phosphodiester bond, such transesterification reactions do not change the number of phosphodiester linkages; rather, there is now a covalent linkage in the hairpin and exposure of another 3′ OH end instead of a phosphodiester linkage between the transposon and the flanking donor DNA. As we see at the end of this chapter, this direct transesterification strategy sharply contrasts with the mechanism of CSSR in which the high energy of DNA phosphodiester bonds is preserved by covalent protein–DNA linkage. If the hairpin forms on the transposon end (Figure 14.12d), transposase opens the hairpin. In all cases, the key outcome is the exposure of the 3′ OH ends of both strands of the transposon.

So what happens once the 3′ OH ends are exposed? The transposase-bound exposed transposon ends engage the target DNA and insertion of the element occurs (Figure 14.10, step 4). The exposed transposon ends are joined to the target DNA by transposase-mediated direct nucleophilic attack of the 3′ OH ends at staggered positions on the top and bottom strands of the target DNA (Figure 14.10, step 5).

The length of this stagger is characteristic of a particular family of transposons. Transposons insert at many sites, although the choice of the insertion site is not entirely random. Instead, the choice of insertion site can be influenced by the DNA sequence that is recognized by the transposase, by non-transposase element-encoded proteins, or by host cell proteins.

The nature of the attack of the 3′ OH ends on the target DNA is very similar to the reaction that occurs during DNA replication or transcription, when the

3′ OH end of the growing nucleic acid strand attacks the high-energy phosphodiester bond of the incoming nucleotide (see Figure 5.6). In both cases, an existing high-energy bond, in the transposon target DNA or within the incoming nucleotide, is converted into a high-energy bond that links the transposon end to the target DNA, or links the incoming nucleotide to another one in the growing chain. It should be noted that there is no covalent linkage between the transposase and the DNA at any step in the cut-and-paste transposition reaction. In other words, formation of a high energy protein-DNA bond does not provide the energy for formation of the phosphodiester that links the transposon end to the target DNA.

As a consequence of the integration reaction, the 3′ ends of the transposon are covalently linked to the target DNA at staggered positions, whereas its 5′ ends are flanked by short gaps that result from the staggered positions of this targeted joining. These gaps are repaired by a host cell polymerase, resulting in the characteristic target-site duplications that flank transposable elements (Figure 14.10, step 5). Thus the regeneration of intact duplex DNA following transposition requires the formation of new phosphodiester bonds by replication and ligation.

Transposases share a common catalytic domain

The release of 3′ OH transposon ends and their direct attack on the target DNA is a highly conserved mechanism, being found in cut-and-paste elements from bacteria and eukaryotes but also in the integration of LTR elements, as we shall see later in the chapter. This mechanistic conservation is reflected in structural similarities in the active site regions of transposases from bacterial elements, eukaryotic DNA-only elements, and the integrases of retroviruses (Figure 14.13).

The central feature of these active sites is the folding of the protein to juxtapose several beta sheet strands and alpha helices in an arrangement that was first identified in the enzyme RNase H; hence, it is called an RNase H fold. Positioned on these beta strands and alpha helices are several acidic amino acids, usually Asp, Asp, and Glu (DDE), to which the essential Mg^{2+} cofactor for transposition reactions is bound. These amino acids can be widely separated in the sequence of the transposase, but become juxtaposed at the active site by the RNase H fold. Moreover, although the amino acid sequences and secondary structures of transposases and other RNase H-related enzymes and polymerases are quite distinct, in all cases Mg^{2+} ions with similar spatial relationships form the active site.

HIV-1 integrase   MuA   Mos1

Figure 14.13 Similarity among transposases and integrases. The structures of the HIV-1 integrase, the bacterial MuA transposase and the eukaryotic Mos1 transposase of the *mariner/ Tc1* family are aligned on their active site regions. The RNase H-like catalytic cores are shown as blue. The conserved acidic amino acids that are juxtaposed in the active site via the RNase H-like fold are shown in red. Hickman *et al.* (2010).

Repair of the gapped donor site resulting from element excision is critical

⊃ We learned about hairpin opening in Section 14.5.

The excision of a cut-and-paste transposon leaves a gap at the donor site that must be repaired. If a hairpin is left on the donor DNA ends upon element excision, an important first step in repair is opening of the hairpin by cellular nucleases. The repair often happens by non-homologous end joining in eukaryotes (Figure 14.14a). Because of the target-site duplication that occurred upon element insertion into what is now the donor site, simple rejoining does not result in the restoration of this site to its original, pre-transposon sequence. Instead, 'footprints' that reflect the element's insertion will persist; the element is said to have undergone 'imprecise excision' (Figure 14.14a, step 2). If transposon insertion occurred within a protein-coding sequence, the proper reading frame and protein function will likely not be restored. 'Precise excision' that restores the donor site to its pre-transposon state may occasionally occur with nucleolytic processing during non-homologous end joining.

⊃ We describe homology-directed DNA repair in more detail in Chapter 13.

Another strategy for repair of the gap at the donor site is to exploit homology-directed DNA repair using a sister chromatid or homolog as templates. As we shall see below, some transposons couple their transposition to DNA replication that may facilitate such repair. If transposon excision occurs and a sister chromatid still containing the transposon is present (Figure 14.14b), the transposon-containing sister chromatid can be used as the template for repair. The consequence of this repair will be to restore a copy of the transposon at the donor site. Note that, in this case, although the transposon moved by a cut-and-paste mechanism, transposition is effectively replicative: a copy of the transposon appears at the new insertion site and persists at the donor site from which excision occurred; it is also present in the sister chromatid.

Alternatively, if repair occurs using as a template a homologous chromosome that lacks a copy of the transposon (Figure 14.14c), the donor chromosome from which the transposon excised will be repaired by copying the sequence from the homolog to the gapped donor site. Such repair effectively restores the donor chromosome to its pre-transposon state and it will appear as if 'precise excision' has occurred.

14.4 MECHANISM OF DNA-ONLY NICK-AND-PASTE TRANSPOSITION REACTIONS

In the previous section, we considered the mechanism of transposition of DNA-only cut-and-paste transposons. Let us now go on to consider a related, but somewhat modified version of the same mechanism, which is exhibited by DNA-only nick-and-paste transposons.

Bacteriophage Mu uses replicative nick-and-paste transposition to replicate its genome

A slight variation of the cut-and-paste mechanism described in the previous section is called the nick-and-paste mechanism, whereby the transposon is actually replicated, resulting in two copies of the transposon. This mechanism has been

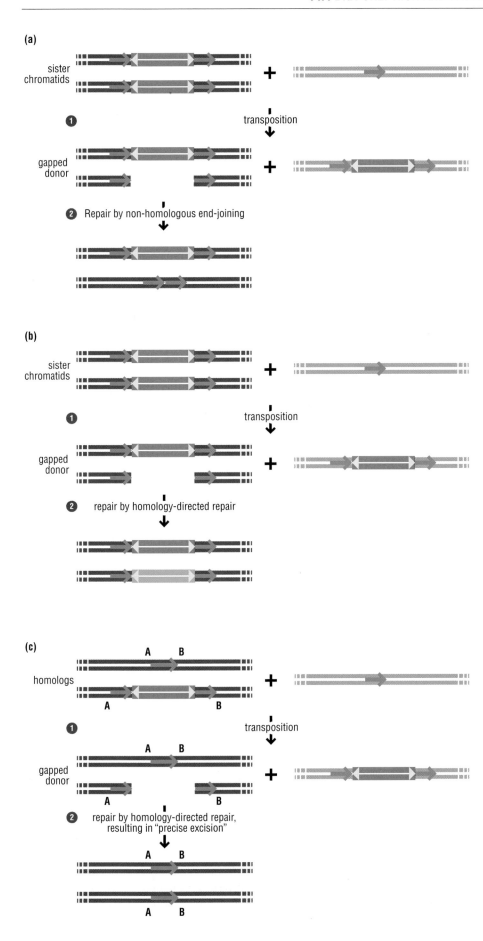

Figure 14.14 Transposition of a DNA cut-and-paste element may or may not change the donor site. Excision of an element from its donor site (gray) and transposition to a new site (light gray) results in a gap that must be repaired. Several different outcomes are possible with any element, depending on the donor substrate and repair pathway used. If transposon excision results in a hairpin on the donor DNA (Figure 14.12a), hairpin opening must occur prior to repair. (a) Rejoining of the gapped donor DNA by non-homologous end joining. Two sister chromatids containing the transposon are present. (1) Excision of the element from one chromatid occurs, with insertion into a target DNA. Because of the target-site duplications (blue arrows) that flanked the transposon in the donor site, simple rejoining by non-homologous end joining, (2) does not restore the donor site to its pre-transposon state with a single copy of the target sequence that was duplicated when the element inserted (not shown). Some nucleolytic processing, however, may occur before joining, resulting in the loss of a few nucleotides. (b) Repair of the donor site by homology-directed repair using a sister chromatid as a template. Two sister chromatids containing the transposon are present. (1) Excision of the element from one chromatid occurs, with insertion into a target DNA. (2) Homology-directed repair of the gapped donor site using the sister chromatid as a template will result in the regeneration of a transposon copy at the donor site (green). Note that, in this case, although transposition occurred by an excision and integration mechanism, there are now three copies of the transposon – one in the target DNA, one formed by homology-directed repair, and one in the sister chromatid from which transposition did not occur. Therefore, transposition is effectively replicative. (c) Repair of the donor site by homology-directed repair using a homolog lacking the transposon as a template. Two homologs, one containing the transposon and one lacking the transposon, are present. The transposon duplicated the target site sequence (blue arrow) upon insertion. (1) Transposon excision and insertion into the target DNA occurs. (2) Homology-directed repair using the homolog as a template results occurs, resulting in the restoration of the donor DNA to its pre-transposon insertion state. This removal of the transposon and one copy of the duplicated target sequence is called "precise excision".

extensively studied for bacteriophage Mu, which can infect *Escherichia coli*. The bacteriophage has a double-stranded DNA genome and uses transposition at several key points in its lifecycle.

Within a virus particle, Mu is a linear DNA flanked at each end by several hundred base pairs of *E. coli* DNA. Upon infection, the Mu genome integrates randomly into the *E. coli* chromosome, using cut-and-paste transposition to excise from the flanking *E. coli* DNA that forms its 'donor site'.

There are two possible outcomes following Mu integration. On the one hand, integrated Mu can remain in a dormant state to form a lysogen. Alternatively, Mu can replicate its DNA for lytic growth by carrying out multiple rounds of replicative transposition, which result in the formation of cointegrates, the mechanism of which we discuss below. In addition to increasing Mu copy number, this replicative transposition destroys the *E. coli* chromosome. Following cointegrate formation, each Mu genome and several hundred base pairs of flanking *E. coli* DNA are packaged into a virus particle and the cell lyses to release the phage particles. By contrast to Mu, other elements, including *IS911*, use an intramolecular nick-and-paste strategy to generate a circularized transposon species that then inserts into a target site.

Nick-and-paste replicative transposition of bacteriophage Mu involves transposon end nicking and joining to the target DNA

Mu replicative transposition involves the introduction of nicks in the donor site at the 3′ ends of the transposon by the MuA transposase (Figure 14.15, step 1), in contrast to the double-strand breaks that excise a transposon from the donor site during cut-and-paste transposition (Figure 14.10). The nick exposes the 3′ OH Mu ends, leaving the 5′ Mu ends still attached to the donor DNA. The 3′ OH Mu ends then attack the target site at staggered positions, mirroring the insertion of an excised transposon into a target DNA (Figure 14.15, step 2). In the resulting fusion product, known as the strand–transfer product, the transposon is covalently linked to the target DNA through its 3′ ends and to the donor DNA at its 5′ ends. These joined molecules can then serve as templates for DNA replication, which initiates at the 3′ OH ends of the single-strand gaps flanking the Mu DNA on the target DNA (Figure 14.15, step 3). Replication then proceeds across the entire element to form the cointegrate, in which the two copies of the transposon link the donor and target DNAs.

Figure 14.15 Mechanism of nick-and-paste replicative transposition to produce a cointegrate. Mu transposition from a donor (gray) plasmid to a target (light gray) plasmid is shown. (1) The MuA transposase initiates transposition by making a nick at the 3′ ends of Mu to expose 3′ OHs, rather than the double-strand breaks that occur in an excision and integration reaction. (2) The Mu 3′ OH ends attack the target DNA at staggered positions (∗), as per cut-and-paste transposition. The transposon is now covalently linked to the donor plasmid DNA (gray) via its uncleaved 5′ ends and to the target plasmid DNA (light gray) via its 3′ ends in a 'strand–transfer product'. The flanking target DNA is gapped because of the staggered positions of target joining, which generate 3′ OHs exposed on this DNA. (3) MuA transposase recruits the host replication machinery to the target 3′ OHs that flank the inserted transposon. The host proteins then fill in the gaps (blue). (4) Filling of the gaps in the strand transfer product and replication of the Mu DNA results in a cointegrate containing both the donor and target DNA linked by two copies of Mu. Target-site duplications (blue) exist at the Mu-target DNA junctions. Note that each Mu contains one 'old' strand from the parent transposon (red) and one newly replicated strand (green).

Thus, although such a nick-and-join transposition reaction does not involve the complete excision of the element from the donor site, the key chemical steps in replicative transposition – exposure of the 3′ OH ends of the transposon and the direct attack of these ends to join the element to the target DNA – are the same as those underlying the transposition of an element that undergoes excision and integration.

Mu transposition proceeds within a series of nucleoprotein complexes

Mu has provided many key insights into the mechanism of transposition and its control. The MuA transposase is converted from an inactive monomer to a tetramer that executes breakage and joining through the multi-step assembly of the transpososome (Figure 14.16). How does this assembly occur? First, monomeric MuA binds to multiple sites at the ends of the Mu genome. Pairing of the ends results in the formation of a MuA tetramer that can execute breakage and joining. Conversion of the MuA monomers to the active tetramer is facilitated by the host DNA-bending proteins HU and IHF, which also bind to the ends of Mu. Conversion is also facilitated by supercoiling, which also likely induces DNA bending. Another Mu-coded protein, MuB, which can also interact with MuA and target DNA, is involved in recruiting the target DNA to the MuA-bound ends (Figure 14.16, step 2).

As discussed in Figure 14.11, Mu transposase executes end cleavage *in trans*: the MuA subunit that cleaves one end of the Mu sequence is bound specifically to the other end of the Mu DNA, as in Figure 14.11. This *trans* assembly strategy ensures that active enzymes can only cleave Mu when the DNA ends are brought

Figure 14.16 Mu transpososomes. Mu transposition proceeds via a series of protein–DNA complexes called transpososomes. The supercoiled donor DNA containing Mu (red) with its terminal TIRs (yellow) is shown. (1) MuA transposase (orange) binds specifically to multiple sites in the transposon ends. In the presence of the DNA-bending proteins HU, IHF, and the Mg^{2+} cofactor (not shown), a stable tetramer of MuA is formed that pairs the ends. MuB (dark blue) binds to the target DNA (light gray). (2) Nicking at each end of Mu exposes the 3′ OH ends of the element and relaxes the flanking donor DNA (blue). (3) Interaction of MuA with MuB juxtaposes the target DNA to the transposon ends and activates the attack of the 3′ OH transposon ends on the target DNA. In the strand transfer product, Mu remains covalently linked to the donor DNA through its 5′ ends and is covalently linked to the target DNA though its 3′ ends.

together in the transpososome. The interaction of MuB with MuA is also important for MuA strand nicking activity and joining of the 3′ Mu ends to the target DNA (Figure 14.16, step 3).

Following the joining of the transposon ends to the target DNA, DNA replication must occur to repair the single-strand gaps that flank newly inserted transposon, and to copy the transposon, generating the cointegrate. Host chaperone proteins convert the highly stable MuA tetramer to a form that recruits the host replication machinery, which mediates this DNA synthesis.

14.5 CELLULAR DOMESTICATION OF A DNA CUT-AND-PASTE TRANSPOSASE IN ADAPTIVE IMMUNITY

The transposition of DNA-only cut-and-paste elements is not just a nuisance event that must be tolerated by the cell, but without any real advantage. Instead, certain cell types have evolved to exploit such transposition reactions for their own benefit. In this section, we consider a particularly important example of this exploitation, the use of transposition systems in the generation of immune system diversity.

Diverse transposition systems have been adopted to mediate a variety of host processes

Transposition proteins have been co-opted to perform a variety of cellular functions. There are, for example, multiple examples of transcription factors derived from DNA cut-and-paste transposases. Perhaps the most spectacular example of transposase domestication is the use of a DNA cut-and-paste transposon derivative in the immune system to mediate the assembly of immunoglobulin superfamily proteins, the antibodies (Figure 14.17) and antigen receptors produced by lymphocytes. This assembly mechanism allows for the generation of a vast diversity of proteins that are able to recognize any disease-causing microorganism.

The generation of diverse antibodies and antigen receptors stems from the way that the genes for these proteins are assembled in the individual immature lymphocytes from multiple gene segments, with each immature lymphocyte assembling a slightly different combination of gene segments. Consequently, each mature lymphocyte synthesizes a slightly different version of the protein.

Gene assembly occurs through recombination reactions that delete the DNA between gene segments and juxtapose the coding regions that remain for subsequent rejoining. The generation of the DNA double-strand breaks that initiate this recombination is very similar in mechanism to the excision process of cut-and-paste DNA transposons. Indeed, the lymphocyte-specific recombinases that mediate this reaction are structurally related to transposases. The gene segments are then joined by the NHEJ system. This important immune-system recombination system probably evolved from a transposable element that was harnessed by the ancestors of vertebrates for this vital role.

Figure 14.17 Schematic representation of an antibody molecule. Antibodies (immunoglobulins) are composed of two identical larger (heavy) chains and two identical smaller (light) chains, attached to each other by disulfide bonds (not shown). Each chain has a variable region (pink, light pink) that is formed by the assembly of different gene segments, and a constant region (gray, light gray). The variable regions form the antigen-binding site that recognizes foreign molecules.

Figure 14.18 Assembly of immunoglobulin gene segments. For simplicity, only the segments encoding a heavy chain are shown. (1) The heavy chain is encoded in four separate gene segments. In the human heavy chain gene locus schematically represented here, there are 65 V segments (V_H1–V_H65), 27 D segments (D_H1–D_H27), and six J segments (J_H1–J_H6) shown in different shades of red along with a constant-region segment shown in gray. (2) In the course of the differentiation of the lymphocytes that make antibodies, one each of the V, D, and J segments become joined by the host NHEJ system to encode a complete variable-region coding domain, here $V_H2D_H2J_H4$. (3) After this DNA rearrangement, the complete gene can be transcribed and the variable region is joined to the constant region by RNA splicing.

Antibody and antigen-receptor genes are formed by joining gene segments together

The antibodies produced by B lymphocytes and the antigen receptors of T lymphocytes feature N-terminal regions called variable regions. These variable regions are encoded by several different loci that are each composed of multiple alternative protein-coding gene segments (Figure 14.18). Both these proteins belong to the larger immunoglobulin superfamily, but they are the only ones whose genes undergo this programmed somatic recombination.

The assembly process is known as **V(D)J recombination** after the names of the gene segments that are joined together, the V (variable), D (diversity) and J (joining) segments. Each locus can include hundreds of different gene segments and spans hundreds of kilobases. Random selection of a gene segment of each type, and deletion of the DNA between them, generates novel combinations of gene segments in each B cell or T cell precursor. The gene segments that are assembled by recombination lie upstream of a C-terminal invariant region, called the constant region.

The immune system recombinase RAG resembles a transposase

Antibody and T cell receptor gene segments are flanked by recombination signal sequences (RSSs), which are inverted repeats that contain conserved heptamer (7 bp) and nonamer (9 bp) segments (Figure 14.19). Thus, they are equivalent to the TIRS of transposable elements. There are two kinds of RSSs, one in which

Figure 14.19 Recombination signal sequences (RSSs). The recombining segments of antibody genes are flanked by specialized recognition sites for the RAG recombinase that executes double-strand breaks at the edges of the gene segments. Each recombination signal sequence (RSS) comprises a seven-nucleotide (heptamer) sequence and a nine-nucleotide (nonamer) sequence, separated by either 23 (green RSS) or 12 (blue RSS) bp. Each type exists in a unique configuration with respect to the coding segments. Recombination requires one RSS with a 12 bp spacer and one RSS with a 23 bp spacer.

the conserved heptamer and nonamer are separated by 23 bp (left) and the other in which the heptamer and nonamer are separated by 12 bp (right); recombination requires one RSS of each type. RSSs are recognized and acted on by the lymphocyte-specific recombinase RAG (Figure 14.20). Thus the RSSs are functionally comparable to the terminal inverted repeats of cut-and-paste transposons whereas RAG is comparable with the transposase.

RAG is composed of two subunits, RAG1 and RAG2; RAG1 is the subunit that carries out the breakage and joining steps at the RSSs and is structurally similar to cut-and-paste transposases. In particular, it appears to have an RNase H fold on which essential conserved acidic amino acids are present. The RAG complex binds specifically to the RSSs (Figure 14.20, step 2) and promotes double-strand breaks that excise the RSS-bounded segment, mirroring the way a transposase recognizes and cleaves the ends of an integrated transposon, excising it from the DNA. This reaction exposes the ends of two gene-coding segments.

The detailed mechanism of double-strand break formation by RAG is mechanistically related to the excision of *hAT* transposons, described in Figure 14.12. RAG introduces a single-strand nick at the junction of the 5′ end of the RSS with the 3′ end of the adjacent coding gene segment (Figure 14.20, step 2). The exposed 3′ OH then intramolecularly attacks the un-nicked strand to covalently link the two complementary strands together to form a hairpin at the end of the coding segment, and leaves a blunt end on the RSS segment, which is now no longer attached to the gene segments it previously sat between (Figure 14.20, step 3).

The similarity of the RAG double-strand break mechanism to that of *hAT* transposons, and the structural similarity of RAG to both the *hAT* family transposases, which are widely present in eukaryotes, and the *Transib* family transposases, which appear in inactive form mostly in insects, suggests that the RAG system evolved from a transposon that invaded a common ancestor of the organisms which today display adaptive immunity.

Gene segments are joined together by the non-homologous end joining DNA repair system

The DNA hairpins on the antibody and T cell receptor gene coding segments exposed by the RAG-promoted double-strand breaks are cleaved by the nuclease Artemis and then joined together to form a functional protein-coding DNA by reactions that involve the universal non-homologous end joining system of DNA repair

Figure 14.20 Breakage and joining by RAG recombinase. The DNA substrate for the joining of a V_H segment to D_H segment is shown. (1) An oligomer of RAG (pale blue) binds to the RSS-23 (green) flanking a coding segment; synapsis then occurs by capture of the RSS-12 (blue) of another coding segment in a reaction facilitated by a DNA-bending protein such as HMG1. (2) RAG then introduces a nick at the 5′ end of the RSS that abuts each coding segment. (3) The exposed 3′ OH ends of the coding segments then attack their complementary strands, forming hairpins at the end of each coding segment; the signal ends containing the RSSs are blunt ended. (4) Joining of the two coding segments (V_H and D_H) forms a coding joint within the chromosome, and joining of the excised RSS-bounded segment creates a circular DNA containing the signal joint. Joining is mediated by RAG in collaboration with universal DNA repair proteins that carry out non-homologous end joining of double-strand breaks. These include Ku, composed of the Ku70 and Ku 80 subunits, which holds the broken ends near to each other and recruits the kinase DNA-PKcs. DNA-PKcs phosphorylates and activates the nuclease Artemis, which then opens the hairpins. Other targets of DNA-PK phosphorylation are not yet understood. Other proteins involved in forming signal joints and coding joints are DNA ligase IV and its partner XRCC4. XLF/Cerunos is involved in coding joint formation.

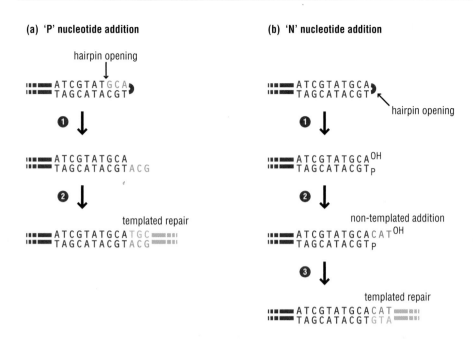

(a) 'P' nucleotide addition

hairpin opening

ATCGTATGCA
TAGCATACGT

❶ ↓

ATCGTATGCA
TAGCATACGTACG

❷ ↓

templated repair

ATCGTATGCATGC
TAGCATACGTACG

(b) 'N' nucleotide addition

ATCGTATGCA
TAGCATACGT

hairpin opening

❶ ↓

ATCGTATGCA OH
TAGCATACGT P

❷ ↓

non-templated addition

ATCGTATGCACAT OH
TAGCATACGT P

❸ ↓

templated repair

ATCGTATGCACAT
TAGCATACGTGTA

Figure 14.21 Diversification of coding joints. Nucleotide addition at the termini of coding segments prior to joining leads to genetic diversity at the coding joints formed during gene assembly by V(D)J recombination. (a) 'P' nucleotide addition. The hairpin end of a coding segment is shown. (1) Instead of opening at the tip of the hairpin by cleavage with Artemis, cleavage may occur at an internal position in the hairpin end. Unfolding of the hairpin creates a single-strand extension (orange). (2) Repair (green) of the extended strand (orange) prior to joining to the next segment represented by green bars adds additional coding capacity to the joint. (b) 'N' nucleotide addition. The hairpin end of a coding segment is shown. (1) Hairpin opening at the end of the hairpin makes a flush-end DNA duplex. (2) Non-templated addition of nucleotides (pink) extends the coding segment. (3) Repair of the gap after segment joining (green) adds diversity to the coding joint.

(see Section 13.2; Figure 14.20, step 4). The joint formed between two coding segments is called a **coding joint**, while the joint formed by the exposed ends of the RSS segment is known as the **signal joint**.

A key step in the formation of the coding joint is the opening of the hairpins at the ends of the coding segments. Opening of the hairpin at its tip generates a blunt end. However, the hairpin is often opened at a position other than its tip, resulting in additional diversity being created in the sequence of the coding joint. Filling-in of the resulting overhangs adds additional nucleotides, called P nucleotides, to the joint (Figure 14.21a). Other nucleotides may also be added directly to the exposed ends of coding segments by the enzyme deoxynucleotidyl terminal transferase. These are called non-templated or N nucleotides (Figure 14.21b). Both blunt and filled ends can also be altered by the addition or deletion of nucleotides before the joint is finally sealed by the XRCC4/DNA ligase IV complex whose activity is modulated by XLF/Cernunnos.

The proteins involved in the repair process include Ku, a protein that binds to DNA ends, DNA-PKcs, a kinase that is apparently recruited to the ends through Ku, the nuclease Artemis that opens DNA hairpins, and a DNA ligase, among other proteins. Mutant organisms lacking functions conferred by these proteins cannot repair other double-strand breaks in DNA, such as those generated by ionizing irradiation, and are thus both immunodeficient and radiation sensitive.

When the protein-coding DNA formed by the coding joint is fully assembled, transcription produces an RNA containing the assembled gene segments and the coding joints between them. RNA splicing then generates an mRNA, which is translated into a functional antibody chain or T cell receptor chain.

The ends of the excised RSS-bounded fragment are joined to form a circular DNA containing the signal joint. This DNA is lost from the cell because it lacks the structures, such as a centromere, that would enable its distribution into daughter cells.

The action of RAG is tightly regulated

For the immune system to function properly, the sequence-specific recombination reactions promoted by RAG must occur only in the right genes in the right cell types at the appropriate developmental stage. One level of control ensures that RAG is only expressed in the precursors of B cells and T cells. Another ensures that recombination at RSS sequences only occurs in the presence of low-level transcription, which itself occurs only in lymphocytes at particular developmental stages. It is likely that this low-level transcription alters chromatin structure in the region of the RSSs, making them accessible to RAG.

Once an intact immunoglobulin protein is generated, an intracellular signaling system halts recombination. This ensures that RAG recombination results in the production of only one type of antibody or T cell receptor protein in each individual lymphocyte. Another signaling system stimulates the growth of B and T cells whose antibodies and antigen receptors interact with an antigen.

V(D)J recombination can be dangerous

Despite the strict controls on RAG action, it can sometimes be misdirected. A significant fraction of lymphoid cancers appears to result from rearrangements of chromosomal DNA that lead to the aberrant and unregulated expression of tumor suppressor or oncogenic proteins, for example, by the fusing of two different genes to encode a novel protein that displays oncogenic activity. These rearrangements are thought to be associated with aberrant RAG-mediated somatic recombination.

14.6 RETROTRANSPOSONS

Having considered the transposition of elements that use only DNA intermediates, we now consider the transposition of elements that have RNA intermediates. Multiple types of transposable elements that move via an RNA intermediate have been found in both bacteria and eukaryotes.

Retrotransposons are numerous and widespread

We note in Section 14.1 how there are three main classes of transposons – the DNA-only transposons, the LTR elements and the non-LTR elements. Two of these classes, LTR elements and non-LTR elements, involve RNA intermediates and are called retrotransposons or retroelements. LTR elements include the retroviruses, which have an extracellular phase and are found only in vertebrates, and retroviral-like elements, the movements of which are strictly intracellular, and which are widespread in fungi, plants and animals. The DNA versions of these elements are called proviruses when integrated into the host genome. The two types of LTR element show distinct similarities that we will discuss below; it is generally thought that retroviruses evolved from the retroviral-like elements by acquiring the ability to exit and then re-enter cells. Non-LTR elements are present in all domains of life.

A notable aspect of retrotransposons is their success in colonizing many eukaryotic genomes. For example, more than 68% of the maize genome is composed of LTR retrotransposons. In some plant genomes, genes have been described as 'small islands surrounded by seas of LTR-retrotransposons'. Non-LTR elements are particularly abundant in vertebrates genomes.

The transposition of both LTR and non-LTR elements uses a reverse transcriptase, which is usually encoded by the mobile element itself. In the case of LTR elements, a DNA copy of the element is generated from an RNA element copy packaged in a nucleoprotein complex containing both the element RNA and reverse transcriptase. This DNA intermediate is then integrated into the host genome by an element-encoded integrase/transposase by the same chemical mechanism described above for DNA-only elements. In the case of non-LTR elements, a DNA copy of the element is generated *in situ* at the insertion site by reverse transcription of an RNA copy of the element.

> ⊝ We discuss the structure of reverse transcriptase and its role in chromosome replication in Chapter 6.

Reverse transcriptase mediates the transposition of LTR and non-LTR elements

Central to the transposition of both LTR and non-LTR retrotransposons is reverse transcriptase, the DNA polymerase that can synthesize DNA using RNA as a template (Figure 14.22). An RNase H activity, which can remove the RNA template strand to facilitate synthesis of the second DNA strand, is also encoded by LTR elements. There is no obvious RNase H domain in non-LTR elements; instead, they are thought to use a host enzyme to remove the RNA strand.

Reverse transcriptases have two distinctive features: they are more error-prone and less processive than replicative polymerases. Both of these properties are reflected in retrotransposon structure and function. The error-prone nature of reverse transcription means that mutations occur at a higher frequency in retro-elements than in the course of chromosomal replication. This property is reflected by the rapid rate at which resistance to antiviral drugs can occur during treatment against the retrovirus commonly known as human immunodeficiency virus (HIV). Another impact of the nature of reverse transcriptase is the truncation of some non-LTR elements during their replication because of the lower processivity of reverse transcriptase. Thus, these truncated elements lack their 5′ end promoters and are not transcribed, and therefore cannot transpose.

14.7 LTR RETROTRANSPOSONS

Having now considered the nature of retrotransposons in overview, let us now turn our attention to the transposition mechanism of LTR elements. We will then discuss the non-LTR elements in the next section.

LTR elements transpose via an RNA intermediate

We note above how LTR elements include both retroviruses, which have an extracellular phase, and retrovirus-like elements, which are purely intracellular. The structures of LTR retroviral-like elements and retroviruses are highly similar, except for the presence of an ENV gene in the retroviruses, which is essential to the extracellular stage in their lifecycle (Figure 14.23).

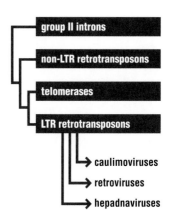

Figure 14.22 Groups of reverse transcriptases. All these elements encode a reverse transcriptase. The three arrows emerging from the LTR retrotransposons represent the independent origins of hepadnaviruses (mammalian viruses that can cause liver infections including hepatitis B in humans), retroviruses and caulimoviruses (plant viruses).

Figure 14.23 The genetic structure of LTR elements. The terminal yellow arrows represent the long terminal repeats (LTRs). *Ty1/copia* and *Ty3/gypsy* are families of LTR retrotransposons of which the prototypes are the yeast *Ty1* and *Ty3* elements and the *Drosophila copia* and *gypsy* elements. Note the different orders of their IN (Integrase) and RT-RH (Reverse transcriptase and RNase H) domains. LTR retrotransposons carry the genes for four proteins required for their transposition: GAG, PRO (PR), RT-RNase H (RT-RH), and integrase (IN) (see text). All retroviruses encode an additional protein, ENV, which enables them to exit cells and infect new ones. MuLV is representative of a minimal retroviral genome. RSV carries an activated SRC gene that results in cancers in its chicken host. HIV includes a number of additional proteins that promote virus multiplication.

The transposition of an LTR element begins with transcription by the host-cell RNA polymerase of an integrated form of the LTR element called the **provirus** (Figure 14.24, step 2). These transcripts are exported from the nucleus (step 3) and initially serve as mRNAs for the synthesis of the element-encoded proteins that are described in detail below (step 4). The LTR element RNAs and proteins assemble into a large complex called a virus-like particle (step 5) in which the RNA is converted to the double-stranded DNA form of the element by reverse transcription (Figure 14.25).

Figure 14.24 LTR element lifecycles. The lifecycles of two LTR elements are compared here: the *S. cerevisiae* LTR retrotransposon *Ty1* (left) and the mammalian retrovirus murine leukemia virus (MuLV) (right). (1) Newly synthesized DNA (green) forms of the elements integrate into the host genome in reactions mediated by integrase (orange balls). (2) Transcription of the element by host RNA pol occurs. (3) The form of the element RNA (blue) is exported to the cytoplasm. (4) Translation of the RNA, and proteolytic processing of the resulting proteins, results in the production of the proteins GAG (red), PRO (dark blue), the integrase IN (orange) and the reverse transcriptase RT-RNaseH (turquoise) from this RNA. These proteins assemble with RNA copies of the element (blue lines) to form a virus-like particle. (5) In the case of the retrovirus, which encodes the ENV protein (gray squares), ENV enters the cell membrane and forms a lipid–protein coat around the RNA virus particle. (6) The retrovirus buds off from the cell. (7) The particle enters a new cell and is uncoated. (8) For both the retrovirus and *Ty1*, the RNAs are reverse-transcribed into cDNA in the cytoplasm, within the particle. (9) The LTR element cDNA, assembled with proteins including Integrase (orange) enters the nuclease and integrates into the cell's genome (1).

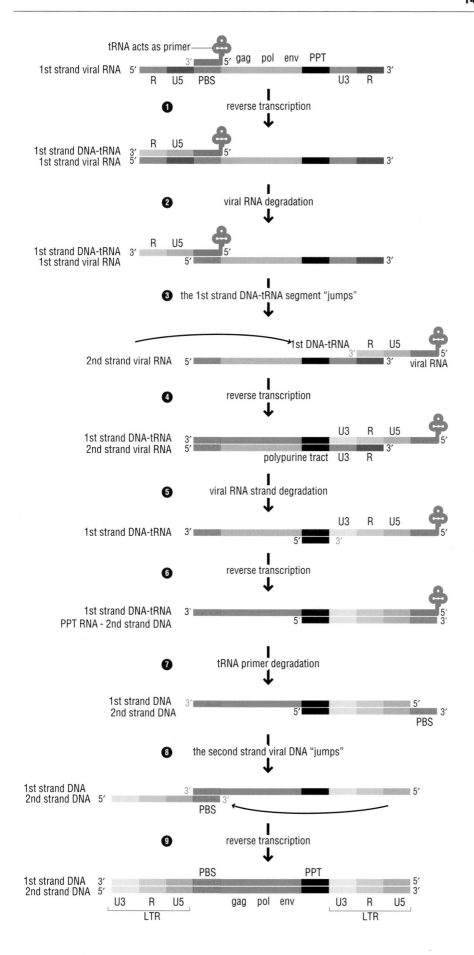

Figure 14.25 Synthesis of long terminal repeat (LTR) element DNA from element RNAs. Host transcription of the integrated provirus generates viral RNA (blue) in which the protein-coding region (GAG, POL, ENV, (red)) is flanked by R, U5, U3 and R segments that play a key role in generating the U3-R-U5 LTRs that bound the DNA form of the element. PBS = Primer Binding Site; PPT = Polypurine Tract. Multiple reverse transcription, RNA degradation and template jumping steps are involved in element DNA production. A primer for the first reverse transcription step is provided by the 3′ end of a tRNA (blue) that binds specifically to the primer binding site (PBS) of the viral RNA. (1) Reverse transcription from the 3′ end of the tRNA synthesizes DNA (green) that extends to the 5′ end of the viral RNA to form the 1st strand DNA-tRNA. (2) The viral RNA of the DNA-RNA hybrid generated above is degraded by the RNaseH domain of RT, exposing the R-U5 DNA of the 1st strand DNA-tRNA segment. (3) The 1st strand DNA-tRNA segment 'jumps' to the other end of the second viral RNA by pairing of the DNA R segment and the RNA R segment. (4) Reverse transcription using the 3′ end of the DNA R segment as a primer extends to the end of the viral RNA. (5) The viral RNA strand is degraded except for the polypurine tract adjacent (PPT) to the U3 region. (6) Reverse transcription using the 3′ end of the polypurine tract as a primer begins the synthesis of the second strand of the viral DNA. (7) The tRNA primer is degraded, exposing the primer binding site in the 3′ end of the second strand of the DNA. (8) The second strand of the viral DNA containing the U3-R-U5-Primer Binding site 'jumps' to the other end of the first DNA strand. (9) Reverse transcription from the 3′ ends of both PBSs completes the synthesis of the viral DNA that is bounded by the U3-R-U5-containing LTRs.

Figure 14.26 Integration mechanism of LTR elements. The proviral DNA of an LTR element in the donor site is shown. (1) The element is transcribed by host RNA polymerase to generate the viral RNA (blue) form of the element. (2) This viral RNA is converted to the double-stranded DNA (green) form of the element by reverse transcription. (3) The 3' OH ends of the viral DNA attack the target DNA at staggered positions just as occurs during target joining of DNA-only elements (Figure 14.10). The newly inserted element is flanked by short gaps. (4) Repair (green) of the gaps flanking the newly inserted element generates the target-site duplications (blue arrow).

The **LTRs** are direct repeats several hundred base pairs long that each contain the U3, R, and U5 segments that play key roles in reverse transcription as sites for the initiation of DNA synthesis. Furthermore, their external 3' termini are the sites at which joining to the target DNA occurs. The DNA form of the virus, which contains two LTRs, is generated through an elaborate series of reverse transcription and RNA degradation steps. Notice how the ends of the LTRs results from the 'jumping' of DNAs to a second viral template strand (steps 3 and 8), underscoring the importance of the packaging of two viral RNAs in each viral-like particle.

Once synthesized, the DNA form of the LTR element is integrated into the host genome by the element-encoded integrase (Figure 14.24, step 1). Integration of the LTR element DNA occurs by the same mechanism as that used by the DNA-only elements (Figure 14.26, see Section 14.3). The 3' OH termini of the LTR element DNA directly attack the target DNA at staggered positions, giving rise to target-site duplications (step 3). For some LTR elements, the actual 3' end of their DNA form joins to the target DNA. For some other elements, endonucleolytic cleavage by the integrase several nucleotides from the 3' ends of the element DNA generates the 3' ends that will actually join to the target DNA.

Like the DNA-only transposons, LTR elements insert at many different sites and can cause insertion mutations or an unwanted activation of adjacent host genes by gene regulatory signals, such as promoters, enhancers and splice sites, which are present in the element.

The LTR transposon lifecycle requires four element-encoded proteins

The lifecycle of an LTR transposon is more complex than that of a DNA-only transposon; this is reflected in the number and type of element-encoded proteins that are required for transposition. An LTR element encodes four proteins: GAG, PRO (protease), RT-RNase H (reverse transcriptase-RNase H) and IN (integrase) (Figure 14.23). These proteins are synthesized by the translation of an mRNA, which is itself transcribed from the integrated element by the transcription machinery of the host cell. Most of the proteins are often first synthesized in the form of a long polyprotein, which is then cleaved to give the individual functional proteins (see Section 11.3).

GAG is itself a polyprotein, which is cleaved after synthesis to give several structural proteins that bind DNA and RNA. PRO is a protease that processes GAG and other polyproteins. RT-RNase H is an enzyme with both reverse transcriptase (RT) and RNase activity; it generates the DNA copy of the element from an RNA copy.

IN is the integrase (transposase) that joins the ends of the retroviral DNA into the new insertion site. IN binds to the extreme ends of the LTR element DNA, which in many LTR elements are composed of very short inverted repeats, like the termini of cut-and-paste transposons (see Figure 14.4). As one would expect from the chemical identity of the LTR element and DNA-only element integration mechanisms (Figure 14.26), the catalytic core of IN is closely related to that of a DNA-only element transposase (see Figure 14.10). Like transposases, the integrase active site contains an RNase H domain that juxtaposes key acidic residues that bind an essential Mg^{2+} cofactor.

LTR element transposition does not involve excision

A very significant difference between the movement of DNA-only cut-and-paste elements and LTR elements (as well as non-LTR elements) is that transposition does not involve the excision of the element from the donor site. Thus, issues of donor site repair (see Figure 14.14) are not part of the LTR element lifecycle. However, changes to the LTR element can occur by homologous recombination between the several hundred base pair long LTRs, deleting the information between them and leaving a solo LTR end in the genome (Figure 14.27).

The solo LTR elements are not mobile but rather serve as markers of the prior presence of an LTR element.

In *Saccharomyces cerevisiae* for example, which contains five families of *Ty* elements, there are about 60 copies of the ~ 6 kb intact elements and about 280 copies of the ~330 bp solo LTR elements, making up about 3% of the genome. The exact number of intact elements and solo LTRs vary considerably between yeast strains. Both LTR elements and solo LTRs are also present in other organisms.

Retroviruses are LTR retrotransposons with an extracellular phase

Retroviruses are LTR retrotransposons that can exit from one cell as an RNA virus, infect another, make a DNA copy of their RNA genome, and integrate it into the DNA of that cell by the transposition mechanism outlined above. Well-studied retroviruses include the murine leukemia virus (MuLV) and HIV (see Figure 14.24).

The feature that makes retroviruses infectious, in contrast to LTR retroviral-like transposons, is that they encode an additional protein called ENV (envelope). This is a glycoprotein that enters the cell membrane and forms a lipid–protein envelope around the retrovirus nucleoprotein core, which enables the particle to be released

⊖ We discuss homologous recombination between repeated sequences in Section 13.7.

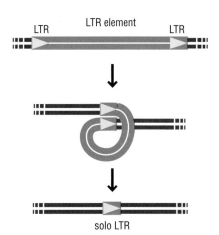

Figure 14.27 Homologous recombination can produce a 'solo LTR' from an LTR element. The proviral DNA of an LTR element with its several hundred bp LTRs is shown. Homologous recombination between the LTRs 'excises' the element, leaving a single LTR copy.

from the cell. Interaction of ENV with specific proteins on cell surfaces also enables the virus particles to infect new host cells. The lifecycles of the retrovirus MuLV and the retroviral-like LTR element *Ty1* of yeast are compared in Figure 14.24. The human genome lacks endogenous active retroviruses; HIV is acquired by infection. However the human genome does contain multiple inactive retroviruses.

The infectious retrovirus particle is called a virion and contains several copies of the RNA genome, and also the RT-RNase H reverse transcriptase. After entry of a virion into a cell and release of the viral RNA from its protein coat, the viral genome is converted into DNA and integrated into the host genome, becoming a permanent component of the genome. This means, for example, that the descendants of cells infected by HIV will contain an HIV provirus that cannot be removed. Retroviruses do not usually actively destroy the cells they infect. Instead, they can both remain in the host genome and continue to generate infectious virus by transcription of the provirus and synthesis of viral proteins. Inhibitors of both the HIV protease and the reverse transcriptase are already used as drugs to block viral multiplication, and inhibitors of integrase that can block integration have been identified, which may also have therapeutic value. In Experimental approach 14.2 we consider how host proteins involved in HIV integration have been identified.

14.8 NON-LTR RETROTRANSPOSONS

Non-LTR retrotransposons are the most common transposable elements in the human genome

Although DNA-only cut-and-paste elements and LTR-elements are widespread, it is **non-LTR transposons** that make up a large proportion of the DNA of many eukaryotic genomes. Strikingly, these elements comprise about 34% of the human genome. While non-LTR elements are particularly prominent in mammalian genomes, they are also present in other organisms, including fungi, *Caenorhabditis elegans*, insects, reptiles, birds, and plants. A distinct group of non-LTR retrotransposons are the **mobile group II introns**, which are found in some bacteria and in mitochondria and chloroplasts of fungi, protists, and plants.

The processes of both cut-and-paste DNA-only and LTR element transposition involve the production of a 'free' DNA form that subsequently engages the target site. By contrast, when a non-LTR element transposes, an RNA copy of the element associates with the target site and is then used as a template for reverse transcription *in situ* at the site of insertion as we discuss below.

> ● We learn more about the mechanism of DNA synthesis in Section 6.1.

LINE elements are autonomous non-LTR retrotransposons

Several groups of non-LTR elements have been identified (Figure 14.28). One type of element found in mammalian genomes is the **LINE** element. These elements form about 21% of the human genome. Intact LINE elements are autonomous retrotransposons about 6 kb in length and encode several transposition proteins (Figure 14.28a): ORF1 encodes an RNA-binding protein that forms a particle with the LINE RNA and is a nucleic acid chaperone; ORF2 encodes a protein with endonuclease and reverse transcriptase activities. These elements also have a 3′ poly(A) tail. Beyond mediating their own movement, these autonomous LINE

Figure 14.28 The structures of non-LTR elements. (a) LINEs (long interspersed elements) and SINEs (short interspersed elements) are transposable elements that move via target-primed reverse transcription. An intact LINE (about 6 kb) is an autonomous element that encodes its own transposition proteins in two open reading frames, ORF1 and ORF2. ORF1 is an RNA chaperone and ORF2 is an endonuclease-reverse transcriptase. Transcription initiates at P (green). SINEs are much shorter than LINES (at 100–300 bp), they do not encode any proteins, and are thought to translocate using LINE-encoded proteins. They do encode A- and B-boxes for RNA pol III. The 3′ ends of both elements consist of poly(A) sequences of variable length. The promoters of both are copied during transcription and subsequent reverse transcription, ensuring that the translocated copy retains the capacity for further rounds of transposition. (b) R2 elements encode a single polypeptide that contains a reverse transcriptase domain and an endonuclease domain. These elements insert virtually exclusively into the 28S rRNA gene. They lack obvious transcription signals in their 5′ untranslated region and may be co-transcribed and translated from the 28S gene. They are prevalent in arthropods and have been observed in some other animal species.

(a)

LINE element

SINE element

(b)

R2 element

elements can also supply transposition proteins to, and promote the movement of, the non-autonomous non-LTR SINE elements (which we discuss below).

Most LINE elements present in the genome are not active, that is, they cannot undergo transposition. Why is this? Most LINE elements are not full-length, but are truncated at their 5′ ends. This truncation reflects the way that these elements are generated by the action of reverse transcriptase, which initiates DNA synthesis at the 3′ end of the LINE element. Unlike the DNA polymerases that carry out chromosomal DNA replication, reverse transcriptase is not highly processive and the enzyme often fails to complete synthesis of the 5′ end of the element. This results in the presence in the genome of truncated copies, which are not transcribed because the promoter located at their 5′ ends has been lost. Taking this one step further, if these 5′ elements are not transcribed, they cannot transpose.

Reverse transcription is also error prone, which means that the LINE genes encoding the transposition machinery are likely to become mutated when they are copied into cDNA. Because of these replication errors, only about 80–100 of the existing 8.7×10^5 human genomic LINEs are intact and still capable of transposition. There are, however, multiple examples where gene disruption by a LINE has caused human disease (Figure 14.2).

Another type of non-LTR element, which is typified by the R2 elements of insects, has only one open reading frame (ORF) (Figure 14.28b). The reverse transcriptase of these elements lies in the middle of the ORF while the C-terminus contains the endonuclease that cuts the target DNA. This endonuclease is highly site-specific and insertion of these elements occurs only into ribosomal RNA genes.

So how does insertion of non-LTR elements occur?

Non-LTR elements insert by reverse transcription at the insertion site

Translocation of a LINE non-LTR retrotransposon begins with transcription of the element from an internal promoter for RNA polymerase II, the addition of a poly(A) tail to the 3′ end of the transcript, and export of the resulting mRNA to the cytoplasm (Figure 14.29, step 1). The LINE ORF1 and ORF2 proteins are synthesized from this mRNA, and preferentially associate with it; the resulting ribonucleoprotein complex then re-enters the nucleus. Nicking of the bottom strand of the target-site DNA by the ORF2 element-encoded endonuclease, which preferentially

14.2 EXPERIMENTAL APPROACH
Hunting for transposition host factors

Transposable elements depend on host-encoded processes and proteins to execute many stages in their lifecycle. These host factors can be identified by the isolation of host mutants that decrease or increase the frequency of the transposition of the element – experiments that, as we shall see here, are facilitated by visual assays for transposition.

To look for *E. coli* host factors that affect transposition of *IS903*, Keith Derbyshire and colleagues first created libraries of *E. coli* mutants containing 2000–5000 different *Tn5* insertions. The *Tn5* transposon inserts into many different sites (i.e. with little sequence specificity) so it was likely that many of the non-essential genes of the approximately 4600 genes in *E. coli* were mutated in the collection of cells represented in the libraries. Because *Tn5* makes mutations by insertion, each mutant gene is marked and the identity

of the mutated gene can easily be determined. Derbyshire and colleagues next transformed the entire library of mutants with a plasmid encoding a specialized mini-*IS903* element and the transposase, and followed the transposition phenotype of thousands of transformants. The mini-*IS903* element contained the terminal inverted repeats (TRs) necessary for transposition flanking a cryptic version of the *lacZ* gene that was not expressed because it did not contain a promoter or translational start site.

Utilization of this *lacZ* gene fragment allowed Derbyshire and colleagues to perform a 'promoter capture' assay: if the *lacZ* fragment is transposed into a gene such that an in-frame translational fusion occurs between the *lacZ* gene fragment and a host gene, *lacZ* will be expressed (and the cells will turn blue in the presence of the indicator dye) (Figure 1a). The frequency of transposition events

Figure 1 Assay for the isolation of *E. coli* mutants with altered *IS903* transposition. (a) The overall scheme of the assay was to look at the expression of a reporter open reading frame (ORF) inside a transposon upon translational fusion to an ORF at a new insertion site. A donor plasmid contained the transposase gene and a transposon containing a promoterless unexpressed *lacZ* reporter whose ORF extends to the end of the transposon terminal inverted repeat (red with yellow triangles). The target gene *X*, with an upstream promoter, is in the chromosome (X can be any non-essential gene). The host is *lacZ⁻*. Transposon insertion into gene *X* results in an in-frame fusion of the transposon *lacZ* reporter with the N-terminus of gene *X*. The X-β-galactosidase fusion protein is synthesized. In the presence of X-gal, a chromogenic substrate for LacZ, in the growth media, transposition is detected as blue papillae (microcolonies) on the white colonies. In wild-type host cells (left), high frequency transposition is observed in a high copy number plasmid background; low frequency transposition is observed in a low copy number plasmid strain (right). Mutants in host genes can increase or decrease the frequency of transposition. (b) Examples of host transposition mutant phenotypes with *IS903* and the unrelated element *Tn552* to determine whether these elements use the same host factors.

Figure 2 Reporter assay for siRNA-mediated screen for the identification of host factors important for HIV replication in human cells. (a) The HIV reporter virus contains a luciferase gene such that replication can be followed by luciferase activity. HIV was packaged in a different viral coat (vsv-g) to facilitate infection. (b) An arrayed library of human cells, each expressing a different siRNA, was infected with several different types of viruses and growth followed. (c) The effect of RNAi against about 2,500 different genes on viral growth is represented using the color scheme below: from blue (strong siRNA inhibition of viral growth) to yellow (weak inhibition of viral growth).

can thus be simply followed as the frequency of blue papillae on otherwise white colonies on a plate. Derbyshire and colleagues used two different hosts, one in which the baseline transposition frequency was high so they could detect host mutants that *decreased* transposition, and another strain in which the baseline level was low so that host mutants that *increased* transposition could be detected. Over 100 different mutants that affected *IS903* transposition were isolated (Figure 1b).

Several mutants exhibited an interesting phenotype in which the *lacZ* color was enhanced in a ring in the colony (Figure 1b). Because colonies grow outward from the edge, this ring of color suggested that events at a particular stage during the growth of the colony were influencing transposition. When the corresponding *Tn5* insertions were cloned and sequenced (remember that the mutagenesis was implemented by a *Tn5* insertional strategy!), they were found in genes involved in guanosine triphosphate (GTP) synthesis, *purD purF*, and *purH*. Moreover, in follow-up experiments, Derbyshire *et al.* showed that the mutant phenotype was complemented (corrected) by plasmids carrying the wild-type versions of these genes. From these studies, it appeared that the GTP synthesis pathway plays a role in *IS903* transposition. Derbyshire and colleagues also showed that this pathway does not influence the transposition of the unrelated *Tn10* transposon.

Why would host mutants in GTP biosynthesis affect transposition and affect specifically one kind of transposon? One possibility is that GTP serves as a cofactor in *IS903* transposition. Such regulation by a cofactor is not unprecedented – the transposon-end pairing activity of the *Drosophila P* element is modulated by GTP, and adenosine triphosphate (ATP) plays a critical role in the assembly of Mu and *Tn7* target DNA-containing complexes, for example. Another interesting possibility is that (p)ppGpp (a global

signal of nutrient availability in bacteria, see Section 10.13) modulates the expression of other host genes involved in transposition. Distinguishing among these models may depend on a fresh experimental approach.

In a conceptually similar study, now in eukaryotes, Sumit Chanda and colleagues screened large siRNA libraries to identify host factors required for HIV integration (König *et al.*, 2008). The lifecycle of retroviruses including HIV is described in Section 14.7.

To follow HIV replication, the virus was engineered to contain a luciferase gene so that the presence of virus could be directly followed by luminescence (Figure 2a). The genome-wide screen was then performed by independently targeting ~20 000 genes with siRNAs, infecting the cells arrayed in ~20 000 wells with the HIV derivative carrying the luciferase reporter, and looking for decreased luciferase signal resulting from reduced integration. As controls, parallel screens were performed on other viral vectors encoding luciferase, including murine leukemia (MuLV) and adeno-associated virus (AAV) constructs; general cellular toxicity was also evaluated (Figure 2b). It was reasoned that comparisons between 'hits' identified for all three viral vectors might prove useful in determining whether the identified host factors might play general or more specific roles in HIV pathogenesis.

As a technical note, siRNA-libraries are becoming more sophisticated, typically carrying more than one siRNA capable of targeting a specific gene in each well, increasing the likelihood of efficient gene knock-down. Moreover, a given gene is typically targeted by several different sets of siRNAs (three in this study) in independent wells, thus allowing reproducibility to be evaluated. While it is possible that an individual siRNA will exhibit 'off target' effects that are misleading in the screen (knocking down another gene in the cell, for example), it is less likely that three different siRNA sets would all have the same secondary targets. Thus, similarities between responses of different siRNAs targeting the same gene should lead to increased confidence in the results.

The effects of specific gene knock-down on replication of the three different viruses is presented in a hierarchical cluster analysis that groups genes with strong inhibition profiles, those with intermediate profiles, and those with no effect (Figure 2c). In this case, the clustering was based on a number of different compiled criteria, not just the siRNA knock-down analysis. As we see, different genes display different effects for the three viral constructs: some genes are essentially HIV-specific, others affect both retroviruses (HIV and MuLV) while some affect all three viruses (HIV, MuLV, and AAV). Genes identified in this siRNA screen were further evaluated in a number of independent ways to determine which particular step in the HIV lifecycle is affected by the identified host factor. Host proteins that uniquely affected reverse transcription included a helicase and other nucleic acid binding proteins and ubiquitin–proteasome proteins. Nuclear importation proteins and proteins that could tether the IN/DNA complex to chromatin were also found to be important.

The use of libraries, whether it is a transposon, siRNA or chemical library, to screen for genes involved in a particular process has been extremely valuable for the identification and characterization of biological pathways such as HIV biology. In all cases, the key to success is a reproducible phenotype or assay in which to screen for the desired effects. The availability of more and more high-throughput approaches with readily available reagents will expand the systems that can be studied and contribute to our understanding of the intricate workings of a cell.

König R, Zhou Y, Elleder D, *et al*. Global analysis of host-pathogen interactions that regulate early-stage HIV-1 replication. *Cell* 2008; **135**:49–60.

Twiss E, Coros AM, Tavakoli NP, Derbyshire KM. Transposition is modulated by a diverse set of host factors in Escherichia coli and is stimulated by nutritional stress. *Molecular Microbiology* 2005;**57**:1593–1607.

cleaves T-rich sequences, then occurs (Figure 14.29, step 2). In humans, the ORF2 endonuclease resembles an apurinic endonuclease (see Section 12.4) and can act at many sites. The role of ORF1 remains to be established.

The poly(A) tail of the LINE mRNA then associates with the target site by base-pairing with the cleaved bottom T-rich strand (Figure 14.29, step 3). The free 3′ OH end of the target DNA generated by endonuclease cleavage then acts as a primer for the synthesis of a new LINE DNA strand and extends to the end of the element RNA this is catalyzed by the ORF2 reverse transcriptase using the LINE mRNA as template. Thus a new LINE DNA strand that is covalently linked to the bottom strand of the insertion site is produced.

This mechanism is called **target-primed reverse transcription (TPRT)** because the target 3′ OH exposed by target nicking provides the primer for reverse transcription of the RNA copy of the non-LTR element. A consequence of TPRT is that a poly(A) sequence always becomes incorporated in the inserted element; this is a characteristic feature of non-LTR retrotransposons; they are sometimes called poly(A) retrotransposons for this reason.

The mechanism by which the second top strand of LINE DNA is synthesized is not yet firmly established. One possibility is that, as the reverse transcriptase enzyme reaches the 5′ end of the LINE RNA template, a second cleavage event exposes a 3′ OH on the top strand of the target DNA (step 4). The newly exposed target 3′OH then acts as a primer for synthesis of a second strand of LINE DNA using as a template the newly synthesized bottom strand LINE DNA. As with other transposons, target-site duplication occurs during non-LTR element insertion (Figure 14.29, step 6).

SINEs and pseudogenes are non-LTR elements generated from cellular RNAs

Another prominent class of non-LTR elements are the **SINEs** (Figure 14.28). These elements derive from cellular RNAs whose initial integration and subsequent cycle of transposition are promoted by the LINE-encoded transposition proteins.

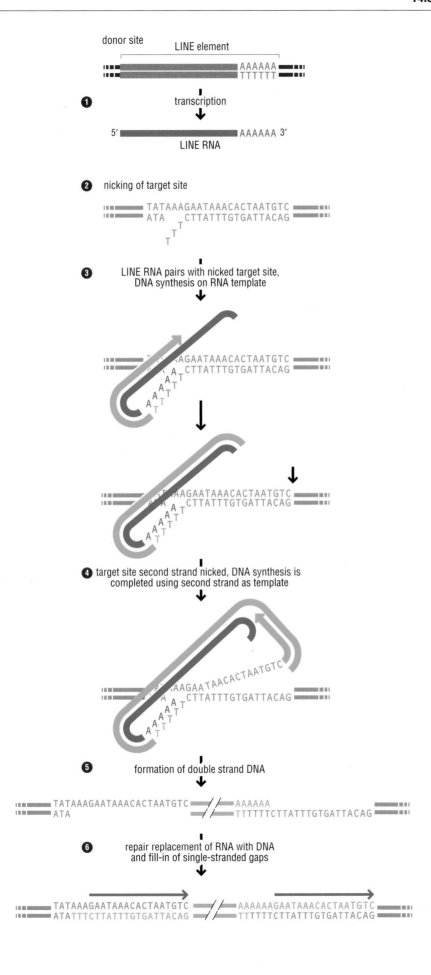

Figure 14.29 The mechanism of transposition via target-primed reverse transcription. A LINE element in a donor DNA is shown. (1) Transposition begins with the transcription by host RNA pol II of the LINE element into an RNA (blue) encoding an RNA-binding protein and a multifunctional protein with endonuclease and reverse transcriptase activity. After RNA export to the cytoplasm, these proteins (not shown) are expressed and associate with the LINE RNA. (2) The endonuclease nicks the DNA at a target site, which contains a poly tract. (3) The target DNA poly tract base-pairs with the poly sequence at the 3′ end of the LINE RNA. Using the target DNA 3′ OH released by nicking, the LINE RNA is then copied by the ORF2 reverse transcriptase into a DNA copy (green), which is covalently attached to the target DNA. The upper target DNA strand is nicked by the endonuclease. (4) A second DNA strand (green) is then synthesized using the newly exposed 3′OH on the top target strand as a primer and the just synthesized first DNA strand of the element as a template. The RNA copy is degraded by RNase H. (5) A double strand DNA copy of the element is now covalently linked to both the top and bottom strand of the target DNA. (6) The single strand gaps at end of the element are repaired to generate the target-site duplications (blue arrows) that flank the newly inserted element.

⊙ We discuss the structure and function of the signal recognition particle in Section 11.2.

Most SINEs are short (100–300 bp long), contain an internal promoter and have a poly(A) tail. SINEs result from the integration of small host RNAs such as tRNAs and 7SL RNA, which is the RNA component of the signal recognition particle involved in targeting proteins to membranes.

The 7SL RNA-derived SINEs are a prominent class of human SINEs. For historical reasons, these 7SL-derived elements are called **Alu elements**, because they include a target site for a restriction endonuclease called Alu.

There are about 1.6×10^6 SINEs in the human genome, forming about 13% of the genome (Figure 14.6) and about 35 examples of human disease resulting from gene disruption by a SINE are known (Figure 14.2).

Pseudogenes are another type of element resulting from LINE protein-mediated target-primed reverse transcription and integration. Pseudogenes are protein-coding sequences that lack the introns associated with their active gene relatives; they also contain poly(A) tails, and are flanked by target-site duplications. Pseudogenes were probably formed by the LINE-protein mediated integration of a spliced mRNA. Pseudogenes are often not transcribed because they lack the promoters that lie upstream of the mRNA start site of their founding mRNA, and have generally been so mutated that they do not encode a gene functional product. The human genome has been estimated to contain about 8×10^3 pseudogenes.

LINEs and SINEs can contribute to genome instability as substrates for homologous recombination

Because LINEs and SINEs are highly repeated sequences that are dispersed throughout the human genome, homologous recombination between these elements can result in intramolecular and intermolecular deletions and duplications by unequal sister chromatid exchange and in translocations, as described in Section 13.9, Figure 13.25. There are more than 25 examples of human diseases that have resulted from homologous recombination between transposable element repeats; some of these rearrangements have occurred multiple times (Figure 14.30).

Mobile group II introns are present in the genomes of bacteria, mitochondria, and chloroplasts

The other major group of non-LTR retrotransposons are the **mobile group II introns**, which are found in bacteria and in mitochondria and chloroplasts of fungi, protists, and plants. Mobile group II introns encode catalytic self-splicing RNAs – that is, they can excise themselves from an RNA transcript without the aid of proteins *in vitro*. Mobile group II introns move by the reverse splicing of their excised intron into the target DNA, followed by reverse transcription.

Group II introns are of particular interest because they are thought to be the progenitors of the nuclear spliceosomal introns of eukaryotic genes. An attractive hypothesis is that they arose in bacteria, and that, after eukaryotic cells acquired mitochondria and chloroplasts (which are of bacterial origin) the mobile group II introns invaded nuclear genomes. Once in the nuclear genome as spliceosomal introns, they can only be excised from mRNA with the aid of spliceosome, the large complex featuring both RNA and proteins (Section 9.5). Bacterial genome sequencing projects have revealed that mobile group II introns are present in a number of bacteria.

Diseases resulting from transposable element-mediated recombination			
	affected protein	element	rearrangement
Alport's syndrome	Collagen type IV	LINE	Deletion
Chronic granulomatous disease	cytochrome b-245, β polypeptide	LINE	Deletion
Glycogen storage disease type II	Phosphorylase kinase (PHKB)	LINE	Deletion
Adenosine deaminase [-] sereve combined immunodeficiency	Adenosine deaminase	Alu	Deletion
Breast cancer	BRCA1	Alu	Deletion
Fabry's disease	α-galactosidase A	Alu	Deletion
Familial hypercholesterolaemia	Low-density lipoprotein receptor	Alu	Deletion
Fanconi's anaemia	FANCA	Alu	Deletion
Glanzmann's thrombastheria	Platelet fibrinogen receptor	Alu	Deletion
Glanzmann's thrombastheria	Glycoprotein Ia	Alu	Deletion
Glioma brain tumours (association to)	Retinoblastoma gene (RB)	Alu	Deletion
Glycogen storage disease type II	Lysosomal acid alpha-glucosidase	Alu	Deletion
Hereditary nonpolyposis colorectal cancer	MutL protein homolog 1	Alu	Deletion
Hereditary persistence of fetal haemoglobin	β-globin cluster	Alu	Deletion
Hypo--lipoproteinaemia	Apolipoprotein B	Alu	Deletion
Insulin-independent diabetes mellitus	Insulin Receptor β	Alu	Deletion
Krabbe's disease	Galactosylceramidase	Alu	Deletion
Maple syrup urine disease	Branched chain acyltransferase	Alu	Deletion
Meningioma	c-sis proto-oncogene	Alu	Deletion
Mucopolysaccharidosis type IVA	Chondroitinase	Alu	Deletion
Sandhoff's disease	β-hexosamindase, β-subunit	Alu	Deletion
Tay-Sach's disease	β-hexosamindase, α-subunit	Alu	Deletion
Thrombophilia	Antithrombin	Alu	Deletion
β- thalassaemia	β-globin cluster	Alu	Deletion
γβδ- thalassaemia	β-globin cluster	Alu	Deletion
α- thalassaemia	α-globin cluster	Alu	Deletion
Von Willebrand's disease type II	Von Willebrand factor	Alu	Deletion
Breast cancer	BRCA1	Alu	Duplication
Ehler-Denlos syndrome	Lsaly hydroxylase	Alu	Duplication
Duchenne muscular dystrophy	Dystrophin	Alu	Duplication
Familial hypercholesterolaemia	Low-density lipoprotein receptor	Alu	Duplication
Lesch-Nyhan syndrome	ypoxanthine phosphoribosyl transferase	Alu	Duplication
Lipoprotein lipase deficiency	Lipoprotein lipase	Alu	Duplication
Glanzmann's thrombastheria	Platelet fibrinogen receptor	Alu	Inversion
Chronic myeloid leukaemia	BCR/ABL	Alu	Translocation
Ewing sarcoma	EWS/FLI1	Alu	Translocation
XX male	XY chromosome	Alu	Complex rearrangement

Figure 14.30 Homologous recombination between non-LTR elements has caused human disease.
From Deininger and Batzer (1999).

Mobile group II introns move by reverse splicing and target-primed reverse transcription of their excised RNA form

Mobile group II introns encode a conserved intron RNA structure that is essential for splicing and mobility; they also encode a protein composed of multiple functional domains including a maturase that assists in element splicing, a reverse transcriptase, and, in some cases, a DNA endonuclease called RME (Figure 14.31). They can participate in two types of mobility reactions: **retrohoming**, in which they insert at high frequency into a specific site with considerable homology to

group II mobile intron

P exon 1 exon 2

RT MT En

1 transcription

exon 1 exon 2

RNA

2 translation of RME

3 splicing assisted by RME

exon 1 exon 2 intron RNA

+

RME

target DNA

4 reverse splicing of intron RNA into target by RME maturase

5 DNA cleavage by endonuclease RME

nick

6 DNA synthesis by reverse transcriptase RME

cDNA

7 DNA processing/repair, RNA degradation by cellular enzymes

8

new copy of intron

Figure 14.31 The mechanism of retrohoming by mobile group II introns. A mobile group II intron located within a gene is shown, flanked by exon 1 and exon 2 (orange). The mobile group II intron encodes a protein RME (dark blue) with reverse transcriptase (RT), maturase (MT) (which facilitates splicing), and endonuclease (En) activity. P is the gene's promoter. (1) Host gene transcription generates an RNA containing the host gene exons separated by the group II intron. (2) Translation of the RNA synthesizes the RME protein (light red). (3) RME facilitates splicing of the mobile group II intron RNA to join exon 1 and exon 2, and forms an RNA-protein complex containing the lariat form of the spliced intron (blue) and RME (4) Assisted by RME, the group II intron RNA (blue) then reverse splices into its cognate homing DNA, i.e. the host gene lacking the mobile group II intron. (5) The RME endonuclease introduces a nick into the DNA strand that lacks the intron at the position of the RNA splice. (6) This cleavage generates a 3′ OH in the target DNA, which then serves as a primer for the third activity of RME, reverse transcription of RNA into DNA (green), forming an RNA–DNA hybrid version on the intron in the target DNA. (7) Further processing results in conversion of the RNA–DNA heteroduplex to double-stranded DNA. (8) The DNA form of the mobile group II intron occupies the new target site.

the intron sequence; and **retrotransposition**, in which they insert at much lower frequency at an ectopic, non-specific site.

The first step in both of the reactions is the splicing of the element from an mRNA to form a lariat (a looped structure), a reaction assisted by the maturase activity (Figure 14.31, step 3). In the retrohoming reaction, the excised RNA reverse splices into one of the DNA strands at a target site that is defined by base-pairing between the intron RNA and the target DNA (Figure 14.31, step 4). The intron-encoded endonuclease then cleaves the DNA strand just outside the insertion site, generating a 3′ OH that on the bottom strand (Figure 14.31, step 5) can act as a primer for the reverse transcriptase to make a DNA copy of the intron RNA. The target site now contains an RNA copy of the element on the top strand and a DNA copy of the element on the bottom strand.

As is suggested for the synthesis of the second strand of LINE elements, the reverse transcriptase can then use this newly inserted intron DNA strand as a template to synthesize the second strand of the element at its new insertion site (Figure 14.31, step 6). The maturase reverse transcriptase, however, lacks an RNase H activity that can degrade the template RNA strand and thus bacterial proteins play an important role in converting the spliced RNA–DNA form into duplex DNA (Figure 14.31 steps 7 and 8).

Mobile group II introns that lack an endonuclease appear to target replication forks at single-strand regions where lagging strand DNA synthesis is occurring; it is thought that the element's reverse transcription is primed by 3′ OH ended fragments on the lagging strand. Again host factors are required to complete the synthesis of duplex DNA.

Despite differences in the origins of the 3′ OHs that prime reverse transcription, and differences in how the RNA template is brought to target DNA – by positioning by a protein or reverse splicing – the mobility of all LINEs, SINEs, and mobile group II introns is based on target-primed reverse transcription.

14.9 CONTROL OF TRANSPOSITION

In the previous sections, we have explored the range of different transposable elements that occur in the genomes of many organisms, and have seen the various molecular processes that underpin transposition events. Very often, it may seem that the host cell is little more than an innocent bystander, powerless to stop transposition events from occurring within its genome. In this section, however, we see how this initial impression is far-removed from reality.

The frequency of transposition is usually tightly controlled

Much remains to be learned about the complex interrelationships between transposons and their hosts. In one view, transposable elements may be considered parasites that infest host genomes. Alternatively, the genetic diversity that can result from transposon mobility may be considered to reflect a more mutualistic relationship between transposons and their hosts. In any case, a high frequency of transposition is generally disadvantageous as it can result in a high frequency of deleterious mutations. An unacceptable burden of mutation (one that would be fatal to the host organism) due to transposition, however, is generally avoided because the transposition of most elements occurs very rarely.

There are many points at which the production of transposition proteins and the formation of protein–nucleic acid intermediates in transposition can be controlled. Although transposon-encoded proteins perform the central catalytic steps in element translocation, host proteins play essential roles in the movement of these elements, providing intimate connections between transposons and their hosts. For example, while element-encoded transposases mediate breakage and joining at the ends of DNA-only cut-and-paste transposons, DNA-bending proteins such as the sequence-specific protein IHF and the sequence non-specific proteins such as HU and HNS, participate directly in several transposition reactions by modulating the assembly of active transpososomes. As we saw in Experimental approach 14.2, many other types of proteins such as those involved in proteasome function and nuclear import are involved in retroviral integration.

The movement of bacterial group II introns (Figure 14.32) and mammalian LINE non-LTR elements, in which element-encoded endonucleases and reverse transcriptases mediate target cleavage and reverse transcription, requires many host proteins such as exonucleases, ribonucleases, and repair polymerases. A recent analysis of cellular factors required for HIV replication has revealed the involvement of even more types of host factors (see Experimental approach 14.2).

Transposase levels can be regulated in a variety of ways

The frequency of transposition is often directly related to the concentration of transposase; the synthesis of transposase, in turn, is frequently subject to elaborate

E. coli host factors involved in group II intron mobility			
host factor	effect	identified function	putative effect on group II intron
Exo III	inhibits mobility	3′-5′ exonuclease	degrades nascent cDNA
ligase	stimulates mobility	DNA ligase	seals DNA nicks
MutD	stimulates mobility	3′-5′ exonuclease ε subunit of Pol III (*dnaQ*)	repairs second-strand cDNA synthesis
Pol I	stimulates mobility	3′-5′ exonuclease; removal of RNA primer from Okazaki fragments	removes intron RNA template
Pol II	stimulates mobility	repair polymerase (*polB*)	repair polymerization across DNA-RNA junctions
Pol III	stimulates mobility	replicative polymerase	second-strand cDNA synthesis
Pol IV	stimulates mobility	repair polymerase (*dinB*)	repair polymerization
Pol V	stimulates mobility	repair polymerase	repair polymerization
RecJ	stimulates mobility	5′-3′ exonuclease	5′-3′ resection of DNA
RNase E	inhibits mobility	ribonuclease; part of RNA degradasome	reduces half-life of intron RNA
RNase H	stimulates mobility	ribonuclease; cleaves RNA strand in RNA/DNA hybrid	removes intron RNA template
RNase I	inhibits mobility	ribonuclease	reduces half-life of intron RNA

Figure 14.32 *E. coli* **host factors involved in mobile group II intron transposition.** Multiple *E. coli.* proteins are involved in the transposition of a mobile group II intron.
From Beauregard *et al.* (2008).

control. For example, the transposition of the bacterial element *IS10* is kept to a low frequency by features of the element itself that result in production of a very low amount of transposase (Figure 14.33). The transposase is translated from an mRNA generated by the promoter P_{in}, which is a very weak promoter (Figure 14.33a). The activity of P_{in} is also decreased by a promoter called P_{out} that promotes transcription from inside the element into the flanking (gray) DNA. The P_{in} transposase mRNA is also poorly translated, not only because of poor translation signals, but also because the P_{out} transcript acts as a regulatory RNA (as seen for the RyhB RNA in Section 9.9); this regulatory RNA forms a duplex RNA with the transposase mRNA that blocks its translation. The inhibition of P_{in} mRNA translation by P_{out} RNA is particularly important when IS10 copy number increases and P_{out} decreases the total amount of transposase synthesized. The translation of transcripts that may enter the element from an adjacent gene also is blocked by the particular secondary structure of these transcripts (Figure 14.33b).

Some elements only undergo transposition in certain cell types. This is the case for the *P* element of *Drosophila*: transposition of the *P* element can occur in germ-line cells but not in somatic cells. Although it may seem paradoxical to allow transposition in the germline because of the potential negative consequences of transposition in terms of mutation, only by transposition in the germline can the element be successfully propagated from generation to generation.

Figure 14.33 Mechanisms by which synthesis of the *IS10* transposase is controlled. The frequency of *IS10/Tn10* transposition is proportional to the amount of transposase present. Multiple levels of regulation act to keep transposase expression at a low level, making transposition infrequent. (a) Transposase is expressed from the weak promoter P_{in}. Transcription from the outward-firing promoter P_{out} decreases P_{in} activity. Moreover, the small antisense P_{out} RNA can base pair with P_{in} mRNA, inhibiting translation of P_{in} mRNA. (b) Expression of transposase from external transcripts is inhibited by the formation of a stem and loop structure including the duplex-forming segments B and A2 in the RNA in the mRNA that occludes the start codon. The role of the A1 loop in transposition remains to be determined. Figure 14.33b from Halling *et al.* (1982).

The production of active *P* element transposase requires the splicing of four exons to form the mRNA (Figure 14.34). A functional mRNA is only produced in the germline, however, because somatic cells contain several factors that specifically inhibit the splicing out of the last intron. A stop codon in this intron results in the synthesis of a truncated transposase that lacks the fourth exon, which encodes the catalytic domain of the protein. Consequently, the truncated transposase can bind to transposon ends, but it cannot promote transposition.

The transposition of some elements is directly influenced by cellular growth conditions. For example, transposition of the yeast *Ty1* transposon increases during filamentous growth, a response to starvation, as a result of increased transcription of *Ty1* and stabilization of the *Ty1* transposase in these conditions. It has been suggested that this increased transposition rate could optimize cell survival in stressful environmental conditions by increasing genetic diversity.

Transposition can be controlled by DNA methylation

We see in Chapter 4 how cells use DNA methylation as a vital means of regulating gene expression. A further important role of DNA methylation is to prevent the movement of transposable elements, often by blocking transposase expression and therefore protecting the host. A particular hazard for DNA-only cut-and-paste transposons is that transposon excision results in a potentially lethal gap at the site of element excision. *IS10* has reduced this potential hazard by coupling its transposition to the passage of the replication fork (Figure 14.35). Thus when *IS10* transposes, one arm of the replication fork will remain intact although

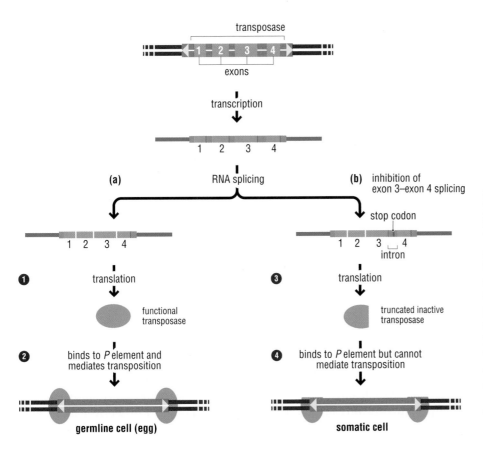

Figure 14.34 Inhibition of splicing of Drosophila *P* element mRNA blocks transposition in somatic cells.
The *P* element transposase is encoded by four exons. Transcription forms the mRNA (blue) that can be alternately spliced. (a) Germline cell (egg). In the germline, splicing can occur, thereby removing all the introns. (1) Translation results in the formation of full length transposase (orange oval). (2) Full length transposase can bind the ends of the transposon and mediate transposition. (b) Somatic cell. In somatic cells, a specific factor inhibits the splicing of exons 3 and 4. (3) A truncated form of transposase is synthesized (truncated orange oval). (4) The truncated transposase can bind the transposon ends but cannot promote transposition.

Figure 14.35 Coupling of IS transposition to chromosomal replication via Dam methylation. The replication fork activates transposition of *IS10*, assuring a single host chromosome does not undergo a potentially lethal double-strand break in the absence of a template for repair. Both ends of *IS10* contain Dam methylation sites (GATC), located within the end sequences required for transposition and at the P$_{in}$ promoter. In unreplicated DNA, both transposon strands are fully methylated and both transposase expression and transposon end activity are low. (1) As the replication fork passes through the element, transient hemi-methylation of the element occurs. Hemi-methylation increases both transposase expression and end activity, resulting in increased transposition. (2) Only one of the elements transposes because the methylated strands are different in each arm of the replication fork and one hemi-methylated element is more active than the other.

the other will have a gap. This gap may be repaired by homology-directed repair using the intact arm as a template.

The coupling of *IS10* transposition to replication is mediated by a Dam (GATC) methylation site in the end of the element. The methylation state of this element can affect transposase expression and the actual activity of the transposon end. As the replication fork passes through the *IS10* element, the two arms contain hemi-methylated *IS10*. Hemi-methylation of the P$_{in}$ promoter stimulates transposase expression and the end activity of one of the hemi-methylated elements, presumably because of differential binding of transposase to the two asymmetric elements.

In eukaryotes, the methylation of transposon DNA also reduces transposition. As described below, this occurs via the RNA silencing mechanisms we learned about in Chapters 8 and 9.

Most eukaryotic organisms suppress transposition globally by gene silencing mechanisms

Most eukaryotic organisms possess a general means of blocking the movement of transposable elements throughout the genome. This suppression depends on mechanisms of gene silencing that are used against repeated DNA sequences throughout the genome. These defense systems act globally to decrease the transposition of all copies of an element by decreasing its transcription and down-regulating the translation of any element mRNA that is synthesized.

⊙ We discuss the production of a double-stranded RNA from the elements in more detail in Chapter 10.

Central to this general control is the production of a double-stranded RNA from the elements.

As multiple copies of the element are located in many different positions in the genome, transcription impinging on the elements from external promoters will generate RNA copies of both strands of the element. These transcripts may then pair to form element-specific double-stranded RNA. Such double-stranded RNA species are specifically degraded to short RNAs by members of the Argonaute family of proteins and can then be targeted to other transposon-derived transcripts to induce mRNA degradation or block translation. Moreover, by recognizing their complementary DNA sequence, these short RNAs can recruit chromatin-remodeling complexes to sites of integrated transposon DNA. These particular chromatin-remodeling complexes modify histones to generate chromatin structure that represses gene expression, thus preventing transcription of the transposase gene (see Chapter 8). The short RNAs can also promote DNA methylation at their target sites by recruiting DNA methylases. It seems likely that these cellular responses to double-stranded RNA arose as specific defenses against transposable elements and viruses.

Although successful transmission of an element to progeny requires element transposition in the germline, it is also important that transposition occurs at a low level to avoid the accumulation of deleterious mutations and maintain genome integrity. While transposition in somatic cells has no heritable effect *per se*, such transposition also is generally blocked to avoid reduced genome integrity, which could result in oncogenic transformation, for example. For some elements, including the *Drosophila P* element, there is an element-specific regulatory mechanism, the inhibition of transposase mRNA splicing, which blocks transposition in somatic cells.

A germline-specific mechanism for inhibition of transposition has recently been discovered in *Drosophila* and some other animals, including mammals. This mechanism uses a specialized class of small RNAs called piRNAs and specialized Argonaute proteins called Piwi, Aubergine, and Ago3. Several *Drosophila* genomic loci that have been long known to encode inhibitors of transposition are now known to encode clusters of inhibitory RNAs, not protein inhibitors, which result from transcription of clusters of elements or fragments of elements no longer capable of transposition. These include *flamenco*, which inhibits the movement of retrotransposons such as *gypsy*, and X-TAS, which inhibits the movement of *P* elements (Figure 14.36). The primary transcripts from these regions are processed in as yet unknown ways to generate short piRNAs that interact with the Piwi, Aubergine, and Ago3 proteins. These piRNA–protein complexes then interact with transcripts from cognate target transposons, leading to the cleavage and inactivation of the transcripts (Figure 14.37). These interactions lead to amplification of the inhibitory piRNAs that can target transposon mRNAs for destruction.

Figure 14.36 Specific loci in *Drosophila* encode inhibitors of transposition. The X-TAS and *flameco* loci on the *Drosophila* X chromosome encode small RNA transposition inhibitors of *P* element and of *gypsy* and several other types of retrotransposons, respectively. These regions do not encode protein inhibitors of transposition. Rather, they encode clusters of inactive transposable element copies whose RNA products are processed to make small RNA (piRNA) inhibitors that act upon the transcripts of active elements to decrease transposition (see Figure 14.37).

Transposition can be controlled through target-site selection

In some cases, the potentially deleterious effects of the movement of transposable elements can be limited by directing transposition away from gene-coding sequences. This means that, even when transposition does occur, the chances of inactivating a gene are small. Although it is true that most elements have the ability to insert into many different sequences, closer inspection reveals that target sites are generally not entirely random.

Target-site choice can be determined exclusively by the interaction of the transposase with the target DNA. In other cases, however, the insertion site is chosen by the interaction of the transposase with host-encoded proteins. The yeast LTR elements *Ty1*, *Ty3*, and *Ty5* have interesting mechanisms of target-site selection, all of which reduce the chance of potentially lethal insertions within

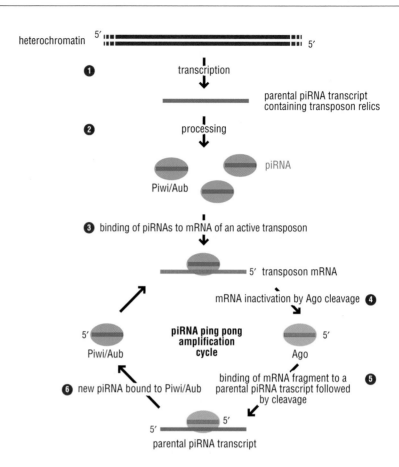

Figure 14.37 Inactivation of transposon mRNAs by piRNAs. A cycle of transcription of long RNAs containing transposon sequences, followed by processing to short piRNAs that can target transposon mRNAs for cleavage can lead to the amplification of piRNAs and increased inactivation of transposon mRNAs. (1) Transcription of regions of heterochromatin containing relics of transposon results in long transcripts (blue) containing transposon sequences. (2) These transcripts are processed in an as yet to be understood pathway by Piwi and Aub proteins (orange) to yield short piRNAs about 23 nts in length. (3) piRNAs bound by Piwi and Aub proteins binding to the mRNA of an active transposon. (4) The transposon RNA is inactivated by cleavage and bound to Ago (green). (5) The mRNA frament binds specifically to another parental piRNA transcript. (6) Another piRNA is generated and bound to Piwi and Aub proteins. From Brennecke *et al.* (2007).

genes. Both *Ty1* and *Ty3* inserts upstream of tRNA genes that are transcribed by RNA pol III by interacting with polymerase-specific transcription factors thereby avoiding the actual coding regions. *Ty5* inserts into silent, heterochromatic regions of the genome, which include the regions around telomeres and the silent mating-type loci. In contrast, certain retroviruses insert preferentially into transcribed regions, a strategy that may favor element expression and, hence, activity.

A feature of many DNA cut-and-paste elements, such as *P* elements and some elements in plants, is that they insert preferentially into DNA close to the donor site, a phenomenon known as **local hopping**. This may reflect limited movement by the excised transposon prior to insertion. Such local hopping is not observed with RNA-based elements that necessarily transit the cytoplasm.

The bacterial transposon *Tn7* is distinguished by the fact that it can transpose to two different types of sites using overlapping but distinct sets of transposition proteins (Figure 14.38). One pathway uses the TnsD targeting protein, which binds to a highly conserved chromosomal sequence, *attTn7*, to insert into a single specific site. This pathway favors the insertion of the transposon into the chromosome and thus its transmission to daughter bacteria; we call this vertical transmission of the transposon. In its other pathway using TnsE, *Tn7* transposes preferentially to plasmids that move between cells by a specialized form of replication called conjugation, likely by recognizing the specialized replication machinery of these plasmids. This strategy favors the horizontal dispersal of *Tn7* throughout bacterial populations.

Despite some restrictions on target-site selection, transposons can insert at many different sites in the genome. In the next sections, we will consider CSSR, a type of DNA movement that can only occur between particular related sites.

14.10 CONSERVATIVE SITE-SPECIFIC RECOMBINATION: OVERVIEW

The previous sections of this chapter have focused on the behavior of transposable elements – discrete portions of DNA that can insert into many different sites in DNA. However, it is not just the movement of defined transposable elements that can lead to the rearrangement of genetic material within the cell. In this section, we explore a quite different way in which DNA rearrangement can occur – the process of **conservative site-specific recombination (CSSR)**. This type of recombination is sometimes called site-specific recombination but as both CSSR and transposition involve specific DNA sites, we will use the term CSSR.

CSSR is required for diverse DNA rearrangements in the cell

CSSR can mediate a variety of different DNA rearrangements such as the integration and excision cycles of bacteriophages, the conversion of circular chromosomal DNA dimers to monomers to allow for accurate chromosome segregation to daughter cells, and the inversion of DNA segments that can control the alternative expression of a gene by changing the orientation of a promoter. Although most known examples of CSSR come from bacteria, it also occurs in eukaryotes and is known, for example, to control the replication of a yeast plasmid. We discuss the use of the bacterial CSSR system cre-*lox* for genome engineering in mammalian cells in Section 16.4. Another role of a CSSR recombinase in eukaryotes is to mediate the integration of a class of eukaryotic retrotransposons; little is known about these retrotransposons other than their structure.

CSSR occurs between specific pairs of special sites that share a short region of homology

Unlike transposition, CSSR is reciprocal and conservative, that is, no DNA sequence is lost or gained during recombination. Instead, DNA is simply exchanged by breakage and joining between the two recombining DNA sites in a short (2–8 bp) region of homology called the spacer that is shared between the sites (Figure 14.39). Note that the spacer has directionality – that is, it does not have an internal repeat structure (Figure 14.40). The spacer region is flanked by recombinase-binding sites in inverted orientation, which position recombinases for action at the spacer region. Once positioned, one recombinase acts on the top strand and one acts on the bottom strand of each parental duplex. The DNA breakage and joining reactions that lead to recombination occur at the edge of these spacers.

Recombination begins by the binding of recombinases to each site, after which the recombinases bind to each other to bring the DNA sites together. Only when

Figure 14.38 The TnsD and TnsE proteins control target-site selection by the *Tn7* transposon. (a) The bacterial transposon *Tn7*. *Tn7* encodes multiple transposition genes, *tnsABCDE*. (b) *Tn7* transposition pathways. TnsABC are required for all transposition reactions: TnsA and TnsB together form the *Tn7* transposase (see also Figure 14.12) and TnsC interacts with and activates the transposase when TnC is recruited to a target DNA by interaction with one of the targeting proteins, TnsD or TnsE, No transposition occurs in the presence of only TnsABC. Two distinct sets of the Tns proteins promote transposition into different types of target DNAs. When bound to its specific binding site *attTn7*, the targeting protein TnsD recruits TnC and thus Tns ABC+D promotes *Tn7* insertion into this specific site which is highly conserved in bacterial chromosomes. When bound to the beta clamp of the replication complex that is involved in lagging strand DNA synthesis, the targeting protein TnsE recruits TnsC and thus TnsABC+E promotes *Tn7* insertion into many different non-*attTn7* sites.

sites for conservative
site-specific recombination

parental

↓

conservative site-specific recombination

recombinant

(a) parental

```
      ATAACTTCGTATAGCATACATTATACGAAGTTAT
      TATTGAAGCATATCGTATGTAATATGCTTCAATA
```

+

```
      ATAACTTCGTATAGCATACATTATACGAAGTTAT
      TATTGAAGCATATCGTATGTAATATGCTTCAATA
```

(b) recombinant

```
      ATAACTTCGTATAGCATACATTATACGAAGTTAT
      TATTGAAGCATATCGTATGTAATATGCTTCAATA
```

+

```
      ATAACTTCGTATAGCATACATTATACGAAGTTAT
      TATTGAAGCATATCGTATGTAATATGCTTCAATA
```

Figure 14.40 Conservative site-specific recombination (CSSR) occurs by DNA breakage and joining within the shared region of homology. The special recombination sites of the Cre-*lox* recombination system of bacteriophage P1 are shown (see also Section 16.4). (a) The parental *lox* sites (blue and pink) share a region of homology (shaded boxes) called the spacer region that is flanked by binding sites for the Cre recombinase in inverted orientation (horizontal blue and pink arrows). The spacer sequence has directionality, i.e. is not a palindrome, as indicated by the green arrows. Each recombinase mediates breakage, exchange, and joining of one strand at each binding site (vertical arrows). (b) Because of the staggered positions of breakage, exchange and joining on each parental duplex, the two recombinant DNA duplexes contain heteroduplex spacer regions. The outcome of CSSR is a precise reciprocal exchange of DNA between the two participating molecules. No DNA sequence is lost or added.

Figure 14.39 Schematic diagram of conservative site-specific recombination (CSSR). Parental duplexes (gray and light gray) contain sites (blue and pink) for conservative site specific recombination that include a region of homology (dark blue) shared between the recombination sites. DNA exchange by CSSR occurs by DNA breakage and joining at the shared regions of homology to generate two recombinant duplexes.

the sites are paired do the recombinases execute DNA breakage and joining so that uncoupled DNA breaks, which may be harmful to the cell, do not occur.

Following pairing, the recombinases then break the parental duplexes of DNA at each spacer at staggered positions (Figure 14.40a). Four recombinases bind to two duplexes and each directs a single cleavage on one strand. Broken strands then are exchanged between the partner duplexes and rejoined to form recombinant duplexes (Figure 14.40b). Because the positions of strand breakage within each spacer are staggered, the DNAs within the exchanged and rejoined spacers are heteroduplex, that is, they contain one DNA strand from each parent between the positions of DNA strand exchange. This requirement for heteroduplex formation helps to enforce the accuracy of the recombination reaction, as only identical DNA sequences can form perfectly paired heteroduplex DNA. The recombinant duplexes contain DNA information from one parent on one side of the spacer and from the other parent on the other side of the spacer.

The outcome of CSSR reactions is determined by the relative positions of the recombining sites

CSSR can promote different kinds of DNA rearrangements, depending on the relative positions and orientations of the two recombining sites (Figure 14.41); the orientation of each recombining site is determined by the directionality of its spacer. When the two recombination sites are located on different DNA molecules – for example, a bacteriophage genome and the host-cell genome – CSSR will integrate the phage genome into the host-cell genome (Figure 14.41a). For this reason, the phage-encoded recombinases that mediate such reactions are called integrases. As we will discuss in detail below, however, these bacteriophage integrases are a different type of enzyme from the integrases (transposases) of retroviruses and other LTR retroelements, and work by a quite different mechanism.

When the recombination sites are present on the same DNA molecule and their spacer sequences are oriented in the same direction, recombination between the two sites will separate the parental substrate DNA into two DNAs (Figure 14.41b). This CSSR reaction is alternatively called 'excision', 'deletion', or 'resolution' depending on the biological process involved.

Finally, when two recombination sites are present on the same DNA molecule and their spacer sequences are in inverted orientation, recombination between them results in the inversion of the DNA segment that lies between them (Figure 14.41c). Inversion of a promoter region by this mechanism can be used to regulate gene expression, a process we explore further in Section 14.13.

It is important to note that not all CSSR systems can carry out all of these reactions. In some cases, they have sophisticated mechanisms to ensure that only particular

Figure 14.41 The outcome of CSSR is determined by the spatial relationship of the two recombination sites. (a) When recombination occurs between CSSR sites (blue boxes with green arrows) on different DNA molecules, the two molecules are joined together or integrated into one larger DNA molecule in which the spacer sequences (green) are oriented in the same direction. (b) When recombination occurs between CSSR sites whose spacer sequences are in direct orientation on the same DNA molecule, two product molecules are formed in a reaction variously termed 'excision', 'deletion', or 'resolution', depending on the biological context. (c) When recombination occurs between CSSR sites whose spacer sequences are in inverted orientation on the same DNA molecule, inversion of the segment between the CSSR sites occurs.

reactions are carried out. For example, in the integration-excision cycles of bacteriophages, once integration occurs, excision cannot take place without the synthesis of additional phage-encoded proteins that are expressed only under excision conditions.

CSSR recombinases are topoisomerases

CSSR occurs by DNA breakage and rejoining. The recombinases that mediate these reactions are sequence-specific topoisomerases. As discussed in Chapter 2, a topoisomerase is a protein that can transiently break DNA, often resulting in a change in DNA topology such as the relaxation of supercoiling. Topoisomerases can break and rejoin DNA in the absence of a high energy cofactor such as ATP because DNA breakage is accompanied by a covalent linking of the topoisomerase to the DNA. This linkage conserves the high energy that was stored in the DNA phosphodiester bond. The high energy of the protein–DNA linkage is then used to rejoin DNA strands by re-forming a phosphodiester bond. Recombination by a topoisomerase activity is therefore quite distinct from transposition, during which DNA synthesis and ligation are required to generate intact duplex DNA.

The topoisomerases that carry out CSSR act specifically at recombination sites because they are sequence-specific DNA binding proteins: they bind to their cognate sites, which flank the spacer regions of CSSR sites. Like transposases, the CSSR recombinases are usually encoded by genes adjacent to the DNA that undergoes recombination.

Two large families of topoisomerase-like recombinases carry out CSSR. Although the structures of these two families are unrelated, they both use the same fundamental mechanism to break DNA: the nucleophilic attack of a hydroxyl group on an amino acid, either a tyrosine or serine, to break DNA, with the concomitant formation of a protein-DNA linkage and release of an OH DNA end. Figure 14.42a shows how one recombinase family breaks DNA by forming a DNA-3′

P-tyrosine linkage and releasing a 5′ OH DNA end. These are called **tyrosine recombinases.** By contrast, recombinases of the other family break DNA by forming a serine–5′ P-DNA linkage and releasing a 3′ OH DNA end, and these are called **serine recombinases** (Figure 14.42b). Intact DNA strands can then be regenerated by the attack of the free DNA 3′ OH or the 5′ OH ends on the cognate covalent protein–DNA linkage, re-forming the high-energy phosphodiester bond in DNA and releasing the recombinase in its original state. Again, in contrast to transposases, CSSR does not require Mg^{2+} as a cofactor.

The serine and tyrosine recombinases differ from each other in that serine recombinases promote CSSR by simultaneously breaking and exchanging both strands in the two parental duplexes. By contrast, tyrosine recombinases first break and exchange one pair of strands from each duplex and then break and exchange the other pair in a subsequent step.

In the next two sections, we explore the application of CSSR in two particular biological contexts. In Section 14.12 we see how CSSR is used in the conversion of DNA dimers to monomers in systems with accessory proteins that regulate the reaction; in the next section, however, we focus on the conversion of DNA dimers to monomers in single-protein systems.

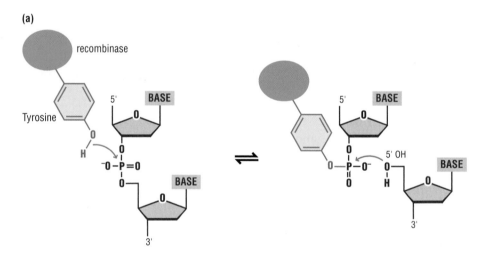

Figure 14.42 The chemistry of serine and tyrosine recombinases. (a) The OH of a tyrosine on a tyrosine recombinase attacks DNA to form a DNA-3′ P-tyrosine–protein linkage and a 5′ OH DNA end. The broken DNA strands can be rejoined by attack of the 5′ OH DNA end on the tyrosine–DNA linkage, regenerating the recombinase. (b) The OH of a serine on a serine recombinase attacks DNA to form a recombinase–serine–5′ P–DNA linkage and a 3′ OH DNA end. The broken DNA strands can be rejoined by attack of the 3′ OH DNA end on the serine–DNA linkage, regenerating the recombinase.
Adapted from Grindley *et al.* (2006).

14.11 CSSR CONVERSION OF DNA DIMERS TO MONOMERS: SINGLE-PROTEIN SYSTEMS

CSSR can covert DNA dimers to monomers

The conversion of a circular DNA to two product DNA circles is part of many cellular processes, and CSSR plays an important role in many such conversions. For example, in the circular genomes that occur in most bacteria and bacteriophages, accurate chromosome segregation to daughter cells is blocked by the formation of dimer chromosomes that result from homologous recombination between two monomer chromosomes (Figure 14.43a). In response, many bacterial genomes, plasmids and some viruses encode CSSR systems that can convert such dimers to monomers (Figure 14.43b). Examples include a segregation system called Cre encoded by the bacteriophage P1 as well as the bacterial chromosomal monomerization system Xer.

Another situation in which a CSSR reaction converts a single DNA circle into two product circles is the processing of some transposition intermediates to their final products. The very similar bacterial transposons γδ and *Tn3* encode a CSSR site

Figure 14.43 CSSR contributes to chromosome segregation by resolving dimeric chromosomes. (a) Dimeric chromosomes cannot be properly segregated to daughter cells. A cell containing two copies of a circular chromosome with an origin of replication (green box) and terminus of replication (red box) is shown. (1) Homologous recombination results in formation of a dimeric chromosome. (2) Attempted segregation can lead to either misegregation that results in a cell without a chromosome or chromosome breakage (not shown). (b) Chromosome monomerization by CSSR. Bacterial chromosomes encode CSSR systems that include a recombinase called Xer that acts at special site called *dif* (yellow arrow), and converts dimers to monomers. (1) Homologous recombination results in formation of a dimeric chromosome in which the *dif* sites are directly oriented. (2) CSSR at *dif* mediated by the recombinase Xer converts a dimer chromosome to two intact and functional monomer chromosomes. (3) The monomer chromosomes can be correctly segregated to daughter cells.

called *res* and a serine recombinase called resolvase, in addition to a transposase and the terminal TIRs (Figure 14.44a). Nick-and-paste replicative transposition of γδ from a donor plasmid to a target plasmid yields a circular cointegrate DNA containing two transposon copies, the donor DNA backbone, and the target DNA (Figure 14.44b, step 1). Since each transposon contains a *res* site, this product circle contains two *res* sites in direct orientation. Resolvase-promoted CSSR between the two *res* sites results in the regeneration of donor plasmid with its transposon copy and the target plasmid, which contains a copy of the transposon (Figure 14.44b, step 2).

We will first consider two single-protein recombination systems in detail: one using transposon γδ resolvase, a serine recombinase, and the other using bacteriophage P1 Cre, a tyrosine recombinase. In Section 14.12 we then discuss CSSR systems in which the activity of a tyrosine recombinase is highly regulated by accessory proteins: the chromosomal Xer system and the bacteriophage lambda integration and excision (resolution) cycle.

Resolvase, a serine recombinase, promotes CSSR via double-strand breaks in DNA

Nick-and-paste replicative transposition of γδ from a donor plasmid to a target plasmid yields a circular cointegrate DNA containing two transposon copies (Figure 14.44). Separation of the donor and target plasmids requires the element-encoded resolvase, which promotes recombination between the *res* sites in the two transposon copies.

The *res* site is large (114 bp) and complicated, containing six resolvase binding sites organized in pairs of binding sites at Sites I–III (Figure 14.45a). The spacer distance between each pair of binding sites is different: 2 bp in Site I, 10 bp in Site II and 1 bp in Site III. Only the spacing at Site I is appropriate for breakage and joining to occur; Sites II and III are auxiliary resolvase-binding sites that organize the DNA and juxtapose the Site I-resolvase complexes for pairing and strand exchange (Figure 14.45b). Two resolvases separated by a 2bp spacer bind at Site

Figure 14.44 Resolution of γδ cointegrates.
(a) The bacterial transposon γδ (and the closely related element *Tn3*) encodes a transposition system and a CSSR resolvase and *res* site resolution system. (b) γδ transposition and resolution. A donor molecule (gray) containing γδ and a target molecule (light gray) are shown. (1) Transposition forms a cointegrate containing both the donor plasmid and target plasmid, (see Figure 14.19) linked by two copies of γδ. Attempted segregation to two different daughter cells could result in chromosome breakage. (2) CSSR mediated by the serine recombinase resolvase at the directly oriented *res* sites in the cointegrate results in two plasmids, each containing a copy of the transposon. Note that the target plasmid containing the transposon is a simple insertion product which results from replicative transposition and CSSR that is identical to the simple insertion generated by cut-and-paste transposition.

(a) transposon γδ

transposase   *res*   serine recombinase resolvase   amp^R

(b)

① transposition   ② resolution

Figure 14.45 Resolvase-mediated CSSR.
(a) The $\gamma\delta\,res$ site. The *res* site contains three pairs of resolvase binding sites, each pair separated by a different number of base pairs. Breakage and joining occurs only in Site I because of the spacing between its resolvase binding sites. Directionality of the strand exchange region in Site I is indicated by the green arrow. The other sites facilitate the synapsis of Site I and are essential for recombination. (b) *res* site synapsis to form the synaptosome in which Site I pairing occurs. A model of a plasmid containing two directly repeated *res* sites is shown with spheres representing resolvase. In one *res* site, the resolvases at Site I are shown in blue; in the other *res* site, the resolvases at Site I are shown in pink. The resolvases at Sites II, III, II' and III' are dark blue. The *res* sites pair by interwrapping of Sites II and III, which results in Site I apposition and strand exchange.
From Li *et al.* (2005).

I and protein–protein interactions result in the pairing of the two Site I segments by four resolvases.

Once paired, the resolvase molecules make a double-strand break at Site I of each *res* site (Figure 14.46, step 1). In each strand, a serine from resolvase attacks the phosphodiester backbone to form a protein-serine-5′ P-DNA linkage and exposes a free 3′ OH DNA end. The two resolvases on each parental duplex thus introduce a double-strand break into each duplex generating four 3′ OH DNA termini and four DNA-5′ P-serine recombinase termini.

Crystallographic studies of complexes containing resolvase bound to cleaved Site I *res* DNAs have revealed the formation of a tetramer of resolvases, 2 *res* Site

Figure 14.46 The serine recombinase resolvase mediates double-strand breakage, exchange, and rejoining at the *res* site. Four molecules of resolvase (blue and pink) bind to their recognition sequences in Site I of the two *res* sites and pair the parental duplexes as in Figure 14.45. (1) The OHs of the resolvase active site serines (circle S) attack the DNA, one per strand, making a covalent link between the recombinases and the DNA, and making a double-strand break in each duplex. (2) The rotation of one pair of resolvases (left blue + pink) relative to the other juxtaposes the broken ends from one parental duplex to the broken ends of the other parental duplex. (3) The free 3′ OHs on the DNAs attack the 5′ phosphoserine linkages, rejoining the DNA backbones and releasing the resolvases. These breakage, exchange, and rejoining events are highly coupled such that no DNA ends or resolvases are released from the complex during this CSSR reaction.

I DNA segments bound on the outside surface of this protein core (Figure 14.47). The two resolvases on one side of the spacer region (left blue, pink) interact with the other two resolvases (right blue, green, pink) across flat hydrophobic protein surfaces. Rotation of the pairs of resolvases on this flat surface is thought to result in the juxtaposition of the DNA ends from one parental DNA duplex with the broken ends from the other parental duplex (Figure 14.46, step 2). Rejoining of the broken DNAs then occurs to form two recombinant duplexes (Figure 14.46, step 3). This rejoining in each hybrid spacer region occurs by the attack of the two 3′ OH DNA ends on the two DNA 5′ P–serine termini, regenerating an intact DNA strand containing a region of heteroduplex and releasing the serine recombinases in their original state.

Figure 14.47 The interacting surfaces of resolvase dimers bound to *res* DNA. The structure of a complex containing four γδ resolvases bound to two cleaved Site I *res* sites is shown. As in Figure 14.46, upper two subunits are blue and bottom two subunits are pink. Note the flat hydrophobic surfaces of the resolvases at the interface. It is likely that these flat surfaces allow rotation of the proteins relative to each other to effect strand exchange without dissociation and re-association PDB 1ZR4. From Li *et al.*, *Science* 2005;**309**:1210–1215.

Figure 14.48 The Cre recombinase promotes pairing of *lox* recombination sites. Four Cre protomers bound to two parental duplexes containing *lox* sites are shown. The circled Ys represent the active site tyrosine in Cre. Note that the spacer regions (green arrows) are arranged in antiparallel fashion. Adapted from Chen and Rice (2003).

➲ We discuss the structure of the Holliday junction in Section 13.7.

Cre, a tyrosine recombinase, promotes CSSR by two cycles of single-strand exchange

As described above, CSSR by a serine recombinase occurs by making and rejoining concerted double-strand breaks in each parental duplex. By contrast, CSSR promoted by tyrosine recombinases occurs by two cycles of single-strand breakage, exchange and rejoining between the two parental duplexes. A well-studied example is the bacteriophage P1 Cre system, which acts at special DNA sites called *lox* sites. As in the resolvase system, recombination begins with the binding of four Cre molecules to inverted recognition sequences that flank the spacer region of two parental *lox* sites (Figure 14.40). Pairing of the *lox* sites in antiparallel alignment occurs by binding of the DNAs to a tetramer of Cre formed by Cre–Cre interactions (Figure 14.48).

Instead of the concerted breakage of all four DNA strands seen with the serine recombinases, tyrosine recombinases break and exchange only two strands at a time, one from each parental duplex. The sequence of the spacer region and the binding properties of Cre at each of its sites determines which pair of Cre molecules will execute the first strand exchanges. Following Cre binding and pairing, which juxtaposes the recombination sites (Figure 14.49) a tyrosine from one Cre protein on each duplex attacks the phosphodiester backbone of one strand to form a protein–tyrosine–3′ P–DNA linkage and expose a free 5′ OH DNA end (Figure 14.49, step 1). The 5′ OH ends from each duplex are then juxtaposed to the protein–tyrosine–3′ P–DNA linkages from the other duplex (Figure 14.49, step 2). DNA rejoining then occurs by the attacks of the 5′ OH DNA ends on the protein–tyrosine–3′ P–DNA linkages. This attack re-forms the DNA phosphodiester linkage and releases the protein (Figure 14.49, step 3).

The resulting CSSR intermediate formed, in which a single pair of strands is exchanged between the parental duplexes, is identical to the Holliday junction that is generated during homology-directed repair and homologous recombination.

Figure 14.50a shows the protein-DNA structure of a Cre complex in which the parental duplexes are paired. The DNA in the Holliday junction intermediate in shown in Figure 14.50b. Following the indicated isomerization of this intermediate, which positions the other non-exchanged strands to be cleaved (Figure 14.50b), CSSR recombination is then completed by a second cycle of Cre-mediated DNA breakage and joining on the other side of the spacer region, using the recombinases on the other pair of strands (Figure 14.49, steps 5–7).

nick one strand on each duplex
to give 3′ P-tyr and 5′OH

exchange 5′ OH ends
between duplexes

rejoin DNA by 5′ OH
attack on 3′ P-tyr

isomerization to
reposition DNA to be
cleaved and recombinase
active sites

heteroduplex DNA at
regions of strand exchange

rejoin DNA by attack
of 5′OH on 3′P

exchange 5′ OH ends
between duplexes

nick non-exchanged strands
to give 3′ P-tyr and 5′OH

Figure 14.49 Cre recombinase promotes recombination between *lox* sites. Four Cre protomers (blue, light blue, pink, and light pink) bind and pair the *lox* sites of two parental duplexes blue and pink in antiparallel fashion (see Figures 14.48, 14.50). The darker shade circles denote an active protomer. (1) In the diagonally opposed Cre molecules (top right blue and bottom left pink circles), the active site tyrosine (circle Y) attacks the phosphodiester backbone of one strand of each parental duplex to form a tyrosine–3′ P-DNA linkage, exposing a free 5′ OH DNA end. (2) The 5′ OH ends are then exchanged between the duplexes such that the tyrosine–3′ P-DNA linkage from one duplex is juxtaposed to the free 5′ OH ends from the other duplex. (3) DNA rejoining then occurs by the attack of the 5′ OHs on the tyrosine–3′ P-DNA linkages. This attack re-forms the DNA phosphodiester linkage and releases the protein. The result of this reciprocal exchange of DNA strands between duplexes is a Holliday junction (see Figure 14.50). (4) The nucleoprotein structure now isomerizes such that the other pair of diagonally positioned Cre proteins are active (note that the darker shade blue and pink circles are now on the top left and bottom right, respectively) (see Figure 14.50); the DNA structure changes slightly (not shown here) to reposition the DNA strands and proteins for the second cycle of strand exchange. (5) Another cycle of single-strand exchange begins, using the other pair of DNA strands. In the diagonally opposed Cre molecules (blue and pink circles), the active tyrosine (circle Y) protein attacks the phosphodiester backbone of one strand to form a tyrosine–3′ P-DNA linkage and expose a free 5′ OH end. (6) The broken 5′ OH ends are then exchanged between the duplexes such that the tyrosine–3′ P-DNA linkage from one duplex is juxtaposed to the free 5′ OH from the other duplex. (7) DNA rejoining then occurs by the attacks of the 5′ OH ends on the tyrosine–3′ P-DNA linkages, releasing the proteins. The recombinant duplexes contain heteroduplex DNA in the spacer regions between the positions of strand exchange.
Adapted from Chen and Rice (2003).

As will be discussed in Chapter 16, the Cre-*lox* system has been extensively used in genome engineering to allow the deletion of defined chromosomal regions at particular times.

14.12 CSSR CONVERSION OF DNA DIMERS TO MONOMERS: SYSTEMS REGULATED BY ACCESSORY PROTEINS

Accessory proteins can control CSSR reactions

In contrast to the resolvase and Cre recombinase systems described above, other CSSR reactions are dependent on proteins in addition to the recombinase, allowing for the regulation of recombination reactions by host factors and by particular

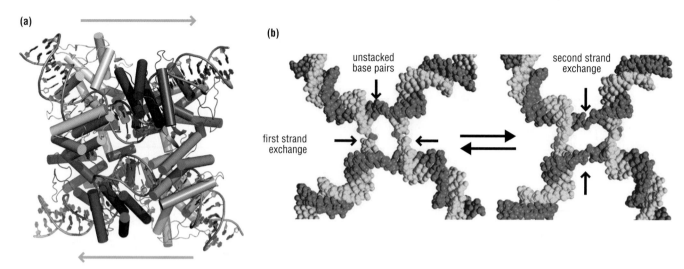

Figure 14.50 The structure of Cre-*lox* recombination complexes. (a) Cre-*lox* complex just before the initiation of DNA breakage and joining. A tetramer of Cre pairs two *lox* sites (gray) in antiparallel orientation as indicated by the green arrows. The darker subunits will perform the first set of strand exchanges (step 1 in Figure 14.49). (b) Isomerization of the Holliday junction intermediate. Reciprocal exchange of one pair of strands results in a Holliday junction (step 3 in Figure 14.49) with several unstacked base pairs at the junction. In the structure on the left, the strand exchange positions (red) in the yellow strands are most closely juxtaposed. During isomerization of the junction (step 4 of Figure 14.49), a modest local change in the structure of the junction DNA more closely juxtaposes the strand exchange positions (red) in the green strands and this second pair of strands can undergo strand exchange. PDBs 2HOI, 3CRK. Figure 14.50a from Ghosh *et al.* (2007); figure 14.50b from Gopaul *et al.* (1998).

environmental conditions. We now discuss two different tyrosine CSSR systems that are regulated by accessory proteins.

Resolution of dimeric bacterial chromosomes is linked to DNA translocation into the daughter cell

⊃ We discuss bacterial chromosome segregation in Section 7.8.

Proper bacterial chromosome segregation and cell division involves the anchoring of the two daughter chromosomes on the two sides of a partition, or septum, that develops in the growing cell prior to replication. A site near the replication origin *oriC*, anchors the chromsome at the distal ends of the dividing cells. DNA is actively pumped into each daughter cell by FtsK, which is positioned at a region of the septum that divides the cells. DNA is pumped in a particular direction (clockwise in Figure 14.31) because particular chromosomal sequences recognized by FtsK are directional.

If the chromosome is a dimer rather than a monomer, one daughter cell will lack a chromosome or the chromosome will be broken when the septum eventually divides the cell into two (see Figure 14.43). Many bacteria, including *E. coli*, have a CSSR system to resolve these hazardous chromosome dimers. The *E. coli* system consists of the tyrosine recombinase Xer, which is composed of two closely related proteins, XerC and XerD, and a recombination site called *dif*, which is located in the region of the chromosome where replication terminates. Under normal circumstances, this termination region is the last region of the replicating chromosome to translocate through the septum into the new daughter cell. Xer acting at the *dif* sites in the replication termination region converts dimeric chromosomes to two single chromosomes, one anchored in each daughter cell. The Xer system is activated by the FtsK DNA translocase. This allows dimer resolution to be coordinated with the late steps of bacterial DNA segregation.

Figure 14.51 Conversion of chromosome dimers to monomers by CSSR at *dif* sites mediated by Xer and FtsK. A cell containing two copies of a circular chromosome is shown. The *oriC* replication origin (green), *ter* replication terminator (red) and *dif* sites (blue with yellow arrow) are shown. (1) Homologous recombination promoted by RecA between monomer chromosomes results in the formation of dimers that cannot be accurately segregated (see Figure 14.43). (2) As cell division proceeds, the DNA translocase FtsK (pink triangles) which is located in two orientations at the septum pumps the DNA into each daughter cell. The replication origins begin to segregate to the opposite poles of the dividing cell. (3) When the replication origins are positioned at the distal ends of the cells, the *dif* sites, which are located in the region of the terminus of replication along with FtsK, are juxtaposed. (4) CSSR at *dif* by FtsK-activated Xer resolves the dimer into two monomers that can be successfully segregated.
Adapted from Val *et al.* (2008).

Bacteriophage lambda integration and excision are highly regulated CSSR events

The CSSR Cre-*lox* system described in the previous section is simple both in its components and directionality. Recombination requires only the recombinase Cre and the *lox* sites, comprising the spacer region and two flanking Cre binding sites. Integration, excision, and inversion can all occur with DNA substrates that possess appropriately oriented *lox* sites. By contrast the directionality of some other CSSR reactions is highly regulated. The integration and excision cycle of the bacteriophage lambda is an example of a highly regulated CSSR system that is mediated by elaborate protein–DNA complexes not unlike those that occur in DNA replication and transcription.

The lambda recombination sites on the phage and bacterial chromosomes are related but distinct

When the bacteriophage lambda infects *E. coli* and enters the lysogenic pathway as described in Section 8.2, CSSR occurs between a special site in the phage genome called *attachment site phage* (*attP*) and a special site in the bacterial genome called *attB* (Figure 14.52). This reaction is mediated by the lambda-encoded tyrosine recombinase Int (integrase), and several accessory proteins. Like the Cre system, *attP* and *attB* share a spacer region called the core region (*O*) in which recombination occurs and which is flanked by Int binding sites *C* and *C'* on *att*P and *B'* and *B* on *att*B (Figure 14.52). The arms of *attP* beyond the Int binding sites flanking the core contain additional protein binding sites and these regions are called the *P* and *P'* arms.

Recombination between *attP* and *attB* generates *attL* and *attR* which flank the integrated prophage; *attL* and *attR* are hybrid sites that contain sequences from both *attP* and *attB*. Prophage excision results from recombination between *attL* and *attR*, which regenerates *attP* and *attB*.

Figure 14.52 Schematic of the bacteriophage lambda integration and excision reactions. The viral chromosome (gray) contains *attP* (blue) which recombines with *attB* (red) in the bacterial chromosome (light gray) during lambda integration. This recombination event forms the prophage, which is flanked by *attL* and *attR*. Excision occurs between *attL* and *attR* and regenerates *attP* and *attB* when the prophage enters the lytic cycle. The *att* sites share a region of homology called the core (*O*) (gray with green arrow) which is flanked by the Int core binding sites *C* and *C′* (dark blue) in *attP* and *B′* and *B* (red) in *attB*. *attP* contains additional recombination protein binding sites in the *P* and *P′* arms (blue); *attB* does not extend beyond the Int core *B′* and *B* sites. The cores are aligned here in antiparallel fashion (green arrows) as in Figure 14.48.

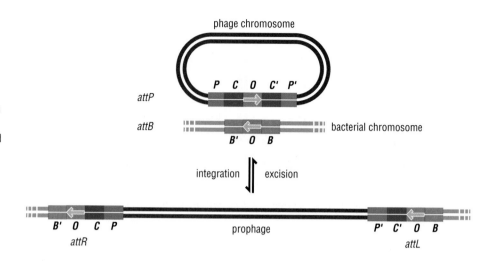

*att*P is large and complex and contains multiple binding sites for additional recombination proteins beyond the Int *C* and *C′* sites that flank the core, *O* (Figure 14.53); no additional protein-binding sites flank the Int *attB* *B* and *B′* sites. Thus each *att* site – *attP*, *attR* and *attL* – is distinct and has different protein binding sites flanking the core.

The unique structure of each *att* site results in particular protein requirements for recombination with these different DNA substrates (Figure 14.53). Integration between *attP* and *attB* requires the phage-encoded integrase and IHF, a host-encoded protein. Excision between *attL* and *attR*, however, requires Int, IHF and two additional proteins, one encoded by the phage, excisionase (Xis) and another host-encoded protein, Factor for Inversion Stimulation (FIS). (As discussed in the last section of this chapter, FIS is an essential protein in a DNA inversion reaction, hence its name). Xis is not present during phage infection when integration occurs.

Multiple recombination proteins are required for the assembly of active intasomes

Like Cre recombination at *lox* sites, Int recombination at *att* sites involves the assembly of a tetramer of Int that binds to the four Int binding sites flanking the regions of homology, pairs the parental duplexes, and carries out strand exchange. The affinity between Cre and *lox* sites and between Cre protomers is sufficiently high that Cre alone can execute recombination. By contrast, the affinities between Int monomers and the *att* core sites, as well as between the Int monomers, are too low to allow the formation of a stable tetramer that can pair the *att* DNA substrates in the presence of only Int; additional proteins are required to form active '**intasomes**'. These additional proteins promote DNA bending to facilitate the binding of the att sites to Int. Furthermore, the different arrangement of the accessory protein binding sites on different *att* sites means that different proteins are required for recombination between different *att* sites. These differential requirements allow for the directional control of recombination, most notably that recombination between *attL* and *attR*, i.e. excision, requires Xis. Xis is only synthesized when the phage lytic functions are derepressed when cellular conditions prompt the phage to shift from the lysogenic state to lytic growth.

Figure 14.53 Structure of the bacteriophage lambda *att* sites. The viral *attP* site contains the core (*O*) flanked by the Int binding sites *C* and *C'* (dark blue) where DNA breakage and joining occur. The *P* arm that flanks the Int *C* site and the *P'* arm that flanks the *C'* site contain multiple binding sites for accessory proteins. Both the *P* and *P'* arms contain Int arm binding sites (blue), whose sequences are distinct from the core *C* and *C'* sites and which are recognized by a different domain of Int (see Figure 14.54). The *P* and *P'* arms also contain binding sites for the host sequence-specific DNA-bending protein integration host factor (IHF) (pink). (The *P* arm contains additional binding sites for proteins important for excision – see below.) The bacterial *attB* site contains the core (*O*), which is flanked by Int-binding sites *B'* and *B* (red) at which that breakage and joining occur. Integration requires Int and IHF, and occurs between *attP* and *attB* to generate *attR* and *attL*. *attR* contains the *B' O C* Int core sites and the *P* arm flanking *C*. In addition to binding sites for Int and IHF, the *P* arm contians binding sites for two proteins required uniquely for excision, the phage-encoded Xis (excisionase) (orange) and host-encoded FIS (green) (factor for inversion stimulation – see Figure 14.58). *attL* contains the *B O C'* sites and the *P'* arm that contains Int arm-binding sites and a binding site for IHF. Excision requires Int, IHF, Xis, and FIS, and occurs between *attR* and *attL* to regenerate *attP* and *attB*.
Adapted from Grindley *et al.* (2006).

Assembly of the *attP* intasome for integration requires IHF and DNA supercoiling

The *attP* *P* and *P'* contains multiple binding sites beyond the core *C* and *C'* sites. Why is the *attP* substrate so elaborate? We now know that the *P* and *P'* arms are required as an assembly platform to position an Int tetramer at the core sites of strand exchange of *attP* and *attB*. The active Int tetramer that mediates integration is initially assembled on *attP* alone and naked *attB* DNA is the recruited into this complex.

Three features are key to the assembly of the Int tetramer in the *attP* intasome. First, Int has two distinct DNA binding domains that recognize two different DNA sequences – an N-terminal domain that binds to the Int arm sites and a C-terminal domain that binds the Int core sites; the C-terminal domain also contains the active site topoisomerase (Figure 14.54a). Int actually binds with higher affinity to the Int arm sites. Thus the binding of 4 Int protomers to 4 Int arm binding sites in *attP* facilitates the recruitment of their C-terminal domains to the core regions.

(a) Int domains

NH₂ — specific binding to Int arm site

specific binding to Int core site

COOH — catalytic tyrosine

(b)

Int arm site   Int arm site

IHF

Int core site   Int core site

Figure 14.54 Int can bind to two different DNA sequences. (a) The domain structure of lambda integrase. Lambda integrase contains two domains: the NH2-terminal domain, which binds to specific sequences in the *P* and *P′* arms of *attP* and the COOH-terminal domain, which binds specifically to the Int sites that flank the core (*O*), *C*, *C′*, *B*, and *B′*, and contains the catalytic topoisomerase-like domain. (b) Int can bind to two DNA segments. Because of its two different sequence-specific DNA-binding domains, the simultaneous binding of one Int protomer to two regions on a single DNA molecule can occur. The bending of the *att* DNA necessary for such bridging is facilitated by DNA bending proteins IHF (as shown), Xis and FIS.

Second, considerable bending of the *P* and *P′* arm DNA is required to allow the simultaneous binding of Int to the arm and core sites of *attP* (Figure 14.54b). The DNA-bending protein IHF, is essential to the bending of *attP*. Third, DNA supercoiling also plays a critical role in promoting the bending and wrapping of the *attP* DNA around the proteins of the *attP* intasome.

Once the *attP* intasome with its four catalytic domains is formed (Figure 14.55), *attB* DNA is thought to enter the reaction as naked DNA, with the *B′* and *B* core sites binding to the pair of unoccupied Int catalytic domains in the already assembled Int tetramer. The Int catalytic domains then act on the *C*, *C′*, *B* and *B′* sites of strand exchange in the paired *attP* and *attB* DNA duplexes, by the mechanism described earlier for Cre (see Figure 14.49). In Experimental approach 14.3 we consider how small molecules have been used as recombination inhibitors to identify intermediates in lambda recombination.

Assembly of the *attL* and *attR* intasomes for excision requires additional accessory proteins

Excisive recombination requires the assembly of *attL* and *attR* intasomes and pairing between them. However, Int and IHF are not sufficient for the formation and interaction of these intasomes. Two additional accessory DNA-bending proteins are required: Xis and FIS. Neither *attL* nor *attR* have Int-arm sites on both sides of the Int core sites to promote assembly of the tetramer on just one of these *att* sites (Figure 14.53). Instead, Xis and FIS binding is required to promote the DNA bending necessary for Int to be able to bind to *both* core and arm sites on attL and attR (Figure 14.56).

As in integration, excision requires the assembly of an active Int tetramer and the wrapping of *attL* and *attR* around that tetramer. In contrast to integration, however, the *P* and *P′* arms that provide the platform for Int tetramer assembly do not lie on the same piece of DNA. The additional bending proteins Xis and FIS

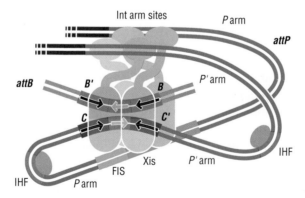

Int arm sites   *P* arm

attP

attB   *B′*   *B*   *P′* arm

C   *C′*

IHF

Xis   *P′* arm

IHF   FIS

P arm

Figure 14.55 Structure of the bacteriophage lambda *attP-attB* intasome required for integration. Integrative recombination occurs in an elaborate nucleoprotein complex. The core regions of *attP* (dark blue) and *attB* (red) bind to a tetramer of Int COOH domains (large blue ovals bound to *C* and *C′*, and large pink ovals bound to *B* and *B′*). The cores of *attB* and *attP* (green arrows) are oriented in antiparallel orientation. Formation of the Int tetramer bound to the core sites is facilitated by the Int NH2 domains (small blue and pink ovals) interacting with Int arm sites in the *P* and *P′* arms. Bending of the *attP* DNA is facilitated by the binding of the DNA bending protein IHF (pink) to the *P* and *P′* arms of *attP*. (Which Int arm binding domains interact with which *attL* and *attR* Int arm binding sequences is not yet known in detail; thus this schematic diagram is meant to indicate only the general path of the *P* and *P′* arms.) Adapted from Van Duyne (2005).

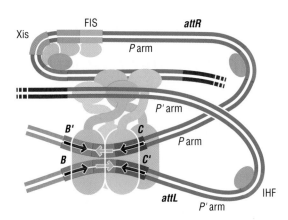

Figure 14.56 Structure of the bacteriophage lambda *attR-attL* intasome required for excision. Excision between *attL* and *attR* occurs within an elaborate nucleoprotein complex in which *attR* and *attL* are paired. The core sites, B'OC in *attR* and BOC' in *attL*, interact with the COOH domains (large blue and pink ovals) of a tetramer of Int. Formation of the tetramer is facilitated by the interaction of the four Int NH2 domains (small blue and pink ovals) with the *P* arm of *attR* and the *P'* arm of *attL*. The bending of the DNAs required for assembly of this complex is facilitated by the binding of the DNA bending proteins IHF (pink), Xis (orange) and FIS (green) to the *P* arm of *attR* and of IHF to the *P'* arm of *attL*. Protein-protein interactions between Ints bound at both the core sites and the arm sites are important to pairing of the *attL* and *attR* DNAs. (Which Int arm binding domains interact with which *attL* and *attR* Int arm binding sequences is not yet known in detail; thus this schematic diagram is meant to indicate only the general path of the *P* and *P'* arms.) Adapted from Van Duyne (2005).

are required to appropriately orient the *P* arm in *attR* such that it can provide a platform for the correct assembly of the Int tetramer. Following the assembly of the intasome containing both *attR* and *attL*, in which the core Int binding sites are appropriately juxtaposed, strand exchange occurs.

The central feature of the lambda recombination directionality control is that recombination cannot occur between *attL* and *attR* in the absence of Xis. Xis (and FIS) are required to promote the bending of *attR*, which provides a DNA platform for the binding of the Int arm domains and, hence, facilitates the assembly of the active tetramer at the core regions. Thus although the same DNA-binding sites are present in the *P* and *P'* arms of *attP*, the separation of the *P* and *P'* arms to the *attR* and *attL* DNAs imposes the requirement for Xis for the assembly of the paired, active *attR-attL* intasomes.

14.13 CSSR SYSTEMS THAT CONTROL GENE EXPRESSION

Throughout this book, we have seen many ingenious ways in which biological systems mediate the regulation of gene expression. We conclude this chapter by considering how CSSR mechanisms provide a further means of controlling the function of the genome.

CSSR deletion reactions can lead to the assembly of new genes

We discussed in Section 14.5 how a transposase has been adapted to mediate the DNA deletion reactions that initiate the combinatorial assembly of gene segments. This combinatorial assembly allows the expression of very diverse

14.3 EXPERIMENTAL APPROACH

Identification of small molecule inhibitors of CSSR

The dissection of a reaction pathway requires identification and analysis of reaction intermediates as well as knowledge of the initial substrates and reaction products. Mechanistic understanding of a pathway can come from the analysis of mutant enzymes (that can only complete certain reaction steps) or the use of modified substrates (that stall the reaction at an intermediate stage). Another powerful strategy for studying reactions is the isolation of small molecule inhibitors that inhibit a process. For many years, chemists and chemical biologists have been developing methods to synthesize large libraries of molecules that can be used in screens to identify inhibitors of a specific biological process. The basic idea behind the use of so-called combinatorial libraries (called this because they are composed of chemical moieties in different combinations) is to apply complex mixtures of compounds to an assay that reports on a biologically interesting event. For example, a screen for antibacterial compounds might start with a growth inhibition assay and a screen for specific biochemical inhibitors might start with purified proteins performing a specific molecular event. The ultimate goal is to identify from within these complex mixtures a small molecule 'lead' compound that inhibits a specific molecular process. The inevitable challenge of such approaches is that the complex mixtures must be 'de-convoluted' to identify specific active molecular species within them, and this requires a careful synthetic plan.

A chemical screening approach was employed by Anca Segall and colleagues to study CSSR mediated by the integrase of bacteriophage lambda. Understanding of the integrase mechanism is incomplete in large part because several of the pathway intermediates are hard to isolate. The Segall group first established an assay that would allow them to screen complex mixtures of small peptides (hexapeptides) to identify inhibitors of the process. By focusing on peptide inhibitors, the researchers reasoned that they might target protein surfaces that are involved in specific molecular interactions, and thus a specific molecular process. As an assay, the CSSR reaction was visualized in a minimal system with radioactively labeled DNA substrate and two purified proteins, Int (the integrase) and IHF (a DNA-binding protein that promotes the bending of DNA), and gel electrophoretic analysis where reaction substrates, products and intermediates have characteristic mobilities (Figure 1). For example, Holliday junction complexes that are reaction intermediates (Figure 14.49) migrate more slowly than the starting *attL* substrate or the recombinant reaction products.

The next step was to ask whether compounds could be identified that resulted in the inhibition of the progress of the reaction or the accumulation of reaction intermediates, such as the Holliday

Figure 1 A biochemical assay for *attL* recombination for screening peptide libraries. A small radioactively labeled *attL* DNA substrate containing the *attL* sequence of a bacteriophage lambda lysogen and a larger unlabeled *attL* DNA substrate are incubated with purified Int and IHF proteins and the products of the reaction resolved on a polyacrylamide gel. The recombinant products, Holliday junction intermediates and covalent Int–DNA complexes are readily seen.
Photo courtesy of Anca Segall (San Diego State University), of reactions like those in Cassell *et al.*, *Journal of Molecular Biology* 2000; **299**:1193–1202.

junction. In a first pass, 120 combinatorial libraries were prepared; in each library one of the 20 amino acids was fixed in a given position in the hexapeptide (6 positions × 20 amino acids = 120) and each contained about 2.5 million distinct molecular species (1 × 19 × 19 × 19 × 19 × 19, excluding cysteine from the randomized amino acid mixtures). Two different concentrations of the library were added to the CSSR reaction, and the progress of the reaction was followed by gel electrophoresis. In Figure 2 we see the analysis of seven different compound mixtures, most of which result in an increase in the amount of Holliday junction intermediate and a decrease in the amount of product (mixture A, by contrast,

Figure 2 Effects of peptide libraries on a CSSR reaction. Peptide mixtures A, C, D, E, F, G, and H refer to the amino acid 'fixed' at position one of the hexamer (OXXXXX where O refers to a specific amino acid and X to random amino acids). Two different concentrations ($1 \times$ and $0.3 \times$) of the peptide mixture were added to the CSSR reaction. All mixtures except A had readily discernible effects on the reaction, increasing the amount of Holliday junction, and decreasing the yield of products.

Figure 3 Inhibition of recombination by peptide libraries with fixed position 1 or 2. The amino acid residue in the fixed position is denoted along the x-axis. Values of percentage recombination were normalized to the extent of recombination in untreated reactions. (a) The most potent libraries carry K, R, P, G, or W at the first position. (b) The most potent libraries carry either W, R, or K at the second position. Subsequent libraries were designed with these positions fixed as indicated by these data.

has little effect). Promising mixtures were subjected to more careful dose-response titrations to determine which mixtures are the most potent for subsequent analysis.

Following this initial screen, other positions in the hexapeptide were 'fixed' according to these initial leads, and new, less complex libraries synthesized around these fixed positions. In Figure 3, we see how the potency of a mixture in the assay can be compared to determine how to 'fix' a given position for synthesis of the next complex mixtures. As such a process is reiterated, the number of molecular species in the synthesized mixture becomes increasingly small, eventually allowing the scientist to identify the most promising leads in the library. The chemical structure of these lead compounds can then be systematically altered to increase the specificity and potency of the compound.

The isolation of these hexapeptide inhibitors has provided a new tool for studying CSSR and may prove useful as an antimicrobial agent if it shows specificity for the bacterial reactions. Such combinatorial library approaches for drug screening are quite powerful in that they allow millions of compounds to be tested simultaneously, thus decreasing the cost of the screening. By requiring the inhibitory activity to emerge from a complex mixture (where each compound is necessarily present at relatively low concentration), the potency of an initial lead must already be reasonably high, and thus worthy of pursuit.

Cassell G, Klemm M, Pinilla C, Segall A. Dissection of bacteriophage lambda site-specific recombination using synthetic peptide combinatorial libraries. *Journal of Molecular Biology* 2000;**299**: 1193–1202.

populations of antibodies and antigen receptors. There are a few examples of CSSR-mediated deletion reactions in bacteria that lead to the assembly of active genes in specialized cell types. In *Bacillus subtilis,* for example, a serine recombinase CSSR deletion reaction generates the gene for a transcription factor required for sporulation by deleting a DNA segment that interrupts the gene in a specialized 'mother cell' which dies during spore formation. Similarly, a tyrosine recombinase CSSR deletion reaction assembles genes involved in the development of the specialized nitrogen-fixing cells of the cyanobacterium *Anabaena*. However, such deletion reactions are relatively rare, as they are not readily reversible and thus represent terminal differentiation events, a state incompatible with bacterial growth.

Some microorganisms change their surface molecules at high frequency by inversion of a promoter-containing DNA segment

Given that the reactions are reversible, there are more examples of CSSR-mediated inversion reactions that control gene expression either by changing the orientation of a promoter with respect to a structural gene, or by changing the identity of the structural gene that is expressed. This type of alteration is most often seen with cell-surface proteins. Such **'phase variation'** enables a microbe to rapidly respond to changes in its environment – for instance, the presence of antibodies – and thereby evade them. Another important feature of phase-variation mechanisms is that they allow a high frequency of change at particular genes without putting the entire genome at risk by a high level of indiscriminate mutation.

While genes for surface antigens on pathogens are most likely to undergo phase variation, this also occurs in genes encoding virulence factors, as well as catabolic and biosynthetic systems in both pathogenic and symbiotic microorganisms.

Inversion of a promoter segment underlies the alternative expression of flagellar proteins in *Salmonella*

A well-studied example of phase variation by DNA inversion is the alternative expression of two types of flagellar protein by the bacterium *Salmonella typhimurium*. Flagella are long whip-like structures on the surface of the bacterium, that enable it to move. The two flagellar proteins expressed by *Salmonella* are antigenically different; therefore, switching from one to the other helps *Salmonella* to circumvent an immune response against the protein expressed previously.

The two flagellar proteins are encoded by two genes, *fljB* and *fliC*, located some distance apart on the bacterial chromosome (Figure 14.57). Which gene is expressed is determined by the orientation of a promoter whose position can be changed by a chromosomal inversion. The FljB protein and the FljA protein (which represses the *fliC* gene) are encoded together in one region while *fliC* is located in a distal region of the chromosome. The *fljB-fljA* region lies adjacent to an invertible DNA segment containing a promoter. When the invertible DNA is in one orientation, transcription from this promoter leads to expression of the flagellar protein FljB and the FljA repressor, which blocks expression of the *fliC* gene. When the promoter segment is in the other orientation, transcription from the promoter does not enter the *fljB-ljA* region so the flagellar FljB protein is not synthesized. In addition, since the repressor protein FljA is not produced, the flagellar FliC protein is now made.

DNA inversion occurs by CSSR between two sites called *hix* sites that lie in inverted orientations on either side of the DNA segment containing the promoter. Each *hix* site consists of the spacer region flanked by two inverted binding sites

Figure 14.57 CCSR-mediated DNA inversion can change the expression of cell surface proteins. The alternative expression of different flagella surface proteins by inversion that changes the relative orientation of a promoter is shown. In *Salmonella*, two different flagellin proteins, FljB and FliC, are encoded by the *fljB* gene (red) and the distal *fliC* gene (dark blue). FljB and a repressor of the *fliC* gene, FljA (orange) can be expressed from a promoter that lies within an adjacent invertible segment. Thus when the promoter is oriented towards the *fljB* and *fljA* genes, the *fljB* gene is expressed and expression of the distal *fliC* gene is blocked by the binding of the FljA repressor. Inversion can occur by recombination between the *hix* (pink and blue) sites, mediated by the serine recombinase Hin that is encoded within the invertible segment. Inversion reorients the promoter such that the *fljB* and *fljA* genes are not expressed, resulting in *fliC* expression.

for the Hin serine recombinase, which is constitutively expressed from a gene in the invertible segment.

Hin inversion is carried out by a sophisticated nucleoprotein complex whose proper assembly is necessary to activate DNA breakage and joining (Figure 14.58). Each *hix* site binds two monomers of Hin, which, like γδ resolvase, mediate a cycle of double-strand breakage and joining to promote strand exchange. Hin-mediated recombination also requires the accessory protein FIS required for bacteriophage lambda excision (Section 14.11). FIS is a DNA-bending protein and binds to specific sites on the invertible segment and is thought to provide a platform for the assembly of the Hin-bound *hix* sites. Assembly of the complex is also facilitated by the host DNA-bending protein HU. Multiple cycles of inversion can occur since, in contrast to lambda recombination, the structure of the recombination sites is not changed by the reaction.

Now that we have learned about the myriad of ways the sequence of a genome can be changed and how exquisitely gene expression and RNA and protein levels are regulated, we turn to a broader view of the genome as a whole. What do we now know about genomes and what do we have yet to learn? We next consider genome structure and evolution.

Figure 14.58 Mechanism of Hin-mediated recombination. The promoter-containing invertible segment is bounded by *hix* sites and also contains an enhancer, a site that can stimulate recombination when positioned in either orientation or at any position with the invertible segment. Dimers of the recombinase Hin (pink and blue) bind to the *hix* sites and dimers of FIS (green) bind to two sites within the enhancer. The right panel shows an assembled 'invertasome' in which the enhancer bound by FIS forms a platform for the juxtaposition of the *hix* sites bound by the Hin dimers, facilitated by the sequence non-specific DNA-bending protein HU (gray). Inversion results from interaction of FIS with Hin that prompts DNA double-strand cleavage at the *hix* sites. Like *res* recombination (see Figure 14.46), *hix* recombination occurs by the rotation of Hin dimers bound on one side of the cleavage site.

✱ SUMMARY

The nature of transposition and CSSR

- Specialized recombination proteins acting at cognate special sequences can promote the rearrangement of defined DNA segments.

- Transposition promotes the movement of discrete DNA segments from place to place in the genome.

- CSSR promotes recombination between pairs of special sites, rearranging the DNA segments containing the sites.

- The mechanisms of DNA breakage and joining are different in transposition and CSSR.

- The three main classes of transposable elements are DNA-only transposons, which move via a DNA intermediate, and LTR and non-LTR elements, which move via an RNA intermediate.

There are multiple types of transposons

- Transposon insertion is accompanied by a target-site duplication, which results from the joining of the transposon to staggered positions of the top and bottom strands of the target DNA.

- DNA synthesis by host polymerases repairs the junction between the newly inserted transposon and the target DNA.

Transposons in nature

- Transposable elements are present in virtually all genomes.

- Transposons usually comprise a small fraction, <10%, of bacterial genomes.

- The rapid replication of bacteria probably selects against the accumulation of transposable elements.

- In eukaryotes, transposable elements can constitute a significant fraction of a genome, for example about 45% of the human genome.

- Transposable elements have great potential to introduce genetic change.

- Transposable element insertion can result in gene disruption and the juxtaposition of element-encoded regulatory signals such as enhancers to host genes.

- Homologous recombination between transposable element copies can result in deletions, duplications, and translocations.

- There are multiple examples of human disease resulting from transposable element insertion and from chromosome rearrangements due to homologous recombination between element copies.

DNA-only transposons

- Movement of DNA-only transposons involves the direct nucleophilic attack of a 3′ OH of the transposon end on the target DNA.

- The transposases that catalyze these reactions share a common catalytic domain.

- The reactions used by DNA-only elements have been co-opted in the immune system to mediate the assembly of antibody and antibody-receptor genes.

- The RAG recombinase that mediates these reactions is related to transposases.

LTR elements

- LTR retrotransposons include retroviruses, and retroviral-like elements that do not leave the cell.

- LTR element transposition begins with the synthesis of an RNA copy of the element using an integrated DNA form as a template.

- The element RNA is the template for an element-encoded reverse transcriptase that generates a double-stranded DNA copy of the element.

- The 3′ OH ends of the DNA copy of an LTR element integrate into the target DNA by the same mechanism as the DNA-only transposons.

- The catalytic domain of the LTR integrase is related to that of the DNA-only transposase.

Non-LTR elements

- Non-LTR elements insert into a new site by reverse transcription of an RNA copy of the element using a 3′ OH in the target DNA to serve as the target for reverse transcription.

- Active non-LTR elements are present in the human genome and their transposition has resulted in disease-inducing mutations.

Control of transposition

- Given the potentially deleterious effects of transposon movement, transposition is tightly controlled at the levels of transposon expression and target-site selection.

- RNAi and DNA methylation are genome-wide defenses that have evolved to protect the genome against transposable elements.

Conservative site-specific recombination

- CSSR between specific pairs of sequences can mediate a variety of DNA transactions.

- CSSR between sites on different DNAs allows for the joining (integration) of DNA molecules.

- CSSR between sites on the same molecule in direct orientation can result in excisions of the DNA between the sites or the resolution of DNA dimers to monomers.

- CSSR can mediate the integration and excision cycles of viruses.

- CSSR between sites on the same molecule in inverted orientation can result in inversion of the DNA between the sites and the regulation of gene expression by changing the orientation of a promoter.

- CSSR is reciprocal and occurs by DNA breakage and joining in the absence of DNA synthesis.

- CSSR is carried out by topoisomerases through a mechanism in which DNA breakage is accompanied by the covalent linkage between the protein and the DNA via either a serine or a tyrosine residue.

 FURTHER READING

GENERAL

Craig NL, Craigie R, Gellert M, Lambowitz AM, eds. *Mobile DNA II.* Washington DC: American Society for Microbiology, 2002.

14.1 TRANSPOSABLE ELEMENTS: OVERVIEW

Curcio MJ, Derbyshire KM. The outs and ins of transposition: from mu to kangaroo. *Nature Reviews Molecular Cell Biology* 2003;**4**:865–877.

Hedges DJ, Batzer MA. From the margins of the genome: mobile elements shape primate evolution. *Bioessays* 2005;**27**:785–794.

Jones RN. McClintock's controlling elements: the full story. *Cytogenetic and Genome Research* 2005;**109**:90–103.

Kazazian HH Jr. Mobile elements: drivers of genome evolution. *Science* 2004;**303**:1626–1632.

Lander ES, Linton LM, Birren B, *et al.* International Human Genome Sequencing Consortium. Initial sequencing and analysis of the human genome. *Nature* 2001;**409**:860–921.

Volff JN. Turning junk into gold: domestication of transposable elements and the creation of new genes in eukaryotes. *Bioessays* 2006;**28**:913–922.

14.2 DNA-ONLY TRANSPOSONS

Feschotte C, Pritham EJ. DNA transposons and the evolution of eukaryotic genomes. *Annual Review of Genetics* 2007;**41**:331–368.

Ivics Z, Li MA, Mátés L, *et al.* Transposon-mediated genome manipulation in vertebrates. *Nature Methods* 2009;**6**:415–422.

Siguier P, Filée J, Chandler M. Insertion sequences in prokaryotic genomes. *Current Opinion in Microbiology* 2006;**9**:526–531.

William S. Reznikoff Transposon *Tn5. Annual Review of Genetics* 2008;**42**:269–286.

14.3 MECHANISM OF DNA-ONLY CUT-AND-PASTE TRANSPOSITION REACTIONS

Gueguen E, Rousseau P, Duval-Valentin G, Chandler M. The transpososome: control of transposition at the level of catalysis. *Trends in Microbiology* 2005;**13**:543–549.

Haniford DB. Transpososome dynamics and regulation in *Tn10* transposition. *Critical Reviews in Biochemistry and Molecular Biology* 2006;**41**:407–424.

Rice PA, Baker TA. Comparative architecture of transposase and integrase complexes. *Nature Structural and Molecular Biology* 2001;**8**:302–307.

14.4 MECHANISM OF DNA-ONLY NICK-AND-PASTE TRANSPOSITION REACTIONS

Harshey RM, Jayaram M. The mu transpososome through a topological lens. *Critical Reviews in Biochemistry and Molecular Biology* 2006;**41**:387–405.

14.5 CELLULAR DOMESTICATION OF A DNA CUT-AND-PASTE TRANSPOSASE IN ADAPTIVE IMMUNITY

Lieber MR, Yu K, Raghavan SC. Roles of nonhomologous DNA end joining, V(D)J recombination, and class switch recombination in chromosomal translocations. *DNA Repair* 2006;**5**:1234–1245.

Mahowald GK, Baron JM, Sleckman BP. Collateral damage from antigen receptor gene diversification. *Cell* 2008;**135**:1009–1012.

Matthew AG, Oettinger MA. Regulation of RAG transposition. *Adv Exp Med Biol* 2009;**650**:16–31.

Oettinger MA. Molecular biology: hairpins at split ends in DNA. *Nature* 2004;**432**:960–961.

Schatz DG, Spanopoulou E. Biochemistry of V(D)J recombination. *Current Topics in Microbiology and Immunology* 2005;**290**:49–85.

Sinzelle L, Izsvák Z, Ivics Z. Molecular domestication of transposable elements: from detrimental parasites to useful host genes. *Cellular and Molecular Life Sciences* 2009;**66**:1073–1093.

14.6 RETROTRANSPOSONS

Callinan PA, Batzer MA. Retrotransposable elements and human disease. *Genome Dynamics* 2006;**1**:104–115.

Eickbush TH, Jamburuthugoda VK. The diversity of retrotransposons and the properties of their reverse transcriptases. *Virus Research* 2008;**134**:221–234.

Hedges DJ, Batzer MA. From the margins of the genome: mobile elements shape primate evolution. *Bioessays* 2005;**27**:785–794.

Medhekar B, Miller JF. Diversity-generating retroelements. *Current Opinion in Microbiology* 2007;**10**:388–395.

14.7 LTR RETROTRANSPOSONS

Engelman A, Cherepanov P. The lentiviral integrase binding protein LEDGF/p75 and HIV-1 replication. *PLoS Pathogens* 2008;**4**:e1000046.

Hazuda D, Iwamoto M, Wenning L. Emerging pharmacology: inhibitors of human immunodeficiency virus integration. *Annual Review of Pharmacology and Toxicology* 2009;**49**:377–394.

Jern P, Coffin JM. Effects of retroviruses on host genome function. *Annual Review of Genetics* 2008;**42**:709–732.

Lewinski MK, Bushman FD. Retroviral DNA integration–mechanism and consequences. *Advances in Genetics* 2005;**55**:147–181.

Nowotny M. Retroviral integrase superfamily: the structural perspective. *EMBO Reports* 2009;**10**:144–151.

14.8 NON-LTR RETROTRANSPOSONS

Babushok DV, Kazazian HH Jr. Progress in understanding the biology of the human mutagen LINE-1. *Human Mutation* 2007;**28**:527–539.

Belancio VP, Hedges DJ, Deininger P. Mammalian non-LTR retrotransposons: for better or worse, in sickness and in health. *Genome Research* 2008;**18**:343–358.

Callinan PA, Batzer MA. Retrotransposable elements and human disease. *Genome Dynamics* 2006;**1**:104–115.

Cordaux R. The human genome in the LINE of fire. *Proceedings of the National Academy of Sciences of the U S A* 2008;**105**:19033–19034.

Lambowitz AM, Zimmerly S. Mobile group II introns. *Annual Review of Genetics* 2004;**38**:1–35.

Robart AR, Zimmerly S. Group II intron retroelements: function and diversity. *Cytogenetic and Genome Research* 2005;**110**:589–597.

Toro N, Jiménez-Zurdo JI, García-Rodríguez FM. Bacterial group II introns: not just splicing. *FEMS Microbiology Reviews* 2007;**31**:342–358.

14.9 CONTROL OF TRANSPOSITION

Aravin AA, Hannon GJ, Brennecke J. The Piwi-piRNA pathway provides an adaptive defense in the transposon arms race. *Science* 2007;**318**:761–764.

Beauregard A, Curcio MJ, Belfort M. The take and give between retrotransposable elements and their hosts. *Annual Review of Genetics* 2008;**42**:587–617.

Dooner HK, Weil CF. Give-and-take: interactions between DNA transposons and their host plant genomes. *Current Opinion in Genetics and Development* 2007;**17**:486–492.

Ebina H, Levin HL. Stress management: how cells take control of their transposons. *Molecular Cell* 2007;**27**:180–181.

Goodier JL, Kazazian HH Jr. Retrotransposons revisited: the restraint and rehabilitation of parasites. *Cell* 2008;**135**:23–35.

Malone CD, Hannon GJ. Small RNAs as guardians of the genome. *Cell* 2009;**136**:656–668.

Nagy Z, Chandler M. Regulation of transposition in bacteria. *Research in Microbiology* 2004;**155**:387–398.

O'Donnell KA, Boeke JD. Mighty Piwis defend the germline against genome intruders. *Cell* 2007;**129**:37–44.

Weil C, Martienssen R. Epigenetic interactions between transposons and genes: lessons from plants. *Current Opinion in Genetics and Development* 2008;**18**:188–192.

14.10 CONSERVATIVE SITE-SPECIFIC RECOMBINATION: OVERVIEW

Grindley ND, Whiteson KL, Rice PA. Mechanisms of site-specific recombination. *Annual Review of Biochemistry* 2006;**75**:567–605.

Mazel D. Integrons: agents of bacterial evolution. *Nature Reviews Microbiology* 2006;**4**:608–620.

Rajeev L, Malanowska K, Gardner JF. Challenging a paradigm: the role of DNA homology in tyrosine recombinase reactions. *Microbiology and Molecular Biology Reviews* 2009;**73**:300–309.

Segall AM, Craig NL. New wrinkles and folds in site-specific recombination. *Molecular Cell* 2005;**19**:433–435.

Seth-Smith H, Croucher NJ. Genome watch: breaking the ICE. *Nature Reviews Microbiology* 2009;**7**:328–329.

14.11 CSSR CONVERSION OF DNA DIMERS TO MONOMERS: SINGLE-PROTEIN SYSTEMS

Lee L, Sadowski PD. Lox and Cre sandwich. *Nature Chemical Biology* 2005;**1**:246–247.

Mouw KW, Rowland SJ, Gajjar MM, *et al.* Architecture of a serine recombinase-DNA regulatory complex. *Molecular Cell* 2008;**30**:145–155.

Van Duyne GD. Teaching Cre to follow directions. *Proceedings of the National Academy of Sciences of the U S A* 2009;**106**:4–5.

14.12 CSSR CONVERSION OF DNA DIMERS TO MONOMERS: SYSTEMS REGULATED BY ACCESSORY PROTEINS

Lesterlin C, Barre FX, Cornet F. Genetic recombination and the cell cycle: what we have learned from chromosome dimers. *Molecular Microbiology* 2004;**54**:1151–1160.

Radman-Livaja M, Biswas T, Ellenberger T, Landy A, Aihara H. DNA arms do the legwork to ensure the directionality of lambda site-specific recombination. *Current Opinion in Structural Biology* 2006;**16**:42–50.

Van Duyne GD. Lambda integrase: armed for recombination. *Current Biology* 2005;**15**:R658–660.

Van Duyne GD. Teaching Cre to follow directions. *Proceedings of the National Academy of Sciences of the U S A* 2009;**106**:4–5.

14.13 CSSR SYSTEMS IN THE CONTROL OF GENE EXPRESSION

Carrasco CD, Holliday SD, Hansel A, Lindblad P, Golden JW. Heterocyst-specific excision of the Anabaena sp. strain PCC 7120 hupL element requires xisC. *Journal of Bacteriology* 2005; **187**:6031–6038.

Cerdeño-Tarraga AM, Patrick S, Crossman LC, *et al.* Extensive DNA inversions in the B. fragilis genome control variable gene expression. *Science* 2005;**307**:1463–1465.

Dhar G, Heiss JK, Johnson RC. Mechanical constraints on Hin subunit rotation imposed by the Fis/enhancer system and DNA supercoiling during site-specific recombination. *Molecular Cell* 2009;**34**:746–759.

15 GENOMICS AND GENETIC VARIATION

INTRODUCTION

Throughout this book we have explored the individual genes, proteins, functional RNA molecules, and metabolites that form the component parts of an organism, and that work together to bring the genetic information carried by that organism to life. Most of these components are similar in different species. Proteins that repair or replicate DNA and those that transcribe and translate mRNA are found in all species and carry out similar functions. Indeed, some of the most compelling evidence for evolution stems from the high degree of similarity in the molecules found among all species.

In this penultimate chapter, we look beyond the individual molecules, and the way they interact to carry out the instructions of our genetic material, to describe the entire collection of genes, proteins, and other functional molecules that act in concert to make up a living organism. The complete collection of genes in a species is known as its **genome**, so the branch of biology that studies genomes is called genomics. Genomics uses the DNA sequence of an organism's genome as a reference point for experimental analyses.

The field of genomics studies both the similarities and differences between the genomes of living organisms. With this in mind, this chapter has two key themes: the evolutionary conservation of genomes on the one hand and the variation that leads to differences between individuals and species on the other. Our first theme is an exploration of the *unity* of living things, that is, the features of the genome which are evolutionarily conserved. Because so many genes and gene products are shared within and between species, genomics illustrates the unity of living things and the descent of species from common ancestors.

Genomics also introduces us to the *differences* between species. Humans are not the same as fruit flies although as many as three-quarters of human genes have an equivalent or similar gene in fruit flies. We are also not the same as chimpanzees, as although we have nearly the same complement of genes, these genes differ by more than 1% at the nucleotide level. Furthermore, because of sequence variation even individuals within a species are not the same despite virtually all of their component molecules being identical. Variation at the level of the genome drives variation at the level of the molecule – and of the organism of which the molecule is a component part. This variation links genomics and evolution and represents the second key theme of this chapter – the biological diversity between species and within a species that emerges from genetic variation.

We begin our exploration of the field of genomics by considering how we study genomes, and the kinds of information these studies can reveal.

15.1 GENOME SEQUENCES AND SEQUENCING PROJECTS

The functional information in a DNA molecule is carried by its nucleotide sequence. The nucleotide sequence encodes the properties of a functional RNA or the amino acid sequence of a protein product of a gene and provides the regulatory signals that direct gene expression. As it became possible to deduce the function of individual genes by determining their nucleotide sequences, the next step was to extend this study to entire genomes – to establish how all of the genes within an organism operate together as a unified whole.

The genome as a whole encapsulates the functional information for every gene – both the coding region and its regulatory region. If the nucleotide sequence of a single gene could suggest its function, so the nucleotide sequence of the whole genome could provide the 'parts list' for the entire organism, as well as many, if not all, of the assembly instructions for using those parts. One of the major challenges facing researchers today is to decipher this parts list, and establish how the different parts work together to generate the fully functional organism.

Many genomes have been sequenced and analyzed

Thousands of genomes have now been sequenced, mostly from bacteria but many from eukaryotes as well. Indeed, all of the model organisms that are used for research purposes have had their genomes sequenced. Genome sequencing projects are also completed or underway for nearly every species that has medical, agricultural, or other economic importance. Other genome projects are focused on species of evolutionary and ecological importance. For example, the genome of the hemichordate *Ciona* has been sequenced because this species represents an important evolutionary transition from invertebrates to vertebrates.

It should be noted that even for genomes that are considered to be completely sequenced, the sequence is often not 100% complete or accurate and the degree of completeness varies between organisms. A 'completed' sequence can have hundreds of small gaps where it has not been possible to amplify the DNA or determine the sequence accurately. In addition, the completed sequence will contain some errors. For example, the exact sequence in regions containing many repeats is sometimes difficult to decipher. Nevertheless, even when the sequence is not 100% accurate, genome projects provide useful and biologically significant information.

Genome projects have revolutionized molecular biology

Having a complete genome sequence available for reference is an enormous benefit for researchers, since they know the nucleotide and inferred amino acid sequence, the splicing pattern, and much more about any individual gene without having to obtain that information themselves. Experiments can be designed quickly on the basis of genome sequence information without having to perform some of the standard cloning and sequence analyses that were necessary even a few years ago.

⊜ Genome sequencing techniques are described in Section 16.14.

The benefit of having sequenced genomes goes beyond exploring individual genes. It is possible to learn quickly about all of the related genes in an organism and in other organisms. This opens up a range of new experimental possibilities, such as probing the function of the genes that are conserved, and identifying similar regions across different organisms. It allows investigators to analyze the organism as a species, with the knowledge of the full complement of genes potentially involved. This has profoundly changed the focus of molecular biology from the characterization of single genes to systematic multi-gene analysis.

Genome sequence information is also used to answer questions about genome structure. For example, what is the distribution of transposable elements and genes? What is the distribution, structure, and function of the non-coding regions of the genome? Although these questions and many more could be approached by studying the process in individual species, genome projects offer deeper insights because they provide information about what has been conserved across species.

The genome sequences of individuals can be determined

With the rapid advances in technology, it has become possible to obtain the genome sequence of individual organisms. The examination of a collection of these sequences allows the variation among individuals to be studied. How frequent is a mutation and where do mutations occur? What are the rates of mutation in different regions of the genome? At a practical level, the sequence of the human genome is extremely valuable for the understanding, diagnosis, and treatment of human diseases. What mutations are associated with cystic fibrosis or breast cancer? What alleles are associated with detrimental side effects of certain drugs? As such, the availability of the genome sequence is revolutionizing medicine. However, the ability to obtain the genome sequence of an individual also raises ethical questions, as will be discussed in Section 15.9.

Genome projects can be applied to ecological niches

One application of genomics is the field of metagenomics. Metagenomics involves obtaining collective sequence information, or the **metagenome**, for all of the organisms that occupy a particular ecological niche, particularly the microbes. This niche could be a particular pond, a soil habitat, a deep-sea vent, or even the microbes on the surface of human skin. The genome sequence information can be used to compile a profile of the population of a particular niche, similar to that performed with ecological population studies of animals such as bird counts. This approach greatly expands genome analysis from individual organisms to their interactions with each other and their environment. It has been suggested that this approach, applied repeatedly over time, could indicate changes in a habitat perhaps due to climate change, or other ecological or industrial activity.

Metagenome sequencing data for microbes reveal information about many more species and from a wider range of niches than other forms of population sampling used to study microbes. This is because sequence information can be obtained from organisms that are not possible to culture in the laboratory. Metagenomics bypasses the obstacle of culturing cells because it is not the organism itself but its DNA that is obtained and amplified. Organisms can be identified based on the DNA sequences, relationships between organisms can be analyzed, and comparisons between niches can be made through the sampling of whole populations of microbes from different environments.

Beyond telling us about ecological implications, metagenomics can offer new insights into human health and disease. For example, the population profile of the bacterial inhabitants of our digestive tract is large and diverse, but can be compiled by obtaining DNA sequences from these microbes. Studies in mice and humans using these metagenomic profiles have found that some obese and lean individuals have different populations of microbes living in the gut. Furthermore, when the guts of lean mice are colonized with bacteria from the guts of obese mice, the lean mice rapidly gain weight. This indicates that, in addition to the diet and genotype of the individual, microbes in the gut can play a role in obesity. This is a striking example of the application of genomics to a novel biological problem. Genome studies reveal the interacting parts in an organism; metagenome studies give insights into the organism as part of a system of interacting species.

Genomes have many different structural and functional elements

Much attention in genome sequencing projects is focused on the protein-coding genes, that is, the regions transcribed into mRNA and translated into the amino acid sequence of a polypeptide. However, for some species, including humans and plants, the protein-coding sequences are a small minority of the total genomic sequence, as illustrated in Figure 15.1. For example, only around 1.5% of the sequence in the human genome codes for proteins.

A major realization coming from genomic analysis is that a substantial fraction of the genome is in fact transcribed without the RNAs being subsequently translated into protein. For some of these transcripts, the functional end product is the RNA itself. Such functional RNA molecules include stable transcripts such as rRNAs, tRNAs, and X-chromosome silencing RNA, among others. Also included are small regulatory RNA molecules, such as siRNAs, miRNAs, and piRNAs. However, the function of some of the untranslated RNAs is not known. Indeed, many may not even have a function.

➲ We learn about the processing and functions of the various regulatory RNAs in Sections 8.15, 9.9, 9.10, and 10.15.

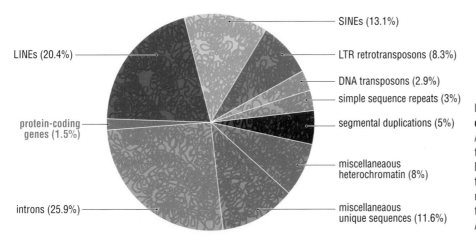

LINEs (20.4%)

protein-coding genes (1.5%)

introns (25.9%)

SINEs (13.1%)

LTR retrotransposons (8.3%)

DNA transposons (2.9%)

simple sequence repeats (3%)

segmental duplications (5%)

miscellaneaous heterochromatin (8%)

miscellaneaous unique sequences (11.6%)

Figure 15.1 Less than 1.5% of the genome corresponds to the protein-coding sequence. As shown by this pie chart, the majority of the human genome sequence consists of DNA encoding intronic sequences and the transposons and long and short interspersed nuclear elements (LINEs and SINEs) elements that are described in Chapter 14. Adapted from Gregory (2005).

In addition to protein-coding and RNA-coding genes, genome sequence analyses have allowed the fine structures of genomes to be examined, including repeated sequences which range from tandemly repeated dinucleotides to repeats several kilobases in length. These repeated sequences vary in copy number from a few to tens of thousands per genome. Included among the repeated sequences are transposable elements or remnants of transposable elements, which were described in detail in Chapter 14. These elements are often found in different locations and in different copy numbers in different individuals.

There are other sequence elements in a genome whose functions are not well understood, and which may not even have an identifiable function. Therefore, a significant challenge in genome analysis is how to identify the sequence elements within a genome that have a true function, the topic we consider next.

15.2 FINDING FUNCTIONS IN A GENOME

Understanding the functions of genes and genomes is a monumental task. Two general approaches are being taken by investigators to elucidate the functions of genes and genomes. In one approach, called **comparative genomics**, insights into potential functions are obtained by comparing sequences. The comparisons can be carried out between individual genes or regions, or even the sequences of whole genomes. The regions of sequence similarity highlight what has been conserved in evolution and can be used to infer function.

The second general approach, termed **functional genomics**, is based on genetic, molecular, biochemical, and cytological experiments, designed to study specific gene products or chromosomal regions to elucidate their functions. Initially, the methods were developed to explore the functions of single genes or single protein complexes. However, with the availability of complete genome sequences, these assays can now be carried out on a genome-wide level.

This chapter will explore what can be learned from both comparative and functional genomics, as together they provide the most complete understanding of the genome.

Genomic analysis begins with genome annotation

The raw sequence of a genome is simply a string of adenosine, thymidine, guanosine, and cytidine nucleotides, albeit a string that is millions or billions of nucleotides in length. How, then, are the different coding and non-coding sequence elements in a genome identified and their functions defined? The process of identifying functional sequence elements in a long sequence string is termed **genome annotation**. Genome annotation uses computer analysis to identify: specific sequence elements such as protein-coding regions or open reading frames (ORFs); sequences that correspond to potential promoter elements or splice sites for genes; or other elements such as tRNA and rRNA genes, all of which have well-established conserved sequences that are easily identified in genomic data.

An increasing number of programs that search a genome sequence for specified features (and particularly the *combination* of features) are being developed.

For example, programs can search for ORFs, canonical promoter sequences or ribosome binding sites, or sequence elements corresponding to splice junctions that define introns. It should be noted, however, that each program has limitations. For example, true protein-coding ORFs become increasingly difficult to distinguish from dubious ORFs as the length decreases. As a result, annotation of protein-coding genes often stops at an arbitrary length, such as 100 amino acids in yeast. Thus programs constantly need to be updated as new information about functional elements becomes available.

No single approach identifies every element, and new functional elements are always being identified and characterized. Thus genome annotation is an open-ended process and online genome databases require frequent updating.

Sequence comparisons allow rapid genome annotation

One of the main themes of this chapter is that the function of a genome sequence element can often be inferred by comparisons with other sequences, either in the same genome or in other sequenced genomes. After all, since the function of a genetic element is encoded in its nucleotide or amino acid sequence, similar sequences are likely to have similar functions. We return to this topic in some more detail below, but for now we will note only the most compelling conclusion: most genes in one species have corresponding genes in other species.

In some cases, the similarity in sequence can be enough to provide specific information about the function of the genes: the amino acid sequence of the histone gene H4 differs by only seven amino acids (of 103 amino acids total) between *Neurospora* and humans, for example (Figure 15.2). This conservation of sequence implies a strong conservation of function. By knowing the function of a gene or gene product in other organisms, we can infer the function of a similar gene or gene product in humans without conducting direct tests. Indeed, in many cases, a given human gene product can substitute for the function of the equivalent gene product in model organisms, allowing studies of the gene product that are not possible in humans.

Homo sapiens	MSGRGKGGKGLGKGGAKRHRKVLRDNIQGITKPAIRRLARRGGVKRISGLIYEETRGVLKVFLENVIRDAVTYTEHAKRKTVTAMDVVYALKRQGRTLYGFGG
Mus musculus	MSGRGKGGKGLGKGGAKRHRKVLRDNIQGITKPAIRRLARRGGVKRISGLIYEETRGVLKVFLENVIRDAVTYTEHAKRKTVTAMDVVYALKRQGRTLYGFGG
Bos taurus	MSGRGKGGKGLGKGGAKRHRKVLRDNIQGITKPAIRRLARRGGVKRISGLIYEETRGVLKVFLENVIRDAVTYTEHAKRKTVTAMDVVYALKRQGRTLYGFGG
Danio rerio	MSGRGKGGKGLGKGGAKRHRKVLRDNIQGITKPAIRRLARRGGVKRISGLIYEETRGVLKVFLENVIRDAVTYTEHAKRKTVTAMDVVYALKRQGRTLYGFGG
Monodelphis domestica	MSGRGKGGKGLGKGGAKRHRKVLRDNIQGITKPAIRRLARRGGVKRISGLIYEETRGVLKVFLENVIRDAVTYTEHAKRKTVTAMDVVYALKRQGRTLYGFGG
Drosophila melanogaster	MTGRGKGGKGLGKGGAKRHRKVLRDNIQGITKPAIRRLARRGGVKRISGLIYEETRGVLKVFLENVIRDAVTYTEHAKRKTVTAMDVVYALKRQGRTLYGFGG
Apis mellifera	MTGRGKGGKGLGKGGAKRHRKVLRDNIQGITKPAIRRLARRGGVKRISGLIYEETRGVLKVFLENVIRDAVTYTEHAKRKTVTAMDVVYALKRQGRTLYGFGG
Anopheles gambiae	MTGRGKGGKGLGKGGAKRHRKVLRDNIQGITKPAIRRLARRGGVKRISGLIYEETRGVLKVFLENVIRDAVTYCEHAKRKTVTAMDVVYALKRQGRTLYGFGG
Caenorhabditis elegans	MSGRGKGGKGLGKGGAKRHRKVLRDNIQGITKPAIRRLARRGGVKRISGLIYEETRGVLKVFLENVIRDAVTYCEHAKRKTVTAMDVVYALKRQGRTLYGFGG
Caenorhabditis briggsae	MSGRGKGGKGLGKGGAKRHRKVLRDNIQGITKPAIRRLARRGGVKRISGLIYEETRGVLKVFLENVIRDAVTYCEHAKRKTVTAMDVVYALKRQGRTLYGFGG
Arabidopsis thaliana	MSGRGKGGKGLGKGGAKRHRKVLRDNIQGITKPAIRRLARRGGVKRISGLIYEETRGVLKIFLENVIRDAVTYTEHARRKTVTAMDVVYALKRQGRTLYGFGG
Oryza sativa	MSGRGKGGKGLGKGGAKRHRKVLRDNIQGITKPAIRRLARRGGVKRISGLIYEETRGVLKIFLENVIRDAVTYTEHARRKTVTAMDVVYALKRQGRTLYGFGG
Zea mays	MSGRGKGGKGLGKGGAKRHRKVLRDNIQGITKPAIRRLARRGGVKRISGLIYEETRGVLKIFLENVIRDAVTYTEHARRKTVTAMDVVYALKRQGRTLYGFGG
Vitis vinifera	MSGRGKGGKGLGKGGAKRHRKVLRDNIQGITKPAIRRLARRGGVKRISGLIYEETRGVLKIFLENVIRDAVTYTEHARRKTVTAMDVVYALKRQGRTLYGFGG
Neurospora crassa	MTGRGKGGKGLGKGGAKRHRKILRDNIQGITKPAIRRLARRGGVKRISAMIYEETRGVLKTFLEGVIRDAVTYTEHAKRKTVTSLDVVYALKRQGRTLYGFGG
Schizosaccharomyces pombe	MSGRGKGGKGLGKGGAKRHRKILRDNIQGITKPAIRRLARRGGVKRISALVYEETRAVLKLFLENVIRDAVTYTEHAKRKTVTSLDVVYSLKRQGRTIYGFGG
Ustilago maydis	MSGRGKGGKGLGKGGAKRHRKILRDNIQGITKPAIRRLARRGGVKRISGLIYDETRGVLKLFLESVIRDSVTYTEHAKRKTVTSLDVVYALKRQGRTLYGFGA
Candida glabrata	MSGRGKGGKGLGKGGAKRHRKILRDNIQGITKPAIRRLARRGGVKRISGLIYEEVRAVLKSFLESVIRDAVTYTEHAKRKTVTSLDVVYALKRQGRTLYGFGG

Figure 15.2 Sequence comparisons allow functions to be inferred. The alignment of the human histone H4 protein sequence with those in various eukaryotes shows the extensive sequence conservation among these proteins; the only differences between them are indicated in pink. The high level of conservation suggests that all of these H4 proteins have similar function.

Frequently, the inferred amino acid sequence can provide some information about the biochemical activity of a protein, such as DNA binding or kinase activity, but not necessarily specific information about the cellular requirements for that activity. In many cases, a domain or amino acid motif in a protein of interest is similar to the sequence of a domain or motif in another protein whose function is known. This helps to predict the function of the protein of interest by extrapolating from known to unknown. For instance, the zinc finger domain binds to nucleic acids (usually DNA) in a sequence-specific fashion. If the amino acid sequence of an untested protein contains zinc finger domains, we can predict from the presence of the domain that the untested protein functions as a DNA-binding protein.

Although sequence comparisons greatly facilitate annotation and can suggest biological functions, additional genetic, molecular, biochemical or cytological experiments are always required to test the suggested functions. Thus, in the example of the zinc finger domain protein, sequence similarity only suggests that the protein binds to DNA; the sequence comparisons do not prove that the protein actually binds nor do they reveal what function the protein performs if in fact it does bind.

We now turn to the many other experimental approaches that are carried out to study gene and genome function. Traditionally, these experiments were carried out on individual genes or gene products, but the availability of completed genome sequences means that they are now being applied on a genome-wide level.

Genetic analysis uses mutant phenotypes to identify functions

Historically, the presence of mutations that lead to a change in phenotype have been used to identify the function of a genetic element. As described in Chapter 1, a phenotype is a trait that can be measured in an organism. For example, the function of the *white* gene for eye color in *Drosophila melanogaster* was inferred from mutations long before DNA was identified as the genetic material, let alone before DNA sequencing was possible. Changes that eliminate the function of the gene change the eye color from red to white – a mutant phenotype (shown in Figure 1.19). Thus, the function of the mutated gene was inferred to be in some aspect of eye pigmentation, although it was not known for many decades that the specific function of the *white* gene is in the transport of pigment granules.

Thousands of genes in dozens of different organisms from all domains of life have been assigned functions by the presence of mutations that disrupt or change function. For many research organisms including humans, the genome project grew directly out of the study of mutations, the creation of genetic maps, and the analysis of individual genes.

The functions of both coding and non-coding sequences are identified by molecular and biochemical assays

The functions of genes can also be elucidated by biochemical and molecular approaches. Much of this book has been devoted to proteins, RNAs, and genomic sequences whose functions we now understand through assays, purification, and molecular cloning. For example, purified proteins with identifiable biochemical activities have been characterized and, ultimately, the genes encoding the proteins identified. Protein interactions have been studied to learn which proteins are found in specific macromolecular complexes, thus helping to elucidate the function of a specific protein through its presence in a complex. In addition, insights into gene function have been obtained from the patterns of their expression and the subcellular location of the gene products.

➲ Many other molecular, genetic, and biochemical assays for function are illustrated in the Experimental approach boxes throughout the book, and are described in Chapter 16.

Functional genomic assays seek to test the function of every gene in the genome

One of the principal changes introduced by genomics is the possibility of genome-wide assays – the ability to examine the function or expression of every gene in a genome. Often genome-wide assays are done by high-throughput techniques involving automated or robotically controlled experimental steps, with thousands of genes identified by their DNA sequence alone. The investigators can effectively test every gene in the genome for its role in a biological process. Genome-wide assays facilitate systematic analysis of the functions of the whole complement of genes and accelerate the discovery of new biological function.

Genome-wide gene disruption assays, in which every suspected gene in an organism has been deleted or disrupted individually, allow the simultaneous analysis of phenotypes associated with defined mutations (Figure 15.3a). These so-called libraries of gene disruptions allow researchers to test the effect of mutations in all genes in their favorite phenotypic assays and immediately know what genes, when disrupted, give the phenotype of interest.

These mutant collections also allow genetic interactions to be examined. Often the deletion of a single gene will have little or no effect. However, when two genes are deleted together, the cells die. Such an interaction is termed **synthetic lethal**: neither deletion alone is lethal but the combination of the deletions is. Synthetic lethal interactions imply that the two genes are functionally redundant or play roles in separate pathways that ultimately affect the same fundamental processes (Figure 15.3b). To assay for synthetic lethal effects, a particular mutation is moved into a **library** of cells that collectively contain single disruptions in all genes such that each cell ends up with two mutations.

In addition to collections of deletion mutants, genome-wide RNAi screens are being carried out to identify gene function, particularly in mammalian cells, in which the generation of gene disruptions is more difficult. In these assays the expression of each gene in an organism is blocked individually by synthetic RNAs that are introduced into the cells. As with the gene disruptions, whole genome RNAi screens have identified proteins whose roles in processes including mitotic spindle formation and RNA processing, and as host factors that affect replication of the human immunodeficiency virus (HIV)-1 (to name but a few examples) were not previously suspected.

➲ See Experimental approach boxes 12.1 and 14.2 for examples of genome-wide disruption and RNAi screens.

Genome-wide RNA expression assays allow the gene expression profile – the levels of all RNAs in a cell – to be determined (Figure 15.4). The population of all RNAs in a cell at a given time is called the **transcriptome** (a fusion of the words transcription and genome). Genome-wide RNA expression assays include approaches such as microarray assays and deep sequencing.

➲ See Experimental approach 8.4 for an example of genome-wide expression analysis.

Figure 15.3 Genome-wide gene disruption assays. (a) By disrupting or replacing each gene by a unique tag or barcode, all mutants in a genome-wide disruption library (for example in *S. cerevisiae*) can be individually followed. If the collection of mutants is grown under some selective condition, those mutants that cannot survive under this condition are lost from the population. By amplifying the set of unique barcodes before and after selective growth and hybridizing the amplified DNA to a microarray, the mutants that are missing after selection can be identified because that unique barcode is missing. (b) In synthetic lethal screens, a particular mutation is moved into a library containing all other gene disruptions strains. If the double mutant strain is unable to grow (a missing spot on the agar plate) while the corresponding single mutant strains are viable, the two mutations are considered to be synthetically lethal.
Figure 15.3a adapted from Boone *et al.* (2007).

Figure 15.4 Genome-wide RNA expression assays. Two general types of gene expression assay can be performed to analyze the transcriptome. (a) RNA or a cDNA copy of the RNA can be hybridized to DNA probes on a microarray. The intensity of hybridization can be determined by labeling the cDNA prior to hybridization or by using labeled RNA or by detecting the DNA–RNA hybrid with labeled antibodies. If the cDNA prepared for two different samples is labeled with different fluorophores, the ratio between the labels can be used to determine whether a gene is up-regulated (green), down-regulated (red) or unchanged (yellow) in one sample compared to the other sample. (b) The cDNA obtained for a sample can also be subject to deep sequencing or NextGen sequencing. All of the sequence reads are then aligned with the genomic region to which the sequence corresponds. The number of times a particular sequence is detected is represented below the reference genomic sequence. Genes that are highly expressed are present in more copies in the cDNA population.

In a microarray assay (Figure 15.4a), each RNA (or a **cDNA** copy of every RNA) from the cell is hybridized to a chip that carries DNA probes for every known gene (or even the entire genome, including sequences not previously known to represent genes) spotted in an array. If the RNA or cDNA corresponding to a particular spot on the array is present it will hybridize to it. The identity of spots on which hybridization has occurred, which represent the RNAs present in the population, can then be determined.

In deep sequencing (Figure 15.4b), also called **NextGen (next generation) sequencing**, the cDNA copy of every RNA molecule in a cell is directly sequenced. The number of identical sequences is proportional to the transcript level of that gene: multiple copies of a cDNA indicate that multiple copies of its corresponding RNA were present in the cell being analyzed, suggesting that the gene to which the RNA corresponds was being actively expressed.

Together these techniques provide a powerful tool to examine what groups of genes are co-regulated in response to specific conditions or in specific tissues. For example, to identify all of the genes whose expression changes when yeast are subjected to a heat shock, we could prepare total mRNA from both non-heat-shocked and heat-shocked yeast and compare the mRNA levels present in the two conditions. Similarly, all of the genes expressed in a liver tumor can be compared with those expressed in normal liver to give an idea of what genes might play a role in cancer growth or progression.

Since all genes are examined simultaneously in these experiments, groups of genes that are regulated together can be defined. Such clusters of co-regulated genes can provide new insights into the function of unknown genes and suggest new functions for known genes based on other genes in the cluster. Genome-wide reporter gene assays, in which the promoters, and sometimes parts of the genes of interest, are fused to a gene encoding an easily assayable protein, also can be used to determine the time and location of gene expression.

Having the complete genome sequence facilitates **proteomics** (a fusion of the words protein and genomics), the large-scale study of proteins, protein complexes, and protein–protein interactions by allowing rapid protein identification and genome-wide protein–protein interaction assays. Rapid identification of proteins made is possible by the availability of completed genome sequences because the amino acid sequences of all of the proteins encoded in the genome can be predicted. (That is, the sequence of nucleotides in a gene can be 'decoded' into the primary sequence of the polypeptides which would be yielded if that gene was to be expressed.) With such genomic information only a short sequence of eight to nine amino acids is needed to uniquely identify a protein from the set of all known proteins sequences. As a result, when a protein complex is isolated, all of the associated proteins are more readily identified.

Understanding stable protein associations is a way of characterizing the function of a protein of interest (and, hence, the function of its corresponding gene) (Figure 15.5a). For example, if a protein of interest is found to interact strongly with a series of proteins that are already known to be subunits of a particular enzyme, we could infer that the protein of interest also forms part of the same complex, and contributes in some way to the function of the enzyme. Many of the proteins in the multisubunit RNA polymerase discussed in Chapter 8 were laboriously identified one by one over a 20-year period before whole genome sequence information was

(a)

tagged protein in complex mixture of other proteins

proteins that bind to tagged protein

PLIDASQ
AAQWERTS
ILSQRCAW

peptide sequences from bound proteins

protein 1
protein 2
protein 3

associated proteins identified from sequence

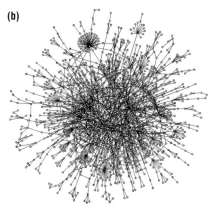

(b)

Figure 15.5 Proteomic approaches facilitated by the availability of genomic sequences. (a) To identify all of the proteins in a protein complex, one protein can be tagged (ref bar) and used to 'pull out' any associated proteins by co-immunoprecipitation or co-purification as described in Chapter 16. All of the proteins isolated are then sequenced. With the availability of a complete genome sequence, a stretch of only eight to nine amino acids is needed to identify the gene that encodes the protein. (b) Using information from a variety of methods that detect protein–protein interactions, protein interaction networks can be displayed as scale-free networks. Such a network shows the protein complexes that are associated with each other and may help point to biochemical pathways that the complex may participate in.

⊙ See Experimental approaches 11.3 and 12.2 for examples of large-scale protein identification.

⊙ ChIP-chip and ChIP-seq are described in more detail in Chapter 16 and Experimental approach 4.1.

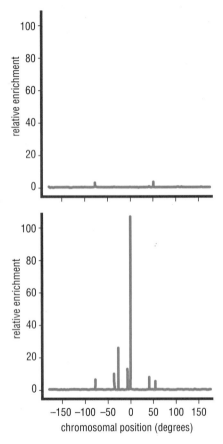

Figure 15.6 **Identification of protein-binding sites across the genome.** Whole genome sequences allow the genome-wide mapping of protein-binding sites. In the example shown here, the binding of the DnaA protein to the *Bacillus* chromosome is compared between cells in exponential growth (top panel) and cells arrested at initiation (bottom). Note how the majority of the DnaA protein is at the origin (position 0) in the initiation-arrested cells. The role of DnaA in the initiation of DNA replication in bacteria is discussed in Chapter 6. From Breier and Grossman (2009).

available. The composition of such a large protein machine could be analyzed in several months today with the availability of whole genome sequences.

In **genome-wide protein–protein interaction assays**, all or most of the proteins that interact with each other in a cell are identified. Each known protein can be individually tagged and purified using the tag, and all of the associated proteins identified. Alternatively, protein-based microarrays can be used to study the interactions. For example, all of the known proteins from the yeast genome have been produced and arrayed on a slide, such that the interactions between a protein of interest and all other proteins in the cell can be examined. The interactions between protein complexes, based on the presence of shared proteins, can be displayed as an interaction network (Figure 15.5b). The interactions between different complexes imply that these complexes may be in shared or related biochemical pathways, and thus give further insight about genome function. The set of protein interactions in a cell type is sometimes referred to as the protein **interactome**.

Genome-wide protein–DNA interaction assays allow protein binding to DNA across the entire genome to be examined (Figure 15.6). As we learned in Chapters 4 and 6–8, protein binding to DNA plays a major role in chromosome function and expression. To understand how sets of genes are co-regulated, whole genome analysis of protein binding can be carried out in addition to whole genome expression analysis. These interactions can be assayed on a genome-wide level using chromatin immunoprecipitation (ChIP). Proteins are crosslinked to genomic DNA, the DNA is fragmented and the fragments bound to the cross-linked proteins are isolated using immunoprecipitation. These DNA fragments are then identified by hybridizing to arrays (termed ChIP-chip), or they can be sequenced directly (termed ChIP-seq). Identification of such a set of binding sites, a global view of all of the genes regulated by a specific transcription factor can be evaluated.

Genome-wide analysis allows the prediction of networks of gene, protein and RNA associations

An important challenge is the need to analyze and synthesize the voluminous amounts of data being obtained from all of the functional genomic assays. When the data from the protein–protein interactions and from synthetic lethal interactions are analyzed as sets, either together or separately, networks of gene product interactions can be assembled. Clusters of interactions detected by these analyses often imply that the gene products function together in a larger complex. Different complexes may share some components and have interactions with each other. Thus the network models can indicate new functional pathways not previously known. The networks also can reveal which gene products are the major nodes, that is, those that have more interactions than others and so are probably the key players in certain biological pathways. For example, these nodes may represent key regulators whose functions should be further explored.

The function of many genes remains unknown

Despite extensive testing, the function of some genes in every organism cannot be determined by any of these assays. For example, dozens or hundreds of genes in every species have no detectable sequence similarity to any gene in another species, so a sequence comparison is not informative. Further, genome-wide mutant screens have found that many genes produce no mutant phenotype, at least under any of the many conditions that have been tested. Instead, many mutant phenotypes are only seen when mutations in two different genes are combined. This lack of phenotype could be because of functional redundancy – that is, whereby two genes carry out the same or similar function – or could be because a particular gene is not needed under the conditions tested.

Genomics has refined our definition of the gene

Comparative and functional genomics have reopened the decades-old debate, 'What is a gene?' When that question was asked during the second half of the twentieth century, a molecular biologist could answer with some confidence: a gene is the nucleotide sequence information needed to make a functional RNA or polypeptide chain. Although that is still a good answer, we now realize that is an incomplete answer. As we have seen, many genes encodes hundreds of proteins due to processing to give alternative mRNA ends, alternative mRNA splicing, and alternative protein processing. In addition, many genes encode regulatory RNA molecules, perhaps even as many as the genes that encode polypeptide chains. Other sequence elements are conserved between species and even transcribed but their functions are still unknown; should these also be classified as genes? Even as we learn more about the structure and function of genomes, the answer to this fundamental question appears to become more elusive.

15.3 EVOLUTIONARY FORCES

The genome-wide comparative and functional genomic studies discussed above enable us to characterize genomes as they currently exist: they give us a snapshot of the current genome sequence and the biological functions it mediates when expressed. However, genomes are not static, unchanging entities; they gradually change over time. Changes to DNA sequence, which will be discussed in subsequent sections, provide the source material from which all evolutionary change arises. These changes can occur in every cell, so effects are observed in both somatic cells and germline cells.

Although mutations in somatic cells are often important in diseases, particularly cancers, the focus in this chapter is on evolutionary change, which involves the transmissible mutations in the germline. However, DNA changes are relatively rare and only a small fraction of the changes are transmitted to subsequent generations. In this section, we will consider different forces that affect the frequency at which DNA changes are transmitted and thus contribute to the genetic diversity of a population. The effects of these forces can be seen whether we are examining populations, individuals, genes, or DNA sequences.

Selection shapes the evolution of a gene and a population

The most powerful force affecting evolution is **natural selection**. In its simplest form, selection describes the process in which some genetic variants contribute more offspring to the next generation than other genetic variants. Selection occurs in many forms and from many sources, but is most usefully characterized according to its effects on the frequency of an allele in the population.

Positive selection occurs when a new genetic variant is favored – when it puts an individual in a stronger position to survive and to pass its genes on to a subsequent

Figure 15.7 Different types of selection. Different kinds of genetic changes are associated with positive and negative selection and neutral mutations. Initially, there may be several different genotypes (represented by the green, red, and blue triangles). If a mutation confers some selected advantage, the genotype (green) will predominate in the population after a period of time (positive selection). If a mutation confers a selected disadvantage, the genotype (red) will be reduced in the population after a period of time (negative selection). For a neutral mutation, the derived population (blue) will be similar to the original population.

generation. In general, positive selection will result in a mutation or genotype that *increases* in frequency in the population over time (Figure 15.7).

Negative selection occurs when a new genetic variant is disfavored, that is, when it puts an individual in a weaker position to survive, such that it is less likely to pass its genes on to a subsequent generation. In general, negative selection will result in a mutation or genotype that *decreases* in frequency in a population over time (Figure 15.7). Most mutations reduce or eliminate the function of a gene, so most newly arisen mutations are subject to negative selection. Since negative selection tends to eliminate the mutation from the population, negative selection is sometimes called **purifying selection**.

Many genetic variants are neither advantageous nor disadvantageous to the organism. These are called **neutral mutations**. One of the best examples observed at the level of DNA sequence lies in the third base of codons, as shown in Figure 15.7. The same amino acid is often specified by more than one codon, as described in Chapter 10. In particular, the third base in a codon may only need to be any purine or pyrimidine in order for the codon to code the correct amino acid (rather than the third base having specifically to be cytosine or uracil, for example). As such, both CAU and CAC specify histidine: the third base must be a pyrimidine, but it does not matter whether that pyrimidine is C or U – the same amino acid is coded for regardless. Such changes in the third base are called **synonymous changes** and are expected to be neutral changes.

The same logic applies at the level of amino acid sequence: serine and threonine are such similar amino acids that substituting one for the other (in evolutionary time) often does not affect the activity of the protein. There are exceptions, of course, but, in general, a mutation that replaces a serine with a threonine or an isoleucine with a valine may well be neutral. The importance of neutral mutations will be developed when we discuss **polymorphisms** and **haplotypes** below.

Genetic drift also affects the evolution of populations and genomes

Another evolutionary force that shapes the genetic structure of a population is known as **genetic drift**. Genetic drift arises because populations (of individuals, cells, or genes) are not infinitely large, so the alleles of a gene that happen to be predominant in a population can change randomly over a period of time. As shown in Figure 15.8, only a subset of possible descendents carrying a random selection of all possible alleles are found in the next generation; thus allele frequencies can change without selection simply due to random sampling in a population.

One particularly important example of genetic drift is the **founder effect**. Imagine that a small group of individuals splits off from the main population to produce a new, isolated population. The new population is likely to have genetic differences from the original population, with some traits being more common and some less common. These differences arise because the new population – the founding population – is much smaller than the original population and does not have all of the same variation as the original population. Imagine selecting ten M&Ms from a packet: while the full packet (your original population) will contain M&Ms displaying a full range of different colors, the ten you have selected (your isolated population) may not represent all of the colors in the original population. Instead, chance dictates that they will probably exhibit less variation and some colors may be missing altogether.

An equivalent occurrence is a *population bottleneck*, when many members of a large population suddenly die (due to a natural disaster for example), and the population re-grows from a much smaller group. The remaining population will probably have only a subset of the alleles that were present in the original population, so subsequent generations will possess a different range of genetic variations.

Although selection and genetic drift are probably the most significant forces that affect the structure of a population, they are not the only ones. The movement of individuals from one population to another (migration) and mating preferences among individuals in a population (non-random mating) also affect the amount and type of genetic variation in a population. All of these forces often work in concert or in opposition with each other. Genetic differences arise as a result of mutation, but they persist or are lost over time largely because of the effects of selection and genetic drift.

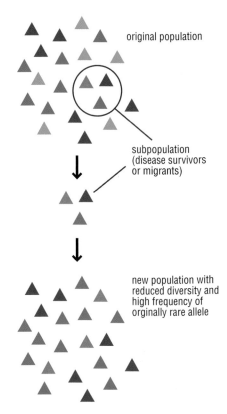

Figure 15.8 Founder effects. There is significant genetic variation in the original large population (represented by the blue, dark blue, orange, and pink triangles). If a subset of the individuals in the original population migrates to a new isolated location or if only a subset survives a disease, the genetic diversity in the new population is reduced.

15.4 THE EVOLVING GENOME

Having learned about the forces that shape evolution, we now review how genomes change, before considering how we can study the changes that have occurred – and what these changes can tell us.

Mutations are a critical source of genetic variation

The essential ingredient for evolution, and for natural selection, is genetic variation – a change in DNA sequence. Without variation at the genetic level, the phenotypic variation that might be selected for would not occur. Changes in DNA sequence (that is, mutations) happen at a low frequency in all cells at all times at all sites in the genome. Small-scale mutational changes in the genome include

wild type	ATG	AAT	ATT	CGA	GAT
point mutation	ATG	AAT	ACT	CGA	GAT
deletion	ATG	AAT	A	CGA	GAT
insertion	ATG	AAT	ATT	ATTCGA	GAT

Figure 15.9 Possible mutations at a nucleotide level. Possible small-scale changes to genome sequence include point mutations as well as the deletion or insertion of one or a few nucleotides.

⊖ We describe the structures of transposable elements and the mechanisms by which they move in Chapter 14.

⊖ We describe the process of recombination in Chapter 13.

the substitution, loss, or gain of one or a few nucleotides as described in Chapter 13 and shown in Figure 15.9. Most of these nucleotide changes occur during DNA replication, although others are generated by damage to DNA (see Chapter 12). Although most nucleotide changes are repaired, some changes do persist; it is these persistent changes that are the basis of genetic variability.

In addition to nucleotide changes, mutations also arise from the activity of transposable elements. Transposable elements do not move at a high frequency; a transposition rate of one movement per 1000 insertion sites per cell generation is probably a realistic estimate. However, there are many transposable elements at different locations in the genome and cells divide many times. Consequently, even a low frequency of transposition has an impact on the genetic structure of a population. Transposable elements frequently produce mutations on both their insertion (Figure 15.10a) and excision (Figure 15.10b) at a site. Not only does the genome of every organism include many transposable elements, but it also harbors evidence of former insertions and excisions.

Transposable elements provide a means to rearrange the genome

In addition to causing mutations by insertion and excision, transposable elements have a major role in rearranging the structure of the genome. Recombination between transposable elements or other repeated sequences is probably one of the principal mechanisms by which genetic regions are lost (Figure 15.10c) or duplicated (Figure 15.10d) on a chromosome. Recombination between transposable elements on two different chromosomes also can produce a translocation (Figure 15.10e), whereby a region of one chromosome is swapped with a region from another chromosome. The transposable element need not be mobile to serve as the substrate for recombination since it is the sequence similarity between elements rather than their movement that provides the basis for rearrangement through recombination.

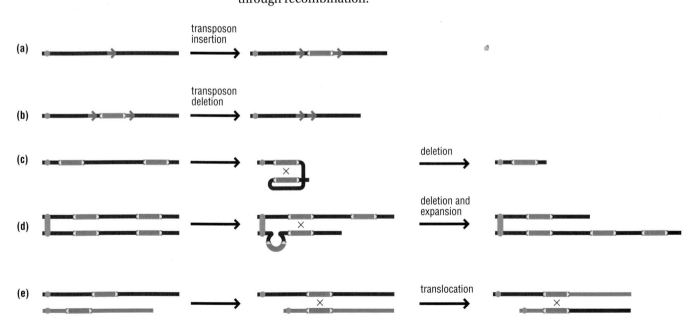

Figure 15.10 Transposons have a major role in rearranging the genome. Transposons can cause changes to the genome sequence by inserting (a) into or excising (b) out of the genome. Recombination between transposable elements at different locations in the genome chromosome can also lead to deletions (c), duplications (d), and rearrangements (e) of regions of the genome.

Genome evolution involves changes in genome size between species

The small-scale changes in nucleotide sequence that we mention above are dwarfed by the large-scale changes that have occurred in the overall size and structure of the genome during evolutionary time. One of the most significant of the large-scale evolutionary changes in genome size is **polyploidy**, an expansion in the number of genome copies (see Figure 1.18). Evidence of evolutionarily ancient polyploidy events is seen in nearly all plant genomes. The presence of multiple copies of the same genome, known as autopolyploidy, results in the possession by an organism of multiple copies of all of the genes in the genome. The frog *Xenopus laevis* has a genome twice the size of that of its close cousin *Xenopus tropicalis*; the banding pattern and structure of the chromosomes suggest that a complete duplication of chromosomes occurred in the evolutionary lineage leading to *X. laevis*. The budding yeast *Saccharomyces cerevisiae* also shows evidence of an evolutionarily ancient genome duplication followed by subsequent loss of many genes. Despite the losses, many genes and regions of the yeast genome are still present in the duplicate copies.

Polyploidy is also a characteristic of domesticated plant genomes. For example, bread wheat is a hexaploid produced from three different ancestral genomes, and bulgur wheat is a tetraploid arising from two different ancestral genomes. Because related genes are not exact duplicates in these cases, they may have different functions. Following polyploidization, some genes may be lost, so not all copies of the genome are complete replicates of the ancestral genomes.

Genome evolution involves rearrangement between existing chromosomes

Even when genomes of related species are approximately the same size, the number and configuration of the chromosomes can change. As we learned in Chapter 13, rearrangement can occur between two chromosomes and within a chromosome, with different consequences.

Within an individual organism, chromosome rearrangements are usually deleterious. However, structural changes in the chromosomes over evolutionary time can help drive the generation of new species. Because pairing of homologous chromosomes is essential for chromosome segregation at meiosis in eukaryotes, two organisms with different chromosomes cannot interbreed, even if they can mate, as a result of the failure of meiosis. Thus, structural changes to chromosomes that prevent pairing can lead to the so-called 'reproductive isolation' of the organisms that carry them, leading eventually to **speciation**, the formation of a new species.

The chromosomal changes that contribute to the speciation of one species from another are caused by many different mechanisms including chromosome fusions, breakages, and rearrangements, as described below. One of the simplest examples of karyotype change is that of chromosome number. There is great diversity in chromosome number among related species; in vertebrates, for example, the haploid chromosome number varies from three to 49 (Figure 15.11a). Strikingly, two species of zebra that look extremely similar have very different chromosome numbers: Grevy's zebra (*Equus grevyi*) has 23 chromosomes and Hartmann's mountain zebra (*Equus zebra hartmannae*) has only 16; however, the oldest

(a) Variation in chromosome number in vertebrates	
species	haploid chromosome number
human	23
mouse	20
chicken	39
carp	49
Indian muntjac	3
Chinese muntjac	23
Mongolian wild horse	33
Grevy's zebra	23
Hartmann's mountain zebra	16

(b) Mongolian wild horse	Grevy's zebra	Hartmann's mountain zebra
haploid chromosome number		
33	23	16

Figure 15.11 Karyotype can evolve through changes in chromosome number. (a) The haploid number of chromosomes varies widely among vertebrates. Even closely related species of zebra that appear remarkably similar have very different chromosome numbers. (b) The most distantly related horse (the Mongolian wild horse) has 33 chromosomes, while two zebras have 23 (Grevy's) and 16 (Hartmann's). (c) The chromosomes of Hartmann's zebra show internal telomere repeats (arrows) indicative of reduction in chromosome number through chromosome end-to-end fusion. (b) Jim Amos/Science Photo Library and John Devries/Science Photo Library. (c) From Santani *et al.* (2002).

(a) chromosome fusion

loss of telomere function and end-to-end fusion

centromere inactivation

chromosome with interstitial telomere repeats

(b) chromosome fission

neocentromere formation

chromosome breakage

new telomere addition

existent species of the same genus, the Mongolian wild horse (*Equus przewalskii*), has 33 (Figure 15.11b). The presence of telomere repeats at the junctions between ancestral blocks of DNA in Hartmann's mountain zebra (Figure 15.11c) suggests that the decrease in chromosome number from the ancestral species occurred through chromosome fusion.

For a stable new chromosome to be generated from two or more ancestral chromosomes, the function of two telomeres and one centromere must have been lost (Figure 15.12a). Chromosome fission – the splitting of a chromosome – also contributes significantly to karyotype evolution. Blocks of sequence that are found on a single chromosome in pigs, cows, and horses are located at the ends of human chromosomes 16 and 19. This implies that breakage of an ancestral chromosome occurred and new telomeres were added, probably by telomerase, in a germ cell early in vertebrate evolution. In addition, a new functional centromere must have been generated to allow segregation of the fragment that did not initially carry one (Figure 15.12b). Thus the ability to inactivate telomere and centromere function in the case of chromosome fusion, and the ability to form new telomeres and new centromeres in the case of chromosome fission, allows karyotype evolution and likely helps drive speciation.

Rearrangements within chromosomes change gene number and location

Three types of rearrangement can occur within a chromosome: deletions, duplications, and inversions. All of these rearrangements change the location

Figure 15.12 Mechanism for change in chromosome number. (a) Chromosome fusion. Two chromosomes can fuse if telomere function is lost. If one centromere also becomes inactivated this chromosome will be a stable larger chromosome and interstitial telomere repeats will be detectable at the point of fusion, as seen for the zebra chromosome in Figure 15.11. (b) Chromosome fission. Neocentromere formation may lead to chromosome breakage directly, or a break may occur at random on a chromosome with a neocentromere. If telomeres are added to the site of breakage by telomerase, two new stable chromosomes may result.

of some genes, and two of them change the copy number of genes, as summarized in Figure 15.13.

Deletions are the loss of regions of the genome (Figure 15.13a). Deletions of coding regions often produce non-viable offspring, and are selected against over the course of time. However, deletions involving non-coding regions may be selectively neutral and can persist. Likewise, deletion following polyploidization or duplication of chromosomal regions is often tolerated and maintained: the multiple copies of genes generated by such a duplication event mean that, even if one copy is rendered inactive through a subsequent deletion event, the organism still has at least one functional copy left. In many, but not all cases, this remaining copy is sufficient for full function. Interestingly, LINE element recombination caused a significant deletion from the chimp genome before divergence to the human genome.

Duplications result in the presence of extra copies of specific regions of the genome (rather than the duplication of the entire genome, as seen with polyploidy). Genome duplications are exceedingly common, and nearly all eukaryotic genomes also show evidence of regional duplication. For example, one species may have the region symbolized A-B-C-D-E-F-G whereas a related species has the same region with the structure A-B-C-D-E-F-D′-E′-F′-G, as the result of a direct duplication of the region (Figure 15.13b). Thus, genes in the regions D, E, and F are present in multiple copies, whereas those in regions A, B, C, and G are not. Duplications can be followed by translocation or transposition, so that not all duplicate copies are in tandem in the genome. The consequences of duplication on genome evolution are discussed below.

Inversions do not change the copy number of genes but do change the order in which they are found. In an inversion, a region of the chromosome symbolized A-B-C-D-E-F-G becomes switched to A-B-C-F-E-D-G (Figure 15.13c). All of the genes located in regions D, E, and F are still present in the same copy number but are in different locations. Inversions can occasionally have the effect of separating a gene from its regulatory region such that the expression of the gene is disrupted. More commonly, however, they have very little effect on an individual.

Translocations and inversions are associated with speciation

Chromosomal translocations and inversions, which change the location but not the number of genes, have important effects on population structure because they affect the ability of individuals to interbreed due to defects in meiosis. Recall that during meiosis, each chromosome synapses along its length with its homologous chromosome.

If one chromosome of the pair has an inversion or translocation, the rearranged homolog is unable to synapse with its unrearranged homolog (or, at best, it synapses very poorly) (Figure 15.14). Inversion heterozygotes have to form a twisted loop in their chromosomes to synapse properly, and these structures are unstable at meiosis, usually resulting in chromosome breakage. Since organisms that cannot properly carry out meiosis are infertile, chromosome rearrangements such as translocations and inversions are often associated with speciation – the formation of a discrete species that is reproductively isolated.

◗ The mechanisms that allow formation of new centromeres, termed neocentromeres, are described in Chapter 4.

Figure 15.13 Deletions, duplications, and inversions affect gene order. The order of the genes A, B, C, D, E, F, and G is affected by chromosomal deletions (a), duplications (b), and inversions (c). Copy number variants are an example of duplications.

◗ The steps in meiosis are described in Chapter 7.

Figure 15.14 Rearranged chromosomes cause problems in meiosis. (a) On the left is the normal chromosome configuration with one X and one Y and two copies of chromosome 16. On the right is the translocation that has occurred between chromosome 16 and the X chromosome denoted 16^X and X^{16}. (b) Types of chromosome arrangements that are observed when cells attempt to synapse the translocated chromosomes during meiosis. Chromosome breaks and chromosome loss can occur when these poorly synapsed chromosomes are pulled apart during chromosome segregation. From Turner *et al.* (2004).

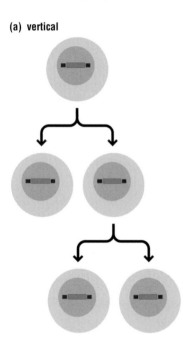

(a) vertical

(b) horizontal

Completely novel sequences are introduced into a genome by horizontal gene transfer

The normal passage of genes from one generation to the next is referred to as **vertical gene transfer**. By contrast, yet one more source of genetic change is **horizontal gene transfer**, the insertion of a segment of the genome from one species into the genome of an unrelated species (Figure 15.15). Horizontal gene transfer produces novel genetic functions in the genome, unlike duplications and changes in polyploidy, which duplicate existing functions. Many of these novel functions are probably detrimental to the recipient organism, so the examples that we observe among living species are likely to be a small fraction of the horizontal gene transfers that have actually occurred during evolution.

Horizontal gene transfer is particularly seen between bacterial genomes but it can occur between bacteria and their eukaryotic hosts. Horizontal gene transfer in bacterial genomes can occur due to the action of transposable elements or other genetic exchange mechanisms such as conjugation (transfer of DNA via cell-to-cell contact), transformation (the uptake of DNA), and transduction (whereby viral particles incorporate up non-viral DNA). Horizontal gene transfer has also been inferred for some regions in plant genomes and in the genomes of some parasitic animals as well.

15.5 DUPLICATION AND DIVERGENCE OF GENE FUNCTION

Increases in the number of copies of a gene by polyploidy or duplication has important consequences for the evolution of gene function. Here we consider these consequences.

Figure 15.15 Vertical and horizontal gene transfer. (a) Normal passage of genes from one generation to the next is referred to as vertical gene transfer. (b) Integration of a segment of DNA from one species into the genome of another species is referred to as horizontal gene transfer.

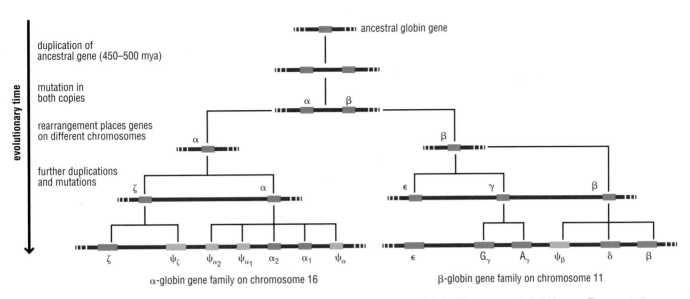

Figure 15.16 α-Globin gene family. The three α-globin and five β-globin genes in humans are all derived from an ancestral globin gene. The ancestral gene was duplicated 450–500 million years ago, and the two copies diverged through different chromosomes. At each of mutation and translocation to these two new chromosomal locations the genes further duplicated and accumulated additional sequence changes. Four copies of the α-gene family and one copy of the β-gene family were inactivated by the sequence changes and are thus pseudogenes shown in green and denoted with a Ψ. From http://www.bio.miami.edu/~cmallery/150/gene/genome.org.htm.

Genome evolution involves changes in the number of genes

Once a gene has been duplicated, the copies of both the regulatory and coding sequences can accumulate different mutations. As a result, the functions of the duplicated genes may diverge over evolutionary time. The duplicate members of a gene or its protein are referred to as **gene or protein families**. The α- and β-globin gene clusters provide an example of a duplicated region and a gene family. Hemoglobin is a dimer of α- and β-globin and is needed for carrying oxygen in the blood. In humans, there are three protein-coding genes at the α-globin locus and five at the β-globin locus (Figure 15.16). In this case, the protein-coding regions of the genes have diverged only slightly in sequence and activity. However, significant divergence is seen in the time of expression. During human embryogenesis, gestation, and postnatal growth, the expression of both the α- and the β-globin genes switch, such that a different gene is expressed at different times in development. The different globin proteins are not just **markers** of developmental stage, but have specialized functions in meeting differing oxygen transfer requirements (for example, placenta versus lungs) as the fetus matures and is born.

For other gene families, the divergence in function occurs when different family members are expressed in different tissues. The individual genes (like the globin genes) retain similar functions but the spatial pattern of expression changes. For example, the *hedgehog* gene family encodes important signaling molecules involved in spatial and developmental patterning during animal development.

Mammals have three members of the *hedgehog* gene family (Figure 15.17), whose expression patterns differ in different tissues. *Sonic hedgehog* has a widespread function in providing developmental patterning information for the

⊖ We describe the maturation of the Hedgehog protein in Chapter 11 and Experimental approach 11.1.

Figure 15.17 *Hedgehog* gene family. The *hedgehog* gene family in humans and *Drosophila* illustrates the difference between paralogs and orthologs. Within an organism the similar copies from the common ancestor are called paralogs and between two different organisms (human and *Drosophila*) the two related genes are called orthologs. One ancestral *hedgehog* gene diverged between humans and *Drosophila*. In humans the gene was duplicated and each copy then further diverged to generate three genes with slightly different function.

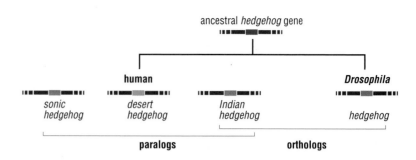

brain, limbs, spinal cord, and many other tissues. By contrast, *desert hedgehog* is involved in patterning the gonad and *Indian hedgehog* affects cartilage and bone specification. The proteins are similar but not identical in sequence, but the most important divergence has occurred in the location and time of gene expression.

Gene families can be compared both within and between species

Gene families arising from duplication and divergence events are found in most eukaryotic genomes. Indeed, there are often many related copies of a gene in a genome. Related copies of a gene in the same genome are referred to as **paralogs** (Figure 15.17). For example, the *sonic hedgehog* and *desert hedgehog* genes in humans are paralogs of one another. The two genes have similar functions but are expressed in different tissues.

Gene families can also be compared between species. Genes that are found in two different species and have arisen from a common ancestral gene are known as **orthologs**. For example, the first member of the *hedgehog* gene family was found as a single gene in *Drosophila melanogaster*, and the gene is named for the mutant phenotype it produces in flies. However, all three genes in mammals are orthologs of the Drosophila *hedgehog* gene; both mice and humans have orthologs of the *sonic hedgehog*, *Indian hedgehog*, and *desert hedgehog* genes. In other words, the single gene found in insects became duplicated in mammals with the concomitant diversification of expression and function.

Both orthologs and paralogs are examples of the more general evolutionary principle of **homology**. Homology describes the way that structures or genes are descended from a common ancestor. It is a qualitative description of an historical event, so two genes are either **homologs**, and are descended from a common ancestor, or they are not. Homology is inferred from the similarity between the two extant genes and the corresponding gene products. The degree of sequence similarity can be quantitatively defined, which means that our inference about homology can be tested statistically. Thus, strictly speaking, we are inferring that the *Drosophila hedgehog* gene and the mammalian *sonic hedgehog* genes are homologs because of the high similarity of the amino acid sequences encoded by these genes. Likewise, we are inferring that *sonic hedgehog*, *desert hedgehog*, and *Indian hedgehog* genes are homologs of one another because of their similarity to each other and to the apparent ancestral *hedgehog* gene.

Duplicate genes experience different evolutionary pressures

Whenever a gene duplication occurs, the evolutionary pressure on the two genes changes. After all, in a previous generation, one gene was sufficient for survival but, following the duplication, there are now two genes. Positive selection can occur for one or both duplicate copies of the gene so mutations that alter the time or amount of gene expression, or that slightly alter the activity of the protein, could be advantageous to the organism. This is a plausible explanation for the divergence in function of the three mammalian *hedgehog* paralogs.

Mutations usually reduce or eliminate the function of the gene. However, because the selective pressure on the function of the gene has been relaxed following duplication (because the organism carrying the duplicated gene now has a 'back-up' if the activity of one of the two genes should become disrupted as a the result of a mutation event), negative selection might not occur, and both the functional and non-functional copies of the gene may be successfully transmitted to the next generation. Thus, many gene families include non-functional gene sequences. These non-functional members of a gene family are termed **pseudogenes**. In pseudogenes, mutations may have altered the promoter region, changed a splice junction, or introduced a stop codon that has eliminated the function of the gene. Once the gene has been rendered non-functional, further mutational events may occur, so that the pseudogene accumulates multiple additional changes. Thus, although the nucleotide sequence of the region resembles other genes in the family, the gene itself is non-functional.

Pseudogenes are common. For example, there are four of them in the α-globin and two of them in the β-globin gene cluster (Figure 15.16). These probably arose by duplication and mutation to non-functional copies of each gene. Indeed, the genomes of all eukaryotes are littered with non-functional versions of once-useful genes. Pseudogenes can also arise from reverse transcription and the subsequent integration of the DNA copy of the processed mRNA, as in the case of non-long terminal repeat (LTR)-mediated transposition described in Chapter 14. These 'processed' pseudogenes are easily recognized because they lack the introns that are present in the original gene.

Conserved sequences in a gene family arise from several different mechanisms

The process of mutation is continually at work to produce changes in the gene family members, and both the resulting similarities and differences between paralogs within a species or orthologs between species are informative. The similarity between two genes can be attributed to one or more of three possible events. First, gene family members that are highly similar in sequence might represent a very recent duplication event. Second, gene family members may be highly similar in sequence because, even if they were duplicated a long time ago, there may be strong selection to maintain the sequence and function of the gene and protein. (That is, changes to the sequence are not tolerated because they lead to loss of function.) Third, gene conversion may be occurring between gene family members, which will result in homogenizing the sequences.

> ⊙ Gene conversion is discussed in Chapter 13.

As described more fully in Chapter 13, gene conversion is a repair process that uses one gene as a template to alter the sequence of a similar gene. By using one gene sequence to repair the sequence of a paralogous gene, gene conversion keeps the sequences highly similar.

15.6 SEQUENCE COMPARISONS

We have now explored some of the changes that happen at the molecular level to drive the evolution of the genome. In this section we look more deeply at the evolution of genomes by looking at how we can study the changes that occur.

Sequence comparisons are a powerful method to infer gene functions

As described in Section 15.2, sequence similarity between genes makes it possible to infer the activity of a gene or its product even in the absence of mutations or biochemical tests. Sequence comparisons can be based on the nucleotide sequence of the DNA or on the amino acid sequence of the polypeptide the gene encodes. Furthermore, each position in one sequence can be compared to a corresponding position in the other sequence, or group of sequences, in what is called an **alignment**. The sequence comparison can be done as a pair-wise alignment between two sequences or as a multiple alignment among multiple sequences. Each of these comparisons provides different information about the relationships and functions of the sequences under analysis, and involves different assumptions and computational approaches. Alignments can be done over the entire length of the sequences, known as a global alignment, or over a section of the sequences, known as a local alignment (Figures 15.18 and 15.19). Most, but not all, alignments allow gaps in the sequences as well, so that a position in one sequence can be aligned with an empty space in the other to accommodate differences in the length of the sequences.

```
                                                                                 Walker A
Walker A
NSF        220   INPDWDFQQMGIGGLDTEFSHIFRRAFASRVFPPEFIEQLGMKHV-RGILLFGPPGTGKTLMARQIGKMLNAREPKIVNGPQILDKYVGESESNVRKLFADAEE
Msp1p       86   INITFQ----DIGGLDPLISDL-HESVIYPLMMPEVYSNSPLLQAPSGVLLYGPPGTGKTMLAKALAKES-GANFISIRMSSIMDKWYGESNKIVDAMFSLANK
AAA ATPase 119   KKAKFS----EIIGQEEAKKKC-RIIMKY-LENPKLF---GEWAP-KNVLFYGPPGTGKTLMARALATET-NSSFILVKAPELIGEHVGDASKMIRELYQRASE
VPS4b      127   PNVKWS----DVAGLEGAKEAL-KEAVILPIKFPHLFT--GKRTPWRGILLFGPPGTGKSYLAKAVATEANNSTFFSISSSDLVSKWLGESEKLVKNLFQLARE
PEX6       653   PNVKWD----DVGGLEDVKTSI-LDTVQLPLLHKDLFSS-GLRKR-SGVLLYGPPGTGKTLLAKAVATEC-SLNFLSVKGPELINMYIGESEKNVRDIFEKARS
FtsH       147   IKTTFA----DVAGCDEAKEEV-AELVEY-LREPSRFQKLGGKIP-KGVLMVGPPGTGKTLLAKAIAGEA-KVPFFTISGSDFVEMFVGVGASRVRDMFEQAKK
Afg3p      283   IKISFK----NVAGCDEAKQEI-MEFVHF-LKNPGKYTKLGAKIP-RGAILSGPPGTGKTLLAKATAGEA-NVPFLSVSGSEFVEMFVGVGASRVRDLFTQARS
VCP        462   PNVHWD----DIGGLEDVKREI-KETVELPLLKPDVFKRLGIRPS-KGFLLYGPPGTGKTLLAKAVATES-NANFISIKGPEVLSKWVGESEKAIREIFKKAKQ
Rpt3       155   PDVSYA----DIGGMDMQKQEI-REAVELPLTHFELYKQIGIDPP-RGVLMYGPPGTGKTMLAKAVAHHT-TASFIRVVGSEFVQKYLGEGPRMVRDVFRLAKE

                                Walker B
Walker B
NSF        EWRRCGANSGLHIIIFDEIDAICKQRGS---MAGSSSVHD--TVVNQLLSKMDGV-EQLNNILVIGMTNRRDMIDEALLRPGRLEVQMEVSL        409(762)
Msp1p      --------LQPCIIFIDEIDSFLRER-----SSTDHEVTA--TLKAEFMTLWDGL-LNNGRVMIIGATNRINDIDDAFLR--RLPKRFLVSL        258(362)
AAA ATPase --------SAPCIVFIDEIDAIGLSRE----YQSLRGDVS--EVVNALLTELDGI-KENEGVVTIAATNNPAMLDPAIRS--RFEEEIEFKL        286(371)
VPS4b      --------NKPSIIFIDEIDSLCGSR-----SENESEAAR--RIKTEFLVQMGVGVDNDGILVLGATNIPWVLDSAIRR--RFEKRIYIPL        300(444)
PEX6       --------ARPCVIFFDEIDSLAPARGA---SGDSGGVMD--RVVSQMLAEIDGLSDSSQDLFIIGASNRPDLIDPALLRPGRFDKLLYVGV        828(941)
FtsH       --------AAPCIIFIDEIDAVGRQRGAG--LGGGHDERE--QTLNQMLVEMDGF-EGNEGIIVIAATNRPDVLDPALLRPGRFDRQVVVGL        321(644)
Afg3p      --------MAPSIIFIDEIDAIGKERGKGGALGGANDERE--ATLNQLLVEMDGF-TTSDQVVVLAGTNRPDVLDNALMRPGRFDRHIQIDS        459(761)
p97        --------VAPAIVFLDEIDSIAPRRG----TTSDSGVTE--RIVNQLLTSLDGI-EVMNGVVVIGATNRPDIMDPALLRAGRFDKLIYIPP        635(745)
Rpt3       --------NAPAIIFIDEIDAIATKRF----DAQTGADREVQRILLELLNQMDGF-DQTTNVKVIMATNRADTLDPALLRPGRLDRKIEFPL        330(413)
```

Figure 15.18 Protein sequence alignment. Amino acid sequence alignment for the ATPase domain of several distinct AAA+ ATPase proteins from different species (described in Section 6.6). The alignment shows how amino acids corresponding to the Walker A and Walker B boxes are conserved from bacteria and yeast to plants, flies, and mice, while much of the other sequence varies.

```
Escherichia coli          TTTGCAA-AAAGTGTTGGACAAGTGCGAATGAGAATGATTATTATTGT-CTCGCG-ATCAGGAAGA----CCCTCGCGGAGAACCT------GAA
Salmonella enteric        TTTGCAA-AAAAAGCTAGACAAGTGCGAATGAGAATGATTATTATTGC-TTTGCG-TTCAGGG-GA----CCCTGACGGAAAACCT------GAA
Klebsiella pneumonia      TTTGCAA-AAAAAAGTAGACAAGTGCGAATGAGAATGATTATTATTGT-CCTGCA-TTCAGGGAGA----CCCTTGCGGAAAGCCT------GAA
Yersinia enterocolitica   TTCTCAACAAAGTGTTTG-CAACTGCTATTGATAACTATTATCATCTG-TTTGCTCTTCGAGACAAAGCTAGTTGTTTCAGGCTTTTGACAGAAA
Erwinia carotovora        TTCTCAACAAAGTACTTTACAGATGATACTGATAACTATTATCATTATTCTCGCACTTCA-GCAGTGG-CTCCTCGC--AGACACT--GAAGGAG
Invariant bases           TT..CAA.AAA....T...CA..TG..A.TGA.AA..ATTAT.AT.......GC...TC..G.............T.....A.....T........A.
```

```
Escherichia coli          AGCACGACATTGCTCACATTGCTTCCAGTATTACTT-AGCCAGC-CGGGTGCTGGCTTTTTTTTTTGATC
Salmonella enteric        AGCACGACATTGCTCACATTGCTTCCAGTATTACTTTAGCCAGC-CGGGTGCTGGCTTTTTTTTTTGCTG
Klebsiella pneumonia      AGCACGACATTGCTCACATTGCTTCCAGTATTACTTTAGCCAGC-CGGGTGCTGGCTTTTTTTTTTGCTG
Yersinia enterocolitica   AGCACGACATTGCTCACATTGCTTCCAGTGTTTTTTAAGCCAGCTCGGGTGCTGGCTTTTTTTTTTGCTT
Erwinia carotovora        AGCACGACATTGCTCACATTGCTTCCAGTATTATTTTAGCCAGCCTACGTGCTGGCTTTTTTTTT-GCCC
Invariant bases           AGCACGACATTGCTCACATTGCTTCCAGT.TT..TT.AGCCAGC....GTGCTGGCTTTTTTTTT.G...
```

Figure 15.19 Nucleotide sequence alignment. DNA sequence alignment for the gene encoding the RyhB non-coding RNA (described in Section 9.9) for several different bacterial species is shown. The sequences encoding the non-coding RNA are pink. The regions encompassing the promoter (underlined), parts of the non-coding RNA, and the terminator (underlined) have a higher number of invariant residues.

Amino acid sequence comparisons can probe genes from evolutionarily distant species

The quickest method to infer the function of a gene is to compare its sequence with that of known genes of other species, commonly done using a **BLAST** program.

If the gene encodes a polypeptide, the most informative comparison is based on the inferred amino acid sequence encoded by the gene rather than the nucleotide sequence of that gene. Why is this? Comparison of amino acids is the preferred method because an amino acid sequence comparison can ignore the changes in synonymous codons (Figure 15.18). In addition, an amino acid sequence alignment can account for the nuances of the biochemical similarities and differences among the 20 amino acids – that is, it can regard a substitution of serine for threonine as more likely to be evolutionarily acceptable than a substitution of lysine for threonine because serine and threonine have similar chemical properties.

Alignments can identify conserved domains and sequence motifs, even when the functions of these domains and motifs are not yet known. They allow comparisons of amino acid sequences from proteins of two very distantly related organisms, such that the functions of proteins in humans can be inferred from comparisons with proteins in fruit flies, worms, budding yeast, and even *Escherichia coli*.

One of the most important features of BLAST alignments is that the program reports a score for each alignment and the probability that the alignment score arose by random similarity between unrelated sequences, called an E value. A very low E value means that there is very little chance that the two sequences arose independently, and implies that they are very likely to be derived from a common ancestor, that is, they are homologs and may have similar activity.

Nucleotide sequence comparisons can be used for closely related species

Amino acid sequence comparisons are nearly always more informative than nucleotide sequence comparisons for the protein-coding regions of the genome. However, nucleotide sequence comparisons are helpful when the functional element is the nucleotide sequence itself rather than its translated product. Among the functional elements of the genome that could be detected using nucleotide sequence comparisons are binding sites for transcription factors, structural elements of chromosomes, and the genes for regulatory RNAs (Figure 15.19).

➲ The method of sequence comparison used in the BLAST program is described in Section 16.14.

Although it may sound easy, defining regulatory regions and other functional DNA sequences can be challenging. Confounding factors include the shortness of many regulatory regions (on the order of eight to 12 nucleotides) and the requirements for conservation of a regulatory sequence often being less stringent than the conservation required to maintain protein function. If a protein sequence is changed there is a strong probability of a functional consequence. By contrast, the binding site for a transcriptional regulator can be degenerate, or multiple binding sites can replace one strong binding site.

Unlike amino acid sequences, in which informative comparisons can be made between two sequences in distantly related species, the most useful comparisons for functional nucleotide sequences require multiple sequences from more closely related species. For example, the genome sequences for 12 species of *Drosophila* could be aligned for protein-coding genes using a local alignment based on the inferred amino acid sequences; the upstream regions of the genes were then compared by a multiple alignment for conserved nucleotide sequences, as shown in Experimental approach 9.3. In the example, we saw how the comparison of *Drosophila* DNA sequences was used to predict RNA editing sites. The assumption is that the conserved regions represent sequences where selection has maintained sequence similarity or identity, possibly because the conserved sequence is the binding site for a regulatory protein. However, we noted previously that conserved sequences may also be evidence of a sequence found in an ancestral species that has not yet diverged. Thus, not all conserved nucleotide sequences between two species are functionally important, but those conserved among multiple species are *more* likely to have a functional role. As more genome sequences from closely related species are available, the methods to detect functional sequence elements are becoming more sophisticated and successful.

Distinguishing the effects of selection from the effects of genetic drift is one of the principal challenges in using nucleotide sequence comparisons to infer function. For example, a sequence repeat at a particular site in the ancestral species may be resident in two different species, even if it confers no selective advantage. In addition, unlike the 20 amino acids with their well-known chemical properties, there are only four nucleotides in only two categories, purines and pyrimidines. Thus, an adenine that is conserved between two nucleotide sequences may be the same because it is functionally significant or because it has been mutated to another base and back to adenine many times.

As mentioned above, many regulatory regions such as protein binding sites are short, and often only some of these individual bases are essential for protein binding. This means that a potential binding site can be found many times in a complex genome simply by chance rather than due to any functional importance. This difficulty is compounded because the regulatory sequences for eukaryotic genes can be located in many different positions with respect to the coding region, and may not be the same distance away from the coding region for different genes. In addition, the activity of potentially functional sites as actual regulatory sites depends on additional factors, such as nucleosome density, which are not easily predicted from the primary structure of the gene.

All of these factors present challenges in inferring the functions of nucleotide sequences using computational comparisons of the sequences. Much current evidence for functional nucleotide sequences comes from molecular assays, such as the chromatin immunoprecipitation (ChIP) assays described in Section 15.2. Sequence data from closely related species have been important in refining computational approaches, so their availability is likely to improve our predictive ability.

Structural changes provide clues to genome history

As discussed above, chromosome rearrangements move entire regions of the genome from one location to another and are often involved in speciation. As a result, regions of a chromosome from one species can be aligned with the chromosomes of another species by using conserved genetic elements as landmarks for the alignment of the nucleotide sequences of two diverged genomes. Landmark genetic elements can include the relative orientation of two genes, the position of a repeated sequence or a transposable element, or some other defined sequence. This conservation of a block of sequence between chromosomes of different species is called **conserved synteny**, and the conserved regions are referred to as **syntenic blocks**. Examples of syntenic blocks between the human and the mouse genomes are shown in Figures 15.20 and 15.21. For human chromosome 22, an autosome, note that large blocks of sequence, much larger than a single gene, are together on different mouse chromosomes (Figure 15.21a). This may be due to the evolutionary history, potential evolutionary pressure, and the amount of time elapsed since the rearrangement as discussed below.

Structural comparisons of genomes involve identifying as many syntenic blocks as possible, and then reassembling the jigsaw puzzle of blocks to observe the results of genome evolution. Specific sequences within syntenic blocks can be in the same orientation or an inverted orientation in different species. The orientation, size, and location of the syntenic regions allow an approximate reconstruction of the rearrangements that shaped the genomes. However, the reconstruction is approximate since there are usually hundreds if not thousands of syntenic blocks, and many of them are the product of more than one rearrangement event. The greater the evolutionary distance, the fewer the number of syntenic regions that are shared between species. A comparison of syntenic maps of humans,

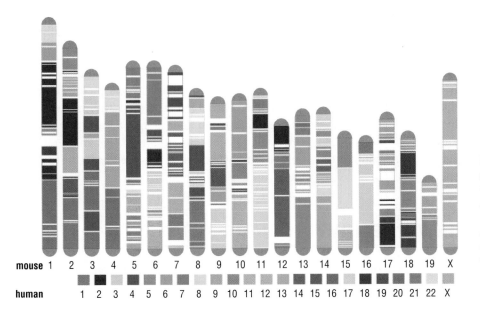

Figure 15.20 Mouse and human genomes show regions of synteny. The regions of the human genome that are homologous to mouse chromosomes are shown mapped onto the 20 mouse chromosomes. Each region of color corresponds to a block of human chromosome that is homologous to that region in the mouse. The colors representing each of the 23 human chromosomes are shown at the bottom. From Waterston *et al.* (2002).

(a)   **(b)**

Figure 15.21 Synteny between autosomes is different from synteny between X chromosomes. (a) Human chromosome 22 has synteny with multiple mouse chromosomes. Blocks of mouse sequence that are homologous to sequences on human chromosome 22 are found on the seven different mouse chromosomes (indicated by the different colored blocks). (b) Blocks of mouse sequence that are homologous to sequences on human X chromosome are all found on the mouse X chromosome although the order is changed.

chimpanzees and mice shows that there is greater similarity between humans and chimpanzees as compared to humans and mice.

The mammalian X chromosome is a special case of synteny

The most notable exception to the general rule about syntenic regions and divergence is the X chromosome, a **sex chromosome**, in placental mammals. The X chromosome is almost completely syntenic between all mammalian species, although intrachromosomal rearrangements such as duplications, deletions, and inversions are seen (Figure 15.21b). The synteny of the mammalian X chromosome was originally recognized because genes that exhibit X-linked inheritance in one mammalian species (that is, those genes that are most likely always to be found together on the X chromosome) are nearly always X-linked in other mammals. This conservation is probably a consequence of X-chromosome inactivation, the mechanism of **dosage compensation** in female mammals.

Dosage compensation occurs because there are two X chromosomes in females and only one X chromosome in males. Different organisms have different ways to compensate for the unequal number of X chromosomes and have evolved mechanisms to equalize gene expression in the two sexes. In mammals, one X chromosome is inactivated so that both males and females only express genes from one X chromosome. For this reason there is evolutionary pressure for the genes on the X chromosome to stay on this chromosome. If a rearrangement was to occur and a region of X chromosome landed on an **autosome**, the transferred genes would be expressed in two doses, which is not tolerated for some X chromosome genes. Conversely if an autosomal region were placed on the X, only one copy would be expressed.

The two examples of synteny described above illustrate the two principal evolutionary forces shaping the large-scale structure of genomes. In the case of the X chromosome, synteny has been preserved by selection, such that any change that disrupted X-linkage also likely disrupted dosage. In the case of the syntenic blocks for human chromosome 22, the extent of synteny correlates with the time elapsed since the divergence of the species. Certainly the arrangement of genes on human chromosome 22 is not the ancestral mammalian linkage arrangement for this chromosome; in fact, no extant mammal has the ancestral linkage arrangement for any chromosome. Nonetheless, we observe that the relative positions of gene elements have remained the same for millions of years. It is possible that this syntenic arrangement has functional significance, like the X-linkage.

It is also possible, and more likely, that the syntenic blocks reveal ancestral regions that have not yet had time to become rearranged. The ancestral mammal, the hypothetical founder of the entire mammalian lineage, had a particular linkage arrangement of its genes on its chromosomes. Remnants of this ancestral linkage arrangement are still observed simply because it was present in the founder species.

15.7 PHENOTYPIC VARIATION

So far, this chapter has considered sequence comparisons between species rather than between individuals of the same species. The sequence differences between individuals within a species are significantly less than between species. Yet, clearly individuals within a species are not identical; this is what allows us to recognize different people from each other or our dog from our neighbor's dog. The comparison of individuals and genome sequences between multiple individuals of the same species has also provided insights into gene function.

Genome comparisons between individuals within a species allows mapping of genes

Although two individuals of a certain species could differ in the large-scale structure of their genomes (for example, in the positions of transposable elements) most genetic variation between two individuals is found as simple DNA sequence changes. The most obvious of these sequence changes are single base changes. For humans, a few of these single nucleotide changes are unique to each individual or family, but many more of them are characteristic of a population group. The human population has DNA sequence variation at many different sites. For example, some individuals have a C at a certain position while others have a T, as shown in Figure 15.22a. A site that has more than one common sequence is known as a **polymorphism**. For sites that differ at only one position, these locations are known as **single nucleotide polymorphisms** (**SNPs** – pronounced 'snips'). On average, a SNP is found at one base in 1000 in the human genome, although some regions are far more polymorphic than others.

Other polymorphic markers are **microsatellite repeats**. These are regions where two or a few nucleotides are repeated in tandem, such as CTTCTTCTTCTT (Figure 15.22b). The number of repeats present at a specific location is variable between individuals. For this reason these microsatellite repeats are frequently used as genetic markers to identify individuals.

Since less than 2% of a mammalian genome consists of protein-coding sequences, most of the sequence variation occurs in non-coding regions. Sequence variation

(a)

```
GATCCAACTTCTGGGTGTCACTTGTCTCTC
GATCCAACTTCTGGGTGTCACTTGTCTCTC
GATCCAACTTCTGGGTGTCACTTGTCTCTC
GATCCAACTCCTGGGTGTCACTTGTCTCTC
GATCCAACTCCTGGGTGTCACTTGTCTCTC
GATCCAACTCCTGGGTGTCACTTGTCTCTC
GATCCAACTTCTGGGTGTCACTTGTCTCTC
GATCCAACTTCTGGGTGTCACTTGTCTCTC
GATCCAACTTCTGGGTGTCACTTGTCTCTC
GATCCAACTTCTGGGTGTCACTTGTCTCTC
GATCCAACTTCTGGGTGTCACTTGTCTCTC
GATCCAACTTCTGGGTGTCACTTGTCTCTC
GATCCAACTTCTGGGTGTCACTTGTCTCTC
GATCCAACTTCTGGGTGTCACTTGTCTCTC
GATCCAACTTCTGGGTGTCACTTGTCTCTC
GATCCAACTTCTGGGTGTCACTTGTCTCTC
GATCCAACTCCTGGGTGTCACTTGTCTCTC
```

(b)

foward primer   8 repeats

...GCTCCAGGCTTAGACTTCTTCTTCTTCTTCTTCTTCTTCGCACTTTAACGATACGG...
...CGAGGTCCGAATCTGAAGAAGAAGAAGAAGAAGAAGAAGCGTGAAATTGCTATGCC...

reverse primer

foward primer   7 repeats

...GCTCCAGGCTTAGACTTCTTCTTCTTCTTCTTCTTCGCACTTTAACGATACGG...
...CGAGGTCCGAATCTGAAGAAGAAGAAGAAGAAGAAGCGTGAAATTGCTATGCC...

reverse primer

foward primer   9 repeats

...GCTCCAGGCTTAGACTTCTTCTTCTTCTTCTTCTTCTTCTTCGCACTTTAACGATACGG...
...CGAGGTCCGAATCTGAAGAAGAAGAAGAAGAAGAAGAAGAAGCGTGAAATTGCTATGCC...

reverse primer

Figure 15.22 Polymorphisms used for gene mapping. Gene mapping is simplified by the presence of various polymorphisms in a genome including (a) single nucleotide polymorphisms and (b) microsatellite repeats. These variable length repeat tracts can be amplified with specific forward and reverse primers as shown in the figure. The different length of the PCR product represents the specific allele of the microsatellite.

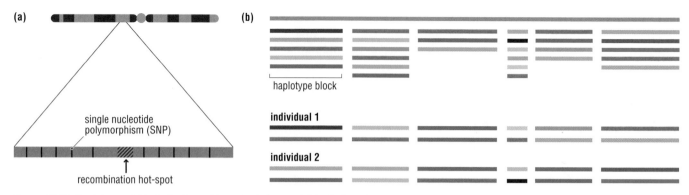

Figure 15.23 Haplotypes. A block of polymorphisms that is frequently inherited together is called a haplotype. (a) Haplotypes arise because of the presence of recombination hotspots, positions in a genome where recombination is more likely to occur. (b) Knowing the structure of haplotype blocks reduces the number of markers needed to associate a phenotype with a particular DNA region. In the top part of the figure the gray line represents the DNA and the colored bars below represent the set of known haplotypes that are present for the six haplotype blocks found on the theoretical chromosome shown here. Below is a representation of two individuals who each have two chromosomes with different combinations of the possible haplotype blocks.

in non-coding regions might not contribute to the phenotype of an organism, and thus could be selectively neutral. Despite the potential lack of biological function, however, SNPs and microsatellite repeats in non-coding regions have become useful and important markers for landmark positions in the human genome. These genetic differences are inherited like any other allele and are thus indicative of the genetic history of a population.

Polymorphisms that lie close together on a chromosome are genetically linked and inherited together, as is the case for any other genetic marker. A block of SNPs and other sequence variations that is frequently inherited together is called a **haplotype** (Figure 15.23). The importance of haplotypes is discussed in Section 15.8.

Copy number variants alter gene expression and phenotype

In addition to SNPs and microsatellites, sequence variation occurs in the form of differences in the copy number of some segments of DNA. Recent analysis of the human genome has found that these **copy number variants (CNVs)** are extremely common. CNVs are one type of chromosome duplication as described in Figure 15.13. The length of the repeated unit in a CNV ranges from less than 1 kb to more than several megabases. Likewise, the number of CNV sites is variable, with the best current estimate being that individuals vary at an average of 11 CNV sites. Although there is no known consequence associated with CNVs at many sites, in some cases, they are functionally significant. For example, the diploid human genome has from two to 15 copies arrayed in tandem of the gene *AMY1*, which encodes the amylase enzyme that breaks down starch. Interestingly, populations with a high starch diet have a high copy number on average and more amylase enzyme expressed in saliva than populations with a low starch diet. This suggests that the copy number of the amylase gene has evolved in concert with cultural differences in diet, or that the diet in part was shaped by the genetic polymorphism. CNVs are also found in association with a number of human diseases, suggesting that having the optimal copy

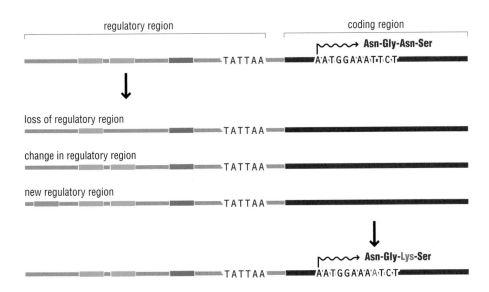

Figure 15.24 Phenotypic variation can result from many different kinds of genetic change. Different phenotypes can arise from mutations in regulatory sequences that affect where, when and how much of a gene product is expressed and from mutations that affect the coding sequences and alter what gene product is made.

number and thus appropriate levels of expression of some genes is important for the organism.

Phenotypic variation arises from regulatory and functional variation

Although we have been describing genetic elements in terms of their sequence and structure, it is also important to consider them in terms of function. For many genetic elements, their function is observed by an effect on gene expression. Gene expression can be measured by variation in RNA levels, in protein levels and activity, as well as in cell morphology. Variation can also be detected at many different stages in the flow of genetic information, such as at transcriptional initiation or splicing, or by measuring mRNA levels. Changes in gene expression at the transcriptional levels may reflect changes in the underlying DNA sequence or genome structure, but the connection between sequence changes and phenotypic variation is often difficult to make.

In general, as depicted in Figure 15.24, variation in *when, how much,* or *where* a gene is expressed can be explained by changes in nucleotide sequence of regulatory elements. This type of variation could reflect base changes in promoters sequences, for instance. A variation in *what* is expressed can be explained by changes in the coding sequences, which affect the nucleotide sequence of the RNA product and/or the amino acid sequence of the protein product (which of course also requires a change in the DNA sequence). As an example, this type of variation could reflect amino acid substitutions that change protein activity. Variations in either regulatory or coding sequences could result in phenotypic changes, so both types of change are important for understanding evolutionary change and population diversity.

Until recently, most of the attention in understanding evolutionary change has been given to variations in the coding sequences, primarily changes in amino acid sequence. However, it is now recognized that changes in the regulatory sequences may be even more important than variation in coding sequences in understanding evolutionary change: changes in regulatory sequences can produce more rapid evolutionary divergence than changes in coding sequences.

Figure 15.25 Phenotypic differences can be due to changes in single genes or in multiple genes. Some human diseases are caused by mutations in one specific gene (a) or by mutations in one of several different genes (b) (monogenic). The inheritance of these phenotypes follows classical mendelian genetics. (c) Other human diseases are the outcome of genetic differences in multiple different genes (polygenic). Often these traits reflect both genetic differences and environmental variation (multifactorial).

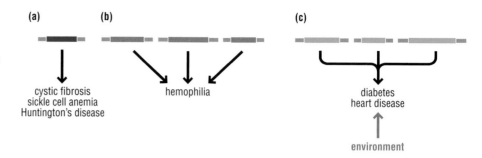

Phenotypes can arise from the effects of one or many genes

'Phenotype' is a very general term for any trait that can be measured in an organism. The connection between the phenotype and the genotype can take many forms, as summarized in Figure 15.25. Some genetic traits are monogenic, arising from changes in single genes. In humans, many of these recognized single gene traits are genetic diseases (Figure 15.25a). Examples include cystic fibrosis, sickle cell anemia, and Huntington's disease. The concepts familiar from mendelian genetics, such as recessive or dominant alleles and autosomal or X-linked inheritance, are relevant to understanding these single gene traits.

Other disease phenotypes arise from a change in a single gene, but changes in many different genes can give rise to the *same* phenotype (Figure 15.25b). For example, there are many different genes involved in blood clotting in humans. A mutation in any one of these genes gives rise to hemophilia, often with very similar clinical symptoms, so it is difficult to know which gene has been affected. For instance, the Russian czars and many descendants of England's Queen Victoria had X-linked recessive hemophilia; there are several different X-linked genes involved in clotting, so that the possible causal mutation is only now being identified through extensive sequencing.

Many other medically and economically important phenotypes are the outcome of several or many genes, each contributing a fraction to the total phenotypic variation in a population (Figure 15.25c). For example, height variation within human populations can be largely (although not entirely) attributed to multiple genetic differences between individuals, rather than a difference at a single locus. A phenotype that arises from the effects of many different genes is known as a **polygenic** trait. Each individual gene is inherited in simple mendelian fashion, but the phenotype being measured is the outcome of the concerted action of many different genes. It is challenging to identify the individual genes that contribute to polygenic traits or to determine how much of the variation can be explained by any one gene.

Often traits reflect both genetic differences and environmental variation, as well as interactions between these two influences. We refer to these traits as **multifactorial traits**. For any given multifactorial trait, it is often difficult to determine how much of the variation can be attributed to genetic differences and how much can be attributed to environmental variation. Many human disease traits and many behavioral traits fall into this category. For example, variations in blood pressure or blood sugar levels arise from both genetic differences and environmental (especially dietary) differences. Some people consuming a high sugar diet develop diabetes, whereas others who consume the same diet do not; this is because different individuals have different alleles of many different genes, all of which have some influence on an individual's risk of developing the condition. The risks for developing many types of cancer are also multifactorial traits, with both genetic and environmental factors playing a significant, but often unknown, role.

In addition, interactions between the genetic composition and environment of an individual impact on the efficacy of potential therapies. We discuss the link between human genomics and clinical interventions in human disease further in Section 15.9.

Heritable variations that do not involve DNA sequence changes can also affect phenotypes

Our discussion has focused on changes in the DNA sequence that affect phenotype. However, sequence changes are not the only source of heritable variation. Epigenetic changes can alter phenotype and gene expression without alterations to the DNA sequence. Epigenetic changes typically are due to the methylation of DNA and the modification of histones; such modifications are heritable between cells and even between generations of individuals, and can affect the final phenotype that arises.

➲ The molecular bases for the inheritance of epigenetic changes are described in Chapter 4 and the consequences to gene expression are described in Chapter 8.

One type of epigenetic change involves methylation of cytosines and is widespread among mammals (see Chapter 4). Most mammalian genes have clusters of CG dinucleotides (known as CpG islands) in their upstream regions, and methylation of cytosines is associated with transcriptional inactivation. The inactive X chromosome in female mammals is highly methylated, as are many individual genes that are transcriptionally silent on other chromosomes. A failure to methylate CpG islands, and the inappropriate gene activation that results, is the underlying molecular defect in the neurological disorder Rett syndrome. Loss of methylation is also frequently found in a variety of human cancers.

Another important epigenetic modification occurs in chromatin, and appears to be important in all eukaryotic organisms. The histone proteins that comprise the structural core of a chromosome are subject to post-translational modifications, such as phosphorylation, methylation, and acetylation, among others. These modifications occur on specific amino acids on specific histones, and different modifications produce different effects on chromosome structure and gene expression.

➲ We discuss the post-translational modification of histones in Chapter 4.

Significantly, DNA methylation or a histone modification is stable and can persist through many cell divisions through the distributive inheritance of old marked histones (as described in Chapter 6). Thus, these changes in the chromatin structure produce changes in the heritable state of genes without actually altering the gene sequence or structure.

➲ We discuss genetic imprinting in Chapter 4.

Genetic imprinting is a special case of an epigenetic effect. A classic example of genetic imprinting is Prader–Willi syndrome. This disease, which results in behavioral changes and obesity, is associated with abnormal expression of a block of genes on chromosome 15 (Figure 15.26). These genes are in an imprinted locus and are normally turned off or silenced by chromatin changes in the chromosome inherited from the mother. However, they are normally expressed from the chromosome inherited from the father. The disease can arise if the genes are deleted or abnormally turned off in the chromosome inherited from the father, resulting in no expression from either gene in the offspring inheriting these chromosomes.

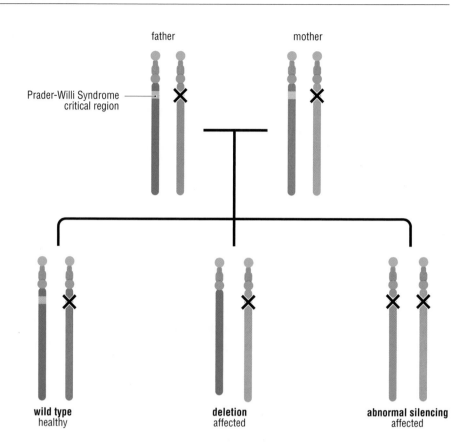

Figure 15.26 Prader–Willi syndrome can be caused by changes in either DNA sequence or chromatin structure. Prader–Willi syndrome is associated with defective expression of a block of genes on human chromosome 15. These genes are normally turned off or silenced by chromatin changes (indicated by the X) in the copy of the chromosome inherited from the mother (pink chromosomes). However, this region is expressed normally in the copy received from the father (blue chromosomes). The disease can arise if the genes are deleted or abnormally silenced in the father, resulting in no gene expression from either chromosome in the offspring. In the parental generation, the block of genes on one of the father's and one of the mother's chromosomes is silenced. The imprinting is normally re-established in the offspring such that the genes will be silenced in any copy from the mother while the copy received from the father will be expressed.

Thus, we see how Prader–Willi syndrome can be caused by a change in chromatin structure rather than a change in the DNA sequence.

Biological complexity comes from many different sources

The diversity and complexity of natural populations arises from many different activities of the genome. Before describing some examples in more detail, let us summarize some of what we have considered about the origins of biological complexity and genetic variation.

First, complexity arises from variation in the **sequences** of the component parts of the genome, both the regulatory sequences and the coding sequences. The number of genes in an organism's genome and the extent of regulation generally increase as we move up a phylogenetic tree. Thus, complexity arises from the number of genes, the sequences of the genes, and regulation of the expression of genes.

Second, complexity and variation arise from **modifications** of the gene products, both RNA and protein. Because of alternative splicing of mRNA, one protein-coding gene can produce many different transcripts and thus different polypeptides. In addition, polypeptides are further modified post-translationally, for instance by phosphorylation or glycosylation, as described in Chapter 11. This produces an even greater range of activities from the same number of protein-coding genes. As a result of these post-transcriptional changes, one structural gene can produce many different products, resulting in more variation and complexity.

Third, complexity and variation arises from **interactions between macromolecules**. For example, as a matter of arithmetic, the number of ways proteins can interact with each other increases greatly with the number of proteins in a cell. This then results in even more biological complexity and variation.

Finally, greater complexity and variation arise from the interaction of the organism with its **environment**. The environment, broadly defined to include nutrition, climate, and all other non-heritable effects, clearly has a profound impact on many phenotypes. In addition, the interaction between the genotype and the environment selects for certain phenotypes and against others. Selection works not only on individual organisms and species but on all of the species that occupy a particular niche.

15.8 EXPLORING EVOLUTION THROUGH GENOMICS: EXAMPLES FROM MAMMALIAN GENOMES

We will now turn to two well-studied and familiar mammals, dogs and humans, to illustrate the effects of genome evolution and the information that can be revealed through genomics. Many of the genetic and genomic approaches described earlier were developed to understand the inheritance of human disease. Here we will see how these approaches can be used to identify the genes responsible for single as well as multigene traits in both dog and human populations. Identifying canine genes responsible for different traits is of both scientific and veterinary interest, and can shed light on the process our ancestors used to domesticate the species. Furthermore, dogs have many of the same diseases as humans, and what has already been learned from dog genetics may have direct applications to human disease.

The dog genome provides an informative parallel to the human genome

Dogs are a single species with many different breeds, and exhibit tremendous phenotypic diversity: a naïve observer would be slow to recognize that a Chihuahua and a St Bernard belong to the same species. The diversity between breeds is largely the outcome of artificial selection by humans. There is little genetic diversity within a breed because purebred dogs are the result of extensive inbreeding and selection from small initial founding populations.

The distinctive morphology and behaviors of different breeds are familiar to us. For example, border collies are incessant herders, especially of sheep and cattle, but do not readily swim. Labrador retrievers attempt to swim in any body of water but do not herd. The appearance and many of the behaviors that are characteristic of the breeds are the products of their genomes, and differences between breeds highlight differences in their genomes.

The extensive genetic variation between breeds, and the lack of genetic variation within a breed, allows investigators to map and define both single gene and multifactorial traits. The map location of a gene can be determined by using the genetic polymorphisms between two or more breeds. Breeds that differ in a phenotypic trait can be crossed, and the subsequent generations bred to find dogs that do and do not share the trait. This represents a classic genetic **linkage analysis**, where

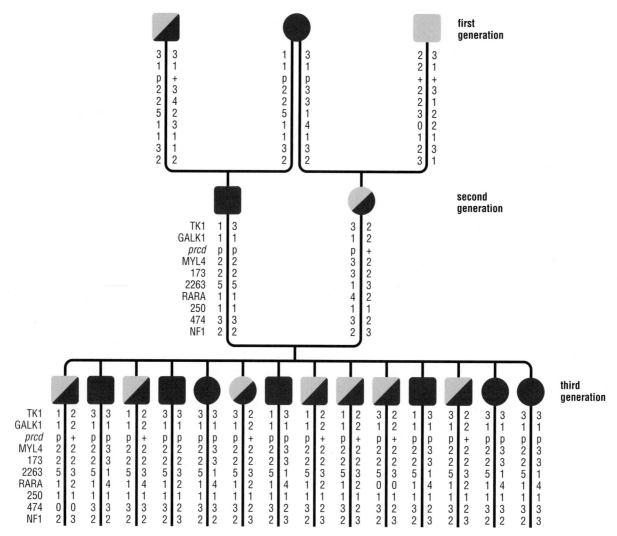

Figure 15.27 Linkage analysis allowed the mapping of the *PRCD* gene in dogs. Analysis of the microsatellite markers in several different dog pedigrees has led to the identification of a gene associated with progressive rod-cone degeneration (*PRCD*), a widespread hereditary retinal disease leading to blindness in dogs. Affected purebred miniature poodles were bred to homozygous normal unrelated miniature poodles and beagles and beagle cross-bred dogs. The heterozygous offspring from these crosses were then backcrossed to affected dogs and the offspring analyzed. Circles = females; squares = males; open symbols = homozygous normal at the *PRCD* locus; solid symbols = *PRCD*-affected; half-filled symbols = *PRCD* heterozygous. The haplotype determined by single nucleotide polymorphism analysis at a number of different loci is shown beneath each individual. The numbers at each position represent the specific length of the microsatellite found at that position; each position has two different alleles as indicated on the left and right side of the line. The haplotypes shared by those dogs that have the disease (TK1 and GALK1) help to identify the location of the disease-associated gene *PRCD*. See Figure 16.79 for a more detailed description of linkage analysis. From Acland *et al.* (1998).

a trait is followed in a specific family (Figure 15.27; see Figure 16.79 for a more detailed description of how linkage analysis is performed). Having a large number of SNP markers, which have been identified by comparing the genome sequence of dogs of multiple breeds, gives power to this linkage analysis. Hundreds or thousands of SNP or other markers can be analyzed for each family member that does or does not have the trait. The marker that tracks faithfully with those members that have the trait is then said to be linked to the trait and can be used to identify the difference in DNA sequence responsible for the trait.

The availability of a dense set of SNP markers also makes possible another type of genetic analysis termed **association studies** (Figure 15.28). In association studies, a large number of unrelated individual dogs from breeds that have one phenotypic

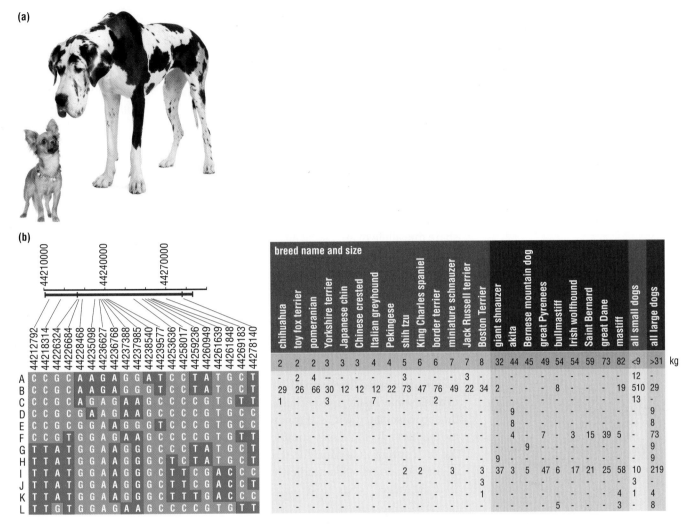

Figure 15.28 Association studies linked differences in dog size with different alleles of *IGF1*. (a) Phenotypic differences between a small dog (Chihuahua) and a giant dog (Great dane) are obvious. (b) Observed haplotypes (A–L) for the 20 single nucleotide polymorphism (SNP) markers spanning the genome near *IGF1* (encoding insulin-like growth factor 1) are shown on the left. Marker alleles shown in yellow correspond to the alleles found in the golden jackal (*Canis aureus*) assumed to be the ancestral allele for each SNP. The number of haplotypes found for specific breeds is given on the right. It is clear that haplotype B is prevalent in small dogs and haplotypes F and I are prevalent in large dogs. This analysis showed that one allele of *IGF1* is present in small breeds whereas a different allele is present in large breeds.
From Sutter *et al.* (2007).

trait (such as short stature) are compared with a large number of dogs from breeds that do not have that trait (such as normal stature). As described above for linkage studies, association studies analyze thousands of SNPs and determine whether one particular SNP is more frequently associated with all of the dogs that have the trait and not with those that do not have the trait. The goal is to identify a specific region of the genome where the gene responsible for the trait is located. As such, association studies do not require breeding and screening for the trait of interest.

With the availability of complete genome sequences and the identification of many SNPs, it is feasible to carry out extensive **genome-wide association studies**, which analyze SNPs from across the genome. In practice, genome-wide association studies are simplified by the presence of blocks of DNA sequences where recombination is very infrequent and all of the SNPs in that region are inherited together. This set of linked DNA sequences defines a haplotype as introduced in Section 15.7 and Figure 15.23. Any SNP in that region will always be linked to the other SNPs in the same region. Thus one SNP is chosen as a 'TagSNP' that marks a particular **haplotype block**. This greatly simplifies the number of variable regions that must be followed in the association study. Because the locations of thousands of SNPs and TagSNPs are now known, genome-wide association studies have been used to find the genes associated with many different traits in many organisms, including humans.

The power of genome-wide association studies is that they can be used whether the trait is controlled by a single gene or by many genes. (In fact, such studies can be used when it is not known at the start of the study how many genes control a trait.) The results of these analyses can be quite striking. For example, an association study found that many small breeds such Chihuahuas, Pomeranians, and Pekingese share one allele of an insulin-like growth factor gene, whereas many large breeds such as St Bernards, Mastiffs, and Great Danes share a different allele of this gene. The ability to identify one gene as being responsible for the variation in a complicated phenotype such as body size between these breeds shows the power of using genomics.

Genome-wide association studies are also being used for studying multifactorial traits such as behaviors and disease risk. The strategy is exactly the same as for single gene traits, but the contribution of any one gene to the variation in phenotype may be relatively small. As a result, there may be several overlapping sets of alleles that result in a particular phenotype for a complex trait. Some alleles that are associated with the trait will only be present in a subset of the individuals that are affected. For a complex trait, a gene that accounts for 10% or 15% of the total variation is considered to have a significant effect. Very large numbers of individuals must be studied to draw meaningful conclusions about multifactorial traits.

Human history is told in the sequences of our genomes

All of the processes of genome evolution that we have described for other species, including dogs, have also affected the human genome. Our genome is approximately the same size and complexity as that of the dog genome, at around three billion base pairs. One major difference is that dog breeds are highly inbred whereas human populations are outbred (with a few exceptions of groups that breed in small populations, such as the Amish population in America). On the other hand, for much of human history, we lived in small populations, so genetic drift has exerted significant effects on human genetic diversity. Although we have recognizable genetic diversity between populations from different geographical regions, most of our genetic variation is found among individuals of the same population. Thus for example, Swedes and Italians (to take two groups) do not look the same. However, the amount of genetic variation between a Swede and an Italian may be much less than the amount of variation found by comparing two Swedes or by comparing two Italians.

As our ancestors lived in small populations, haplotypes are a powerful way to trace our peripatetic past. For example, the native populations of the Americas have haplotypes that are also found in populations that arose in east-central Asia, consistent with migration across Siberia and the Bering Strait and south along the west

Figure 15.29 Haplotype analyses confirm maps of human migration. Because our ancestors lived in small populations, genetic analyses (here shown for the Y chromosome inherited from the father (blue) and mitochondria DNA which is inherited from the mother (pink)) are a powerful way to trace human migration. The native populations of the Americas have haplotypes that are also found in populations that arose in east-central Asia, consistent with multiple waves of migration across Siberia and the Bering Strait and south along the west coast of the Americas.
From Rinaldi A, *EMBO Rep* 2007;**2**:121–125.

coast of the Americas (Figure 15.29). Multiple prehistoric waves of migration can be inferred, the first being found in South America and the most recent in western and northern Canada. These waves of migration are consistent with a regular southward and eastward movement of the immigrants. The prehistoric migrations across the islands of Polynesia and the general northwesterly settlement of Europe can also be reconstructed from examining the patterns of haplotypes.

Genome analysis supports what we learn from recorded history as well. For example, the presence of the Vikings in Ireland, Phoenicians in port cities around the Mediterranean Sea, and Spanish soldiers in Colombia are also evident from analysis of DNA polymorphisms and haplotypes.

The ancestral nature of some mutations and the effects of population migration are also apparent from some genetic diseases. One example is cystic fibrosis, the most common genetic disease among people of European descent. The gene responsible for cystic fibrosis is named *CFTR* and the disease is inherited as an autosomal recessive trait. Although more than 125 different mutations in the gene *CFTR* have been identified, more than 75% of all cystic fibrosis cases are due to the same molecular lesion, a 3 bp deletion of a phenylalanine at position 508 of the *CFTR* protein (referred to as ΔF508). Analyses of polymorphisms near to the ΔF508 mutation have revealed a common haplotype or a group of closely related haplotypes, suggesting that the causative mutation in almost 75% of cystic fibrosis cases arose in, and spread from, one particular individual.

Many other human genetic diseases show a similar phenomenon whereby one particular molecular lesion on a given haplotype accounts for a large percentage of the disease cases today. These results and the haplotype structure of the human genome have led to the 'common disease–common variant' hypothesis. The tenet of this hypothesis is that other relatively common genetic diseases shared among a particular population or geographical group will stem from the same molecular lesion. However, genome-wide association studies that have recently indicated this may not be the case for all diseases.

The human genome also shows evidence for natural selection

We noted earlier that genetic drift and selection shape the sequence of a genome. This is also true for humans, although some indications of natural selection are subtle, and are occasionally best seen in heterozygotes. Cystic fibrosis provides one example of the past effects of natural selection. Heterozygotes for the ΔF508 mutation are resistant to several water-borne pathogens, including the causative agent of cholera. Thus, although there is strong selection against being homozygous for the ΔF508 mutation, being a heterozygous individual confers a selective advantage. As a result, the mutation and its surrounding haplotype spread among European populations.

Another example of selection in favor of the heterozygote is seen with sickle cell anemia. The causative mutation in the β-globin gene is common in populations where malaria has been a significant health threat: heterozygotes for the mutation are resistant to malaria. This mutation has been found in association with several different haplotypes, indicating that the same mutation arose independently in different populations. In these examples, the selective agent is a pathogen and the most favored genotype is the heterozygote. Similar pathogen-resistance effects have been postulated for heterozygotes for other genetic diseases.

Selection is also evident from non-disease traits in humans. Most humans (and mammals generally) shut off transcription of the gene encoding the enzyme lactase once adulthood is reached. Indeed, recent evidence indicates that Neandertals also shut off lactase expression. Thus, most humans cannot digest milk as adults. Although we refer to this as 'lactose intolerance' as if it were the unusual situation, the ability to continue to drink milk, known as lactase persistence, is the derived or mutant condition. Lactase persistence is common among populations that were traditionally herders. Thus, a mutation that allowed the population to utilize the milk produced by their herds would provide an additional nutritional source.

At least five different lactase-persistence mutations are known, all in the regulatory region of the gene (Figure 15.30). In fact, analysis of the haplotypes for lactase persistence indicates that these mutations arose separately in different geographical locations. These mutations conferred no selective advantage to non-herding populations and were probably lost once they occurred. In this case, as with our previous example for CNV of the amylase gene, the selective agent appears to have been cultural differences, including the sources of nutrition.

15.9 HUMAN GENETIC DISEASES

Genome variation among populations is an indicator of our species' past, but we may be even more interested in the genetic variation among current individuals.

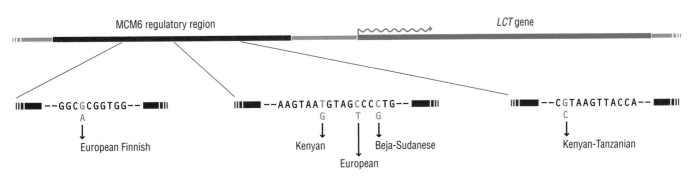

Figure 15.30 Lactase persistence mutations. Five mutations that confer the ability to continue to drink milk into adulthood, known as lactase persistence, have been identified in the regulatory region of the *LCT* gene on human chromosome 2. Haplotype analysis suggests that these mutations arose separately in different geographical locations.

In particular, what genes contribute to the different incidences of inherited diseases among different families? This has become a significant focus of modern medicine, using information obtained from the Human Genome Project.

Diseases arising from single genes are the best-researched areas of human genetics

Thousands of genetic diseases that arise from a mutation in a single gene are known (Figure 15.31). Although some genetic diseases are common in certain populations, most genetic diseases are rare in all populations. But the incidence of the diseases in the population matters little to the couple who worries about their own child, so individual cases of human genetic diseases are carefully recorded. An authoritative database of human genetics, known as the Online Mendelian Inheritance in Man or OMIM, is maintained by the National Center for Biotechnology Information (http://www.ncbi.nlm.nih.gov/omim). This database is the best source of information about any genetic disease, including inheritance, gene identification and function, animal models, known disease variants, and even links to support groups for afflicted families.

Nearly all of the genetic diseases in OMIM are the result of changes in the protein-coding portion of the gene. Included are mutations that result in amino acid

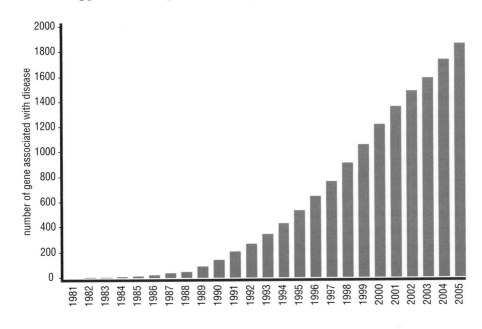

Figure 15.31 Number of mapped human disease genes. As shown by this graph from the Online Mendelian Inheritance in Man (OMIM) database, the number of mapped human disease genes has increased significantly.

substitution or loss, and ones that generate stop codons, mutations in splice sites, and small deletions of the gene. Few of them are due to changes in regulatory regions, although such mutations are common in other species and must certainly occur in humans as well.

As these genetic diseases arise from mutations in a single gene, simple mendelian traits such as hemophilia or adenosine deaminase deficiency are the best candidates for treatment using gene therapy, that is, using a normal copy of a gene to correct the defect or supply the function lost from the mutated version. Ideally, a normal copy of the gene with its regulatory region will be stably introduced into a person who has a mutation; the introduced normal copy then complements the defective function of the chromosomal copies, restoring the phenotype to the 'wild type'. But how is the normal copy introduced? Many investigators have engineered viruses that could be used as vectors to introduce the gene. However, the number of patients who have been successfully treated by gene therapy is small. Insertional mutagenesis, in which the introduced gene copy inappropriately inactivates an endogenous gene when it integrates, is a hazard associated with gene therapy. Thus, for many individuals with single gene diseases, therapies using drugs and nutrition are still preferred.

Multifactorial genetic traits include our susceptibility to many adult-onset diseases

Most of the single gene diseases in OMIM exhibit their phenotypes early in life. Many other traits and diseases 'run in families' but are not inherited in a simple mendelian fashion. For those of us who do not exhibit any of the single gene diseases, the more significant inherited diseases are multifactorial, with phenotypes that arise at different times in our life, and arise from the action of many genes and their interaction with the environment.

The risk for many types of cancers, for heart disease, for adult-onset diabetes, and many other medically significant traits, are familial multifactorial traits. Recent research has involved using genome-wide association studies to identify genes that affect our susceptibility for these traits. A quick survey of the scientific literature for human genetic analysis finds reports of genome-wide association studies to identify genes affecting our risk for schizophrenia, bipolar illness, colorectal cancer, type 1 and type 2 diabetes, multiple sclerosis, and coronary heart disease, among many others. Other papers report genes associated with traits such as obesity and body mass index, liver function, skin color, baldness, bone mineral density, and many more. All of these traits have both genetic and environmental factors that contribute to our susceptibility or phenotype and so are multifactorial.

Identifying the genes responsible for these traits may help to predict risk, improve diagnostics and suggest new therapies and treatments but the multifactorial basis of these traits complicates such prospects. Many of the genetic variants being identified appear to increase our susceptibility by as little as 20%, which is below the impact of some known environmental risk factors. There is a complex interplay between our genetic variation and our environment, which collectively determines our overall risk for developing any particular genetic disease or disease-related trait. Identification of the genetic susceptibility factors is a beginning but not the end point.

Individualized medicine is possible with genomic sequence

As we identify genes associated with many human traits and diseases, there is increased interest in analyzing individual genomes to help plan medical treatment on an individual level. **Pharmacogenomics** integrates information about a person's reaction and response to a particular drug with an understanding of their unique genetic makeup (Figure 15.32), and has had some striking successes in recent years. About 25% of primary breast cancers tumors are sensitive to the drug trastuzumab (Herceptin), such that pretreatment testing can be used to determine if this drug is likely to be effective. Determining a patient's genotype is also useful in avoiding treatment with a drug that has potential risk. For example, about 5% of Caucasians have at least one copy of a specific variant of an immune response protein Human Leukocyte Antigen (HLA)-B that results in significant risk for a fatal hypersensitivity to the antiretroviral drug abacavir (Ziagen), which is used in the treatment of individuals with human immunodeficiency virus (HIV) infection. Pretreatment genetic testing can determine whether an individual carries this specific variant, and therefore indicate whether this drug should be avoided.

Proponents of individualized medicine suggest that side effects of other medications could be reduced through the use of genetic information. For example, the blood-thinning drug warfarin (Coumadin) is prescribed hundreds of thousands of times annually in the United States for people with blood clots or at risk of clotting after surgery. However, approximately 1% of the patients develop serious life-threatening side effects from warfarin treatment due to known genetic variants at one of two different loci, and these side effects could be avoided if patients were tested for these alleles, and the dose more carefully monitored for those individuals carrying the known genetic variant. In addition, better treatments may be identified for these individuals.

As another example, the drug gefitinib (Iressa) is the most effective therapy for about 10% of people with lung cancer who have a particular polymorphism, but is much less effective for most other cases. A genetic test could identify those individuals who would benefit from this treatment.

Obtaining a personal genome sequence is powerful and controversial

The examples described above illustrate the value of individualized medicine. A newer related area for genomics is personal genomics, where companies worldwide are offering services to analyze the whole transcribed genome or a subset of genes that affect susceptibility to multifactorial disease. These direct-to-consumer companies are marketing genome sequencing to individuals, outside a medical setting. In principle, such personal genomic information can allow an individual to recognize a susceptibility to a multifactorial disease, such as diabetes or some cancers, before the symptoms are manifested. The person can then make informed choices and lifestyle changes.

However, many people have raised doubts about the efficacy and ethics of personal genomic testing, especially outside the medical setting. Studies indicate that many individual consumers have a poor understanding of the concept

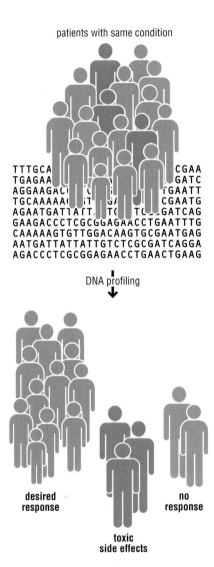

Figure 15.32 Individualized medicine is possible with genomic sequence. Genomic sequence information for a population of individuals with the same disease can aid in the development of a treatment regimen by identifying individuals who may not respond to the drug or even worse may experience a toxic side effect of the drug.

of risk factors so that follow-up counseling is often needed but not available. Furthermore, the personal genome testing industry is largely unregulated by government agencies. Consequently, genetic testing companies in the United States and most of Europe do not currently have to be certified or licensed. This means that the information on which people are basing life decisions or lifestyle changes could be flawed.

Ethical issues arise with the availability of individual genome sequences

Prenatal genetic testing is already used to allow parents to make more informed reproductive decisions. Genetic testing can be sought when there is increased risk of genetic disease because a previous child was affected with a disease, other family members have a disease, or there is an increased risk of disease because of environment, such as maternal age. There are, however, significant ethical issues about prenatal testing and 'selective reproduction'. Is it ethical for parents to use genetic testing to 'select for' children who may be more likely to have a 'desired trait'? Is it ethical to 'select against' children who may have an increased risk of an 'undesirable trait', rather than a significantly debilitating disease?

Privacy is also a consideration. There is considerable disagreement about who should have the ability to request the genetic information gathered for individuals. Many observers have raised the possibility of genetic discrimination in health insurance as well as in the life insurance arena. Recent regulations in the United States ban the use of genetic information for discrimination for insurance purposes, but many are skeptical of the effectiveness of such legislation.

Genetic privacy considerations extend to ethical issues related to inheritance and consent since information derived from genetic testing can affect family members who do not want to be tested or do not have a legal right to consent (for example, minor children). For example, some young women have asked to have their *BRCA1* and *BRCA2* loci tested to determine if they may be at increased risk for susceptibility to breast and ovarian cancer. What if a daughter is found to have an elevated risk due to an allele inherited from her mother, although her mother has declined to be tested? The issue is even more complex in cases where the presence of a particular gene allele always results in severe disease, for example with the neurodegenerative disorder Huntington disease whose symptoms usually only begin at middle age. The impact of genomics is felt in many areas of biology, and in fields that touch the essence of who we are.

✱ SUMMARY

In the book *'On the Origin of Species'* Darwin ended by musing on the 'endless forms most beautiful' as he contemplated the diversity of living things. Darwin saw the life of organisms and species in their environments; he knew little or nothing of cells, genomes, genes, RNA, and proteins. We now recognize that the unity and diversity of all living things that inspired Darwin is apparent at all levels of biology from ecosystems to molecules.

Genomics both focuses and reflects all of biology. We may not yet have the insights necessary to understand the subtle interrelationships between molecules or between species, but genomics provides both the tools and perspective that we need to approach these investigations. It gives us molecular data to explore human history and our interaction with our environment. It can shape our understanding of disease, lifespan, and behavior. Yet the power of genomics also brings responsibility. We recognize that humans are but one species among millions, and each of us is but one individual among billions. We understand how we are different, but we also recognize how we are similar.

Throughout this book we have explored the molecular components that cooperate in a series of processes to bring the information in our genome to life – what we call 'molecular biology'. However, while these components and processes remain remarkably conserved throughout the natural world, the end results of genome function – the organisms from which the natural world is formed – are remarkably diverse. Molecular biology paints only part of the picture of the intricacies of life. Only by considering molecular biology in a wider context – alongside fields of biology such as ecology, physiology, and evolution – can we really begin to appreciate just how intricate – and spectacular – the process of life really is.

Principles of genomics

- **The ability to obtain the complete genome sequence of thousands of different species, different individuals of one species (genomics) or a collection of species (metagenomics) has revolutionized biology.**

- **The genome includes protein-coding genes, RNA-coding genes, regulatory sequences, transposable elements, other repeat sequences, and many sequences of unknown function.**

- **Genome annotation involves using sequence information and knowledge obtained from other experimental approaches to assign functions to coding and regulatory sequences.**

- **Despite a wealth of information from molecular biology, the function of many genes and much of the intergenic sequence remains unknown.**

Evolutionary forces

- **Selection and genetic drift work on genetic variation to shape natural populations.**

- **Evolutionary differences between species involved changes in size of the genomes, the structure of genomes, and the number of genes, as well as the sequence of genes.**

- **Translocations and inversions, and large-scale chromosome rearrangements including fusions are associated with speciation.**

Analyzing genomic sequence

- **Many genes are members of gene families with related sequences and functions.**

- **Amino acid sequence alignments provide functional information about genes from distantly related species, whereas nucleotide sequence alignments provide functional information about sequences from more closely related species.**

- **Sequence polymorphisms are inherited in linkage blocks known as haplotypes. These blocks, which can be followed by SNPs or microsatellite repeats, are extremely important for the mapping of mammalian genes.**

Phenotypic diversity

- **Biological complexity arises from both genetic and environmental sources, as well as the interactions between these sources.**

Human genomics

- **The rapidly expanding knowledge of the sequence and haplotype structure of the human genome allows us to trace our history, as well as to identify genes associated with many different traits and important diseases, but also carries with it the responsibility to use the information wisely.**

FURTHER READING

15.1 GENOME SEQUENCES AND SEQUENCING PROJECTS

Berg P. Origins of the human genome project: why sequence the human genome when 96% of it is junk? *American Journal of Human Genetics* 2006;**79**:603–605.

Collins FS, Morgan M, Patrinos A. The Human Genome Project: lessons from large-scale biology. *Science* 2003;**300**:286–290.

Olson MV. The Human Genome Project: a player's perspective. *Journal of Molecular Biology* 2002;**319**:931–942.

Shendure J, Mitra RD, Varma C, Church GM. Advanced sequencing technologies: methods and goals. *Nature Reviews Genetics* 2004;**5**: 335–344.

Tringe SG, Rubin EM. Metagenomics: DNA sequencing of environmental samples. *Nature Reviews Genetics* 2005;**6**:805–814.

15.2 FINDING FUNCTIONS IN A GENOME

Boffelli D, Nobrega MA, Rubin EM. Comparative genomics at the vertebrate extremes. *Nature Reviews Genetics* 2004;**5**:456–465.

Brent MR. Steady progress and recent breakthroughs in the accuracy of automated genome annotation. *Nature Reviews Genetics* 2008; **9**:62–73.

Friedman A, Perrimon N. Genome-wide high-throughput screens in functional genomics. *Current Opinion in Genetics and Development* 2004;**14**:470–476.

Gerstein MB, Bruce C, Rozowsky JS, *et al.* What is a gene, post-ENCODE? History and updated definition. *Genome Research* 2007;**17**:669–681.

Mardis ER. Next-generation DNA sequencing methods. *Annual Review of Genomics and Human Genetics* 2008;**9**:387–402.

Piskur J, Langkjaer RB. Yeast genome sequencing: the power of comparative genomics. *Molecular Microbiology* 2004;**53**:381–389.

Weinstock GM. ENCODE: more genomic empowerment. *Genome Research* 2007;**17**:667–668.

Wold B, Myers RM. Sequence census methods for functional genomics. *Nature Methods* 2008;**5**:19–21.

15.3 EVOLUTIONARY FORCES AND
15.4 THE EVOLVING GENOME

Bohne A, Brunet F, Galiana-Arnoux D, Schultheis C, Volff JN. Transposable elements as drivers of genomic and biological diversity in vertebrates. *Chromosome Research* 2008;**16**:203–215.

Dujon B. Yeasts illustrate the molecular mechanisms of eukaryotic genome evolution. *Trends in Genetics* 2006;**22**:375–387.

Eichler EE, Sankoff D. Structural dynamics of eukaryotic chromosome evolution. *Science* 2003;**301**:793–797.

Frazer KA, Murray SS, Schork NJ, Topol EJ. Human genetic variation and its contribution to complex traits. *Nature Reviews Genetics* 2009;**10**:241–251.

McCarthy MI, Abecasis GR, Cardon LR, *et al.* Genome-wide association studies for complex traits: consensus, uncertainty and challenges. *Nature Reviews Genetics* 2008;**9**:356–369.

Tishkoff SA, Verrelli BC. Patterns of human genetic diversity: implications for human evolutionary history and disease. *Annual Review of Genomics and Human Genetics* 2003;**4**:293–340.

Vinogradov AE. Evolution of genome size: multilevel selection, mutation bias or dynamical chaos? *Current Opinion in Genetics and Development* 2004;**14**:620–626.

15.5 DUPLICATION AND DIVERGENCE OF
GENE FUNCTION

Conrad B, Antonarakis SE. Gene duplication: a drive for phenotypic diversity and cause of human disease. *Annual Review of Genomics and Human Genetics* 2007;**8**:17–35.

Fryxell KJ. The coevolution of gene family trees. *Trends in Genetics* 1996;**12**:364–369.

Gogarten JP, Olendzenski L. Orthologs, paralogs and genome comparisons. *Current Opinion in Genetics and Development* 1999;**9**: 630–636.

15.6 GENOME AND SEQUENCE COMPARISONS

Ferguson-Smith MA, Trifonov V. Mammalian karyotype evolution. *Nature Reviews Genetics* 2007;**8**:950–962.

Kumar S, Filipski A. Multiple sequence alignment: in pursuit of homologous DNA positions. *Genome Research* 2007;**17**:127–135.

Payer B, Lee JT. X chromosome dosage compensation: how mammals keep the balance. *Annual Review of Genetics* 2008;**42**:733–772.

Pei J. Multiple protein sequence alignment. *Current Opinion in Structural Biology* 2008;**18**:382–386.

15.7 PHENOTYPIC VARIATION

Cooper GM, Nickerson DA, Eichler EE. Mutational and selective effects on copy-number variants in the human genome. *Nature Genetics* 2007;**39**:S22–29.

Gu CC, Yu K, Rao DC. Characterization of LD structures and the utility of HapMap in genetic association studies. *Advances in Genetics* 2008;**60**:407–435.

15.8 EXAMPLES FROM MAMMALIAN GENOMES

Sutter NB, Ostrander EA. Dog star rising: the canine genetic system. *Nature Reviews Genetics* 2004;**5**:900–910.

Wayne RK, Ostrander EA. Lessons learned from the dog genome. *Trends in Genetics* 2007;**23**:557–567.

15.9 HUMAN GENETIC DISEASES

Altshuler D, Daly MJ, Lander ES. Genetic mapping in human disease. *Science* 2008;**322**:881–888.

Evans WE, Relling MV. Moving towards individualized medicine with pharmacogenomics. *Nature* 2004;**429**:464–468.

Kaye J. The regulation of direct-to-consumer genetic tests. *Human Molecular Genetics* 2008;**17**:R180–183.

Lunshof JE, Chadwick R, Vorhaus DB, Church GM. From genetic privacy to open consent. *Nature Reviews Genetics* 2008;**9**:406–411.

McCarthy MI, Abecasis GR, Cardon LR, *et al.* Genome-wide association studies for complex traits: consensus, uncertainty and challenges. *Nature Reviews Genetics* 2008;**9**:356–369.

Renegar G, Webster CJ, Stuerzebecher S, *et al.* Returning genetic research results to individuals: points-to-consider. *Bioethics* 2006; **20**:24–36.

Tian C, Gregersen, PK, Seldin MF. Accounting for ancestry: population substructure and genome-wide association studies. *Human Molecular Genetics* 2008;**17**:R143–150.

Weinshilboum RM, Wang L. Pharmacogenetics and pharmacogenomics: development, science, and translation. *Annual Review of Genomics and Human Genetics* 2006;**7**:223–245.

➤ USEFUL WEBSITES

- National Human Genome Research Institute: http://www.genome.gov/Education

- OMIM: http://www.ncbi.nlm.nih.gov/omim

16 TOOLS AND TECHNIQUES IN MOLECULAR BIOLOGY

INTRODUCTION

Scientific discovery is driven by individuals who are able to look at a problem in a new way and have the right tools at their disposal. The Experimental approach panels throughout this book describe the process of discovery that has illuminated our understanding of some of the molecular processes and components described in the preceding chapters. Here we describe the tools and experimental methods that are used by biologists today in their continuing quest to understand how molecules in each cell carry out their many amazing functions.

Molecular biology became defined as a discipline around the middle of the twentieth century, when genetics and biochemistry were converging to provide an explanation for how genetic information was stored and harnessed by the cell. Although the use of model organisms in genetics was long established, we have since learned that many fundamental processes are so highly conserved from single-celled bacteria to multicellular eukaryotes that study of these model organisms yields information relevant to all domains of life. With the development of techniques to manipulate genetic material and reintroduce it into cells, it has become possible to study biochemical pathways as never before.

The advent of whole genome sequencing, and the resulting explosion of information about the nucleic acids and proteins in each organism, has made possible new types of experiments that exploit both cutting-edge techniques and the availability of vast amounts of information from thousands of organisms. The process of discovery has also been aided by technological advances, including the use of robotics and the printing of slides containing thousands of different DNA or protein molecules. These methods have greatly increased the pace of discovery and have yielded profound insights into the inner workings of the cell. It is hard to believe that it was less than 70 years ago that DNA was first shown to be the carrier of genetic information in cells.

The term 'molecular biology' was coined to place emphasis on the molecules that carry out biological processes. To understand how something takes place in the cell, one must identify the molecules that participate in the process and determine their functions. A powerful approach to determining the components required for a given process, as well as the function of each macromolecule, is to replicate the process with purified components *in vitro*, meaning in an artificial vessel (literally 'within the glass'). An mRNA copy of a gene, for example, can be readily synthesized in the test tube by mixing the DNA with purified RNA polymerase, nucleoside triphosphates, magnesium, and the appropriate buffers. The fact that RNA synthesis will now occur without any other cellular components present tells us that the molecules in the test tube mixture comprise the minimal requirements for RNA synthesis in the cell. *In vitro* RNA synthesis also makes possible quantitative studies of binding constants, reaction rates, and the way in which particular mutations affect one or more steps in the process.

However, *in vitro* studies have their limitations: when a particular process is reconstituted in the test tube, some of the proteins that help to regulate the process in the cell might still be missing, or an important post-translational modification might not be found in a recombinant protein used in the experiment. Despite these caveats, much of what is known in molecular biology derives from *in vitro* experiments.

We begin this chapter by describing the organisms that have served as models for the study of biology in all types of cells and explore how cells and viruses can be grown in culture. From there, we explain **recombinant DNA** technology and how it is used to both identify new genes and construct modified genes and chromosomes. We then present a large array of techniques used to study individual molecules, beginning with their purification from whole cells, through modern methods for identifying particular proteins and nucleic acids, before considering how we study their interactions with one another. The techniques used to image whole cells and study processes *in vivo* (in the living organism) on the one hand, and individual macromolecules on the other hand, are then explained. The chapter concludes with an explanation of how whole genomes are sequenced and how that information can be used to identify genes implicated in human disease.

16.1 MODEL ORGANISMS

Although millions of species exist in nature, only a few dozen, called model organisms, are extensively used in research. Many model organisms are studied not for their intrinsic importance to human health or the economy, but because they are easy to propagate and can provide general insights into key molecular processes and components – insights which can be extrapolated to other organisms. This is because so many molecules and molecular processes are evolutionarily conserved. For example, *Arabidopsis* is not a horticulturally important flowering plant and *Caenorhabditis elegans* is not an agriculturally or medically important nematode worm, but both organisms serve as laboratory models for flowers and nematodes that have greater economical or biological significance. Most strains of *Escherichia coli* are not human pathogens, yet they are similar in many ways to bacteria that are.

Model organisms have certain shared properties

The most widely used model organisms share several biological properties that have made them the standards for investigation. A model organism must have fairly simple and, ideally, very defined nutritional requirements, as well as a well-understood life cycle that allows it to be easily and safely grown in large quantities in the laboratory. A short generation time, small size and the ability to store or maintain stocks of the organisms for years are also advantages.

Nearly all important model organisms have been extensively studied by genetic analysis. This means that mutations, which reduce or alter the function of a gene, can be readily introduced, propagated, identified, and characterized. Since the advent of recombinant DNA technology, a further requirement is the ability to introduce modified genes into the organism, so that the effect of any type of mutant can be studied. The genomes of most model organisms have been sequenced, and the large communities of investigators studying these organisms have established databases containing genome sequence information, descriptions of strains and phenotypes, and the latest information about gene expression, enzyme activity, and protein–protein interactions. The availability of these data has greatly accelerated the pace of discovery and made possible new approaches to studying biology, as we will see throughout this chapter.

It is important to note that the nomenclature adopted when presenting gene and protein names varies from organism to organism. Refer back to p ii for a summary of the nomenclature rules adopted for different model organisms.

Different model organisms have distinct advantages

Model organisms come from different kingdoms and phyla, and represent different evolutionary and biological properties. Well-studied models include bacteria, single-celled eukaryotes, small animals, flowering plants, and lower vertebrates and mammals. How, then, does a scientist choose an experimental model in which to study a biological process? The answer depends on the question being asked, as well as the experimental tools that are available at the time. For example, much of our knowledge about DNA replication, transcription, and translation was originally gained from studies of bacteria. In part, this is because bacteria were among the first model organisms available, since it is easy to obtain large quantities of bacterial cells and they are easy to mutate. Many aspects of the processes characterized in bacteria are universal to all organisms, validating the use of this model.

Phenomena that are unique to eukaryotes must, of course, be studied in eukaryotic cells. The inner workings of the nucleus or a cell can be readily studied in single-celled eukaryotes such as yeast, while the mechanisms by which a multicellular organism develops from a fertilized egg must be studied in a higher eukaryote, such as the fruit fly or the nematode. Similarly, phenomena unique to vertebrates or to mammals must be studied in a representative of that class of organisms. Often, the choice of the appropriate model organism has been instrumental in making rapid experimental progress. The ones mentioned below are just a selected subset of the important and informative model organisms that have been used.

Escherichia coli, *Caulobacter crescentus*, and *Bacillus subtilis* are commonly studied bacteria

The gut bacterium, *E. coli,* is the best-studied bacterium, and quite possibly the best-studied species. Many of the basic principles of molecular biology that are

(a)

(b)

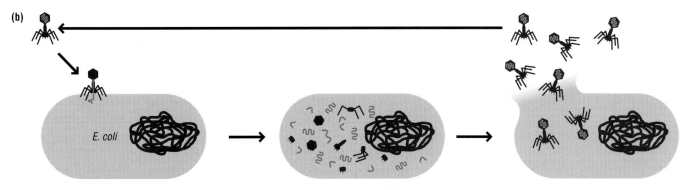

Figure 16.1 Bacterial cell with plasmid and phage. (a) Bacterial cell harboring a plasmid, a circular double-stranded DNA molecule that is separate from the bacterial chromosome. Plasmids are typically much smaller than the chromosome. (b) The lifecycle of bacteriophage T2: infecting an *E. coli* bacterium by injecting its DNA (pink) into the bacterium, then expressing its genes to make multiple copies of itself, and finally being released by cell lysis.

described throughout this book were first investigated in *E. coli.* This bacterium can be grown easily in the laboratory, and divides about every 20–30 minutes under optimal conditions. Most importantly, genes can be readily introduced into *E. coli* using extrachromosomal DNA elements known as **plasmids** (Figure 16.1a). These double-stranded, circular DNA molecules harbor multiple genes and can be readily introduced into bacterial cells. Plasmids can be easily manipulated by genetic engineering, making it straightforward to introduce mutant genes into bacteria and subsequently to study their effects. *E. coli* cells can also be infected by a variety of viruses such as bacteriophage T2 and lambda (Figure 16.1b). These bacteriophages provide another means by which to introduce genes into the cell. The dependence of bacteriophages on host cell function for their propagation has helped provide insights into basic cellular processes.

Many other bacteria have interesting or important biological properties that make them useful as models. The widely used model, *Bacillus subtilis,* can form stress-resistant spores, and its sporulation program is a thoroughly studied example of a biological switch. Sporulation involves transcriptional activation of certain key genes, with a change in the sigma factor subunit of RNA polymerase. Thus, *B. subtilis* has given us insights into transcription initiation. The bacterium *Caulobacter crescentus* can adopt either of two different morphological forms at different points in a lifecycle. One form is the highly motile swarming cell that moves by means of a flagellum. However, a swarming cell can also anchor itself to a surface and differentiate into a stalk – that is, a stationary cell with a very different morphology. The switch between swarming and stalk formation is an example of a bi-stable biological switch. *Caulobacter* has also been used to study bacterial cell division.

◉ We provide examples of what has been learned from studying bacteriophage lambda in Sections 8.10 and 14.12.

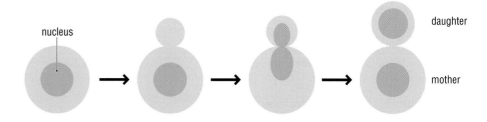

Figure 16.2 Saccharomyces *cerevisiae* budding. Diagram of asymmetrical division of budding yeast.

Saccharomyces cerevisiae, Schizosaccharomyces pombe and Dictyostelium discoideum are commonly studied single-celled eukaryotes

Despite the enormous differences evident between highly complex eukaryotes, such as humans and single-celled eukaryotes, most fundamental pathways are remarkably conserved across all eukaryotes. Single-celled eukaryotes are therefore very attractive model systems for the study of eukaryotic molecular biology because of the ease with which they can be grown in culture and genetically manipulated. In addition, their short doubling time – about 90 minutes per generation for brewers' yeast – greatly enhances the rate at which studies of the effects of mutations can be performed.

The yeast *Saccharomyces cerevisiae*, which is of widespread practical use in brewing, baking, and production of biofuels, is also the best-understood model for a eukaryotic cell. *S. cerevisiae* has been a key model organism for understanding the fundamental processes of transcription, meiosis, recombination, cell division, chromosome structure and segregation, and intracellular signaling. Yeast has a fairly compact genome, with about 6000 genes in a genome of 12 million base pairs. This type of yeast is distinct from other eukaryotes, and even from some other yeasts, in its manner of cell division: rather than simply dividing into two cells of equal size, *S. cerevisiae* reproduces by budding (Figure 16.2), in which the daughter cell first appears as a small bud on the mother cell. This small bud eventually grows into a daughter cell of equal size. As a result, it is possible to distinguish mother and daughter cells through subsequent generations, which facilitates studying pedigrees and inheritance in yeast.

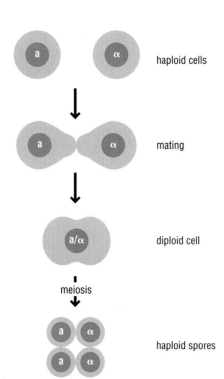

Figure 16.3 Life cycle of budding yeast. Yeast can exist as either haploid or diploid cells. Two haploid cells of opposite mating type, **a** and α, can mate with one another and fuse to form a diploid **a**/α cell. Under conditions of starvation, the diploid cell undergoes meiosis to produce four haploid spores: two **a** and two α.

The nature of the budding yeast life cycle lends itself to rapid genetic manipulation. As we learned in Chapter 8, this yeast can exist in both haploid and diploid form (Figure 16.3). There are two haploid yeast cell types: **a** cells and α cells. A haploid yeast cell can mate with a cell of opposite mating type, that is an **a** can only mate with an α to produce a diploid, **a**/α cell. Diploid cells can be propagated indefinitely but, under conditions of starvation, they undergo meiosis to form four haploid cells: two **a** cells and two α cells. This life cycle can be easily exploited in genetic complementation studies, since it is straightforward to produce two different haploid spores, each with a different mutation, and then mate the two cells to produce a diploid cell containing both mutations.

Some phenomena common to higher eukaryotes that are absent from budding yeast can be studied in its evolutionarily distant cousin, *Schizosaccharomyces pombe*. This yeast reproduces by symmetrical cell division and is therefore called a fission yeast. The DNA silencing and DNA structures of the centromeres in *S. pombe* are more similar to those found in multicellular eukaryotes than those of *S. cerevisiae*. Therefore, the similarities and contrasts between the two yeasts are often used to provide a broader perspective on the fundamental properties of eukaryotic cells.

The single-celled yeasts do not have extensive extracellular signaling or multicellular organization (although they do have some). Therefore another organism, the slime mold, *Dictyostelium discoidum*, is used to provide insights into such intercellular communication and signaling. *Dictyostelium* lives as motile solitary cells for much of its life cycle. However, in response to a secreted signal from one cell, the nearby single cells stream together to form a colony of functionally differentiated cells called a slug. The slug moves by the coordinated action of individual cells and can attach itself to a surface, whereupon it differentiates into a complex cellular structure comprising stalk and mushroom-like features, in which the cells have undergone functional differentiation (Figure 16.4). The secretion of the signal and the response of the cells to a signaling **gradient** for slug and stalk formation have been important in understanding the role of calcium in extracellular signaling in a relatively simple eukaryote.

Figure 16.4 Slime mold *Dictyostelium discoidum*. Picture of stalks formed by a colony of *D. discoidum* cells.
From Science Musings blog. http://www.sciencemusings.com/uploaded_images/slime-784557.jpeg

Drosophila melanogaster, *Caenorhabditis elegans* and *Arabidopsis thaliana* are commonly studied multicellular eukaryotes

Studies of single-celled eukaryotes, while tremendously important for understanding the inner workings of all eukaryotic cells, provide limited insight into how multicellular organisms develop from a single fertilized egg, with distinct tissue types and morphology. The common fruit fly, *D. melanogaster*, has been one of the most important experimental models for studying development since it was first established as a model organism over 100 years ago. Because *Drosophila* was the principal model for genetic analysis for the first half of the twentieth century, *Drosophila* mutants have been instrumental in studying almost every process in animals.

The entire development cycle in *Drosophila* is very rapid, taking just about a week to develop from a fertilized egg into a mature fly (Figure 16.5). A single female can lay thousands of eggs, thereby producing an enormous number of progeny. Random mutations can be induced by feeding flies chemicals that induce mutations or by subjecting flies to ionizing radiation, after which the many progeny can be screened for new and interesting phenotypes. Indeed, modern recombinant DNA techniques make it possible to engineer specific mutations into flies. By analyzing the effects of mutations on the mature organism, as well as on the different stages of development, genes that control development and differentiation as well as behavior can be identified. A great number of genes that control development in all metazoans were first identified in *Drosophila*.

The nematode worm, *Caenorhabditis elegans*, was developed as a genetic model organism because nematodes carry out complex biological processes and behaviors but have many fewer cells than other multicellular animals. An adult worm has only 909 cells (excluding the germline and the gonad), of which 302 are neurons. In addition, these cells arise from a pattern of precise and largely invariant cell divisions, which can be tracked by virtue of the worm's transparency. The cell lineage patterns, which were established by the early 1980s, were instrumental in understanding the important role of apoptosis in organismal growth and development. The *C. elegans* lifecycle, which takes about three days, also has features that differ from *Drosophila* and are useful for genetic studies: the worms are either hermaphrodites or males and can reproduce by self-fertilization (hermaphrodites) or by mating (a hermaphrodite with a male).

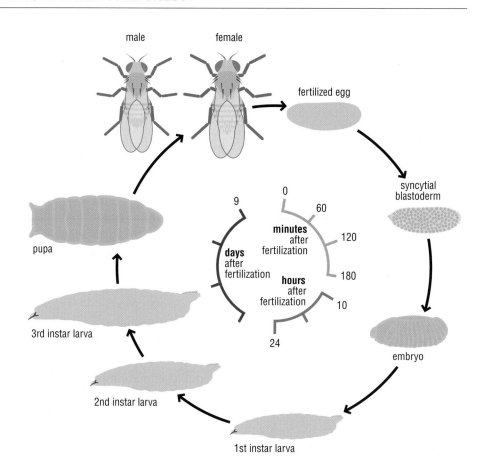

Figure 16.5 Life cycle of the fruit fly *Drosophila melanogaster.* It takes 24 hours for a fertilized egg to develop into an embryo, which then proceeds through several discrete larval stages over the next few days. The final stage of development is the formation of the pupa, from which the fully developed fly emerges. From FlyMove. http://flymove.uni-muenster.de/

● See Experimental approach 9.4 to learn more about the discovery of RNAi.

An unanticipated benefit of developing *C. elegans* as an experimental model was the relatively recent discovery of the RNA interference (RNAi) machinery, which grew out of attempts to manipulate gene expression in worms using antisense RNA. Once it was realized that double-stranded RNA was causing targeted genes to be turned off, standard methods of worm genetics were rapidly used to identify components of the RNAi machinery.

The most prominent model organism from the plant world is the flowering plant, *Arabidopsis thaliana*. Many other plants have served as model organisms, including maize, pea, corn, tobacco, petunias, and snapdragons, but *Arabidopsis* is much smaller and has a shorter generation time. It also has a much smaller genome than agriculturally or horticulturally important plants. As we will see below, genetic manipulation of *Arabidopsis* also is quite straightforward since foreign genes can be introduced by infecting the plants with a soil bacterium, *Agrobacterium tumefaciens*. This soil bacterium harbors a tumor-inducing (Ti) plasmid, into which can be cloned the gene to be introduced into the plant cells. Through the process of infection, a segment of the Ti plasmid into which the gene was cloned is ultimately incorporated into the *Arabidopsis* genome.

The zebrafish, frog, and mouse are model vertebrates

Several organisms serve as model vertebrates—that is, animals having a backbone and spinal column—which include reptiles, birds, and mammals. A small fish commonly found in pet stores known as the zebrafish (*Danio rero*) has many features of other model organisms that makes it amenable to study. Zebrafish are easy to raise and breed in the laboratory, with each female producing hundreds of eggs in a single clutch. Although the generation time of zebrafish is similar to the mouse (three to four months), the embryos develop in a matter of days. Importantly, embryo development occurs externally to the mother and the embryos are transparent, making it possible to observe the different stages of vertebrate embryogenesis and organ formation. A variety of techniques are available for genetic manipulations in zebrafish, making it possible to conduct large-scale screens.

Another vertebrate, *Xenopus laevis*, the African clawed frog, is very widely used in molecular biology and biochemical studies. *X. laevis* is relatively large, so it is simpler to get large quantities of material from a single frog. Since its oocytes and eggs are very large, it is possible to inject DNA and other molecules (and even whole nuclei) into them. *Xenopus* oocytes are often used as a vertebrate cellular system for testing the expression and function of cloned genes. In addition, *Xenopus* eggs are a rich source for biochemical studies because the egg extracts can be highly concentrated. This concentrated nature of the extracts makes it feasible to carry out many biochemical reactions. The extracts can also then be fractionated to find the important components for a given reaction. Cell cycle and DNA replication studies, in particular, have used such *Xenopus* extracts to identify many biological molecules. *Xenopus* is not, however, extensively used for genetic analysis because it is a partial tetraploid. The genome of the closely related, but diploid, *Xenopus tropicalis* has been sequenced and will allow genetic approaches to be used in the study of this important vertebrate model system.

Some biological processes in humans can only be analyzed in other mammals. The house mouse, *Mus musculus*, provides invaluable insights into human biology and disease. It is often the experimental organism of choice for comparisons with humans because it, too, is a mammal, and yet it can be cultivated for laboratory research (unlike humans). Our understanding of the complexities of the mammalian immune system, for example, was largely derived from experimental studies in mice. The mouse genome is similar to the human genome in size, and there is extensive sequence homology between genes. Mouse genetic analysis is very well established, and many studies on human diseases are greatly accelerated by studies in mice. Mice offer an excellent system to study the effects of genes associated with human disease; single gene traits and, increasingly, the complex traits can be studied in mice.

Despite the complexity of the mouse, it is possible to generate genetically modified mice by manipulating the genes in mouse embryos. The techniques for generating genetically modified mice are discussed in Section 16.4. The ability to disrupt single genes, thereby creating a 'knockout mouse', has been an invaluable tool in studying gene function in mammals. In addition, knockout mice serve as models for those human diseases that result from the inactivation of a particular gene.

Some genes cannot be disrupted in this way because embryos containing the knockout fail to complete development and reach adulthood. In many of these cases, **conditional** mutants are used. These mutants make it possible to delete genes that are only deleted in certain tissues in the adult. As the gene is not deleted in the entire mouse, embryonic development can occur. It is also possible to produce strains of **transgenic mice**, that is, mice whose cells express a foreign or mutant form of a gene, and thus are also useful for studying disease.

Tetrahymena is an example of a model organism studied for its special properties

Many of the model organisms described above are representatives of particular phylogenic branches and together give a general overview of biological systems. Some model organisms, however, are chosen not for their common properties, but rather for some unique trait that can give insights into specific biological processes. A prime example is the ciliated protozoan, *Tetrahymena thermophila*. *Tetrahymena* is a single-celled eukaryote that is found in fresh water ponds (Figure 16.6a). It is classified as a ciliate because it is covered with cilia, small appendages that allow it to move through the water. This single-celled organism is unusual in that it has two nuclei: a diploid micronucleus, which serves as a germline 'vessel', and a polyploid somatic macronucleus, from which gene expression occurs.

The macronucleus is derived from a copy of the diploid micronucleus. This takes place in a developmental process that involves both gene amplification and the rearrangement of some regions of DNA and the elimination of other sections (Figure 16.6b). As a result of the amplification, there is a very high copy number of small chromosomes, with many additional copies of both ribosomal RNA genes. These features led to the discovery of both self-splicing rRNA and telomeres. The presence of both a transcriptionally silent micronucleus and a transcriptionally active macronucleus in the same cell allowed investigation of the role of histone modification in transcriptional regulation. These discoveries are described in Experimental approaches 4.2 and 9.1.

Even though its unique genome gymnastics makes it appear unusual, studies in *Tetrahymena* have turned out to be applicable across phyla. Given the conservation of fundamental mechanisms in biology, what is learned in *Tetrahymena* and other unusual organisms is applicable across biological systems. For this reason, using diverse organisms to study specific biological processes makes it possible to gain new fundamental insights.

(a) **(b)**

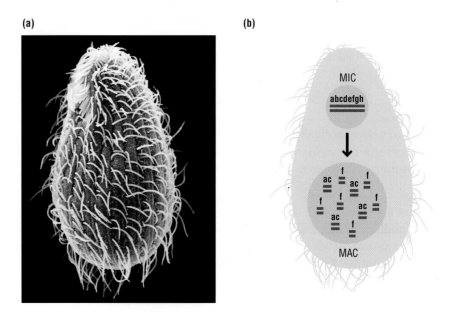

Figure 16.6 The ciliated protozoan *Tetrahymena thermophila*. (a) Electron micrograph of *Tetrahymena thermophila*. (b) *Tetrahymena* has two nuclei, a diploid micronucleus (MIC) that serves as a germline and a polyploid somatic macronucleus (MAC). Gene expression occurs from the macronucleus, which is derived from a copy of the diploid micronucleus in a developmental process involving multiple DNA rearrangements. (a) Courtesy of Aaron J. Bell.

16.2 CULTURED CELLS AND VIRUSES

Detailed biochemical studies require samples of individual macromolecules that have been isolated from a cell. It is generally important to purify the molecule or molecules of interest from a homogeneous population of cells, in order to avoid variations in sequence or post-translational modifications that may be found among different strains or tissue types. Although proteins and nucleic acids can be isolated from relatively homogeneous large organs, such as a bovine heart or liver, most of the experiments underlying the information in this book were carried out with proteins and nucleic acids isolated from a single type of cell grown while suspended in liquid or immobilized on agar, plastic, or glass surfaces. We refer to this type of cell growth as **cell culture**. Some of the different types of cells that are grown in culture and the advantages and limitations of cell culture are the focus of this section.

Single-celled organisms can be grown in liquid culture or on agar plates

Single-celled organisms are most readily grown in culture. Model organisms such as the bacterium *E. coli*, and the budding yeast, *S. cerevisiae*, can be grown in liquid medium consisting of a mixture of nutrients that is buffered at a pH necessary for cell growth (Figure 16.7a). An advantage of culturing cells is that conditions under which the cells are growing can be controlled. For example, the cells can be continuously supplied with a continual supply of air by bubbling in gas or by vigorous stirring, or they can be grown completely in the absence of oxygen. Similarly, the cells can be maintained at different temperatures or pressures. Many species of bacteria, algae, and some archaea can also be grown in liquid culture, or on solid medium such as agar. (Agar is a substance added to culture medium that is liquid at high temperatures, but which solidifies at lower temperatures to form a gel on which microorganisms can grow.) When cells are spread on agar at low density, each cell grows and divides, giving rise to single colonies as seen in Figure 16.7b. When cells are spread at high density, the proliferating cells coat the surface of the agar, giving rise to a lawn. Some organisms, such as those that live in unusual environments, cannot be grown in the laboratory because we do not as yet know how to create culture conditions under which they will grow and thrive. Our detailed knowledge of biochemical processes within cells has therefore been limited so far to those processes occurring in organisms that can be grown in culture.

Many differentiated cells from multicellular organisms need to undergo modifications to be maintained in culture

Eukaryotic cells from multicellular organisms can also be grown in culture. These cells, which can be derived by homogenizing tissues and are known as **primary cells**, are typically grown in a plastic culture bottle or on a plate under conditions that allow the cell to adhere to the surface (Figure 16.8). There are some cell types, such as blood cells, that can be grown for short periods while suspended in liquid culture.

(a)

(b)

Figure 16.7 Culturing of single-celled organisms. (a) Growth of the yeast, *S. cerevisiae* in a glass tube. The solution is turbid because of the presence of yeast cells at about 10^8 cells/mL. The nutrient solution itself is yellow. (b) Growth of the yeast *S. cerevisiae* on an agar plate. A solution containing several hundred yeast cells was spread on the surface of a nutrient-containing agar plate; the agar is yellow because of the nutrients. On incubation, each cell gives rise to a single colony containing several million progeny cells.
Photos courtesy of Susan Michaelis, Johns Hopkins University School of Medicine.

Figure 16.8 Eukaryotic cells grown in tissue culture. (a) Plastic flask containing mammalian cells in culture. The growth media is pink because of a pH indicator dye. (b) Layer of single tissue culture cells, as viewed by light microscopy. (b) Kindly provided by Sin Urban, Johns Hopkins University.

The growth of cells in tissue culture is often limited by a number of factors. Contact between cells may trigger cell cycle arrest so that cells cease to grow, a phenomenon known as **contact inhibition**. In addition, somatic cells can only divide a limited number of times due to growth inhibition mechanisms. For example, growth is limited in many cases by a lack of telomerase expression. Since telomeres shorten at every division, the short telomerases eventually cause **senescence**, which cell division ceases. To avoid these problems, researchers often use special strains of cells that have been developed to grow in cell culture without undergoing senescence. Called **cell lines**, these cells are often derived from tumors that actively express telomerase, and have undergone other genetic changes that remove the normal inhibitions to continued cell division. Cells can also be made to grow continuously in culture by infection with a cancer-causing virus.

Some of the most commonly used cell lines are listed in Figure 16.9. While these cell lines have been central to the study of eukaryotic cells, the very changes that permit their continuous growth in culture also mean that these cells differ in fundamental ways from somatic cells. These differences can affect the outcome of certain experiments.

Stem cells are the precursor cells to all cell types

Stem cells are a type of cell that can give rise to many different cell lineages. **Stem cells** such as **embryonic stem cells** (ES cells) are derived from early embryos

Some commonly used cell lines		
cell line	**cell type**	**origin**
3T3	fibroblast	mouse
BHK21	fibroblast	Syrian hamster
MDCK	epithelial cell	dog
HeLa	epithelial cell	human
PtK1	epithelial cell	rat kangaroo
L6	myoblast	rat
PC12	adrenal gland cell	rat
SP2	plasma cell	mouse
COS	kidney cell	monkey
293	kidney cell, transformed with adenovirus	human
CHO	ovary	chinese hamster
DT40	lymphoma cell	chick
R1	embryonic stem cell	mouse
E14.1	embryonic stem cell	mouse
H1, H9	embryonic stem cell	human
S2	late embryonic cell	*Drosophila*
BY2	undifferentiated meristematic cell	tobacco

Figure 16.9 Commonly used cell lines. Adapted from *The Molecular Biology of the Cell*, 4th edition.

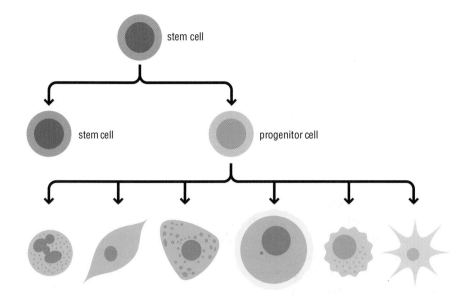

Figure 16.10 Stem cells are capable of self-renewal and differentiation. Stem cells divide in a special manner yielding one daughter cell that is a stem cell and another daughter that can differentiate into a particular cell type. In some cases, the second daughter cell is a progenitor cell, which can give rise to multiple cell types.

and are considered **totipotent cells**, meaning that they have the potential to give rise to every cell type of the body. Other stem cells can generate a more limited repertoire of cell types; these are called **pluripotent cells**, as they give rise to a restricted set of cell lineages. Such pluripotent stem cells are present during development and are also found in fully developed animals. Adult stem cells, which are stem cells that can be derived from a fully mature adult organism, are usually pluripotent, and the number of different cell types they can generate depends on the specific cell type. Stem cells from skin, hair follicles, and the intestine have been studied, but the best characterized stem cells are those from blood.

Stem cells typically have an unlimited capacity to divide, in stark contrast to differentiated cells, which have a limited cell division capacity. When stem cells divide, they do so in a special manner: one of the two daughters forms a new stem cell while the other daughter may differentiate into a particular cell type. Under some circumstances, the differentiated daughter can also give rise to other cell types and is referred to as a progenitor cell (Figure 16.10). Progenitor cells may have a more restricted set of cell fates than stem cells, and may also have a more limited cell division capacity.

The best way to demonstrate that a cell is indeed a stem cell is to isolate the cell and implant it into a recipient animal. A stem cell will give rise to the entire set of differentiated cells in this new setting. This type of transplantation is also the basis of bone marrow transplants that are carried out for the treatment of blood diseases and cancer. In bone marrow transplants, stem cells that generate blood cells are isolated from the bone marrow, where they normally are found, and transferred to another individual. These blood precursor cells are referred to as **hematopoietic stem cells**, as **hematopoiesis** is the process of regenerating blood cells that occurs every day in adults. Blood cells have a limited lifespan, with some types lasting only a day. Blood cells must therefore be continually renewed throughout the life of the organism. The differentiation of hematopoietic stem cells into specific blood cell lineages is one of the best-understood hierarchies of stem cell differentiation (Figure 16.11).

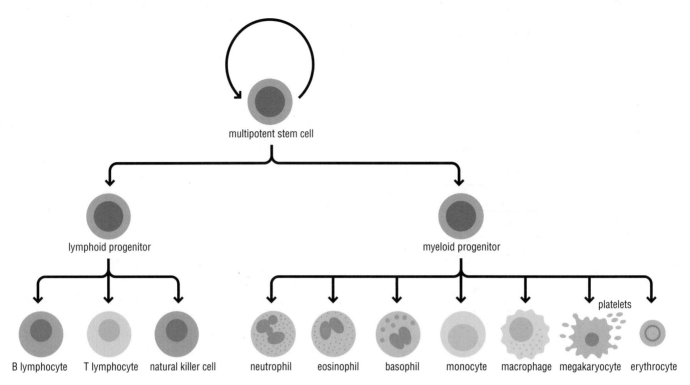

Figure 16.11 The differentiation of hemapoietic stem cells into specific blood cell lineages. Hemapoietic stem cells are isolated from blood or bone marrow. They are capable of self-renewal (that is, they can produce more stem cells) and also produce progenitor cells that can differentiate into many other cell types.

Embryonic stem cells can give rise to cell types needed to treat disease

ES cells have received a great deal of attention for their potential medical usage. These totipotent cells have been manipulated in the laboratory to generate specific cells lineages that might be used to treat specific diseases. For example, ES cell treated with certain growth factors in culture can cause >99% of the cells to differentiate into neurons. It is hoped that these neurons could be transplanted into patients with Parkinson disease as a way to treat the loss of function of specific neurons, which is a characteristic of this disease. One of the dangers of the use of ES cells, however, is the potential to generate tumors. This can occur in the event that less than 100% of the cells fully differentiate, making it possible that a few cells will continue to grow as ES cells and later differentiate randomly to form a tumor. Indeed, direct injection of ES cells into mice gives rise to a tumor known as a teratoma, which has a mixture of many different differentiated cell types.

iPS cells are a special form of stem cells derived from differentiated cells

In recent years, a new method has been developed to form totipotent cells for medical research, termed iPS for induced pluripotent stem cells. Researchers found that introducing four specific transcription factors into fully differentiated skin cells would 'reprogram' the skin cells, turning them into stem cells; that is, the transcription factors induce the expression of a set of genes that is needed in early embryonic development. iPS cells seem to be in many ways equivalent to ES cells.

Importantly, this new approach makes it possible to treat an individual with differentiated cells derived from their own tissue. Called autologous transplants, these cells are less likely to be rejected by the host. The ability to form iPS cells in this way bypasses the ethical issues surrounding the use of ES cells derived from human embryos.

Viruses must be cultured in living cells

As we saw in earlier chapters, studies of viruses that infect bacteria or eukaryotic cells have provided important insights into basic cellular mechanisms. Viruses are also a valuable means for introducing DNA into cells. However, since viruses are not free-living life forms, they must be propagated within other types of cells. Most viruses can infect just a few types of cells at most, so the ability to grow a particular virus depends on the ability to culture the type of cell it grows in. Perhaps the most important advance in developing a vaccine against the crippling disease, polio, was the development of methods to grow poliovirus in cultured human cells.

Viruses that have been propagated in cultured cells can be isolated in several ways. Since viruses infect a cell and harness the cell's machinery to produce many virus particles, one isolation method is to infect cultured cells with virus and harvest the virus-producing cells that result. The cell membrane can then be disrupted with detergents to release the virus particles, which can be separated from the cell debris by a variety of methods. With lytic viruses, which cause the cell membrane to burst in order to release the virus particles, cells can be infected and grown until the cells burst and release virus particles into the culture medium, or onto the surface of a Petri dish, where lysis of cells in a lawn produces a clear circle called a plaque. The virus particles present in a plaque or shed into the culture medium can then be purified.

16.3 CLONING AND AMPLIFICATION OF DNA AND RNA SEQUENCES

The ability to manipulate DNA has been central to the advancement of molecular biology in the past few decades. Thanks to these methods, hundreds of gene functions have been studied by creating specific mutations; proteins have been purified or localized by creating gene fusions to other proteins or tag sequences; and protein activities and structures have been determined by over-producing the proteins in specialized cells. Most of these experimental approaches require that the DNA of interest be isolated and altered. In this section, we discuss ways to manipulate and amplify DNA and RNA sequences.

Specific DNA fragments can be amplified and modified by cloning into plasmid vectors

Recombinant DNA techniques make it possible to isolate and manipulate particular DNA fragments. The DNA is introduced into an appropriate **vector**, which makes it possible to propagate the desired DNA fragment to be propagated independently of the host cell chromosome. The process of isolating and propagating specific DNA regions is called **DNA cloning** because many copies of the same

DNA sequence are often generated from one original. Vectors for cloning can be plasmids, circular DNA that can replicate independently of the chromosome, or viruses. These vectors usually carry a gene that allows the cells with that vector to grow selectively, that is, culture conditions are used in which cells that do *not* carry the vector cannot grow. For example, vectors can carry an antibiotic-resistance gene that allows only the vector-containing cells to grow in the presence of an antibiotic drug, or a gene encoding a metabolic enzyme that allows the cells to grow in specific medium. The gene that allows the cells to grow under these special conditions is known as a **selectable marker**, since it makes it possible to select cells containing the marker from a larger population of cells. DNA cloning is often carried out using bacteria, although other host organisms such as yeast or cultured cells are also used.

In addition to providing multiple copies of a DNA fragment of interest, DNA clones can be used to test for the ability of a particular gene to complement a mutation (that is, to restore a particular gene activity lost by a given mutation), to fuse the gene of interest to a **reporter gene** or tag that facilitates detection or purification, or to provide for high levels of a gene product. (Various types of reporters and tags will be described in subsequent sections.) The cloned DNA fragment can also be more easily manipulated than the equivalent DNA sequence situated in its 'native' chromosomal environment. Such manipulation may be used to introduce specific mutations.

There are two main strategies for cloning a DNA fragment:

- ligation – where the DNA fragment is excised with restriction enzymes and is then 'glued' to the vector DNA by a DNA ligase
- recombination – where the fragment DNA is amplified by **polymerase chain reaction** (**PCR**) and then inserted into the vector by recombination enzymes by virtue of sequence homology.

Both these approaches are discussed below.

Cloning by ligation requires restriction enzymes and DNA ligase

The first step in DNA cloning by ligation is to isolate the desired fragment of DNA (Figure 16.12a). The fragment is excised from a larger region of DNA using restriction enzymes. These are the bacterial enzymes that cleave double-stranded DNA at specific recognition sites, which are usually 4–8 bp long. The cut DNA often has a small region of 5′ or 3′ overhanging bases available to base pair with a complementary sequence (Figure 16.12b). The DNA fragments are then isolated by **gel electrophoresis**. At the same time, the vector is prepared by digesting it with restriction enzymes to generate overhanging ends that are complementary to the overhangs of the DNA fragment (Figure 16.12b). The DNA insert and vector DNA are then incubated together so that the compatible ends of the fragment and plasmid can base pair with one another. A DNA ligase is used to covalently join the adjacent 5′ and 3′ ends thereby, connecting the DNA backbone (Figure 16.12a). Finally, the ligated plasmid is introduced into bacteria or another host organism, a process called **transformation** (Figure 16.12a). The receiving cells usually need to be treated with salts or exposed to an electric field before they are competent to take up the DNA. (The cells able to take up DNA are referred to as competent cells.)

Figure 16.12 Cloning using restriction endonucleases. (a) In this cloning scheme, a plasmid is cut with a restriction endonuclease to create two ends. The fragment DNA to be cloned is cut with the same restriction endonuclease, making the single-stranded ends of the plasmid and the DNA fragment complementary, such that they can anneal. To clone the fragment DNA into the plasmid, the cut plasmid and the DNA fragment are mixed, allowed to anneal, and then the nicks are sealed by DNA ligase. The new plasmid is then transformed into a host, most commonly *E. coli*, where it can be propagated. (b) An example of DNA annealing following cleavage by the restriction endonuclease EcoRI.
From Stem Cell Information, 2001. http://stemcells.nih.gov/info/scireport/chapter5.asp

E. coli is usually the organism of choice for most routine cloning. A wide range of vectors can be used to propagate genes in *E. coli*; the vector chosen depends on the experiment to be undertaken. Vectors can accommodate inserts of different size; typical bacterial plasmids might have inserts of up to 15 kb, whereas a specialized kind of bacterial vector, called a bacterial artificial chromosome (BAC), can have inserts that are hundreds of kilobases in size. Most bacterial vectors contain selectable markers that confer resistance to a particular antibiotic, so the cells that contain the plasmid can be selected for by growing the transformed bacteria in the presence of the appropriate antibiotic. Some vectors also contain a promoter and additional regulatory sequences that make it possible to regulate expression of inserted genes.

Many vectors are capable of replicating in several different organisms, and are therefore called shuttle vectors. These vectors contain two or more origins of replication: one for replicating in *E. coli* and the others for replicating in a different organism of choice. These vectors may also include two or more selectable markers, one for *E. coli* and the other(s) for another organism. In this way, the cloning steps can be done in *E. coli*, after which the vector then introduced into cells of

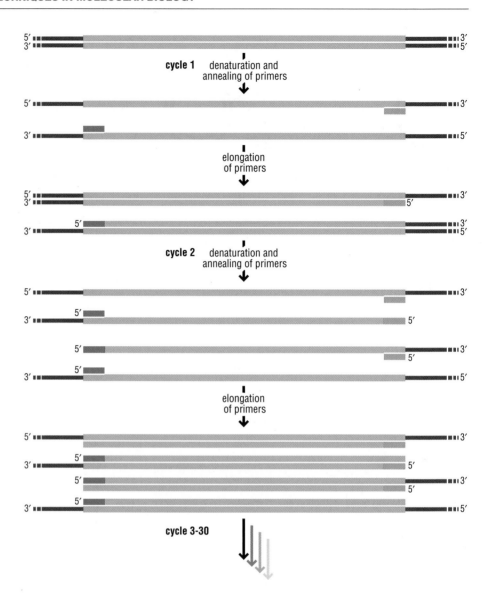

Figure 16.13 The use of PCR to amplify a specific DNA fragment, using multiple rounds of primer annealing and extension. Oligonucleotides (in blue and green) are designed to be complementary to each end of the fragment that is to be amplified. Such nucleotides are typically around 20 bases in length. When the DNA is denatured, these oligonucleotides anneal to the complementary site on the single-stranded DNA. They then serve as primers for DNA replication, usually by a heat-resistant DNA polymerase from the bacterium, *Thermus aquaticus*. After the DNA is fully replicated, it is again denatured and used in a second round of primer annealing and extension. With every round, the amount of new DNA present in the reaction mixture is doubled.

the organism of choice. The copy number of the gene in the specific organism is determined by the specific type of vector used. Some vectors are present in one or two copies per cell, whereas others are present in more than 50 copies.

PCR can be used to amplify a specific DNA sequence

We described above how a piece of DNA can be isolated using restriction enzymes. A limitation of this method is that it depends on the presence of restriction sites – the DNA sequences that are cleaved by particular restriction enzymes – in convenient locations flanking the gene. Using the Polymerase Chain Reaction (PCR) bypasses this process and makes it possible to make many copies of a DNA fragment of choice. Virtually any region of DNA can be amplified by PCR, after which the DNA fragment can be cloned into a vector of choice.

In the PCR method (Figure 16.13), the two strands of a DNA template are first separated by heat denaturation. Two synthetic oligonucleotides are then allowed to hybridize to the denatured DNA, each complementary to the 5' end of one strand of the sequence to be amplified. These oligonucleotides serve as primers for a DNA polymerase, which makes a copy of the complementary strand extending beyond the 3′ ends of the primers. This procedure will produce two new copies of the desired DNA sequence. The cycle is then repeated; the DNA is denatured and primers again can hybridize and be extended. In this second cycle four template strands are now available. With each further cycle, the number of template strands doubles. It is this amplification that makes PCR such a powerful experimental tool: it allows over one million copies of a single double-stranded template to be generated in just 20 cycles. PCR is usually done in an automated manner in a single test tube using a DNA polymerase that is stable at high temperatures, so that repeated cycles of heat denaturation do not inactivate the enzyme.

It is possible to introduce new restriction enzyme sites into the 5′ ends of amplified fragments. This is done by adding sequences to the 5′ end of each primer that are not complementary to the DNA sequence to be amplified, but instead contain a sequence that can be cleaved by a restriction enzyme of choice. In successive rounds of PCR, the bases complementary to this sequence will be added as the complementary strand is synthesized. The resulting DNA fragment that is amplified can then be digested with the restriction enzymes and cloned into a vector as described above.

Although the PCR method is extremely powerful, a few caveats should be noted. Oligonucleotide primers can sometimes hybridize to each other or to sites other than those intended due to partial sequence complementarity, which causes the wrong sequence to be amplified. Thus, primers need to be designed with care. (In practice, investigators often use computer programs that can help in primer design.) In addition, errors by the polymerase can introduce random mutations during DNA synthesis. If such a mutation is introduced in an early cycle, the mutation will be amplified at each cycle. (Although mutations are typically not desirable, the introduction of mutations by PCR can be exploited to generate randomly mutated fragments.) Finally, the efficiency of amplification diminishes beyond a certain DNA length (typically ~4000–5000 bp).

Cloning by recombination requires regions of DNA homology at the ends of the DNA fragment

The ability to amplify DNA by PCR and to introduce desired additional sequences to the ends of the DNA fragment has allowed investigators to introduce fragments into vectors by recombination (Figure 16.14). Using the same approach described for introducing restriction sites at the ends of the amplified DNA fragment, it is also possible to add DNA sequences that can recombine with sequences in the target vector. This method was initially very popular among yeast researchers, where the yeast cell carries out the recombination reaction *in vivo* if supplied with a linear vector and an insert that both had regions of homology at their ends. More recently, methods have been developed to conduct this type of cloning in bacteria. Indeed, there are now ways to carry out this recombination reaction in a test tube, using purified recombination enzymes. The advantage of this method is that it bypasses the need for restriction enzymes and ligases.

A library of clones can be made from genomic sequences or cDNA copies of expressed mRNAs

Sometimes, the sequence encoding a gene of interest is not known. For example, one could isolate a mutant organism with an interesting phenotype, but not know which mutated gene is responsible for this phenotype. Thus is it often desirable to

(a)

amplification by PCR

(b)

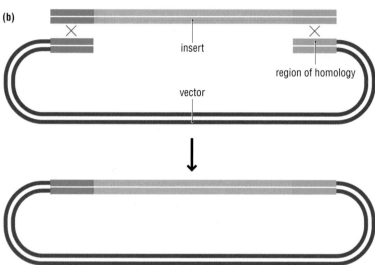

insert

region of homology

vector

Figure 16.14 Cloning by recombination.
(a) To clone a piece of DNA by recombination, the DNA can be amplified with a set of primers that contain, at their 5′ end, a sequence that is homologous to the vector (in turquoise and green). These primers also contain sequences that are complementary to the end of the DNA that is amplified (orange), such that after multiple rounds of amplification by PCR the amplified DNA contains at its end vector sequences.
(b) The recombination reaction can take place in cells, such as yeast cells or bacteria. A linearized vector and the fragment (for simplicity, the two DNA strands are shown as one) are transformed into cells and the recombination reaction occurs *in vivo*, to produce a recombinant plasmid. Certain types of recombination reactions can also be performed in a test tube.

have a collection or **library** of many different clones from which the sequence of interest can be isolated. A library can be made of total genomic DNA, of expressed genes only, or of a subset of expressed genes. For a genomic library, a genome is fragmented and each fragment is independently ligated into the plasmid vector so that each plasmid carries a different fragment of the genome. In the example above, the desired clone can then be identified by its ability to suppress a particular mutant phenotype. The suppressing plasmid is then isolated, from which the gene of interest can be cloned.

A library can also be constructed from just the expressed mRNAs. A complementary DNA fragment of each mRNA, known as cDNA, is first made using the reverse transcriptase enzyme described in Chapter 6. This enzyme uses RNA as a template to generate a complementary DNA strand. The single-stranded DNA is then copied by a DNA polymerase to generate a double-stranded DNA fragment that can be cloned into a vector. A **cDNA library** featuring all mRNAs expressed in a certain cell type or tissue can be produced; such libraries make it possible to identify and study genes expressed in certain biological contexts, for example, from a particular human tissue.

Specific mutations can be introduced by site-directed mutagenesis

Although we frequently want to make exact copies of a region of DNA, it is also often helpful to change the function of a gene or inactivate it by introducing specific mutations. **Site-directed mutagenesis** can be used to change a specific

DNA sequence in a region of cloned DNA, thereby introducing a mutation (Figure 16.15). Several methods can be used to make these specific changes in a DNA sequence. In some of the most common methods, an oligonucleotide that is mostly complementary to the DNA sequence of interest, but has one or a few nucleotide changes, is hybridized to single-stranded DNA. Under appropriate

Figure 16.15 Site-directed mutagenesis. (a) In this scheme, oligonucleotides are synthesized whose sequence contains the desired mutations (depicted as an inverted V in the DNA) but are otherwise complementary to the flanking codons. These oligonucleotides are annealed to the plasmid and serve as primers for DNA replication, which extends the primers and replicates the entire plasmid. The plasmid is then transformed into bacteria, and the colonies that arise are tested for plasmids with the desired mutations. (b) In this scheme, the mutagenesis reaction takes place as part of a PCR amplification process. The first step of the mutagenesis reaction takes place in two PCR reactions, one amplifying the left half of the DNA fragment using a terminal (pink) primer and an interior mutagenic primer (green) and the other amplifying the right half of the DNA fragment using an interior mutagenic primer (dark blue) and terminal (orange) primer. (the process is not shown but the product, from light blue to green with the mutation in red, is shown below). By incorporating the mutation into the oligonucleotide used in the PCR reaction, the majority of the amplified products will, eventually, carry the mutation in both strands. Once the two half reactions are completed, the produces are used in a 'sewing' PCR reaction, as shown, using the terminal primers. The end result is a DNA fragment that is like the starting material, except that it carries a mutation at the desired position.

conditions, such a mismatched oligonucleotide can still hybridize and serve as a primer for DNA polymerase. The annealed primer is then extended by the DNA polymerase as usual. The products, in which 50% of the clones will contain the base changes, can be transformed into bacteria. These can be screened for the desired mutation or the original template destroyed by a variety of means, so that nearly100% of the clones will have the mutagenized sequence.

PCR can also be used to introduce a specific change into the amplified DNA. In this instance, the PCR primer carries the desired mutation such that the products of the PCR will also predominantly carry the mutation. The resulting product can be used in a subsequent PCR reaction in which two PCR fragments are 'sewn' together by annealing the ends to each other (Figure 16.15b). The products can then be introduced into a vector or a specific chromosomal locus by recombination. In all cases the resulting clones are sequenced to verify that the desired change has been introduced. We discuss methods for sequencing below.

16.4 GENOME MANIPULATION

In the preceding section, we learned how mutations are introduced into cloned regions of DNA. However, scientists also often want to introduce mutations into a genome, particularly the genomes of model organisms, to eliminate or modify a gene. In general terms, two basic strategies are employed when using a genetic approach to study a problem in biology. One is **forward genetics**, in which the researcher generates a mutant that is detective in a particular process (or uses an existing mutant) and then seeks to identify and understand the gene(s) involved in causing the observed phenotype or disease. The other approach is **reverse genetics**, in which a gene is already in hand and the researcher seeks to understand that gene by altering its sequence and analyzing the phenotypic changes that result when the sequence is expressed *in vivo*.

The introduction of random mutations throughout the chromosome frequently forms the basis of forward genetic analysis. For reverse genetic analysis (as well as for other assays of specific genes or regions of DNA), it is desirable to introduce a specific gene or specific mutations into a genome in a non-random fashion. The introduction of specific genes into the genome of an organism is denoted **transgenesis** and results in the generation of a **transgenic organism**.

In this section, we will discuss several different methods for genome manipulation including chemical mutagenesis, transposition, homologous recombination, and conservative site-specific recombination (CSSR). For some organisms, however, it is difficult or even impossible to introduce specific mutations into the genome. In these organisms, scientists can turn to approaches such as RNAi and chemical libraries to block the expression or function of the gene products, as will be described at the end of the section.

Mutations can be introduced by radiation or chemicals that damage DNA

As discussed in Chapter 12, mutations can be introduced more or less at random by exposure to radiation or DNA-damaging chemicals. Thus, one of the simplest ways of introducing mutations is to expose organisms to agents such as ethyl methanesulfonate, ethylnitrosourea, or irradiation wih ultraviolet (UV) light. The advantage of these types of mutagens is that they often introduce point mutations that can change the function of the gene product instead of simply inactivating the gene. However, this type of mutagenesis brings with it the challenge of identifying the DNA sequence change in the context of a large genome (for example, that of the mouse). This limitation is rapidly being overcome by advances in DNA sequencing technology, making it likely that chemical mutagenesis will remain a widely used tool in genetic screens.

Transposons can be exploited for mutagenesis and transgenesis

Transposons were discovered because of the way their movement resulted in mutagenesis, since the insertion of a transposon into a gene disrupts gene function. This property of transposons has been exploited in every genetically tractable model organism to carry out insertional mutagenesis, whereby genes are disrupted by transposons.

Transposon mutagenesis is most often carried out using a two component system where one component is a regulated source of transposase and the other component is a 'mini' transposon consisting of a selectable marker gene flanked by the end sequences that are necessary for transposition (Figure 16.16). This separation is useful because, once the element inserts into the target DNA, it is stable and cannot transpose again as it lacks transposase. Cells containing the transposon are then isolated by screening for the selectable marker, which may be an antibiotic resistance gene or a visual marker such as a fluorescent protein.

The plasmids or viruses containing the transposon are introduced into the cell by transformation, injection or infection, followed by selection or screening for a selectable marker encoded by the transposon. The donor plasmids or viruses are unable to replicate, which means that they will not be propagated as cells grow and divide. This helps to ensure that the original vector will eventually be lost or inactivated, so that only transposons that have been incorporated into the genome will be selected for and propagated.

Transposons are a particularly useful experimental tool because the DNA that flanks the transposon can be isolated and sequenced, thereby providing a molecular tag to identify the site of the insertion mutation. This strategy is far less laborious than, for example, determining the position of a single base pair change that results from chemical mutagenesis. Some of the transposons that are the most widely used for mutagenesis and transgenesis are listed in Figure 16.17. One class of transposon called 'DNA-only cut-and-paste transposons' are most often employed for insertional mutagenesis, although retroviruses have also been used occasionally for mutagenesis in vertebrates. Transposons have also been exploited as vectors for expressing foreign genes, generating transgenic organisms.

Transposon insertion can be used to obtain information about the activity of specific promoters or enhancers, or to artificially activate a gene. For example, a transposon can contain a reporter gene, whose expression can easily be assayed (Figure 16.18). When the transposon is inserted in the vicinity of a promoter or

Figure 16.16 Two-component transposon system. (a) A transposon as found in nature is shown with its terminal recombination sequences (yellow triangles) and transposase gene (orange). (b) In a two-component system, one component is a 'mini' transposon in which the transposon ends flank a cargo DNA segment, here NeoR (which provides resistance to the drug neomycin). The other component is the transposase gene under the control of a regulatable, heterologous promoter. When a cell contains both components, induction of transposase expression mobilizes the mini transposon.

⊖ We explore mutagenesis of the human genome by an endogenous transposable element in Experimental approach 14.2.

Some transposons commonly used as tools	
organism	**commonly used transposons**
bacteria	*Tn5*
plants	*Ac*
insects	P element, *piggyBac, hobo*
fish	*Tol2*
mammals	*Sleeping Beauty, Tol2, piggyBac, Minos*

Figure 16.17 Transposons commonly used as tools.

Figure 16.18 An enhancer-sniffer transposon. The transposon contains a selectable marker and a reporter gene (*lacZ*) with only a weak basal promoter such that it is not expressed in the absence of an enhancer. If the transposon inserts at a chromosomal region lacking a nearby enhancer, the *lacZ* reporter will not be expressed. If, however, the transposon inserts adjacent to the enhancer of a host gene, the reporter is expressed in the same pattern as the host gene.

enhancer element, the level at which the reporter gene is expressed reflects the expression of the host gene. Another strategy is to put a strong promoter in the transposon, such that expression of a host gene nearby the inserted transposon becomes regulated by the inserted promoter (rather than its own endogenous promoter).

Transposons are also being used as vehicles for so-called 'gene therapy', in which a wild-type copy of a protein is introduced into the host chromosome in an effort to cure disease. We are some way from using gene therapy as a generally applicable therapeutic option in humans, however: nearly all such studies to date have been performed in model organisms.

Plant genomes can also be modified

Mobile DNA elements have been the major tool for manipulating the genome of the plant, *Arabidopsis thaliana*. *Arabidopsis* can be infected by the bacterium *A. tumefaciens*, resulting in the transfer of a DNA segment called the T-DNA from a specialized Ti plasmid into the plant genome (Figure 16.19). The virulence

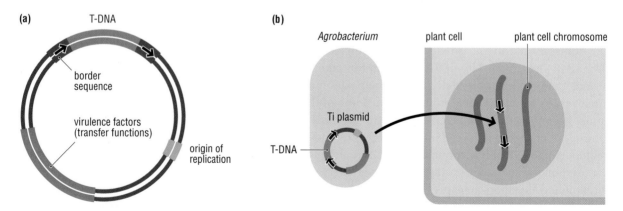

Figure 16.19 T-DNA for modification of plant genomes. (a) The Ti plasmid from *A. tumefaciens* includes a bacterial origin of replication (green), the T-DNA, which is composed of two border sequences (dark blue) and the DNA that lies between (blue), and the virulence region (pink) that codes for proteins that copy and move the T-DNA to the plant cell. (b) Insertion of the T-DNA segment into the plant genome occurs when the protein coded by the bacterial virulence region copies the T-DNA segment and moves it to the plant cell, where it is integrated into the genome. When used for creating transgenic plants, the T-DNA segment is where the DNA of interest is inserted.

region of the Ti plasmid encodes the proteins that transfer the T-DNA segment from the bacterium to the plant; the T-DNA is bound by special sequences that are recognized and acted on by these proteins. In nature, the T-segment encodes plant hormones that stimulate the plant cells to grow in unregulated fashion as 'crown gall tumors'; these tumors produce compounds that the bacterium can metabolize for growth. The frequency of T-DNA transfer is so high that it can be exploited for mutagenesis.

For genome engineering, the tumor-inducing genes inside the T-DNA are replaced with a selectable marker (for example, an antibiotic-resistance gene) and any other DNA of interest (for example, a reporter gene whose expression will depend on an *Arabidopsis* gene adjacent to the position of T-DNA insertion, or a gene whose properties are to be studied). Another useful strategy is to use the T-DNA to deliver a transposon into the plant genomic DNA, which can be subsequently remobilized for further mutagenesis. siRNAs described below can also be expressed from T-DNAs to inhibit targeted genes.

Specific sequences or gene disruptions can be introduced by homologous recombination

In addition to randomly mutagenizing a genome, researchers often want to either replace or substitute an exogenously supplied piece of DNA for a segment within the genome or add a new DNA segment to a genome. This can be accomplished by homologous recombination; several strategies for this are depicted in Figure 16.20.

Homologous recombination has been used to modify the genomes of several bacterial species, of yeast, and of the mouse, among others. In addition to simply knocking out a gene, this approach can be used to introduce reporter genes as described for transposons, as well as to introduce tags that allow gene products to be more easily detected or purified.

To alter a genome using this method, a linear fragment of DNA is introduced into cells and then becomes incorporated into the host cell genome by homologous recombination. This process may be mediated by proteins that are normally found in the host cell, or the cell may be engineered to increase the efficiency of the recombination reaction. In some organisms such as *B. subtilis*, *S. cerevisiae*, and mammalian cells, the ends of the linear targeting DNA are sufficiently recombinogenic, so that recombination can occur in the wild-type strain. By contrast, bacteria such as *E. coli* need to be modified, for example by decreasing the nuclease activity of RecBCD and introducing another recombination system from bacteriophage lambda to increase the efficiency of recombination.

In some organisms, it is necessary to introduce a double-strand break in the recipient chromosome to stimulate recombination. This can be done with engineered enzymes that generate breaks at specific locations in the chromosome, promoting homologous recombination at these sites.

The incorporated DNA fragment can be engineered to contain unique sequences that make it easy to identify cells containing that particular gene replacement. In yeast, for example, it has been possible to make a substitution of each gene with a DNA fragment containing a selectable marker gene flanked on either side by 50 base pairs that overlap with an endogenous gene located upstream and

(a) replacement

(b) addition

Figure 16.20 Gene replacement and addition by homologous recombination. (a) Gene replacement by homologous recombination. Homologous recombination between the a and d regions, which flank the *NeoR* gene on the targeting DNA ('construct'), and the genome results in the substitution of the genomic b and c regions with the *NeoR* gene from the targeting construct. (b) Gene addition by homologous recombination. Homologous recombination between the b and c regions of a targeting DNA ('construct'), in which the b and c regions flank a *NeoR* gene, results in the incorporation of the *NeoR* gene between the b and c regions of the genomic DNA.

➲ We discuss homologous recombination in more detail in Chapter 13.

➲ RecBCD is discussed in Chapter 13 and in Experimental approach 13.1.

⊖ The generation and use of knockout strain collections is described in Section 15.2.

■ primer for kan^R gene
molecular barcodes
■ universal primer
gene-specific sequences

upstream primer   downstream primer

collection of primer pairs that includes each gene

+

kan^R

PCR amplification of kan^R gene using each primer pair

↓

collection of insertion cassettes

Figure 16.21 Gene disruptions (knockouts) of *S. cerevisiae* genes. Homologous recombination between the gene-specific sequences on each end of the gene disruption cassette results in the replacement of the gene with a kanR fragment (conferring kanamycin resistance) flanked by universal primers and two barcodes specific to the targeted gene. In this way, it is possible to carry out experiments with pools of mutants, each of which contains an insertion at a different locus. The strains of interest (for example, those that are able to survive a particular treatment) can then be identified by virtue of their unique barcode associated with the kanR cassette. From Meneely (2009).

Figure 16.22 Characterization of gene integration in mammalian cells. The insert used in this example contains neoR, which confers resistance to neomycin, and *HSV-tk*, which confers sensitivity to the drug ganciclovir. (a) When the construct is not integrated, cells are sensitive to neomycin. (b) Random integration of the targeting vector by non-homologous end joining results in cells that are neomycin resistant and sensitive to ganciclovir. (c) Homologous integration results in cells that are resistant to neomycin and insensitive to ganciclovir. From Meneely (2009).

downstream, respectively, of the targeted gene. The 50 base pairs of homology at either end of the fragment allow homologous recombination, replacing the resident gene with the selectable marker. In the resulting knockout strain collections, there are two short unique sequences (bar codes) upstream and downstream of the selectable marker (Figure 16.21). These barcodes make it possible to identify the different knockout strains uniquely in a pool of mutants, either by DNA sequencing or by hybridizing the chromosomal DNA to microarrays containing the corresponding probes.

Homologous gene replacements have also been extremely useful in mammalian cells in culture. However, the sequence must not include any repetitive DNA, which may target integration to a unwanted locus. In mammalian cells, DNA often integrates by non-homologous end joining (NHEJ) and thus it is important to distinguish between cells in which targeted integration has occurred and those in which NHEJ has occurred. A useful strategy is to include two markers on the targeting DNA, one that will be retained when homologous recombination occurs and another that will be lost when homologous recombination does not occur (Figure 16.22).

(a) no integration

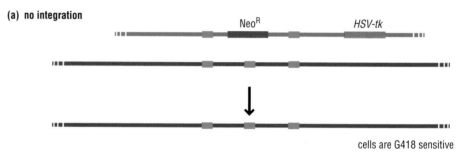

NeoR   HSV-tk

cells are G418 sensitive

(b) random integration

NeoR   HSV-tk

NeoR   HSV-tk

cells are G418 resistant and ganciclovir sensitive

(c) targeted integration

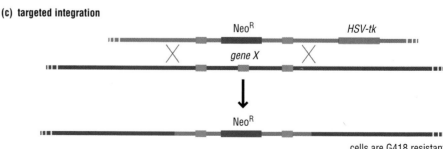

NeoR   HSV-tk

gene X

NeoR

cells are G418 resistant and ganciclovir insensitive

Knockout mice are generated using homologous recombination

Knockout mice, in which one or more genes have been disrupted, are also generated through a procedure that relies on homologous recombination to re-engineer mouse embryos (Figure 16.23a). The gene of interest in a mouse ES cell in culture is first replaced by homologous recombination with a selectable marker as described above (Figure 16.23a, step 2). For this example, we will assume the ES cell is derived from a gray mouse. The ES cells are placed in a selective medium so that only the cells containing a correctly targeted selectable marker will survive (Figure 16.23a, step 3). The transformed cells are then transplanted into a mouse embryo from a white-coated mouse, which is at the blastocyst stage of development (Figure 16.23a, step 4) and contains several hundred cells. The ES cells become incorporated into the blastocyst and will contribute to the tissues of the animal that develops. The blastocyst is implanted in the uterus of a female mouse (Figure 16.23a, step 5). The coat color of the recipient blastocyst (white) is distinct from that of the ES cell (gray) to facilitate the identification of progeny mice in which the ES cells contribute to the tissues. Since the implanted embryo contains a mixture of engineered ES cells

Figure 16.23 Generation of a transgenic mouse. (a) Scheme for generating a transgenic mouse: (1) Embryonic stem (ES) cells from a mouse with gray coat color are grown in culture. (2) The targeting vector is introduced into the ES cells by selection for a marker. (3) Cells in which the targeting vector has been correctly incorporated are grown as a pure population. (4) The gray coat color targeted ES cells are introduced into a blastocyst from a mouse with gray coat color. (5) The blastocyst containing the targeted ES cells is injected into a female surrogate black mouse with black coat color. (6) Chimeric pups that contain gray fur from the ES cells and white coat color from the mother are born. (7) Chimeric mice are backcrossed to a mouse with white fur. (8) Gray coat color mice derived entirely from the ES cells are obtained. (b) Photograph of a chimeric mouse, with patches of both white and brown fur. From Meneely (2009).

and normal cells, the resulting offspring will be chimeric (Figure 16.23a, step 6): some of their cells will derive from the ES cell (gray coat color) and some from the recipient embryo (brown coat color). This is readily apparent from the appearance of a chimeric mouse (Figure 16.23b), which has patches of brown and white fur.

The chimeric mice must be further bred in order to obtain mice in which all cells lack an active copy of the gene of interest. This means identifying those chimeric mice in which the germline cells – the sperm and ova – harbor the disrupted gene. It is not possible to know ahead of time what tissues the ES cells contributed to in the embryo. Therefore, the chimeric mice are bred with white mice and, if the germ cells came from the gray ES cells, all gray progeny result (Figure 16.23a, step 8).

CSSR allows the expression of mutations in specific cells

If a gene plays a critical role during development, it may not be possible to study its function in the mature animal by generating a knockout in an ES cell. An example is a gene knockout that causes the embryo to die or cease developing, thereby making it impossible to assess the function of the gene in the mature organism. In this case, a useful alternative strategy is to make a **conditional** allele of the gene of interest, which allows the gene to be deleted at a chosen time or in a specific cell type. A conditional allele can also be useful in studying the effects of a knockout in just a subset of tissue types, rather than in the whole organism.

A common method for producing conditional knockouts relies upon the site-specific Cre recombinase, which mediates recombination between DNA sequences called *lox* sites (Figure 16.24a). Cre promotes the deletion of DNA that lies between *lox* sites that are in a direct orientation (meaning in the same orientation in the DNA). A *lox*-Cre construct can be used to make a conditional deletion allele in a mouse (Figure 16.24b). First, a targeting DNA fragment is constructed that contains *lox* sites at positions chosen to flank the gene or exon to be deleted (Figure 16.24b, step 1). These constructs, called floxed for flanked by lox, are then recombined into an ES cell, and a knockout mouse is made (Figure 16.24b, step 2). Cre can be supplied to the construct by crossing the strain to another mouse strain that has been engineered to express Cre only in certain cell types or under certain conditions. This is done by having tissue-specific or -regulatable (for example, heat-shock) promoters that control the expression of the Cre gene (Figure 16.24b, step 3). These promoters can be engineered to respond to heat shock or to the addition of a specific drug.

Another genome modification strategy takes advantage of the phage *attP-attB* recombination system from the phage, phiC31, different genes at specific chromosomal locations without disrupting host gene function. This takes advantage of naturally occurring short sequences in the mammalian genome that resemble recombination sites.

◗ We describe the regulation of transcription though the binding of specific promoter sites in Chapter 8.

The expression of genes can be disrupted by siRNAs

Rather than disrupting a gene to study its function, the RNAi pathway can be exploited to target particular mRNAs for degradation, thereby reducing expression of the targeted gene.

(a)

(b)

Figure 16.24 Using Cre-*lox* recombination.
(a) The principles of Cre-*lox* recombination. The substrate DNA contains a *lox* site (pink arrows) in the intron between exons 1 and 2 and another *lox* site in the intron between exons 2 and 3. Cre-mediated recombination results in deletion of exon. (b) Scheme for making a conditional deletion using Cre-*lox* recombination. The targeting DNA contains *lox* sites flanking a *neoR* gene in the intron between exons 2 and 3 and another *lox* site flanking exon 3. The genomic DNA to be targeted contains five exons. (1) Targeted integration is performed in ES cells, selecting for neomycin resistance, conferred by the *neoR* gene. (2) Following transient expression of Cre in ES cells (for example, from a plasmid), cells are screened for those that have deleted the neoR marker by recombination at the *lox* sites flanking the *neoR* cassette, and have retained exon 3 and the *lox* site between exons 3 and 4. This step is done to eliminate any contribution by the presence of the *neoR* gene itself. (3) A mouse containing the gene with the targeted DNA is generated by blastocyst injection and backcrossing, as described previously, and then crossed to a mouse in which Cre expression is regulated, for example by the insulin regulatory region. Cre is expressed in the pancreas and exon 3 of the targeted gene is deleted. In non-pancreas cells. Cre is not expressed and exon 3 is not deleted. This approach allows the study of gene function in a particular tissue.
From Meneely (2009).

➲ We learn about the double-stranded RNA fragments called siRNAs (21–22 bp), which can trigger the degradation of mRNA containing a region of complementarity to the siRNA, in Chapter 10.

In order to reduce expression of a particular gene, a double-stranded RNA fragment that matches the sequence of the target gene is introduced into the cell. The duplex RNA is then cleaved into siRNA fragments by Dicer. In many cases, RNA oligonucleotides corresponding to the cleaved siRNA can be introduced directly into the cell. The siRNA then binds to the RISC complex, which targets the corresponding mRNA for degradation. This approach can be used in large-scale screens, since different RNA molecules can be readily synthesized and there is no need to mutate individual genes. Indeed, siRNAs are now commercially available for every gene in the human genome.

A limitation to the RNAi approach is the phenomenon of off-target effects, in which the expression of genes whose sequence is a near-match to the siRNA is sometimes also suppressed. This can be overcome by evaluating the effects of multiple independent siRNAs that hybridize to different regions of the gene for each gene being analyzed, as seen in Experimental approach 14.2.

A variety of methods are available to introduce siRNAs into a cell. Annealed synthetic RNA oligonucleotides can be transfected into cells in culture and their effects evaluated. However, because the concentration of the RNA will decrease over time, especially as cells divide, a more useful strategy is to introduce plasmids or viruses encoding the siRNA precursors (typically a long double-stranded RNA molecules, or a single transcript that forms a hairpin) into cells so that siRNAs can be continuously produced. In the case of *C. elegans*, the worms can be fed bacteria-harboring plasmids that express the double-stranded RNA corresponding to a particular gene, which is then cleaved to produce the siRNA.

Protein activity can be changed by small molecules

An alternative approach to making mutations takes advantage of small molecules that can alter or inhibit a protein's activity. Technological advances in making large combinatorial libraries of peptides, nucleic acids, and many other compounds provide rich sources of small molecules that can be screened for compounds that alter protein activity. Instead of screening for mutations or engineering mutants with a desired phenotype, small molecules are used to achieve the same result.

In a 'forward chemical' approach, a particular phenotype can be monitored after adding a library (or collection) of small molecules to a cell or organism. To identify the small molecule(s) in the large library that causes the particular phenotype, the original mixture of molecules is divided into smaller pools, each containing a subset of the molecules. The phenotypic screen is then repeated with each pool. This process of subdividing the set of molecules, followed by phenotypic screening, is reiterated until the molecule that gives rise to the phenotype is isolated. The small molecules can then be used to isolate the protein to which the small molecule binds by taking advantage of the affinity of the protein for the molecule in question. This process thus can be used to link a particular protein to a cellular process.

In a 'reverse chemical' approach, the goal is to identify a small molecule that interferes with a particular biochemical pathway. For example, in Experimental approach 14.2, a combinatorial library of peptides was added to an *in vitro* recombination system to look for small molecules that alter recombination by inhibiting the recombination protein, causing reaction intermediates to accumulate. This type of screen is done by adding a mixture of small molecules to a protein sample or to cells and then looking for a defect in a particular pathway. If the mixture produces the desired effect, the set of molecules is subdivided into smaller pools and the activity assay is carried out again. By reiterating this process with pools containing fewer and fewer different compounds, the small molecule that produces the desired effect can eventually be identified. This molecule can then be used to study the downstream effects of defects in the function of a particular protein, without the need to mutate the protein directly.

16.5 DETECTION OF BIOLOGICAL MOLECULES

In order to purify and study cellular components, methods are needed to detect the presence of biological molecules and distinguish them from others in the mixture. Some detection methods take advantage of the intrinsic properties of a molecule, such as its biochemical activity or its intrinsic spectroscopic properties. Molecules can also be detected by derivatizing them with (or incorporating within them) moieties with particular spectral, chemical or fluorescent properties that are then readily detectable. Many of these detection methods can also be used to determine the *amount* of a molecule that is present. Such measurements are essential for quantitative studies of the thermodynamic and kinetic properties of the molecules. In this section, we review some of the basic methods currently in use.

We note that molecules are most generally studied and followed as a population – in 'bulk' – and less often as single molecules. Thus when we refer to the properties of 'a protein' or 'a DNA fragment', what we most typically mean is the properties of a whole population of proteins or DNA fragments.

Cellular components can be monitored by their spectroscopic properties

Biological molecules absorb electromagnetic radiation over a range of wavelengths specific to the molecule of interest; this property can be exploited to identify and quantify molecules in solution. The amount of radiation absorbed as a function of wavelength is called the **absorption spectrum**. Several key biological molecules have spectroscopic properties that allow their presence to be readily monitored by virtue of the characteristic absorption spectra they generate. For example. DNA, RNA, and proteins have distinct UV absorption spectra (Figure 16.25a). A spectrum typically has one or more peaks of absorbance that can be used as a chemical 'signature' to monitor the presence of a particular species. DNA and RNA have an absorbance maximum at ~260 nm that reflects the presence of heterocyclic bases, whereas a typical protein with aromatic side chains (tryptophan, tyrosine, or phenylalanine) usually has an absorbance peak at ~280 nm. Other peaks in the DNA, RNA, and protein absorption spectra at shorter wavelengths reflect the properties of the peptide and nucleic acid backbones.

Many small molecules in the cell also have characteristic absorption spectra that make them relatively easy to visualize (and identify). As shown in Figure 16.25b, the nucleotide cofactor flavin adenine dinucleotide (FAD) has a characteristic absorption spectrum with peaks at 380 and 450 nm. Proteins that are complexed with FAD

(a)

(b)

Figure 16.25 Representative absorption spectra. (a) Absorption spectra of DNA and protein at equal concentrations. At A260, a 50 μg/mL solution of double-stranded DNA or a 40 μg/mL solution of single-stranded RNA has an absorbance of ~1.0 $m^{-1}cm^{-1}$. At A280, a 1 mg/mL of an 'average' protein has an absorbance of ~1.0 $m^{-1}cm^{-1}$, although this value depends on amino acid composition. (b) The absorption spectrum of FAD/FADH$_2$. Both FAD and FADH$_2$ exhibit two absorption peaks, at 380 and 450 nm. From Harm (1980).

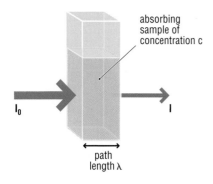

Figure 16.26 Absorption spectroscopy. The relative amount of light absorbed by a sample (I/Io) is determined by the type of molecule it contains, the concentration of the molecule, and the distance the light traverses through the liquid. These parameters are related to one another according to Beer's Law:

$A_\lambda = \varepsilon_\lambda \, l \, c$ where A is the measured absorbance at wavelength λ, E is the molar absorptivity or extinction coefficient, I is the path length and c is the concentration. The key term is ε, which varies with wavelength and is a property of the particular molecule being evaluated. The extinction coefficient of a protein at a wavelength of 280 nm, for example, is dictated primarily by its total number of tryptophan and tyrosine residues.

(or its reduced derivative, $FADH_2$) can therefore be directly monitored by measuring their absorbance at these wavelengths. Cofactors that cause proteins to absorb light in the visible portion of the electromagnetic spectrum (from 400 nm to 700 nm wavelength) cause otherwise colorless proteins to appear colored, and are therefore called **chromophores** (from the Greek 'chromo', which means color, and 'phore', which means carrying). For example, the red color of hemoglobin comes from the oxygenated heme prosthetic group complexed with the globin polypeptide chains.

Optical absorbance is measured in an instrument called a **spectrophotometer**. Typically, a liquid sample is placed in a transparent vessel called a cuvette (Figure 16.26), which is placed in the spectrophotometer. The intensity of a light beam that has passed through the sample is compared to the intensity of a beam that has passed through air or through a sample containing water or buffer. The relative amount of light absorbed by the sample is determined by three things: the concentration of the molecule, the distance the light traverses through the liquid, and the type of molecule the sample contains. Each molecule has intrinsic properties which affect the absorbance of light and are circumscribed by a parameter called the **extinction coefficient**. The extinction coefficient of a particular protein at 280 nm, for example, is denoted E and is dictated primarily by the total number of tryptophan and tyrosine residues in the protein. The light absorbed by the sample at a given wavelength, denoted A, will then be the product of the extinction coefficient, the concentration of the molecule, and the distance the light traverses through the fluid: $A = \varepsilon cl$ (see Figure 16.26).

When the extinction coefficient of the molecules under study is known, the absorbance of a sample can be used directly to determine the concentration of a given molecule (particularly if no other species are present that could contribute to the overall absorbance at that wavelength). In addition, the purity of a sample can sometimes be evaluated by asking whether the ratios of absorbance at different wavelengths match those of a pure sample. For example, the ratio of the absorbance at 260 nm and 280 nm is typically ~2 for a DNA population, whereas the ratio is ~0.6 for a typical protein preparation.

Cellular components can be monitored by binding or reaction with fluorochromes

In addition to using their intrinsic spectral properties, molecules can be detected by the binding or attachment of a specific molecule that either fluoresces or has a distinct color or spectroscopic property. A method for determining protein concentration known as the Bradford assay relies on incubating the protein with a dye known as Coomassie Brilliant Blue, which binds to proteins. The color of the protein sample is compared to the signal for different concentrations of a standard protein and used to generate a titration curve. This standard curve is then used to estimate the protein concentration in the sample (Figure 16.27). A drawback to this assay, and others of its type, is that it assumes that the unknown protein binds about as much dye per mg of protein as the protein used as the standard. Coomassie Blue is also commonly used to stain proteins immobilized in gels after seperation by electrophoresis. Silver staining can also be used as a more sensitive way to detect proteins that are similarly separated on a gel. In this case, silver ions bind to the amino acid side chains (in particular the sulfhydryl and carboxyl groups) and are then chemically reduced to metallic silver, which is easily visualized.

A commonly used molecule for the detection of DNA, and sometimes RNA, is ethidium bromide. This molecule inserts (intercalates) between the DNA or RNA bases, and the intercalated molecules fluoresce when irradiated with ultraviolet light. Methylene blue dye is also commonly used to stain nucleic acids. Although not as sensitive as ethidium bromide, methylene blue is less toxic and is less likely to interfere with subsequent hybridization steps because it does not intercalate in the nucleic acid. Both reagents are commonly used to stain nucleic acid samples that have been resolved on acrylamide or agarose gels.

We discussed above how the biochemical activity of a molecular species can often be directly monitored, even in a complex mixture. A common technique used to analyze many enzymes is to derivatize a substrate molecule with a chromophore whose spectroscopic properties change once the reaction occurs – such approaches are routinely used for the analysis of a wide range of proteases.

Particular nucleic acid species and proteins can be detected by exploiting their ability to bind to other large molecules. Because of their base-pairing potential, specific DNAs or RNAs can be detected through hybridization to labeled nucleic acid probes; these approaches include **Southern blot** analysis and **northern blot** analysis. These methods are discussed in Sections 16.8 and 16.9, respectively. By contrast, specific proteins can be detected through their binding to antibodies; these approaches will be discussed in Section 16.10. The methods described here focus on the relatively non-specific detection of proteins and nucleic acids.

Radioactive labeling can be used to detect molecules

A powerful method for detecting very small amounts of a molecule is to incorporate radioactive atoms into them. Isotopes such as ^{32}P or ^{33}P, ^{35}S, ^{14}C, or ^{3}H are unstable, unlike their more common isotopes, ^{31}P, ^{32}S, ^{12}C, or ^{1}H. This instability means that they break down over time, emitting ionizing radiation that can be detected. Various methods are available for incorporating radioactive isotopes into larger molecules, including proteins, nucleic acids, sugars, or lipids. For example, ATP synthesized with ^{32}P at the γ phosphate (the outermost phosphate) can be used as a donor to transfer the radioactive phosphate to a different molecule. To do this, T4 polynucleotide kinase is used to transfer the γ phosphate of ATP to a nucleotide substrate (for example, the 5′ end of a strand of DNA or RNA). In an analogous fashion, protein kinases typically transfer the same γ phosphate of ATP to certain serine, threonine, or tyrosine residues in a protein. Radioactively labeled molecules can be detected by a range of techniques (Figure 16.28), many of which can be used to measure the amount of radioactivity and, hence, the quantity of radioactively labeled molecular species present in a sample.

Cellular components can be monitored by activity assays

Cellular components that have particular binding or enzymatic activities can be detected with assays that monitor the particular activity of the protein. For example, if the molecule of interest is a protein that binds to a particular sequence of DNA, the presence of the protein can be detected using one of the DNA-binding assays that we shall learn about later in this chapter. The presence of a protein with kinase activity can be assayed, for example, by measuring the addition of a radioactive phosphate to the target of the kinase. The presence of radioactivity

(a)

(b)

Figure 16.27 Typical standard curve for a Bradford assay. (a) A standard curve is constructed by performing the assay with known quantities of protein. Note that the assay becomes somewhat non-linear at higher protein concentrations. Samples are typically diluted until they fall within the linear range of the assay. (b) Visual examples of Bradford assay tubes with increasing concentrations of protein. BSA, bovine serum albumin.
From www.eiroforum.org/media/gallery/embl.php

Methods of detection of radioactivity
Geiger-Muller survey meter (Geiger counter)
Typically handheld gas-filled monitor that detects ions formed upon radioactive decay.
liquid scintillation counting
A scintillation counter converts radioactive emissions into visible light. Liquid scintillation fluid is an organic solvent containing a fluor (a compound that fluoresces in the presence of radioactivity). Scintillation fluid is mixed with the radioactive sample (which may be a small amount of fluid, a gel slice or a small piece of filter paper) in a vial which is then placed inside the counter in the dark so that the light emissions can be detected and counted.
film autoradiography
Photographic emulsions contain silver halide crystals that form silver atoms in the presence of radioactivity, leaving an image on the film after processing. X-ray film containing silver halide crystals is the format usually used in the laboratory and is exposed by pressing it flat and tight against a gel or paper that contains the radioactivity.
phosphorimager autoradiography
A phosphorimager screen is coated with a phosphor compound whose atoms are excited when they encounter radioactivity. When the screen is exposed to radioactivity, for example by being pressed against a gel, a latent image of excited atoms forms in the screen. When the screen is scanned with a laser, the stored energy in the excited atoms is released in the form of light, which can be measured.

Figure 16.28 Methods for detection of radioactivity.

Figure 16.29 Measurement of enzymatic activity by coupling several reactions.
ATP hydrolysis by DNA topoisomerase can be followed by coupling the hydrolysis reaction to the oxidation of NADH to NAD$^+$. (a) The reaction scheme. Topoisomerase II hydrolyzes ATP, releasing phosphate. Phosphate release can be measured using two subsequent coupled reactions. First, pyruvate kinase dephosphorylates phosphoenolpyruvate (PEP) to generate pyruvate, utilizing the phosphate to convert ADP in to ATP. Second, lactate dehydrogenase converts pyruvate in to lactate, using NADH as a cofactor. Thus the NADH concentration is directly correlated with phosphate production. (b) Absorption spectra of NAD$^+$ and NADH. The spectra of these molecules are markedly different: NADH absorbs light at 340 nm whereas NAD$^+$ does not. Thus, when phosphate is released, the concentration of NADH decreases and absorption at 340 nm decreases in tandem.

⊘ Experimental approach 11.2 provides an example of an activity assay.

can be followed in a variety of ways (as described below). The advantage of a direct assay for activity is that it can, in many cases, be sensitive enough to detect minute quantities of the species that might not be detectable by any other means, especially if the sample in question is impure.

In other cases, the products of the reaction cannot be easily monitored directly, but the reaction can be coupled to a second reaction whose products can be monitored more readily. In the example shown in Figure 16.29, the amount of phosphate released during the topoisomerase-catalyzed hydrolysis of adenosine triphosphate (ATP) is measured by coupling the reaction to one in which nicotinamide adenosine dinucleotide (reduced form) (NADH) is oxidized to NAD⁺. NADH is readily detected by its spectroscopic properties, as described below. Therefore, the oxidation of NADH to NAD⁺ can be followed in a **spectrophotometer**: the conversion of NADH to NAD⁺ is marked by a decrease in absorbance at around 330 nm. This change in absorbance gives an indirect measure of the extent of ATP hydrolysis.

Activity assays also make it possible to look for and isolate a biological molecule whose existence is suspected, but whose identity is unknown. For example, many of the proteins required to initiate RNA polymerase II transcription were identified by fractionating nuclear extracts and assaying the different fractions for their ability to promote transcription *in vitro*. Once a sufficiently pure sample with the desired activity is obtained, the proteins and other molecules can be identified through a variety of methods.

16.6 SEPARATION AND ISOLATION OF BIOLOGICAL MOLECULES

A great deal of information in this book derives from experiments done in the test tube using purified components. In order to study the behavior and properties of a specific cellular component or complex in solution, one must be able to separate the molecules of interest from other cellular components. This can be done by taking advantage of the different physical and chemical characteristics that distinguish one molecule from another – for example, mass, shape or overall charge. It is also possible to introduce a tag into a protein or nucleic acid molecule that is then used to isolate the tagged molecule. Molecular tags can also be used to locate a molecule within a cell, as we will see later in this chapter.

Organelles and molecules can be separated by centrifugation

The first step in purifying a biological molecules from a cell involves separating different organelles, or cytosolic and membrane fractions, a process known as **cell fractionation**. This can be done with whole tissues, or with cells grown in culture. To begin this process, the intact cells must be disrupted to release their contents. At the same time, the treatment must not be so harsh as to disrupt the organelle or the molecules being studied. A variety of methods can be used to disrupt cells; these include the use of a device similar to a mortar and pestle, or the use of ultrasonic waves, **high pressure**, or detergents (Figure 16.30).

Once the cells have been disrupted, the individual components can be separated. This is commonly done using a **centrifuge**, an instrument used to spin vessels at high speed. The centrifuge creates centripetal force perpendicular to the rotation axis. Test tubes (or other vessels) are placed in a central holder in the centrifuge called a **rotor**, which either allows the tubes to swing outward as high speeds are achieved, or holds the test tubes at a fixed angle (Figure 16.31). In either case, the centripetal force causes organelles or molecular complexes to move towards the bottom of the tube and form a pellet. The rate at which a particular organelle

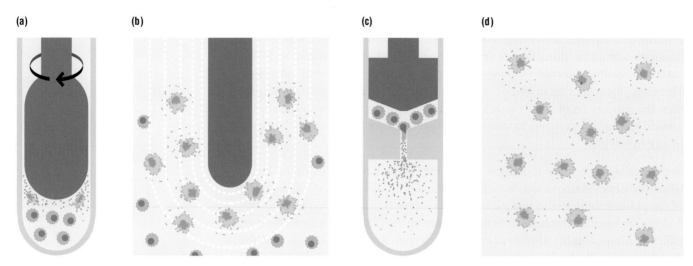

Figure 16.30 Methods for disrupting cells. Various methods are used to disrupt cells, including: (a) grinding of cells with a pestle that fits tightly in a test tube, (b) subjecting cells to ultrasonic waves using an instrument called a sonicator (the tip of the sonicator is shown), (c) passing cells through a small opening by applying very high pressure, and (d) treatment of cells with mild detergents that generate holes in membranes.

descends through the tube depends upon its size and the speed at which the centrifuge rotates, since the force exerted on the organelle increases with increasing speed of rotation. By choosing a particular speed and amount of time, the desired organelle can be collected at the bottom of a test tube (Figure 16.31). For example, the cytosolic components of a cell that has been broken or lysed can be separated from the heavier membrane components by centrifugation. At the appropriate speed, the membrane components will settle at the bottom of the tube while the cytosolic components, commonly referred to as the soluble components, will stay in solution.

Better separation between different subcellular fractions can be achieved by centrifuging the fractions through a gradient of different sugar or salt concentrations. When sucrose (a sugar) or cesium chloride (CsCl; a salt) is dissolved in water, the resulting solution is significantly more viscous than water or dilute buffer. One can then create a **gradient** of concentrations in which the most dense and concentrated

Figure 16.31 Centrifugation. Solutions can be subjected to centripetal force in a centrifuge rotor, which is spun at very high speeds. This causes larger complexes and organelles to fall to the bottom of the test tube. (a) In a swinging bucket rotor, the test tubes are placed in metal buckets that can swing outward as the rotor begins to spin. The centripetal force causes large complexes and organelles to sediment and collect in the bottom of the tube. (b) In a fixed-angle rotor, the text tube bottoms slant away from the central axis of rotation and remain at a fixed angle. The centripetal force acts perpendicular to the axis of rotation. As the rotor slows and stops, the layers of increasing density reorient so that they vary along the vertical axis.

solution is on the bottom of the centrifuge tube and the least dense (and least concentrated) solution is on the top. The gradient can be established prior to adding the sample, or it can be allowed to develop during centrifugation as a result of the centripetal force on the solute molecules (Figure 16.32).

A gradient can separate different cellular components in one of two ways. In velocity sedimentation, the sample is layered on the top of the gradient and the sample is centrifuged for a fixed amount of time. The rates at which the molecules or organelles migrate to the bottom of the tube depend on both their size and shape. After centrifuging the sample for a fixed amount of time, the sample is collected from the bottom of the tube (Figure 16.31). (This can be done by punching a hole in the bottom of the tube and allowing the sample to drip out.) The rate at which a molecule or complex of molecules sediments is denoted by a unit called a Svedberg (or S), as mentioned for the ribosomal subunits in Chapter 10. In equilibrium sedimentation, the sample is spun for a long period of time until the molecules reach equilibrium and are no longer moving within the gradient (Figure 16.32b). A molecule or complex will stop moving through the gradient when it reaches a position in the tube where the density of the gradient equals the density of the molecule.

Proteins and protein complexes can be purified by column chromatography

Once cells have been disrupted and fractionated, the individual molecules within a given sample can be further separated from one another. The most common method for purifying proteins is column chromatography, which uses a glass or plastic tube filled with material that separates proteins based on their physical properties as they flow through the column. The protein sample is first applied to the top of the column (Figure 16.33). As a buffer solution is flowed continuously through the column, proteins in the sample migrate through the column at different rates, depending on the nature of the column matrix and the physical and chemical properties of the proteins. Different types of column are used to separate proteins based on size, charge, hydrophobicity, and other properties.

In **size exclusion chromatography**, protein molecules and complexes can be separated on the basis of size and shape on a **gel filtration** column (Figure 16.33a). The column matrix consists of microscopic beads of inert material, each of which contains many tiny pores. In these columns, small molecules are more likely to go through the pores of the matrix and thus travel through the column more slowly. Larger molecules move through the column more quickly because they are excluded from entering the pores in the beads and can therefore only pass through the spaces between the beads; they travel a shorter distance overall.

Ion exchange chromatography columns separate proteins on the basis of surface charge (Figure 16.33b). These columns contain a matrix bearing either positively or negatively charged chemical groups. In low-salt solutions, proteins with a negative surface charge will bind more strongly to positively charged **anion exchange** columns, while proteins with a positive surface charge will bind to negatively charged **cation exchange** columns. The bound proteins can then be washed off (or 'eluted') in buffer containing higher salt concentrations, which competes for the charged chemical groups on the matrix. Other properties of proteins, such as hydrophobicity or their binding to specific molecules, can also be used to separate proteins on

Figure 16.32 Density gradient sedimentation. Cellular components can be separated from one another by centrifugation through a solution whose density is highest at the bottom of the test tube and lowest at the top. (a) In a sedimentation velocity experiment, the gradient (for example, of sucrose) is first created by layering solutions of decreasing density in the test tube and then introducing the sample at the top of the tube. The rate at which particles sediment through this gradient is determined by their mass and shape. In the example shown, the red particles are larger and denser than the dark blue and orange particles and thus, during centrifugation, they move faster through the gradient and end up at a more dense part of the gradient. (b) In an equilibrium sedimentation experiment, particles are separated by density. A solution containing the sample and a high concentration of a substance such as the salt CsCl is centrifuged at high speeds for long periods. Eventually, a density gradient forms under the centripetal force. The different particles in the sample will be found at a position in the tube where the solution density matches the particle density.

⊘ An example of the uses of CsCl gradient centrifugation to separate DNA of different densities is given in Experimental approach 6.1.

(a)

(b)

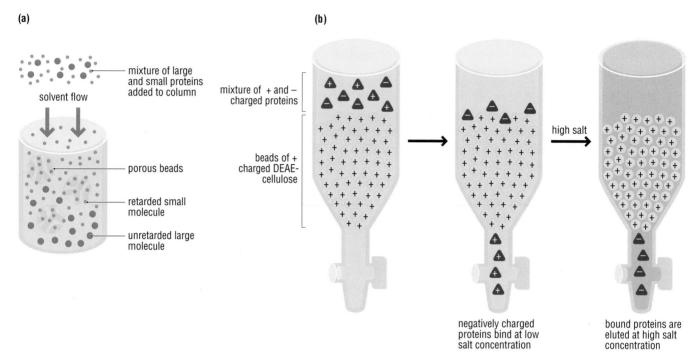

Figure 16.33 Chromatographic separation by size or charge. (a) Size exclusion chromatography, also known as gel filtration, separates molecules based on their size and shape. Large molecules that cannot fit into the pores in the beads pass quickly between the beads, while smaller molecules that can enter the beads take more time to pass through the column. (b) Ion exchange chromatography separates molecules based on charge. If the beads in the column are positively charged, negatively charged proteins will bind to the column matrix at low salt concentrations, as a result of ionic interactions. The proteins can be induced to dissociate from the beads by increasing the salt concentration.

columns. For example, the heparin polymer mimics the structure of a long DNA molecule and, when attached to a matrix, can be used to purify DNA-binding proteins.

Sometimes, the binding of a protein to a specific molecule can be used in purification. For example, a DNA fragment with a particular DNA sequence can be attached to a column matrix and used to purify a protein that specifically binds to that sequence. This approach is called **affinity chromatography**, since it relies on the protein binding to a molecule for which it has specific affinity.

Proteins can be engineered to contain tags that facilitate purification

Instead of simply relying on the native properties of a protein to separate it from other proteins in the cell, the recombinant DNA technology described in Section 16.3 can be used to add an affinity tag that can be exploited to purify a protein using affinity chromatography. The tag consists of additional amino acids added to the N- or C- terminus, thus extending the polypeptide chain (see Figure 16.33). Some commonly used tags are a histidine-rich peptide that binds tightly to a matrix containing nickel ions; a peptide that binds to a specific antibody immobilized on a column; or an entire protein domain such as glutathione-*S*-transferase (GST) that binds to a matrix containing immobilized glutathione (see also Section 16.3). More than one affinity tag can be incorporated into a single polypeptide chain, making it possible to use different types of affinity purification sequentially.

How are tagged proteins purified? The engineered fusion protein containing the protein of interest plus the tag is expressed from a plasmid or the chromosome. The

(a)

tag

bead   protein

(b)

mixture of tagged and untagged proteins

tagged proteins bind while other proteins flow through

tagged protein is eluted with a competitor molecule

Figure 16.34 Using tags to purify proteins. (a) A fusion protein containing an additional peptide or domain can be engineered into a protein. (b) Affinity purification of a tagged protein. The tag binds to a molecule (pink) that is immobilized on a column (orange beads), causing the fusion protein (dark blue) to be retained while other proteins (green) flow through. The protein can be eluted by adding a molecule that competes with the fusion domain for binding to the column matrix.

fusion protein is then purified by lysing the cells and resolving the resulting mixture on the appropriate affinity column (Figure 16.34). The fusion protein is then eluted from the column with a molecule that reverses the interaction with the matrix. Imidazole, for example, is used to elute histidine-tagged proteins from a nickel column (since it competes for metal binding with the imidazole ring of histidine side chains), whereas glutathione is used to elute proteins from a GST column.

An important caveat is that affinity tags have the potential to alter the activity of a protein. For systems where the activity of the protein can be assayed *in vivo* or *in vitro*, it is important to check that the tag has not altered the activity. Tagged proteins can also be engineered to contain intein sequences, which we learned about in Chapter 11, or a protease recognition site between the tag and the protein sequence of interest. These can then be exploited to remove the tag, either by changing the solution conditions to promote intein self-splicing or by adding a protease.

High-performance liquid chromatography and thin layer chromatography can be used to separate small molecules

The method most commonly used to separate small peptides, nucleic acids and small molecules is denoted **high-pressure** or **high-performance liquid chromatography** (**HPLC**). In a common application of this method known as reverse-phase chromatography, the column matrix consists of small beads derivatized with hydrophobic alkyl chains (such as $C_{18}H_{37}$ or $C_{18}H_{17}$). Small molecules bind to the beads by forming hydrophobic interactions, and are then eluted with increasing concentrations of organic solvent. The liquid is passed through the column at high pressure, which increases the resolution.

Instead of using a column matrix, small molecules can also be separated on a flat support covered by a thin layer of an absorbant such as silica or cellulose. Termed **thin layer chromatography** (**TLC**), this method of separation is also based on the principle that different molecules will travel at different rates through the absorbant. The sample is applied to the bottom of the plate, whose end is then dipped in a suitable solvent containing organic compounds. The different molecules in the sample migrate through the absorbant by capillary action; the distance they travel in a given time is determined by their chemistry and thus their interaction with the plate and with the solvent. Different types of molecule can then be distinguished by their different locations on the TLC plate.

RNA and DNA molecules can be separated on the basis of size on agarose and acrylamide gels

As we recall from Chapter 2, DNA and RNA are negatively charged. Thus, when nucleic acids are subjected to an electrical field, they migrate to the positive pole. This property is used to separate DNA and RNA molecules of different sizes by a method called **gel electrophoresis** (Figure 16.35). This method relies on a porous matrix made of agarose or polyacrylamide through which nucleic acids migrate when the gel slab is immersed in an aqueous solution and an electric field is applied. A gel slab is prepared by heating a solution containing agarose and then allowing it to solidify in a mold, or by polymerization of acrylamide, which can be induced by adding a cross-linking chemical. The pore size in the gel can be adjusted by altering the concentration of agarose or acrylamide. The rate at which nucleic acids migrate through the gel depends on their length, as well as the pore size.

Figure 16.35 Slab gel electrophoresis. (a) Apparatus showing an agarose slab gel immersed in buffer and connected to a power supply. The DNA is introduced into the slots (gray) at the negative end of the gel and a voltage is applied, causing the fragments to migrate towards the positive pole. (b) Example of DNA fragments separated on an agarose gels. After electrophoresis, the gel was immersed in a solution containing a dye that binds to DNA (ethidium bromide). The DNA bands (in white) were detected by exposing the gel to UV light.

To separate nucleic acid fragments of different sizes, the DNA or RNA is applied to a well at one end of the gel (the negative pole; Figure 16.35a) and induced to migrate through the gel to the other end (the positive pole) by applying a voltage. As short fragments migrate quickly and long fragments more slowly, the nucleic acid polymers separate into discrete bands according to size. The separated DNA or RNA molecules can be visualized by a variety of methods, including the intercalation of dyes such as ethidium bromide (Figure 16.35b) and radioactive labeling (which must be detected by autoradiography, Figure 16.28). Polyacrylamide gels have a smaller pore size than agarose beads and are best for separating nucleotides that are between two and several hundred bases in length, while agarose gels are best for separating nucleic acids ranging from several hundred to 250 000 bases in length. The window of optimal separation can be adjusted by changing the thickness of the gel and the acrylamide or agarose concentration.

DNA molecules of longer than approximately 100 000 bases are not well-resolved or separated on agarose gels when subjected to an electric field. Instead, separation of chromosomes and large DNA fragments requires a variant of the gel method, termed a **pulsed-field gel**. In these gels, electric fields are applied in a series of orthogonally oriented pulses. Every time the orientation of the electric field is changed, the DNA molecules are reoriented. Larger DNA molecules move more slowly through the gel because they take longer to reorient each time the electrical field is shifted, allowing for separation of very large fragments on the basis of size. A variant of this method is the **field inversion gel** (**FIG**), in which the charge at the two ends of the gel is periodically reversed in pulses of several seconds. This switching or inversion of the electric field serves the same purpose as the pulsed field gels: to reorient the large DNA molecules. A FIG gel is simpler to manufacture and use, although the separation of FIG with typically not as good as it is for orthogonally oriented pulsed field gels.

Proteins can be separated by size and charge on one- and two-dimensional acrylamide gels

Unlike nucleic acids, different proteins vary in charge and secondary structure. Although proteins can also be separated by molecular weight using gel electrophoresis, they must first be denatured and coated with an ionic detergent that is negatively charged. This is done by adding the strong ionic detergent, sodium dodecyl sulfate (SDS), and usually by also heating the sample to induce the protein to unfold and allow the SDS molecules to bind to the hydrophobic regions of the protein. In the presence of SDS, proteins behave as unstructured polymers coated with a negative charge, especially if all disulfide bonds are eliminated by adding a reducing agent. Proteins that have been coated with SDS can be separated similarly to nucleic acids by SDS-polyacrylamide gel electrophoresis (**SDS-PAGE**; Figure 16.36).

SDS-PAGE gels differ from the gels used to separate nucleic acids in that they are usually discontinuous. The sample first migrates through a small, slightly acidic 'stacking' gel with a low acrylamide concentration, followed by a longer, mildly basic 'resolving' gel containing a higher concentration of acrylamide (Figure 16.36a). The running buffer contains glycine, a zwitterion that is neutral in the acidic stacking gel and becomes negatively charged in the basic resolving gel. This change in ionization, along with the differing acrylamide concentrations in the stacking and resolving gel, helps the protein sample to become focused into

Figure 16.36 Sodium dodecyl sulfate-polyacrylamide gel electrophoresis (SDS-PAGE). (a) This type of gel is used to separate proteins by molecular weight after they have first been denatured with the detergent, SDS. The gel is discontinuous, with the sample first migrating through a stacking gel, which helps to compact the sample into a single, tight band, followed by separation on a resolving gel, which separates proteins according to their size. (b) Depiction of an SDS-PAGE gel that has been stained with Coomassie Brilliant Blue to visualize the different proteins. The right lane contains a mixture of proteins of known mass, to indicate the approximate position of proteins of similar mass on the gel (numbers are in kilo Daltons).

a tight band in the stacking gel. Once the proteins reach the resolving gel, they migrate according to their molecular weights, with small proteins separating into quickly migrating bands and larger proteins migrating more slowly. The mixture of proteins is thus separated into discrete bands according to size. The percentage of acrylamide in the resolving gel can be adjusted to maximize the resolution of individual bands over the desired molecular weight range. The proteins can be visualized by a variety of means, including the binding of the dye, Coomassie Brilliant Blue (Figure 16.36b), or silver nitrate as described in Section 16.5.

Sometimes it is useful to separate proteins on the basis of other properties, such as isoelectric point. The isoelectric point, pI, is the pH at which a protein has no overall charge. An **isoelectric focusing (IEF) gel** is a type of polyacrylamide gel that has a pH gradient across the gel (Figure 16.37a). When a native protein is loaded onto the high pH end of the gel, the protein will begin to migrate towards the negative pole of the gel because the protein has an overall negative charge (assuming the pI is lower than the highest pH in the IEF gel, which is typically around 9 or 10). The protein will continue to migrate through the gradient of decreasing pH until it reaches a zone in the gel whose pH corresponds to the pI of the protein. At this point, the protein is no longer charged and ceases to migrate in the gel.

IEF can be used in combination with SDS-PAGE to separate proteins by both pI and molecular weight. This method is called **two-dimensional gel electrophoresis** and is depicted in Figure 16.37b. Two-dimensional gels are particularly useful for separating proteins that have a similar pI (whereby isoelectric focusing alone will fail to achieve a clean separation): proteins can first be separated on the basis of pI using IEF, and separated subsequently on the basis of size using SDS-PAGE (Figure 16.37c).

Figure 16.37 Separation by isoelectric point. (a) An isoelectric focusing gel contains a pH gradient. Proteins (pink) migrate through the gel until they reach a pH that matches their pI, at which point they are no longer charged and therefore stop migrating. (b) Two-dimensional gel electrophoresis. Proteins are first separated along one dimension by their pI, and along the second dimension by molecular weight. (c) Example of a two-dimensional gel of *D. melanogaster* sperm. The purified sperm proteins are run on a pH 4–7 isoelectric focusing strip and run in the second dimension on a 12.5% SDS-PAGE gel. This gel demonstrates the resulting separation of proteins, which allows one to excise selected proteins from the gel (circled) for further analysis. From Karr (2008).

16.7 IDENTIFICATION OF BIOLOGICAL MOLECULES

Once a cellular component of interest has been purified away from the other components, it is usually desirable to determine its identity. Many biological molecules are linear polymers, and thus can be identified by determining the sequence of nucleotides or amino acids. Alternatively, the identity of the molecule may be determined by measuring its exact mass.

The sequence of RNA polymers can be determined by reaction with specific nucleases or chemicals

The sequence of an RNA molecule can be determined with the aid of enzymes or chemical reagents that cleave RNA molecules after particular nucleotides. The first step in such an experiment is to label either the 5′ or 3′ end of the unknown nucleic acid species. RNA can be easily labeled at the 5′ end with radioactive phosphate using T4 polynucleotide kinase and γ-[^{32}P]-ATP. Alternatively, the 3′ end can be labeled with [^{32}P]-pCp using T4 RNA ligase. The end-labeled samples are then exposed to sequence specific nucleases or certain chemical modification reagents, which result in strand cleavage or a physical block that prevents a primer extension reaction from proceeding (see below). These experiments are done under conditions in which a partial reaction is performed such that each molecule in the population is cleaved or modified once on average. These fragments can then be resolved on denaturing polyacrylamide gels to directly read out the sequence of the polymer. By comparing the length of all of the singly cleaved molecules, it is possible to obtain information about the entire sequence. For example, the ribonuclease T1 preferentially cleaves after G residues, whereas ribonuclease U2 cleaves after A residues. Thus, for example, the positions of G residues can be determined by partially digesting a labeled RNA molecule with ribonuclease T1, which produces a variety of fragments each ending in a G. The positions of each of these fragments on a gel is then compared with the positions of the cleaved fragments resulting from a partial base (OH⁻) digestion of the same labeled RNA fragment, where a band is seen for each consecutive nucleotide (Figure 16.38a).

Another approach is to use chemicals that modify bases at particular positions, and then to detect the location of the modified base. Dimethylsulfate, for example, modifies a number of positions on nucleic acids: the Watson–Crick face of A and C residues (meaning chemical groups involved in Watson–Crick base-pairing interactions), and the N7 position of G residues. When a G residue is modified at the N7 position, treatment of the RNA strand with the chemical aniline will cause strand cleavage at the modified base. This can be detected by polyacrylamide

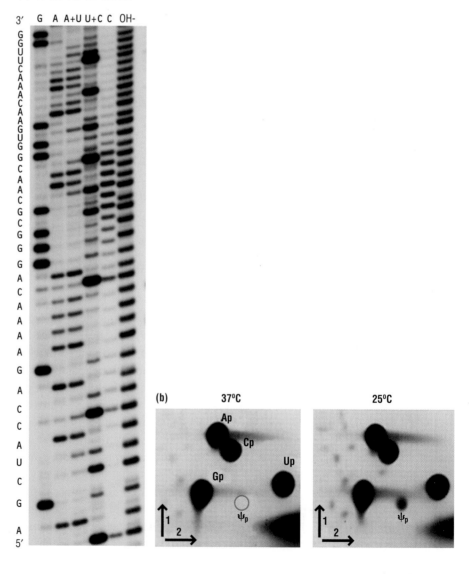

Figure 16.38 Sequencing RNA polymers.
(a) Nuclease and chemical digestion of RNA. 5′ end-labeled RNA was prepared by *in vitro* transcription, subjected to digestion with ribonuclease T1 or with base (OH⁻), and analyzed by denaturing polyacrylamide gel electrophoresis. The treatment with base generates a ladder of products resulting from cleavage at each nucleotide position, whereas T1 only cleaves after guanosine (thus the incomplete banding pattern). (b) Detection of a modified nucleotide by thin layer chromatography. *In vivo* labeled ³⁵S pre-rRNA was isolated at 37°C or 25°C from yeast containing a temperature-sensitive allele in the pathway for RNA pseudouridylation. The RNA was digested with RNase T2 and analyzed by two-dimensional cellulose thin layer chromatography. Arrows indicate the two directions of chromatography. Spots corresponding to the Ap, Gp, Cp, and Up residues are indicated. Pseudouridine 3′-monophosphate, whose expected position is indicated in each panel (Ψp), is not detectable in the heat-treated population (37°C, on left). Weak spots represent 2′-O-methylated dinucleotides resistant to RNase T2 digestion.
Figure 16.38a from Barrick *et al.* (2005); figure 16.38b from Bousquet-Antonelli *et al.* (1997).

gel electrophoresis as described in Figure 16.38a for enzymatically cleaved DNA. Positions that are modified at the Watson–Crick face of the nucleotide bases can be used for determining sequence as well, but the location of the modification must be identified using a different method described below, known as a primer extension reaction. Chemical reagents typically used for RNA modification include DMS (A+C residues), kethoxal (G specific), and 1-cyclohexyl-(2-morpholinoethyl) carbodiimide metho-p-toluene sulfonate (CMCT) (U specific).

Because post-transcriptionally modified nucleotides often behave like standard nucleotides in typical sequencing methods, it is sometimes useful to use other methods such as TLC to resolve single nucleotides after the RNA is degraded with various nucleases. This approach, for example, can be used to detect pseudouridine in an RNA molecule (Figure 16.38b).

⊙ The application of chemical probing to ribosome structural analysis is described in Experimental approach 10.1.

The sequence of both RNA and DNA molecules can be determined by synthesizing a cDNA copy

RNA and DNA molecules can both be sequenced by generating DNA copies of the molecules of interest in the presence of chain-terminating nucleotides in a procedure referred to as **chain-termination dideoxy sequencing**. In this approach, an oligonucleotide primer that specifically hybridizes to the RNA or DNA is annealed, and the primer is extended with an enzyme (a DNA polymerase in the case of DNA or reverse transcriptase in the case of RNA). If all four deoxynucleotide triphosphates (dNTPs) are supplied, a full-length DNA copy of the template strand will be synthesized. If, however, a small amount of nucleotide lacking a 3′ hydroxyl group is added, chain elongation terminates when that nucleotide is incorporated. This occurs because the chain cannot be extended by polymerase when there is no 3′OH group. For the sequencing reactions, four independent chain elongation reactions are run, each containing one of the four dNTPs together with a small amount of a 2′,3′-dideoxynucleotide (ddA, ddC, ddG, and ddT) that lacks the 3′-hydroxyl group on the sugar. Each analog terminates chain extension at one of the four bases. When supplied in an appropriate amount, the analogs will be incorporated at random, generating a pool of extension products terminating at each position that a given nucleotide occurs. The sequence of the DNA being made can be read by comparing the pattern of extension products of the four different reactions. Figure 16.39 outlines such an extension protocol.

Typically, radioactively or fluorescently labeled dideoxynucleotides are included in the reactions to make it possible to detect the extension products when they are resolved by gel electrophoresis. As we will see in Section 16.14, automated DNA sequencing and new techniques are continuously being developed, allowing the complete genome sequences of thousands of organisms and even individuals to be obtained with decreasing cost and increasing speed.

The N-terminal sequence of a protein can be determined by derivatization and cleavage

The N-terminal sequences of a protein can be determined by sequential derivatization, release, and identification of the N-terminal amino acid in a process called **Edman degradation** (Figure 16.40). In this procedure, the most N-terminal amino

❶

primer + template
+
DNA Pol I
+
dATP, dCTP, dGTP, dTTP
+
limiting amounts of fluorescent
ddATP, ddCTP, **ddGTP**, ddTTP

❷

primer
5′   3′
TGCATGCTCGA
TGCATGCTCGA**G**
TGCATGCTCGAGC
TGCATGCTCGAGC**G**
TGCATGCTCGAGCG**G**
TGCATGCTCGAGCGGC
TGCATGCTCGAGCGGCC
TGCATGCTCGAGCGGCC**G**
TGCATGCTCGAGCGGCCGC
TGCATGCTCGAGCGGCCGCC
TGCATGCTCGAGCGGCCGCCA
TGCATGCTCGAGCGGCCGCCA**G**
TGCATGCTCGAGCGGCCGCCAGT
TGCATGCTCGAGCGGCCGCCAGT**G**
TGCATGCTCGAGCGGCCGCCAGTGT
TGCATGCTCGAGCGGCCGCCAGTGT**G**
TGCATGCTCGAGCGGCCGCCAGTGTGA
TGCATGCTCGAGCGGCCGCCAGTGTGAT
TGCATGCTCGAGCGGCCGCCAGTGTGAT**G**
TGCATGCTCGAGCGGCCGCCAGTGTGATG**G**
TGCATGCTCGAGCGGCCGCCAGTGTGATGGA
AGCTACCTGCTCGCCGGCGGTCACACTACCTATGGACGTCTTAAG
3′   40   50   60   5′
                        template

❸

dye-labeled segments
are applied to a capillary
gel and subjected to
electrophoresis

large

DNA migration

small

laser beam

laser

detector

❹

ATGCATGCTCGAGCGGCCGCCAGTGTGATGGATATCTGCAGAATTC

Figure 16.39 Chain-termination DNA sequencing using dideoxynucleotides. (1) A DNA primer (turquoise) is annealed to the DNA (pink) whose sequence is being determined; DNA synthesis is carried out by DNA Pol I (or reverse transcriptase) + dNTPs in the presence of limiting amounts of dideoxynucleotides (ddNTPs), which terminate DNA synthesis when they are incorporated because they lack a 3′OH. Each ddNTP is uniquely fluorescently labeled. (2) Products of DNA synthesis, each terminated by one of the labeled dideoxynucleotides. (3) The products of DNA synthesis are separated by capillary gel electrophoresis, with the color of the fluorescent label at the 3′ end corresponding to the terminating nucleotide. (4) The results are displayed on a chromatogram that reveals the sequence of the newly synthesized strand. The arrow marks the position of the first nucleotide incorporated beyond the primer.

Figure 16.40 Steps in Edman protein degradation. The peptide is adsorbed to a solid support via its C-terminus and sequenced using sequential Edman degradation steps, in which one residue is removed from the N-terminus per cycle. Each cycle has the following steps: (1) Phenylisothiocyanate (PITC) is reacted with the primary amine function of the peptide. (2) Under mildly alkaline conditions, a phenylthiocarbamoyl derivative is formed. (3) Mild acid treatment causes the terminal amino acid is released, freeing the N-terminus of the peptide for another cycle of PITC reaction. (4) The released amino acid isomerizes to form a phenylthiohydantoin species that can be identified by chromatography or mass spectrometry.

acid is specifically modified by phenylisothiocyanate (PITC). This modified amino acid is then specifically cleaved from the protein by mild acid treatment (which does not affect the remainder of the protein). The released derivatized amino acid is then identified by column chromatography. The cycle is then repeated with the next amino acid, and so on.

Two limitations of the method are: some residues are resistant to PITC modification or acid hydrolysis, and the ability to identify an amino acid decreases with each round of cleavage (due to limitations in efficiency as the process is iterated). In practice, only a limited number of N-terminal amino acids can be identified through this approach, though in many cases the number of residues correctly identified may be sufficient for protein identification. There are also methods for determining the C-terminal sequence of a protein based on the same principles as the N-terminal sequencing, but these approaches are more problematic and are much less widely used.

Mass spectrometry can be used to identify most molecules

Mass spectrometry allows the identification of the widest spectrum of molecules, even if the molecules of interest are only present in small amounts. Mass spectrometry is used extensively to identify proteins following their separation in and excision from a polyacrylamide gel (Figure 16.41); it can also be used to identify carbohydrates, nucleic acids, and small molecules.

The method of mass spectrometry makes it possible to determine the molecular weight of a molecule with very high accuracy by measuring its mass-to-charge ratio. A small amount of the material of interest must first be converted to gas-phase ions. This can be achieved by bombarding a sample embedded in a specific chemical matrix with a laser (matrix-assisted laser desorption/ionization or MALDI) or by generating a spray of the sample in the presence of a strong electric field (electrospray ionization or ESI). The ionized molecules are separated according to their mass-to-charge ratios, and the mass of the separated ions is then measured upon striking a detector (Figure 16.41a). The mass of small biological molecules can be determined very accurately to within a single Dalton, while the mass of larger biological molecules can be determined with somewhat less accuracy.

Often, particularly in the case of larger molecules, the mass resolution is not sufficient to allow for unambiguous identification of a molecular species. In these cases, the molecules are usually digested or hydrolyzed into small fragments whose individual masses can be determined with greater accuracy. For example, proteins can be digested with specific proteases such as trypsin, and the masses of each of the resulting fragments determined (Figure 16.41b). For organisms for which the entire genomic sequence is known, this information is usually sufficient to identify the protein, based on predicted coding regions as described in Chapter 15.

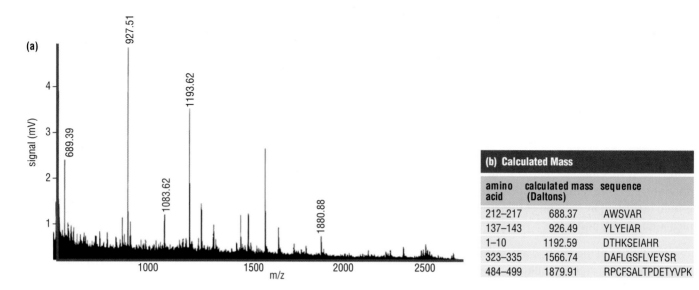

Figure 16.41 Representative mass spectrometric analysis. (a) Bovine serum albumin (BSA) was digested with trypsin and a MALDI spectrum was generated. The mass/charge (m/z) ratios are indicated on the x-axis and relative intensity is plotted on the y-axis. (b) Table showing the tryptic fragments of BSA that were detected and their calculated mass (accounting for the relative abundance of different. From Shevchenko *et al.* (1996).

If identification is not possible based on the fragment masses alone, however, the proteolytic fragments can be broken down further to the individual amino acids, whose masses can then be determined. This is often referred to as tandem mass spectrometry (or MS/MS): first the mass of the proteolytic fragment is determined, then the masses of the amino acids in the fragment are determined.

There are several different approaches for the ionization, separation (for example gas chromatography or liquid chromatography), and detection of molecules by mass spectrometry; the combination of approaches suitable for a particular application depends on the complexity, amount and type of sample.

16.8 DETECTION OF SPECIFIC DNA SEQUENCES

The base-pairing ability inherent in the DNA sequence, which allows two complementary strands of single-stranded DNA to hybridize, can be exploited to detect a specific DNA sequence amidst a vast excess of unrelated sequences. In this section, we describe a number of methods by which unique regions of DNA can be specifically identified and distinguished from other DNA sequences in the context of many DNA fragments, a pool of DNA clones, or even chromosomes within a cell.

Specific DNA fragments can be detected by Southern blot hybridization

One method to detect a specific DNA fragment among many DNA fragments is called **Southern blot** analysis (illustrated in Figure 16.42). This method is named after its inventor, Edwin Southern, and hence the name carries an uppercase 'S'. (Note that in the terms 'northern blot' and '**western blot**', the 'n' and 'w' are not

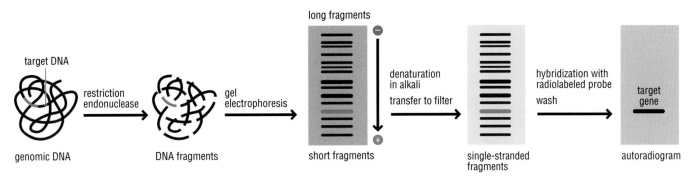

Figure 16.42 Southern blot analysis. The DNA is first digested with restriction enzymes and separated by gel electrophoresis. The DNA is then denatured and transferred to a filter (in blue). In parallel, a DNA fragment complementary to the region of interest is labeled, for example, with a radioactive isotope. The radioactive DNA probe is then hybridized to the filter. The probe will hybridize to its complementary DNA on the filter, which can be detected by exposing the filter to radiation sensitive film.

capitalized because neither term is derived from a person's name; rather, they are whimsical names given to these techniques as they are similar to Southern blot.) In overview, Southern blot analysis proceeds as follows. DNA fragments cut by specific restriction enzymes are separated on an acrylamide, agarose, or pulse-field gel as described in Section 16.6. Without fragmentation, the DNA does not enter the gel, nor would it be possible to analyze specific genomic regions. The DNA fragments are then denatured with alkali to expose the single strands for hybridization and transferred to a membrane. The membrane is then probed with a radioactively labeled single-stranded DNA that is complementary to the sequence to be identified. The labeled DNA anneals to its complementary strand and the location in the gel of the target DNA fragment is identified by exposing the membrane to an x-ray film. Other methods of DNA labeling such as chemically or fluorescently labeled probes, which can be detected by other means, are sometimes also used.

Using this method, one can determine whether a particular chromosomal locus is altered by the insertion or deletion of sequences, which will be reflected in a DNA fragment size that is either larger or smaller than expected. Experimental approach 14.1 describes how Southern blot analysis was used in the detection of a transposon inserted into the factor VIII gene that causes hemophilia.

Specific DNA clones can be identified in libraries by colony or plaque hybridization

A similar hybridization method can be used to identify a specific DNA clone in a library of many different clones. If the library is on a bacterial plasmid or BAC vector, colony hybridization can be used. Alternatively, if the library is represented in a bacteriophage vector, plaque hybridization can be used.

For colony hybridization, a collection of bacterial colonies, which each arise from a single cell containing one DNA clone, is transferred to the positively charged membrane, generating a copy of the plate on the membrane (Figure 16.43). The cells are then lysed *in situ* on the membrane. The DNA from the colony is denatured and probed in the same way as for the Southern blots described above. In an analogous method called plaque hybridization, the bacteriophages that formed plaques on bacterial lawns are transferred to the membrane and the DNA is denatured. Hybridization to the DNA probe is then performed as for

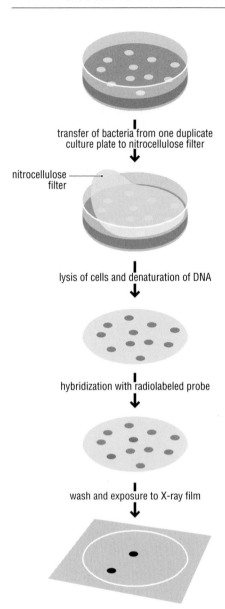

transfer of bacteria from one duplicate culture plate to nitrocellulose filter

nitrocellulose filter

lysis of cells and denaturation of DNA

hybridization with radiolabeled probe

wash and exposure to X-ray film

colonies where probe hybridized exposes the film in spots

Figure 16.43 Colony hybridization. The figure outlines the method for probing bacterial colonies, each of which contains a unique plasmid, for the presence of a plasmid containing a DNA sequence of interest. In the first step colonies are transferred onto a filter that is pressed on the surface of the plate. The filter is processed in much the same way as that described for Southern blot analysis, then hybridized with a DNA probe to the DNA region of interest. If the colony carrying the relevant plasmid was present on the original plate, there will be a spot corresponding to its location on the filter. The investigator can then isolate that colony from the original plate. Copyright Pearson Education, publishing as Benjamin Cummings.

Southern blots. Because the membrane is a copy of the original plate, the researcher can go back to the plate once a positive signal is obtained and isolate the bacterial colony or bacteriophage that gave rise to the positive signal. These hybridization approaches make it possible to identify a single clone that contains the specific sequence of interest from among thousands of clones in a library.

Chromosome abnormalities can be detected by chromosome spreads

Since each metaphase chromosome in multicellular eukaryotes has a characteristic banding pattern (Figure 16.44), chromosomes can be stained with a dye such as Giemsa and examined directly under the microscope to detect the presence of additional chromosomes, missing chromosomes, and gross deletions or rearrangements. This approach, by which regions of a chromosome are detected visually, is denoted karyotype analysis, and is frequently used in medical diagnostics.

Aneuploidy (an extra or missing chromosome) and other chromosomal aberrations, such as deletions and translocations, are frequently the underlying cause of birth defects, mental retardation, miscarriages, and other syndromes. Thus, direct visualization of chromosomes is routinely used in prenatal screening procedures, such as amniocentesis, a medical procedure in which fluid is removed from the amniotic sac surrounding the fetus. This fluid contains cells shed from the fetus, which can be cultured and used for karyotypic analysis. Chromosome abnormalities are similarly diagnosed in the case of malignancies and other diseases that may involve chromosomal rearrangements.

DNA sequences can be localized along a chromosome by fluorescent *in situ* hybridization

While examining the overall chromosome structure makes it possible to detect gross chromosomal changes, it is often helpful to locate a specific DNA sequence within a chromosome. The method used to accomplish this is called **fluorescent** *in situ* **hybridization** (**FISH**), (Figure 16.45). This approach is also useful in diagnosing genetic diseases associated with chromosome abnormalities.

Figure 16.44 The karyotype of a human stem cell showing trisomy 12 and trisomy 17. The karyotype is determined by staining chromosomes of metaphase cells. In the example shown, the cell has one X chromosome, one Y chromosome, and two copies of each autosome, with the exception of chromosomes 12 and 17, which are present in three copies each (marked by red arrows). Adapted from Meisner and Johnson (2008).

healthy   leukaemic

Figure 16.45 Fluorescent *in situ* hybridization (FISH). This assay was used to detect a reciprocal translocation between chromosomes 4 and 11, using probes close to the break point. (a) In normal individuals, each probe gives rise to two spots, one for each homologous chromosome. (b) In a leukemia patient, however, there was a reciprocal translocation between one chromosome 4 and one chromosome 11. The unaffected chromosomes still give green or red signals, but the probes are very close to each other on the chromosomes that experience the translocation, giving rise to merged signals (yellow). Photo courtesy of Sabine Strehl, Children's Cancer Research Institute.

In this method, cells are fixed to a slide and their DNA is hybridized with differently colored, fluorescently labeled DNA probes in a manner similar to the Southern blots described above. The chromosomes are also stained with a fluorescent dye, 4′-6′-diamidino-2-phenylindole (DAPI), described in Section 16.12, which binds specifically to DNA. The image that appears in a fluorescent microscope shows the chromosomes (DAPI blue), with the regions of interest highlighted in different colors by the fluorescent probes. In Figure 16.45, the probes used were against regions on two different chromosomes, as shown in the example from a healthy individual (Figure 16.45a). In a certain type of leukemia (Figure 16.45b), a translocation occurs that bring these two regions in close proximity. As a result, the two probes overlap. The use of FISH to study chromosome segregation was described in Experimental approach 7.1.

Two modifications of the FISH procedure, called **chromosome painting** and **spectral karyotyping** (**SKY**), can be used to examine whole chromosomes. These methods are used to identify specific chromosomes or chromosomal rearrangements. For chromosome painting, the DNA for one entire chromosome is purified and labeled with a fluorescent dye. SKY is similar to chromosome painting except that, in this case, every chromosome is separately purified, and each is labeled with a different and spectrally distinct ratio of five different fluorescent dyes. This set of labeled chromosome probes is then pooled, denatured, and used to probe a metaphase spread of chromosomes. In a normal karyotype, each pair of chromosomes has a unique color. If chromosome translocations or rearrangements have occurred, an abnormal karyotype will be seen (Figure 16.46). The specific chromosome that has been translocated and the partner it was translocated to are then known because the color uniquely identifies each chromosome.

Alterations in DNA from patients can be detected by array comparative genomic hybridization

Although FISH and SKY are very useful, they are not sensitive enough to detect small deletions or amplifications. In the case of FISH, we also need to have prior knowledge of which DNA region is to be analyzed. Array comparative genomic hybridization (aCGH) is a quantitative method for detecting changes in copy number (deletion,

(a) (b) (c)

Figure 16.46 Spectral karyotyping (SKY). (a) DNA probes are generated for each individual chromosome by creating a unique combination of fluorophores, making each chromosome appear as a different color when these probes are hybridized to mitotic chromosome spreads. (b) In parallel, the chromosomes are also stained with a DNA dye that reveals the specific banding pattern of each chromosome. (c) An arranged SKY karyotype of the chromosomes can be generated from the chromosomes shown in (a) Photo kindly provided by Margaret Strong, Johns Hopkins School of Medicine, unpublished.

amplification) on a genome-wide basis (Figure 16.47). This method involves labeling genomic DNA from a patient with a fluorophore (for example, fluorescein, which emits green light when excited). Instead of hybridizing to chromosomes, however, the labeled DNA is hybridized to a **DNA microarray** that contains DNA fragments from the entire genome, spotted in rows and column onto a microchip. (The use of DNA microarrays to detect RNAs on a genome-wide level will be discussed in Section 16.9.) At the same time, DNA from normal cells, which is labeled with a different fluorophore (for example, rhodamine, which emits red light when excited), is also hybridized to the microarray. If the copy number of a particular region is the same in the patient and in the control, the signal from the two fluorophores will be equal (and will appear yellow because the equal amounts of red and green fluorochromes generate a yellow light). However, if the patient has a deletion or duplication, the

Figure 16.47 Array comparative genome hybridization (aCGH). (a) DNA from the sample to be tested is labeled with a green fluorescent dye and reference DNA is labeled with a red fluorescent dye. The two samples are mixed and co-hybridized to an array containing genomic DNA fragments that have been spotted on a glass slide. (b) The resulting ratio of fluorescence intensities is proportional to the ratio of the copy numbers of DNA sequences in the test and reference genomes. The green spots indicate extra chromosomal material (duplication) in the test sample at that particular region, while red spots indicate relatively less test DNA (deletion). The black spots are areas where no hybridization was observed.
From Shinawi and Cheung (2008).

signal will be red or green, respectively. (In this instance, the excess of one fluorophore will generate a signal for the color that is in excess.) The microarray is scanned by a special fluorescent scanner that measures the signal from each microarray spot and plots it with respect to each chromosome. In the example shown (Figure 16.47), the DNA from the patient is missing a particular region, and this is detected by the reduced green fluorescence in a number of microarray spots.

16.9 DETECTION OF SPECIFIC RNA MOLECULES

As we learned in Chapter 8, many genes are regulated at the level of transcription. It is therefore often desirable to study the expression of an individual gene or a collection of genes by assaying the amount of RNA that is present. As described in this section, many methods used to examine RNA levels rely on hybridization to specific probes, although the use of reporter genes can also provide information about timing and patterns of expression.

Northern blot analysis, qRT-PCR, RNase protection, and deep sequencing can be used to evaluate the expression of individual genes

Specific RNA transcripts are generally detected by hybridization to specific DNA or RNA probes that have been labeled in some way. Just as we learned in Section 16.8 that specific DNA fragments can be detected by Southern blots, we can detect RNA through an analogous approach known as a northern blot.

So how is a northern blot performed? First, RNAs are separated by size on denaturing acrylamide or agarose gels. (The denaturing step eliminates secondary structures in the RNA that can alter its migration in the gel (Figure 16.48a).) The RNA is then transferred directly to a positively charged membrane. Unlike a Southern blot, where the DNA is fragmented prior to electrophoresis, RNA molecules are relatively small and thus will enter the gel and transfer efficiently to the filter without prior fragmentation. Also, RNA can be probed directly because it is single-stranded, and therefore does not need to be denatured (unlike double-stranded DNA). At this stage, the membrane is 'probed' with a radioactively or fluorescently labeled DNA (or RNA) probe and the hybridized labeled probe is detected. Northern blot analysis is the least biased approach for RNA analysis as it directly follows the signal of the actual transcript in the sample, in contrast to other techniques described below. As such, the size of a full-length transcript can be determined by comparing the position of the signal in the membrane to the positions of marker bands of known size. Moreover, if there are multiple RNA species that hybridize to the selected probe (for example, alternatively processed versions of the gene), each will be observed as an independent signal on the blot. Northern blot analysis can also be used to determine the relative abundance of a specific transcript under different growth conditions by comparing the intensity of the signals from samples isolated under different conditions.

The PCR method described in Section 16.3 can also be exploited to determine the relative quantities of specific RNAs in what is commonly called quantitative reverse transcription-polymerase chain reaction (**qRT-PCR**). In this method,

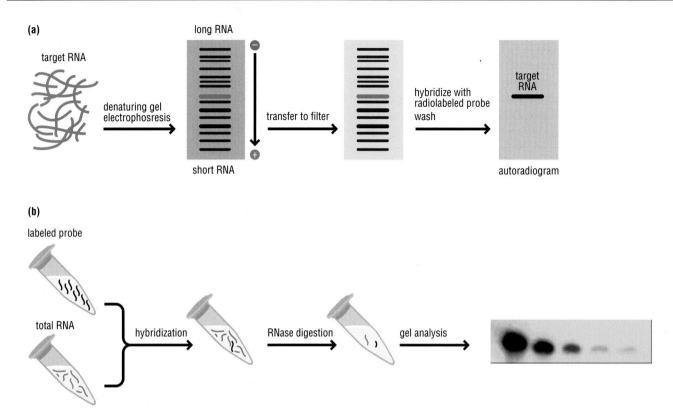

Figure 16.48 Analyzing RNA abundance. (a) Northern blot analysis. RNA is isolated from the cell and separated by size on a denaturing gel. The RNA is then transferred to a filter (light blue) and probed with a DNA or RNA probe that is coupled to a detectable molecule (for example, a radiolabeled isotope). The filter is then analyzed to determine the positions on the gel where the probe has hybridized (black band). In the case of a radioactive probe, the filter can be exposed to an x-ray film or scanned on a phosphorimager. (b) Nuclease protection assay. RNA from the cell is hybridized with a radioactive probe (DNA or RNA) targeting a specific sequence. The mixture is then treated with nucleases that digest single-stranded RNA (or DNA) but not double-stranded RNA or DNA–RNA hybrids. Both the probe and the RNA are protected from nuclease digestion wherever the two are complementary to one another. The resulting mixture is subjected to gel electrophoresis; the amount of undigested probe visible in the gel is proportional to the abundance of the RNA of interest.

a cDNA copy of the RNA of interest is first synthesized by reverse transcription. The cDNA products are then amplified by PCR, utilizing a second primer within the proposed transcript. This step is designed to yield a conveniently sized PCR product for analysis on an agarose gel.

The amount of product obtained by qRT-PCR will be proportional to the amount of RNA present in the sample as long as every sample being compared is within the exponential phase of PCR amplification. Machines that automatically quantitate the amount of PCR products after each cycle in 'real time' (through binding to a dye) have greatly simplified these assays, though proper controls must always be included to ensure that the amount of the PCR product is correlated with that of the input RNA. A distinct advantage of qRT-PCR is that its extreme sensitivity allows very small amounts of RNA to be detected. A disadvantage of this approach is that only a limited segment of the RNA molecule corresponding to the region between the chosen primers will be detected. As a result, for example, if the RNA exists in multiple forms within the sample, this type of information may be missed in the analysis.

Another approach that has traditionally been used to determine the amount of an RNA is a **nuclease protection assay** (Figure 16.48b). As with a northern blot, a radioactively labeled DNA or RNA probe of a defined size is hybridized to the transcript of interest, though in this case the RNA of interest is found in solution rather

Figure 16.49 Methods to determine the identity of the ends of RNA molecules. (a) Primer extension. A radioactive oligonucleotide (green) is hybridized to the RNA (blue) and then extended using reverse transcriptase to produce single-stranded cDNA (black). The resulting cDNA is subsequently analyzed on a denaturing polyacrylamide gel (PAGE) (bottom panel). A manual sequencing reaction is run next to the primer extension product to determine the size (in this case is it 46 bases long). (b) 3′ RACE. (1) An oligonucleotide of known sequence (pink) is ligated to the 3′ end of the RNA. (2) The primer is extended by reverse transcriptase to form cDNA. (3) The DNA near the tagged end is then amplified by PCR using one oligonucleotide complementary to the tag sequence (green) and a second oligonucleotide complementary to a known segment of the RNA (light gray). (4) The duplex products of the PCR amplification are then sequenced to determine the unknown 3′ end of the RNA species.
Figure 16.49(a) from Opdyke *et al.* (2004).

than immobilized on a membrane. The probe is often deliberately designed to have some number of nucleotides that are not complementary to the molecule of interest and thus will remain single-stranded even after hybridization. Nucleases, such as S1 nuclease, that cleave single-stranded nucleic acids but not double-stranded DNA–RNA or RNA–RNA hybrids, are then used to digest unhybridized probes and any overhanging ends of the hybridized probes. These digestion products are then resolved by gel electrophoresis.

A final approach of increasing importance is the simple direct sequencing of cDNAs generated from an RNA population of interest. As sequencing methods have improved, it is now possible to sequence vast numbers of cDNAs, and then simply sort through the many sequences to identify those of interest. If enough sequences are read, then the data from such an experiment can be used to evaluate actual amounts of a species in a population, in particular when directly compared with a related population. The 'next generation' (NextGen) sequencing methods used for this approach are described in Chapter 15 and in more detail below. Direct sequencing approaches have been extensively used to characterize small RNA populations in a variety of organisms. Because certain high throughput sequencing approaches only obtain sequence information on about 40 contiguous nucleotides at a time, the small RNA populations are ideally suited to such an approach. A caveat is that the need to generate a cDNA copy of the RNA can constrain the population that is recovered for sequencing: for example, an RNA will be missed if reverse transcriptase cannot make a copy of the transcript due to inhibitory secondary structures or nucleotide modifications.

The specific ends of RNA molecules can be determined by nuclease protection, primer extension, or 5′- or 3′-RACE

Sometimes it is desirable to determine the 5′ or 3′ end of an RNA molecule – for example, to identify the promoter of a gene. All methods for identifying end regions rely on the hybridization of DNA or RNA probes in the vicinity of the ends. For example, the nuclease protection assays described above can be used to map the ends of an RNA, since comparing the sizes of the undigested and digested probes indicates the position of the original 5′ or 3′ end.

Primer extension assays can also be used to identify the 5′ end of an RNA molecule. As shown in Figure 16.49a, a labeled oligonucleotide complementary to sequences near the 5′ end of an RNA is extended by reverse transcriptase. Since the reverse transcriptase enzyme falls off when it reaches the 5′ end of the transcript, it is possible to determine the 5′ end precisely by comparing the extended product with a sequencing ladder.

In another approach, called **5'- or 3'-RACE** (rapid amplification of complementary DNA ends), an oligonucleotide of known sequence is ligated to the 5' (or 3' end) of the RNA before the RNA is reverse transcribed (see Figure 16.49b). The resulting cDNA product will then have an additional string of nucleotides (whose sequence is known) at one end. The DNA near the tagged end is then amplified by PCR using two oligonucleotides; one that is complementary to the tag sequence and one that is complementary to a known segment of the RNA. The products of PCR amplification are sequenced, revealing the complete sequence of the original RNA molecule through the unknown 5' (or 3') portion.

DNA microarray analysis is used to determine genome-wide expression patterns

The availability of whole genome sequences has made it possible to develop chips containing nucleotide fragments for all or much of the genome, which can then be used to detect specific DNA or RNA sequences on a genome-wide scale. A DNA microarray contains many different DNA fragments that are immobilized on the surface of a chip (Figure 16.50). Each spot contains a unique DNA segment, of

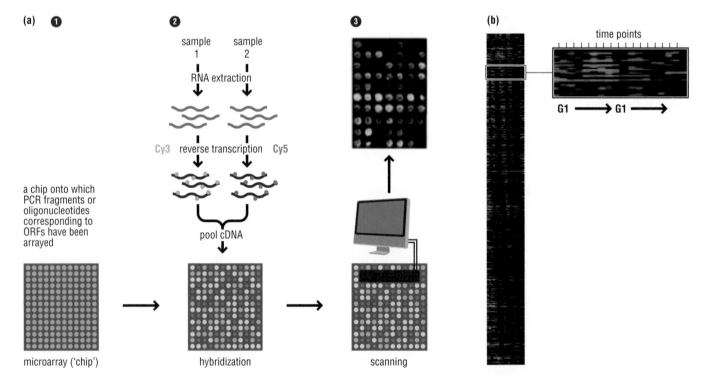

Figure 16.50 Microarray analysis. (a) Steps involved in microarray analysis. (1) A slide or membrane is prepared containing sequences representative of a select group of genes or an entire genome. (2) RNA is made from two cultures, one containing control cells and the other non-control cells. RNA is reversed transcribed to generate cDNA, and each cDNA pool is labeled with a different fluorophore (in this case, Cy3, which emits green light, and Cy5, which emits red light). The labeled RNAs are pooled and hybridized to the microarray. (3) Each spot on the chip is excited with a laser (at the Cy3 and Cy5 excitation wavelengths), and the emission is captured by a camera. If a particular transcript is expressed more robustly in the control than the non-control cells, the spot will appear green, while higher expression in the non-control culture will yield a red spot. Equal expression gives a yellow signal. (b) Genome-wide microarray analysis of cell cycle dependent gene expression in *S. cerevisiae*. Cells progressed synchronously through the cell cycle and the expression of every gene was examined at fixed time intervals. The data are displayed in the form of clusters, meaning that genes that have a similar expression pattern are placed next to each other in the presentation. Each row represents the expression pattern of a single gene and each column represents a time point. At the left most time point, the cells are found in G1, and the cells proceed through two full cell cycles as the time course progresses. Red indicates genes with increased transcription relative to the control (RNA from asynchronous cells) while green indicates genes with decreased transcription. The right panel is an enlargement of a small section from the left panel. ORF, open reading frame.
Adapted from Spellman *et al.* (1998).

which thousands can be fit onto a single chip. The DNA molecules can be either synthetic oligonucleotides or larger fragments that have been generated by PCR. Chips can be constructed, for example, so that each spot contains a DNA fragment corresponding to a portion of an open reading frame. The chips can then be used to monitor transcription in a population of cells by isolating the pool of RNA, generating fluorescently labeled cDNA copies of this population of RNA molecules, and then hybridizing the mixture of cDNAs to the microarray. A labeled cDNA will hybridize to a spot in the microarray that contains a complementary DNA fragment. The relative intensity of each spot on the microarray reflects the level of transcription of the corresponding gene.

Expression of the same set of genes under different conditions (say, in the wild-type organism as compared with a mutant) can be compared by labeling each cDNA pool with a different fluorescent dye. Both samples are then used to probe the same microarray and the relative intensities of each dye on each spot can be determined as described above. If red and green dyes are used, genes that are expressed at equal levels under both conditions will yield a yellow spot on the array, whereas differences in the ratio of expression will give rise to a spot that is either green or red in color. This approach can be used to assess relative gene expression under two different conditions, such as in the presence and absence of a stress such as heat shock, as discussed in Experimental approach 8.4. To circumvent possible artifacts introduced by the generation of a cDNA copy, RNA can also be hybridized directly to the chip and the RNA–DNA hybrids can be detected by specific antibodies, which in turn can be detected by fluorescence.

Reporter genes can simplify the detection of RNAs

A common method for assaying the expression of a particular gene is to fuse its regulatory region to a **reporter gene**, which encodes a gene product that can be readily detected. Products of some commonly used reporter genes include β-galactosidase (encoded by the *E. coli lacZ* gene), which breaks down galactose derivatives to give a colored product; luciferase, which catalyzes a chemical reaction that emits light; and the green fluorescent protein from jellyfish, which has intrinsic fluorescence. The availability of easily detected indicators of expression makes it possible to screen thousands of microbial colonies (as seen in Experimental approach 14.2).

Reporter genes can be quickly and repeatedly assayed, making it possible to study the way gene expression changes with time. Reporter genes also can be used to monitor gene expression that might be limited to certain tissues within the whole organism. The reporter can be fused to the promoter of the gene to reflect the transcriptional regulation in operation within the organism (Figure 16.51). Alternatively, the reporter can be fused directly to an open reading frame of interest such that the reporter reflects both the transcriptional and translational regulation of the gene.

16.10 DETECTION OF SPECIFIC PROTEINS

It is often desirable to follow the fate of a specific protein. Two main approaches for detecting proteins, the use of antibodies directed against the protein and the use of tags added to the protein by cloning, are described in this section. In addition we consider an approach for evaluating the relative levels of specific proteins in two different samples.

(a)

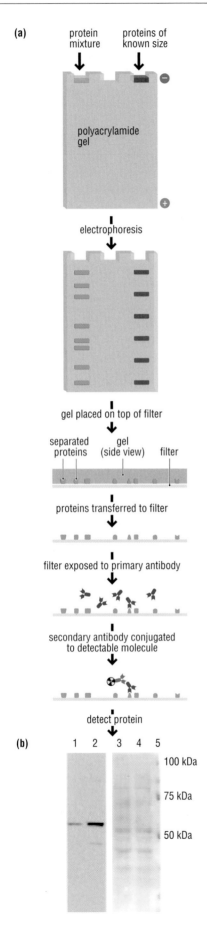

protein mixture

proteins of known size

polyacrylamide gel

electrophoresis

gel placed on top of filter

separated proteins

gel (side view)

filter

proteins transferred to filter

filter exposed to primary antibody

secondary antibody conjugated to detectable molecule

detect protein

(b)

1   2   3   4   5

100 kDa

75 kDa

50 kDa

Figure 16.51 **The use of reporter genes** *in vivo.* Expression of a cyclin E:GFP reporter in *C. elegans.* GFP fused to the promoter of the *cyclin E* gene (cye-1::GFP) reports on its transcriptional activity. Shown is a larval stage (L1) animal with strong expression of the cye-1 reporter gene in a row of dividing cells (P lineage descendents) in the ventral nerve cord (arrow), as well as in other tissues. From Brodigan *et al.* (2003).

Western blots rely on antibodies to detect proteins

One powerful method for detecting specific proteins takes advantage of the mammalian immune system's ability to generate antibodies that bind to proteins with very high affinity and specificity. The antibodies can be obtained by exposing the appropriate animal (often mouse, rabbit, or goat), to either a purified protein or just a short polypeptide corresponding to a region of the protein. Generally, two classes of antibodies are used: monoclonal antibodies, which are obtained from a clonal population of cells and only react with only one epitope or region of the protein, and polyclonal antibodies, which are a collection of antibodies that react with different regions of the same protein.

The western blot, or **immunoblot**, detects a specific protein immobilized on a membrane. In this method, proteins are separated by SDS-PAGE (Figure 16.52), and transferred from the gel to a hydrophobic or positively charged membrane. The membrane is then incubated with specific antibodies in the presence of a protein-rich blocking agent that reduces non-specific binding of the antibody, such as powdered milk, or a solution containing bovine serum albumin, which. After the membrane is washed, the bound antibody can be detected in a variety of ways, most of which rely on 'secondary' antibodies that bind to the 'primary' antibody. These 'secondary' antibodies themselves can be directly labeled with

Figure 16.52 **Western blot analysis.** (a) A schematic illustration of western blot analysis. (1) A protein mixture is loaded on a polyacrylamide gel and separated by electrophoresis, as described in Figure 16.36. Proteins of known size that serve as molecular weight markers can also be loaded. (2) The gel is placed on a specialized filter that can bind proteins. (3) The proteins are then transferred to the filter in a special chamber, using an electric field. The filter can then be stained with dye, such as Ponceau S Red, which reveals the proteins on the filter and shows the positions of the molecular weight markers (see (b)). (4) The filter is exposed to the primary antibody that recognizes the protein of interest. This reaction takes place in a special buffer that minimizes non-specific binding of the antibody to irrelevant proteins (known as cross-reactivity). (5) The filter with the bound primary antibody is washed several times to remove unbound antibody and then exposed to a secondary antibody that is conjugated to a detectable molecule, such as a radioactive isotope or a fluorophore. (6) The filter is again washed to remove unbound secondary antibody and then processed to detect where the secondary antibody using autoradiography or a phosphorimager. For fluorescent probes, the filter can be scanned by a machine that detects fluorescence. (b) An example of western blot analysis. Two samples containing a mixture of proteins were run on an SDS-PAGE gel and transferred to a filter. The filter was then stained with Ponceau S Red, revealing the proteins in both lanes that were transferred to the filter (lanes 3 and 4: note the presence of many protein bands). The Ponceau S staining also shows the locations of the molecular weight markers (lane 5). The same filter was then treated with antibodies against one protein of interest followed by secondary antibodies that recognize the first. The resulting bands in lanes 1 and 2 correspond to the protein of interest.

radioactivity, coupled to a fluorescent molecule, or linked to an enzyme such as horseradish peroxidase, which produces a colored or fluorescent band when incubated with the appropriate substrate. The accuracy of this method depends upon the specificity of the primary antibody; some antibodies may also react with other proteins besides the intended target.

The addition of specific tags can be used to detect proteins

It is not always possible to use antibodies to detect proteins. For some proteins, it is extremely difficult to generate antibodies. In addition, one may want to detect a protein whose gene sequence is known, but which has not been purified. An alternative approach is to introduce a protein tag that is fused to either the N- or C-terminus of the protein of interest, and then use an antibody that binds specifically to that tag. A summary of commonly used tags is given in Figure 16.53. Some of the tags comprise only a few amino acids, while other tags correspond to an entire small protein. Very specific antibodies have been developed for most of the tags, allowing for sensitive detection of proteins carrying these tags. As described in Section 16.6, several of the tags also bind to specific matrices and can be used to purifiy the tagged proteins.

The relative levels of proteins in two different samples can be determined by isotope labeling followed by mass spectroscopy

A recent innovation in the area of mass spectroscopy has made it possible to compare the relative amounts of various protein species in two samples (collected under different conditions). This quantitative proteomic approach is referred to as stable isotope labeling with amino acids in culture (SILAC; Figure 16.54).

Commonly used protein tags		
tag name	**size**	**source and comments**
Myc	10 amino acids (EQKLISEEDL)	derived from the human c-Myc proteins
hemagglutinin epitope (HA)	9 amino acids (YPYDVPDYA)	derived from the influenza hemagglutinin glycoprotein
FLAG	8 amino acids (DYKDDDDK)	peptide sequence that can is recognized by a specific antibody and can also be cleaved by enteropeptidase.
hexa histidine-tag (6 × His)	6 (or more) amino acids (HHHHHH)	binds to resin containing nickel (Ni^{2+}); can be eluted with EDTA, is also recognized by specific antibodies
vesicular stomatitis virus glycoprotein (VSV-G)	11 amino acids (YTDIEMNRLGK)	derived from a viral glycoprotein needed for the budding of vesicular stomatitis virus
V5	14 amino acids (GKPIPNPLLGLDST)	derived from a small epitope present on the P and V proteins of the paramyxovirus of simian virus 5 (SV5)
calmodulin binding peptide (CBP)	26 amino acids	derived from muscle myosin light-chain kinase. Binds to resin containing calmodulin in the presence of Ca^{2+} and is eluted with EGTA
glutathione-S-transferase (GST)	220 amino acids	naturally occurring protein that binds to resin containing glutathione and is eluted with free glutathione
green fluorescent protein (GFP)	238 amino acids	a naturally fluorescent protein from the jellyfish, *Aequorea victoria*, commonly used in applications such as light microscopy to visualize directly the location of the tagged protein
maltose binding protein (MBP)	371 amino acids	*E. coli* protein, binds to resin containing amylose and is eluted with maltose

Figure 16.53 Commonly used protein tags. For short tags, the amino acid sequence is shown. A 'designed tag' means that it was devised by scientists, rather than being derived from a naturally occurring protein.

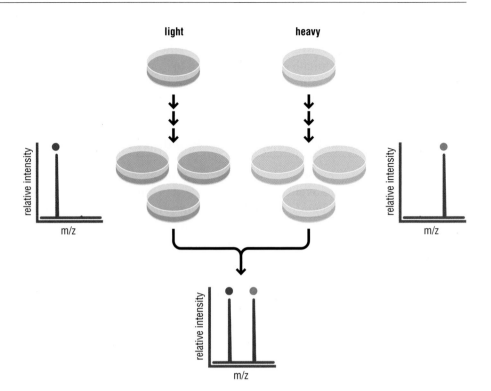

Figure 16.54 The principle of SILAC. Cells are differentially labeled by growing them in medium containing two different isotopes, for example normal arginine (purple) or heavy arginine (light red). The incorporation of the amino acids into the proteins results in a mass shift of the corresponding peptides, which can be detected by a mass spectrometer. In the example shown, a particular peptide has an increased mass following growth in the heavy arginine medium, as compared to the normal arginine medium. The samples are combined and the relative amounts of each peptide are compared. In the example shown, the peptide is present at similar levels in both samples.

The basic idea behind the approach is to grow cells in either normal media, or media containing isotopically heavy amino acids (such as arginine labeled with ^{13}C atoms instead of the usual ^{12}C). One isotope is used under one set of conditions while the other isotope is used in a population of cells subjected to different conditions. (For example, one population of cells might be treated with heat to stimulate a heat-shock response.) As a result of this isotopic labeling, any protein fragment containing an arginine will have a mass 6 Da heavier than an equivalent fragment from the culture grown with normal amino acids. Following the whole cell labeling, protein extracts are prepared and the two samples digested with proteases. The samples are then injected simultaneously into the mass spectrometer, where the ratio of peak intensities can provide information on the relative abundance of the peptides under different conditions.

Although not all proteins in the cell can yet be resolved and identified with this approach, a substantial number can be followed in a single experiment, yielding significant insights into global protein levels in the face of a given insult to the cell. As the technology is further developed, this method will undoubtedly become increasingly sophisticated.

16.11 DETECTION OF INTERACTIONS BETWEEN MOLECULES

The goal of the many techniques we have discussed above is to isolate and identify a single biological molecule of interest, often a specific RNA or a specific protein. However, since no biological molecule acts alone, it is frequently of interest to examine the interaction between molecules – especially interactions between proteins, and proteins and nucleic acids. Thus we next consider some of the commonly used approaches to study binding between biological molecules.

Co-purification and co-immunoprecipitation can be used to identify and study interacting molecules

Among the simplest methods used to study interactions between macromolecules are co-purification and co-immunoprecipitation. If certain molecules of interest remain associated with one protein (for instance, during many purification steps) the molecules are said to co-purify. Thus, co-purification can be exploited to identify molecules that interact with one another in the cell.

Co-immunopurification relies upon antibodies that bind to specific proteins. An antibody against a protein of interest is added to a mixture of proteins or to a cell extract. The antibody is then precipitated from solution by adding microscopic beads that are covered with proteins that bind tightly to any antibodies. The beads can be readily precipitated, bringing with it the protein that binds to the immobilized antibody. If any other proteins are bound to the protein of interest, they, too, will precipitate along with the rest of the complex. The binding partners that are precipitated in this way can then be identified by the method of mass spectrometry as described in Section 16.7.

The same approach can be used to isolate DNA or RNA fragments that bind to the protein precipitated by the antibody. Further variations of these methods can be carried out with fusion proteins containing tags, which we learned about in Sections 16.6 and 16.10. For example, a fusion protein containing GST can be precipitated by beads coated with glutathione. In all of these experiments, care must be taken to wash off loosely bound proteins that may adhere non-specifically to the protein or the beads. The interaction can be confirmed by performing in reverse, in which the binding partner is precipitated and the association of the original protein is examined.

Protein binding sites in the chromosome can be identified by chromatin immunoprecipitation

An application of immunoprecipitation can be used to identify proteins that are bound to a particular region of chromosomal DNA in the cell. In this method, called **chromatin immunoprecipitation (ChIP)** (Figure 16.55), cells are treated for a limited period with a chemical reagent that covalently cross-links proteins to DNA. Once the bound proteins are immobilized in this way, the chromosomal DNA is sheared into relatively small fragments. Antibodies against a particular DNA-binding protein are then added to the mix and used to immunoprecipitate the protein, along with the piece of DNA to which it is cross-linked. A reagent is then added that reverses the cross-links, releasing the DNA into solution. The DNA fragments can then be identified by PCR using primers for DNA regions of interest or by probing the DNA microarrays, as described in Section 16.8. Chromatin immunoprecipitation followed by hybridization to a microarray, sometimes referred to as 'ChIP on chip' or 'ChIP-chip', can be used to identify all of the sites in a genome to which a particular protein binds. Alternatively, the precipiated DNA fragments can be individually seque nced. This method is known as '**Chip-seq**'.

Two-hybrid analysis is a genetic approach for detecting interactions between proteins

Co-purification and co-immunoprecipitation are convenient methods for identifying new interaction partners *in vitro*. A common method for studying protein interactions that is carried out *in vivo* and lends itself to large-scale screening

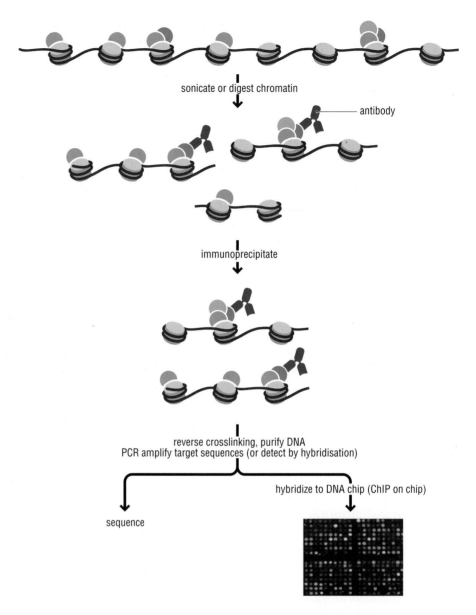

Figure 16.55 Chromatin immunoprecipitation (ChIP). Chromatin is treated to chemically cross-link proteins to the DNA, after which the DNA is digested by restriction enzymes or mechanically sheared. The DNA fragments and their associated proteins are then subjected to immunoprecipitation, using an antibody against a protein of interest (pink), which isolates only those DNA fragments that are bound to this protein. The cross-links are then removed, and the DNA is amplified by PCR and then analyzed by one of a variety of methods, including sequencing (ChIP-seq) and hybridization to a DNA chip (ChIP on chip).

is called the **two-hybrid assay**. This method is generally carried out in budding yeast, although it can be done in bacteria (*E. coli*) as well. The approach often relies on the modular nature of many transcriptional activators, which contain separate DNA-binding and activation domains.

Typically, the two-hybrid assay is carried out using a specially constructed strain of yeast containing a reporter gene. Instead of relying on a single protein, such as GAL4, to bind DNA and activate transcription of the reporter gene, the DNA-binding domain and the activation domain are separately engineered to be part of two separate polypeptides. The reporter gene is only turned on when the two polypeptides bind to one another. Recombinant DNA techniques are used to construct a hybrid gene containing the protein of interest fused to a DNA-binding domain (Figure 16.56). This fusion is the 'bait' that binds upstream of a reporter gene. To search for a protein that interacts with the bait, a second set of constructs referred to as the 'prey' are made. The prey consists of a different protein (or protein fragment) fused to a transcription activation domain. A construct containing the prey is introduced into the strain containing the bait. If the bait and prey proteins bind to one another, the activation domain attached to the prey will stimulate transcription of the reporter gene.

This method lends itself to a genome-wide screen for interacting partners, since it is relatively straightforward to construct a library of 'prey' strains, each containing a different protein fused to an activation domain. Variations of the two-hybrid approach also can be used to detect protein–RNA and RNA–RNA interactions.

A shift in DNA mobility in gels can be used to measure binding affinity

The mobility of molecules in an electrophoretic gel depends upon their size. If the electrophoresis is carried out under conditions where interactions between molecules are maintained, a complex between molecules will migrate more slowly in the gel than the individual molecules alone. This approach, termed an **electrophoretic mobility shift assay** (**EMSA**), is frequently used to measure the strength of protein binding to a DNA fragment, an example of which is given in Experimental approach 13.2. In an EMSA (sometimes referred to as a gel shift assay) of protein binding to DNA, a radiolabeled DNA fragment is incubated with a protein and then loaded onto a gel (Figure 16.57). DNA with bound protein will migrate more slowly in the gel than the free DNA fragment and the relative amounts of bound and unbound DNA in each resulting band can be quantified. If this is done for a series of samples with different amounts of protein, a binding curve can be constructed, from which the equilibrium binding constant can be determined.

Precise binding sites on DNA and RNA can be identified by footprinting

The precise binding site of a protein on a DNA fragment or RNA molecule can be identified using **footprinting**. The underlying principle of this method

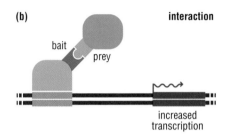

Figure 16.56 Yeast two-hybrid assay. A protein or domain of interest (the 'bait', in blue) is fused to the GAL4 DNA-binding domain (orange), while a different protein fragment, or a collection of protein fragments, the 'prey' (in green), is fused to the activation domain (orange). (a) If the bait and prey do not interact, the reporter gene will not be expressed. (b) Transcription of the reporter gene is only activated if the 'bait' and 'prey' bind to one another is.

⮕ We discuss binding equilibria in detail in Chapter 3.

Figure 16.57 Electrophoretic mobility shift assay (EMSA). A non-denaturing gel is used to assay binding of a protein to a DNA fragment. The first lane on the left shows the migration of the DNA fragment on its own. The middle shows the migration of the DNA fragment when protein is present; the protein–DNA complex is larger than the DNA fragment alone and therefore migrates more slowly through the gel than the free DNA. The right lane shows the migration of the DNA fragment in the presence of 10 times more protein than in the middle lane. This allows more DNA to associate with the protein. As a result of this association, more of the DNA fragment migrates more slowly. (b) An example of an EMSA, showing how a radiolabeled DNA fragment bound to the OxyR protein is shifted to a more slowly migrating form as increasing amounts of the OxyR protein are added.
Figure 16.57b adapted from Tartaglia *et al.* (1991).

Figure 16.58 DNA footprinting assay. (a) DNA is labeled at one end and subjected to limited digestion, producing a ladder of fragments when electrophoresed on a gel. If a protein is bound to the DNA during the digestion, it will protect a portion of the DNA from cleavage. See text for more complete description. (b) An example of footprinting in a whole cell lysate. The region that is spanned by the black bar contains an *S. cerevisiae* origin of replication. When the cells are in G2/M the origin is not bound by the pre-replicative complex (pre-RC, see Chapter 6), and is thus exposed to DNase I digestion, generating the observed bands. In G1, however, the origin is bound by the pre-RC, protecting it from DNase I activity and generating a footprint that reveals the protected region. As seen here, positions of enhanced digestion sometimes occur in regions of protein-DNA interaction.
Adapted from Labib *et al.* (2001).

(Figure 16.58) is that a protein bound to DNA or RNA can protect the region of contact from chemicals or enzymes that cleave the nucleic acid. To carry out a footprinting experiment, many copies of a particular DNA or RNA molecule that has been radioactively or fluorescently labeled is incubated in solution with the protein of interest. The binding reaction is then treated with an endonuclease (a chemical that can cleave the nucleic acid), under conditions in which each DNA or RNA molecule in the solution is cleaved only once on average. The resulting fragments are separated on a polyacrylamide gel and the gel is autoradiographed.

If no protein is bound, a ladder of bands corresponding to the collection of cleavage products will appear. If, however, a protein is bound to the DNA or RNA before it is incubated with the enzyme, the nucleic acid will be cleaved everywhere except where it was protected by the protein. The resulting ladder on the gel will show fainter bands or no bands in the protected region. Base-pairing interactions between two different RNA molecules and interactions between proteins can also be examined by variations of the footprinting technique.

Spectroscopic signals provide sensitive approaches for detecting molecular interactions

Many biological molecules are fluorescent, meaning that they absorb light at one wavelength but emit it at a different (longer) wavelength. The chemical environment of a fluorescent moiety can substantially affect its absorption and emission properties: tryptophan side chains, for example, have different emission spectra depending on whether they are buried in the hydrophobic core of a protein or lie on the surface of the protein, where they are exposed to solvent. Tryptophan fluorescence can therefore be monitored to follow protein folding and unfolding.

Similarly, the fluorescence of a molecule can either increase or decrease when another species binds to it. Fluorescence changes can therefore be monitored to measure the binding of one molecule to another. Even if a molecule lacks intrinsic fluorescence, a fluorescent dye can be covalently attached and its fluorescence can then be monitored. Proteins can also be engineered to have tryptophan side chains on the surface, whose fluorescence can then be monitored to follow a binding interaction. The instrument used to measure fluorescence is called a fluorimeter.

Another approach that also takes advantage of fluorescently labeled molecules is **Förster resonance energy transfer** (also called **fluorescence resonance energy transfer** or **FRET**). This method can be used to monitor interactions between two molecules, as well as to measure the distance between them. A FRET experiment requires two different fluorophores with distinct absorption and emission spectra (Figure 16.59). The emission spectrum of one molecule (the donor) overlaps with the absorption spectrum of the second molecule (the acceptor). If the molecules are sufficiently close to one another (within 5 nm), the energy emitted by the donor can be absorbed by the acceptor, causing the acceptor to fluoresce. By exciting the donor molecule and then measuring the fluorescence of the acceptor, it is possible to monitor interactions with a high degree of accuracy. Since the efficiency of the energy transfer depends strongly on distance (varying as $1/r^6$, where r is the distance between the donor and acceptor), the intensity of the emitted light can even be used to measure the distance between the donor and acceptor. We shall see in Section 16.12 how FRET measurements can be used in whole cells to detect molecular interactions *in vivo*.

Another sensitive technique for studying interactions between two molecules relies on surface plasmon resonance (SPR)-based technology. This method is used to measure binding interactions where one of the interacting species has been immobilized on a surface. Like the fluorescence experiments described above, this approach can yield quantitative information on molecular interactions.

As illustrated in Figure 16.60, one of the molecular species of interest (the 'bait') is immobilized on a thin gold film that lies on top of a glass surface. Buffer containing the other molecular species of interest (the 'prey') is then flowed across the surface. If the prey has an affinity for the immobilized bait, more and more prey will bind to the bait, until the surface is saturated. If fresh buffer is then flowed across the surface, the prey molecules will dissociate at a characteristic rate until no more molecules are bound. The binding of molecules to the bait immobilized on the surface, and their subsequent dissociation, is monitored by detecting changes in the refractive index in the immediate vicinity of the surface layer (based on

Figure 16.59 Principle of Förster resonance energy transfer (FRET). (a) If two fluorophores are present in the same environment but are not close to each other, the excitation of one (CFP) will not affect the other (YFP), and therefore no fluorescence will be detected in the YFP emission wavelength. (b) If, however, the two are very close to each other, excitation of CFP will result in the transfer of energy to the YFP, which will become excited and emit light in the YFP emission wavelength. Thus, the appearance of YFP emission in response to CFP excitation indicates that the two fluorophores are in close proximity.

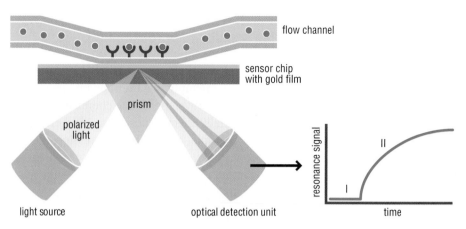

Figure 16.60 Surface plasmon resonance. A glass sensor chip contains a thin layer of gold on which proteins or other molecules (shown in dark blue) can be immobilized. A solution containing a binding partner (pink spheres) flows across the upper surface. At the same time, a light is shone at an angle on the lower glass surface and the refracted light is detected. As molecules in the solution bind to the immobilized protein, the refractive index of the glass changes (due to a physical phenomenon known as surface plasmon resonance). This alters the angle at which the refracted light exits the sensor chip, which is then recorded by the optical detection unit. The changes can be monitored in real time. From Cooper (2002).

physical principles that are beyond the scope of this book). A plot of the change in the resonance angle as a function of time, known as a sensorgram (Figure 16.60), is then used to evaluate binding. Because the interactions between the prey and bait can be monitored in real time, the surface plasmon resonance method, like the fluorescence methods described above, can allow researchers to obtain quantitative information on the equilibrium dissociation constants (the K_ds), as well as the rate constants (k_{on} and k_{off}) that define the equilibria (see Section 3.2). Taken together, these approaches make possible a very detailed analysis of the binding interactions between molecular species that lie at the heart of their function in the cell.

16.12 IMAGING CELLS AND MOLECULES

The ability to visualize, or image, cells and to localize specific cellular components has been fundamental to our understanding of many aspects of biology. In this section we discuss a range of microscopy techniques that are used to image cells and molecules. Microscopy makes it possible to determine the subcellular localization of particular proteins or cellular structures, to follow the dynamics of cellular components over time, to examine the expression patterns of a gene of interest, and to determine whether two proteins interact *in vivo*. In addition, the analysis of single cells reveals information that is masked when examining an entire population. For example, Figure 16.61 shows the variation in gene expression observed between different *E. coli* cells, all of which are genetically identical. These cell-to-cell differences cannot be observed when assessing the protein levels in a population by western blot analysis, for example. Cells can be visualized live or treated with chemical cross-linkers (a process known as fixation) to preserve intracellular structures. Moreover, cells can be observed whole or they

Figure 16.61 Cell-to-cell variation in gene expression. Each of these *E. coli* bacteria is expressing two proteins, one fused to a cyan fluorescent protein (CFP) and one to a yellow fluorescent protein (YFP), both expressed from the same promoters. Although these are genetically identical and they are all growing in the same environment, different cells exhibit different levels of CFP (in green) and YFP (in red). This observation was made possible because the expression pattern was monitored on a cell-by-cell basis.
From Elowitz *et al.* (2002).

can be sectioned. Importantly, the method of choice is dictated by the biological question at hand.

Many cell features can be visualized by light microscopy

Light microscopy is so called because it uses light refraction to detect various cellular features. These can be visualized and magnified because light passing through the sample will be absorbed differently by cell pigments or by the differing thicknesses of parts of the cell. However, because most biological structures are similar in density to the cytoplasm, various methods have been developed to increase contrast. These usually involve special lenses and filters that alter the light path within the microscope. Two common types of light microscopy are **phase contrast** and **differential interference contrast** (**DIC**) (Figure 16.62). Phase contrast often provides a penetrating image of the cell while the DIC method provides a three dimensional-like image of the cell. Light microscopy can be used in combination with various strains or dyes that further increase the contrast and highlight particular cellular structures.

Intracellular structures can be seen by fluorescence microscopy in both fixed and live samples

Fluorescence microscopy makes use of the light emitted by certain dyes or proteins when they are illuminated with light of a shorter wavelength (known as the excitation wavelength, Figure 16.63a). The characteristic excitation and emission wavelengths depend upon the chemical structure of the fluorescent molecule, and scientists often use several different fluorescent molecules concomitantly, to highlight multiple cellular structures at once. A variety of approaches have been developed to visualize specific cellular components using fluorescent molecules. These fall into two main categories: the use of fluorescent dyes or antibodies conjugated to fluorophores that are added to the cells, and the expression of fluorescent protein tags, such as **green fluorescent proteins** (**GFP**), fused to the protein of interest. The former approach is usually used on fixed samples, because cells often need to be permeabilized to allow the dyes or antibodies to penetrate the cell. The fluorescent protein fusion approach can be used with either live or fixed cells.

Figure 16.62 Phase contrast and differential interference contrast (DIC) images of metaphase chromosomes. (a) A phase image of metaphase chromosomes from a flattened endosperm cell of the African blood lily *Haemanthus katherinae*. (b) A DIC image of metaphase chromosomes from newt lung cells. Note that in the phase image the chromosomes have light and dark regions, while DIC creates a three-dimensional image, but the chromosomes look uniform in composition.
Figure 16.62a from Inoué S and Oldenbourg R (1998); figure 16.62b from Conly Rieder's website, http://www.wadsworth.org/bms/SCBlinks/web_mit2/res_mit.htm.

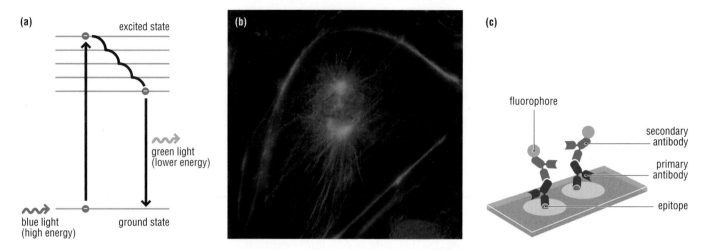

Figure 16.63 Fluorescence microscopy. (a) Each fluorophore or fluorescent protein has a specific excitation wavelength and emission wavelength. When the fluorophore is illuminated with high energy light at the excitation wavelength, an electron moves to an excited state. On its return to the ground state, energy is released in the form of light at the emission wavelength. In fluorescence microscopy, the specimen is placed on a slide on the microscope stage and is illuminated at the excitation wavelength. The light emitted by the fluorescent dye or protein is then emitted and can be visualized through the microscope's eyepiece or captured by a specialized digital camera. (b) Multiple fluorescent dyes or fluorophores can be used at once. In the image shown, the DNA is shown in blue, microtubules in green and actin filaments are in red. These structures can be distinguished from one another because they are labeled with fluorescent reagents that have different excitation and emission wavelengths. (c) Indirect immunofluorescence. To detect proteins for which antibodies are available, the cells are placed on a slide, permeabilized and treated with a the primary antibody that recognizes the protein of interest (in red). The sample is then treated with a secondary antibody that is conjugated to a fluorophore and which recognizes the primary antibody. Thus, on exposure to the appropriate excitation wavelength, the fluorophore associated with a secondary antibody will reveal the location of the protein of interest. This method is called indirect immunofluorescence because the fluorophore is associated indirectly with the protein.
Figure 16.63b from the Wadsworth center website, http://www.wadsworth.org.

A common example of a fluorescent dye is DAPI, which associates with double-stranded DNA (Figure 16.63b). Other fluorescent dyes bind to different subcellular structures, such as actin filaments, membranes or mitochondria. However, not every cellular protein or structure can be stained with a fluorescent chemical, and instead specific antibodies are used. A fluorescent signal that emanates from the antibody indicates the location of the target protein. We encountered the use of antibodies to label target proteins for western blot analysis in Section 16.10. The method commonly used for detecting proteins with antibodies within the cell is called **indirect immunofluorescence**, and it involves fixing and permeabilizing the cell, treating it with an antibody against the protein of interest, and then treating the sample with a fluorescently tagged secondary antibody that binds to the first antibody. (For example, if the primary antibody was generated in mouse, the secondary antibody could be a fluorescently tagged antibody against mouse immunoglobulin generated in goat, see Figure 16.63c.)

Since a highly specific antibody against the protein of interest is not always available, scientists sometimes create fusion proteins (as described in Section 16.10) between the protein and an epitope for which there are good antibodies. The fusion protein is then expressed in the cell so that its location can be detected by indirect immunofluorescence. While this is a powerful method, a potential pitfall is that the fusion protein may sometimes localize to places where the endogenous protein does not usually reside. Moreover, the process of fixation and permeabilization can sometimes alter cellular structures.

While significant information about the localization of a molecule can be obtained by immunofluorescence of fixed cells, it is often useful to be able to monitor the localization of a molecule for a period of time in live cells. The use of the jellyfish **green fluorescent protein (GFP)** (Figure 16.64a), as well as other naturally fluorescent proteins that emit and fluoresce at different wavelengths (Figure 16.64b), has greatly facilitated these studies. The localization of proteins tagged with these fluorescent proteins can be followed over time by recording a series of images (Figure 16.64c). In these types of experiment, researchers generate a gene fusion between their protein of interest and a fluorescent protein, such as GFP, and then express them *in vivo*. As is the case with immunofluorescence, these cells are then illuminated on the microscope stage with the appropriate excitation wavelength and images are acquired using a sensitive digital camera.

Co-localization between two proteins can be detected by FRET

When several different proteins or other components are detected by distinct dyes, antibodies, or tags, we can determine if all of these components are found in the same location, a situation referred to as **co-localization**. While this is a good indication that the proteins are in the same cellular compartment, it does not necessarily mean that they are in the exact same spot (for example, that they are part of the same complex). FRET, which we described in Section 16.11, can also be applied to determine the distance between molecules within the cell. To do so, two proteins are expressed as a fusion to a FRET-compatible pair of fluorescent proteins, such as yellow fluorescent protein (YFP) and cyan fluorescent protein (CFP) (Figure 16.65). If the two proteins are close to each other (namely 10–100 Å), the energy generated by the excitation of the CFP will be transferred to the YFP, which will get excited and emit light that is detectable by microscopy. If the proteins are further away from each other, no such energy transfer will occur. Using FRET, it is also possible to sort out the organization of a multiprotein complex. For example, the nuclear pore complex consists of dozen of proteins. To determine how these proteins are organized relative to each other, researchers have measured the distances between pairs of proteins (Figure 16.65) and created a map of interactions among the individual components of the nuclear pore.

The movement of molecules can be followed using photobleaching

Photobleaching occurs when a fluorophore or a fluorescent protein is exposed to light of sufficiently high intensity that it destroys the ability of the fluorescent molecues to fluoresce. This can be a problem when taking repeated images of a cell, such as when acquiring time-lapse images, since the resulting photobleaching causes the fluorescence of a fluorescent protein to diminish over time.

(a)

absorbs blue and ultraviolet light

emits green light

(b)

Figure 16.64 Fluorescent proteins. (a) The structure of green fluorescent protein (GFP). PDB 1EMA. (b) A series of fluorescent proteins, engineered from naturally occurring proteins, that emit at different wavelengths. (c) A series of images (fluorescence on the left, differential interference contrast on the right) taken of a *C. elegans* embryo synthesizing both YFP-lamin-, which outlines the nuclear envelope, seen exclusively in time point 9:00) and GFP-β-tubulin. which highlights the spindle, seen exclusively in time point 0:40. The times (minutes:seconds) are relative to anaphase onset. The times (minutes:seconds) are relative to anaphase onset.
Figure 16.64b from http://www.tsienlab.ucsd.edu/Images.htm; figure 16.64c Franz *et al.* (2005).

(a) Nup82-ECFP   Nup120-EYFP

Nup82-ECFP   Nup49-EYFP

(b) Nup82/Nup120   Nup82/Nup49

Figure 16.65 Analysis of yeast nuclear pore complex architecture by FRET. (a) Yeast cells co-expressing the nuclear pore complex protein Nup82 fused to enhanced cyan fluorescence protein (ECFP) and Nup120 fused to enhanced yellow fluorescent protein (EYFP) (top panel) or yeast cells expressing Nup82-ECFP and Nup49-EYFP (bottom panel) were excited with either the ECFP or EYFP excitation wave lengths and the emission of either protein was recorded. The fluorescence images are superimposed on DIC images of the same cells. (b) When these cells were analyzed by FRET, namely they were excited with only the ECFP wavelength and the EYFP emission was recorded, a FRET signal could be seen between Nup82 and Nup120 but not between Nup82 and Nup49, indicating that in the nuclear pore complex Nup120 is relatively close to Nup82 but that Nup49 is not. Adapted from Damelin and Silver (2002).

However, this irreversible loss of fluorescence can be exploited to assess the movement or diffusion of a fluorescent protein. In this approach, called **fluorescence recovery after photobleaching (FRAP)**, a small region of interest (smaller than the total area occupied by the fluorescent protein) is illuminated repeatedly with high intensity light to bleach the fluorescent protein in that region (Figure 16.66). If the protein is mobile, it will diffuse away from the bleached region and be replaced by undamaged fluorescent protein molecules, resulting in the recovery of fluorescence signal in that illuminated zone. However, if the protein is immobile, it will neither leave the bleached area nor be replaced by unaffected protein. Consequently, the bleached area will remain dark. By measuring the rate of recovery of fluorescent signal, we can estimate the rate of protein movement or diffusion in the cell.

A complementary method is **fluorescence loss in photobleaching (FLIP)**, where a small region is repeatedly bleached. This method is useful, for example, when a fluorescent protein occupies two distinct cellular compartments and the investigator wants to determine if the proteins can exchange freely between the two compartments. If there is an exchange, repeated photobleaching of an area within one pool will lead to loss of fluorescence in the other pool. If, however, there is no exchange, the protein in the compartment that is not being photobleached will retain its fluorescence.

Electron microscopy produces images at very high magnification

The resolution of light microscopy images is limited by the wavelength of light. The use of electrons rather than light allows for significantly better resolution and thus can be viewed at higher levels of magnification. Samples must generally be

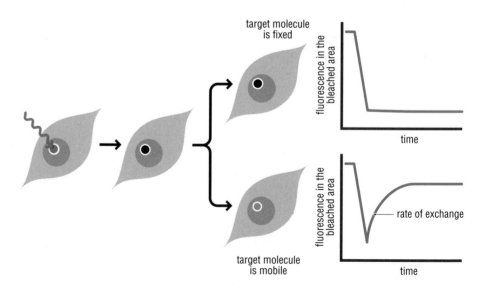

Figure 16.66 Fluorescence recovery after photobleaching (FRAP). The cells in the diagram are expressing a fluorescent protein (for example GFP) in the area indicated in pink. In the FRAP method, a small area (shown in red) is rapidly bleached such that the fluorescent proteins in this area lose the ability to fluoresce. The recovery of fluorescence in this region is monitored over time. If the fluorescent proteins are not mobile (i.e. fixed to their location and not free to diffuse), there will be no fluorescence recovery in the bleached area. If, however, the fluorescent proteins are mobile, the fluorescence in the bleached area will recover, and the rate of recovery will be indicative of the protein's mobility.

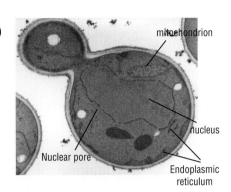

Figure 16.67 Electron microscopy. Budding yeast were examined by (a) scanning electron microscopy (SEM) and (b) transmission electron microscopy (TEM). Note that SEM is a powerful method for examining shape while TEM reveals intracellular structures.

Figure 16.67a from www.micron.ac.uk/organisms/sacch.html; figure 16.67b from www.medresearch.utoronto.ca/mil_transmission.html.

fixed, either by embedding in a resin or by flash freezing. Subcellular structures are often visualized by exposing the sample to chemicals that differentially bind to or stain components of the cell. Transmission electron microscopy (TEM) operates on the same basic principles as the light microscope, but electrons rather than light pass through the sample. In this case, the sample has to be cut into thin slices to allow the electrons to pass through. Objects whose dimensions are on the order of a few Ångstroms (10^{-10} m) can be detected, such that details can be visualized at near atomic levels. In a second type of electron microscopy, scanning electron microscope (SEM), the scattering of an electron beam from a surface is monitored. The advantage of this method is the large depth of field, which allows much of the sample to be in focus at one time. This method mostly reveals the shape of the object of interest. Yeast cells imaged by both types of electron microscopy are shown in Figure 16.67.

Electron tomography is a method for generating a three-dimensional image using electron microscopy. The images for electron tomography are collected while tilting the specimen by small increments around an axis. The images are then aligned to give a three-dimensional view of the object. This method is well suited for studying cellular structures (Figure 16.68).

Atomic force microscopy can reveal the contour of a cell surface or molecule

Atomic force microscopy (AFM) is a method designed to characterize and analyze the surface area of a specimen at very high resolution. It utilizes a fine tip, situated at the end of a flexible cantilever, which tracks very closely to the specimen's surface (Figure 16.69a). The tip rises and falls as the cantilever moves across the specimen's surface. The surface either attracts or repels the tip due to various forces acting between the surface atoms and those of the tip, and changes in the tip's position are recorded by a laser that is reflected off the cantilever. The result is a topological representation of the specimen, which not only provides a detailed image of the specimen's shape but also allows precise measurements of the specimen's dimensions. AFM is used to analyze the surface of polymers and semiconductors, but also of complex biological molecules and structures, such as proteins bound to fragments of DNA, the topography of membranes and membrane bound proteins, and the structure of mitotic chromosomes, as shown in Figure 16.69b.

Figure 16.68 Electron tomography. An electron tomograph of a fission yeast cell, in which the three-dimensional structure of mitochondria was reconstructed (in yellow). The image also shows microtubules in green, and the reconstruction reveals that mitochondria align with microtubules. The cell surface is shown in green and the nucleus in purple.

Adapted from Hoog *et al.* (2007).

(a)

(b)

Figure 16.69 Atomic force microscopy (AFM). (a) General principles of AFM. The tip, which is attached to a cantilever, is moved across the surface of a sample while the laser is used to measure the relative displacement of the tip from the surface. See text for details. (b) Structure of a metaphase chromosome as determined by AFM. The chromosomes were isolated form the human cell line BALL-1. The profile (top right) is produced by calculating the average of the height in the region indicated by the box in the micrograph at left. The bottom left panel shows a three-dimensional representation of the chromosome, while the bottom right panel is a closer view of a part of the chromosome showing an aggregation of globular or fibrous structures.
Adapted from Ushiki and Hoshi (2008).

16.13 MOLECULAR STRUCTURE DETERMINATION

Much insight into biological processes has come from information on the three-dimensional structures of biological molecules. As will be discussed in this section, there are several techniques used for structure determination, though nearly all of the macromolecular structures depicted in this book were determined by a method known as **x-ray crystallography**. This technique can be used to define the structure of a molecule by examining the way in which a crystal of the molecule of interest scatters x-rays. As compared with other techniques, x-ray crystallography can produce the most accurate and detailed model of a macromolecule. A key limitation to this method, however, is that it can only be used for molecules or complexes that can be induced to form a crystal. Thus, other approaches, such as **nuclear magnetic resonance spectroscopy** and cryoelectron microscopy, are used to obtain structural information about biological molecules and complexes that crystallize less readily.

A macromolecular structure can be determined from the diffraction of x-rays

X-ray crystallography takes advantage of the fact that the way in which molecules scatter x-rays is determined by molecular shape. X-rays, like visible light, are part of the electromagnetic spectrum that also includes infrared radiation, ultraviolet light, and microwaves. The wavelength of x-rays used in the laboratory, around ~1.5 Å, is most appropriate for visualizing biological molecules because it is on the order of the distance between covalently bonded atoms (1.2–1.5 Å). There are fundamental technical barriers to constructing an x-ray microscope that could produce a magnified image of a molecule using x-rays as the 'light', so the technique of x-ray crystallography was developed instead.

Instead of scattering x-ray waves from a single molecule, x-ray crystallography uses scattering from a crystal. A crystal consists of molecules lined up in a repeating three-dimensional array (Figure 16.70). Proteins and nucleic acids can be induced to form a crystal by adjusting solution conditions to reduce the solubility of the molecule, by adding certain salts or organic solvents, for example. While most solution conditions that reduce protein solubility simply cause the protein to precipitate as an amorphous substance, certain conditions can cause the protein to form crystals. A trial-and-error approach is used to screen hundreds of conditions that can potentially induce a molecule to crystallize. It is important to note that the crystal is simply a means to an end, namely that of getting the molecule of interest in a form in which its atomic structure can be determined.

When a crystal is placed in an x-ray beam, the x-ray waves are scattered by the electrons of each atom in the molecule (Figure 16.71). We refer to this scattering as **diffraction**. When the scattering is recorded on an x-ray detector or a piece of film, a pattern of light and dark spots results, with the intensity of each spot determined by the three-dimensional structure of the crystallized molecule. It is from many pictures like this, taken from different angles, that the information about the molecular structure is extracted.

While information about the molecule's structure is contained in the intensity of each spot in the diffraction pattern, one more piece of information is needed to calculate the structure. Each diffraction spot results from the x-ray waves scattered from the molecule, which sum to yield a wave. Each wave has peaks and crests with characteristic amplitude – its maximum height – and relative phase, which gives a measure of how much the peaks (or crests) are shifted relative to the other waves scattered by the crystal. If the amplitudes and phases of all the scattered waves are known, it is possible to reconstruct mathematically the molecular shape of the molecule that has been crystallized. While the amplitude of the waves can be obtained directly from the intensity of each diffraction spot, additional approaches must be used to obtain the phase information.

Crystallography yields an electron density map that must be interpreted

The result of experiments to record x-ray diffraction intensities and obtain phase information is an experimental **electron density map** (Figure 16.71b), which shows the outline of all the visible atoms. The electrons surrounding each atom give each atom its shape, and the electron density map depicts the surface formed by the covalently joined atoms within a molecule. For nearly all protein and nucleic acid structures, the hydrogen atoms cannot be visualized, and crystal structures in the Protein Data Bank (and throughout this book) typically lack hydrogen atoms. The three-dimensional electron density map is calculated with a computer, and a model of the protein is then constructed to best fit the electron density map. The accuracy of the model will depend on the resolution of the data, which is a function of how far from the center of the image diffraction spots can be recorded (Figure 16.71a). A structure determined at 1.8 Å resolution shows

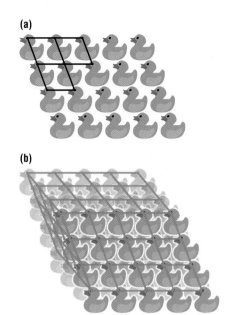

Figure 16.70 A crystal lattice. (a) A two-dimensional lattice of ducks. The parallelogram is the fundamental repeat unit of the crystal, called the unit cell (shown are three repeats of the unit). The lattice is generated by adding many identical copies of the unit cell, as indicated, along each of the two unit cell axes. Note that each unit cell contains a complete duck, although it is made up of bits and pieces of several intact ducks. (b) A three-dimensional crystal lattice contains many repeats of the unit cell, replicated along the three unit cell axes. (The use of ducks to illustrate crystal lattices is an established, albeit obscure, tradition in protein crystallography.)

Figure 16.71 X-ray crystal structure determination. (a) An x-ray source is used to irradiate a crystal with x-rays and the scattered x-rays are recorded on a detector. The intensity of each individual spot contains information about the structure of the molecule that has been crystallized. These intensities, together with additional information about the phase of the scattered wave, can be used in computer programs to calculate an electron density map, which is then used to construct a model of the macromolecules. The protein structure shown is a ribbons depiction of TATA-binding protein (TBP). (b) An electron density map (blue lattice) of Ubp8, a subunit of the yeast SAGA deubiquitinating module. The molecular structure that was constructed to fit the electron density map is shown in green (carbon), blue (nitrogen) and red (oxygen). The ribbons model on the right shows four proteins in the complex: Ubp8 (green), Sgf11 (pink), Sus1 (blue) and Sgf73 (salmon); orange spheres are bound zinc atoms. PDBs 1TBP, 3MHH.
Figure 16.71b courtesey of CW.

the positions of atoms with high accuracy, with all side chains clearly defined, whereas the side chains in a structure determined at 3.5 Å resolution are not well determined. Experimental approach 2.1 described the application of x-ray crystallography to determine the structure of the DNA double helix.

The model for the protein (or nucleic acid) and its bound ligands is stored in a data file as a list of atoms along with their corresponding positions in space, which are given by x, y, and z coordinates. A single macromolecular structure may have several thousand atoms. The coordinates for protein and nucleic acid structures, including those depicted throughout this book, can be obtained from the Protein Data Bank website and displayed on a personal computer.

Nuclear magnetic resonance spectroscopy gives structural information on macromolecules in solution

Protein and RNA structures can be determined in solution by **nuclear magnetic resonance spectroscopy**, commonly referred to as **NMR**. The distinct advantage of this technique is that it does not require a crystal; only a concentrated solution of soluble protein. In addition to yielding information on the three-dimensional structure of a macromolecule, it is also possible to study structural fluctuations – known as protein dynamics – by NMR. For this reason, NMR spectroscopy can provide information on proteins that complements the static picture of a protein structure provided by x-ray crystallography. Although NMR was originally best suited to determining structures of small proteins, recent advances have made it possible to extend the technique to the study of proteins containing hundreds of amino acids.

NMR takes advantage of the fact that certain atomic nuclei behave as small magnets and can be aligned in a magnetic field. Any nucleus with an odd number of nuclear particles – protons plus neutrons – has a property known as spin that causes it to behave like a small magnet. Thus, hydrogen atoms, ^1H, have spin, as do the isotopes of carbon and nitrogen, ^{13}C and ^{15}N. When placed in a magnetic field, these nuclei behave like small gyroscopes, with their axes precessing around the direction of the magnetic field. If a pulse of electromagnetic radiation at radio frequencies is applied in a direction perpendicular to the magnetic field, the spinning nuclei are realigned in the perpendicular direction. Once the pulse is removed, the nuclei gradually return to their original orientations along the magnetic field, emitting energy as they do so. The energy emitted by individual nuclei and the rate at which they return to their original orientation are influenced by neighboring atoms in the molecule. By measuring these parameters, one can derive information about the bonding arrangement of the atoms with spin and their relationship to their neighbors. By applying a sequence of radiofrequency pulses of varying duration and frequency, different properties of individual atoms can be probed. NMR spectroscopists have developed numerous pulse sequences that give information on interatomic distances and bond angles.

Figure 16.72 Structure determined by NMR spectroscopy. Shown is a set of superimposed models for the structure of a bromodomain of the transcriptional co-activator CBP (CREB binding protein) as determined by solution NMR spectroscopy. The models shown are the ten best fits to the distance and angle restraints derived from the NMR measurements. The regions shown in green, orange and red are the same in all models and their structure fixed, whereas the other regions are more mobile. (PDB code 1JSP).

In contrast with x-ray crystallography, which yields a set of atomic coordinates that constitute the best fit to an x-ray data set, the result of an NMR structure determination is a set of constraints on the interatomic distances and bond angles in a polypeptide chain. Together with the amino acid sequence of the protein, one can construct a model that provides a best fit to these constraints. It is generally possible to construct a number of different models that are consistent with the NMR data yet differ from one another in some way. Figure 16.72 shows an example of the set of models that provide a best fit to an NMR data set on a particular protein. Portions of the protein that are well ordered and have many constraints are nearly identical in the different models, while regions that are either mobile or have few distance constraints vary in the different models.

Cryoelectron microscopy can be used to image large macromolecular assemblies

Some macromolecular complexes are so large and flexible that their structures cannot be determined either by NMR spectroscopy or by x-ray crystallography. In these cases, it is possible to use electron microscopy to determine their structures. A homogeneous preparation of the complex of interest is laid down on a special grid and flash-frozen. An electron micrograph is then recorded, yielding an image of the randomly oriented particles (Figure 16.73a). The image of each individual particle represents a view down a particular axis: a projection of the structure along that axis. Thousands of these projections are collected, combined, and averaged, yielding an image of the particle with all its contours, in three dimensions (Figure 16.73b). A structure determined by this method typically does not provide sufficient detail to distinguish individual atoms or residues. However, it provides the outlines of the complex to which is often possible to fit models of individual proteins and nucleic acid structures (Figure 16.73b).

Figure 16.73 Cryoelectron microscopy: single molecule reconstruction. (a) The arrows symbolize the electron beam impinging on the specimen, which contains many randomly oriented molecules embedded in a thin layer of ice. The resulting images show the different projections of the particles. (b) Structure of the *E. coli* signal recognition particle bound to the ribosome. Ribosomal subunits are shown in yellow (30S) and blue (50S), SRP is shown in red. The surface depiction shows the structure obtained by cryo electron microscopy. Crystal structures of the individual subunits are superimposed on the cryoelectron microscopy structure.
Figure 16.73a from Frank (2002); figure 16.73b from Schaffitzel *et al.* (2006).

16.14 OBTAINING AND ANALYZING A COMPLETE GENOME SEQUENCE

The availability of complete genome sequences, a prerequisite for systematic genome-wide studies, relies equally on rapid methods of DNA sequencing and bioinformatic analyses of whole genome sequences. The first genome to be completely decoded, that of *Haemophilus influenzae*, was completed in 1995. Fifteen years later, at the time of writing this book, more than 1000 completed genomes have been reported. The pace of sequence generation continues to accelerate, driven by advances in DNA sequencing as well as in the methods for their analysis summarized in this section.

There are two general approaches for generating libraries of genome fragments for sequencing

How can one obtain the sequence of a very large genome-sized piece of DNA, when only about 500–700 bp of sequence can be generated from one primer using the standard sequencing approaches described in Section 16.7? Historically, two approaches were used to obtain whole genome sequences, including the initial human genome sequence. Both strategies ultimately are based on obtaining complete DNA libraries of the genomic sequence. In the one strategy, known as 'shotgun' sequencing, the genomic DNA is randomly sheared into small fragments (approximately 1 kb), cloned into plasmids that are propagated in *E. coli*, and then sequenced (Figure 16.74a). Since the genomic DNA is fragmented in a random manner, this approach generates many DNA sequences that overlap partially with other fragments. By comparing the sequences of overlapping regions between many different clones, the complete sequence of the genome can be assembled from these many fragments with the aid of a computer.

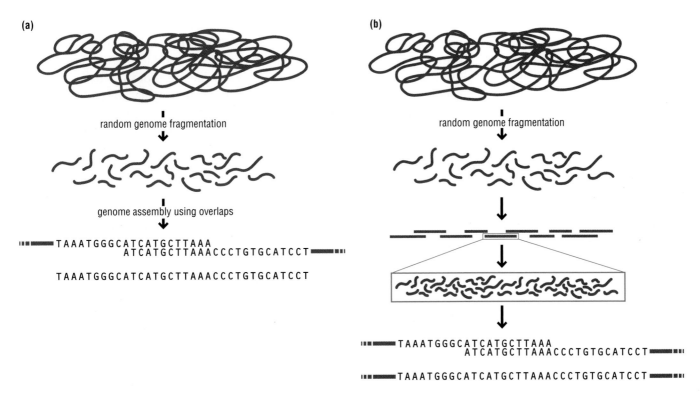

Figure 16.74 Assembly of a genome sequence from the sequences of multiple overlapping fragments. (a) The sequence of a genome can be obtained by shearing the DNA into many random fragments, and then obtaining the sequences of these small fragments. The genome sequence can then be assembled using computer programs that detect overlaps between these fragments. The sequencing of such random fragments is termed 'shotgun sequencing'. (b) The genome can also first be fragmented into larger pieces, each of which is then subjected to shotgun sequencing.

The second strategy involves an additional step: first a library of large DNA fragments (100–3000 kb) is made from the genome and propagated, for example, as 100–3000 kb fragments in yeast artificial chromosomes (YACs) or as 150–350 kb fragments in bacterial artificial chromosomes (BACs). These are then mapped to assemble a complete set for each chromosome. The sequence of each large segment is then subjected to shotgun sequencing, after which the entire genome sequence is assembled from the sequences of the DNA fragments (Figure 16.74b). Since the large fragments that are sequenced were mapped to the whole genome, the assembly into a contiguous genome sequence is easier.

One technical issue regarding the sequencing of small clones is that it is impossible to sequence some regions of the genome because high copy number plasmids containing those fragments are toxic to *E. coli*. Since bacteria containing these subclones do not survive, these regions are not represented in the genome library. Thus the sequences of these regions must be obtained in other ways, such as by the more labor-intensive approach of sequencing low copy number plasmids.

Genome sequencing requires oversequencing

In order to obtain a complete genome sequence, it is necessary to sequence many more base pairs than the total length of the genome. There are two reasons for this. First, since DNA inserts are cloned and sequenced randomly, many different overlapping, and sometimes duplicate, fragments must be sequenced to ensure that all regions of the genome have been sequenced at least once. Theoretically, clones corresponding to eight genome equivalents need to be sequenced to ensure coverage of 99.9% of the genome. Thus, for example, to obtain a nearly complete sequence of the 4.6 million nucleotide (4.6 Mb) *E. coli* chromosome using shotgun sequencing, it is necessary to sequence about 37 Mb of DNA (eight genome equivalents). Such higher coverage sequencing has the additional benefit of increasing the accuracy of the final sequence. This is because an error in the sequence of a single fragment can be recognized and corrected when sequences are obtained from independent clones corresponding to the same region.

The second reason that oversequencing is needed to generate a complete genome sequence is that enough random overlapping DNA segments must be sequenced so that the genome can be assembled by aligning overlapping sequences. The presence of repetitive DNA in the sequence complicates the assembly of short sequence fragments because repetitive sequences from non-contiguous portions of the genome may align with each other inappropriately. These issues are, of course, substantially greater with whole genome shotgun assemblies than with assembly of smaller clones (like BAC clones, for example). However, increasingly sophisticated assembly algorithms, and the inclusion of linkage data (giving the physical distance between two sequences), are facilitating assembly of regions containing repetitive DNA sequences.

DNA sequences can be obtained by multiple methods

For many years, the primary method used in whole genome sequencing was chain-termination sequencing using dideoxynucleotides as chain terminators, as described in Section 16.7. Originally this DNA sequencing method was performed by hand and analyzed by gel electrophoresis. Later machines were developed that automated the sequencing and detection process.

A single high-throughput automated chain-termination sequencing machine can process about 1000 reactions each day, generating about 0.5 Mb of total sequence (each sequencing reaction, or 'read', provides about 500–700 bp of sequence). Using this approach, sequencing eight genome equivalents of the 4.6 Mb *E. coli* genome (37 Mb) would require about 75 sequencing runs. To sequence the human genome (3×10^9 bp), about 70 million 500 bp chain-termination sequencing reactions were needed to generate 35×10^9 bp (ten genome equivalents) of sequence. This immense effort required 100 automatic sequencing machines generating 1000 sequencing reactions per day for about two years.

Recent advances in sequencing technology have greatly increased the speed of DNA sequencing, while at the same time reducing the cost. As described in Chapter 15, these new methods are collectively called deep sequencing, or NextGen sequencing because of the new technologies that these methods employ. NextGen sequencing methods share certain features, but most importantly, they are 'massively parallel' methods, meaning that huge numbers of unique DNA molecules can be sequenced simultaneously. In addition, these methods do not involve the propagation of clones in *E. coli*, eliminating the bias against sequences that are deleterious to the host. Rather, in current NextGen methods, DNA for sequencing is amplified by PCR directly from genomic DNA and then sequenced. Methods are currently being developed that will make it possible to sequence individual single DNA molecules directly, eliminating the need for PCR and thereby avoiding the errors that can be introduced during amplification.

While the pace of innovation in sequencing technology exceeds the pace of text-book writing, we will describe the principles behind two NextGen sequencing methods called **454 sequencing** and **Solexa sequencing**, named after the first commercial instruments. In the 454 NextGen sequencing method (Figure 16.75),

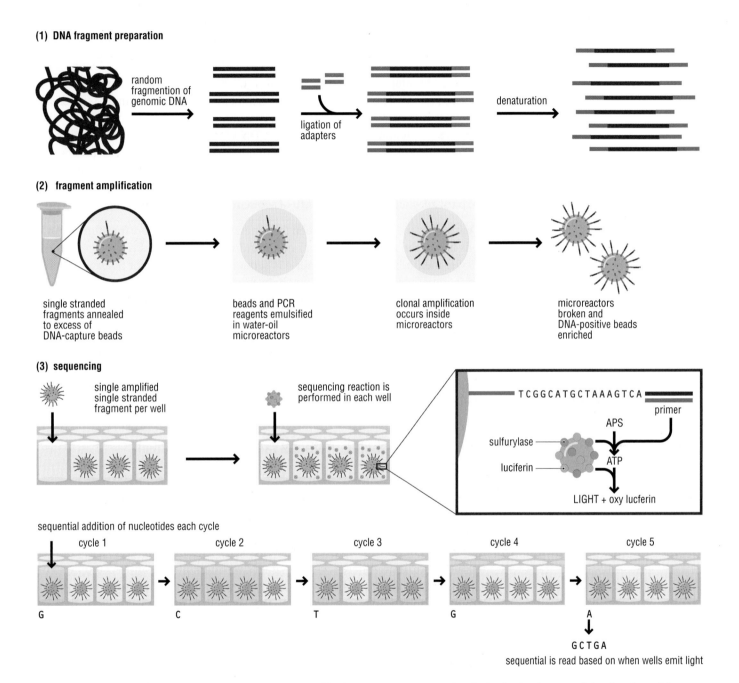

Figure 16.75 454 sequencing. (1) DNA fragment preparation. The DNA fragments to be sequenced are broken by shearing or restriction digestion and the adapters A and B ligated to the ends of the fragments. (2) Fragment amplification by emulsion PCR. Single-strand fragments are then captured in oil droplets containing PCR reactants and a bead with complementary oligonucleotides. Millions of copies of fragment are generated by multiple rounds of PCR regions. (3) Sequencing. The DNA-containing beads are distributed into individual wells for sequencing during which nucleotides flow in a precise order across the wells. When a nucleotide is added, inorganic PPi is released which reacts with adenylyl sulfate in the presence of sulfurylase to yield ATP, which reacts with luciferin in the presence of luciferase to emit light that it detected by a special camera. It is the parallel sequencing of the many identical fragments in a well that yields the detectable signal.
Figure 16.75, step 3 from Mardis (2008); figure 16.75, step 4 from Roche.

oligonucleotide adapter segments are ligated to the ends of the DNA fragments that are to be sequenced. Single DNA fragments containing the adapter are then trapped in an oil emulsion and individually amplified by PCR. In the sequencing step, each nucleotide is added to the reaction chamber in turn and nucleotide addition is scored by a light signal that results from each round of nucleotide addition. Hundreds of thousands of individual sequences can be sequenced in parallel using this method, with a single run yielding about 100 Mb of sequence information.

In the Solexa method (Figure 16.76), as with 454 sequencing, specific oligonucleotide adapter segments are added to the ends of the DNA fragments to be sequenced. The DNA strands are then separated and the single strands are allowed to hybridize to a surface coated with immobilized oligonucleotides that are complementary to the adapter. The DNA fragments bound to the surface are then amplified by PCR. During sequencing a camera monitors incorporation of fluorescently labeled nucleotides, with A, T, G, and C bases each producing a unique signal. In this method, tens of millions of individual sequences are scored in parallel, yielding up to 1300 Mb of sequence information in a single run.

One important difference between these new methods and more traditional methods is that the read length obtained for individual sequences is shorter with the newer methods: about 250 bp for 454 sequencing and 40 bp for Solexa sequencing compared with 500–700 bp for chain-termination capillary sequencing. Recall that, in whole genome shotgun sequencing, the genome sequence is assembled by aligning overlapping regions among individual sequence reads. The shorter reads obtained with NextGen sequencing compound the problem of assembly. One strategy to counter this problem is to increase the depth of coverage, that is, to sequence 20–25 genome equivalents rather than six to eight genome equivalents; in 2008, 454 sequencing was used to determine the genome sequence of a single human individual in several months.

NextGen sequencing methods facilitate new kinds of experiments

The ability to quickly generate large amounts of DNA sequencing information has prompted development of the kinds of experiments that were not feasible with previous approaches. NextGen sequencing can be used to monitor gene expression in particular cells or tissues because, with this method, it is possible to sequence many different cDNAs generated from the total mRNA complement. The same approach can be used to identify non-coding RNAs. NextGen sequencing can also be used to identify the DNA sequences bound by proteins that are isolated in chromatin immunoprecipitation experiments (see Sections 15.2 and 16.11 as well as Experimental approach 3.2).

In an application to human genetics, NextGen sequencing is a powerful means of screening for sequence variation in candidate disease genes that have been identified by any of a variety of methods. NextGen sequencing can also be used to rapidly re-sequence entire genomes and thereby identify naturally occurring genetic variation. In such applications, the original genome sequence is used as a template to guide alignment of short sequence reads into a complete genome, greatly reducing the difficulties in assembly. As sequencing becomes even cheaper, it is likely that whole genome sequencing of large numbers of individuals will become a powerful means of identifying disease-associated mutations.

Figure 16.76 Solexa sequencing. (1) DNA fragment preparation. The DNA fragments to be sequenced are broken by shearing or restriction digestion and adapters ligated to the ends of the fragments. (2) Fragment capture and amplification. The fragments are attached to a surface decorated with a dense lawn of primers complementary to the adapters. The individual fragments are then amplified by unlabeled nucleotides using a nearby primer to form a double-strand molecular bridge. Finally, the DNAs are denatured to yield clusters of identical fragments that are sequenced in parallel. (3) Sequencing. A specific primer and four 3′ blocked nucleotides, each with a different fluorescent label, are added. After washing away the unincorporated nucleotides, laser excitation and optical detection identify the added nucleotide at each cluster position. The 3′ blocks are removed and the cycle is repeated. From Mardis (2008).

NextGen sequencing also facilitates metagenomics, the study of the genomes of the mixtures of organisms that inhabit a particular niche, for example in the soil or in the human gut. Since many of these organisms cannot be grown in laboratory cultures, the very small amounts of DNA needed for NextGen sequencing, as compared with older sequencing methods, make it possible to sequence and identify genomic DNA from these organisms. Further, the ability to rapidly sequence very large amounts of DNA makes it easy to identify organisms that may be present in only small amounts in mixed populations.

➲ We discuss metagenomics in Chapter 15.

Genome annotation assigns meaning to sequence information

Having obtained and assembled the sequence of a new genome, how it is analyzed? As we discussed in Chapter 15, a genome must be annotated to make sense of the many millions of base pairs of DNA sequence that are obtained. There are two levels of annotation. On one level, annotation involves identifying coding sequences, the intron and exon structure of genes, transcriptional start sites and polyadenylation sites, genes encoding functional RNAs, elements of chromosome structure such as origins of replication or centromeres, and the location of transposable elements or endogenous viral genomes. The next level of annotation is the identification of the biological function of annotated genes or non-coding sequences, which can often be inferred from sequence similarities to genes whose functions are known.

How are genes that encode proteins identified? A number of different features of the DNA sequence must be examined to determine whether a given segment of the chromosome is likely to encode a protein. Each protein-coding gene should contain an initiation codon and a stop codon. Protein-coding genes are generally identified based on the fact that stop codons are found by chance at a fairly high frequency in non-coding parts of the genome but are excluded in the coding frame from coding sequences. Since the appropriate reading frame is not known, the nucleotide sequence must be translated in all three possible reading frames for each DNA strand, making a total of six potential reading frames. In the case of eukaryotic genes, introns and splice junctions must also be identified and excluded when translating the protein sequence. There are limitations to this approach, for example it is difficult to distinguish small protein genes from non-functional small open reading frames that occur at random.

Additional information can be included to assist in identifying gene sequences that serve important functions. Information about protein-coding genes can be obtained by constructing and sequencing cDNA libraries for the organism of interest, thereby making it possible to identify the transcription unit, including intron and exon junctions, transcription start sites and polyadenylation sites.

Computational approaches can be used to identify other DNA sequences that are conserved between different species, suggesting that they may contain functionally important genes. Intragenic, non-coding sequences diverge rapidly between species, so areas of high conservation that do not encode obvious open reading frames often encode functional elements such as non-coding RNAs or structural components of the chromosome, and these can be identified computationally by comparing divergent sequenced genomes.

Sequence alignment programs are used to assign functions to genes

The second critical step in annotating a genome sequence is to infer the functions of the specific sequences or genes. One of the most common and powerful methods used to infer the function of an unknown gene is to compare its sequence to that of genes with known functions. The comparison of two or more sequences is called an alignment, in which the nucleotide or amino acid at each position in one sequence is compared with the nucleotide or amino acid at the corresponding position in another sequence. Each position is then assigned a score, based on whether the residues are identical, similar, or different. The comparisons can be made across the sequences of whole proteins or RNAs. Such alignments make it possible to identify other proteins or RNA molecules that are homologs with a common evolutionary origin, and therefore are likely to carry out similar functions. The region of sequence similarity may encompass an entire coding sequence or it may be restricted to one or more protein domains (for example, kinase domains, zinc finger DNA-binding domains, or carbohydrate binding domains). The presence of an identifiable domain whose function in other proteins is known often provides clues about the function of an uncharacterized protein.

The most widely used method for sequence alignments is a suite of related programs known collectively as **BLAST** (Basic Local Alignment Sequence Tool), which are available from the National Center for Biotechnology Information (NCBI) website and on many other web servers worldwide. The BLAST programs carry out pair-wise alignments – that is, all of the comparisons are done between a pair of sequences – between the query sequence and each sequence in a database. BLAST algorithms can align either DNA sequences (BLAST) or protein sequences (BLASTP). Each alignment is given a score that provides a measure of how closely two sequences match one another. The score depends on the amount of similarity between the two sequences and the length of the similar sequence: a high score indicates a high degree of sequence identity. Gaps may be introduced into one or another protein sequence to maximize the alignment of two sequences, although a penalty (namely a decrease in the score) is assigned for each gap. The extent of sequence similarity is also scored statistically with the E value, which is a measure of the probability that the sequence identity observed could occur by chance, that is, it indicates the likelihood that the query sequence and target sequence are in fact unrelated.

An example of the output from a BLASTP search is given in Figure 16.77, which shows the results of a search for proteins similar in sequence to the Hermes transposase from the house fly, *Musca domestica* (the query sequence). The upper portion of the results page (Figure 16.77a) shows putative conserved domains present in the query protein. The actual sequence alignment comparing the query domains to the consensus sequences of the conserved domains is shown in Figure 16.77b. The alignment of Hermes to proteins in the searched database is shown graphically in the lower portion of Figure 16.77a and the names and scores of the related proteins are shown below. For the top hit, the E value is 0, indicating that there is no possibility that this protein is unrelated to the query sequence. The query sequence is already present in the database and the search has simply found itself (Figure 16.77c); the likelihood of this is $<10^{-200,}$ which in the BLAST display is reported as 0. The third protein in the list is a truncated form of Hermes used to determine its crystal structure (PDB code 2BW3). The next best matches

Figure 16.77 BLASTP search using Hermes transposase as a query. See text for details. (a) Upper portion of the BLASTP results page shows the presence within the Hermes transposase of putative conserved domains indicated in colored balloons, along with the domain names. In the middle portion, each line represents a protein that is similar in sequence across the indicated portion of Hermes. The lower portion shows a ranked table of related proteins, each with a score that reflects the degree of sequence similarity and length of polypeptide over which the sequence similarity extends, and the E value, which gives the probability that this alignment represents homology compared with the chance that the alignment is obtained by chance. For top hits, the E value is 0. (b) Conserved domains found by BLASTP. Alignment of the Hermes transposase query sequence to the consensus sequences of the conserved domains is shown. (c) Alignment of Hermes transposase with the closely-related Hobo transposase from *Drosophila*.

are orthologous proteins from *Drosophila*, the *hobo* transposases (Figure 16.77c), followed by a number of high-scoring hits from other insects.

While BLAST searches are the most commonly used algorithms for conducting sequence alignments, other programs are also used, and the identification of homologs, domains, and gene functions is becoming more and more sophisticated with programs that incorporate multiple parameters including, for example, structural information.

16.15 MAPPING HUMAN DISEASE GENES

A critical benefit of having the complete and annotated human genome sequence is that this information has greatly accelerated the identification of new disease genes. The strategies used to map and identify these genes are summarized in this section.

Genetic markers are of great benefit in mapping genes

The principles of mapping human disease genes are the same as those used to map a locus in any organism: markers, which are positions of sequence variation that can be used to distinguish DNA segments from different individuals, are identified all over the genome. The frequency at which each of these markers is found associated with the trait of interest is then scored. The higher the frequency of association, the more likely it is that the gene of interest is close to the marker in the chromosome. The lower the frequency of association, the more likely it is that the gene and the marker are far apart. The variation in association frequency is due to recombination between homologous chromosomes, which can separate two positions on a chromosome and therefore can separate a marker from a trait, or from another marker. The probability of recombination occurring between the two sites is proportional to the distance between them: the farther apart they are, the more likely it is that recombination will occur.

The availability of the human genome sequence has greatly facilitated identification of genetic markers. Using these markers, human geneticists can map the location of genes for various traits, including those related to disease. It is important to appreciate that these markers do not themselves correspond to mutations that cause disease, but rather are genetically linked, that is, their position on the chromosome is in close proximity to the disease-causing mutation. In effect, the markers act in an analogous fashion to street names (which identify the geographical region in which a particular house may be found).

If we consider two loci and assume that they are inherited independently of one another (that is, they are unlinked genetically), then after a large amount of time alleles of these two loci in a given population will be at 'equilibrium' in terms of how frequently they are inherited together. Being in equilibrium means that all combinations of alleles for the two loci will be found in the population at frequencies that would be expected if the alleles were inherited independently of one another. Two alleles that are found together more frequently in the population than expected based on random assortment are said to be in linkage disequilibrium; the term 'disequilibrium' simply means that two alleles are not randomly associated as they would be if they were unlinked. Linkage disequilibrium can, of course, be assessed for any two given loci. Linkage disequilibrium can also be assessed for one marker (an allele at a particular locus) and the presence of disease in an individual. If a marker is genetically linked to a disease-causing mutation, people with the disease will co-inherit the linked marker more frequently than if the mutation were inherited independently of the disease. By genotyping individuals with and without a particular disease, we can use this logic to deduce where the disease-causing gene is located in the genome.

Sequence tagged sites make useful markers in the human genome

We discuss above that markers are often used to map a disease-associated gene. The DNA markers that are used to follow disease genes represent specific sequences that can be uniquely identified in the human genome. There are two general classes of markers that are used to locate genes, **sequence tagged sites (STSs)** and single nucleotide polymorphisms (SNPs). STS is a general term referring to a unique location in the human genome for which sequence variants occur in the population. One frequently used STS is a microsatellite repeat (more formally called a **simple sequence length polymorphism (SSLP)**). As discussed in Chapter 15, microsatellites represent known regions of the genome where tracts of simple repeats exist, for example, CACACACA. The number of repeats of the unit (in this case, CA) at a given microsatellite locus varies between individuals in the human population. Since the locations of these variable sequence repeats are known, primer sequences that can be used for PCR amplification across the repeat region are listed in the genome database UniSTS, maintained and updated by the NCBI (http://www.ncbi.nlm.nih.gov/genome/sts/). To determine the length of a particular microsatellite repeat for an individual, a PCR reaction is performed and the size of the product analyzed on a gel (see Figure 15.22b). The length of the product in a particular person's genome can be used to follow inheritance of that region of the chromosome when carrying out the **linkage studies** described below. Microsatellites are easy to identify by PCR and standard electrophoresis, and are thus used for many kinds of genetic studies.

SNPs are the most frequent markers in the genome

For studies in which a large number of markers must be used, SNPs provide a more powerful tool. As described in Chapter 15, SNPs are sequences in the genome at which a single base differs in some individuals. For example, the reference sequence of the human genome has the sequence TTGACGTCA at a particular position, yet in some people it may read TTGACCTCA as a result of a mutation at that specific site. SNP loci typically have two alleles (out of the possible four), corresponding to the two variant bases. To be useful, a SNP should be present at a frequency of 1% in the population. Below that, so few people would have base variations at that locus that the utility of the SNP would be too limited for most studies. As of May 2009, there were 79 485 538 verified SNPs. The genomic location of SNPs are known and are listed in the dbSNP database at the NCBI website (http://www.ncbi.nlm.nih.gov/SNP/index.html).

Known SNPs can be used as genetic markers to map disease genes, as described in Chapter 15 and summarized below. To do this, the sequence at SNP loci must be determined for individuals with and without the disease. Because the technology for SNP detection is relatively new, many different methods have been developed over the past several years and continue to be developed. Currently, the method used most commonly to detect SNPs is high-density microarray analysis (referred to as SNP Chips). The SNP Chip has arrayed on it thousands of different 25-base oligonucleotides, corresponding to locations of individual SNPs. Because a SNP has two possible sequences at a given position (as described above), both variants are represented on the SNP Chip. To analyze the SNPs present in an individual, genomic DNA is fragmented and amplified by PCR to generate enough material for the analysis. The small fragments generated by PCR amplification are then denatured and hybridized to the SNP Chip under conditions in which only a perfect match will hybridize to a given spot; consequently, hybridization to a spot on the chip uniquely identifies which of two possible sequences is present (Figure 16.78).

Newer methods for SNP detection are on the horizon. With development of rapid, very high throughput sequencing (see Section 16.14), it may become more efficient to simply sequence all SNPs, or even the whole genome of an individual to identify, at one stroke, the entire complement of SNPs and other sequence variants.

Linkage analysis and association studies are two major approaches to disease gene identification

In Chapter 15 we describe the principles of gene mapping used to identify disease genes in dogs and in humans. Here we will briefly describe the methods currently used to carry out this mapping. Look at Chapter 15 for a discussion of some of the nuances of gene–gene and gene–environment interactions that can complicate the interpretation of mapping studies. To identify specific genes that are associated with human disease, one needs to start with a population of individuals who have the disease and compare their DNA to a matched population of people who do not.

DNA is collected from each individual and each person's DNA is analyzed for the presence of specific markers – either STSs or SNPs – as described in the previous section. We will discuss two different approaches to mapping. **Linkage analysis** can be used to trace specific disease alleles within families. By contrast, **association studies** are used to identify disease genes by studying large groups of unrelated individuals. The method chosen to identify a disease-related gene will depend in part on the population to be studied, and the availability of samples from people who have the disease of interest. If the disease is known to segregate in families, and DNA is available from many individuals in several such families, then linkage analysis will be an effective way to identify a potential disease gene. If DNA samples from related individuals are not available, association studies on large numbers of unrelated people sharing the disease can be employed.

Linkage analysis can determine disease genes in families

Linkage analysis is the method of choice for identifying single gene mendelian traits, if DNA from individuals in large families is available for study. Linkage analysis examines the way in which alleles are passed from one generation to the next, and uses the pattern of inheritance to determine the association of a marker with disease. The principle is the same as discussed above: if a marker is close to a disease gene on a chromosome, the marker and phenotype will typically be inherited

(a)

(b)

Figure 16.78 SNP Chips allow identification of single nucleotide polymorphisms (SNPs) in DNA samples. SNP Chips are used to detect single base changes at specific loci in the human genome. (a) A chip is synthesized that has many thousand different 25-nucleotide single-stranded DNA molecules arrayed in a known pattern. DNA from one individual is then hybridized to the chip under conditions where a single base mismatch will *not* hybridize. (b) The DNA is labeled with a fluorescent probe so that hybridization to a particular spot on the chip can be detected. More than 500 000 different SNP sites from a single individual can be examined simultaneously.

together. By contrast, if a marker is relatively distant, there is an increased likelihood of recombination, reducing the chance of co-inheritance.

As closely related family members inherit large blocks of the genome from a common ancestor, fewer recombinations will have occurred that might separate a disease allele from a given marker within a family, as compared with the number of recombinations that would occur in the population in general. For this reason, a marker that is reasonably close to the disease gene has a higher probability of being present in all family members who have the disease (Figure 16.79) but absent in family members without the disease. Since members of the family

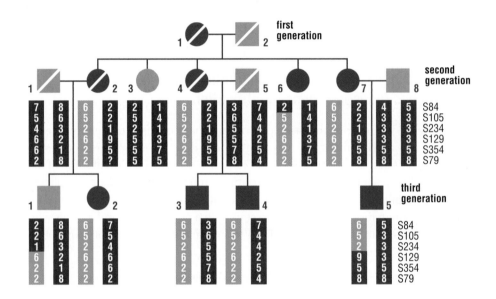

Figure 16.79 Linkage analysis follows the association of markers with a trait in families. This pedigree shows the inheritance of a dominantly inherited skin disorder, Darier–White disease. The pedigree is drawn using standard nomenclature: square = male; circle = female; blue = affected individual; slash = deceased. In the first generation there are two grandparents labeled 1 and 2, the second generation has eight people labeled 1–8, and the third generation has five people labeled 1-5. Each individual is referred to by the generation number followed by the individual number. For example the grandfather would be I-2. The two vertical bars below each individual represent the two homologous chromosomes inherited from each parent. Each chromosome contains six STS markers that segregate with the disease (shown on far right: S84, S105, S234, S129, S353, S279). By examining how a particular version of each marker correlated with disease, it was possible to map the location of the disease gene between markers S84 and S129. There are different versions of each STS marker that can be found at each position; these are numbered 1-9. For example individual II-1 has the 7 and 8 versions of the S84 marker, while individual II-2 has the 6 and 2 version of this marker. Affected individuals II-1, II-4 II-7, III-2, III-3, and III-4 all contain the linked set of alleles, '6,5,2,6,6,2' (this block is colored orange). But where is the gene responsible for the disease? To find out, one looks at chromosomes that have recombined and sees which of the "partial sets" of disease markers still cause the disease. Affected individual II-6 carries the '2' version of the D84 marker but carries the '5,2,6,6,2' version of the rest of the region. Thus, the disease gene must lie below the S84 marker. Individual III-1 has only the bottom half of the disease markers and does not have the disease, while individual III-5 has only the top half and does have the disease. Since individual III-1 *does not* have the disease, the disease gene must be above the S129 marker. Together with the information from individual II-6 and III-1, we can conclude that the disease gene must lie between S84 and S129.
Adapted from Strachan and Read (2004).

without disease can be assumed to have inherited the non-disease allele, information from these unaffected individuals is informative about the potential linkage of a given marker to the disease locus. Given that the pedigree of all individuals in a family is known, one can accurately follow the co-inheritance of a given marker and the disease-related gene to determine whether the association of a given marker with the disease is higher than expected if the marker and disease locus were unlinked.

To determine how tightly a marker is linked to a particular trait, statistical analysis is used. Below we describe the **LOD score** (log of the odds score), which is the score assigned to a marker to describe the likelihood that a marker is linked to a trait. The larger the family that can be analyzed, the more powerful is the approach of linkage analysis to identifying disease genes. For example, for a dominant mutation, an affected parent and affected child share half their genomes. Simply by genotyping these two, approximately half the genome (that which is found in the child but not the parent) can be excluded as being a candidate for containing the disease locus. A second affected sibling in the same family would exclude a further half of the remaining half, narrowing the potential location of the disease gene to one fourth of the genome. In this way, each child of an affected parent narrows the genomic region in which the potential disease gene may lie. Such family transmission information makes linkage analysis very powerful, with the result that fewer people and fewer markers are needed to identify a gene than for association studies: typically 500–6000 markers will be used in linkage studies compared with 500 000–1 000 000 markers that may be needed for some association studies.

Association studies can link markers to disease in unrelated individuals

Often large families that exhibit a strong disease phenotype are not available for study. In this case, association studies can be a powerful way to identify genes associated with disease. Association studies have also become the method of choice for identifying complex traits, diseases, or phenotypes that are influenced by many different genes as well as by the environment. Each of these genes may have only a small effect when considered in isolation, but their concerted action can be significant. Association studies use the same conceptual framework as linkage analysis, and statistical analysis is the tool used to determine the probability that a given marker associates with a trait. The association of markers in a collection of people who do have the trait is compared with the markers in a collection of people who do have the trait. If a marker is enriched in the pool of people with the trait, it may be linked to the trait (Figure 16.80). Because the people used in the studies are unrelated, however, the likelihood of recombination occurring between a given marker and the disease is higher and so a much larger number of people and many more markers must be used to find disease-associated loci by this method.

In Chapter 15 we describe an association study that was done with dogs, which correlated the size of a dog with the allele of the insulin-like growth factor gene *IGF1* it carries: large dogs have an allele that is associated with a higher level of expression of this growth factor (Figure 15.28). In that study, a large number of dogs within a single breed were first genotyped using microsatellites, and a

Figure 16.80 Association studies compare marker frequencies in populations that do or do not have a particular trait. frequency of markers corresponding with a trait.
(a) The allele frequency for any single nucleotide polymorphism (SNP) in the population reflects the distribution that developed over evolutionary time as described in Chapter 15. (b) Given the initial known allele frequency in the general population, the frequency in the disease population (cases) and in a control population is compared. In the example shown, the pink allele is more frequent than the gray allele in the cases, while the control population reflects the known allele distribution. This particular SNP (shown in pink) is thus more frequent in the disease population and may be linked to the disease. This kind of analysis is done for all 500 000 SNPs to find the one or several that are most highly enriched in the disease population. GWAS, genome-wide association studies.

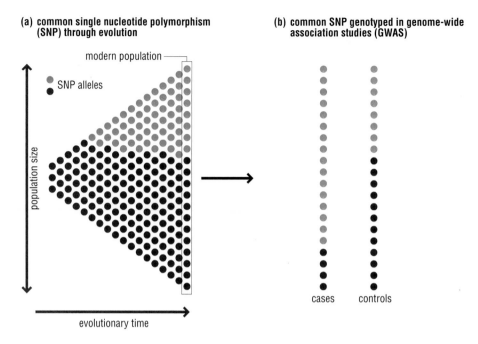

(a) common single nucleotide polymorphism (SNP) through evolution

modern population

SNP alleles

population size

evolutionary time

(b) common SNP genotyped in genome-wide association studies (GWAS)

cases   controls

chromosomal region was identified that associated with the size of the dogs. (That is, the study identified a particular chromosomal region whose sequence varied in a characteristic and conserved way between dogs of different sizes: small dogs exhibited particular STSs in this region, while larger dog breeds exhibited different STSs.) This region was further analyzed in a second set of experiments using a SNP Chip and those SNPs that were most frequently associated with large dogs, but not small dogs, were identified. The SNPs that correlated with size were near the *IGF1* gene. When the gene itself was sequenced, a polymorphism was found that did indeed correlate with size.

Similar to the approach used to identify the genetic difference between large and small dogs, association studies or linkage analysis can be used in an initial screen to locate the large region where a gene might be, and then followed up with additional screens to further narrow the region of interest (Figure 16.81).

In the dog example, a single dominant locus was found. However, in many such studies multiple independent regions of the genome are identified, each of which may contribute to only a small percentage of the phenotype. Thus, to assign a region of the genome as being linked to a phenotype, robust statistical tools are required.

The odds ratio is a statistical method to establish linkage

As seen from the examples above, identifying a marker that is associated with a genetic trait using both linkage analysis and association studies requires statistical analysis to assess the significance of the observed linkage. This analysis will mathematically define the probability of the marker being associated with the trait, telling us whether the association is significant or whether it is merely a chance occurrence. For statistical analysis to be meaningful in practice, one needs to examine the association of a large number of markers with a trait of interest in an individual, and then do this independently for many individuals. For example, if a SNP Chip is used, DNA from each person with or without the disease would be

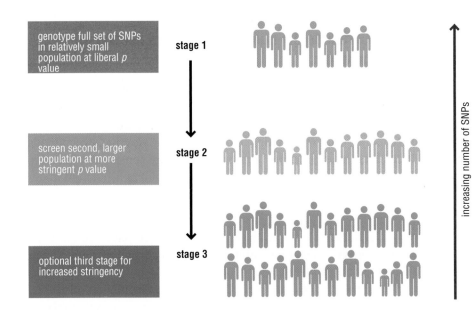

Figure 16.81 Association studies are often carried out in sequential steps. In the first stage, low-density coverage in a small population may be used in an association study to find an initial region where a disease gene might lie. Alternatively, linkage analysis may be initially used. (2) Once candidate regions of the genome are identified, a large number of SNPs that are specific to the potential genomic region or regions are interrogated. (3) A third step may be used if the region identified has too many genes to be sure which one is right or if several potential regions were identified and the investigators need more statistics to determine which region is real.

prepared and hybridized to the chip, and the presence or absence of every one of over 500,000 SNPs would then be determined for each person. This large amount of data must then be analyzed to determine if one of the SNP markers is always present in those people with the disease.

To evaluate the data, a statistical calculation is done to determine the probability of a marker being linked to the disease. This so-called **odds ratio** is the ratio of the probability of a given marker being linked to a given disease to the probability of it not being linked. The odds ratio is expressed on a $\log_{10}$ scale and called LOD score. The higher the LOD score, the more likely a marker is associated with the disease. For linkage analysis, a LOD score above 3 is typically considered to represent linkage, while markers with a score near 1 are unlinked. Scores higher than 5 represent tight linkage (Figure 16.82).

Figure 16.82 LOD scores are a quantitative description of the likelihood that a marker is associated with a trait. A whole genome scan was carried out using SNP Chips to identify genes associated with prostate cancer. The LOD score representing the likelihood of each SNP association with disease is plotted on the y-axis. The location of each of the SNPs assayed in the human genome is plotted on the x-axis. This analysis revealed that there were a set of SNPs located on chromosome 8 at the map position 8q24 that were tightly associated with disease. From Freedman *et al.* (2006).

Haplotype blocks reduce the number of markers needed to find a gene association

As SNPs from many individuals were characterized, it was discovered that SNPs are found across many unrelated individuals in characteristic sets called haplotype blocks. Of course, in related individuals, SNPs that are linked will tend to be inherited together; what was striking about the finding was that, even among unrelated individuals in large populations, some SNPs are consistently linked to other nearby SNPs. The characteristic size of haplotype blocks is different in different populations. For example, the blocks are about 7.3 kb in African populations and 16.3 kb in Europeans.

The presence of haplotype blocks primarily reflects the fact that there has been insufficient evolutionary time for recombination to bring the genome into equilibrium; in this regard, the difference in block size between different populations reflects the different number of generations (and therefore recombination events) since the founding of the population. Another contributing factor is that recombination occurs predominantly at certain characteristic spots in the genome, and so markers between those recombination spots tend to be inherited together (see Figure 15.23). Practically, this has enormous consequences for gene mapping: if you follow one marker in this block, the others will almost always move with it.

We can think of the human genome as a series of independent blocks. Knowing about these haplotype blocks in the genome simplifies both linkage analysis and association studies because one needs only to look at one defining marker within the haplotype block to identify it. This is the basis of the human **HapMap**, which defines particular SNPs that best identify each haplotype block, hence making mapping easier. There is no need to use ten markers that are already known to associate with each other within a block, as just one informative SNP is sufficient to define a specific haplotype block. These particularly informative SNPs, each of which acts to tag a specific known haplotype block, are known as tagSNPs. Using tagSNPs in an association study greatly reduces the number of markers that need to be analyzed.

⊜ We discuss the concept of haplotypes in Chapter 15.

✲ SUMMARY

Advances in our understanding of molecular biology and genomics rely on ongoing experimentation, drawing on the kinds of tools and techniques we have introduced in this chapter. It is not only our understanding of molecular biology as a discipline that continues to move forward, but also the techniques used to probe genome function. As existing techniques become enhanced, and new methodologies become available, new and exciting approaches can be adopted to probe areas of interest, illuminating our understanding in new – and often unexpected – ways.

The principles of genome function presented throughout this book will doubtless continue to be refined as investigators around the world continue their research; so, too, will the tools and techniques used to explore these principles. It is this dynamic interplay between what we know and how we know what we know that makes molecular biology fascinating, continually uncovering new questions that remain to be explored.

An overview of molecular biology methods

- Molecular biology methods range from genome-wide analysis to the dissection of interactions between isolated, single molecules.

- A key aspect of molecular biology is the study of processes both *in vivo* and *in vitro*.

- Most molecular components and processes are studied in a relatively small range of model organisms.

- Studies of the molecular components of a wide variety of genetically tractable organisms have proved Jacques Monod's prediction: 'What is true for *E. coli* is true for the elephant'.

Cell culture

- Both bacterial and eukaryotic cells can be grown in culture.

- Individual molecules can be purified from cultures of these relatively homogenous populations of cells.

- Embryonic stem cells and iPS stem cells can be grown in culture and manipulated to give rise to differentiated cells.

Gene and genome manipulation

- The ability to manipulate genes and genomes is central to molecular biology.

- DNA cloning, whereby specific DNA regions are isolated and propagated, is often carried out in bacteria.

- A library of clones may represent total genomic DNA, expressed genes only, or just a subset of expressed genes.

- The targeted introduction of mutations through DNA engineering (mutagenesis) allows for the generation of variant DNAs, RNAs, and proteins for study.

- Transposons and recombination are commonly used to introduce specific genes or mutations into the genome of an organism.

- Increasing use is being made of siRNAs and small molecules for genome manipulation and analysis.

- The availability of genome sequences has provided new tools to assess transcription and translation on a genome-wide basis.

The isolation and characterization of biological molecules

- The isolation and characterization of molecules *in vitro* is a key method in molecular biology.

- A variety of methods can be used to both identify and quantify molecules that are present in a sample.

- Many such detection techniques are spectroscopic in nature, and exploit the characteristic ways in which molecules interact with electromagnetic radiation.

- Chromatography techniques exploit the different physical characteristics of molecules to separate them from one another.

- Various approaches exist for determining the sequence (primary structure) of a biological molecule.

- Specific DNA sequences and RNA molecules can be detected using hybridization-based methods.

- Proteins can be detected by virtue of their specific interaction with antibodies.

- Genome sequence information and DNA manipulation allow for the addition of tags to specific proteins and RNA, facilitating their isolation without prior knowledge of their function.

- Our understanding of protein or RNA function can be enhanced by identifying components that interact with the molecule.

Cellular imaging and molecular structure determination

- Advances in imaging methods, including the use of reporter proteins such as GFP, have provided important new tools for studying cells in both live and fixed samples.

- Imaging can reveal information about both molecular localization and movement within the cellular environment.

- The three-dimensional structure of molecules is most widely determined by x-ray crystallography.

- The study of human molecular biology has been facilitated and expanded by the availability of complete human genome sequences.

Genome sequencing

- Advances in genome sequencing have enabled us to obtain genome sequence information with increased speed and accuracy, and at lower cost.

➤ ONLINE RESOURCES FOR GENOMICS AND MODEL ORGANISMS

- National Institutes for Health
  http://www.genome.gov/
  http://www.ncbi.nlm.nih.gov/Genomes/

- US Department of Energy Office of Science
  http://genomics.energy.gov

- Major genomics centers
  J. Craig Venter Institute: http://www.jcvi.org/
  The Sanger Institute: http://www.sanger.ac.uk/
  Broad Institute of MIT and Harvard: http://www.broadinstitute.org/
  The Genome Center at Washington University: http://genome.wustl.edu/
  Baylor College of Medicine, Human Genome Sequencing Center: http://www.hgsc.bcm.tmc.edu/

- Microbes
  http://www.microbes.info/resources/General_Microbiology/Databases/Genetic/index.htm

- *Escherichia coli*
  http://www.ecolicommunity.org/

- *Saccharomyces cerevisiae* (budding yeast)
  http://www.yeastgenome.org/

- *Schizosaccharomyces pombe* (fission yeast)
  http://www.genedb.org/genedb/pombe/

- *Dictyostelium discoideum*
  http://dictybase.org/

- *Caenorhabditis elegans* (worm)
  http://wormbase.org/
  http://www.ncbi.nlm.nih.gov/projects/genome/guide/nematode/

- *Drosophila* (fruit fly)
  http://flybase.org/
  http://www.ncbi.nlm.nih.gov/projects/genome/guide/fly/

- Arabidopsis
  http://www.arabidopsis.org/

- Xenopus (frog)
  http://www.ncbi.nlm.nih.gov/genome/guide/frog/

- *Danio rerio* (zebrafish)
  http://zfin.org/cgi-bin/webdriver?MIval=aa-ZDB_home.apg
  http://www.ncbi.nlm.nih.gov/projects/genome/guide/zebrafish/

- Mouse
  http://www.informatics.jax.org/

- Human
  http://www.genome.ucsc.edu/

- The Human Genome Project
  http://www.genome.gov/10001772

GLOSSARY

(p)ppGpp ('magic spot'): a pentaphosphate guanosine analog that plays an important role in signaling the stringent response in bacterial systems.

–10: promoter sequence located ten nucleotides to the 5′ of the start of transcription and comprising part of the sequence directing bacterial RNA polymerase to bind and initiate transcription; also called the Pribnow box.

3′ CCA tail: the three terminal nucleotides of all tRNAs and the site of attachment for amino acids.

3′ splice site: consensus dinucleotide (AG) sequence found at the 3′ end of the intron that signals the intron–3′ exon boundary.

–35: promoter sequence located 35 nucleotides to the 5′ side of the start of transcription and comprising part of the sequence directing bacterial RNA polymerase to bind and initiate transcription.

454 sequencing: massively parallel sequencing technique in which individual sequence reads can be up to 250 bp.

4E-BP: class of protein that bind directly to the cap-binding protein eIF-4E thus preventing functional interactions with eIF-4G.

5′ cap: a special structure at the 5′ end of eukaryotic mRNAs containing a 7-methylguanosine residue attached via an unusual 5′–5′ triphosphate linkage.

5′ capping: process by which an mRNA is modified with the cap structure.

5′ splice site: consensus dinucleotide (GU) sequence found at the 5′ end of the intron that signals the 5′ exon–intron boundary.

β protein: subunit of the sliding clamp protein in bacteria; it dimerizes to generate a symmetrical ring structure with six similar domains.

A

AAA+ family (ATPases associated with a variety of cellular activities): family of proteins with a conserved region that includes the Walker A and Walker B ATPase motifs. Many replication proteins are members of this family, including the clamp loader and helicases. AAA+ family members are also involved in cellular activities other than DNA replication.

abasic site: site in DNA at which a base has been lost from a nucleotide residue.

abortive initiation: non-productive initiation of transcription leading to the production and release of short RNAs rather than elongation of the RNA transcript.

absorption spectrum: the amount of light absorbed by a sample as a function of wavelength.

acceptor stem: one of four helical elements in the tRNA that brings together the 5′ and the 3′ end of the molecule via base-pairing interactions.

accommodation: the step in tRNA selection that follows guanosine triphosphate (GTP) hydrolysis, which occurs when a tRNA is finally accepted by the ribosome and is thus committed to peptide bond formation.

acetylation: covalent addition of an acetyl group from acetyl-CoA to a nitrogen atom at either the N-terminus of a polypeptide chain or in a lysine side chain. The reaction is catalyzed by *N*-acetyltransferase. In the case of histone acetylation, acetyl groups are added to lysine residues in the histone tails.

acetyl transferase: enzyme that adds acetyl groups to side chains on proteins. A subgroup is the histone acetyl transferases (HATs).

acid: molecule that releases H⁺ into solution.

activate: in transcription, to increase the number of RNA transcripts of a given gene.

activation energy: the energy required to bring a species in a chemical reaction from the ground state to a state of higher free energy, from which it can transform spontaneously to another low-energy species.

activator: sequence-specific DNA-binding protein that leads to increased transcription.

active site: pocket on or near the surface of a macromolecule that promotes chemical catalysis when the appropriate ligand (substrate) binds.

actomyosin ring: The contractile ring involved in cytokinesis. This ring attaches to the cell membrane at the cleavage furrow and its contraction leads to the formation of a membrane that separates the cytoplasms of the daughter cells.

ADAR: Acronym for 'adenosine deaminase that acts on RNA'.

A-DNA: right-handed DNA double helix with 11 bp per turn; A-DNA has a deeper and narrower major groove than B-DNA.

affinity chromatography: purification of a molecule, for example, a protein, based on its ability to bind to a particular compound immobilized on a support.

alignment: comparison of each position in one sequence with the corresponding position in another sequence, or group of sequences.

alkylating agent: mutation-causing chemical that modifies bases in DNA with a methyl or ethyl group.

allele: different version of a gene, such as a wild-type or a mutant version.

allosteric effector: molecule that bind to a protein and induces a conformational change, thereby changing the activity of the protein.

alpha helix: an element of protein secondary structure in which a portion of the polypeptide chain forms a right-handed spiral, with the backbone NH group of every residue *i* forming a hydrogen bond with the C=O group of every residue *i*+4.

alpha satellite: highly repetitive DNA found at the centromere and elsewhere on human chromosomes. It consists of tandem repeats of a 171 bp sequence.

alternative lengthening of telomeres (ALT): break-induced replication (BIR) mechanism for the addition of telomeres to chromosome ends that have lost telomere sequences, usually because of the inactivation of telomerase.

alternative sigma (σ) factor: sigma factor that recognizes promoters of specialized sets of genes and is expressed or activated under specific conditions, when it associates with the core RNA polymerase and enables the specialized genes to be expressed.

Alu element: genomic sequence derived from 7SL RNA, the RNA component of the signal recognition particle that targets protein to membranes. There are more than one million copies of this sequence in the human genome.

amino acid: chemical building block of proteins containing an amino group, a carboxyl group, and one of 20 possible side chains.

amino acid residue: the atoms that remain after an amino acid is incorporated into a polypeptide chain.

aminoacyl site: position on the small and large ribosomal subunits of the ribosome where the aminoacyl-tRNA binds.

aminoacylation: process by which an amino acid is covalently attached to the terminal adenosine of a tRNA species.

aminoacyl-tRNA: a tRNA species carrying an esterified amino acid covalently bound to its terminal adenosine.

aminoacyl-tRNA synthetase: one of ~20 different ubiquitous enzymes, each of which is specific for one type of amino acid, which catalyzes the covalent attachment of an amino acid to all cognate tRNA species.

A-minor motif: a commonly seen structural motif in RNA, where adenosines found in distant regions interact via their minor-groove face with the minor groove of a neighboring helix, often with their 2′ OH groups. Various versions of the A-minor motif function similarly to bring together distant regions of the RNA.

amphipathic: having both polar and non-polar character. Synonym for amphiphilic.

amphipathic alpha helix: alpha helix with a hydrophilic side and a hydrophobic side.

amphitelic attachment: *See* **bi-orientation**.

anaphase: the stage of mitosis in which the sister chromatids are segregated by the mitotic spindle. In most species, anaphase is divided into anaphase A, the initial movement of chromosomes to the spindle poles, and anaphase B, the movement of spindle poles away from each other.

anaphase-promoting complex (APC): large, multisubunit ubiquitin-protein ligase (also known as an E3 enzyme) that catalyzes the attachment of ubiquitin to mitotic proteins such as M cyclin and securin, thus promoting their ubiquitin-dependent proteolysis. It is also called the cyclosome or APC/C.

anomer: cyclic sugar isomer with a characteristic orientation of a hydroxyl group relative to the carbon atom.

antibiotics: small molecules that kill or slow the growth of bacteria or fungi by targeting essential cellular processes.

anticodon: three nucleotides in the anticodon loop of the tRNA that are responsible for directly interacting with the mRNA and decoding the genetic information.

anticodon loop: the single-stranded loop region of the tRNA that presents the anticodon nucleotides to interact with the mRNA.

antiparallel: arrangement of two DNA strands with the 5′ to 3′ polarity of one strand opposite to that of the complementary strand.

antiparallel beta sheet: a beta sheet, often formed from contiguous regions of the polypeptide chain, in which each strand runs in the opposite direction from its immediate neighbors.

anti-Shine–Dalgarno sequence: polypyrimidine sequence at the 3′ end of the 16S rRNA that directly interacts by base-pairing with the upstream leader sequence of mRNAs to position the AUG start codon in the P site.

anti-sigma factor: protein that binds a sigma factor and inhibits its function.

anti-termination: promotion of elongation over termination.

apoptosis: programmed cell death using a special cell machinery that occurs in response to unrepaired DNA damage.

apurinic/apyrimidinic endonuclease (AP endonuclease): an enzyme involved in base excision repair that recognizes the absence of a base (an abasic site) in DNA.

architectural DNA-binding protein: protein whose role is to induce a significant bend in the DNA, thereby either promoting or inhibiting interactions between proteins bound on either side of the bend.

Argonaute: key protein component of complexes (RISC and miRNP) that bind small RNAs and mediate RNAi or miRNA-mediated repression.

association study: method used to find the DNA marker closely associated with a disease gene or other trait when DNA is available from a large number of unrelated people who have the disease and a large number who do not.

ATM (ataxia telangiectasia mutated): a regulator kinase of the DNA damage response of multicellular eukaryotes first identified in patients with ataxia telangiectasia that interacts with MRN, which is a senor of DNA damage.

Atomic force microscopy (AFM): a high-resolution technique for examining the surface of a specimen. It uses a nanometer-sized tip associated with a cantilever that scans the specimen surface and records changes in height that result from forces between the tip and surface features.

ATP-dependent nucleosome-remodeling complex: protein complex that uses the energy released by ATP hydrolysis to alter the positions of nucleosomes on the DNA.

ATR (ATM related): a regulator kinase of the DNA damage response of multicellular eukaryotes identified by its homology to ATM.

ATRIP (ATR interacting protein): a regulator of the eukaryotic DNA damage response that interacts with RPA bound to single-stranded DNA.

attenuation: mechanism for regulating transcription termination that exploits competition between ribosome binding and formation of a stem–loop structure in the transcribed RNA.

autocrine loop: when a secreted mitogen acts directly back on the cell that secreted it, increasing cell division of the secreting cell.

autonomous element: transposable element that encodes the transposase necessary for its mobility and also retains the sequences at which the transposase acts.

autoregulation: process by which a protein controls its own synthesis by activating or repressing transcription of the gene that encodes it.

autosome: a chromosome that is not a sex chromosome. Males and females have equal numbers inherited from the mother and the father.

B

backtrack: when an RNA polymerase enzyme stalls and then slides backwards along the template, causing the most recently added ribonucleotides to dissociate from the template.

bacteriophage: a virus that infects bacteria. Also referred to as phage.

base: molecule that reduces the concentration of H+ in solution.

base (in a nucleic acid): aromatic purine or pyrimidine group attached to the sugar of a nucleotide.

base excision repair: mechanism of DNA repair in which a nucleotide with a damaged or missing base is excised from the DNA and replaced with an undamaged one, using the undamaged DNA strand as template.

base stacking: the stacking of bases in parallel planes along a nucleic acid strand. Base stacking, which is stabilized by van der Waals interactions, can occur in double- or single-stranded DNA and RNA.

base triple: three interacting, coplanar bases typically consisting of a Watson–Crick base pair that hydrogen bonds with a third base via its remaining unpaired chemical groups.

B-DNA: right-handed DNA double helix that predominates in cells, with 10.5 bp per turn and a rise of 3.4 Å per base pair.

beta sheet: secondary-structure element composed of multiple beta strands that interact through backbone hydrogen bonding interactions.

beta strand: an individual single strand of extended polypeptide chain in a beta sheet.

binding curve: a curve in which the amount of complex formed is plotted as a function of the concentration of the other component.

bi-orientation: the correct binding of kinetochores of sister chromatids to microtubules emanating from opposite spindle poles. Also known as **amphitelic attachment**.

BLAST: Basic Local Alignment Sequence Tool, a sequence alignment algorithm available on the web (http://blast.ncbi.nlm.nih.gov/Blast.cgi) for aligning either DNA or protein, in which sequence comparisons are based on local, that is regional, rather than global, comparisons.

boundary elements: DNA sequences that specify the transition between euchromatin and heterochromatin. Moving a boundary element sequence moves the transition point of chromatin structure.

branch migration: a process by which Holliday junctions move along DNA, increasing or decreasing the region of heteroduplex.

branch-point nucleotide: the adenosine residue that is directly involved in the nucleophilic attack on the exon–intron boundary and that is thus located at the branch site of the intron lariat structure.

break-induced replication (BIR): replication from the site of a double-strand break to the end of the chromosome using a homologous chromosome as a template.

C

carbohydrate: organic molecule composed of one or more monosaccharides.

catalyze: to increase the rate of a chemical reaction by decreasing the activation energy.

catastrophe (in microtubules): sudden shrinkage that occurs when all the tubulin at the microtubule end is in a GDP-bound form.

Cdk-activating kinase (CAK): protein kinase that activates cyclin-dependent kinases by phosphorylating a threonine residue (Thr 160 in human Cdk2).

cDNA: complementary DNA that is synthesized from mature (i.e. spliced) mRNA template by reverse transcriptase.

cDNA library: collection of clones carrying sequences of spliced mRNAs expressed under specific conditions.

cell culture: the process by which cells are grown under controlled conditions.

cell cycle: the highly regulated sequence of events that take a cell from its formation to its own division. In eukaryotic cells it is divided into four main phases: G1, S, G2, and mitosis (M).

cell fractionation: the process of breaking open cells and separating distinct organelles.

cell line: a particular strain of eukaryotic cells that will grow continuously in culture.

centrifuge: an instrument that uses the centripetal force generated by rotation at high speed to cause cells or molecules suspended in liquid to sediment to the bottom of the tube.

centromere: a region of heterochromatin in each eukaryotic chromosome through which the chromosome becomes attached to the mitotic spindle, allowing daughter chromosomes to segregate correctly at each cell division.

centromere inactivation: an epigenetic process by which a centromere loses its ability to promote chromosome segregation, presumably due to changes in chromatin structure.

chain-termination dideoxy sequencing: a method for sequencing DNA in which dideoxynucleotides, which cannot prime further DNA synthesis during primer extension because they lack a 3′ OH, are present in limiting amounts such that termination occurs at some frequency at each nucleotide position. Reaction products are analyzed by display following denaturing electrophoresis.

chaperone: a protein that aids in the folding of another protein by preventing the unwanted association of the unfolded or partially folded forms of that protein with itself or with others.

checkpoints: regulatory pathways that monitor certain cellular processes and can arrest cell cycle progression at specific transition points in the cell cycle when these processes fail.

chiasma (*plural* chiasmata): X-shaped chromosomal structure that is seen in the microscope at sites of crossing-over between homologous chromosomes at the end of meiotic prophase.

Chi (χ) sequence: asymmetric, 8 bp DNA sequence that changes the activity of the bacterial RecBCD–protein complex.

chloroplast: organelle found in plant cells that houses the machinery for harvesting energy from sunlight for the production of sugars within the cell.

cholesterol: a lipid containing conjugated hydrocarbon rings that is a steroid.

chromatin: the complex of DNA bound by proteins and assembled into the higher-order structure known as a chromosome.

chromatin assembly factor (Caf1): the protein that binds the H3–H4 tetramer and facilitates its incorporation into nucleosomes. Caf1 interacts with the sliding clamp protein, PCNA, to facilitate histone loading.

chromatin immunoprecipitation (ChIP): a method for identifying DNA sequences bound to a particular protein by cross-linking the DNA to the protein, shearing the DNA to produce fragments, and then immunoprecipitating particular protein-DNA complexes with a specific antibody against the protein. The associated DNA is then analyzed by microarray hybridization or sequencing (ChIP-seq).

chromophore: small molecule bound to a protein that causes it to absorb visible light and therefore have a color.

chromosome: a structure composed of DNA complexed with specific proteins, which is the form in which the genetic traits encoded in DNA are maintained and transmitted from one generation to the next. Generally, organisms have multiple, independent linear chromosomes or a single, circular chromosome.

chromosome condensation: the structural changes that result in compaction of the chromosomes into smaller, more compact structures before cell division.

chromosome cycle: variation in chromosome condensation state during the cell cycle, which changes from relatively decondensed in interphase to highly condensed during mitosis.

chromosome painting: method used to detect one unique chromosome in the set of all metaphase chromosomes.

chromosome segregation: the distribution of chromosomes to daughter cells at mitosis or meiosis.

chromosome territory: the unique space occupied by a chromosome in the nucleus.

cis splicing: the standard form of splicing in which exons from the same RNA transcript are joined in a series of events coordinated most typically by the spliceosome.

clamp loader: five-subunit protein that loads the sliding clamp onto the DNA during DNA replication.

class 1 release factor: any of the protein factors responsible for recognizing the three stop codons in the A site of the ribosome and for facilitating the hydrolytic reaction that releases the growing peptide chain from the peptidyl-tRNA.

class 2 release factor: the GTPase protein factor involved in promoting different facets of the termination process in bacteria and eukaryotes.

Class I transposable element: an element whose transposition pathway contains an RNA intermediate.

Class II transposable element: an element whose transposition pathway contains only DNA intermediates.

cleavage furrow: during cytokinesis, the ingression of the cell membrane that is associated with the contractile ring and that eventually divides the cell in two.

closed complex: RNA polymerase bound to promoter DNA before the DNA strands are separated.

closed loop complex: complex of mRNA and eukaryotic translation initiation factors (minimally eIF4E, eIF4G, and PABP) thought to form a closed loop that is competent for translation initiation.

co-activator: protein or complex of proteins recruited to DNA by sequence-specific DNA-binding proteins that activates transcription.

coaxial stacking: end-to-end stacking of RNA helices commonly seen in folded RNA molecules.

coding joint: chromosomal fusion between various V, D, and J coding segments that occurs during RAG-mediated V(D)J recombination and results in the assembly of diverse antibody and antigen receptor genes in lymphocytes.

coding, non-template, or sense strand: DNA strand that bears the same sequence as the RNA.

codon: three consecutive nucleotides (i.e. a triplet) in an mRNA that correspond to an amino acid or stop and are decoded by tRNAs, release factors, and the ribosome.

coenzyme: a cofactor that is an organic or organometallic molecule that assists in enzyme catalysis.

cognate interaction: when the codon of the mRNA and the anticodon of the tRNA are perfectly matched – these pairing interactions involve Watson–Crick interactions at the first and second positions of the codon and either a Watson–Crick or 'wobble' at the third position (typically a G–U interaction).

cohesin: protein complex containing a heterodimer of structural maintenance of chromatin (SMC) proteins, a kleisin, and other subunits, that links sister chromatids together after S phase.

cohesion: association between sister chromatids that is established immediately after DNA replication and is dissolved during M phase.

coiled coil: pair of alpha helices that coil around one another and associate via hydrophobic interactions to form a dimerization interaction.

cointegrate: DNA product generated by the replicative transposition of some DNA-only transposons, in which the transposon has been replicated to produce two new copies that link the donor and target DNA.

co-localization: degree to which distinct signals obtained from two or more fluorescent signals, each emitting at a different wavelength, overlap in the same spatial position within cells examined by microscopy.

combinatorial control: the use of several different proteins to regulate a single gene.

comparative genomics: comparison of genomic sequences to gain insights into genes and genome function.

competitive inhibitor: species that competes with substrate for binding to the active site of an enzyme and thus inhibits catalytic activity (affecting the apparent K_M of the reaction).

complementary (of DNA): containing the base pair partner(s) predicted by the Watson–Crick base pairs.

compound transposon: bacterial transposon in which two insertion sequence (IS) elements flank a segment of DNA. Coordinated movement of the IS elements results in the movement of the interior segment.

condensation: compaction of chromosomes.

condensin: protein complex containing a heterodimer of structural maintenance of chromatin (SMC) proteins and other subunits that mediates chromosome condensation.

conformation: three-dimensional arrangement of the atoms in a molecule.

conservative site-specific recombination (CSSR): exchange of DNA at specific sites with short regions of homology. It can result in integration, excision, recombination, or inversion depending on whether the sites are on different DNA molecules or the same one, and on their relative orientations.

conserved synteny: the regions of each chromosome that are very similar in gene composition and order to regions in the other species.

contact inhibition: contact between cells that stimulates cell cycle arrest, causing cells to cease dividing.

contractile ring: ring of proteins, including contractile assemblies of actin and myosin, which forms at the site of cleavage in dividing animal cells. Its gradual contraction pinches the cell in two.

convergent evolution: evolution of a common structure or function in macromolecules or organisms that do not share a common ancestor.

cooperative binding: occurs when binding of one ligand alters the affinity for the other ligand.

co-protease: protein that stimulates the self-cleavage activity of another protein.

copy number variants (CNV): a DNA sequence variation where the copy number of some segments of DNA at a specific position in the genome differs between members of a species.

core enzyme: RNA polymerase enzyme that is capable of DNA-dependent RNA synthesis *in vitro* but is incapable of specific promoter recognition. Consists of β, β', and two α subunits in bacteria, and more than ten subunits in eukaryotes and archaea.

co-repressor: protein or complex of proteins recruited to DNA by sequence-specific DNA-binding proteins that represses transcription.

core promoter elements: short DNA sequences found within eukaryotic promoter regions that serve as binding sites for proteins in the pre-initiation complex. Also called promoter proximal elements.

covalent bond: a chemical bond that results from sharing of electrons between two atoms.

covariation: concerted changes in sequences observed throughout evolution that provide evidence for interaction between two elements.

CpG and CpXpG: nomenclature designating the sequence that reads 5′ to 3′ CG or C (any nucleotide) G. These are the sequences in mammalian DNA that most often contain methyl cytosine.

cryptic splice site: sequence elements that are similar to the canonical splice sites but do not define biologically utilized exons.

CTD: term used for both the carboxy (C)-terminus domain of the bacterial α subunit and the carboxyl (C)-terminus domain of the eukaryotic RNA polymerase II Rpb1 subunit. There is no structural similarity between the bacterial and eukaryotic CTDs.

C-terminus (carboxyl terminus, also called carboxy terminus): end of a polypeptide chain with an exposed carboxylic acid.

cullin: a protein found in a core subunit of some ubiquitin-protein ligases, including SCF and the APC.

cyclin: regulatory subunit that binds and activates a cyclin-dependent kinase (Cdk). Cyclin levels oscillate in the cell cycle and different cyclins are present at different cell cycle stages.

cyclin box: a stretch of about 100 amino acids that is conserved between cyclin proteins and that identifies these proteins as members of the cyclin family.

cyclin-dependent kinase (Cdk): protein kinase whose catalytic activity depends on an associated cyclin subunit. Cdks are key components of the cell cycle machinery.

cyclin kinase inhibitor (CKI): protein that interacts with Cdks or Cdk–cyclin complexes to block activity, usually during G1 or in response to inhibitory signals from the environment or damaged DNA.

cysteine desulfurase: an enzyme belonging to the family of transferases which transfer sulfur-containing groups from one species to another.

cytokinesis: process following the completion of chromosome segregation in mitosis by which the parent cell divides, giving rise to two daughter cells.

cytoplasm (meaning 'cell substance'): contents of the cell contained within the cell membrane and, in eukaryotic cells, outside of the nucleus.

D

deacetylase: enzyme that specifically removes acetyl groups from the side chains on proteins. Histone deacetylases (HDACs) represent a subgroup.

deamination: loss of an amino group. If this occurs on cytosine the base is now uracil; if it occurs on 5′-methyl cytosine the base is now thymine.

decatenation: separation and unlinking of the two DNA molecules that result from replication of a circular DNA molecule.

decoding: process of interpreting the genetic code by pairing aminoacyl-tRNAs with cognate codons; this process is facilitated by an elongation factor known as EFTu (bacteria) or eEF1A (eukaryotes).

degradosome: complex of proteins in bacteria containing the endonuclease RNase E, which initiates the degradation of many bacterial RNAs, as well as a helicase and a 3′ to 5′ exonuclease.

deletion: in relation to chromosomes, the loss of part of a chromosome or of a whole chromosome.

deoxyribonucleotide: deoxyribonucleoside with attached phosphates.

deoxyribose: five-carbon sugar with H in place of OH at the 2′ position of ribose that is a component of the nucleic acid DNA.

depurination: removal of a purine base from a nucleotide in DNA.

depyrimidation: removal of a pyrimidine base from a nucleotide in DNA.

Dicer: RNAse III-type enzyme that carries out the cleavage of double-stranded RNA to generate siRNAs.

differential interference contrast (DIC): illumination technique in microscopy that increases the contrast of the specimen. Sometimes also referred to as Nomarski microscopy, which is a particular type of DIC. This illumination highlights the edges of the specimen, generating an illusion of a three-dimensional image.

diffraction: scattering of waves (such as X-rays) in different directions when they encounter a group of atoms or molecules.

diploid: containing two sets of chromosomes.

dipole: separation of positive and negative charge.

directed gene conversion: targeting of gene conversion to a specific position by a double-strand break in DNA.

disaccharide: two sugars covalently linked by a glycosidic bond.

distributive: refers to a type of DNA replication in which the polymerase adds only a small number of nucleotides to the growing DNA strand before dissociating from the template. The template is fully copied by repeated rounds of polymerase association, elongation, dissociation, and reassociation.

disulfide bond: a covalent bond joining two sulfur atoms, which can form by a chemical reaction between the sulfhydryl groups of two cysteine side chains.

divergent evolution: evolution from a common ancestor.

D loop (DNA): a three-stranded DNA structure in which one segment of a duplex is displaced by pairing with another complementary strand.

D loop (RNA): single-stranded region of tRNA that contains conserved dihydrouridine residues and participates in critical interactions with the T-loop to stabilize the elbow region of the tRNA.

DnaB: replicative DNA in *Escherichia coli* comprising six identical subunits.

DNA-binding domain: protein domain that binds to DNA.

DNA-binding motif: protein fold that binds to DNA. It may be a discrete, globular domain or a portion of a protein domain.

DNA cloning: generation of multiple copies of a particular DNA fragment. Often refers to the insertion of a DNA region into a plasmid or virus, which can then be propagated and amplified.

DNA damage response: cellular response to extensive DNA damage, in which cell cycle progression is delayed and DNA repair is promoted.

DNA glycosylase: enzyme involved in base excision repair of DNA that removes damaged bases by hydrolyzing the linkage between the base and the sugar. Different glycosylases are specific for different types of damaged bases.

DNA helicase: protein that moves along DNA, unwinding the double helix as it goes. DNA helicases use the energy of ATP hydrolysis to move along the DNA. The helicases involved in DNA replication in bacteria and eukaryotes are six-subunit ring structures.

DNA ligase: enzyme that links the ends of two strands of DNA by catalyzing the formation of a phosphodiester bond between the 3′ hydroxyl group at the end of one fragment and the 5′ phosphate at the end of the other.

DNA methylases: another name for DNA methyltransferases.

DNA methylation: addition of a methyl group to a base in DNA. Most often cytosine is methylated, although adenine is methylated in some organisms.

DNA methyltransferases: enzymes that catalyze the addition of methyl groups to bases in DNA.

DNA microarray: also known as a chip, because of its resemblance to a computer chip, a microarray contains a series of microscopic dots, each containing oligonucleotide sequence that corresponds to a specific chromosomal locus or gene. The DNA (or cDNA) under study is labeled with a fluorophore and hybridized to the microarray. A special scanner is used to detect the fluorescence at positions in the array to which labeled DNA has hybridized.

DNA-only cut-and-paste transposable element: DNA-only transposable element that moves by DNA breakage and joining reactions in which the transposon is cut out of the donor site and then inserted into a new site.

DNA-only transposon: transposable element whose transposition involves DNA intermediates only.

DNA polymerase: enzyme that synthesizes new DNA by copying a single-stranded DNA template. The polymerase moves along the template and synthesizes a new strand of complementary DNA sequence by adding nucleotides, one at a time, to the 3′ OH end of the new strand.

DNA repair: biochemical process that reconstitutes an intact DNA molecule after DNA damage.

domain: compact unit of protein structure that is usually capable of folding stably as an independent entity in solution. Domains do not need to comprise a contiguous segment of peptide chain, although this is often the case.

dominant mutation: change in DNA sequence that has phenotypic consequences even in the presence of the wild-type gene.

donor site: the site of a transposable element before transposition.

double bond: two covalent bonds joining two atoms, the result of sharing two pairs of electrons.

double helix: the helix formed by two base-paired DNA strands that wind around one another in a helical fashion.

double-stranded RNA-binding domain (dsRBD): RNA-binding motif consisting of a fold containing alpha helices and beta sheets, which can contact the major and minor grooves of the RNA helix.

downstream: to the 3′ side of any given reference point, such as a transcription start site or a coding sequence.

duplication: in relation to chromosomes, a duplication of part of a chromosome.

dynamic instability: tendency of microtubules to switch between states of rapid growth and rapid shrinkage.

E

editing: process by which an aminoacyl-tRNA synthetase catalyzes the removal from the tRNA of amino acids that have been inappropriately loaded onto a given tRNA species.

editing site: site on the aminoacyl-tRNA synthetase that catalyzes a hydrolytic reaction on misactivated or aminoacylated amino acids.

Edman degradation: method for protein sequencing in which the N-terminal amino acid residue is removed by repeated cycles of degradation, each of which generates a derivatized amino acid that can be identified by comparison to known standards.

electron density map: depiction in three dimensions of the possible locations of all the electrons in a molecule, as is produced using the method of x-ray crystallography. The electron density map shows the outlines of the molecule, which can be used to construct a three-dimensional atomic model.

electronegative: partial negative charge on an atom that results from its power to attract electrons to itself.

electrophile: reagent that is attracted to centers of negative charge and can accept electrons to form a chemical bond.

electrophoretic mobility shift assay (EMSA): method for examining the interaction of a protein with a nucleic acid by comparing the mobility in an electrophoretic gel of a DNA fragment on its own and in complex with a protein.

electropositive: partial positive charge on an atom that results from its tendency to donate electrons to other atoms.

elongation (transcription): iterative process of addition of ribonucleotides complementary to the DNA template to the nascent transcript; catalyzed by RNA polymerase.

elongation (translation): iterative process of addition of an individual amino acid to the growing polypeptide chain and the advancement of the mRNA–tRNA complex in the 3′ direction; catalyzed by the ribosome.

elongation factor (transcription): protein that promotes transcription elongation over pausing or termination.

elongation factor (translation): group of proteins that collectively is important for facilitating the various steps of the elongation cycle; best known are EFTu (eEF1A) and EFG (eEF2).

embryonic stem cells: stem cells isolated from very early stage embryos.

endonuclease: enzyme that cleaves a nucleic acid internally, rather than degrading it from the ends as an exonuclease does. There are both DNA endonucleases and RNA endonucleases.

endoplasmic reticulum: organelle found in eukaryotic cells comprising a network of tubules wherein lipids are synthesized and secreted proteins are translated, folded, modified and then sent on their way.

endosymbiont: organism living within the cell or body of another organism.

end-replication problem: describes the problem of fully replicating the end of a linear DNA molecule, which would lead to progressive sequence loss without the addition of telomeric DNA by telomerase.

enhanceosome: complex consisting of a variety of proteins bound to an enhancer sequence.

enhancer: DNA sequences recognized by gene-specific regulators that can function efficiently at large distances from the promoter elements, both upstream and downstream of the promoter and in either orientation.

enthalpy: form of energy that can be released or absorbed as heat at constant pressure.

entropy: a measure of the disorder or randomness of a molecule or system.

enzymatic terminator: terminator sequence in the DNA that requires the action of additional enzymes or other proteins to cause RNA polymerase to stop transcribing and dissociate from the DNA.

enzyme: biological molecule that catalyzes a chemical reaction.

enzyme kinetics: rates of the different steps in an enzymatic reaction.

epidermal growth factor (EGF): small peptide that is a mitogen and stimulates cell division in response to receptor binding.

equilibrium: state in which the rate of the forward reaction and the rate of the reverse reaction in a chemical transformation (or binding reaction) are equal. At equilibrium, the relative concentrations of reactants and products no longer change, although the reaction continues to proceed in both directions.

equilibrium association constant (K_a): quantitative measure of binding affinity for the association reaction $A + B \rightleftharpoons AB$. K_a is equal to a ratio of the concentration of free components and complex at equilibrium ($[AB]/[A][B]$) or of the rate constants for the forward and reverse reactions (k_{on}/k_{off}). The higher the association constant, the higher the affinity of the interaction.

equilibrium constant (K_{eq}): ratio of the forward and reverse rate constants for a reaction; also equal to the ratio of the product of the concentrations of reaction products to the product of the concentration of the reactants. For the reaction $A + B \rightleftharpoons C + D$, $K_{eq} = [C][D]/[A][B]$.

equilibrium dissociation constant (K_d): quantitative measure of binding affinity for the dissociation reaction $AB \rightleftharpoons A + B$; K_d is equal to the inverse of the equilibrium constant for the association reaction ($1/K_a$ or $[A][B]/[AB]$ or k_{off}/k_{on}). The lower the dissociation constant, the higher the affinity of the interaction.

euchromatin: active chromatin. The DNA that is packaged as euchromatin is accessible to many DNA-binding proteins that are involved in transcription, replication, recombination, and repair.

eukaryote (meaning 'well kernel'): organism having a nuclear compartment that contains its genetic information.

exit (E) site: the position on the small and large ribosomal subunits which the deacylated tRNA occupies before completely exiting the ribosome following involvement in a round of elongation.

exon: RNA sequence that is interrupted by introns in a pre-RNA and are brought together by splicing.

exon definition: the hypothesis that exons, rather than introns, are defined by proteins that bind to the exon sequence and by interactions between the 3′ and 5′ splice site spliceosome components across the defined exon.

exonic enhancer/silencer sequence: sequences in the exons that either increase or decrease, respectively, the likelihood of that exon being included in the spliced mRNA product.

exon junction complex (EJC): group of proteins deposited at exon–exon junctions during the splicing reaction that 'marks' the transcript as processed at that site.

exon shuffling: process by which the DNA sequences encoding exons are exchange and reordered through genetic recombination between the DNA sequences encoding introns.

exonuclease: enzyme activity that cleaves nucleotides from the ends of a nucleic acid molecule. There are DNA exonucleases and RNA exonucleases.

exosome: complex of proteins in eukaryotes and archaea, related to the bacterial degradosome, and responsible for the 3′ to 5′ degradation of many RNAs.

extein: the sequences flanking an intein, which are religated after intein excision to form the functional protein.

extended −10 region: additional bases to the 5′ side of some −10 sequences that are bound by domain 3 of the sigma subunit of bacterial RNA polymerase.

extinction coefficient: coefficient that indicates how much light is absorbed by a particular molecular species at a given wavelength.

F

fatty acid: hydrocarbon chain with a carboxylic acid at one end.

fidelity: the accuracy of synthesis (by DNA or RNA polymerases or the ribosome). All polymerases occasionally make an error and add the wrong building block during chain extension. A polymerase that makes very few errors is said to have high fidelity.

first law of thermodynamics: states that energy can be converted from one form to another, but is neither created nor destroyed.

fluorescence loss in photobleaching (FLIP): a method for examining the movement of fluorescent molecules in a cell. It is based on the repeated photobleaching of a specific locus of the cell and determining whether the fluorescence in other loci is diminished.

fluorescence recovery after photobleaching (FRAP): a method for quantifying the diffusion rate of fluorescent molecules, such as green fluorescent protein (GFP)–protein fusions, in a cell, based on the rate at which fluorescence is recovered in a particular region that was previously photobleached.

fluorescent *in situ* hybridization (FISH): method for detecting a specific DNA sequence in a chromosome *in situ*. FISH typically involves placing cells on a slide, denaturing the DNA and hybridization to a DNA probe that can be detected by fluorescence microscopy.

fold: three-dimensional shape adopted by an RNA molecule or protein.

footprinting: method for identifying interacting macromolecular surfaces, for example the region of DNA that is bound by a specific transcription factor, by comparison of the consequences of treating the DNA or the protein–DNA complex with a reagent that modifies the DNA. The reactions are typically analyzed by gel electrophoresis.

forward genetics: random mutagenesis of a population of organisms followed by isolation of organisms with a given phenotype. An example is the screening for mutant yeast that cannot synthesize uracil, with the goal of identifying the genes involved in uracil biosynthesis.

founder effect: type of genetic drift that occurs when a new population is established by a very small number of individuals.

frameshifting: the process by which a ribosome shifts along the mRNA in a forward or reverse direction – thus changing the 'frame' of reading the triplet genetic code.

functional genomics: genetic, molecular, biochemical, and cytological studies carried out on a genome-wide scale to gain insights into gene and genome function.

furanose: five-membered ring that forms when the C-2 keto group of a sugar reacts with a C-5 hydroxyl group.

G

G0: a prolonged non-dividing state that is reached from G1 when cells are exposed to extracellular conditions that arrest cell proliferation.

G1: first gap phase of the cell cycle. It is a growth stage that follows cell division (M phase) and precedes DNA replication (S phase).

G2: second gap phase of the cell cycle. This stage follows DNA replication (S phase) and precedes mitosis and cell division (M phase).

gamete: specialized haploid reproductive cell, for example, egg and sperm in animals, which is produced through meiosis.

GATC sites: DNA sequences that are usually methylated on the A residue in both strands in bacterial DNA.

gel electrophoresis: separation of protein or nucleic acid mixtures using an electric field to promote their movement in liquid through an agarose or polyacrylamide gel.

gel filtration: chromatographic method for separating molecules based on their size, as well as shape. Also known as size exclusion chromatography.

gene: region of DNA that controls a discrete hereditary characteristic.

gene conversion: replacement of a gene sequence with another version of that gene.

gene expression: utilization of the information provided by a gene to generate a functional product.

gene family: duplicate members of a gene.

general transcription factors: proteins required for promoter recognition and transcription initiation in eukaryotes and archaea. Also known as basal transcription factors.

gene regulation: mechanisms through which the expression of a gene product is controlled in time and space.

genetic code: set of three nucleotide 'words' that allow for the translation of the DNA sequence into proteins composed of amino acids.

genetic drift: change in the frequency with which a gene occurs in a population, resulting from the fact that genes in offspring are a random sample of those in the parents, and because chance determines whether a given individual survives and reproduces.

genome: the complete DNA sequence of an organism, which serves as its instructional blueprint. It should be noted that the genome of some viruses consists of RNA.

genome annotation: the process of annotating the functional sequence elements in a genome sequence.

genome-wide association study: an examination of how a particular trait is linked to various markers throughout the genomes of many unrelated individuals.

genotype: the collective DNA sequence of an organism.

germline cells: cells in multicellular eukaryotes that are involved in the production of the progeny.

Gibbs free energy (*G*): the total energy change resulting from a chemical process, which is related to the entropy and enthalpy changes.

global genomic repair (GGR): the basic pathway of nucleotide excision repair that can act on many different types of damage at any site in the genome.

globular: roughly spherical in shape.

glycan: a complex carbohydrate constructed from monosaccharide units in either a linear or branched arrangement.

glycolipid: lipid molecule with covalently attached carbohydrate.

glycoprotein: protein with covalently attached carbohydrate.

glycosidic bond: chemical bond that joins a carbohydrate to another group. This term is used to describe the linkage between the heterocyclic bases of nucleic acids and the ribose or deoxyribose sugars.

glycosylation: the post-translational covalent addition of carbohydrate molecules to asparagine, serine, or threonine residues on a protein molecule. Glycosylation can add a single sugar or a chain of sugars at a given site and is usually catalyzed by an enzyme.

glycosylphosphatidylinositol (GPI) anchor: complex structure involving both lipids and carbohydrate molecules that is reversibly attached to some proteins to target them to the cell membrane.

Golgi apparatus: organelle found in most eukaryotic cells that processes proteins and modifies them with sugars before packaging them for export.

gradient: solution in which the concentration of a solute varies continuously from one value to another.

green fluorescent protein (GFP): protein normally found in the jellyfish *Aequorea victoria* that exhibits green fluorescence when excited by blue light. Derivatives of this protein have been made that fluoresce at other wavelengths.

G-rich strand: the DNA strand in telomere DNA that is rich in G residues; the end of this strand forms a short single-stranded extension at the end of the telomere.

group I intron: class of introns with a conserved secondary and tertiary structure, which catalyzes two sequential transesterification reactions initiated by the 3′ OH of an exogenous guanosine nucleotide.

group II intron: class of introns with a conserved secondary and tertiary structure that catalyzes two sequential transesterification reactions initiated by the 2′ OH of an internal adenosine nucleotide (the branch site).

group II intron transposable element: a mobile element that excises by RNA splicing from an mRNA transcript and then reverse-splices into its target DNA and is converted to DNA by an element-encoded reverse transcriptase.

growth factor: extracellular factor that stimulates cell growth (an increase in cell mass). This term is sometimes used incorrectly to describe a factor that stimulates cell division, for which the correct term is mitogen.

GTPase-activating protein (GAP): proteins that interact with G-proteins (GTPases) and accelerate their rate of GTP hydrolysis.

guanine-nucleotide exchange factor (GEF): proteins that catalyze the exchange of a GDP bound to a specific G-protein (GTPase) for GTP.

guide RNA: RNAs that direct modifications of RNAs by base-pairing with the target RNAs.

H

hairpin: RNA strand that folds over on itself to form a double helix separated by a short single-stranded region referred to as a loop.

haploid: containing one copy of each chromosome.

haplotype: a block of single nucleotide polymorphisms, copy number variation, and other sequence variations that are inherited together.

haplotype block: block of DNA along a chromosome that is usually linked together within which recombination rarely occurs.

HapMap: map of all of the known haplotype blocks in the human genome. The generation of the HapMap is ongoing as more and more individual human genomes are sequenced.

head group: hydrophilic end of a lipid molecule.

helix-turn-helix: DNA-binding motif consisting of a pair of α helices, the second of which fits in the major groove of the DNA (the recognition helix).

hemi-methylated: DNA sequence that is methylated on one strand. Usually a site of methylation is symmetrical and both strands are methylated at the equivalent position.

heterochromatin: tightly packed chromatin that is less accessible to nuclease digestion than euchromatin and is associated with different chromosomal proteins. Regions packaged as heterochromatin are usually silenced.

heteroduplex: duplex containing complementary DNA strands from homologous parental duplexes, but which may not themselves be identical.

hexose: six-carbon monosaccharide.

high-performance or high-pressure liquid chromatography (HPLC): form of column chromatography in which the liquid phase, usually at high pressure, is pumped through the column.

histone acetyltransferases (HATs): enzymes that covalently attach acetyl groups to lysine residues in histones, as well as other proteins. Also known as histone acetylases or lysine acetyltransferases.

histone deacetylases (HDACs): enzymes that remove acetyl groups from lysine residues in histones and proteins.

histone methyltransferase (HMT): enzyme that catalyzes the covalent attachment of methyl groups to lysine or arginine residues in histones.

histone modification: covalent chemical alteration of amino acid side chains in histone proteins.

histone octamer: complex of eight histone proteins made up of two copies each of the histones H2A, H2B, H3, and H4.

histone protein: any one of the highly conserved histone family of proteins that bind and package DNA in eukaryotes. The family comprises the very highly conserved proteins of the histone octamer, H2A, H2B, H3, and H4, a number of histone variants, and linker histones such as H1.

histone variants: proteins that are related in sequence and in structure to the highly core histones, and that play specific roles in chromosome function.

holocentric: chromosome that lacks a defined centromere; instead, the mitotic spindle attaches to multiple points along the length of the chromosome.

holoenzyme: bacterial RNA polymerase consisting of the core enzyme (the β, β′, and two α subunits), together with a σ subunit that is capable of specific promoter recognition.

homeodomain: eukaryotic member of the helix-turn-helix family of DNA-binding proteins.

homologous chromosomes: two chromosomes with identical, or similar, gene composition and organization, one inherited from each parent. Homologous chromosomes are very similar, but not identical, in sequence.

homologous recombination: exchange of large DNA segments between homologous chromosomes that is mediated by strand exchange recombinases such as RecA and Rad51.

homolog (gene): gene descended from a common ancestor. Two different types of homolog are orthologs and paralogs.

homology: sequence similarity reflecting descent from a common ancestor.

homology-directed repair: mechanism for accurately repairing double-strand breaks in DNA in which a homologous DNA molecule provides the template for repair synthesis.

horizontal gene transfer: insertion of a segment of the genome from one species into the genome of an unrelated species.

hybrid states model: model for the elongation cycle wherein the tRNAs move distinctly with respect to the large and small subunits of the ribosome following peptidyl transfer.

hydrogen bond: non-covalent interaction between the donor atom, which is bound to a positively polarized hydrogen atom, and the acceptor atom, which is negatively polarized.

hydrogen bond acceptor: partially negatively charged atom participating in a hydrogen bond.

hydrogen bond donor: partially positively charged hydrogen atom in a hydrogen bond.

hydrophilic: tending to interact with water. Hydrophilic molecules are polar or charged and, as a consequence, are very soluble in water.

hydrophobic: tending to avoid water. Hydrophobic molecules are non-polar and uncharged and, as a consequence, are relatively insoluble in water.

hyperacetylated: containing many acetylated side chains.

hypoacetylated: containing few acetylated side chains.

I

identity element: the group of nucleotides in a given tRNA species that confers specificity for the appropriate aminoacyl-tRNA synthetase.

immunoblot: an analytical method in which a protein is detected by the addition of a specific antibody.

imprinting: the inactivation of a gene or group of genes on one chromosome of a homologous pair, but not the other. The imprinted gene is typically methylated and is packaged as heterochromatin.

indirect immunofluorescence: method for detecting specific cellular molecules (for example, a specific protein) by microscopy using two antibodies: a primary antibody that recognizes the molecule of interest, and a secondary antibody that is fused to a fluorescent molecule and recognizes the primary one.

initial selection: early steps in tRNA selection including codon recognition and subsequent GTP hydrolysis.

initiation (transcription): steps by which RNA polymerase begins transcription.

initiation (translation): process by which a ribosome is loaded with the initiating methionyl-tRNA species in preparation for the elongation phase of protein synthesis.

initiation codon: AUG codon (or sometimes UUG or GUG) that is the starting point for an open reading frame that encodes a protein product.

initiation factor: any one of a number of proteins that is critical in the initiation process.

initiator protein: any of a complex of proteins that bind to origins of replication in DNA and initiate the unwinding of the double helix in preparation for DNA replication.

initiator tRNA: a distinct class of tRNAs with particular structural and functional features that allow them to be specifically deposited in the P site to decode the initiation codon.

insertion sequence (IS element): small bacterial transposable element encoding only the functions required for its transposition.

insulator: thought to be a boundary element that helps define the limits of specific forms of chromatin, including preventing the spreading of heterochromatin. An insulator inserted between a gene and a gene-regulatory element preceding it can also block activation of the gene.

intasome: elaborate nucleoprotein complexes in which CSSR integrases carry out recombination.

integral membrane protein: membrane protein in which the protein chain is largely embedded in the membrane.

integrase: protein that mediates the integration of a retroviral provirus or long terminal repeat (LTR) retrotransposon into DNA.

intein: internal portion of a protein sequence that is post-translationally excised in an autocatalytic reaction while the flanking regions are spliced together to making a new protein product; a protein intron (intervening sequence).

interactome: complete network of the protein–protein interactions in a cell type.

intercalating agent: chemical such as ethidium bromide that becomes inserted between adjacent base pairs in DNA and sometimes gives rise to mutations.

internal ribosome entry site (IRES): RNA element found in the 5′ untranslated region (UTR) of viral (and cellular) transcripts (in eukaryotes) that identifies the AUG start site for translation initiation independent of the standard cap-dependent scanning mechanism.

interphase: period between two M phases that includes G1, S, and G2.

intrinsic or simple terminator: DNA sequence that, when transcribed into mRNA, causes RNA polymerase to terminate transcription and dissociate from the DNA template.

intron: portion of RNA that is included in the primary transcript but is removed by splicing.

intron definition: hypothesis that introns, rather than exons, are defined through interactions between sequential 5′ and 3′ splice site bound factors.

intronic enhancer/silencer sequence: sequences in the introns that either increase or decrease, respectively, the likelihood that neighboring exons will be included in the spliced RNA product.

in vitro: in an artificial environment, such as a test tube (literally 'in glass').

in vivo: in the living organism.

ion: an atom or molecule that is positively or negatively charged.

ion exchange chromatography: type of chromatography in which proteins and other molecules are separated according to charge.

ionized: the charged state of a previously uncharged atom or molecule.

IRES-transacting factor (ITAF): protein factor that interacts with IRESs to facilitate the non-standard translation initiation pathway in eukaryotes.

K

karyotype: set of condensed metaphase chromosomes in eukaryotes stained so that each chromosome can be uniquely identified. The chromosomes are usually displayed by arranging them in order from the largest, designated chromosome 1, to the smallest.

kinetochore: a complex of proteins that attaches to the centromere and mediates the attachment of the chromosome to the spindle during mitosis.

K_M: the Michaelis constant, which is equal to the substrate concentration at which the reaction rate is 1/2 of V_{max}.

Kozak sequence: the weak consensus sequence of nucleotides surrounding the AUG start site in eukaryotic genes that favors efficient start site recognition.

L

lagging strand: in the replication of DNA, the new DNA strand that is synthesized on the template strand that runs from 5′ to 3′ into a replication fork. It is synthesized discontinuously as a series of short DNA fragments that are subsequently joined together to form a continuous DNA strand.

lariat: the circular intron structure generated during the process of splicing by the spliceosome or the group II introns.

leading strand: in the replication of DNA, the new DNA strand that is synthesized on the template strand that runs from 3′ to 5′ into a replication fork. It is synthesized continuously as the fork moves forward.

lectin: protein that binds specifically to particular polysaccharides or other carbohydrate structures.

library: collection of clones, plasmids, or viruses that contain different DNA inserts, often representing a particular cell type. This could be a genomic library representing the entire genome or a cDNA library representing the set of expressed genes.

linkage analysis: method used to find disease genes in family pedigrees when DNA samples from a number of people in several large families is available (also denoted linkage study).

lipid: term used to describe a range of molecules that are all strongly hydrophobic, including fatty acids and cholesterol.

lipid bilayer: the structure of cellular membranes, formed when the hydrophobic tails from two sheets of lipid molecules pack against each other to form the interior of the sandwich while their polar head groups cover the outside.

locus control region: eukaryotic regulatory regions containing a combination of enhancer and insulator elements.

LOD score: a score that indicates the probability that two genetic markers or loci are linked to one another, and therefore likely to be inherited together; given by the logarithm of the odds ratio for linkage.

long interspersed element (LINE): genomic sequence derived from the duplication and transposition of retrotransposons that move by target-primed reverse transcription.

LTR elements: class of transposable elements that are characterized by LTRs and move via an RNA intermediate. They include the LTR retrotransposons and the retroviruses.

long terminal repeats (LTRs): direct sequence repeats of hundreds of base pairs at the ends of retroviral genomes and some types of retrotransposons.

loss of heterozygosity (LOH): loss of one allele of a gene in a cell that is heterozygous for that gene (Aa), causing the cell to become effectively homozygous for the other allele (by virtue of becoming haploid). The term is usually used in the context of genes containing deleterious mutations, when the loss of the normal copy results in effective homozygosity for a deleterious recessive mutation.

LTR retrotransposon: type of long terminal repeat (LTR) element that does not form infectious particles and thus can only move within a single cell.

lysogen: bacterial cell containing a copy of the bacteriophage chromosome integrated into the host cell chromosome.

lysogeny: type of viral growth in which the bacteriophage chromosome is integrated into the host cell chromosome.

lytic growth: type of viral growth in which the bacteriophage reproduces within the infected host cell, ending with lysis (bursting) of the host cell to release the phage particles.

M

macromolecule: large molecule; it is usually also a polymer.

major groove: one of two grooves running along the helix of double-stranded DNA, in which the N7 of purine bases is exposed. In B-DNA, this is the wider of the two grooves.

marker: a unique DNA sequence that can be followed in individual people to determine if it is linked to a disease gene or other trait. Sequence tagged sites (STSs) and single nucleotide polymorphisms (SNPs) are two common kinds of DNA marker.

mass spectrometry: analytical method for determining the composition of a molecule by generating ions and determining their mass to charge ratio.

mating: in yeast, the process by which two haploid cells, one **a** and one α cell, fuse to form a diploid **a**/α cell.

mating type: cell type of a yeast cell **a** or α for the yeast, *Schizosaccharomyces cerevisiae*.

MAT locus: location in *S. cerevisiae* chromosome III of the genes that are expressed in particular cell types. **a** cells contain the **a** cassette at this location, which encodes the **a**1 protein, while α cells contain the α cassette encoding α1 and α2.

mature RNA: fully processed and functional RNA.

MCM: replicative DNA helicase in eukaryotes and archaea comprising six different proteins.

MEC1: the ATR regulator kinase of *S. cerevisiae*.

Mediator: complex composed of 22 proteins required by RNA polymerase II for appropriate transcription and response to transcriptional activators.

meiosis: specialized division that occurs during formation of the gametes through two successive round of chromosome segregation. The products of meiosis have half the ploidy of the parent cell.

metagenome: genomic sequence information for all of the organisms that occupy a particular ecological niche.

metaphase: stage in mitosis in which the sister chromatids are fully attached to the spindle and await the signal to separate.

methylation: the covalent attachment of a methyl group to a molecule. In the case of histone methylation, methyl groups are added to lysine and arginine residues. Some bases on DNA and RNA can also be methylated.

methyl transferase: enzyme that adds methyl groups to side chains on proteins (lysine or arginine methyl transferase) or to DNA bases (DNA methyl transferase).

Michaelis–Menten equation: equation that expresses the rate of an enzyme-catalyzed reaction as a function of substrate concentration and is characterized by constants, V_{max} and K_M.

microsatellite repeats: short repeated stretches of DNA, such as CACACACA, which are variable in length between individuals and can be used as markers in genetic linkage and association studies.

microtubule: long hollow polymer of tubulin heterodimers with two distinct ends, a plus end and a minus end, which display different polymerization behaviors.

microtubule flux: flow of tubulin within microtubules from the microtubules plus end to the spindle poles, as a result of loss of tubulin from the minus ends at the poles. It is usually accompanied by addition of tubulin at the plus ends.

minor groove: one of two grooves running along the helix of double-stranded DNA, in which the N3 of purine bases is exposed. In B-DNA, this is the narrower of the two grooves.

miRNAs (microRNA): 21–24 nucleotide RNAs, encoded by distinct genes and specifically excised from characteristic hairpin structures, which generally modulate translation or RNA degradation.

miscoding agent: small molecule antibiotic that binds in the decoding region of the small subunit of the ribosome and induces the misincorporation of amino acids during the tRNA selection process.

mismatch repair: repair of nucleotide mismatches that are generated during DNA replication.

missense mutation: single nucleotide change in the DNA sequence that changes a codon encoding a particular amino acid to one encoding a different one.

mitochondrion: organelle found in eukaryotic cells that contains the macromolecular machinery responsible for deriving chemical energy from food precursors (known as the power house of the cell).

mitogen: extracellular molecule that stimulates cell proliferation.

mitosis: process by which the duplicated chromosomes, the sister chromatids, are segregated by the mitotic spindle and packaged into daughter nuclei.

mobile group II intron: type of group II intron that, in addition to catalyzing its own excision from RNA, encodes a maturase protein with reverse transcriptase activity that enables the element to insert itself at other sites.

model organisms: organisms that are extensively studied to gain insights into biological processes that occur in many other organisms, and that are easy to manipulate under laboratory conditions.

molecule: two or more atoms joined by covalent bonds.

monogenic: phenotypes or diseases that result from variation in a single gene.

monosaccharide: simple sugar with formula $(CH_2O)_n$; the building block of carbohydrates.

M phase: mitosis, the cell cycle phase during which the duplicated chromosomes are segregated and packaged into daughter nuclei.

MRN: complex in multicellular eukaryotes containing MRE11-RAD50-NBS1 that interacts with DNA and plays several important roles in the DNA damage response.

MRX: complex in *S. cerevisiae* containing Mre11-Rad50-Xrs2 that plays an important role in the DNA damage response and in DNA transaction in homology-dependent repair, homologous recombination and non-homologous end-joining.

multifactorial trait: phenotype determined by both the genetic composition of an individual and environmental influences.

mutator phenotype: phenotype characterized by an increased mutation rate as the result of genetic defects in DNA repair processes.

N

natural selection: process by which a heritable trait that causes an organism to produce more offspring in each generation eventually becomes more common than other genetic variants after multiple generations.

near-cognate interaction: when the codon of the mRNA and the anticodon of the tRNA are only appropriately matched at two of the three positions – these tRNAs are usually rejected by the ribosome.

negative selection: situation where a new genetic variant is disfavored in a population over time (also called purifying selection).

neocentromere: new centromere that forms at any position on the chromosome other than the normal centromere position.

neutral mutation: genetic variant that is neither favored nor disfavored.

NextGen (next generation) sequencing: massively parallel sequencing methods that have the capacity to generate hundred of megabases of sequence information.

nitrosylation: modification of a cysteine or tyrosine residue by addition of nitric oxide (NO). The nitrosylation of cysteine is called *S*-nitrosylation.

***N*-linked carbohydrate:** sugar covalently linked to the nitrogen of an asparagine side chain.

NMR: nuclear magnetic resonance, a property of atomic nuclei in a high magnetic field that causes them to absorb energy when reoriented by an electromagnetic pulse, and then emit energy again as they return to their original orientation (a process called relaxation). This phenomenon is exploited by the technique of NMR spectroscopy, which can be used to obtain information on the structure and dynamics of biological molecules.

non-autonomous element: transposable element lacking a transposase gene, but which can be mobilized by a transposase from another similar element.

non-canonical base pair: base pair that does not form a standard Watson–Crick base-pairing interaction, but which generally is stabilized by some number of hydrogen bonds.

non-coding, template, or antisense strand: DNA strand complementary to the coding strand and which is the template for synthesizing an RNA copy of the coding strand.

non-cognate: a term used to describe a lack of correspondence between an amino acid and a tRNA isoacceptor species (or a codon) that will result in faithful interpretation of the genetic code.

non-cognate interaction: when the codon of the mRNA and the anticodon of the tRNA are mismatched at two or three of the potential three positions – these tRNAs are very effectively rejected by the ribosome.

non-competitive inhibitor: molecule that lowers the apparent maximal rate (V_{max}) of an enzymatic reaction by binding to the enzyme-substrate complex.

non-covalent interactions: interactions between atoms or molecules that do not depend on formation of covalent bonds.

non-homologous end joining (NHEJ): mechanism for repairing double-stranded breaks in DNA in which the broken ends are rejoined directly, usually with the loss of nucleotides at the joint.

non-LTR retrotransposon: transposable element that has an RNA intermediate and inserts into a new insertion site via target-primed reverse transcription. LTR, long terminal repeat.

non-polar: (of a bond or molecule) containing an even distribution of electric charge.

non-recombinant: (section of DNA) produced by a homologous-recombination event between two DNA duplexes in which the DNA segments flanking the position of interaction between the duplexes that initiated recombination have not been exchanged.

nonsense mediated decay (NMD): process in eukaryotic cells that identifies RNA transcripts containing premature termination codons and targets them for degradation (thus preventing the production of truncated proteins).

nonsense mutation: change in the DNA sequence that introduces a premature stop signal such that only truncated gene products are made.

non-specific binding: lower affinity binding that occurs at random as opposed to relatively tight binding between correct partners.

northern blot: method for analyzing RNA in which RNA is extracted from the cell, fractionated by size by gel electrophoresis, and detected by hybridization to a labeled complementary nucleic acid probe (DNA or RNA).

N-terminus: end of a polypeptide chain with an exposed amino group.

nuclear receptor protein: member of a conserved family of proteins containing characteristic DNA-binding and ligand-binding domains. These proteins regulate transcription in response to the binding of hormones and other small molecules.

nuclease sensitivity: a way of differentiating euchromatin from heterochromatin experimentally by its greater resistance to digestion by the endonuclease, DNase I.

nucleic acid: polymer consisting of nucleotides joined by phosphodiester linkages.

nucleoid: region of bacterial and archaeal cells in which the chromosomal DNA is sequestered and packaged with protein.

nucleoid occlusion: a regulatory mechanism that prevents cell division in bacteria when the nucleoid lies near the division ring.

nucleophile: a reagent that is attracted to centers of positive charge and can donate electrons to form a chemical bond.

nucleoside: base joined to ribose or deoxyribose by a glycosidic bond.

nucleosome: basic unit of DNA packaging in eukaryotes, consisting of a protein core composed of a histone octamer, with DNA wrapped around the outside.

nucleotide: base joined to ribose or deoxyribose and one or more phosphate groups. Nucleotides form the repeating unit of the nucleic-acid polymers DNA and RNA.

nucleotide excision repair: DNA repair pathway in which major structural defects, such as thymine dimers, are excised along with a stretch of 10–30 nucleotides surrounding the site of damage. The now excised strand is resynthesized using the undamaged strand as the template.

nucleus: membrane-bound compartment found in eukaryotic cells where the chromosomes are contained.

null or loss of function mutation: change in DNA sequence that eliminates the function of a gene. These types of mutation typically are associated either with a change in the product that completely eliminates function (for example, by unfolding the protein) or from a large disruption in the DNA sequence that grossly perturbs gene expression.

O

odds ratio: odds of linkage of a marker to disease in individuals who have the disease divided by the odds in controls who do not. If the odds ratio is significantly greater than one, then the marker or genetic variant is associated with the disease.

Okazaki fragment maturation: the process by which the short DNA fragments, the Okazaki fragments, which are synthesized on the lagging strand during DNA replication, have their primers removed and are joined together to make a continuous DNA strand.

Okazaki fragments: short fragments of DNA that are made during lagging strand synthesis. They are initiated with RNA primers at the replication fork, synthesized in the direction away from the direction of fork movement and subsequently joined together by DNA ligase after removal of the RNA.

oligosaccharide: polymer composed of multiple (up to around ten) monosaccharide subunits.

***O*-linked carbohydrate:** sugar covalently linked to the oxygen of a serine or threonine side chain.

oncogene: a gene whose protein product promotes cell division that leads to cancer. Generally mutations or rearrangements in the normal gene have resulted in the oncogene version of the protein that is either overactive or overproduced.

open complex: complex formed by RNA polymerase and the promoter DNA in which the two DNA strands are separated.

operator: bacterial regulatory DNA sequence recognized by a repressor protein.

operon: a set of genes with a common promoter and operator that are transcribed as a single, polycistronic mRNA.

organic: containing the element carbon.

origin of replication (*ori*): a site on a chromosome at which DNA replication can begin. It contains a DNA sequence that is recognized by proteins that unwind the DNA and initiate DNA replication.

origin recognition complex (ORC): protein complex that binds to replication origins in eukaryotes and recruits other proteins to unwind DNA and initiate DNA replication.

ortholog: related copy of a gene in the genome of different species.

oxidation: loss of electrons from an atom or molecule.

P

p53: gene regulatory protein that orchestrates the long-term DNA damage response in multicellular organisms by activating a cell cycle checkpoint and/or apoptosis in response to DNA damage or to DNA replication occurring in an incorrect phase of the cell cycle.

parallel beta sheet: beta sheet formed from non-contiguous regions of the polypeptide chain in which each strand runs in the same direction.

paralog: related copies of a gene in the same genome.

parental histone segregation: the random redistribution of the old modified histones on the parental strand onto the two new daughter strands. Half the old nucleosomes go to one daughter and half to the other.

PCNA (proliferative cell nuclear antigen): subunit of the sliding clamp protein in eukaryotes; it forms a trimer to make a symmetrical ring structure with six similar domains.

penetrance: the percentage of individuals carrying a particular mutant genotype that exhibit the mutant phenotype.

pentose: five-carbon monosaccharide.

peptide backbone: the regularly repeating atoms in a polypeptide, typically referring to the atoms NH, Cα, and C=O, and excluding the side chain R group.

peptide bond: another name for amide bond, a chemical bond formed when a carboxylic acid condenses with an amino group with the loss of a water molecule. The term peptide bond is used only when the reactive groups come from amino acids.

peptidoglycan: rigid polymer of polysaccharides joined by peptide cross-links that is a major structural element of most types of bacterial cell walls.

peptidyl (P) site: position on the small and large ribosomal subunits which the peptidyl-tRNA occupies prior to peptide bond formation.

peptidyl transferase: activity of the enzyme responsible for catalyzing the formation of a peptide bond between the growing peptide chain and the incoming aminoacyl-tRNA.

peripheral membrane protein: protein associated with membranes but which does not contain a complete transmembrane region.

pH: negative log (base 10) of the H^+ ion concentration.

pharmacogenomics: the use of an individual's genetic and genomic information to predict their response to specific drugs.

phase contrast: illumination technique in microscopy that amplifies small changes in the phase of light waves as they pass through a specimen, effectively increasing contrast.

phase variation: alternative expression of a cell surface protein on a microbe that facilitates evasion of recognition by the immune system.

phenotype: visual features and properties of an organism.

phospholipid: lipid containing a phosphate group on the head group.

phosphorylation: covalent attachment of a phosphate group to a molecule. In the case of proteins, the phosphate is added to a serine, threonine, tyrosine, or histidine residue.

photolyase: a DNA repair enzyme present in organisms other than placental mammals, which splits pyrimidine dimers formed in DNA by UV light.

phragmoplast: an organelle in a dividing plant cell on which the new cell membranes and cell walls between the two daughter cells are constructed.

phylogenetic analysis: a method for predicting interacting regions in RNA molecules by comparing homologous RNA molecules from different species and identifying covariations indicative of conserved interactions.

phylogenetic tree: representation in a quantitative branched model of the extent of relatedness between various species.

PIKK: ATM, ATR and DNA-PK are phosphoinositide-3-kinase-related protein kinases.

piRNAs (Piwi-interacting RNAs): 26–31 nucleotide long RNAs that associate with the Piwi subfamily of Argonaute proteins in RISC complexes and primarily act to silence transposons in the germlines of animals.

pK_a: negative log (base 10) of the acid dissociation constant (K_a) for an ionizable group.

plasma membrane: membrane that encloses a cell.

plasmid: DNA molecule, usually circular and containing an origin or replication, distinct from the chromosomal genome of the host.

ploidy: number of homologous sets of chromosomes in an organism.

P loop motif: weak amino acid consensus motif found in a variety of NTP binding proteins; gross conformational changes are observed in this motif following triphosphate hydrolysis.

pluripotent cells: stem cells that can give rise to a limited set of cell lineages.

point centromere: type of centromere that covers a very small region of chromosome (~100 bp of DNA).

polar: bond or molecule containing an uneven distribution of electric charge, with positive charge at one end and negative at the other.

polyadenosine (poly(A)) tail: tail of approximately 200 adenosine residues added to the 3′ ends of translated mRNAs.

polyadenylation: process by which a poly(A) tail is added to mRNAs.

polyadenylation site: specific sequences responsible for recruiting the polyadenylation machinery.

polycistronic message: RNA transcript containing the coding sequences for several genes; typically found in bacteria.

polygenic: phenotype or disease in which multiple independent genes play a role.

polymer: molecule composed of many copies of a simple building block covalently linked to one another.

polymerase chain reaction (PCR): method to amplify a specific region of DNA with known sequence primers. The DNA is denatured and copied using primers and a DNA polymerase. Many cycles of denaturation and primer extension are carried out resulting in an exponential amplification of the sequence between the primers.

polymerase III holoenzyme: complex of proteins that carries out the elongation phase of DNA replication in bacteria, moving along with the replication fork and synthesizing new DNA on both template strands simultaneously. It is composed of two molecules of DNA polymerase III and the clamp loader with a bound sliding clamp. Also called the replisome.

polymerase switching: in eukaryotes, the switch that occurs from DNA synthesis by the polymerase α–primase complex to DNA synthesis by the processive replicative polymerase that carries out the bulk of DNA elongation.

polymorphism: a site in a genome that has more than one common sequence in a population.

polypeptide: a polymer of amino acids joined together by peptide bonds.

polyploidy: expansion in the number of genome copies.

polyprotein: expressed polypeptide containing several discrete protein domains that must be separated by proteolytic cleavage before they become active.

polypyrimidine tract: pyrimidine-rich sequence found just upstream of the 3′ splice site recognized by U2AF65 during the early steps of splicing.

polysaccharide: large carbohydrate composed of a large number of sugars; can be a linear or branched polymer.

polytene chromosome: giant chromosomes that are formed by multiple rounds of DNA replication without separation of sister chromatids.

position-effect variegation: variation in expression of a gene as a result of changes in its relative position on the chromosome. The variegation refers to the visible effects of different degrees of silencing of a given gene by encroachment of heterochromatin when the gene is moved into a region adjacent to heterochromatin.

positive cooperativity: cooperative interaction that makes binding of more than one ligand energetically more favorable.

positive selection: situation where a new genetic variant is favored in a population over time.

post-transcriptional: refers to events that take place following the process known as transcription.

post-transcriptional gene silencing (PTGS): mechanism by which short fragments of double-stranded RNA lead to mRNA degradation and thus lead to the silencing of RNA derived from foreign genes.

post-translational modifications: changes such as cleavage, or the addition of lipids, sugars or small molecules, to a polypeptide chain that occur after protein synthesis.

precursor RNA (pre-RNA): full-length unprocessed RNA transcript synthesized by RNA polymerase.

pre-initiation complex: complex formed by the general transcription factors and the RNA polymerase core enzyme, which is capable of transcription initiation.

premature termination codon (PTC): stop (nonsense) codon that is found in the transcript 5′ to the authentic stop codon for the gene.

prenylation: the irreversible attachment of either a farnesyl or geranylgeranyl group to a protein via a thioether linkage.

prereplicative complex (pre-RC): large complex of proteins, including the origin recognition complex and its associated proteins, which assembles at replication origins in late mitosis and G1 and is activated to initiate DNA replication at the beginning of S phase.

presynaptic filament: nucleoprotein complex formed by the binding of RecA/Rad51-like protein to single-stranded DNA and that can initiate a search for homology in duplex DNA.

primary cells: cells derived from differentiated tissue.

primary sigma factor: sigma factor that is present in all normal growth conditions and associates with the core RNA polymerase to bind the promoters of many highly expressed genes. Each bacterial species has only one primary sigma factor.

primary structure: the amino-acid sequence of a polypeptide chain.

primase: specialized RNA polymerase that synthesizes a short stretch of primer RNA on the template at the beginning of DNA replication, thus generating a 3′ end to which nucleotides can be added by DNA polymerases.

primer: in DNA synthesis, a short stretch of RNA synthesized by a primase on a DNA template, and from which the new DNA strand is elongated by DNA polymerase.

processivity: ability of a DNA polymerase to remain associated with the template while synthesizing DNA. A highly processive enzyme will stay bound to the template and polymerize thousands of nucleotides before it dissociates.

progenote: a putative common original life form from which the entire diversity of modern-day life derives, often referred to as the last universal common ancestor (LUCA).

prokaryote: (meaning 'before the kernel' or pre-nuclear) unicellular species, including members of the bacterial and the archaeal kingdoms, which do not have internal membranes within their cells.

prometaphase: second stage of mitosis, in which the nuclear envelope breaks down and the sister chromatids become attached to the spindle.

promoter: DNA sequence needed for RNA polymerase to bind to the DNA and initiate transcription.

promoter clearance, promoter escape: step in transcription in which RNA polymerase moves past the promoter and enters into productive elongation.

promoter proximal pausing: pausing of RNA polymerase II somewhat downstream of the promoter after transcribing around 50 base pairs.

proofreading (DNA synthesis): when DNA polymerase recognizes that an incorrect nucleotide has been added and resynthesizes the region following the targeted action of an associated exonuclease function.

proofreading (protein synthesis): a phase in the tRNA selection process (a requisite step in elongation) that follows GTP hydrolysis and includes the alternative pathways of accommodation and rejection.

prophage: copy of the phage chromosome that is integrated into the bacterial host cell chromosome.

prophase: first stage of mitosis, when chromosome condensation, centrosome separation, and spindle assembly begin.

proteasome: large multisubunit enzyme complex that degrades proteins tagged with polyubiquitin chains into short peptides.

protein family: proteins that are related in structure or amino acid sequence and thus are presumably descended from a common ancestor.

protein fold: overall structure of a protein, as defined by its shape, secondary structure elements and topology.

protein folding: process by which a polypeptide chain collapses to form a compact, folded structure.

protein kinase: enzyme that adds phosphate groups to side chains on proteins.

protein phosphatase: enzyme that removes phosphate groups from phosphorylated serines, threonines or tyrosines on proteins.

proteomics: the study of proteins, protein complexes and protein interactions that is aided by having complete genome sequence of all predicted proteins.

protofilament: a protein structure formed by the head to tail attachments of tubulin dimers. Microtubules are made of 13 protofilaments that are aligned side by side in a cylindrical arrangement.

provirus: DNA copy of a retrovirus genome integrated into the cellular DNA.

pseudogene: gene that has a sequence similar to an active gene, but that is not expressed or does not encode a functional product.

pseudoknot: RNA structure minimally composed of two helical segments connected by single-stranded regions or loops; typically, the bases in the loop of a hairpin form another stem through interactions with a distinct set of bases.

pseudouridylation: conversion of uridine to pseudouridine.

purifying selection: *See* **negative selection**.

purine: nitrogenous base found in nucleosides and nucleotides and consisting of a double ring structure.

pyranose: six-membered ring formed when the C-1 aldehyde of a sugar is attacked by C-5 hydroxyl.

pyrimidine: nitrogenous base found in nucleosides and nucleotides and consisting of single ring structure.

pyrimidine dimer: structure formed by the covalent linkage of two adjacent pyrimidine bases in a DNA strand, typically caused by UV light. It cannot be replicated correctly and thus causes a mutation.

pyrophosphate: two covalently linked inorganic phosphate groups, also denoted PP_i.

Q

quaternary structure: subunit structure of a protein.

R

rad3+: the ATR regulator kinase of *S. pombe*.

Rad51: eukaryotic strand-exchange recombinase.

Ramachandran plot: a two-dimensional plot of the values of the backbone torsion angles phi (φ) and psi (ψ), with allowed regions indicated for conformations where there is no steric interference between neighboring atoms.

rate constant: proportionality constant that describes how the rate of a reaction varies as a function of reactant concentrations.

reaction intermediate: chemical species that forms transiently in the course of a chemical reaction.

RecA: bacterial strand-exchange recombinase.

recessive mutation: change in DNA sequence that is compensated at the phenotypic level by the wild-type version of the gene.

recoding: process by which the genetic code is reinterpreted to allow for the encoding of different information.

recognition helix: the α helix in DNA-binding proteins that is used to recognize specific DNA sequences by forming side chain contacts with base pairs.

recombinant: DNA produced by a homologous-recombination event between two parental DNA duplexes in which the DNA segments flanking the position of interaction between the duplexes that initiated recombination have been exchanged.

recombinant DNA: DNA molecule that does not occur naturally but is a product of artificial manipulation, such as insertion of DNA sequences into plasmid or virus DNA.

recombinase: enzyme that recognizes specific DNA sequences and catalyzes breakage and joining reactions that result in their rearrangement.

recombination crossover: in meiosis, a homologous recombination event that results in the reciprocal exchange of DNA between two homologs.

redox reaction: reaction in which both oxidation and reduction occur.

reduction: gain of electrons by an atom or molecule.

regional centromere: a type of centromere that covers a large area of the chromosome (from hundreds to many thousands of base pairs of DNA).

regulatory sequence: DNA sequence to which one or more transcriptional regulators bind and regulate a gene or set of genes.

rejection: the final step during the tRNA selection process by which an aminoacyl-tRNA can be discarded based on inherent binding stabilities for the ribosome.

replication: the process wherein a complementary copy of each strand of DNA is generated by DNA polymerase.

replication bubble: open structure formed in DNA undergoing bidirectional replication from an origin of replication. The bubble is generated after the DNA at the replication origin has been unwound and the two replication forks at either end move away from each other.

replication complex: large complex of proteins assembled at the eukaryotic replication fork and functionally equivalent to the bacterial polymerase III holoenzyme.

replication-dependent histone (S phase histone): histones H2A, H2B, H3, and H4, which are synthesized in large amounts at the same time as DNA is synthesized (during the S phase of the cell cycle) so that new nucleosomes can be assembled on the new DNA.

replication factor C (RFC): clamp loader in eukaryotes consisting of five different protein subunits.

replication fork: site at which DNA strands are separated and new DNA is synthesized. It is a Y-shaped structure and moves away from the site of replication initiation. Both strands of the DNA are copied at the replication fork.

replication fork barrier: site in eukaryotic DNA that prevents the progression of a replication fork coming from one direction by means of proteins that bind specifically to the site.

replication protein A (RPA): the single-stranded DNA-binding protein in eukaryotes which is a trimer of three proteins Rpal, Rpa2, and Rpa3.

replication restart: resumption of DNA replication at a site where the fork has collapsed because of a DNA lesion.

replicative transposition: transposition pathway in which DNA-only transposons are copied during transposition to form a cointegrate.

replisome: alternative name for the polymerase III holoenzyme.

reporter gene: gene that is used to report on the expression of another gene by making a transcription fusion. The reporter gene product may produce a change in color or fluorescence, as in the case of green fluorescent protein (GFP), have an enzymatic activity that is easily detected, or may be detected by existing reagents such as nucleic acid probes or antibodies.

repressor: sequence-specific DNA-binding protein that leads to decreased transcription.

rescue (in microtubules): sudden shift from shrinkage to growth that occurs when a GTP cap forms at the microtubule end.

resolution: parameter that governs the limit of accuracy of a molecular structure and determines how well individual atoms can be distinguished from one another; related to how a crystal diffracts x-rays.

resonance: delocalization of bonding electrons over more than one chemical bond in a molecule. Resonance greatly increases the stability of a molecule. It can be represented, conceptually, as if the properties of the molecule were an average of several structures in which the chemical bonds differ.

response regulator: protein that transduces a specific response on phosphate transfer from a sensor kinase to a specific aspartate residue.

restriction endonuclease: enzyme that recognizes a specific sequence in DNA and cleaves both DNA strands at defined locations.

restriction point: term used in mammalian cells to describe the same regulatory points as Start. It is the point in the cell cycle beyond which cells are committed to complete the cell cycle.

restrictive: term for a type of ribosome mutation that results in high-fidelity protein synthesis.

retrotransposon: transposable element whose transposition involves an RNA intermediate and a reverse transcription step.

retrovirus: RNA virus whose genome is reverse transcribed into DNA and integrated into the cell's DNA as part of the viral life cycle. The viral genome is characterized by long terminal repeats (LTRs) similar to those present in the LTR retrotransposons.

reverse genetics: analysis of the consequences to the organism of directly inactivating a particular gene.

reverse transcriptase: DNA polymerase that synthesizes DNA from an RNA template.

rho-dependent terminator: terminator sequence in bacteria that requires the additional action of Rho protein for transcription termination.

rho-independent terminator: intrinsic terminator sequence in bacteria that does not require the Rho protein for transcription termination.

ribonuclease: nuclease enzymes that act on RNA.

ribonucleoprotein (RNP): complex of RNAs and proteins.

ribonucleoprotein (RNP) domain: *See* **RNA-recognition motif (RRM) domain**.

ribonucleoside: base covalently linked to ribose via a glycosidic bond with the C1′ atom of the sugar.

ribonucleotide: ribonucleoside with one or more attached phosphates.

ribose: five-carbon sugar that is a component of the nucleic acid RNA.

ribose 2′-O-methylation: methylation of the 2′ oxygen of ribose.

ribosomal ambiguity (ram): term for a type of ribosome mutation that results in decreased fidelity during protein synthesis.

ribosomal protein: one of the ~55 (bacteria) or ~80 (eukaryotes) different types of protein present in the corresponding ribosome.

ribosomal RNA (rRNA): one of the three (bacteria) or four (eukaryotes) species of RNA present in the corresponding ribosome.

ribosome: macromolecular ribonucleoprotein machine responsible for the translation of the genetic information in the mRNA into the encoded polypeptide.

ribosome recycling: process by which the small and large subunits of the ribosome are separated and the tRNAs and mRNA released after termination of translation at stop codons in preparation for the next cycle of protein synthesis.

riboswitch: RNA element typically found in the 5′ untranslated region (UTR) of transcripts that, on binding to small molecule metabolites, change their conformation in a way that affects transcription or translation.

ribozyme: enzyme composed of RNA.

RISC (RNA-induced silencing complex): molecular complex which through the specificity of bound siRNAs acts as the effector for RNA interference or silencing.

RNA-binding domain (RBD): *See* **RNA-recognition motif (RRM) domain**.

RNA-binding motif: protein fold that binds to RNA.

RNA editing: process by which individual nucleotides in an RNA are modified, removed, or inserted.

RNA half-life: time in which half of the initial amount of an RNA is degraded.

RNA interference (RNAi): mechanism by which short fragments of double-stranded RNA lead to mRNA degradation.

RNA polymerase: the molecular machinery responsible for copying the DNA into RNA.

RNA processing: modifications that can occur to RNA including cleavage, chemical modification of individual nucleotides, as well as the addition and deletion of nucleotides.

RNA-recognition motif (RRM) domain: RNA-binding motif consisting of an alpha-beta sandwich structure in which the beta strands form a platform for RNA recognition. Also called RNA-binding domain (RBD) or ribonucleoprotein (RNP) domain.

RNA splicing: process by which introns are excised from a pre-RNA.

S

S-acylation: reversible attachment of a fatty-acid group to a protein via a thioester linkage; palmitoylation is an example of *S*-acylation.

salt bridge: a non-covalent interaction in which both donor and acceptor atoms are fully charged.

saturated (fatty acid): containing no double bonds in the hydrocarbon chain.

SDS-PAGE: sodium dodecyl sulfate – polyacrylamide gel electrophoresis, a method for separating proteins based primarily on their molecular weight by denaturing proteins in the presence of SDS and then comparing their electrophoretic mobility in a polyacrylamide gel.

secondary structure: folded segments of a polypeptide chain with repeating, characteristic phi (φ) and psi (ψ) backbone torsion angles, that are stabilized by a regular pattern of hydrogen bonds between the peptide –N-H and C=O groups of different residues.

second law of thermodynamics: states that the entropy of an isolated system can only increase.

securin: inhibitor of separase protease activity. Securin is inactivated by degradation following ubiquitination by the anaphase-promoting complex (APC).

segregation: process through which the replicated DNA is divided into two chromosome clusters, in order for progeny cells to obtain a full complement of chromosomes.

selenocysteine insertion sequence (SECIS): RNA sequence element that specifies the insertion of selenocysteine at UGA stop codons.

self-splicing intron: intron that is able to catalyze its own excision from the primary RNA transcript in the absence of any protein cofactors; such an intron is typically referred to as a ribozyme.

semi-conservative: a term applied to DNA replication to indicate that a new DNA double helix has one old strand of DNA and one new strand. Half of the parental DNA is conserved in each daughter helix.

senescence: when cells lose the ability to divide.

sense codon: a nucleotide triplet sequence that specifies a particular amino acid.

sensor kinase: protein that detects a signal which causes the protein to autophosphorylate a histidine residue.

separase: protease necessary for sister chromatid separation. Separase cleaves the kleisin subunit of cohesin.

sequence tagged site (STS): site that is uniquely identified in the human genome and for which PCR primer sequence are available to allow detection of this site.

sex chromosome: a chromosome that determines sex in a given species. The number of sex chromosomes often varies between males and females. In humans, females have two X chromosomes and males have one X and one Y chromosome.

Shine–Dalgarno sequence (ribosome binding site): polypurine sequence in the upstream leader of an mRNA that base pairs with a polypyrimidine tract in the 16S rRNA to help in the identification of the start site of the open reading frame.

short interspersed element (SINE): genomic sequence derived from small cellular RNAs such as tRNAs or 7SL RNA that spreads non-autonomously in the genome by target-primed reverse transcription mediated by LINE-encoded recombination proteins.

sigma factor: bacterial RNA polymerase subunit required for promoter recognition.

sigmoidal curve: s-shaped binding curve characteristic of cooperative binding interactions.

signal joint: fusion between the ends of the nonamer and heptamer bounded segments that are recognized by RAG and excised when the V, D, and J coding segments joined during RAG-mediated V(D)J recombination.

silencer sequence: *see* **intronic enhancer** and **exonic enhancer**.

silencing: establishment of a heritable state of chromatin – heterochromatin – in which genes are not transcribed, origins of replication are inactivated or replicate late, and recombination does not take place.

silent mutation: change in the DNA sequence that does not result in obvious phenotypic change.

simple sequence length polymorphism (SSLP): sites in the DNA where there are repeated sequences, usually di- or tri- nucleotide repeats, in which the number of repeats at a given site varies from person to person. Thus a region of the genome can be marked and followed by determining which version of the SSLP is present in association with disease.

single bond: one covalent bond joining two atoms, the result of sharing a pair of electrons.

single nucleotide polymorphism (SNP): DNA sequence variation where a single nucleotide at a specific position in the genome differs between members of a species.

single-stranded binding protein (SSB protein): protein that binds specifically to single-stranded DNA that is exposed when the double helix is unwound or damaged.

siRNAs (small interfering RNA): 21–24 nucleotide RNAs typically generated through non-specific excision from double-stranded RNA of either external or endogenous origin, which direct RNA interference and transcriptional gene silencing.

sister chromatid: one of the two daughter chromosomes generated after DNA replication. The two chromatids contain identical DNA sequences and remain attached to each other until chromosome segregation takes place.

sister chromatid cohesion: linkages that hold sister chromatids together between S phase and anaphase.

site-directed mutagenesis: generation of one or more mutations at a particular location in a gene.

size exclusion chromatography: method of chromatography that separates molecules based on size, as well as shape.

sliding clamp: ring-shaped protein that moves along the DNA with the DNA polymerase in DNA replication and thus increases its processivity. The ring is made of two copies of the β protein in *Escherichia coli* and three copies of PCNA in eukaryotes.

slipped-strand misreplication: errors in DNA replication that can occur in repetitive sequence regions, when intrastrand base-pairing of either the template strand or newly synthesized strand causes the replication machinery to slip on the template.

SL RNA: short, defined RNA sequences carrying a methylated cap structure, which are spliced onto the 5′ end of distinct transcripts in a *trans*-splicing reaction.

snoRNA: small nucleolar RNA that guides ribose methylation and pseudouridylation.

snoRNP: snoRNA-containing RNP that catalyzes ribose methylation and pseudouridylation.

snRNA: RNA component of a snRNP.

snRNP: small nuclear ribonucleoprotein; ribonucleoprotein complexes composed of an snRNA and a variety of proteins that assemble to form the spliceosome.

Solexa sequencing: massively parallel sequencing technique in which individual sequence reads can be up to 40 bp in length.

somatic cells: cells in multicellular eukaryotes that constitute the building blocks of the body of an organism (such as liver cells) but are not involved in the production of progeny.

sorting motif: sequence identifiers within proteins that target the proteins to specific subcellular compartments (also denoted sorting signals or signal sequences in some cases).

SOS response: response of bacterial cells to maintain viability in cases of extensive DNA damage.

Southern blot analysis: method for detecting specific DNA sequences. The DNA fragments are typically separated on a gel, denatured, transferred to a filter, and probed by hybridization with a radioactive or otherwise labeled complementary DNA fragment.

species: reproductively isolated group of organisms.

specific binding: relatively tight binding between correct partners as opposed to lower affinity 'non-specific binding' that occurs at random.

spectral karyotyping (SKY): a method for detecting abnormalities in chromosome structure, by hybridizing chromosome-specific probes, each with a different fluorescent signature.

spectrophotometer: instrument for measuring the amount of light absorbed by a sample.

S phase: the stage of the cell cycle during which the DNA is replicated, resulting in chromosome duplication.

start: major regulatory transition at the entry into the cell cycle in mid to late G1, also called the restriction point (in animal cells). It is the point in the cell cycle beyond which cells are committed to complete the cell cycle.

stem cells: cells that have an unlimited capacity to divide and can give rise to many different lineages.

stop codon (nonsense codon): a nucleotide triplet sequence that specifies a stop in the translation of the genetic information; there are usually three such triplets (UAA, UAG, and UGA).

strand-exchange recombinase: recombination protein that identifies regions of homology shared between a single-stranded DNA and double-stranded DNA and promotes the formation of heteroduplex DNA containing the parental single strand and its complement in the DNA duplex.

structural maintenance of chromatin (SMC) protein: structural protein involved in cohesion, condensation, and DNA repair.

substrate: molecule that is transformed in a reaction.

sugar–phosphate backbone: the regularly repeating part of the DNA polymer, which consists of deoxyribose sugar joined by phosphodiester linkages between the 3′ oxygen of one sugar and the 5′ oxygen of the next.

sugar pucker: conformation assumed by the non-planar sugar in nucleic acid (ribose in RNA and deoxyribose in DNA).

sulfhydryl: –SH group, which is found in the side chain of cysteine. Also known as a thiol.

sumoylation: attachment of SUMO (small ubiquitin-like modifier) to a protein.

synapsis: recognition of regions of homology between a single strand of one DNA molecule and an intact duplex, usually promoted by a strand exchange protein.

synaptonemal complex: protein structure that links a pair of homologous chromosomes along their length in early meiosis.

synonymous changes: changes to a coding sequence that do not change the encoded amino acid due to the redundancy of the genetic code.

syntenic blocks: conserved regions that contain the same genes in the same order between chromosomes of different species.

synthesis-dependent strand annealing (SDSA): a pathway for double-stranded DNA break repair using non-conservative DNA replication in which the newly synthesized DNA that restores a break is displaced from the template duplex and transferred to the broken duplex.

synthetic lethal: situation when the combination of two gene disruptions or mutations is lethal while the individual disruptions or mutations do not affect viability.

T

TAF (TBP-associated factor): polypeptide that associates with TBP to form the TFIID complex and includes proteins that are required for activation of transcription.

TagSNP: specific single nucleotide polymorphism that uniquely identifies a **haplotype block**.

T antigen: protein encoded by the mammalian SV40 virus. SV40 is a tumor-causing virus and T antigen stands for tumor antigen. T antigen binds the origin of replication in viral DNA to initiate replication and also serves as the helicase that unwinds the DNA ahead of the replication fork as well as a transcription factor.

target-primed reverse transcription (TPRT): insertion mechanism by which a DNA copy of a RNA copy of a non-long terminal repeat (LTR) element is copied *in situ* at the new insertion site.

target site: the site in DNA into which a moving transposable element is inserted.

target-site duplication: short direct duplication of the target-site sequence that is generated on either side of an inserted transposon by the insertion process.

TATA box: AT-rich core promoter element located approximately 25–30 nucleotides upstream of the transcription start of many archaeal and eukaryotic promoters. It is bound by the general transcription factor TATA-binding protein (TBP).

tautomer: structural isomer of a molecule that can change spontaneously from one form to the other, usually by a change in position of a particular hydrogen atom or proton, and exists in equilibrium with alternative isomers.

TBP (TATA-binding protein): present in the TFIID fraction. It binds to the TATA box and directs transcription.

tel1⁺: the ATM regulator kinase of *S. pombe*.

TEL1: the ATM regulator kinase of *S. cerevisiae*.

telomerase: enzyme that synthesizes the terminal telomere repeats on chromosome ends.

telomere: specialized structure at the end of a eukaryotic chromosome that enables the DNA to be fully replicated and also maintains the integrity of the chromosome. It contains repetitive DNA sequences packaged into heterochromatin.

telophase: final stage of mitosis, in which the spindle is disassembled, the chromosomes decondense, and the nuclear envelope re-forms.

template: parental nucleic acid strand that is copied by a polymerase, for example during replication or transcription. Each base is read by the appropriate polymerase and the complementary nucleotide is added to the new nucleic acid strand. For example, a G in the template specifies that a C be added to the new strand.

termination (transcription): steps by which RNA polymerase stops transcription.

termination (translation): process by which a stop codon in the mRNA is recognized and the growing polypeptide chain is released from the ribosome.

terminator: DNA sequence that signals RNA polymerase to stop transcription and be released from the DNA along with the RNA transcript.

terminus region: region on the *Escherichia coli* chromosome at which DNA replication terminates and the two interlinked daughter DNAs are resolved.

tertiary structure: folded conformation of a protein.

tetraploid: containing four complete sets of chromosomes.

thin layer chromatography (TLC): type of chromatography in which a sheet of glass or plastic is coated with an absorbent material (such as polyethyleneimine or silica) and a solvent containing a mixture of molecules moves across the plate by capillary action, separating the components of the mixture based on their varying interactions with the absorbent material.

three domains of life: three distinct groups (the bacteria, archaea, and eukaryotes), or branches, into which all organisms can be divided.

TIR: the terminal inverted repeats at the ends of transposons that interact with transposase.

t-loop (telomere): structure formed at human telomeres when the G-rich single-stranded region loops around and base-pairs with the nearby double-stranded telomeric DNA.

T loop (tRNA): single-stranded region of tRNA that contains ribothymidine and participates in critical interactions with the D loop to stabilize the elbow region of the tRNA.

tmRNA: bacterial RNA species known to function both as a tRNA and an mRNA to recover stalled ribosomes.

TOR: protein kinase that regulates cell growth through the stimulation of protein synthesis.

totipotent cells: stem cells that can give rise to every cell type of the body.

transamidase: the enzyme responsible for catalyzing the amination of a carboxylic acid, for example to transform aspartate and glutamate into asparagine and glutamine.

transcript: RNA copy of a gene.

transcript cleavage factor: protein that promotes cleavage of the most recently added ribonucleotides that have separated from the template strand in a backtracked RNA polymerase complex.

transcription: process of copying the DNA sequence of interest into the corresponding RNA.

transcriptional arrest: cessation of RNA transcription.

transcriptional pausing: a temporary cessation of transcript elongation.

transcription bubble: unpaired region of DNA associated with RNA polymerase.

transcription-coupled repair (TCR): nucleotide excision repair that is preferentially targeted to DNA damage in genes that are being actively transcribed.

transcription unit: the RNA copy of a region of the genome, which may include one or more genes along with non-coding regulatory sequences.

transcriptome: levels of all of the transcripts detected for a genome.

transesterification: energetically neutral reaction involving the exchange of phosphodiester linkages.

transformation: process of introducing DNA into cells. Cells are often treated with chemicals to make them competent to take up the DNA.

transgene: specific gene that has been added to a genome to make a transgenic organism.

transgenic organism: organism containing a transgene.

transition mutation: when a purine base, A or G, is replaced by the other purine base, A or G, or a pyrimidine base, C or T, is replaced by the other pyrimidine base, C or T.

transition state: species of highest free energy either in a reaction or a step of a reaction.

translation: process wherein the nucleotide sequence of the mRNA is deciphered in three-base triplets into the corresponding amino acids to form the encoded protein via the actions of the ribosome and auxiliary protein factors.

translational fusion: direct fusion of the open reading frame (ORF) of a reporter gene to the ORF of another gene.

translation factors: host of proteins that facilitate the various steps in the translation cycle.

translocation (chromosome): movement of a segment of a chromosome to another chromosome.

translocation (translation): step in elongation that follows peptidyl transfer that is defined by the directional 3-nucleotide movement of the mRNA–tRNA complex.

transmembrane helix: alpha helix spanning a cell membrane.

transposable element: a discrete mobile DNA segment that can move between non-homologous positions in a genome.

transposase: type of recombinase required for the movement of a transposable element and which is usually encoded by the element. It usually recognizes sequences at the termini of the transposon and executes the DNA cleavage and target-joining reactions that enable the element to move.

transposition: movement of a **transposable element**.

transposon: *See* **transposable element**.

transpososome: elaborate nucleoprotein complexes in which transposition occurs.

***trans* splicing:** form of splicing in which exons from two separate RNA transcripts are joined in a reaction that is otherwise analogous to that of *cis* splicing mediated by the spliceosome.

transversion mutation: when a purine base, A or G, is replaced by a pyrimidine base, C or T, or a pyrimidine base, C or T, is replaced by a purine base, A or G.

tRNA: bifunctional adaptor molecule that interprets the genetic code and participates in the synthesis of proteins by the ribosome.

tumor suppressor gene: gene that encodes a protein that normally restrains cell proliferation or tumorigenesis, such that loss of the gene increases the likelihood of cancer formation.

two-component signal transduction pathways: signal transduction cascades common in bacteria that signal extracellular changes and in the simplest cases comprise two proteins, a histidine kinase and a response regulator.

two-dimensional gel electrophoresis: separation method in which gel electrophoresis is carried out first in one direction, separating molecules based on one property, such as size, and then subjecting the gel to a second electrophoretic separation at a 90° angle to the first, using a different basis of separation such as pH.

two-hybrid assay: an *in vivo* technique for identifying proteins or protein domains that interact with one another.

U

ubiquitin: small protein that is covalently attached to lysine side chains in substrate proteins, either singly or in multiple copies as a polyubiquitin chain. Ubiquitin modification serves multiple signaling roles, notably targeting proteins to the proteasome for degradation, as well as for transcriptional regulation, the DNA damage response, and endocytosis.

ubiquitination: attachment of ubiquitin to a protein; also known as ubiquitylation.

unequal crossing over: homologous recombination between non-allelic repeat sequences of homologous chromosomes that result in chromosomal deletions or expansions.

unfolded protein response: multilayered response in eukaryotic cells when unfolded proteins are detected in the endoplasmic reticulum.

unsaturated (fatty acid): fatty acid containing one or more double bonds linking carbon atoms.

untranslated region (UTR) (5′ or 3′): region of an mRNA that precedes the start site of translation is known as the 5′ UTR whereas the region that follows the termination site is known as the 3′ UTR.

UP element: AT-rich sequence located just to the 5′ side of the –35 sequence and bound by the α subunit of RNA polymerase.

UPF: a group of proteins initially identified through genetic approaches for their ability, when mutated, to increase frameshifting (up frameshifting). These proteins are involved in recognizing exon-junction complexes and in signalling to the translation machinery that a particular stop codon is premature.

upstream: to the 5′ side of any given reference point, such as a transcription start site or a coding sequence.

upstream ORFs (uORFs): small open reading frames (ORFs) located 5′ of the principal large ORF in the transcript that in eukaryotes can play important roles in regulating translation.

V

van der Waals interaction: a weak attractive force between two atoms or groups of atoms, arising from the fluctuations in electron distribution around the nuclei.

van der Waals radius: effective radius of an atom or chemical group.

V(D)J recombination: cellular process for producing antibodies and immune system receptors by recombination of independent gene segments, selected at random from the V, D, and J clusters, to assemble a gene encoding an immune system protein. This process generates a diverse repertoire of antibodies and receptors that can recognize a variety of antigens.

vector: DNA molecule that is used in cloning. Often contains an origin of replication, a multiple cloning site and a selectable marker. A DNA fragment of interest is cloned into the vector, which can then be introduced into cells and propagated.

vertical gene transfer: normal passage of genes from one generation to the next.

viruses: particles composed of a nucleic acid-based genome surrounded by a protein coat. Although viruses contain their own genomes, they cannot replicate on their own and must replicate within another organism, where they usurp components of the host cell machinery.

W

Watson–Crick base pairs: characteristic pairing of guanine with cytosine and adenine with thymine that is stabilized by hydrogen bonds.

western blot: an analytical method in which an antibody is used to detect a protein of interest that has been separated by gel electrophoresis and transferred to a filter.

wild-type: the typical or most common form of an organism or gene in nature.

wobble pairing: non-optimal non-Watson–Crick interactions (most commonly G–U interactions); such pairings are best known for being accepted as cognate at the third nucleotide of the codon (and the first nucleotide of the anticodon) by the translational machinery.

X

X inactivation: the process in mammals that generates equal expression of genes on the X chromosome in males and females. In females one of the two X chromosomes is permanently inactivated.

x-ray crystallography: a method for determining three-dimensional molecular structures from x-rays scattered from crystals.

Y

Y family polymerases: class of specialized DNA polymerases that can replicate DNA in an error-prone fashion past a site of DNA damage.

Z

Z-DNA: left-handed DNA double helix that can form in alternating purine–pyrimidine sequences under conditions of high salt or torsional stress.

zinc finger: a protein domain whose single alpha helix and two antiparallel beta strands are held together by a central zinc ion.

Z ring: the contractile ring in bacteria that is made of the tubulin-like protein, FtsZ. The contraction of the Z ring in the driving force of bacterial cell division.

INDEX

Page numbers in *italic* indicate figures.